Cultivation of Comfrey

Medicinal and Food Uses for People and Livestock

Comfrey as Food and Medicine for Livestock
Comfrey Meal, Pellets, Hay and Silage
Livestock / Pet Species and Comfrey

Medicinal Comfrey Overview
Medical Research about Comfrey and Healing
Personal / Clinical Observations of Healing
Comfrey Heals: Allantoin
Making and Using Comfrey Medicine
Warnings and Negative Reactions to Comfrey
Alkaloids in Comfrey
Some Uses of Comfrey are Restricted by Governments
Humans Eating Comfrey
Miscellaneous Uses of Comfrey

Care of Comfrey Plant: Overview and How to Propagate
Planting, Soil, Fertilization, Water, Disease
Productivity and Farm Economics of Comfrey
Harvesting Comfrey Leaves
How to Get Rid of Unwanted Comfrey

Volume 2
Comfrey Book Series

CAUCASIAN PRICKLY COMFREY.

Caucasian Prickly Comfrey 1883
'Ensilage: A System for the Preservation in Pits of Forage Plants and Grasses,
Independent of Weather: A Collection of Facts and Statistics on the
Cheapest Mode of Providing Winter Food for Dairy Cattle, Sheep, Horses, Etc.'
by Thomas Christy, London, England, 1883, page 62.

©Copyright 2022 Nancy Shirley. All rights reserved. Published in the United States.
Nantahala Farm and Garden www.nantahala-farm.com ncfarmgarden@gmail.com
ISBN 978-0-9890851-3-7

Basic Table of Contents for Volume 2

Preface for Volume 1 and 2 . . . . . . page 8

PART D: LIVESTOCK, PETS AND COMFREY
Chapter 20: Comfrey as Food & Medicine for Livestock . . . . page 10
Chapter 21: Comfrey Meal, Pellets, Hay and Silage . . . . page 30
Chapter 22: Livestock / Pet Species and Comfrey . . . . page 46

PART E: COMFREY AS MEDICINE AND HUMAN FOOD
Chapter 23: Medicinal Comfrey Overview . . . . page 81
Chapter 24: Medical Research about Comfrey & Healing . . . . . . page 101
Chapter 25: Personal / Clinical Observations of Healing . . . . . . page 127
Chapter 26: Comfrey Heals: Allantoin . . . . page 155
Chapter 27: Making and Using Comfrey Medicine . . . . page 166
Chapter 28: Warnings and Negative Reactions to Comfrey . . . . page 199
Chapter 29: Alkaloids in Comfrey: Overview, Types, Detection, Amounts, Food . . page 216
Chapter 30: Alkaloids in Comfrey: Scientific Studies and Various Perspectives . . page 247
Chapter 31: Some Uses of Comfrey are Restricted by Governments . . . . page 291
Chapter 32: Humans Eating Comfrey . . . . page 316
Chapter 33: Miscellaneous Uses of Comfrey . . . . page 338

PART F: CULTIVATION AND PRODUCTIVITY OF COMFREY
Chapter 34: Care of Comfrey Plant: Overview and How to Propagate . . page 353
Chapter 35: Planting, Soil, Fertilization, Water, Disease . . . . page 374
Chapter 36: Productivity and Farm Economics of Comfrey . . . . page 410
Chapter 37: Harvesting Comfrey Leaves . . . . page 430
Chapter 38: How to Get Rid of Unwanted Comfrey . . . . page 436

Artwork

Apothecary Tin and Jar: page 198

Comfrey Chemistry: page 215

Common Comfrey: pages 126, 165, 246, 337, 352

Culinary Vegetables: page 337

First Aid Kit: page 100

Medicinal Comfrey for Sale: page 100

Pigs: page 80

Prickly Comfrey: pages 1, 2, 29, 373, 436

Sheep: page 80

Sickles and Scythes: page 436

Silage Pits: page 45

Tobacco Drying: page 352

Tuberous Comfrey: pages 290, 409

Unknown Comfrey: pages 154, 315

Book Cover by: Baker Vail Design

FLOWER OF THE TRUE VARIETY—THE BUD IS RED AND THE EXPANDED FLOWER IS BLUE.

Prickly Comfrey 1880
"The bud is red and the expanded flower is blue."

'New Commercial Plants with Directions
How to Grow Them to the Best Advantage, No. 3, 1880'
by Thomas Christy, Fellow of Linnaean Society, London, England.
Number 3 includes Prickly Comfrey pages 11-14. Includes 'Caucasian Prickly Comfrey' and 'Russian Comfrey' live root advertisements.

Expanded Table of Contents for Volume 2

Preface for Volume 1 and 2......page 8
Complex, Fascinating Comfrey
Definitions of Words
Finding Botanical Literature
Fair Use and Registration Marks
Legal Disclaimers

PART D: LIVESTOCK, PETS AND COMFREY

Chapter 20: ....page 10
Comfrey as Food & Medicine for Livestock
Comfrey Leaves and Livestock Nutrition
    Feeding Comfrey to Livestock 1800s
    Feeding Comfrey to Livestock 1900s
Advantages of Russian Comfrey for Livestock
Livestock and Safety of Comfrey
Tethered Stock and Rotational Grazing
Livestock Food in Early Spring
Prickly Comfrey as a Forage Crop
    1871, 1876, 1877, 1880, 1881, 1882, 1896
    1910, 1911, 1976, 1978
Comfrey as Livestock Medicine
    Overview
    Broken Bones in Animals
    Foot-and-Mouth Disease
    Mastitis (Breast Tissue Infection)
    Muscle Strain in Animals
    Parasites in Animals
    Pregnancy Problems
    Scours (Diarrhea)
    Skin Problems
    Sprains and Strains
    Ticks
    Walking Problems
    Wounds

Chapter 21: ....page 30
Comfrey Meal, Pellets, Hay and Silage
Comfrey Bristles
Dried Comfrey Meal Use Shredder and Drier
Comfrey Pellets
Comfrey as Hay
    Drying Small Amounts of Comfrey Leaf
    Drying on Ground, then Sometimes Use Drier
    Drying Using Tripods
    Hanging in Shed
    Drying on Racks and on Roof
    Comfrey as Silage
    Comfrey Silage Overview
    Wrongs Ways to Make Comfrey Silage
    Correct Ways to Make Comfrey Silage
        Mix Comfrey with Cereal Grains
        Add Molasses to Comfrey
        Or Add Sugar and Yeast
        Or Add Lactic Acid Bacteria
        Or Add Chemical Acid
    Silo Should Have Drainage
    Tank Storge System 1877

Chapter 22: ....page 46
Livestock / Pet Species and Comfrey
Cats and Dogs
Cattle: Beef and Dairy
    Cattle Love Comfrey
    Amount of Comfrey for Cattle
    Comfrey Early Spring and Late Fall
    Use Comfrey with Other Feeds for Cattle
    Cows Eating Comfrey Produce Good Milk
    Feeding Comfrey Silage to Cows
    Comfrey and Cattle Health
    Feeding Calves Comfrey
Chickens and Ducks (see Poultry)
Dogs (see Cats and Dogs)
Donkeys (see Zoo Animals)
Ducks (see Poultry)
Fish
Geese (see Poultry)
Goats
    Comfrey and Goat Milk
    Comfrey as Goat Medicine
Guinea Pigs
Hares (see Rabbits)
Hogs (see Pigs)
Horses
    Pounds of Comfrey Per Day
    Comfrey as Horse Medicine
Pigs and Hogs
    Pigs, Comfrey and Land Use
    Amount of Comfrey for Pigs
    Comfrey Diet and Meat Quality
    Problems with Comfrey and Pigs (Nitrate)
    Pigs and Vitamin B12
    Timing Pig Rearing with Comfrey Growth
    Breeding Sows and Piglets
    Comfrey as Pig Medicine
Poultry (Chickens, Duck, Geese, Turkey)
    Poultry Love Comfrey
    Comfrey, Vitamin A, and Egg Yolk Color
    Comfrey Extracts and Poultry Diet
    Feed Comfrey to Chicks and Poults
    Ducks and Geese
    Turkey
Rabbits and Hares
    Rabbits Eat Comfrey
    Rabbits are Resistant to Alkaloids
    Comfrey as Rabbit Medicine
Rats as Pets
Sheep
Snails as Food
Swine (see Pigs)
Turkey (see Poultry)
Zoo Animals and Donkeys

PART E:
COMFREY AS MEDICINE AND HUMAN FOOD

Chapter 23: ....page 81
Medicinal Comfrey Overview
Types of Herbal Preparations
Commercial Preparations of Comfrey
Defining Medicinal Effects of Comfrey
Combination of Ingredients in Comfrey Plant
Old Remedies and Current Medicine Overview
Herbalist Recommendations
    Overview
    1800s, 1900s, 2000s
Chinese Medicine and Comfrey
Ayurvedic Medicine and Comfrey
Physician Recommendations

Chapter 24: ....page 101
Medical Research about Comfrey & Healing
More Research Needed
Overview of Topical (Skin) Applications
Positive Medical Research about Comfrey
    Anti-Bacterial (see Infection)
    Antifungal
    Anti-Inflammatory (see Inflammation)
    Anti-Microbial (see Infection)
    Antioxidant Properties
    Blood Pressure and Blood Flow
    Blunt Trauma / Injury (see Bruises or see Sprains)
    Broken Bones
    Bruises (Contusions, Blunt Injury)
    Burns
    Cancer (also see Tumors)
    Eczema and Psorasis
    Infection
    Inflammation
    Joint Pain
    Muscle Soreness
    Muscle Tension
    Nerve Injury
    Osteoarthritis
    Pain in General
    Pain in Back and Neck
    Pressure Sores
    Rheumatism
    Skin Health
    Skin Redness
    Skin- Ulcers
    Sports Fatigue
    Sprains and Blunt Injury
    Tumors Reduced (also see Cancer)
    Ulcer: Gastric / Dueodenal
    Varicose Veins
    Wounds- Abrasion (scrapes)

Chapter 25: ....page 127
Personal / Clinical Observations of Healing
Overview of Comfrey as Healer
Abscess or Boil (skin)
Anal Itching
Anemia
Arthritis
Athletes Foot (fungus)
Bedsores
Bites (see Stings)
Bladder and Urinary Problems (also see Kidneys)
Blood and Circulation; Blood Pressure
Boil (see Abscess)
Breast (see Mastitis)
Bones (see Broken Bones and see Osteoporosis)
Broken Bones (also see Osteoporosis)
Bruises (see Cuts)
Burns (heat, sun)
Bursitis
Calcium Deficiency
Cancer and Tumors
Chicken Pox Itching
Cold (see Cough)
Constipation (see Diarrhea)
Cough, Cold, Sore Throat
Cuts / Bruises / Sprains
Dental (see Mouth)
Diarrhea and Constipation
Eczema and Psorasis
Eyes
Feet
Female Problems (see Mastitis or Vagina)
Hair
Headache
Hemorrhoids
Hernia
Inflammation (see Swelling)
Kidneys (also see Bladder)
Liver
Lungs (also see Pleurisy; see Cough / Cold / Sore Throat)
Mastitis and Breast Problems
Mouth / Dental
Nails
Neuralgia, Nerve Pain
Oral (see Mouth)
Osteoporosis (also see Broken Bones)
Pleurisy (also see Lungs)
Rash (Skin)
Rheumatism
Skin Health (also see Rash)
Skin Injuries
Skin: Stretch Marks
Skin Ulcer
Sore Throat (see Cough)
Sprains (see Cuts)
Stings and Bites (insects, nettle plant)
Swelling

Teeth (see Mouth)
Tonsillitis
Tumor (see Cancer)
Ulcer: Gastric or Duodenal (Stomach & Intestines)
Urinary (see Bladder)
Vagina: Dry, Healing/Tearing, Discharge
Varicose Veins and Inflammation of Veins
Wounds (also see Skin Injuries or see Cuts)

Chapter 26: ....page 155
Comfrey Heals: Allantoin
What is Allantoin?
Where Allantoin is Found in Comfrey
Seasonal Variations of Allantoin
Amounts of Allantoin in Comfrey Leaves
    Russian Comfrey Leaf Allantoin
    Symphytum Officinale Leaf Allantoin
Amounts of Allantoin in Comfrey Root
    Russian Comfrey Root Allantoin
    Symphytum Officinale Root Allantoin
Health Benefits of Allantoin
Allantoin is Not Antiseptic/Antibiotic
Allantoin Increases Phagocytosis and Leucocytosis
Solubility, Stability, Heat Sensitivity of Allantoin

Chapter 27: ....page 166
Making and Using Comfrey Medicine
Medicinal Gathering of Comfrey Leaves and Roots
Drying and Processing Comfrey Roots
    (Drying Comfrey Leaves: see 'Comfrey Tea', Chapter 32)
Making Comfrey Flour from Leaves and Roots
Storing Dried Comfrey
Dosages of Comfrey
Comfrey Medicine Overview
Alkaloids in Comfrey Medicine Overview
    includes Harvest Time
Comfrey Powder
Comfrey Extract to Make Tinctures and Powder
    Definition of Extract and Tincture
    How to Make Comfrey Extract as Tincture
    Extracting Constituents by Alcohol, Water & Others
    Comfrey Extract Formulas
Comfrey Fermentation
Comfrey Syrup
Comfrey Lozenges or Cough Drops
Comfrey Bolus or Suppository
Comfrey Oil Infusion
Comfrey Ointment, Cream or Salve
    Ointment or Cream from Comfrey Root
    Ointment, Cream or Salve from Comfrey Leaves
Comfrey Compress or Fomentation
Comfrey Poultices
    Comfrey Flour Poultice
    Chopped Leaf Poultice
    Comfrey Root Poultice
Comfrey Baths and Soaks
    Drawing Bath
    Sitz / Hip Bath

Homeopathic Comfrey
Comfrey Flower Essences and Remedies

Chapter 28: ....page 199
Warnings & Negative Reactions to Comfrey
Negative Reactions to Comfrey (Side Effects)
Individual Differences in Reactions to Alkaloids
Age and Alkaloid Susceptibility
Kidney Function and Alkaloid Susceptibility
Liver Function and Alkaloid Susceptibility
Genetic Variation & Alkaloid Susceptibility: Cytochrome
    Drugs & Herbs that Increase Cytochrome P450 Levels
    Drugs & Herbs that Decrease Cytochrome P450 Levels
    Drugs Metabolized by Cytochrome P450
Diet and Alkaloid Susceptibility: Glutathione
    Food with Glutathione
    Low Protein Diet and Glutathione Levels
    Alcoholism and Low Glutathione Levels
    Sulfur and Selenium Needed to Make Glutathione
Environment and Alkaloid Susceptibility
Drug and Herb Interactions with Comfrey
    Caution When Using These Herbs with Comfrey
    Do Not Use with Drug/Herb with Alkaloids or Harm Liver
    Other Drugs Not to be Used with Comfrey
    Comfrey Affecting Lab Test Results
    Heavy Metals and Comfrey
Diagnosis of Herbal Hepatotoxicity

Chapter 29: ....page 216 **Alkaloids in Comfrey:**
    Overview, Types, Detection, Amounts, Food
Warning about Alkaloids
Definition of Alkaloid and Pyrrolizidine Alkaloids
Some Alkaloids are More Toxic Than Others
Metabolic Pathways of Pyrrolizidine Alkaloids
Alkaloids Become Toxic when Metabolised by Liver
Solubility, Stability, Heat Sensitivity of Alkaloids
Detecting Pyrrolizidine Alkaloids in Herbal Products
Detecting Pyrrolizidine Alkaloid in Humans
Symptoms of Pyrrolizidine Alkaloid Poisoning
Alkaloids in Comfrey Develop in Roots and Flowers
Research 1970s-1980s Types/Percent Alkaloids in Comfrey
Research 1990s on Types & Percent Alkaloids in Comfrey
Research 2000s on Types & Percent Alkaloids in Comfrey
Symphytum officinale and Echimidine
Human Foods with Alkaloids
    Honey and Alkaloids
    Eggs and Alkaloids
    Milk, Meat, Grain and Alkaloids

Chapter 30: ....page 247 **Alkaloids in Comfrey:**
    Scientific Studies and Various Perspectives
Scientific Studies Showing Dangers of Alkaloids in Comfrey
    with some Rebuttals such as:
    -'Toxicity of Pyrrolizidine Alkaloids' by Mattocks, 1968.
    -'An Outbreak of Hepatic Veno-Occlusive Disease in
       North-Western Afghanistan' by Mohabbat, 1976.
    -'Carcinogenic Activity of Symphytum Officinale' by
       Hirono, 1978.

- *'Comfrey and Liver Damage'* by Roitman, 1981.
- *'Comfrey: Assessing Low-Dose Health Risk'* by Abbott, 1988.
- *'Hepatic Veno-Occlusive Disease Associated with Comfrey Ingestion'* by Yeong, 1990.
- *'Determination of Pyrrolizidine Alkaloids in Commercial Comfrey Products'* by Betz, 1994.
- *'Analysis of Herbal Teas Made from the Leaves of Comfrey'* by Oberlies, 2004.
- *'Toxicity of Pyrrolizidine Alkaloids to Humans and Ruminants'* by Wiedenfeld, 2011.

Scientific Studies and Indirect Evidence of Low Risk
More Alkaloid Research is Needed
Proper Botanical Identification is Needed
   Examples of Wrong Identification
Flawed Experimental Design
Inaccurate Statements: Poor Understanding of Research
Putting Alkaloids in Perspective
Comfrey with Low or No Alkaloids

Chapter 31: ....page 291
Some Uses of Comfrey are Restricted by Governments
Overview of Legal Restrictions by Governments
Australia Bans Sale of Comfrey for Internal Use
   1984, 1986, 1988, 2011
   Response to Decision: Comfrey on Poison Schedule
Canada Bans Sale of Some Comfrey for Internal Use
   1982, 1984, 1986-1989, 1993, 1997, 2003
   2004, 2007-2009, 2011
United States Regulates Sale of Comfrey for Internal Use
   USA 1990: FDA Tests Comfrey Products
   USA 1994: Dietary Supplement & Health Education Act
   USA 1996: AHPA Recommends Restricting Comfrey
   USA 1998: USP Discourages Internal Use of Comfrey
   USA 2001: FDA Bans Sale of Comfrey for Internal Use
   USA 2006: Adverse Effects from Dietary Supplements
   USA 2007: Good Manufacturing Practices
   USA 2007, 2008: Regulations & Pyrrolizidine Alkaloids
   USA 2011: New Dietary Ingredients
   USA 2011: Comfrey Sold Under 'Dietary Supplement Act'
   USA 2013: Comfrey Extracts and Safety
   USA 2015, 2016: Supplement Claims and Safety
United Kingdom Bans Some Uses of Comfrey
   1984, 1992, 1993, 2004, 2005, 2013, 2016
Germany Bans Some Uses of Comfrey
   1992, 1996, 2004
   German Commission E Monographs
European Union: Comfrey Medicinal Regulations
   1992, 2001, 2004, 2011, 2015

Chapter 32: ....page 316
Humans Eating Comfrey
Humans Eating Comfrey Overview
Overview of Many Ways to Eat Comfrey
Reasonable Eating of Comfrey
Eat Small or Large Leaves
   Eat Small Leaves Because Tender
   Eat Large Leaves Because Less Alkaloids

Ways to Eat Comfrey
   Raw Comfrey Leaves and Flowers in Salad
   Comfrey Cooked Alone
   Comfrey Soup and Stew
   Baked Comfrey
   Comfrey Bread, Pasta, and Flour
   Vegetables/Fruit and Comfrey
   Frying and Dry Roasting Comfrey
   Green Drinks
   Root with Sweeteners
   Root Wine
   Beer and Ale
Comfrey Tea (Infusion or Decoction)
   Overview of Extracting Substances from Comfrey
   Drying Leaves for Comfrey Tea
      (see 'Comfrey as Hay', Chapter 21)
   Making Comfrey Tea
Comfrey Protein Concentrates

Chapter 33: ....page 338
Miscellaneous Uses of Comfrey
Tannins in Comfrey
Leather and Tanning
Spinning
Comfrey is a Natural Dye
Comfrey Soap
Comfrey Toiletries and Cosmetics
Smoking and Vaporizing Comfrey
Making Paper with Comfrey
Wildlife and Fire Protection
Feed Worms
As Energy Source (Biofuel)
Protection During a Journey
Improve Finances
Astrology
Increase Fertility and Help Pregnancy
Comfrey, Minnesota
Comfrey Postage Stamps

PART F: CULTIVATION AND PRODUCTIVITY OF COMFREY

Chapter 34: ....page 353
Care of Comfrey Plant: Overview & How to Propagate
General Care of Plant
General Cultivation of Prickly Comfrey
Collecting Pods and Seeds
   Drying Comfrey Pods and Seeds
   Number Dry Seeds Per Gram
   Storing Seeds
Propagation by Seeds
   Starting Seeds Indoors or in Greenhouse / Cold Frame
   Direct Sowing Seeds in Fall
   Direct Sowing Seeds in Spring
   Cold, Moist Stratification

Propagation by Stem Cuttings
Propagation by Root Cuttings
   Overview
      Rapid Increase Area of Land Planted
      Plant Hormones and Comfrey Root Cuttings
   Methods of Digging Root Cuttings
      Definitions and Overview
      Garden Method of Digging Roots
      Farm Method of Digging Roots
   Size of Root Cuttings (Offsets)
   Care and Storing of Root Cuttings

Chapter 35: ....page 374
Planting, Soil, Fertilization, Water, Disease
Location of Planting
Timing of Digging and Planting
   Best Time to Dig Up Comfrey Roots
   Best to Plant in Spring or Early Fall
   Latest Planting is Early Fall
   Dig Roots Starting Third Year
Temperature
   Frost Resistance of Leaves
   Cold Hardiness of Entire Plant
   Heat Sensitivity of Comfrey
Photoperiod (Day Length)
Soil Type and Soil Depth for Root Growth
   Soil Depth Needed by Comfrey
How to Plant Comfrey Roots
   Overview
   Spacing Between Roots, and Number Per Acre
   Depth of Planting Roots; Vertical/Horizontal
Fertilization of Comfrey Plant
   Manure
   Fishmeal or Dried Blood Fertilizers
   Urine as Fertilizer
   Lime Fertilization
   Chemical Fertilizers
   Mineral Fertilization
   Rock Powders and Seaweed
   Timing of Fertilization
Water
   How Much Water Needed (Right Amount vs Too Much)
   Very Drought Tolerant
Weeds
   Mechanical and Chemical Control of Weeds
   Fertilization Helps Comfrey Outgrow Weeds
Pests and Disease
   Rust (Fungus)
   Other Fungus Disease
   Pyrethrum Eelworm (Nematode)
   Bacterial Disease
   Insect Damage
   Virus Disease
   Unwanted Deer, Rabbits & Groundhogs Eat Comfrey
   Gopher Problems

Chapter 36: ....page 410
Productivity and Farm Economics of Comfrey
Rate of Growth
   First, Second, Third and Fourth Years
Production Per Plant or Small Area (Yield)
   Pounds Fresh Comfrey Leaves Per Plant or Small Area
   Pounds Dried Comfrey Leaves & Roots Per Part of Acre
Production Per Acre (Yield)
   Tons or Pounds of Fresh Comfrey Leaves Per Acre
   Tons or Pounds of Dried Comfrey Leaves Per Acre
   Tons or Pounds of Fresh Comfrey Roots Per Acre
   Tons or Pounds of Dried Comfrey Roots Per Acre
Comparing Comfrey to Other Crops
   Comfrey versus Alfalfa: Yield and Land Use
   Comfrey versus Amaranth
   Comfrey versus Timothy Yield
Comfrey and Farm Economics
   International Comfrey Production and Marketing
   Usefulness Varies with Number of Tons Per Acre
   Usefulness Varies with Type of Environment
   Best Farm Comfrey is Solid Stem & Magenta Flowers
Intercropping (Growing Comfrey with Other Crops)
Pasture and Comfrey
Age of Comfrey Plot and Yield
Older Comfrey Plots and How to Restore Them

Chapter 37: ....page 430
Harvesting Comfrey Leaves
(For Drying Comfrey leaves, see 'Comfrey as Hay', Chapter 21)
Overview of Harvesting Leaves
Tools Used for Harvesting Comfrey Leaves
Timing and Frequency of Harvests
   Timing According to First Cut of Season
   Timing According to Height
   Timing According to Month
      First Cut for Comfrey Planted that Spring
      First Cut for Established Comfrey Plants
      Cutting in June or July
      Last Cut in September or October
   Cutting Prebloom versus Full Bloom
   Number of Cuts Per Year (or Days Between Cuts)

Chapter 38: ....page 436
How to Get Rid of Unwanted Comfrey
Invasiveness of Comfrey
   S. Asperum, S. Officinale, S. Tuberosum
   Symphytum x Uplandicum Invasiveness
Plow Field in Winter, Then Grow Grass
Let Pigs Dig Up and Eat Comfrey Roots
Let Chickens or Other Poultry Kill the Comfrey
Kill by Sheet Mulching or Covering with Plastic
Dig, Dig, Dig
Use Salt to Kill Comfrey Roots
Rot Roots with Too Much Water
Use Herbicides to Get Rid of Comfrey

Preface for Volume 1 and 2

Some references to 'Sections' and 'Subsections' are in Volume 1.

Complex, Fascinating Comfrey
When I started writing this book I had no idea there was so much information available about Comfrey.

Volume 1
The botany is complex and still has areas of uncertainty. Botanical research continues with updates about taxonomy, chromosomes and species.

The history section is an overview of medical references, taxonomical investigations, livestock feeding, garden use, and international trade. Comfrey has been used medicinally from at least 400 BC in Greece.

Comfrey can be used to help your garden grow. It is beneficial for compost, fertilization, and permaculture. The nutritional value is examined.

Volume 2
This volume shows how Comfrey is used as a forage plant for livestock. Also how to successfully make comfrey hay and silage.

The healing and potentially dangerous properties of Comfrey are investigated from both the scientific/research and the everyday/practical points of view.

Then how to take care of your Comfrey plants including propagation, fertilization and harvesting. Yields are given to help determine your expected productivity. Last is how to get rid of unwanted Comfrey plants.

Definitions of Words
I have included many definitions of words for those who are not botanical, historical, geographical, medical or livestock experts. As much as possible I have given the complete words for abbreviations various authors have used.

In writing this book on a broad range of topics about Comfrey, it is surprising and sometimes frustrating how many abbreviations are used by expert writers. In many cases they are writing for other experts, so they assume everyone knows what the abbreviations mean.

At times I had to do a lot of searching to find out what they meant. At one point I could not figure out the meaning of a particular botanical phrase, so I contacted one of the well-known plant organizations about the meaning. I was put in contact with a botanical classification expert with a Ph.D. who works in the Department of Botany at Harvard University, Cambridge, Massachusetts.

I have included definitions of some words that may be unfamiliar to people who do not speak English as their first language. This book is relevant to people all over the world. This is also useful for clarifying the meaning of what the author is trying to say for those who are not experts in that field.

Many authors wrote research reports about a particular country or region and expected the only readers to be people in those areas. So towns, cities and other geographic regions are frequently mentioned with no country or other identifier for the reader to know where they are located. I defined as many as I could. Some ancient writings are harder to figure out.

In certain situations I include more detailed descriptions about the region because it is helpful to understand the botany about Comfrey both from a taxonomical and cultivation point of view.

If an author uses measurements in metric, then I added the English/Imperial measurements, and vice-versa. This way people in all countries understand the length, weight or distance.

Finding Botanical Literature
It was a fun mystery to find some of the more obscure literature. In some cases, I contacted the authors who were able to send me pdfs of their research reports.

Some articles are written in languages other than English with no translations available. I tried online translators but in most cases they were not very helpful. For people who are able to translate reports into English, I am hoping you will send me translations. My goal is to write an update to this book with the new information I receive.

For all research articles in this book, some may not be relevant to the average person. However, I like to include them because there are researchers and other scientists who are interested in them. These scientists can then find these articles and read the entire report.

For an unknown reason, sometimes the search function for a particular term works better online at the original source such as www.biodiversitylibrary.org or https://www.hathitrust.org, than it does on a pdf that I downloaded from the same site.

Fair Use and Registration Marks
Fair use is a doctrine in United States copyright law that allows limited use of copyrighted material without requiring permission from the rights holders. It provides for the legal, non-licensed citation or incorporation of copyrighted material in another author's work.

"Title 17 United States Copyright Act, Section 107. Chapter 1 - Subject Matter and Scope of Copyright.
Limitations on exclusive rights: Fair use.
Notwithstanding the provisions of sections 106 and 106A, the fair use of a copyrighted work, including such use by reproduction in copies or phonorecords or by any other means specified by that section, for purposes such as criticism, comment, news reporting, teaching, scholarship, or research, is not an infringement of copyright."

All the contents of this book are under the protection of Section 107 of the United States Copyright Law. Under the fair use rule of copyright law, an author may make limited use of another author's work for purposes of criticism and comment such as quoting or excerpting a work in review or criticism for purposes of illustration or comment. There are also provisions for education, teaching and research, among others.

As much as possible I put '®' after a company name or registered product if they used it on their website or literature. The '®' symbol means it is federally registered with the United States Patent and Trademark Office, Alexandria, Virginia, either on the Principal Register or Supplemental Register. Please consider all company names and products as potentially trademarked or registered.

Legal Disclaimers
This book is for entertainment purposes only. This is especially true for the medical and medicinal information provided. Statements about health and healing for humans, pets or livestock are not a substitute for professional medical advice, diagnosis and treatment. If you, your family, pets or livestock have health issues, please contact your physician, veterinarian or other health care provider.

This book includes ancient, folk, traditional and various historical accounts of healing that may not be accurate or safe to follow. And there is great controversy about the safety of Comfrey for internal use. I have quoted all of the information I could find about that, but it is up to you and your health care provider to decide what is right. I can not make that decision, especially because each person or animal has unique health needs.

This book is the result of extensive investigation from many sources. Some quotes are from scientific research or years of practical experience. Others are opinions, folklore or grapevine/gossip. The information may not be accurate or appropriate for you to use in your situation. Therefore, I assume no responsibility if you decide to follow the quoted articles, reports, books, etc. Every quote is referenced with author, publication and year. It is up to you to do your own research to see if it applies to you.

Agricultural information may or may not be useful for your environment or type of farm/garden. Check with your local horticultural, agricultural or livestock department/agency to see what is right for you. Go online to various websites, forums and blogs to see what is good agriculture and ranching practice to achieve your goals. Some jurisdictions consider Comfrey to be invasive and unwanted.

Researchers, farmers and historical records sometimes have very different ideas about the safety of Comfrey as food for animals. Contact your veterinarian or other livestock specialist to see what is right for you and your animals.

I mention many companies, organizations, products, resources, services, journals, books, etc. Except for 'Nantahala Farm and Garden' and 'Western North Carolina Farm and Garden Calendar', I do not have a connection with any of them. I am not recommending or endorsing them. I am providing references so you can do your own research about them.

PART D

LIVESTOCK, PETS AND COMFREY

Chapter 20

Comfrey as Food and Medicine for Livestock

See subsection 'Timing and Frequency of Harvests' in section 'Harvesting Comfrey Leaves' (Chapter 37).

<u>Comfrey Leaves and Livestock Nutrition</u>

*"**Comfrey is the fastest vegetable protein builder we have found so far, and it is in theory the best cheap source as food for stock to replace concentrates.**"*
-Comfrey: Fodder, Food and Remedy by Lawrence D. Hills. New York: Rizzoli Universe Books: **1976**, page 83.
    (Feed Concentrates are mixes of ingredients that have a high amount of cereal grains. Mixes of only grains are also called concentrates.)

*"If you want fat cows, you plant rye grass and clover, but you will still get cows with worms and cows with deficiency symptoms. **Newman Turner recommends a whole lot of perennial herbs that should be put along hedgerows**. We know, for instance, that when cows can just browse along hazel tips and buds, the butterfat content in milk increases, and the cows are healthier. **Cows will always eat some Comfrey, though it is not a preferred plant.**"*
-'Introduction to Permaculture: Permaculture Design Course Series, Pamphlets 1 to 14' by Bill Mollison at 'The Rural Education Center', Wilton, New Hamshire, **1981**; published by Yankee Permaculture, Sparr, Florida.
    (F. Newman Turner wrote 'Fertility Farming', 'Fertility Pastures', and 'Herdsmanship', all classics of practical organic husbandry.)   (A hedgerow is a line of closely planted shrubs and trees, usually bordering a road or field.)

"So you have three strategies, then, with these cattle and deer and goats and sheep.
One is, instead of just relying on annual pastures, have areas of permanent, high-mineral mobilization herbs throughout all your pastures: dandelion, chicory, Comfrey.
Have evergreens, standing, high-nutrition tree crop within forage range that the cattle will coppice.
Have high-sugar summer pods that will carry cattle through the semi-arid seasons. This group is critically important to range capacity. Also, you must have a winter high carbohydrate source: large nuts and acorns."
-'Introduction to Permaculture: Permaculture Design Course Series, Pamphlets 1 to 14' by Bill Mollison at 'The Rural Education Center', Wilton, New Hamshire, **1981**; published by Yankee Permaculture, Sparr, Florida.
    (Coppice means to periodically cut back a tree or shrub to ground level to stimulate growth.)

*"One can justifiably claim in summary that **there is no fodder plant so versatile as Comfrey in its application to the farm need for a farm-grown food supply.**"*
-Comfrey: Nature's Healing Herb & Health Food by Andrew Hughes. Japan: Sanyusha Publishing Co., Ltd, **1992**, page 178.

*"**Newman Turner in 'Fertility Farming':***
'The thing we must do is to get back into our dairy pastures as many herbs as possible to assist the health of the cattle grazing the leys and to benefit the topsoil in a way any amount of chemical dressing can never do.
Hedgerows should contain Comfrey, *garlic, raspberry, hazelnut, docks and cleavers, etc.'"*
-'Benefits of Biodiverse Forage' by Jerry Brunetti, Acres USA: A Voice for Eco-Agriculture, Austin, Texas, Volume 33, No. 10, October **2003**. (Ley farming is growing grass or legumes in rotation with grain or tilled crops as a soil conservation measure.)

*"**The importance of pasture plant species diversity, creating healing fields on farm:***
*This can be achieved relatively easily by growing wide ranges of suitable plants on farm and allowing animals free access to them by, for example, **incorporating them into pasture or planting them along cattle track fence lines.***
I saw this on a number of farms in Australia and New Zealand with Comfrey, *flax, alder, elder, willows, chicory, yarrow, burnet, alsike, lucerne, sweet clovers and similar adding both aesthetic appeal as well as wildlife habitat, shelter and shade*

advantages besides."
-'Improving Pasture Quality for Animal and Ultimately Human Nutrition and Health' by Ben Mead, N.Sch., Nuffield Farming Scholarships Trust Report, Cornwall, England, date unknown but around **2006**, page 23.

"Comfrey (Symphvtum oficinale), a prolific producer of fodder, is very rich in protein, calcium, potassium, magnesium, and trace minerals including copper, zinc, manganese, and boron. *It contains moderate levels of sulfur and phosphorus and has impressive TDN (Total Digestible Nutrients), ADF (Acid Detergent Fiber), and NDF (Neutral Detergent Fiber) values. Rich in vitamins B-complex, C, E, and beta-carotene,* ***this plant is especially attractive as forage in its early growth stages."***
-Alternative Health Practices for Livestock edited by Thomas F. Morris and Michael T. Keilty. Ames, Iowa: Blackwell Publishing, **2006**, page 98.
(For TDN, ADF, NDF, see subsection 'Digestibility of Comfrey Leaf' in section 'Nutritional Value of Comfrey' {Vol. 1, Chapter 19}.)

*"**Forage Crops in Russia:***
Highly productive forage crops and plants for improving natural hayfields and pastures are very important for stock-raising and farming. ***In Russia only 25 main forage species are widely used and botanical research is aimed at the selection of special forage species and cultivars.***
In a number of Russian botanic gardens research is being undertaken on Heracleum, Lathyrus, Vicia, Polygonum, Astragalus, Oxytropis, Hedysarum and others.
The search for new forage crops in the north of Russia is one of the main tasks of the 'Botanic Garden of the Komi Institute of Biology' in Syktyvkar. *This garden has been successful in introducing Heracleum sosnovskii, Polygonum weirichii, Rhaphanus sativus,* ***Symphytum spp.*** *and many others.*
In the Novosibirsk region some crops selected by the 'Central Siberian Botanic Garden' for a high-protein content have been are being tested on farms. These include about 50 species of Bromus, Medicago, Melilotus, Astragalus and others."
-'The Role of Russian Botanic Gardens in the Study and Development of Economic Plants' by Yu. N. Gorbunov; Botanic Gardens Conservation International (BGCI), www.bgci.org, Volume 3, Number 7, February **2001**. BGCI provides a global voice for all botanic gardens, championing and celebrating their work. We are the world's largest plant conservation network, open to all.

*"**Forage production: Many species of legumes and other herbs make excellent forage, either in conjunction (at same time) with grasses, or when grown alone or in mixtures.***
Red clover is a particularly good choice for silage production with high yields and feed quality, with up to 20% crude protein, without the need for nitrogenous fertilisers. It is highly palatable to stock and results in higher milk yields and liveweight gains than ryegrass alone silage.
When grown as a forage crop it can be combined with Italian and hybrid ryegrasses, providing them with all the nitrogen they need for maximum production. Vetches and other winter-hardy legumes can also be used as a companion crop in a mix with cereals like maize (corn), sorghum or oats to be cut for silage. ***Other species to consider, either alone or in mixtures, include ribwort plantain, chicory, lucerne (alfalfa), sainfoin, phacelia and Comfrey."***
-'Organic Pastoral Resource Guide' by the 'Organic Dairy and Pastoral Group' and 'BioGro New Zealand'. Wellington, New Zealand: 'Soil & Health Association of New Zealand' and 'BioDynamic Farming and Gardening Association in New Zealand', 144 pages, **2010**, page 67.

*"**I mineralise all of the animals on the farm*** *so that their manures are of a very high quality which all goes through the pastures, through the animal pens. They also get composted and turned into highly oxygenated compost tea that mineralises the landscape.*
We do it by spreading minerals throughout the whole farm so that the herbs we give them as forage (pigeon pea, acacia, Comfrey, mugwort, arrowroot, and bamboo leaves), whole mixtures of forage that we cut for the animals, are already mineralised. *They have a very high quality health, they have very vital energy. -Geoff Lawton, England"*
-Permaculture Pioneers: Stories from the New Frontier edited by Kerry Dawborn and Caroline Smith. Hepburn, Victoria, Australia: Melliodora Publishing, holmgren.com.au, **2011**.

-'Comfrey Miracle or Mirage?' by R.H. Hart, Crop Soils, Volume 29, No. 1, pages 12-14, 1976.
(I could not find this report. If you have a copy, could you please send it to me.)

Feeding Comfrey to Livestock in the 1800s

*"In **1873** Kinard Edwards from Leicestershire, England wrote 'An Acre of Land and How to Make the Most of It':*
He gives clear directions for the cultivation of the Comfrey plant as a 'soiling crop' for stall-fed cows, *stressing the importance of keeping the land clean and well manured, and with adequate cutting.*
*He gives yields of 10 to 15 tons per acre per cut; totals of 50 tons the first year, 80 the second and after that 100 to 120 tons. For the other crops mentioned in his work, yields such as 40 tons an acre for mangolds are normal on good land today. In his 'The Amateur's and Cottager's Cow', date about 1875, he quotes **a neighbor of his who has fed three cows and two horses off the produce of a quarter acre of Comfrey**, and gives a normal annual yield of 80 to 90 tons."*
-Russian Comfrey: A Hundred Tons an Acre of Stock or Compost for Farm, Garden or Smallholding by Lawrence D. Hills. London England: Faber and Faber, Limited, 1953, page 32.

(A 'soiling crop' is cut green and fed to livestock immediately without further curing or processing.)
(A mangold is a beet with a large yellow or yellow-orange root that is grown primarily as cattle feed. It is also called Beta vulgaris vulgaris, mangel-wurzel, mangold-wurzel, field beet and fodder beet. Fodder is a type of animal feed for domesticated livestock. It refers to food given to animals including plants cut and carried to them, rather than what they forage or graze for themselves.)

Henry Doubleday in 1876 fed 160 pounds of Comfrey leaves a day to his 3 cows and 1 pony.

"In **1878** in 'Kew Annual Report' of the Royal Botanic Gardens (London, England) about **Russian Comfrey: 'In England it has been found very useful for winter fodder**, as it forms large tufts of root leaves which start into growth early in the year and bear several cuttings. It is greedily eaten by animals which refuse ordinary Comfrey, the habit and appearance of which is not very dissimilar.' "
-Russian Comfrey: A Hundred Tons an Acre of Stock or Compost for Farm, Garden or Smallholding by Lawrence D. Hills. London England: Faber and Faber, Limited, 1953, page 29.

From "Catalogue of Messrs. Suttons" of Reading, Berkshire, England in **1878**: "**It (Russian Comfrey) is very hardy and gives an early cutting**, supplies a constant succession of green food, and, when once planted, is permanent.
It is much relished by all kinds of stock, either cut up and mixed with chaff or separately. For milch (milk) cows it is most valuable and it is much relished by domestic poultry."
-Russian Comfrey: A Hundred Tons an Acre of Stock or Compost for Farm, Garden or Smallholding by Lawrence D. Hills. London England: Faber and Faber, Limited, 1953, page 33.
(Chaff is the husks of corn, wheat or other seed/grain separated by winnowing and threshing.)

In **1882** "The Reverend F. Gilbert White (of Ashburton, Devonshire County, England) **considered Comfrey splendid fed chaffed with hay, and producing very high-quality butter**. It did not taint or cause hove (bloat)."
-Russian Comfrey: A Hundred Tons an Acre of Stock or Compost for Farm, Garden or Smallholding by Lawrence D. Hills. London England: Faber and Faber, Limited, 1953, page 35.
(A goat's rumen {one of its stomachs} ferments food and produces carbon dioxide and methane gas that needs to be expelled. If they cannot expel the gas, pressure builds up and the goat bloats.)

"The Reverend E. Highton, of Bude, Cornwall, England, writes:
'**The Prickly Comfrey crop furnished green food from May to November for my horse,** which only required a very little chaffed hay and straw besides, with half the corn usually allowed, i.e., three quarts (2.8 liters) instead of six quarts (5.6 liters) a day.
The cows eat the Prickly Comfrey after being cut like other green food, but next season I intend to put it through the chaff-cutter mixed with hay and straw, and then add crushed oats and bran, as I think this will be a more economical mode of utilisation.' "
-'The Journal of the Royal Agricultural Society of England, Second Series, Volume 18', London, England, **1882**. Includes article: 'On Green or Fodder Crops Not Commonly Grown, which Have Been Found Serviceable for Stock-Feeding' by Joseph Darby.

"**After flowering, Prickly Comfrey (Symphytum asperrimum) well deserves its name of prickly,** and if then offered to cows or animals for the first time, they will, in nine cases out of ten, refuse to have anything to do with it, merely because it hurts their mouths. **If required as green forage, cut it just as it shows for flower, when the prickles are scarcely perceptible, and it is exceedingly succulent.**
If by chance some plants have gone ahead, flowered, and got a little too tough and prickly, cut them down with your sickle, and leave them on the ground for a couple of days. The prickles become flaccid (limp), and cattle eat the food readily."
-'Ensilage: A System for the Preservation in Pits of Forage Plants and Grasses, Independent of Weather: A Collection of Facts and Statistics on the Cheapest Mode of Providing Winter Food for Dairy Cattle, Sheep, Horses, Etc.', by Thomas Christy, F.L.S. (Fellow of Linneaen Society), London, England, **1883**. Caucasian Prickly Comfrey, pages 57-63.
(A sickle is a tool for cutting grain, grass, etc. that consists of a curved, hooklike blade mounted in a short handle.)

"**Compound Ensilage and Soiling Crops: I am in the retail milk business; keep 25 cows;** sell at 7 cents per quart the year round. My ration is found on page 1 of 'Feeding Animals', i.e., 60 pounds (27.2 kg) corn ensilage, 5 pounds (2.2 kg) hay, 2 pounds (0.9kg) linseed meal, and 4 pounds (1.8 kg) bran. This gives me good results. **I don't know what to plant, or how much. Answer from editors: We think you would do well to try the Prickly Comfrey (Symphytum asperrimum),** as there have been a sufficient number of favorable reports to show that cattle are fond of it, and one analysis made by the New-York Experiment Station which we will show as compared with turnips, mangolds, carrots and fodder corn. Percent of water, ash, albuminoids, carbohydrates, fat, and nutritive ratio:

	Water	Ash	Albuminoids	Carbohydrates	Fats	Nutritive Ratio
Prickly Comfrey	84.36	2.45	1.63	5.77	0.27	3.9
Turnips	92.40	0.72	1.25	5.40	0.23	4.6
Mangold Beets	86.00	0.80	1.15	9.80	0.15	8.5
Carrots	90.00	1.00	1.10	6.60	0.29	6.6
Maize Fodder	80.98	1.13	1.19	10.87	0.31	9.9

This shows Prickly Comfrey in a very favorable light. *It will be seen that it is superior to fodder corn, superior to the roots (turnips, mangolds, carrots) in albuminoids, that **it has a narrower nutritive ratio, and therefore would be an improvement to fodder corn if ensilaged or fed with it**, and as it is a perennial plant, when once well established in a deep, rich soil, would produce very large green crops for years without much labor except harvesting.*
***Answer from reader:** I have been raising Prickly Comfrey for at least a dozen years, and consider it a forage plant of much value, especially for calves. Pigs eat it readily and seem to do well on it. -W.P.W., Wilmington, Ohio."*
-'The Cultivator and Country Gentleman: Devoted to the Practice and Science of Agriculture and Horticulture at Large' edited by Luther Tucker and Son, and John J. Thomas, Volume 56, Albany, New York, **1891**. Comfrey pages 47, 916.

Feeding Comfrey to Livestock in the 1900s

*"**In recent years Comfrey (Symphytum asperrimum and several other species, S. officinale, etc.) have been cultivated as soiling crops, especially for swine.** These are Oriental herbs, some of them naturalized in the Middle States (of Germany). They are very succulent, contain little crude fiber, and are rich in protein. The crude protein content is 9.5 percent, starch value 37.5 percent. **They are usually greedily eaten by swine (pigs), although cattle care less for them on account of their rough hairy character. Comfrey is also used as feed for horses, goats, rabbits, geese and ducks.***
The plant grows well under a variety of indifferent conditions, is not easily exterminated (killed) and yields well."
-'Scientific Feeding of the Domestic Animals' by Dr. Martin Klimmer, Ph.D., D.M.V., Professor of Hygiene and Feeding, Director of the Hygienic Institute of the Veterinary College of Dresden, Germany, **1923**, page 56.
   (A 'soiling crop' is cut green and fed to livestock immediately without further curing or processing.)

*"**Cultivated Comfrey often referred to as Russian or Quaker Comfrey, was introduced in Uganda (east Africa) in 1950s.** Comfrey could become a major source of fresh herbage for most of the year in Uganda where there is plenty of sunshine and rainfall. Comfrey is also the only known land plant that extracts vitamin B12 from the soil (Odwongo et al, 1987).*
The upland soils of Kabanyolo are classified as ferralitic soils. The area has a moist tropical climate with mean maximum temperatures varying from 28.5 C (83.3 F) in January to 26.0 C (78.8 F) in July and minimum temperatures ranging from 17.4 C (63.3 F) in April to 15.9 C (60.6 F) in July.
Mean annual rainfall is about 1300 mm (51 inches), with peaks in April and November and two periods of low rainfall in January and July when the monthly mean drops to 60 mm (2.36 inches). Comfrey was harvested at weekly intervals from eight to 12 weeks after planting. Harvesting was stopped at 12 weeks of growth because the plants had become very coarse. They were also heavily infested with bacterial blight caused by Cladosporium sp. (species).
*There was a steady increase in Crude Protein (CP) yield during both growth and regrowth, but CP content stayed constant at about 16%. **This is comparable to Crude Protein values of good to medium quality forage legumes but higher than for most pasture grasses** (Soneji et al, 1971; Sabiiti and Mugerwa, 1990).*
***During all harvests, the Comfrey plants had higher contents of both macro-and micro-minerals** than are usually observed in many fodder crops (Kabaija and Smith, 1988)."*
-'Potential of Russian Comfrey as an Animal Feedstuff in Uganda' by Bareeba, Odwongo and Mugerwa, Department of Animal Science, Faculty of Agriculture and Forestry, Proceedings of the First Uganda Pasture Network Workshop, Makerere University, Kampala, Uganda, Africa, 1987.
   (Ferralitic soils include red-yellow, red, laterite, ferralitic-gley, and other soils. They are widespread in south and central America, central Africa, south and southeast Asia, and northern Australia.)
   (Cladosporium is a genus of fungi of some of the most common indoor and outdoor molds. They produce olive-green to brown or black colonies.)

*"Comfrey was once widely cultivated as a fodder plant. **Sheep and cows seek it in the ditchside and eat it greedily. English gypsies say that a handful of Comfrey roots,** cleaned and fed daily to horses and cows in the spring, will rid them of all winter torpor and put them into fine bloom in one week."*
-The Complete Herbal Handbook for Farm and Stable by Juliette de Bairacli Levy. London, England: Faber and Faber, **1952**.
   (Torpor is a state of decreased physiological activity in an animal. Bloom is a glow or flush that shows health; it
   is the glossy, healthy appearance of the coat of an animal.)

In **1953**: *"My Comfrey grows simply huge, I cut about a barrow-load most days, both goats love it, and the pigs!"*
-Russian Comfrey: A Hundred Tons an Acre of Stock or Compost for Farm, Garden or Smallholding by Lawrence D. Hills. London England: Faber and Faber, Limited, **1953**, page 51.

*"Another Australian, who valued 'Henry Doubleday Research Association' research, was Foster Savage from Nambour (Queensland, Australia) in **1954**. Wilted Comfrey was fed to his animals in large amounts. Why did he allow it to wilt? He told me that **animals could eat much more , each day, when it was wilted.**"*
-How Can I Use Herbs in My Daily Life: Over 500 Herbs, Spices and Edible Plants: An Australian Practical Guide to Growing Culinary and Medicinal Herbs' by Isabell Shipard. David Stewart Publisher: 2003.

*"**In recent years scattered interest has been shown in Quaker (Russian) Comfrey.** The plant is said to be higher yielding than Prickly Comfrey.*

'Cooperative Extension Experiment Station' plantings have been made at several points in the United States. The cost of planting, limitations with respect to grazing spaced plants, and its unsuitability for hay or silage would seem to limit its use to soiling. **The 'United States Department of Agriculture' cannot recommend Quaker Comfrey as a forage plant on the basis of available information.** This recommendation may be modified, however, as more information becomes available from existing experiments."
-'Quaker Comfrey' by Agronomy Notes, AP 198, University of California Cooperative Extension, 2 pages, August **1957**. Based on reports of the 'Crops Research Division', Agricultural Research Service, United States Department of Agriculture.
    (You can make silage with Comfrey. See subsection 'Comfrey as Silage' in section 'Comfrey Meal, Pellets, Hay and Silage' {Chapter 21}.)

"In 1968 fifty percent of green leaf Comfrey was fed to pigs, and broilers were fed 50% of their total food needs on dried leaf mixed in their feed. Ishii Animal Husbandry Institute at Ohito Machi (Japan) took up this raising method, and they report that stockfeed costs are reduced to nearly one third. Compared with other stockfood, labor is saved in cultivation management."
-Comfrey: Nature's Healing Herb and Health Food by Andrew Hughes. Japan: Sanyusha Publishing Co., Ltd, 1992, page 28.

Advantages of Russian Comfrey for Livestock

Thomas Christy in **1876** in "Forage Crops" wrote: *"The Solid Stem Variety (Russian Comfrey) is far more palatable and in every way they have proved superior to anything grown in this country as Prickly Comfrey.*
On good land, Russian Comfrey is fully equal to giving 120 to 150 tons an acre from plants placed one yard (3 feet = 0.9 meter) apart each way.
David Wemyss of St. Andrews (Fife County, Scotland), in 'The Cultivation of Prickly Comfrey and Its Uses as a Fodder Plant' (Transactions of the Highland and Agricultural Society of Scotland, **1881**) refers to the **'even better Solid-Stemmed Variety' (Russian Comfrey)'** and gives a yield of 100 to 120 tons of good fodder per acre per annum.
He recommends feeding wilted until the stock are used to it, and cutting when half-grown; the plant should never be allowed to become hard and woody. He also stresses the richness of the butter."
-Russian Comfrey: A Hundred Tons an Acre of Stock or Compost for Farm, Garden or Smallholding by Lawrence D. Hills. London England: Faber and Faber, Limited, 1953, pages 31, 35.

Livestock and Safety of Comfrey
For nitrates/nitrites in Comfrey, see sub-subsection 'Problems with Comfrey and Pigs (Nitrate)' in subsection 'Pigs and Hogs' in section 'Livestock / Pet Species and Comfrey' (Chapter 22).

"The use of Comfrey leaves as stock feed appears to present a large margin of safety because of the apparent detoxification of alkaloids in the rumen (one of the stomachs) of cattle, sheep and goats, and the apparent resistance of rabbits and horses to pyrrolizidine alkaloids.
Pigs are kept for such a relatively short time that there is no chance of them assimilating the large quantity of Comfrey required to give them a toxic dose of alkaloids.
It appears probable that the absence of veterinary records of alkaloid poisoning is because such poisoning has never happened, due to the large quantities of Comfrey required, rather than because stockmen or veterinary surgeons have failed to correlate alkaloid poisoning with the consumption of Comfrey, as has been naively suggested in some quarters.
I cannot believe that an animal could consume possibly forty times its own bodyweight of Comfrey without the owner or stockman being aware of the fact."
-'The Safety of Comfrey: Report from the Henry Doubleday Research Association' by J.A. Pembery B.Sc., Special Advisor, Research Chemist, Aldgate Press: London, England, 20 pages, **1983**, page 14.

"Large species differences exist in susceptibility to Pyrrolizidine Alakaloid (PA) toxicosis.
Small herbivores such as sheep, goats, rabbits, guinea pigs and other herbivorous laboratory animals are highly resistant to PA toxicity, associated with a low rate of hepatic production of reactive metabolites (pyrroles) and (or) a high rate of activity of detoxifying enzymes. **The PA and their metabolites are secreted in the milk of lactating animals, but this probably does not represent a significant human health hazard."**
-'Toxicity and Metabolism of Pyrrolizidine Alkaloids' by P.R. Cheeke, Oregon State University, Corvallis, Oregon; Journal of Animal Science: American Society of Animal Science, Volume 66, No. 9, pages 2343-2350, October **1988**.

"The susceptibility of animals species to the toxic effects of pyrrolizidine alkaloids varies widely, exhibiting differences of more than 2 orders of magnitude.
Swine (pigs) are reported to be exceptionally sensitive, followed by chicken (less sensitive by a factor 5), cattle and horses (factor 14), rats and mice (factor 50 and 150) and sheep (factor 200, meaning not very sensitive).
Ruminal microflora is assumed to detoxify pyrrolizidine alkaloids to a significant extent.
However, numerous outbreaks of poisonings partly connected to extensive morbidity and mortality have been recorded in animals grazing pastures contaminated with pyrrolizidine alkaloids containing plants (not necessarily Comfrey)."
-'European Medicines Agency- Symphyti Radix: Committee for Veterinary Medicinal Products' by European Agency for the

Evaluation of Medicinal Products, Veterinary Medicines Evaluation Unit, London, England, EMEA/MRL/649/99-FINAL, 5 pages, August **1999**. (Rumen is one of the stomachs of ruminants such as cattle, sheep and goats.)

*"**Excretion of Pyrrolizidine Alkaloids:** There are no data on the pharmacokinetics of Pyrrolizidine Alkaloids in man; in experimental animal studies, however, both the parent alkaloids and their nontoxic metabolites are largely excreted within 24 hours after exposure. As much as 80% of the pyrrolizidine ring is excreted unchanged in the urine.*
In ruminants, blood levels of Pyrrolizidine Alkaloids decrease rapidly to zero within 1 hour, with only 10% to 15% of the total dose appearing in urine after 48 hours. This is because bacteria in the rumen detoxify diesters, but not cyclic esters, to 1-methylene derivatives.
Pyrrolizidine alkaloids excreted in urine can be auto-oxidized to N-oxides. Ignorance of this fact has given rise to misinterpretation of metabolic pathways."
-'Pyrrolizidine Poisoning: A Neglected Area in Human Toxicology' by Michael J. Stewart and Vanessa Steenkamp, University of the Witwatersrand Medical School, Johannesburg, South Africa; Therapeutic Drug Monitoring Journal, New York, Volume 23, No. 6, pages 698-708, **2001**.
    (Pharmacokinetics is the branch of pharmacology about the movement of drugs in the body.)
    (There are three main categories of Pyrrolizidine Alkaloids: macrocyclic diesters, branched diesters, and monoesters. Pyrrolizidine Alkaloids can be free bases or N-oxides.)

*"**Types of Feed Ingredients for Animals: Includes 'nutraceuticals' and unapproved substances (herbal and botanical products and dietary supplements such as Comfrey, kava).***
According to Association of American Feed Control Officials (AAFCO), many undefined or unrecognized ingredients are being marketed for use in animal feed, or they are being marketed for unapproved purposes (AAFCO EMSI Working Group, undated). (EMSI = Enforcement Strategy for Marketed Ingredients).
    Recently (2002) AAFCO has recommended nationwide 'enforcement events' to crack down on two botanical products, the herb Comfrey*, *and kava, a plant in pepper family."*
-'Feed for Food-Producing Animals: A Resource on Ingredients, the Industry and Regulation' by Lisa Y. Lefferts, Margaret Kucharski, Shawn McKenzie and Polly Walker, Center for a Livable Future, Johns Hopkins Bloomberg School of Public Health, www.jhsph.edu/clf, Baltimore, Maryland, Fall **2007**.
(* -'Target Ingredient Announcement: AAFCO Recommended Enforcement Event', Association of American Feed Control Officials, Inc., Champaign, Illinois, 2003. The Association is a group of local, state and federal agencies that regulate the sale and distribution of animal feeds and animal drug remedies. **As of 2019 there is no information on the AAFCO site about 'Comfrey' or 'Symphytum'**.)
    (A nutraceutical is a pharmaceutical that is a standardized nutrient. In the United States they are regulated as dietary supplements and food additives by the Food and Drug Administration rules in 'Federal Food, Drug, and Cosmetic Act'.)

*"**The following release was posted 2002 on American Association of Feed Control Officials (AAFCO) web site: During the AAFCO Annual Meeting in August 2002, the Enforcement Strategy for Marketed Ingredients (ESMI) Working Group announced the target ingredient recommended for a future regulatory enforcement event. The target ingredient, Comfrey, has been shown to be a health and safety concern for animals and humans,*** *prompting regulatory action by the United States, Canada, and Germany. Comfrey does not meet any of the recognized criteria for use as an animal feed ingredient or animal feed.*
Comfrey was identified by the ESMI Working Group based on the following published scientific information provided by the United States Food and Drug Administration (FDA), Center for Veterinary Medicine with references provided at the end of this document:
    1. *Leaf and root of Comfrey (Symphytum officinale) plant have been used in supplements. Supplement use has been orally for ulcers, diarrhea, cough, bronchitis and rheumatism, or topically for treatment of inflammation, arthritis, wounds, and bruises. This supplement has been banned in Germany and Canada due to safety concerns.*
    2. *Comfrey has shown to be hepatotoxic (relating to or causing injury to the liver) in both humans and rats. The toxic compounds found are pyrrolizidine alkaloids (8 have been identified), which include lasiocarpine and symphtine. Highest content of these substances were found in products containing bulk Comfrey root or leaf.*
    3. *Pyrrolizidine alkaloids have been associated with lung and liver cancer. The primary liver ailment associated with comfrey consumption is veno-occlusive disease (form of Budd-Chiari syndrome), a non-thrombotic destruction of small hepatic veins leading to cirrhosis and eventual liver failure. Consumption of 85 mg pyrrolizidine alkaloids (15mg/kg body weight/day) for 6 months resulted in venocclusive disease in a 49-year-old woman.*
    4. *Signs of liver toxicity have been seen in rats consuming low doses of Comfrey (50mg/kg body weight three times a week for three weeks), which included loss of sinusoidal lining cells, sinusoids filled with cellular debris (hepatocyte organelles and red blood cells), and narrowing of terminal hepatic venules.*
    5. *Pyrrolizidine alkaloids extracted from Comfrey were shown to damage chromosomes when administered to human lymphocytes at concentrations of 140 mg/mL and 1400 mg/mL. At these concentrations, sister-chromatid exchange and chromosome aberrations were observed. On July 6, 2001, the FDA advised dietary supplement manufacturers that comfrey should not be used in dietary supplements due to safety concerns.*
AAFCO recommended to feed control officials that an enforcement event occur to clarify the regulatory status of ingredients sold for consumption by animals as animal feed, including livestock feed and pet food. All feed ingredients must be shown to be safe and efficacious for their intended use prior to distribution.

Feed manufacturers have several methods for meeting this requirement that are summarized in an ingredient fact sheet entitled, 'Options Available for Acceptance of a Proposed Feed Ingredient,' available on the AAFCO website (www.AAFCO.org). Feed ingredients not recognized or acceptable for their intended purpose may be subject to regulatory action by the feed control official and the FDA."
-'Safety of Comfrey' by The Horse Staff, www.thehorse.com, The Horse: Your Guide To Equine Health Care is a monthly equine publication providing the latest news and information on the health, care, welfare, and management of all equids, Lexington, Kentucky, (January 17 2003).

(This article did not include the references that AAFCO referred to. However, they are cited and reviewed in the section 'Alkaloids in Comfrey: Scientific Studies and Various Perspectives' {Chapter 30}.)

"Based on these data it is not possible to establish a concrete lethal dose of particular PAs (Pyrrolizidine Alkaloids) for each ruminant species, the dosages and especially the duration of administration are too variable.
On the other hand these experiments give clear evidence for the fact that indeed **the susceptibility for PA intoxication is different in individual species: in case of ruminants it is decreasing from cattle (which seem to show the highest sensitivity) to goats and sheep which are the most resistant animals** discussed in this overview presumably due in large part to their different rumen microflora."
-'Toxicity of Pyrrolizidine Alkaloids to Humans and Ruminants' by Helmut Wiedenfeld and John Edgar, Phytochemistry Reviews: Proceedings of the Phytochemical Society of Europe, Volume 10, Issue 1, pages 137-151, March **2011**.

Tethered Stock and Rotational Grazing

"Tethered beasts will eat Comfrey readily, but the tether should be moved before they crop it within three inches (7.6 cm) of the crown (plant at soil level), and if stock is on the land in winter great damage will be done.
Folding Comfrey on an 'on and off system' with electric fencing (rotational grazing) would be an ideal method of cashing its yield, possibly with poultry to follow and clear up the waste plus harrowing."
-Russian Comfrey: A Hundred Tons an Acre of Stock or Compost for Farm, Garden or Smallholding by Lawrence D. Hills. London England: Faber and Faber, Limited, 1953, page 74.

(Rotational grazing is moving livestock between pastures/paddocks on a regular basis. Land is divided using permanent or temporary fencing.) (A harrow is dragged behind a tractor to break up dirt clods, remove weeds, and cover seed.)

Livestock Food in Early Spring

"The produce (Prickly Comfrey) may, in times of great scarcity for other cattle food, be found of peculiar (unusual) value, as the Reverend F. Gilbert White, of Lensdon Vicarage, Ashburton (Devonshire, England), experienced it to be last spring, the subjoined (added) statement from whom was made public on April 30th, 1881:

'**This spring,** in spite of my having neglected to give the **Prickly Comfrey** plants good mulching of stable litter to protect them from the frost, they again **came up early and strong**, and gave promise of a most abundant yield.
**The end of the mangolds (fodder beets), which we were pulping to mix with our chaff, loomed close at hand; the coming in of vetches or any other green food looked a long way off in the cold, dry, easterly winds.
Then a new idea struck us. We brought two large hand-cart loads of the luxuriant young Comfrey leaves up into the hay-loft. We laid them in the trough of the chaff-cutter, with about equal quantities of hay and of forage** (i.e., of oats cut before the corn is ripe enough to be threshed out), and we cut up all together; then we left the large heap to welter (ferment) for two or three days upon the floor.
The result is that we now have an abundant supply of sweet moist food, which every cow, calf, and horse eats with the utmost greediness, literally licking out their mangers (troughs) lest a fragment of the leaf should escape them; and this, with the aid of a little decorticated (de-husked) cotton-cake, will render us independent of all extraneous (outside) aid till summer is fully come. I may observe that the cook, who knew nothing about the cows' change of food, at once remarked upon the **improvement of the butter, both in colour and in texture.**'

The plant is very much liked by Lord Moreton's farm manager, Mr. John Watts, of Whitfield Farm, Gloucestershire (England), who writes to me as follows:

'**One of the chief advantages derived from this valuable forage plant (Prickly Comfrey) is, that it is so much more forward (more grown) in spring than any other green food,** and therefore greedily relished by any kind of livestock. **The practice here is to cut it up with straw into chaff, which, when roots become scarce and straw dry and husky,** enables us to keep the store stock in the yards a week or two longer, **to give the grass a good start (before grazing it).** At other periods, through the summer, it is used for weaning calves, pigs, horses, or any other kind of stock that may happen to be about the homestead.' "
-'The Journal of the Royal Agricultural Society of England, Second Series, Volume 18', London, England, **1882**. Includes article: 'On Green or Fodder Crops Not Commonly Grown, which Have Been Found Serviceable for Stock-Feeding' by Joseph Darby.

(Chaff is the husks of corn, wheat or other seed/grain separated by winnowing and threshing.)

"A mild winter will start Comfrey early, and it gets 'rich' more rapidly than grasses because it is growing from stored nutriment in the root."

-Russian Comfrey: A Hundred Tons an Acre of Stock or Compost for Farm, Garden or Smallholding by Lawrence D. Hills. London England: Faber and Faber, Limited, **1953**, page 86.

Prickly Comfrey as a Forage Crop, 1871

"**Prickly Comfrey:** *We cannot do better than give you, in answer to your questions, the following remarks by a correspondent of the 'Field Magazine':*
> '*I procured in the year around 1832, 25 sets of the Prickly Comfrey. I remember the advertisement of it prophesying that it would soon become as commonly sold in London, England for green fodder as Vetches. This was not a prediction likely to be fulfilled, for* **the leaf is extremely marcescent when plucked,** *and does not give the idea of a food calculated to make a green store of. It would probably ferment and become unwholesome in this state.*
> **But, as taken immediately from the plant grown on the spot, I know of nothing at all comparable to it to set before cattle for nourishment.** *If they do not take to it at first, they soon come round to it. Where cows are kept,* **there is nothing probably that can be so advantageously placed before them when they are tied up for milking.**' "

-'The Gardeners Chronicle and Agricultural Gazette for 1871' published in Covent Garden, WC, London, England, 1269 pages, January 7 to December 30 **1871**. Article: 'Notices to Correspondents', January 21 1871, pages 59-60.
(Vetches are the genus Vicia of flowering plants in the legume family. It is cultivated for forage and silage.)
(Marcescent means the leaves wither but remain attached to the stem.)

Prickly Comfrey as a Forage Crop, 1876

"**If a quarter of an acre of ground is set apart in a suburban garden, sufficient forage can be obtained to maintain a cow and a horse all the year round,** and keep the family in good, rich, wholesome milk and butter of superior quality to what is obtained from the grocer.
Gentlemen who have grown Comfrey for twenty years in England, say they get better and richer milk from their cows, and a higher price for their butter.
A writer in the 'The Field' of December 17th, 1870, states: 'It is nearly forty years since my attention was directed to **Prickly Comfrey; its excellence as a food for cattle,** its specialties the chief of which is its rapid and continuous reproduction.
I procured (obtained) in the year I spoke of (about 1832) twenty-five (root) sets of the Prickly Comfrey taken immediately from the plant grown on the spot.
I know nothing at all comparable to it to set before cattle for nourishment. The weight of each leaf will be somewhat more than an ounce (28 grams).
When the whole crown (top roots with leaf buds) was taken at once, it reproduced in the short space of ten or twelve days in the summer time, and in a fortnight (2 weeks) or little more in late spring and autumn.
Mr. W. Doubleday 'thinks **the Symphytum asperrimum (Prickly Comfrey) will prove a highly valuable plant for India,** quite a mine of wealth for the farmers who keep cattle. Its extreme luxuriance, and its power of withstanding a long drought, mark it out for such a climate'.
If this opinion is worth anything, and I submit that it is, **what a valuable plant Comfrey will be in New South Wales (Australia).** A few acres on a large run will keep cattle and sheep from starving, but if grown in sufficient quantities, will prevent them losing condition."
-'Prickly Comfrey: Its History, Cultivation, Extraordinary Production, and Uses: A Letter Addressed to His Excellency Sir Hercules Robinson, President of the Agricultural Society of New South Wales' by Arthur T. Holroyd, Sydney, Australia, **1876**, pages 5-6.
('The Field' is the world's oldest country and field sports magazine, having been published continuously in England since 1853.)

"**Symphytum Asperrimum: Experiencing the great want in this state (Virginia) of some good fodder plant, I was induced to place myself in communication with a gentleman in Europe who has devoted many years to this subject,** and who has, after considerable expense of time and money, succeeded in introducing the above named plant into pretty general notice in France, Germany and England, where it is daily becoming more appreciated and valued.
The common name of this plant is Prickly Comfrey. There are several varieties of Symphytum indigenous to Great Britain, but none of them are of much value for feeding purposes; but the true Symphytum asperrimum is a native of the Caucasus, and **produces enormous crops of the best fodder, which, both in the green and dry state, is greedily eaten by horses, cows, sheep, pigs and poultry.**"
-'The Southern Planter and Farmer, Devoted to Agriculture, Horticulture and Rural Affairs' by L.R. Dickinson, Richmond, Virginia, No. 1, January **1876**. Article: 'Symphytum Asperrimum' by C.E. Ashburner, an Englishman in Henrico County, Virginia, page 55.

Prickly Comfrey As a Forage Crop, 1877

"**New Forage Plant: A good deal of attention is being attracted in agricultural circles by the introduction into Victoria**

(Australia) of a new forage plant from France and England. This is the 'Prickly Comfrey' or 'Solid Stem Comfrey', a plant which is described as yielding as much as sixty tons to the acre of a fodder to which horses and cattle take alike, and which has great fattening properties.
The specialty of the plant is its success in withstanding droughts, and the accounts received from the Southern States of America are very satisfactory in this respect. In Virginia last year no rain fell for over two months, the thermometer was seldom below 95 F degrees (35 C) in the shade, and frequently over 105 F degrees (40 C), and the Comfrey grew well on dry ground, furnishing abundant green fodder, while every other crop was parched up. Mr. W.R. Church, of Camberwell, has imported the plant, and is now showing it, so that agriculturists can judge for themselves."
-'The Mercury' Newspaper, Hobart, Tasmania, Australia, April 18 **1877**, page 3. 'The Mercury' is still printing newspapers.

Prickly Comfrey As a Forage Crop, 1877 (This section is from one author, Thomas Christy.)

"**There is no plant yet discovered which possesses such milk-producing qualities as the Symphytum asperrimum (Prickly Comfrey).** As soon as the roots have taken hold of the ground, cattle and sheep may at once be turned on, but it will be as well to move them off when the foliage is eaten down, and if flower stems remain they should be cut down. This plan ensures a continuous crop." -page vi

"This plant both blooms and seeds, but the seed will not germinate in this country, except in a very small per-centage, and the plants raised by artificial means are not found to possess the properties of the true Caucasian variety." -page 2

"A little over 30 years ago, the English Agricultural press contained some notes upon the species (Symphytum asperrimum) as one deserving of cultivation for its cattle feeding properties. **No sooner did the cows (especially milch/milk cows), horses, sheep and pigs begin to understand it, than they eat it most greedily, and our report upon it was that while all creatures seemed to thrive upon the Comfrey, yet in no instance could we find the slightest evidence of any evil effects.**
The crop was enormous, and this too upon land of very medium quality; but we have this year been trying its growth on light sandy soil, and can report that all through the season of drought the thick deep roots of the Comfrey have drawn up the moisture which rises hygrometrically in our sand bed, and the result has been a succession of green leaves when surface plants were an utter failure." -pages 4, 5

"Lately, the Messrs. Christy & Co. have **introduced a form of Prickly Comfrey, which is distinguished from the ordinary one (Symphytum officinale) by a solid stem,** and they claim for it an advantage over the common form, inasmuch as it contains a greater amount of mucilage and yields a larger produce." -page 6

"Now it should be stated that in all probability our earlier experiments were conducted with the hollow-stemmed sort (of Symphytum asperrimum), but **we are now growing the solid-stemmed kind, which we have no doubt will prove superior, not only as yielding a heavier crop,** but from the mucilaginous nature, especially of the stem." -page 7

"Professor J. A. Barral, Perpetual Secretary of the 'Central Society of Agriculture in France' ('Societe Centrale de l'Agriculture en France') writes in the 'Journal de l'Agriculture', of 7th October, 1876 (No. 391):
The quantity of green forage obtained per hectare, in a soil deep and moist enough for vegetation, appears to rise as high as 300,000 kilogs (kilograms) (661,386 pounds) per hectare, and perhaps more, **it is certainly a plant to be tried, particularly for pickling purposes in tanks or pits, (especially from the standpoint of silage) in order to provide green food for cattle throughout the winter.** -page 9
    (A hectare is a square 100 meters on each side, also 100 ares {10,000 m2} or 1 square hectometre {hm2}.
    An acre is 0.405 hectare, and one hectare is 2.47 acres.)

" 'The Solid-Stem Symphytum Asperrimum or Caucasian Prickly Comfrey' by Henry Doubleday, 1876:
'It is a very ornamental plant, and **bears a profusion of blossom very rich in honey. This imparts a sweetness to the plant which animals are very fond of, and when dried the stem and leaves have a most agreeable perfume.**
This Solid Stem Prickly Comfrey branches out much more than the old varieties, so that though the plants be placed three feet (0.9 meter) from each other they soon cover the ground. The crowns and stem cuttings blossom the first year, but as a rule the root cuttings generally not till the second year, but the latter produce an amazing quantity of leaves forming a head of great beauty from the graceful wavy curve of its long leaves. **No other cultivated plant produces the enormous weight per acre of such valuable food.**' " -page 11

" 'The Solid-Stem Symphytum Asperrimum or Caucasian Prickly Comfrey' by Henry Doubleday, 1876:
'There is another great advantage which this Solid Stem variety has over the Hollow Stem in propagating. **The stem has the nature of a succulent root, and if pieces of the stem are cut with two eyes, and planted in the ground and kept moist, they strike (grow roots) very quickly** and flower during the summer, throwing down large solid roots, producing six or eight crowns the same year. A plant of a year's growth will produce in root and stems fifty-fold, if taken up and replanted by dividing its crowns and roots, besides the cuttings taken from each flower stem during the year.' " -page 13

" 'The Solid-Stem Symphytum Asperrimum or Caucasian Prickly Comfrey' by Henry Doubleday, 1876:

*An impression prevails in some quarters that animals will not eat it- this is quite a mistake; it may have arisen from persons confounding (confusing) some of the wild species with the cultivated ones.
It sometimes happens that the first time an animal is offered Comfrey it does not take the new food as readily as that it has always been fed upon, but if persevered with for a few days* **Comfrey never fails to become the favourite feed of any animal, and on none do they show so quickly such an improvement in condition.** *Lambs and pigs were extremely fond of it green, and did well. They are also very fond of it in the dry state."* -pages 13, 14

" *'The Solid-Stem Symphytum Asperrimum or Caucasian Prickly Comfrey' by Henry Doubleday, 1876:
I think the Symphytum Asperrimum will prove a highly valuable plant for India—quite a mine of wealth for the farmers who keep cattle.* **Its extreme luxuriance and its power of withstanding a long drought** *mark it out for such a climate.
I had a proof of the fondness of goats for the plant, having been obliged to part with a favorite one because I could not keep it away from my Prickly Comfrey, though it had plenty of other good food."* -page 14

*"A writer in 'The Field' of December 17th, 1870, states ('The Field' magazine in the United Kingdom is the world's oldest country and field sports magazine in publication since 1853.):
'The leaves begin to show themselves in April, and have lasted to the end of October.
As to the place where Comfrey is to be planted, any soil seems to suit it. I have two spots where it is planted: one, an open sunny place on almost a sand; the other, a stiff clay against a north wall; and they seem to do nearly equally well in both.
A root will divide into six or seven cuttings, and may be planted at any time of the year. I have been in the habit of giving horses that were kept in the stables in the summer a daily allowance of 8 pounds (3.6 kg) with very good effect. They become very fond of it. It keeps them cool, and is a diuretic.' "* -page 17

"Extract from the 'The Times of India', April 16, 1876: Plant Symphytum Asperrimum, or Caucasian Prickly Comfrey, throughout the length and breadth of India; it may have much to do with the progress of India, probably not less than Cotton has in times past.
Being deep-rooted, it is independent of the weather and climate, *it is only cut down by severe frost; being perennial, once planted all expense is at an end, and the crowns or plants increase in size each year.*
Any soil but chalk suits it. *It ought to be tried in every Collectorate, in every Government farm and garden from Cape Comorin (India, south) to the Khybur (now Khyber Pass, Pakistan, north) from Sylhet (now Jalalabad, Bangladesh, east) to Kurrachee (now Karachi, Pakistan, west).
The roots can be shipped in boxes as ordinary merchandise in the hold (where cargo is stored in a ship)."* -page 22
  ('The Times of India' is an Indian English-language daily newspaper that is the fourth-largest newspaper in India and the largest selling English-language daily in the world. It is the oldest English-language newspaper in India with its first edition published in 1838.)
  (Chalky soil has high alkalinity and pH due to the lime and calcium carbonate. It is stony, shallow and well draining that holds little water.)

In the 'Ceylon Times Overland Summary', December 7th, 1876:
Prickly Comfrey is a plant specially adapted to the soil and climate of this country. *There should be no reason why it might not be grown successfully on most of the poor waste lands of the low country. With this plant at their disposal the natives of the poorer districts of the maritime (by the sea) provinces, might turn to good account many an unproductive field, many a barren waste."* -page 23   (Ceylon is now Sri Lanka in south Asia.)

-'**Forage Plants and Their Economic Conservation by the New System of Ensilage**: Part I: Caucasian Prickly Comfrey' by **Thomas Christy**, Jun., F.L.S. (Fellow of the Linnean Society), Christy & Co., London, England, **1877**.

Prickly Comfrey As a Forage Crop, 1880

*"The Comfrey is much relished by cattle, particularly by cows, and affords a large crop of herbage, which would prove profitable on moist rich land as a fodder plant. Being perennial it lasts several years without renewal.
In moist situations it grows very freely, dying down in winter, but if cut before the flowers quite expand, many crops might be gathered each season. Owing to the successful growth of the plant being only possible on rich and moist land, our agriculturists have paid it but little attention.*
Professor Buckman, however, tells us that a species of Symphytum (Symphytum asperrimum, Prickly Comfrey) greatly resembling our Common Comfrey was introduced into England from the Caucasus in 1811, *and was recommended at the time chiefly as an ornamental plant in shrubberies and large gardens.
Recently, however, it has been tried as a green 'soiling plant', with very good results.*
From an analysis of the plant made by Professor Voelcker, it appears to be equal to some of our more important green food crops; and certainly, if we take into consideration the quantity of its produce, there are few plants capable of yielding so much of green food as the Comfrey.
Dr. Voelcker says 'that the amount of flesh-forming substances is considerable. The juice of this plant contains much green and mucilage, and but little sugar.' "
-'**English Botany (Sowerby's)**; or Coloured Figures of British Plants: Volume 7' by John T. Boswell and John Edward Sowerby, London, England, **1880**, page 116.

Prickly Comfrey as a Forage Crop, 1881

"*An analysis of the Prickly Comfrey, made by Voelcher, gave the following results. The plant taken in green state and also dried at 212 F (100 C).*

	Natural State	Dry
Water	90.66%	
Nitrogenous or flesh forming matters (containing nitrogen)	2.72	29.12
Non-nitrogenous or heat & fat producing compounds	4.78	51.28
Mineral matter (ash)	1.84	19.60
	100.00%	100.00%

Like many other forage plants, Comfrey, in a green state, contains a very large percent of water.
It is extremely rich in mucilage and contains the essentials for forming flesh and milk in abundance with little increase of oil or butter. *Hence when milk is too rich in oil or butter to be wholesome, as often occurs in the Jersey cow, feeding Comfrey would reduce the excess of oily matters in the milk and increase the quantity of the latter.*
I find it excellent for nursing sows (female pigs), and indeed for all kinds of hogs, cattle, horses, mules, etc. Pigs for slaughtering, fed freely with Comfrey and sweet potatoes with a little corn or meal, furnish probably the most deliciously nice pork that can be produced by any feed whatever.
Most animals require some training to learn the value of this plant and to acquire a relish for it. *But when they do eat, and it requires but little effort to induce them to try it,* ***they become excessively fond of it.***
I have never found it necessary, as practised by some persons, to confine animals to make them eat it, nor to mix it with other food. If hungry, the animal may be more ready to taste; but even when full, they have been induced to test it.
With a hand full of leaves, go among your animals; if one will take a leaf, others from jealousy will come and try one. If this does not succeed, have with your Comfrey, some other green plant that will be readily taken, only enough to give one animal a mouthful. Others seeing the one eating will come and try the Comfrey. A few trials will get up a lively competition for what they soon regard as a choice luxury. They may at first nibble daintily; but soon eat greedily.
Poultry also may be taught to eat it with great benefit."
-'The Farmers Book of Grasses and Other Forage Plants for the Southern United States' by D.L. Phares, A.M. (Master of Arts), M.D., Professor of Biology, A&M College of Mississippi, Starkville, Misssissippi, **1881**, pages 22-24.

"*Prickly Comfrey:*
C.N. Merriwether, Esquire, of **St. Bethlehem, Montgomery county, Tennessee** *writes me that he has been experimenting on a small scale with Comfrey since 1876, but that he has been 'deterred from doing anything with it by* ***the impossibility of keeping it in a dry state and by the distaste which stock have for it.*** *These difficulties may be overcome by ensilage and habit.'*
The scientific name of the plant is Symphytum asperrimum. Several of the species of this genus have escaped from cultivation and are now common road side weeds, of coarse and unattractive appearance."
-'Experimental Work of the Agricultural Department of the University of Tennessee (Report of the Experimental and Other Work of the School of Agriculture and Botany of the University of Tennessee for Session of 1880-1881)' by John M. McBryde, Professor of Agriculture, Horticulture and Botany, Knoxville, Tennessee, **1881**.

Prickly Comfrey as a Forage Crop, 1882

"*Never surely were there so many conflicting opinions as to the value of any crop for agricultural purposes as of Prickly Comfrey.*
> *Mr. H. Doubleday, of Coggeshall, Essex (southeast England), has written of it, that 'four or five cuts of 20 tons each to the acre may be taken when the plants are fully established, and they will last for twenty years if the ground is kept clean and occasionally stirred'.*
> *But against this the following statement of Mr. T.R. Hulbert, of North Cerney (Gloucestershire county, southwest England), may be set in opposition: 'I believe Prickly Comfrey to be quite a delusion. It wants very good land, plenty of dung and attention, and no stock will eat it if they can get plenty of other food'.*

In seasons of scarcity, at other periods of the year besides winter, *it is an immense advantage to be able to fall back upon thousand-headed kale, purple-sprouting broccoli, green rye, or the early cuttings of sainfoin, lucerne (alfalfa), and Italian rye-grass in spring; on cabbages, vetches,* ***Prickly Comfrey*** *and repeated cuttings of lucerne and sainfoin during summer;* ***and as autumn advances,*** *on the breadths of thousand-headed kale, cabbages, and kohl-rabi which were spring-sown, together with green maize and* ***still further supplies from the plots or fields giving lucerne or Prickly Comfrey.***
Even Prickly Comfrey is not new, as it has the reputation of having been ***introduced into England as early as 1790, and grown in Russia and Circassia (north Caucasus) long before;*** *but, so far as Great Britain is concerned, its use was very much confined to medicinal purposes until within the past twenty years.*"
-'The Journal of the Royal Agricultural Society of England, Second Series, Volume 18', London, England, **1882**. Includes article: 'On Green or Fodder Crops Not Commonly Grown, which Have Been Found Serviceable for Stock-Feeding' by Joseph Darby, pages 114, 141. (Sainfoin, also known as onobrychis, is a perennial in the legume family.)

"**Prickly Comfrey (Symphytum asperrimum) is relished by all animals, horses being frequently kept through the winter on it, without hay, and only a moderate allowance of oats, which has the effect of giving them fine coats of hair. When given to milk cows, it not only produces a full yield of milk, but the butter made possesses the quality and natural rich colour of the best summer butter.** Young cattle also thrive well upon it, with or without turnips.
Sheep do not eat it so freely, unless being compelled by much snow being on the ground.
Should cattle at first be found not to take it readily, their dislike will probably be caused by roughness of leaves; for this reason they should be withered before being given. But this does not often happen, as **all animals, as a rule, eat Comfrey readily.**"
-'Transactions of the Highland and Agricultural Society of Scotland, Fourth Series, Volume 14', Edinburgh, Scotland, **1882**. With the article: 'The Cultivation of Prickly Comfrey, and Its Use as a Fodder Plant' by David W. Wemyss, Newton Bank, St. Andrews, Fife Council, Scotland, pages 264-267.

Prickly Comfrey As a Forage Crop, 1896

"*Prickly Comfrey, Symphytum officinale:*
Cut and fed in the green state the leaves and stalks of this plant are a valuable food for stock. In some portions of Europe it has been profitably grown for many years. It is also much prized by a few growers in New York and in other parts of America. But it is little known anywhere in this country.
    In Michigan I had some experience with it as a plant for sandy lands. Its good qualities led me to desire to try it in Florida. **The plant does very well on light lands, and grows in great luxuriance on fertile soils.**
At first animals do not like the leaves, but a little 'education' makes them extravagantly fond of them. The plant is not to be pastured. The leaves should be cut and fed in the yard or stable.
In the spring of 1895 I procured a few roots from Michigan, and a few from New York, which were **planted at Lake City, Florida.** All the roots that came to hand in good condition began at once to grow, and grew well throughout the summer. It continued to grow well throughout the fall. **As the cooler weather of December and January came on it grew less luxuriantly, but has been green all winter.** The outer leaves died, but the center leaves have kept fresh, and now, in the middle of February, are growing well. Work with this plant will go on another year. Roots can be procured of A.S. Cotton, Clifton Springs, New York."
-'Cassava, the Velvet Bean, Prickly Comfrey, Taro, Tropical Yam, Canaigre, Alfalfa, Flat Pea, Sachaline' by O. Clute, M.S., LL.D., Station Director, Lake City, Florida; Florida Agricultural Experiment Station, Bulletin No. 35, Jacksonville, Florida, pages 345-346, April **1896**.
    (Prickly Comfrey is not Symphytum officinale which is Common Comfrey.)
    (Lake City is the county seat of Columbia County, Florida. The county is in northern Florida on the border with Georgia. It is 60 miles west of Jacksonville.)

Prickly Comfrey As a Forage Crop, 1910

"**Prickly Comfrey (Symphytum asperrimum Donn)** is a perennial herbaceous plant, a native of the Caucasus region of Europe, which was introduced into England as early as 1801.
Apparently it was first grown in the United States near Richmond, Virginia in 1876. The only recorded importation of this plant by the 'Department of Agriculture' was made in February, 1899, from France.
In 1830 it attracted attention in England as a forage plant, and from that date until 1876 or later some little interest was exhibited in its dissemination (spreading) by agriculturists.
Thomas Christy, Jr., of London (England), was especially prominent in it's advertisement and published a lengthy article descriptive of its value as a food for hogs, sheep, and dairy cows, especially as a soiling crop and in the form of ensilage.
Although Prickly Comfrey was grown rather extensively years ago in Europe and to some extent in the United States, it has never attained any considerable importance in either country as a forage crop.
At the present time it is probably grown more generally in Germany than in any other country, and its success there may be ascribed to the intensive methods of cultivation employed on small farms, a practice which calls for some crop that will respond with yields to heavy applications of fertilizer. Only under such methods can the yields of forage mentioned in reports from Germany be expected. None of the government experiment stations in European countries have seen fit to commend Prickly Comfrey in their reports far as noted.
Propagation of the Plants:
    Although the Prickly Comfrey produces large crops of seed, only a small percentage of this seed will germinate, so **it is generally found more practicable to plant new fields by division of the roots than by seed.** These root cuttings may be either crown cuttings or transverse (cross) sections of the lower taproots, and they may be quite small, so that the number secured from a single plant will be considerable even in one year.
    They are planted in rows, usually about 3 feet (0.9 meters) apart each way, or 3 feet between the rows and 1 1/2 to 2 feet (0.45-0.60 meters) apart in the row, the distance depending on the fertility of soil.
    When first planted, young sets (roots) must be given frequent and thorough cultivation. Sets made from crown cuttings usually bloom the first year, while those made from pieces of taproots will not bloom as a rule until the second season.
Culture of the Crop:
    Cultivation should be continued after each cutting until the plants are large enough to shade the ground, and a light top-dressing of manure should be given the field after each cutting if large and frequent crops are to be expected.

The cuttings should always be made before seed has formed. From three to six crops a year may be obtained, and in good soil a field is supposed to last from fifteen to twenty years without replanting, returning a yield of 10 to 40 tons of green feed per acre each year.

Value of Prickly Comfrey as a Soiling Crop for Dairy Cows:

(A 'soiling crop' is cut green and fed to livestock immediately without further curing or processing.)

It is as a soiling crop for dairy cows that Comfrey has proved of most value. Dr. Henry Foster, of Clifton Springs, New York, has been in the past the most enthusiastic advocate (promoter) of Comfrey for this purpose.

Doctor Foster top-dressed his fields with manure after each cutting and cultivated thoroughly. In this way he claimed to have secured a yield of 50 tons per acre in five cuttings. According to his statement **the cows ate it greedily, and no other crop equaled it in producing quantity and quality of milk.**

At the 'New York Agricultural Experiment Station' dairy cows at first refused to eat green Comfrey. Corn meal was then sprinkled over the Comfrey in the manger, but it was knocked off and licked up from the bottom of the feed boxes. As a last resource, salt was scattered over the Comfrey and the animals were thus induced to eat it. They soon became fond of it and afterwards ate it readily without salting.

Crop Yields:

From 14 to 16 tons of green matter per acre are reported by 'New York Agricultural Experiment Station'. Sixteen tons by the Vermont station. 6 1/2 to to 17 1/2 tons by the North Carolina Station, and **33 1/2 tons by the Wisconsin station**.

In dry matter the 'Wisconsin Agricultural Experiment Station' reports a yield of 6,475 pounds (2937 kg) of Comfrey to the acre, compared to 7,987 pounds (3622 kg) of red clover.

'Pennsylvania Agricultural Experiment Station' reports yield per acre of digestible material in Comfrey to be: Fat 30.5 pounds (13.8 kg), Crude Fiber 35.3 pounds (16.0 kg), Nitrogen-Free Extract 623.1 pounds (282.6 kg), Protein 221.0 pounds (100.2 kg), Total Digestible Matter 909.9 pounds (412.7 kg), Total Green Matter 16,500.0 pounds (7484 kg).

Diseases of Prickly Comfrey and Insect Enemies:

Comfrey has been grown at the 'North Carolina Agricultural Experiment Station' since 1899. It grew well but was **injured by both caterpillars and a fungous disease**, which reduced the crop to two or three cuttings each year.

Effect of Prickly Comfrey on the Soil:

The soil in an old Comfrey field is usually left in good condition, owing to frequent cultivations and to the top-dressings of barnyard manure. **The large, fleshy roots of the Comfrey also penetrate to a considerable depth and add humus to the subsoil.**

Assuming 20 tons of green material to the acre as an average crop of Comfrey, there would be removed from the soil 165 pounds (74.8 kg) of nitrogen, 65 pounds (29.4 kg) of phosphoric acid, and 74 pounds (33.5 kg) of potash (potassium).

Conclusion:

From the present knowledge of Prickly Comfrey, it is advisable to experiment with it only on a small scale as a soiling crop. There seems little to justify its extended use in a region where alfalfa or red clover will succeed. **Large yields have not been obtained without heavy applications of fertilizer."**

-**'Prickly Comfrey as a Forage Crop'** by United States Department of Agriculture, Bureau of Plant Industry, Office of Forage Crop Investigations, Washington, DC, Circular No. 47, 10 pages, January 26 **1910**.

"Prickly Comfrey, Symphytum asperrimum: **This plant, occasionally exploited by advertisers, has little merit in comparison with standard forage plants.** When carefully cultivated it gives quite large returns of forage which at first is not relished by cattle. Woll of the Wisconsin Station found that red clover returned 23 percent more dry matter and 25 percent more crude protein than the same area of carefully cultivated Prickly Comfrey."

-**Feeds and Feeding: A Handbook for the Student and Stockman** by W.A. Henry, D.Sc., D.Agr. and F.B. Morrison, B.S. University of Wisconsin, Madison, 10th Edition: **1910**, page 196.

Prickly Comfrey As a Forage Crop, 1911

"**Prickly Comfrey has been grown successfully as a soiling food in Great Britain and other countries of Europe for many years. Some experiments have been made in growing it in the United States, but the reports from these are conflicting.**

These reports agree first in regard to the productiveness of the plants, and second in regard to the ability of the same to grow on light lands not possessed of high fertility. They also agree in speaking of the little relish which live stock manifest for Prickly Comfrey when it is first fed to them.

But they do not agree as to its value for soiling uses. When fed to live stock at the Ontario, Canada agricultural college farm, the live stock did not manifest any fondness for it. Some other experiment stations have reported similarly. **It may be that domestic animals may be educated to eat it, so that ultimately they will manifest a fondness for it.** Were it otherwise there would seem to be no good reasons for growing it to the considerable extent to which it is grown in several of countries of Europe.

This Comfrey plant is not likely to be grown as a soiling food, at least to any great extent, on the arable soils of the northern and central United States, where other and better soiling plants are or may be grown so numerously.

It may be different however in the southern United States where cultivated grasses of the better yielding varieties grow but shyly.

A plant that has rendered service in providing soiling food even in England where soiling foods grow in such variety, is at least well worthy of a fair trial in all those sections of the United States which are possessed of fair adaptation for producing it."

-'Soiling Crops and the Silo: How to Cultivate and Harvest the Crops; How to Build and Fill the Silo; and How to Use Silage' book by Thomas Shaw, Professor of Animal Husbandry, University of Minnesota; published in New York by Orange Judd Company, **1911**, pages 226-230.   (A 'soiling crop' is cut green and fed to livestock immediately without further curing or processing.)

Prickly Comfrey As a Forage Crop, 1976

"Prickly Comfrey, which is very similar to Quaker Comfrey, even considered to be identical by some botanists, received considerable publicity as a forage plant around the turn of the century (late 1800s- early 1900s). It was studied by the 'United States Department of Agriculture' and several state experiment stations, with the general conclusion that it did not have a place as a forage plant in the United States (USDA Circular AP 198). **Prickly Comfrey produced high green matter yields under favorable conditions, but the quality and dry weight yields were inferior to those of common forage plants such as alfalfa and red clover.**"
-'Comfrey as a Forage' by Lester R. Vough, Extension-Research Agronomist, Oregon State University, Corvallis, Agronomic Crop Science Report, August 1976, page 2. (I was unable to get 'USDA Circular AP 198'. If anyone has a copy, could you send it.)
(I found statements in this report that I know are not true. For instance, the quote above of: 'even considered to be identical by some botanists'. Prickly Comfrey {Symphytum asperriumum} and Quaker Comfrey {Symphytum peregrinum} are not identical.)

"According to Richard H. Hart, United States Department of Agriculture Agronomist, **Comfrey has consistently been inferior to the more common forage crops both in yield and in nutritional value** (Richard H. Hart, 'Comfrey Yields and Forage Value', United States Department of Agriculture, Agricultural Research Service Plant Physiology Institute, Beltsville, Maryland, CA-NE-2, December 1972, 21 pages)."
-'Comfrey as a Forage' by Lester R. Vough, Extension-Research Agronomist, Oregon State University, Corvallis, Agronomic Crop Science Report, August 1976, page 2.
(I was unable to get 'Comfrey Yields and Forage Value'. If anyone has it, could you send it to me. The type of Comfrey in Hart's report is not mentioned by Mr. Vough except results from an English study included 8 'strains' of Comfrey.)

Prickly Comfrey As a Forage Crop, 1978

"Symphytum asperum, as a fodder crop, yielded 46.2 tonnes/hectare. Uncut plants, which started to bloom in late May, flowered for 32 days; sugar production was 242 kg/hectare. If the plant was left to flower after the first cut, the flowering period lasted 38 days, with a sugar production of 198 kg/hectare.
The crop is recommended both for silage and for bee forage."
-'Is Prickly Comfrey Worth Cultivating?' by J. Straigis and V. Marciulionis, Lithuanian Research Institute for Agriculture; Lietuvos Zemdirbystes Mokslinio Tyrimo Instituto Darbai (Works of the Lithuanian Research Institute for Agricultural Research), Volume 22, pages 103-107, 1978. (I was not able to get this report. If you have a copy, could you please send it to me.)

Comfrey as Livestock Medicine
See sections 'Medicinal Comfrey Overview' and 'Making and Using Comfrey Medicine' (Chapter 27).

Overview

"**Prickly Comfrey (Symphytum asperrimum) is a diuretic, cooling, and strengthening, and its good effect in increasing and improving the supply of milk (in livestock) is beyond question.**"
-'New Commercial Plants with Directions How to Grow Them to the Best Advantage, No. 1 (**1878**) and No. 3 (**1880**)' by Thomas Christy, F.L.S. (Fellow of Linnaean Society), London, England. Number 1 includes Caucasian Prickly Comfrey pages 14-16. Number 3 includes Prickly Comfrey pages 11-14. Includes 'Caucasian Prickly Comfrey' and 'Russian Comfrey' advertisements.
(A diuretic increases the output of urine.)

"**Use of Comfrey:** Cure of all internal haemorrhages (bleeding), including uterine. To aid the reunion of wounds and the knitting of fractured and broken bones. Since ancient times Comfrey has been famed as a bone-mender, especially valued in limb fractures. It encourages the natural healing process and speeds up the formation of new bone cells.
Such powers are probably due to the presence of allantoin in Comfrey, for this substance is known to promote granulaton and the making of epithelial cells. **Another important constituent of Comfrey is cholin (choline), also known to be a powerful healing agent.**
Treatment of internal ulcers, also ruptures. The mucilaginous content of leaves and roots makes this herb a valuable remedy for pulmonary (lung) ailments. Also a proved effective remedy for rheumatism and arthritis. Externally the leaves or bruised roots make an important poultice for all types of swellings and fistulas."
-**The Complete Herbal Handbook for Farm and Stable** by Juliette de Bairacli Levy. London, England: Faber and Faber, Inc., **1952**.
(Choline is a water-soluble vitamin-like essential nutrient. It is a constituent of lecithin, which is present in many plants and animal organs.)   (A fistula is an abnormal connection between an organ and another body structure.)

"A good drench is made from one pound (0.45 kg) of Comfrey, boiled slowly in 1 1/2 quarts (1.4 liters) of water, boiling for one hour. When boiled, a handful of ground-ivy plant should be added, and two ounces of Spanish liquorice: brew well. A half-pint (1 cup= 0.23 liter) drench should be given three times daily. For treatment of internal ulcers, molasses should be added, one dessertspoon per half-pint (1 cup), and the drench given mixed with an equal part of raw, skimmed milk.
For bone knitting, *feed two handfuls of well-bruised roots daily. The gypsies bruise their roots by stamping upon them. When bone breaks are set with splints and bandages, pour over the bandages three times daily a strong brew of Comfrey. Also aids binding. The foliage also possesses powerful knitting properties.*
For poultice *take a good handful of leaves, cut finely, mix with bran, place upon a square of flannel, fold, boil in water for five minutes, wring well, and apply hot. Also for swellings, make a pulp form the fresh leaves spread over the area and hold in place with cold, wet bandages."*
-The Complete Herbal Handbook for Farm and Stable by Juliette de Bairacli Levy. London, England: Faber and Faber, Inc., **1952**.
    (A drench is giving an animal a liquid oral medication.)
    (Glechoma hederacea is an aromatic, perennial, evergreen creeper in the mint family Lamiaceae. It is known as ground-ivy, gill-over-the-ground, creeping charlie, alehoof, tunhoof, catsfoot, field balm, and run-away-robin.)
    (Liquorice is a constituent of the root of the shrub that grows wild around the Mediterranean, including Spain. The Spanish root is sweet enough to be eaten raw. However, liquorice root used in candy is from the Eastern Mediterranean plant and has to be mixed with sugar to make it edible.)
    (A dessertspoon is between the size of a tablespoon and a teaspoon.)

"Feeding a profitable fodder crop (Comfrey) which provides free precautionary medicine is good farm economy, even at the expense of the chemical manufacturer."
-Russian Comfrey: A Hundred Tons an Acre of Stock or Compost for Farm, Garden or Smallholding by Lawrence D. Hills. London England: Faber and Faber, Limited, **1953**, page 141.

"(Comfrey Bocking No. 1, 16, 17) was supplied by F. Newman Turner, then of Ferne Farm, Shaftesbury, Dorset County, England. It dates from when his farm was a racing stable. Mr. F. Newman Turner used his bed to provide a kind of 'nature cure' treatment for pedigree cattle, including one Supreme Dairy Show Champion. **Jersey cows fed on concentrates to force their gallonage higher and higher can suffer from sterility, and a diet of Comfrey and water only could solve this problem, perhaps with the combination of allantoin and Vitamin E that Comfrey contains.**
Those who wish to follow Mr. Newman Turner's system of herbal veterinary treatments and organic farming stock rearing methods will find them in 'Herdsmanship' (Faber and Faber, 1954)."
Comfrey: Fodder, Food and Remedy by Lawrence D. Hills. New York: Rizzoli Universe Books: **1976**, pages 52, 74.

"Symphyti radix (root) of Symphytum officinale and preparations thereof are contained in 2 veterinary medicinal products used in phytotherapy.
1. One preparation is a solution containing 5 active ingredients, of which 1 is a tincture prepared with 60% isopropanol from 4 different plants, including 1 gram Symphyti radix.
    This tincture constitutes 53.8% (w/w = weight/weight) of the finished product and additionally 22% (w/w) of a decoct of Symphyti radix (in total extracts from 2.1 grams Symphyti radix in 100 grams of the finished product). **It is used for topical (skin) administration to all food producing species, particularly cows, horses, sheep and goats, to treat strains and ruptures of muscles and tendons, and swollen joints.** *The usual dose is up to 10 ml of the product, depending on size of lesion. The product is only intended for use on intact skin and duration of treatment is 2 to 3 days.*
2. The other preparation is the pulverised (powdered) dried Symphytum roots **(17.5% (w/w) and 3 other ingredients, part of an anti-diarrhoeic powder for cattle, horses, sheep, pigs and poultry.** *The recommended daily oral dose is 30 to 90 grams of the powder for cattle and horses, and 15 to 30 grams for calves, foals, pigs and sheep, given for 3 days."*
-'European Medicines Agency- Symphyti Radix: Committee for Veterinary Medicinal Products' by European Agency for Evaluation of Medicinal Products, Veterinary Medicines Evaluation Unit, London, England, EMEA/MRL/649/99-FINAL, 5 pages, August **1999**.

"Comfrey is available in combination products in veterinary medicine for topical (skin) treatment of muscle strains and ruptures and for oral administration as an antidiarrheal."
-Professional's Handbook of Complementary and Alternative Medicines, 3rd Edition' by Charles H. Fetrow, Pharmd (Doctor of Pharmacy) and Juan R. Avila, Pharmd. Pennsylvania: Lippincott Williams & Wilkins (imprint of Wolters Kluwer), **2003**, page 236.

"Poultices (for animals) can be made easily. Use oatmeal, cornflour, or any other starches you have in your pantry to mix with the herbs to aid drawing and adhesion (so it stays in place).
Linseed (Flax): *Stir ground linseed into sufficient boiling water to make smooth dough. Add olive oil to keep it pliable. Spread on warm cotton or muslin (cotton fabric of plain weave), wrapped up and applied to the area needed. Ensure poultice is not too hot.*
Herbal poultices: Cook Comfrey root until it is soft enough to mash. Add other infused herbs to the mashed Comfrey roots. Apply directly to wound. *Consult a herbal book for appropriate herbs."*
-'Organic Pastoral Resource Guide' by the 'Organic Dairy and Pastoral Group' and 'BioGro New Zealand'. Wellington, New Zealand: 'Soil & Health Association of New Zealand' and 'BioDynamic Farming and Gardening Association in New Zealand', 144 pages, **2010**, page 90.

"Comfrey has also been used in veterinary medicine. ('Medicinal Plants in the Veterinary Medicine' by M.I. Rabinovich,

Russagricultural Publishing House: Moscow, Russia; pages 49-50, published 1981.)"
-'Comfrey: A Clinical Overview' by Christiane Staiger in Germany, Phytotherapy Research, Volume 26, Issue 10, pages 1441-1448, February **2012**. (I was unable to get 'Medicinal Plants'. If you have a copy of the Comfrey pages, could you send it.)

"Comfrey (Symphytum officinale L., Symphyti radix, Symphyti herba):
The farmers prepared ointments and tinctures from Comfrey roots, or used the leaves directly to treat sprains, contusions, swollen joints, or indigestion. *Comparable results were found in Swiss Safiental (Graubunden Canton). In veterinary medicine, Comfrey is used in topical application to treat contusions, sprains, and pulled muscles.*
Allantoine and hydroxycinnamic acid derivatives are considered to be responsible for the analgesic, antiphlogistic (anti-inflammation) properties of Comfrey."
-'Traditional Use of Herbal Remedies in Livestock by Farmers in 3 Swiss Cantons (Aargau, Zurich, Schaffhausen)' by Kathrin Schmid, Silvia Ivemeyer, Christian Vogl, Franziska Klarer, Beat Meier, Matthias Hamburger & Michael Walkenhorst (Austria and Switzerland); Forsch Komplementmed (Research in Complementary Medicine), Switzerland, Vol. 19, No. 3, pages 125-136, **2012**.

"During the years 2011 and 2012, 80 farmers on 64 farms in seven cantons (states) of Northern Switzerland (Aargau, Zürich, Schaffhausen, St. Gallen, Thurgau, Appenzell Innerhoden and Appenzell Ausserrhoden) were interviewed.
More than 500 homemade herbal remedies (HMHR) were documented regarding the used plant species, modes of preparation, dosage, routes of administration, category of use and origin of knowledge.
The 34 selected HMHR contained twelve plant species from 8 families.
The most frequently used plant species were from the family of Asteraceae, and flowers were the most often used plant parts. The processing of herbs included mostly extraction with oil/fat or water, but also maceration (soaking to soften) with ethanol (alcohol) of varying percentage. **In contrast, fresh Comfrey roots were grated and administered directly to skin (compress).**
The formulations were used in 49 different applications for treatment of wounds and other skin alterations in livestock, mainly in cattle. Most of the documented concentrations were in a lower range compared to literature."
-'Traditional Homemade Herbal Remedies Used by Farmers of Northern Switzerland to Treat Skin Alterations and Wounds in Livestock' by M. Disler, K. Schmid, S. Ivemeyer, M. Hamburger and M. Walkenhorst, (Dept of Pharmaceutical Sciences, Switzerland) (Department of Farm Animal Behaviour and Husbandry, Germany) (Research Institute of Organic Agriculture, Switzerland), Planta Medica, 79, PL24, **2013**.

"Comfrey (Symphytum officinale L.): **Preparations made from Comfrey root were externally applied in case of injuries of the musculoskeletal system, skin afflictions, and sores and mastitis (breast infection). Roots were either used freshly crushed, or as extracts prepared with alcohol, oil or fat.**
Leaves were applied directly onto skin to treat injuries of the musculoskeletal system.
Comparable uses have been previously reported from Switzerland and Austria.
Phytoveterinarian literature recommends topical (skin) use of Comfrey preparations for treatment of contusions (bruises), sprains, and pulled muscles. *The documented concentrations in gram dry plant equivalent per 100 grams finished product were lower than the recommended concentration in literature."*
-'Ethnoveterinary Herbal Remedies Used by Farmers in Four North-Eastern Swiss Cantons (St. Gallen, Thurgau, Appenzell Innerrhoden and Appenzell Ausserrhoden)' by M. Disler, S. Ivemeyer, M. Hamburger, C.R. Vogl, A. Tesic, F. Klarer, B. Meier and M. Walkenhorst (Switzerland and Austria), Journal of Ethnobiology and Ethnomedicine. Volume 10, No. 32, 22 pages, **2014**.
(Ethnoveterinary medicine {EVM} considers traditional practices of veterinary medicine legitimate. There are many non-Western traditions of veterinary medicine such as acupuncture and herbal medicine in China, Tibetan veterinary medicine, Ayurveda in India, etc. Phytoveterinary is the use of plants and plant materials to heal animals.)

Broken Bones in Animals

"Broken Legs in Livestock: The farther a leg is broken from the joint, the better; fractures in the hip are seldom cured. **Cure: Take solomon seal root, buck horn and Comfrey roots**, *each a handful, to be boiled in tar for a knitting plaster to be placed next the leg; then splinter it in the proper place, and with your narrow bandage bind it up, let it remain till it is well. It is sometimes necessary to sling the beast, that he may not misplace the leg by standing."*
-'The New-England Farrier, or A Compendium wherein Most of the Diseases to which Horses, Cattle, Sheep and Swine are Incident' by Paul Jewett of Rowley, Massachusetts, published in Newburyport, Massachusetts, **1795**.

"Broken Leg in Poultry: This is quite a common occurrence in poultry, especially when they are confined and are not free ranging. The broken legs of birds mend easily.
Treatment: Make splints, preferably of elder wood. Bandage firmly. **Stand the leg in a cold brew of Comfrey leaves, three times daily.** *The water should come well above the place of breakage.* **Also give one teaspoonful of the Comfrey brew internally morning and night."**
-Herbal Handbook for Farm and Stable by Juliette de Bairacli Levy. London, England: Faber and Faber, Inc., **1952**. First edition that later was called 'The Complete Herbal Handbook for Farm and Stable'.

"Broken Limbs in Sheep: **Sheep, being great wall and crag climbers, frequently suffer from broken limbs, especially forelegs. They can be repaired readily.** *I still consider splint and bandage to be the best method; especially the gipsy splint*

made from hollowed elder bough, and encased around the limb; lined internally with sphagnum moss, and kept in place by bandaging. Stand the limb in a bucketful of cold water thrice (3 times) daily.
Internally to aid the joining of the broken bone-parts, give daily a brew of chopped holly leaves and/or Comfrey leaves, *a handful of herb or herbs to 11 pints (= 5.5 quarts = 5.2 liters) of water.*
In breakages, to allow for the usual early swelling of the limb, splinting should not be used for twenty four hours, only light bandaging. Fast for one to two days to promote speedy healing."
-Herbal Handbook for Farm and Stable by Juliette de Bairacli Levy. London, England: Faber and Faber, Inc., **1952**. Later called 'The Complete Herbal Handbook for Farm and Stable'.

Foot-and-Mouth Disease

Foot-and-mouth disease or hoof-and-mouth disease (Aphthae epizooticae) is an infectious and sometimes fatal viral disease that affects cloven-hoofed animals (cattle, goats, sheep). The virus causes a high fever for two to six days, followed by blisters in the mouth and on the feet that may rupture and cause lameness.

"If it is not in all cases a complete preventative of foot and mouth disease, it is a palliative (provides relief), and further experiments may probably prove that it may be applied succssfulllly to Pleuro-pneumonia.
The true Symphytum asperrimum (Prickly Comfrey) is diuretic. 'Hence,' says Mr. Thomas Christy, 'it prevents fever, and **cattle fed upon this plant are free from the ravages of lung and foot-and-mouth diseases,** *and are strongly fortified against infection; its curative properties have been long known.' "*
-'Prickly Comfrey: Its History, Cultivation, Extraordinary Production, and Uses: A Letter Addressed to His Excellency Sir Hercules Robinson, President of the Agricultural Society of New South Wales' by Arthur T. Holroyd, Sydney, Australia, **1876**, page 11.
(Thomas Christy from England promoted Prickly & Russian Comfrey in mid to late 1800s. See 'History' section in Volume 1.)
(Pleuro-pneumonia is inflammation of the lungs and pleura. Pleurisy is the inflammation of only the pleura. Pleura is a membrane that covers the lungs.)

"Foot-and-Mouth Disease:
   I have eight dairy cows; one was affected with foot-and-mouth disease on the 8th September, 1875. Two of the eight have escaped altogether (of the disease), and under circumstances which lead me to believe that the distemper is not infectious. I attribute the mildness of the attack in my animals to their **having been drenched as soon as the complaint appeared, and also to the use of Prickly Comfrey, with which they have been fed at midday for some months. As Comfrey is said to be valuable as a febrifuge (reduce fever), and as it also affords much mucilaginous food. I have no doubt that its influence is great in lessening the violence of this complaint,** *and keeping up a healthy condition in all animals fed upon it. -W.L.C.*
In confirmation of the remarks of W.L.C, I may state that **I lately turned (put) a cow, which I have occasionally fed with Comfrey during the past two summers, amongst a herd of eighteen or twenty other cows, all of which were bad with the foot-and-mouth disease, and although she has now been feeding in company with these cows for six weeks, she continues perfectly healthy.**
There is I think, no plant that grows so luxuriantly, or that can be more rapidly multiplied. A very large quantity of spurious (not genuine) and utterly worthless varieties have been sold in this country under the name of Comfrey; hence the importance of securing the correct thing. -Kinard B. Edwards, Burbage Hall, Hinckley, Leicestershire, England."
-'Forage Plants and Their Economic Conservation by the New System of Ensilage: Part I: Caucasian Prickly Comfrey' by Thomas Christy, Jun., F.L.S. (Fellow of the Linnean Society), Christy & Co., London, England, **1877**, page 18.
   (A drench is giving an animal a liquid oral medication.)

Mastitis (Breast Tissue Infection)

*"***Bovine (cattle) mastitis, the inflammation of the mammary (breast) gland associated with bacterial infection, continues to be the most frequent and costly disease of dairy cattle all over the world.** *Based on informal interview, ethnoveterinary information about plants used in the prevention and control of bovine mastitis in Southern Brazil were obtained.*
Symphytum officinale L. ('confrei', Common Comfrey) and 10 other plants were cited as helpful. In order to validate the traditional practice, the decoctions obtained with the plants were analyzed for the 'in vitro' anti-microbial activity against Staphylococcus aureus and Salmonella choleraesuis by the agar dilution method.
Aqueous (water) plant extracts were obtained boiling (15 minutes) 100 grams (3.5 ounces) of the dried and powdered plant materials with 1.000 mL of distilled water. The volume was completed to 1.000 mL (0.033 fluid ounces) with water in order to compensate the evaporation.
Alternanthera brasiliana, Achillea millefolium, Baccharis trimera and Solidago chilensis extracts were active against S. aureus while **Symphytum officinale,** *Sambucus nigra, Mentha sp., Ocimum basilicum, Parapiptadenia rigida and Cuphea carthagenensis* **extracts were active against both microorganisms (Staphylococcus aureus and Salmonella choleraesuis)**.
For all the cited species, scientific data were reviewed aiming to establish a correlation between popular use and biological

*properties. **The data found in literature for several of these plants could justify the use in the bovine mastitis treatment for anti-microbial, anti-inflammatory and wound healing activities.**"*
-'Antimicrobial Activity of Plants Used in the Prevention and Control of Bovine Mastitis in Southern Brazil' by C. Avancini, J. Wiest, R. Dall'agnol, J.S. Haas and G.L. von Poser, Universidade Federal do Rio Grande do Sul, Porto Alegre, RS, Brazil; Latin American Journal of Pharmacy (formerly Acta Farmaceutica Bonaerense), Volume 27, No. 6, pages 894-899, November 2008.
('In Vitro' means in laboratory equipment.)
(Staphylococcus aureus is a Gram-positive, round-shaped bacterium that is a member of Firmicutes phylum. It is part of the normal flora of the body, frequently found in the nose, respiratory tract, and on skin. It is a common cause of skin infections including abscesses, respiratory infections such as sinusitis, and food poisoning.)
(Salmonella enterica, formerly called Salmonella choleraesuis, is a rod-shaped, flagellate, facultative anaerobic, gram-negative bacterium.)

Muscle Strain in Animals

"*Receipts For the Cure of Cattle:* **A remedy for an Ox that is backstrained. Take Comfrey,** archangel (angelica), knot grass, and plantain, a handful of each; boil these, tied up in bunches, in about five pints (10 cups = 2.3 liter) of ale-worth, or, for want of that, in middling beer, free from yeast, till the liquor is strong of the herbs; then add an ounce of anniseeds, and about quarter of a pound of bole ammoniac finely powdered; when these are boiled, and strained through a seive, **give half the liquor to an ox in the morning, and the other half the morning following,** not suffering him to drink till the afternoon."
-'The Family Physician and the Farmer's Companion', author unknown, New York, **1840**, page 22.
(An ox is a bovine trained as a draft animal. Oxen are usually castrated adult male cattle.)
(I think by 'bole ammoniac' he meant Armenian bole, also known as bolus armenus or bole armoniac, that is an earthy clay, usually red, native to Armenia. It is red due to the presence of iron oxide.)

Parasites in Animals

"**Plants used for treating endo- and ectoparasites of rabbits and poultry in British Columbia included** Arctium lappa (burdock), Artemisia sp. (wormwood), Chenopodium album (lambsquarters) and C. ambrosioides (epazote), Cirsium arvense (Canada thistle), Juniperus spp. (juniper), Mentha piperita (peppermint), Nicotiana sp. (tobacco), Papaver somniferum (opium poppy), Rubus spp. (blackberry and raspberry relatives), **Symphytum officinale (Comfrey),** Taraxacum officinale (common dandelion), Thuja plicata (western redcedar) and Urtica dioica (stinging nettle).
Ethnoveterinary medicine used for poultry and rabbits in British Columbia: **Symphytum officinale Comfrey fed fresh or dried leaves were used for diarrhoea and endoparasites in poultry. The plant is mucilaginous and high in protein. Self-medicating birds apparently did not ingest enough pyrrolizidine alkaloids to be harmed and the content of these alkaloids varies from plant to plant.**"
-'Organic Parasite Control for Poultry and Rabbits in British Columbia, Canada' by Cheryl Lans and Nancy Turner, Journal of Ethnobiology and Ethnomedicine, BioMed Central, 7:21, pages 2-9, **2011**.
(An endoparasite lives in the internal organs or tissues of its host. An ectoparasite lives on the exterior of its host.)

Pregnancy Problems

"**The times when a cow is most likely to part (premature birth) with her calf** are at eighteen, twelve, nine, and six weeks from her regular time of calving. Give the drink at each of these times except the nine weeks: 4 ounces of Bole, 2 do. Irish Slate, 1 do. Tormentil Root or **Comfrey Root,** 1 do. Madder, and 1 do. Prepared Chalk or Crabs' Claws.
Powder all these, mix them in a quart of water, and give it fasting, and to fast three hours after. If proper care be taken as advised, and the drink be repeated as directed, you will soon get rid of the disorder, as I am of opinion that there is not a better drink found out than the above, therefore more recipes are needless."
-'Knowlson's Complete Cow Doctor: The Complete Cattle Doctor: A Treatise on the Diseases of Horned Cattle and Calves' by John C. Knowlson, New York, **1847**. (I do not know what 'do.' means. If you do, please contact me.)

Scours (Diarrhea): Preventing or Curing

Historically, Comfrey was used by pig farmers to cure scour (diarrhea). Currently, cattle, pig and horse farmers use it to prevent and cure intestinal and digestive problems.

"**For the Bloody Flux (Diarrhea): Comfrey, under Capricorn, flowers in June and July.** The Oak is under Jupiter; and its leaves, bark, and acorn cups, are all in use for this disorder, either in powder or made into tea."
-'Knowlson's Complete Cow Doctor: The Complete Cattle Doctor: A Treatise on the Diseases of Horned Cattle and Calves' by John C. Knowlson, New York, **1847**

Skin Problems

*"Put the animal (cattle) on to a diet of green food only- whatever quantity is normal for the age of the animal. With ringworm dress the parts with veterinary iodine initially and then follow with **daily dressing of a herbal healing cream which I have evolved from the most effective cleansing and healing herbs: garlic, marshmallow, Comfrey and chickweed.**
For skin troubles other than ringworm, no iodine should be used. The affected parts should be bathed in warm water, then daily dressed with the herbal healing cream, until the trouble has nearly subsided when it should be allowed to dry up and be left without further attention."*
-Herdsmanship: A Guide for the Herd Owner, Herdsman and Cowman by Newman Turner. Austin, Texas: Acres USA, **1952**, page 129.   (This formula should be good for most animals.)

Sprains and Strains

*"**Herbal Treatments for Horses: Strained Tendons:**
Rest for at least seven days. Apply frequent cold water applications and loose, water-soaked bandages. Later massage well into the area a liniment made from half a pint of strong brew of seaweed, to which two dessertspoonfuls of vinegar have been added. A Spanish gipsy remedy is standing the limb in hot salt water for a good half-hour. Then friction with warm olive oil. Bandage with cold water-soaked cloths. **The English gipsy specific is massage with a strong brew of Comfrey; also bandaging with cold Comfrey-soaked flannel.**"*
-Herbal Handbook for Farm and Stable by Juliette de Bairacli Levy. London, England: Faber and Faber, Inc., **1952**. Later called 'The Complete Herbal Handbook for Farm and Stable'.   (A dessertspoon is between the size of a tablespoon and a teaspoon.)

Ticks

*"**Topical application of crude ethanol extracts (CEs) of the leaves of 43 of 51 Jamaican plants produced varying degrees of multiple acaricidal effects on engorged Boophilus microplus Canst.**, including mortality (M), inhibition of oviposition (IO) and inhibition of embryogenesis (IE).*
    Fresh green leaves of each plant were macerated and a 10 gram (0.35 ounce) sample was extracted with 95% ethanol for 5 days in a darkened room with occasional shaking. The extract was filtered, concentrated to 0.5 ml in a rotary evaporator and re-dissolved in acetone to provide a 10% stock solution (Williams and Mansingh, 1993). The ticks were held in Petri dishes under laboratory conditions of 55-60% Relative Humidity and 27-29 C (80.6-84.2 F) for 12 days during which mortality was recorded daily.
Acaricidal indices (AI) ranged from 50 to 100 for the CEs of 29 plants, 38 to 47 for 9 plants and from 0 to < 25 for 13 plants. The most active CEs, in decreasing order of activity (AI values in parentheses) were those of: Simarouba glauca (100), Symphytum officinale (99), *Nicotiana tabacum (95), Hibiscus rosa-sinensis (93), Ervalamia divaricala = Ricinus communis (82), Salvia serotina (80), Stachytarphetajamaicensis (79), Blighia sapida = Ocimum micranthum (76), Spigelia anthelmia (75), Cycloptis semicordata (74), Mormordica charantia (71), Bontia daphnoides (69), Azadirachta indica (68), Capsicum annum = Catharantus roseus = Peliveria alliacea (66), Gliricida sepium (64), Lippia alba (62), Cuscuta americana = Erythrina corallodendrum (61), Piper amalago (60), Cannabis sativa = Cecropia peltata (58), Dioscorea polygonoides (56), Artocarpus altilis (53), Crotalaria retusa (51), Citrus aitrantium (50).*
    Differences in the bioactivity of the CEs were significant (P< 0.05); 36-hour mortality (M) ranged from 21 to 50%, inhibition of oviposition (IO) from 10 to 86% and inhibition of embryogenesis (IE) from 1 to 100%.
Symphytum officinale L.:
    % 36-Hour Mortality: 30.0 (high death rate).
    % Inhibition of Oviposition: 56.5 (high).   % Inhibition of Embryogenesis: 95.7 (high).
    Relative Reproductive Success (RRS): 1 (very low).
    Acaricidal Index (AI): 99 (very high). *Acaricidal index was calculated as: 100 minus RRS."*
-'Pesticial Potential of Tropical Plants II: Acaricidal Activity of Crude Extracts of Several Jamaican Plants' by
Ajai Mansingh and Lawrence A.D. Williams, Pesticide and Pest Research Group, Department of Life Sciences, University of the West Indies, Jamaica; Insect Science and Its Application: The International Journal of Tropical Insect Science, Volume 18, No. 2, pages 149-155, **1998**.
    (Acaricide is a substance poisonous to species of the arachnid subclass Acari such as mites and ticks.)
    (Rhipicephalus microplus, previously called Boophilus microplus, is also known as the Asian blue tick,
    Australian cattle tick, southern cattle tick, Cuban tick, Madagascar blue tick, and Porto Rican Texas fever tick.)

*"**An enumeration (list) of plant species that have been documented in literature to contain compounds and/or active ingredients that have effects on livestock ticks worldwide:**
Symphytum officinale L. (Comfrey) Boraginaceae (borage family): **Fresh leaf.** Topical application of crude ethanol extracts. Used in Jamaica, United States. Being toxic and inhabitant of oviposition and embryogenesis of Boophilus microplus Canst."*
-'Potential of Traditional Knowledge of Plants in the Management of Arthropods in Livestock Industry with Focus on (Acari) Ticks' by Wycliffe Wanzala, Kenya, Africa; Hindawi: Evidence-Based Complementary and Alternative Medicine, Cairo, Egypt, Volume

Walking Problems

"Hiped and Half Hiped (Lameness): When the bones of the hip fall so low as to be called hiped, the horse becomes useless; but when they are only half hiped, or hip-shot (dislocated hip or one hip lower than the other), the hip may be strengthened, and the horse (though disfigured) may perform much labour.
Cure: Take white-oak bark, elm and whitepine bark; roots, Solomon-seal, buck horn and Comfrey; boil them all together, and frequently bath the hip with it. This in a little time will strengthen the hip and fit the horse for business."
-'The New-England Farrier, or A Compendium wherein Most of the Diseases to which Horses, Cattle, Sheep and Swine are Incident' by Paul Jewett of Rowley, Massachusetts, published in Newburyport, Massachusetts, **1795**

Wounds

"Herbal Treatments for Cows: Wounds: Wash the surface very clean with cold water, slightly salted, then bathe with a cleansing and healing herbal brew, possessing disinfectant properties. Use preferably Comfrey plant or elder leaves and/or blossoms. If knapweed herb is available, add some of this to the brew; it is an excellent astringent and tonic. The common daisy plant is also astringent and tonic.
Approximate quantities for the brew would be two large handfuls of finely chopped Comfrey and elder, to one quart (0.94 liter) of water. If knapweed or daisy are added, take half a handful, and mix with the Comfrey or elder.
When bathing the wound, apply fairly hot for the first treatments; the subsequent treatments should receive cold applications. Bathe at least three times daily.
Rub also into the wound some of the green leafage residue mixed with a small amount of vinegar, as there is generally also local bruising of tissues if the wounds be sizable. One or two teaspoonfuls of vinegar to one cupful (236 ml) of green herb pulp. Plantain leaves yield a very soothing mucilage, valuable in treating inflamed surroundings of wounds, and adder's tongue ointment is highly curative in old wounds."
-Herbal Handbook for Farm and Stable by Juliette de Bairacli Levy. London, England: Faber and Faber, Inc., **1952**. First edition that later was called 'The Complete Herbal Handbook for Farm and Stable'.

Prickly Comfrey 1910

'Prickly Comfrey as a Forage Crop' by United States Department of Agriculture, Bureau of Plant Industry, Office of Forage Crop Investigations, Washington, DC. Circular No. 47, 10 pages, January 26, 1910. Comfrey page 5.

A. Flowering stem of Prickly Comfrey.
B. Crown cutting.
C. Taproot cutting.

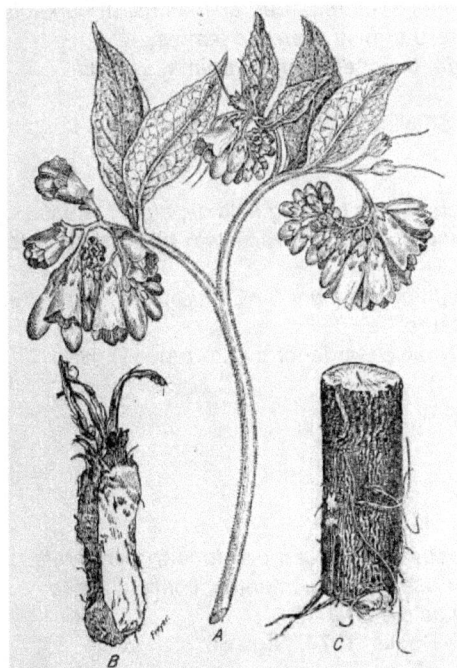

Symphytum asperrimum (Prickly Comfrey) 1877

'Gartenflora', Volume 26, by Eduard August von Regel, Stuttgart, Germany, 1877. Page 151.

It was a monthly illustrated botanical magazine published 1852-1940. Described as a 'General Monthly Magazine for German, Russian and Swiss Horticulture and Botany'.

Regel was Director of the Russian Imperial Botanical Garden of Saint Petersburg, Russia from 1855-1892.

Chapter 21

Comfrey Meal, Pellets, Hay and Silage

Comfrey Bristles

Bristles on Comfrey leaves: *"**These are very light and flimsy structures made largely of silica,** securing fragile stiffness on the same principle as wheat straw and **explaining the relatively high proportion of this element in the ash.** They are rigid because they are filled with water which exudes through their tops at times.*
***When Comfrey is first cut it is still transpiring at its normal speed, and the first wilting is extremely fast; in an hour 88 percent moisture has become 78 percent. The water drains out of these bristles back into the leaf,** and the empty shells are so frail that they are not even capable of repelling tame rabbits once these tender-mouthed animals are used to the fodder."*
-Russian Comfrey: A Hundred Tons an Acre of Stock or Compost for Farm, Garden or Smallholding by Lawrence D. Hills. London England: Faber and Faber, Limited, **1953**, page 76.
>(Ash residue is various types of minerals, i.e., inorganic matter. The main components of ash are usually phosphorous and calcium, but it also contains iron, zinc and other minerals. Anything in food that won't burn is counted as ash.)
>(Transpiring is giving off water vapor through the stomata, i.e., leaf pores.)

*"**Commercial grass drier: With Comfrey the dustiness reported would be due to the remains of the finely divided silica of the bristles** and that on the way up the stem to the leaves as yet ungrown at the time of cutting."*
-Russian Comfrey: A Hundred Tons an Acre of Stock or Compost for Farm, Garden or Smallholding by Lawrence D. Hills. London England: Faber and Faber, Limited, **1953**, page 106.

*"**The hairs on Comfrey leaves** limit its use for pasture. The fresh leaves are eaten by sheep, pigs, and poultry but are often not palatable for cattle and rabbits. **The leaf hairs collapse when the forage is wilted or ensiled (made into silage);** consequently, cattle and rabbits will eat the wilted forage. The forage is also fed to horses, goats, chinchillas, and caged birds. Daily harvesting and feeding as green chop is an effective use of large plants; apparently, cattle do not object to the hairiness after the plants are chopped."*
-'Comfrey: A Controversial Crop' by Robert G. Robinson, University of Minnesota, Agricultural Experiment Station, Minnesota Report MR-191, Item No. AD-MR-2210, **1983**, page 4.

Dried Comfrey Meal Using Shredder and Drier

*"Early in 1950 a trial was made at Messrs. Wissington Estates Nurseries at Southery (Norfolk County, England), on the possibilities of **Symphytum peregrinum (Russian Comfrey) as a crop for the grass drier.***
It was discovered that the meal produced, though high in protein and low in fibre, was so dark as to be unsaleable.
*In the trial, however, the experience of the Drier staff enabled them to overcome what has been regarded as the unsurmountable obstacle to conserving the crop by this method: **the thick, fleshy stems which dry more slowly than the leaves.***
***The crop was wilted and put through a 'Robust' shredder, in the same way as kale, brussels sprout stems, unsold cabbage or sugar-beet tops,** and proved no more difficult to dry."*
-Russian Comfrey: A Hundred Tons an Acre of Stock or Compost for Farm, Garden or Smallholding by Lawrence D. Hills. London England: Faber and Faber, Limited, **1953**, page 45.

*"My father designed many driers. We tried a Calor® gas burner, miniature aeroplane propellers blowing cold air, electric heaters, and chicken brooder lamps. **The final drier, which served us for years, was a cabinet of eight drawers with perforated zinc bottoms, one above the other, fitted above one of my father's early heaters.***
*We dried leaves only, leaving them on wire netting to wilt in the sun first, and crushing them when dry for Comfrey tea. **Comfrey is far more difficult to dry than any herb, because of its high protein and thick stems.**"*
-Fighting Like the Flowers: An Autobiography: The Life Story of Britain's Best-Known Organic Gardener by Lawrence D. Hills. Bideford, Devon, England: Green Books, **1989**, page 118.
>(Calor Gas company, formed 1935, sells bottled butane and propane in Britain and Ireland.)

Comfrey Pellets

*"There are processes under development for **direct pelleting of Comfrey under pressure which can produce the equivalent to pig and dairy nuts (concentrate)** for conservation through the winter as well as for use as supplementary concentrate to with high fibre foods for starch equivalent, or to balance high carbohydrate foods such as pig swill."*
-Comfrey: Fodder, Food and Remedy by Lawrence D. Hills. New York: Rizzoli Universe Books: **1976**, page 86.

(Feed Concentrates are mixes of ingredients with a high amount of cereal grains. Mixes of only grains are also called concentrates.) (Swill is food for animals in liquid or partially liquid form. It includes kitchen and food processing scraps.)

"Besides processing the Comfrey leaf for the greenish powder, and harvesting the underground root for drying, **there is an opportunity to produce cattle feed alternatives as a 60% Comfrey / 40% Alfalfa pellet.** *This constitutes a 'whole food' for cattle. While this might seem rather simple,* **Comfrey is actually very difficult to handle, requiring some rather sophisticated management techniques.**
Most details beyond this first overview are considered proprietary (exclusive rights), and will probably require an outside consultant to assist in the harvest, drying, and eventual processing details for ease into the marketplace.
Further, the alternative cattle food supplement market has yet to be developed, requiring further professional help. This is how I currently make a living, working as an outside consultant."
-'Comfrey Yields, Drying' by Richard Alan Miller, Commercial Herb Production and Marketing, Richters Herbs, Goodwood, Ontario, Canada, November 12 **1999**.

*"***Comfrey was cultivated in Saskatchewan (Canada)** *in the 1950s and has long been used as a forage crop for livestock giving exceptionally high yields of fodder (50-80 tons per acre). Comfrey was shown to effectively prevent digestive disorders in poultry and livestock, especially scours (diarrhea) in calves.*
Drying Comfrey Pellets: *The transplants of S. officinale and S. uplandicum (Bocking No. 4 and Bocking No. 14) were planted in the first week of June 1996 at three sites: Saskatoon, Shellbrook and Tisdale (all in Saskatchewan province in western Canada). Comfrey samples were collected 3 times throughout the season.* **The steady state moisture content (6%) of pellet was achieved after 30 hours of drying at 40 C (104 F), as compared to 11 hours at 70 C (158 F).** *No visible differences in colour of pellets dried at 40 C and 70 C could be observed.*
Alfalfa vs Comfrey Pellets: *Preliminary pilot-scale pelleting trials using 4 kg (8.8 pounds) of dry Comfrey grind were performed using California Pellet Mill (CPM CL-5) with a 3/16" (4.76 mm) die under similar conditions used for alfalfa pellets. In comparison to alfalfa pellets,* **Comfrey pellets were found less durable and less hard. However, in their appearance (nice green colour) and physical properties, Comfrey pellets met the alfalfa industry standards.** *Cattle feed market in Japan seems to be the largest and the most attractive.*
Pyrrolizidine Alkaloids: *The quantities of PAs vary significantly between plant populations and some success has been claimed in growing alkaloid-free comfrey (PAs <5 ppm) (parts per million).* **The difficulty was experienced in developing reliable assay for PAs, since these are unstable compounds typically present in very low concentration, at 0.1-400 ppm.** *The lack of standards prevented identification of PAs known to occur in Symphytumm spp. (species); total PA content could only be determined semi-quantitatively.*
Comfrey Feed Analysis, 100% Dry: *Protein 17.02%, Sodium 0.04%, Phosphorus 0.27%, Potassium 5.54%, Sulphur 0.22%, Calcium 1.75%, Magnesium 0.36%, Copper 15.20 ppm, Iron 551.00 ppm, Manganese 57.30 ppm, Zinc 48.60 ppm, Molybdenum co.50 ppm, Selenium co.20 ppm, Nitrate 0.40%, A.D. (Acid Detergent) Fibre 42.39%."*
-'Feasibility of Producing Comfrey (Symphytum spp.) Pellet as a Feed Supplement' by Barl, Gibson, Crerar, Shao, and Sokhansanj, Dept of Plant Sciences and Department of Agricultural & Bioresource Engineering, University of Saskatchewan, Saskatoon, Canada at Soils & Crops Conference, February **1999**. (I do not know what 'co.' means in front of Molybdenum and Selenium.)

Comfrey as Hay

Overview
See section 'Harvesting Comfrey Leaves' (Chapter 37).
See sub-subsection 'Drying Leaves for Comfrey Tea' in subsection 'Comfrey Tea' in section 'Humans Eating Comfrey' (Chapter 32).

"The aroma of dried Comfrey is superior to that of the finest clover hay. *Where a large enough plantation exists, if a small stack of dried Comfrey be used to flavour any inferior quality hay from bad pasture or badly cured, I doubt if ever again the Comfrey stack will be absent. A handful of it will scent a large quantity of otherwise flavourless hay."*
-'Ensilage: A System for the Preservation in Pits of Forage Plants and Grasses, Independent of Weather: A Collection of Facts and Statistics on the Cheapest Mode of Providing Winter Food for Dairy Cattle, Sheep, Horses, Etc.', by Thomas Christy, F.L.S. (Fellow of Linneaen Society), London, England, **1883**. Caucasian Prickly Comfrey, pages 57-63.

In 1916 'The Agricultural Notebook' by Primrose McConnell, B.Sc. records Comfrey hay with 18.5% crude protein versus 13% for the very best meadow hay.
(The first edition was compiled by Primrose McConnell in 1883, and it is now in its 20th edition in the United Kingdom.)

"Various miscellaneous hays, percent dry matter by Kellner (1880, 'Centralbl f. Agrikulturchem'):

	Crude Protein	**Fibre**	**Starch Equivalent**	**Protein Equivalent**
Buckwheat	12.2%	36.5%	32.2%	6.4%
Rape (Brassica napus)	18.6	24.3	42.4	11.6
White Mustard	13.1	31.1	32.6	6.1
Comfrey	24.4	13.5	44.1	12.7
Nettle	20.7	12.0	54.2	12.5 "

-'The Science and Practice of Conservation: Grass and Forage Crops, Volume 1' by S.J. Watson, D.Sc., F.I.C., London, England, pages 425, 471, 472, **1939**.

"The main problem with Comfrey hay is the thick mid-ribs of the leaves and stems which may be an inch (2.5 cm) in diameter. This is partly overcome by cutting at a very leafy stage so that the main stems are few, but though the first 20 percent of moisture vanishes rapidly, and the next 30 percent with almost the same speed, reducing the plant to the stage at which it is used for compost, the further drying to make it into hay is more difficult.
Unlike short grass and other crop-drier raw materials available in summer, **Comfrey can be wilted without loss of nutriment** and still be picked up cleanly by mechanical handlers, where grass blades crisp and break."
-Russian Comfrey: A Hundred Tons an Acre of Stock or Compost for Farm, Garden or Smallholding by Lawrence D. Hills. London England: Faber and Faber, Limited, **1953**, pages 79, 84.

"Russian Comfrey, Symphytum asperum. Lepech.
Percentage composition on a dry matter basis:
    T.D.N. = 50%; Protein = 15-18%: Digestible protein =11%; Fibre =12%; Ash =20%.
    The ash content is very high and this accounts for a relatively low T.D.N. value.
Russian Comfrey is a leafy perennial which grows to a height of 2 to 5 feet (0.6-1.5 meter). The leaves are high in protein, palatable and are excellent feed for cattle and sheep. Also they are a suitable green feed for poultry.
Russian Comfrey Hay: Russian Comfrey hay is a good quality roughage with a high protein content (13 per cent) and a correspondingly low fibre content (13 percent)."
-'Animal Feeds of the Federation' by John H. Topps, Occasional Paper No. 1, Department of Agriculture, University College of Rhodesia, Salisbury, Rhodesia (now Zimbabwe), page 43, **1961**.
    (Total Digestible Nutrients or TDN measures available energy of feeds and energy requirements of animals.)

"One aspect of drying Comfrey should be kept in mind. It is that slow drying will reduce the quantity of Beta Carotene for conversion to Vitamin A in the liver. One fresh weight (green leaf) analysis of Comfrey done in England revealed 77 mg per kg of leaf of Beta Carotene. In dried Comfrey leaf the highest figure known was 400 mg per kg. The New Zealand analysis shows 170 mg per kg, which is a good average taken over a year. **Spring growth is highest in Carotene.**
Surplus Comfrey should be dried as fast as possible, compressed into bales, and stored in a cool dry dark place to be fed out when needed in the winter."
-Comfrey: Nature's Healing Herb & Health Food by Andrew Hughes. Japan: Sanyusha Publishing Co., Ltd, **1992**, page 161.
    (One kg or kilogram = 2.2 pounds)

"Comfrey, Quaker (also called Russian Comfrey, Symphytum peregrinum):
Yield: Hay, 3-5 tons per acre.   pH Range: 6.0-6.5.
Fertilizer: 60 pounds (27 kg) Nitrogen. Apply 60 pounds P205 and 60 pounds K20 at medium soil test levels."
-'Agronomy Handbook' by Virginia Cooperative Extension, Virginia PolyTechnic Institute, and Virginia State University, Publication 424-100, **2009**.

Drying Small Amounts of Comfrey Leaf
See sub-subsection 'Drying Leaves for Comfrey Tea' in subsection 'Comfrey Tea' in section 'Humans Eating Comfrey' (Chapter 32).

"After the passing of this grand lady (Comfrey grower Mrs. Peggy Greer of England) in 1970 at the age of eighty-nine, Brickwall Farm (Halstead, Essex, England) was left to her nephew, Mr. Bowen-Colthurst.
During my visit to the farm during the summer of 1972, I was most impressed with Comfrey farming and preparation. Two farm labourers were in the process of sorting freshly cut Comfrey leaves and stems for drying.
Suitable leaves are spread out in special trays arranged on shelves under glass in a large building resembling a greenhouse-cum-barn. In this manner natural sunshine allows the leaves to dry to the ideal texture for crumbling to make Comfrey tea. *The crumbling process is carried out by rubbing leaves on special sieves. Comfrey stems are also cut up to produce stalk tea mixture.*
The procedure for obtaining these ingredients is somewhat lengthy for the drying of leaves or stems by artificial heat does not have the same effect. **There are six Comfrey preparations from crops grown on the farm:** *Comfrey leaf tea, Comfrey stem tea, Comfrey root powder, Comfrey ointment, Comfrey tablets, Comfrey stem powder."*
-About Comfrey: The Forgotten Herb by G.J. Binding, M.B.E., F.R.H.S. Wellingborough, Northhamptonshire, England: Weatherby Woolnough, **1974**, pages 53, 54.

"Dry the Comfrey leaves either by suspending the whole leaf stalks from the ceiling of your attic or cellar, or by placing them loosely next to or over your warm oil burner. If you're using the floor of an unoccupied (but preferably warm) room, **be sure to stir the leaves once a day.** They may also be dried by suspension in the kitchen hallway or in the garage.
In all cases, **be sure to keep the drying herbs away from direct sunlight.** Above all don't crowd the material, to prevent possible mold. **Unlike aromatic herbs which are best gathered between 10 A.M. and 12 noon, Comfrey leaves (and roots) may be taken any time of the day providing the leaves are dry."**
-Comfrey: What You Need to Know by Ben Charles Harris. New Canaan, Connecticut: Keats Publishing, Inc., **1982**, page 95.

Making Comfrey Hay by Drying on Ground, then Sometimes Use Drier

"We also dry as much Comfrey as we can to provide the goats with their favorite treat during the winter.
At the end of a sunny, non-humid day (when food value in the leaf is at its peak), we sickle our way through the patch. On such occasions, the garden cart becomes our hay wagon to convey the cuttings to their drying spot on the grass.
Since Comfrey leaves are so high in moisture and protein, **we spread them out well to avoid the heating and spoilage that would take place if the foliage were heaped up.** Two days of good clear weather does the job, and we pile the result in big cartons and store it in the garage.
You'd best finish harvesting your winter's supply by mid-August, or the heavy dews that appear later in the summer (here in the eastern United States) will hinder the process. **A rack or wire netting screen that holds the drying Comfrey up off the ground can considerably extend your 'haying' season for the plant.**"
-'Comfrey for the Homestead' by Nancy Bubel, Mother Earth News, Kansas, May/June **1974**.
 (A sickle is a tool for cutting grain, grass, etc. that consists of a curved, hooklike blade mounted in a short handle.)

"Mr. Willing (Saskatoon, Saskatchewan, Canada in 1956) attempted curing the Comfrey leaves for hay in a small way. **The forage was cut and spread out on the ground and allowed to dry. The dried leaves were gathered and stored with mixed grass hay in the barn. The cows ate the dry Comfrey leaves readily along with the other cured forage.**"
-Comfrey: Fodder, Food and Remedy by Lawrence D. Hills. New York: Rizzoli Universe Books: **1976**, page 66.

"**Leaving the Comfrey crop to wilt almost to hay in the field and then putting it through a shredder and a grass drier will reduce the fuel cost considerably,** but it is the last 40 percent of moisture that is hardest to take out."
-Comfrey: Fodder, Food and Remedy by Lawrence D. Hills. New York: Rizzoli Universe Books: **1976**, page 120.

"In 1973 Elmer Jesky of Aurora, Oregon **made Comfrey hay with his third cut, leaving it three days to dry in windrows and taking it up with an ordinary pick-up bailer.** It dries black, rather like 'rooftop hay' but is palatable (agreeable taste) for stock, and it is fed as a supplement in the early spring.
Making Comfrey hay for baling, like all hay making, is a matter of watching the weather and knowing when it is ready to bale."
-Comfrey: Fodder, Food and Remedy by Lawrence D. Hills. New York: Rizzoli Universe Books: **1976**, page 222.
 (A windrow is a row of cut hay or small grain. It is formed by a hay rake that puts it into a row after it has been
 cut by a mower or scythe. It is dried before being baled or rolled.)

"**Let the cut Comfrey leaf (Symphytum officinale) come to a 50% sun-cure wilt,** and then pick it up with a flail-chop to be taken to a drying facility (i.e., Hop Kiln, Corn Dryer, etc.). Tobacco dryers and plywood kilns are other alternatives for dryers. **Comfrey is easy to grow, but the key to success with this crop lies in proper dehydration and handling.**
 Both Comfrey leaf and dried root should be kept in average room temperature and humidity for long-term
 storage. Store to eliminate the potential for insects, rodent contamination, and yeast and molds."
-'A Farm Project: Comfrey Leaf and Root' by Richard Alan Miller, Oak Publishing Inc., Grants Pass, Oregon, and Northwest Botanicals, www.nwbotanicals.org, **1999**.

"**To dry it from a field, one normally puts a light wilt on the Comfrey before picking it up with a flail chop and then taking it to a dehydrator, hop kiln, or tobacco dryer.** All will work. If you try to dry it in the field it will rot. If you do not put a light wilt on it, it will probably rot in a dehydrator. **It is not an easy handling herb,** as the indol alkaloids will cause it to blacken and then rot before it can properly dry. This is due to the high muscilage in the leaf."
-'Comfrey Running Rampant, Can I Sell It?' by Richard Alan Miller, Commercial Herb Production and Marketing, Richters Herbs, Goodwood, Ontario, Canada, July 15, **2006**.
 (A kiln is thermally insulated chamber/oven that produces temperatures to harden, dry or chemically change the product.)
 (An oast, oast house or hop kiln is a building designed for drying hops as part of the alcohol brewing process.)
 (Flue-cured tobacco was originally strung onto tobacco sticks, which were hung from tier-poles in curing barns or kilns.
 The barns have flues run from externally fed fire boxes, heat-curing the tobacco without exposing it to smoke. It slowly
 raises the temperature over the course of the curing which is about a week.)

Making Comfrey Hay Using Tripods

"**If you find difficulty in drying the Comfrey tops for hay,** without crumbling, heating or rotting, after drying by sun as much as convenient, **put the tops into small piles or stacks of 40 to 100 pounds (18-45 kg) each,** according to the then stage of drying; and, if occasion require, before stowing it in the barn, pitch it over lightly, each into another, putting top for bottom, and then let them remain out till sufficiently dry for preserving."
-'The New England Farmer, and Horticultural Register' by Joseph Breck, Vol. 23, New Series Vol. 13, Boston, Massachusetts, **1845**. ('Symphytum or Comfrey, as Food for Men and Cattle' by Ezekiel Rich, Troy, New Hampshire, page 10, July 10 1844.)

"The quality of Comfrey hay is so good that it is worth using the Highland crofter's system of tripoding. **These tripods of stout**

stakes about four feet (1.2 meter) long are used to make hollow-centered haycocks, heaping wilted material round them to complete drying by the natural draught (draft) up the middle.
While this system is costly in labour and is **popular only where heavy rainfall makes it the only way of making hay at all.** The small 1952 plot was cut by hand and good weather enabled the crop to be dried completely in the field by normal methods."
-Russian Comfrey: A Hundred Tons an Acre of Stock or Compost for Farm, Garden or Smallholding by Lawrence D. Hills. London England: Faber and Faber, Limited, **1953**, page 79.
    (Crofting is a traditional system in Scotland of community small-scale food production including forage and food production.)   (A haycock is a cone-shaped pile of hay in a field.)

Making Comfrey Hay by Hanging in Shed

"It is easy to make good hay from Comfrey on this scale (small farming).
The crop is cut, wilted for twenty-four hours, tied in bundles and hung in a dry shed until the leaves are crisp and the stems break like sticks. The wilting is to prevent the strings coming loose as the crop shrinks in drying, and when this is complete the fodder is stacked and the next cut is ready to hang. **It should be cut each time when the first flowers are just opening,** and analysis will show that both the protein and carotene are far richer than normal meadow hay."
-Russian Comfrey: A Hundred Tons an Acre of Stock or Compost for Farm, Garden or Smallholding by Lawrence D. Hills. London England: Faber and Faber, Limited, **1953**, page 88.

Making Comfrey Hay by Drying on Racks

"**Drying Comfrey on wire racks can be satisfactory.** Stacked on wire racks inside a large shed is also possible. **Free air circulation is necessary,** and drying should take not more than three weeks. In a hot, dry climate it can be done in four days. It would appear that autumn in Japan would be the best time, when the humidity is down from summer. Or over even late spring before the summer humidity rises."
-Comfrey: Nature's Healing Herb & Health Food by Andrew Hughes. Japan: Sanyusha Publishing Co., Ltd, **1992**, page 161.
    (Japan's islands are a temperate climate like United States and Great Britain. Climate south of Tokyo is warm and mild, with weather resembling southeastern United States. In northern Japan {Hokkaido island} it is comparable to New England, United States. Southern Honshu {mainland} through Kyushu {southwest island} is mild enough for crops all year.)

Making Comfrey Hay by Drying on Roof

"Another Comfrey drying method, which needs hot but not necessarily rainless weather, and is very much less time-consuming, is as follows: **Cut and spread thinly on top of a black-painted corrugated iron roof, and leave the Comfrey there until it is dry,** before replacing with a further cutting. A black surface absorbs sun-heat, metal conducts this until the inside is uncomfortably hot, and if the roof is over a pigsty or poultry-run, this is of considerable additional benefit to those inside (keeping them cooler). **The quality is better than for normal hay, and in a wet summer there is an advantage from more sun-heat in proportion, and the fact that the rain runs quickly off the roof.**
With the corregated iron slope covered about two inches (5 cm) deep with Comfrey and a single turning, the product can go straight into the hammermill when the stems are dry enough to snap like sticks. The result is a meal resembling normal vegetable meal, but with low fibre to suit pigs and poultry."
-Russian Comfrey: A Hundred Tons an Acre of Stock or Compost for Farm, Garden or Smallholding by Lawrence D. Hills. London England: Faber and Faber, Limited, **1953**, pages 84, 88.
(Hammermill is a mill that shreds, grinds or crushes material into smaller pieces by repeated blows of rotating little hammers.)

"One of the Holbrook (England) 'Young Farmers Club' boys had the bright idea of **using the football (United States soccer) goal netting in summer to hold their hay on the roof, and this was quite effective.**
With a net made of nylon thick enough and with large enough meshes to hold the hay down in a windy summer, large quantities can be dried with little trouble, but in batches which must stay on the roof for a week, the leaves go black before the stems dry. **As the basic variation of the Stephenson Strain is Bocking No. 14, Mr. Stephenson dried better hay on this system than growers with the thick stemmed Webster variations (Bocking No. 4).**
When the stems are dry enough to snap like sticks, this material can be put through a hammermill to add to the pig ration. This method is recommended for goat keepers who feed it in winter as a source of protein and minerals."
-Comfrey: Fodder, Food and Remedy by Lawrence D. Hills. New York: Rizzoli Universe Books: **1976**, page 119.
    ('National Federation of Young Farmers Clubs' is a rural youth organisation in England and Wales.)

Comfrey as Silage

Definition of Silage

*"**Ensilage is the packing of green crops in air and water-tight pits called silos.** If the packing is rapidly performed, the forage evenly spread, covered with planks, and heavily weighted so as to drive out the air, **any green vegetable matter will keep fresh for at least a year in a silo**. As the mass becomes compacted together, fermentation is arrested, and **on the pit being opened after some months, it is found full of excellent and appetizing food,** a kind of sauer-kraut for cattle which can be made available at any period of the year.*

The practice of ensilage is very ancient. It was known centuries before our era, and probably had its origin among warlike nations, who, in the first instance, concealed their crops in this way to prevent pillage. It is mentioned by Euripedes and nearly all the Latin writers on agriculture. 'There is nothing new except the forgotten'."
-'Journal of the Society of Arts', Volume 31, November 17 1882 to November 16 1883; George Bell and Sons, London, England. Comfrey, February 2 **1883**, page 238.

Silage is fermented, high-moisture fodder that is fed to cattle, goats, sheep and other ruminants (cud-chewing animals). It is fermented in a process called ensilage or silaging, and is usually made from grasses such as corn, sorghum and other cereals, using all of the green plant (not just grain). **Three ways silage is made:**
    1. Place cut green plants in a silo or pit.
    2. Place plants in a large pile and compress to remove oxygen, then cover with plastic sheeting.
    3. Wrap plants tightly in plastic so it forms a large round bale.
A silo is a structure for storing bulk materials such as coal, cement, food, grain, sawdust and woodchips. It is used in agriculture to store grain or fermented feed known as silage. Three types of silos are used today: tower silos, bunker silos, and bag silos.

*"**Silage is the term used for the product formed when any green plant material is put in a place where it can ferment in the absence of air** (Narayanan and Dabadghao, 1972).*

The major changes which occur during ensiling are the fermentation of sugars to form acids and the breakdown of some of the forage proteins to simpler compounds, including ammonia (Wilkins and Wilson, 1970). This fermentation occurs during the first two to three months; after that the silage, remains practically unchanged for another 12 to 18 months.
***The process of making silage is called ensiling,** and the container for keeping silage is a silo."*
-'Silage and Hay Making' by B.N. Chatterjee and S. Maiti, Dept of Agronomy, Bidhan Chandra Krishi Vishwa Vidyalaya (Bidhan Chandra Agricultural University), Kalyani, West Bengal, India; Indian Council of Agricultural Research, New Delhi, India, **1978**.

*"**Forage which has been grown while still green and nutritious can be conserved through a natural 'pickling' process.** Lactic acid is produced when the sugars in the forage plants are fermented by bacteria in a sealed container (silo) with no air. Forage conserved this way is known as 'ensiled forage' or 'silage' and will keep for up to three years without deteriorating. Silage is very palatable to livestock and can be fed at any time."*
-'Silage Making for Small Scale Farmers' by Food and Agriculture Organization of United Nations, date unknown but around **2013**.

Comfrey Silage Overview

*"**Ensilage and Prickly Comfrey:** Mr. Henry Doubleday, writes to The Chelmsford Chronicle, Essex, England:*
*'In the book on Forage Plants by T. Christy, mention is made of the plan adopted by the French for preserving green food for winter use, and which they style, ensilage. Although I had read this account I did nothing in it till quite lately, and then only commenced by accident. **I was cutting a larger quantity of the tops of the solid stem Comfrey than my stock could consume fresh and green, aud the surplus was placed in cocks.** These soon heated, and after standing some time I gave these heated leaves to my stock, and found they were extremely fond of them.*
*I then made a further trial, and put into a wood tank about half a ton of green Comfrey leaves and trod them well down. These also soon heated and settled down into a mucilagiuous pulp, and **I have been succesful with my new food and find that cattle take to it most kindly.***
I am now trying the fattening properties ot the green Comfrey leaves and this sour keep (sour fodder), by feeding a bullock upon them with an occasional bit of the sort of which he is very fond. **I am quite astonished at the result of my trial of Prickly Comfrey as a fattening food.** I give my bullock some cut straw and 2 pounds (0.9 kg) of barley meal daily. Some farmers were looking at him last week, and could hardly believe but that he had been fed with oilcake."
-The Farmers Magazine and Monthly Journal of the Agricultural Interest', London, England, July-Dec **1877**. Comfrey page 399.
    (Cocks: Haystacks are sometimes called haycocks. These are small piles of cut and gathered hay to be later put into larger stacks. One loose hay stacking technique in Britain is to put freshly cut hay into smaller mounds called foot cocks, hay coles, kyles, hayshocks or haycocks, to help initial curing.) (Trod is walking on something to press it down.) (Oilcake is compressed plant material left after oil has been extracted, used as fodder or fertilizer.)

*"**Prickly Comfrey, Symphytum asperrimum, a plant that has been rather slow in finding a place in our agriculture (United States), is in Europe commended (praised) as one useful for ensilage, especially to mix with fodder com in the silo.** We have seen no definite accounts of experiments with this plant in the form of ensilage.*
*In some localities in Virginia, and on some dairy farms in New England (eastern United States), it has been cultivated to some extent. The chief merit claimed for it is its abilty to furnish green fodder very early in spring and late in the fall, and we enumerate it as **one of those plants that may possibly be of value in the silo**."*
-'Silos and Ensilage: The Preservation of Fodder Corn and Other Green Fodder Crops' by Dr. George Thurber of 'The American

Agriculturalist', New York, **1881**.

" '*I ensilaged*', *says a New York farmer,* '*1,400 tons in 1881, comprising clover, Hungarian grass, maize (corn), sorghum, dourra (millet), rye, oats, with peas and Prickly Comfrey.* Some of these were mixed in the silos, and some of each were ensiled separately. **All of it did well, and the cattle liked it all, except the Comfrey, which the cattle would not eat.** Other cultivators speak more favourably of their success with Prickly Comfrey. Mr. Remington, the farmer whom I have quoted, found that he could preserve this forage in as good condition as other kinds, but that probably from contrast with other more agreeable kinds of food, it was distasteful to his cattle."
-'Ensilage in America: Its Prospects in English Agriculture' by James E. Thorold Rogers, M.P., London, England, **1883**. Page 48.

"**Roofed Silos with Portable Weights: Silage Unchopped:** Mr. A. Copley, East Cowton, Northallerton, England.
The original silo was 12 feet (3.6 meter) long, 7 feet (2.1 meter) wide, and 8 feet (2.4 meter) deep, but it is now 2 feet (0.6 meter) deeper. A second silo, built this year, is 15 feet (4.5 meter) long, 7 feet wide, and 10 feet (3.0 meter) deep. Both are 8 feet below and 2 feet above the level of the ground.
The first silo was built of bricks and mortar, faced with 1 inch (2.5 cm) of cement lining, and was covered for one year with a low movable span roof. The second one was built of concrete, faced with 1/4 inch (0.6 cm) of cement lining.
Filling is done at various times, when the crops are ready and other things convenient, but we endeavour to catch the material at that stage of growth when it contains most nutriment. **Meadow-grass, aftermath, Comfrey, oats, tares (vetch), and clover have been pitted, all in a whole state.** The filling will be completed in September.
A small sprinkling of salt has been added- but not more than 1 pound (0.45 kg) to 2 1/2 cwts (280 pounds = 127 kg) of grass, etc. -with the idea of destroying noxious life-germs, such as liverfluke, etc.
The material is covered with transverse boards, 2 inches (5.0 cm) thick, and weighted with iron blocks of 1 cwt. each, so as to give a pressure of 1 1/2 cwt (168 pounds = 76 kg) per square foot (0.09 square meter).
The effect of using the pitted fodder has been that the milch (milk) cows improved in condition, and the yield of milk, cream, and butter increased in quantity and improved in quality.
Each cow received in previous years daily 2 pounds (0.9 kg) of oats, 10 pounds (4.5 kg) of hay, and 45 pounds (20.4 kg) of pitted fodder, but this year the use of hay has been discontinued. **The fodder can be consumed without deterioration, if taken out by degrees, during a period extending over several months.**"
-'Report on the Practice of Ensilage at Home and Abroad' by H.M. Jenkins, F.G.S.; The Journal of the Royal Agricultural Society of England, Volume 20, Second Series, pages 126-246, **1884**.
   (Aftermath is new grass growing after mowing or harvesting.)
   ('cwt' is the abbreviation for hundredweight. In the United Kingdom one hundredweight is 112 pounds. In the United States and Canada it is 100 pounds. There are 20 hundredweight in a ton, producing a 'short ton' of 2000 pounds and a 'long ton' of 2240 pounds.)

" '**With regard to ensilage (and Comfrey), George Fry's*, of Chobham (Surrey, England), system of sweet ensilage must, I think, be adopted.** I have a small silo now filled with it on that principle not yet opened, but it appears to be sweet and promising' by Henry Doubleday, Coggleshall (Essex, England)."
-'The Gardeners Chronicle: A Weekly Illustrated Journal: Horticulture and Allied Subjects', London, England, October 24 **1885**.
(* -'The Theory and Practice of Sweet Ensilage' by George Fry. London, England: The Agricultural Press, 66 pages, 1885. It does not specifically mention Comfrey or Symphytum.)

"**Corn for Ensilage: I want to seed my land down to some crop for ensilage, that will be permanent, and that will stand the climate of this province (New Brunswick, Canada);** the thermometer goes down to 3 degrees below zero some times, but we have plenty of snow. The crop would be top dressed heavily every fall with good stable manure. I want to cut two good crops every year from it and in return would manure heavily.
A heavy crop of meadow hay will make good ensilage, but corn on rich land will give three or four times as much to the acre, and its superiority and heavy product would probably more than pay for the annual labor.
Our correspondent wishes a continuous series of crops, like that from Prickly Comfrey, which, if benefited by passing through the silo, and animals could be induced to eat it, might possibly be the thing wanted. **It appears to have succeeded best at the South, but some cultivators at the North have tried it successfully, and others have failed and rejected it.**
An advertisement in the same issue:
   Prickly Comfrey Roots for Sale. Address: W.C. Townsend, Locust Valley, Long Island, New York."
-'The Cultivator and Country Gentleman: Devoted to the Practice and Science of Agriculture and Horticulture at Large' edited by Luther Tucker and Son, and John J. Thomas, Volume 56, Albany, New York, **1891**. Comfrey pages 167, 178.

"**At 'Jealott's Hill Research Station' (Bracknell, Berkshire, England), a bag of Prickly Comfrey (Symphytum asperrimum), treated by the A.I.V. process, was buried in one of the silos, weighed, and analyzed before and after ensilage.**
Finnish or Virtanen (A.I.V.) Process: The final establishment of the underlying principles of acidification in the making of silage and their application to practice is due to Virtanen and his co-workers in Finland (1929). The process has been called the A.I.V. process after A.I. Virtanen, though Braccini (1936) claims it for early Italian workers under name of Italo Giglioli (I.G.) process.
Composition and losses involved in the ensilage of Prickly Comfrey, based on one bag buried in a silo filled with grassland herbage:

	In	*Out*
Weight of bag contents	1.87 pounds (0.84 kg)	1.59 pounds (0.72 kg)
Loss percent of dry matter	-	14.9%
Composition percent of dry matter		
Ether extract	1.82%	2.72%
Fibre	17.96%	19.85%
Crude protein	17.70%	22.42%
Ash (Minerals)	14.69%	12.33%
N-free extractives	47.83%	42.68%
Organic matter	85.31%	87.67%
True protein	14.37%	16.65%
Ratio of true protein to crude protein	0.81	0.74
Moisture in fresh material percent	89.30%	79.20%

The Prickly Comfrey made a good silage of high protein content.
The break-down of protein was not excessive, and the material was eaten readily. *Usual changes in composition were noted."*
-'The Science and Practice of Conservation: Grass and Forage Crops, Volume 1' by S.J. Watson, D.Sc., F.I.C., London, England, pages 425, 471, 472, **1939**.

"At Jealott's Hill good silage has been made from Symphytum asperrimum (Prickly Comfrey)."
-The Conservation of Grass and Forage Crops by Stephen John Watson and Michael J. Nash, University of Edinburgh. Edinburgh, Scotland: Oliver and Boyd, 1960, pages 382, 408. First published 1939.

"Nettles are one of the richest known sources of protein in nature, and for this reason of all weeds the nettle offers probably the best possibilities for development as a commercial crop. **Comfrey, the greatest yielder of protein, is already accepted as a farm crop,** *thanks to the recent research of Mr. Lawrence D. Hills and the 'Henry Doubleday Research Association' (England). Nettle hay is well known as a food for goats, and* **I have made excellent silage for cows from a mixture of nettles and Comfrey.** *I am developing Russian Comfrey now for cattle silage and compost material."*
-Fertility Pastures: Herbal Leys as the Basis of Soil Fertility and Animal Health by Newman Turner. Austin, Texas: Acres USA, **1955**, page 58.

"I.I. Astakhov (Northwest Scientific Research Institute of Agriculture, Settlement of Siverskaya, Leningrad Oblast) reported on studies of the Comfrey. **Many years of testing at the Siverskaya Station and at other scientific and test establishments of the scabrous (rough) Comfrey points to the large promise of this ensilage feed plant.**
Especially sizeable successes in the study and introduction into cultivation of new ensilage plants have been achieved in the Komi ASSR (Autonomous Soviet Socialist Republic) Branch of the Academy of Sciences USSR.
Here the following plants are being widely tested and also used: mallow, oil-bearing radish, wild mustard, Soznovskiy cowparsnip, Veyrikha buckwheat, **scabrous Comfrey,** *butterbur.*
Scabrous Comfrey is devoured quite voraciously (eagerly) by hogs in the fresh and ensilaged form. *The use of the Comfrey green mass begins from the second year of life and continues for 10 or more years. It yields 40-60 tons of green mass per hectare."*
-'Seminar on New Ensilage Forage Plants' by V.S. Sokolov and P.F. Medvedev, Botanical Institute V.L. Komarov, Academy of Sciences, Leningrad; Botanicheskiy Zhurnal (Botanical Journal), Moscow, Russia, No 9, pages 1404-1406, **1963**.

*"***Silage crops suggested for north-east European parts of the USSR*** are mallow, sugar beet, oil radish, white mustard, green maize, beans, Sosnov cow parsnip (Heracleum),* **Comfrey** *and sunchoke (Helianthus annuus x H. tuberosus)."*
-'New Silage Plants in the North' by P.P. Vavilov and K. Moiseev from Russia, Nutrition Abstracts and Reviews, No. 6, pages 43-44, **1963**. (This is all I have from this report. If you have a copy, could you please send it to me.)
    (U.S.S.R. also known as Soviet Union was a state in Eurasia from 1922 to 1991. Now Russia and 12 independent nations.)

*"***Silage plants which are not at present widely cultivated in the USSR, but are of potential value, are briefly appraised.***
These include mallow (Malva sp.), Jerusalem artichoke (Plelianthus tuberosus), sunchoke (H. annum x H. tuberosus), white sweet-clover (melilotus alba),* **Comfrey (Symphytum asperum, Russian Comfrey),** *cow parsnip (heracleum sosnowskyi), Polygonum weyrichii and a Silphium sp. (a new silage plant under study in the west Ukraine)."*
-'Seminar on New Silage Plants, February **1963**, under auspices of the V.L. Komarov Botanical Institute, Academy of Sciences, USSR', Botanicheskii Zhurnal (Botanical Journal), No. 9, 1964. (This is all I have. If you have a copy, could you send it.)

*"***It was recommended that Symphytum asperum plants for silage should be cut at the end of July when they give high herbage yield and have a high sugar content.***"*
-'Okopnik na Sakhaline' {Comfrey in Sakhalin, Russia} by N.G. Khrushkova and A.A. Odegova; Trudy Sakhalinskogo Kompleksnogo Nauchno-Issledovatel'skogo Instituta, Russia, Volume 23, pages 175-179, **1971**.

*"***The main interest in Symphytum spp. (species) worldwide has been, and will probably continue to be, as a silage crop. It ensiles well*** (*Watson and Nash, 1960; Mikhkiev, Rozenberg and Il'in, 1970; Raman, 1970; Krushkova and Odegova, 1971),* **although serious nutrient losses occur if the herbage is not sufficiently wilted before ensiling** *(**Van der Zweerde, 1965).*

Even after wilting, Comfrey silage may have such a high moisture content that dry matter intake by animals is seriously restricted. Dry matter contents of fresh Comfrey range from about 110 to 200 grams/kg, similar to those of green forage crops such as cabbage, kale and rape, or the leaves of sugar beet and turnips, but somewhat lower than that of grazed grass.
The dry matter content of Comfrey silage depends on the length of time for which the crop is wilted before ensiling, the method of ensilage and whether or not molasses or other materials are added to raise the carbohydrate level for fermentation. *It may, however, be as low as 135 grams/kg (Hills, 1976), considerably below what is normally regarded as a desirable level for grass silage.*
Growing Comfrey together with grass to provide herbage with a higher dry matter content for ensiling appears not to be successful, because the Comfrey is intolerant of competition from grass *(Saunders, 1977; North of Scotland College of Agriculture, 1978)."*
-'Comfrey Symphytum spp. as a Forage Crop' by J.C. Forbes, A.D. McKelvie, and P.J.C. Saunders, North of Scotland College of Agriculture, Aberdeen, United Kingdom; Herbage Abstracts, Volume 49, No. 12, pages 523-539, **1979**.
(* -The Conservation of Grass and Forage Crops by Stephen John Watson. Edinburgh, Scotland: Oliver and Boyd, 1960.)
(** -'Verslag van een Proef met het Gewas Kaukasische Smeerwortel Uuitgevoerd in de Jaren 1953-1960' {Report on a Trial with Russian Comfrey in 1953-1960} by H. Van Der Zweerde, Verslagen Instituut voor Biologisch en Scheikundig Onderzoek van Landbouwgewassen {Institute for Biological and Chemical Research on Field Crops and Herbage}, Netherlands, No. 35, 12 pages, 1965. All in Dutch. If you have an English translation of this, please send it to me.)
(Grass and Comfrey can be grown in different fields so that argument against silage does not make sense.)

"**Silage is now the most common way for grass to be conserved as winter fodder.** *It has become so only within the last twenty years, but this is the culmination of a process which has been going on since about 1880 in Britain. The technique was introduced into this country from continental Europe in early 1880s, and generated much interest in wet summers of that decade.*
In July 1875 *the farm bailiff on Earl Cathcart's farm near Thirsk in North Yorkshire, England, recorded in the farm diary*:*
'Finished leading Grass to make it into pickeled Hay.' And in that year, or the one after, Mr. Arthur Scott of Rotherfield Park, Alton, in Hampshire, England, began to **experiment with ensilage of** *vetches, clover, ryegrass, oats and meadow grass, which were successful, and* **mangold leaves, cabbages, Comfrey and artichoke stalks, which were not successful.**
The agricultural press began to give their attention to silage, and early in 1883 James Howard MP suggested to the Journal Committee of the Royal Agricultural Society that the society should commission an investigation into ensilage and its suitability for English conditions.
Most of those who replied had covered silos in which the ensiled material was compressed by portable weights, and there was a roughly equal split between those who used chopped and those who used unchopped material.
Many different crops were ensiled: *vetches, oats, clover, ryegrass, meadow grass, rye, lucerne, maize, tares (vetch), trefoil, coarse grass from the orchard, sainfoin,* **Prickly Comfrey,** *beans, peas - in short, just about anything green was ensiled by one or another of Jenkins' correspondents.*
After his exhaustive account of the experiences of a **relatively small sample of silage producers,** *Jenkins set out his conclusions.* **Maize was the best crop for silage, grass and clover would do well if cut earlier than for hay, and green oats and rye, possibly buckwheat, but never Prickly Comfrey."**
-'Silage in Britain 1880-1990: The Delayed Adoption of an Innovation' by Paul Brassley, Agricultural History Review, Volume 44, No. 1, pages 63-87, **1996**.
(* -'Report on the Practice of Ensilage at Home and Abroad' by H.M. Jenkins, F.G.S.; The Royal Agricultural Society of England, Volume 20, Second Series, pages 126-246, 1884.)

"*Small-Scale Silage: Comfrey:*
A problem I see is that you will be making a harvest about every week or 2, and trying to add to the pile? This is not the best way to produce silage. The whole 'silo' should be filled and left alone for 3 weeks. Contantly adding to it will introduce O2 (oxygen), disturb the good bugs, and create multiple lines of poor 'top' silage in your bunker silo.
An important safety warning: A well functioning silo will produce gases that are heavier than air, and displace all oxygen. *As well as probably producing some gases that make your lungs want to foam (farmer's lung). If your 'hole in the ground' is actually a deeper pit, be very careful of this, especially in those first 3 weeks. Many, many farmers have needed hospital visits or coroner visits from entering a silo too soon and running out of oxygen. I'll guess it's not too serious on such a small scale, but I wouldn't feel right in not mentioning this."*
-'Small-Scale Silage: Comfrey' by Homesteading Today, www.homesteadingtoday.com, Forums, articles, media. February 3 **2004**.

"**Nicolaus Remer in writing his biodynamic work 'Laws of Life in Agriculture'* suggests that high yielding bulk fodder plants (15 tons of dry matter per hectare; or 6.7 tons per acre) could consist of the following forages:**
Sorghum (fodder millet); Jerusalem artichokes; timothy at 2-3 cuts per year; Phafuris tuberosa (reed grass); meadow fescue and tall fescue; **Comfrey (for silage);** *and lucerne. All these examples provide a much better amino acid profile than corn."*
-Alternative Health Practices for Livestock edited by Thomas F. Morris and Michael T. Keilty. Ames, Iowa: Blackwell Publishing, **2006**, page 98.
(* -Laws of Life in Agriculture by Nicolas Remer. Forest Row, East Sussex, England: Rudolf Steiner Press, 1995.)

"**Pyrrolizidine Alkaloids (PAs) are greatly reduced by making silage:** *In this context the question occurs whether there is a decrease in PAs during hay production/storage or during silaging. PA contents are shown of hay and silage produced from*

Jacobaea vulgaris (Senecio jacobaea, stinking willie). The data show that in case of hay no reduction of the PA level (compared with dried plants) can be observed. Contrary to this, **the results for silage show a decrease in the PA level down to 10%. It can be assumed that this is due to an enzymatic decomposition.** *Therefore, it can be concluded that in case of hay possible toxic risk is equal to original plant material (dried), whereas* **feeding silage to animals seems to be without any toxic risk.**"
-'Plants Containing Pyrrolizidine Alkaloids: Toxicity and Problems' by Helmut Wiedenfeld, Pharmaceutical Institute, University of Bonn, Germany; Food Additives and Contaminants, Volume 28, No. 3, pages 282-292, **2011**.

    (The PA reduction in Jacobaea vulgaris would most likely also happen with Comfrey silage. The hay comparison was made against the dried plant. I'm not sure what difference there is between hay and 'dried plant'. It would be interesting to know the difference between hay and fresh plant.)

Wrong Ways to Make Comfrey Silage

"**Thomas Christy's 'Ensilage' (*1877 publication)** *on the new process of ensilage advocated it strongly, and so did the following letter in 'The Times' (London, England newspaper) for 24 October* **1882**, *from J. Bailey Denton, a famous authority on agriculture: '***a valuable foreign forage crop (Symphytum asperrimum) which appears especially suited for the silo.** *In fact the extraordinary bulk of cattle food which may be gathered from it, would suggest its special growth as an ingredient of ensilage.'* **Unfortunately, the methods of silage-making then known were particularly unsuited to Comfrey, a high protein and moisture crop relatively low in carbohydrates.**
To quote Professor Stephen J. Watson's 'Silage' (1951): 'Then, in **1885**, George Fry published his book, **'Sweet Ensilage'**, and this, more than anything else, sounded the death knell of silage for the time being, and the process was set back fifty years.'
This process involved rapid heating, and like those used earlier did not introduce molasses. **A crop which is naturally high in carbohydrates had the best chance of producing sufficient lactic acid to arrest decay by acidity; one higher in protein (for example, Comfrey) swamps this effect with the evil-smelling and complex by-products of their decay.**
Russian and Prickly Comfrey were the worst possible subjects for bad silage methods, and it is to this period that all reports of the poor quality and unpalatibility of silage made from the crop can be traced. The few modern experiments have been almost entirely successful."
-Russian Comfrey: A Hundred Tons an Acre of Stock or Compost for Farm, Garden or Smallholding by Lawrence D. Hills. London England: Faber and Faber, Limited, 1953, page 36.
(* -'Forage Plants and Their Economic Conservation by the New System of Ensilage: Part I: Caucasian Prickly Comfrey' by Thomas Christy, Jun., F.L.S. {Fellow of the Linnean Society}, Christy & Co., London, England, 1877.)

"**Prickly Comfrey is a little known forage plant** although it is occasionally exploited (used) in some districts (Ottawa, Canada). The species commonly cultivated for forage is the gigantic Prickly Comfrey. **Only one experiment was made with this crop, the silage being an absolute failure. The moisture content of the crop at the time of ensiling was 92 percent.** The material rotted in the silo, had a very disagreeable putrid odour and was entirely unfit for feed. While one experiment is not sufficient on which to base a definite conclusion, there seems little reason to believe that Prickly Comfrey will become a popular silage crop."
-'Silage Production' by E.S. Hopkins and P.O. Ripley, Field Husbandry Division, Experimental Farms Service, Ottawa, Department of Agriculture, Dominion of Canada, Publication 525, Farmers' Bulletin 13, July **1944**, page 39.
    (One mistake they made is that Comfrey needs to be wilted before being used as silage.)

"A 10 square meter (107 square feet) area of Comfrey (Symphytum officinale) was chopped and ensiled in duplicate laboratory silos, either unwilted or following a 24 hour wilt, to test the hypothesis that the crop might be suitable for ensiling as animal feed. **Both crops were very difficult to chop due to the mucilaginous nature of the material.** On chopping, the unwilted crop formed a sticky mass, which did not pass through the chopper without considerable physical assistance from the operator. After a 24 hour wilt, the crop still required physical assistance to aid its passage through the chopper.
The laboratory silos did not allow liquid effluent (waste) to escape, and the DM (dry matter) concentrations were slightly, though not significantly, lower in the silages than in the fresh crop before ensiling. The presence of liquid within the silos, coupled with the relatively low DM concentrations may have influenced fermentation.
The concentrations of water-soluble carbohydrates, ash and nitrogen in the crops at ensiling averaged 107.0 and 118.0, 143.0 and 148.5, 32.2 and 34.7 g/kg DM for the unwilted and wilted materials, respectively. Buffering capacity averaged 488 and 472 mE/kg DM for the unwilted and wilted crops at harvest. Mean pH values of the silages were 5.43 and 5.16 for unwilted and wilted materials.
Concentrations of soluble and ammonia nitrogen (NH3-N) were, respectively, 708 and 470, and 238 and 179 g/kg total nitrogen in the unwilted and wilted silages, indicating **extensive proteolysis, particularly in the unwilted crop.**
The **contents of total nitrogen** in the crops at harvest (32.2 and 34.7 g/kg DM for unwilted and wilted materials, respectively), and in the silages (28.4 and 30.5 g/kg DM for unwilted and wilted materials, respectively), were relatively high compared to grass silages, and similar to concentrations for lucerne (alfalfa) silage (31.0 g/kg DM), and whole crop pea silage (28.6 g/kg DM) made in the United Kingdom ('Ministry of Agriculture, Fisheries and Food', Feed Composition, Second edition, Chalcombe Publications, Canterbury, United Kingdom, 1992).
The fermentation quality of the silages was poor, with relatively low concentrations of lactic acid and mean concentrations of n-butyric acid of 44.1 and 29.4 g/kg DM in the unwilted and wilted silages, respectively.
The epiphytic microflora may have been inappropriate for ensilage, in which case treatment of the crop with an effective additive might have produced beneficial results. **Without additive treatment and extensive wilting, the fermentation of Comfrey**

leads to poor quality silage."
-'A Laboratory Evaluation of Comfrey (Symphytum officinale L.) as a Forage Crop for Ensilage' by J.M. Wilkinson, School of Biology, University of Leeds, United Kingdom, Animal Feed Science and Technology, Volume 104, Issues 1-4, pages 227–233, February **2003**.   (Proteolysis is the breakdown of proteins into smaller polypeptides or amino acids.)

Correct Ways to Make Comfrey Silage

*"***Ensilage and Prickly Comfrey:*** Mr. Henry Doubleday, writes to 'The Chelmsford Chronicle' (England):*
*In the book on 'Forage Plants' by T. Christy, mention is made of **the plan adopted by the French for preserving green food for winter use, and which they style, 'ensilage'**. Although I had read this account I did nothing in it till quite lately, and then only commenced by accident. I was cutting a larger quantity of the **tops of the solid stem Comfrey** than my livestock could consume fresh and green, and the surplus was placed in cocks. These soon heated, and after standing some time **I gave these heated leaves to my stock,** and found they were extremely fond of them.*
*I then made a further trial, and **put into a wood tank about half a ton of green Comfrey leaves and trod them well down**. These also soon heated and settled down into a mucilaginous pulp, and I have been most succesful with my new food and find that cattle take to it most kindly.*
*I am now trying **the fattening properties of the green Comfrey leaves and their sour keep,** by feeding a bullock upon them with an occasional bit of the sort of which he is very fond. I am quite astonished at the result of my trial of Prickly Comfrey as a fattening food. I give my bullock some cut straw and 2 pounds (0.9 kg) of barley meal daily. Some farmers wre looking at him last week, and could hardly believe but that he had been fed with oilcake."*
-'The British Farmers Magazine', New Series, Volume 74; Rogerson and Tuxford, London, England, **1877**. Comfrey page 399.
(A cock or handcock is when grass and other plants cut to make hay are made into a small pile by hand. After a few days all the handcocks are put together in one pile to make a trampcock. Then someone gets on top of it to tramp it down. Then all the trampcocks are gathered together to make a haystack.)
(Oilcake is the solid residue left after oily seeds, such as cottonseed and linseed, have been pressed free of their oil. It is ground and used as livestock feed or fertilizer.)

*"**Storage of Comfrey by Ensilage:** The acquirement of Prickly Comfrey plants having now been brought within the reach of all farmers, either as regards convenience or cost, there is naturally more anxiety than heretofore, for the knowledge of what to do with it when grown; and as we promised a fortnight (2 weeks) since, **we give some few practical hints and instructions from Mr. Christy's useful pamphlet,** after reading which every grower of this prolific and useful forage plant will know how to store and utilise it to the greatest advantage.*
***With respect to the construction of M. Goffart's silage pits at Chateau Burtin, Department Loire et Cher, France,** the largest -36 feet (10.9 meter) in length by 6 feet (1.8 meter) in depth and width- has well-pointed stone wall sides, and paved bottom, bonded with cement. A second, adjoining it, is not walled round, but has, like the one just alluded to, vertical sides, so as to facilitate the pressing down of its contents. The largest frequently contains forty tons of green maize, mixed with about one-fifth of its weight of rye-straw chaff.*
These pits are worked as follows: *As fast as the fodder falls into them from the steam chaff cutter, it is spread out and firmly trodden by men. When the pit is full, some salt is sprinkled on the uppermost layer, then comes a coating of long straw, and finally a covering of planks well weighed down with logs or stone, not earth for it filters through to the fodder.*
***For some time after the completion of the above work, the pits need to be carefully examined every day,** as owing to the settling down of the fodder, cracks are apt to form in the roof, and these, if not closed, would admit a quantity of air, and injure the fodder by turning it mouldy.*

> *The same danger is incurred to some extent when a portion of the provender (food for livestock) is taken out for feeding purposes, and on these occasions the precaution of covering over the exposed parts again as quickly and completely as possible has to be observed.*

The greater 'proofiiness' of the sour food, in the case of sainfoin, lucerne, and vetches, he ascribes to the fact of the crop being allowed to remain standing longer than it would if fed green or made into hay.

> *The coarser woody portions, which, under ordinary circumstances, the cattle would refuse to eat, or which would pass through their bodies undigested, are rendered soft and assimilable by the process of fermentation; and the somewhat acid taste and peculiar penetrating aroma which the fodder acquires cause it to be consumed with avidity (eagerly).*

*This is especially applicable to India, where forage is so liable to become dry and woody. It has been tried and found to answer in England, but **too much salt was often employed,** and now some farmers are opposed to this common practice of sprinkling salt on each layer in the pit, maintaining that the salt does more harm than good, as it retards fermentation, and renders the lower portions oversalt.*
Opinion differs also as to the degree of ripeness the crop should be allowed to attain before being cut; but the leaves and flowers (particularly in the case of Comfrey and clover) are never allowed to become so withered and dry that they drop off during transport.
A very succulent crop will heat in the pit, and acquire a darkish colour and disagreeably pungent (sharply strong) smell.
It was supposed there was less danger, however, to be apprehended from over succulency than from over-wetness; and careful farmers avoided carrying and housing their green food as long as the latter was saturated with rain or dew.

> *But from experiments tried this season when the forage was wet, and gathered in October 1870, during storms of rain, and cut up quickly and placed in a deep pit in the open air, it has turned out in first rate order and quite equal to that put*

up in dry weather, opened last week December 1876. It is of great importance that farmers in Ireland, Scotland, and north of England should note this.
Generally speaking, it requires to remain at least six weeks in the pit before it is in a state fit for consumption. As the presence of an excess of lactic acid -the acid to which the sharp taste is due- may cause the sourkeep to act as a purgative, and even bring on diarrhoea, a certain amount of caution is necessary in feeding with it. When, occuring only to the ordinary extent, and the cattle are in a healthy state, no fear need be entertained; the acid then appears to serve the usual purpose of promoting digestion. **Experience shows that the deeper the pit, and consequently the greater the pressure on the forage, the better is the quality."**
-'Storage of Comfrey by Ensilage', Australian Town and Country Journal, Sydney, NSW, Australia, page 21, January 26 **1878**.

"Ensilage in the United States by Professor James E. Thorold Rogers, M.P.:
That corn or maize is the commonest, and will remain the most frequently ensilaged crop, cannot be doubted. But it is an error to suppose that maize is the only article stored; green oats, clover, trifolium (clover), grass, all kinds of leguminous plants, sorghum, dourra (Indian millet), **Prickly Comfrey,** and even apples, are stored in silos."
-'Journal of the Society of Arts', Volume 31, November 17 1882 to November 16 1883; George Bell and Sons, London, England. Comfrey, February 2 **1883**, page 232.

"**Mr. Thomas Christy**, F.L.S., of Malvern House, Sydenham (England), called in; and Examined, **April 22 1885**:
Chairman: "As to the actual feeding values of the different types of fermented foods: now, **you have tried Comfrey?** Do you grow it on a large scale at all?"
    Mr. Christy: "Yes and yes, in Essex, near Coggeshall (England)."
Chairman: "**The ordinary Comfrey, with the solid stem?**"
    Mr. Christy: "Yes. This is an instance of a plant holding a large quantity of liquid; if this plant is dried up like hay, you will hardly find anything when you come to take it out of a stack, because the liquid has almost disappeared, and to such an extent, if the Comfrey is cut green, that the leaves won't stand upright, but **if this Comfrey is cut and partially fermented, and then put into a silo,** it fattens the cattle more rapidly than almost anything else. That is, the heat just started as it is put into the silo. Some people start their heat by putting it outside the silo in heaps, so that they can spread it over, and others spread it at the bottom of the silo and so allowing it to get up a heat."
Mr. Michael Henry: "Do you think that with the fermented Comfrey the fermentation adds something to the nutritive value of the Comfrey itself; **that the fermented Comfrey is more nutritious than the fresh Comfrey?**"
    Mr. Christy: "It is an opinion of myself and some others that it does, but perhaps not to the extent of many other crops. There is no sugar in the sap of Comfrey, if you do not allow it to go into the full flowering stage."
Chairman: "Have you any analysis of Comfrey?"
    Mr. Christy: "It has been analyzed by a gentleman connected with the Cirencester College (England), Professor Wrightson and Professor Voelcker."
Chairman: "On what sort of soils?"
    Mr. Christy: "On clay soils, and on steep fields that it is difficult to plough."
Mr. Harris: "**Have you made much ensilage from Comfrey? Was it good?**"
    Mr. Christy: "**I think we made 20 to 30 tons last year (1884). Yes, it was very good indeed.**"
Chairman: "How many tons can you grow to the acre? And you cut it how many times?"
    Mr. Christy: "Twenty tons of green Comfrey at a cutting in certain seasons. That depends on the season; if it is very dry the Comfrey does not grow so well."
Chairman: "And therefore you would not recommend it for light sandy soils?"
    Mr. Christy: Without it is allowed to get a good head first, because at Kew we dug 8 feet (2.4 meter) and found the roots of Comfrey; the garden of Kew (Royal Botanic Garden, London, England) was the bed of the river originally."
Mr. Harris: "**Would you recommend siloing Comfrey alone, or with grass put in layers?**"
    Mr. Christy: "**It is a very good thing in any silo to mix crops;** I am quite sure that for food for cattle the more the crops can be mixed the better. I have been trying some experiments with poultry and pigeons for getting eggs for food supply, and getting weight on fowls; and I find the more I mix the foods the better."
Mr. Harris: "In making silage would it improve the mechanical property of ensilage by mixing layers with grass rather than grain?"
    Mr. Christy: "Yes."
Mr. Henry: "**Would you chop the Comfrey?**"
    Mr. Christy: "**I would not; it is not necessary;** if you have long stems, then it is advisable to cut it like grass, but it is unnecessary."
Mr. Henry: "Have you used Comfrey for food, say for feeding poultry?"
    Mr. Christy: "Yes, I could not keep the poultry away from one variety that was sent to me by the director of the Botanical Gardens at Saint Petersburg (Russia); but I had to wire them in, and from the small cuttings I have propagated many acres; along with the other varieties of Comfrey that I had grown, I found that the cattle preferred this one variety to any other. **I adopted the solid stem Comfrey.**"
Chairman: "Is that Caucasian Comfrey?"
    Mr. Christy: "No, it is a new variety."
'The Sessional Papers Printed by Order of the House of Lords, or Presented by Royal Command in the Session 1884-1885, 48 & 49 Victoria', Volume 12, London, England. Comfrey, April 22 **1885**, page 45; pdf page 230.

Mix Comfrey with Cereral Grains

"Mr. A. Goffart, in his work on the 'Cultivation and System of Ensilage of Maize and other Green Forage Plants' (Manuel de la Culture et de L'Ensilage des Mais et Autres Fourrages Verts), writes, Paris, France, 1877:
'Maize (corn) barely contains 1.20 to 1.25 percent of azotic (nitrogen) matter, whereas a recent analysis gives 2.70 percent to Comfrey, or more than double. Thus these two plants, instead of rivalling one another, are rendered (made) each one by the other more complete, to the great advantage of agriculture."
-'New Commercial Plants with Directions How to Grow Them to the Best Advantage, No. 1 (**1878**) and No. 3 (**1880**)' by Thomas Christy, F.L.S. (Fellow of Linnaean Society), London, England. Number 1 includes Caucasian Prickly Comfrey pages 14-16. Number 3 includes Prickly Comfrey pages 11-14. Includes 'Caucasian Prickly Comfrey' and 'Russian Comfrey' advertisements. (If you have an English translation of the Comfrey section of Mr. Goffart's work, could you please send it to me.)

*"Mr. Goffart finds that he can ripen maize quite sufficiently for purposes of Ensilage, and it must be borne in mind that **any crops like rye or anything with much straw, want something of the nature of clover or Comfrey to help them to solidify and form a mass.***
When intended for the silo, Comfrey should be allowed to come into full flower before cutting."
-'Ensilage: A System for the Preservation in Pits of Forage Plants and Grasses, Independent of Weather: A Collection of Facts and Statistics on the Cheapest Mode of Providing Winter Food for Dairy Cattle, Sheep, Horses, Etc.', by Thomas Christy, F.L.S. (Fellow of Linneaen Society), London, England, **1883**. Caucasian Prickly Comfrey, pages 57-63.

"One of the chief objects of the farmer, so Mr. Bernard Dyer once wrote, is to make, during the season of active growth, as much provision as possible for feeding his stock during the winter, when vegetation is comparatively dormant.
The condition of dryness in which the fodder is packed for ensilage is also a matter calling for careful attention. Dr. Voelcker once said that his experience of ensilage is such as to show that no ordinary green English farm crop can be successfully stored without an **admixture (addition) of a considerable portion of fine straw chaff, to absorb the superfluous (extra) moisture**, and so retard (reduce) fermentation and putrefaction.
This precaution being found frequently necessary even with the green maize (corn plant) pitted (put in pit or silo) in America which is less succulent (juicy) than rye grass, clover, tares (vetch), or Prickly Comfrey.
The most promising use of ensilage is in connection with dairy farms, for milch cows love moist and succulent food, and, when properly and judiciously (wisely) supplied, always repay a good allowance of such food by a good flow of milIk."
-'Live Stock Journal, Vol. 87', Vinton & Company Limited, London, England, May 10 **1918**. Weekly newspaper established 1874.

"**The most successful silage maker, a Surrey County, (England) farmer who makes 300 tons of cereal-legume mixture silage a year, merely mixes wilted Comfrey from his acre in with the rest; the high carbohydrate of the mixture levels up the high protein in Comfrey**, and the result is more silage of a slightly better nutritional value which is much appreciated by cattle. The easiest way of using unwanted summer cuts of Comfrey to help winter keep is as an addition where silage is already made. **Up to 25 percent of the fill with grass, cereal-legume mixture or maize (corn) can be Comfrey, either chaffed or as cut,** and the result is excellent, with no special precautions or variation in technique. The carbohydrates in the crop balance the excess protein, and the finished product is balanced at a higher level of quality."
-Russian Comfrey: A Hundred Tons an Acre of Stock or Compost for Farm, Garden or Smallholding by Lawrence D. Hills. London England: Faber and Faber, Limited, **1953**, page 51, 81.

"The late Mr. J.W. Hobbs of the Great Glen Cattle Ranch, Fort William (Highland, Scotland), used to **put the production of his 1 1/2 acre Comfrey bed through a chopper blower and mix it with 'mashlum', which is Scots for oats and vetches cut at the milky high carbohydrate stage, and made silage with 14.57 percent protein including Comfrey,** and 10.08 percent for mashlum alone, according to the 'North of Scotland Agricultural College' analysis."
-Comfrey: Fodder, Food and Remedy by Lawrence D. Hills. New York: Rizzoli Universe Books: **1976**, page 116.

*"J.W. Hobbs of the Great Glen Cattle Ranch, Fort William uses **25% Comfrey with 75% of a cereal-legume mixture, cut with the oats at the milky stage**. The gain in protein from the Comfrey has more than justified his 3-acre crop."*
-Comfrey: Nature's Healing Herb & Health Food by Andrew Hughes. Japan: Sanyusha Publishing, 1992, pages 21, 159.

"**For silage, the Comfrey crop must be cut and allowed to wilt for at least 24 hours, and carbohydrate preservatives such as molasses or grain are needed. Probably the most economical preservation is to ensile a forage mixture by blending up to 25 percent Comfrey with small grain or corn forage.**"
-'Comfrey: A Controversial Crop' by Robert G. Robinson, University of Minnesota, Agricultural Experiment Station, Minnesota Report MR-191, Item No. AD-MR-2210, **1983**, page 4.   (Corn forage is the entire plant.)

"**Comfrey must be cut and allowed to wilt for a minimum of 24 hours when used as silage. Additives such as molasses or grain are sometimes helpful, and mixing up to 25% Comfrey with small grain or corn forage serves as an economical method to make high quality silage.**"
-'Comfrey' by Teynor, Putnam, Doll, Kelling, Oelke, Undersander and Oplinger, 'Alternative Field Crops Manual', University of Wisconsin: Cooperative Extension, University of Minnesota: Center for Alternative Plant & Animal Products, and Minnesota

Cooperative Extension, February 1992, updated November 1997, page 4.

Add Molasses to the Comfrey

"Twenty-four-hour wilting is advised to bring the cut down by 30 to 50 percent of the water content. *The N.A.A.S. (National Agricultural Advisory Service of England) advise that **two gallons (7.5 liters) of molasses mixed with a further two gallons of water is the minimum; three gallons (11.3 liters) of molasses was rather better to make sure of a good lactic acid fermentation** and the fall in pH value after heating that keeps the crop in a state of arrested decay. Comfrey, with the protein higher than any (other crop), but low in carbohydrates, needs the addition; if too little (molasses) is added the proteins will not be preserved and their decay will make the whole product ill-smelling and distasteful to stock, apart from lost nutritional value.*"
-Russian Comfrey: A Hundred Tons an Acre of Stock or Compost for Farm, Garden or Smallholding by Lawrence D. Hills. London England: Faber and Faber, Limited, **1953**, page 80.

Or Add Sugar and Yeast

"*H.D.R.A. (Henry Doubleday Research Association, England) member Mr. B. Hickey of Ascot, Berks (Bershire County, England), **invented a variation on silage for poultry** (12 hens):*
'*October 1967 stuffed Comfrey into 25 gallon (94.6 liters) barrels, filled to top with water and two pounds (0.9 kg) of sugar, added one ounce (28.3 grams) dry yeast. Fed hens on three barrels during winter, lost no hens to date, average five or six eggs, eat about 10 ounces (283 grams) each Comfrey daily, moulted like wild birds, fed wheat and a little range mash, cost about two pence (pennies) a day.' The Comfrey was packed into the barrels which were covered but not airtight. There is scope for experiment with this unusual mixture which appears to have been nearer Comfrey chutney for chickens than silage, for the yeast could well have produced a vinegar fermentation.*"
-Comfrey Report: The Story of the World's Fastest Protein Builder and Herbal Healer, Conservation Gardening and Farming Series: Series C by Lawrence D. Hills. England: Henry Doubleday Research Association, **1975**, page 71.
(English-style chutney includes sugar and vinegar to give a long shelf life so that autumn fruit can be preserved.)

Or Add Lactic Acid Bacteria

"*The making of silage is used widely in dairying, from the standpoint of cost and high food value; 'Nihon University Farm' (Japan) tested Comfrey by this method:*
Comfrey (leaf and stem) was dried 2-3 hours under the sun, cut before the first frosts (middle to end of October); it was then cut and mixed with 10% of bran, with a little lactic acid bacteria added and put into the silo. Lactic acid bacteria was used to promote lactic acid fermentation.
Two kinds of lactic acid bacteria, Lactobacillus plantarum Orla-Jensen and Lacto bacillus brevis Orla-Jensen, were used. Mix 800 grams (28.2 ounce) of bacteria in 40 liters (10.5 gallons) of water, and culture for 3 days at the temperature of 30 degrees C (86 F).
Ten liters (2.6 gallons) of the culture was used on 3.7 tons of Comfrey plus bran, and then it was packed into the silo. **The silage was matured for 2 months** *and proved to be very good in smell and color and in no way inferior to Dent Corn silage. Cows and pork pigs liked it very much.*
The analysis of Comfrey silage done by the 'Agriculture and Veterinary Food Research Department':
Water 9.76%, Protein 9.25%, Crude Fat 7.77%, Crude Fiber 13.39%, Crude Ash 23.51%, Soluble Non-Nitrogenous 36.32%. The volatile fatty acid analysis, an important factor in determining silage value, shows that **it is very suitable for stockfeed.**"
-Comfrey: Nature's Healing Herb & Health Food by Andrew Hughes. Japan: Sanyusha Publishing Co., Ltd, **1992**, page 188.
(Nihon University is a private research university in Tokyo, Japan.)
(Lactic acid is an organic compound with the formula $CH_3CH(OH)CO_2H$. As a solid, it is white and water-soluble. As a liquid, it is colorless. It is produced both naturally and synthetically. Lactic acid fermentation is performed by lactic acid bacteria, which convert simple carbohydrates such as glucose and sucrose to lactic acid.)

Or Add Chemical Acid

"**The acid compound IB-2 was superior to sodium nitrite as an additive for ensiling herbage of** *Helianthus tuberosus,* **Symphytum asperum (Prickly Comfrey)** *and Silphium perfoliatum; no additive was suitable for ensiling herbage of Heracleum sosnowskyi, Helianthus tuberosus and Rhaponticum carthamoides.*"
-'Chemical Conservation of Fresh Matter of New Silage Plants' by A.I. Mikhkiev, V.M. Rozenberg and H.G. Il'm, Institut Bilogii Karel'skogo Filliala Akademii Nauk SSSR (Biology Karelian Branch of the Academy of Sciences of the USSR), Petrozavodsk, USSR; Fifth Symposium on New Silage Plants: Part 1, Lenningrad, Russia, pages 63-64, **1970** or 1972.
(I prefer natural ways of fermenting Comfrey rather than this method. Nutrition is better with old-fashioned methods.)

Silo Should Have Drainage

"The silo should be filled loosely, and the first four feet (1.2 meters) should go in totally unconsolidated and allowed to warm to blood heat before the rest of the crop is filled in.
***There should be some drainage provision** as there will be loss of liquid, but not on any great scale if the crop is well wilted.*
There is no need to chaff it, as unlike other stemmy materials the fibre is low and the stems break up."
-Russian Comfrey: A Hundred Tons an Acre of Stock or Compost for Farm, Garden or Smallholding by Lawrence D. Hills. London England: Faber and Faber, Limited, **1953**, page 80.

*"**The effluent (liquid waste) from Comfrey silage is very rich in potassium and therefore should be run into a slurry tank, or distributed straight back to the land.**"*
-Comfrey: Fodder, Food and Remedy by Lawrence D. Hills. New York: Rizzoli Universe Books: **1976**, page 116.
    (A slurry tank, slurry pit or slurry lagoon is a hole or dam where farmers gather animal waste with other organic matter, such as hay/straw and water runoff to convert it to fertilizer that can be reused.)

Tank Storage System for Comfrey Silage from 1877

*"**Prickly Comfrey as Silage:**
    The Storage of Fodder has for some time past been exciting much interest, especially in France. Mr. Barral's report to the 'Agricultural Journal of France' (in 1876) **speaks highly in favor of Symphytum Asperrimum (Prickly Comfrey), particulary with a view to its conservation (ensilage).** It is perennial and a very heavy crop producer. It contains an unusually large percentage of Azote or Nitrogen, and gives an early, or I may say the earliest, spring crop on the farm. **When other crops are ready for cutting and green food becomes abundant, the Symphytum Asperrimum should still be regularly cut and stored in Tanks or Pits (for silage)**, as a provision against a long summer drought or for winter use.
    Opinion differs also as to the degree of ripeness the crop should be allowed to attain before being cut; but **the leaves and flowers (particularly in the case of Comfrey and clover) are never allowed to become so withered and dry that they drop off during transport. A very succulent (juicy) crop will heat in the pit,** and acquire a darkish colour and disagreeably pungent smell.
    It was supposed there was less danger, however, to be apprehended (had) from over-succulency (moist, juicy in the plant) than from over-wetness (outside the plant); and careful farmers avoided carrying and housing their green food as long as the latter was saturated with rain or dew.
    But from experiments tried this season when the forage was wet, and gathered in October 1876, during storms of rain, and cut up quickly and placed in a deep pit in the open air, it has turned out in first rate order and quite equal to that put up in dry weather, opened last week December 1876. It is of great importance that farmers in Ireland, Scotland, and North of England should note this.
Bury Green Fodder:
    As the **Tank Storage System** is so imperfectly understood and almost unknown in Great Britain and the Colonies, I will describe it in detail. When the French first went to Algeria (north Africa), they experienced great difficulty in securing forage in the dry season, but by bribing some of the chiefs they learnt the secret of the Arab system of ensilage or **simply burying the green fodder in trenches, carefully concealed by replacing the earth**, and they soon learnt how to trace them out.
    The lesson thus learnt in Algeria has not been forgotton, and M. Goffart, of Chateau Burtin, Dept Loire et Cher (France) has of late devoted much time to experiments of this new method of preserving green food and with great success.
How to Build Tank or Pit:
    With respect to the construction of M. Goffart's pits at Chateau Burtin, the largest- 36 feet (10.97 meters) in length by 6 feet (1.8 meters) in depth and width- **has well-pointed stone wall sides, and paved bottom, bonded with cement.** A second, adjoining it, is not walled round, but has, like the one just alluded (referred) to, vertical sides, so as to facilitate the pressing down of its contents.
    M. Lecouffe in a communication to the 'Lille Agricultural Society' (city in northern France), dated August 31, 1876 states that **his pits are constructed of brick, built on the slope of a hill to facilitate (allow) the escape of moisture, and made broader above than below to prevent the formation of hollow spaces when the food settles down.**
    The exact dimensions of his pits are 4 1/2 feet (1.37 meters) in depth by 9 3/4 feet (2.97 meters) in width, increasing at the top to 10 1/2 feet (3.2 meters).
    Experience shows that the deeper the pit, and consequently the greater the pressure on the forage, the better is the quality. A thin partition of stone or brick may be made between the tanks at a small cost, with cross sections or supports and arches in them. **There ought to be no filtration of water from the outside.**
How to Fill Tank or Pit:
    These pits are worked as follows: As fast as the fodder falls into them from the Steam Chaff Cutter, it is spread out and firmly trodden by men. When the pit is full, some salt is sprinkled on the uppermost layer, then comes a coating of long straw, and finally a covering of planks well weighted down with logs or stone, not earth for it filters through to the fodder.

Another method is used by M. Lecouffe: **To exclude the air more effectually, a layer of beetroot pulp is placed above the contents of the pit.** *This method has been tried and found to answer (work) in England, but* **too much salt was often employed,** *and now some farmers are opposed to this common practice of sprinkling salt on each layer in the pit, maintaining that salt does more harm than good, as it retards fermentation, and renders the lower portions oversalt.*

Care of Pit During Fermentation:

For some time after the completion of the above work, the pits need to be carefully examined every day, as owing to the settling down of the fodder, cracks are apt to form in the roof, and these, if not closed, would admit a quantity of air, and injure the fodder by turning it mouldy. *The same danger is incurred to some extent when a portion of the provender (food for livestock) is taken out for feeding purposes, and on these occasions the precaution of covering over the exposed parts again as quickly and completely as possible has to be observed.*

Time in Pit Before It is Ready:

Generally speaking, it requires to remain at least six weeks in the pit before it is in a state fit for consumption.

Feeding to Livestock:

As the presence of an excess of lactic acid (the acid to which the sharp taste is due) may cause the sour-keep to act as a purgative (laxative), and even bring on diarrhoea, *a certain amount of caution is necessary in feeding with it. When, occuring only to the ordinary extent, and the cattle are in a healthy state, no fear need be entertained; the acid then appears to serve the usual purpose of promoting digestion.*

Conclusions:

M. Lecouffe's experience with pitted or preserved fodder, both with and without the admixture (addition) of straw chaff, has been highly satisfactory, *and he strongly recommends its employment in the cattle stall in conjunction with linseed (flaxseed) cake, turnips, etc.*

Monsieur Sottom writing to the 'Journal d'Agriculture Practique' ('Journal of Agriculture Practice' in publication in Paris, France from 1853 to 1936) from Gers, in the south of France, fully indorses (endorses) the estimate formed of the sour-keep by Monsieur Lecouffe and Goffart, and has found it excellent food for working oxen.

Two no slight advantages connected with the pitting system are, **the economy of labour, and the comparative independence of weather,** *the farmer adopting it enjoys in harvesting his forage and herbage crops as well as change of diet. As remarked by Professor Wrightson in his recent 'Report on the Agriculture of Austria-Hungary', this method of storing green food is one well worth the attention of English agriculturists."*

-**'Forage Plants and Their Economic Conservation by the New System of Ensilage: Part I: Caucasian Prickly Comfrey' by Thomas Christy,** Jun., F.L.S. (Fellow of the Linnean Society), Christy & Co., London, England, **1877**, pages 27-31.

(Professor J. A. Barral, Perpetual Secretary of the 'Central Society of Agriculture in France' wrote in the Journal de l'Agriculture, of 7th October, 1876.)

(Lactic acid is found in sour milk products such as cheese, clabbered milk, kefir, koumiss, laban and yogurt, and in sauerkraut and soy sauce. It gives the sour taste of sourdough bread. The casein in fermented milk is coagulated/curdled by lactic acid.)

(Austria-Hungary or the Austro-Hungarian Empire was a constitutional union of Austria and the Kingdom of Hungary from 1867 to 1918, when it collapsed due to defeat in World War I.)

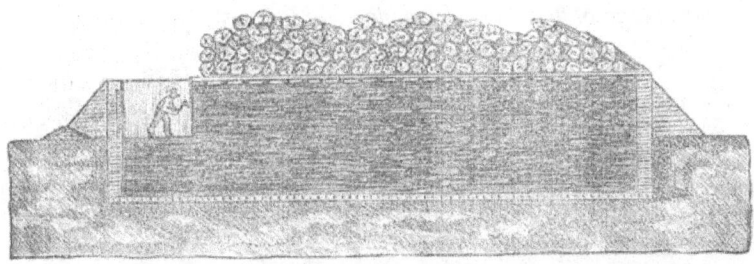

PLAN OF PITS AT CHATEAU BURTIN.

1877 Plan of Silage Pits at Chateau Burtin Department Loire et Cher, France

'Forage Plants and Their Economic Conservation by the New System of Ensilage:
Part I: Caucasian Prickly Comfrey'
by Thomas Christy, Fellow of the Linnean Society, Christy & Co., London, England, 1877. Page 27.

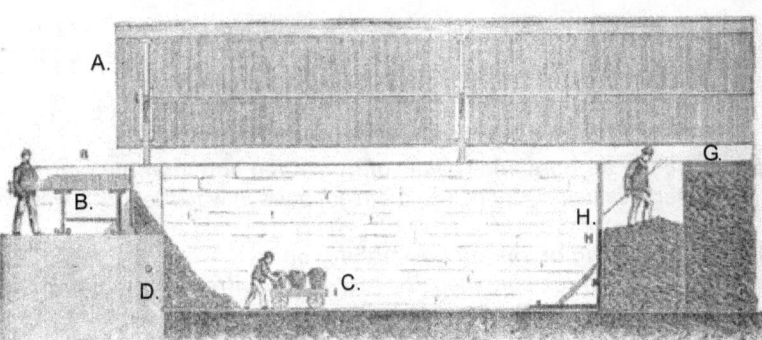

A, the roof; B, chaff-cutting machine; C, trolly on rails; D, the outside walls; E, centre wall; F, filled silo; G, first section filled; H, moveable partition.

1883 Silo: Silage Pit

'Ensilage: A System for the Preservation in Pits of Forage Plants and Grasses, Independent of Weather', by Thomas Christy, London, England, 1883.
Silage pit, page 20. Prickly Comfrey, pages 57-63.

A. The roof
B. Chaff-cutting machine
C. Trolly on rails
D. The outside walls
E. Centre wall (in another drawing)
F. Filled silo (in another drawing)
G. First section filled
H. Moveable partitions

Chapter 22

Livestock / Pet Species and Comfrey

See sections 'Medicinal Comfrey Overview' and 'Making and Using Comfrey Medicine' (Chapter 27).

Consult with your veterinarian or farm professional about your livestock and pet health needs.

Cats and Dogs

"*Therapeutic veterinary use of the Comfrey (Symphytum officinale) plant has been reported. Preparations, sometimes called Kytta®, made from Comfrey roots have been applied topically to promote healing of tongue ulcers in cats*,**.*"
-Poisonous Plants in Britain and Their Effects on Animals and Man by Marion Cooper and Anthony W. Johnson for 'Ministry of Agriculture, Fisheries, and Food'. London, England: H.M. The Stationery Office, **1984**.
(* -'Tongue Ulcers in Cats' by J.A. Rohrbach, The Veterinary Record, Volume 101, No. 14, page 292, October or November 1977. I was unable to get this. If you have it, could you please send it to me.)
(** -'Kytta® Preparations Used by Veterinary Practitioners' by H. Gonnermann, Tierarztl Umsch Journal, Volume 31, pages 402-404, 1976. All in German. I was unable to get this. If you have it, could you please send it to me.)

"**Comfrey poultices are an excellent first aid measure in the treatment of cuts, burns, abrasions and other injuries,** for Comfrey contains allantoin, a cell growth stimulant that speeds healing. Practically everyone who works with herbs and dogs, cats, birds or rabbits has a Comfrey story, and most involve its direct application.
A widely published warning about Comfrey is that it should not be applied to puncture wounds or infected cuts because its rapid healing action may cause the wound to close, trapping infection beneath it.
    To guard against infection, clean the wound first with topical disinfectants such as grapefruit seed extract, lavender essential oil or echinacea tea or tincture before applying the Comfrey.
    Another technique is to alternate Comfrey and other herb poultices before the wound closes.
Comfrey dramatically speeds the healing of wounds and even broken bones.
    In fact, some herbalists have fed Comfrey to dogs after surgical pins were placed in their broken bones following traffic accidents, and the bones healed so quickly that they forced the pins out, requiring additional surgery. (Comfrey's common name is 'knit bone.')
Limit Internal Use of Comfrey:
Veterinarian Beverly Cappel-King cautions that she has seen dogs who began vomiting after taking several Comfrey capsules daily for three to five weeks. When she tested the blood of these dogs, Cappel-King found elevated liver counts.
    She therefore recommends giving Comfrey in food with caution, if at all. PA-free (Pyrrolizidine Alkaloid) Comfrey tinctures are available for those who wish to take advantage of the plant's exceptional properties without the risk of liver disease, and small quantities of fresh or dried Comfrey are unlikely to cause problems."
-The Encyclopedia of Natural Pet Care by C.J. Puotinen. New Canaan, Connecticut: Keats Publishing, Inc., **1998**.
    (See subsection 'Comfrey Poultices' in section 'Making and Using Comfrey Medicine' {Chapter 27}.)
    (Tea tree oil can be toxic to cats and dogs.)

"**Comfrey is one of those herbs which is high in nutritional value.** It contains also chlorophyll and allantoin- which is a 'cell proliferant- encouraging the multiplication of cells more rapidly than usual to heal wounds, burns, or tissue irritations such as frequently occur in the intestinal tract from prolonged use of processed carbohydrate foods.
Comfrey can be purchased at most health food stores in bulk as a tea, then blended into the live-food supplemental drink, or can be simmered in the broth during fasting.
The Comfrey root is more potent, but must be ground to a powder before being added to soups or blender so it is better utilized."
-'Perfect Health for Dogs and Cats' by Kit Cain, Soulful Stories Publishing, Yarmouth, Nova Scotia, Canada, 36 pages, **2006**.
    (Limit internal use of Comfrey. Comfrey root contains more toxic Pyrrolizidine Alkaloids than Comfrey leaf.
    Fasting may or may not be good for your pet depending on his/her health. Please contact your veterinarian first.)

"**Interview of dog owners and traditional healers, experience of authors and review of published and unpublished studies were employed to collect data.** Preparation of the herbs is usually by boiling the plant parts in water for 15-20 minutes, strained and cooled before use (decoction).
    This is then applied on the skin of the dog through bathing or smudging (rubbing on).
The herbs used for wound management are the mesophyll of aloe (Aloe barbadensis), **roots of Comfrey (Symphytum officinale)**, leaves of the eggplant (Solanum melongena), leaves of guava (Psidium guajava), and dye of atchuete (Bixa orellana).
    Comfrey Active ingredient: allantoin, tannin, mucous substances.

> **Part of plant used:** Comfrey roots.
> **Preparation of herb:** bruise roots and immediately apply topically, or apply leaf poultice overnight.
> **Indications:** wounds, abscess, cuts."

-'Commonly Used Herbs for Canine Dermatologic Problems in the Philippines' by Marianne Leila Santiago-Flores, Marco F. Reyes, Marian C. Jaro, Michael Remil G. Amparo, and Frances C. Recuenco, Philippines and Japan; Presented at Third Academic Conference of Asian Society of Traditional Veterinary Medicine, Nihon Veterinary Life Science University, Tokyo, Japan, July 31 to August 1 **2010**.

"**Comfrey for Animals- extract and powder:**
Comfrey is anti-inflammatory and a cell proliferator. It is useful for topical (skin) treatment of wounds, insect bites, nicks, scrapes, burns. It is a healing aid in contusions, sprains and fractures when applied directly to the closed wound as a poultice.
External Use of Comfrey:
> **Comfrey contains allantoin, a cell proliferator, and topical application may heal so fast as to trap infectious microbes.** It is recommended that 'Wound Aid for Animals'®, a potent antimicrobial, be used with powdered Comfrey herb on open wounds to disinfect.

Internal Use of Comfrey:
> **Internally Comfrey should be used with caution due to its alleged toxicity.** Use one cc per 50 pounds (22.6 kg) weight twice daily, simultaneously with 400 mg ground seed of milk thistle. Unlikely or not it is advisable to simultaneously use milk thistle with Comfrey extract to provide for hepatoprotection (liver protection).

Safety of Comfrey:
> Comfrey is contraindicated during pregnancy, lactation or with a pre-existing liver disorder.
> The argument that Comfrey is not safe for internal use is unconvincing since it is based on dosing rodents with high levels of purified pyrrolizidine alkoloids. It is my opinion that Comfrey has received an undeserved bad rap. Clinical trials are in order to separate fact from fiction and hysteria.

Comfrey Root versus Leaf:
> The root is about two orders of magnitude higher in concentration of pyrrolizidine alkoloids than is the herb. Therefore hepatotoxicity likelihood resultant from herb extract is minimal.

Buck Mountain Botanicals is a certified organic grower and producer of veterinary botanical products for all creatures."
-'Buck Mountain Animal Health: Wholesale Catalog' by Buck Mountain Botanicals, Inc., Miles City, Montana, **2010**.
(I do not have any connection with this company or the products that they sell.)

"**Symphytum officinale:**
> **Parts Used:** All aboveground parts.
> **Primary Medicinal Activities:** Heals wounds; anti-inflammatory; astringent; lubricates, soothes and protects internal mucous membranes; expectorant.
> **Strongest Affinities:** Skin, digestive tract, respiratory tract.
> **Preparation:** Water or oil infusion, poultice, salve, fomentation.

For treatment of colitis, stomach ulcers, or just about any other inflammation of the digestive tract: For dogs and cats, 1/2 to 1 teaspoon of the dried Comfrey herb for each pound (0.45 kg) of food fed should be of therapeutic benefit.
Comfrey is traditionally used for bronchitis and other respiratory ailments that are relieved by its soothing anti-inflammatory nature. A cooled tea serves this purpose well. Administer 1 tablespoon per 20 pounds (9 kg) of the animal's body weight, twice daily.
> In all cases of internal use, feeding of Comfrey should be limited to occasional short-term therapies. For the same reason, highly concentrated preparations such as tinctures or strong decoctions should not be employed internally.

Cautions:
> Don't use Comfrey in pregnant or lactating animals, or in animals with pre-existing liver disease.
> Common sense dictates that anything ingested in over-abundance is potentially toxic."

-Herbs for Pets: The Natural Way to Enhance Your Pet's Life by Mary L. Wulff and Greg L. Tilford. Los Angeles, California: i5 Publishing, **2011**, pages 86-89.

"**The leaves from many other types of plants,** e.g. violet (Viola odorata), nasturtium (Nasturtium officinale), **Comfrey** (Symphyturn officinale) and nettle (Urtica dioica), **also have healing effects on animal skin. They not only revitalise the skin and make animals' coats shine but also have anti-itching, disinfectant, and star-healing (energy) effects.**"
-'Useful Plants for Animal Therapy' by M. Laudato and R. Capasso, Department of Pharmacy, University of Naples Federico II, Italy; OA Alternative Medicine, Open Access, London, England, Volume 1, 6 pages, February **2013**.

"**Healing Broken Bones in Dogs: The study was undertaken to evaluate the efficacy of Symphytum officinale as a stimulating agent in the process of long bone fracture healing in canines (dogs).**
Twelve clinical cases of long bone fracture in dogs presented at 'Teaching Veterinary Clinical Services Complex', Nagpur (India) reported during February 2014 to July 2014, were divided into two equal groups and subjected to the treatment as per schedule.
The method adopted for the administration of the Symphytum officinale homoeopathy drug of 30 c potency for oral administration in the dogs with fracture of long bones during the present study was found suitable and easy and the quantity of globules administered **was found satisfactory as an osteoinducer.**
In group II, in addition to immobilization of fractured bone, Symphytum Officinale homoeopathy drug was administered

at four globules twice a day for days for 21 days postoperatively.
>   ***The animals of group II with long bone fracture required less time for partial weight bearing** 3.25 +- 0.25 postoperative days and complete weight bearing 25.75 +- 2.17 postoperative days as compared to group I and reduction in weight bearing was statistically significant.*

Radiographs (X-ray or similar) taken at schedule intervals showed accelerated fracture healing with complete bridging of fragment, bony deposition along with periosteal and medullary continuity across the fracture site on animals of treated group II whereas radiographic union and periosteal bridging at fracture site were relatively slow and incomplete in dogs of group I at schedule interval.

Chhavi et al. (*2011) reported that **Symphytum officinale removes the inflammation surrounding the fracture.** It induced union of affected bones, and contains allantoin, a crystallizable substance used in orthodox medicine to encourage epithelial formation in ulcer and wound."
-'Role of Symphytum Officinale as an Osteoinducer in Long Bone Fracture Repair in Canine' thesis by Swarop Chandel for Master of Veterinary Science degree, Department of Veterinary Surgery and Radiology, Nagpur Veterinary College, India, **2014**.
(* -'Recent Update on Proficient Bone Fracture Revivifying Herbs' by S.S. Chhavi, A. Drabu, R. Verma, A. Dhiman and A. Sharma, all from India; International Research Journal of Pharmacy, Rampur, India, Volume 2, No. 11, pages 3-5, 2011.)
>   (Osteoinduction is the process by which osteogenesis is induced for bone healing to take place. Immature cells are stimulated to develop into pre-osteoblasts so that bone starts growing.)

"**Ellie's Dog Biscuits:**
1 cup (236 ml) bran, 1 1/2 cups (354 ml) whole meal flour
1/2 cup (118 ml) olive oil or sunflower or SoyaOlive is great for their coat
1/2 cup sunflower seeds, 1 cup oatmeal, 1 egg, 1 cup milk or water, 1 teaspoon brewers yeast, 1/2 teaspoon salt or kelp,
1/2 cup coconut , **1 Comfrey leaf, finely chopped** and can add parsley etc.
Mix everything together and form balls (or shapes) with your hands. Place on baking tray and flatten with a fork. Bake slowly at 150 degrees C (302 F) until hard, about 40-45 minutes. I double the recipe, and it makes heaps - about 2 trays."
-'152 Homemade Dog Food Recipes' by Fury Joy, **2015**.

"**Comfrey and Calendula Skin Balm for Pets:**
>   Organically produced ingredients are used to make this lovely soothing balm that's used to **calm and soften dry, scaly and damaged skin.** This should be part of every pet owner's pet care kit and is ideal for use on skin folds, pressure calluses, to protect sore pads and on areas of roughened but unbroken skin.
>   **Active ingredients:** Organically produced sunflower oil, beeswax, **Comfrey,** calendula, lavender and rosemary.
>   **Administration:** Apply regularly until all signs of soreness, dryness or irritation have disappeared. Care should be taken when using near the eye area.
>   Use short and long term: Yes.   Use during pregnancy and lactation: Yes.
>   Minimum age for use: 8 weeks.
>   **Interactions:** No known interactions. Can be given in conjunction with any medication.

Homeopathic Symphytum 15C for Pets:
>   Being prepared from Comfrey herb, whose old country name is Knitbone, it's not surprising that this remedy **supports healthy ligaments and tendons, particularly following a strain or sprain.**
>   **Administration:** Give one 3 times per day for 10 days. Repeat after 10 days if necessary.

Remember we are not veterinarians and do not diagnose. You should always consult a veterinarian if you are concerned about your pet's health."
-'Dorwest Veterinary: Experts in Herbal Pet Care' by Dorwest Herbs Ltd., Dorchester, Dorset, England, Autumn **2017**.
>   (I do not have any connection with this company or their products. This is to show you some Comfrey products available. Research online for companies you like. Contact your veterinarian for advice on what your pet needs.)
>   (See subsection 'Homeopathic Comfrey' in section 'Making and Using Comfrey Medicine' (Chapter 27).)

"**Liver Disease in Dogs: Symptoms:** Early in the course of liver disease dogs may not have any clinical signs.
>   **The earliest clinical signs seen in dogs with liver disease are often non-specific** and include: vomiting, diarrhea, weight-loss, polyuria/polydipsia, and hyporexia. More liver specific signs such as icterus, ascites, and encephalopathy occur in late in the progression of chronic hepatitis. When any of these clinical signs are present, they warrant further investigation in an attempt to determine their cause.

Causes: There are many causes of increased liver enzymes activities, so it very important for clinicians to go from a list of all the possible causes to a list of all the causes that are 'probable for that patient on that day'.
>   **When taking a history, it is very important to ask specifically about exposure to hepatotoxins** such as cycads, blue green algae, amanita mushrooms, aflatoxins, heavy metals, xylitol, or chlorinated compounds.
>   **A variety of drugs can also be hepatotoxic, these include:** ketoconazole, various antimicrobial agents, azathioprine, carprofen, lomustine, acetaminophen, ketoconazole, mitotane, and phenobarbital.
>   **It is important to specifically ask about any herbal remedies that the dog is receiving as many of these have been reported to be hepatotoxic, including:** herbal teas, pennyroyal oil, and Comfrey."
-'Approach to Gastrointestinal Bleeding in Dogs and Cats' & 'Increased Liver Enzymes in Dogs: What Do I Biopsy?' by Jonathan Lidbury, BVMS, MRCVS, PhD, DACVIM, DECVIM, Texas A&M University, College Station, Texas, no date but **2017**-2018.

*"**Muscle Magic for Dogs: A soothing warming massage lotion, rich in Arnica and Comfrey tincture with aromatic relaxing Lavender essential oil.** Just what your dog needs to help relax tired and sore muscles.*
Easy to apply and quickly absorbed, the lotion can be massaged in gently to any area as a general massage lotion, therapeutically after exercise, or in conjunction with physiotherapy, hydrotherapy or chiropractic treatment. Your dog will love you for it! As with all products containing arnica, avoid open wounds."
-'Hilton Herbs: Natural Supplements for Animals', Europe & United States, www.hiltonherbs.com, date unknown, **2018** or earlier.
    (I do not have any connection with this company or their products. This is to show you some Comfrey products available for cats and dogs. Please contact your veterinarian for advice.)

*"**Your Dog's Feet in Winter:** A dog's foot pads are composed of several layers of keratin, a harder form of skin cells. **You can actually build up the toughness of your dog's pads.***
If you have a big trip or hike planned, you can treat your dog's pads in the weeks before with a product called Pad-Tough®. It is a botanical product with aloe and Comfrey. It comes in a spray form and simply coat your dog's paw pads liberally before any rigorous activity."
-'Grab that Leash and Hit the Trail: A Primer for Hiking with Your Dog' by Doug Gelbert, Doggin' America, Great Vacation Ideas for You and Your Dog, western North Carolina, no date but **2018** or earlier.   (I do not have any connection with this product.)

*"**Herbal Legal Regulations in the United States: Recent market statistics indicate that sales of supplements for companion animals are on the rise.** Tinctures, tablets, and powders designed to supplement the diets and treat the ailments of humankind's furry friends are increasingly available, despite on-going regulatory challenges.*
Regulation of Supplements for Pets
Supplements for pets have been able to achieve impressive strides in the marketplace despite a complicated legal status and a lack of formal regulations. The 'Dietary Supplement Health and Education Act' of 1994 (DSHEA) established regulations for human dietary supplements at a time when there were few similar products for pets on the market.
    *DSHEA did not specifically address the topic of supplements for pets, but a posting in the 'Federal Register' later specified that the **United States Food and Drug Administration (FDA) does not consider such supplements to be covered under DSHEA.***
Supplements marketed for dogs, cats, and horses have therefore been left with two possible legal categories under United States law: they may be defined as animal foods/feeds or animal drugs.
    1. Most supplement products for pets are classified by the manufacturers as nutritional or feed supplements. *However, hundreds of ingredients commonly used for human and pet dietary supplements are not technically approved for use in animal feed products.*
    2. The other regulatory option for pet supplements, as animal drugs, *requires that supplement marketers submit a 'New Animal Drug Application' to the FDA's 'Center for Veterinary Medicine' (CVM), demonstrating the product's safety and efficacy according to certain criteria. According to Bookout of NASC (National Animal Supplement Council), these applications, and the testing that they require, are expensive and problematic."*
All 'National Animal Supplement Council' (NASC, Arizona) member companies also employ warning or cautionary statements in labeling for products that contain certain herbal ingredients, *as recommended by the NASC Scientific Advisory Committee.*
Some herbal ingredients are banned from member companies' products. *Bookout explained that the FDA-CVM (Food and Drug Administration - Center for Veterinary Medicine) has recommended that kava (Piper methysticum, Piperaceae), **Comfrey (Symphytum spp., Boraginaceae),** and pennyroyal (Mentha pulegium, Lamiaceae) oil not be used in pet food products, and the NASC has prohibited these herbal ingredients in supplements due to the agency's recommendation. Comfrey contains hepatotoxic pyrrolizidine alkaloids."*
-'The Expanding Market and Regulatory Challenges of Supplements for Pets in the United States' by Courtney Cavaliere, HerbalGram, American Botanical Council, Austin, Texas, Volume 82, pages 34-41, **2009**.

Cattle: Beef and Dairy

Definitions

Bovine or Bovinae subfamily includes cattle, bison, African buffalo, water buffalo, yak, and antelopes.
Beef cattle are raised for meat production.   **Dairy cattle** are raised for milk production.
Cow: A female who has had at least one calf.
Heifer: A female who has never had a calf.
Bull: A mature male, not neutered/castrated, who is used for breeding.
Steer: A mature male that has been neutered/castrated.
Bullock: Definitions vary but always male.
Calf: A male or female under the age of one year.
Ox: Usually a castrated male used for pulling equipment. The plural is oxen.

Cattle Love Comfrey

*"**Prickly Comfrey, from the 'Journal of the Royal Agricultural Society': I know of no plant that would answer so well, if cattle would like it, and thrive upon it.** I have no doubt that on good land it would produce 40 tons per acre per annum (year), with little or no expense in the culture; but I should like to know how cattle would do upon it before giving up much ground to it. **Cattle do not appear to be fond of it, but that may be owing to not having enough to give it a fair trial; as many sorts of food are not eaten by cattle readily at first, which they are fond of when used to.**
There appears to be a large quantity of mucilage in the plant, from which I should suppose it would be nutritious."*
-'The Farmers Register: A Monthly Publication Devoted to the Improvement of the Practice and Support of the Interests of Agriculture' edited by Edmund Ruffin, Proprietor, Petersburg, Virginia, 741 pages, Volume 9, **1841**. April 30 1841, page 248.

*"**The Symphytum officinale, or Common Comfrey**, is a hardy perennial, found growing wild in wet places, by the side of ditches and sluggish streams. The plant pushes its vegetation very forward in the spring, and thus produces a great quantity of tender succulent shoots and leaves, which are **readily eaten by cattle; cows particularly seem to do well on it.**
In all cases the plants (S. officinale and S. asperrimum) should not be allowed to get too much growth before they are given to cattle, as otherwise the stems get hard and woody, and are neither palatable nor so digestible when eaten."*
-'Our Farm Crops: Being a Popular Scientific Description of the Cultivation, Chemistry, Diseases, Remedies, Etc., of the Various Crops Cultivated in Great Britain and Ireland', Volume 2, Part 3: Gorse, Ryegrass, Chicory, Lentil, Melilot, Birdsfoot-Trefoil, Comfrey, Flax' by John Wilson, London, England, **1859**, page 252.

*"Mr. Robert Ashburner, of San Mateo County, California, has a field of Prickly Comfrey (Symphytum asperrimum), of which he has a high opinion as summer feed for dairy cows. **He has no trouble in getting his cows to eat the Comfrey, if it is given to them when green and succulent, before the 'prickles' become firm and sharp.**
He cuts the Comfrey with a reaping hook when the lower leaves begin to turn yellow. The leaves are then about 18 inches (0.45 meter) in length. He likes the Comfrey so well that he will double his planting this winter."*
-'Pacific Rural Press', Volumes 17-18, San Francisco, California, 843 pages, every other week from January 4 to December 27 **1879**. Article: 'A Visit to Baden Farm', San Mateo County, California, October 11, 1879.

*"**Prickly Comfrey plant has attracted attention even in South Australia.** A correspondent to 'Darling Downs Gazette' says :
'I planted some roots of Prickly Comfrey in January 1881, four feet (1.2 meter) apart. They grew a dense mass of leaves, and ultimately covered the intervening space. Three weeks ago we had a hailstorm that riddled all the leaves.
I cut them, and gave a quantity to an old cow that was used to hand-feeding- they were new to her. She left several times, but always came back again, and ultimately she ate the whole. **I gave the rest of the leaves to a lot of yearling heifers; they seemed to relish them,** for they gobbled them up in a very short time.' "*
-'The Journal of the Royal Agricultural Society of England, Second Series, Volume 18', London, England, **1882**. Includes article: 'On Green or Fodder Crops Not Commonly Grown, which Have Been Found Serviceable for Stock-Feeding' by Joseph Darby.
   (The 'Darling Downs Gazette' was a newspaper published from 1848 to 1922 in Drayton and Toowoomba in Queensland, Australia.)   (A heifer is a young female cow that has not given birth yet.)

*"In a letter of Mr. Thomas Christy's of the 6th of November 1875, he calls attention to the fact, that 'it is stated that cattle reject Prickly Comfrey owing to the prickly nature of the matured leaves. This is quite possible, but it can be easily remedied by **offering at first a few of the younger and more tender portions of the foliages. The animals soon learn to discriminate (judge), and no other food will be held in such esteem.**"*
-'Prickly Comfrey: Its History, Cultivation, Extraordinary Production, and Uses: A Letter Addressed to His Excellency Sir Hercules Robinson, President of the Agricultural Society of New South Wales' by Arthur T. Holroyd, Sydney, Australia, **1876**, page 10.
(Thomas Christy from England promoted Prickly and Russian Comfrey in mid to late 1800s. See 'History' section, Chapter 15.)

*"**I have at this time three cows and a splendid Russian cob which are fed entirely upon Prickly Comfrey (Symphytum asperrimum), with the addition of a small quantity of bran, and a few grains for the cows.**
I only give one bushel of these latter per day for the three animals. The cows are in splendid condition, and yield a large quantity of milk, and are quite a match for those fed in pastures and at much greater cost.
And as for the Russian cob, several good judges have said they never saw a finer specimen of his breed, or in better condition; and this is as it should be, for as the pedigree of this cob is Russian so it has been fed with Russian Prickly Comfrey."*
-'Ensilage: A System for the Preservation in Pits of Forage Plants and Grasses, Independent of Weather: A Collection of Facts and Statistics on the Cheapest Mode of Providing Winter Food for Dairy Cattle, Sheep, Horses, Etc.', by Thomas Christy, F.L.S. (Fellow of Linneaen Society), London, England, **1883**. Caucasian Prickly Comfrey, pages 57-63.
     (A cob is a stocky short-legged riding horse.)
     (A United States bushel: equal to 64 United States pints {1 pint = 2 cups} equivalent to 35.2 liters, used for dry goods.
     A British bushel: equal to 8 Imperial gallons equivalent to 36.4 liters, used for dry goods and liquids.)

*"At New Bells Farm, Haughley, Suffolk, (England), **the 'Soil Association' use Russian Comfrey for cattle food,** and recommend it both as fodder and compost material. The leaves are on the rough side. It is distinct from our native Comfrey, Symphytum officinale, which is a weed, and useless as fodder, but **once they get the taste for the crop, stock will leave even grass to eat it.**"*
-Fertility Farming by Newman Turner. Austin, Texas: Acres USA, **1951**, page 54.
     (The 'Soil Association', founded in 1946, is a charity in the United Kingdom. Its does campaign work on issues such as

opposition to intensive farming, support for local purchasing, public education on nutrition, and certification of organic food. Today it certifies over 80% of organic produce in the United Kingdom.)

*"Steers were put to graze on an old test site. The first species grazed was reed canary grass (Phalaris amndinaced) followed by brome grass / orchardgrass / creeping red fescue mixture, then orchardgrass (Daclylis glomerata), **then sainfoin (Onobrychis sativa) and dandelion growing with Russian Comfrey (Symphytum peregrinum), then lucerne (alfalfa) growing with Comfrey, and finally the Comfrey itself.** Shepherd's purse (Capsella bursa-pastoris) was left ungrazed."*
-'Grazing Preferences of Hereford Steers' by D.B. Wilson, Lethbridge, Alberta, Canada; Forage Notes, Volume 7, No. 3, page 41, **1961**.

(Steer is a male bovine animal and especially a domestic ox castrated before sexual maturity.)
(**In this field the Comfrey was not the favorite.** I do not have the rest of the report. Please send a copy if you have it.)

*"**When the farmer cut the long deep green leaves of Comfrey and called his cows, they came running to get their daily ration.** They were not a hungry half-starved herd; they were really fat stock. But clearly they knew and relished their Comfrey rations. **The free range method of grazing grass and Comfrey selectively has shown that cows will eat out the Comfrey first**, right down to the ground, and pigs will not only eat the leaves but root out the plants as deep as their noses can take them if allowed to do so."*
-Miracle Grass, Comfrey by Andrew Hughes published in **1966** in Japanese. Part of it is in: 'Comfrey: Nature's Healing Herb & Health Food' by Andrew Hughes, Japan: Sanyusha Publishing Co., Ltd, 1992. And part is in 'Comfrey Report No. 1: The 1954 Research Results' by Lawrence Hills published in 1955. If you have a translation in English, I would appreciate having a copy.

Amount of Comfrey for Cattle

*"**My cows consume daily about 130 pounds (58.9 kg) each of the (fresh) Russian Comfrey, so that the consumption per week is about 1 1/4 tons for my three cows and pony.** -H. Doubleday, Coggebhall, Essex, England, 4th July, 1882."*
-'Ensilage: A System for the Preservation in Pits of Forage Plants and Grasses, Independent of Weather: A Collection of Facts and Statistics on the Cheapest Mode of Providing Winter Food for Dairy Cattle, Sheep, Horses, Etc.', by Thomas Christy, F.L.S. (Fellow of Linneaen Society), London, England, **1883**. Caucasian Prickly Comfrey, pages 57-63.

*"**The late Kenneth Crawley estimated that the ideal ration of Comfrey for dairy cows in milk was half-cwt (56 pounds = 25.4 kg) per day as wilted fodder or as silage in winter,** to provide for maintenance and first two gallons (7.5 liters). This was from his own experience with Ayrshire cattle, and he calculated from Henry Doubleday's figures that the 80 to 120 ton yield an acre of Symphytum peregrinum (Russian Comfrey) should produce the greater part of the feed for nine cows."*
-Russian Comfrey: A Hundred Tons an Acre of Stock or Compost for Farm, Garden or Smallholding by Lawrence D. Hills. London England: Faber and Faber, Limited, **1953**, page 85.
('cwt' is abbreviation for hundredweight. In United Kingdom one hundredweight is 112 pounds. In the United States and Canada it is 100 pounds. There are 20 hundredweight in a ton, producing a 'short ton' of 2000 pounds and 'long ton' of 2240 pounds.)

*"**Twenty kg (44 pounds) of Comfrey per day was supplied to each of 6 cows. Each cow took 5 kg (11 pounds) of Comfrey in 20 minutes without hesitation.**
Comfrey was supplied to a calf suffering from scours (diarrhea) and as a result its droppings became firm."*
-'Comfrey Report No. 3: Feeding Dairy Cattle in Japan' by Meiji Milk Producing Co., Tokyo, Japan for Henry Doubleday Research Association, Braintree, Essex, England, 31 pages, July **1964**.

*"Mr. Willing (Saskatoon, Saskatchewan, Canada in 1956) cut and hauled **fresh Comfrey from the field each day to his milking cows. He estimated that each cow received approximately 30 pounds (13.6 kg) per day.**
The Comfrey was fed just before milking, and there was no taint to the milk and no bloat occurred. When first presented with the fresh forage, the cows did not appear to relish it. However, by slightly wilting the forage, the cows would eat it, and after a short time they ate the fresh forage with relish."*
-Comfrey: Fodder, Food and Remedy by Lawrence D. Hills. New York: Rizzoli Universe Books: **1976**, page 66.

*"**Cutting Comfrey for cows and horses should be done at a more mature stage than for chickens and pigs, between 28 and 35 days being appropriate according to the season. This means higher fiber.**"*
-Comfrey: Nature's Healing Herb & Health Food by Andrew Hughes. Japan: Sanyusha Publishing Co., Ltd, **1992**, page 175.

Comfrey Early Spring and Late Fall

*"The question for the dairy farmer is one of profit per cow, and it is certain that the most cheaply produced feed is the good grass that the digestion of the animal is designed to take.
An economic milk yield needs reinforcement by concentrated feeding-stuffs at both ends of the season, and it is here that the (Comfrey) crop can produce savings by the replacement of part of the bought or home-grown feeding-stuffs with more cheaply produced fodder.

We have abundant evidence from the past that there is enough to increase the butter fats in the early spring. There is also complete agreement between past and present that fresh Comfrey has never caused bloat or hoven even when eaten in quantity."
-Russian Comfrey: A Hundred Tons an Acre of Stock or Compost for Farm, Garden or Smallholding by Lawrence D. Hills. London England: Faber and Faber, Limited, **1953**, page 85.
> (Hoven means having the disease hove. In ruminants it is inflammation of the stomach with gas, usually caused by eating too much fresh, green food.)

Use Comfrey with Other Feeds for Cattle

"The reason for the need of cows for good hay along with the Comfrey is the low fiber content of Comfrey. Cows being ruminants, they must have high fiber. But Comfrey is the richest food source to go with the hay. It is rich in calcium, another must for cows; they need calcium in considerable quantities naturally in their food rather than as a supplement."
-Comfrey: Nature's Healing Herb & Health Food by Andrew Hughes. Japan: Sanyusha Publishing Co., Ltd, **1992**, page 171.

"Recent studies with Comfrey leaf have shown that it contains several essential amino acids missing in alfalfa. When it is combined with alfalfa in a 60/40% ratio, it constitutes a 'whole food' for feedlot cattle. It is metabolized into harmless proteins, including two essential amino acids missing in alfalfa (lysine and alanine)."
-'Comfrey Leaf: A New Animal Food Supplement' in The Encyclopedia of Alternative Agriculture by Dr. Richard Alan Miller, Agricultural Consultant and Researcher, USA, **1992**.

Cows Eating Comfrey Produce Good Milk

"In a Clergyman's plantation where Prickly Comfrey did well, he gave it to his cow, a fine young animal of large size, which had not long calved. **The cow when fed on Symphytum asperrimum gave 36 quarts (34 liters) of milk every 24 hours (was thrice {3 times day} milked), being an increase of rather more than 12 quarts (11.3 liters) on what she had given previously, from which a good quantity of splendid butter was made twice a week,** *and new milk was used freely in the family. The Clergyman's horses, donkey, pigs, rabbits, etc. all fed on it and did well.*
A writer in the 'Country Gentleman's Magazine' mentions that:
> *'his cows ate the Prickly Comfrey greedily and used to follow us about the fields as it were begging for it, and it seemed that it was not only highly relished by these stall cattle and pigs, but with all kinds of stock it seemed to act most favourably.'*

The writer of 'The Field' (magazine):
> *'knows nothing at all comparable to Prickly Comfrey to set before cattle for nourishment. If they do not take to it at first, they soon come round to it. Where cows are kept there is nothing probably that can be so advantageiously placed before them when they are tied up for milking.' "*

-'Prickly Comfrey: Its History, Cultivation, Extraordinary Production, and Uses: A Letter Addressed to His Excellency Sir Hercules Robinson, President of the Agricultural Society of New South Wales' by Arthur T. Holroyd, Sydney, Australia, **1876**, page 10.
> (The 'Country Gentleman', 1831-1955, was an agricultural magazine published in Rochester, New York. It covered the business side of farming.) ('The Field' is the world's oldest country and field sports magazine, having been published continuously in England since 1853.)

"During a late call on Mr. G. Hunziker, at **Cloverdale, California,** *to whose interesting experiments we have already had occasion to refer, we were shown a patch of* **Prickly Comfrey,** *which was set out last April, and has been cut eight times since then.* **Mr. Hunziker informs us that after feeding it to his cows there was a perceptible increase in the flow of milk, to the extent of a pint (= 2 cups = 0.47 liter) or more each.**
The habit and appearance of the plant is something like the mullen, and, as it sends down its roots to a depth of six or seven feet (1.8 to 2.1 meter), it seems as though with a good start it ought to do very well without irrigation.
Where it has had a fair trial it is especially recommended for hogs."
-'Pacific Rural Press', Volumes 15 and 16, San Francisco, California, 853 pages, every other week from January 5 to December 28 **1878**. Article: 'Prickly Comfrey, Russian River Flag', January 19 1878.

"Dr. Walter Moore, of Bourton-on-the-Water (Gloucestershire County, England), has recently informed me that, within his recollection, **plots of Comfrey were grown as a food for cattle, on account of its reputation for producing milk, rich both in quality and quantity."**
-The Medicinal Uses of Comfrey by Dr. Charles MacAlister, M.D., F.R.C.P., **1935**.

"Louis Bromfield at Malabar (his famous organic farm in Ohio, United States) says:
'With regard to the Vitamin B12 content of milk (from Comfrey fed cows), *the experiments conducted at Missouri University have so far been nothing short of startling, the content rising on an average upon the feeding of trace elements from 0.006 content to a content of 0.043."*
-Comfrey: Nature's Healing Herb & Health Food by Andrew Hughes. Japan: Sanyusha Publishing Co., Ltd, **1992**, page 171.

(No reference was given as to the source of this study. If you have it, please send it to me.)

Feeding Comfrey Silage to Cows

Silage is fermented, high-moisture fodder that is fed to cattle, goats, sheep and other ruminants (cud-chewing animals). It is fermented in a process called ensilage or silaging, and is usually made from grasses such as corn, sorghum and other cereals, using all of the green plant (not just grain).

"Feeding Comfrey silage over 55 days at a rate of 20 or 30 kg/day (44-66 pounds) in a balanced diet (oat straw, fodder beet, carrots, concentrates) *to Black Pied Lowland cows (3 groups of 10) in months 2-3 of lactation **did not have any appreciable effects on the composition of their milk and on the quality** of 18% fat tvorog, 30% fat smetana (cultured cream) and soured milk made from it. **Cows given Comfrey silage at 30 kg/day produced most milk."***
-'Quality of Cultured Milk Products Made from Milk of Cows Fed Comfrey Silage' by A.V. Beloborodov, Moskovskay Vet Akademiya, Moscow, USSR; Sbornik Nauchnykh Trudov Moskovskaya Ordena Trudovogo Krasnogo Znameni Veterinarnaya Akademiya (Collection of Scientific Proceedings of the Moscow Order of the Red Banner of Labor Veterinary Academy), Volume 91, pages 63-66, **1977**.
    (Feed Concentrates are mixes of ingredients that have a high amount of cereal grains. Mixes of only grains are also called concentrates.) (Tvorgo, kwark, quark or quarg is a fresh dairy product made by warming soured milk until the desired amount of curdling, then straining it.)

"Three matched groups of 10 Black Pied cows received in a preliminary period a common ration supplying daily per cow 20 kg (44 pounds) maize (corn) silage. In the experimental period, the control group continued on the common ration.
For the 2 experimental groups the maize silage was replaced by Comfrey silage *made from plants cut in the flowering stage when 120-130 cm (47-51 inches) high, respectively at 20 and 30 kg (44-66 pounds) daily per cow.*
*Differences in main constituents and protein fractions of milks expressed as percent gains or losses in relation to the controls are presented and discussed. **No difference was found in organoleptic quality of Yaroslavl' cheeses made from milks of the 3 groups. Use of Comfrey silage in dairy cow feeding is recommended."***
-'Quality of Milk and Yaroslavl' Cheese in Relation to Feeding to Cows of Comfrey Silage' by A.S. Shuvarikov, Intensifikatsiya Protsessov Proizvodstva Natural'nykh Syrov i Sovershenstvovanie ikh Tekhnologii (Intensification of the Processes for the Production of Natural Cheeses and Improvement of their Technology), USSR, pages 140-143, **1977**.
    (Organoleptic properties are the sensory aspects of food, water or other substances. It is taste, look, smell and touch.)

*"During the adjustment period, cows were fed on a complete diet of clover and timothy silage at 25 kg/cow (55 pounds) daily. In a test period control cows continued to eat the adjustment diet, while the test cows were **given daily 25 kg of Comfrey (Symphytum) silage** in place of the clover and timothy silage.*
There was no difference between the groups in milk colour, odour or consistency or in DM (Dry Matter) and vitamin contents. Milk fat was 3.33% with clover and timothy silage and 3.30% with the Comfrey silage."
-'Some Indicators of Quality of Milk from Cows Fed on Mixed Comfrey and Grass Silage' by A.V. Beloborodov, Moskovskay Vet Akademiya, Moscow, USSR; Sbornik Nauchnykh Trudov Moskovskaya Ordena Trudovogo Krasnogo Znameni Veterinarnaya Akademiya (Collection of Scientific Proceedings of the Moscow Order of the Red Banner of Labor Veterinary Academy), No. 123, pages 45-47, **1981**.

Comfrey and Cattle Health

*"Prickly Comfrey's (Symphytum asperrimum) medicinal properties also cannot be overlooked, it having been found to **be a curative as well as a preventative in cases of foot-and-mouth disease;** and it has been affirmed that cows fed on Comfrey have escaped this disease, while others in the same dairy not fed on it had the disease badly."*
-'Transactions of the Highland and Agricultural Society of Scotland, Fourth Series, Volume 14', Edinburgh, Scotland, **1882**. With the article: 'The Cultivation of Prickly Comfrey, and Its Use as a Fodder Plant' by David W. Wemyss, Newton Bank, St. Andrews, Fife Council, Scotland, pages 264-267.
    (See sub-subsection 'Foot-and-Mouth Disease' in subsection 'Comfrey as Livestock Medicine' in this section.)

*"**Cows fed regularly on Comfrey have been found to keep much freer from mastitis (udder infection) than cows without Comfrey.** This is understandable because of Comfrey's power to make animals (including man) more strongly resistant to infection."*
-Comfrey: Nature's Healing Herb & Health Food by Andrew Hughes. Japan: Sanyusha Publishing Co., Ltd, **1992**, page 173.

"Palatability (good taste) to a cow is influenced by the degree of production of suitable microflora in the rumen of the cow. When cows are fed the same kind of food every day only those microflora which are suitable to break down and ferment the food are propagated.
A sudden change of stockfood means that there are not enough microflora in the rumen to deal with the new food.
Therefore in changing the diet, a small quantity must be given for upwards of a week, *otherwise cows have no appetite for*

the new food, and they may suffer scours (diarrhea)."
-Comfrey: Nature's Healing Herb & Health Food by Andrew Hughes. Japan: Sanyusha Publishing Co., Ltd, **1992**, page 192.
    (The rumen is the first chamber in the alimentary canal of ruminant animals such as cattle, goats and sheep.
    It is the primary site for microbial fermentation of food.)

"**Another Australian, who valued HDRA (Henry Doubleday Research Association) research, was Foster Savage** who I had the opportunity to know him, personally, when he settled in Nambour (Queensland, Australia).
I also fed Comfrey to my farm animals.
    Knowing the power of Comfrey to restore a worn out animal quickly, and make her milk again, I once bought an old cow at the Dandenong Market, when farming in Victoria (Australia). It had been discarded by some farmer, as worn out.
    I put her on Comfrey, giving her 90 pounds (40.8 kg) of wilted Comfrey (wilted to increase the cow's intake of Comfrey's extraordinary nutrients), and 90 pounds made a pretty big heap, about 4 feet (1.2 meter) high.
    This poor, old, creature took to the Comfrey, without hesitation. She was starving for minerals and her instincts gave her a craving for Comfrey. When she began to eat, she would eat off the heap of leaves for a couple of hours, then sit down for an hour or so. Later, she would continue eating, until every leaf was gone.
    She doubled her milk output, within a week, and in a fortnight (2 weeks), trebled (tripled) it. The remarkable thing, was that the cream that settled overnight, was some 3/4 inch (1.9 cm) thick and the separation of cream from the milk was so perfect, that the cream could be lifted off, with none remaining.
I fed Comfrey to calves, as much as they could eat, again with only gratifying results.
I fed pigs, entirely on Comfrey and grain, as much Comfrey as they could eat, and the quality of those pigs was legendary in the district. The fame of Comfrey spread far and wide, for my farm was visited by 6,000 farmers from around Australia and from overseas.
Finally, I well remember the enthusiastic remarks of the butcher who regularly killed our Comfrey-fed calves. He told us that he had never before, seen such healthy livers, that, mind you, after being reared on a herb that was supposed to cause liver diseases!' "
-How Can I Use Herbs in My Daily Life: Over 500 Herbs, Spices and Edible Plants: An Australian Practical Guide to Growing Culinary and Medicinal Herbs by Isabell Shipard. New York, NY: David Stewart Publisher / Simon & Schuster, **2003**.

"**The aim of this study was to survey the current day ethnoveterinary practices of ethnic Hungarian (Szekely) settlements situated in the Erdovidek commune in Covasna County, Transylvania, Romania,** and to compare them with earlier works on this topic in Romania and other European countries.
Plant taxa used in ethnoveterinary medicine of the selected villages:
    Symphytum officinale L., Boraginaceae: fekete nadaly, nadaly, forrasztofu.
    Comfrey root as fodder with bran for rumination of cattle."
-'Ethnoveterinary Practices of Covasna County, Transylvania, Romania' by Samuel Gergely Bartha, Cassandra L. Quave, Lajos Balogh and Nora Papp (Hungary and Atlanta, Georgia); Journal of Ethnobiology and Ethnomedicine, Volume 11, No. 35, 22 pages, **May 2015**.

"**Comfrey Drench for Cattle: Take one pound (0.45 kg) of Comfrey leaves,** and boil slowly in two litres (2.1 quarts) of water for one hour. Add handful of ground ivy and two ounces (226 gram) of wild licorice root, and simmer for one more hour. Give eight ounces (236 ml) up to three times daily.
For internal ulcers, add one tablespoon of molasses to equal parts of above and raw skim milk."
-'Comfrey' by Glenn Axford, Our Common Roots: An Exploration of the Healing Power of Nature, Canada, www.ourcommonroots.com, date unknown but **2018** or earlier.

    Feeding Calves Comfrey

"**By all accounts Comfrey is an excellent calf feed, chopped at first and then as whole wilted leaves or hay.** The softness and low proportion of fibre make it good introduction to normal diet.
At Little Weighton (Yorkshire, England) it was used in the 1940's for rearing calves on calf gruel with a very small milk allowance. It was from this work that its value as a scour (diarrhea) preventative for calves was established."
-Russian Comfrey: A Hundred Tons an Acre of Stock or Compost for Farm, Garden or Smallholding by Lawrence D. Hills. London England: Faber and Faber, Limited, **1953**, page 87.

"As in every country, the good fields are always owned by the man who is feeding his stock and making a profit, and one of these is Elmer Jesky of Aurora, Oregon (in the 1970s). He feeds pedigree (purebred) Aberdeen Angus calves to sell at eight months old, **supplementing their grazing with a daily Comfrey ration which gives him an average live weight gain of 100 pounds (45.3 kg) more in eight months than he would get on grass alone.**
The Comfrey high mineral and high protein feed gives perfect condition of coat, vigour and bone structure, which counts when better prices are paid on market days. The calves are brought in, and **they fit the growing period of Comfrey perfectly, with their maximum appetites in the autumn,** and no stock in the winter, which avoids need for conservation (drying or ensiling Comfrey). If 100 head of calves can be fed off of six acres, then sixty acres should feed 1,000."
-Comfrey: Fodder, Food and Remedy by Lawrence D. Hills. New York: Rizzoli Universe Books: **1976**, pages 221, 222.

"From the very beginning, Comfrey can be fed to calves.
One method is to crush the fresh leaves and add the juice to the milk at first for baby calves. From that stage they can go on to chopped up leaves in the milk, and many calves will start eating the fresh or wilted leaves direct from 10 or so days old. **Comfrey will cure scours (diarrhea), but calves raised on Comfrey in this way never get scours**, which basically is caused by deficiency."
-Comfrey: Nature's Healing Herb & Health Food by Andrew Hughes. Japan: Sanyusha Publishing Co., Ltd, **1992**, page 172.

Chickens and Ducks (see Poultry)

Dogs (see Cats and Dogs)

Donkeys (see Zoo Animals)

Ducks (see Poultry)

Fish

"National Science Foundation sponsored **closed-system aquaculture** project:
Feeds: Leafy plants: **Comfrey plants** (35170 protein), vetch, purslane and alfalfa, among others. Comfrey (Symphytum species) can also be ground and dried into pellets for storable feeds."
-'Summary of Fish Culture Techniques in Solar-Algae Ponds' by John Wolfe and Ron Zweig, Journal of the New Alchemists, New Alchemy Institute, Falmouth, Massachusetts, Volume 6, 1977, 1980, pages 97-99.
    (Aquaculture is raising aquatic animals or cultivating aquatic plants for food. I do not know what '35170 protein' is.)

"Crop variety: **Crops that are palatable to the fish**, rich in nutrition, resistant to disease, easy to manage and have well developed roots to protect the slope should be used. If the crop serves as a straw manure, it should decompose easily.
Fodder Crops: **Symphytum peregrinum**, Yield 7500-12500 kg/mu, Sowing Time March-April, Seed 2500 kg/mu, Transplanting Type, Harvesting Time May-October, Leaves as food, Food Conversion Rate 40-45."
-'Integrated Fish Farming in China' by Asian-Pacific Regional Research and Training Center in Integrated Fish Farming, Wuxi, China. Published by Network of Aquaculture Centres in Asia and the Pacific, Technical Manual 7, Bangkok, Thailand, 1989. Chapter 7: Introduction of Chinese Integrated Fish Farming and Some Major Models by Yang Huazhu and Hu Baotong.
(Mu or Mou is a Chinese unit of land measurement that varies but is commonly the same as 806.65 square yards = 0.165 acre = 666.5 square meters. However, the mou is defined by customs treaty as 920.417 square yards = 769.59 square meters.
(Feed Conversion Ratio or Feed Conversion Rate measures the efficiency that livestock convert animal feed into milk or meat. The lower the number, the more efficient the use of feed.)

"Food Security Technology's strategy on **fish and shellfish feeds** for its Residential and Commercial Aquafarm systems is to develop our own proprietary blend with growers around the world supplying the bulk raw sources, such as Moringa Olifera, **Comfrey,** and Spirulina to provide the fish and shellfish feed foundation."
-'Fish Feed' by Food Security Technology, Alabama, www.foodsecuritytech.com/fish-feed.html, 2019.

Geese (see Poultry)

Goats (also see 'Sheep' section)

"**Rationing is a matter of experience and can be judged by normal green-fodder allowances** from standard textbooks. This can then be stepped up to an ad lib (as much as they want) feeding with a watch on the milk yield of extra-greedy nannies (female goats). The midsummer cuts can go for hay and with roots alone can make up a complete winter diet if sufficient is available. The early cuts are of special value because rough pasture is later starting than improved grasses."
-Russian Comfrey: A Hundred Tons an Acre of Stock or Compost for Farm, Garden or Smallholding by Lawrence D. Hills. London England: Faber and Faber, Limited, **1953**, page 95.

"*The Principles of Goat Feeding:*
Provided the goat-keeper knows the approximate composition of the mixture he is feeding, and has gathered some understanding of the mechanism of mineral balance in the goat, he has a fair chance of juggling successfully with a selection of mixtures.
The best that can be done here is to consider the broad categories of trouble:
1. The 'High Calcium Diet' consists almost entirely of high-quality fodder crops, clover, lucerne (alfalfa), oats and tares (vetch), Comfrey, etc., with perhaps a more or less generous allowance of sugar-beet pulp and 2 or 3 pounds (4.4 or 6.6 kg) of concentrate.
The danger here is twofold. The wealth of calcium may block the uptake of phosphorus from the diet; excess calcium in the

blood may 'brake' the effect of thyroxine, preventing the release of skeletal phosphorus, and putting a strain on resources of iodine. Symptoms are most likely to arise with the introduction of winter feeding legume hay, etc., and take the form of lowered production and miscellaneous breeding problems."
-Goat Husbandry by David Mackenzie. London, England: Faber and Faber Ltd., **1967**, page 157.
    (Legume hay species include alfalfa/lucerne, clover and birdsfoot trefoil.)

"The Principles of Goat Feeding:
Hay made on a weed patch with the maximum of nettles and docks is always preferable. Dried branches, pure nettle hay, and **dried Comfrey leaves are luxuries for which the minority of goat-keepers can find the time, space and sunshine."**
-Goat Husbandry by David Mackenzie. London, England: Faber and Faber Ltd., **1967**, page 167.

"**Cropping for Goats:** *The following table shows the productivity of the major goat crops:*

Comfrey	Weight of Crop Per Pole	Starch Equivalent Per Pole	Notes
	300 pounds (136 kg)	15 pounds (6.8 kg)	Protein rich and very early."

-Goat Husbandry by David Mackenzie. London, England: Faber and Faber Ltd., **1967**, page 313.
    ('Pole' or 'perch' or 'rod' is 30.5 square yards = 3 yards x 10 yards = 25.5 square meters.
    One acre = 160 square poles or square rods. One ton per acre = 1 stone per pole. One stone = 14 pounds = 6.3 kg.)

"**Goats:** *Our four dairy does relish dried Comfrey and prefer it to alfalfa when they have a choice.*
They also eat the herb green: a total of five to eight full grown plants a day from May to October.
Since Comfrey doesn't cause bloat, we could just tether (tie) the herd in the patch, but **we find it easier to cut the leaves with a sickle and put them in the manger (hay rack). We start kids (baby goats) on this vegetable before offering them other greens,** *both to form good habits and to prevent scours or diarrhea."*
-'Comfrey for the Homestead' by Nancy Bubel, Mother Earth News, Kansas, May/June **1974**.
    (A goat's rumen {one of its stomachs} ferments food and produces carbon dioxide and methane gas that needs to be
    expelled. If they cannot expel the gas, pressure builds up and the goat bloats.)
    (A sickle is a tool for cutting grain, grass, etc. that consists of a curved, hooklike blade mounted in a short handle.)

"**The plants were cut for the goats who liked it very much as an extra, It was fed in their hay rack.** *Through the summer of 1966 both Comfrey plots grew well, and Plot A was fed to the goats,* **roughly two plants a day for 3 goats**.
It is cheaper to supply the minerals that goats need in Comfrey *than to grow trees for them to bark, and to feed their roughage as a high cellulose material such as barley straw."*
-Comfrey: Fodder, Food and Remedy by Lawrence D. Hills. New York: Rizzoli Universe Books: **1976**, pages 137, 152.

"In many instances it has been observed that when animals were in the greatest need of minerals they turned to Comfrey instead of chewing so much of the mineral-rich bark and twigs off the trees around their pasture."
-Enchanted Garden: Alan Chadwick's Organic Method of Gardening by Tom Cuthbertson. London, England: Rider & Company / Hutchinson & Co. Publishers Ltd, **1978**, page 121.
    (When I had goats, if they got out of their pasture, they liked to eat the bark off of trees, especially fruit trees. So giving
    them a good supply of Comfrey will reduce their urge to eat bark that might then kill the tree.)

"The best way to feed Comfrey is to offer a few leaves once or twice a week to goats that are stall fed, **they appear to find it very palatable (agreeable)."**
-Natural Goat Care by Pat Coleby. Austin, Texas: Acres USA, **2001**, page 188.
    (I had goats for ten years and fed them armloads of Comfrey leaves and stems. They loved it more than their pasture.)

"The digestion of goats is similar to cattle and sheep but they are better able to ingest and digest forage rich in cell wall and poor in nitrogen (Morand-Fehr 2005:31). They need large amounts of carbohydrate with some protein.
In the last few hundred years green crops including lucerne, Comfrey, oats, vetches, and root vegetables have been fed, *with pea and bean plant waste, bolting lettuce, spinach, apples and brassicas. Kale and mustard are particularly useful for over-wintering and cereal chaff provides roughage (Halliday and Halliday 1988:83; Salmon 1981:59)."*
-'Livestock and Landscape: Exploring Animal Exploitation in Later Prehistory in the South West of Britain' by C.E. Randall, **2010**. Chapter 2: Farming Animals: Pastoral Farming, Possibilities and Strategies. Doctorate Thesis, Bournemouth University, England.

    Comfrey and Goat Milk

"**Goat-keepers have always been the most successful small-scale Comfrey growers.**
Though the modern milking goat is as productive in proportion to body weight as a Jersey cow, this yield cannot be secured on roughage. In Comfrey wilted to roughly lucerne (alfalfa) value he has found a very useful cheap balancer protein source.
Goats will graze Comfrey. They have been folded (rotational grazing) on the crop, but where it is their sole feed, falls in milk yield have been noticed. **Comfrey's role should be that of a low-labour-cost 'concentrate' to balance the maintenance ration from poor grass and roughage."**
-Russian Comfrey: A Hundred Tons an Acre of Stock or Compost for Farm, Garden or Smallholding by Lawrence D. Hills. London

England: Faber and Faber, Limited, **1953**, page 94.
> (Rotational grazing is moving livestock between pastures/paddocks on a regular basis. Land is divided using permanent or temporary fencing.)

"**The amount that can be fed each day depends on the individual goat, but it has been observed that there is a curve that can be plotted on the milk chart.** Up to a limit there is an increase in yield as the amount of Comfrey fed goes up, then it falls sharply, because Comfrey is high protein and low fibre, and goats need that fibre.
Comfrey is not a complete diet for anything, and attempts to use it as a sole source of protein will fail, but it is a useful bulk source of cheap protein which can be substituted for more expensive foods, or used to upgrade a poor diet."
-Comfrey: Fodder, Food and Remedy by Lawrence D. Hills. New York: Rizzoli Universe Books: **1976**, page 152.

"**Comfrey:** One particular plant deserves special attention, because so many people are interested in it and because **it's controversial (differences of opinion).** That's Comfrey, also known as boneset.
Several years ago there was a rash (lack of careful consideration) of statements from 'County Extension' agents and state Departments of Agriculture knocking (negative statements) Comfrey. Some of their reasons for not growing it are practical for large farmers, not homesteaders. And some of their information is just plain wrong.
It is true that a study conducted in Australia some years ago suggested that Comfrey might be carcinogenic when fed excessively or over an extended period. That study didn't involve goats. **And since then goats have consumed tons of Comfrey, with no problems showing up in the scientific veterinary literature.**
Even aside from that, Comfrey should be in every goat owner's garden for at least limited use. **Many goat and rabbit raisers swear by Comfrey as a feed, a tonic, and medication for certain conditions such as scours (diarrhea).**
It's high in protein, ranking with alfalfa (lucerne), although there is some question about the digestibility of the protein. But it is easier to grow and harvest than alfalfa, using hand methods."
-Storey's Guide to Raising Dairy Goats: Breeds, Care, Dairying, Marketing, 4th Edition by Jerry Belanger and Sara Thomson Bredesen. North Adams, Massachusetts: Storey Publishing, **2010**.
('County Extension' also known as 'Cooperative State Research, Education and Extension Service' is an agency within United States Department of Agriculture. Its mission is to 'advance agriculture, the environment, human health and well-being, and communities' by supporting research, education, and extension programs at land-grant universities and other organizations.)

Comfrey as Goat Medicine

"Because it helped heal broken bones, knit-bone was the old folk name for Comfrey. **Comfrey is of great assistance when used internally or topically (on skin) for bone problems, including breaks.**
Comfrey may be used in poultices and will often reduce bony swellings in a matter of days. It may be made into an **ointment or used as a liquid** obtained by boiling the leaves. **Distilled Comfrey oil** is the best source of the plant, if obtainable.
Like many plants it has a poison (in this case alkaloid) constituent which if separated from the plant could be dangerous, however fed as a plant it is safe. In spite of much publicity to the contrary, it is completely safe both internally and externally."
-Natural Goat Care by Pat Coleby. Austin, Texas: Acres USA, **2001**, page 188.

"**An ordinary bone break where the skin is unbroken is fairly easy to heal in a light animal like a goat.** If the fracture is compound (where the bone protrudes), call the veterinarian quickly. A healthy goat will be healed in ten days or less. In both cases, compound and ordinary fractures, add one 500 IU capsule of vitamin E, and three to five grams (0.10-0.17 ounces) of oral vitamin C to the diet, plus one teaspoon of cod liver oil a week. An extra teaspoon dolomite daily should also be given.
Those who have Comfrey growing should feed three to four leaves a day to a kid, and double or more to an adult goat. If you can find Comfrey tablets, give two a day for a kid, and four for an adult until healing is complete- goats will often chew the Comfrey tablets up- otherwise crush them in their feed."
-Natural Goat Care by Pat Coleby. Austin, Texas: Acres USA, **2001**, page 222.
> (Dolomite differs from limestone because it has magnesium as well as calcium. It is sourced from rock and is 8-12% magnesium and 18-22% calcium. Limestone is calcium carbonate with up to 40% calcium.)
> (I would feed a lot more Comfrey leaves than that. A small armful or more for an adult is good.)

"*Symphytum officinale* L. (Boraginaceae)
1. Goats are treated for proud flesh with several herbs:
    Wound-knitting herbs Comfrey (*Symphytum officinalis*), goldenseal (*Hydrastis canadenis*) or calendula (*Calendula officinalis*) are not used on fresh wounds since they are thought to close the wound too quickly, before it has healed underneath. Proud flesh is dealt with by scrubbing until it bleeds twice a day with a stiff scrub brush. Then hydrogen peroxide is applied using a syringe. A purchased product called 'Wonder Dust' antifungal powder is sprinkled on the wound. Once the wound is healed vitamin E, and infused oil or salve of St. John's Wort (*Hypericum perforatum*) or essential oil of lavender (*Lavandula officinalis*) is put on the area.
    Another treatment involves a Comfrey poultice (*Symphytum officinalis*) made with 1 teaspoon curcumin or fresh grated turmeric and bromelain (crush 1 or 2 purchased pineapple or papaya enzyme tablets for papain).
2. **Deep wounds, broken horn, shearing cut, wire cut:**
    Another treatment consists of a wash made with an infusion (tea) of 2 teaspoon dried aerial (above ground) parts of self

heal (Prunella vulgaris) steeped in 1 cup of boiling water and allowed to cool.
Ample fresh or dried Comfrey aerial parts are fed. *To boost the immune system and fight infection, Echinacea or Oregon grape teas are given for seven days.* **Comfrey (Symphytum officinale) and calendula (Calendula officinalis) are used on injuries only after the threat of infection has passed.**

3. Flystrike (maggot infestation):
All ruminants are treated for flystrike with **Comfrey salve,** *if the wound is partially healed or if it is not deep. Pine tar is applied if it is warm weather (corresponding to the fly season).*

4. Diarrhoea, scours:
A combination of fresh plantain leaves (Plantago sp.), flower heads of calendula (Calendula officinale), tops of nettles (Urtica dioica) and **leaves of Comfrey (Symphytum officinale) was given.**

5. Milk production:
Pregnant and lactating goats and cows are allowed access to fresh nettles or wilted cut nettles. **Armfuls of Comfrey (Symphytum officinale) are reputed to increase butterfat and act as a laxative.** *A handful of fresh or dried leaves of thornless raspberry (Rubus sp.) is given free choice (as much as they want).*

6. Udder edema (swelling):
A handful of dandelions (Taraxacum officinale) leaves and/or cornsilk (Zea mays) are fed as diuretics. Both can be dried (on a cookie sheet on low heat: 100 to 200 degrees in the oven) and used in the winter.
Fresh or dried Comfrey (Symphytum officinalis) leaves and/or stems are also fed.

7. Milk reduction (drying off):
A couple of stalks of Comfrey (Symphytum officinale) are given every couple of days during the lactation (milk production) period.

8. Pregnancy toxaemia - ketosis:
Animals are hand fed all and any tasty forest browse, e.g. salal (Gaultheria shallon), huckleberry (Vaccinium sp) or **armfuls of Comfrey (Symphytum officinale).**

9. Various injuries – abscess:
A root decoction of Oregon grape (Berberis aquifolium/Mahonia aquifolium) or root decoction of Echinacea (Echinacea spp.) is given as the drinking water for seven to ten days.
Ruminants are also feed ample amounts of fresh or dried Comfrey (Symphytum officinale)."

-'Ethnoveterinary Medicines Used for Ruminants in British Columbia, Canada' by C. Lans, N. Turner, T. Khan (DVM), G. Brauer and W. Boepple, (University of Victoria, British Columbia) (Canadian Liaison National Saanen Breeders); Journal of Ethnobiology and Ethnomedicine, Volume 3, No. 11, 22 pages, **2007**.

(Proud flesh is granulation tissue which is new connective tissue and microscopic blood vessels on surface of wound.)
(Flystrike or myiasis is parasitic infestation of an animal by fly larvae {maggots} that grow inside and feed on its tissue.)

"Recipe for daily tonic/drench for dairy goats and/or sheep:
An excellent daily tonic to maintain health and assist with worm management.
10 cloves garlic- no need to peel. Handful fresh fennel leaves or 1 tablespoon fennel seeds.
5 whole cloves or 1/4 teaspoon ground cloves. Handful of fresh thyme leaves.
1 dessertspoon seaweed/kelp granules or 1 cup (236 ml) liquid seaweed tonic.
1/2 cup (118 ml) molasses. 2 teaspoons cod liver oil. 4 cups (946 ml) cider vinegar. 4 cups hot water.
If available also add some Comfrey leaves and/or stinging nettles to the mix. *Place dried and fresh herbs in food processor with 1/2 cup cider vinegar and process until shredded. Put into bowl and add 2 cups (473 ml) hot water, leave to steep for 4-6 hours, then strain out solids. Add rest of water, cider vinegar, molasses and cod liver oil to make 2 litres (2.1 quarts) of drench. Mix well and give 10-15 ml as a drench per goat daily or mix into food. The solids can be added to food rations."*

-'Organic Pastoral Resource Guide' by the 'Organic Dairy and Pastoral Group' and 'BioGro New Zealand'. Wellington, New Zealand: 'Soil & Health Association of New Zealand' and 'BioDynamic Farming and Gardening Association in New Zealand', 144 pages, **2010**, page 91.

(A drench is giving an animal a liquid oral medication. A dessertspoon is between size of a tablespoon and teaspoon.)

<u>Guinea Pigs</u>

"The third trial was with growing guinea pigs fed fresh Comfrey (Symphytum peregrinum) foliage free-choice (as much as they want) and either:
(i) King grass and 50 grams/day of a cereal-rich concentrate (Control).
(ii) Chopped sugar cane stalk (from which the rind had been removed previously), and 15 grams/day of a protein supplement (40% protein) formulated for growing pigs.
(iii) Chopped derinded cane stalk, and leaves from the legume tree Erythrina poeppigiana.
(iv) Sugar cane juice, and 15 grams/day of the protein concentrate used in the second treatment.
There were no digestive problems and no differences in growth rates (5 to 8 grams/day) among the treatments. *But it tended to be less for the derinded cane stalk treatments than for the control and cane juice treatments. Growth rate was significantly related with intake of protein ($r_5 = 0.83$) but not with dry matter ($r_5 = 0.21$)."*

-'Development of Feeding Systems for Rabbits and Guinea Pigs, Based on Sugar Cane Juice and Tree Foliages' by A. Solarte, Food and Agricultural Organization (FAO) of the United Nations, Livestock Research for Rural Development, Volume 1, Number

1, CIPAV, Cali, Colombia, November **1989**.

Hares (see Rabbits)

Hogs (see Pigs)

Horses

"The most rapid effect was from feeding several handfuls a day of the washed (Comfrey) roots, and very probably their higher proportion of allantoin, removed all 'winter torpor' after the season when wayside grazing would have been poorest and brought the beasts into splendid 'bloom' in only a week."
-Russian Comfrey: A Hundred Tons an Acre of Stock or Compost for Farm, Garden or Smallholding by Lawrence D. Hills. London England: Faber and Faber, Limited, **1953**, page 89.
(Bloom is a glow or flush that shows health; it is the glossy, healthy appearance of the coat of an animal.)

"The Webster strain (Bocking No. 4) and its modern-day reselections are low potash, which means that chickens eat them, they are better feed for stock or as vegetables.
Curiously enough, horses like a high potash Comfrey (Bocking No. 14), *and pigs eat any one."*
-Compost, Comfrey and Green-Manure: 'Henry Doubleday Research Association' First Gardeners Report by Lawrence D. Hills. Braintree, Essex, England: Henry Doubleday Research Association, **1959**.

*"**Green Comfrey plus wheat cavings (pieces) (to provide balancing starch equivalent as digestible fiber) is a complete substitute for hay quality fed in racing stables.**"*
-Comfrey Report Number Two: For Gardeners, Farmers and All Comfrey Growers in All Countries, with Analysis, Yields and Cultivation Methods, and the Results of Seven Years More Work Since Our First Report by Henry Doubleday Research Association, Braintree, Essex, England, 45 pages, March **1963**.

"I once led a mare (female horse) called Lyon's Lass for a mile and a half along a country lane and let her graze as she would.
She took yarrow, chicory and all the deep rooting herbs rather than the lushest grasses which were at their best in the full green tide of summer.
Just like the cows at Southery (Norfolk County, England) leaving the pedigree (purebred) Aberystwyth grasses for cut Comfrey, **she enjoyed her Comfrey ration every day**, *for she knew what she wanted when she had a chance to get it."*
-Comfrey: Fodder, Food and Remedy by Lawrence D. Hills. New York: Rizzoli Universe Books: **1976**, page 142.

"Horses need minerals to build their bones and give them speed. Mares in foal (pregnant) need calcium and phosphorus to build up the bones of wobbly-legged foals (baby horse) which must soon follow the herd.
Cattle take mineral licks; mineral supplements can be added to pig feed, but horses seem to take their minerals only from plants, and do best on organically-grown grazing and feed, not articially-made substitutes."
-Fighting Like the Flowers: An Autobiography: The Life Story of Britain's Best-Known Organic Gardener by Lawrence D. Hills. Bideford, Devon, England: Green Books, **1989**, page 96.

*"**Cutting Comfrey for cows and horses should be done at a more mature stage than for chickens and pigs, between 28 and 35 days being appropriate according to the season. This means higher fiber.**"*
-Comfrey: Nature's Healing Herb & Health Food by Andrew Hughes. Japan: Sanyusha Publishing Co., Ltd, **1992**, page 175.

There is a horse named 'Comfrey'. He was born February 9, 2002 to breeder Jayeff 'B' Stables in the United States. He was exported to Japan in August 2006. His great grandfather is the famous horse, Secretariat. Horses named 'Comfrey' were also born in 1894 (Great Britain), 1964 (United States), 1973 (Great Britain), and 1976 (Great Britain).
-Pedigree Online Thoroughbred Database, www.pedigreequery.com, **2018**.

Pounds of Comfrey Per Day for Horses

*"The writer in 'The Field' (magazine) has been in the habit of **giving horses that were kept in the stable in the summer a daily allowance of 8 pounds (3.6 kg) with very good effect.** They become very fond of it. It keeps them cool, and is a diuretic."*
-'Prickly Comfrey: Its History, Cultivation, Extraordinary Production, and Uses: A Letter Addressed to His Excellency Sir Hercules Robinson, President of the Agricultural Society of New South Wales' by Arthur T. Holroyd, Sydney, Australia, **1876**, page 11.
('The Field' is the world's oldest country and field sports magazine, having been published continuously in England since 1853.) (A diuretic increases the excretion of urine.)

*"The experiment of the Reverend E. Highton in 1875, was repeated in 1953 by Mr. Stephenson with Tom Atom, a four year old gelding and a stallion. **Comfrey ad lib (as much as they want) turned out to be roughly 40 pounds (18 kg) per beast a day fresh weight,** and the oats were reduced from 8 pounds (3.6 kg) a day to 4 pounds (1.8 kg). With this as the sole diet both*

beasts throve."
-Russian Comfrey: A Hundred Tons an Acre of Stock or Compost for Farm, Garden or Smallholding by Lawrence D. Hills. London England: Faber and Faber, Limited, **1953**, page 91.
    (A gelding is a castrated male horse. A stallion is a male horse that is not castrated.)

"At Little Weighton (East Riding of Yorkshire County, England) Symphytum peregrinum is fed ad lib to working stallions but when it has to be cut down because of the needs of mares, **the regular allowance for adult stock is approximately fourteen pounds 96.3 kg) per head per day.**
It is part of the regular diet from early April to the end of November, being replaced in winter with bought carrots."
-Russian Comfrey: A Hundred Tons an Acre of Stock or Compost for Farm, Garden or Smallholding by Lawrence D. Hills. London England: Faber and Faber, Limited, **1953**, page 89.

"Mr. E.V. Stephenson of Hunsley House Stud, Little Weighton, York, England, owns the oldest continuously cultivated Comfrey plot in Britain. He obtained his stock from the late Kenneth Crawly of Lockerbie (southwest Scotland) in 1938. His plot has now completed its 10th cutting season on 3/4 acre.
By feeding Comfrey first, Mr. Stephenson established that a 14 pound (6.3 kg) a day ration to mares with foals ruled out all risk of scour (diarrhea). Now a portion of the plot is covered with Chase barn cloches in February as though it were early lettuce to bring on the ration in time for the 'luxury market' of the first mares to foal."
-Comfrey Report Number Two: For Gardeners, Farmers and All Comfrey Growers in All Countries, with Analysis, Yields and Cultivation Methods, and the Results of Seven Years More Work Since Our First Report by Henry Doubleday Research Association, Braintree, Essex, England, 45 pages, March **1963**.
    (A cloche is a long, low cover made of glass or clear plastic that is put over plants to protect them from cold. Traditional barn cloches of 1930's, developed by Chase, have largely been replaced by plastic tunnel cloches. The wire framework holds 3 mm {0.118 inch} horticultural glass.)

"In 1952 E.V. Stephenson of Hunsley House Stud, Little Weighton, near Hull (Yorkshire, England) gave each racehorse a daily ration of about 14 pounds of Comfrey that was cut by each groom with a sickle and gathered on canvas sheets to carry in and feed to them when they were in their looseboxes."
-Fighting Like the Flowers: An Autobiography: The Life Story of Britain's Best-Known Organic Gardener by Lawrence D. Hills. Bideford, Devon, England: Green Books, **1989**, page 96.
    (A groom is responsible for the care of horses and their stable.) (A sickle is a tool for cutting grain, grass, etc. that consists of a curved, hooklike blade mounted in a short handle.)

Comfrey as Horse Medicine

"Mr. Shilling's remarks upon **the medicinal properties of Prickly Comfrey** are important, and will be read with interest:
    '**For horses and cattle in cases of constipation or incipient fret, known as colic or gripes** (which some horses are very subject to after having worked very hard or been a long time without food or water), **it has been used with the best results,** acting as a solvent on the system, as well as on the bowels and urinary passages.
    No establishment whatever with the slightest pretentions (claims) to economy and real management should be without a good plantation of this plant near the homestead.
    After three or four consecutive feeds, the animals will not fail (the horses particularly) to remind you that its feeds are deficient in quality should you fail to replenish it with the green herbage. He will eat it with the same avidity (eagerness) as green clover or tares (vetch). Cattle of all classes will thrive and fatten on it.
    For horses with broken wind (COPD), or when they have had a chill in the blood after been much heated by hard drivng, which often produces a surfeit (skin problem) or scurf disease (flakes) on the skin, very hard to cure as well as being irritating to the horse as soon as harnessed.
    The writer has known more good result by feeding with this herbage, and mucilage extract in the feeds of corn, than all the nostrums (medicine of doubtful usefulness) devised in the shape of powders with nitre (niter= potassium nitrate), etc., and given as diuretics and skin cleansers, which are ruinous (bad) to the horse's constitution (health).' "
-'Prickly Comfrey: Its History, Cultivation, Extraordinary Production, and Uses: A Letter Addressed to His Excellency Sir Hercules Robinson, President of the Agricultural Society of New South Wales' by Arthur T. Holroyd, Sydney, Australia, **1876**, page 11.
    (Colic in horses is abdominal pain. It is a clinical sign rather than a diagnosis. It includes all forms of gastrointestinal distress which cause pain as well as other causes of abdominal pain.)
    (Broken wind or COPD is 'Chronic Obstructive Pulmonary Disease' which is the constriction of the airways with difficulty breathing.) (Diuretics increase the production of urine.)

"Dublin, Ireland, February 29th, 1876 from W. Brabazon: **I write you an account of my experience of the value of Prickly Comfrey (Symphytum asperrimum).** I had a valuable hunter horse (used in fox and deer hunting), who after a severe cold, broke out with button farcy (a form of the bacterial disease glanders affecting skin and lymphatic vessels), and nearly died. He was cured at a veterinary hospital, and sent home a miserable object, scarcely able to walk. At first the horse would hardly eat the Comfrey, but in a couple of weeks, he would come neighing to his stable-door at feeding time.
At the end of two months, on this food, he was perfectly recovered, and in splendid condition. I find this is first-rate

food for carriage horses- that is, to give two good feeds of it daily, along with a very small quantity of hay, and their usual measure of oats, as it never purges (causes diarrhea), and gives them a very fine coat. I also found the cows that were fed on it gave much more milk than the others, although they were all on good pasture."
-'Forage Plants and Their Economic Conservation by the New System of Ensilage: Part I: Caucasian Prickly Comfrey' by Thomas Christy, Jun., F.L.S. (Fellow of the Linnean Society), Christy & Co., London, England, **1877**, page 16.

"Herbal Treatments for Horses: Capped Hock:
This is most frequently caused by a bruise *obtained in the many ways possible when the animal is confined in the stable; commonly by a kick. However, many horse owners uphold that the trouble is purely constitutional.*
Treatment: *The swollen area should be treated with a cold brew of seaweed or of elder leaves, and then some oil, preferably linseed or almond, rubbed well into the area. The animal must be given thorough rest from work, and placed on a laxative diet, which should include a daily quarter-pint dose of linseed oil.*
Comfrey is an important remedy for bruises*, also the common daisy plant."*
-Herbal Handbook for Farm and Stable by Juliette de Bairacli Levy. London, England: Faber and Faber, Inc., **1952**. First edition that later was called 'The Complete Herbal Handbook for Farm and Stable'.
> (Hock or gambrel is the joint between the tarsal bones and tibia of a mammal such as a horse, cat or dog. It is the anatomical equivalent of the human ankle. A capped hock means the joint is inflamed.)

*"**When horses are fed concentrates they become extra frisky, hence the phrases 'full of beans' and 'feeling his oats',** while extra green food has a certain tranquilizing effect.*
It is certain that a good, well kept Comfrey plot would be an asset for anyone concerned with horses."
-Comfrey: Fodder, Food and Remedy by Lawrence D. Hills. New York: Rizzoli Universe Books: **1976**, pages 143, 145.
> ("Feel one's oats" is a reference to the behavior of a horse that has eaten a lot of oats. It means to feel energetic and frisky. "Full of beans" means to be lively and in high spirits.)

*"**Osteoarthritis is a degenerative disease of the joints that affects a great number of horses**, and accounts for a considerable economic burden on the industry. The condition is typically treated with NSAIDs (NonSteroidal Anti-Inflammatory Drugs) and steroids, but these treatments may elicit (create) negative side-effects when used over the long-term.*
*Herbal remedies have increased in popularity over the past 3 years, but little scientific documentation exists to further the knowledge of these products for horses. **This study investigated the effects of an herbal mixture 'Mobility®' on equine osteoarthritis.***
*'**Mobility®' is an herbal product developed specifically for horses by Selected Bioproducts (Guelph, Ontario). It is formulated to improve musculoskeletal function by reducing the inflammatory response in chronically inflamed tissue. The supplement contains a mixture of five herbs with historic application to the treatment of inflammation and pain: dandelion, devil's claw, stinging nettle, burdock, and Comfrey.** The primary active phytochemicals in Mobility are glycosides, flavonoids, allantoin, mucilage, and sesquiterpenes.*
*Supplementation with Mobility in arthritic horses in the current study did not have any effect on hyaluronic acid levels in synovial fluid, or on any of the hematological or biochemical blood parameters tested. However, **Mobility supplementation did result in a reduction of PGE2 production in arthritic joints in horses during the first two weeks of supplementation.***
This observation could explain the anecdotal (story) testimonials to the efficacy (effectiveness) of Mobility in the management of osteoarthritis in horses. The trends observed in SeGAG and SynGAG levels should be further investigated with a larger sample size to confirm their artifact nature, or to determine if the increase is an anabolic or catabolic process."
-'Effect of a Propietary Herbal Product on Equine Joint Disease' by W. Pearson, S. McKee and A.F. Clarke, Equine Research Centre, Guelph, Ontario, Canada; Journal of Nutraceuticals, Functional & Medical Foods, Volume 2, No. 2, pages 31-46, **1999**.
> (I do not have any connection with the product 'Mobility®'.)
> (Osteoarthritis is a joint disease from breakdown of joint cartilage and underlying bone. Symptoms are joint pain and stiffness. Initially, symptoms occur following exercise, but over time may become constant. It creates joint swelling, decreased range of motion, and weakness/numbness of arms and legs.)
> (A steroid is any of a large group of fat-soluble organic compounds such as sterols, bile acids, and sex hormones. They are classified by function:
>> Corticosteroids: Glucocorticoids such as cortisol. Mineralocorticoids such as aldosterone.
>> Sex steroids such as progestogens {progesterone}, androgens {testosterone}, estrogens {estradiol}.
>> Neurosteroids such as DHEA and allopregnanolone.)
> (The prostaglandins are physiologically active lipid compounds with various hormone-like effects in animals.)

*"**A famous show jumper,** who was in Scotland for some horse trials, had an unfortunate accident when the horse slipped. First I attended to the rider and prescribed Comfrey for external use. Afterwards I examined and treated the horse.*
***Both made a remarkable recovery. Comfrey gel is a wonderful remedy.**"*
-Traditional Home and Herbal Remedies by Jan De Vries. Edinburgh, Scotland: Mainstream Publishing Company Limited, **2004**.

"1. Plants used for abscesses and wounds in horses:
> ***An external treatment for abscesses consisted of a wash of Comfrey tea*** *(Symphytum officinalis). This tea could include an infusion of Oregon grape (Mahonia aquifolium).*
> *Compresses were made of powdered aerial (above ground) parts of: betony (Stachys officinalis), figwort (Scrophularia*

nodosa) and motherwort (Leonorus cardiaca).

Comfrey (Symphytum officinalis) root was added. Equal amounts of the herbs were made into a paste with water, applied onto a gauze pad and placed onto the wound.

2. Plants used for hoof problems in horses:

One or two leaves of Comfrey were crushed and applied to cracks on the hoof and then bandaged.

3. Plants used for exercise induced pulmonary haemorrhage (EIPH) in horses:

One breeder used a commercial herbal product containing lungwort (Pulmonaria officinalis) compounds, bioflavonoids and vitamin K for EIPH. That breeder also used alfalfa hay (or soaked alfalfa pellets) in a 1 : 4 ratio with the regular hay. Furosemide (Lasix®), a diuretic often used in the treatment of EIPH, was thought to dehydrate the horse. To reduce this effect, a tea was given with 1 part each of the following: licorice (Glycyrrhiza glabra) root, aerial parts of mullein (Verbascum thapsus) or mallow (Althea sp.), and **Comfrey (Symphytum officinalis) root.**

4. Plants used for eye problems, eye infections in horses:

An infusion with saline (salt) solution was made with equal parts of the following: eyebright (Euphrasia officinalis) fresh or dry leaves, calendula (Calendula officinalis) flowers, and **Comfrey (Symphytum officinalis) leaves. The infusion was strained carefully and used as an eyewash.** The infusion was weakened as the condition improved."

-'Ethnoveterinary Medicines Used for Horses in Trinidad and in British Columbia, Canada' by C. Lans, N. Turner, G. Brauer, G. Lourenco and K. Georges (University of Victoria, British Columbia) (Trinidad and Tobago Racing Authority and University of the West Indies), Journal of Ethnobiology and Ethnomedicine, Volume 2, No. 31, 20 pages, **2006**.

(An abscess is an infected, swollen area with an accumulation of pus.)

"**Symphytum officinale: Parts Used:** All aboveground parts.

Primary Medicinal Activities: Heals wounds; anti-inflammatory; astringent; lubricates, soothes and protects internal mucous membranes; expectorant.

Strongest Affinities: Skin, digestive tract, respiratory tract.

Preparation: Water or oil infusion, poultice, salve, fomentation.

For treatment of colitis, stomach ulcers, or just about any other inflammation of the digestive tract, a handful of the fresh leaves or 2-3 ounces (56-85 grams) of the dried Comfrey leaves can be fed directly to horses and other large herbivores on a daily basis for up to two weeks.

Comfrey is traditionally used for bronchitis and other respiratory ailments that are relieved by its soothing anti-inflammatory nature. A cooled tea serves this purpose well. Administer 1 tablespoon per 20 pounds (9 kg) of the animal's body weight, twice daily.

In all cases of internal use, feeding of Comfrey should be limited to occasional short-term therapies. For the same reason, highly concentrated preparations such as tinctures or strong decoctions should not be employed internally.

For external treatment of closed injuries, skin ulcers or mastitis, you can make a poultice or fomentation from the fresh or dried Comfrey leaves and apply to the affected area.

With horses and other large animals, this might require you to secure the compress onto a swollen limb with a strip of gauze or a clean piece of fabric torn from some old clothing. Leave the compress on for as long as possible- preferably eight hours or more.

Pyrrolizidine Alkaloids:

Given the fact that the Pyrrolizidine Alkaloid content in a fresh Comfrey leaf amounts to only about 0.3 percent of its total chemistry, **the average horse would need to eat several hundred pounds (200 pounds = 90 kg) of the leaves each day before any toxic effects would be observed.**"

-Herbs for Pets: The Natural Way to Enhance Your Pet's Life by Mary L. Wulff and Greg L. Tilford. Los Angeles, California: i5 Publishing, **2011**, pages 86-89.

(How make poultice/fomentation, see subsection 'Comfrey Poultices' in section 'Making and Using Comfrey Medicine' {Chapter 27}.)

"In spite of much publicity to the contrary, the whole Comfrey plant is completely safe, both internally and externally. In many parts of Germany and also Japan, Comfrey is used exclusively for diary cattle fodder during the summer months.

Comfrey is one of the few plants that contains natural vitamn B12. This may be one of the reasons why it is so useful in cases of sickness.

It is highly nutritious and of great assistance used internally or topically (on skin) for bone problems, including breaks.

It may be used in poultices and will often reduce bony swellings like splints of recent origin in a matter of days.

It may be made into an ointmnent or used as a liquid obtained by boiling the leaves; distilled Comfrey oil is the best if obtainable. All forms are useful at some time or other The plant also has the reputation as an inhibitor of cancer.

The best way to feed Comfrey is to offer a few leaves to horses once or twice a week; they appear to find it very palatable (agreeable taste). In United Kingdom, most racing stables have a bed of it by the barns and feed each horse a few leaves a week."

-Natural Horse Care by Pat Coleby. Austin, Texas: Acres USA, **2013**.

(Splints is a hard, bony swelling, usually on inside of horse's front leg, between splint and cannon bone or on splint bone itself.)

Prevent and Cure Diarrhea in Horses

"**The greatest gain is in avoiding digestive disturbances, urinary complaints and above all, foal scouring (diarrhea), and this benefit is incalculable.** Though Comfrey is no 'cure-all' and its limitations can only be determined by veterinary research,

its use as a feed reduces the number of things that can go wrong with highly-geared and sensitive race-horses. One racing trainer is using Comfrey, which he wilts, chaffs, and mixes by hand into a previously boiled mash and feeds twice daily."
-Russian Comfrey: A Hundred Tons an Acre, page 90

*"E.V. Stephenson of Hunsley House Stud, Little Weighton, near Hull (Yorkshire, England) found that **Comfrey prevented 'scour' (diarrhoea) among the foals,** and when mares and foals are returned to a customer scouring there is just cause for complaint. Middle Piece Farm, which he also used for grazing, had no Comfrey, and constant scour problems. So he ended this problem by establishing a second Comfrey plot."*
-Fighting Like the Flowers: An Autobiography: The Life Story of Britain's Best-Known Organic Gardener by Lawrence D. Hills. Bideford, Devon, England: Green Books, 1989, page 96.

Pigs and Hogs

Definitions

Based on Age and Sex
Barrow: (shortened to bar) male pig castrated before puberty.
Boar: mature male hog (of breeding age).
Gilt: female that has not reproduced.
Hog: any age, status or gender of animal.
Pig: unweaned baby hog.
Piglet: very young baby hog.
Runt: unusually small and weak piglet.
Shoat (Shote): young hog that has been weaned.
Sow: female that has reproduced.
Stag: male pig castrated after puberty.
Sucker: pig between birth and weaning.
Swine: any age, status or gender of animal.
Weaner: young pig recently separated from the sow.

Pig Slaughter Definitions
Suckling Pig: piglet slaughtered for its tender meat.
Feeder Pig: weaned pig between 18-37 kg (40-82 pounds) 6-10 weeks old to be finished for slaughter.
Grower Pig: pig between 18-100 kg (40-220 pounds).
Porker: market pig between 30-54 kg (66-119 pounds) dressed weight.
Baconer: market pig between 65-80 kg (143-180 pounds) dressed weight.
Finisher: grower pig over 70 kg (150 pounds) live weight.
Butcher Hog: pig of approximately 100 kg (220 pounds), ready for the market.
Backfatter: cull (unwanted) breeding age pig sold for meat.
(Dressed weight, dead weight or carcass weight is the weight of an animal after being partially butchered, removing all the internal organs and often the head as well as inedible portions of the tail and legs.)

Overview of Pigs and Comfrey

*"Mr. Thomas Christy says: '**Pigs fatten by Comfrey leaves being mixed with their food, either in the green or dry state, or when boiled with the ordinary food.'** "*
-'Prickly Comfrey: Its History, Cultivation, Extraordinary Production, and Uses: A Letter Addressed to His Excellency Sir Hercules Robinson, President of the Agricultural Society of New South Wales' by Arthur T. Holroyd, Sydney, Australia, **1876**, page 10.
 (Thomas Christy from England promoted Prickly and Russian Comfrey in the mid to late 1800s.)

*"Two lots of younger pigs, in fair condition, **fed during three weeks all the Prickly Comfrey they could eat** and a little corn meal, failed to make any gain. The pigs in Lot A averaging 64.3 pounds weight consumed each an average per week of 23.2 pounds of Comfrey and 2.3 pounds corn meal and drank 35.3 pounds of water. Those in Lot B averaging 64.6 pounds weight consumed each an average of 22.8 pounds Comfrey and 2.3 pounds corn meal and drank 34.7 pounds of water. The average loss of weight in Lot A being 0.9 pounds per week and of Lot B 1.6 pounds per week.*
Younger pigs fed up Prickly Comfrey, chopped and dusted with corn meal, would not eat enough for any increase in weight."
-New York Agricultural Experiment Station, Bulletin No. 22 New Series: Pig Feeding Experiments Without Milk, Geneva, New York, August **1890**, page 292. Includes 'Feeding Prickly Comfrey'.
 (Comfrey is meant to be a valuable addition to the pig diet, not the only thing or the main thing fed.)
 (The New York Agricultural Experiment Station article below has more details about the above pig feeding.)

*"For a short time in July and August , two pens of pigs (Cheshires) each containing two sows and a barrow...were **fed all the**

Prickly Comfrey they could eat *and a little corn meal. The Comfrey formed over 90 percent of the total food consumed in both pens, and less than 58 percent of the water-free food, the moisture being 88 percent in the fresh plant. Neither would eat enough to make any gain or even hold their weight."*
-New York Agricultural Experiment Station, Bulletin No. 28 New Series: Pig Feeding Experiments With Coarse Food, Geneva, New York, April **1891**, page 438. Includes 'Results with Prickly Comfrey'.

*"**Prickly Comfrey is a fine thing for pigs of all ages and descriptions.** It takes them a little time to get used to it, but when once they do take to it, they are fond of it, and it is a fine thing to keep them in good health. I have had both sows and pigs when unwell eat a few leaves of it in preference to anything else. On all farms where a quantity of pigs are bred a rood, or even a few poles, will be found extremely handy."*
-'Practical Pig-Keeping: A Manual for Amateurs, Based on Personal Experience in Breeding, Feeding and Fattening; Also in Buying and Selling Pigs at Market Prices' by R.D. Garratt, London, England, **1897**. page 41.
　　(Rood is an English unit of area equal to 1/4 acre or 10,890 square feet or 1,012 square meters.)
　　(Pole or perch is a unit of area. A square perch is equal to 30.25 square yards or 25.29 square meters or 1/160 acre.)

*"**The additional vegetable protein from Comfrey might enable cheaper meals with higher carbohydrates, or processed swill, to be used with profit.** It should be recognized that the pig is an omnivorous animal and non-vegetable protein is also required, as fish or meat meal or in a compound form.*
The Vitamin A and minerals in Comfrey, especially the calcium and iron in the ash, would be supplied more easily than in an artificial mixture, apart from the value of the allantoin."
-Russian Comfrey: A Hundred Tons an Acre of Stock or Compost for Farm, Garden or Smallholding by Lawrence D. Hills. London England: Faber and Faber, Limited, **1953**, page 101.
　　(An omnivore is an animal or person who eats both plants and animals.)
　　(Swill is food for animals in liquid or partially liquid form. It includes kitchen and food processing scraps.)

*"**The most profitable stock to feed with Comfrey in Britain today is the pig,** and pig feeding pays even with yields of 7,500 kg (16,535 pounds) per 1/4 acre, when over 85% of the production costs for pigs is bought meal.*
The high ratio of protein to fiber fits the digestion of the pig, and their great appetite for it makes it highly suitable for a modern version of the Lehman feeding system, *without the troubles from excess fiber in lucerne (alfalfa) usually fed under this system.*
Comfrey should not be compared with fodder beet (mangel), which contains about four times as much carbohydrate, and only one-third of the protein, and slightly less fiber. The root crop (beet) is very much higher in labor cost than green Comfrey after the first year.
The biggest live weight gain from vegetable protein fed as much as they want comes in the early stages, and small weaners quickly level up to the bigger ones. Feed them all they will eat, and reduce the usual supply of meal to one half.
The procedure is to take a daily cut of Comfrey, near the piggery or pig pasture, and throw it to them. ***No wilting is necessary.*** *The moisture, roughly the same as that of the mangel-wurzel (fodder beet), merely means that they drink a bit less than when on a dry meal diet.*
With **baconers** *there comes a stage, usually at about 60 kg (132 pounds) live weight, when carbohydrate needs increase, and the meal proportion should go up by as much as 500 grams (1.1 pounds) per day, finishing at about three-quarters of the previous amount used. If growth slows after a good start, increase the meal a bit.*
The yardstick is profit per pig, not maximum live weight increase in the minimum period, and small store pigs bought cheaply, so long as they are healthy, mean the most return for the small farmer on this system."
-Comfrey Report No. 1: The 1954 Research Results by Lawrence D. Hills, Henry Doubleday Research Association, Braintree, Essex, England, published **1955**.
(A weaner is a young pig recently separated from the sow. A baconer is a market pig between 65-80 kg or 143-180 pounds dressed weight. Dressed weight is after being partially butchered, removing all internal organs, parts of the tail/legs, and head.)

> *"**The well-known Lehman ration (for pigs)** specifies two and two-tenths pounds (2.2 pounds = 0.99 kg) of a mixture (70 percent cereals, often rye, and 30 percent protein concentrates, often 20 percent meat meal and 10 percent fish meal) per head per day supplemented by potatoes to the amount animals desire.*
> *Pigs weighing a hundred and fifty pounds (68 kg) or more will usually eat at least 14 pounds (6.3 kg) of potatoes daily. As potatoes do not constitute a satisfactory food for young pigs, only the basal ration of cereals and protein is given until appetites exceed two and two-tenths pounds of concentrates often at the age of twelve to fourteen weeks."*
> -'Potato Fed Swine in Germany' by Earl Shaw; Economic Geography, Clark University, Worcester, Massachusetts, Volume 18, No. 3, pages 287-297, July **1942**.
> 　　(A basal ration has the needed energy but lacks one or more food substances such as vitamins.)

"An initial crude palatability (taste) test using two 150-pound (68 kg) hogs indicated that they would consume reasonable quantities of a ration containing up to 40% dehydrated (dried) Comfrey. Higher levels were not fed.
The swine digestion trial indicated that the Comfrey used had a dry matter content of 12.1% digestible Crude Protein, and 52.7% TDN (Total Digestible Nutrients) content.
The ratings for Crude Protein, Total Digestible Nutrient content and Nitrogen-free extract were all lower for Comfrey than for the control ration. *The control ration was ground barley, cottonseed meal (41% of ration's Crude Protein), plus meat*

and bone meal (50% of ration's Crude Protein)."
-'Comfrey as a Feed for Swine' by Hubert Heitman, Jr. (University of California) and Sergio E. Oyarzun (University of Chile); California Agriculture (Hilgardia: Journal of Agricultural Science by the California Agricultural Experiment Station), Volume 25, No. 1, pages 7-8, January **1971**.

"The fact that it is possible to feed pigs successfully on Comfrey and cereals alone has long been known to growers. Every writer since 1810 has stressed the real pleasure pigs have in eating Comfrey; *there is never any question of their getting accustomed to it by degrees.*
Comfrey supplies minerals in digestible form as well as the high protein and low fibre that suits their needs."
-Comfrey: Fodder, Food and Remedy by Lawrence D. Hills. New York: Rizzoli Universe Books: **1976**, pages 89, 121.

"Pigs are eager to eat feeds containing garlic, juniper fruits, Comfrey root, *dandelion, nettle, yarrow, St John's wort, knotgrass, peppermint leaves and couch grass rhizome. They are not very willing to consume calamus rhizome (sweet flag), creeping thyme and wormwood.*
An important property of herbs is that they improve the palatability (good taste) of feeds. Herbs stimulate gustatory (taste) and olfactory (smell) receptors in pigs, making them more willing to eat feed.
Stimulation of the appetite plays a significant role, especially during the period when piglets are adjusted to solid feed and weaned from their mothers *(*Rekiel, 1998)."*
-'Modulating Gastrointestinal Microflora of Pigs through Nutrition Using Feed Additives' by Dorota Bederska-Lojewska and Marek Pieszka, Department of Animal Nutrition and Feed Science, National Research Institute of Animal Production, Krakow, Poland; Annals of Animal Science, Volume 11, No. 3, pages 333-355, **2011**.
(* -'Effectiveness of the Use of Herbal Mixture in Rearing of Piglets' by A. Rekiel, Medycyna Weterynaryjna: Veterinary Medicine-Science and Practice, Lublin, Poland, Volume 54, No. 8, pages 545-549, 1998. In Polish with Summary in English.)

"In smallholder agriculture, the fast-growing and perennial accumulator plant Comfrey (Symphytum spp.) was used to supply pigs with protein and minerals. *Comfrey leaves show similar values in dry matter as soybean or blue lupine in crude protein content, but much higher levels of calcium and phosphorus. However, in terms of increased efficiency in animal husbandry, Comfrey has been displaced by mainly soybean and cereals. Due to its profile of macro- and micronutrients the use of Comfrey could have the potential to re-establish local resource cycles and help remediate over-fertilized soils.*
The aim of the study was to evaluate whether a modern pig breed accepts a continuous feed supplement of dried Comfrey leaves. *After an initial adaptation period post-weaning,* **German Landrace piglets** *were subjected to either a standard control diet or* **a diet supplemented with 15% dried Comfrey leaves for 4 weeks**.
Body weight was reduced in Comfrey-supplemented piglets *compared to controls, which might be attributed to reduced palatability in the experimental setting. Nevertheless, Comfrey supplemented piglets exhibited adequate bone mineralization and intestinal integrity.*
The microbiome profile in feces and digesta revealed higher diversity in Comfrey-supplemented piglets compared to controls, with pronounced effects on the abundances of Treponema and Prevotella. This may be due to described bio-positive components of the Comfrey plant, as data suggest that the use of Comfrey leaves may **promote intestinal health**. *Digestive tract phosphorus levels were reduced in piglets receiving Comfrey supplementation, which may ultimately affect phosphorus levels in manure.*
Results indicate that Comfrey leaves could serve as a feed component in integrated agricultural systems to establish regional nutrient cycles. *The trial provides a basis for further work on Comfrey as a regionally grown protein source and effective replacement for rock mineral supplements."*
-'Comfrey (Symphytum spp) as a Feed Supplement in Pig Nutrition Contributes to Regional Resource Cycles' by Oster, Reyer, Keiler, Ball, Mulvenna, Ponsuksili, Wimmers (Germany and United Kingdom); Science of the Total Environment, Volume 796, Article 148988, November **2021**.

Pigs, Comfrey and Land Use

"Comfrey has greater weight of nutritional value per acre, over twice as much carbohydrate on a 75-ton yield as from an 18-ton-an-acre lucerne (alfalfa) crop. This means rather **more than twice the number of pigs are fed off the same area of Comfrey, from a fodder that is suitable for a wider range of soils."**
-Russian Comfrey: A Hundred Tons an Acre of Stock or Compost for Farm, Garden or Smallholding by Lawrence D. Hills. London England: Faber and Faber, Limited, **1953**, page 101.

"In his view (Mr. George Halling of Stevenage, Hertfordshire, England) **the second most important aspect of Comfrey on the modern piggery was as a disposal for slurry,** *for he found that it was possible to continue irrigating (watering) with this right through the winter without harming the crop."*
-Comfrey: Fodder, Food and Remedy by Lawrence D. Hills. New York: Rizzoli Universe Books: **1976**, page 124.
> (Slurry is a semi-liquid mixture. It is created in livestock systems where little or no bedding is added to the manure and urine. Slurry manure is 5-15% solids.)

Amount of Comfrey for Pigs

*"**The late Kenneth Crawley fattened pigs on Comfrey; he states that one can feed it ad lib and save half the normal meal.** Wilted to 75 percent moisture, which is very easily achieved, the crude protein is 4.1 percent, carbohydrates 9.9 percent, the fibre is 7.2 percent, and the ash 2.4 percent. Pigs digest some fibre, but cellulose digestion is always incomplete."*
-Russian Comfrey: A Hundred Tons an Acre of Stock or Compost for Farm, Garden or Smallholding by Lawrence D. Hills. London England: Faber and Faber, Limited, **1953**, page 99.
(Ad lib = Ad libitum is Latin for 'at one's pleasure' or 'as you desire'. It means free feeding or giving animals as much as they want.)

*"**The Russian or Quaker Comfrey (Symphytum peregrinum),** from both plots and harvested in the spring of 1967, was dehydrated in a forage dryer. The relatively high level of crude protein and moderate level of crude fiber are noteworthy, but so is the high level (over 23%) of ash.*
***An initial crude palatability (taste) study was made utilizing two 150-pound (68 kg) hogs. They consumed reasonable quantities of feed containing as much as 40% of dehydrated Comfrey.** Higher levels were not tested."*
-'Plant Leaf Protein with Emphasis on Comfrey' by Hubert Heitman and Milton D. Miller with Edward Johnson (agronomic phases of study), Sergio E. Gyarzun of University of Chile (swine digestion trial), Bob D. Wilson (rat study) and James T. Elings (Extension Animal Scientist); California Experiment Station and Agricultural Extension Service, University of California, **1969**.

*"The cut Comfrey for each tethered (tied) pig was thrown down outside the pigloo in varied quantity, **building up to a maximum of 30 pounds (13.6 kg) a day for a gilt (young female pig that never gave birth) just before sale, and to fit the appetite of a sow (adult female pig that has given birth) with a growing litter.***
As a general average he found he could save 25 percent of the meal that would be fed to pigs on this grazing system.
*A trial was undertaken at the Devon Farm Institute at Bicton (Devonshire County, England), with a twentieth of an acre of Comfrey planted in September 1958. Further trials showed that **1 pound (0.45 kg) fattening meal could be replaced with 5 pounds (2.26 kg) of fresh Comfrey.***
As a basic rule, a pig should eat up to 10 percent of its body weight in addition to fattening meal.
*A test at the Fujisawa Farm of the Nihon Agricultural University in Japan showed that **adult pigs do well on 8 to 9 kg (17 to 19 pounds) of green Comfrey leaf per day.** This means replacing 30 percent of the meal with cut Comfrey. The University report states 'a noticeable result was the improved health of the pigs fed on Comfrey' not only from allantoin, which banished scouring (diarrhea), but better mineral balance. This was also observed at the Devon farm institute."*
-Comfrey: Fodder, Food and Remedy by Lawrence D. Hills. New York: Rizzoli Universe Books: **1976**, pages 121, 122.
 (In 1958, Cargill develped the 'Pigloo'. It was a 12-sided house for a sow and her piglets. It allowed sows to lie with their backs against the wall and udders facing a metal guard. Newborn piglets had an electric heat lamp for warmth, distanced from their mother who could not accidentally crush them.)

*"**Studies were done on substitution of fresh Comfrey for up to 25% of the concentrates or of Comfrey meal for up to 20% of the concentrates in diets for fattening pigs.** Effects on weight gain, carcass weight, dressed weight, dressing percentage and fat thickness (at the shoulder, back and loin) were investigated.*
The results show that the Comfrey-containing diet gave carcass weight and quality closely resembling those achieve with the control diet."
-'Studies on the Suitability of Comfrey as Feed for Pigs' by G. Nakanishi, M. Akahori, T. Ohmi and Y. Niwa, College of Agriculture and Veterinary Medicine, Tokyo, Japan; Bulletin of the College of Agriculture and Veterinary Medicine,   Nihon University, Japan, No. 35, pages 271-281, **1978**.

*"**In some respects Comfrey is regarded as having its greatest economy of use with pigs.***
Pigs can be raised very successfully and economically on a diet almost wholly of Comfrey, as high as 80-90%.
In pork and bacon raised on 80% Comfrey, the ratio of flesh and fat will be balanced. The porkers will come to market with a bloom that marks all first class pigs properly fed, and the flesh-fat ratio will be such that one can eat it all.
***Comfrey diet will help also after castration (testicles removed) of piglets,** who recover after the operation without any check to their growth.*
In addition to the basic meal ration, a pig seems capable of eating a maximum quantity of Comfrey per day equal to approximately 10% of its live weight."
-Comfrey: Nature's Healing Herb & Health Food by Andrew Hughes. Japan: Sanyusha Publishing, **1992**, pages 165, 168, 169.
 (A porker is a market pig between 30-54 kg or 66-119 pounds dressed weight.)
 (Bloom is a glow or flush that shows health; it is the glossy, healthy appearance of the coat of an animal.)

Comfrey Diet and Meat Quality

*"In the summer (of the 1950s), with no mares and foals, Vernon Stephenson (of Hunsley House Stud, Little Weighton, near Hull, England) fed Comfrey to Essex pigs in harness which he tethered (tied) on his constantly-used horse pastures to prevent the risk of a build-up of parasitic worms. This cut down the food requirement, and **because he was feeding more protein, produced leaner porkers, which he sold at a special price to butchers who wanted leaner pork.**"*
-Fighting Like the Flowers: An Autobiography: The Life Story of Britain's Best-Known Organic Gardener by Lawrence D. Hills.

Bideford, Devon, England: Green Books, **1989**, page 97.

"A total of 22 Large White pigs were fed based on live weight, either a common growing-finishing diet (A) or the same diet supplemented with 10% leaves from Comfrey (B).
Pigs were slaughtered at an average live weight of 105 kg (231 pounds). Tissue samples of backfat and m. longissimus dorsi (muscles of the back) were collected 24 hours after slaughter.
Animals of treatment B (Comfrey) had lower stearic and palmitic, but higher oleic and linolenic acid concentration in the adipose (fat) tissue than those of treatment A. The differences between treatments were not evident in the muscle lipids. Drip and cooking losses as well as both average pH values measured 45 minutes and 24 hours post mortem were not affected by the Comfrey supplementation.
With respect to the colour measurements, the redness and chroma (purity of color) tended to be higher in treatment B (Comfrey) compared to A.
Furthermore, the taste panel evaluation did not reveal any treatment differences. The data of the present study suggest that **a moderate Comfrey supplementation has an impact on the composition of backfat lipids without affecting other meat quality traits.**"
-'Effect of Comfrey (Symphytum Peregrinum) Fed to Pigs on Meat Quality Traits' by G. Bee, G.J.S. Lotscher and P.A. Dufey, Swiss Federal Research Station for Animal Production; Conference Paper, Proceedings of the Joint Session of EAAP Commissions on Pig Production, Animal Genetics and Animal Nutrition, Zurich, Switzerland, pages 237-424, **2000**.
(Backfat is the well-developed panniculus adiposus {subcutaneious tissue} along back of the pig that is well-formed and firm.)
(Essential fatty acids can not be synthesized by the body so they must be eaten. Only two fatty acids are known to be essential for humans: alpha-linolenic acid which is an omega-3 fatty acid, and linoleic acid which is an omega-6 fatty acid.)

"*Three treatments consisting of control (ingredients of animal origin), vegetarian, and Comfrey (Symphytum peregrinum) diets were examined using 36 Large White pigs.*
The inclusion of 10% Comfrey leaves in the diet reduced feed intake. Replacement of ingredients of animal origin by ingredients of plant origin had no negative effects on finishing performance and carcass traits."
-'**Comphrey (Comfrey) and Vegetarian Diets for Fattening Pigs**' by P. Stoll and A. Gutzwiller, RAP, Swiss Federal Research Station for Animal Production; IFOAM 2000: The World Grows Organic, Proceedings 13th International IFOAM Scientific Conference, Basel, Switzerland, 28 to 31 August **2000**, page 369.
(IFOAM= International Federation of Organic Agriculture Movements {Organics International} is the worldwide umbrella organization for the organic agriculture movement, representing 800 affiliates in 117 countries.)

Problems with Comfrey and Pigs (Nitrate) If your pig or other animal is sick, I recommend contacting a veterinarian.

"**Nitrate nitrogen in feed crops has an important effect upon the physiological condition of livestock** and causes public discussion usually.
Russian Comfrey contains 0.3-0.4% of nitrate, and its content is within the limit of concentration of nitrate in many other feed crops, though some differences in nitrate concentration are recognized in different growth periods.
Usually feed crops contain nitrate, and the nitrate content of the crops has a direct dependence on rates of nitrogenous fertilization.
In rye grass the maximum value is 14.96 mg nitrate nitrogen/gram dry matter. The value for fodder rye is 10.67, for white mustard 25.6, and for Liho rape 16.11. The mean nitrate content of 133 samples of meadow hay is 0.12% nitrate on dry matter (minimum 0.02, Maximum 0.43% nitrate).
In 12 samples of silage (corn, clover, and mixture) the mean nitrate content is 0.05%. And in an other study the average nitrate nitrogen of 48 samples of grass is 0.02%.
Kageyama and others examined a method using NO_3-N (nitrate nitrogen) content in leaves as an indicator to establish a suitable rapid method for diagnosing the nitrogen-nutritional status of vegetables.
The concentration of nitrate in forages is influenced greatly by species, part of the plant, stage of maturity, level of nitrogenous fertilization, and light intensity, time of season, daily low temperature and variations between daily low and high temperatures.
  It is least affected by closely related species, varieties, time of nitrogenous fertilization, kind and placement of nitrogenous fertilizer, and lack of certain plant nutrients.
On the kinds of nitrogenous fertilizer and nitrate in forage:
  Griffith reported that when nitrogen was applied as $(NH_4)_2SO_4$ (ammonium sulfate) at increasing rates up to 12 hundredweight/acre, nitrate content of grass mixtures, also increased, reaching 2,000 ppm (parts per million) 6 weeks after the heaviest treatment, and rape contained high levels of nitrate.
  There was no difference between Nitrochalk® and $(NH_4)_2SO_4$ (ammonium sulfate) in respect of nitrate accumulation.
  Eriksson observed that fertilizing with $Ca(NO_3)_2$ (calcium nitrate) had a relatively strong influence on nitrate content."
-'On the Nitrogen Content of Russian Comfrey' by Minoru Ikeda, Itaru Kunisaki and Hiroko Matsumura, Department of Animal Husbandry, Fisheries and Animal Husbandry, Hiroshima University, Fukuyama, Japan; Journal of Faculty of Fisheries and Animal Husbandry, Volume 5, No. 1, pages 165-173, **1963**.
  ('cwt' is the abbreviation for hundredweight. In the United Kingdom one hundredweight is 112 pounds. In the United
  States and Canada it is 100 pounds. There are 20 hundredweight in a ton, producing a 'short ton' of 2000 pounds and a

'long ton' of 2240 pounds.)
(Nitrochalk® is a trademarked chemical fertilizer that contains calcium carbonate and ammonium nitrate. It is 26% nitrogen, half is in the ammonical form and half is nitrate. Some nitrogen is immediately available and some later. It is also used under other trade names.)

"**Nitrate and Nitrite Poisoning in Pigs:**
Comfrey (Symphytum asperrimum, Prickly Comfrey), *producing a high yield of forage with little labour, is a traditional feed in parts of the German Democratic Republic, especially for pigs, which seems to like it.*
From May to September 1974 Comfrey had nitrate 5.4 to 32.0 gram/kg (0.19-1.12 ounce/2.2 pounds) dry matter.

1974	grams nitrate/kg Fresh Weight	grams nitrate/kg Dry Weight
May 26	4.4	-
June 10	3.5	32.0
June 25	2.5	18.0
July 25	1.8	14.6
August 25	2.8	22.2
September 2	0.7	5.4

Pigs given Comfrey in large amounts, after 2 to 4 days had signs of poisoning, *usually affecting general condition, heart rate, gait, mucosae, blood and faeces. Pigs were thirsty and refused feed, but responded to medication.*
It was suggested that Comfrey should be replaced by an alternative forage plant, or failing that, be grown in unshaded areas, picked in the evening, and supplied in small amounts with feed rich in carbohydrate, minerals and vitamins, and devoid of nitrate, and with water low in both nitrate and nitrite."
-'Nitrate and Nitrite Poisoning in Pigs Caused by Eating Comfrey' by A. Keindorf and H.J. Keindorf, STGP Dromling-Allertal, Germany; Monatshefte fur Veterinarmedizin (...for Veterinary Medicine), Volume 33, No. 11, pages 425-427, **1978**.

(A high source of inorganic nitrate comes from diets rich in leafy green foods when nitrates are used as fertilizers. The main nitrate fertilizers are ammonium, sodium, potassium, and calcium salts. The lethal dose of nitrites in humans is about 22 milligrams per kilogram (2.2 pounds) of body weight. The maximum allowed nitrite concentration in meat products is 200 parts per million. About 80 to 90% of nitrite in the average United States diet is from nitrite production in vegetables. Nitrates are in soil, nitrogen-based fertilizer, water, and nitrogen in the atmosphere. Nitrates are converted to nitrites when they are eaten.)

(During drought, nitrate in the soil increases greatly because of a lack of leaching {washing away}, reduced nitrate uptake by plants, and decomposition of organic matter. When it rains again, nitrate uptake by plants is high, especially the first week after rain. If animals eat these plants, stock losses from nitrate/nitrite poisoning may be disastrous. Factors that increase uptake of nitrate by plants are use of nitrogen-containing fertilizers, certain herbicides, cloudy/cold weather, low soil sulphur or molybdenum, and areas where stock have urinated/defecated.)

"**Clinical Symptoms of Nitrate and Nitrite Poisoning in Pigs:** *The intoxications occur 2 to 4 days after the start of Comfrey feeding. Small Comfrey rations do not make pigs sick. As with other intoxications, not all animals that have taken up the nitrate-rich Comfrey become sick. Pigs of all ages are affected but older animals are less susceptible (less likely to be harmed). Clinical symptoms vary as with any poisoning. Individual symptoms vary from animal to animal.*
General examination:
General condition of animals changes to poor. Animals are apathetic (indifferent). They hide in the straw and lie in prone position (lying flat on stomach). After flapping, they lie down quickly. The body temperature is usually greatly increased (41 C = 105.8 F).
Special examination:
Pigs show dyspnea (difficulty breathing). Heartbeat is throbbing, and heart rate is accelerated (fast). The mucous membranes are cyanotic (bluish color). The blood has a brownish or dark red hue. Also, in some cases dark colored urine is excreted.
The sick animals refuse food, but not water (increased thirst). In a few cases, mild tympanic (eardrum membrane) and colic (abdominal pain) symptoms occur. The feces have a thin pulpy consistency and may have a light to dark green color depending on the other feed components. Pigs show an uncertain, partly fluctuating (irregular) course (movement) of the hindquarters. Bigger pigs cannot keep up on their hind legs; they collapse immediately after standing up.
Therapy for Poisoned Pigs:
Administration of 2% methylene blue solution at a rate of 30 to 50 ml has been proven effective. Animals with circulatory insufficiency should be treated with circulation-supporting preparations. The application of mild laxatives is indicated. Withdrawal of feed for 1-2 days with sufficient water intake promotes the healing process.
Targeted therapy with methylene blue solution leads to a rapid normalization of the health status. After surviving the disease and taking into account feeding measures, the further development of the animals is not affected.
Reducing Nitrite/Nitrate Risk:
Certain factors may favor nitrate accumulation in Comfrey, such as growing on nutrient-poor sandy soils, poor water supply, and insufficient radiation (sun) in shady locations. **Cultivation of Comfrey in areas without shade is recommended.**
Feed Comfrey harvest in the evening (low nitrate content). *Feeding of small amounts of Comfrey. Other feeds in rations must be nitrate-free, thus no additional feeding of nitrate-storing weeds. Administration of low-nitrate drinking water. Give a carbohydrate, mineral- and vitamin-rich diet.*"
-'Nitrate and Nitrite Poisoning in Pigs Caused by Eating Comfrey' by A. Keindorf and H.J. Keindorf, STGP Dromling-Allertal,

Germany; Monatshefte fur Veterinarmedizin (...for Veterinary Medicine), Volume 33, No. 11, pages 425-427, **1978**.
(This is translated from German to English so there are some words or sentences that are awkward.)
    (If your pig or other animal is sick, I recommend contacting a veterinarian.)
    (Normal rectal temperature of a healthy pig is 101.5 to 102.5 F = 38.6 to 39.1 C.)
    (Methylene blue or methylthioninium chloride is a medication used to treat methemoglobinemia. It is used when methemoglobin levels are greater than 30%. Methemoglobinemia is caused by elevated levels of methemoglobin in the blood. Methemoglobin is a form of hemoglobin that has the ferric form of iron. The affinity for oxygen of ferric iron is damaged. This leads to a reduced ability of the red blood cell to release oxygen to tissues.)

*"**Nitrite Levels and Harvest Time: The highest levels of nitrite contained in spinach and mustard were during the afternoon and evening.** Spinach and mustard are better harvested in the morning because it contains nitrite less than in spinach picked afternoon and evening. Level of nitrite increases from morning to afternoon and decreases from afternoon to evening.*
***Nitrate Levels and Harvest Time: The highest levels of nitrate contained in spinach and mustard were during the morning.** The level of nitrate decreases from morning to afternoon and increases from afternoon to evening."*
-'Effect of Harvesting Time at Morning, Afternoon, and Evening on Nitrate and Nitrite Level in Spinach (Amaranthus Tricolor L.) and Mustard (Brassica Rapa L.)' by Nahitma Ginting, Jansen Silalahi, Tuty Roida Pardede, Sudarmi Sudarmi and Nerdy Nerdy, Department of Pharmaceutical Chemistry, Universitas Sumatera Utara, Medan, Indonesia; Asian Journal of Pharmaceutical and Clinical Research, Volume 11, Special Issue 1, pages 110-113, 2018.

    Pigs and Vitamin B12 See subsection 'Vitamin B12' in section 'Nutritional Value of Comfrey' (Chapter 19) in Volume 1.

*"**Pigs, poultry and human beings are all omnivorous**, and though it is possible with skill enough to live on an all vegetable diet, pigs have few principles and without what used to be called '**the animal protein factor**' they failed to thrive. This was not an amino-acid, but a vitamin which is present in such small traces that it is measured in nanograms per 100 grams- thousandths of a millionth of a gram- **vitamin B12**.*
*One of the **Comfrey growers** in the 1954 Comfrey Race was Paul Weir, who was perfectly satisfied with his 22 ton an acre yield because **it enabled him to rear what may have been Britain's first veganic (vegetarian) pigs**.*
*In defiance of theory, the pigs throve, for **a pig as heavy as a man can easily take its entire bodily needs of vitamin B12 in about four pounds (1.8 kg) of fresh Comfrey a day**. This applies also to chickens and to any animal which does not have the digestive bacteria to synthesize its own supply, like the ruminants such as cattle and sheep."*
-Comfrey: Fodder, Food and Remedy by Lawrence D. Hills. New York: Rizzoli Universe Books: **1976**, page 96.
    (An omnivore is an animal that eats both plants and animals.)

    Timing Pig Rearing with Comfrey Growth

*"**The longer period in cut (of harvest) of Comfrey (than lucerne/alfalfa) gives a chance to put through two batches of store pigs in a year.** Starting off with a really early cut of young Comfrey when the need was only roughly 60 pounds (27.2 kg) per pen, rising to about 225 pounds (102 kg) in time for the June growth peak.*
The second batch would start off with a low Comfrey need, but a cut for hay or silage then would put something in the barn towards winter keep for breeding sows. The final batch would finish on the last cuts.
***The general average should work out at about 80 pigs off an acre of Russian Comfrey.** The Germans allow 25 square metres per pig, roughly 30 square yards. As a trial on a first-year planting on reasonably rich land, a quarter of an acre should run four pigs on this system in a single batch."*
-Russian Comfrey: A Hundred Tons an Acre of Stock or Compost for Farm, Garden or Smallholding by Lawrence D. Hills. London England: Faber and Faber, Limited, **1953**, pages 99, 101, 102.

    Feed Comfrey to Breeding Sows and Piglets

*"**No branch of the subject of pigs is of more importance than the management of breeding-sows.***
*After being served (bred by a boar), they should have plenty of exercise in a roomy yard or paddock; they should be fed with a little bran or pollard mixed with slops or wash (leftovers) from the house; **in summer they should have green food: grass, tares (common vetch), or Prickly Comfrey.***
*I may mention that well-fed pigs, getting up for show, are **exceedingly fond of Comfrey- it is first rate food for any kind of pig**- it grows luxuriantly, and yields a constant supply throughout the summer."*
-'The Journal of the Royal Agricultural Society of England, Second Series, Volume 17', London, England, **1881**. Includes article: 'Pigs; and Experience in their Breeding and Management' by James Howard, M.P., Clapham Park, Bedfordshire, England.
    (White flour from wheat is made from the endosperm or inside part of the grain. The rest is bran and germ and is called wheat middlings. Wheat pollard is a finely milled blend of middlings.)

*"**The best small scale application (of feeding Comfrey) is for breeders, who feed after farrowing (giving birth) as a*

supplement to normal rations, increasing the quantity as the piglets take to it.
This takes the strain of a large litter off her (the sow's) milk and pulls the runts (small piglets) ahead with the live weight gain of high vegetable protein and low fibre.
It removes the scour (diarrhea) risk from both sow and litter, *including baccilus coli (gram-positive bacteria), as well as digestive troubles. It has been fed green to piglets still suckling as a scour preventative and cure, with complete success."*
-Russian Comfrey: A Hundred Tons an Acre of Stock or Compost for Farm, Garden or Smallholding by Lawrence D. Hills. London England: Faber and Faber, Limited, **1953**, pages 101, 103.

The Comfrey fed to the piglets was always young leaves without any coarse stems full of fibre from allowing plants to run to flower. *Adult pigs will eat any Comfrey, but it pays to cut frequently and with care for piglets, just as it does with poultry.*
Perhaps the greatest profits have been made by feeding unlimited Comfrey to the runts, nisgards, wrecklings, caddies, and all the other names for the small pig in the litter.
The late Charles Rogers, a retired pigman of St. Austell (Cornwall County, England), made a good living by buying up all these variously named rejects on Cornish markets, and rearing them for sale as stores (store pigs) after this Comfrey treatment.
His Comfrey yield was only 20 tons an acre, but with runts, and buying up any pigs that were scouring (diarrhea) so the price was down, to sell cured with Comfrey."
-Comfrey: Fodder, Food and Remedy by Lawrence D. Hills. New York: Rizzoli Universe Books: **1976**, pages 123, 124.
('Store pig' is an old-fashioned term. It is similar to what is now called 'Feeder pig', around live weight 40-100 pounds =18-45 kg.)

"The traditional use for Comfrey is as a scour (diarrhea) preventative in pigs, *and it is still in use casually on some farms where there is no knowledge of it other than tradition.*
The modern system is to feed 2-3 kg (4.4-6.6 pounds) of fresh Comfrey leaf to the sow while she is in pig (pregnant), *not altering her diet greatly, saving a little on meal. Her mineral needs are high at this time, and she likes the Comfrey.*
The ration is continued and increased after farrowing (giving birth), *giving her plenty of young Comfrey leaves.* **A sow will eat Comfrey at any stage of growth, and very small piglets will eat fresh Comfrey as their first solid food (even as young as 1 week).**
This reduces their demand on the sow's milk and with a big litter, the runts get a better chance at the teats while their bigger competitors are eating Comfrey. The result is to even up the batch, giving a quick early live weight gain from suitable vegetable protein with the minimum fiber, the leaves supplementing the sow's milk."
-Comfrey: Nature's Healing Herb & Health Food by Andrew Hughes. Japan: Sanyusha Publishing Co., Ltd, **1992**, page 167.

"**Three hundred piglets** of (Polish Large White x Polish Landrace) x (Duroc x Hampshire) crossbreeds (150 gilts and 150 barrow) were allotted to 5 feeding groups.
In Group V (control), the animals were fed with the standard PW (post weaning) mixture for piglets. Feed and water were available ad libitum. The experiment lasted 40 days; blood samples were taken afterwards.
In Group II there were leaves of great nettle, herb of common yarrow and knotgrass (Polygonum aviculare), capitula of wild chamomile (Matricaria chamomilla) (Chamomilla recutita), lyophilized garlic bulbs, **root of Common Comfrey (Symphytum officinale)** and the seeds of echinacea (Echinacea purpurea).
The herb supplements to the diets allowed to achieve the gains even better, in comparison with the antibiotic additive. The best daily gains, feed utilization and blood indices were found for the piglets fed the mixture containing great nettle and plaintain leaves, common yarrow herb, wheat grass rhizomes, garlic bulbs and juniper fruits (Group I)."
-'Effectiveness of Herbs Additive in Weaning Piglets' by E.R. Grela, A. Czech and M. Baranowka, Instytut Zywienia Zwierat Wydzialu Biologii i Hodowli Zwierzat AR (Institute of Animal Nutrition of the Faculty of Biology and Animal Breeding of UR), Lublin, Poland; Annales Universitatis Mariae Curie-Sklodowska, Sectio EEE, Horticultura, Volume 9, No. Supplementum, pages 249-254, **2001**.   (Ad libitum means as much as they want.)

Comfrey as Pig Medicine

"**Comfrey has veterinary uses, too. In some parts of Ireland, swine fever is treated by boiling Comfrey roots in milk, and adding everything, roots and all, to the pig's food.** *This has to be kept up for some weeks (*Logan).*
But Norfolk, England pigkeepers added Comfrey leaves to the pig's feed just to keep them in good health, but also to make sure they could not be bewitched.
Pigweed is a Wiltshire county (England) name for the Comfrey plant, though it must have had a much wider significance."
-Dictionary of Plant Lore by Donald Watts, BA, MIL, Bath, England. London, England: Academic Press, **2007**.
(* -Irish Country Cures by Patrick Logan. Dublin, Ireland: Talbot Press, 1972.)
  (Classical swine fever or hog cholera is a highly contagious disease of pigs. It causes fever, skin lesions, convulsions and sometimes death.)   (Bewitch means to cast a spell on someone or an animal.)

Poultry (Chicken, Duck, Goose, Turkey)

"**Poultry Feeding: Finally, a good supply of green feed is at all times desirable. Chickens kept in pens should be given green stuff like** *nettles, notchweed, dandelions, sweet grasses, clover, alfalfa, serradella, vetches, spinach, mangels (fodder*

beet), spurry, rape, buckwheat, corn, artichokes, **Comfrey,** *beet and cabbage leaves."*
-'Scientific Feeding of the Domestic Animals' by Dr. Martin Klimmer, Ph.D., D.M.V., Professor of Hygiene and Feeding, Director of the Hygienic Institute of the Veterinary College of Dresden, Germany, **1923**, page 209.

*"****The digestion of the fowl is even less fibre-tolerant than that of the pig.**** It is simple and contains no bacteria able to digest cellulose, and the stomach secretes no enzyme which can deal with it.*
When the crude fibre of any ration reached 10 percent, 10 pound in every 100, there may be a reduction in the digestibility of the carbohydrates; at higher rates this is certain to produce a heavy fall in egg yields and delayed maturity, however rich the rest of the diet.
The crude fibre should vary between 5 percent and 8 percent of the total diet, and the nearer the lower limit the better. Those who rear their own poultry replacements may well find that it pays to take fewer eggs per bird and keep more birds, as poultry food increases in price, by supplying maximum of easily-grown green food to save bought concentrates."
-Russian Comfrey: A Hundred Tons an Acre of Stock or Compost for Farm, Garden or Smallholding by Lawrence D. Hills. London England: Faber and Faber, Limited, **1953**, pages 104, 128.
(Feed Concentrates are mixes of ingredients with high amount of cereal grains. Mixes of only grains are also called concentrates.)

*"****In Japan there are about 28,000 recorded growers of Comfrey for poultry feed.**** Comfrey is rich in carotene (precursor to Vitamin A) and produces a yellow colour in poultry meat.*
It also reduces the soft fat in geese, which could have commercial significance."
-World Protein Resources by Allen Jones. NY: John Wiley & Sons, **1974**. (Chapter 22: 'Green Leaf Protein', pages 209-216.)
(There are 2 types of body fat: subcutaneous or under the skin 'soft fat', and visceral 'hard fat' that is deeper inside the body.)

"The Chingola poultry system was tried in Japan by a Mr. Suzuki of Yokosuka. ***He fed equal parts of the chopped Comfrey foliage and his normal meal ration*** *to 2,000 mature birds and 1,000 pullets (young female chickens) from April 1963 to November 1963, when the crop goes dormant for the winter in Japan, just as it does in England.*
The first results were a saving of 790,000 yen (about 790 British pounds at 1964 exchange rates) on feed costs, and a rise in production to 70 to 75 percent (100 percent = one egg a day for every bird) which fell when the Comfrey stopped in November, to 60 to 65 percent, the same as the ordinary meal fed birds.
There was also improved egg quality shown in yolk colour, and better growth of the younger birds, with pullets maturing to egg production 15 to 20 days sooner than when no Comfrey was fed.
Comfrey was also fed to table birds and was found to produce rather yellower flesh than normal because of the extra Vitamin A."
-Comfrey: Fodder, Food and Remedy by Lawrence D. Hills. New York: Rizzoli Universe Books: **1976**, page 131.

"As the weeks go by into summer one finds a mid-morning feed followed by a later afternoon one of Comfrey can be given. ***About one third of their feed capacity now consists of Comfrey.*** *Egg output is retained all the time. For me at least,* ***Comfrey and poultry or vice-versa are symbiotic. Comfrey to feed Poultry. Poultry Manure to feed Comfrey."***
-Comfrey: Fodder, Food and Remedy by Lawrence D. Hills. New York: Rizzoli Universe Books: **1976**, page 133.

*"****Local cultivars of Comfrey (Symphytum spp.) have been used to cover protein and mineral requirements of farm animals in low-input systems.**** Due to its known health-promoting (e.g. allantoin), but also anti-nutritive ingredients (e.g. pyrrolizidine alkaloids), multidisciplinary approaches are essential in order to quantify the nutritional value and the potential of its use in poultry and farm animals in terms of meeting animal needs, using local resources as well as remediating over-fertilized soils.*
Focusing on animal effects, here ***one-day old sexed Cobb500 broiler chickens*** *were subjected to either a standard control diet or* ***a standard diet supplemented with 4% dried Comfrey leaves for 32 days.*** *Performance traits indicate good acceptance of supplementation with Comfrey leaves.*
Parameters for liver function, mineral homeostasis, bone mineral density as well as intestinal microanatomy revealed no signs of impairment. Quantified pyrrolizidine alkaloids were below the detection limit in liver and breast muscle (<5 microgram / kg tissue). Comfrey supplemented male broiler chickens showed higher ash content in breast muscle and revealed altered gene expression profiles for metabolic pathways in blood cells. In healthy broiler chickens, the transcriptome analyses revealed no aberrations in the immune-related pathways due to Comfrey supplementation.
The results imply that the use of Comfrey leaves in a high-performance broiler line seems feasible and offers the potential for closed nutrient cycles *in site-adapted local agricultural systems. Further analyses need to focus on possible growth-promoting and health-improving components of Comfrey and the safe use of chicken products for human consumption."*
-'Comfrey (Symphytum spp.) as an Alternative Field Crop Contributing to Closed Agricultural Cycles in Chicken Feeding' by Oster, Reyer, Keiler, Ball, Mulvenna, Murani, Ponsuksili, Wimmers (Germany and United Kingdom), Science of the Total Environment, Volume 742, Article 140490, June **2020**.

Poultry Love Comfrey

*"****Prickly Comfrey (Symphytum asperrimum):*** *The variety I have found to be the best has been raised from one plant out of a number of others sent me by the Director-General of the Botanical Gardens at St. Petersburg (Russia).*
Upon growing a quantity of it, Mr. H. Doubleday noticed that the animals preferred it to all others growing on the same

land. *Poultry would pass over all the other beds of Comfrey, but stop at this one and greedily devour the leaves;* so much was this the case that it had to be protected to enable Mr. Doubleday to save sufficient for propagation."
-'New Commercial Plants with Directions How to Grow Them to the Best Advantage, No. 1 (1878) and No. 3 (**1880**)' by Thomas Christy, F.L.S. (Fellow of Linnaean Society), London, England. Number 1 includes Caucasian Prickly Comfrey pages 14-16. Number 3 includes Prickly Comfrey pages 11-14. Includes 'Caucasian Prickly Comfrey' and 'Russian Comfrey' advertisements.

*"It is possible that **Russian Comfrey, cut at leafy stage, would provide a special poultry meal**, with 22.7 percent protein to 10.9 percent fibre, and 38.2% carbohydrates (Hannah Dairy Institute figures). Russian Comfrey with 400 mg per kg carotene."*
-Russian Comfrey: A Hundred Tons an Acre of Stock or Compost for Farm, Garden or Smallholding by Lawrence D. Hills. London England: Faber and Faber, Limited, **1953**, page 105.

*"Common chickweed (Stellaria media), dead nettle (Lamium spp.), dock (Rumex spp.) and **Comfrey (Symphytum officinale L.) are mentioned as fodder with good feeding value for chicken and chicks.***
They are good quality in a general alimentary way, variation in diet, and eaten with pleasure (when fed fresh). Rearing fodder for chicks, good fodder quality (palatable and good nutritive value only if harvested properly), appetising and easily digestible."
-'**Local Knowledge Held by Farmers in Eastern Tyrol (Austria) about the Use of Plants to Maintain and Improve Animal Health and Welfare**' by Christian R. Vogl, Brigitte Vogl-Lukasser and Michael Walkenhorst, Department for Sustainable Agricultural Systems, University of Natural Resources and Life Sciences (BOKU), Vienna, Austria; Journal of Ethnobiology and Ethnomedicine, Volume 12, No. 40, 17 pages, **2016**.

Comfrey, Vitamin A and Egg Yolk Color

*"**Vitamin A is a fat-soluble vitamin** present in large quantities in certain fish oils. Supplied in concentrated form to the poultry mash it is extremely expensive, but a deficiency produces disastrous results.*
***The normal supply is from green food such as Comfrey, in which it is present as carotene**, which is built up by the bird into the vitamin. As green fodder contains not only this essential substance, but the whole Vitamin B complex."*
-Russian Comfrey: A Hundred Tons an Acre of Stock or Compost for Farm, Garden or Smallholding by Lawrence D. Hills. London England: Faber and Faber, Limited, **1953**, page 105.

*"The same feeding system was tried by Mr. E.V. Stephenson with **his laying battery (caged chickens), feeding 2 1/2 to 3 ounces (70.8-85.0 grams) a head of wilted Comfrey per day**, with no fall in egg yield, and a saving in concentrates amounting to an ounce (28.3 grams) a day per bird.*
He states that his birds were in perfect health, and if the daily Comfrey ration was missed, they became unsettled and unhappy.
The Comfrey was wilted down to approximately 75 percent moisture, before being put through a chaff-cutter and incorporated with the mash (ground feed). As this stage its carotene would be nearly the same as that of fresh grass in spring, and ample Vitamin A would mean good colour in the egg yolks for the whole period.
Comfrey is the cheapest of all sources of Vitamin A in terms of labour."
-Russian Comfrey: A Hundred Tons an Acre of Stock or Compost for Farm, Garden or Smallholding by Lawrence D. Hills. London England: Faber and Faber, Limited, **1953**, pages 107, 108.
    (A chaff cutter is a machine that cuts straw, hay or other fodder into small pieces.)

*"**It seems best to keep the Comfrey in the patch and cut it fresh for the hens.** The birds pick each day's offering to shreds (leaving only the stem), and **all that vitamin A goes a long way toward giving us nutritious eggs with deep yellow yolks."***
-'Comfrey for the Homestead' by Nancy Bubel, Mother Earth News, Kansas, May/June **1974**.
    (My ducks eat most of the stems that chickens find too big.)

*"Comfrey keeps hens healthier and happier, providing them with much of the protein they miss when they eat feed without worms or insects. **The hens lay eggs with stronger shells and darker yolks, too."***
-Enchanted Garden: Alan Chadwick's Organic Method of Gardening by Tom Cuthbertson. London, England: Rider & Company / Hutchinson & Co. Publishers Ltd, **1978**, page 122.

*"**A normal diet without or with 8% ground Comfrey (Symphytum officinale, Common Comfrey) was given to laying Leghorn hens for up to 44 days.***
A growing diet was given to 137 chickens initially 3 weeks old for 56 days without or with 8% Comfrey or with Comfrey as a 2% decoction (tea) in the drinking water.
Egg yolks of hens given Comfrey was of a more intense yellow than that of the controls; values on the La Roche scale were 9.3 and 6.6, respectively.
*Feed with 8% Comfrey was palatable and well accepted. Chickens given 8% Comfrey showed no difference in weight gain from the controls but **those given 2% Comfrey decoction in their water showed an increase in gain**. With 8% Comfrey in the diet, calcium and phosphorus balances were positive and bone mineralization was normal.*
***When up to 80% Comfrey was given in the feed or 20% decoction in the water,** some chickens had diarrhoea, and there were slight signs of congestion and haemorrhage in internal organs."*
-'Use of Symphytum Officinale as a Feed for Poultry, as a Palatability Factor, and as a Coloring Agent of Egg Yolk of Laying

Hens' by C. Statescu, V. Martin, M. Crivineanu and M. Paraschivescu, Instiutul Agronomic (Agronomic Institute) Nicolae Balcescu, Romania, Volume 28, pages 77-82, **1985**.

('La Roche Yolk Color Fan®', now called 'DSM Yolk Color Fan®' is an industry standard, 15 scale color index to distinguish the yolk color density. It helps determine the carotenoid-based pigmentation in egg yolks, broilers and salmon.)

"High Protein – Low Fibre Comfrey:
This is where Comfrey offers a solution to the home and small scale producer. Ideally chicken feed should contain between 5% and 8% maximum indigestible fibre. Even oats at 10% fibre should form a limited portion of the diet.
Comfrey being low in fibre but high in protein is the ideal green to feed chickens. *The allantoin is potentially health promoting and **the high levels of Vitamins A and B12 results in both a rich yellow yolk to the eggs and a yellow tinge to the flesh** as with the expensive corn-fed hens."*
-'Feeding Comfrey to Poultry and Other Livestock' by John Harrison, Allotment and Gardens: Grow Your Own - Allotment - Gardening Help, https://www.allotment-garden.org, Penygroes, Gwynedd, Wales, **2018**.

Comfrey Extracts and Poultry Diet

*"**Consumers are increasingly demanding in their desire for broiler (chicken) meat grown without antimicrobials**. Consequently, nutritionists have been seeking alternative methods of promoting growth that have no effect on animals and humans and do not affect broiler performance.*
In this context, the addition of aloe vera and Symphytum officinale extracts alone or in combination with symbiotics (synbiotics) can improve broiler performance and replace chemical antimicrobials in broiler chicken diets.
This study aimed to test the effects of dietary aloe vera and Symphytum officinale (Common Comfrey) extracts added separately or in combination with symbiotics on the performance, nutrient utilization, serum (blood) biochemical parameters, biometrics, and intestinal histomorfometry (form/shape of tissue) of broilers.
*The experiment had a randomized block design with five treatments and six replicates of ten broilers each. Treatments were as follows: negative control and positive control (diet without and with antibiotic, respectively); 0.2% aloe vera (AV); **0.2% Symphytum officinale (S);** 0.2% functional supplement, composed of symbiotics fermented in aloe vera and **Comfrey plant extracts (S+PE)**.*
At seven days of age, FI (Feed Intake) of birds fed the aloe vera extracts diets were lower than that observed for birds consuming the diet with Symphytum officinale extract and S+PE. *Broiler performance remained unaffected by treatments at other ages evaluated.*
*At 10 to 14 days of age the lowest ADCDM (Apparent Digestibility Coefficients of Dry Matter) and **ADCCP (Apparent Digestibility Coefficients of Crude Protein)** was shown in group feed NC (Negative Control). **The highest ADCCP was observed in PC (Positive Control) group and in diets supplemented with Aloe vera and S+PE (Comfrey Plant Extracts).***
*Serum levels of cholesterol, triglycerides, and phosphorus were affected by addition of extracts at seven, 21, and 35 days of age. **The longest duodenal villi (intestinal projections) were observed in broilers fed S+PE diets at seven days of age.** Aloe vera and Symphytum officinale extracts and symbiotics can be used in broiler diets as an alternative to growth-promoting antibiotics."*
-'Symbiotics and Aloe Vera and Symphytum Officinale Extracts in Broiler Feed' by P.R. Oliveira, F.R. dos Santos, E.F. Duarte, G.S. Guimaraes, N.S.C. Mattos, C.S. Minafra, Semina: Ciencias Agrarias (Agricultural Sciences), Londrina, Brazil, Volume 37, No. 4, Supplement 1, pages 2677-2690, **2016**.

(Histomorfometry is the quantitative study of the microscopic organization and structure of a tissue especially by computer-assisted analysis of images.)

*"**When Gibson introduced the concept of prebiotics he speculated as to the additional benefits if prebiotics were combined with probiotics to form what he termed as Synbiotics**. A synbiotic product beneficially affects the host in improving the survival and implantation of live microbial dietary supplements in the gastrointestinal tract by selectively stimulating the growth and/or activating the metabolism of one or a limited number of health promoting bacteria.*
The health benefits imparted by probiotics and prebiotics as well as synbiotics *have been the subject of extensive research in the past few decades. These food supplements termed as functional foods have been demonstrated to alter, modify and reinstate the pre-existing intestinal flora. They also facilitate smooth functions of the intestinal environment. Most commonly used probiotic strains are: Bifidobacterium, Lactobacilli, S. boulardii, B. coagulans.*
Prebiotics like FOS, GOS, XOS, Inulin; fructans are the most commonly used fibers which when used together with probiotics are termed synbiotics and are able to improve the viability of the probiotics."
-'Probiotics, Prebiotics and Synbiotics: A Review' by Kavita R. Pandey, Suresh R. Naik and Babu V. Vakil, Guru Nanak Khalsa College of Arts Science and Commerce, Mumbai, India; Journal of Food Science and Technology, Association of Food Scientists and Technologists, India, Volume 52, No. 12, pages 7577-7587, December 2015.

Feeding Comfrey to Chicks and Poults (Baby Turkey)

*"**Though Comfrey seems of considerable value to growing poultry, before 8 weeks old it should be fed chaffed (cut)**. The*

fibres are few but stringy, and there is a risk of crop trouble with young chicks. This is avoided by making sure it is short lengths."
-Russian Comfrey: A Hundred Tons an Acre of Stock or Compost for Farm, Garden or Smallholding by Lawrence D. Hills. London England: Faber and Faber, Limited, **1953**, page 108.

"The protein requirements of poultry vary with age; *the need is highest for young chicks, at 20 percent up to 8 weeks, and 16 percent for the next 10 (weeks); laying or breeding birds need roughly 15 percent."*
-Russian Comfrey: A Hundred Tons an Acre of Stock or Compost for Farm, Garden or Smallholding by Lawrence D. Hills. London England: Faber and Faber, Limited, **1953**, page 104.

"In order to investigate energy utilization of defatted rice bran and **Comfrey (Symphytum peregrinum) meal,** hatched single comb White Leghorn male chicks were fed for 10 days a commercial chick mash and for the subsequent 10 days on diets containing 17.0% of cellulose, wheat bran, defatted rice bran and **Comfrey meal,** respectively.
During experimental feeding period, daily body weight gain did not show significant differences among treatments though **birds fed defatted rice bran diet consumed more diet than those fed Comfrey diet. And protein retention did not show significant differences while birds fed Comfrey diet retained lower lipids than those fed other diets."**
-'Energy Utilization of Defatted Rice Bran and Comfrey (Syphytum Peregrinum) in Chicks' by I.S. Shin, K.T. Nam, S.K. and T.S. Koh, (College of Animal Husbandry, Konkuk University, Seoul, Korea); Korean Journal of Animal Sciences, Volume 28, No. 9, pages 607-611, **1986**.

Ducks and Geese

"Feeding Ducks: *Beginning with the sixth week young ducks receive crumbs of soft feed composed of 3 parts corn meal, 5 parts bran, 4 parts feed meal and 1/2 part fish meal.*
Curds (coagulated milk), sour milk, earthworms, crushed snails or May beetles, lettuce and Comfrey are favorite by-feeds. *When feathers begin to develop, soaked ground oats, followed a little later with dry ground oats and whole grains, are given. The fattening proper of ducks begins about the fifth week. The feed then consists of wheat bran and barley meal or rye meal, 1 part of each; steamed mashed potatoes and greens (Comfrey), fish meal and coarse sand, 2 parts.*
Beginning with the eighth week the greens are displaced with cracklings."
-'Scientific Feeding of the Domestic Animals' by Dr. Martin Klimmer, Ph.D., D.M.V., Professor of Hygiene and Feeding, Director of the Hygienic Institute of the Veterinary College of Dresden, Germany, **1923**, page 212.
 (Cracklings are rendered pieces of pork or poultry fat. 'Rendering fat' means slowly cooking until it melts and then straining it. The solid pieces are cracklings.)

"Geese are grazing birds and will graze Comfrey but leave the midribs. *They should therefore be confined with hurdles (moveable light fence) on an area they will clear in two days, and then be moved to another run.*
'Henry Doubleday Research Association' member Paul N. Griesenaur of the United States found that **geese could be reared on an almost all Comfrey diet.** *On April 28th, 1958, I received two dozen Emden goslings two days old. All were fed nothing but 'mushed' Comfrey for four weeks. At this end, one dozen was put out on range where they had* **free access to Comfrey plants in the field** *and a large area of Ladino Clover."*
-Comfrey: Fodder, Food and Remedy by Lawrence D. Hills. New York: Rizzoli Universe Books: **1976**, page 139.

"Geese in particular are grass eating birds, and do very well on Comfrey, but other water or semi-water birds also consume large quantities of grass, and here is a field for the goose and duck farmer. *Chopped up and fed with the usual mash, Comfrey can save the farmer much of his outlay for food by* **supplying up to 90% of the bird's needs."**
-Comfrey: Nature's Healing Herb & Health Food by Andrew Hughes. Japan: Sanyusha Publishing Co., Ltd, **1992**, page 176.

"Geese are perhaps the most ideally suited to Comfrey.
Starting at a few days old they consume the herb most readily when chopped up or mushed. The goslings will thrive on very little else, after a week or two they can be put on **free range Comfrey to eat ad lib** *(as much as they want)."*
-Comfrey: Symphuo Symphytum: A Multi-Purpose Herb by Philip Clarke. Edinburgh, England: Pentland Press, **1997**, page 15.

"Feeding Comfrey to Geese: *Geese are better equipped to handle Comfrey than chickens as they already eat grass and herbs.*
Allowing the flock access to a portion of the Comfrey bed, *perhaps controlled with electric fencing, can improve weight gain and health without additional cost."*
-'Feeding Comfrey to Poultry and Other Livestock' by John Harrison, Allotment and Gardens: Grow Your Own - Allotment - Gardening Help, https://www.allotment-garden.org, Penygroes, Gwynedd, Wales, **2018**.

Turkey

"The turkey is far less of a seed-eater than the fowl (chicken) and is therefore capable of taking more bulk in the diet.
The main difficulty with turkey raising is the high mortality and disease risk in the early stages. Large quantities of cut and wilted Comfrey were supplied to a neighboring turkey raiser as a feed for chicks (poults) and growing birds.

The material was chaffed and mixed with the mash supplied to turkey chicks from a fortnight (2 weeks) old, in the proportion of half by volume. **Now there was a considerable fall in mortality, and the chicks thrived on the new Comfrey diet.** *It is possible that the small percentage of allantoin in the leaf prevents or cures the minor gastric irritations which are among the enemies of the turkey."*
-Russian Comfrey: A Hundred Tons an Acre of Stock or Compost for Farm, Garden or Smallholding by Lawrence D. Hills. London England: Faber and Faber, Limited, **1953**, page 110.

"On turkeys, one report published in Japan says:
'The problems of raising young stock in both categories (chickens and turkeys) and especially turkeys, which are so hard to bring to maturity, can be largely overcome with Comfrey because of its high mineral and medicinal value. All such stock have shown a marked preference for Comfrey when given free choice, *refusing to eat young grass until all Comfrey was consumed. Comfrey is favored over brassicas both by the stock and the farmer, by the stock for its flavor and food value, and the farmer for its reduced labor. Once Comfrey is established, it is there for a lifetime."*
-Comfrey: Nature's Healing Herb & Health Food by Andrew Hughes. Japan: Sanyusha Publishing Co., Ltd, **1992**, page 177.
    (Forage brassicas include cabbage, cauliflower, broccoli, brussel sprouts, kale, radishes, rape, swedes and turnips.)

Rabbits and Hares

Rabbits Eat Comfrey

"Mr. E.J. Barnes, of Rose Cottage, Leytonstone, east London, England, writes to the 'Standard':
I should like to say one word more upon another subject which has been mentioned from time to time in your columns, and this is the **Prickly Comfrey. I use it frequently and regularly, both for my fowls (poultry) and for a lot of Belgian hares which I breed.**
The fondness of the hares for it is most extraordinary. They will always leave cabbage for Prickly Comfrey; *but its value to me for their keep is that, however much they eat of it, and no matter how frequently I give it to them,* **Prickly Comfrey never has the pernicious (harmful) effect upon them that too much green food of any other kind produce.**
For the fowls I hang up a fresh cut bunch for them to peck at, and much they enjoy it. Prickly Comfrey answers admirably with me, but of two purchases of it which I made one has turned out worthless, and is, I suppose the wrong sort."
-'The Farmers Magazine and Monthly Journal of the Agricultural Interest', London, England, July to December **1879**. Comfrey page 331.
(Rabbits, jackrabbits and hares are in the family Leporidae. Jackrabbits and hares are in the genus Lepus. Rabbits are in the these genera: Pentalagus, Bunolagus, Nesolagus, Romerolagus, Brachylagus, Sylvilagus, Oryctolagus and Poelagus.
Hares are similar in size/form to rabbits and have similar herbivore diets, but have longer ears and live solitarily or in pairs.)

"In reference to Prickly Comfrey's action upon animals, a letter appeared in the 'Standard' (London newspaper) in October (1880): 'I should like to say one word on another subject which has been mentioned from time to time in your columns, and this is the **Prickly Comfrey. I use it frequently and regularly, both for my fowls and for a number of Belgian hares (rabbits) which I breed.**
The fondness of the hares for Comfrey is most extraordinary. They will always leave cabbage for Prickly Comfrey; *but its value to me for their keep is that, however much they eat of it, and no matter how frequently I give it them,* **Prickly Comfrey never has the pernicious (bad) effect upon them that too much green food of any other kind produces.**
For the fowls I hang up a fresh-cut bunch for them to peck at and much they enjoy it.
Prickly Comfrey answers admirably with me.' -from Rose Cottage, High Street, E.J. Barnes, Leytonstone, Essex, England."
-'New Commercial Plants with Directions How to Grow Them to the Best Advantage, No. 1 (1878) and No. 3 (**1880**)' by Thomas Christy, F.L.S. (Fellow of Linnaean Society), London, England. Number 1 includes Caucasian Prickly Comfrey pages 14-16. Number 3 includes Prickly Comfrey pages 11-14. Includes 'Caucasian Prickly Comfrey' and 'Russian Comfrey' advertisements.
    (For most greens, if rabbits are given too much, they get diarrhea.)

"Rabbits may be given Comfrey ad libitum, and never suffer from a surfeit (too much) of this splendid green food."
-'Ensilage: A System for the Preservation in Pits of Forage Plants and Grasses, Independent of Weather: A Collection of Facts and Statistics on the Cheapest Mode of Providing Winter Food for Dairy Cattle, Sheep, Horses, Etc.', by Thomas Christy, F.L.S. (Fellow of Linneaen Society), London, England, **1883**. Caucasian Prickly Comfrey, pages 57-63.
(Ad lib = Ad libitum is Latin for 'at one's pleasure' or 'as you desire'. It means free feeding or giving animals as much as they want.)

"Though wild rabbits and tame will not eat the plant growing, it is excellent fed wilted, *and once they are used to it this is a very useful green food for tame rabbits. This is from modern rabbit-keepers who consider it superior to cabbage."*
-Russian Comfrey: A Hundred Tons an Acre of Stock or Compost for Farm, Garden or Smallholding by Lawrence D. Hills. London England: Faber and Faber, Limited, **1953**, page 128.   (My rabbits would eat unwilted Comfrey.)

"Nutritive values of Russian Comfrey (Symphytum peregrinum) and kale, recently introduced into Korea, were estimated with 30 weaned male New Zealand White rabbits.
Contents of crude protein and ether extract in kale were higher than in Russian Comfrey. Russian Comfrey had more N-free

extract and crude fibre. The plants did not differ in contents of Ca (Calcium) or P (Phosphorous) or energy value.
When rabbits were fed on the 2 crops, there was no significant difference in average daily gain in weight or metabolic size. Although the same amount of roughage dry matter was eaten, rabbits given kale ate significantly less concentrate.
Amount of Russian Comfrey or kale and concentrate required per unit gain was not affected by source of roughage. Kale was more palatable than Comfrey.
Digestibility of protein was higher for kale. Digestibility of ether extract was lower in kale than in Russian Comfrey and that of N-free extract was significantly higher. The plants did not differ in digestibilities of crude fibre or energy.
Amount of nitrogen retained by rabbits was similar. There was no difference in total digestible nutrients but kale had more digestible crude protein."
-'Comparative Studies of the Nutritive Value of Russian Comfrey (Symphytum peregrium) and Kale (Brassica oleracea var acephala) for Growing Rabbits' by I.K. Han, K.I. Kim and K.S. Lee, College of Agriculture, Seoul National University, Suwon, Korea; Nong-sa Si-hem Yen-ku Po-Ko = Research Report to Office of Rural Development, Suwon, Korea, Volume 11, No. 4, pages 89-95, **1968**.   (Palatable means acceptable or agreeable in taste.)

"We could give these smaller Comfrey leaves to the poultry and rabbits. The high protein content of Comfrey, and its concentration of potassium and other minerals, make it an ideal fodder. **The high content of potassium has the drawbacks of making Comfrey unattractive to rabbits and poultry when it is growing,** and the fibrous character of the old flower stems makes them hard for the chickens to digest; **but if the leaves and stems are picked while still young, and are wilted for a day before being fed to the livestock, they will be both attractive and nutritious.**
When Comfrey is fed regularly to livestock, as an additive to fodder, **it maintains the health of the animals in a more general way. It definitely improves the hair, the meat, and the friskiness of rabbits.**"
-Enchanted Garden: Alan Chadwick's Organic Method of Gardening by Tom Cuthbertson. London, England: Rider & Company / Hutchinson & Co. Publishers Ltd, **1978**, page 120.

"**The 14 New Zealand White rabbits initially 3 months old were caged singly and given for 4 hours each day, on 14 successive days, a different green feed. On average, out of the 100 grams (3.5 ounce) offered, the rabbits ate** sunflower leaves 98.4, green bean vines 95.4, red clover 94.5, carrot tops 92.9, cauliflower leaves 92.5, dandelions 83.3, white clover 80.8, Swiss chard 78.0, maize (corn) leaves 66.4, amaranthus 65.6, cocksfoot grass 59.7, grape vine leaves 56.8, black locust leaves 42.2, and **Comfrey (Symphytum) 35.4 grams.**"
-'Feed Preference Studies with Rabbits Fed Fourteen Different Fresh Greens' by D.J. Harris, P.R. Cheeke and N.M. Patton, Rabbit Research Center, Oregon State University, Corvallis, Oregon; Journal of Applied Rabbit Research, Volume 6, No. 4, pages 120-122, **1983**.
    (The Comfrey was the least favorite in this study. I would like to know what type of Comfrey they were fed.
    I was unable to get this report. If you have a copy, could you please send it to me.)

"**Although it is recognised that the rabbit is not partial to foods with high potash (potassium) content, experiments now being conducted show that Comfrey can reduce the feed bill and increase profit by preventing disease.**
The rabbit can take a good percentage of its food in Comfrey, and if protein percentage is considered, it will soon be seen that considerable reduction of costs can be effected."
-Comfrey: Nature's Healing Herb & Health Food by Andrew Hughes. Japan: Sanyusha Publishing Co., Ltd, **1992**, page 106.

"**Rabbits can be fed 100% on Comfrey.** One report on our files, prepared in Japan, points out that rabbits so fed put on weight faster, and the fur improves in quality."
-Comfrey: Nature's Healing Herb & Health Food by Andrew Hughes. Japan: Sanyusha Publishing Co., Ltd, **1992**, page 177.

"**Comfrey (Symphytum officinalis): Comfrey, a leafy forage plant, may be fed as a fresh green during the growing months or dried and fed as a hay during the winter months.** Comfrey is a good protein source, comparable to alfalfa (lucerne); however, the digestibility of protein in Comfrey is only about 70% of the digestibility of alfalfa protein because it is bound to fiber. Comfrey is lower in fiber compared to alfalfa and has about double the ash (mineral) content."
-Raising Healthy Rabbits by Dr. W. Sheldon Bivin and Dr. William W. King. Seattle, Washington: Christian Veterinary Mission, **1994**, page 60.

"I know that my rabbits love the stuff just by their reaction when I'm harvesting Comfrey. The rabbits hear me cutting the leaves in the Comfrey beds, and you can hear them running around knowing their healthy tonic is on the way.
When entering the rabbitry with a basketful, the whole herd comes alive waiting for their treat. I highly recommend Comfrey for rabbits. **It is a great digestive aid and will help with wool block. Do not overfeed, as that may cause diarrhea.**
Comfrey gets all of its nitrogen from the soil, so some type of regularly added organic matter to the soil is needed. Of course I cannot think of anything better to use than 'Bunny Berries' (rabbit manure)! I top dress my Comfrey plants every spring and fall. I feed Comfrey to all the livestock on my homestead. The chickens love it. When the free ranging chickens get to run in the Comfrey beds they will eat it to the ground. The pigs go crazy when they see you carrying in to them, grunting and doing their happy dance.
I have fed it to my rabbits for 30+ years, and they love it! So far, I have had no adverse effects on feed Comfrey to any of the livestock I raise on the homestead!"
-Beyond the Pellet: Feeding Rabbits Naturally by Boyd Craven Jr (The Urban Rabbit Project) and Rick Worden (Rise and Shine

Rabbitry), **2013**.
> (Rabbits groom themselves and eat small amounts of hair. Usually, the hair passes through the digestive system and is excreted. Sometimes a hairball is trapped. Wool block occurs more during a molt. It is more of a problem with wool breeds. If the digestive track is blocked completely, the rabbit will die.)

"**Adult Flemish rabbits, non-breeding does and all bucks from 9 months of age**, get 1 1/2 cups of pellets a day plus hay. Greens and fruits are given 1-2 times a week (usually 1/3 cup as a treat), and can include **fresh Comfrey**, parsley, clover, dandelion leaves, apples, oranges, fresh papaya, mango, kale, spinach, and carrot tops (once in a while carrots). I have almost no issues with stasis."
-'Rabbit Nutrition and Nutritional Healing' by Lucile Moore, Ph.D. College Station, Texas: VBW Publishing, **2017**.
> (Gastrointestinal stasis is when the digestive system slows down or stops completely. This can kill a rabbit.)

"When I lived near Joussard (Alberta, Canada), on Lesser Slave Lake, I raised rabbits.
Fresh Comfrey leaves in summer, and dried forage in winter was a treat quickly polished off."
-'Comfrey' by Glenn Axford, Our Common Roots: An Exploration of the Healing Power of Nature, Canada, www.ourcommonroots.com, date unknown but **2018** or earlier.

Rabbits are Resistant to Alkaloids

"Rabbits were fed Senecio jacobaea (ragwort, groundsel) for 263 days. Total intake of the plant averaged 112.5% of initial body weight. No gross lesions, or changes in serum (blood) protein and albumin occurred. Microscopic changes in liver tissue were seen. Two rabbits injected with 150 mg Senecio pyrrolizidine alkaloid/kg (2.2 pounds) body weight died in less than 24 hours. **The resistance of the rabbit to chronic Senecio intoxication, but not to injected alkaloid, suggests that alkaloid absorption may be low in this species**."
-'Resistance of the Rabbit to Dietary Pyrrolizidine (Senecio) Alkaloid' by Pierson, Cheeke and Dickinson, Research Communications in Chemical Pathology and Pharmacology, 16(3):561-4, April **1977**.

"**Rabbits were given diets containing Comfrey for 6-12 months without affect on health or liver pathology. Unlike other livestock and man, rabbits seemed to be resistant to the toxic alkaloids and probably can be given Comfrey with safety.** Previous studies suggest that dried Comfrey is slightly inferior to lucerne (alfalfa) for rabbits."
-'Comfrey: An Excellent Forage or a Poisonous Plant?' by P.R. Cheeke, OSU Rabbit Research Center, Oregon State University, Corvallis, Oregon; Journal of Applied Rabbit Research, Volume 2, No. 3, pages 7-11, **1979**.
> (I was not able to get this report. If you have a copy, could you please send it to me.)

"**The pyrrolizidine alkaloids symphytine and sinkirkine, as the laboratory tests reveal, are carcinogenic for rodents, including rabbits.** Lasiocarpine in moderate doses produces damage to the liver, cirrhosis, fibrosis and malign tumours and is also responsible for the mutagenic and teratogenic effects. It has been proven that children, fetus and young animals are the most sensitive, developing veno-occulusive disease after an exposure of less than one week."
-'Monitoring of Some Hepatic Cytolysis Indicators at Rabbits Treated with Symphytum Officinale Phytopreparate' by C Prisacaru, V. Sandu and C. Tulcan, Lucrai Scedilla tiinifice- Medicina Veterinara, Universitatea de Stilinte Agricole si Medicina Veterinara (Cedilla Scientific Works- Veterinary Medicine, University of Agricultural Sciences and Veterinary Medicine), Romania, Volume 48, No. 7, pages 771-774, **2005**. (I do not have this report. If you have a copy, could you please send it to me. I would like more details to fully understand it. **Most studies show rabbits are resistance to alkaloids.**)
> (About 40% of all mammal species are rodents. Common rodents include beavers, chipmunks, gerbils, guinea pigs, hamsters, hares, mice, porcupines, prairie dogs, rabbits, rats and squirrels.)

"**Fortunately, rabbits are very resistant to pyrrizolidine alkaloids so Comfrey can be safely fed.**"
-Rabbit Production by Steven D. Lukefahr, Peter Robert Cheeke and Nephi M. Patton. Prairie Village, Kansas: Interstate Publishers, **2013**.

Comfrey as Rabbit Medicine

"In our pre-Comfrey days we lost several litters in a row to scours (diarrhea) by feeding weeds and garden greens too early. Now we have green Comfrey available for babies to nibble on as soon as they leave the nest and, for about 2 weeks, this is their only fresh vegetable. They grow up liking it, and do well (we haven't lost a litter since we started this program)."
-'Comfrey for the Homestead' by Nancy Bubel, Mother Earth News, Kansas, May/June **1974**.

"**Comfrey also contains allantoin and has been reported to alleviate diarrhea in young rabbits.**"
-Raising Healthy Rabbits by Dr. W. Sheldon Bivin and Dr. William W. King. Seattle, Washington: Christian Veterinary Mission, **1994**, page 60.

"**Comfrey added in our rabbit feed fills the gap in vitamins, minerals and has a great general medicinal effect on them.**

I (Rick) have been using Comfrey for a long time in my rabbitry. When I was young I always talked to all the old-timer rabbit breeders around my neighborhood and on my paper route. **They all agreed and swore by Comfrey as the best rabbit tonic and said every rabbit breeder should grow it.** *They would give a little each day, and told me this is why their rabbits never got sick.* **They told me it would prevent everything from snuffles (upper respiratory disease) to premature kindling (birth) and help nursing does (females) produce milk.** *Comfrey is a great source of Vitamin A and good for pregnant and nursing does as it also supports the immune system.* **When I do feed, it makes up about 25% of the basket of greens/weeds I pick every other day or so in the spring / summer / fall."**
-Beyond the Pellet: Feeding Rabbits Naturally by Boyd Craven Jr (The Urban Rabbit Project) and Rick Worden (Rise and Shine Rabbitry), **2013**.

Rats as Pets

"Rat (21-day old Sprague Dawley males) Feeding Trial:
Even when only half of the crude protein came from dehydrated Russian or Quaker Comfrey (the other half came from casein), there were drastic reductions in feed consumption *(about 24%),* **gain** *(about 62%)* **and feed conversion** *(about 50%). On the 40% Comfrey ration all 10 rats lost weight continuously through the 21-day trial.* **The palatability (taste) of Comfrey for rats was apparently less than for swine** *as indicated by the preliminary trial with swine."*
-'Plant Leaf Protein with Emphasis on Comfrey' by Hubert Heitman and Milton D. Miller with Edward Johnson (agronomic phases of study), Sergio E. Gyarzun of University of Chile (swine digestion trial), Bob D. Wilson (rat study) and James T. Elings (Extension Animal Scientist); California Experiment Station and Agricultural Extension Service, University of California, **1969**.
(Rats are omnivores, eating fruit, nuts, seeds, grain, flowers, leaves, insects, birds, reptiles, fish and eggs. If you are going to feed your pet rats Comfrey, give it to them separately from their regular food. Experiment with fresh & dried.)

Sheep (also see 'Goats' section)

"The Rot: Few disorders have been more fatal to sheep, *or have more frequently exercised the attention of graziers and breeders than the rot; for the origin of which various causes have been assigned.*
Various medicines have been recommended to the attention of farmers and breeders; though we conceive, they can only be employed with probability of success in incipient (early) cases.
Ellis recommends a peck (2 dry gallons = 7.5 liters) of malt, or more, to be mashed and brewed into twelve gallons (45 liters) of wort, in which a quantity of bloodwort, **Comfrey,** *pennyroyal, plantain, sage, shepherd's purse, and wormwood, are to be boiled; the liquor to be worked with yeast, some common salt to be added, when it is to be put into a cask for use.*
Of this medicated beer seven or eight spoonsful are to be given to each sheep, *once in the course of a week during wet weather; but with longer intervals in dry seasons."*
-'The Complete Grazier; or Farmers and Cattle-Breeder and Dealers Assistant' by a Lincolnshire (England) Grazier, London, **1816**, page 221.   (Sheep rot is a disease of the liver in sheep and cattle caused by liver flukes.)
(Malt is barley or other grain that has been steeped, germinated and dried. It is used for brewing alcohol and making vinegar.)   (Wort is the liquid extracted during mashing when brewing beer or whisky.)

*"***Sheep were stall-fed on Symphytum asperrimum (Prickly Comfrey) in Ireland in the nineteenth century.*** It was also the sole diet of working donkeys on the Carnew Castle farm (Ireland). As with cattle, the value is at both ends of the season, with hay or silage to take the summer excess if grass is available and more profitably fed.*
There is abundant evidence that sheep never suffer from a 'surfeit' (too much) on it as they will from other lush forage crops. *But their close biting means damage to the crowns, which should not be grazed closer than three inches (7.6 cm) from the ground."*
-Russian Comfrey: A Hundred Tons an Acre of Stock or Compost for Farm, Garden or Smallholding by Lawrence D. Hills. London England: Faber and Faber, Limited, **1953**, page 92.

*"***Queensland, Australia Trials on Russian Comfrey:*** On a dry basis it was found to contain up to 24 percent of protein.*
Some of the harvested material was used in a palatability (taste) trial with sheep. The tests showed that shade drying for 24 hours approximately doubled it palatability."
-'Queensland Agricultural Journal' edited by C.W. Winders, B.Sc.Agr., Queensland Department of Primary Industries, Brisbane, Australia, Volume 83, July to December 1957, No. 7 to No. 12. Article: 'Queensland Trials on Russian Comfrey', Volume 83, No. 11, page 624, November 1 **1957**.

"An interesting experiment was recently conducted by the 'Department of Agriculture' in south Australia, **feeding sheep on a mixture of Comfrey and straw, not hay,** *be it noted, just dry wheaten straw. The sheep were fed in the ratio of 1 kg (2.2 pounds) of dry weight Comfrey to 1.5 kg (3.3 pounds) of straw. This was found preferable to the usual 1 kg of oats with straw. The Comfrey was cut and wilted before feeding, and showed that* **Comfrey promoted digestion of the straw, and is a valuable contribution when sheep are grazed on wheat stubble,** *as they often are."*
-Comfrey: Nature's Healing Herb & Health Food by Andrew Hughes. Japan: Sanyusha Publishing Co., Ltd, **1992**, page 177.

Snails as Food

Snail farming, heliciculture or heliculture is the raising of land snails for human use, either to eat (escargot) or for snail slime to use in cosmetics. The best known species in the Western world are Helix pomatia (Roman snail, Burgundy snail) and Cornu aspersum (Helix aspersa). Worldwide sales from snail farms are around $12 billion a year.

"We investigated the influence of Armoracia rusticana (Horseradish) and Symphytum officinale (Common Comfrey) upon the development of one of the most farmed and consumed terrestrial snails Helix aspersa Muller (Cornu aspersum), an excellent animal protein source for humans.

The experiments took place in town of Baisoara, Cluj County, Romania in the June-August 2010 interval. The adult Helix aspersa Muller snail populations in the witness lots display a survival mean rate under extreme heat wave and drought conditions of just 6.50 +- 1.29 while **the population in the experimental lots with Armoracia rusticana and Symphytum officinale presented higher survival mean rate of 14.00 +- 0.82 and 16.75 +- 1.50 respectively.**

As far as prolificacy (breeding), the witness enclosure batches accounted for 100 +- 5.72 while the batches in experimental enclosures accounted for 248.00 +- 8.29 and 140.00 +- 5.10 respectively.

We included the fitness component test, as an indicative of snail's vitality and energetic condition, measuring their crawling mean speed and the mean heart bit rate of new born baby snails that have similar body weights.

While the heart bit rate differences between the different batches were not significant, **the mean speed presented significant differences directing us to the conclusion that plants such as Armoracia rusticana and Symphytum officinale not only represent valuable nutritive foods to Helix aspersa Muller adults and youngsters but also have protective values against stress conditions such as severe drought and prolonged heat waves.**

Such plants can represent an additional assurance for **an efficient and ecological snail farming technology** as an important source of quality animal protein for human consumption, capable to deliver a high quantity of animal protein from a given land surface, representing a venue towards sustainable agro-economical development."

-'The Influence of Horseradish (Armoracia Rusticana) and Common Comfrey (Symphytum Officinale) Upon the Edible Terrestrial Snails Helix Aspersa Muller (Cornu Aspersum) During Heat Wave and Drought as Means to Improve Snail Farming Technologies' by Adrian Toader-Williams (Romania) and Roberto F. Nespolo (Chile); Abstracts of the International Congress 'Natural Cataclysms and Global Problems of the Modern Civilization', Istanbul, Turkey, September 19-21 **2011**. (I was only able to get the abstract. If you have the full paper, could you please send it to me.)

"Two species of edible terrestrial snails are studied, namely Helix pomatia and Cornu aspersum.
The experiments took place at Craiesti and Baisoara, Cluj County, Romania. During experiments, the weather condition was extreme, the temperatures were high.

Many variations and repetitions were in different pens. In each pen of a square meter (10.7 square feet) each 40 snails were introduced. There were times when were used 160 snails per square meter, both Helix pomatia and Cornu aspersum of different ages.

The plants that are used as fodder are: Taraxacum officinale, Sonchus oleraceus, Equisetum arvense, Atriplex hortensis, Lupinus polyphyllus (perenis), Rumex acetosa, Rheum officinale (Rheum rhaponticum), Armoracia rusticana (Armoracia lapathifolia), Arctium lappa, Thymus vulgaris, Lavender officinalis, Foeniculum vulgare, Hyssopus officinalis, **and Symphytum officinale. The best results were obtained in batches containing predominantly Rumex acetosa (sorrel), Armoracia rusticana (horseradish), and Symphytum officinale.**

Helix pomatia snail meat grown in batch CV-3(E) in the presence of Symphytum officinale showed 14.7530% protein, the highest level for Helix pomatia. In the group CV-2(E) with Rumex acetosa was obtained 12.0981% protein, compared to the control CV-1(M) with 10.3213% protein.

The meat of Cornu aspersum snail raised in BV-3(E) in the presence of Symphytum officinale showed highest level of protein for Cornu aspersum, namely 15.4662% protein.

The BV-2(E), Armoracia rusticana had a level of 14.0970% protein and the control BV-1(M) with 12.3110% protein.

Feeding the Helix pomatia snails with Rumex acetosa and with Symphytum officinale leads to considerable increases in weight. The snails contain much higher protein and fat as opposed to the lots where these plants are absent. Regarding the Cornu aspersum snails, the protein content follows the same pattern.

Breeding: The degree of prolificacy (breeding) of Cornu aspersum snails is far superior in the lots with Armoracia rusticana followed by the lots with Symphytum officinale.

Survival: Degree of survival is over two times higher in both experimental lots as opposed to the control lots.

Health: Speed of the snail is an indicator of metabolic rate and health. Digital video of snails moving on graph paper of the method is at the base of the method. **The index of energy metabolism (speed) of the Cornu aspersum juvenile is about 30% higher in the variant where Armoracia rusticana is present as opposed to control lot and almost 20% higher in the lot where Symphytum officinale is present.**

Snail Farming: The use in snail farming activity of perennials that have a high organic nitrogen productivity, combined with use of land for solar energy capture, confer synergistic opportunities for bioeconomic exploitation of agricultural potential of land."

-'The Influence of Feed Upon the Helix (Sp) Edible Snails Production Performance Under the Bioeconomic Aspect' by Adrian Toader, Eng., M.Sc. for Ph.D. Thesis, University of Agricultural Science and Veterinary Medicine Cluj-Napoca, Doctoral School, Faculty of Animal Sciences and Biotechnologies, Romania, **2012**.

(Bioeconomics is the study of the dynamics of living resources using economic models.)

Swine (see Pigs)

Turkey (see Poultry)

Zoo Animals and Donkeys (also see 'Horses')

"At Carnew Castle (Ireland), **Symphytum asperrimum (Prickly Comfrey) was the sole diet of the working donkeys,** and an H.D.R.A. (Henry Doubleday Research Association) member, Major Linton (in Peru), who farmed on the east-facing slopes of the Andes (mountains in South America), used Bocking No. 14 which he smuggled out in the early days (1954) of our trial ground (in England), as food for both **mules and donkeys**. Their animals were given a Comfrey supper and breakfast before they started their long journey via mountain paths with a barrel (of wine) on each side of their pack saddles (going to Argentina)."
-Comfrey: Fodder, Food and Remedy by Lawrence D. Hills. New York: Rizzoli Universe Books: **1976**, page 152.

"In 1962, Whipsnade Zoo (in Bedfordshire, England) bought enough **Comfrey** for about 2 1/2 acres from the trial ground, and Mr. H. Tong, the then curator, **fed it to elephants, hippopotami, rhinoceroses and giraffes,** which were the animals that broke in and stole it most often from the farms of growers in Kenya and Rhodesia (now Zimbabwe, Africa). **An elephant will eat at least a hundredweight (100 to 112 pounds) of Comfrey a day**, because it is also a highly suitable diet."
-Comfrey: Fodder, Food and Remedy by Lawrence D. Hills. New York: Rizzoli Universe Books: **1976**, pages 152, 153.

"**Those who run zoos are often short of space, and a Comfrey plot could be the answer to the summer green feed problem for animals with digestions that fit Comfrey.**
As with chinchillas and racehorses, however, the object is not money-saving but giving valued stock the best possible diet."
-Comfrey: Fodder, Food and Remedy by Lawrence D. Hills. New York: Rizzoli Universe Books: **1976**, page 154.

"Herbs do work, sometimes remarkably well. Here are three that have been used effectively for centuries: **Comfrey:** Extensive European studies show that its main constituent, allentoin, effectively promotes the growth of connective tissue, bone, and cartilage, and it's easily absorbed through the skin.
Fresh or dried Comfrey leaves can be used in poultices, fomentations, salves, and oils to treat wounds, bruises, sores, minor burns, and insect bites (in donkeys).
Modern science confirms what herbalists have claimed for hundreds of years: Comfrey poultices effectively ease pain, reduce inflammation, and support the healing of minor fractures. Comfrey salve or oil rubbed into arthritic joints, muscle strains, and bowed tendons bring healing relief.
In emergency situations, powdered Comfrey leaves and flowers quickly staunch (stop) bleeding.
Comfrey poultice is a classic treatment for wounds, bruises, and swellings of all types."
-The Donkey Companion: Selecting, Training, Breeding, Enjoying and Caring for Donkeys by Sue Weaver. North Adams, Massachusetts: Storey Publishing, **2008**.

The Black-Faced Sheep 1839

'A Cyclopaedia of Practical Husbandry and Rural Affairs in General' by Martin Doyle, Dublin, Ireland, 1839. Sheep page 430. Comfrey page 108.

'A Cyclopaedia of Practical Husbandry and Rural Affairs in General' by Martin Doyle, London, England, 1851. Comfrey pages 151-152.

Pigs Love Comfrey 1897

'Practical Pig-Keeping: A Manual for Amateurs, Based on Personal Experience in Breeding, Feeding and Fattening; Also in Buying and Selling Pigs at Market Prices' by R.D. Garratt, London, England, 1897.

This is the book cover. Comfrey is on page 41.

PART E

COMFREY AS MEDICINE AND HUMAN FOOD

Chapter 23

Medicinal Comfrey Overview

Always consult with your healthcare practitioner when you need medical advice.
Each individual is different so some herbs are not good for everyone. The information here is for entertainment purposes only.

Types of Herbal Preparations

Balm: A fragrant cream, ointment or liquid used to heal and soothe skin.
Bolus: A suppository in the vagina or rectum that draws poisons or brings healing.
Capsule: Used to swallow herbal medicines.
Compress: Used to improve circulation of blood or lymph. Applied warm or hot. When cold, it reduces pain and swelling.
Cream: See ointment.
Decoction: Used to extract mineral salts and bitter principles from hard material such as roots, bark, seeds and wood. It is boiled for at least 10 minutes and then is steeped for a few hours.
Essential Oil: It is made by distilling an herb to a highly concentrated extraction. It is used for external aromatherapy in drops.
Extract: Made by extracting a part of the herb, usually using a solvent such as ethanol or water. May be tincture or powder.
Fomentation: It is the application of a hot moist substance to the skin. See compress.
Flower Essence: It gives an energetic signature similar to homeopathy. It is made from an infusion. Then the plant parts are removed, and it is preserved with brandy.
Granule: Used in Chinese medicine. Made by dehydrating herbal decoctions into concentrated granulated forms added to water.
Homeopathy: Treats disease by giving tiny doses of a remedy that in larger amounts produce symptoms similar to that disease.
Infusion: When a tea bag or ball is put in a cup of hot water and allowed to steep for a few minutes before drinking. It extracts vitamins and volatile ingredients from leaves and flowers.
Liniment: A medicated preparation applied to skin. It has similar or less stickiness than lotions. It is rubbed in to create friction.
Lotion: It is a low-viscosity (not sticky, not thick) preparation applied to skin.
Lozenge: It is a small, flavored tablet made with sugar or syrup, usually medicated.
Ointmint: A semi-solid substance for external application to skin or mucous membranes. Uses almond oil, beeswax, lanolin, lard or petroleum jelly. Ointments have a higher concentration of oil compared to creams. A salve is an ointment, cream, or balm.
Oil: Instead of extracting oil from the herb, a vegetable-based carrier oil is infused with the herb.
Pill: Easy way to swallow herbal medicines.
Poultice: It is a soft, moist mass of meal, herbs, etc. applied warm or hot to the skin.
Powder: It is made from dried parts of plants that are pulverized until it is fine particles.
Salve: See ointment.
Soak: Soaks and Sitz baths use herbal teas, decoctions, or essential oils to immerse the skin. Sitz baths are for hips and legs, whereas an herbal soak can be any part of the body.
Suppository: See bolus.
Syrup: An herbal syrup is a concentrated decoction with either honey or sugar.
Tea: An herbal infusion is like tea, except an infusion is steeped longer and uses a larger amount of herb.
Tincture: It is an alcoholic extract or solution. The extract is 25-60% ethanol (50-120 United States proof). See extract.
Wine and Cordials: An herbal wine or cordial is similar to a tincture except the herbs are infused in brandy (for a cordial) or wine, rather than the more highly-alcoholic tincture.

"Suggested methods of extraction for Comfrey: Bolus/suppository, decoction, fomentation, infusion, lotion, oil, poultice, salve, syrup, and tincture."
-The Herbal Medicine-Maker's Handbook: A Home Manual by James Green, Herbalist. NY, NY: Random House, 2000, page 38.

Commercial Preparations of Comfrey

"Lacking any official standards of quality, the identity and purity of many herbal preparations (all herbs, not just Comfrey) may not always be those specified on the label. This is a complicating factor when attempting to evaluate the safety and

efficacy of a particular plant product.
The only recourse (choice) the pharmacist has is to purchase the best-quality products, preferably standardized herbal materials, from producers with impeccable (respectable) reputations. *For now, experience is the pharmacist's best guide. Comfrey has traditionally been used as a wound-healing agent. Common Comfrey is the rhizome (underground stem), roots, and leaves of Symphytum officinale L. Related species include Russian Comfrey, derived from S. x uplandicum Nym., and Prickly Comfrey from S. asperum Lepech.* **Comfrey species tend to hybridize, and commercial samples are often misidentified.** *Unfortunately, the various species of Comfrey also contain pyrrolizidine alkaloids with 1,2-unsaturated necine moieties; these are highly carcinogenic.* **Pharmacists would be wise to discourage the use of the toxic Comfrey plant entirely."**
-'What Pharmacists Should Know about Herbal Remedies: Pharmacists Can Help Patients Differentiate the Useful Herbs from the Harmful Ones' by Varro E. Tyler, Ph.D., Purdue University School of Pharmacy and Pharmacal Sciences, West Lafayette, Indiana; Journal of the American Pharmaceutical Association, Volume NS36, No. 1, pages 29-37, January **1996**.

*"***Common Trade Names for Comfrey Products (all ®):** *Wise Woman Comfrey Salve®.*
Several combination products are available, including Alticort, Atri-Res, Black Ointment, C&F Formula, Comfrey/Aloe Capsules, Comfrey and Fenugreek, EB5 Footcare Formula, EB5 Toning Formula, #483 Oxox Cell Activator, Goldenseal Salve, H-Complex, Heal-All Salve, Kytta-Plasma f, Kytta-Salbe F, Liniment Virtue, Muco-Plex, Mucoplex, Mustard Salve, Pain-Less Rub, Plantain Salve, Procomfrin, Respa-Herb, Simicort, Super Salve, T-ANEM, T-ASMA, T-BC, T-BF, Traumaplant, T-SLC, and T-ULC.®
Comfrey Common Forms: *Available as a blended plant extract also known as 'green drink', homeopathic preparations, a poultice or liniment, a tea (dried leaf and whole root), and a topical cream or ointment and in bulk roots or leaves, capsules, elixir, mucilaginous decoctions, powder, and tincture.*
Pyrrolizidine alkaloid (PA)-free Comfrey preparations are also available.
Commercial root preparations are available, but they are not recommended for internal or external use because of their high concentration of PAs. Comfrey is available in combination products in veterinary medicine for topical treatment of muscle strains and ruptures and for oral administration as an antidiarrheal."
-Professional's Handbook of Complementary and Alternative Medicines, 3rd Edition' by Charles H. Fetrow, Pharmd (Doctor of Pharmacy) and Juan R. Avila, Pharmd. Pennsylvania: Lippincott Williams & Wilkins (imprint of Wolters Kluwer), **2003**, page 235-238.   (All are registered or trademarked products. I do not have a connection with any of these products.)

"Seventy of the 100 most commonly prescribed phytomedicines in 2002 in Germany were single-herb products. These can be reduced to a total of 34 herbs or active constituents, which are ranked below in order of sales.

Rank	Herb or active constituent	Products	Indication	Sales in thousand Euros
1	Gingko leaves: extract	6	A	73,239
2	St.John's wort: extract	10	A	36,772
3	Mistletoe: extract	3	F	32,729
4	Ivy leaves: extract	4	B	18,563
5	Hawthorn leaves/flowers: extract	3	C	15,299
6	Saw palmetto berries: extract	3	O	13,933
7	Saccharomyces: powder	2	E	13,550
8	Nettle root: extract	1	O	10,619
9	Pelargonium root: extract	1	B	8,715
10	Horse chestnut seed: extract	2	C	8,605
11	Thyme: extract	5	B	5,795
12	Eucalyptus oil and cineol	2	B	5,476
13	Milk thistle fruits: extract	1	E	4,535
14	Petasite rhizome: extract	1	A	4,108
15	Pumpkin seeds: extract	1	O	4,084
16	Chaste berry: extract	2	H	3,792
17	Bromelain	1	G	3,469
18	Black cohosh rhizome: extract	1	H	3,344
19	Chamomile flowers: extract	3	G	3,309
20	Valerian root: extract	2	A	2,860
21	Colchicum: extract	1	G	2,801
22	Grass pollens: extract	1	D	2,586
23	Artichoke leaves: extract	1	E	2,282
24	Goldenrod: extract	1	D	2,072
25	Nettle leaves: extract	1	G	1,995
26	Devil's claw: extract	1	G	1,956
27	Sage: extract	1	G	1,847
28	**Comfrey root: extract**	1	G	**1,507**

A *Central nervous system disorders ,* **B** *Respiratory disorders,* **C** *Cardiovascular disorders,* **D** *Urinary tract disorders*
E *Disorders of the stomach, bowel/liver, or biliary tract,* **F** *Increasing resistance to diseases*
G Skin and connective tissue disorders, **H** *Gynecologic disorders."*

-Rational Phytotherapy: A Physician's Guide to Herbal Medicine by V. Schulz, R. Hansel and V.E. Tyler. Heidelberg, Germany: Springer Verlag, **2004**.

"Proprietary single-ingredient preparations:
   Austria: Traumaplant®.   Czech Republic: Traumaplant®.   Germany: Kytta-Plasma f®; Kytta-Salbe f®; Traumaplant®.
   Switzerland: Kytta Pommade®.   Venezuela: Traumaplant®.
Proprietary multi-ingredient preparations:
   Czech Republic: Dr Theiss Beinwell Salbe®; Stomatosan®.   Germany: Kytta-Balsam f®; Rhus-Rheuma-Gel N®.
   Israel: Comfrey Plus®.   Switzerland: Gel a la consoude®; Keppur; Keppur®; Kytta Baume®.
   United States: MSM with Glucosamine Creme®."
-Herbal Medicines by Joanne Barnes, Linda A. Anderson and J. David Phillipson. London, England: Pharmaceutical Press, third edition **2007**, page 190.   (I do not have any connection with these companies or their products.)
(Proprietary means made and/or sold only by the particular company that has legal right to do so. Some of these Comfrey preparations plus others are mentioned in the section 'Medical Research about Comfrey and Healing' {Chapter 24}.

Defining Medicinal Effects of Comfrey
Historically, Comfrey has been used as analgesic (pain killer), antidiarrhetic, antiphlogistic (anti-inflammation), astringent (tightens tissues), cicatrizant (wound healer), demulcent (soothes), diuretic (increase urine), epispastic (apply to skin), expectorant (help expel lung mucus), hemostat (anti-hemorrhagic), sedative (tranquilizer), stimulant, sudorific (sweat enhancer) & more.

Alterative: Purifies the blood. Increases health and vitality.
Anodyne and Analgesic: Reduces pain by reducing sensitivity of the nervous system.
Anti-diarrhea: Stops loose bowels.
Anti-exudative: Reduces fluids coming from an injury or infection.
Anti-inflammatory: Reduces inflammation and swelling.
Astringent: Causes contraction of body tissues and stops discharges.
Cell Proliferant: Helps wounds heal by increasing growth of cells.
Demulcent: Helps soothe inflamed tissues.
Emollient: Softens and relaxes, especially skin.
Expectorant: Makes coughing up mucous easier.
Hemostatic: Reduces bleeding.
Mucilagenous: Slippery substance that promotes healing.
Pectoral: Good for chest infections.
Styptic: Reduces bleeding.
Tonic: Invigorates the body. Improves health.
Vulnerary: Promotes healing of wounds by applying directly to a cleaned area.

"Comfrey (Knitbone) is vulnerary and demulcent, having unparalleled wound, ulcer, and fracture healing action. It is anti-inflammatory and soothing to dry inflamed digestive tract; astringent, able to allay (reduce) hemorrhaging whereever it occurs; and expectorant as an age-old remedy for dry irritable coughs, especially when accompanied by blood streaked mucus."
-The Herbal Medicine-Maker's Handbook: A Home Manual by James Green, Herbalist. NY, NY: Random House, **2000**, page 31.

"Comfrey (Symphytum spp. {species}):
Most contributors cited Symphytum officinale but few studies are vouchered, and the species are difficult to determine.
***Activities (Comfrey)- Alterative** (f; CRC); **Analgesic** (1; CAN); **Antiaging** (f; CRC); **Antihemorrhagic** (f; CAN); **Antiinflammatory** (2; APA; KOM; PH2; WAM); **Antileukocyte** (1; PH2); **Antimitotic** (1; PHR; PIP); **Antimutagenic** (1; PNC); **Antipsoriatic** (1; PNC); **Antitumor** (1; FAD); **Astringent** (1; APA; FAD; FEL; PNC); **Callus-Promoter** (1; PHR); **Carcinogenic** (1; APA; CRC); **Demulcent** (1; CAN; FEL; PH2; WAM); **Emollient** (1; CRC; WAM); **Expectorant** (f; CRC; MAD); **Hemostat** (f; CRC); **Hepatotoxic** (1; APA); **Hypotensive** (1; PH2); **Tonic** (f; FAD); **Uterotonic** (1; CAN); **Vulnerary** (1; APA; CAN; WAM)."*
-Handbook of Medicinal Herbs, second edition by James A. Duke with Mary Jo Bogenschutz-Godwin, Judi duCellier and Peggy-Ann K. Duke. Boca Raton, Florida: CRC Press, **2002**, page 214.
   (Vouchered means the Comfrey is from an authenticated botanical voucher system. The specimen is identified by an expert in the field and is stored at a facility where researchers can obtain it for research. An example is: 'Voucher Specimen No. CANB 286704 Australian National Herbarium, Canberra'.)
   (APA Peirce: 1999; CAN Newall, Anderson, and Phillipson: 1996; CRC Duke: 1985; FAD Foster and Duke: 1990; FEL Felter and Lloyd: 1898; KOM Blumenthal et al.: 1998, Commission E; MAD Madaus: 1976; PH2 Gruenwald et al.: 2000; PHR PDR for Herbal Medicine, 1st ed., Fleming, et al.: 1998; PIP Schilcher: 1997; PNC Williamson and Evans: 1989; WAM White and Mavor: 1998.)

*"**Comfrey Uses and Properties:** Mainly used externally for the treatment of inflammation, bruises, sprains, dislocations, pulled ligaments and muscles, arthritis, glandular swellings, slow healing wounds, and boils. Traditionally, roots or leaves were taken internally against lung disorders, gastritis, stomach ulcers, and bleeding."*
-Medicinal Plants of the World: An Illustrated Scientific Guide to Important Medicinal Plants and Their Uses by Ben-Erik Van Wyk and Michael Wink. Portland, Oregon: Timber Press, **2004**, page 314.

*"**Energetics of Ancient Western Herbalism: The origins of the four elements lie far back in the mists of time,** but it was Empedocles (495-444 BC) who introduced them into Greek philosophy. Plato (428-348 BC) lent his view to this but Aristotle (384-322 BC) felt that the four elements were derivative (secondary).*
The classification of plants according to the Greek four qualities (hot, cold, damp, dry) often agrees with modern ideas about their use, but there are sometimes area of disagreement.
For instance, mucilages, which we would consider moistening, are classified by ancient Greeks as drying. This is because they glue together the lips of a wound and prevent bleeding or fluid loss.
Thus, Comfrey and plantain are considered to be drying when we would consider them to be mucilaginous and moistening. *Practitioners abandoned Greek medicine in the late seventeenth century. It was replaced by the 'theory of irritation', which led to the tissue state model of the nineteenth century."*
-The Practice of Traditional Western Herbalism: Basic Doctrine, Energetics and Classification by Matthew Wood. Berkeley, California: North Atlantic Books, **2004**, pages 41-43.

*"**Comfrey: Energetics: Cooling, moistening, and slightly constricting.** Properties: Emollient and vulnerary."*
-The Modern Herbal Dispensatory: A Medicine-Making Guide by Thomas Easley and Steven Horne. Berkeley, California: North Atlantic Books, **2016**, page 217.

Combination of Ingredients in Comfrey Plant
For more about the constituents and nutrients in Comfrey, see section 'Nutritional Value of Comfrey' (Chapter 19, Volume 1).

*"**Allantoin and rosmarinic acid are nowadays often considered as the only active agents in preparations based on Symphyturn officinale L. (Comfrey).** However, it seems that no broad systematic studies on the correlation between efficacy and composition have been done, **despite the fact that preparations containing the pure compounds mentioned above are less potent than those based on plant extracts.**"*
-'Relating Antiphiogistic Efficacy of Dermatics Containing Extracts of Symphytum Officinale to Chemical Profiles' by B. Andres, R. Brenneisen and J.T. Clerc, Institute of Pharmacy, University of Berne, Switzerland; Planta Medica: Society for Medicinal Plant and Natural Product Research by Thieme Medical Publishers, Volume 55, pages 643-644, **1989**.

*"Let it be stressed again that **the other elements in Comfrey, the high ratio of high grade protein, complete with all amino acids, and its high level and range of vitamins, plus its exclusive two elements: Vitamin B12 and Allantoin, put it in a category by itself.** It can be considered probable that the low level of alkaloids even help its curative and preventive functions so long known in its history of effective cure. The experts have never studied this aspect."*
-Comfrey: Nature's Healing Herb & Health Food by Andrew Hughes. Japan: Sanyusha Publishing Co., Ltd, **1992**, page 63.

*"Pyrrolizidine alkaloids have been linked to liver and lung cancers and a range of other deleterious effects. **As with many natural toxicants, major problems arise in determining the effects of the different members of the class and the importance of various forms of ingestion (eating or drinking).***
In this study we have investigated the levels of pyrrolizidine alkaloids in Comfrey (Symphytum officinale), *determined the levels in different parts of the plant and in herbal remedies, separated the alkaloids into **two main groups (the principal parent alkaloids and the corresponding N-oxides)** and, finally, carried out a simple bioassay based upon the mutagenic (changes genetic material) capability of the separated compounds in a human cell line.*
We conclude that the part of the plant ingested is important in terms of alkaloid challenge and that the effect of two of the major groups of alkaloids individually is different from that of alkaloids in the whole plant extract.
*The precise nature of the structure/activity relationships of the alkaloids remains unclear. In addition, **the possibility of synergistic effects must be considered in any assessment of the potential risk due to the presence of specific alkaloids.***
Work at this laboratory has resulted in the development of a bioassay for carcinogens (causes cancer) based upon their ability to cause a specific mutation in the p53 tumour suppressor gene*.
The bioassay of the separated alkaloids and the whole Comfrey extract suggested that the genotoxic effect results from a number of components in the Comfrey and not simply symphytine/symlandine. *This confirms previously published animal studies and offers the possibility of the development of a rapid screening method for natural toxins based upon the effect upon the p53 tumour suppressor gene."*
-'Analysis Separation and Bioassay of Pyrrolizidine Alkaloids from Comfrey (Symphytum Officinale)' by C.E. Couet, C. Crews & A.B. Hanley, CSL Food Science Laboratory, Norwich, Norfolk, United Kingdom; Natural Toxins, Vol 4, No. 4, page 163-167, **1996**.
(* -'The Biological Assay of Natural Mutagens Using the P53 Gene' by C.E. Couet, A.B. Hanley, S. MacDonald and L. Mayes, in 'Food and Cancer Prevention' edited by K.W. Waldron, I.T. Johnston and G.R. Fenwick, London, United Kingdom: Royal Society of Chemistry, 1993.)
    (A bioassay determines the concentration or potency of a substance by its effect on living cells or tissues.)
    (Synergistic means the interaction of two or more substances that create a combined effect greater than the sum of their separate effects. In other words, the whole is greater than the sum of its parts.)

*"**The alkaloids found in Comfrey assist in the healing action achieved together with its other properties.** These alkaloids are small in proportions and acting with the other bland ingredients from the plant give a favourable result in the body.*

*But using the alkaloids in large doses without the accompanying natural properties of the Comfrey plant, the substance becomes a destroying agent, as does any other concentrate. This is where the chemist finds himself in trouble, for he thinks that the active ingredient is the only thing of worth, and that if extracted and used separated from its other minerals, it will work better in stronger concentrated form. Then if there is what is considered to be good effects, the alkaloid is synthesized to make it more profitable dollar-wise, but **here is failure for they have made a dead product** that will in time, bring disaster."*
-'Comfrey: School of Natural Science' by Dr. Frederick John Steed. Ph.D. (Natural Science), NSc., D.N., Australia, 2 pages, **1997**. Founder of Natural Science correspondence course about healthy food and herbal medicine.

"Please keep in mind, that **while alkaloids and their salts have distinctive therapeutic properties of their own, they do not fully nor exactly represent the action of the whole plant from which they are derived.**"
-The Herbal Medicine-Maker's Handbook: A Home Manual by James Green, Herbalist. New York, New York: Random House, Inc., **2000**, page 93.

"**So, why is there this need for isolating the 'active ingredients'?** I can understand the need for the scientific process of establishing the fact that particular medicinal herbs work on a particular disease, pathogen or whatever.
Rather than trying to isolate the active ingredient(s), why not test these medicinal herbs, utilizing the knowledge of professional herbalists, on patients in vivo (living), using the myriad of technology available to researchers and medical diagnosticians to see how and why these medicinal herbs work in living, breathing patients, rather than in a test tube or on laboratory rats and mice (which, by the way, are not humans and have a different physiology).
Big Pharma is not really interested in effects of medicinal herbs as a whole, but rather in whether they can isolate a therapeutic substance which can then be manufactured cheaply and marketed as a new drug! **The problem with this approach is however, that medicinal herbs like Comfrey, dandelion and other medicinal herbs usually contain hundreds if not thousands of chemical compounds that interact, yet many of which are not yet understood and cannot be manufactured.** This is why the manufactured drugs, based on so-called active ingredients, often do not work or produce side effects."
-A Guide to Understanding Herbal Medicines and Surviving the Coming Pharmaceutical Monopoly by Dr. Michael Farley, N.D. and Ty M. Bollinger. Infinity 510 Squared Partners, **2011**, page 14.

"Herbal medicine needs to be defined because it is different from the type of medicine found in the official pharmacopoeias and the British National Formulary. **The practitioners of herbal medicine use the whole plant or parts of it in treatment, for example the dried leaves of Comfrey. They do not accept that the active principle in the plant should be extracted from it, believing that the many compounds in a plant all contribute to its therapeutic value and diminish toxicity.**
Herbal medicines are not authenticated (found to be effective) by the strict trials to which pharmaceutical medicines are subjected. A herbal medicine could be said to be a preparation of a plant that has not been properly tested. In the United States there are over 20,000 herbal preparations."
-'John Gerard, Physic Gardens and Medicinal Plants' by Arthur Hollman, Sussex, United Kingdom, Journal of Medical Biography, Volume 19, pages 47-48, May **2011**.

Old Remedies and Current Medicine Overview

"***Comfrey is a very important medicinal plant. The truly great herb, Comfrey has recently been condemned by scientists following unnatural experiments. Ignore their baneful (destructive) findings.***"
-The Complete Herbal Handbook for Farm and Stable by Juliette de Bairacli Levy. London, England: Faber and Faber, **1952**.

"**Comfrey: Seldom does one encounter the degree of enthusiasm about anything which the modern herbalists display for Common Comfrey, the rhizome and roots as well as the leaves of Symphytum officinale L. (family Boraginaceae).**
    *Aikman's attractive compilation of folk medicine features photographs of a Virginia woman brewing up an oversized
    kettle of Comfrey leaves to make 'a sweet tea for calming coughs and stomach ulcers'.*
Although Comfrey is presently one of the most common herbs sold to the American public, there is reason to believe that using it internally is definitely hazardous to the health."
-The Honest Herbal: A Sensible Guide to the Use of Herbs and Herbal Products by V.E. Tyler, Ph.D. Philadelphia, Pennsylvania: G.F. Stickley Co., **1981**. Now called 'Tyler's Honest Herbal'.
(* -'Natures Healing Arts' by L. Aikman, National Geographic Society, Washington, DC, pages 30-31, 1977.)

"**Control of Pyrrolizidine Exposure: In North America, where crop weeds are controlled by herbicides and foods are inspected, the biggest risk of exposure to pyrrolizidines comes from herbs, 'health foods', 'natural foods', and so-called food supplements.** Risk is increased and increasing because of the developing anti-rationalism and anti-authoritarianism of our society, where everybody's opinion is as good as that of his physician.
There is an urgent need for the public to become more educated concerning the use and abuse of herbs. Unfortunately, there is a dearth (lack) of good sources of information on the subject. Much of the vast popular literature available is valueless. Herbalists quote from one another, and errors can be traced for centuries.
Typically, numerous claims are made for plants. Three characteristics of such claims should be noted, which appear throughout the vast herbal literature:
    1. The ancient medical terminology (e.g., phthisis, gravel, gleet).

2. The emphasis on treatment of symptoms rather than causes (e.g., leukorrhea, enlarged glands).
3. Use of vague, all-encompassing descriptors (e.g., gynecological disorders, liver malfunction, disturbances of the bloodstream).
These characteristics are indications both of the persistence of herbal lore and of its derivation from accepted medical practice in an era when etiologies (causes) were uncertain, diseases with similar symptoms could not be differentiated, and the physician had little to aid him.
Any ethnobotanic value such claims as those listed may hold are obviated (made useless) by the lack of information as to the origin of the claim, the cultural groups that used the plant for the claimed purpose, the extent of such use, and the historic period of such use. Even books by academic writers are not exempt from the above strictures (critical remarks). There are, of course, exceptions. **One of the finest of the books addressed to the lay reader is that of Tyler's*.** This examines the evidence underlying claims made for plants and draws rational conclusions."
-Toxicants of Plant Origin, Volume I: Alkaloids edited by Peter R. Cheeke. Boca Raton, Florida: CRC Press, **1989**. Chapter 3: 'Human Health Implications of Pyrrolizidine Alkaloids and Herbs Containing Them' by Ryan J. Huxtable, University of Arizona.
(* -The Honest Herbal: A Sensible Guide to the Use of Herbs and Herbal Products by V.E. Tyler. Philadelphia, Pennsylvania: G.F. Stickley Co., 1981. Now called 'Tyler's Honest Herbal'.)

"**Time and again science is confirming what earlier generations believed on the basis of folk-lore.**
All over the modern world the search is going on to find effective plants and herbs to be used for health, and old remedies are being tested by new scientific methods. **This is true of Comfrey.** Modern science is revealing the wonderful chemical structure of Comfrey. One main use of Comfrey should be to prevent sickness. While it is true that it can be used to cure many ills, the approach we want to emphasize is prevention, which is so much better than cure."
-Comfrey: Nature's Healing Herb & Health Food by Andrew Hughes. Japan: Sanyusha Publishing Co., Ltd, **1992**, page 45.

"**With the recent return of the public to interest in herbs, it was only to be expected that the FDA (United States Food and Drug Administration) would protect the drug companies' finances by zeroing in on the most popular herbs. Comfrey's content of pyrrolizidine alkaloids was exploited to excite fear, so that international disapproval of this wonderful herb surfaced and rapidly destroyed common usage of it.**
As with a number of other plants containing such alkaloids (chocolate, coffee, tobacco), addiction is possible. If a man or animal lives on just one item, such addiction does indeed bring trouble, in this case, liver damage, even to the point of death. How sensible is it to fuel such an addiction? **Well balanced diet prevents such foolishness. That should not condemn the plant!** How often one hears of persons who drink coffee or tea all day and are unable to function without the daily morning dose of alkaloids!"
-The Organic Method Primer: A Practical Explanation: The How and Why for the Beginner and the Experienced by Bargyla and Gylver Rateaver. San Diego, California: The Rateavers, **1993**, page 163.

"**In view of the large number of conditions and disorders to which preparations of S. officinale are therapeutically applied, this plant can almost be regarded as a panacea (universal remedy).** Initial pharmacological experiments seem to justify its application to wounds to stimulate healing but unfortunately clinical studies are lacking.
Furthermore, there is evidence that S. officinale extracts possess healing properties toward diseases in which the immune system is involved, e.g. rheumatic arthritis. It is tempting to connect an action on such a central physiological system to broad medicinal use, particularly since S. officinale is applied in many conditions, e.g. wound healing, ulcers, and rheumatic diseases, which have at least an inflammatory component.
In conclusion, it can be stated that the healing properties of S. officinale claimed by folk medicine are poorly supported by scientific data. Additional research will be needed to rationalize its ethano-medical (ethnic medicine) use."
-Saponins Used in Traditional and Modern Medicine by Manuel F. Balandrin with editors George R. Waller and Kazuo Yamasaki. Part of 'Advances in Experimental Medicine and Biology Series 404'. New York: Springer, **1996**. Chapter: Phyto-Pharmacology of Saponins from Symphytum Officinale L. by Khalid Aftab, Fehmeena Shaheen, Faryal Vali Mohammad, Mushtaq Noorwala and Viqar Uddin Ahmad; H.E.J. Research Institute of Chemistry, University of Karachi, Pakistan, pages 429-442.

"**Comfrey (Symphytum officinale) Medicinal virtues: Although Russian official medicine is still conducting research on Comfrey and warns against taking it in large doses internally, Russian folk healers use Comfrey extensively.**
As official research continues, many folk healers believe that Comfrey preparations will become an authorized herb for Russian physicians to prescribe for their patients.
Traditional folk healers have long used Comfrey root preparations to speed the healing of wounds. They have also used it extensively to promote the mending of broken bones and reduce inflammation around the site of the break.
Folk herbalists have also found that Comfrey preparations reduce pain of the intestinal tract, improve digestive processes, and heal damaged mucous membranes of the stomach and intestines. They also prescribe Comfrey decoctions for treating chronic inflammation of the stomach and intestines, as well as to relieve symptoms of dysentery, gastritis, and stomach ulcers.
Russian official medicine:
Official medicine uses Comfrey externally in compresses and washes for treating burns and skin grafts, as well as to facilitate the healing of broken bones. They are also used to help relieve the swelling and pain of hemorrhoids.
Fresh sliced roots and their juice are also used by physicians to help heal wounds and to stop external bleeding."
-A Russian Herbal: Traditional Remedies for Health and Healing by Igor Vilevich Zevin, Nathaniel Altman and Lilia Vasilevna Zevin. Rochester, Vermont: Healing Arts Press, **1997**.

"Comfrey has been praised throughout history as a premier healing plant used extensively in folkloric herbalism internally and externally for the repair of innumerable body wounds and illnesses.
However, in the past few years reductionist science has proclaimed Comfrey (in particular, the root and the early spring leaves) to be the possessor and conveyer of certain toxic components called pyrrolizidine alkaloids which are said to cause damage to liver of human beings. Many herblists have accepted this as truth and a number of us have not."
-The Herbal Medicine-Maker's Handbook: A Home Manual by James Green, Herbalist. New York, New York: Random House, Inc., **2000**, page 31.
> (Scientific reductionism reduces complex interactions and substances to the sum of their constituent parts to make them easier to study. The opposite view is the holistic approach of systems biology.)

"Comfrey (Symphytum officinale): Comfrey root and leaf have similar properties; the root is stronger, but the leaf is more palatable (better tasting). *Use them both in salves and ointments. When served as a tea, Comfrey will soothe inflammation in the tissues.* ***The root is decocted, the leaf infused.*** *Comfrey can also be administered via capsules.*
Comfrey was widely used by herbalists in the 1960s and 1970s, but studies a few years ago found traces of PAs (Pyrrolizidine Alkaloids) in the plant. However, the studies were never conclusive.
I'm so absolutely convinced that Comfrey is safe that I continue to use it personally, though I don't recommend it to others. You must make the choice for yourself."
-Rosemary Gladstar's Herbal Recipes for Vibrant Health: 175 Teas, Tonics, Oils, Salves, Tinctures and Other Natural Remedies by Rosemary Gladstar. North Adams, Massachusetts: Storey Publishing, **2008**, page 325.
> (An infusion is when a tea bag or ball is put in a cup of hot water and allowed to steep for a few minutes before drinking. It extracts vitamins and volatile ingredients from soft ingredients like leaves, flowers, etc. A decoction is used to extract the mineral salts and bitter principles from hard material such as roots, bark, seeds and wood. They need boiling for at least 10 minutes and then are allowed to steep for a number of hours.)

"The construction of risk (forming attitudes/beliefs about what is dangerous) in media reporting about Herbal Medicine *is a specific phenomenon that has received little attention within the discipline that has become known as the* **Sociology of Complementary and Alternative Medicine (CAM)**.
> *As a marginalised (pushed to the side) form of medicine in the context of mainstream Australian healthcare, Herbal Medicine media representations highlight the numerous tensions that exist between lay (regular people) and expert knowledges and biomedicine and CAM, as well as the relationship between these forms of knowledge and the multiple ways in which they become mediated (discussed/resolved) socially and politically.*

Evans* (2008) makes the point that what has been defined by Bensoussan and Myers (1996) as 'predictable risk' is what herbalists have for centuries understood as 'danger': 'Herbalists have long understood that some plants are dangerous, and should be avoided or handled with extreme care'. **She notes the frustration for herbalists who are not allowed access to plant medicines categorised as 'registered', such as Comfrey, yet other mainstream health professionals such as pharmacists and doctors, who may have no training in the use of these Herbal Medicines, are able to access them. As Mendel (**2001) points out in her research, the 'promotion of science as the ultimate authority' is a key rhetorical (persuasion) tool in elite medical discourse in Australia.**
> *It is important to note that the nature of this promotion of scientific ideology in relation to Herbal Medicines in Australian elite medical discourse may not necessarily be in keeping with what is deemed 'sound' scientific theory and method according to the principles of Evidence-Based Medicine (EBM)."*

-'Herbal Medicine and Risk Constructions: Representations in Australian Print Media' by Monique Renae Lewis, Media Studies Ph.D. Thesis, Southern Cross University, Lismore, New South Wales, Australia, December **2011**, pages 6, 85.
(* -'Challenge, Tension and Possiblity: An Exploration into Contemporary Western Herbal Medicine in Australia' by S. Evans, Ph.D. thesis, School of Arts and Social Sciences, Lismore, Southern Cross University, Australia, 2008.)
(** -'Risk and Evidence: Political and Philosophical Hegemonies in Australian Health Care' by J. Mendel, School of Social and Workplace Development, Southern Cross University, Lismore, Australia, 2001.)

> **"Complementary and Alternative Medicine (CAM) or 'complementary health approaches' are a group of diverse medical and health care practices and products that are not presently considered to be part of conventional medicine (National Center for Complementary and Integrative Health, https://nccih.nih.gov).**
> *The 'National Center for Complementary and Integrative Health' (NCCIH) classifies most complementary health approaches into one of two subgroups:*
> *1) Natural products, including herbs, vitamins, minerals, and probiotics, often sold to consumers as dietary supplements.*
> *2) Mind and body practices, including a large and diverse group of procedures or techniques administered or taught by a trained practitioner or teacher. These include but are not limited to yoga, chiropractic and osteopathic manipulation, meditation, massage therapy, acupuncture, relaxation techniques, tai chi, qi gong, healing touch, hypnotherapy and movement therapies. Other complementary health approaches include traditional healers, Ayurvedic medicine, traditional Chinese medicine, homeopathy and naturopathy.*
> **CAM therapies are termed as Alternative when used in place of conventional treatments and Complementary when used together with conventional treatments.** *Integrative medicine combines mainstream medical therapies and CAM therapies in a coordinated way."*
> -United States National Library of Medicine®, National Institutes of Health®, Bethesda, Maryland, www.nlm.nih.gov, 2018.

"*Comfrey has been used for a wide variety of medicinal purposes for over 2000 years* (Rode, 2002), and it continues to be consumed as an herbal tea or a vegetable in many countries (Mei et al., 2010). **More recently, controlled studies have found multiple Comfrey based topical treatments to be beneficial for treatment of a variety of muscle and joint pains** (Koll et al., 2004; Staiger 2012). The anti-inflammatory and analgesic (pain killing) effects of Comfrey are thought to be as a result of the imidazolidinylurea allantoin and the phenylpropanoid rosmarinic acid (Staiger, 2012)."
-'The Comparative Toxicity of a Reduced, Crude Comfrey (Symphytum officinale) Alkaloid Extract and the Pure, Comfrey-derived Pyrrolizidine Alkaloids, Lycopsamine and Intermedine in Chicks' by Browna, Stegelmeiera, Colegatea, Gardnera, Pantera, Knoppela and Hallc, Journal of Applied Toxicology, Volume 36: 716-725, May 1 **2016**.

"**Comfrey is renowned since ancient times as a wound healer and a choice for a first aid kit, and it is one of the easiest herbs to grow.** It is necessary that the active herbal ingredients should be effectively transported into skin and blood circulation for their therapeutic actions. However, many active herbal compounds have limited solubility and therefore it creates a hurdle (problem) in formulating patient compliant transdermal patches.
For the systemic delivery of drugs, Transdermal Drug Delivery System (TDDS) is one of the suitable routes. TDDS has the ability to deliver medications more convenient and effective way than conventional systems.
Nano-technology can be utilized to process the herbal extracts and thereby solves the issue of poor aqueous (water) solubility and toxicity of active ingredients in the herbal formula. **The traditional topical herbal paste can be further reformulated into a form of nano emulsion** by homogenization process and design into a perfect shaped matrix-type transdermal patch. Delivery of drugs to the skin is the potential application of nanotechnology."
-'Possibility and Scope of Transdermal Comfrey Multipurpose Nano Patch: For the Treatment of Broken Bones, Wounds and External Ulcers' by Akhilesh Dubey, N.G.S.M. Institute of Pharmaceutical Sciences, Nitte University, India; Pharmacy and Pharmacology International Journal, Volume 5 Issue 1, **2017**.
  (Transdermal is applying a medicine or drug through the skin, usually with an adhesive patch, so it is absorbed slowly.)
  (Nanotechnology deals with dimensions of less than 100 nanometers, especially the manipulation of individual atoms and molecules. A nanometer is one-billionth of a meter.)

Herbalists Recommendations for Comfrey

Overview of Herbalists

"**Symphytum officinale L. root and other parts of the herb have been valued medicinally for more than 2,000 years.** The specific name 'officinale' designates its inclusion in early lists of official medicinal herbs. **Comfrey has been prepared as a poultice or compress with healing properties** for blunt injuries, fractures, swollen bruises, boils, carbuncles, varicose ulcers, and burns. Poultices were also applied to ease breast pain in breastfeeding women. The hot, pulped root, applied externally, was used to treat bronchitis, pleurisy, and to reduce pain and inflammation of sprains*.
Comfrey, taken internally as a tea or expressed juice, has been used to soothe ulcers, hernias, colitis, and to stop internal bleeding. As a gargle it has been used to treat mouth sores and bleeding gums. The herbal tea has also been used to treat nasal congestion and inflammation, diarrhoea, and to quiet coughing.
As the popularity of Comfrey grew over the centuries so did its indications for use.
Comfrey has been used to treat respiratory problems (bronchitis, catarrh, haemoptysis, pleurisy, whooping cough), gastrointestinal diseases (cholecystitis, colitis, dysentery, diarrhoea, ulcers, hematemesis), metrorrhagia, phlebitis, and tonsillitis."
-'European Medicines Agency- Assessment Report on Symphytum Officinale L., Radix' by European Medicines Agency, Committee on Herbal Medicinal Products (HMPC), London, England, 27 pages, May 5 2015.
(* -'The Gale Encyclopedia of Alternative Medicine, Volume 1' edited by J.L. Longe, Thomson Gale, Michigan: Gale Group, pages 526-527, 2005.)
  (Carbuncle is inflammation of the subcutaneous tissue. Catarrh is inflammation of the mucous membranes in the airways, usually throat and paranasal sinuses. Haemoptysis = hemoptysis coughing up blood. Pleurisy is inflammation of the membranes that surround the lungs and line the chest cavity. Cholecystitis is inflammation of the gallbaldder. Colitis is inflammation of the colon. Hematemesis is vomiting blood. Metrorrhagia is uterine bleeding between periods. Phlebitis is inflammation of a vein.)

"**Symphytum officinale has a long tradition and is still applied nowadays** as an external treatment for inflammatory disorders of joints, wounds, gout, bone fractures, distortions, haematomas and thrombophlebitis. It is also applied as a decoction (tea) for oral and pharyngeal (throat) gargle.
For internal application, Comfrey is claimed to benefit gastritis and gastroduodenal ulcers, though its effects have never been demonstrated in controlled investigations. In addition, herbal practitioners recommend Comfrey capsules for the treatment of rheumatoid arthritis, bronchitis, various allergies and for diarrhoea, regardless of the pathogenic (micro-organism) cause*."
-'European Medicines Agency- Assessment Report on Symphytum Officinale L., Radix' by European Medicines Agency, Committee on Herbal Medicinal Products (HMPC), London, England, 27 pages, May 5 2015.
(* -'The Efficacy and Safety of Comfrey' by F. Stickel and H.K. Seitz, Public Health Nutrition Journal of The Nutrition Society, London, United Kingdom, 3:501-508, published 2000.)
  (Gout is inflammatory arthritis. Haematoma is a localized swelling filled with blood. Thrombophlebitis is a phlebitis {inflammation of a vein} related to a blood clot.)

"Plants in the eighteenth to twentieth century (1700s to 1900s): Comfrey: Symphytum officinale:
Mild tea for flooding after childbirth; for flux (diarrhea) or dysentary; roots in water for gonorrhea; for pregnant women with heartburn costiveness (constipation); for sprains and bruises."
-Cherokee Plants: Their Uses- A 400 Year History by Paul B. Hamel and Mary U. Chiltoskey. 1975, page 30.
> (Flooding is heavy bleeding from the uterus, especially after childbirth or in severe cases of menorrhagia {heavy or too long menstruation}.)

Herbalists in 1800s

*"**Comfrey is a plant which possesses considerable medical properties,** though they are but little regarded. The root abounds in a pure, tasteless mucilage, and is useful in irritations of the throat, intestines, and, above all, the bladder.*
***A conserve (food made by cooking fruit with sugar) of the roots** cures the whites, and a decoction of them is excellent in coughs and soreness of the breast.*
***Dried and powdered,** they are good against fluxes of the belly, attended with griping pains and bloody stools. It is serviceable in defluxions (copious discharge of fluid matter as in catarrh) on the lungs, spitting of blood, and other disorders of the breast.*
***Bruised and applied** to foul ulcers, it cleanses and disposes them to heal. It removes the inflammation, eases the pain, and stops the bleeding of the piles (hemorrhoids), and is of considerable efficacy in ulcerations of the kidneys and urinary passage, particularly if occasioned by the use of cantharides or Spanish flies (blister beetles)."*
-'**The Universal Herbal; or, Botanical, Medical, and Agricultural Dictionary;** Containing an Account of All the Known Plants in the World, Arranged According to the Linnean system. Specifying the uses to which they are or may be applied, whether as food, as medicine, or in the arts and manufactures, with the best methods of propagation, and the most recent agricultural improvements', Volume 2, by Thomas Green, London, England, **1824**, page 641.

*"Pulmonary (lung) Balsam (aromatic resin): Take of spikenard (nardin, muskroot) root one pound and a half (680 grams), hoarhound tops, elecampane root, and **Comfrey root, one pound each (453 grams);** add a suitable quantity of water. Boil, and pour off the infusion repeatedly; until the strength is all extracted; then strain and reduce the whole of the liquid down to about twelve porter (brand) bottles; then add of white sugar twelve pounds (5.4 kg), and good honey six pounds (2.7 kg); clarify it with the white of eggs. Let it stand twenty-four hours, in order that it may settle: add one quart (0.94 liter) of spirits (alcohol), and finally bottle for use. Dose, a wine glass full three or four times a day.*
***Use: This preparation is highly useful in the treatment of pulmonary affections (problems),** and coughs of long standing. It is admirably calculated to relieve that constricted state of the lungs which is often met with in consumption (tuberculosis), and to assist expectoration."*
-'**The Botanic Family Physician,** or, The Secret of Curing All Diseases on Improved Hygeian Principles, Fully Disclosed: Containing Also Formulas or Recipes for the Cure of Every Disease Incidental to Human Nature' by L. Meeker Day, Botanic Physician, New York, **1833**, page 11.

"The leaves and flowers of Comfrey are rarely employed in medicine. The root, which is most frequently used, is inodorous (no odor), insipid (no flavor), sweetish, viscid (sticky), and glutinous (like glue).
It contains an abundance of mucilage, more tenacious (firm) than that of marsh-mallow, accompanied with gallic acid in sufficient quantity to render the aqueous decoction quite black by the addition of sulphate of iron (ferrous sulfate).
***Dioscorides** used it in haemoptysis, and **Pfann** states that a small quantity of the powdered root snuffed up the nostrils will stop bleeding at the nose. **Simon Pauli** recommended the application of it to fractured bones, and **Houston**, in hernia.*
***Tachenius** advises the bruised root to be applied in the form of a cataplasm (poultice) in gout, and a case is mentioned by **Camerarius** in which the pain was removed by this application, but an eruption of pustules was the consequence, which at length spread over the whole body."*
-'**The British Flora Medica, or, History of the Medicinal plants of Great Britain,** Volume 1' by Benjamin Herbert Barton and Thomas Castle, London, England, **1838**. (Symphytum officinale, pages 211-215.)
> (Gallic acid is a trihydroxybenzoic acid, a type of phenolic acid, found in gallnuts, sumac, witch hazel, oak bark and other plants. It is found both free and as part of tannins.)
> (The hydrated form of ferrous sulfate is used medically to treat iron deficiency. Other forms are not safe to be consumed. These old-time writings are for entertainment purposes only.)

"For Sciatic or Sciatica Arteria:
***Syrup:** Spikenard, **Comfrey,** white Solomen's Seal, Johnswort (St. Johns Wort), Sweet Agrimony, Prince's Feather (Amaranthus hypochondriacus), or what is called 'love lies a bleeding', (grows in the garden), Swamp Brake roots (a plenty), one pound of raisins, two ounces of saffron. Put all into an earthen pot, adding a layer of sugar between each layer of roots and herbs; cover the pot or jar with a rye dough or paste, put it in a hot oven, and, when it is sufficiently digested, wring out the liquor; add one-third rum or brandy. The syrup is to be taken a wine glassful two or three times a day. A young lady, who had been a long time afflicted and a cripple, with this complaint, was cured by the above. 'In all instances of sciatica during my practice', says Dr. Seely, 'I found this medicine a sovereign specific (broad curative); and I well know that the generality of mankind afflicted with sciatica call it rheumatism.' "*
-'**The American Practice Condensed, or The Family Physician:** Being the Scientific System of Medicine' by Wooster Beach,

New York, 1847, page 403.
(Sciatica is pain going down the leg from the lower back usually from a spinal disc herniation pressing on one of the lumbar or sacral nerve roots. Rheumatism causes chronic pain in joints and/or connective tissue.)

*"**An extremely useful syrup for coughs and colds,** and of occasional benefit in consumption (tuberculosis), is made by **taking Elecampane, Comfrey, and Slippery-elm bark,** of each one ounce (28.3 grams), and pouring on them three pints (6 cups = 1.4 liters) of hot water, boiling down to a quart (0.94 liter), straining, and adding white sugar enough to make a syrup. A teaspoonful of this taken whenever necessary, is a dose."*
-'**The Book of Herbs Giving Descriptions of Medical Plants and Directions for Gathering and Preserving Them**' by John B. Newman, M.D., New York, **1847**, pages 33, 43.

"Comfrey (Consolida) (oo-ster-oo-ste-lur-e-stee in Cherokee):
Of this plant there are two kinds, the wild and garden Comfrey. **Of the two species the garden Comfrey is some the best, owing to its containing more mucilage or jelly, and not being quite so hard and tough as the wild.**
A handful of the roots boiled in new milk and drank freely, is good for flooding after child-birth.
A gill (1/4 of pint= 1/2 cup = 118 ml) of the milk in which Comfrey root has been boiled, given every half hour, is amongst the best remedies for flux (diarrhea) or dysentery.
The root sliced and steeped in water, and used as a common drink, *is good in clap (gonorrhoea), also for strictures (narrowing of body duct or passage), or heat in making water, it is excellent. The root infused in cold water, and made a constant drink, is valuable for pregnant women who are troubled with heart-burn, costiveness (constipation), etc. It is also excellent for such females as from sexual weakness are troubled with menstrual discharges and other symptoms of abortion during pregnancy.*
A poultice made by bruising and boiling the root *is valuable to reduce inflammation and prevent mortification (necrosis = tissue death). The writer can bear testimony of the efficacy of this poultice, it having removed the inflammation from a wound for him, after it had thrown him into high fever, without the aid of other remedies. The poultice is made by pounding or bruising the root fine, boiling it in new milk or water, and thickening it with wheat bran or corn meal. The bruised root wet with vinegar is excellent applied to sprains, bruises and healings; it will often drive back the worst of healings when other applications fail."*
-'**The Cherokee Physician or Indian Guide to Health** as Given by Richard Foreman, A Cherokee Doctor', New York, **1857**

*"**The Comfrey** is slightly astringent, and was formerly regarded as a vulnerary (wound healer), but its styptic (blood stopping) qualities are very slight.* **A mucilage abounding in the leaves, stems, and particularly the root, renders the plant more valuable as an emollient,** *and has been found useful in cases of irritation of intestinal canal, bronchitis and some disorders of the bladder; it is probably a good substitute for marsh mallow. It is now but little employed in medicine in this country."*
-'**The Useful Plants of Great Britain:** Part I, August' by John E. Sowerby & C. Pierpoint Johnson, London, England, **1862**, p.182.

*"**United States Patent No. 56,072** by Lucinda Marmaduke, Shelbyville, Missouri, **Medical Compound, July 3, 1866. For the cure of pulmonary and cognate disorders.** Composed of ground ivy, cohosh, hoarhound, polypody, valerian, Iceland moss, licorice root, spikenard, balm of Gilead, elecampane, Indian turnip, liverwort, **Comfrey,** striped elder, water ash, blood root. Boil, sweeten, add tincture of lobelia and spirits to preserve it from fermentation."*
-'Annual Report of the Commissioner of Patents for the Year **1866**', Volume 2, Washington Government Printing Office, D.C., 1867. Page 919.

*" '**The Solid-Stem Symphytum Asperrimum or Caucasian Prickly Comfrey**' by Henry Doubleday, 1876:*
I have myself frequently proved its efficacy in **healing up the flesh of a wound or cut, by rubbing in the sap from the leaf or stalk,** *and have found the place heal very quickly under this treatment.*
There is no doubt that these healing properties extend to animals suffering from foot or mouth disease, foot rot, etc., and if fed regularly upon this food, we have good reason for hoping its use would entirely prevent such diseases."
-'**Forage Plants and Their Economic Conservation by the New System of Ensilage:** Part I: Caucasian Prickly Comfrey' by Thomas Christy, Jun., F.L.S. (Fellow of the Linnean Society), Christy & Co., London, England, **1877**, page 13.

*"**United States Patent Office, Patent No. 280,281: Medicine for Scrofula, June 26, 1883:***
Be it known that I, Gallahill Atkins, a citizen of the United States, and residing at Pikeville, in the county of Pike, and State of Kentucky, **have invented certain new and useful Improvements in Medical Compounds;**
My discovery has relation to that class of medical compounds which are administered in the treatment of scrofula or King's Evil; and it consists in a purely vegetable compound of the ingredients hereinafter named:
I take one ounce each of black-pine buds (Germina pinus nigra), tulip-tree bark (Cortex liriodendron), sarsaparilla, (Sarsaparilla), **wild Comfrey root, (Radix symphyti oflicinalis),** *golden seal (Hydrastis canadensis), wild silkweed (Asclepias syriaca), magnolia-bark (Cortex magnoliae), and dandelion-root (Radix taraxaci dens leonis). These several ingredients, in a dry state, are first comminuted or reduced to a* **coarse powder,** *after which 1 1/2 pint of alcohol is added. After steeping thoroughly, this is filtered, and 2 gallons of water is added, which by evaporation is reduced to 1 gallon, to which I add 8 pounds of refined sugar. If desired,* **a sirup** *may be formed of the sugar and water, evaporating half the quantity of the water, and then adding the* **tincture** *to the sirup, which should be again subjected to the action of heat until all the alcohol is vaporized, after which the sirup is filtered, when it is ready for use.*
I prefer to use this compound in doses of one tea-spoonful for children and one table-spoonful for adults, three times a day."
(Scrofula or Tuberculous lymphadenitis was known in Europe in 1700s and earlier as 'King's Evil', where the royal touch was

believed to cure the disease. It is a skin disease that is a form of tuberculosis affecting the lymph nodes, especially the neck.)

"This thesis reports on the research of 25 plants used as herbal remedies **since the 1800s by the author's Native American ancestors (the Day family) and the Cherokee tribe.** The plants were identified in four state parks in southwestern Indiana. Information sources included the research literature, articles on Cherokee herbal remedies, and interviews (oral traditions) with Cherokee elders and medicine men from Cherokee Village, North Carolina, and Tahlequah, Oklahoma.
The early settlers used many of the remedies that they had learned from the Cherokees. An astonishing number of Native American drugs and treatments have proved to be of enormous value."
"**Comfrey (Symphytum officinale) Cherokee Uses:**
The Cherokees drank Comfrey root tea for 'bad memory', cancer, genital infections, milky urine, and internal bleeding. Fresh leaves were poulticed for sprains, tendinitis, bruises, and pulled muscles (Foster & Duke, 'Eastern/Central Medicinal Plants', 1990). **The dried leaves were used like tobacco and smoked. They harvested the leaves during May through June when flowers had begun to bud. Leaves were highest in toxic alkaloids during this time.**
The roots were gathered in autumn, cut into slices, and dried (J. Olds, 'The Encyclopeida of Organic Gardening', 1975). The dried roots were crushed, powdered, and placed on cuts, burns, and wounds to promote healing.
Day Family Remedies (1800s and 1900s):
Comfrey tea and another common root, Sassafras tea, were drunk by the Day family as spring tonics.The grandparents also mixed a small amount of baking soda in a cup of hot Comfrey tea for 'settling an upset stomach'."
"Sources Consulted: Mr. Walkingstick (1993) stated **a syrup made by boiling Comfrey roots** has been long accepted as a medicine of great value for coughs and problems of the lungs. It has also been used for the purpose of cleansing the body of impurities and for use as a tonic. A poultice of fresh or powdered Comfrey dried leaves has helped heal fresh wounds, swellings, burns, and bruises. Defoe, Langdon, and Rock ('Flowering Plants of the Great Smoky Mountains', 1989) indicated Comfrey was found occasionally and was well distributed, but nowhere abundant in the Smoky Mountains."
-'**A Manual of Cherokee Herbal Remedies:** History, Information, Identification, Medicinal Healing' by Patricia D. Schafer, Master's Thesis, Indiana State University, March 1993.
    (The Great Smoky Mountains are a mountain range along the Tennessee- North Carolina border in the southeastern United States. They are a subrange of the Appalachian Mountains.)

"**Traditional Local Medicinal Uses for Symphytum caucasicum M. Bieb. Boraginaceae:**
Armenia:
The Comfrey roots are raw material for pharmacology and traditional medicine. It's a good remedy for sciatic nerve inflammation and for fractures (Amirdovlat 1927; Isotova et al. 2010; Gabrielyan 2001; Gammarman and Grom 1976; Grossheim 1952; Gubanov et al. 1976; Harutyunyan 1990; Mardjanyan 2008; Nosal and Nosal 1991; Tsaturyan and Gevorgyan 2014; Turova and Sapojnikova 1982; Vardanyan 1979; Zolotnitskaya 1958–1965).
The Comfrey roots contain tannins, essential oils, choline, allantoin, and asparagine acid (Grossheim 1952; Budantseva 1994–1996; Sokolov 1984–1993; Tsaturyan and Gevorgyan 2007; Zolotnitskaya 1958–1965).
Azerbaijan:
A water infusion of the Comfrey root is used for diarrhea and other gastroenteric diseases (Grossheim 1943).
An infusion of the Comfrey root is used for baths, compress for fractures, dislocations, injury and aches of joints, and also for different skin diseases, and old and bad healing wounds and ulcers. An ointment from the roots is also used for the same purposes (Damirov et al. 1988). The water infusion and decoction of the Comfrey roots has anti-inflammatory, antiseptic and hemostatic property and is used to stop hemorrhage (Damirov et al. 1988).
Georgia (country):
Comfrey leaves and roots are used in salves and to cure fractures. The leaves are also applied to furuncles (carbuncles, boils), and the root is prepared as tea to treat gastro intestinal problems (Bussmann et al. 2014, 2016, 2017)."
-**Ethnobotany of Caucasus** by Ketevan Batsatsashvili, et. al. (Chapter: Symphytum caucasicum M. Bieb. Boraginaceae, pages 683-688). Cham, Switzerland: Springer, 2017.
    (Azerbaijan is a country in south Caucasus region of Eurasia at the crossroads of eastern Europe and western Asia.)

Herbalists in 1900s

"Allantoin is a fresh instance of the good judgment of our rustics (country people), especially of old times, with regard to the virtues of plants. **The great Comfrey or consound,** though it was official with us down to the middle of the eighteenth century, never had a very prominent place in professional practice; but **our herbalists were loud in its praise and the country culler (harvester) of simples (single herbs) held it almost infallible (always effective) as a remedy for both external and internal wounds bruises, and ulcers, for phlegm, for spitting of blood, ruptures, haemorrhoids, etc.**
For ulcers of the stomach and liver especially, the root (the part used) was regarded as of sovereign (great) virtue. It is precisely for such complaints as these that Allantoin, obtained from the rhizome (root) of the plant, is now prescribed.
One old '**Syrupus de Symphyto**' (Spanish) was a rather complicated preparation.
Gerard has a better formula, also a compound, which he highly recommends for ulcers of the lungs.
The old Edinburgh (Scotland) formula is the simplest and probably the best: 'Fresh Comfrey leaves and fresh plantain leaves, of each lb.ss.; bruise them and well squeeze out the juice, add to the dregs (remnants) spring water lb.ij.; boil to half, and mix the strained liquor with the expressed juice; add an equal quantity of white sugar and boil to a syrup.' "

-**The Chemist and Druggist newspaper** published weekly, London, England, August 13, **1921**, No. 2168, page 53
(There was an advertisement for Comfrey roots for sale in this issue on page xxii {pdf page 96}. The newspaper was established in 1859 and is still in publication as the leading trade journal for pharmacists in the United Kingdom.)
('Herbal Simple' means any at-home natural remedy consisting of one ingredient only.)
(I do not know what 'lb.ss.' or 'lb.ij.' mean. If you know, please contact me.)

"**Comfrey Medicinal Action and Uses:**
Demulcent, mildly astringent and expectorant. As the plant abounds in mucilage, it is frequently given whenever **a mucilaginous medicine** is required and has been used like Marshmallow (Althaea officinalis) for intestinal troubles. It is very similar in its emollient action to Marshmallow, but in many cases is even preferred to it and is an ingredient in a large number of herbal preparations.
It forms a gentle remedy in cases of **diarrhoea and dysentery**. A decoction is made by boiling 1/2 to 1 ounce (14.17-28.34 grams) of crushed root in 1 quart (0.94 liter) of water or milk, which is taken in wineglassful doses, frequently.
For its **demulcent action** it has long been employed domestically in lung troubles and also for quinsy (inflammation of tonsils) and whooping-cough. **The root is more effectual than the leaves and is the part usually used in cases of coughs.** It is highly esteemed for all pulmonary (lung) complaints, consumption (tuberculosis) and bleeding of the lungs.
A strong decoction, or tea, is recommended in cases of internal haemorrhage (bleeding), whether from the lungs, stomach, bowels or from bleeding piles (hemorrhoids) -to be taken every two hours till the haemorrhage ceases, in severe cases, a teaspoonful of Witch Hazel extract being added to the Comfrey root tea.
A modern medicinal tincture, employed by homoeopaths, is made from the root with spirits of wine, 10 drops in a tablespoonful of water being administered several times a day.
Comfrey leaves are of much value as an external remedy, both in the form of fomentations (application of hot moist substance to body), for sprains, swellings and bruises, and as a poultice, to severe cuts, to promote suppuration (formation of pus) of boils and abscesses, and gangrenous (dead tissue) and ill-conditioned ulcers.
The whole Comfrey plant, beaten to a cataplasm (soft, moist mass) and applied hot as a poultice, has always been deemed excellent for soothing pain in any tender, inflamed or suppurating part.
It was formerly applied to raw, indolent (slow to heal) ulcers as a glutinous (sticky, adhesive) astringent. It is useful in any kind of inflammatory swelling.
Internally, the leaves are taken in the form of an infusion, 1 ounce of the leaves to 1 pint (2 cups = 0.47 liters) of boiling water. Fluid extract: dose 1/2 to 2 drachms (drachms= 60 grains= 64.79891 milligrams).
The reputation of Comfrey as a vulnerary (wound healer) has been considered due partly to the fact of its reducing the swollen parts in the immediate neighbourhood of fractures, causing union to take place with greater facility. Gerard affirmed: 'A salve concocted from the fresh herb will certainly tend to promote the healing of bruised and broken parts.'
Surgeons have declared that the powdered root, if dissolved in water to a mucilage, is far from contemptible (it is good) for bleedings and fractures, whilst it hastens the callus of bones under repair.
Its virtues as a vulnerary are now attributed to the Allantoin it contains. According to MacAlister (British Medical Journal, January 6, 1912), Allantoin in aqueous (water) solution in strengths of 0.3 percent has a powerful action in strengthening epithelial formations, and is a valuable remedy not only in external ulceration, but also in ulcers of the stomach and duodenum (small intestine). Comfrey Root is used as a source of this cell proliferant Allantoin, employed in the dealing of chronic wounds, burns, ulcers, etc., though Allantoin is also made artificially."
A Modern Herbal: The Medicinal, Culinary, Cosmetic and Economic Properties, Cultivation and Folk-Lore of Herbs, Grasses, Fungi, Shrubs and Trees with their Scientific Uses by Mrs. M. Grieve. NY: Dover Publications, 1971. First published in **1931**.
(An infusion is the steeping of a substance in water to obtain its soluble properties.)
(As an apothecary measure, wineglass / wineglassful / cyathus vinarius was defined as 1/8 pint or 2 fluid ounces or 59 ml {2 1/2 fluid ounces in the imperial system}. Before 1800 it was 1 1/2 fluid ounces or 44 ml. These measurements are no longer relevant to current capacity of wineglasses.)

Mausert's Formulas with Comfrey mixed with other herbs:
'Formula No. 3, Powder for Asthma', 'Formula No. 36 for Bronchial Cough', 'Formula No. 36 for Bronchial Cough in Powder Form', 'Formula No. 81, Tea for Coughs', 'Formula No. 165 for Hoarseness', and 'Formula No. 249, Tea for Pleurisy' use Comfrey root that 'acts soothing to respiratory tract'.
'Formula No. 117, Ointment for Athlete's Foot', 'Formula No. 243, Rectal Wash for Piles' (hemorrhoids), 'Formula No. 252 for Poison Oak and Poison Ivy', and 'Formula No. 309 for Gargle and Mouth Wash'.
'Formula No. 297 for Ulcers of the Stomach and Duodenum No. 1' and 'Formula No. 300 for Ulcers of the Stomach and Duodenum No. 2' **use Comfrey root that is 'soothing and healing to mucous membranes'.**
'**Formula No. 81, Tea for Coughs**': (drachms= 60 grains= 64.79891 milligrams)
1. Thyme Leaves, 4 drachms, Acts quieting and soothing on the mucous membrane.
2. Quillaya, 2 drachms, Aids expectoration.
3. Couch Grass, 6 drachms, Loosens mucous accumulations.
4. Lobelia Herb, 1/2 drachm, Stimulates the respiratory centers.
5. Lungwort Herb, 5 drachms, A valuable expectorant.
6. Chondrus, 2 drachms, Very useful in chronic pectoral (chest) affections.
7. Elecampane Root, 5 1/2 drachms, Facilitates expectoration.
8. Licorice Root, 9 drachms, Relieves irritation, loosens phlegm.

9. Anise Seed, 6 drachms, Allays (reduces) irritation in the air passages.
10. Comfrey Root, 6 drachms, Acts Soothing to respiratory tract.
Mix well and divide into 20 doses using Herbs especially cut for Tea.
Directions: Add one dose to three cups (0.7 liters) of boiling water, cover, boil slowly for about 2 to 3 minutes, let it stand for ten minutes, then strain and take one third in the morning, noon and at night before or after meals. If boiling water is not available, use hot water and allow to stand for half hour. It may be sweetened with honey, rock candy, etc., to suit taste.
This is a very good formula for the treatment of affections of the Bronchial tubes and lungs. Coughs and colds settled in these organs, and the tickling and irritation in the throat are quickly relieved by its quieting and soothing effect.
-**Mausert's Formulae from 'Herbs for Health:** A Concise Treatise on Medicinal Herbs, Their Usefulness and Correct Combinations in the Treatment of Disease. A Guide to Health by Natural Means', by Otto Mausert, San Francisco, California, **1932**

(Pleurisy is inflammation of the membranes that surround the lungs and line the chest cavity.)
(Otto Mausert, N.D. {Naturopathic Doctor} was a German Naturopath who practiced herbalism in San Francisco before World War II. Naturopathy then was a populist {common people} medical-anarchism {independent thinking}. Today Naturopathy is alternative health care taught in post-graduate medical schools and board-licensed with emphasis on anatomy/physiology, nutrition, body adjustments, phytopharmacy {pharmaceutical substances of plant origin}, Traditional Chinese Medicine, Homeopathy, etc.
Mausert's formulas were right before the American 'Herbal Renaissance' of the 1960s and 1970s. Together with the book 'Back to Eden' {1939} by Jethro Kloss {a self-taught populist}, they are almost the sole published remnants of 'The Herbal Dark Ages' {1930s to early 1960s in the United States}. Up until about 1970, Mausert's formulas were sold by 'Nature's Herb Company' in San Francisco, 'Wide World of Herbs' (Montreal, Canada) and other companies. Herb formulas similar to these were used by John Christopher and other herbalists. Dr. Christopher {1909-1983} was an American herbalist and Utah Naturopathic Physician who wrote the book 'School of Natural Healing'.)

"**The action of Comfrey is similar to that of Marsh Mallow, and consequently it is a popular cough remedy.**
It is also used as a fomentation in strained and inflammatory conditions of the muscles, and will promote suppuration (pus formation) of boils and other skin eruptions.
A decoction is made by boiling 1/2 to 1 ounce of the crushed root in 1 quart of water, reducing to 1 1/2 pints (3 cups), and is taken in wineglass doses. Dr. Coffin tells us the root of the plant is also '**a good tonic medicine, and acts friendly on the stomach;** very useful in cases where, from maltreatment, the mouth, the throat and stomach have become sore.' "
-'**Herbal Manual:** The Medicinal, Toilet, Culinary and Other Uses of 130 of the Most Commonly Used Herbs' by Harold Ward; L.N. Fowler & Co. Ltd., London, England, **1936**, page 37.

(American Samuel Thomson, 1769-1843, helped form botanic societies to have Physio-Medicalism {Thomsonian System} accepted at medical colleges. This system was brought to England in 1838 by the natural healer and herbalist A.I. Coffin, 1798-1866. He worked in Leeds, Manchester, and London.)

"**Comfrey (Symphytum officinale) is a powerful remedy** in coughs, catarrh (mucous buildup), ulcerations or inflammation of the lungs, consumption (tuberculosis), hemorrhage (bleeding) and excessive expectoration (phlegm) in asthma and tuberculosis. Very valuable in ulceration or soreness of the kidneys, stomach, or bowels. The best remedy for bloody urine.
Apply a fomentation (compress) wrung out of the strong hot Comfrey tea for bad bruises, swellings, sprains, fractures: it will greatly reduce the swelling and relieve the pain. Also use as a fomentation on boils.
A poultice of the fresh Comfrey leaves is excellent for ruptures, sore breasts, fresh wounds, ulcers, burns, bruises, gangrenous sores, insect bites, and pimples.
The tea taken internally is useful in scrofula, anemia, dysentery, diarrhea, leukorrhea, sores and pains.
Take one or two capsules daily for one or two weeks, then take a week's rest. Boil one ounce of the root in one quart of water and take several wineglassfuls a day as a decoction."
-**Back to Eden:** The Classic Guide to Herbal Medicine, Natural Foods and Home Remedies by Jethro Kloss. Wisconsin: Lotus Press, **1939**-1999, page 121.

(A fomentation is the application of a warm or hot, moist substance to the body to ease pain and promote healing.)
(Scrofula is lymphadenopathy of the neck, usually from an infection of the lymph nodes, known as lymphadenitis. It can be caused by tuberculous or nontuberculous mycobacteria.)
(Anemia is when there are not enough healthy red blood cells to carry adequate oxygen to the body's tissues.)
(Leukorrhea is a thick, whitish or yellowish vaginal discharge. The most common cause is estrogen imbalance. Discharge may increase due to vaginal infection or sexually transmitted diseases.)

The herbalist, Dorothy Hall, wrote in 1975:
"**Russian Comfrey and garlic could together almost halve the present ills of western civilization.**"
In 1991 she wrote: 'Creating Your Herbal Profile: How and Where to Find the Herbs that Match Your Personality Traits and Health Needs'. In 1994 she wrote 'The Herb Tea Book'. In 1998 she wrote 'Dorothy Hall's Herbal Medicine' book.

"**I first met Dorothy in 1986, during the campaign to reverse the government's prohibition of Comfrey.**
I saw first-hand how she touched the lives of thousands of students, teaching herbal medicine at both the clinical and philosophical levels. The teachings were unique, ahead of their time and emphasised the essence of healing.
She was the matriarch of Australian herbal medicine, the founder of 'Australian Traditional-Medicine Society', and an author of seven books.

*Once upon a time, not so very long ago, before Coles sold vitamins, before there was a 'Diploma of Advanced Western Herbal Medicine', before there was a 'Complementary and Alternative Medicine' Industry, there was the **'Traditional Herbalist'**. **The practitioner of the garden and the hedgerows and the waste lands and the bush was the keeper of the old ways, the old knowledge, the ancient wisdoms, the elder, the sage. That is where we find Dorothy Hall, walking the well worn paths of ancestral knowledge and bringing that knowledge to modern times."*
-'Vale Dorothy Hall: Dorothy Hall, the Founder of ATMS, Passed Away on March 24th, 2012. Si Monumentum Requiris, Circumspice', Journal of the Australian Traditional-Medicine Society, New South Wales, June 1 2012.

"**Most external ulcers will respond satisfactorily if a fresh macerated Comfrey leaf is applied to them 3 times a day.** *Be sure to bruise the leaf's veins which will come into direct contact with the affected part, to better heal it.*
Leaves are similarly employed by British veterinarians to open wounds and stubborn sores and ulcers. The bruised wet leaf is also a safe and effective styptic- it stops all kinds of minor bleedings."
Comfrey: What You Need to Know by Ben Charles Harris. New Canaan, Connecticut: Keats Publishing, Inc., **1982**, page 5.
  (An external ulcer is a sore on skin or mucous membranes with disintegration of tissue and sometimes formation of pus. Macerate means to soften or separate into parts. Styptic means to contract or bind, an astringent that stops bleeding.)

"**Comfrey is one of the finest healers for the respiratory system,** *especially where there is hemorrhage of the lungs; it has saved thousands of lives. The root has been used reputably as both a tonic and a vulnerary from very ancient times up to the present. The root and leaves are most beneficial as a poultice in healing any obstinate (hard to heal) or ulcerous wound.*
Comfrey forms an ingredient in a large number of herbal preparations, and it may be given wherever a mucilaginous or demulcent medicine is required."
-**School of Natural Healing:** Herbal Reference Guide by John R. Christopher. Colorado: Nutri Books Corp, **1996**.

Herbalists in 2000s

"**There is provided a topical composition for treating and/or preventing skin irritation induced by a retinoid.** *The composition has Comfrey. There is also provided a method for treating and/or preventing skin irritation induced by a retinoid.* **An effective amount of Comfrey is applied to the skin.** *The Comfrey of the present invention may be derived from any part of the Comfrey plant (preferably,* **Symphytum officinale***) These parts include the flowers, leaves, roots, seeds and stems.* **Preferably, the Comfrey is extracted from the leaves and roots of the plant.**
The **Comfrey extract** *may be in the form of a liquid or powder. A suitable example of a Comfrey extract is available from Cosmetochem® under the tradename Herbasol™. It is an extract obtained from Comfrey herbs in approximately 60:40 v:v water in propylene glycol solution. It is about 3.5% to 5.5% active.*
Comfrey should be present in the topical composition of the invention in an amount, on an active basis, of at least about 0.005 wt % based on the total weight of the composition to treat retinoid-induced skin irritation. Most preferably about 0.01 wt. % to about 1.0 wt. % based on the total weight of the composition.
As a guideline, it is preferred that when the retinoid is retinol, the amount of **Comfrey to retinol ratio** *is from about 1:5 to about 10:1, more preferably from about 1:4 to about 5:1, or most preferably from about 1:3 to about 5:1.*
Assignee: Avon® Products, Inc. (New York, NY), Inventor: Michele C. Duggan (Middletown, NY)."
-'United States Patent for Compositions Having Comfrey and Methods for Reducing Retinoid-Induced Skin Irritation', Patent #6,583,184, Justia Patents, https://patents.justia.com, June 24 **2003**. Justia provides open and free access to the law.
(Cosmetochem® International AG has headquarters in Steinhausen, Switzerland. It is a raw material producer for cosmetic and toiletry products. Customised botanical extracts and actives are available in liquid {Herbasol®} and dry powder.)

"**Comfrey Availability:** *Capsules: 50 mg, 100 mg, and 225 mg, along with other natural products.*
    Compounded Oil: Contains multiple herbal components along with Comfrey.
    Cream, Ointment: Contains 5% to 20% leaf or root extract, for external use.
    Leaf Extract: 1:1.5 strength; 25 mg/ml with 35% to 40% grain alcohol.
Comfrey For External Use: *Ointments and other products containing 5% to 20% dried herb should be applied topically to intact skin with the daily amount not exceeding 100 mcg (microgram) of pyrrolizidine alkaloids with 1,2 unsaturated necine structure. The roots contain more allantoin (0.6% to 0.7%) than the leaves (0.3%).*
Comfrey For Internal Use: *Herb isn't recommended.*
Comfrey Adverse Reactions: *Gastrointestinal (GI): pancreatic islet cell tumors. Genitourinary (GU): urinary bladder tumors.*
    Hepatic (Liver): hepatotoxicity, liver damage, veno-occlusive disease (abdominal distension, anorexia, lethargy, dull ache in upper-right abdomen). May increase liver function test values.
Interactions: Herb-Drug. *Disulfiram (used to treat chronic alcoholism): Comfrey may cause a disulfiram reaction if the herbal product contains alcohol. Discourage use together.*
Comfrey Cautions: *Comfrey shouldn't be taken internally. Pregnant women, women planning to become pregnant, breastfeeding women, and those with a history of alcohol abuse or liver disease should avoid this herb. Infants may be more susceptible to hepatic veno-occlusive disease. Use of Comfrey should be limited to 4 to 6 weeks yearly, to prevent exposure to large amounts of pyrrolizidine alkaloids. Inform the patient that manufacturers need not specify the amount of pyrrolizidine alkaloids their products contain, making it impossible to determine the amount of toxic alkaloids being ingested."*
-**Nursing Herbal Medicine Handbook** by Springhouse. Netherlands: Lippincott Williams & Wilkins, an imprint of Wolters Kluwer

Health, February **2005**.

"**Symphytum- Comfrey. The root and leaves are astringent, mucilaginous and contain allantoin, useful externally on cuts and burns, and internally as an expectorant and demulcent.** The astringency makes Comfrey useful for stopping bleeding and healing ulcers, while the mucilage soothes the irritated tissues. Comfrey contains pyrroilizidine alkaloids which are toxic to the liver tissues. Toxicity is variable from species to species."
-**Botany in a Day:** The Patterns Method of Plant Identification: Herbal Guide to Plant Families of North America by Thomas J. Elpel, written **2008**, page 144. (This book is excellent.)
    (An astringent tends to shrink or constrict body tissues. Mucilaginous means moist, gluey, gummy and sticky. An expectorant helps expel lung mucus. A demulcent forms a soothing, protective film on a mucous membrane surface.)

"**As a medical and naturopathic doctor who sees patients with a wide range of conditions, I am truly delighted to be telling them about the healing powers of Comfrey.** Now that this ancient herb is in a form that is groundbreaking- Traumaplant®. After being used for millennia (thousands of years) as a literal lifesaver and then being tossed aside in favor of 'modern medicine', Comfrey is roaring back into action because its healing powers are unique, potent and sorely needed.
Thanks to the rigorous studies that show that the Comfrey extract in Traumaplant® works: Heals sprained ankles in half the time. Reduces pain and immobility in injured knees by three-quarters in just seven days. Shaves three days off full recovery from abrasions (scrapes). Shrinks abrasions in adults and children in half the time. Reduces by at least three days rehabilitation for a frozen shoulder (painful and stiff). Cuts the immobility of acute back pain by two-thirds within three days. Relieves pain by 90 to 100 percent in locomotor (movement) problems with strong muscular compoments, and can even ease pain in some degenerative conditions. Is so safe that it accelerates the healing process for children as young as four without any reactions or adverse effects."
-**The Healing Power of Trauma Comfrey** by Holly Lucille, N.D., R.N. Brevard, North Carolina: To Your Health Books, **2013**.
    (I do not have any connection with the company or product Traumaplant®.)
    (Degenerative disorders have progressive impairment of both the structure and function of part of the body such as osteoarthritis and rheumatoid arthritis.)
    (Acute disease means it has an abrupt/rapid onset. It usually means an illness of short duration, rapidly progressive, and in need of urgent care. Acute is in reference to time as opposed to subacute or chronic. Chronic means lasting a long time, usually 3 months or more. Subacute is between acute and chronic.)

"**The active ingredients contained in Traumaplant® cream are taken from freshly harvested Symphytum blossoms and leaves, a specially cultivated medical Comfrey (Symphytum x uplandicum Nyman): 'Trauma-Comfrey'.** By means of a particularly careful process, the pharmacologically active, high dosage and standardized ingredients in Traumaplant® are isolated from the freshly pressed plant sap.
Choline, rosmarinic acid derivates and allantoin provide the anti-phlogistic (anti-inflammatory), anti-exudative and wound-healing effects.
    1. Choline: In intermediary metabolism, choline is the central building block for various biosynthetic processes and achieves vasodilatation through the parasympathetic nerve endings and therefore, improves perfusion in the inflamed tissue. Inflammatory exudates and metabolic products are removed. Acting as a vasoactive substance with capillary membrane-sealing properties, choline is the most important anti-exudative ingredient among the active components in Traumaplant®.
    2. Rosmarinic Acid: Besides anti-exudative properties, rosmarinic acid derivates possesses an excellent anti-phlogistic effect and inhibits the formation of inflammation mediators.
    3. Allantoin, an active component in Traumaplant®, stimulates cell proliferation and promotes regeneration of damaged tissue. It is, therefore, responsible for the effects which promote granulation and the wound-healing action of Traumaplant® cream.
The various pharmacological properties of the ingredients in Traumaplant®: anti-phlogistic, anti-exudative, wound healing, and **analgetic (relieves pain)** are responsible for the **extensive and reliable effects of this anti-traumatic medication in cases of painful diseases of joints, sports injuries, sprains, bruises, slow-healing wounds and inflammatory symptoms associated with disorders of the muscles, tendons and joints.**
Traumaplant® acts directly on superficial wounds and, after percutaneous (through the skin) absorption, on deeper lying injuries - especially in cases of deeper inflammations and exudates resulting from blunt injuries.
Successful treatment using Traumaplant® provides rapid decrease in swelling and inflammation, pain relief, reduces feeling of tension, additional wound healing, thus, Traumaplant® is **reliable and effective in cases of both open and blunt injuries.**
Indications: Bruises and sprains (sports and accidental injuries), joint and muscle pain, slow healing wounds.
Application and Dosage: For external applicaton. If not otherwise instructed, the cream should be applied up to several times daily to the afflicted region. Especially suitable for cream bandages.
Contra-Indications: No contra-indications are known.
Traumaplant® is free of problematic Pyrrolizidine alkaloids (detection limit 0.1 ppm, officially fixed safety limit 10 mg Pyrrolizidine alkaloids / day)
Side effects: In rare cases erythema (reddening of the skin) may occur, especially in patients with extremely sensitive skin or with an allergic tendency to a component of Traumaplant® cream.
Interaction with other drugs: No interactions with other drugs are known."

-'**Traumaplant® Cream:** The Percutaneous Anti-Traumatic' by Harras Pharma Curarina, Arzneimittel GmbH, Munchen, Germany, **2018**.   (I do not have any connection with the company or product Traumaplant®.)
(An exudate is any fluid that forms at lesions or areas of inflammation. It can be pus or clear. The fluid is composed of serum, fibrin, and white blood cells. A lesion is damaged tissue such as a wound, ulcer, abscess, tumor, etc.)
(Vasoactive means it influences the diameter of blood vessels and thereby blood pressure.)

*"**Although internal use of Comfrey was not advised (by 'European Medicines Agency'), dosages for oral administration for traditional uses were recommended in older standard herbal reference texts.** The recommended oral (unless otherwise stated) doses of 'The British Herbal Pharmacopoeia' (1974 and 1983) for the treatment of gastric and duodenal ulcer, colitis and hematemesis were as follows:*
- ***Dried root/rhizome:** 2-4 grams (0.07- 0.14 ounces) (in the 1983 edition) or 2-8 grams (0.07- 0.28 ounces) (in the 1974 edition) as a decoction (tea) three times daily.*
- ***Root, liquid extract:** 2-4 ml (in the 1983 edition) or 2-8 ml (in the 1974 edition) (1:1 in 25% alcohol) three times daily."*

-'European Medicines Agency- Assessment Report on Symphytum Officinale L., Radix' by European Medicines Agency, Committee on Herbal Medicinal Products (HMPC), London, England, 27 pages, May 5 **2015**.

*"**Symphytum spp. Comfrey:** The internal use of Comfrey leaf can only be contentious (leading to agrument) and excite conflicting views! PA (Pyrrolizidine Alkaloids) in Comfrey have been shown to be genotoxic (gene mutations) and carcinogenic (cancer causing). **I argued that there was less risk because the PA are monoesters or open diesters and thus more likely to be metabolised before the CYP (Cytochrome P450) route comes into play. The argument against this is that in the end a pyrrole is a pyrrole, and the risks associated with ingestion of PA override the value of the plants.**
But what if the Comfrey was prescribed for a life-threatening disease, for example for bleeding associated with gastric or duodenal ulcers, or where the patient had survived oesophageal (esophageal) varices (abnormal veins)?
Finally, my argument depended on dosage, and I argued that mature leaves should be used, and that S. officinale (Common Comfrey) should be used, not S. x uplandicum (Russian Comfrey) as the level of PA is lower*.
This highlights the need for further studies into wild and cultivated plant materials to investigate this point!"*
-'**Pyrrolizidine Alkaloids: Key Points for Herbal Practitioners**' by Helen Phillips M.Sc. Herbal Medicine, Diploma of Phytotherapy, MNIMH and Alison Denham, M.A., FNIMH (National Institute of Medical Herbalists); The Forager's Path, LLC, School of Botanical Studies, Flagstaff, Arizona, August **2016**.
(* -'Using Herbs that Contain Pyrrolizidine Alkaloids' by Alison Denham, B.A., MNIMH {National Institute of Medical Herbalists}, University of Central Lancashire, England; The European Journal of Herbal Medicine, Volume 2, No. 3, pages 27-38, 1996.)
(For more about Cytochrome P450, see subsection 'Individual Diffferences in Reactions to Alkaloids' in section 'Warnings and Negative Reactions to Comfrey' {Chapter 28}.)

Chinese Medicine and Comfrey
For more about Chinese medicine, see year '1830' in subsection 'Age of Revolutions' in section 'History' (Chapter 15, Volume 1}.

*"**Comfrey (Symphytum officinale): Will it one day be added to the materia medica of Traditional Chinese Medicine?** One must acknowledge the possibility that Comfrey may commend itself to contemporary practitioners of Traditional Chinese Medicine, whether in China or abroad.
In adopting an herb deriving from the Euro-American herbal literature, the physician needs to realize that the herbal literature is in significant ways unreliable and misleading. This is demonstrably so for Comfrey.*
> *'A very wide chasm now exists between the scientific study of plant drugs --a part of the discipline known as pharmacognosy-- and the field of popular herbal medicine. The later is composed of varying parts of outdated information, folklore, superstition, wishful thinking, hokum (nonsense), and even hoax', one *authority has written."*

-'Comfrey in the Chinese Materia Medica' by Robert Anderson, Mills College, Oakland, California; Asian Medicine Newsletter, International Association for the Study of Traditional Asian Medicine, New Series, Volume 2, pages 7-11, July **1992**.
(* -Examining Holistic Medicine by Douglas Stalker. Buffalo, New York: Prometheus Books, 1985, page 323.)

"**Meridian Herbal Chart:**
- **Fire:** small intestine, heart, circulation/sex, **triple warmer (Comfrey)**.
- **Earth:** spleen, **stomach (Comfrey)**.
- **Metal: lung (Comfrey)**, large intestine.
- **Water:** bladder, kidney.
- **Wood:** gall bladder, liver.

-'Meridian Herbal Chart' by Kinesiology College of Canada: Discover the Fundamentals of Energy Medicine, Lake Country, British Columbia, Canada, **2001**. Approved by the 'International College of Professional Kinesiology Practice'.
(In Chinese medicine 'Triple Warmer' is one of the 12 vital organs. It regulates the flow of energy through the organs.)

*"**Comfrey, Symphytum officinale:Energy, Taste and Organs Affected:** Cool, bitter, sweet; Lungs, stomach, bones, muscles.
   Actions: Tonify Yin."*
-Healing with the Herbs of Life: Hundreds of Herbal Remedies, Therapies, and Preparations by Lesley Tierra, L.Ac., Herbalist, A.H.G. New York: Crossing Press / Crown Publishing Group, **2003**.

(In Traditional Chinese Medicine, health is a balance of yin and yang. These forces are opposite, interconnected and complementary. Without yin, there would be no yang, and without yang, no yin. Yin is negative, dark, cold, moon, contracting, water, receptive and feminine. Yang is positive, bright, warm/hot, sun, expanding, fire, active, masculine.)

"**Comfrey Energetics:** *bitter, sweet, cool.* **Meridians/Organs Affected:** *lungs, stomach (and kidneys with the Comfrey root)."*
-'Herbalpedia®: Comfrey' by Maureen Rogers, The Herb Growing & Marketing Network, Silver Spring, Pennsylvania, www.herbalpedia.com, **2006**.   (Meridians are paths through which life-energy known as 'qi' flows through the body.)

"*Comfrey's nickname, knitbone, is highly appropriate as one of its constituents (allantoin) actually causes cellular proliferation, quickly healing broken bones, fractures, torn skin (try it on torn perineums after childbirth, using the fresh herb poultice daily), and strengthening tendons, bones and ligaments (take internally and apply externally).*
I have found Comfrey, along with perhaps plantain and echinacea, to be incomparable in drawing out the poison from spider bites, healing them quickly and painlessly.
A wonderful herb for the lungs (tonifies Lung Yin), Comfrey's cooling moistening effect *heals bronchitis, tonsillitis, pharyngitis (inflammation of the pharynx of throat), pleurisy, pneumonia, pulmonary tuberculosis, coughs (including whooping cough), expels phlegm, soothes the throat, lowers fevers and overall, rejuvenates the lungs and mucous membranes.*
It helps the pancreas regulate blood sugar levels and promotes the secretion of pepsin, thus aiding digestion.
The root can be used as well as the leaf, and is stronger in tonic properties for healing lungs and mucous membranes, especially in cases of Dryness, Heat, Deficient Yin and inflammation.
Leaves are more astringent and anti-inflammatory. Comfrey energy is cool. The taste is bitter and sweet. It tonifies Yin."
-'Comfrey Comfort' by East West School of Planetary Herbology®, Santa Cruz, California, probably **2014**.
     (Certain spider bites such as from a Brown Recluse can be deadly. If in doubt, see a doctor right away.)

"*Yin and Yang Herbal Tonics: There are specific Yin and Yang herbal tonics.* **On the one hand, Yin-tonics feed the organs, providing them with necessary nutrients. Tierra (*1998) asserts that the most valuable Yin-tonics are the 'seaweeds (kelp and Irish Moss), alfalfa, Comfrey and dandelion leaf'.**
On the other hand, Yang-tonics balance and stimulate the life-energy of the organs, thereby, improving their assimilation and utilization of nutrients. According to Tierra (1998), Chinese root tonic herbs, burdock, dandelion root, parsley, Oregon grape root and goldenseal root are considered to be among the more efficacious (effective) Yang-tonics."
-'Traditional Chinese Medicine: Part II: Tonifying the Body' by M.G. Garko, Ph.D., Health and Wellness Monthly, Let's Talk Nutrition®, Tampa, Florida, www.letstalknutrition.com, April **2014**.
(* -The Way of Herbs: Fully Updated with the Latest Developments in Herbal Science by Michael Tierra, L.Ac., O.M.D.; New York: Pocket Books, 1998.)

"*One of the biggest obstacles to progress in* **Traditional Chinese Medicine (TCM)** *development in Western countries is the difficulty of applying the traditional concepts to the Western medicinal plants, which are not traditionally described in ancient literature.*
During recent years, new advances in the field of understanding Yin/Yang aspects from a modern bioscientific point of view have led to the conclusion that antioxidation-oxidation concepts might mirror a Yin-Yang relationship.
Trolox Equivalent Antioxidant Capacity (TEAC) of **six vegetal aqueous (water) extracts (Symphytum officinalae (radix = root)- SYM,** *Inula helenium (radix)- INU, Calendula officinalis (flores = flowers)- CAL, Angelica arhanghelica (folium = leaf)- ANG(F), Angelica arhanghelica (radix)- ANG(R), Ecbalium Elaterium (fruits)- ECB) and luminol-enhanced chemi-luminescence of PMNL (polimorphonuclear leukocytes) on addition of these vegetal extracts were measured.*
The most potent Yin herb was found to be Symphytum officinalae (SYM). In Western phytotherapy, SYM is characterized as being cold (Yin) in the first degree, earthy and having heat clearing action (anti-inflammatory, anti-ulcerative, etc.) (Culpeper, 1995).
The nature of European herbs, which is described by certain Western authors (Culpeper, 1995), in terms of four qualities (cold/warm, moist/dry) might be partially correlated with the Yin-Yang nature described in Traditional Chinese Medicine system.
In our study SYM showed the biggest TEAC values (Yin), and it also had a significant inhibitory (reducing) effect of the CL (chemi-luminescence) signal (cold-Yin).
It is worthy to mention that SYM contains vitamin B12 (*EMA Assessment Report, 2009), which is uncommon for herbs, and allantoin. Vitamin B12 is required for the red blood cell production (Blood is Yin) and allantoin stimulates the cell growth and proliferation (Yin) *(Araujo et al, 2010; Ahn et al, 2013).*
We also suggest that the herbs with higher antioxidant capacity, stronger inhibitory effect on CL signal, and which are slower in action **(Yin active herbs)** *might be useful in the treatment of those diseases associated with oxidative stress ('excess of Yang' or 'deficiency of Yin'). While those herbs with lower antioxidant capacity, stronger stimulatory effect on the respiratory burst, and which are faster in action (Yang active herbs) might be beneficial in those pathologies associated with reductive stress ('excess of Yin' or 'deficiency of Yang').*"
-'Estimating the Yin-Yang Nature of Western Herbs: A Potential Tool Based on Antioxidation-Oxidation Theory' by Marilena Gilca, Laura Gaman, Daniela Lixandru and Irina Stoian, Bucharest, Romania; African Journal of Traditional, Complementary and Alternative Medicines, Volume 11, No. 3, pages 210-216, **2014**.
(* -'European Medicines Agency- Assessment Report on Symphytum Officinale L., Radix' by European Medicines Agency, Committee on Herbal Medicinal Products -HMPC, London, England, 27 pages, May 5 2015.)

"The concept of Yin and Yang balance existed in ancient Traditional Chinese Medicine for more than 2000 years. The author of Chinese medical treatise Su Wen writes 'The imbalance of Yin and Yang was the cause of all diseases.' This Yin and Yang balance sought in Traditional Chinese Medicine seems to have been reflected in modern western medicine as the balance between antioxidation and oxidation in the human body.
Numerous research show that the balance between antioxidation and oxidation is critical for maintaining a healthy biological system. Human body systemically generates oxidative reactive oxygen species that are critical for cellular signaling process. At the same time, these oxidative reactive oxygen species are extremely unstable that cause oxidation and damages of biological structures such as cells."
-'Antioxidation and Oxidation: Reflection of Yin-Yang Theory?' by Jin Ji, Ph.D., Brunswick Laboratories: BioAnalytical Testing and Research Laboratories, Southborough, Massachusetts, https://brunswicklabs.com, September 30 **2014**.

"Comfrey root (Symphytum officinale) is an herb traditionally used by herbalists throughout the Western hemisphere as a Yin tonic that promotes growth and maintenance of bones and muscles. It contains an abundance of allantoin which is a recognized cell proliferent. Comfrey root has all indications of a Yin tonic, having a cool, moist, nutritive energy. It is used not only to strengthen bones but also to counteract inflammation and arthritic conditions. Recent findings of trace amounts of pyrolizidine alkaloids in certain species of Comfrey have made many herbalists question its long term use. In Traditional Chinese Medicine there are many Yin or blood tonic herbs that have some of properties of Comfrey and could be substituted."
-'Integrating the Traditional Chinese Understanding of the Kidneys into Western Herbalism' by Michael Tierra, L.Ac., O.M.D., East West School of Herbology, Santa Cruz, California, www.Planetherbs.com, date unknown but around **2015**.

"Energetics: Cooling, moistening, and slightly constricting. Properties: Emollient and vulnerary."
-The Modern Herbal Dispensary: A Medicine-Making Guide by Thomas Easley and Steven Horne. Berkeley, California: North Atlantic Books, **2016**, page 217.

Ayurvedic Medicine and Comfrey

Ayurveda or Ayurvedic medicine has been used in India for thousands of years. It is the traditional system of medicine of India and seeks to treat and integrate body, mind, and spirit using a comprehensive holistic approach especially by emphasizing diet, herbal remedies, exercise, meditation, breathing, and physical therapy.

There are 3 energy forces (Vata, Pitta and Kapha) in every person in different proportions.
Vata has the active nature of Wind energy = Air + Ether.
Pitta has the transformative nature of Fire energy = Fire + Water.
Kapha has the binding nature of Water energy = Water + Earth.

*"**Samana Karma: Stambhana Therapy: Stambhana therapies are primarily a treatment for pitta,** emphasising moistening, cooling and salty foods, sufficient water, electrolytes, bathing in cool water, residing next to water, and exposure to moonlight. Stambhana karma tends to have constipating action and is thus used in paittika diseases such as diarrhoea and dysentery. The qualities of stambhana karma are sita (cold), manda (slow), sandra (solidifying) and sthira (stabilising). Herbal treatment in stambhana therapy are primarily madhura (sweet), tikta (bitter), kasaya (astringent) in rasa.*
***Useful non-Indian herbs include** astringents such as Blackberry root (Rubus discolor), Cranesbill Geranium root (Geranium maculatum), White Pond Lily root (Nymphaea odorata); **demulcents such as Comfrey leaf (Symphytum officinalis)** and Marshmallow root (Althaea officinalis); and bitter herbs such as Gentian root (Gentiana spp.), Dandelion root (Taraxacum officinalis), and Calendula flower (Calendula officinalis)."*
-Ayurveda: The Divine Science of Life by Todd Caldecott. Philadelphia, Pennsylvania: Elsevier Ltd, **2006**, pages 155-156.

"A skin scrub should be mild enough to be used every day. Scrubs stimulate circulation and cleanse the pores, preventing blackheads by efficiently removing the dead scaly outermost layer of the skin. It helps bring lustre to your complexion and stimulates new skin growth.
***Homemade Ayurvedic Skin Scrubs: Normal Skin-** Oat flour plus a little Basil (tulsi), sandalwood (chandan) powder and rose (gulab) petals. **Mix to a paste with water, milk or Comfrey tea.**"*
-'Ayurveda and All: The Complete Ayurvedic Quarterly Magazine', New Delhi, India, Volume 14, No. 4, page 32, October-December **2017**.

*"**Vata and Pitta: Favor Comfrey herbal tea. Kapha: Comfrey herbal tea is OK in moderation.**"*
-'Food Guidelines for Basic Constitutional Types' by The Ayurvedic Institute from 'Ayurvedic Cooking for Self Healing' by Usha and Vasant Lad, MASc., Albuquerque, New Mexico: The Ayurvedic Press, www.ayurveda.com, **2018**.

Physician Recommendations for Comfrey

*"**Common Comfrey in the Treatment of Stomach Cough:** A lady of 75 years developed a cough which steadily increased in severity until there appeared to be considerable risk of cardiac and general exhaustion. By process of exclusion I diagnosed

stomach cough and prescribed several different stomachics, but neither these nor careful dieting had the desired effect.
I accordingly decided to try Symphytum officinale, or Common Comfrey, and ordered a simple infusion (tea) of the root to be given in teaspoonful doses every half-hour. The effect was certainly remarkable, the cough beginning to subside in a few hours, and within a week had completely ceased.
The cell proliferating properties of Comfrey, which have been investigated by Dr. MacAlister, will probably account for the repair of what seems likely to have been a patchy denudation of the gastric or oesophageal epithelium, its beneficial effects being unretarded by any untoward influence of the gastric secretions, as would have happened in the stomach."
-'Letters, Notes and Answers' by **William Bramwell, M.A., M.D., B.Ch.**, The British Medical Journal, London, England, page 748, April 24 **1915**.
(Infusion is extracting compounds from plants in solvent such as water, oil or alcohol by keeping it in solvent over time; steeping.)

"Comfrey-Knitbone (Symphytum officinale): The root contains a crystalline solid, that stimulates the growth of epithelium on ulcerated surfaces. It may be administered internally in the treatment of gastric and duodenal ulcers; injuries to sinews, tendons and the periosteum. Acts on joints generally.
Neuralgia of knee: Of great use in wounds penetrating to perineum and bones, and in non-union of fractures; irritable stump after amputation, irritable bone at point of fracture. Psoas abscess. Pricking pain and soreness of periosteum.
Head: Pain in occiput (back of head), top and forehead; changing places. Pain comes down bone of nose. Inflammation of inferior maxillary (upper jaw) bone, hard, red, swelling.
Eye: Pain in eye after a blow of an obtuse (blunt) body. For traumatic injuries of the eyes, no remedy equals this.
Relationship-: Compare: Arn (Arnica montana); Calc phos (Calcarea phos, calcium phosphate).
Dose:Tincture. Externally as a dressing for sores and ulcers and pruritus ani (itchy skin)."
-The Pocket Manual of Homeopathic Materia Medica by **Dr. William Boericke, 1927.**

(Dr. William Boericke was born in Austria in 1849 and later moved to the United States. He graduated from Hahnemann Medical College in 1880. He was co-owner of homeopathic pharmaceutical company 'Boericke & Tafel' in Philadelphia, Pennsylvania. He helped form Hahnemann College of San Francisco, California where he was professor of 'Materia Medica and Therapeutics'. He compiled and edited 'The Pocket Manual of Homeopathic Materia Medica' with the 1st edition in 1901. The last and 9th edition was 1927. It is an encyclopedia of therapeutic properties of each medication.)
(Periosteum is a membrane that covers the outer surface of all bones, except at the joints of long bones. Neuralgia is intense burning or stabbing pain caused by irritation or damage to a nerve. The perineum is the area between the anus and scrotum in a male and between the anus and vulva in a female. Non-union of fractures means bones that will not fuse or heal.) (Psoas abscess is a lesion, usually tuberculous, originating in tuberculous spondylitis and extending through the iliopsoas muscle to the inguinal region. Iliopsoas is a large muscle group going from the inside of the back wall of the pelvis and lower abdomen. Inguinal is in the groin area or in the lowest lateral regions of the abdomen.)

"The true method of healing the sick is to tell them of the herbs that grow for the benefit of man.
As a medical doctor with sixty years experience, **in my practice the herb Comfrey Symphytum has a large place among those simple yet effective remedies.**"
-Nature's Healing Grasses by **H.E. Kirschner, M.D.** California: H.C. White Publications. First printing **1960**, sixteenth printing 1980, page 118

"Comfrey (Symphytum officinale): The medicinal parts are the fresh root and the leaves.
Compounds: Allantoin, Mucilages (Fructans), Triterpene saponins including symphytoxide A, Tannins, Silicic acid (to some extent water-soluble). Pyrrolizidine alkaloids (0.03% in the leaves) including echinatine, lycopsamine, 7-acetyl lycoposamine, echimidine, lasiocarpine, symphytine, intermedine, symveridine.
Anti-inflammatory Effect: Comfrey suppresses leukocyte infiltration during the inflammation process (Shipochliev, 1981).
Demultant Effect: The mucilages act as demultants for a soothing and irritation reduction effect.
Hypotensive Effect (lowers blood pressure):
Symphytoxide A, a triterpene saponin, exhibited hypotensive activity in anesthetized rats (Ahmad, 4P 1993).
Tissue/Nerve Stimulation:
Allantoin, a component in Comfrey, stimulates tissue repair and wound healing through cell proliferation (Rieth, 1968).
Allantoin has had significant effect on cellular multiplication in degenerating & regenerating peripheral nerves (Loots, 1979).
Clinical Trials:
The anti-inflammatory effects of Comfrey were studied in musculoskeletal disorders. Forty-one patients with musculoskeletal rheumatism were treated with either a pyrrolizidine alkaloid-free ointment or placebo for 4 weeks. The patient illnesses consisted of epicondylitis, tendovaginitis, and periarthritis. Efficacy was determined by evaluation of different **pain parameters** (tenderness on pressure, pain at rest, pain on exercise). There was significant improvement with the ointment compared to placebo at weeks 1, 2 and 4 in patients with epicondylitis. There was improvement with M. tendovaginitis at week 1 and 2, but not at week 4 with the ointment compared to placebo. There was no improvement in the peri-arthritis patients in either of the two treatment groups (Petersen, 1993).
Indications and Usage:
Approved by Commission E: Blunt injuries externally. Comfrey used for bruises, sprains and promotion of bone healing.
Unproven Uses: The root has been used externally as a mouthwash and gargle for gum disease, pharyngitis, and strep throat. Internally, the root has been used for gastritis and gastrointestinal ulcers. In Folk medicine, the root of the plant has been used for rheumatism, pleuritis, and as an antidiarrheal agent.

Contraindications: Comfrey is contra-indicated in pregnancy and in nursing mothers.

Precautions and Adverse Reactions:

Hepatotoxicity: Internal administration of the drug, due to the presence of pyrrolizidine alkaloids, has resulted in hepatocyte membrane injury with hemorrhagic necrosis (tissue death) and loss of microvilli (Yeong, 1993). Hepatic veno-occlusive disease and severe portal hypertension has been associated with Comfrey ingestion, and in one case report, death resulted by liver failure (Ridker, 1989; Yeong, 1990).

Carcinogenic / Mutagenic Effects: Mutagenic effects are associated with aqueous (water) extracts of the alkaloid fractions (Furmanowa, 1983). Hepatocelluar adenomas have been reported in animal models receiving diets containing Comfrey roots and leaves (Hirono, 1978).

Comfrey also has chromosome-damaging effects in human lymphocytes (Behninger, 1989).

Gastrointestinal / Kidney / Pancreas Effects: Comfrey, through the pyrrolizidine alkaloids, has been shown to produce lesions in the gastrointestinal tract, pancreas, and renal glomeruli in animal models (Winship, 1991).

Respiratory Effects: Pulmonary endothelial hyperplasia from the pyrrolizidine alkaloids has been seen in animal models (Miskely, 1992).

Dosage:

Mode of Administration: The crushed root, extracts, and pressed juice of the fresh plant are used as semi-solid preparations and poultices for external use. The drug is a component of standardized preparations of analgesics, antirheumatic agents, antiphlogistics, antitussives, and expectorants.

How Supplied: Cream- 1.25 ounce, 2 ounce.

Preparation: To make an infusion, pour boiling water over 5 to 10 grams (0.17-0.35 ounces) comminuted (small particles) or powdered drug, steep 10 to 15 minutes, then strain (1 teaspoonful = 4 grams drug). For external application, a decoction of 1:10 is used, or the fresh roots are mashed.

External Use: The daily dosage should not exceed 1 mg of pyrrolizidine alkaloids for external preparations calculated with 5 to 7% drug, maximum 1 ppm/gm (parts per million / gram) for commercial pharmaceutical preparations. The drug should be used for a maximum of 4 weeks.

Tea: When using the infusion, take 1 cup, 2 to 3 times daily, but not for a long duration."

-PDR for Herbal Medicines (Physicians Desk Reference), 4th Edition. Montvale, New Jersey: Medical Economics Company, Inc., **2000**, page 463. (Lists 600 herbs. It is mostly based on findings of the German Regulatory Authority for herbal medicine, called Commission E.)

J. FLEMONS & SONS
Wholesale Herb Merchants,
DUNSTABLE & SOHAM.

Special lines: English Herbs, Roots and Flowers, fresh and dried.

LEAVES & HERBS.
Digitalis, Belladonna, Broom Tops, Dandelion Tops, Coltsfoot, Mugwort, Ragwort, Wild Carrot, Meadowsweet, etc.

ROOTS.
Belladonna, Burdock, Bryony White and Black, Comfrey, Dandelion, Male Fern, Doggrass, etc.

Full List on Application.

Advertisement: Wholesale Herb Merchants 1859

Roots: Belladonna, Burdock, Bryony White and Black, Comfrey, Dandelion, Male Fern, Doggrass, etc.

'The Chemist and Druggist' newspaper published weekly, London, England, August 13, 1921, No. 2168, page xxii.

The newspaper was established in 1859 and is still in publication as the leading trade journal for pharmacists in the United Kingdom.

World War 1 First Aid Kit with Comfrey

"Symphytum offinale Lin. No. 86, Parke, Davis & Co., Detroit, Mich. U.S.A."

"Properties: Demulcent, slightly astringent, and tonic. It is also used in bronchial irritation, diarrhea, dysentary and leucorrhea. Dose: 1/2 to 2 fluid ounces (15 to 60 mils) of an infusion, one ounce to the pint."

The box is 2 inches by 2 inches. World War 1 was from 1914 to 1918.

Founded in 1866, Parke, Davis & Co. was America's oldest and largest drug maker. It had the first modern pharmaceutical laboratory and developed systematic methods of performing clinical trials on medications.

"Russian soldiers have used Comfrey as a handy field remedy since Czarist times. Handwritten journals describe cavalry soldiers who 'sliced and slew Comfrey roots to mash' and applied them to each other's wounds, holding them on 'until the blood stopped and the groans of pain were gone'. During those times, Comfrey was known as 'knit-together herb'."
-A Russian Herbal: Traditional Remedies for Health and Healing by Igor Vilevich Zevin, Nathaniel Altman and Lilia Vasilevna Zevin. Rochester, Vermont: Healing Arts Press, 1997.

Chapter 24

Medical Research about Comfrey and Healing

More Research Needed about Comfrey

"Research into the medicinal possibilities of Comfrey suffers not only from the shortage of money- it is further handicapped by the absence of acceptable clinical data. Though I could fill perhaps four chapters with reported successes with Comfrey as ointment, tea, tablets, or fresh, none of this evidence would be accepted medically. It is mere anecdote (story) to any doctor in every country, and because Comfrey is to them folk medicine or herbalism, no research has been done."
-Comfrey: Fodder, Food and Remedy by Lawrence D. Hills. New York: Rizzoli Universe Books: **1976**, page 178.

*"***Comfrey has no mighty pharmaceutical combine behind it able to make generous grants towards research in University medical departments***, for 'double blind' tests and all the statistical refinements of science. This book is written to be read by those who are interested in Comfrey, nature cure practitioners, herbalists, nurses, and housewives in all countries."*
-Comfrey: The Herbal Healer by Henry Doubleday Research Association, **1975**, 41 pages. (in 'Comfrey Report' book), page 4.

"Clinical studies: There is a lack of clinical research assessing the effects of Comfrey, and rigorous randomised controlled clinical trials are required."
-Herbal Medicines by Joanne Barnes, Linda A. Anderson and J. David Phillipson. London, England: Pharmaceutical Press, third edition **2007**, page 189.

"The therapeutic properties of Comfrey are based on its anti-inflammatory and analgesic effects. Comfrey also stimulates granulation and tissue regeneration, and supports callus formation*.
However, **the key activity-determining constituents of Comfrey extracts and its molecular mechanisms of action have not been completely elucidated (determined)***. Allantoin and rosmarinic acid are probably of central importance to its pharmacodynamic effects**.*
No clinical-pharmacokinetic investigation results in humans have been published so far on the absorption, distribution and elimination of the constituents of Comfrey extracts.*"*
-'Comfrey: A Clinical Overview' by Christiane Staiger in Germany, Phytotherapy Research, Volume 26, Issue 10, pages 1441-1448, February **2012**, page 1.
(* -'Commission E Monograph Phytotherapy, Comfrey Herb and Leaf, Symphytii Herba-Folium', published July 27, 1990 by the 'Bundesanzeiger', the official publication of the Federal Republic of Germany.)
(** -'Relating Antiphlogistic Efficacy of Dermatics Containing Extracts of Symphytum Officinale to Chemical Profiles' by P. Andres, R. Brenneisen and J.T. Clerc, Planta Medica, Volume 55, pages 66–67, 1989.)
   (Pharmacokinetics is study of time course of drug absorption, distribution, metabolism and excretion. Clinical pharmacokinetics is application of pharmacokinetic principles to safe and effective therapeutic management of drugs.)

*"***Symphytum species*** *belongs to the Boraginaceae family and have been used for centuries for bone breakages, sprains and rheumatism, liver problems, gastritis, ulcers, skin problems, joint pain and contusions, wounds, gout, hematomas and thrombophlebitis.* **The present review aims at summarizing the main data on the therapeutic indications of the Symphytum species based on the current evidence, also emphasizing data on both the efficacy and adverse effects.**
   The present review was carried out by consulting PubMed® (Medline®), Web of Science®, Embase®, Scopus®, Cochrane Database®, Science Direct® and Google Scholar® (as a search engine) databases to retrieve the most updated articles on this topic. All articles were carefully analyzed by the authors to assess their strengths and weaknesses, and to select the most useful ones for purpose of review, **prioritizing articles published from 1956 to 2018***.*
The pharmacological effects of the Symphytum species are attributed to several chemical compounds, among them allantoin, phenolic compounds, glycopeptides, polysaccharides and some toxic pyrrolizidine alkaloids. Not less important to highlight are the risks associated with its use. In fact, there is increasing consumption of over-the-counter drugs, which when associated with conventional drugs can cause serious and even fatal adverse events.
Although clinical trials sustain the folk topical application of Symphytum species in musculoskeletal and blunt injuries, with minor adverse effects, its antimicrobial potency was still poorly investigated. *Further studies are needed to assess the antimicrobial spectrum of Symphytum species and to characterize the active molecules both in vitro and in vivo."*
-'Symphytum Species: A Comprehensive Review on Chemical Composition, Food Applications and Phytopharmacology' by Bahare Salehi, et al.; Molecules: A Journal of Synthetic Chemistry and Natural Product Chemistry, Basel, Switzerland, Volume 24, No. 12, 2272, June **2019**.

Overview of Topical (Skin) Applications of Comfrey (in chronological order)

"*Aim:* To analyze the anti-inflammatory and analgetic (pain killing) properties of the topical (skin) Comfrey preparations Kytta-Salbe® f, Kytta-Plasma® f and Kytta-Balsam® f applied to bruises, sprains and distortions and painful conditions of the muscles and joints.
Method: A prospective open multicentric observational study complying with paragraph 67(6) of the AMG (Arzneimittelgesetz: Medicinal Products Act) and involving 162 general practitioners. **During the two-week period of observation, the patients received an average of one to three applications of the Comfrey preparation per day.** All 492 questionnaires were evaluated. Efficacy and tolerability were assessed by both physician and patient.
Results: **Pain at rest and on movement, as also tenderness, improved in the overall observation group by an average of 45-47%. The duration of morning joint stiffness decreased from 20 minutes initially to 3 minutes.**
During the course of treatment with Comfrey, more than two-thirds of the patients were able to reduce or even discontinue their intake of non-steroidal anti-inflammatory drugs (NAIDs) and other specific concomitant (at same time) medication. In most of the cases, both effectiveness and tolerability were assessed to be excellent or good.
Conclusion: The results of the study confirm the effectiveness and tolerability of the topical Comfrey preparation investigated in the treatment of bruises, sprains and distortions as well as painful conditions affecting muscles and joints."
-'Therapeutic Properties and Tolerability of Topical Comfrey Preparations: Results of an Observational Study' by R. Koll and S. Klingenburg, Richard-Zanders-Strasse, Bergisch Gladbach, Germany; MMW Fortschritte der Medizin (Advances in Medicine), Volume 144, pages 1-9, April **2002**. (I do not have any connection with these topical Comfrey preparations.)

> **Kytta-Salbe® f:** Comfrey root topical salve by Merck for sore muscles, 'beat-up' joints, tennis elbow, and lower back pain. Contains butyl-, ethyl-, methyl-, 2-methylpropyl-and propyl-4-hydroxybenzoate (parabens), peanut oil, cetostearyl alcohol, benzyl benzoate and bergamot.
> **Kytta-Plasma® f:** Comfrey root topical ointment by Merck for painful muscles and joints, bruises, strains and sprains.
> **Kytta-Balsam® f:** Comfrey root extract topical cream by Merck. Active ingredients: Comfrey and Methyl Nicotinate. For joint and muscle pain, knee arthrosis (osteoarthritis), and cartilage wear. Methyl Nicotinate is a nicotinic acid methyl ester, used as rubefacient in cosmetics to produce a redness or inflammation for a short period of time, often in lip plumpers or other lip products.
> **Merck & Company, Inc.**, doing business as Merck Sharp & Dohme (MSD) outside the United States and Canada, is an American pharmaceutical company and one of the largest pharmaceutical companies in the world. It was established in 1891 as the United States subsidiary of the German company Merck, which was founded in 1668.

"**Topical preparations containing Comfrey extracts are used for the treatment of various muscle and joint complaints.** There are numerous recent clinical studies and postmarketing surveillance studies available for these preparations.
They confirm the analgesic and anti-inflammatory as well as the anti-exudative (reduce fluid) spectrum of action of the medicinal plant and also prove its high degree of efficacy and tolerance. The findings are presented in this overview."
-'Comfrey A Modern Medicinal Plant' by C. Staiger, Merck Selbstmedikation GmbH, Darmstadt, Germany; Zeitschrift fur Phytotherapie (Journal of Phytotherapy), Germany, Volume 26, No. 4, pages 169-173, **2005**.
(I have this report in German with the above abstract in English. If you have an English translation, could you send it to me.)

"In the past already, **clinical trials have confirmed the use of topical Comfrey extract preparations for treatment of various muscle and joint complaints in traditional medicine.**
In recent years further clinical research was conducted, the latest results of which are presented in this overview."
-'Comfrey: State of Clinical Research' by C. Staiger, Merck Selbstmedikation GmbH, Darmstadt, Germany; Zeitschrift fur Phytotherapie (Journal of Phytotherapy), Volume 28, No. 3, pages 110-114, **2007**.
(I have this report in German with the above abstract in English. If you have an English translation, could you send it to me.)

"**The herbal monograph (detailed study) selects and summarises scientific studies and textbooks regarding efficacy, dosage and safety supporting the therapeutic uses of Comfrey root. This herbal drug by definition consists of the dried rhizomes and roots of Symphytum officinale L.** Studies with its main characteristic constituents allantoin, mucilage polysaccharides, phenolic acids, glycopeptides and triterpene saponins are included.
The therapeutic indications are pain and swelling related to muscle and joints; acute myalgia (muscle pain) in the back, strains, contusions (bruises) and distortions (sprains), epicondylitis (one type is tennis elbow), tendovaginitis (tendon inflammation) and periarthritis (one type is frozen shoulder).
Administration of Comfrey root addresses posology (pharmacology dosages); its duration of use; contra-indications; special warnings; special precautions for use; interactions with other medicinal products; other forms of interaction; in pregnancy and lactation; its effects on ability to drive; undesirable effects; overdose.
> In vitro experiments with Comfrey root extracts demonstrate **anti-inflammatory, wound healing and antibacterial properties.** In vivo experiments with animals indicate anti-inflammatory and wound healing activities. A pharmacological study in humans concerned effects related to wound healing.

Several controlled clinical studies with topically (skin) applied Comfrey root extract demonstrated its therapeutic efficacy in patients with sprained ankles, painful joint complaints and osteoarthritis of the knee.
Preclinical safety data for Comfrey root were assessed in toxicity studies and indicated that 99% of pyrrolizidine alkaloids should be removed from its extracts. Safety data were assessed in human studies. Adverse effects of topically used Comfrey root extracts were limited to skin reactions.
Medical knowledge is ever-changing. As new research and clinical experience broaden our knowledge, changes in treatment may be required.

In their efforts to provide information on the efficacy and safety of herbal drugs and herbal preparations, presented as a substantial overview together with summaries of relevant data, the authors of the material herein have consulted comprehensive sources believed to be reliable.
The material complies with the monograph of the 'Deutscher Arzneimittel-Codex: Beinwellwurzel -Symphyti radix' ('German Drug Codex: Comfrey Root'), 1979 or 'British Herbal Pharmacopoeia: Comfrey Root -Symphyti radix', 1996."
-'Symphyti Radix (Comfrey Root)' by ESCOP Monographs: The Scientific Foundations for Herbal Medicinal Products, European Scientific Cooperative on Phytotherapy, Exeter, United Kingdom, 16 pages, **2012**.

"*Posology (pharmaceutical dosage):*
External use only. Adults and children from 3 years: ointments or other preparations containing up to 35% of Comfrey root extract (1:2, ethanol 60% V/V {volume of solute per volume of solvent}), applied 3-4 times daily (Koll 2002; Koll 2004; Predel 2005; Staiger 2005; Staiger 2007; Grube 2007; Staiger 2008; Giannetti 2010).
Some national authorities restrict the external use of Comfrey root. The German health authorities, for example, limit the content of pyrrolizidine alkaloids (PA) (including N-oxides) with 1,2-unsaturated necine structures in the amount of a preparation to be applied daily to not more than 100 microgram. There are no restrictions for preparations containing less than 10 microgram PA (Schilcher 2010).
Method of administration: For topical (skin) application only.
Duration of use: The German health authorities recommend not more than 4-6 weeks per year if the daily application contains between 10 microgram and 100 microgram of pyrrolizidine alkaloids (including their N-oxides) with 1,2-unsaturated necine structures (Teuscher 2009; Schilcher 2010).
Contraindications: None known.
Special warnings and special precautions for use: In cases of wound healing and skin inflammation, PA-free (pyrrolizidine alkaloids) Comfrey root preparations should be used.
Interaction with other medicaments and other forms of interaction: None reported.
Pregnancy and lactation: As no human data are available, the potential risk is unknown. In accordance with general medical practice the product should not be used during pregnancy and lactation without medical advice."
-'Symphyti Radix (Comfrey Root)' by ESCOP Monographs: The Scientific Foundations for Herbal Medicinal Products, European Scientific Cooperative on Phytotherapy, Exeter, United Kingdom, 16 pages, **2012**.

"Comfrey (Symphytum officinale L.) has been used over many centuries as a medicinal plant. In particular, the use of the root has a longstanding tradition.
**Today, several randomised controlled trials have demonstrated the efficacy and safety. Comfrey root extract has been used for the topical treatment of painful muscle and joint complaints.
It is clinically proven to relieve pain, inflammation and swelling of muscles and joints in the case of degenerative arthritis, acute myalgia (muscle pain) in the back, sprains, contusions (bruises) and strains after sports injuries and accidents**, also in children aged 3 years and older. This paper provides information on clinical trials, non-interventional studies and further literature published on Comfrey root till date."
-'Comfrey Root from Tradition to Modern Clinical Trials' by Christine Staiger, Wiener Medizinische Wochenschrift (Vienna Medical Weekly, Austria), 163, pages 58–64, **2013**. (A contusion is a blow from a blunt instrument where the skin is not broken.)

"*Background:* **External preparations of the herb Comfrey (most commonly Symphytum officinale L.)** are widely available for over-the-counter, practitioner and healthcare professional usage.
Traditional practice suggests Comfrey can be used to treat musculoskeletal disorders, wounds and various other conditions; however a full and critical coverage of the evidence base has not yet been undertaken.
Methods: A critical scoping review was undertaken. Six bibliographic databases, 10 grey literature databases and nine trials registers were searched plus reference lists of included studies.
Results: Of 1348 identified records, 64 full texts were screened for inclusion and 26 were included in the review- 13 RCTs, 5 non-randomised controlled trials and 8 observational studies evaluating treatments for ankle distortion (sprained ankle), back pain, abrasion wounds (scrapes), venous leg ulcers and osteoarthritis. **The majority of included trials had an overall unclear risk of bias due to poor quality of reporting. Few adverse events were reported.**
Conclusions: **Individual clinical trials showed evidence of benefit for ankle distortion, back pain, abrasion wounds and osteoarthritis. Topical (skin) application appears to be safe but further rigorous assessment is needed.**"
-"A Critical Scoping Review of External Uses of Comfrey (Symphtum spp.)' by R. Frost, H. MacPherson and S. O'Meara, Complementary Therapies in Medicine, University of York, United Kingdom, Volume 21, Issue 6, pages 724-745, December **2013**.
(Grey literature are materials and research produced by organizations outside of the traditional commercial or academic publishing and distribution channels. These organizations include government departments and agencies, civil society or non-governmental organisations, academic centers and departments, and private companies and consultants.)

"**Cosmetic products: Symphytum officinale extract is used as a soothing agent, as an emollient and against skin impurities, in concentrations of 0.5-4%** (Council of Europe 2008). **The defined functions of Comfrey according to 'CosIng' are: skin conditioning, abrasive (tough cleaning), antidandruff, soothing.**
Several products (>100) were identified containing Symphytum officinale in the databases EWG's Skin Deep® and Codecheck®. These comprise the following product categories: -Cream -Moisturizer -Facial moisturizer -Facial cleanser -Shampoo -Hair colour and bleaching -Mask -Body wash/cleanser -Toners."

-'Risk Profile Symphytum Officinale Extracts', CAS No. 84696-05-9, www.mattilsynet.no, Statens tilsyn for planter, fisk, dyr og naeringsmidler (State supervision of plants, fish, animals and food), Brumunddal, Norway, March 11 **2013**.
  ('CosIng' {Cosmetic Ingredient database} is the European Commission database for information on cosmetic substances and ingredients.)

"The pharmacologic effects and clinical efficacy of topical (skin) Comfrey preparations are supported by several studies that evaluated its effects in the treatment of pain, inflammatory conditions, and cutaneous (slkin) wounds.
In these trials, S. officinale (Common Comfrey) proved efficacious (effective) as an anti-inflammatory and analgesic (pain killing) wound healer and promoter of granulation.
Phytochemical analyses have isolated several compounds, including allantoin, choline, triterpenoids, saponins, rosmarinic acid (RA) derivatives, tannins, and essential oils, that may be partly responsible for the pharmacologic properties of Comfrey due to their known biologic activities.
Rosmarinic acid may be responsible for the anti-inflammatory properties of Comfrey, though its wound-healing action is attributed to allantoin. Other components such as saponins possess antibacterial and anti-edematogenic properties, and choline causes vasodilation.
Comfrey seems to be a promising strategy for treating various skin problems based on its traditional use in ethnopharmacology as well as the biologic effects of its compounds; *however, preclinical and clinical trials still need to be developed to increase knowledge of the effectiveness and safety of the topical application of Comfrey."*
-'Botanical Briefs: Comfrey (Symphytum Officinale)' by Cíntia D.S. Horinouchi, M.Sc. and Michel F. Otuki, Ph.D., Cutis (clinical journal for the dermatologist, allergist and general practitioner), Volume 91, No. 5, pages 225-228, May **2013**.

<u>**Positive Medical Research about Comfrey**</u> (For Allantoin research see section 'Comfrey Heals: Allantoin' {Chapter 26}.)

  Antibacterial / Antimicrobial (see Infection)

  Antifungal

"Antifungal activity was determined by the tube dilution method on **different extracts of Symphytum sylvaticum Boiss. subsp. sepulcrale (Boiss. & Bal.) Greuter & Burdet var. sepulcrale**. *The antifungal activity of the isolated compound from the root alkaloid fraction (Echimidine-N-Oxide) was also tested against ten fungal cultures.*
The antifungal activity was found to be mainly due to Echimidine-N-Oxide (ENO).
The quantitative determination of ENO was also carried out with capillary gas chromatography in the roots and the aerial (above ground) parts of the plant.
In each extract some activity was observed for certain fungi. **However, evaluation of the extracts for antifungal property clearly indicated that the fungicidal activity was maximum in root alkaloid extract (ARF). This may lead to the conclusion that the alkaloids of this plant are responsible for this activity.**
Traditionally, the aqueous extract of Heliotropium (Boraginaceae) species are used topically for fungal infections of the feet in Turkey. This usage is generally attributed to pyrrolizidine alkaloids.
Test microorganisms: *Anti-fungal studies were carried out against Epidermophyton floccosum, Microsporum canis, Nigrospora oryzae, Allefsheria boydii, Pleuretus ostreatus, Stachbotrys atra, Curvularia lunata, Drechslera rostrata, Aspergillus niger and Candida albicans.*
As estimated from the results of the extracts, strong inhibitory activity for nine fungal cultures out of ten (except C. albicans) was demonstrated for ENO at 200 microgram/ml.
Dried samples of Symphytum sylvaticum Boiss. subsp. sepulcrale (Boiss. & Bal.) Greuter & Burdet var. sepulcrale plant gave a yield of total alkaloids of 0.199% in the roots and 0.093% in the aerial parts.
The major alkaloid in the roots was Echimidine-N-oxide (ENO). The ENO content was higher in the roots (0.1078%) than in the aerial parts of the plant (0.0061%).
The yield of ENO was 55.96-57.54% in the root and 6.66-6.86% in the aerial part crude alkaloid fraction. This explains why the alkaloid fraction from the root was more active than the alkaloid fraction from the aerial part compared with the fungal cultures used. Consequently, the activity was found to be mainly due to Echimidine-N-Oxide (ENO)."
-'Antifungal Activities of Different Extracts and Echimidine-N-oxide from Symphytum Sylvaticum Boiss. subsp. sepulcrale (Boiss. & Bal.) Greuter & Burdet var. Sepulcrale' by Murat Kartal (Turkey), Semra Kurucu (Turkey) and M. Iqbal Choudary (Pakistan), Turkish Journal of Medical Sciences, Ankara, Turkey, Volume 31, pages 487-492, **2001**.

Anti-Inflammatory (see Inflammation)

Antioxidant (Free Radical Scavenger) Properties

Antioxidants are molecules that inhibit oxidation of other molecules. Oxidation is a chemical reaction that produces free radicals. Free radicals are highly reactive and unstable compounds that damage cell components such as DNA, proteins and lipids.

"**Symphytum officinalis 8% and 10%** (mass concentration) hydro-alcoholic extracts in 50% ethylic alcohol and 50% methylic alcohol were obtained. They were further purified and concentrated through membrane procedures (ultrafiltration).
In all types of extracts were assessed: the content of total polyphenols, of flavones, as well as the antioxidant capacity through two procedures (with ABTS and DPPH). The raw polysaccharides were extracted from roots of **Symphytum asperum and S. caucasicum,** and fractionated by ultrafiltration (UF), using filtering membranes of different cut-offs.

A higher amount of **flavones** was extracted in ethanol extracts than in methanol ones, an amount which increased as increased the ethylic alcohol concentration. The extract with highest amounts of flavones was the 10% hydro-alcoholic (mass concentration) extract obtained in 50% ethylic alcohol. More optimum flavones concentrations were achieved with 5000 Da Millipore than PSF membranes.

The ethanol extracts had higher **polyphenol** amounts than the methanol ones- similar to flavones. The richest extract in polyphenols was that of 10% plant mass in 50% ethanol, as in the flavones case. In case of polyphenols, a better concentration was realized on PSF membrane.

By both tested methods, the highest radical scavenging activity and the highest DPPH inhibition percent had the 10% mass concentration ethanol extracts in 50% ethylic alcohol.

A correlation between the amount of tested biologic active principles- flavones and polyphenols -and the radical scavenging activity of the analyzed extracts exists, the same ethanol extracts of 10% mass concentration and 50% ethylic alcohol also having the highest amount of such substances. The obtained inhibition values of DPPH varied from 62.45% for the initial extract 8% EtOH of **Symphytum officinalis,** till 95% for 10% EtOH extract UF2 concentrated.

It was concluded that by use of the membranare processes to concentrate the biologic active principles (polyphenols and flavones) Symphytum officinalis concentrated extracts were obtained with very high radical scavenging activity - above 90% DPPH inhibition."
-'Antioxidant Capacity of some Symphytum Officinalis Extracts Processed by Ultrafiltration' by Elena Neagu, Gabriela Paun Roman, Gabriel Lucian Radu, University of Bucharest, Romania, Romanian Biotechnological Letters, Volume 15, No. 4, pages 5505-5511, **2010**.

(Flavones are a class of flavonoids found in spices, and red-purple fruits/vegetables. Flavonoids, polyphenolic compounds in plants, are antioxidants that may reduce the risk of disease. They are anti-inflammatory. Polyphenols are antioxidant phytochemicals.)

"**Symphytum officinale L.,** is a perennial plant from Boraginaceae family that has been used in medicine for treatment of painful joints and muscles, menstruation pain and bronchial problems in addition to stimulating healing of wounds, among other effects. On the other hand, **the presence of polyphenols, triterpenoids and tannins in this specie allows us to think that it could be a promising source of natural compounds with high antioxidant activity.**
We evaluate the antioxidant potential of the extracts and the essential oil from S. officinale.
Dry and powdered S. officinale leaves underwent successive extractions with ethanol and hexane and the essential oil was obtained by hydrodistillation. The antioxidant potential was determined on the trapping capacity of stable radical 2-2-diphenyl-1-picrylhydrazyl (DPPH assay). The samples showed EC50 values from 0.06 to 0.44 gram sample/micromol DPPH, comparable to natural extracts like the essential oils from oregano and rosemary and to standard substances like ascorbic acid.
S. officinale leaves showed an important antioxidant effect, and this result support the medicinal properties of the plant in tissue recovery and therapy for muscular pains and others pathologies."
-'In Vitro Antioxidant Capacity of Comfrey (Symphytum Officinale L.)' by M.A. Puertas-Mejia, J.F. Zuleta-Montoya and F. Rivera-Echeverry, Universidad de Antioquia, Medellin, Colombia; Revista Cubana de Plantas Medicinales (Cuban Journal of Medicinal Plants), Volume 17, No. 1, pages 30-36, **2012**.
(This report is in Spanish with an abstract in English.) (EC50 is concentration of a drug that gives half-maximal response.)

"**This paper shows the antioxidant activity of different extracts of marigold (Calendula officinalis L., Asteraceae), Comfrey (Symphytum officinale L., Boraginaceae) and yarrow (Achillea millefolium L., Asteraceae).**
As the activity and availability of antioxidants are directly linked to their chemical structure and solubility, the effectiveness of n-butanolic, ethyl acetate, ether, chloroform and water extracts of marigold, Comfrey and yarrow **against lipid peroxidation (LP), hydroxyl (OH) and 2, 2-diphenyl-1- picrylhydrazyl (DPPH) radicals** were investigated, as well as the effectiveness of water, n-butanolic and ethanolic extracts against nitric oxide (NO) radical.
The extracts had greatly inhibited LP (more than 70%) in liposome. The most successful were 10% n-butanolic extract of Comfrey, 0.5% chloroform extract of Comfrey and 10% n-butanolic extract of yarrow. We noted that scavenging activity against hydroxyl radical in liposome was almost twice as weak as the effectiveness against LP.
The most successful were 2.5% ethyl acetate extract of marigold, 10% water extract of Comfrey and 10% chloroform extract of marigold. Ethyl acetate extract of yarrow was most effective against DPPH radical, followed by n-butanolic extract of yarrow and **ethyl acetate extract of Comfrey.** However, water extract of marigold was the most effective against NO radical, followed by ethanolic extract of yarrow and **water extracts of yarrow and Comfrey.**
It can be concluded that all three investigated plant species contain compounds that produce considerable antioxidant activity, and because of their structural diversity, antioxidant activity of investigated extracts was strongly affected by the solvent used for the extraction and antioxidant test applied."
-'Antioxidant Activity of Different Extracts of Marigold, Comfrey and Yarrow' by S. Boskovic and N. Mimica-Dukic, Faculty of Environmental Protection, Educons University, Serbia; Fresenius Environmental Bulletin, Germany, Volume 22, No. 6, pages 1731-1735, January **2013**.

(A liposome is a tiny round sac of phospholipid molecules enclosing a water drop, especially as formed artificially to carry drugs into tissues.)

"**The antioxidant activity of plant extracts is of particular interest both of their beneficial physiological activity on human cells and the potential they have to replace synthetic antioxidants used in foodstuff** (Amarowicz, 1999).
Most of the methods of determination of total antioxidant activity characterize the ability of the tested compound or product to scavenge free radicals and to complex metal ions driving the oxidation process (Tirzitis and Bartosz, 2010).
The aim of this study to demonstrate the antioxidant potential of methanolic, ethanolic and aqueous (water) extracts of plants of Symphytum L. species. The DPPH radical scavenging effect was assessed by the discoloration of methanol solution of 2.2-diphenyl-1-picrylhydrazyl after 10 minutes according to Brand-Williams et al. (1995).
Total Antioxidant Activity (TAA) of above-ground part of Symphytum asperum Lepech. during vegetation was from 73.59% (spring vegetation stage) to 79.65% (fruiatage) in methanolic extracts, from 17.53% (budding stage) to 36.31% (spring vegetation stage)- in ethanolic extracts, from 59.64% (spring vegetation stage) to 76.15% (stage of stem growth) - in water extracts.
Methanolic **extracts of Symphytum caucasicum Bieb.** showed the TAA from 75.48% (budding stage) to 80.81% (spring vegetation stage), ethanolic extracts - from 22.61% (blossoming stage) to 73.99% (spring vegetation stage), water extracts - from 67.46% (budding stage) to 74.45% (blossoming stage).
In plant raw material of **Symphytum x uplandicum Nyman** TAA was from 16.24% (blossoming stage) to 79.18% (budding stage) in methanolic extracts, from 5.64% (blossoming stage) to 21.06% (budding stage)– in ethanolic extracts, from 17.23% (stage of stem growth) to 60.58% (budding stage)- in water extracts.
Maximal sign of radical inhibition was noticed in methanolic extracts of Symphytum caucasicum in stage of spring vegetation and minimal in ethanolic extracts of Symphytum x uplandicum in blossoming stage. Obtained data allow to use these plants as plant raw material with antioxidant potential."
-'Total Antioxidant Activity Of Plants of Symphytum L. Species' by V. Olena, B. Jan and R. Djamal, from Gryshko National Botanical Garden of NAS of Ukraine, and Slovak University of Agriculture in Nitra; Agrobiodiversity, pages 488-492, November **2017**.   (A complex ion has a metal ion at its center with other molecules or ions surrounding it.)

"Comfrey root preparations are used for the external treatment of joint distortions and myalgia, due to its analgesic and anti-inflammatory properties. **Up to date, key activity-determining constituents of Comfrey root extracts have not been completely elucidated (determined). Therefore, we applied different approaches to further characterize a Comfrey root extract (65% ethanol).**
The phenolic profile of Comfrey root sample was characterized by HPLC-DAD-QTOF-MS/MS. **Rosmarinic acid was identified as main phenolic constituent (7.55 mg/g extract).**
Moreover, trimers and tetramers of caffeic acid (isomers of salvianolic acid A, B and C) were identified and quantified for the first time in Comfrey root.
In addition, pyrrolizidine alkaloids were evaluated by HPLC-QQQ-MS/MS and acetylintermedine, acetyllycopsamine and their N-oxides were determined as major pyrrolizidine alkaloids in the Comfrey root sample.
Lastly, the antioxidant activity was determined using four assays (tests): DPPH and ABTS radicals scavenging assays, reducing power assay and 15-lipoxygenase inhibition assay.
Comfrey root extract exhibited significant antioxidant activities when compared to known antioxidants. Thus, Comfrey root is an important source of phenolic compounds endowed with antioxidant activity which may contribute to the overall bioactivity of Symphytum preparations."
-'Is Comfrey Root More Than Toxic Pyrrolizidine Alkaloids? Salvianolic Acids Among Antioxidant Polyphenols in Comfrey (Symphytum officinale L.) Roots' by Adriana Trifan, et al., Food and Chemical Toxicology, Volume 112, pages 178-187, February **2018**.
    (In organic chemistry, phenols or phenolics, are a class of chemical compounds of a hydroxyl group {-OH} bonded directly to an aromatic hydrocarbon group. Phenolic compounds are classified as simple phenols or polyphenols based on the number of phenol units in the molecule.)
    (Rosmarinic acid is a plant chemical that is an anti-oxidant. It is caffeic acid ester of 3-(3,4-dihydroxyphenyl)lactic acid.)
    (Caffeic acid is an organic compound classified as a hydroxycinnamic acid that consists of both phenolic and acrylic functional groups. It is found in all plants because it is an intermediate in the biosynthesis of lignin, one of the principal components of woody plant biomass.)
    (Due to their polyphenolic structure, salvianolic acids are thought to be free radical scavengers.)

"**The root of Symphytum officinale L. is commonly used in folk medicine to promote the wound healing, reduce the inflammation and in the treatment of broken bones.** The objective of our investigation was to analyse the extract from S. officinale in terms of its antioxidant activity and the effect on cell viability and proliferation of Human Skin Fibroblast (HSF). Moreover, the quantification of main phenolics and allantoin was conducted using HPLC–DAD method.
Five compounds were found: rosmarinic, p-hydroxybenzoic, caffeic, chlorogenic and p-coumaric acid. DPPH, FRAP and TPC assay showed the **high antioxidant activity of the extract.** MTT test proved the stimulatory effect on cell metabolism and viability of HSF cells. Moreover, no changes in cytoskeleton structure and cells shape were observed.
The obtained results indicate that non-toxic extract from S. officinale root has strong antioxidant potential and a beneficial effect on human skin fibroblasts."
-'Proliferative and Antioxidant Activity of Symphytum Officinale Root Extract' by Ireneusz Sowa, et al., Lublin, Poland; Natural Product Research, Volume 32, No. 5, pages 605-609, **2018**.
    (A fibroblast is a cell in connective tissue that produces collagen and other fibers.)

Blood Pressure and Blood Flow

*"During a preliminary screening for hypotensive (lower blood pressure) activity in anesthetized rats, the ethanolic extract of Symphytum officinale was found active. Consequently, bioassay-directed fractionation of the ethanolic extract of S. officinale resulted in the isolation of **symphytoxide-A, a new triterpenoidal saponin, and two known saponins, leontoside-A and -B**. The structures of these saponins were established on the basis of ID and 2D NMR spectroscopy. In anesthetized rats, the ethanolic extract, fractions, and saponins caused a **fall in blood pressure in a dose-dependent manner.***

In isolated guinea-pig atria (heart cavity), saponins produced inhibitory (reducing) effects on force and rate of contractions while in smooth muscle preparations such as guinea-pig ileum (part of small intestine) and rat uterus, they induced stimulant responses. These results suggest that Symphytum officinale contains active principles (saponins) which may explain the hypotensive effects observed in the in vivo studies."

-Saponins Used in Traditional and Modern Medicine by Manuel F. Balandrin with editors George R. Waller and Kazuo Yamasaki. Part of 'Advances in Experimental Medicine and Biology Series 404'. New York: Springer, **1996**. Chapter: Phyto-Pharmacology of Saponins from Symphytum Officinale L. by Khalid Aftab, Fehmeena Shaheen, Faryal Vali Mohammad, Mushtaq Noorwala and Viqar Uddin Ahmad; H.E.J. Research Institute of Chemistry, University of Karachi, Pakistan, pages 429-442.

(Saponins are amphipathic glycosides grouped by the soap-like foam they produce when shaken in water. A glycoside is a molecule where a sugar is bound to another functional group by a glycosidic bond.)

*"**Diseases Related to Blood Flow and Nerve Problems:***

***The involvement of sodium-hydrogen exchangers (NHE) has been described in the pathophysiology of diseases including ischemic heart and brain diseases, cardiomyopathy, congestive heart failure, epilepsy, dementia, and neuropathic (nerve) pain.** Synthetic NHE inhibitors have not achieved much clinical success; therefore, plant-derived phytoconstituents may be explored as NHE inhibitors.*

Methods of Experiment:

*In the present study, **the NHE inhibitory potential of hydroalcoholic and alkaloidal fractions of Malus domestica, Musa x paradisiaca, Daucus carota, and Symphytum officinale was evaluated.** The different concentrations of hydroalcoholic and alkaloidal extracts of the selected plants were evaluated for their **NHE inhibitory activity in the (blood) platelets** using the optical swelling assay.*

Results:

*Among the hydroalcoholic extracts, the highest NHE inhibitory activity was shown by M. domestica followed by Musa x paradisiaca, D. carota, and S. officinale. Among the alkaloidal fractions, the highest NHE inhibitory activity was shown by the alkaloidal fraction of Musa x paradisiacal followed by D. carota, M. domestica, and S. officinale. The IC50 of alkaloidal fractions was comparable to the IC50 of synthetic NHE inhibitor, EIPA [5-(N-ethyl-N-isopropyl)amiloride]. **It may be concluded that the alkaloidal fractions of these plants possess potent NHE inhibitory activity and may be exploited for their therapeutic potential in NHE activation-related pathological complications."***

-'Sodium-Hydrogen Exchanger Inhibitory Potential of Malus domestica, Musa x paradisiaca, Daucus carota, and Symphytum officinale' by Vivek Verma, Nirmal Singh and Amteshwar Singh Jaggi, Journal of Basic and Clinical Physiology and Pharmacology, Germany, Volume 25, No. 1, pages 99-108, **2014**.

(Ischemia is not enough blood supply to an organ or part of the body, especially heart muscles.)

Blunt Trauma / Injury (see Bruises or see Sprains)

Broken Bones (Fracture)

*"**Symphytum officinale, Comfrey: Osseous (Bone) Fractures: Compresses applied three times daily or 3-4 ml tincture three times daily (for no more than six weeks).** If sufficient area can be left exposed during the casting process, topical (skin) applications of Symphytum officinale (Common Comfrey) poultices may help speed the healing process. Sometimes, it is possible to leave a window in the cast to allow such access; other times, it is not.*

Homeopathic Symphytum is used internally to similar ends. Internal use of Comfrey can also be beneficial for speeding fracture repair. Unfortunately, internal Comfrey use has been accused in the literature of being a cause of hepatic (liver) venoocclusive disease. A close analysis of this information reveals that patients either took large doses for long periods of time or had preexisting liver disease.

Short-term Comfrey use (6 weeks or less) of moderate doses (3-4 ml three times a day of tincture) carries little to no risk of causing hepatic damage in patients with no concomitant (at the same time) liver disease or not on hepatotoxic (liver toxic) drugs."

-'Botanical Medicine and Sports Injuries' by Eric Yarnell, M.D., Alternative and Complementary Therapies, 3(3), pages 183-186, June **1997**.

(A compress is a cloth or other material applied with pressure to skin and held in place. It can be any temperature and can be dry or wet. It can have medication or an herbal remedy.)

(A poultice is a soft moist mass, often heated and medicated, that is spread on cloth over skin to treat injury or disease.)

"Symphytum officinale Linn. (Knit Bone) removes the inflammation surrounding the fracture.
It induces (causes) union of affected parts and contains allantoin, a crystallizable substance used in orthodox (approved modern) medicine to encourage epithelial formation in ulcer and wound."
-'Recent Update on Proficient Bone Fracture Revivifying Herbs' by S.S. Chhavi, A. Drabu, R. Verma, A. Dhiman and A. Sharma, all from India; International Research Journal of Pharmacy, Rampur, India, Volume 2, No. 11, pages 3-5, **2011**.
(Epithelial tissues are sheets of cells covering all the surfaces of the body exposed to the outside world and lining the outside of organs.)

*"**The study was undertaken to evaluate the efficacy of Symphytum officinale as a stimulating agent in the process of long bone fracture healing.** Twelve clinical cases of long bone fracture in dogs presented at 'Teaching Veterinary Clinical Services Complex', Nagpur (India) reported during February 2014 to July 2014, were divided into two equal groups and subjected to the treatment as per schedule.*
The method adopted for the administration of the Symphytum officinale homoeopathy drug of 30 c potency for oral administration in the dogs with fracture of long bones** during the present study was found suitable and easy and the quantity of globules administered **was found satisfactory as an osteoinducer.
***In group II, in addition to immobilization of fractured bone, the Symphytum Officinale homoeopathy drug was administered** at four globules twice a day for days for 21 days postoperatively.*
***The animals of group II with long bone fracture required less time for partial weight bearing** 3.25 +- 0.25 postoperative days and complete weight bearing 25.75 +- 2.17 postoperative days as compared to group I and reduction in weight bearing was statistically significant.*
***Radiographs (X-ray or similar) taken at schedule intervals showed accelerated fracture healing with complete bridging of fragment, bony deposition** along with periosteal and medullary continuity across the fracture site on animals of treated group II whereas radiographic union and periosteal bridging at fracture site were relatively slow and incomplete in dogs of group I at schedule interval.*
***The animals of group I and group II revealed non-significant changes of hematological values** such as hemoglobin (iron in red blood cells), total erythrocyte (red blood cell) count, total leucocyte (white blood cell) count, differential leucocyte count which does not affect the bone healing process. The serum (blood) alkaline phosphatase level revealed decreasing trend in both the group I and II which was statistically significant.*
*Chhavi et al. (*2011) reported that **Symphytum officinale removes the inflammation surrounding the fracture.** It induced union of affected bones, and contains allantoin, a crystallizable substance used in orthodox medicine to encourage epithelial formation in ulcer and wound."*
-'Role of Symphytum Officinale as an Osteoinducer in Long Bone Fracture Repair in Canine' thesis by Swarop Chandel for Master of Veterinary Science degree, Department of Veterinary Surgery and Radiology, Nagpur Veterinary College, India, **2014**.
(* -'Recent Update on Proficient Bone Fracture Revivifying Herbs' by S.S. Chhavi, A. Drabu, R. Verma, A. Dhiman and A. Sharma, all from India; International Research Journal of Pharmacy, Rampur, India, Volume 2, No. 11, pages 3-5, 2011.)
(Osteoinduction is the process by which osteogenesis is induced for bone healing to take place. Immature cells are stimulated to develop into pre-osteoblasts so that bone starts growing.)

Bruises (Contusions, Blunt Injury)

*"**Comfrey extract for treating blunt injuries in children:** In a post-marketing surveillance study the tolerability and efficacy of an ointment containing a fluid extract of Comfrey root were investigated in 306 children aged 3 to 12 years.*
All assessed clinical symptoms improved considerably during the treatment. The topical application of the investigated medicinal product is therefore considered to be effective and safe also for children."
-'Beinwellinder Therapie: Stumpfer Traumen Anwendung bei Kindern' (Comfrey in Blunt Trauma Therapy Use in Children) by Christiane Staiger and Tankred Wegener, Zeitschrift fur Phytotherapie (Journal of Phytotherapy), 29, pages 58-64, in German, **2008**. (All of the report is in German except for this English translation of the summary. I tried using an online translator but it was not clear. If you have an English translation, could you please send it to me.)

*"In an open observational study the therapeutic applicability and safety of **application of a topical (skin) cream preparation from the aerial (above ground) plant parts of the Comfrey cultivar 'Symphytum x uplandicum Nyman Harras'** was tested in 196 children in the age of 4 to 12 years with respect to the **paediatric treatment of acute blunt trauma (contusions, strains and distortions).***
*The extent of symptom improvement observed in this study (pain on palpitation/touch, pain in motion, functional impairment, oedema and haematoma) was found in the range of 84.5 to 100% for the single parameters ($p < 0.001$) with at the same time an excellent tolerability and compliance. **Topical Comfrey cream was therefore confirmed as a remedy of choice for the treatment of acute blunt trauma in children.**"*
-'Application and Safety of Comfrey Cream in Paediatric Treatment of Acute Blunt Trauma' by Grunwald, Bitterlich, Nauert and Schmidt, Zeitschrift fur Phytotherapie (Journal of Phytotherapy), Volume 31, No. 01, pages 61-66, November **2009**.

*"**Comfrey root has a strong historical record in the treatment of blunt injuries due to its anti-inflammatory, de-swelling***

and pain-relieving properties. The efficacy granted by the Commission E has been further substantiated with clinical data."
-'Comfrey Root from Tradition to Modern Clinical Trials' by Christine Staiger, Wiener Medizinische Wochenschrift (Vienna Medical Weekly, Austria), 163, pages 58–64, **2013**.

>(The German Commission E is a scientific advisory board of the 'Bundesinstitut für Arzneimittel und Medizinprodukte' {German equivalent of the USA 'Food and Drug Administration' FDA}. For more about Commission E see the subsection 'Germany Bans Some Uses of Comfrey' in section 'Some Uses of Comfrey Banned'.)

"The use of preparations made from **Comfrey (Symphytum officinale L.) for medicinal purposes** has been documented ever since antiquity. A first concentration of the formerly broad reported uses was made by experts of **the official German 'Commission E', which acknowledged an external use in cases of contusions, bruises and sprains**.
Meanwhile several randomised, placebo- and also reference-controlled studies **confirm the efficacy and safety of topically used Comfrey root extract for osteoarthritis of the knee, back pain and sprains, contusions and strains after sports injuries and accidents both in adults and in children above 3 years.**
These studies were exclusively obtained for a 60% ethanolic **extract of the fresh roots of Comfrey where almost all of toxic pyrrolizidine alkaloids had been removed by a special method.**
The reported controlled studies show that when using a special extract from the fresh root drug from Symphytum officinale L. reliable effects when applied to arthritic disorders of the knee joint, in back pain as well as in sport and accident injuries are to be expected.
Many years experience from the postmarketing as well as from the controlled ones and also the open studies indicate a high application of security as well as compatibility. This also applies to the use in children from 3 years."
-'Comfrey Root, Symphytum Officinale L., Radix: An Update' by T. Wegener and B. Deitelhoff, Bruckstr, Weinheim, Germany; Zeitschrift fur Phytotherapie (Journal of Phytotherapy), Volume 39, No. 1, pages 5-13, **2018**.
(I have this report in German that has this English translation of the abstract. I used an online translator for some of this. If you have an English translation of it or relevant parts of it, please send it to me.)

>(A placebo is a harmless and useless substance in the form of a pill/capsule or liquid that is given for the psychological benefit of the patient rather than for any medicinal effect. In clinical trials it is used as a control or baseline to compare against a similar substance given that does have medicine.)

Burns

"1. **Pharmacological study of burn wound healing activity of ointment containing high-molecular polysaccharide (and novel biopolymer) fractions from Comfrey roots (Symphytum asperum and S. caucasicum)**, free of allantoin and pyrrolyzidine alkaloids, revealed that by efficiency it is superior to allantoin ointment.
2. The obtained results allow to assume that established burn healing activity of the composition is associated with synergistic action of its constituents due to the shortening of the inflammatory phase of wound repair.
3. The established pharmacological action of ointment allows **recommending it for treatment of second and third-degree burns.**"
-'Burn Healing Compositions from Caucasian Species of Comfrey (Symphytum L.)' by Mulkijanyan, Barbakadze, Novikova, Sulakvelidze, Gogilashvili, Amiranashvili, and Merlani, Bulletin of the Georgian National Academy of Sciences, Tbilisi, Georgia (the country in Caucasus region of Eurasia), Volume 3, No 3, pages 114-117, **2009**.

>(Synergistic means the interaction of two or more substances that create a combined effect greater than the sum of their separate effects. In other words, the whole is greater than the sum of its parts.)

"**Survey Results: The experience with B & W Ointment® and LT® (Burdock Leaf Therapy) treatment of mild to severe burns** was related by 32 respondents from Anabaptist communities residing in eight states in the United States (Illinois, Indiana, Kentucky, Missouri, New York, Tennessee, Pennsylvania, and Wisconsin) and one Canadian province (Ontario). The 32 respondents reporting included both males and females, with an age range of 26-79 years. The respondents represented different roles in the community such as caregivers, household members, and community leaders.
The respondents reported positive results with the use of the B & W Ointment® and LT®, including minimal scarring, limited pain, and rapid healing. Three of the respondents reported the appearance of new skin growth or healing within 5-17 days; 5 respondents reported healing within 3-6 weeks even with severe burns.
They described positive experiences with dressing changes as the gauze did not stick to the new skin when B & W Ointment® and LT® were used. They reported that pain was also limited in children who were treated with this nonconventional method, with the children ceasing to cry when the dressing was changed. One respondent reported that the pain was 'gone' within 29 minutes after the application of treatment following several hours of pain associated with 'shaking'."
-'Treatment of Burns with Burns & Wounds (B & W) Ointment® and Leaf Therapy' by Maria E. Main, Deborah Williams and M. Susan Jones, Western Kentucky University, Nursing Faculty Publications, Journal of Alternative and Complementary Medicine', Volume 18, No. 2, pages 109-111, **2012**.

"**RegenaDerm® ointment has been shown to be of benefit in the management of burns and wounds.** The majority ingredients in RegenaDerm® include: lanolin, honey, beeswax, glycerin, aloe vera, wheat germ oil and olive oil. **Other herbal therapeutic ingredients constituting less than 1% of the product include: Comfrey, wormwood, Indian tobacco, white oak bark, marshmallow root, and myrrh.**

Comfrey root or Symphytum officinale: Two extraction methods result in a medicinal product from the Comfrey root: oleic and hydrolytic. When the oleic method is used, the oleic acid extraction formulates a hemangoendothelial-stimulating product. Oleic acid is an omega-9 fatty acid, which is considered one of the healtheir sources of fat in the diet. Topically applied, no toxicity has been reported with the oleic formulation, which is used in RegenaDerm®.

The author hypothesizes that an action of the oleic extract of Symphytum officinale is the creation of an epithelial growth factor that contributes to the accelerated re-epithelialization observed clinically. The reasoning is as follows: vascular endothelium is both embryonically and histologically nearly identical to dermal epithelium. Vascular (blood vessel) and dermal (skin) endothelium regenerates through 'cell movement'.

Dermal epithelial cell growth requires 'margin convergence' in a skin lesion healing by secondary intention. Vascular epithelial cells 'slide forward' vertically at first, forming epithelial 'nests' at the surface of the wound. From these 'nests' cells then begin to migrate radially in a circumferential fashion.

It has been observed in individuals treated with RegenaDerm® that epithelial 'nests' rapidly and visibly develop as the wound heals. *These epithelial 'nests' scattered throughout the wound bed are white, so they can be mistaken for infection. It has also been observed that while the wounds heal from the edges in, they also heal from the center of the nests out. Cessation (stopping) use of RegenaDerm® once the wound is re-epithelialized is important to prevent excessive scarring.*

The hydrolytic extraction method *results in a N-oxide pyrrolizidine alkaloid: riddelline. Oral ingestion of massive amounts of the hydrolytic extract has been associated with excessive internal vascularization and tumors.* **This extract is not used in RegenaDerm®.**"

-'Summary Review of Pharmacological Makeup of RegenaDerm®' by Mark Finneran, M.D., FAADEP, Ohio Agricultural Research and Development Center, Ohio State University, Wooster, Ohio, unpublished monograph, **2012**.

(I do not have any connection with this author or the product.)

(Oleic acid is a fatty acid in animal and vegetable fats/oils. It is a monounsaturated omega-9 fatty acid that is common in human diet. Eating monounsaturated fat is associated with decreased low-density lipoprotein {LDL} cholesterol.)

(Histology is the study of the microscopic anatomy of cells and tissues of plants and animals.)

"*Comfrey root (Symphytum officinale) has oleic extract that is believed to accelerate re-epithelialization needed to heal damaged dermal and epidermal tissues* (*Finneran, 2012).

Of note is the presence of Comfrey root extract in the B&W® ointment. Because of the potential for side effects, **it is recommended that creams containing Comfrey root extract be used less than 10 days in a row and no more than 6 weeks per year.** Average healing times for the burns observed in this study (8 days) fell within the recommended time."

-'The Effect of Burns and Wounds (B&W®) Burdock Leaf Therapy® on Burn-Injured Amish Patients: A Pilot Study Measuring Pain Levels, Infection Rates, and Healing Times' by Amish Burn Study Group, Nicole M. Kolacz, Mark T. Jaroch, Monica L. Bear and Rosanna F. Hess; Journal of Holistic Nursing, Volume 32, No. 4, pages 327-340, March **2014**.

(-'Summary Review of Pharmacological Makeup of RegenaDerm®' by Mark Finneran, M.D., FAADEP, Ohio Agricultural Research and Development Center, Ohio State University, Wooster, Ohio, unpublished monograph, 2012.)

"This case report describes the phases of an 'Amish Burn Care Project' and the lessons learned throughout the process. Amish, distinct by their horse and buggy travel, plain dress, and German dialect commonly known as Pennsylvania Dutch, number over 250,000 across the United States and Ontario, Canada.

In 2008, a group of Amish in this region of Ohio, known locally as **burn dressers, initiated a project with health care professionals to document the outcomes of their burn care using the Burns and Wounds (B&W) Ointment®** and Burdock Leaf Therapy® (Amish Burn Study Group, Kolacz, Jaroch, Bear, & Hess, 2014).

They had observed that patients under their care seldom used analgesics (pain killers) and rarely experienced pain during dressing changes in sharp contrast to conventional burn care (Kornhaber & Wilson, 2011; Malloy & Milling, 2010; Morris, Louw, & Grimmer-Somers, 2009; Smith, Murray, McBride, & McBride-Henry, 2011).

The B&W Burdock Leaf Therapy® originated with an Amish man named John Keim (Keim, 1999). This herbal-based burn care technique was born out of Keim's intense desire to alleviate suffering associated with hospital-based care using opiates for pain management and scrub tanks and skin grafts for wound care.

After discovering the efficacy of the B&W Ointment®, whose ingredients include honey, Comfrey, beeswax, lanolin, aloe, lobelia, olive oil, wheat germ oil, and white oak bark (Amish Burn Study Group et al., 2014), **and the burdock leaf for pain reduction** (Chan et al., 2011) during dressing changes, Keim trained hundreds of his fellow Amish to use this method ('Burn Certification List' 2008)."

-'Amish-Initiated Burn Care Project: Case Report and Lessons Learned in Participatory Research' by Rosanna F. Hess, D.N.P., R.N. from 'Research for Health, Inc.', Ohio, Journal of Transcultural Nursing, 28(2), pages 1-8, November **2015**.

Cancer (also see 'Tumors Reduced')

"**Overview of Development of Liver Cancer: The only known sequence of tissue changes seen during liver cancer development** involves microscopic foci or islands of altered hepatocytes (liver cells), hepatocyte nodules (group of cells), a subset of these nodules, the persistent nodules, nodules in nodules, and ultimately hepatocellular carcinoma (cancer).

The nodules show an array of architectural, fine ultrastructural, vascular (blood vessel), biochemical and physiological properties characteristic of this new population of hepatocytes.

Despite their origin following initiation with a chemical carcinogen, the vast majority (98-99%) of nodules undergo a

complex process of remodeling or redifferentiation to normal looking mature liver. A very small minority persist, continue to grow slowly and ultimately may act as a site of origin for new later pre-cancerous nodules and metastasizing (spreading) hepatocellular carcinoma.
The basis for the different behaviour patterns, remodelling of the majority and persistence of the minority is not understood."
-'The Biology of Carcinogen-Induced Hepatocyte Nodules and Related Liver Lesions in the Rats' by Emmanuel Farber, M.D., Ph.D., F.R.S.C., Departments of Pathology and Biochemistry, University of Toronto, Ontario, Canada; Toxicologic Pathology: Society of Toxicologic Pathologists, Volume 10, No. 2, pages 197-201, **1982**.

"In the present study, we investigated the efficacy of a novel phytochemical poly[3-(3, 4-dihydroxyphenyl) glyceric acid] (p-DGA) from Caucasian species of Comfrey (Symphytum caucasicum) and its synthetic derivative syn-2, 3-dihydroxy-3-(3, 4-dihydroxyphenyl) propionic acid (m-DGA) against PCA LNCaP and 22Rv1 cells.
We found that both p-DGA and m-DGA suppressed the growth and induced (caused) death in PCA (prostate cancer) cells, with comparatively lesser cytotoxicity (cell damage) towards non-neoplastic human prostate epithelial cells.
Furthermore, we also found that both p-DGA and m-DGA caused G1 arrest in PCA cells through modulating the expression of cell cycle regulators, especially an increase in CDKIs (p21 and p27). In addition, p-DGA and m-DGA induced apoptotic death by activating caspases, and also strongly decreased AR and PSA expression.
Consistent with in vitro results, our in vivo study showed that **p-DGA feeding strongly inhibited 22Rv1 tumors growth by 76% and 88% at 2.5 and 5 mg/kg body weight doses, respectively, without any toxicity, together with a strong decrease in PSA (Prostate-Specific Antigen) level in plasma;** *and a decrease in PCNA, AR and PSA expression but increase in p21/p27 expression and apoptosis (cell death) in tumor tissues from p-DGA-fed mice.*
Overall, present study identifies p-DGA as a potent agent against PCA (prostate cancer) without any toxicity, and supports its clinical application."
-'Poly3-34-DihydroxyPhenyl Glyceric Acid from Comfrey Exerts Anti-Cancer Efficacy Against Human Prostate Cancer via Targeting Androgen Receptor, Cell Cycle Arrest and Apoptosis' by Sangeeta Shrotriya, et al., Aurora Colorado, Tbilisi Georgia, and Athens Greece; Carcinogenesis: Integrative Cancer Research, Volume 33, No. 8, pages 1572-1580, **2012**.

*"***This review article was carried out by searching studies in PubMed, Medline, Web of Science, and IranMedex databases.*** The initial search strategy identified about 102 references. In this study, 44 studies was accepted for further screening and met all our inclusion criteria [in English, full text, therapeutic effects of Symphytum officinale L. and dated mainly from the year 1992 to 2016. The search terms were 'Symphytum officinale L.', 'therapeutic properties', 'pharmacological effects'.*
Based on the result, Symphytum officinale L. was shown to be effective in carcinoma treatment *as well as being effective in relieving pain and stiffness and in improving physical functioning and were superior to placebo in those with primary osteoarthritis of the knee without serious adverse effects.*
> *The efficacy of a novel phytochemical (p-DGA) from Symphytum officinale L. was investigated. It was found that both p-DGA and m-DGA suppressed the growth and induced death in PCA cells. So, p-DGA with no toxicity confirmed to have anti-cancer activity."*

-'Anti-Cancer and Osteoarthritic Pain Activity of Symphytum Officinale L.' by Mansoureh Masoudi and Milad Saiedi, Iran; Der Pharmacia Lettre: International Journal of Pharmaceutical Sciences, Volume 9, No. 3, pages 68-73, **2017**.
> (Carcinoma is cancer in the epithelial tissue of the skin or of the lining of the internal organs.)

*"**Recently, biologically active polymer has been isolated from Caucasian species of Comfrey Symphytum asperum Lepech. and Symphytum caucasicum Bieb.***
This polymer - poly[3-(3,4-dihydroxyphenyl)glyceric acid] (p-DGA) is a representative of a new class of natural polyethers with a residue of 3-(3,4-dihydroxyphenyl)-glyceric acid (DGA) as the repeating unit.
P-DGA exhibits high antioxidant, anti-inflammatory, wound healing and anticancer activities."
-'New Biopolymer from Comfrey: Chemistry and Biological Activity' by Maia Merlani, Vakhtang Barbakadze, Lela Amiranashvili and Lali Gogilashvili, Tbilisi State Medical University, Georgia; Joint Event on 5th International Conference on Bioplastics and 6th World Congress on Biopolymers, Paris, France, September 7-9 2017; Journal of Chemical Engineering and Process Technology, Volume 8, No. 4, Supplement, **2017**.

Eczema and Psoriasis

"It has been shown that plant extracts containing large amounts of allantoin produced healing of intractable (nonhealing) ulcers, as did pure solutions of allantoin.
Though it might seem that this action would not have any direct application in the treatment of psoriasis, allantoin has been shown to possess a keratin-dispersing activity, which may explain its beneficial activity in psoriasis. *Clinical trials in the United States of America have shown that allantoin appears of value in the management of psoriasis. From 70% to 88% of patients were helped, and 50% or more had their lesions cleared completely by the treatment. Clearing tended to be slow and treatment had to be maintained for relatively long periods of time, while any recurrence had to be treated immediately.*
It is also suggested that allantoin is of value in the treatment of chronic scaling dermatoses other than psoriasis, but not enough has been written about its use in such conditions to enable any assessment to be made.
Allantoin has not been reported as causing any skin reactions and phototoxicity has not been observed. *The preparations are acceptable to the patients. The cream can be applied under polyethylene (plastic) occlusion (closure, cover, wrap) with*

increased effect.
It seems that allantoin may well be of value in the management of psoriasis. It certainly avoids the hazards of antimetabolites and local and systemic corticosteroids, and it does not have the disadvantages of staining, burning, and discoloration that occur with other topical preparations."
-'Allantoin' in 'Todays Drug' section by expert contributors, British Medical Journal, Vol 4, No. 5578, page 535, December 2 **1967**.

Infection (Antibacterial / Antimicrobial)

"Dr. H. Wagner*, Institute of Pharmaceutical Biology, University of Munich, Germany, is probably the best known researcher on plants with immunomodulating activity. *Wagner also explained the types of immunomodulating activity and seven methods for evaluating such effects. The immunostimulatory effects can be specific (enhancing reaction to specific antigens) or nonspecific (generally stimulating immune function), with the latter category possibly serving to counteract every form of immune suppression.*
Since 'immunostimulant' is not a term found in older literature, it is difficult to find leads to this type of activity. Because of this, references to **antibacterial and antiviral agents** *were evaluated as well as those which may have been used for fever or other illnesses generally related to bacterial or viral infection.*
The methods employed in testing these plants include in vitro phagocytosis, in vivo carbon clearance test, in vitro T-lymphocyte transformation, and four others.
Dr. Wagner elaborated on some of the compounds and plants other than Echinacea which have been found active, including *Acanthopanax or Eleutherococcus, Chamomile,* **Comfrey,** *Aristolochia, and the compound vincristine from the Madagascar Periwinkle."*
-'HerbalGram, Issue 15' by American Botanical Council, Austin, Texas, 24 pages, Winter **1988**.
(* -'Immunostimulants from Medicinal Plants' by H. Wagner in 'Advances in Chinese Medicinal Materials Research: An International Symposium Held in Meridien Hotel Hong Kong June 12-14, 1984' by H.M. Chang, Henry Wai-Chung Yeung, W. W. Tso and A. Koo, Chinese University of Hong Kong. Chinese Medicinal Research Centre. Singapore: World Scientific Publishing Co. Inc., 1985. **"A marked immunstimulating activity was also observed with mucilages from Altheaea, Abelmoschus, Plantago, and Symphytum."**)
(Immunomodulator is a chemical agent that modifies the immune response or the functioning of the immune system such as by stimulation of antibody formation or inhibition of white blood cells.)

"The results show that of the 68 extracts, 64 (94.1%) possessed an **activity against one or more microorganisms** *whereas only four samples (Humulus lupulus, Agrostemma githago, Satureja hortensis and Althaea officinalis) were completely inactive. The most promising plants are those exhibiting a broad spectrum of activity. Among these, Grossheimia macrocephala and Origanum vulgare showed activity against all the tested* **gram-positive and gram-negative bacterial microorganisms**.
Cichorium intybum, Silybum marianum, Marrubium vulgare were active against seven of the eight bacterial species. Fumaria officinalis against six bacterial species. Sambucus nigra, Tanacetum vulgare and Allium cepa against five bacteria.
Symphytum officinale (roots and leaves) and Thymus serpyllum were active against four microorganisms.
Six plant species were active against Salmonella typhi (Typhoid Fever) *(Allium cepa, Cichorium intybum and Origanum vulgare) (MIC< 4 microgram), Sambucus nigra,* **Symphytum officinale** *and Tanacetum vulgare (MIC between 6 & 9 microgram).*
Three were active against Bacillus subtilis (Hay/Grass Bacillus) *(Origanum vulgare (MIC<4 microgram), Fumaria officinalis (MIC between 6 and 9 microgram),* **Symphytum officinale** *(MIC between 9 and 12 microgram).*
Three were active against Proteus mirabilis *(Origanum vulgare (MIC <4 microgram),* **Symphytum officinale (leaves,** *MIC between 6 and 9 microgram), Silybum marianum (MIC between 9 and 12 microgram).*
A search of the literature showed that antibacterial properties were attributed to *the phenols pyrocatechol in Allium cepa (Paris and Moyse, 1981),* **chlorogenic and caffeic acids in** *Sambucus nigra (Luckner et al., 1969) and* **Symphytum officinale (Paris and Moyse, 1981), to the alkaloids** *protopine and corydaline in Fumaria officinalis and* **consolidine and consolicine in Symphytum officinale (Paris and Moyse, 1981)."*
-'Biological Screening of Italian Medicinal Plants for Antibacterial Activity' by Izzo, Carlo, Fusco, Mascolo, Borrelli, Capasso, Fasulo and Autore, Phytotherapy Research, Volume 9, Issue 4, pages 281-286, June **1995**. (I'm not sure what 'MIC' stands for.)

"Comfrey is promoted in Ayurvedic and other herbal systems, with claims for benefit in disorders such as peptic ulcer. Comfrey also has been commonly used as a topical anti-inflammatory healing agent.
The present study was carried out **antibacterial activity of Symphytum officinale L.** *by Streak plate method using different solvent such as ethanol, methanol, ethyl acetate and chloroform.*
Both gram positive and gram negative bacterial strains were found to be sensitivity to the Comfrey leaf extracts of all the solvents except chloroform at higher concentration (75% and 100%)."
-'Antibacterial Activity of the Plant Extract of Symphytum Officinale L. Against Selected Pathogenic Bacteria' by Sumathi, Kumar, Bharathi and Sathish, International Journal of Research in Pharmaceutical Sciences 2, page 92, January **2011**. (I was unable to get this report. If you have one, could you please send it to me.)
(Ayurvedic / Ayurveda is the traditional system of medicine of India and seeks to treat and integrate body, mind, and spirit using a comprehensive holistic approach especially by emphasizing diet, herbal remedies, exercise, meditation, breathing, and physical therapy.)

"The aim of this study was to evaluate the antimicrobial effect of dried plant extracts and propolis against strains of Staphylococcus aureus isolated from bulk raw milk of Rio Pomba, Minas Gerais, Brazil. **The methodology involved the assessment of antimicrobial activity of extracts of garlic, pepper tree, Comfrey, rosemary, yucca, terramycin, marigold, parsley, oregano and propolis against 29 strains of Staphylococcus aureus** by Mueller-Hinton Agar diffusion.
Paper discs were soaked into extracts and transferred to plates previously inoculated with strains of S. aureus. It was used the Pairwise Clustering Method Using Arithmetic Averages and Scott-Knott tests at 5% probability to evaluation of obtained results. **Extracts of pepper tree and oregano were effective against all strains of S. auereus. The other extracts showed inhibition zone diameter reduced for most of the isolates tested.** Therefore, more studies are needed to determine the mechanisms of action and effective dose of pepper tree and oregano extracts in order to develop appropriated technologies to control bovine mastitis."
-'Antimicrobial Activity of Plant Extracts and Propolis Against Strains of Staphylococcus Aureus Isolated from Bulk Raw Milk' by L.R. da Silva, et al., Instituto Federal de Educacao, Clencia e Tecnologia do Sudeste (Federal Institute of Education, Science and Technology of the Southeast), Minas Gerais, Brazil; Higiene Alimentar (Food Hygiene), Volume 29, No. 248 / 249, pages 67-71, **2015**. (I was unable to get this report except the abstract in English. If you have it, could you please send me a copy.)

"**Allantoin, rosmarinic acid and ellagic acid were identified as major bioactive compounds (in Symphytum officinale root). The results obtained by the determination of the antimicrobial activity showed that Escherichia coli ATCC 8739 and Salmonela typhimirium ATCC 6538 were most sensitive to the aqueous (water) extract of Comfrey root.** The results showed that allantoin did not express the antimicrobial activity on all the investigated bacteria species, and based on this it can be concluded that **allantoin is not responsible for the antimicrobial activity of the aqueous extract of Comfrey root.**"
-'The Identification and Quantification of Bioactive Compounds from the Aqueous Extract of Comfrey Root by UHPLC–DAD–HESI–MS Method and its Microbial Activity' by Vesna Lj. Savic, Sasa R. Savic, Vesna D. Nikolic, Ljubisa B. Nikolic, Stevo J. Najman, Jelena S. Lazarevic and Aleksandra S. Dordevic, University of Nis, Serbia; Hemijska Industrija (Chemical Industry), Volume 69, No. 1, pages 1-8, **2015**.

Inflammation

"**Plants of 27 families, encompassing 75 species, have been selected on the basis of medicinal folklore. They are being studied in a broad screening programme for their anti-inflammatory activity,** using carrageenin foot oedema (edema/swelling) in rats. **Only 4 species were very active (this included Comfrey),** inhibiting carrageenin foot oedema by 42 to 74%, but overall 72% exhibited some anti-inflammatory activity.
With respect to the 45% inhibition by indomethacin (non-steroidal anti-inflammatory drug), **the following plants seem of special interest:** Agrostemma githago (38%), Anthriscus cerefolium (31%), Bryonia dioica (33%), Calendula arvensis (32%), Coriandrum sativum (31%), Foeniculum vulgare (36%), Grindelia robusta (41%), Juniperus communis (60%), Prunus cocumilia (74%), **Symphytum offcinale (reducing swelling 42%).**
These plants reduced the swelling induced in the rat's paw by carrageenin on each of the five occasions it was measured. **We also observed that the anti-inflammatory activity of** Prunus cocumilia, **Symphytum oficinale,** Juniperus communis and Grindelia robusta extracts is **dose-dependent, with 200 mg/kg** resulting in 89%, 60%, 79% and 63% **inhibition** respectively.
These plants are interesting for further studies, in particular Prunus cocumilia which is the most active, Symphytum officinale, Grindelia robusta and Juniperus communis.
Symphytum officinale is interesting for an additional reason since it also possesses anti-ulcer activity (British Herbal Pharmacopoeia; Stamford and Tavares, 1983) whereas most anti-inflammatory agents are ulcerogenic (creates ulcers).
The toxicity studies, carried out on Prunus cocumilia, Symphytum officinale, Juniperus communis and Grindelia robustu, showed that the rats tolerated the oral dose of 2.5 gram/kg of the tested extracts. No mortality occurred in any of the animals, and no side effects were recorded except mild diarrhoea in 2 animals receiving 2.5 g/kg of extract of Prunus cocumilia."
-'Biological Screening of Italian Medicinal Plants for Anti-Inflammatory Activity' by Mascolo, Autore, Capasso, Menghini and Fasulo, Phytotherapy Research, Volume 1, No. 1, pages 28-31, **1987**.

"**Inflammation is part of various pathological conditions: e.g., arthritis, atherosclerosis, the metabolic syndrome, allergies and other autoimmune diseases as well as cancer.** For most of these conditions no satisfying treatment of the associated inflammation is available.
Many inflammatory processes in different cell types are mediated by... Important molecular players inhibiting NF-kB responses are nuclear receptors, like the glucocorticoid receptor, the peroxisome proliferator-activated receptors (PPARs), the farnesoid X receptor (FXR), the liver X receptor (LXR) and the orphan nuclear receptors class 4A (NR4A).
The present analyses revealed potency of the leaf and root extracts prepared from Symphytum officinale to activate PPARs and to downregulate IL-8 and E-selectin mRNA."
-'Ethnopharmacological In Vitro Studies on Austria's Folk Medicine: An Unexplored Lore In Vitro Anti-Inflammatory Activities of 71 Austrian Traditional Herbal Drugs' by Sylvia Vogl, et al., Vienna, Austria; Journal of Ethnopharmacology, Volume 149, pages 750-771, **2013**.   (**In other words, Comfrey reduces inflammation.**)

"**Acute inflammation is a short-term process,** usually appearing within a few minutes or hours and ceasing upon the removal of the injurious stimulus. It is characterized by five cardinal signs: Dolor: pain), Calor: heat, Rubor: redness, Tumor: swelling and Functio laesa: loss of function.

The present study was carried out to assess the possible anti-inflammatory effect of methanolic extract of roots of Symphytum officinale Linn (MERSO) using egg white induced edema (swelling) in rats (acute model of inflammation).

The anti-inflammatory effect of MERSO was evaluated in acute inflammation model using 30 Wistar albino rats and divided into five groups including normal saline 10 ml/kg orally, diclofenac 5 mg/kg IP (intraperitoneal = injected into abdomen), and MERSO 250, 500 and 700mg/kg bodyweight orally.

Suppression of paw inflammation by either diclofenac or MERSO and the percentage of inhibition of paw edema were assessed. **The data obtained from this study reported that oral administration of Comfrey root extract significantly ($P<0.05$, $P<0.01$) inhibited raw egg albumin- induced rat paw edema** *as compared to control group. Maximum inhibitory effect (33.53%) was observed at a dose of 750 mg/kg, at the end of 240 minutes when compared to control group."*

-'Anti Inflammatory Activity of Symphytum Officinale Linn Root on Wistar Albino Rats' by D. Yashwanth Kumar, D.S.S.N. Neelima, Pradeep Kumar Choda and Namani Srilatha, Telangana, India; International Journal of Research in Pharmacy and Life Sciences, India, Volume 4, No. 1, pages 47-50, **2016**.

"**Chronic joint inflammatory disorders such as osteoarthritis and rheumatoid arthritis have in common an upsurge of inflammation, and oxidative stress,** *resulting in progressive histological (tissue) alterations and disabling symptoms. Currently used conventional medication (ranging from pain-killers to biological agents) is potent, but frequently associated with serious, even life-threatening side effects. Used for millennia in traditional herbalism, medicinal plants are a promising alternative, with lower rate of adverse events and efficiency frequently comparable with that of conventional drugs. Nevertheless, their mechanism of action is in many cases elusive and/or uncertain. Even though many of them have been proven effective in studies done in vitro (in laboratory equipment) or on animal models, there is a scarcity of human clinical evidence."*

-'Phytomedicine in Joint Disorders' by Dragos, Gilca, Gaman, Vlad, Iosif, Stoian and Lupescu, Journal of Human Nutrition, Basel, Switzerland, Volume 9, Issue 1, page 70+, January **2017**.

"**Symphytum officinalis, family Boraginaceae:**
   Traditional knowledge:
Symphytum officinalis, also known as Comfrey, is a medicinal plant traditionally used in Europe **for the treatment of inflammatory disorders.** *[Cavero, R.Y.; Calvo, M.I. Medicinal plants used for musculoskeletal disorders in Navarra and their pharmacological validation. J. Ethnopharmacol. 2015, 168, 255-259.] [Di Lorenzo, C.; Dell'Agli, M.; Badea, M.; Dima, L.; Colombo, E.; Sangiovanni, E.; Restani, P.; Bosisio, E. Plant food supplements with anti-inflammatory properties: A systematic review (II). Crit. Rev. Food Sci. Nutr. 2013, 53, 507-516.]*
   In vitro studies:
An **extract of Comfrey** *significantly inhibited the respiratory burst of polymorphonuclear leukocytes, suggesting an* **anti-inflammatory potential**. *[Gilca, M.; Gaman, L.; Lixandru, D.; Stoian, I. Estimating the yin-yang nature of Western herbs: A potential tool based on antioxidation-oxidation theory. Afr. J. Tradit. Complement. Altern. Med. 2014, 11, 210-216.]*
   Animal studies:
Comfrey extracts showed anti-inflammatory activity, *by inhibiting carrageenan-induced rat paw oedema (edema, fluid swelling). [Hiermann, A.; Writzel, M. Antiphlogistic glycopeptide from the roots of Symphytum officinale. Pharm. Pharmacol. Lett. 1998, 8, 154-157.] [Mascolo, N.; Autore, G.; Capasso, F.; Menghini, A.; Fasulo, M.P. Biological screening of Italian medicinal plants for anti-inflammatory activity. Phyther. Res. 1987, 1, 28-31.]*
   Human clinical studies:
A study on people aged 50–80 with **osteoarthritis (OA) of the knee proved that topically applied Comfrey preparation decreased pain**, *although was unable to decrease the burden of inflammatory molecules or the rate of cartilage breakdown, the only noticeable adverse effect being local rash. [Laslett, L.L.; Quinn, S.J.; Darian-Smith, E.; Kwok, M.; Fedorova, T.; Korner, H.; Steels, E.; March, L.; Jones, G. Treatment with 4Jointz reduces knee pain over 12 weeks of treatment in patients with clinical knee osteoarthritis: A randomised controlled trial. Osteoarthr. Cartil. 2012, 20, 1209-1216.]*
Similar results yielded another study on a similar population of years-long sufferers from OA of the knee: a **Comfrey-containing ointment improved the quality of life by decreasing pain and increasing knee-mobility**. *[Grube, B.; Grunwald, J.; Krug, L.; Staiger, C. Efficacy of a comfrey root (Symphyti offic. radix) extract ointment in the treatment of patients with painful osteoarthritis of the knee: Results of a double-blind, randomised, bicenter, placebo-controlled trial. Phytomedicine 2007, 14, 2-10.]*
   Active phytochemicals:
Phenolic acids (e.g., rosmarinic acid), glycopeptides and amino acids are considered to be, at least in part, responsible for the anti-inflammatory potential of Comfrey root extracts, *in various vitro models [Gracza, L.; Koch, H.; Loffler, E. 'Biochemical-pharmacologic studies of medicinal plants. 1. Isolation of rosmarinic acid from Symphytum officinale L. and its anti-inflammatory activity in an in vitro model'. Arch. Pharm. 1985, 318, 1090-1095].*
Rosmarinic acid inhibited prostaglandin (a hormone) synthesis, and carrageenan (collloid)- and gelatine-induced erythrocyte (red blood cell) aggregation (coming together). ['Testing the Membrane-Sealing Action of a Phytopharmaceutical and Its Agents' by L. Gracza, Z Phytother 8: 78-81, published 1987.]

-'Phytomedicine in Joint Disorders' by Dragos, Gilca, Gaman, Vlad, Iosif, Stoian and Lupescu, Journal of Human Nutrition, Basel, Switzerland, Volume 9, Issue 1, page 70+, January **2017**.

"**Regarding the mechanism of anti-inflammatory action, it has been found that a glycopeptide isolated from the Symphytum officinale aqueous (water) extract inhibits the release of prostaglandins and leukotrienes** *by decreasing expression of phospholipase A2."*

-'Medicinal Herbs as Possible Sources of Anti-Inflammatory Products' by Andreia Corciov, Daniela Matei and Bianca Ivanescu,

University of Medicine and Pharmacy, Iasi, Romania; Balneo Research Journal, Bucharest, Romania, Volume 8, No. 4, pages 231-241, December **2017**.

"Today, **Symphytum officinale topical use is based on its analgesic and anti-inflammatory effects,** which have been substantiated by modern clinical trials. However, the molecular basis of its action remained elusive.
Here, we show that a hydroalcoholic extract of Comfrey root impairs the development of a pro-inflammatory scenario in primary human endothelial cells in a dose-dependent manner.
The extract, and especially its mucilage depleted fraction, impair the interleukin-1 (IL-1) induced expression of pro-inflammatory markers including E-selectin, VCAM1, ICAM1, and COX-2. Both preparations inhibit the activation of NF-kB, a transcription factor of central importance for the expression of these and other pro-inflammatory genes.
Furthermore, our biochemical studies provide evidence that Comfrey inhibits NF-kB signaling at two stages: it dampens not only the activation of IKK1/2 and the subsequent IkBa degradation, but also interferes with NF-kB p65 nucleo-cytoplasmatic shuttling and transactivation."
-'A Symphytum Officinale Root Extract Exerts Anti-Inflammatory Properties by Affecting Two Distinct Steps of NF-kB Signaling' by Jacqueline Seigner, Marc Junker-Samek, Alberto Plaza, Gilda D'Urso, Milena Masullo, Sonia Piacente, Yvonne M. Holper-Schichl and Rainer de Martin, Medical University of Vienna, Austria, and R&D, Procter & Gamble Health®, Darmstadt, Germany; Frontiers in Pharmacology: Ethnopharmacology, Volume 10, Article 289, April **2019**.

Joint Pain (also see Rheumatism, and see Osteoarthritis)

"**A dermatological (skin) preparation from the aerial (above ground) parts of Comfrey (Symphytum x uplandicum Nyman, Russian Comfrey) was investigated** in an open, non-interventional study with 24 patients complaining about **moderate arthralgia (joint pain)** in a university hospital. **Twice daily application for one week decreased the pain score from 1.2 to 0.5, while hydratization (hydration) of the upper skin layers improved significantly.**
One of the main active components of Symphytum, allantoin, showed **anti-inflammatory effects in experimental assays with dermal cell cultures,** i.e. primary fibroblasts and keratinocytes.
Allantoin as well as Diclofenac® (non-steroidal anti-inflammatory drug) reduced the release of pro-inflammatory interleukin IL-6 induced by TNF-alpah or solar simulated radiation, with effective concentrations of about 50 micrometer/ml. Allantoin was not cytotoxic, had no effects on cell metabolism and proliferation."
-'Comfrey Ointment: Clinical Relevance and Mode of Action' by F. Casetti, U. Wolfle, G. Seelinger and C.M. Schempp, Forschungszentrum (Research Center) Skinitial Universitats, Freiburg, Germany; Zeitschrift fur Phytotherapie (Journal of Phytotherapy), Volume 35, No. 6, pages 268-272, 2014.   (The report is in German with the above English abstract.)

Muscle Soreness

"An improvement of pain on movement was already shown in the comparison of groups after 15 minutes in 66.7% versus 16.7% of subjects (p=0.036). Assessments of strength of effect led to the demonstration of a trend towards a superiority of verum (Symphytum cream) over placebo after 30 minutes, with statistical significance reached after 120 minutes (39.0+- 21.2 vs 19.6+- 22.8% improvement, p=0.060).
The quick onset of pain-relieving effects of the topical Comfrey herb extract cream already known from other studies could also be observed when applied against muscle pain related to overload induced muscle soreness."
-'Topical Comfrey Cream for the Pain Relief of Exercise-Induced Muscle Soreness: A Randomized, Placebo Controlled Study' by R. Uebelhack and M. Shaudt, Journal Pharmkol Therapy, Volume 23, No. 1, page 3, **2014**.

Muscle Tension

"Authors in the last two years have been continuously informing in specialised seminars about planned or carried out monitoring of comparative study performed in cooperation of two workplaces. The monitoring concerns the **therapeutic possibilities in cervical spine functional disorders** - one of the most common disorders among patients who enter rehabilitation workplace or spa. Since the patients, who enter ambulatory rehabilitation workplaces differ a little bit from the patients at spa, the authors decided to carry out the study in two workplaces. 50 patients were included into project in ambulatory (walking) workplaces and 300 patients at spa.
Results: Rehabilitation in medication coverage with the help of Traumaplant® proved to be effective for release of shortened muscles (muscle relaxation) and for improvement of fascia movement."
-'Final Report of Monitoring of Rehabilitation of Cervical Spine Functional Disorders in Medication Coverage with the Help of Traumaplant® in Bratislava and Bojnice (Slovakia)' by Guth, Mercekova, Hrdy and Ritomsky; Lekarsky Obzor (Medical Horizon), Slovakia, Volume 57, pages 409-412, January **2008**. (I do not have any connection with the medication mentioned in this report.)
> (Fascia is a band or sheet of connective tissue, primarily collagen, beneath the skin that attaches, stabilizes, encloses, and separates muscles and internal organs.)

Nerve Injury

*"**Injury to a peripheral nerve** is, among other things, followed by degeneration of axons and myelin, as well as by a sharp increase in the number of cells (especially Schwann cells) in the part distal to the injury. **The effect of allantoin - a cell proliferant - was tested** on the above mentioned reactions in the sciatic nerve of rats. This paper describes an investigation of the effects of allantoin on both cellular multiplication and removal of products of degeneration.*
***The results showed that allantoin had a statistically significant effect on the cellular multiplication seen in the nerve 7 and 14 days after the injury.** Myelin degeneration was also found to be more advanced in the (Group A) allantoin-treated nerve preparations examined 14 and 21 days postoperatively than in the control (no allantoin) nerve preparations. **Various workers observed that allantoin facilitated the removal of scales, crusts and necrotic (dead) tissue.** In a similar way, the removal of the products of degeneration was accelerated in the injured nerves of the animals in Group A (with allantoin)."*
-'The Effect of Allantoin on Cellular Multiplication in Degenerating and Regenerating Nerves' by J.M. Loots, G.P. Loots and W.S. Joubert, University of Pretoria, South Africa; South African Medical Journal, Volume 55, pages 53-56, January 13 **1979**.
(The peripheral nervous system is 43 pairs of motor and sensory nerves that connect brain and spinal cord {central nervous system} to the rest of the body.) (Distal means situated away from the center of the body or from the point of attachment.)

Osteoarthritis (inflammation of joints; also see Rheumatism)

*"Trials undertaken by 'Henry Doubleday Research Association' members, also showed that **Comfrey is a valuable plant for pain relief.** I have reports submitted by Dr. S.J.L. Mount, Berkshire, United Kingdom, who supervised the trials in 1983, testing 90 members with **osteoarthritis and rheumatoid arthritis**. Members took Comfrey, as either 4 cups (0.94 liter) of tea or 9 tablets, daily. Dr. Mount reported there were no side effects from this dosage, whatsoever, and no reports of any symptoms, which could be construed as liver syptomatology.*
Patients reported improvement in well-being, with 23-35% pain relief and mobility."
-How Can I Use Herbs in My Daily Life: Over 500 Herbs, Spices and Edible Plants: An Australian Practical Guide to Growing Culinary and Medicinal Herbs by Isabell Shipard. New York, New York: David Stewart Publisher / Simon & Schuster, Inc., **2003**.

"This randomised, double-blind, bicenter, placebo-controlled clinical trial investigated **the effect of a daily application of 6 grams Kytta-Salbe® f (Comfrey Salve by Merck®) (3x2 grams) over a 3 week period with patients suffering from painful osteoarthritis of the knee.**
In the course of the trial, the Visual Analog Scale (VAS) total score (primary target value) in the verum (Comfrey salve) group dropped by 51.6 mm (54.7%) and in the placebo group by 10.1 mm (10.7%). The average difference between the groups of 41.5 mm (95% confidence interval=34.8 to 48.2 mm) or 44.0% is significant (p<0.001).
The WOMAC (Western Ontario and McMaster Universities) total score (secondary target value) also improved similar to the VAS total score. At the end of the trial, a reduction by 60.4 mm (58.0%) was recorded for the verum group and a reduction of 14.7 mm (14.1%) for the placebo group. The average group difference of 45.7 mm (95% confidence interval=37.1 to 54.3 mm) or 43.9% is significant (p<0.001).
The difference between the treatment groups increased systematically and significantly, in parallel with the duration of the treatment. Thus, the superiority of the treatment with Kytta-Salbe® f over that with the placebo is proven, even by means of the multi-factorial multivariate (multiple variables) analysis for repetitive measurements.
In respect of the explorative secondary target values SF-36 (quality of life), angle measurement (mobility of the knee), CGI (Clinical Global Impression) and global assessment of efficacy by the physician and the patient, a significant superiority (p<0.001 each) of the verum (Comfrey salve) group over the placebo group was also proven.
The results suggest that the Comfrey root extract ointment is well suited for the treatment of osteoarthritis of the knee. Pain is reduced, mobility of the knee improved and quality of life increased."
-'Efficacy of Comfrey Root (Symphti Officinale Radix) Extract Ointment in the Treatment of Patients with Painful Osteoarthritis of the Knee' by Grube, Grunwald and Staiger, Phytomedicine, Volume 14, Issue 1, pages 2-10, January 10 **2007**.
(Kytta-Salbe® by Merck & Company, Inc.: *"Sore muscles, beat-up joints, tennis elbow, lower back pain- these things happen to just about everyone at some point. The remedy of choice for many has been ibuprofen or other over-the-counter pain medication. But there is a non-pharmaceutical alternative- Kytta Salbe® {Comfrey Salve}.*
Comfrey Root has a long history as a medicinal herb. It's been used for centuries in Chinese medicine to treat wounds and reduce arthritis pain. This herbal salve is often used by customers with arthritic and rheumatic pain.
It was featured in an article posted on WebMD®. The article praises Kytta Salbe® as an excellent natural cream to help relieve acute back pain not caused by a identifiable source, such as a bulging disc or trauma. Based on the study, participants using the Comfrey Salve experienced 95% reduction in pain - very convincing results.")
(I have no connection with this product or the company.)
(A placebo is a harmless and useless substance in the form of a pill/capsule or liquid that is given for the psychological benefit of the patient rather than for any medicinal effect. In clinical trials it is used as a control or baseline to compare against a similar substance given that does have medicine.)

*"**The purpose of this study was to determine the effect of 2 concentrations of topical (skin), Comfrey-based botanical creams containing a blend of tannic acid and eucalyptus to a eucalyptus reference cream on pain, stiffness, and physical functioning in those with primary osteoarthritis of the knee.***

Forty-three male and female subjects (45-83 years old) with diagnosed primary osteoarthritis of the knee who met the inclusion criteria were entered into the study. Participants applied the cream 3 times a day for 6 weeks and were evaluated every 2 weeks during the treatment.
Repeated-measures analyses of variance yielded significant differences in all of the Western Ontario and MacMaster Universities Osteoarthritis Index categories (pain P b .01, stiffness P b .01, daily function P b .01), confirming that **the 10% and 20% Comfrey-based creams were superior to the reference cream.**
The active groups each had 2 participants who had temporary and minor adverse reactions of skin rash and itching, which were rapidly resolved by modifying applications. **Both active topical Comfrey formulations were effective in relieving pain and stiffness and in improving physical functioning and were superior to placebo in those with primary osteoarthritis of the knee without serious adverse effects.**
Comfrey (Symphytum officinale L.), also known as knit bone, has long been advocated in folk medicine for the treatment of wounds, sores, sprains, and bone fractures. **In Germany, Comfrey has been used in medicine since 1920 for the treatment of musculoskeletal conditions. It has been suggested that the efficacy of Comfrey is primarily due to its anti-inflammatory, analgesic, granulating promoting, and antiexudative properties.**
Comfrey pharmacological components include rosmarinic acid and tannin. Rosmarinic acid is a natural polyphenol antioxidant, and both rosmarinic acid and tannin are considered anti-inflammatory agents."
-'Effect of a Blend of Comfrey Root Extract (Symphytum officinale L.) and Tannic Acid Creams in the Treatment of Osteoarthritis of the Knee' by Doug B. Smith Ph.D., Bert H. Jacobson Ed.D., Journal of Chiropractic Medicine, Vol 10, pages 147-156, **2011**.

"**Objective:** To assess the efficacy of thrice (3 times) daily topical (skin) 4Jointz® utilizing Acteev technology (**a combination of a standardized Comfrey extract and a pharmaceutical grade tannic acid**, 3.5 grams/day) on osteoarthritic knee pain, markers of inflammation and cartilage breakdown over 12 weeks.
Patients and methods: Adults aged 50-80 years (n = 133) with clinical knee OA (Osteoarthritis) were randomised to receive 4Jointz® or placebo in addition to existing medications.
Pain and function were measured using a Visual Analogue Scale (VAS) and the Knee Injury and Osteoarthritis Outcome Score (KOOS) scale at baseline, 4, 8 and 12 weeks. Inflammation was measured analysing IL-6 expression and CTX-2 presence as representative for cartilage breakdown using ELISA, at baseline and 12 weeks.
Results: Pain scores significantly reduced in the group who received 4Jointz® compared to the group who received placebo after 12 weeks using both the VAS (-9.9 mm, $P = 0.034$) and the KOOS pain scale (+5.7, $P = 0.047$). Changes in IL-6 and CTX-2 were not significant (-0.04, $P = 0.5$; -0.01, $P = 0.68$).
Post-hoc analyses suggested that treatment may be most effective in women (VAS -16.8 mm, $P = 0.008$) and those with milder radiographic (X-ray or similar) osteoarthritis (OA) (VAS -16.1 mm, $P = 0.009$).
Rates of adverse events were similar in both groups, excepting local rash that was more common amongst participants receiving 4Jointz® (21% vs 1.6%, IRR 13.2, $P = 0.013$), but only 26% (n = 4) of participants with rashes discontinued treatment. There were no changes in systemic blood results.
Conclusions: Topical treatment using 4Jointz® reduced pain but had no effect on inflammation or cartilage breakdown over 12 weeks of treatment."
-'Treatment with 4Jointz® Reduces Knee Pain Over Twelve Weeks of Treatment in Patients with Clinical Knee Osteoarthritis: A Randomised Controlled Trial' by L.L. Laslett, et al., Australia; Osteoarthritis and Cartilage: Osteoarthritis Research Society International, Volume 20, pages 1209-1216, **2012**.   (I do not have any connection with this product.)

"**This review article was carried out by searching studies in PubMed, Medline, Web of Science, and IranMedex databases.** The initial search strategy identified about 102 references. In this study, 44 studies was accepted for further screening and met all our inclusion criteria [in English, full text, therapeutic effects of Symphytum officinale L. and dated mainly from the year 1992 to 2016. The search terms were 'Symphytum officinale L.', 'therapeutic properties', 'pharmacological effects'.
Based on the result, Symphytum officinale L. was shown to be effective in carcinoma treatment as well as being effective in relieving pain and stiffness and in improving physical functioning and were superior to placebo in those with primary osteoarthritis of the knee without serious adverse effects."
-'Anti-Cancer and Osteoarthritic Pain Activity of Symphytum Officinale L.' by Mansoureh Masoudi and Milad Saiedi, Iran; Der Pharmacia Lettre: International Journal of Pharmaceutical Sciences, Volume 9, No. 3, pages 68-73, **2017**.

Pain in General

"In an open, uncontrolled study, 105 patients with locomotor (movement) system symptoms were **treated twice daily with an ointment containing a Symphytum active substance complex.** A clear therapeutic effect was noted on chronic and subacute (between chronic and acute) symptoms that were accompanied mainly by functional disturbances and pain in the musculature. **The preparation was most effective against muscle pain, swelling and overstrain, arthralgia (pain in joint) / distortions (sprains), enthesopathy (tendon/ligament problem), and vertebral syndrome.** Activity was weaker against degenerative conditions, for which the ointment may have an adjuvant role with the aim of improving muscular dysfunction and alleviating (reducing) pain."
-'Effects of Symphytum Ointment on Muscular Symptoms and Functional Locomotor Disturbances' by Kucera, Kalal, and Polesna, Advances in Therapy Journal, Volume 17, No. 4, pages 204-210, July-August **2000**.
      (An adjuvant is a pharmacological or immunological agent that modifies the effect of other agents.)

"Comfrey has been used in folk medicine as a poultice for treating burns and wounds. The roots of the Comfrey is used in case of pulmonary (lung) complaints. The leaves of the Comfrey have been used for the treatment of rheumatism and gout.
The roots of Symphytum officinale were purchased from Ghaziabad (state of Uttar Pradesh, India) in April 2013. The effect of petroleum ether, chloroform and ethanol extracts (400 mg/kg body weight) (kg= 2.2 pounds) of roots of Symphytum officinale were examined in Swiss albino mice to evaluate analgesic (pain killing) activity.
In conclusion, the present study demonstrate that the Symphytum officinale roots has significant analgesic (pain killing) activity. The one, two or more constituents of Symphytum officinale are responsible for the activity which needs to be investigated further."
-'Evaluation of Analgesic Activity of Various Extracts of Roots of Symphytum Officinale' by Kaur, Kaur and Sekhon, International Journal of Medicine and Pharmaceutical Sciences, Volume 4, Issue 3, pages 59-62, June **2014**.

Pain in Back and Neck

*"**The objective was to show the superiority of Comfrey root extract ointment to placebo ointment in patients with acute upper or lower back pain.** The study was conducted as a double-blind, multicentre, randomised clinical trial with parallel group design over a period of 5 days. The patients (number=120, mean age 36.9 years) were treated with verum (true/actual substance) or placebo ointment three times a day, 4 gram (0.14 ounce) ointment per application. The trial included four visits.*
There was a significant treatment difference between Comfrey extract and placebo regarding the primary variable. In the course of the trial the pain intensity on active standardised movement decreased on average (median) approximately 95.2% in the verum group and 37.8% in the placebo group.
***The results of this clinical trial were clear-cut and consistent across all primary and secondary efficacy variables. Comfrey root extract showed a remarkably potent and clinically relevant effect in reducing acute back pain.** For the first time a fast-acting effect of the ointment was also witnessed."*
-'Efficacy and Safety of Comfrey Root Extract Ointment in the Treatment of Acute Upper or Lower Back Pain' by Giannetti, Staiger, Bulitta, and Predel, British Journal of Sports Medicine, Volume 44, Issue 9, July **2010**.

"Comfrey has a centuries-old tradition as a medicinal plant. Today, multiple randomized controlled trials have demonstrated **the efficacy and safety of Comfrey preparations for the topical (skin) treatment of pain, inflammation and swelling of muscles and joints in degenerative arthritis, acute myalgia (muscle pain) in the back, sprains, contusions (bruises) and strains after sports injuries and accidents**, also in children aged 3 or 4 and over.
This paper provides information on clinical trials and non-interventional studies published on Comfrey to date and further literature, substantiating the fact that topical Comfrey preparations are a valuable therapy option for the treatment of painful muscle and joint complaints."
-'Comfrey: A Clinical Overview' by Christiane Staiger in Germany, Phytotherapy Research, Volume 26, Issue 10, pages 1441-1448, February **2012**.

*"The efficacy and tolerability of a topical preparation with **an active substance concentrate made from the aerial (above ground) parts of medicinal Comfrey (Symphytum x uplandicum Nyman) was tested in 215 patients with acute or chronic myalgia (muscle pain) of the upper and lower back** (active substance n=104, reference n=111) (Kucera et al. 2005).
A low dose but otherwise identical preparation (1% versus 10% active substance) was used as a reference. **The result was a significantly better and clinically relevant reduction in pain at rest and on movement after 4-5 days with the high active substance concentration.** The superiority of active substance compared to reference was clinically relevant in both subgroups.
Topical Comfrey extract has a fast acting analgesic (pain killing) effect in myalgia of the back caused by chronic strain as well as acute blunt injury."*
-'Topical Comfrey Extract: Study Confirms Rapid Effects on Myalgia Due to Strain or Acue Blunt Trauma' by M. Kucera and M. Hladfkova, Journal Pharmkol Therapy Volume 21, No. 4, pages 112-117, **2012**.
   (Acute disease means it has an abrupt/rapid onset. It usually means an illness of short duration, rapidly progressive, and in need of urgent care. Acute is in reference to time as opposed to subacute or chronic. Chronic means lasting a long time, usually 3 months or more. Subacute is between acute and chronic.)

*"This randomised, multicentre, double-blind, three-arm, placebo-controlled trial **compared a topical (skin) combination of 35% Comfrey root extract plus 1.2% methyl nicotinate versus a single preparation of methyl nicotinate or placebo cream for relief of acute upper or low back pain.***
*379 patients were randomly assigned to three groups (combination, n = 163; methyl nicotinate, n = 164; placebo, n = 52). They applied a 12 cm (4.7 inch) layer of cream three times daily for 5 days. The primary efficacy variable was the 'Area Under the Curve' (AUC) of the 'Visual Analogue Scale' (VAS) on active standardised movement values at visits 1 to 4.
Secondary measures included back pain at rest, pressure algometry (pain), consumption of analgesic (pain killing) medication, functional impairment measured with Oswestry Disability Index, and global assessment of response.
The AUC of the VAS on active standardised movement was markedly smaller in the combination treatment group than in the methyl nicotinate and in the placebo group (ANOVA: $p<0.0001$). **The combination demonstrated superiority to the two other treatment arms, while methyl nicotinate displayed a considerable effect as well. The clinical trial at hand confirms the topical combination is an effective and well-tolerated treatment option for acute back pain.**"*

-'Combination of Comfrey Root Extract Plus Methyl Nicotinate in Patients with Conditions of Acute Upper or Low Back Pain: A Multicentre Randomised Controlled Trial' by Helmut Pabst, Axel Schaefer, Christiane Staiger, Marc Junker-Samek and Hans-Georg Predel, Germany; Phytotherapy Research, Volume 27, pages 811-817, **2013**. (The species of Comfrey was not given.)
(Methyl Nicotinate is a nicotinic acid methyl ester, used as rubefacient in cosmetics to produce a redness or inflammation for a short period of time, often in lip products. A rubefacient dilates capillaries and increases blood circulation.)

Pressure Sores

"In an open, prospective (future) use study, 161 patients with 198 **decubitus ulcers (pressure ulcers**, ITT population) in stages II and III were treated with the **topical (skin) preparation Symphytum herb extract cream**. The bandages with the cream were changed every 2–3 days.
Complete healing of the pressure sores within 4 weeks was observed in 85.9 % (PP population)/79.8 % (ITT population) of the treated ulcers. Over a treatment duration of 25-30 days, a 89.2% reduction of the total decubitus area was observed. The same result was found for the depth of the pressure ulcer with a reduction of 88%.
The overall treatment success was from both the perspective of the physician and the patient considered successful in 90.4% (5-point scale) of cases and 87.9% (100 mm VAS, PP population).
Two cases of local irritation were observed after 25/30 days (1.2% of the patients with exposure), thus showing very good skin compatibility. **The efficacy of Symphytum herb extract cream is surprisingly good in the treatment of pressure ulcers.**"
-'Efficacy and Safety of Topical Symphytum Cream in the Treatment of Pressure Ulcers' by J. Stepan, J. Ehrlichova and M. Hladikova, Zeitschrift fur Gerontologie und Geriatrie (Journal of Gerontology and Geriatrics), Vol 47, No. 3, page 228-235, **2014**.

"**Objectives**: The purpose of this work was development, formulation and testing of new herbal ointment for the treatment of pressure ulcers.
Patients and methods: 50 patients (27 males and 23 females) with total 84 ulcers of stage II and III were treated 28 days (twice a day) with the **ointment containing the following ingredients: Symphytum officinale,** Plantago major, Calendula officinalis, Matricaria chamomilla, Bellis perennis, Achillea millefolium, Salvia officinalis, Hypericum perforatum, Olea europaea, Lavandula officinalis, Melaleuca alternifolia, Cympobogon martini, Origanum vulgare, Eugenia caryophyllata, Thymus vulgaris ct. thymol, Cera alba, honey, and glycerol. The healing process was assessed by Pressure Ulcer Scale for Healing (PUSH) tool version 3.0.
Results: Prior to the therapy mean value and standard deviation of the PUSH score for ulcer surface area, quantity of exudate (fluids secreted), type of tissue and the total score were 8.39+-0.79, 1.35+-0.84, 2.81+-0.40 and 12.5+-1.94, respectively.
All the mentioned values decreased significantly after only seven days of the treatment ($p<0.00001$). Further treatment resulted in linear decrease of PUSH parameters reaching zero values after 28 days of the therapy.
Slough (removal of dead tissue) disappeared after 14 days of the therapy and epithelial tissue was obtained on the edge of 67.86% of the ulcers. Following the 21 day of the treatment 17.86% of the ulcers were completely closed while after 28 days all the ulcers healed completely.
Conclusion: Four weeks of the topical treatment with Bioapifit® herbal wound healing ointment resulted in complete closure of all ulcers with mean healing time of 26.4 days. Such excellent results could be attributed to the ointment's formulation containing the ingredients with strong wound healing, anti-inflammatory and antimicrobial potential."
-'Treatment of Pressure Ulcers with Bioapifit® Wound Healing Herbal Ointment: A Preliminary Study' by Visnja Orescanin, Orescanin Ltd., Laboratory for Herbal Drugs Development, Croatia; IJRDO: Journal of Biological Science, Volume 2, Issue 10, October **2016**. (I do not have any connection with this product.)

Rheumatism (also see Osteoarthritis)

"**Phytopharmaceuticals are successfully administered externally and internally in the treatment of rheumatic diseases.** Topically (on skin) used medicines are Capsicum (Capsicum annum), Arnica (Arnica montana) and **Comfrey (Symphytum officinale)** which have a similar therapeutic value compared to Non-Steroidal Anti-Inflammatory Drugs (NSAIDs).
In comparison to conventional NSAID (Non-Steroidal Anti-Inflammatory Drugs), these drugs (herbs) have a better compatibility and cost efficiency as well as reduced adverse effects.
Comfrey (Symphytum officinale) promotes wound healing and is anti-inflammatory in external application. In interaction with other ingredients allantoin is effective. It is anti-inflammatory, anti-exsudative (reduces oozing of fluid at injury), stimulates cell proliferation and promotes the regeneration processes in injured tissue.
From toxicological view, in Austria the upper limit of 0.1 ppm (parts per million) pyrrolizidine alkaloids which are called carcinogenic (cancer causing) and are classified hepatotoxic (liver toxic).
The medicinal specialty Traumaplant® ointment met these requirements. It contains an extract from fresh cultured cut herb (Dev 2-3 : 1; ethanol 30%), taking this preparation a maximum duration of treatment of 4 weeks.
This in Germany is an approved product: Kytta® ointment that contains Comfrey root fluid extract (Dev 1 : 2, ethanol 60%) and is at a level of <0.35 ppm pyrrolizidine alkaloids."
-'Phytotherapy in Rheumatology' by F. Schullner and E. Mur, Krankenhausapotheke (Dispensary) LKH Innsbruck, Austria; Zeitschrift fur Phytotherapie (Journal of Phytotherapy), Volume 33, No. 4, pages 158-167, **2012**.
(I have this report in German with an abstract in English. Then I used an online translator for the Comfrey part of the report. I do not know what 'Dev' means. If you know, please contact me.) (I do not have any connection with the ointments described.)

Skin Health

"***Comfrey (Symphytum officinale leaves) Suggested Cosmetic Uses*** *(External use only. Not for drug use.):*
Actiphyte® of Comfrey can be used in creams, lotions, ointments, salves, douches, hair rinses, shampoos, and massage / body oils.
Actiphyte® are basic botanical extracts *consisting of a single herb that present a wide range of different performance traits. According to the product claims being sought, each extract can be incorporated into specific cosmetic formulas.*
    Usage Level Recommended: 5 - 10% in skin and hair care products.
    pH: 4.0 - 6.5 at 25 C (77 F).   Solubility: Soluble in any proportion in water.
    Specific Gravity: 1.02 - 1.05 at 25 C.   Refractive Index: 1.3860 - 1.3950 at 25 C.
    Microbial Plate Count: Less than 100 organisms per gram."
-'Actiphyte® of Comfrey' by Active Organics, Lewisville, Texas, **2002**. Now part of Lipotec®, Barcelona, Spain, www.lipotec.com. (Cosmetics are products that enhance the appearance of the face or fragrance / texture of the body.)

"*Improving Skin Renewal during the Aging Process with Plant Stem Cells:*
PhytoCellTec® Symphytum is a powder based plant stem cell extract of Comfrey (Symphytum) roots.
PhytoCellTec® Symphytum has been shown to clearly restore the regenerative capacity of epidermal stem cells cultivated in an aging environment. *Clinical studies performed with twenty women demonstrated a significant increase of the skin renewal rate as well as an 11% improvement in skin smoothness. PhytoCellTec® Symphytum can thus boost and improve the skin renewal from the deepest layers of the epidermis during aging.*
It prevents the age-related thinning of the epidermis and ensures that the skin remains smooth.
Ingredients: 50% Symphytum Officinale Callus Culture Extract, *Isomalt, Lecithin, Sodium Benzoate and Water.*"
-'PhytoCellTec® Symphytum: Speed Up Your Skin Renewal through Stem Cell Activation' by Mibelle Group Biochemistry, Buchs, Switzerland, www.mibellebiochemistry.com, December **2013**. (I do not have any connection with this company or the product.)

"*In zebrafish, UV (Ultra-Violet radiation) exposure leads to fin malformation phenotypes (fin deformity) including fin reduction or absence.* **The present study evaluated UV-protective activities of Comfrey leaves extracts** *in a zebrafish model by recording fin morphological (form / structure) changes. Inflammation, oxidative stress and DNA damage are caused by exposure to UV radiation.*
Comfrey leave extracts had UV-absorbance abilities and significantly reduced ROS (Reactive Oxygen Species) production *in UV-exposed zebrafish embryos, which may attenuate (reduce) UV-mediated apoptosis (cell death). Generation of Reactive Oxygen Species (ROS) is considered the most important adverse (bad) effect after UV exposure.*
Comfrey extracts increased the rate of fin repair.
In conclusion, Comfrey leaves extracts may have the potential to be developed as UV-protective agents *to protect zebrafish embryos from UV-induced damage. Taken together, we propose that the UV-protective ability of Comfrey extract may mostly come from its photochemical properties, which can isolate UV. That is,* **Comfrey leave extracts may act just like a sunscreen,** *providing protection against UV-induced fin damage from the extracellular level.*
In summary, this study suggests that Comfrey can be used to protect zebrafish fins from UV-induced damage, implying that it may be applied to **aquaculture** *to enhance the survival of juvenile fish.*"
-'Protective Role of Comfrey Leaves Extracts on UV-Induced Zebrafish Fin Damage' by Chien-Chung Cheng, Chi-Yuan Chou, Yao-Chin Chang, Hsuan-Wen Wang, Chi-Chung Wen and Yau-Hung Chen, Taiwan; Journal of Toxicologic Pathology: The Japanese Society of Toxicologic Pathology, Tokyo, Japan, Volume 27, pages 115-121, **2014**.

> "*Solar UV and other ionizing radiations cause a generation of Reactive Oxygen Species (ROS), induce cellular DNA damage and alter skin homeostasis (balance).* **The use of exogenous (external) antioxidants is increasingly frequent. We attempt to demonstrate that a Rosmarinic Acid extract acts as photo-protector; both free radical scavenger as an inducer (promoter) of the body's own endogenous (internal) defence mechanisms** *by regulating tyrosinase activity and stimulating melanin production.*"
> -'Rosmarinic Acid: A Photo-Protective Agent Against UV and Other Ionizing Radiations' by M. Sanchez-Campillo, J.A. Gabaldon, J. Castillo, O. Benavente-García, M.J. Del Bano, M. Alcaraz, V. Vicente, N. Alvarez and J.A. Lozano; Murcia, Spain; Food and Chemical Toxicology, Volume 47, pages 386-392, **2009**.
> > (For more about Rosmarinic Acid in Comfrey, see sub-subsection 'Rosmarinic Acid' in subsection 'Other Constituents' in section 'Nutritional Value of Comfrey' {Chapter 19, Volume 1}.)

Skin Redness (Erythema)

"**Symphyti radix- Comfrey root:** *In vitro, rosmarinic acid and glycopeptides isolated from Comfrey root have been shown to possess the anti-inflammatory activity. Rosmarinic acid inhibits synthesis of the prostaglandins, and glycopeptides, dose-dependently, inhibit release of the prostaglandins (PGE2 and PGI2), and also of 12-HETE and the arachidonic acid.*
In pharmacological studies in healthy humans, dermatological (skin) preparations containing 5% or 10% of Comfrey root extract were effective in reducing UV-B-induced erythema (skin redness).

Anti-inflammatory potency of the extracts was comparable, or even greater than that of diclofenac (Non-Steroidal Anti-Inflammatory synthetic Drug- NSAID).
A positive correlation could be demonstrated between the efficacy and the concentration of the caffeic acid."
-'Medicinal Plants Used in Treatment of Inflammatory Skin Diseases' by Renata Dawid-Pac, Department of Medicinal and Cosmetics Natural Products, Poznan University of Medical Sciences, Poland; Postepy Dermatologii i Alergologii (Advances in Dermatology and Allergology); Volume XXX, No. 3, pages 170-177, **2013**.
(Erythmea is inflammation from injury or irritation that causes reddening of the skin.)

Sports Fatigue

"**This paper studied the influence of Comfrey extract on alleviating (reducing) sport fatigue of athletes after long distance running training of 1 month.** The results showed that the average heart rate and maximum heart rate of the observation group (Comfrey users) after training appeared significantly enhanced over the control groups ($P<0.05$), and the morning heart rate of the observation group significantly declined more than the control groups ($P<0.05$). The sleep quality, diet condition and body weight were significantly higher (better) than the control group ($P<0.05$).
The average exercise power, maximal anaerobic power, and maximum aerobic power of the observation group (Comfrey users) were significantly higher than the control group ($P<0.05$).
The creatine kinase and blood urea nitrogen of the observation group were significantly higher than the control group ($P<0.05$) and hemoglobin and testosterone of the observation group were significantly decline (less) than the control groups ($P<0.05$).
The feeling of fatigue of the observation group (Comfrey users) were obviously lower than the control group. In all of the study there did not occur any adverse reactions which showed that Comfrey extract had obvious good effect on alleviating sports fatigue and was safe."
-'Research in the Effect of Comfrey Extract on Alleviating Sports Fatigue' by Zhao XiaoKun, Tangshan University, Hebei, China; Food Research and Development, Volume 38, No. 4, pages 172-175, 2017.   (This was translated from Chinese to English.)

Sprains and Blunt Injury (also see Bruises)

"**Treatment with a Symphytum peregrinum (synonymous with Symphytum x uplandicum Nyman) ointment was carried out for an average of 8 days on 40 patients suffering from recent knee joint injuries, distortions (sprains), and contusions (bruises) which did not require surgery.** Except for physical measures such as compression bandages, ice packs, and directions for independent exercise therapy, no other additional treatment was carried out.
Assessment criteria were pain, swelling, and restriction of movement. **Of the patients, 34 (85%) rated the efficacy of the preparation as good to very good**, and only 9 (15%) rated the therapeutic result as less satisfactory.
**All 40 patients assessed the tolerance of the ointment positively. No side effects occurred in any of the cases.
It can be concluded from the results that the 10% Symphytum ointment is very suitable for the local treatment of knee joint injuries.**"
-'Effect of a Symphytum Ointment with Sports Injuries of the Knee Joint' by H. Hess, German Journal of Sports Medicine, Volume 42, No. 4, pages156-162, **1991**.
(A contusion is an injury from a blow with a blunt instrument where the subsurface tissue is injured but the skin is not broken; i.e., a bruise.)

"In a controlled, double blind, randomized multicentre study, **the efficacy and safety of the topical Comfrey product Traumaplant® (10% active ingredient of a 2.5 : 1 aqueous ethanolic pressed juice of freshly harvested, cultivated Comfrey herb {Symphytum x uplandicum Nyman}), corresponding to 25 grams of fresh herb per 100 grams of cream**; n = 104).
It was tested against a 1% product (corresponding to 2.5 g of fresh Comfrey herb in 100 g of cream; n = 99) in 203 **patients with acute ankle distortion (sprain)**.
With the high concentration, decrease of the scores for pain on active motion, pain at rest and functional impairment was highly significant and clinically relevant on days T3-4 as well as T7 ($p < 0.001$).
**Amelioration (reduction) of swellings as compared to reference was also significant on day 3-4 ($p < 0.01$). Efficacy was judged good to excellent in 85.6% of cases with verum (Symphytum) and in 65.7% of cases with reference on day 3-4.
Overall tolerability was excellent.**"
-'Efficacy and Safety of Topically Applied Symphytum Herb Extract Cream in the Treatment of Ankle Distortion: Results of a Randomized Controlled Clinical Double Blind Study' by Kucera, Barna, Horacek, Kovarikova and Kucera, Wiener Medizinische Wochenschrift (Vienna Medical Weekly, Austria), Volume 154, No. 21-22, pages 498-507, November **2004**.

"**Comfrey (Symphytum officinale L.) is a medicinal plant with anti-inflammatory, analgesic (pain killing) and tissue regenerating properties.** In Germany, Comfrey products have been used as medicine since the 1920s. External Comfrey preparations have been applied particularly often in the fields of sports and casualty medicine.
In a double-blind, multicenter, randomized, placebo-controlled, group comparison study on patients suffering from **unilateral acute ankle sprains** (number = 142; mean age 31.8 years, 78.9% male), the percutaneous (through skin) efficacy of **an ointment of Comfrey extract (Kytta-Salbes f®, four treatments per day for 8 days) was confirmed decisively.
Compared to placebo, the active treatment was clearly superior** regarding the reduction of pain (tonometric measurement,

p<0.0001; as the primary efficacy variable) and ankle edema (swelling) (figure-of-eight method, p=0.0001). Statistically significant differences between active treatment and placebo could also be shown for ankle mobility (neutral zero method), and global efficacy. **Under active treatment, no adverse drug reactions were reported. The good local and global tolerance of the trial medication could also be confirmed.** *The study results are consistent with the known pre-clinical and clinical data concerning Comfrey."*
-'Efficacy and Tolerance of a Comfrey Root Extract (Extr. Rad. Symphyti) in the Treatment of Ankle Distortions: Results of a Multicenter, Randomized, Placebo-Controlled, Double-Blind Study' by R. Koll, et al., Phytomedicine, Vol 11, page 470-477, **2004**. (This article is based on a study first reported in German in Zeitschrift fur Phytotherapie {Journal of Phytotherapy}, Volume 21, pages 127-134, 2000.)   (I have no connection with the trademarked ointment used in this study.)

    (A placebo is a harmless and useless substance in the form of a pill/capsule or liquid that is given for the psychological benefit of the patient rather than for any medicinal effect. In clinical trials it is used as a control or baseline to compare against a similar substance given that does have medicine.)

"Herbal Medicine Review of:
'Efficacy of a Comfrey Root Extract in Comparison to a Diclofenac Gel in the Treatment of Ankle Distortions: Results of an Observer-Blind, Randomized Multicenter Study' by Predel, Giannetti, Koll, Bulitta and Staiger, Phytomedicine12, pages 707-714, 2005. The purpose of this German study was to compare the efficacy of Comfrey ointment against the pharmaceutical drug Voltaren® (brand name for Diclofenac) gel in the treatment of unilateral ankle sprains.
In terms of ankle pain, the researchers found that the Comfrey ointment group had a statistically better clinical outcome than the Voltaren® group. *In regards to pain sensation at rest and during movement, the results showed that participants applying Comfrey ointment experienced less pain at both rest and movement than the Voltaren group.*
The results for ankle swelling showed that Comfrey ointment produced a better result than the application of Voltaren® gel. *The results of this study should also be of interest to massage therapists, who see more cases of ankle sprain than herbalists."*
-'Comfrey Ointment Gives Better Results in Ankle Sprains than Voltaren Gel' by Raymond Khoury, Journal of the Australian Traditional-Medicine Society, Volume 11, Issue 2, page 91, June **2005**.

"The ankle sprain is one of the most frequent sports injuries. Amongst the many treatments used, ointments containing for example a Comfrey extract are an interesting alternative to Non-Steroid Anti-Inflammatory Drugs (NSAIDs).
A comparative clinical study, comparing an ointment made from Comfrey extract and a diclofenac gel, showed very convincing results: the results of the Comfrey extract ointment (Kytta-Salbe® f) equalled and even surpassed those of the gel, achieving significant pain and inflammation reduction in a week."
-'The Sprained Ankle: Comparative Study of a Comfrey Root Extract Ointment Versus a Diclofenac Gel' by S.K. Coulibaly, S. Courau and Christiane Staiger, Merck in France and Germany; Phytotherapie (Phytotherapy), Volume 7, No. 3, pages 147-149, June **2009**.   (All in French except abstract is in English. If you have an English translation, I would appreciate a copy of it.)

    (I do not have any connection with these patented topical Comfrey preparations.)
    (Kytta-Salbe® f: Comfrey root topical salve by Merck for sore muscles, 'beat-up' joints, tennis elbow, and lower back pain. Contains butyl-, ethyl-, methyl-, 2-methylpropyl-and propyl-4-hydroxybenzoate {parabens}, peanut oil, cetostearyl alcohol, benzyl benzoate and bergamot.
    Merck & Company, Inc., doing business as Merck Sharp & Dohme outside the United States and Canada, is an American pharmaceutical company and one of the largest pharmaceutical companies in the world. It was established in 1891 as the United States subsidiary of the German company Merck, which was founded in 1668.)

"Harras cultivar/variety of Comfrey: A specifically selected Comfrey herb extract for phytotherapy Comfrey (Symphytum officinale sensu lato)is an outstanding example for the rational use of medicinal plants against blunt traumas such as contusions (bruises), strains and sprains. The use against muscle and joint complaints, e.g. against muscle pain in the back, is newly added to these well-established indications.
A high performance Comfrey cultivar was specifically selected and cultivated for the use in medicinal products. **It was registered at the 'European Plant Variety Office' with the denomination Symphytum x uplandicum Nyman 'Harras'.**
In an exemplary (best) way this variety fulfils the requirements for quality, efficacy and safety. Among other factors, quality is given by the reproducible cultivation, the -within narrow margins- constant phytochemical composition and the non-application of herbicides and pesticides. The efficacy against blunt traumas with and without abrasions (scrapes) has been demonstrated in clinical trials. It also includes the application in children.
Safety is concluded from the observation of an exceptionally good clinical tolerability and from the **lack of potentially toxic pyrrolizidine alkaloids in the harvested plant material**.
From the point of view of phytotherapy the fact that all clinical trials with the preparation from the above ground plant parts have been performed with exactly the same plant variety and quality is also unique: It allows a direct transferability between studies of clinical findings with respect to efficacy and safety."
-'Comfrey Herbal Extract in Phytotherapy' by M. Schmidt, Zeitschrift fur Phytotherapie (Journal of Phytotherapy), Germany, Volume 33, No. 3, pages 114-117, **2012**.   (sensu lato= Latin 'in the wide sense')

    (Harras Pharma Curarina Arzneimittel GmbH, Munich, Germany, www.harraspharma.de. For over 60 years, it has 'developed, manufactured, and marketed herbal medicine products. Harras Pharma Curarina commissions experienced scientists to carry out multicentre, placebo-controlled trials.')

See subsection 'Hybrids, Hybrid Swarms, Introgression' in section 'Symphytum Species Overview' (Chapter 6, Volume 1).

"Comfrey (Symphytum officinale sensu lato = Latin 'in the wide sense') is a complex of hybrids of Symphytum officinale L. (sensu stricto = Latin 'in the strict sense') and Symphytum asperum Lepechin (Prickly Comfrey) and/or other species of Symphytum. **The various hybrids, summarized under the botanical name Symphytum x uplandicum Nyman, differ in their phytochemical properties.**
The cultivation of a high performance cultivar especially selected for the absence of pyrrolidine alkaloids may therefore be considered an important contribution to efficacy and safety of comfrey preparations.
Based on selection, the cultivar 'Harras' (Symphytum x uplandicum Nyman) was introduced into cultivation. The cultivar excels in the absense of detectable quantities of pyrrolidine alkaloids in the aerial (above ground) plant organs.
A preparation using herbal drug substance obtained from the cultivar 'Harras' was amply and successfully tested in clinical trials for its applicability in medicinal products. The absence of pyrrolidine alkaloids allows the application on broken skin, e.g., abrasions.
-'High Performance Cultivar Harras as a Contribution to Quality, Efficacy and Safety of Comfrey (Symphytum x uplandicum Nyman)' by M. Schmidt, Journal of Medicinal and Spice Plants, Germany, Volume 13, No. 4, pages 182-184, December **2008**.

"Growing Harras Cultivar of Comfrey:
    In southern Germany under biological conditions, i.e., without use of herbicides or pesticides, 'Symphytum x uplandicum Nyman Harras' is grown.
    The harvest takes place at the time of full flowering, *the climatic conditions allow two harvests per year in Bavaria. It will be with annual harvest in the range of 16-20 tons/hectare of fresh Comfrey.*
Pyrrolizidine Alkaloids in Above-Ground Harras Comfrey:
    The analysis techniques have with the introduction of new procedures, **regular reductions in detection limits for pyrrolizidine alkaloids.** *Such were the first, thin-layer chromatographic methods from today's perspective still relatively insensitive.*
    Today's method by means of LC-MS (Liquid Chromatography Electrospray Ionization Mass Spectrometry) or GC-MS (Gas Chromatography-Mass Spectrometry) allowed the **detection of pyrrolizidine alkaloids far below 0.1 ppm (parts per million) and thus in irrelevant low quantities.**
Traumaplant® Ointment Production and Testing:
    The Comfrey variety 'Harras' came from the exclusively vegetatively propagated plant material with the resulting manufactured Topikum (Traumaplant®). Systematic investigations of clinical features of the Comfrey species 'Harras' have been documented **since 1989.**
    The production of the ointment begins with the freshly harvested, flowering above-ground parts of plants, from those immediately it becomes a concentrate. *This liquid ingredient will be checked through appropriate analysis methods in terms of defined marker compounds and the absence of contaminants like heavy metals, pesticides or aflatoxins, but also of pyrrolizidine alkaloids.*
    It will then go into an emulsion base incorporated in the finished medicinal product: **10 grams of the aqueous-ethanolic concentrate from the variety Comfrey 'Harras' (DEV = 2.5: 1) in 100 grams of ointment."**
-'High Performance Cultivar Harras as a Contribution to Quality, Efficacy and Safety of Comfrey (Symphytum x uplandicum Nyman)' by M. Schmidt, Journal of Medicinal and Spice Plants, Germany, Volume 13, No. 4, pages 182-184, December **2008**. (This was translated online from German to English. If you have a better translation, please send it.)

Tumors Reduced (see Cancer in this section)

"A cold water extract of leaves of the plant Symphytum officiale (officinale) was given to mice bearing spontaneous or transplanted tumors. *The tests involved 73 control and 52 experimental mice bearing spontaneous mammary tumors of various initial sizes, and 71 control and 63 experimental animals bearing transplanted mammary tumors.*
Survival time of the spontaneous tumor mice that received Comfrey extract was increased an average of 59% as compared with the controls.
In the transplanted tumor tests, tumor weight at autopsy averaged 24% less in the Comfrey extract mice as compared with the controls. These differences between control and experimental groups are statistically significant.
Spontaneous tumors in 2 mice receiving the plant extract regressed completely within 2 weeks of the treatment. The extract was not associated with any evidence of toxic effects.
Most of the work on the problem of cancer therapy has been carried out from the standpoint of finding drugs which directly interfere with the growth of tumor tissue. But there is the possibility that host resistance to the disease can be so increased that the tumor is secondarily affected.
In the present investigation, the experimental mice appeared to be more lively and have a generally healthier appearance, even when the tumor mass was as large or larger than in the control animals."
-'Protective Effect of Symphytum Officiale (Officinale) on Mice Bearing Spontaneous and Transplant Tumors' by Alfred Taylor and Nell Carmichael Taylor, Clayton Foundation Biochemical Institute, University of Texas, Austin, Texas; Proceedings of The Society for Experimental Biology and Medicine, Washington, D.C., Volume 114, No. 3, pages 772-774, January **1964**.

"Report About Anti-Mitotic Effects of Comfrey:
The crude watery extract of Symphytum officinale *and certain proteic (protein) and carbohydrate components isolated from it were studied for their effect upon the in vivo and in vitro proliferation of Ehrlich ascites cells, EL-4 cell line and of human T lymphocytes (white blood cells) and upon the respiratory burst of human PMN granulocytes stimulated via Fc receptors.*
The results indicate that the Comfrey crude extract and its proteic (protein) fraction stimulate the in vivo proliferation (increase) of the studied neoplastic (tumor) cells and exert an anti-mitotic (reduces cell division) effect on human T lymphocytes in vitro stimulated with PHA (Phytohemagglutinin: a toxic plant protein).
The vegetal preparations have remarkable effects on the respiratory burst of the granulocytes non-stimulated and stimulated via Fc receptors. ***The data underline the necessity to study thoroughly the effects of different phyto compounds through both pharmacological and immunological methods."***
-'Action of Some Proteic and Carbohydrate Components of Symphytum Officinale upon Normal and Neoplastic Cells' by A. Olinescu, G. Manda, M. Neagu, S. Hristescu and C. Dasanu, Cantacuzino Institute, Bucharest, Romania; Roumanian Archives of Microbiology and Immunology, Volume 52, No. 2, pages 73-80, April-June **1993**.
(I only have the English abstract of this report. If you have an English translation, I would appreciate a copy.)

"Previous Two-Month Comfrey Experiments and Tumor Formation:
Comfrey or Symphytum officinale L. *(Boraginaceae) is a very popular plant used for therapeutic purposes. Since the 1980s, its effects have been studied in 'long-term' carcinogenesis (cancer causing) studies, in which Comfrey extract is administered at high doses during several months and* ***the neoplastic (uncontrolled growth of abnormal tissue) hepatic (liver) lesions (damage) are evaluated.***
However, the literature on this topic is very poor considering the studies performed under short-term carcinogenesis protocols (official procedures), such as the 'Resistant Hepatocyte Model' (RHM). In these studies, it is possible to observe easily the phenomena related to the ***early phases of tumor development,*** *since Pre-Neoplastic Lesions (PNLs) rise in about 1-2 months of chemical induction.*
Current Comfrey Experiment:
Herein (in this report), the effects of chronic oral treatment of rats with ***10% Comfrey ethanolic extract were evaluated in a RHM.*** *Wistar rats were sequentially treated with N-nitrosodiethylamine (intraperitoneal) and 2-acetilaminofluorene (po), and submitted to hepatectomy (remove all or part of liver) to induce carcinogenesis promotion. Macroscopic/microscopic quantitative analysis of PNL was performed. The level of significance was set at $P<0.05$.*
Comfrey treatment reduced the number of Pre-Neoplastic macroscopic Lesions up to 1 mm *($P<0.05$), the percentage of oval cells ($P=0.0001$) and mitotic figures ($P=0.007$), as well as the number of Proliferating Cell Nuclear Antigen (PCNA) positive cells ($P=0.0001$) and acidophilic Pre-Neoplastic nodules ($P=0.05$).*
On the other hand, the percentage of cells presenting megalocytosis ($P=0.0001$) and vacuolar degeneration ($P=0.0001$) was increased. Scores of fibrosis, glycogen stores and the number of nucleolus organizing regions were not altered.
Anti-Mitotic (stops cell division) versus Cancer-Causing Effects of Comfrey:
The study indicated that oral treatment of rats with 10% Comfrey alcoholic extract reduced cell proliferation (growth) in this model. This convergence (joining or meeting) of results is according the previous studies in which the anti-mitotic (reduces cell division) effects of pyrrolizidine alkaloids are preponderant (dominant) over the carcinogenic ones in short-term experimental procedures.
Conclusion: Comfrey Reduces Pre-Tumor Damage:
The study of Comfrey and other herbs largely used as complementary (helping) medicine deserves more ***specific studies, since plant extracts are chemically diverse and may be able to modulate (control) several metabolism steps simultaneously.***
In conclusion, the present findings can contribute to understand a very particular and under explored Comfrey pharmacological activity, since ***the treatment of rats with its extract seems to protect them from early development of pre-neoplastic (pre-tumor) liver lesions, by inhibiting (reducing) cell proliferation (growth) and modulating atypical (not typical) phenotype (observable characteristics)."***
-'Comfrey (Symphytum Officinale. L.) and Experimental Hepatic Carcinogenesis: A Short-Term Carcinogenesis Model Study' by Gomes, Massoco, Xavier and Bonamin, Evidence-Based Complementary and Alternative Medicine, Volume 7, No. 2, pages 197-202, **2010**.   (**An anti-mitotic inhibits mitosis or cell division, which reduces tumor growth.**)
(Different types of studies about the tumor-causing or tumor-reducing effects of Comfrey extract give opposite results. **This study is different from most research because it compares the anti-mitotic effects of Comfrey to the carcinogenic effects. Most studies have ignored the anti-mitotic effects.**)

Ulcer: Gastric / Duodenal

*"**Symphytum officinale: A strong mucilaginous Infusion of the root. Synonym: Mucilago Symphyti.***
Internal use of the Infusion satisfactory in gastralgia, eases pain of gastric ulcer, *indeed the internal use is more wonderful than its effect in external application. It promotes the growth of new mucous tissue. -W. Bramwell, *B.M.J. i./12,12.*
This plant and Symphytum tuberosum have both the same properties, *resembling the Common Borage, Borago Officinalis. Tisanes (teas), in many respects, both yielding Allantoin-containing infusions."*

-'Extra Pharmacopoeia, 18th Edition' by Martindale, Volume 1. London, England: H.K. Lewis & Co., **1924**. Symphytum page 862.
(* -'The New Cell Proliferant: A Note on the Symphytum Officinale or Common Comfrey' by William Bramwell, M.A., M.D., B.Ch., The British Medical Journal, London, England, pages 12-13, January 6 1912.)
(Gastralgia is pain in the stomach or epigastrium especially of a neuralgic {nerve} type.)

"*Dried Comfrey leaves (Symphytum officinale) gathered as a mature crop on a herb farm in Germany, were homogenized in Krebs solution, filtered and a stock solution made equivalent to 10 mg dried leaves in 1 ml (0.0338 fluid ounces).*
The bioassay results demonstrate that an aqueous (Krebs solution) extract of Comfrey (concentrations equivalent to 50 microgram/ 5 mg dried leaf) increases the release of prostaglandin-like material from rat gastric corpus and antrum.
Radioassay and h.p.l.c. indicated greater outputs of PGF 2alpha, and 6-keto-PGF 1alpha.
Since various prostaglandins can protect the gastric mucosa (Robert 1977) this might explain the use of Comfrey leaves in gastric upsets. *Comfrey leaves contain protein, pyrrolizidine alkaloids (Hirono et al 1978), allantoin, tannin and mucilage, but it is not known to what extent these contribute to the increased prostaglandin synthesis by rat stomach.*"
-'The Effect of an Aqueous Extract of Comfrey on Prostaglandin Synthesis by Rat Isolated Stomach' by I. F. Stamford and I. A. Tavares, King's College Hospital Medical School, London, England; Journal of Pharmacy and Pharmacology, Volume 35, pages 816-817, December **1983**.
(Krebs–Henseleit solution contains sodium {Na}, potassium {K}, chloride {Cl}, calcium {Ca}, magnesium sulfate {MgSO4}, bicarbonate {HCO3}, phosphate {PO4}, glucose, albumin, and tromethamine {THAM}.)
(Prostaglandins are physiologically active lipid compounds that have hormone-like effects in animals.)

"*As numerous prostaglandins have been found to protect the gastric mucosa (mucous memberane), there may be a biologic basis for use of Comfrey as a treatment for peptic ulcers.*"
-Toxicology and Clinical Pharmacology of Herbal Products edited by Melanie Johns Cupp, PharmD, BCPS. Totowa, New Jersey: Humana Press, **2000**. Chapter 18: Comfrey by David Burch and Melanie Johns Cupp, page 205.

Wounds: Abrasion (scrapes)

"***The wound-healing efficacy of a 10% Symphytum peregrinum (synonymous with Symphytum x uplandicum Nyman, Russian Comfrey) ointment*** *was tested on experimentally produced open shallow wounds with an intact columnar layer in healthy volunteers. The ointment base and a polyacrylamide agar gel without active substances served as controls (no Comfrey in ointment).*
After application of the Symphytum peregrinum ointment, the time taken for the wounds to heal completely was clearly reduced in comparison to the two control preparations, and the difference with respect to the active substance-free ointment vehicle was statistically significant *(p < 0.05).*"
-'Effect of an Active Substance Complex from Symphytum on Epithelialization' by R. Niedner, Acta Therapeutica: Journal of Pharmaceutical Medicine, Brussels, Belgium, Volume 15, pages 289-297, **1989**.

"*Wound healing effects of a topically (skin) applied preparation (Traumaplant®) containing a concentrate (10% active ingredient) from the aerial (above ground) parts of medicinal Comfrey (Symphytum x uplandicum Nyman) were examined in a randomized clinical double-blind study including 278 patients with fresh abrasions (scrapes) (verum: number = 137), among them 64 patients of up to 20 years of age (verum n = 29, reference product n = 35).*
After 2-3 days of application of the study medication a highly significantly and clinically relevantly faster initial reduction of wound size *of 49 + or - 19% versus 29 + or - 13% per day in favour of verum (Comfrey) (p < 5x10(-21)) was found. The physicians rated efficacy as good to very good in 93.4% of cases, as compared to 61.7% in the group treated with the reference product (p = 2 x 10(-11)).*
No adverse effects or problems with drug tolerability occurred. *Specifically, cutaneous (skin) reactions were observed in none of the patients throughout the 10 day observation phase.* ***Symphytum herb extract can be attributed distinct wound healing effects, effects that can explicitly (clearly) be used in paediatry (children).*"
-'Wound Healing Effects of a Symphytum Herb Extract Cream (Symphytum x uplandicum Nyman): Results of a Randomized, Controlled Double-Blind Study' by M. Barna, A. Kucera, M. Hladicova and M. Kucera, Wiener Medizinische Wochenschrift (Vienna Medical Weekly, Austria), Volume 157, No. 21-22, pages 569-574, **2007**.
(Also see 'Randomized Double-Blind Study: Wound-Healing Effects of a Symphytum Herb Extract Cream {Symphytum x uplandicum Nyman} in Children' by M. Barna, A. Kucera, M. Hladicova and M. Kucera, Prague, Czech Republic; Arzneimittel-Forschung (Drug Research), Germany, Volume 62, No. 6, pages 285-289, June 2012.)

"*Healing is a physiological process with the objective of repairing damaged tissue. Synthetically, this mechanism has three stages – inflammatory, (cell) proliferation and remodeling – that occur gradually and dynamically (Mondolin and Bevilacqua 1985; Serhan et al. 2008).* ***Comfrey was cited as one of the most used plants to heal wounds and to treat external skin problems, according to the studies performed on the Brazilian population*** *(Parente and Rosa 2001; Luz 2001; Ritter et al. 2002; Champs et al. 2003; Souza and Felfili 2006).*
The present work evaluates wound healing activity of leaves extracts of Symphytum officinale L. (Common Comfrey) incorporated in three pharmaceutical formulations. *Wound healing activity of Comfrey was determined by qualitative and quantitative histological (microscopic) analysis of open wound in rat model, using allantoin as positive control.*

Three topical (skin) formulations, carbomer gel, glycero-alcoholic solution, and O/W (Oil/Water) emulsion (soft lotion) were compared. The histological analysis of the healing process shows significant differences in treatment, particularly on its intensity and rate. **The results indicate that emulsion (fine mixture) containing both extracts, commercial and prepared, induced the largest and furthest repair of damaged tissue.**
This could be evidenced from day 3 to 28 by increase in collagen deposition from 40% to 240% and reduction on cellular inflammatory infiltrate from 3% to 46%.
However, 8% prepared Comfrey extract in emulsion presented the best efficacy. This work clearly demonstrates that Comfrey leaves have a wound healing activity. *The O/W (Oil/Water) emulsion showed to be the vehicle most effective to induce healing activity, particularly with extracts obtained from Comfrey leaves collected in Minas Gerais state in Brazil.* **It shows the best efficacy to control the inflammatory process and to induce collagen deposition at 8% concentration."**
-'In Vivo Healing Effects of Symphytum Officinale L. Leaves Extract in Different Topical Formulations' by L.U. Araujo, et al., Departamento de Farmacia, Unversidade Federal (Dept of Pharmacy, Federal University) dos Vales do Jequitinhonha e Mucuri, Brazil; Pharmazie (Pharmacy), Volume 67, No. 4, pages 355-360, **2012**.
(Histology is the study of the microscopic anatomy of cells and tissues of plants and animals.)
(Collagen is a protein that is major component of connective tissues that are part of tendons, ligaments, skin, muscles.)

"Wound and Burn Ointment: The authors present and evaluate an original product in the armamentarium of the preparations for burns and wounds treatment, made solely on the basis of medicinal plants and natural ingredients.
The ointment formulation comprises olive oil extract from a mixture of **nine medicinal plants with wound healing activity** *(Calendula officinalis L., atricariachamomilla L.,* **Symphytum officinale L.,** *Hypericumperforatum L., Achilleamillefolium L., Arctiumlappa L., Plantago major L., Althaea officinalis L., Quercusrobur L.), sea buckthorn oil, lavender essential oil and as thickening agents, coconut oil, beeswax and conifer resin.*
The LC-MS (Liquid Chromatography-Mass Spectrometry) analyses of the ethanolic extracts from the plant mixture and of the ethanolic re-extracts from oil plant mixture extract and from ointment have evidenced **high levels of polyphenols** like caffeic, chlorogenic, gallic and ferulic acids, as well as quercetin and rutin, **all of which being known compounds with good wound healing activity.** The neutral red assay has shown **no cytotoxic (toxic to cells) effect** on fibroblast NCTC cell line exposed to herbal extracts. **Finally, the wound healing action of the submitted ointment has been clinically confirmed and highlighted by some case reports."**
-'An Innovative Ointment Made of Natural Ingredients with Increased Wound Healing Activity' by Andrei Zbuchea, Liliana Lungu, Claudia-Valentina Popa, Victorita Tecuceanu, Valentina Alexandru and Rodica Tatia, Romanian Biotechnological Letters, University of Bucharest, Romania, Volume 21, No. 2, pages 11427-11437, January **2016**.
(Armamentarium is the medicines, equipment, and techniques available to a medical practitioner.)

FIG. 4.—COMFREY (SYMPHYTUM OFFICINALE).

Comfrey (Symphytum Officinale) 1907

'American Root Drugs: Bulletin No. 107'
by Alice Henkel, Drug-Plant Investigations,
Bureau of Plant Industry, United States Department
of Agriculture, Washington, DC, October 25 1907.

Symphytum pages 57-58.
Plate VI, Figure 4: *"Comfrey (Symphytum officinale) showing the thick, rough leaves, the clusters of flowers, lower portion of plant with root, and sections of root."*

"More than half of the root drugs recognized in the Eighth Decennial Revision of the United States Pharmacopoeia occur in this country, some native and not growing elsewhere and others introduced. All of the official root drugs found in the United States have been included in this bulletin." -page 9

Chapter 25

Personal / Clinical Observations of Healing

If you are not feeling well, please consult your health care advisor.
These medicinal observations with Comfrey are anecdotal / testimonial that are for entertainment purposes only.

Overview of Comfrey as Healer

"**Thomson had witnessed the good done by the widow Benton with the simple herbs,** in which he so much delighted, and **when suffering in the ankle from severe laceration, he applied Comfrey root** and turpentine, and cured it.
He had cured numberless forms of disease in himself, wife, children, and neighbours. He must find some thing like a system to build his practice upon. The herbs he felt convinced were the gift of God to the human family."
-'A Plea for the Botanic Practice of Medicine' by John Skelton, Lecturer and Professor of the Botanic Practice of Medicine, Leeds and London, England, **1853**. A Thomsonian herbal practitioner (Physio-Medicalism).

"**Comfrey, Symphytum officinale Properties and Uses: The plant is demulcent and slightly astringent. All mucilaginous agents exert an influence on mucous tissues,** hence the cure of many pulmonary and other affections in which these tissues have been chiefly implicated, by their internal use. Mucilaginous agents are always beneficial in scrofulous and anemic habits.
Physicians must not expect a serous disease to yield to remedies which act on mucous membranes only. To determine the true value of a medicinal agent, they must first ascertain true character of the affection, as well as of tissues involved.
Comfrey root is very useful in diarrhoea, dysentery, coughs, hemoptysis or bleeding of the lungs, and other pulmonary affections; also in leucorrhoea and female debility: **all these being principally affections of mucous membranes.**"
-'The Complete Herbalist, or the People Their Own Physicians by the Use of Natural Remedies' by Dr. O. Phelps Brown, Jersey City, New Jersey, **1897**, page 74.   (Scrofulous is bacterial infection of the lymph nodes such as tuberculosis.)
(Serous fluid is body fluid resembling serum. It is a pale yellow, transparent liquid.)

"**An interesting point is the fact that in country districts Comfrey is still valued by agricultural and other workers on account of its curative properties.** When visiting a farm at Tarvin in Cheshire (County, England) many years ago, I was interested to find that its owner always kept a bed of Comfrey in order that he might provide villagers with it when occasions arose. Mr. Edwin Green of Cheltenham (Gloucestershire County, England) in 1912 wrote that this plant (Comfrey) when cooked the same way as spinach is used very largely by many people during the spring.
He stated that **Comfrey is well known there as a blood purifier** not only for human beings but also for cattle and horses, the effect on the latter being to produce a wonderful glossy coat."
-The Medicinal Uses of Comfrey by Dr. Charles MacAlister, M.D., F.R.C.P., written **1935**. (in 'Comfrey: Fodder, Food & Remedy')

"**Ever since our first Newsletter (Henry Doubleday Research Association) was issued in 1958, we have been receiving reports from Members and others of the benefits of using Comfrey medicinally, and we have reproduced these reports in our Newsletter.** They are not in any sense medical reports, for none of these appear to be available, except of course from the work of the late Dr. Charles MacAlister in the 1930s, and from the late Dr. H.B. Kirschner in the United States.
They are merely unsolicited testimonials to the healing powers of Comfrey which have been known since 400 A.D."
-Comfrey: The Herbal Healer by Henry Doubleday Research Association, **1975**, 41 pages, page 3. (in 'Comfrey Report' book.)
(I think the writer may have meant Dr. H.E. Kirschner who wrote 'Live Food Juices: For Vim, Vigor, Vitality' in 1972, and 'Natures Healing Grasses' in 1975.)

"**Comfrey isn't a cure-all, of course. It's just one of the many health-sustaining plants in the garden.** We see it more as an example of strength, as a balancer, a provider of many qualities and substances for the diet and for the garden soil that are not supplied by other plants."
-Enchanted Garden: Alan Chadwick's Organic Method of Gardening by Tom Cuthbertson. London, England: Rider & Company / Hutchinson & Co. Publishers Ltd, **1978**, page 122.

"**Apart from its high protein content, the four major components of Comfrey implicated in biological activity are allantoin (5-ureidohydantoin), mucilage (mucopolysaccharide of fructose and glucose), tannin, and pyrrolizidine alkaloids.**"
-'Comfrey' or 'Herbal Medicine: Comfrey' by D.V.C. Awang, Health Protection Branch, Health and Welfare Canada; Canadian Pharmaceutical Journal, Volume 120, pages 101-104, February **1987**.

"**It is not a cure-all, but Comfrey will help to build up our defenses against all forms of infection and the effects of pollution of the air, water and food.** It is a protective addition to help balance our needs and strengthen our resistance."
-Comfrey: Nature's Healing Herb & Health Food by Andrew Hughes. Japan: Sanyusha Publishing Co., Ltd, **1992**, page 9.

"Comfrey is probably used for more different health purposes than any other herb."
-Modern Encyclopedia of Herbs by Joseph M Kadans, N.D., Ph.D. New York, New York: Simon & Schuster, Inc., **1993**, page 98.

Abscess (Skin), Boil or Furuncle (swollen skin with pus)

Do not use Comfrey externally on deep wounds. See 'Wounds' in this section for more information about it.

"An Impostume (abscess): Comfrey root, dock root, valerian root, butter, old lard, and sulphur, pounded well together, and expressed through a cloth, are useful for an impostume."
-'The Physicians of Myddvai; Meddygon Myddvai, or The Medical Practice of the Celebrated Riwallon and His Sons, of Myddvai (Myddfai), in Caemarthenshire (Wales), Physicians to Rhys Gryg, Lord of Dynevor and Ystrad Towy, About the Middle of the Thirteenth Century', English translation by John Pughe, edited by Reverend John Williams Ab Ithel, for Welsh MSS Society, published in Llandovery, Wales and London, England, **1861**, page 46.

Anal Itching (Pruritis Ani) (also see Hemorrhoids)

*"One patient, an old gentleman of gouty diathesis (susceptibile to attacks of gout), who will not diet or take medicine, **finds the nightly application of the mucilage of Symphytum the only remedy of the many he has tried which gives him any relief for pruritis ani**. His sufferings at one time were so intense that he says he could not sleep for weeks together. I judge the explanation of its efficacy in this case to be very simple.*
When the mucilage is taken into the mouth, there is a sense of dryness, due to some astringent principle, probably tannin; at the same time, if the air be drawn in there is a sensation of coldness and numbness much like that felt after peppermint, though to a lesser degree, but without the previous burning sensation, and hence it is slightly anaesthetic.
*In pruritus ani these two principles acting together, the one driving back the acid-laden blood, the other **soothing the irritated nerve endings**, will, I think, account for the relief which is felt in this troublesome complaint."*
-'The New Cell Proliferant: A Note on the Symphytum Officinale or Common Comfrey' by William Bramwell, M.A., M.D., B.Ch., The British Medical Journal, London, England, pages 12-13, January 6 **1912**.
(Gout is a type of arthritis that causes inflammation of joints due to excess uric acid.)

*"**Pruritus Ani: This is a condition of itching in the region of the anus or rectum. The direct application of the Comfrey powdered leaf or root will usually result in a healing of the tissue.**"*
-Modern Encyclopedia of Herbs by Joseph M Kadans, N.D., Ph.D. New York, New York: Simon & Schuster, Inc., **1993**, page 100.

Anemia (not enough red blood cells, or hemoglobin in blood is low)

*"**Comfrey has been found valuable for cases of anemia, increasing appetite and providing nutrients** needed by the body. It may be taken with unsweetened pineapple juice or as a tea. It may also be swallowed in the form of gelatin capsules."*
-Modern Encyclopedia of Herbs by Joseph M Kadans, N.D., Ph.D. New York, New York: Simon & Schuster, Inc., **1993**, page 102.

*"**Anemia:** 1 ounce (28.3 gram) Comfrey root (Symphytum officinale)*
1 tablespoonful Garlic, fresh juice (Allium sativum)
Preparation: *Simmer the mixture slowly for 20 minutes in 1 quart (0.94 liter) of water. Strain, bottle and keep in a cool place.*
Dosage: *1 wineglassful, every 4 hours."*
-'Beatrice® Nutrition Recipes: Comfrey' by Beatrice® Nutrition and Health Sciences, Division of Beatrice® Foods Co., Phoenix, Arizona, **2017**.
(As an apothecary measure, wineglass / wineglassful / cyathus vinarius was defined as 1/8 pint or 2 fluid ounces or 59 ml {2 1/2 fluid ounces in imperial system}. Before 1800 it was 1 1/2 fluid ounces or 44 ml. These measurements are no longer relevant to current capacity of wineglasses.)

Arthritis

*"**The following is Dr. Thomson's poultice for 'White Swelling': Take fresh Comfrey roots,** scrape or grate them fine, until you obtain half a pint (1 cup = 236 ml), add to this the white of three eggs, and a gill (1/2 cup = 118 ml) of brandy, or sufficient to make it into a proper consistence, for a poultice; spread and bind it on the affected part as firmly as it can be borne, renewing or wetting it with brandy, as often as it becomes dry. Bear in mind that the object to keep in view is, the restoration of a healthy circulation to the parts, for there is no healing power apart from the blood."*
-'Family Medical Adviser', Fifth Edition (First Edition 1852), by Dr. John Skelton, Senior, Lecturer and Professor of Medicine, Leeds, England, **1857**. A Thomsonian herbal practitioner (Physio-Medicalism).
('White swelling' is a localized enlargement in the knees, ankles or other joints. It can occur in tuberculous arthritis that is caused by the bacteria, Mycobacterium tuberculosis.)

"Report from Mrs. T. of Godalming (Surrey County, England): 'You will be pleased to hear that in my case the **Comfrey tea was a positive help and cure. I am relatively young to have arthritis, had pain in my hands** on and off for the last three years but the 1969 winter was the worst. More than that the pain stayed with me through the spring and summer of 1970. I bought a 1/4 pound (113 gram) packet of tea. In early September I tried it for my hands taking approximately half a cup (118 ml) twice a day.
For about a month there was no change at all, my hands were aching, then one day the ache just stopped. I kept on drinking the tea which lasted until about the end of November. I had no pains in my hands for 3-4 weeks after I stopped taking Comfrey tea. Since Christmas the pain came back though not so strongly, but gradually got worse."
-Comfrey: The Herbal Healer by Henry Doubleday Research Association, **1975**, 41 pages, page 29. (in 'Comfrey Report' book.)

"In the economically less developed rural areas of Ukraine and Russia, the use of NWFPs (Non-Wood Forest Products) continues to be an important part of livelihoods, both as a source of income and for domestic use as food and medicine.
In Sweden the collection of wild food has become mainly a recreational activity, and the use of medicinal plants is no longer prevalent among our respondents. **Symphytum officinale L., Comfrey: aerial parts (leaves) are used as liquids with alcohol (tincture) for Arthritis 'when hands hurt'.**"
-'From Economic Survival to Recreation: Contemporary Uses of Wild Food and Medicine in Rural Sweden, Ukraine and NW Russia' by N. Stryamets, M. Elbakidze, M. Ceuterick, P. Angelstam and R. Axelsson, Journal of Ethnobiology and Ethnomedicine, Volume 11, No. 53, 18 pages, **2015**.

Athlete's Foot (a fungus)

"Mrs. R.P.E. Dorking (Surrey County, England) reports: '**I have completely cured Athlete's Foot** which I picked up in a nursing home when my daughter was born 21 years ago. I have tried countless remedies over the years, but the **Comfrey ointment cleared it up in about three weeks**, and it has stayed quite clear."
-Comfrey: The Herbal Healer by Henry Doubleday Research Association, **1975**, 41 pages, page 5. (in 'Comfrey Report' book.)

"**There's one gardener who even uses Comfrey against athlete's foot.
Whenever her toes are invaded by fungus, she breaks off a leaf of Comfrey at the stem and rubs the mucilaginous juice over the cracks that form betweeen her toes. The cracks close up, and within a week, she says, the itching is gone.** I wouldn't claim that Comfrey will cure all fungus infections, or even that it can control all cases of athlete's foot, but this is a good example of how you can make use of the strength of a plant that grows right in the garden instead of going to the drugstore to buy some cream or another that you know nothing about."
-Enchanted Garden: Alan Chadwick's Organic Method of Gardening by Tom Cuthbertson. London, England: Rider & Company / Hutchinson & Co. Publishers Ltd, **1978**, page 122.

"**Fungal Skin Infections, Including Athlete's Foot: Make a Comfrey poultice** and firmly apply to the affected area for 1-2 hours each day. Caution: Do not use Comfrey on broken skin."
-Encyclopedia of Herbal Medicine: The Definitive Reference to 550 Herbs and Remedies for Common Ailments by Andrew Chevallier, FNIMH. London, England: DK (Dorling Kindersley), **2000**.
     (A poultice is a soft, moist mass of meal, herbs, etc. applied warm or hot to the skin.)

Bedsores (pressure sores)

"G.M. of London SW15 (England) reports on **Comfrey cream**:
'My mother has **cured my stepfather's bedsores with Comfrey** and finds it invaluable for the other rubbed places and miseries of the semi-paralysed. Both the district nurses now visiting him are completely converted."
-Comfrey: The Herbal Healer by Henry Doubleday Research Association, **1975**, 41 pages, page 6. (in 'Comfrey Report' book.)

"**Bed Sores: Make a salve of six parts pulverized Comfrey leaves** with one part each pulverized calendula flowers, horsechestnut leaves, and plantain leaves."
-'A Mini-Course in Medical Botany' by Dr. James A. Duke, Ethnobiologist, Fulton, Maryland, Dr. Duke's Phytochemical and Ethnobotanical Databases, date unknown but around **1996**, page 33.   https://phytochem.nal.usda.gov/phytochem/search

Bites (see Stings)

Bladder and Urinary Problems (also see Kidneys)

"**An experienced remedy for bloody water:** Take waters of the black alder, of mallows, of each three ounces, **syrup of Comfrey** one ounce: mix them, and let the patient take four spoonfuls immediately; and four or five times a day.

An excellent medicine, though not curative, for those that are tormented with the stone in the bladder:
Take powder of comfrey-roots *an ounce and half, marsh-mallow-roots three ounces, liquorice-powder two drams, seeds of daucus of Crete two drams, seeds of purslane, of winter-cherries, of each half a dram, nutmegs two drams, saffron one dram (drachm = 60 grains = 64.79 milligrams), the species diamargariton (a composition of pearls) frigid six drams, syrup of marsh-mallows four ounces; mix and make a soft electuary (medicinal substance mixed with honey or other sweet), of which let the sick daily take the quantity of a walnut. It is profitable against the stone in the reins (kidneys) and bladder, but chiefly against the latter; as also against the strangury, dysuria, etc."*
-'The Works of the Honourable Robert Boyle in Six Volumes, Volume 5, London, England, **1772**, page 368, 379. Robert Boyle, 1627-1691, was one of the most influential scientists and philosophers of the 1600s. These books were compiled after his death.
(Robert Boyle was born in Ireland, part of the Anglo-Irish. He was founder of modern chemistry. He helped change it from an occult science into a logical method of experiments and observation. His research on properties of gases and his mechanistic theory of matter are forerunners of modern theories. He is known for 'Boyle Lectures' and 'Boyle's Law' that pressure and volume of a gas are inversely proportional. A founding fellow of the 'Royal Society of London'.)
(Strangury is blockage or irritation of the bladder, with severe pain and a strong desire to urinate. Dysuria is painful or difficult urination.)

*"**Inflammation of the Bladder,** a medicine made and given as follows:*
Comfrey root, an ounce (29 ml). *Clivers, an ounce. Burdock root, an ounce and a half. And tansy, an ounce. Bruise and boil the roots in two quarts (1.89 liter) of water down to three pints (6 cups = 1.4 liter), with the cover on, add the herbs, and boil for five minutes more, strain, sweeten with honey, add two ounces of powdered gum arabic, and take a wine glassful four times a day, or oftener for the first two or three days."*
-'Family Medical Adviser', Fifth Edition (First Edition 1852), by Dr. John Skelton, Senior, Lecturer and Professor of Medicine, Leeds, England, **1857**. A Thomsonian herbal practitioner (Physio-Medicalism).

*"Besides other herbs useful and helpful in **management of UTIs (Urinary Tract Infections)**, there are:*
First, the herbs that alkalize urine *are sarsaparilla, peppermint, marshmallow,* ***Comfrey root,*** *plantain, and ginger."*
-'Urinary Tract infection (UTI): Western and Ayurvedic Diagnosis and Treatment Approaches' by Mahsa Ranjbarian, California College of Ayurveda, Nevada City, California, www.ayurvedacollege.com, date unknown but around **2016**.

Blood and Circulation; Blood Pressure

*"**In some parts of Ireland, Comfrey is eaten as a cure for defective circulation and poverty of blood,** being regarded as a perfectly safe and harmless remedy."*
-A Modern Herbal: The Medicinal, Culinary, Cosmetic and Economic Properties, Cultivation and Folk-Lore of Herbs, Grasses, Fungi, Shrubs and Trees with their Scientific Uses by Mrs. M. Grieve. New York: Dover Publications, 1971. First published in **1931**.

"In 'The Herbs of Siberia Used for Cardiovascular Diseases'*, Dr. I.M. Krasnoborov and Dr. S.B. Kaznacheev report that ***Comfrey can reduce blood pressure and facilitate breathing.***
Although some physicians caution that large doses of Comfrey taken internally can injure the central nervous system and the liver, ***testimonies to the value of this herb can be found in nearly all Russian herbals."***
-A Russian Herbal: Traditional Remedies for Health and Healing by Igor Vilevich Zevin, Nathaniel Altman and Lilia Vasilevna Zevin. Rochester, Vermont: Healing Arts Press, **1997**.
(* -The Herbs of Siberia Used for Cardiovascular Diseases by Dr. I.M. Krasnoborov and Dr. S.B. Kaznacheev. Novosebirsk, Russia: Sibirskoe, Otdelenie, 1991.)

Boil (see Abscess)

Bones (see Broken Bones and also see Osteoporosis)

Breast (see Mastitis)

Broken Bones (also see Osteoporosis)

*"**To promote the union of bone. Take Comfrey,** and bruise with wine, pepper and honey, drinking it daily for nine days, and they will unite compactly."*
-'The Physicians of Myddvai; Meddygon Myddvai, or The Medical Practice of the Celebrated Riwallon and His Sons, of Myddvai (Myddfai), in Caemarthenshire (Wales), Physicians to Rhys Gryg, Lord of Dynevor and Ystrad Towy, About the Middle of the Thirteenth Century', English translation by John Pughe, edited by Reverend John Williams Ab Ithel, for Welsh MSS Society,

published in Llandovery, Wales and London, England, **1861**, page 78.

*"Mr. Cockayne relates that the locksman at Teddington (south London, England) informed him how **the bone of his little finger being broken,** was grinding and grunching so sadly for two months, that sometimes he felt quite wrong in his head. One day he saw a doctor go by, and told him about the distress. **The doctor said: 'You see that Comfrey growing there? Take a piece of its root, and champ (chop) it, and put it about your finger, and wrap it up.' The man did so, and in four days his finger was well."***
-'Herbal Simples Approved for Modern Uses of Cure' by W.T. Fernie, M.D., Bristol, England, **1897**.

The same story about the broken finger was told in this book with more added to it:
*"Perhaps herbs are more really effectual than we shall easily believe. This story struck me the more since **Comfrey is the 'confirma' of the middle ages, and the 'symfyto' of the Greeks, both which names seem to attribute to the plant the same consolidating virtue.** Besides the instances in the medical treatises which survive, and which are the less characteristic as they are borrowed, **we find the healing power of worts** spoken of as a thing of course."*
-'Leechdoms, Wortcunning and Starcraft: Being a Collection of Documents Illustrating the History of Science Before the Norman Conquest, Volumes 1-3' also called 'Chronicles and Memorials of Great Britain and Ireland During the Middle Ages' edited by Thomas Oswald Cockayne, London, England, **1864**. Volume 1, preface liii.
(The Norman Conquest of England was an 11th-century invasion and occupation of England by Norman, Breton, Flemish and French soldiers.)
(Worts are plants used as medicine. Examples include Bellwort, Dragonwort, Figwort, Honeywort, Lungwort and Saint Johns Wort. Most of these names come from the 1600s or earlier.)

"Symphyti Radix (root), Common Comfrey root, Symphytum officinale: The black rind (outside of the root) is scraped off, and the mucilaginous root is then scraped carefully into a nice even pulp.
This spread to the thickness of a crownpiece (4 mm or 1/8 inch) upon cambric or old muslin (cloth), is wrapped round the limb and bandaged over; it soon stiffens, and forms a casing (cast) superior to starch, giving great support and strength to the part. The late author knew a bone-setter who practised more than fifty years ago, and rendered himself famous by treating fractures after this method, which he kept secret, **the bandage not being removed until the limb was well."**
-'Squires Companion to the Latest Edition of the British Pharmacopoeia: Comparing the Strength of Its Various Preparations with those of the United States and Other Foreign Pharmacopoeias, 17th edition' by Peter Squire, London, England, **1899**, page 619.

"Some idea of the traditional therapeutic virtues of Comfrey may be gathered from the names by which it was popularly known. For instance **'Knitbacke'** (Gerard, 1597), **'Comfort Knitbene'** (Scotland), in Aberdeen (Scotland) it was called **'Comfer Knitbeen',** and a prepartion made by boiling the root in oil or lard was extolled by old women for hardening and strengthening fractures. This property also accounted for its being called **'Bone-set' or 'Knit Bone'** in Lancashire (County, England).
It appears to have been used both internally and externally in fractures in all districts."
-The Medicinal Uses of Comfrey by Dr. Charles MacAlister, M.D., F.R.C.P., **1935**. (in 'Comfrey: Fodder, Food & Remedy')

"Comfrey, Symphytum officinale: also called Knitbone, Nipbone and Blackwort. **Is used by many Gypsies to bind broken bones, and was known in this connection as long ago as the time of Pliny (23-79 AD)."**
-'Gypsies of Britain: An Introduction to Their History' by Brian Vesey-Fitzgerald, FLS, FRES, MBOU. London, England; Chapman & Hall, Ltd., **1944**, page 141.

"Mrs. C. reports: 'I fractured small bone in foot about 18 months ago and was in plaster (cast) for three weeks and gradually restored to normal function.
About a month ago I slipped and badly strained the previously damaged foot which became so painful I had difficulty walking. **I took some Comfrey ointment which I applied daily, and the pain disappeared in about three days."**
-Comfrey: The Herbal Healer by Henry Doubleday Research Association, **1975**, 41 pages, page 15. (Included in 'Comfrey Report' book.)

"Horsetail is famous among herbalists as a source of silicon (Si). Comfrey contains from 50-80% of the silicic acid content of horsetail. Silicon was found to be essential in the 1970s for normal development of the connective tissues, mucopolysaccharides, cartilage, elastin, and bone (Carlisle). It is an important rate-limiting enzyme cofactor in the formation of the collagen matrix of bone, and **its presence facilitates bone repair and the uptake of other minerals into bone.**
Silicic acid, one form in which silicon exists in equisetum (horsetail) and Comfrey (Symphytun spp.), is readily soluble in water, readily absorbed in the digestive tract once dissolved, and readily diffuses to the extracellular fluid reservoir and connective tissues.
Thus a small amount of infused (tea) herbal material containing soluble silicic acid may provide more physiologically available silicon than much larger amounts of food in which the silicon is bound by fiber or fails to be extracted into solution in the small volume of fluid in the stomach and intestine."
-'Equisetum: Silicon in Horsetail and Comfrey' by Paul Bergner, Medical Herbalism: Journal for the Clinical Practitioner, North American Institute of Medical Herbalism Inc., Volume 10, No. 4, page 10, **2001**.

"One interesting story is that of a registered nurse in Provo, Utah. Her **14-year-old boy broke his arm,** so she rushed him to the

'Dugway Proving Grounds Hospital' to be taken care of.
When the doctor x-rayed the arm, he told them the bone was clean broken, so clean that he would have to put the boy into a brace for a few days until knitting started, and then into a cast. He put on the brace and told them to come back in five days. The arm was bare so on arriving home **she put Comfrey poultices and fomentations around the arm,** *and she gave him Comfrey tea, Comfrey green drink, Comfrey tablets and capsules, and put Comfrey into salads and steamed Comfrey as a vegetable.*
In five days she took him back to Dugway to get the cast on and when the doctor came out of the dark room with the new x-ray he said, 'What have you done to this boy? **You're a registered nurse and this boy's arm is completely healed, and the bone knit together without a hairline crack-- it is perfect in five days- what did you use?'** *"*
-'Comfrey: Heaven's Gift to Man' by Dr. John R. Christopher, M.H., The Herbalist, Volume 1, Number 5, 1976 as reported in 'Comfrey: Tomorrow's Food, Today's Medicine' by Ingri Cassel, Idaho Observer Newspaper, July **2002**.
    (I do not recommend using as much Comfrey as the mother did in this story.)
    (A poultice is a soft, moist mass of meal, herbs, etc. applied warm or hot to the skin.)

"Comfrey is so good at 'knitting' that it must not be used on broken bones until they have been set, or it will start bonding them together in the wrong position. Once a bone has been set by a qualified person, apply a fresh Comfrey poultice. *If the fracture is in plaster, take the Comfrey up to the edges of the plaster.*
In addition, use homeopathic Comfrey (Symphytum 6x) internally as directed by a homeopath.
Or - as long as you are not pregnant or breastfeeding - you can drink a couple of cups of Comfrey leaf tea a day until the bone heals. Use a leaf or half a large leaf per cup of tea, infusing for 5 minutes."
-Backyard Medicine: Harvest and Make Your Own Herbal Remedies by Julie Bruton-Seal and Matthew Seal. New York, New York: Skyhorse Publishing, **2009**. First published in Great Britain by Merlin Unwin Books, 2008 as 'Hedgerow Medicine'.

"Homeopathy, reiki, qi gong, polarity therapy, healing touch, acupuncture, and massage are all non-conventional energy healing modalities with applications for **fracture healing. Common over-the-counter homeopathic remedies** *include arnica as an anti-trauma remedy for immediately after the fracture (not to be used if the person is unconscious),* **Symphytum (Comfrey) for pain relief and the joining of set bones,** *and Calcarea phosphorica for fractures that are difficult to heal.*
Low-potency homeopathic remedies (6x, 6c to 30x, 30c) are often used for self-help, as detailed in 'Homeopathic Self-Care' by Robert Ullman and Judyth Reichenberg-Ullman (Prima Publishing, 1997). Homeopathy is a powerful medicine and when possible the best policy is to seek the advice of a professional homeopath."
-'How to Speed Fracture Healing' by Dr. Susan E. Brown, Ph.D., Center for Better Bones, East Syracuse, NY, 11 pages, **2016**.

*"***Comfrey Root** *(Symphytum officinale, S. x uplandicum): This controversial but easy-to-grow garden plant is slimy, soothing, and promotes rapid healing of tissue and bones when applied internally or externally.* **You would want to be sure a bone was set properly before taking Comfrey."**
-'Autumn Roots, Barks and Berries: Day-Long Field Workshop' by Maria Noel Groves, Clinical Herbalist, Wintergreen Botanicals LLC, Allenstown, New Hampshire, www.WintergreenBotanicals.com, **2016**.

"The original Carpathian herbal remedies were either substituted by knowledge gathered from classical medicinal monographs or were identical to them.
Medieval Celtic physicians (Scottish, Irish, and Welsh) often received education at the University of Padua, Italy, similar to their Ruthenian and Polish counterparts (Scottish Historical review, 1906). **Subsequent handwritten manuscripts** *and surveys reported both smaller numbers of total medicinal plants, as well as those used for wound healing.*
Comfrey root (Symphytum officinale) was used on multiple occasions to heal wounds associated with bone fractures, minced in animal fat.
    The type of fat used in this ointment varied depending on the location:
        in Silesian Beskids, pork fat was preffered,
        in Zywiec Beskids, goose lard was believed to possess the best curative properties,
        while in Sacz Beskids they were both considered as effective."
-'Botanical Provenance of Traditional Medicines From Carpathian Mountains at the Ukrainian-Polish Border' by Weronika Kozlowska, Charles Wagner, Erin M. Moore, Adam Matkowski and Slavko Komarnytsky from Wroclaw, Poland and North Carolina; Frontiers in Pharmacology, Lausanne, Switzerland, Volume 9, pages 295-311, April **2018**.
(The Carpathian Mountains or Carpathians form an arc across central and eastern Europe. It includes parts of Austria, Czech Republic, Hungary, Poland, Serbia, Slovakia, Ukraine and Romania.)
(Ruthenians and Ruthenes are Latin names formerly used in Western Europe for ancestors of modern East Slavic peoples, especially the Rus' people with Ruthenian Greek Catholic religious background and Orthodox believers who lived outside Rus'.)
(Beskids or Beskid Mountains are mountain ranges in the Carpathians, going from Czech Republic in the west along the border of Poland with Slovakia up to Ukraine in the east.)

<u>**Bruises**</u> (see Cuts / Bruises / Sprains)

<u>**Burns**</u> (heat/fire, sun, chemicals) (also see Wounds)

See a medical practitioner if you are seriously burned.

"I might quote many cases of various kinds which clearly confirmed **the cell-proliferative qualities of allantoin**. Among these were several cases of varicose ulcers, but it was found that **burns and scalds of the lesser degrees** were very useful fields for observation. **This was because islets of epithelium (skin), many of them at first invisible to the naked eye, formed centres from which new epithelial growths could be seen spreading from day to day with remarkable rapidity.**
In a letter to the 'British Medical Journal' (13 January, 1912) Mr. R.W. Murray (Honorable Surgeon Liverpool Northern Hospital, England) confirmed the value of allantoin as a cell-proliferant as follows:

> 'I can confirm Dr. MacAlister's remarks upon the value of allantoin as a cell proliferant. We were called upon to treat a large number of men who were severely burnt. The results were so satisfactory and so convincing to house-surgeons, dressers and nurses, that **dressing with allantoin solution soon became general. It not only stimulates epithelial growth, but cleans up sloughing surfaces in a most remarkable fashion.**' "

-The Medicinal Uses of Comfrey by Dr. Charles MacAlister, M.D., F.R.C.P., **1935**. (in 'Comfrey: Fodder, Food & Remedy')
   (Comfrey contains allantoin.)

"**The Comfrey burn paste is made with equal parts of Comfrey leaf or powder, wheat germ oil and honey.** It is applied directly to the wound or burn and more is added to the skin when the original is absorbed. There is no need to remove the original application when adding more of the mixture.
We have many wonderful success stories about using the Comfrey burn paste. Most recently, one of our 'School of Natural Healing' staff members accidently **scalded his hand with burning olive oil during a kitchen fire. He acquired Third Degree burns** from this misfortune.
There was no one around to properly dress the wound, so he went to the local emergency room to have the toasted, dead flesh cut away. They cleaned the burn and informed him that he would need skin grafting if he wished to regain the use of his hand. He said, "No, thank you. Not unless there is a donor for the skin. I don't want flesh cut from anywhere on my body!" The hospital assured him that the skin grafting was the only route to go. He still refused.
But when he got home, he asked a friend to 1) take pictures of the hand for documentation, and 2) make up the Comfrey burn paste and apply it to the hand. **After a few weeks of using the Comfrey paste, he was able to move his hand.**
The hand is still scarred somewhat, but new flesh has grown in and he has total use of the hand. With dry skin brushing and use of the Cayenne and BF & C Ointments®, he began improving the circulation as soon as the skin grew back. The Comfrey paste had turned an almost mummified-looking hand to one that is now living again."
-School of Natural Healing: Herbal Reference Guide by John R. Christopher. Colorado: Nutri Books Corp, **1996**.

"**Burns can be caused by fire, sunlight, or chemicals.**
First- and second-degree burns can generally be treated effectively at home, but you must be certain to keep the area clean to avoid infection. If infection should occur, seek medical advice. Always seek medical attention for third-degree burns.
To treat a burn, first cool the area, thus 'putting out the fire'. Immerse the area in ice water or apply diluted apple cider vinegar compress to the damaged area for at least 30 mintues.
St.-Johns-Wort salve or oil applied topically is especially helpful for healing burns.
This particular salve is also excellent for rashes, cuts and wounds: 1 part calendula flower, 1 part Comfrey leaf, 1 part St.-Johns-Wort leaf and flower. Apply to the affected area two or three times daily."
-Rosemary Gladstar's Herbal Recipes for Vibrant Health: 175 Teas, Tonics, Oils, Salves, Tinctures and Other Natural Remedies by Rosemary Gladstar. North Adams, Massachusetts: Storey Publishing, **2008**, page 74-75.
   (For how to make a salve, see subsection 'Comfrey Ointment, Cream or Salve' in section 'Making and Using Comfrey Medicine' {Chapter 27}.)

"Calendula salve, grated carrot, **Comfrey poultice**, cucumber slices, raw honey, tofu, wheat grass, plantain poultice, raw potato or potato juice, vinegar, yogurt, or cooled damp black tea bags can all be **applied topically (to skin) to first-degree burns after the heat has been soaked out of them. These all will have a cooling and anti-inflammatory effect.**
Cures from Grandma's Kitchen: Treat second- and third-degree burns by blending 1/2 cup (120 ml) wheat germ oil with 1/2 cup (170 grams) raw honey and 1/2 teaspoon lobelia powder. Store in a clean glass jar in a cool place.
When needed, add enough chopped or blended Comfrey leaves to make a paste. Apply gently with a new clean paintbrush. Don't clean off the paste, just keep painting on additional layers 2 or 3 times daily to regenerate new skin. **Comfrey contains allantoin, which stimulates tissue regeneration.**"
-The Country Almanac of Home Remedies: Time-Tested & Almost Forgotten Wisdom for Treating Hundreds of Common Ailments, Aches & Pains Quickly and Naturally by Brigitte Mars and Chrystle Fiedler. Minnesota: Quayside Publishing Group, **2011**.

"**Natural Sun Blockers: Allantoin is a nucleotide that naturally occurs in the body and absorbs the spectrum of Ultraviolet (UV) radiation which damages the cell's fragile DNA.**
   Allantoin is an extract of the Comfrey plant and is used for its healing, soothing, and anti-irritating properties. **This extract can be found in anti-acne products, sun care products, and clarifying lotions** because of its ability to help heal minor wounds and promote healthy skin. Some clinical studies confirm that allantoin enhance skin repair. From avocado oil to botanicals such as rosemary and Comfrey, these ingredients soothe and protect the skin."
-'Potential of Herbs in Skin Protection from Ultraviolet Radiation' by Radava R. Korac (Belgrade, Serbia) and Kapil M. Khambholja (Gujarat, India), Pharmacognosy Reviews, Mumbai, India, Volume 5, Issue 10, pages 164-173, July-December **2011**).

(A clarifying lotion exfoliates skin, meaning it cleans away dirt, dead skin and pollution.)

"Burn Paste (burn, sprains, wounds, etc.):
   3 parts Comfrey root (Symphytum officinale)
   1 part Lobelia, powder (Lobelia inflata)
   Base of Honey and wheat germ oil (equal parts)
Preparation: *Mix the base of honey and wheat germ oil in the blender, gradually adding first the Comfrey, then the lobelia until reaching a paste consistency. Cover and keep cool. The paste may be preserved by*
adding a little glycerine but better results are obtained when freshly-made.
Administration: *Apply the burn paste externally on the afflicted area.*
Internally drink a tea of Comfrey, or a tea combined of pineapple and Comfrey."
-'Beatrice® Nutrition Recipes: Comfrey' by Beatrice® Nutrition and Health Sciences, Division of Beatrice® Foods Co., Phoenix, Arizona, **2017**.

Bursitis (inflammation of fluid in joints)

"In total, 140 different gathered plant species were listed by respondents.
Herbal tea is the most frequently mentioned use.
The most frequently listed plant species gathered in the wild at the 'Biosphere Reserve Grosses Walsertal, Austria *(36 plant species are on the list):*
   Symphytum officinale L. leaf, *Beinwell, Boraginaceae, gathered growing wild.*
      Medicinal, application to skin, raw (smashed) for bursitis.
      Nutrition: food salad, food omelette, drink tea.
   Symphytum officinale L. root
      Medicinal, Veterinary: grease/ointment for bursitis, *feet, joints, violent pressure, wounds.*
      Medicinal, Veterinary: bathing tea for wounds."
-'Gathering Tea: From Necessity to Connectedness with Nature. Local Knowledge about Wild Plant Gathering in the Biosphere Reserve Grosses Walsertal (Austria)' by Susanne Grasser, Christoph Schunko and Christian R. Vogl, Division of Organic Farming, University of Natural Resources and Life Sciences, Vienna, Austria; Journal of Ethnobiology and Ethnomedicine, Volume 8, No. 31, 23 pages, **2012**.

Calcium Deficiency

See subsection 'Minerals' in section 'Nutritional Value of Comfrey' (Chappter 19, Volume 1).

*"**It is a possibility that the increasing use and benefit from Comfrey is that it is very rich in easily assimilated organic calcium.**"*
-Natures Healing Agents: The Medicines of Nature or the Natura System by R. Swinburne Clymer, M.D. Philadelphia, Pennsylvania: Dorrance and Company, **1963**, page 123.

"Comfrey has been reported as being very rich in easily assimilated organic calcium."
-Modern Encyclopedia of Herbs by Joseph M Kadans, N.D., Ph.D. New York, New York: Simon & Schuster, Inc., **1993**, page 99.

"Bone health depends not so much on calcium intake, but rather on its metabolism and utilization.
While calcium is clearly important, there are at least 19 other key nutrients that each play a vital role in the structural integrity and overall health of our bones. The major players in this regard are vitamin D, vitamin K, and magnesium.
Good dairy-free sources of dietary calcium: *Whole wheat, Brassica family vegetables, dark leafy greens, canned fish with bones, beans, nuts, seeds, and mineral water.* **Herbal teas and infusions, e.g., oatstraw, nettle, red clover, Comfrey."**
-'The Calcium Myth: Bone Health' by Susan E. Brown, Ph.D., Women's Health Network, Portland, Maine, www.womenshealthnetwork.com/osteoporosisandbonehealth/calciummyth.aspx, January 21 **2009**.

Cancer and Tumors

*"**Mr. William Thomson, President of the Royal Academy of Medicine, in Ireland, delivered an address before that body,** on 'Some Surprises and Mistakes in Medicine':*
   He relates a case of malignant (cancer) tumors of the antrum *confirmed by consultation with other eminent members of the profession, where a radical operation was refused, but an exploratory one allowed only to confirm the diagnosis, and a microscopical examination showing, without doubt, the growth to be a **round-cell sarcoma**. The growth returned, the patient then consulted Dr. Semon, of London, England, who advised immediate **removal of the jaw**. This was done after the usual methods, and the whole base of the skull was found infiltrated. All that possibly could be, was removed.*

*After a month the growth returned, bulging upon the face, almost closing the right eye. This was in **June** last and we all agreed up on a speedily **fatal prognosis**.*

Healing with Comfrey Root:
Early in October he walked into my office looking better than I had ever seen him, and I was not able to identify a trace of the former trouble. **He told me he had applied poultices of Comfrey root and that the swelling had gradually disappeared.**

Now this was a case of undoubted diagnosis by all the best known methods, in which the disease recurred twice and the second time in an extreme degree; and yet this recurrent tumor has disappeared.

*I do not know the cause, but **I do believe that Comfrey root can remove a sarcomatous tumor. The fact that this recurrent tumor has not sloughed away, but simply with unbroken covering disappeared, is to me one of the greatest surprises and puzzles that I have met with.**"*

-'The Atlantic Medical Weekly: A Journal of Reform and Progress in the Medical Sciences, Volume VII', Atlantic Medical Publishing Company, Providence, Rhode Island, January-June **1897**. Article from January 23 1897, pages 62-63.

(Antrum is a natural cavity in a bone or other anatomical structure.)

(Sarcoma is a type of cancer in bones or connective tissue such as fat and muscle.)

(Slough means dead tissue separating from living tissue.)

"Comfrey and allantoin do not appear to have any capacity for producing a somatic (body) cell from a malignant (cancerous) one, or a carinomatous (cancerous) or sarcomatous cell from a somatic one, and no reasonable explanation can be afforded for the occasional cures which have followed their employment.

From its influences on normal cells it can only be inferred that since allantoin leads to their proliferation, it may possibly have something to do with the activities of the nucleic acid."

-The Medicinal Uses of Comfrey by Dr. Charles MacAlister, M.D., F.R.C.P., written **1935**. (in 'Comfrey: Fodder, Food & Remedy')

(Nucleic acids are biopolymers or small biomolecules essential to life. They are composed of nucleotides made of: 5-carbon sugar, phosphate group, and nitrogenous base. If the sugar is a compound ribose, the polymer is RNA- ribonucleic acid; if the sugar is derived from ribose as deoxyribose, the polymer is DNA- deoxyribonucleic acid.)

*"Comfrey, Symphytum officinale: These are numerous uncontradicted reports of **lung cancer cured** where all other means had failed and in which the sole treatment consisted of infusions made from the whole green Comfrey plant and, even in some instances, of infusion made from the powder of the entire plant."*

-Natures Healing Agents: The Medicines of Nature or the Natura System by R. Swinburne Clymer, M.D. Philadelphia, Pennsylvania: Dorrance and Company, **1963**, page 124.

Chicken Pox Itching

*"Mrs. G.M. of SW5 (southwest London, England) reports: '**Chicken pox struck this summer** (1973) and every child for miles around was spotty. The pattern with my two was one isolated spot for a day and then millions.*
***I smeared Comfrey ointment on fairly liberally, and there was no itching.** Everybody else's children had at least one day of intolerable itching and misery."*
-Comfrey: The Herbal Healer by Henry Doubleday Research Association, **1975**, 41 pages, page 3. (in 'Comfrey Report' book.)

Constipation (see Diarrhea)

Cough, Cold, Sore Throat (also see Lungs) (also see Tonsillitis)

"For a dry cough take elecampane and Comfrey; let the patient eat them in virgin honey."
-'Leechdoms, Wortcunning and Starcraft: Being a Collection of Documents Illustrating the History of Science Before the Norman Conquest' also called 'Chronicles and Memorials of Great Britain and Ireland During the Middle Ages' edited by Thomas Oswald Cockayne, London, England, 1864. Volume 2, page 59.

(The Norman Conquest of England was a **1066 AD** invasion and occupation of England by Norman, Breton, Flemish and French soldiers.)

*"**Should there be a cough or tightness upon the chest,** give a tea spoonful of the following syrup, or less or more, six or eight times a day, as the case may require :*

Hyssop, an ounce (29 ml). **Comfrey root, half an ounce.** Ginger root, half an ounce.

Liquorice root, half an ounce. Sweet fennel seed, quarter of an ounce. And two ounces of the best raisins.

Bruise and boil the roots in a quart (0.9 liter) of water down to a pint and a half (3 cups = 0.7 liter), with the cover on, strain off the liquid boiling upon the herb, sweeten well with honey."
-'Family Medical Adviser', Fifth Edition (First Edition 1852), by Dr. John Skelton, Senior, Lecturer and Professor of Medicine, Leeds, England, **1857**. A Thomsonian herbal practitioner (Physio-Medicalism).

"**Comfrey, Symphytum officinale: Useful in colds and coughs,** it should be combined with other agents such as Wild Cherry, Horehound and Racemosa (black cohosh).
In domestic use one ounce (28 gram) of Comfrey was boiled for a few minutes, strained, and either sugar or honey (preferred) added to form a syrup. **If to be kept for a period of time, good brandy was added.**"
-Natures Healing Agents: The Medicines of Nature or the Natura System by R. Swinburne Clymer, M.D. Philadelphia, Pennsylvania: Dorrance and Company, **1963**, page 123.

"Mrs. J.L. who lives near Newbury (Berkshire County, England) reports:
'**The sinusitis was giving me a stuffy and swollen nose**, pain under my eyes and on the temples and a persistent throbbing at the back of my head. I could only use **Comfrey ointment** on my face at night because of discoloration, but **I inserted it into my nostrils several times a day. The facial pain eased in about a week.**"
-Comfrey: The Herbal Healer by Henry Doubleday Research Association, **1975**, 41 pages, page 13. (in 'Comfrey Report' book)

"Mr. L.M. of Fakenham (Norfolk, England) reports: '**We use Comfrey pills for colds. My wife was a sinus sufferer and has not been one after taking Comfrey pills whenever a cold threatened.**
The most positive result in our experience is a friend in the next village, a farm worker, who suffered badly every dusty season from **sinus pain and general nasal inflammation**. We put him on Comfrey pills, 3 per day, and he is enormously improved.' "
-Comfrey: The Herbal Healer by Henry Doubleday Research Association, **1975**, 41 pages, page 24. (in 'Comfrey Report' book)

"**Comfrey, Symphytum officinale Indications: Lungs, bronchitis, tonsillitis, pharyngitis, pleurisy, pneumonia and pulmonary consumption (tuberculosis), coughs including whooping cough, expels phlegm, soothes sore throat, lowers fever. A wonderful herb for the lungs (tonifies Lung Yin). Comfrey's cooling, moistening effect rejuvenates the lungs and mucous membranes.**
The root can be used as well as the leaf, and is stronger in tonic properties for healing lungs and mucous membranes, especially in cases of Dryness, Heat, Deficient Yin and inflammation. **The leaves** are more astringent and anti-inflammatory."
-Healing with the Herbs of Life: Hundreds of Herbal Remedies, Therapies, and Preparations by Lesley Tierra, L.Ac., Herbalist, A.H.G. New York: Crossing Press / Crown Publishing Group, **2003**.

"**To a striking extent the main uses to which mallows have been put and the relative frequencies of those uses parallel those recorded for Comfrey (Symphytum officinale).** This strongly suggests that the two have served as alternatives, the mallows standing in for Comfrey in areas where that much less generally distributed plant is rare or absent.
Not only have both been valued for treating swellings (pre-eminently for sprains in the case of Comfrey), but they have both been widely used as well, if to nothing like the same extent, for two other purposes. The more important of these, accounting for the 40 mallow records, is **as a demulcent for coughs, colds, sore throats, asthma and chest troubles**—chewed or sucked or infused and either drunk or gargled in the case of mallows.
The other is for easing rheumatism, stiff joints or backache, though that category of complaints might equally well be subsumed within the main one of poulticed inflammation."
-Medicinal Plants in Folk Tradition: An Ethnobotany of Britain and Ireland by David Allen and Gabrielle Hatfield. Portland, Oregon: Timber Press, **2004**, page 109.
    (A demulcent forms a soothing film over mucous membranes, relieving pain and inflammation.)

"**A cough syrup is made from baked onion juice, Comfrey tea, and honey.** Drink it daily to get relief from a dry cough."
-'Cough Suppressant Herbal Drugs: A Review' by Shahnaz Sultana, Andleeb Khan, Mohammed M. Safhi and Hassan A. Alhazmi, College Of Pharmacy, Jazan University, Saudi Arabia; International Journal of Pharmaceutical Science Invention, Volume 5, Issue 5, pages 15-28, August **2016**.

"**Mucilage of Comfrey Root Formula: 2 ounces (56.6 grams) Comfrey root, cut (Symphytum officinale)**
        1 quart (0.94 liter) distilled water, 6 ounces (170 grams) Honey, 2 ounces Glycerine
Preparation: Soak the Comfrey root in water for 12 hours. Bring to a boil, cover and simmer for 30 minutes. Strain, then filter and squeeze through muslin or linen cloth. Return liquor (liquid) to the cleansed vessel, add
the honey and glycerine, simmer for 5 minutes and set aside to cool. Placing in a wide-mouthed bottle and keep in a cool place.
Dosage: Coughs, raw or sore throat, slight hemorrhage: 1 wineglassful every hour until tissues are healed & coughing stops.
Note: Do not give food for at least twelve hours while the treatment is being administered in order for the mucilage to reach the blood and lungs without interference and loss of energy (it is very nutritive itself)."
-'Beatrice® Nutrition Recipes: Comfrey' by Beatrice® Nutrition and Health Sciences, Division of Beatrice® Foods Co., Phoenix, Arizona, **2017**.

<u>Cuts, Bruises and Sprains</u> (also see Wounds)

Do not use Comfrey externally on deep wounds. See 'Wounds' in this section for more information about it.

"**An effectual medicine for a strain: Take Comfrey-roots (beating them to a pulp)** half a pound (8 ounces = 226 grams), powder of Japan Earth four ounces (113 gram), spirit of wine a sufficient quantity; mix and **apply it to the part.**"

-'The Works of the Honourable Robert Boyle in Six Volumes, Volume 5, London, England, **1772**, page 390. Robert Boyle, 1627-1691, was one of the most influential scientists and philosophers of the 1600s. These books were compiled after his death.
> ('Japan Earth' is catechu which is an extract containing tannin, usually from an Indian acacia tree, used mostly for tanning and dyeing.)

"Beaumont and Fletcher knew of its healing quality and mention it in *'The Knight of the Burning Pestle' thus:
> Jasper after beating Humphrey unmercifully says to him in derision:
>> "Go, get to your night-cap and the diet to cure your broken bones."
> Jasper's sister sighs: "**Alas, poor Humphrey! Get thee some wholesome broth, with Sage and Comfrey,**
>> a little oil of roses and a feather to 'noint (annoint) thy back withal."

-'Old Time Herbs for Northern Gardens' by Minnie W. Kamm. Boston, Massachusetts: Little, Brown & Co, **1938**. Symphytum pages 115-116.
(* -The Knight of the Burning Pestle is a play in five acts by Francis Beaumont and John Fletcher, first performed at Blackfriars Theatre, London, England, in 1607 and published in a quarto in 1613.)

"Mrs. J.L. who lives near Newbury (Berkshire County, England) reports: 'I have had a fall resulting in a cut hand and badly bruised knee- **with the aid of Comfrey, the hand had healed quickly, and the bruises have given me hardly any pain at all- nor did the knee stiffen- in spite of being all colours from black to puce!** (reddish purple brown)' "
-Comfrey: The Herbal Healer by Henry Doubleday Research Association, **1975**, 41 pages, page 14. (in 'Comfrey Report' book)

"A mother reports on a sore finger: 'Our younger child aged 3 caught the two smallest fingers on his left hand in a door which was slammed by the wind. The skin was broken, and he cried bitterly for about 10 minutes. **We wrapped the fingers in Comfrey-smeared lint and put a bandage on the hand. He seemed to have no more pain and wanted to have the bandage off later in the day. There seemed to be very little swelling, and the fingers healed in a few days.**' "
-Comfrey: The Herbal Healer by Henry Doubleday Research Association, **1975**, 41 pages, page 14. (in 'Comfrey Report' book)

"**Unless a wound is severe, the direct application of the powdered Comfrey leaf or root** will result in the formation of a scab immediately, and healing of the injury will usually take place much more rapidly than ordinarily.
In any severe wound, a physician should be called to stitch the parts together."
-Modern Encyclopedia of Herbs by Joseph M Kadans, N.D., Ph.D. New York, New York: Simon & Schuster, **1993**, page 101.

"**By far the commonest use of Comfrey, recorded from most parts of the British Isles- except apparently the southern half of Wales- has been for treating injuries to limbs and ligaments, in particular, sprains** (twice as often mentioned as fractures). Identified with a herb mentioned by Dioscorides (herbalist 40-90 A.D.), whose name for it passed into Latin as Symphytum, grow-together plant. The plant is rich in allantoin, which promotes healing in connective tissues through the proliferation (growth) of new cells. Not for nothing was it widely known as 'knitbone', a name which still lingers on in places. **Various parts of the plant yield a strongly astringent oily juice, but for treating injuries the roots are most often preferred.** The usual process is to clean, peel, pound or grate and boil these, in order to extract a thick paste which is then applied like plaster of Paris.
Alternatively, the leaves and/or stem are heated and put on as a poultice.
A third, much rarer method is to mix the juice with lard and rub the ointment in."
-Medicinal Plants in Folk Tradition: An Ethnobotany of Britain and Ireland by David Allen and Gabrielle Hatfield. Portland, Oregon: Timber Press, **2004**, page 208.
> (Plaster of Paris is a white powder that is hydrated calcium sulfate made from gypsum. Water is added to it to make casts and molds.)

"**Comfrey, Symphytum Officinale Medicinal Properties:**
Cell proliferators, heals wounds, encourages bone / cartilage / muscle cell growth, speeds healing when applied to injured limb.
Comfrey leaves and flowering tops are used in ointments and oils for sprains, arthritic joints.
Comfrey root used for varicose ulcers, as wrap around joint for sprains and arthritis, make tea, relieves insect stings, stops burns instantaneously, faster healer.
> **Comfrey's got gel in it like this, and you can use the leaf. You make a poultice with paper towel.** You wet the paper towel, and you put the tuft (clump of Comfrey) in it. **If you have anything like a bad knee or anything like that you wrap it around with this, and it's amazing how fast it will heal.** My mom had some problem, and her dad wrapped her legs with the Comfrey leaf, and the next couple of days it was gone. She had, **her lower leg was all blue**, and by the time she left a couple of days later it was normal. It's amazing fun."

-'Metis Nation of Ontario: Southern Ontario Metis Traditional Plant Use, Traditional Ecological Knowledge Study' by Metis Nation of Ontario, Ottawa, Canada, 40 pages, Spring / Summer **2010**.

"Tormentil can be of exceptional value as an astringent in heavy periods, and, although this action is not explained, **it is one of those actions exerted by medicinal plants, of which perhaps the action of Comfrey, Symphytum officinale, in bruising is the archetype, where once seen, always believed. 'Probatus est' as the old writers called it.**"
-The Western Herbal Tradition: 2000 Years of Medicinal Plant Knowledge by Graeme Tobyn, Alison Denham and Margaret Whitelegg. London, England: Churchill Livingstone, page 241, **2011**.   (Probatus est: Latin for 'the approved'.)

Dental (see Mouth / Teeth)

Diarrhea and Constipation

"Astringent drink extolled (praised) by the celebrated Robert Boyle is thus prepared: Take of Comfrey-root...two drachms; Catechu...two drachms; Water one pint (2 cups= 0.47 liters). Reduce by boiling to six ounces (0.17 liters), then add Syrup of Clove-pink...one ounce; Cinnamon water...half an ounce.
A desert spoonful to be taken every hour in chronic diarrhoea and dysentery."
-'The British Flora Medica, or, History of the Medicinal plants of Great Britain, Volume 1' by Benjamin Herbert Barton and Thomas Castle, London, England, **1838**. (Symphytum officinale, pages 211-215.)
    (drachms= 60 grains= 64.79 milligrams)
    (Catechu is an extract containing tannin, usually from an Indian acacia tree, used chiefly for tanning and dyeing.)

"Protective Colloids Found in Ancient Remedies:
*The 'scraped-apple' diet used by German peasants in **the treatment of infantile diarrhea and constipation is an interesting example of the use of gelatin-like, moisture-absorbing substances called 'hydrophilic colloids'** in human nutrition. These hydrophilic colloids are just now becoming known to investigators of science, but they have long been known to the homemaker for another reason; they are the substance which gives jelly its quivery firmness or 'set'; they are 'pectin'.*
The okra and Comfrey plants are other examples of hydrophilic colloids. *Comfrey is of particular interest. The word "Comfrey" is attributed to the old French word 'to preserve'.*
    Early discoveries in nutrition were concerned with only the missing elements caused by 'indiscretions in the diet,' but today we must consider the factors which come within the realm of enzymes, hydrophilic colloids and other activators.
Substances such as pectin, Comfrey and mineral-earths formerly regarded as virtually inert biologically, now are being considered in terms of nutrient value, *not because they contribute calories or weight, but because they possess activities which have heretofore been unsuspected or ignored in spite of practical evidence to the contrary."*
-'Protective Colloids Found in Ancient Remedies' by Dr. Royal Lee; Let's Live Magazine, The Dr. Royal Lee Historical Archive Collection from Selene River Press, www.seleneriverpress.com, Loveland, Colorado, **1958**. A publishing company specializing in holistic nutrition education.
    (A hydrophilic colloid or water-loving colloid is colloid particles that are hydrophilic polymers dispersed in water. Gelatin and agar agar are hydrophillic colloids.)

"An ounce of powdered Comfrey root or leaves is placed in a container, over which is poured a pint (2 cups = 0.47 liters) of hot water. The mixture is then stirred and allowed to soak for about fifteen minutes. One-half cup (0.11 liters) of this infusion may then be taken four times a day.
***The infusion is found to be helpful in relieving cases of diarrhea, as well as in relieving inflammation of the membrane of the large intestine, referred to as dysentary**."*
-Modern Encyclopedia of Herbs by Joseph M Kadans, N.D., Ph.D. New York, New York: Simon & Schuster, Inc., **1993**, page 99.

Eczema and Psoriasis (inflamed and/or itchy, dry skin)

*"Mr. W.R. of Plymouth (Massachusetts) reports **success in treating Eczema, which he had for over 20 years,** covering an area about 2 inches wide and 6 inches (5-15 cm) long on the inside of each wrist.*
*'When I received the **Comfrey ointment** in May 1966, I put it on each evening at bedtime, and when the tube was used up, then I looked at my arms, and they looked a little different. In 3 days I looked at it, it was almost gone, and about a week later it was all gone, and to this day April 1967 it's still cleared up."*
-Comfrey: The Herbal Healer by Henry Doubleday Research Association, **1975**, 41 pages, page 10. (in 'Comfrey Report' book)

"Used Formulas for Skin Disorders:

Skin disorder	Herbal extract/oil	Quantity	Ingredients in cream
Eczema	Comfrey leaf tincture	10 ml	5%
Seborrheic dermatitis	Comfrey leaf tincture	10 ml	5%"

-Aromatic and Medicinal Plants: Back to Nature edited by Hany El-Shemy. London, England: IntechOpen Limited, **2017**. Chapter 1: 'Medicinal Plants to Calm and Treat Psoriasis Disease' by Azadeh Izadyari Aghmiuni and Azim Akbarzadeh Khiavi.
    (Seborrheic dermatitis is scaly patches, red skin and dandruff on the scalp.)

Eyes (If you have an eye injury or problem, see a medical practitioner.)

"An opthalmic (eye) surgeon employed allantoin dressings for a patient having a very extensive and deep burn of the eyeball.

He wrote: 'The disappearance of the chemosis, the firm healing of the deeper layers and the formation of new tissue have been most marked and have even astonished the nurses who nothing about the stuff."
-The Medicinal Uses of Comfrey by Dr. Charles MacAlister, M.D., F.R.C.P., **1935**. (in 'Comfrey: Fodder, Food & Remedy')
(Chemosis is swelling or edema of the conjunctiva. Conjunctiva lines the eyeball and the inner eyelid.)

"Mrs. J.C. Catania, Sicily: 'For the past couple of weeks I had been suffering from something between **a sty and conjunctivitis (pinkeye) in one eye.** I decided to try the Comfrey cream. I rubbed it round the rim of the eyelid at night and found the eye improved by morning. On application the next night, it was completely cured."
-Comfrey: The Herbal Healer by Henry Doubleday Research Association, **1975**, 41 pages, page 7. (in 'Comfrey Report' book)
(A sty is a red, painful lump at the edge of the eyelid that looks like a boil. Pink eye or conjunctivitis is inflammation or infection of the transparent membrane of the eyelid and white part of the eyeball.)

Feet

"**For foot ache,** betony, germen leaves, that is mallow, fennel, ribwort, of all equal quantities; mingle milk with water, and bathe the swollen limb, from the upper part of it, with that, lest swelling go inwards; **then take sodden (soaked) Comfrey, lay on it.**"
-'Leechdoms, Wortcunning and Starcraft: Being a Collection of Documents Illustrating the History of Science Before the Norman Conquest' also called 'Chronicles and Memorials of Great Britain and Ireland During the Middle Ages' edited by Thomas Oswald Cockayne, London, England, 1864. Volume 2, page 69.
(The Norman Conquest of England was a **1066 AD** invasion and occupation of England by Norman, Breton, Flemish and French soldiers.)

"**To treat tired feet,** *Gardner recommends a soothing footbath with mugwort, Comfrey (Symphytum spp., Boraginaceae) and mint (Mentha spp., Lamiaceae)."
'Aromatic Herbal Baths of the Ancients' by Farid Alakbarov, Ph.D., HerbalGram: The Journal of the American Botanical Council, Austin, Texas, Issue 57, pages 40-49, 2003.
(* -'The Art of Artemisias' by Jo Ann Gardner, The Herb Companion, Volume 12, No. 4, pages 28-33, 2000.)

Female Problems (see Mastitis or see Vagina)

Hair

"**Dry Hair:** Chamomile, Red Clover, Quince Seed, Horsetail, **Comfrey Root,** Elderflower, Orange Blossom, Peach Leaves, Rosemary, Sage, Basil, Southern Wood.
Hair, Split Ends: Horstail, **Comfrey,** Fenugreek, Quince, Rosemary, Echinacea, Lavender, Olive, Basil."
-'Botanicals in Cosmetics: Reference Chart' by Vege-Tech Botanicals: Nature and Science, certified organic specialties, www.vegetch.com, Glendale, California, **2011**.

"**The use of bioactive materials from the natural system stimulates the biology of dermis and hair for usual growth that presents healthful hair and epidermis.** Mainly natural system provides much nutrition, antioxidants, various oils, proteins, terpenoids, and many most important oils.
List of 100 plants having hair growth promoting, nutritional support, and antidandruff activity: includes Symphytum officinale, Boraginaceae, Comfrey."
-'The Wonder of Herbs to Treat Alopecia' by Pushpendra Kumar Jain and Debajyoti Das, India; Innovare Journal of Medical Science, Mandsaur, Madhya Pradesh, India, Volume 4, Issue 5, pages 1-6, October **2016**.
(Dermis is the thick layer of tissue below the epidermis, containing blood capillaries, nerve endings, sweat glands, hair follicles and other structures.)
(Alopecia is the partial or complete loss of hair from areas of the body where it normally grows; baldness.)

"**Comfrey has many hair care benefits as well. Comfrey leaf can be added to hair conditioners** to help repair any over-processed or damaged hair. It can also help to control and possibly reverse hair loss, and it brings the hair volume and shine. Comfrey is also a great way to naturally nourish your hair, as well as fight dandruff."
-'Comfrey Class' by Natures Garden: Wholesale Candles, Soaps and Cosmetics, Wellington, Ohio, www.naturesgardencandles.com, (date unknown but **2018** or earlier).

"**Copper (Cu) peptides such as allantoin from Comfrey root have been shown to improve strand thickness and speed growth when applied topically (top of skin) or subdermally (under the skin).** Long term effects are not known to this author currently. These compounds actually stimulate cell growth from the undifferentiated human cells. Stimulating hair growth using GLH-Cu Complexes or Glycine-Histidine-Lysine complexed with a metal are having excellent lab results.
While there are patents issued for the use of Comfrey for hair regrowth, you can buy oil infused with Comfrey, Comfrey leaves or roots, or even grow your own Comfrey from root cuttings for poultices.

Growing is easy from roots, just place the root section horizontally about 2-3 inches (5.08-7.62 cm) deep and water when dry for a week to get growth."
-'Hair Regrowth FAQ: Aimed at the Best of What Works and Only What Works' by Hair Growth and Regrowth, http://plainthoughtworks.com/Health/Hair/Hair_Regrowth_faq.htm, (date unknown but **2018** or earlier).
(Copper peptides are naturally occurring complexes used in skin and hair care products. They are a combination of the element copper and three amino acids. Copper peptides are found in trace amounts in blood plasma, saliva and urine.)

Headache

*"**A poultice of Comfrey is recommended for headache pains** occurring in the lower back part of the head, the top of the head, and the forehead."*
-Modern Encyclopedia of Herbs by Joseph M Kadans, N.D., Ph.D. New York, New York: Simon & Schuster, Inc., **1993**, page 101.
(A poultice is a soft, moist mass of meal, herbs, etc. applied warm or hot to the skin.)

Hemorrhoids (Piles) (also see Anal Itching)

*"**Pile Ointment:** Pinus canadensis, pulverized, 1/2 ounce (14 grams). **Comfrey root, pulverized, 1/2 ounce.** Cut up an ounce of crane's bill, fine. Mix in a clean saucepan, and gently simmer for one hour, with ten ounces (283 gram) of lard and two of mutton (sheep) suet (hard white fat). Strain first through a coarse cloth, and squeeze or press out the strength as much as possible. Now pass it through a cloth, rather finer, and add two ounces of olive oil.*
This is a most excellent ointment for piles, chaps upon the hands, lips, etc. It is also good for sore nipples in females."
-'Family Medical Adviser', Fifth Edition (First Edition 1852), by Dr. John Skelton, Senior, Lecturer and Professor of Medicine, Leeds, England, **1857**. A Thomsonian herbal practitioner (Physio-Medicalism).

*"**Application of the powdered Comfrey leaf or root will repress (reduce) bleeding of hemorrhoids.**"*
-Modern Encyclopedia of Herbs by Joseph M Kadans, N.D., Ph.D. New York, New York: Simon & Schuster, Inc., **1993**, page 101.
(Hemorrhoids are swollen and inflamed veins in the lower part of the rectum and anus.)

*"**Soothing practices for hemorrhoids:** Though it takes a bit of effort, a sitz bath (hip bath) can bring great relief to hemorrhoids. **Simply fill two baby bathtubs—one with a hot tea of witch hazel, white oak bark, Comfrey, and plantain, the other with ice water.** Sit your bottom into the hot tea bath for three minutes. Then go immediately to the ice cold bath for one minute. Alternate from one to the other. Always start with the hot bath and end with the cold one.*
Dry off, apply some healing salve to the inflamed hemorrhoids, and then dress comfortably."
-The Country Almanac of Home Remedies: Time-Tested & Almost Forgotten Wisdom for Treating Hundreds of Common Ailments, Aches & Pains Quickly and Naturally by Brigitte Mars and Chrystle Fiedler: Minnesota: Quayside Publishing Group, **2011**.

*"**Haemorrhoids/piles:** Bulging veins in the anus which can be very painful when going to the toilet.*
    Comfrey: Apply poultice of 1 teaspoon fresh crushed leaf."
-'Healthy Harvest: A Training Manual for Community Workers in Good Nutrition and the Growing, Preparing and Processing of Healthy Food' by 'Food and Nutrition Council of Zimbabwe', 'Food and Agriculture Organization' of the United Nations, and 'United Nations Children's Fund', 123 pages, **2015**.

*"**Hemorrhoidal disease is a benign (not cancerous) perianal (around the anus) disease**, which is basically caused by vasodilation (blood vessel dilation) on 'pleux haemorrhoidalis' vein. It is believed that the number of cases is much higher than the actual reported number. The majority of the patients' complaints consist of pain, itching, bleeding and feeling of discomfort.*

Latin Name	Common Name	Turkish Name	Used Part	Application Method
Symphytum officinale	Common Comfrey	Karakafesotu	Root, Leaf	Internal (Systemic)* "

-'Plants Used in Anatolian Traditional Medicine for the Treatment of Hemorrhoid' by Ufuk Koca Caliskan, Ceylan Aka and Mehmet Goker Oz, (Ankara, Turkey and Rotterdam, The Netherlands); Records of Natural Products, Volume 11, No. 3, pages 235-250, **2017**.
(* -'Investigation of Vasoactive Ion Content of Herbs Used in Hemorrhoid Treatment' by Mahir Gulec, Recai Ogur, Husamettin Gul, Ahmet Korkmaz and Bilal Bakir, Gulhane Medical School, Ankara, Turkey; Pakistan Journal of Pharmaceutical Sciences, Volume 22, No. 2, pages 187-192, April 2009.) (Systemic means affecting the body generally. Not locally.)

Hernia

A hernia is when an organ or tissue squeezes through a weak spot in muscle or connective tissue. The most common are inguinal (inner groin), incisional (from an incision), femoral (outer groin), umbilical (belly button), and hiatal (upper stomach). If you think you have a hernia, see your medical practitioner.

*"**Reduce hernia by the use of blackroot (Comfrey) which should be held on the hernia until warm.** It should be then

planted, and if it grows the hernia will be cured. References: Berks County, Schuylkill County, York County, Pennsylvania."
-'Beliefs and Superstitions of the Pennsylvania Germans' by Edwin Miller Fogel, PhD. Philadelphia, Pennsylvania: American Germanica Press, **1915**, page 288.

"**Hernia** often result from heavy lifting or straining, or because the tissue or muscle at the lower end of the abdominal cavity is either congenitally (from birth) weak or becomes weak. Take care to avoid constipation.
Oatstraw and Horsetail Hernia Remedy:
4 parts Comfrey leaf, 3 parts raspberry, 2 parts lemon balm, 2 parts nettle, 2 parts white oak bark, 1 part horsetail, 1 part oatstraw. Make an infusion of the herbs. Drink 3 to 4 cups daily. **There is a current, as yet unresolved controversy surrounding the use of Comfrey for internal purposes. Generally, I no longer recommend Comfrey internally for others, though I continue to use it myself abundantly.**
In the case of hernias, because of extreme usefulness in healing torn or damaged tissue, I am braving the wrath of the herbal community and including it in my formula. Please educate yourself as to the controversy, and make up your own mind."
-Rosemary Gladstar's Herbal Recipes for Vibrant Health: 175 Teas, Tonics, Oils, Salves, Tinctures and Other Natural Remedies by Rosemary Gladstar. North Adams, Massachusetts: Storey Publishing, **2008**, page 259.
(An infusion is when a tea bag or ball is put in a cup of hot water and allowed to steep for a few minutes before drinking. It extracts vitamins and volatile ingredients from soft ingredients like leaves, flowers, etc.)

Inflammation (see Swelling)

Kidneys (also see Bladder)

"And from Pittsburgh, Pennsylvania, this word from a gentleman who reports spectacular relief: 'Thanks to your book and **the use of fresh Comfrey leaf tea, I have been cured of the following ailments: kidney stones, gall stones, sour stomach.** For 13 1/2 years I suffered with kidney stones. I am now free from this bothersome disability and am as healthy as can be.' "
-Nature's Healing Grasses by H.E. Kirschner, M.D. California: H.C. White Publications. First printing **1960**, sixteenth printing 1980, page 124.

"**Congenial (favorable) combinations with Comfrey: For an inflamed kidney or urinary condition, Comfrey in combination with gravel root** (Eupatorium puupureum) **will relieve the inflamed kidney or urinary condition.**"
-'Beatrice® Nutrition Recipes: Comfrey' by Beatrice® Nutrition and Health Sciences, Division of Beatrice® Foods Co., Phoenix, Arizona, **2017**.

Liver

"**Inflammation of the Liver:** There is generally a severe pain in the right side, just below the ribs, often with pain between the shoulders. **As soon as all the painful symptoms are removed, make up the following medicine:**
Barberry bark, one ounce. **Comfrey root, one ounce (29 ml).**
Dandelion root, one ounce. Agrimony, one ounce.
Bruise and boil the roots in three pints (6 cups = 1.4 liter) of water down to a quart (4 cups = 0.9 liter); add the herbs, and half an ounce of the curative powder; mix well, let it settle, strain, add half an ounce of Spanish juice. When cold, take half a wine glassful four or six times a day."
-'Family Medical Adviser', Fifth Edition (First Edition 1852), by Dr. John Skelton, Senior, Lecturer and Professor of Medicine, Leeds, England, **1857**. A Thomsonian herbal practitioner (Physio-Medicalism).
(If you are in pain, please see your medical practitioner.)

"**Liver Herbs: This recipe contains herbs traditionally used to help liver function.**
6 parts Comfrey root (Symphytum officinale), 6 parts tanner's / white oak bark (Quercus alba), 3 parts gravel root (Eupatorium purpureum), 3 parts Jacob's staff / mullein herb (Verbascum thapsus), 2 parts licorice root (Glycyrrhiza glabra), 2 parts wild yam root (Dioscorea villosa), 2 parts milk thistle herb (Silybum marianum), 3 parts black walnut bark (Juglans nigra), 3 parts white marshmallow root (Althea officinalis), 1 part lobelia plant (Lobelia inflata), 1 part skullcap (Scutellaria lateriflora).
Mix all the herbs. Add 1/2 cup (118 ml) of the mixture to 2 quarts (1.89 liters) of water. Bring to a boil. Put lid on. Let sit for six hours. Strain and drink 1 1/2 cups (354 ml) per day. Put the strained herbs in the freezer, and use them one more time."
-The Cure for All Diseases with Many Case Histories by Hulda Clark. Chula Vista, CA: New Century Press, **1995**, page 552.

Lungs (also see 'Cough / Cold / Sore Throat') (also see Pleurisy)

"Syrup of Comfrey extolled (praised) by the celebrated Robert Boyle in spitting of blood is thus prepared:
Take of Comfrey-root.....six ounces; Plantain-leaves.....three ounces. Bruise them together in a marble mortar, and express the juice; then, after clearing the liquid, add of white sugar an equal quantity. Boil to the consistence of a syrup.

Two or three spoonsful may be taken at a dose. ***If for coughs,*** *a small quantity of liquorice-root may be added before the sugar."*
-'The British Flora Medica, or, History of the Medicinal plants of Great Britain, Volume 1' by Benjamin Herbert Barton and Thomas Castle, London, England, **1838**. (Symphytum officinale, pages 211-215.)

*"****Prickly Comfrey has been held in high esteem as a simple and efficacious (effective) remedy for chest and lung complaints.*** *The simple extract is administered to children and adults as a tea, or in the form of a jelly or drops made from the gum (created) by boiling the roots."*
-'Prickly Comfrey: Its History, Cultivation, Extraordinary Production, and Uses: A Letter Addressed to His Excellency Sir Hercules Robinson, President of the Agricultural Society of New South Wales' by Arthur T. Holroyd, Sydney, Australia, **1876**, page 11.

*"****Pulmonary (Lung) Remedy: Take of the roots of spikenard, elecampane, Comfrey and blood-root;*** *of the leaves and flowers of hoarhound, and of the bark of wild cherry, each one pound (0.45 kg).* ***These may all be ground and tinctured,*** *by adding alcohol, water, and sugar sufficient to make three gallons (11.3 liters) of syrup.*
> *Or any portion of the above compound may be tinctured in sufficient alcohol to cover them, when the herbs may be boiled until their strength is obtained, and the tincture and watery infusion may be mixed, and a sufficient amount of refined sugar added to make a thick syrup.*

For coughs and colds, *to be taken in teaspoonful doses as required."*
-'The Complete Herbalist, or the People Their Own Physicians by the Use of Natural Remedies' by Dr. O. Phelps Brown, Jersey City, New Jersey, **1897**, page 460.

"A letter from a lady living in Rockville, Connecticut, reports on her husband's condition: 'I truly believe the ***Comfrey root tea*** *has at last begun to show results in my husband's case.* ***He can breathe better,*** *and it is easier for him to get about; his appetite is wonderful; and his kidneys and bowels are okay. His broken rib from coughing seems better.' "*
-Nature's Healing Grasses by H.E. Kirschner, M.D. California: H.C. White Publications. First printing **1960**, sixteenth printing 1980, page 124.

"I became interested about Comfrey back in 1956 when an acquaintance of mine, who had suffered from bleeding of the lungs for many years, started using ***Comfrey root tea.***
To the marvel of everyone, including the doctors that had taken care of her, ***her lung bleeding stopped in about three months after she had started drinking the tea,*** *and she has never had any bleeding since then. Maybe this is pure coincidence, but after reading all of the literature about Comfrey and how it has been considered both an internal and an external healer, not only by the common people, but also by medical men, I believe that it was an authentic case."*
-Comfrey and Chlorophyll: A Report About the Medicinal Value of Comfrey and Chlorophyll as Found in Old and Modern Literature by Vincent Licata. California: Continental Health Research, **1971**, page 4.

*"****There are traditional herbs for helping lungs. Grow your own Comfrey and garlic.*** *Make mullein tea from the dried herb. Read herb books for more help. Dry some for winter use, being careful to do it right and not let it mold."*
-The Cure for All Diseases with Many Case Histories by Hulda Clark. Chula Vista, CA: New Century Press, **1995**, page 137.

*"****In treating bronchial illnesses:*** *Without medical tests it is difficult to determine how much damage the lungs have endured. We must assume the worst and act accordingly. Damaged lungs will be ripe for re-infection.* ***The use of Comfrey in three forms can help repair damage and strengthen the lungs.*** *Mullein can be used, but tends to best as a preventative.*
> *1. Remember please, Comfrey is transdermal. It will go through the skin.* ***Comfrey oil, made with the fresh leaves, olive oil, vitamin B capsules as a preservative, and perhaps a bit of almond oil, can be rubbed on the chest to ease the battered muscles of the lungs.*** *This should be done two to three times a day. It can be used warm, but it is not necessary to heat the oil to the point of discomfort. A linen covering can be placed on the chest to keep the oil in place and to prevent staining bed clothes.*
> *2.* ***A Comfrey tincture should be given,*** *starting at 90 drops twice a day.*
> *3. In between tinctures a warm cup or two of* ***Comfrey tea*** *can be taken. The tea works best with dry leaves."*

-'Curandero Articles: Notes on Steam and Folklore' by California School of Traditional Hispanic Herbalism, Director Charles R. Garcia, Richmond, California, June 30, **2013**. To teach and preserve healing traditions of the Hispanic curanderos and curanderas (folk healers). www.hispanicherbs.com/articles-hiatus.html

*"****Comfrey Therapeutic action:*** *Demulcent, cell proliferant, pectoral, astringent, nutritive, tonic, expectorant, hemostatic, alterative, vulnerary, mucilage, and styptic.* ***Comfrey is one of the finest healers for the respiratory system, especially where there is hemorrhage of the lungs; it has saved thousands of lives.***
The root has been used reputably as both a tonic and a vulnerary (wound healer) from very ancient times up to the present.
> ***Lung Tonic Recipe:***
> *1/2 ounce (14.1 gram) Comfrey root (Symphytum officinale)*
> *1/2 ounce Horehound (Marubium vulgare),   1/2 ounce Elecampane root (Inula helenium)*
> *1/2 ounce Ground ivy (Glechoma hederacea),   1/2 ounce Ginger root (Zingiber officinale)*
> *1/2 ounce teaspoonful Cayenne (Capsicum minimum; C. fastigiatum)*
> *1 1/2 pounds (0.68 kg) Yellow D® dark brown sugar*
> *Preparation: Simmer the first 4 herbs slowly in 3 pints (6 cups = 1.4 liters) of water for 20 minutes. Add the nutmeg,*

cover and simmer 4 minutes longer. Strain over ginger and cayenne and add sugar while hot. Allow to cool and bottle. Dosage: 1 - 2 tablespoonfulls every 2 hours."
-'Beatrice® Nutrition Recipes: Comfrey' by Beatrice® Nutrition and Health Sciences, Division of Beatrice® Foods Co., Phoenix, Arizona, **2017**.

Mastitis and Breast Problems
Mastitis is infection of the breast resulting in pain, swelling, warmth and redness.

"French nurses treat cracked nipples by applying a hollow section of the fresh Comfrey root over the sore caruncle (red swelling)."
-'Herbal Simples Approved for Modern Uses of Cure, by W.T. Fernie, M.D., Bristol, England, **1897**.

"Breasts: In case of swollen breasts due to excess milk, the application of a poultice of Comfrey is reported to be very beneficial."
-Modern Encyclopedia of Herbs by Joseph M Kadans, N.D., Ph.D. New York, New York: Simon & Schuster, Inc., **1993**, page 101.

"Soothe sore nipples from breastfeeding by applying a salve of calendula and Comfrey."
-The Country Almanac of Home Remedies: Time-Tested & Almost Forgotten Wisdom for Treating Hundreds of Common Ailments, Aches & Pains Quickly and Naturally by Brigitte Mars and Chrystle Fiedler. Minnesota: Quayside Publishing Group, **2011**.

Mouth / Dental

*"**Nontoxic dental care herbal formulation for preventing dental plaque and gingivitis.** Wolf et al. (United States Patent # 5,989,604 and 6,159,508) suggested use of neem oil in their xylitol based **formulation for dogs (non-human animals) for reducing the incidence of dental carries (cavities).***
*The invention is based on xylitol, however as a suggestion **they have listed several plants such as** gum arabic, beef broth, chicken broth or distilled water, lecithin, Coenzyme Q10, folic acid, aloe vera, **Comfrey,** rosemary, goldenseal, horsetail, arnica, calendula, barley grass, chamomile, bloodroot, siwak-miswak, pullulan, horse chestnut, neem, peelu, propolis, green tea, myrrh, birch bark, white oak bark, tea tree oil, grape seed extract, wheat germ, bromelain, papain and quercetincan **be added to promote health of hard and soft dental tissue."***
-'Nontoxic Dental Care Herbal Formulation: European Patent Specification, European Patent Office' by Council of Scientific and Industrial Research, New Delhi, India, **2008**.

*"Periodontal diseases, if left unchecked, can lead to major health problems. **There are a number of traditional herbal remedies for the treatment and management of diseases related to teeth, gum and oral hygiene.** Use of clove oil is an age old remedy still practiced for periodontal problems.*
Symphytum officinale (Comfrey) Root, Leaf: Gargle."
-'Herbal Remedies for the Treatment of Periodontal Disease: A Patent Review' by Pramod Kumar, Shahid H. Ansari and Javed Ali, Faculty of Pharmacy, Jamia Hamdard, Hamdard Nagar, New Delhi, India; Recent Patents on Drug Delivery and Formulation, United Arab Emirates, Volume 3, No. 3, pages 221-228, November **2009**.

"Regrowing your teeth is simple. You'll just need two things: Comfrey root and organic eggshells.
Eggshells:
  Eggshells are used because they contain 27 minerals and loads of calcium, so they contain the ideal building materials to regrow your teeth. When you're regrowing your teeth, aim to eat one organic eggshell each day. Blending your shells into fresh fruit smoothies is a great way to consume them because blending breaks them into tiny particles that are easily consumed.
Comfrey:
  ***Comfrey root is used because it accelerates bone, teeth and tissue growth.** In fact, another name for Comfrey root is knitbone, primarily because of its ability to knit - or regrow - bone together so quickly.*
  ***You'll want to use Comfrey root on your teeth and gums.** Either fresh or dried Comfrey root will do the trick, but if it's dried, boil the root lightly for ten minutes to rehydrate it. Then, blend a square inch (6.4 square cm) of the root with a few tablespoons of water to make a liquid - and **swish the liquid in your mouth and between your teeth for about 20 minutes.** When you're finished, just spit it out.*
  Using Comfrey in this manner is best done once a day, and you'll likely see progress within a few weeks. Many cavities can be completely regrown within a month or two with regular use.
Caution, Limit Use:
  Comfrey root can be a little hard on your liver so if you have liver problems, you'll want to avoid using Comfrey. After your teeth have regrown, you'll also want to end the use of Comfrey so as not to over do it."
-'Forget Filling Cavities: Regrow Your Teeth Instead' by Kim Evans, Natural News Network, independent natural health advocacy organization, www.naturalnews.com, July 1 2010.
  (This is obviously not endorsed by the 'American Dental Association'. I include this information to show what some

natural healers believe. Comfrey does help with bone repair. For teeth to regrow there would have to be something for the Comfrey to mend/meld together like it does with a regular bone fracture. It could not regrow a tooth completely from just a hole in the mouth. I do not know anything about how well it might work. For entertainment purposes only.)

"Since herbal therapies aids in effectiveness, safety, accessibility and control over treatment, hence can be tried in Dentistry as they are used in medical disorders.
Comfrey (Symphytum asperum): Soak a washcloth in warm Comfrey tea and use as a compress to ease jaw tension and relieve the pain of jaw and tooth fractures or adjustments to braces."
-'Herbal Therapy in Dentistry: A Review' by Shivayogi Charantimath and Rakesh Oswal, K.L.E.V.K. Institute of Dental Science, Belgaum, India; Innovative Journal of Medical and Health Science, Bhopal, India, Volume 1, No. 1, pages 1-4, **2011**.

"**Herbal Tooth and Gum Powder Recipe: The following recipe makes a fabulous tooth powder.** I purchase herbs at the local herb shop and grind them in a coffee grinder. The recipe is from the late master herbalist Dr. John Christopher.
3 parts oak bark, **6 parts Comfrey root**
3 parts horsetail grass, 1 part lobelia, 1 part cloves, 3 parts peppermint
People need to be aware of the herbs they use, and use them wisely. With prolonged and exclusive use, this tooth powder may cause some slight tooth discoloration that can be cleaned off. I alternate using this herbal formula with the baking soda formula. The tooth and gum powder can also be placed on gums for gum healing."
-Cure Tooth Decay: Remineralize Cavities and Repair Your Teeth Naturally with Good Food by Ramiel Nagel. Los Gatos, California: Golden Child Publishing, **2011**.
(Tooth powder was used before there was toothpaste. Wet toothbrush, then dip it into powder. After brushing, rinse mouth.)

"A topical application of clove oil, for instance, will stop the pain of toothache. **Moist herbal wraps, either hot or cold, can be used on specific affected parts of the body.** These wraps are especially effective for sore, tense muscles such as those in the neck, shoulders, back, or jaw when **temporomandibular joint syndrome (TMJ)** is present. **Comfrey: Use as a compress to ease jaw tension and relieve the pain of jaw and tooth fractures or adjustments to braces.**"
-'Herbs in Dentistry' by Jamile B. Taheri, Somayyeh Azimi, Nasrin Rafieian and Hosein Akhavan Zanjani in Iran; International Dental Journal, London, England, Volume 61, No. 6, pages 287-296, December **2011**.
(Temporomandibular joint syndrome is pain and restricted movement of the jaw joint and surrounding muscles.)

"**Data Source:** A literature review was performed in PubMed Central and Cochrane library using MeSH (Medical Subject Heading) terms: 'herbal medicine', 'periodontitis', and 'dentistry': **Symphytum officinale (Comfrey) root, leaf: Gargle.**"
-'Dentistry Meets Nature: Role of Herbs in Periodontal Care: A Systematic Review' by Venisha Pandita, Basavaraj Patthi, Ashish Singla, Shipli Singh, Ravneet Malhi and Vaibhav Vashishtha, Department of Public Health Dentistry, D.J. College of Dental Sciences and Research, Ghaziabad, Uttar Pradesh, India; Journal of Indian Association of Public Health Dentistry, Volume 12, Issue 3, July-September **2014**.

"**Herbs have been used for centuries to prevent and control dental disease.** Herbs may be good alternatives to current preventive and curative treatments for oral health problems, but it is clear that we need more research.
Comfrey: Used as a compress to ease jaw tension and relieve the pain of jaw and tooth fractures."
-'Herbs: A Good Alternatives to Current Treatments for Oral Health Problems' by M. Anushri, R. Yashoda and Manjunath P. Puranik, Government Dental College and Research Institute, Bangalore, Karnataka, India; International Journal of Advanced Health Sciences, Maharashtra, India, Volume 1, Issue 12, pages 26-32, April **2015**.

"**Comfrey Promotes Healthy Teeth: Swirling an infusion (tea of the fresh or dried leaves steeped for a few hours or overnight) can strengthen weak teeth.** I have an old roommate who couldn't afford a dentist when he had a cavity, so he put chewed up Comfrey leaves on the tooth for a few days. The pain went away, and he didn't have any further problems after that."
-'Why You Should Have Comfrey in Your Garden' by Gabe Garms, Raven's Roots Naturalist School: Wilderness Instruction, http://www.ravensroots.com/blog, Sedro-Woolley, Washington, August 1, **2016**.

Nails

"**Nails which were always thin and easily torn now firm and hard (from Comfrey leaf tea).**"
-Comfrey: The Herbal Healer by Henry Doubleday Research Association, **1975**, 41 pages, page 33. ('Comfrey Report' book)

"**Aromatic miscellany:** Gentle shaping and moisturizing encourage healthy growth and strengthen nails.
Herbal tea soaks or herb-infused oil treatments of Comfrey, oat straw and horsetail can strengthen nails and cuticles."
-Horticultural Sciences: Aromatic Plants, Volume 7 edited by Professor K.V. Peter. Volume 7 by Baby P. Skaria, P.P. Joy, Samuel Mathew, Gracy Mathew, Ancy Joseph, and Regina Joseph. Aromatic and Medicinal Plants Research Station, Kerala Agricultural University, Kerala, India, **2007**.

"**Nails, to Strengthen:** Horsetail, Calendula, Aloe, **Comfrey.**"
-'Botanicals in Cosmetics: Reference Chart' by Vege-Tech Botanicals: Nature and Science, certified organic specialties,

www.vegetch.com, Glendale, California, **2011**.

Neuralgia (nerve pain that feels burning or sharp)

"Facial neuralgia is pain that occurs along the path of a facial nerve. Some of these treatments mentioned are a result of analysis of the herbal literature from many countries.
1. Poultice of freshly grated Apple, Lettuce and Comfrey leaf. *Use as needed in the evening. This is an old-time recommendation.*
2. Facial Massage *whenever wanted.* ***Use an infused Comfrey root oil.*** *Use 2 ounces of the Comfrey root infused oil to which you have added 2 ounces of Calophyllum oil and a pinch (1/2 teaspoon) of Golden Seal Root. Mix thoroughly. Use regularly."*
-'Facial Neuralgia' by Jeanne Rose, Aromatherapy & All Things Herbal, www.jeannerose.net, San Francisco, California, **2003**. Principal tutor of 'Herbal Studies Course' and 'Aromatherapy Studies Course'.

*"**Comfrey, Symphytum officinale, is highly effective in curing Facial Neuralgia.** Take Comfrey extract. Put 4 to 5 drops on cotton pad. Keep in that ear which is at the painful side of the face."*
-'Facial Neuralgia Herbal Treatment, Prevention, Symptoms, Causes, Cured By' by Herbpathy: Make Life Healthy, https://herbpathy.com, **2018**.

Oral (see Mouth / Dental)

Osteoporosis (reduced bone density) (also see Calcium Deficiency)

*"**Extracting Minerals from Herbs and Food:** From the Wise Woman perspective, the perfect way to maintain bone health, bone flexibility, and resistance to fracture is to use mineral-rich herbs and foods.*
To extract minerals, we need heat, time, and generous quantities of plant material. I prefer to extract minerals into water or vinegar. ***My favorite nourishing herbal infusions (teas) are made from oatstraw (Avena sativa) or nettle (Urtica dioica) or red clover (Trifolium pratense) or Comfrey leaves (Symphytum uplandica x).***"
-'Healthy Bones the Wise Woman Way' by Susun S. Weed, www.susunweed.com, Woodstock, New York, **2000**. Author of the books 'Wise Woman Herbal for the Childbearing Year', 'Healing Wise', 'Menopausal Years the Wise Woman Way', 'Breast Cancer? Breast Health!', and 'Down There: Sexual and Reproductive Health the Wise Woman Way', Ash Tree Publishing.

"Osteoporosis is considered a weakness of the 'Kidney Essence' (neuroendocrine and endocrine system) with physical manifestations of not feeling supported in life. Intervention using botanicals that also build inward anabolic 'Essence' and mediate the neuroendocrine and endocrine systems is vital for bone health (strength and density).
An herbal tea made from nutrient dense herbs provides additional bone-strengthening minerals *in an easily absorbable and tasty beverage. Drink 2-3 cups (0.47-0.70 liters) of this tea daily.*
Donnie Yance's Simple Nutrient-Rich Mineralizing Bone Tea: Combine equal parts of horsetail, nettles, Comfrey leaf, and hibiscus flowers. *Pour freshly boiled water over herbs. Use 1-2 heaping tablespoons of herbal mixture to one cup (0.23 liter) of water. Steep, covered, for 20 minutes. Enjoy hot or chilled."*
-'Building Healthy Bones' by Donnie Yance, Clinical Master Herbalist and Certified Nutritionist, www.donnieyance.com, The Mederi Center: Holistic Health and Healing, Ashland, Oregon, December 11 **2012**.

*"**There is a group of herbs that can be taken daily as a tonic for building general bone health, including strength and flexibility. Key herbs include** Urtica dioica (nettles), Medicago sativa (alfalfa), Trifolium pratense (red clover), Equisetum arvense (horsetail),* ***Symphytum officinale (Comfrey),*** *and Avena sativa (oat straw, or oat pods).*
This is best prepared as a long infusion (tea), which allows for the extraction of the sought-after minerals. The typical dose is 1 to 2 cups daily; the herbs can be rotated or combined."
-'Maintaining Healthy Bones: Embracing the Plants' by Robin DiPasquale, ND, RH (AHG), Madison, Wisconsin; Naturopathic Doctor News and Review, https://ndnr.com, August 9 **2013**.

*"**Drinking Comfrey leaf infusion (tea) has many benefits. It will keep your bones strong for those who experience thinning bones.** It strengthens digestion as well as elimination and can be used as a bulk laxative.*
For those with inflammation from top to bottom including sore throat, ulcers or inflamed intestines, think of it as a cool drink of water soothing and healing the tissues as it goes down.
Comfrey leaves are rich in proteins including ones needed for the formation of short-term memory cells. They are a great source of folic acid, many vitamins, and every mineral and trace mineral we need for a strong immune system."
-'The Thymekeeper: A Sea of Comfrey' by Mari Marques-Worden, Certified Herbalist; UTE Country News: Putting the Unity Back in Community, Divide, Colorado, Volume 10, No. 9, September **2018**.

Pleurisy (also see Lungs)

Pleurisy is inflammation of the membranes that surround the lungs and line the chest cavity.

"Pleurisy: 3 parts Comfrey root (Symphytum officinale)
1 ounce (28.3 gram) Vervain (Verbena officinalis, V. Hastata); 1 ounce Pleurisy root (Asclepias tuberosa)
1/2 ounce (14.1 gram) Hyssop (Hyssopus officinalis); 1 teaspoonful Cayenne (Capsicum minimum, C. fastigiatum)
Preparation: *Boil the first 4 herbs slowly in 3 pints (6 cups = 1.4 liters) of water down to 1 1/4 pints (2 1/2 cups = 0.59 liters). Strain over the cayenne, set aside to cool, bottle and keep in a cool place.* **Dosage:** *3 tablespoonfuls every 2-3 hours.*
Administration: *Give slippery elm gruel and also a strong stinging nettle tea (Urtica dioica) freely. Relieve any constipation with a catnip (Nepeta cataria) injection (suppository)."*
-'Beatrice® Nutrition Recipes: Comfrey' by Beatrice® Nutrition and Health Sciences, Division of Beatrice® Foods Co., Phoenix, Arizona, **2017**.

Rash (Skin)

*"**Comfrey or Healing Herb (Symphytum officinale):** Truly a healing herb, Comfrey is used in many ways from wounds of the skin to internal problems. The Cherokee used it with boneset as a healing agent for broken bones.*
I use it externally only, as it contains allantoin, which is good for skin rashes.
Comfrey mixed with the juice of a broadleaf plantain leaf makes a good skin ointment for enhancing the healing process of wounds on skin; *the plantain is a good natural antiseptic. Comfrey is always in my Medicine Bag. I do caution you to be sure that in the wild you don't choose foxglove, thinking it is Comfrey, since they are similiar-looking plants."*
-'Medicine of the Cherokee: The Way of Right Relationship' by J.T. Garrett, Ed.D., M.P.H, and Michael Tlanusta Garrett, Ph.D., Eastern Band of Cherokee. Rochester, Vermont: Bear & Company Publishing, **1996**.

*"**Inflamed Skin Rashes:** Calendula (Calendula officinalis) and Comfrey (Symphytum officinale). Apply calendula or* **Comfrey ointment, cream, or lotion** *to troubled areas 2-4 times a day.*
For the lotion, make an infusion, strain, cool, and then apply. Caution: Do not apply Comfrey to broken skin."
-Encyclopedia of Herbal Medicine: The Definitive Reference to 550 Herbs and Remedies for Common Ailments by Andrew Chevallier, FNIMH. London, England: DK (Dorling Kindersley), **2000**.

Rheumatism (inflammation / pain in muscles, joints or fibrous tissue) (also see Arthritis)

*"**Treatment of rheumatism in Hungarian ethnomedicine:** Hungarian folk tradition considered cold, constraint and bad blood to be the cause of the rheumatism.* **They placed compress on the painful and injured (sometimes even broken) parts of the body made of freshly collected and pulped herbal parts,** *e.g., nettle leaves, althaea leaves (marshmallow),* **Comfrey root,** *linaria (toadflax), willow bark, or a fabric compress soaked in the decoction of the above plants.*
They released the pain of the limbs by massaging them with an ointment made of some medicinal herbs, e.g., Common Comfrey root, *verbascum (mullein), agrimony root.*
The creams were made of dried and powdered roots mixed with animal fat (lard or goose fat). Having them mixed, the cream had to stand for some days, before usage."
-'Centuries of the Traditional Medicine in Hungary' by Peter Babulka, Budapest, Hungary, **2000**, page 8.
(Ethnomedicine is traditional medicine practiced by various ethnic groups, and especially by indigenous {native} peoples.)

"Medicinal plant species contribute significantly to folk medicine in Colombia (South America). Leaves are used most commonly prepared by decoction or infusion and administrated orally. **Symphytum officinale L. herb (leaves), cultivated in garden. Used for rheumatism as infusion, taken orally.** *Comfrey used for headache as concoction (mixture) with Rosmarinus officinalis (rosemary) in decoction, taken orally. Used for prostate complaints."*
-'Use and Valuation of Native and Introduced Medicinal Plant Species in Campo Hermoso and Zetaquira, Boyaca, Colombia' by Ana Lucía Cadena-Gonzalez, Marten Sorensen and Ida Theilade, University of Copenhagen, Rolighedsvej, Frederiksberg C, Denmark; Journal of Ethnobiology and Ethnomedicine, Volume 9, No. 23, pages 1-34, **2013**.
(An infusion is when a tea bag or ball is put in a cup of hot water and allowed to steep for a few minutes before drinking. It extracts vitamins and volatile ingredients from soft ingredients like leaves, flowers, etc.
A decoction is used to extract mineral salts and bitter principles from hard material such as roots, bark, seeds and wood. They need boiling for at least 10 minutes and then are steeped for a number of hours.)

Skin Health (also see Rash)

*"**Comfrey Cosmetic Uses:** Not only will Comfrey restore the skin's freshness and thoroughly cleanse the pores without overstimulating greasy skin, but it has the beneficial side-effect of relieving laryngitis (voicebox inflammation), catarrh (mucus discharge) and smoker's cough.*
First, clean the face and neck with non-greasy cleanser, rinsing several times afterward with tepid (lukewarm) water and patting the skin dry. Then use the following steam to cleanse: **Steam Facial for Oily Skin / Large Pores**

    1 tablespoon Comfrey leaf, cut, sifted
    2 tablespoon lavender flowers,   3 tablespoon powdered licorice root,   1 tablespoon lemon peel, cut sifted
    1 tablespoon peppermint, cut or whole,   1 tablespoon pansy, cut or whole,   1 tablespoon parsley
    1 tablespoon rose bud or leaf,   1 tablespoon strawberry leaves,   2-3 cups water

Combine all herbs in bowl, mix well. Place 1/4 cup (59 ml) mixture in tall narrow enamel or glass pot. Add water, cover pot. Bring to a boil. Simmer for 3-5 minutes, remove from heat. Steep for several minutes. Cover hair with towel and remove lid from pot. Position face over pot (not too close). Cover pot and sides of face with the towel. Allow the herbal steam to relax, cleanse and medicate your pores. Wipe away dirt and oil with clean washcloth, splash warm water on your face and rinse with cool mineral water to close pores. Pat dry."
-'Herbalpedia®: Comfrey' by Maureen Rogers, The Herb Growing & Marketing Network, Silver Spring, Pennsylvania, www.herbalpedia.com, **2006**.

Skin Injuries (also see Skin Ulcer)

"In addition to Comfrey's tissue regenerative abilities, it seems to be effective in destroying harmful bacteria. Applying Comfrey extract to infections on both farm animals and humans has quickened the healing process."
-Rodale's Illustrated Encyclopedia of Herbs edited by Claire Kowalchik and William H. Hylton. Emmaus, Pennsylvania: Rodale Press, **1998**, page 105.

"Comfrey is a cell proliferant (encouraging only healthy cells to grow, not scar tissue), and possibly the most healing herb ever yet discovered. I (a midwife) include some in the mom's birth kit, and during labor make it into an infusion (strong tea) that the mother can dilute in her peri (perineum) bottle for rinsing each time she uses the toilet. If the bottle is sprayed before she begins to urinate, the urine will not burn, even if she had stitches. It heals from the inside out, reduces healing time significantly, and virtually eliminates infections.
The solids left over from making the infusion I roll into the leftover gauze pads and make peri compresses, *which also go into the refrigerator for the mom to apply a fresh one whenever she needs to. The feedback is that they feel awesome. Several grand-multips (multiparous woman; one who has given birth at least 2 times) have told me that they recovered better from their birth with me than they ever did before, and I am sure that the herbs played a big part in that."*
-Lorri Carr, CPM (Certified Professional Midwife)" -'Comfrey, Symphytum Officinale, Boraginaceae Family' by Sarah Heany, Highland Midwife Birth Services LLC (TM), Yakima, Washington, November **2011**.
    (Perineum is the space between the anus and scrotum in the male and between the anus and the vulva in the female.)

Skin: Stretch Marks

"Stretch marks are a common aftermath of pregnancy. It's important to lubricate your body daily to help prevent them. This is especially important the last three months of pregnancy so your skin will be supple (flexible) and stretch during delivery. You'll want to apply this 'Pregnant Belly Butter'® to the perineal area (perineum), belly, hips, thighs, and breasts, which are all areas where stretch marks can occur. : **Pregnant Belly Butter®**
    2 cups (475 ml) olive oil or coconut oil,   1/2 ounce (15 grams) calendula flowers
    1/2 ounce (15 grams) **Comfrey leaves**
    1 tablespoon (15 ml) vitamin E oil,   5 drops essential oil of lavender

In a dry, clean glass jar, place the crushed herbs and oil. Allow to steep for two weeks. Strain through a clean dry cotton cloth while squeezing the oil out of the herbs. Discard the herbs. Stir in the vitamin E and lavender oil. Bottle. Apply to areas prone to stretching at least twice daily. **Soothe sore nipples from breastfeeding by applying a salve of calendula and Comfrey."**
-The Country Almanac of Home Remedies: Time-Tested & Almost Forgotten Wisdom for Treating Hundreds of Common Ailments, Aches & Pains Quickly and Naturally by Brigitte Mars and Chrystle Fiedler. Minnesota: Quayside Publishing Group, **2011**.
    (Perineum is space between the anus and scrotum in the male and between the anus and the vulva in the female.)

Skin Ulcer (also see Skin Injuries)

Do not use Comfrey externally on deep wounds. See 'Wounds' in this section for more information about it.

Skin ulcers are slow healing, open sores usually with shedding of inflamed tissue.

*"**I have on more than one occasion cured old (skin) ulcers which have resisted other treatments, by the simple extract from the Comfrey root applied on lint to saturation (well soaked).** After a few hours this dressing, in favourable cases, sets quite hard, and can only be removed by the lengthy application of water. This hard setting acts, in some measure, like strapping, drawing the edges of the ulcer towards one another, but probably much more evenly and accurately than the most skillful strapping could effect. This looks as though the cure is wrought similarly to healing under a scab."*
-'The New Cell Proliferant: A Note on the Symphytum Officinale or Common Comfrey' by William Bramwell, M.A., M.D., B.Ch., The British Medical Journal, London, England, pages 12-13, January 6 **1912**.

*"The patient was a woman aged 87. Having resisted all kinds of previous treatment, it was **a very large ulcer** (measuring 4 by 3 inches = 10.1 by 7.6 cm) involving the skin and deeper structures over the upper thorax (chest), and it was slowly spreading. There was some seropurulent (pus) discharge.*
***After being dressed with the mucilaginous (Comfrey root) infusion for about a week, the surface cleaned and a distinct ingrowth of epithelium (skin) could be seen** taking place from some of the marginal points. The epithelium became stronger and closed in to a considerable extent."*
-The Medicinal Uses of Comfrey by Dr. Charles MacAlister, M.D., F.R.C.P., written **1935**. (in 'Comfrey: Fodder, Food & Remedy')
    (Infusion also known as steeping extracts chemical compounds from plant material in a solvent such as water, oil or alcohol, by keeping it in the solvent over time.)

*"Recently a most interesting case came under my observation. A middle-aged woman came to me with a **large ulcer** below the eye and close to the nose. **I prescribed a Comfrey poultice, and the 'Green Drink' containing Comfrey leaves. Soon after the application of the Comfrey leaf poultice, the painful swelling subsided, and rapid improvement was noted.** Only a few months after the initial treatment there was a complete healing over of the infected area, and the ulcer disappeared."*
-Nature's Healing Grasses by H.E. Kirschner, M.D. California: H.C. White Publications, p. 59. First printing **1960**, 16th printing 1980.

*"From a 75-year-old lady living in Mariposa, California: 'God works in mysterious ways. His wonders to perform. I would like to write you about the great benefit to me of your article, **'Comfrey: The Miracle Herb'**. I have been using the fresh Comfrey leaf poultices on an obstinate (stubborn) sore on my neck. The results of this simple treatment are most gratifying.' "*
-Nature's Healing Grasses by H.E. Kirschner, M.D. California: H.C. White Publications, page 124, 1960.

*"**Skin Ulcers:** An infected wound which is often hard to clear up. **Comfrey: Apply poultice of 1 teaspoon fresh crushed leaf."***
-'Healthy Harvest: A Training Manual for Community Workers in Good Nutrition and the Growing, Preparing and Processing of Healthy Food' by 'Food and Nutrition Council of Zimbabwe', 'Food and Agriculture Organization' of the United Nations, and 'United Nations Children's Fund', 123 pages, **2015**.

Sore Throat (see Cough)

Sprains (see Cuts / Bruises / Sprains)

Stings and Bites (Insects, Nettles)

*"Miss W. of Broadtown, near Swindon (Wiltshire County, England) **keeps (honey) bees, and she reports that she has cured some very nasty stings with Comfrey ointment, removing the pain almost at once.**
I suggest that it should be tried for wasps and mosquito bites."*
-Comfrey: The Herbal Healer by Henry Doubleday Research Association, **1975**, 41 pages, page 8. (in 'Comfrey Report' book)

*"R.T. reports: 'A few weeks ago my five year old skidded on his two-wheeler (bicycle) and **fell into a bed of stinging nettles, face first. I rushed him home and applied Comfrey ointment.** He looked a mess but the pain was taken away immediately. Within a short time the spots had gone."*
-Comfrey: The Herbal Healer by Henry Doubleday Research Association, **1975**, 41 pages, page 12. (in 'Comfrey Report' book)

*"**An English beekeeper kept Comfrey ointment on hand for bee stings.**
It is also used to relieve pain of mosquito, wasp and horsefly bites."*
-The Herbalist: A Publication of The Herb Society of America, No. 44. Boston, Massachusetts: **1978**. Article: 'Quaker or Russian Comfrey' by Rosella F. Mathieu.

*"**I have found Comfrey, along with perhaps plantain and echinacea, to be incomparable in drawing out the poison from spider bites, healing them quickly and painlessly.** It is the fastest wound healer around."*
-Healing with the Herbs of Life: Hundreds of Herbal Remedies, Therapies, and Preparations by Lesley Tierra, L.Ac., Herbalist, A.H.G. New York: Crossing Press / Crown Publishing Group, **2003**.

*"A customer came into my health food store and told me about her experience with a **Brown Recluse Spider bite** that she got while pulling weeds. **She snipped off a leaf of Comfrey, rolled it between her hands until the juice began to flow and then bandaged it around the bite.** When she took off the leaf that night it was black and running with poison. She got a fresh leaf and repeated the treatment.
The next morning the leaf was black and running with poison. She repeated the treatment and that night when she took off the leaf it looked fresh and green as if it had just been picked. No loss of flesh, no pain."*
-'Letters to the Editor: Bacon Cures' by Yvonne Hursh, Missouri, Acres USA: The Voice of Eco-Agriculture, Issue #550, page 4, April **2017**.

(A Brown Recluse Spider bite is serious so I recommend you see a doctor immediately.)

Swelling (also see 'Inflammation' and 'Cuts / Bruises / Sprains')

"***Against a sudden swelling,*** *take horehound, beat and mingle it with lard, lay on.*
Again, mingle together the cottony potentilla, commonly called silverweed, groats of malt, smede or fine flour, cress, the white of an egg, bishopwort, helenium, ontre, lupins, 'sigsonte', **Comfrey***. Lay on.*"
-'Leechdoms, Wortcunning and Starcraft: Being a Collection of Documents Illustrating the History of Science Before the Norman Conquest' also called 'Chronicles and Memorials of Great Britain and Ireland During the Middle Ages' edited by Thomas Oswald Cockayne, London, England, 1864. Volume 2, page 75.
    (The Norman Conquest of England was a **1066 AD** invasion and occupation of England by Norman, Breton, Flemish and French soldiers.)

Teeth (see Mouth / Dental)

Tonsillitis (infection of lymph nodes in throat) (also see 'Cough / Cold / Sore Throat')

"***Comfrey, Symphytum officinale: There are reports of acute cases of ulcerated tonsils*** *with a high temperature present in which infusions of the green Comfrey plant given frequently* **-poultices of the dry powder placed about the throat-** *responded to the treatment within a short period of time.*"
-Natures Healing Agents: The Medicines of Nature or the Natura System by R. Swinburne Clymer, M.D. Philadelphia, Pennsylvania: Dorrance and Company, **1963**, page 124.
    (See subsection 'Comfrey Poultices' in section 'Making and Using Comfrey Medicine' {Chapter 27}.)

"*Comfrey is a healing herb and used extensively for cuts, bruises, wounds burns and even abrasions.* **Lately, many studies have proved that Comfrey leaves if used for gargling can cure most of the problems related to tonsils.** *It is recommended that Comfrey should not be consumed but the decoction can be prepared by boiling the leaves and letting it steep for a while and gargle with the herbal Comfrey preparation. Doing this twice every day is sure to cure the troubles of the tonsils.*"
-'Effective Herbs for Tonsils: Comfrey' by Richa, Search Herbal Remedy, www.searchherbalremedy.com, October 25 **2013**.

Tumor (see Cancer)

Ulcers: Gastric and Duodenal (Stomach and Intestines)

"*Whereas there are cases of internal ulcerative conditions in which improvement has been noted, there have been a good many in which the results were doubtful. One can only speak with any certainty when diagnosis has been clear and where beneficial results have not followed other forms of medical treatment.*
One such experience occurred in the case of an elderly man who after **suffering pain after food, vomiting of blood, melaena and other clinical evidences of ulceration which were unrelieved by medical treatment. He was then treated with allantoin dissolved in a Comfrey infusion** *and a purely milk diet was prescribed, with the result that he was free from signs and symptoms in a month and is still alive at 90 odd years of age.*"
-The Medicinal Uses of Comfrey by Dr. Charles MacAlister, M.D., F.R.C.P., written **1935**. (in 'Comfrey: Fodder, Food & Remedy')
    (Melaena is dark black, tarry feces associated with upper gastrointestinal bleeding.)
    (Infusion or steeping extracts chemical compounds from plant material in a solvent such as water, oil or alcohol, by keeping it in the solvent over time.)

"***Gastric and duodenal ulcers: In addition to aiding stomach ulcers, Comfrey is reported as being of value in ulcers involving the small intestines, known as the duodenum.***
The powder, taken in gelatin capsules, will usually relieve any pains in the stomach, a condition known as gastralgia."
-Modern Encyclopedia of Herbs by Joseph M Kadans, N.D., Ph.D. New York, New York: Simon & Schuster, Inc., **1993**, page 100.

"***Although many herbs are available as capsules or tablets, for some conditions the herbs really do need to be taken as powders; they shouldn't be encapsulated. This is especially true in cases of severe stomach ulceration.***
The herbs should be powdered, then mixed with liquid and consumed. This allows the herb to make contact with entire affected area. Many people don't like to do it this way because of the flavor of the herbs- they are often bitter. I just use juice or honey.
If ulceration is in the duodenum, which lies just below the stomach, then capsules should be used:
    The capsules tend to sit at the bottom of the stomach and then drop through into the duodenum where they are needed.
    Duodenal ulcers are often accompanied by painful cramping or spasming. This can be alleviated by the addtion of a few drops of peppermint essential oil to the herbal mixture before encapsulating it.

For duodenal ulcers, you would take the below formula in capsules.
Herbal regimen for an ulcerated stomach:
The herbs should not be in capsules in order to allow them to fully coat the stomach lining.
Ingredients: 4 ounces dried licorice root, 4 ounces dried marshmallow root, **4 ounces dried Comfrey root,** *2 ounces berberine plant tincture, 1 quart wildflower honey.*
Powder the licorice, marshmallow, and Comfrey roots (as finely as possible), and mix together in equal parts.
Take 2 tablespoons, twice a day (morning and evening), mixed in any liquid (e.g., apple juice), for 30 days.
For the next 60 days use just powdered licorice and marshmallow roots (omitting the Comfrey), again mixed in equal parts; take 1 tablespoon in the morning only.
At the same time, as you first begin the treatment with the powder, take 1 teaspoon of the berberine plant tincture three times daily for 15 days. And take 1 tablespoon honey six times daily for 30 days.
Instead of the powder you can take Symphytum (Comfrey) tincture of the dried root:
1:5, 50 percent alcohol. Dosage: 30-60 drops up to 3 times daily for up to 30 days."
-Herbal Antibiotics: Natural Alternatives for Treating Drug-Resistant Bacteria by Stephen Harrod Buhner. North Adams, Massachusetts: Storey Publishing, **2012**, page 354-355, 378.

Urinary (see Bladder)

Vagina: Dry and also Healing Tears

In 'Wounds' subsection below, see 'Healing Stories 3 and 4' about the vagina. There is other information in that section about vaginal tears (such as from childbirth) and repair of the hymen. The hymen is a membrane that partially closes the opening to the vagina. Also there are cautions about using Comfrey on and around the vagina.

"**Dry Vagina:**
1. Comfrey root sitz bath *(two quarts/liters of the infusion) is an old favorite for keeping vaginal tissues flexible, strong, and soft. Sitz for 5 to 10 minutes several times a week.*
2. Comfrey ointment is the ally of choice when skin needs flexible strength. *Rub in the morning and night, and use as a lubricant for love play. The vulva will be noticeably plumper and moister within three weeks."*
-Menopausal Years the Wise Woman Way: Alternative Approaches for Women 30 - 90 by Susun S. Weed. Ashcroft, Canada: Ash Tree Publishing, **2002**. (Vulva is the female external genitals.)

(For information about Comfrey sitz baths, see sub-subsection 'Sitz / Hip Bath' in subsection 'Comfrey Baths and Soaks' in section 'Making and Using Comfrey Medicine' {Chapter 27}.)

(For information about Comfrey ointment, see subsection 'Comfrey Ointment, Cream or Salve' in section 'Making and Using Comfrey Medicine'.)

"**Vulneraries to Heal Vaginal Tissue: The vaginal mucosa (mucous membrane) is receptive to topical moistening. Vaginal salve containing Comfrey oil may increase epithelial cell growth and tissue integrity.**
Other botanicals, such as St. John's wort and calendula, also contribute to cell regeneration and help to fight infection. Topical application of vitamin E (400 IU) included in vaginal salves and suppositories can help reduce irritation, inflammation, and promote healing.
Comfrey moistens, heals, and soothes irritated and inflamed tissue. Its mucilaginous nature lends well to its emollient effects on the vaginal tissue. *Comfrey contains a number of constituents, notably allantoin and rosmarinic acid, as well as tannins and mucilage, which show evidence of anti-inflammatory activity and tissue healing activity.*
Note that PAs (Pyrrolizidine Alkaloids) from Comfrey, associated with venoocclusive liver disease, may be absorbed transvaginally in small amounts; caution should be observed if topical Comfrey products are used daily for prolonged periods of time, although adverse effects are not expected.
The German Commission E approves the topical use of Comfrey on unbroken skin only, not to exceed 4 to 6 weeks in duration, and not to exceed 100 microgram equivalents of PAs daily."
-Botanical Medicine for Women's Health by Aviva Romm, CPM, Registered Herbalist of the American Herbalists Guild. London, England: Churchill Livingstone, **2009**. (Chapter 19: Menopausal Health, pages 513, 514.)

(A vulnerary promotes healing of wounds by being applied directly to a cleaned area.)

"**Vaginal Lubricant: This vaginal lubricant is intended for topical use. It can be inserted vaginally and applied to the labia daily. It also can be applied liberally prior to intercourse.** *Clients can be taught to prepare this product themselves, or it can be prepared by the practitioner and stored in a cool place (i.e., refrigerated) for up to 1 year. Warm to room temperature prior to insertion. The following recipe makes 8 ounces (236 ml). Preparing the salve:*
1. *Sterilize four 2 ounce (59 ml) glass wide mouth jars with plastic caps. Be sure the jars and caps are completely dry before using.*
2. *Place in a double boiler: 0.8 ounce (23 ml) grated beeswax,* **1/4 cup (59 ml) herbal oils made from dried black cohosh and Comfrey roots, calendula flowers, and wild yam root powder.** *Add 1/4 cup coconut oil and 1/4 cup jojoba oil.*

3. After heating the above and allowing cooling, add: 1 tablespoon wheat germ oil, 1 tablespoon liquid vitamin E in the form of d-alpha tocopherol, 1 teaspoon emulsified vitamin A. Add pure certified therapeutic essential oils as follows: 30 drops lavender, 30 drops rose geranium essential oil, 10 drops Rosa Damascena (Bulgarian. Buy Otto, not Absolute)
4. Cap immediately. Keep out of direct sunlight and away from heat accept body heat. Use clean fingers or applicator to rub onto labia and into the vaginal canal."
-Botanical Medicine for Women's Health by Aviva Romm, CPM, Registered Herbalist of the American Herbalists Guild. London, England: Churchill Livingstone, **2009**. (Chapter 19: Menopausal Health, page 515.)

Vaginal Discharge

"**Whites (Leucorrhea):** *This is generally the result of weakness, and is a complaint to which many females are subject; it is both disagreeable and exceedingly debilitating, and on no account should be suffered to continue.*
The following medicine may be taken internally:
    White pond lily root, half an ounce. **Comfrey root, one ounce (29 ml).**
    Crane's bill root, two ounces. Female restorative powder, half an ounce.
Bruise and boil the roots in three pints (6 cups = 1.4 liter) of water down to a quart (4 cups = 0.9 liter), now add the powder, and boil for five minutes more, mix well, let it stand half an hour, strain, and add half a pound (0.22 kg) of lump sugar. Take of this mixture from four to six half wine glassesful daily.
    Female Restorative Powder:
    Balmony, 1/2 ounce (14.7 ml). Bistort root, pulverized, 1/2 ounce.
    White pond lily, pulverized, 1/2 ounce. **Comfrey root, pulverized, 1 ounce (29.5 ml).**
    Cloves, pulverized, 1 ounce. Tormentil, pulverizd, 1 ounce.   Cayenne, as much as will stand upon a shilling.
    Mix well, and take well sweetened with treacle (molasses), or in wine glassful doses.
    This will be found a most excellent remedy for leucorrhoea (whites) and all excess of menstruation, etc."
-'Family Medical Adviser', Fifth Edition (First Edition 1852), by Dr. John Skelton, Senior, Lecturer and Professor of Medicine, Leeds, England, **1857**. A Thomsonian herbal practitioner (Physio-Medicalism).
    (Leucorrhea is an odorless, whitish or yellowish discharge of mucus from the vagina.)
    (A shilling in 1857 was 23.6 mm diameter = 0.9 inch.)

"**Western herbalists call vaginal discharges leucorrhea ('the whites').** *Eastern herbalists call it a 'damp disorder'. Both agree that astringent herbs - usually in the form of sitz baths or suppositories - are useful as treatment. When effective, they end the discharge within five to ten days. My favorites include:* **Comfrey (Symphytum officinale) root and/or leaf infusion, to strengthen, soothe, and heal all the tissues down there. Does not get rid of infections."**
-'Vaginal Distress' by Susun S. Weed, www.susunweed.com, Woodstock, New York, **2007**. Author of the books 'Wise Woman Herbal for the Childbearing Year', 'Healing Wise', 'Menopausal Years the Wise Woman Way', 'Breast Cancer? Breast Health!', and 'Down There: Sexual and Reproductive Health the Wise Woman Way', Ash Tree Publishing.

Varicose Veins, and Inflammation of Veins

"**Herbs for Varicosities (Varicose Veins): External applications: The following herbs can be used as poultices, sitz baths (hip bath), oils, suppositories or fomentations (hot pack): Comfrey,** *Slippery elm, Plantain.*
These plants are mucilaginous and have a demulcent effect thereby soothing and healing irritated tissues.
All three of these plants are well known for their ability to speed the healing process when applied to irritated tissue.
Comfrey, Symphytum officinalis:
Useful externally as a poultice, paste or fomentation for contusions (bruises), sprains, dislocations, wounds, burns, ulcers and all inflammatory skin disorders.
Used to decrease inflammation of thrombophlebitis and phlebitis. *This herb decreases the healing time for all manner of skin wounds and irritations."*
-'Varicose Veins and Hemorrhoids' by Sharol Tilgner, N.D., Herbal Transitions: A Herbal Newsletter From Wise Acres Farm, Pleasant Hill, Oregon, www.herbaltransitions.com, **2005**.
(A suppository is a solid medicine inserted into the rectum or vagina. It dissolves and has both local and systemic effects.)
(Phlebitis is inflammation of a vein, both surface and deep veins. Thrombophlebitis is blood clots in a vein that cause inflammation.)

"**Varicose veins occur when capillaries and veins lose their elasticity and become distended and distorted.** *More than a cosmetic problem, varicose veins can be painful and make walking and sitting uncomfortable, and are a sign that the vascular system needs attention.*
Vitamin C Poultice:
    1 part calendula, **1 part Comfrey,** *1 part yarrow, 5000 I.U. vitamin C, witch hazel extract. Chop and mix together the calendula, Comfrey, and yarrow. Mix in the vitamin C. Add enough witch hazel extract to make a thick paste.* **Apply directly to the varicose veins or wrap in a muslin cloth and place over the veins.** *Leave on for 30 to 45 minutes."*
-Rosemary Gladstar's Herbal Recipes for Vibrant Health: 175 Teas, Tonics, Oils, Salves, Tinctures and Other Natural Remedies by Rosemary Gladstar. North Adams, Massachusetts: Storey Publishing, **2008**, page 300.

(Muslin is cotton fabric of plain weave made in different degrees of fineness, used for sheets and other purposes.)

Wounds (also see Skin Injuries and see Cuts)

Do not use Comfrey externally on deep wounds. The reason is given below.
If you are treating a serious wound, see a healthcare practitioner.

*"In the year 1788 when I was in my nineteenth year, my father purchased a piece of land on Onion river, in the state of Vermont. On the 2nd of December, when **I had the misfortune to cut my ancle (ankle) very badly,** which accident prevented me from doing any labor a long time, and almost deprived me of life.*
> ***The wound was a very bad one, as it split the joint and laid the bone entirely bare,*** *so as to lose the juices of my ancle joint to such a degree as to reduce my strength very much.*

*My father sent for a Dr. Cole, of Jericho, Vermont, who ordered sweet appletree bark to be boiled, and the wound to be washed with it, which caused great pain, and made it much worse, so that in eight days my strength was almost exhausted; the flesh on my leg and thigh was mostly gone, and **my life was despaired of**; the doctor said he could do no more for me.*
*My father was greatly alarmed about me, and said that if Dr. Kittridge, of Walpole, New Hampshire, could be sent for, he thought he might help me; but I told him it would be in vain to send for him, **I could not live so long as it would take to go after him,** without some immediate assistance.*
*He said he did not know what to do; I told him that there was one thing I had thought of, which I wished to have tried, if it could be obtained, that I thought would help me. He anxiously inquired what it was, and **I told him if he could find some Comfrey root, I would try a plaster made of that and turpentine.***
*He immediately went to an old place that was settled before the war, and had the good luck to find some; a **plaster was prepared by my directions and applied to my ancle, the side opposite the wound, and had the desired effect.***
The juices stopped running in about six hours, and I was very much relieved, though the pain continued to be very severe, and the inflammation was great; the juices settled between the skin and bone and caused a suppuration (pus formation), which broke in about three weeks, *during which time I did not have three nights' sleep, nor did I eat any thing."*
-'A Narrative of the Life and Medical Discoveries of Samuel Thomson: Containing an Account of his System of Practice, and the Manner of Curing Disease with Vegetable Medicine' by Samuel Thomson, Columbus, Ohio, **1833-1835**.

> (What I find interesting is that the plaster was put on the side opposite the wound. If you read all of the information here about wounds, you see that it is not good to put Comfrey directly on a deep wound because the top can heal before the inside, causing problems.)

*"**New Cure for Stubborn Wounds Results from Clue Given by Fly:***
> *The insect that gave the clue to this discovery is one of the flies -in the maggot stage- that gained fame as **a medical aid on World War (One) battlefields, where an Army doctor found that wounds infested with maggots healed better and faster than wounds without them.** Since then surgeons all over the world have used maggots in treating deep infections difficult to cure by ordinary surgery.*

Allantoin:
> *Dr. William Robinson, of the 'Bureau of Entomology and Plant Quarantine', now finds that **allantoin, which is given off by the maggots as they work their way through a wound, is responsible for part of this power.** Allantoin, Dr. Robinson says, is not a new discovery. Dr. C.J. Macalister, who used it successfully 23 years ago for ulcers, reported that **Europeon peasants had long applied the roots of Comfrey, which contain allantoin to sores.***

How to Control Healing of Deep Wounds:
> *His recent tests, Dr. Robinson says, show that **allantoin is particularly useful for non-healing wounds, such as chronic ulcers and extensive burns that refuse to mend.***
> *After a few treatments, pinkish granulation tissue begins to grow and soon the tissues are knitting together rapidly.*
> ***A specially promising feature of the new treatment is that it can be made to control healing. Healing from the bottom up can be ensured in a deep wound by applying the allantoin solution in a small packing at the base of the wound and covering the sides with vaseline.***
> *General granulation can be promoted by filling the wound with gauze well saturated with the solution."*

-'Illinois Medical Journal: Official Organ of the Illinois State Medical Society, Volume 70' edited by Charles J. Whalen MD and Henry G. Ohls MD, Oak Park, Illinois, July-December **1936**.

> *"Maggots specifically Licilia sericata (sheep blowfly), were used during and immediately after World War I in the treatment of hard to heal wound infections.*
> ***Allantoin, in both maggot therapy and Comfrey, works by increasing the water content of the extracellular matrix and enhancing the skin peeling or desquamation of upper layers of dead skin cells,*** *increasing the smoothness of the skin, promoting cell proliferation and wound healing."*
> -'Comfrey, Maggot Therapy and Cell Regeneration' by Herbal Goddess Musings: Celebrating Gaia's Herbal Gifts, https://herbalgoddessmedicinals.wordpress.com, March 2 2017.

*"**Comfrey, Symphytum officinale:** Its proponents or converts become so by what it has done for them.*
In chronic sores of various types, a poultice made of either the crushed green plant or the powder of the plant *has*

effected elimination of the sores within a short period of time in which all other agents and procedures had failed."
-Natures Healing Agents: The Medicines of Nature or the Natura System by R. Swinburne Clymer, M.D. Philadelphia, Pennsylvania: Dorrance and Company, **1963**, page 124.

"**Russian soldiers have used Comfrey as a handy field remedy since Czarist times.**
Handwritten journals describe **cavalry (horse) soldiers who 'sliced and slew Comfrey roots to mash' and applied them to each other's wounds, holding them on 'until the blood stopped and the groans of pain were gone'.** During those times, Comfrey was known as 'knit-together herb'.
But the weapons of battle changed, and the name of the herb changed as well. The widespread use of cannons and guns forced soldiers to fight from trenches, and they actually dug up Comfrey from the edge of the trench. Modern Russian herbalists soon adopted the name 'trench herb'. Although Comfrey was no longer needed to treat injuries caused by sabres and swords, its antibacterial and styptic (stops bleeding) properties made it a popular remedy for treating bullet and shrapnel wounds.
Parts used: Comfrey roots, sometimes fresh leaves."
-A Russian Herbal: Traditional Remedies for Health and Healing by Igor Vilevich Zevin, Nathaniel Altman and Lilia Vasilevna Zevin. Rochester, Vermont: Healing Arts Press, **1997**.
(Czar or Tsar means monarch or ruler. The first Czar in Russia was Rurik, Prince of Novgorod, around 862 AD.)
(Shrapnel is fragments of a bomb, shell or other object from an explosion.)

"**Healing Story 1 (Finger):**
Here is an example of incorrect and correct wound treatment that demonstrates how important it is to understand how healing works, rather than to simply throw herbs at a problem blindly.
This is a medicine story told by herbalist 7Song. A man came to see him who had **a deep cut on his finger that had been stitched up by a doctor.**
Someone told him to put Comfrey on the wound. After a few days the wound got putrid (rotting), starting discharging pus and the finger turned blackish. 7Song removed the Comfrey, put on yarrow and the next day the end of finger had 'pinked up.'
What happened here was that **Comfrey, which is well known for encouraging cellular proliferation or growth, caused the cells in the end of the finger to grow more quickly. Yet, they had no blood supply so they started to die off, shown by the blackish color.** The yarrow, on the other hand, stimulated blood flow into the finger to nourish the tissues and out to remove waste products. It is an excellent wound healer, so it started the job of recovery correctly.
Healing Story 2 (Ankle):
A man had gouged out a chunk of flesh on the front side of the ankle on a barb wire fence. The tendons were freely visible in the wound.
It was fortunate that the man did not put the Comfrey on the wound externally since, as we have noted, this herb tends to cause a wound to grow back on the outside, rather than from the inside out. This would have caused the wound to heal on the outside but become rotten on the inside.
However, Comfrey is very good for regenerating tissue and is especially beneficial to the tendons, so the internal use was likely to be very helpful.
Healing Story 3 (Vagina):
One of my students in Pennsylvania reported that after birth she used **Comfrey on a vaginal tear.** The remedy, rather than bringing the lips of the wound together, **caused healing all over the surface of the tear, leaving the gaping wound intact but covered up by a membrane.** In the end the doctors had to sew up the tear surgically.
Healing Story 4 (Vagina):
Another student in New York recounted how an acquaintance of hers used **Comfrey for diaper rash on her baby, only to have it cause the lips of the vagina to grow together.** They had to be surgically separated. Another student in the class realized that exactly the same thing had happened when she used Comfrey for diaper rash on her girl baby."
-'Herbal First Aide: Cuts, Lacerations, Bruises, Burns, Boils, Broken Bones, Bites' by Matthew Wood, Masters of Science (Scottish School of Herbal Medicine), Registered Herbalist (American Herbalists Guild), Sunnyfield Herb Farm, Minnesota, www.woodherbs.com, date unknown but around **2010**.
(The herbalist 7Song created the Northeast School of Botanical Medicine, Ithaca, New York, www.7song.com.)
(For deep or serious wounds see your healthcare practitioner.)

"**Do not apply Comfrey on deep wounds, which can close at the top before the deep part has healed underneath.** Saint John's wort is better for deep puncture wounds.
Virginity: This Comfrey herb had such a reputation for repairing tissue that it was said that **less virtuous brides would bathe in it before their wedding day to restore their virginity!**"
-Backyard Medicine: Harvest and Make Your Own Herbal Remedies by Julie Bruton-Seal and Matthew Seal. New York, New York: Skyhorse Publishing, 2009. Published in Great Britain by Merlin Unwin Books, 2008 as 'Hedgerow Medicine'.

"**Comfrey must be used with caution and respect.**
It is such an excellent and speedy wound-healing remedy that it actually should not be used on deep wounds or lacerations (cuts). It could potentially heal the top layer of skin before the bottom layer, resulting in an abscess.
So please do avoid Comfrey for major skin wounds, and use it only for bruises, sprains and minor cuts and scratches."
-'Comfrey (Symphytum Officinale): A Healer of Wounds, Bruises and Bones' by Steph Zabel, Herbalist, Ethnobotanist and

Educator, Cambridge Naturals, Cambridge, Massachusetts, www.cambridgenaturals.com, April 31 **2016**.
(An abscess is a swollen area with a pocket of pus.)

"Do not apply Comfrey initially to a wound, especially if it is deep or liable to become infected.
Infection can proliferate (grow) underneath the superficial (top) skin growth initiated by Comfrey. Topical (skin) cell proliferants such as Comfrey can be applied if the wound is not too deep and there is little chance of infection.
Anti-inflammatory tea blend: Licorice, Marshmallow, **Comfrey**, Calendula, Yarrow.
Vulnerary (wound) salve:
Calendula, Chickweed, **Comfrey**, St. John's wort (vitamin E is often added to these as a preservative)."
-'Herbal First Aid Wound Care' by Herbalist 7Song, Northeast School of Botanical Medicine, Ithaca, New York, www.7song.com, March **2017**.

"In Comfrey the allantoin repairs, the mucilage soothes, protects and absorbs acid, while the tannins tighten mucous membranes and tissue.
Specifically, it binds cornified (hard) layers together, moistens and prevents drying and has a positive effect on keratin and skin in general by holding moisture.
Comfrey increases epithelial skin tissue growth, increases leukocyte infiltration and reduces growth of necrotic (dead) tissue. Allantoin increases granulation tissue in wounds and reduces inflammation.
It is best to combine an anti-bacterial herb *such as Oregon grape root, St. John's wort or usnea (a lichen) with Comfrey in treating open wounds.* **If sealed too quickly, especially in burns, there is a danger of sealing in bacterial infection.**
Vagina: Midwives make great use of Comfrey to quickly heal vaginal tears from birthing.
In some cultures, the hymen was reputed to repair from repeated use of Comfrey leaf bolus (suppository)."
-'Comfrey' by Glenn Axford, Our Common Roots: An Exploration of the Healing Power of Nature, Canada, www.ourcommonroots.com, date unknown but **2018** or earlier.
(Keratin is a fibrous protein that creates the structural parts of hair, feathers, hoofs, claws, and horns.)
(Hymen is a membrane that partially closes the opening to the vagina.)

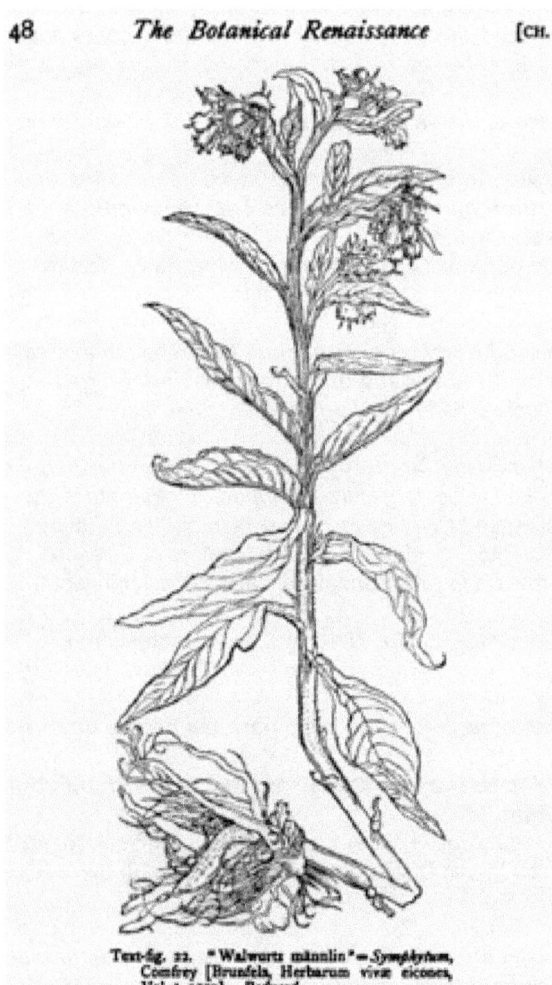

Symphytum 1530 Brunfels Herbarium
Text-fig. 22. "Walwurtz mannlin = Comfrey Brunfels, Herbarum Vivae Eicones, Volume I. 1530".

'Herbals: Their Origin and Evolution: A Chapter in the History of Botany: 1470 to 1670' by Agnes Robertson Arber, University College, London, England; Cambridge University Press, England, 1912.

Art is on page 48 in Chapter 4: 'The Botanical Renaissance of the Sixteenth and Seventeenth Centuries'. The below text, pages 47-50.

"In his 'History of Botany', Kurt Sprengel first used the honoured title, 'The German Fathers of Botany', to describe a group of herbalists -Brunfels, Bock, Fuchs and Cordus- whose work belongs principally to the first half of the sixteenth century. The earliest of these was Otto Brunfels (Otho Brunfelsius), who is said to have been born in 1464.

A new era in the history of the herbal may be said to date from the year 1530, when the first part of Brunfels' work, the 'Herbarum Vivae Eicones', was published by Schott of Strasburg.

In this book, with its beautiful and naturalistic illustrations, there is, as the title indicates, a real return to nature; the plants are represented as they are, and not in the conventionalised aspect which had become traditional in the earlier herbals, through successive copying by one artist from another, without reference to the plants themselves."

Chapter 26

Comfrey Heals: Allantoin

What is Allantoin?

Allantoin is described as a cell proliferant (increases cell production), epithelization stimulant (improves wound healing), and chemical debrider (removal of dead and diseased tissue) in the 'United States Dispensatory and Physicians Pharmacology', 'Merck Index of Chemicals and Drugs', and 'British Pharmaceutical Codex'.
The 'United States Dispensatory' is a collection of monographs on unofficial drugs and drugs recognized by the 'United States Pharmacopiea' (USP), the 'National Formulary' (NF) of the United States, and the 'Pharmacopoeia of Great Britain'.

"Allantoinum, Allantoin: $C_4H_6N_4O_3$, eq. 158.088. White, odourless and tasteless, glistening, prismatic crystals. Chemically it is a Diureide of Glyoxylic Acid. **It is contained in Comfrey Root, from which it may be prepared;** *in vegetable tissues, e.g., the young shoots of plane trees, and in foetal allantois.*
It may be also prepared synthetically by the oxidation of Uric Acid in alkaline solution, by means of Potassium Permanganate. Solubility: 1 in 102 of Water; practically insoluble in Alcohol (90 p.c), and in Ether.
Employed as a cell-proliferant in the form of a 0.4 to 0.5 percent solution in Physiological Salt Solution.
It also affords a useful local application for ulcerated surfaces, slow-healing wounds and sores, applied on gauze or lint and renewing the dressings three or four times daily. *It may be given internally, and is successfully employed in gastralgia, gastric and duodenal ulcers. It is also very useful in pruritus ani (itchy anus).*
Dose: *1.2 to 2 grains = 0.032 to 0.13 gramme.*
Tests: *Allantoin melts at about 226 degrees C (438.8 F), wth partial decomposition. It is only slightly soluble in cold Water, but readily in hot Water. It undergoes decomposition when boiled for some considerable time with Water, and is decomposed by alkali Hydroxides. When oxidised it is converted into Allantoic Acid. It leaves no appreciable ash."*
-'Squires Companion to the Latest Edition of the British Pharmacopoeia: Comparing the Strength of Its Various Preparations with those of the United States and Other Foreign Pharmacopoeias, 19th edition' by Peter Squire, London, England, **1916**, page 1369.

*"**Allantoin has been used as a cell proliferant.** Experiments on plants seem to show that it exerts a perceptibly stimulating action on growth.* **It may replace the decoction of Comfrey root in the treatment of gastric ulcer.** *Proof as to its value in clinical medicine is still lacking.* **Allantoin has no antiseptic properties;** *it has been applied locally to sluggish (slow to heal) wounds, sores and abscesses to promote the formation of epithelial tissue.* **Dose:** *0.03 to 0.12 gramme (1/2 to 2 grains).*
The healing action of Comfrey has been attributed to the presence of allantoin, *which has some reputation as a cell proliferant.* **Comfrey has been used as an application to wounds, sores and ulcers of various kinds,** *a mucilaginous decoction of fresh root, peeled and bruised into a pulp, being applied. A decoction has also been given internally in gastralgia and gastric ulcer."*
-'The British Pharmaceutical Codex: An Imperial Dispensatory for the Use of Medical Practitioners and Pharmacists' by the Council of the Pharmaceutical Society of Great Britain, London, England, **1923**, pages 86-87, 1036.

*"**Allantoin is a white crystalline substance (powder),** melting at about 226 degrees C (438 F), with decomposition.*
It is only slightly soluble in cold water (0.6 percent), but readily in hot water. It is rather more soluble in cold alcohol, but it is quite insoluble in ether. Dry allantoin is quite stable, but if boiled with water for a considerable time it undoubtedly undergoes decomposition to some extent. It is decomposed by alkali, giving a variety of products, the nature of which entirely depend upon the conditions under which the experiment is carried out. It has been thought well to mention it as pointing to the necessity for dissolving crystals in distilled water."
-The Medicinal Uses of Comfrey by Dr. Charles MacAlister, M.D., F.R.C.P., written **1935**. (in 'Comfrey: Fodder, Food & Remedy')

*"**Allantoin is in use for cosmetics and topical pharmaceuticals for over 70 years. It is Food and Drug Administration (FDA) classified as Category I (safe and effective) as an active ingredient for skin protection.** Its keratolytic (softens skin) properties without any adverse effects (e.g. no skin irritation) made it a part of many cosmetics.*
Remarkable for a substance with assumed healing properties that is in wide use for so long there are no studies that quantify the beneficial effects or investigate the specific mechanism. *The first study proving the healing properties of allantoin was published in 2010 (*Araujo et al. 2010).*
> *One explanation for the lack of studies is originated in the production of allantoin. It can be chemically synthesized by uric acid oxidation with permanganate. A method that was invented by Liebig and Wohler (1837) and an important source for allantoin in the 1930s.* **Nowadays a major source of allantoin are extracts of the Comfrey plant (Symphytum officinale), their roots and herbs contain 0.6-1.0% allantoin (**Staiger 2012).**

Most studies and cosmetics use those Comfrey plant extract *and thus observed effects do not solely originate from allantoin and can not be attributed to a single component (***Araujo et al. 2012)."*
-'New Tools for Maggot Debridement Therapy Research: From the Establishment of qRT-PCR to the Characterization of Lucilia

Sericata Urate Oxidase' by M.Sc. Andre Baumann, Inaugural-Dissertation for degree Doctor Rerum Naturalium of Natural Science Department, Justus Liebig University Giessen, Hesse, Germany, **2017**.
(* -'Profile of Wound Healing Process Induced by Allantoin' by L.U. Araujo, A. Grabe-Guimaraes, V.C. Mosqueira, C.M. Carneiro and N.M. Silva-Barcellos, Pharmacy School, Federal University of Ouro Preto, Minas Gerais, Brazil; Acta Cirurgica Brasileira, Volume 25, No. 5, pages 460-466, 2010.)
(** -'Comfrey: A Clinical Overview' by Christiane Staiger in Germany, Phytotherapy Research, Volume 26, Issue 10, pages 1441-1448, February 2012.)
(*** -'In Vivo Wound Healing Effects of Symphytum Officinale L. Leaves Extract in Different Topical Formulations' by L.U. Araujo, et al., Departamento de Farmacia, Unversidade Federal {Dept of Pharmacy, Federal University} dos Vales do Jequitinhonha e Mucuri, Brazil; Pharmazie {Pharmacy}, Volume 67, No. 4, pages 355-360, 2012.)

*"**Allantoin** is a chemical compound with formula $C_4H_6N_4O_3$. It is also called 5-ureidohydantoin or glyoxyldiureide.* **It is a diureide of glyoxylic acid.** *Allantoin is a major metabolic intermediate in most organisms including animals, plants and bacteria. It is produced from uric acid, which is a degradation product of purine nucleobases, by urate oxidase (or uricase). Chemically synthesized bulk allantoin, which is chemically equivalent to natural allantoin, is safe, non-toxic, compatible with cosmetic raw materials and meets CTFA (Cosmetic, Toiletry and Fragrance Association) and JSCI (Japanese Standards of Cosmetic Ingredients) requirements.*
Over 10,000 patents reference allantoin. Manufacturers cite several beneficial effects for allantoin as an active ingredient in over-the-counter cosmetics, including: a moisturizing and keratolytic (helps shed skin) effect, increasing the water content of the extracellular matrix and enhancing the desquamation (skin peeling) of upper layers of dead skin cells, increasing the smoothness of the skin; promoting cell proliferation (growth) and wound healing; and a soothing, anti-irritant, and skin protectant effect by forming complexes with irritant and sensitizing agents."
-Wikipedia®: The Free Encyclopedia, www.wikipedia.org, **2018**.

Where Allantoin is Found in Comfrey

*"**Distribution of Allantoin in the Vegetable Kingdom:**
It is generally found in parts which are related to growth, either active or potential. We have a striking example of this in the Comfrey where it is present in the roots and terminal buds. The leaves have not yet been thoroughly investigated, but the interesting point is the large amount of allantoin in the rhizome (root), greater than in any plants heretofore investigated."*
-The Medicinal Uses of Comfrey by Dr. Charles MacAlister, M.D., F.R.C.P., **1935**. (in 'Comfrey: Fodder, Food & Remedy')
    (A terminal bud is the primary growing point at the top of the stem of a plant.)
    (A rhizome is a modified underground stem of a plant that sends out roots and shoots, also called creeping rootstalks and rootstocks. Rhizomes grow horizontally and can send new shoots upwards.)

*"**The presence of allantoin in Comfrey root was shown by Titherley and Coppin (1912) and by Harvey-Gibson,** who used a solution of mercuric nitrate for its detection. Vogl (1918) described a test for the detection of allantoin in fresh roots of Comfrey."*
-'Comfrey (Symphytum Officinale L.) Root: Its Anatomy and Its Detection in Admixture with Chicory in Dandelion Coffee' by J.M. Rowson, Museum of The Pharmaceutical Society of Great Britain, Transactions of the Society, Journal Royal Microscopical Society, Great Britain, Volume 75, No. 2, pages 119-128, 1955 and/or January **1956**.
    (R.J. Harvey-Gibson wrote 'Notes on the Anatomy and Herbal History of Symphytum Officinale Comfrey', Pharm. and Pharmacist 88, or Pharm. J., London, Vol 34, page 91-92, 1912. I was unable to find it. If you have it, could you send it.)

*"**In 1912 Comfrey was found to be of value in the formation of epithelial tissue in external ulceration and in ulcers of the stomach and duodenum, due to the presence of allantoin** (Bramwell, 1912; Wood and Lawall, 1926).*
The infusion is still used (in 1961) as a fomentation for reducing inflammations *associated with sprains and bruises and as a medicine for chest complaints. It is widely used in country districts, particularly Warwickshire County (England) (Ministry of Agriculture and Fisheries, England, 1941), and is generally available commercially."*
-'The Anatomy of the Leaf of Symphytum Officinale L.', by J.M. Peck and K.R. Fell, Pharmacognosy Research Laboratory, Bradford Institute of Technology, West Yorkshire, England; The Journal of Pharmacy and Pharmacology, Royal Pharmaceutical Society of Great Britain, Volume 13, pages 154-65, March **1961**.

*"**Glycine-2-C14, hypoxanthine-8-C14, and urea-C14 were fed to both leaf disks and sterile root cultures of Symphytum uplandicum (Nyman).** In both tissues, hypoxanthine-8-C14 was largely converted to allantoin, in which distribution of radioactivity suggested its formation by way of a symmetrical intermediate.*
In leaf disks, the conversion of glycine-2-C14 to allantoin was at most very slight. In roots, it was partially converted to allantoin with a more complex pattern of labelling. Urea-C14 was not converted to allantoin in either tissue.
It is concluded that in Comfrey roots allantoin is synthesised from glycine through purine intermediates."
-'The Biosynthesis of Allantoin in Symphytum' by G.W. Butler, J.D. Ferguson and R.M. Allison, Plant Chemistry Division, Palmerston North and Crop Research Division, Lincoln, New Zealand; Physiologia Plantarum: Scandinavian Plant Physiology Society, Volume 14, pages 310-321, **1961**.
    (Glycine is an amino acid. Purine is colorless crystalline compound with basic properties, forming uric acid on oxidation.)

"Allantoin is also present in Comfrey roots, and in the temperate climates when the plant goes dormant in winter, this is returned to the roots for storage until growth begins in the spring.
It is recommended that roots an inch (2.5 cm) in diameter or under should be used for the maximum medicinal value, for elderly roots which can be losing vigour may be less effective."
-Comfrey: The Herbal Healer by Henry Doubleday Research Association, **1975**, 41 pages, page 21. (in 'Comfrey Report' book)
(Temperate climates are those without extremes of temperature and rain/snow. It is between the tropics and the polar regions.)

*"**Symphytum Species: The underground parts of Comfrey contain 0.6 to 0.7% allantoin** and 4 to 6.5% tannin.*
***The leaves are poorer in allantoin, containing only about 0.3%,** but richer in tannin, 8-9%.*
Large amounts of mucilage are present in both roots and leaves."
-The Honest Herbal: A Sensible Guide to the Use of Herbs and Herbal Products by V.E. Tyler, Ph.D. Philadelphia, Pennsylvania: G.F. Stickley Co., **1981**. Now called 'Tyler's Honest Herbal'.

*"**Comfrey's most significant ingredient, however, is allantoin**, a nitrogenous crystalline substance that appears as white crystals on the dried roots. **It is produced mostly by the root system (it constitutes from 0.6 to 1 percent of that part)** and much less by the terminal (end) buds and the large leaves."*
-Comfrey: What You Need to Know by Ben Charles Harris. New Canaan, Connecticut: Keats Publishing, Inc., **1982**, page 21.
    (Nitrogenous means containing nitrogen.)   (The species of Comfrey was not given.)

*"**In Symphytum, allantoin can be found in the whole plant, although higher concentrations are observed in roots and rhizomes.** Roots were considered to be the main site of ureide allantoin synthesis*, even though leaves were also shown to synthesize these compounds when detached and kept in the dark (Hartmann and Arnold, 1974).*
***The allantoin concentration in Comfrey varies according to the season, organ, physiological / developmental stage and nutritional state** (*; Matsumoto et al., 1977; Saito and Oliveira, 1986)."*
-'Influence of Photoperiod on the Initial Growth of Comfrey (Symphytum Officinale L.) Plants' by A.H.F. Castro and A.A. de Alvarenga, Departamento de Biologia / Setor de Fisiologia Vegetal da Universidade Federal De Lavras (Department of Biology / Department of Plant Physiology, Federal University of Lavras), Brazil; Ciencia e Agrotecnologia (Science and Agrotechnology), Volume 26, No. 1, pages 77-86, **2002**.
(* -'Urea, Ureides, and Guanidines in Plants' by H. Reinbothe and K. Mothes, Deutsche Akadeniie der Wissenschaften zu Berlin {German Academy of Sciences Berlin}, Am-Kirchtor 1, Halle/Saale, Germany; Annual Review of Plant Physiology, Volume 13, pages 129-149, June 1962.)

*"**Comfrey, a medicinal herb with healing properties that are attributed to allantoin**, was studied in this work. The accumulation and metabolism of **allantoin (ALN)** and its degradation product, allantoic acid (ALA), were examined.*
***ALN was the main ureide (derivative of urea) in Comfrey leaves and roots, with young leaves showing the highest content.** ALA was the predominant ureide in the xylem sap, and together with ALN represented 27% of the nitrogen (N) transported in the xylem. Amino acids were the most abundant N-compound in the xylem sap with a high proportion of glutamine (an amino acid) (C-14).*
*Xanthine feeding experiments showed that ALN and ALA were actively formed in leaves and roots by degradation of xanthine. Both xanthine and uric acid were rapidly degraded to form ALN and ALA. Enzyme studies showed that calculated V-max/K-m are low for allantoinase and alantoicase, supporting the results from the feeding experiments, and indicating that **accumulation of ALN in Comfrey is due to a low capacity for the enzymatic degradation of ureides.**"*
-'Control of Allantoin Accumulation in Comfrey' by P. Mazzafera, K.V. Goncalves and M.M. Shimizu, Instituto de Biologia da Universidade Estadual de Campinas (State University of Campinas, Institute of Biology), Sao Paulo, Brazil; Natural Product Communications, Volume 3, No. 9, pages 1411-1422, **2008**.
    (Urea is a water-soluble compound that is the major nitrogenous end product of protein metabolism and is the chief nitrogenous component of the urine in mammals.)
    (Xylem is one of the two types of transport tissue in vascular plants, phloem being the other.)
    (Xanthine is a purine base found in most body tissues and fluids. A number of stimulants are derived from xanthine, including caffeine and theobromine.)

<u>Seasonal Variations of Allantoin in Comfrey</u>

*"**Compare this with the Comfrey rhizome (root) which in the earliest months of the year (January to March) contains from 0.6 to 0.8 percent of allantoin. Analysed a couple of months later, it contains about 0.4 percent.***
In July the amount is still further diminished, and when the plant is in full growth practically none is to be found in the rhizome but it is discoverable in the terminal buds, leaves and young shoots.
This important fact may be regarded as evidence that the plant withdraws allantoin from its storehouse in the rhizome and utilizes it for purposes of cell-proliferation."
-The Medicinal Uses of Comfrey by Dr. Charles MacAlister, M.D., F.R.C.P., **1935**. (in 'Comfrey: Fodder, Food & Remedy')
    (The species was not given but it is probably Symphytum officinale.)

*"Mothes and Engelbrecht (*1954) have traced **the variation in allantoin contents of roots and aerial (above ground) parts under different conditions; they find a minimal content during the summer and a maximum in spring.**"*
-'Comfrey (Symphytum Officinale L.) Root: Its Anatomy and Its Detection in Admixture with Chicory in Dandelion Coffee' by J.M. Rowson, Museum of The Pharmaceutical Society of Great Britain, Transactions of the Society, Journal Royal Microscopical Society, Great Britain, Volume 75, No. 2, pages 119-128, 1955 and/or January **1956**.
(* -'About Allantoic Acid and Allantoin II: Their Behavior in the Storage Roots' {Uber Allantoinsaure und Allantoin II} by Professor Dr. K. Mothes and Dr. L. Engelbrecht; Flora oder Allgemeine Botanische Zeitung {Flora Or General Botanical Newspaper}, Volume 141, Issue 2, 1954, pages 356-378. All in German. I tried translating this online but it did not make any sense. If you have an English translation, could you send it to me.)

*"**The total allantoin content in maple (tree) and Comfrey is involved in an annual rhythm. In spring these compounds ascend (go up) chiefly in the xylem, arriving in the leaves where they are consumed in protein synthesis.**
In autumn, the reciprocal (oppositie) process takes place. The leaves are almost depleted of soluble nitrogen and storage organs are filled up, especially with allantoin and allantoic acid.
The conclusions of Mothes have been fully confirmed by Bollard*. Allantoin and allantoic acid provide efficient forms of storage and transport of nitrogen, keeping this readily available for further metabolic events."*
-'Urea, Ureides, and Guanidines in Plants' by H. Reinbothe and K. Mothes, Deutsche Akadeniie der Wissenschaften zu Berlin {German Academy of Sciences Berlin}, Am-Kirchtor 1, Halle/Saale, Germany; Annual Review of Plant Physiology, Volume 13, pages 129-149, June **1962**.
(* -'Nitrogenous Compounds in Plant Xylem Sap' by E.G. Bollard, Fruit Research Station, Department of Scientific and Industrial Research, Auckland, New Zealand; Nature, Volume 178, Issue 4543, pages 1189-1190, November 1956.)

*"**Allantoin is also present in Comfrey roots, and in the temperate climates when the plant goes dormant in winter, this is returned to the roots for storage until growth begins in the spring.**"*
-Comfrey: The Herbal Healer by Henry Doubleday Research Association, **1975**, 41 pages, page 21. (in 'Comfrey Report')

*"**The allantoin content of the Comfrey root is greatest during the winter months. In temperature climates when the plant goes dormant in winter, allantoin is returned to the roots for storage until growth begins in the spring.** It has been reported that during this dormant stage root of S. officinale contains from 0.6-0.8% of allantoin and S. tuberosum up to 0.96%."*
-'Studies on Symphytum Species: HPLC Determination of Allantoin' by R. Dennis, C. Dezelak and J. Grime; Acta Pharmaceutica Hungarica, Budapest, Hungary, Volume 57, No. 6, pages 267-274, November **1987**.

*"**At this stage of early spring the plant is just breaking into growth, and the allantoin in the plant is up in the new-forming buds,** and the new plants will get away to a quick start.
In the spring the plants go into early flower, and the leaves are smaller, with heavy flower stems. **This is good Comfrey. In some respects this is the best medicinal Comfrey becase allantoin content is high in the flowering stage.**"*
-Comfrey: Nature's Healing Herb & Health Food by Andrew Hughes. Japan: Sanyusha Publishing Co., Ltd, **1992**, page 154.

*"**Comfrey Collection: The roots should be unearthed in the spring or autumn when the allantoin levels are the highest.** Split the roots down the middle and dry in moderate temperatures of about 40-60 degrees C (104-140 F)."*
-'Comfrey' by BotanicEye: Herbs and Botanics, https://botanicseye.com, Ireland, (date unknown, **2019** or earlier). BotanicEye aims to be a one-stop information site for all matters herbal.

<u>Amounts of Allantoin in Comfrey Leaves</u>

Russian Comfrey Leaf: 0.10 to 1.60% allantoin
Russian Comfrey No. 4 Leaf: 0.34% allantoin
Russian Comfrey No. 14 Leaf: 0.44% allantoin
Symphytum Officinale (Common Comfrey) Leaf: 0.30%, 0.11-2.00% allantoin

*"**There appears to be more Allantoin in the stems than the leaves**, and chopping these up and putting them through a liquidiser (blender) with water is an emergency method of getting a strong solution.
From the experience of generations of stockmen, it should be said that if too much Comfrey tea is taken, it can act as a laxative, but otherwise it passes harmlessly through the body."*
-Comfrey: The Herbal Healer by Henry Doubleday Research Association, **1975**, 41 pages, page 28. (in 'Comfrey Report')

*"**Allantoin (in Comfrey) has been detected at a concentration of 0.8% in roots and 0.4% in leaves.**"*
-'Comfrey and One of Its Constituent Alkaloids: Symphytine, Review of Toxicological Literature' by Raymond Tice, Ph.D., Integrated Laboratory Systems, Research Triangle Park, North Carolina, October **1997**. **(Species was not given in the report.)**

*"**An optimised method of extraction and detection was developed for Allantoin,** using 80% ethanol as extraction solvent under sonication (sound) to disrupt the cell membrane, followed by filtration and solvent removal. The dry residue was dissolved with 50% aquous methanol for silica gel TLC (Thin-Layer Chromatography) with Butanol-Acetate-Water as mobile phase. Boiling

the extract 15 minutes with 10% KOH (potassium hydroxide) and 4 minutes with 2,4- dinitrophenylhydrazine was the optimum method of quantification by spectrophotometry at 520 nm (nanometer).
Allantoin was found to be higher in Symphytum officinale tissues than in Symphytum x uplandicum."
-'IPSAM Book of Abstracts' by Irish Plant Scientists Association Meeting, University College Cork, Ireland, April 28-29 **2014**. Includes article 'A Comparison of the Pyrrolizidine Alkaloids and Allantoin Content in Symphytum Officinale and Symphytum x Uplandicum, and Optimisation of Extraction and Detection Methods' by M. O'Keeffe and G. Levieille, Department of Biological Sciences, Cork Institute of Technology, Ireland.
(I only have the abstract. If you have the entire article, I would appreciate a copy.) (It did not say whether it was leaves or roots.)

Russian Comfrey Leaf Allantoin

*"Allantoin is present in varied amounts through the season. In 1960 Dr. A.H. Ward discovered a method of **analyzing the dried leaf** despite the gumminess of the proteins, and he measured what the variations contained:*
Russian Comfrey Bocking No. 4 Leaf = 0.34% allantoin
Russian Comfrey Bocking No. 14 Leaf = 0.44% allantoin."
-Comfrey: The Herbal Healer by Henry Doubleday Research Association, **1975**, 41 pages, p. 27. (in 'Comfrey Report')
  (Dr. A.H. Ward worked at Aynsome Laboratories, Grange-over-Sands, Lancashire, England and was Consulting Analyst to the 'Henry Doubleday Research Association'.)

"In 1965 some of the tea (Russian Comfrey leaves) made on the Trail Ground (Henry Doubleday Research Association) was sent to Dr. Ward to test exactly how much Allantoin was being diluted by the two methods of making in common use:
*1. Two heaped teaspoonfuls (4 grams) used to make four cupfuls (600 millimeters) of tea, by pouring on the boiling water and steeping for 5 minutes. **Amount of Allantoin extracted from leaf: 0.045%**.*
2. Four grams of the sample boiled with 600 millimeters of water for 5 mintues.
***Amount of Allantoin extracted from leaf: 0.046%.'* "**
-Comfrey: The Herbal Healer by Henry Doubleday Research Association, **1975**, 41 pages, p. 27. (in 'Comfrey Report')

"Allantoin is the healing principle in Comfrey used by herbalists for 2,000 years.
Allantoin is present in varied amounts throughout the season. *Dr. A.H. Ward discovered a method of analyzing the* ***dried Comfrey leaf*** *despite the gumminess of the proteins, and he measured what the variations contained:*

Bocking No. 2	0.23 % **Allantoin in Leaf**
Bocking No. 4	**0.34%**
Bocking No. 14	**0.44%**
Bocking No. 17	0.23%
Mixed New Zealand	0.18% "

-Comfrey Report: The Story of the World's Fastest Protein Builder and Herbal Healer, Conservation Gardening and Farming Series: Series C by Lawrence D. Hills. England: Henry Doubleday Research Association, **1975**, page 47.

*"In 1965 Dr. A.H. Ward, analyst to the 'Henry Doubleday Research Association', tested (**Russian Comfrey leaf**) **tea made by two different methods (water or alcohol) to determine the allantoin**:*
*1. Allantoin content determined in the usual way after **alcoholic extraction: 0.083 percent**.*
*2. Two heaped teaspoonfuls (4 grams) used to make four cupfuls (600 ml) of **Russian Comfrey leaf tea**, by pouring on the boiling water and steeping for 5 minutes. **Amount of allantoin extracted: 0.045 percent**.*
*3. Four grams of the sample boiled with 600 ml of water for 5 minutes. **Amount of allantoin extracted: 0.046 percent**.*
These were leaf-only samples, for the major portion of the allantoin is in stems and midribs, *but it will be seen that the tea as drunk is approximately the same as the strength recommended for sores and gastric ulcers in the early editions of the British Pharmacopoeia."*
-Comfrey: Fodder, Food and Remedy by Lawrence D. Hills. New York: Rizzoli Universe Books: **1976**, page 175.
  (The first edition of the 'British Pharmacopoeia' was in 1864.)

*"**Hart (1976) reported that dried (Quaker / Russian) Comfrey leaves contain 0.1 to 1.6% allantoin** while dried roots have 0.4 to 1.5% allantoin."*
-'Comfrey: Miracle or Mirage?' by R H. Hart, Crops and Soils, Volume 29, No. 1, pages 12-14, **1976**. (I was unable to find this report. If you have a copy, could you please send it to me.)

Symphytum Officinale (Common Comfrey) Leaf Allantoin

*"**Symphytum Officinale:***

	Low PPM	High PPM	Low Percent	High Percent	Standard Deviation
Allantoin Leaf	1,100	20,000	0.11%	2.00%	1.0
Allantoin Root	6,000	25,500	0.60%	2.55%	1.41 "

-Handbook of Phytochemical Constituents of GRAS Herbs and Other Economic Plants by Dr. James A. Duke. Boca Raton, Florida: CRC Press, **2000**. (GRAS= Generally Regarded As Safe) This information is also at: Dr. Duke's Phytochemical and Ethnobotanical Databases, https://phytochem.nal.usda.gov/phytochem/search.
(PPM is Parts Per Million.) (Standard Deviation is a statistic that measures the dispersion of a dataset relative to its mean and is calculated as the square root of the variance. In other words, Standard Deviation is a measure of how spread out the numbers are.)

"*Plant species with highest amount of Allantoin:*
Symphytum officinale Comfrey: 13,000 ppm in Leaf; 6,000-8,000 ppm in Root."
-'Allantoin: A Safe and Effective Skin Protectant' by Akema Fine Chemicals: Raw Materials for the Cosmetic and Pharmaceutical Industry, Province of Rimini, Italy, www.akema.it, **2006**.
(**Leaf: 13,000 ppm = 1.30%**. Root: 6,000-8,000 ppm = 0.60-0.80%.)

"*The parts of the herb (Symphytum officinale) under the ground, especially the root, encloses around 0.6 to 0.7 percent of allantoin and approximately 4.0 to 6.5 percent tannin. On the other hand, **Comfrey leaves contain lesser amount of allantoin (approximately 0.3 percent)**, but more of tannin (anything between 8.0 to 9.0 percent).*"
-'A Practical Guide for Nutritional and Traditional Health Care' based on the book 'A Modern Herbal' by Mrs. Grieves, www.Herbs2000.com, **2018**.

Amounts of Allantoin in Comfrey Root

Russian Comfrey Root: 0.4-1.5% allantoin
Symphytum officinale (Common Comfrey) Root: 0.6-0.8%, 0.8%, 0.6-4.7%, 0.60-2.55% allantoin
Symphytum tuberosum: up to 0.96% allantoin (during winter)

"*In 1933 Dr. Titherley (head of the Organic Chemistry Department, University of Liverpool, England) and he (Mr. Coppin) found that the **Comfrey root contained about 0.8 percent of this crystalline substance**, and by accurate determination of its carbon, hydrogen and nitrogen contents, showed that is possessed the **same empirical formula as allantoin**, which it greatly resembled in its chemical properties.*"
-The Medicinal Uses of Comfrey by Dr. Charles MacAlister, M.D., F.R.C.P., written **1935**. (in 'Comfrey: Fodder, Food & Remedy')
(**It is unknown which species of Comfrey this is though it is probably Symphytum officinale.**)

"*HPLC Method of Allantoin Determination: A sensitive and accurate High Performance Liquid Chromatography (HPLC) method has been used for the direct determination of allantoin in Comfrey root. **The allantoin in powdered Comfrey root samples** was completely extracted with 70% methanol using a soxhlet continuous extraction apparatus.
Quantitative HPLC analysis of the extract was carried out using external and internal standardisation procedures. The results obtained from the two methods of analysis were in close agreement and **indicated a much higher allantoin content for Comfrey root than had been reported using other procedures**.
Factors that could result in variation of the allantoin content:
Comfrey was obtained from different sources. It could have been grown under different conditions, perhaps with a different soil type. It may have been harvested at another time of year or stage of growth. A further possibility is that the root sample may have been dried and prepared in a different manner.
Allantoin Content of Comfrey Root (%w/w): 3.05 to 4.72%.
It seems that the allantoin content of Comfrey root can be much higher than that generally reported. The reason for this could be that HPLC provides a far better method of assay for allantoin than the spectrophotometric methods commonly employed.*"
-'Studies on Symphytum Species: HPLC Determination of Allantoin' by R. Dennis, C. Dezelak and J. Grime; Acta Pharmaceutica Hungarica, Budapest, Hungary, Volume 57, No. 6, pages 267-274, November **1987**.
(The species was not given though Symphytum officinale and Symphytum tuberosum were mentioned earlier in the article.)

"**Allantoin (in Comfrey) has been detected at a concentration of 0.8% in roots** and 0.4% in leaves."
-'Comfrey and One of Its Constituent Alkaloids: Symphytine, Review of Toxicological Literature' by Raymond Tice, Ph.D., Integrated Laboratory Systems, Research Triangle Park, North Carolina, October **1997**. (**Species was not given in the report.**)

Russian Comfrey Root Allantoin

"*Hart (1976) reported that dried (Quaker / Russian) Comfrey leaves contain 0.1 to 1.6% allantoin while **dried roots have 0.4 to 1.5% allantoin**.*"
-'Comfrey: Miracle or Mirage?' by R H. Hart, Crops and Soils, Volume 29, No. 1, pages12-14, **1976**. (I was unable to find this report. If you have a copy, could you please send it to me.)

Symphytum Officinale (Common Comfrey) Root Allantoin

"*Constituents:* The chief and most important constituent of **Common Comfrey (Symphytum officinale) root** is mucilage, which it contains in great abundance, more even than Marshmallow. **It also contains from 0.6 to 0.8 percent. of Allantoin** and a little tannin."
-A Modern Herbal: The Medicinal, Culinary, Cosmetic and Economic Properties, Cultivation and Folk-Lore of Herbs, Grasses, Fungi, Shrubs and Trees with their Scientific Uses by Mrs. M. Grieve. New York: Dover Publications, 1971. First published in **1931**.

"***Symphyti radix is the fresh or dried root section of Symphytum officinale*** (synonym: Comfrey) which is a plant of the Boraginaceae family. **Symphyti radix contains allantoin (0.6 to 0.8%).**"
-'European Medicines Agency- Symphyti Radix: Committee for Veterinary Medicinal Products' by European Agency for the Evaluation of Medicinal Products, Veterinary Medicines Evaluation Unit, London, England, EMEA/MRL/649/99-FINAL, 5 pages, August **1999**.

"*Symphytum Officinale:*

	Low PPM	High PPM	Low Percent	High Percent	Standard Deviation
Allantoin Leaf	1,100	20,000	0.11%	2.00%	1.0
Allantoin Root	6,000	25,500	**0.60%**	**2.55%**	**1.41** "

-Handbook of Phytochemical Constituents of GRAS Herbs and Other Economic Plants by Dr. James A. Duke. Boca Raton, Florida: CRC Press, **2000**. (GRAS= Generally Regarded As Safe) This information is also at: Dr. Duke's Phytochemical and Ethnobotanical Databases, https://phytochem.nal.usda.gov/phytochem/search.

"*Plant species with highest amount of Allantoin:*
Symphytum officinale Comfrey: 13,000 ppm in Leaf; **6,000-8,000 ppm in Root.**"
-'Allantoin: A Safe and Effective Skin Protectant' by Akema Fine Chemicals: Raw Materials for the Cosmetic and Pharmaceutical Industry, Province of Rimini, Italy, www.akema.it, **2006**.
    (Leaf: 13,000 ppm = 1.30%. **Root: 6,000-8,000 ppm = 0.60-0.80%.**)

"***The constituents of Comfrey (Symphytum officinale) root include 0.6–4.7% allantoin (Dennis et al., 1987);*** abundant mucilage polysaccharides (about 29%) composed of fructose and glucose units (Franz, 1969); phenolic acids such as rosmarinic acid (up to 0.2%), chlorogenic acid (0.012%) as well as caffeic acid (0.004%) and a-hydroxy caffeic acid (Andres, 1991; Grabias and Swiatek, 1998; Teuscher et al., 2009); glycopeptides and amino acids (Hiermann and Writzel, 1998); and triterpene saponins in form of monodesmosidic and bidesmosidic glycosides based on aglycones hederagenin (e.g. symphytoxide A), oleanolic acid (Aftab et al., 1996) and lithospermic acid (Wagner et al., 1970)."
-'Comfrey: A Clinical Overview' by Christiane Staiger in Germany, Phytotherapy Research, Volume 26, Issue 10, pages 1441-1448, February **2012**.

"The herb contains i.a. allantoin, mucilage, tannins, steroidal saponins, pyrrolizidine alkaloids, inulin and proteins (Sigma-Aldrich Co). **Comfrey consists of the dried root and rhizome of Symphytum officinale (Boraginaceae); the leaf has also been used. It contains about 0.7% of allantoin,** large quantities of mucilage, and some tannin."
-'Risk Profile Symphytum Officinale Extracts', CAS No. 84696-05-9, www.mattilsynet.no, Statens tilsyn for planter, fisk, dyr og naeringsmidler (State supervision of plants, fish, animals and food), Brumunddal, Norway, March 11 **2013**.
    (I'm not sure what 'i.a.' means.)

"***Comfrey (Symphytum officinale L.)*** is a perennial half-bushy plant with developed spindle-shaped branched root which is used in traditional medicine. **The main pharmacological activity of the Comfrey root extract is to stimulate the tissue regeneration, anti-inflammatory, analgesic, anti-edematic, and astringent effect.** It is used for the treatment of damaged ligaments, for arthritis, rheumatic pain, and wounds that would not heal easily*. In traditional medicine, Comfrey is known for its effects on bone healing; and because of that, it got the name.
The most important pharmacologically active ingredient of the Comfrey root is allantoin and its proportion is 0.6–0.8%*."
-'Comparative Study of the Biological Activity of Allantoin and Aqueous Extract of the Comfrey Root' by Vesna Lj. Savic, et. al., University of Nis, Serbia; Phytotherapy Research, Volume 29, No. 8, pages 1117-1122, April **2015**.
(* -'In Vivo Wound Healing Effects of Symphytum Officinale L. Leaves Extract in Different Topical Formulations' by L.U. Araujo, et al., Departamento de Farmacia, Unversidade Federal {Dept of Pharmacy, Federal University} dos Vales do Jequitinhonha e Mucuri, Brazil; Pharmazie {Pharmacy}, Volume 67, No. 4, pages 355-360, 2012.)

Health Benefits of Allantoin

"*Allantoin's value as a cell-proliferant in making the edges of the wounds grow together, healing sores, and internally for gastric and duodenal ulcers and intestinal irritations causing diarrhoea, is still recognized in pharmacy.*"
-Russian Comfrey: A Hundred Tons an Acre of Stock or Compost for Farm, Garden or Smallholding by Lawrence D. Hills. London England: Faber and Faber, Limited, **1953**, page 20.

"The 'Journal of American Podiatry' (Volume 56, Number 8, pages 357-364) of August 1966 carries an article ('The Allantoins') by Deray W. Meixell, D.S.C. and S.B. Mecca, B.S. in which is stated that **in dermatologic disturbances of the feet (e.g., inflamed ulcerations, dry skin, bromidrosis {sweat that smells}, and other skin problems of the feet) allantion is 'a valuable cell-proliferant agent which stimulates healthy tissue formation.**
Allantoin is also a chemical debrider (cleaner) of necrotic (dead) tissue and serves to clean up the area to which it is applied. When such an agent can further be described as bland, stable, nontoxic, soothing and nonirritating, it becomes a tool of more than casual interest to the podiatrist (foot doctor).' "
-Comfrey and Chlorophyll: A Report About the Medicinal Value of Comfrey and Chlorophyll as Found in Old and Modern Literature by Vincent Licata. California: Continental Health Research, **1971**, page 20.
    ('Journal of the American Podiatric Medical Association' was founded in 1907 and is located in Bethesda, Maryland.)

"**Medicinal, Pharmaceutical, and Cosmetic Uses:** Comfrey root and leaves and their extracts are used as ingredients in various types of cosmetic preparations such as lotions, creams, ointments, eyedrops, hair products, and others."
-Leung's Encyclopedia of Common Natural Ingredients Used in Food, Drugs and Cosmetics, Third Edition by Ikhlas Khan and Ehab Abourashed. New Jersey: John Wiley & Sons, **2010**, page 226.

"**Allantoin and Wound Healing: To evaluate and characterize the wound healing process profile induced by allantoin incorporated in soft lotion oil/water emulsion** using planimetric and histological methods.
Methods:
    Female Wistar rats (number=60) were randomly assigned to 3 experimental groups: (C) control group-without treatment; (E) group treated with soft lotion O/W (oil/water) emulsion excipients; (EA) group treated with soft lotion O/W emulsion containing allantoin 5%. Emulsions either containing or not allantoin were topically administered for 14 days.
Results:
    The data which were obtained and analyzed innovate by demonstrating, qualitatively and quantitatively, by histological analysis, the profile of healing process induced by allantoin. The results suggest that **the wound healing mechanism induced by allantoin occurs via the regulation of inflammatory response and stimulus to fibroblastic proliferation and extracellular matrix synthesis.**
    This work shows, for the first time, the histological wound healing profile induced by allantoin in rats and demonstrated that **allantoin is able to ameliorate and fasten the reestablishment of the normal skin.**"
-'Profile of Wound Healing Process Induced by Allantoin' by L.U. Araujo, A. Grabe-Guimaraes, V.C. Mosqueira, C.M. Carneiro and N.M. Silva-Barcellos, Pharmacy School, Federal University of Ouro Preto, Minas Gerais, Brazil; Acta Cirurgica Brasileira, Volume 25, No. 5, pages 460-466, **2010**.
    (Planimeter is an instrument that measures the area of a surface by tracing its boundaries.)
    (Histology is the study of the microscopic anatomy of cells and tissues of plants and animals.)

"**Manufacturers cite several beneficial effects for allantoin as an active ingredient in over-the-counter cosmetics, including:**
a moisturizing and keratolytic (remove warts and calluses) effect, increasing the water content of the extracellular matrix and enhancing the desquamation (skin peeling) of upper layers of dead skin cells, increasing the smoothness of the skin; promoting cell proliferation and wound healing; and a soothing, anti-irritant, and skin protectant effect by forming complexes with irritant and sensitizing agents.
An animal study in 2010 found that based on the results from histological analyses, **a soft lotion with 5% allantoin ameliorates (improves) the wound healing process, by modulating the inflammatory response**. The study suggests that quantitative analysis lends support to the idea that allantoin also promotes fibroblast proliferation and synthesis of the extracellular matrix."
-Wikipedia®: The Free Encyclopedia, www.wikipedia.org, **2018**.

Allantoin is Not an Antiseptic or Antibiotic

"On my return to Liverpool (England) Dr. Alfred Adams, working in the Biochemical Laboratories of the University on 1 November, 1911, reported that the Bacillus coli, Staphylococci, Streptococci and Tubercle Bacilli were not retarded in their growth and multiplication by allantoin.
Allantoin, then, is not an antiseptic in the usual acceptation of the term, and its action in this respect must depend upon some influence brought to bear upon the cells, whereby their resistance, stability and immunity are established and their proliferation promoted."
-The Medicinal Uses of Comfrey by Dr. Charles MacAlister, M.D., F.R.C.P., written **1935**. (in 'Comfrey: Fodder, Food & Remedy')

"**Allantoin, rosmarinic acid and ellagic acid were identified as major bioactive compounds (in Symphytum officinale root)**. The results obtained by the determination of the antimicrobial activity showed that Escherichia coli ATCC 8739 and Salmonela typhimirium ATCC 6538 were most sensitive to the aqueous (water) extract of Comfrey root.
The results showed that allantoin did not express the antimicrobial activity on all the investigated bacteria species, and based on this it can be concluded that allantoin is not responsible for the antimicrobial activity of the aqueous extract of Comfrey root."

-'The Identification and Quantification of Bioactive Compounds from the Aqueous Extract of Comfrey Root by UHPLC–DAD–HESI–MS Method and its Microbial Activity' by Vesna Lj. Savic, Sasa R. Savic, Vesna D. Nikolic, Ljubisa B. Nikolic, Stevo J. Najman, Jelena S. Lazarevic and Aleksandra S. Dordevic, University of Nis, Serbia; Hemijska Industrija (Chemical Industry), Volume 69, No. 1, pages 1-8, **2015**.

Allantoin Increases Phagocytosis and Leucocytosis

Phagocytosis is when a cell engulfs a soid particle. In the immune system, phagocytosis is a major mechanism used to remove pathogens and cell debris.
Leukocytosis is when white cells or the leukocyte count is above the normal range in the blood. It is a sign of an inflammatory response, usually from infection.

"*In connection with the local application of allantoin to septic (infected) and sloughing (shedding) surfaces,* the cleaning-up was originally ascribed (attributed) only to an increase in the vitality or resistance in the cells themselves, but the point arose as to whether this might not be brought about, in part at any rate, by **some change in the cellular environment.**
That the cleaning-up did not result from a chemical antiseptic itself, and by the circumstances that many of the ulcers had failed to clean up or heal when treated by various antiseptics before the solution of allantoin had been applied.
The question thus arose as to whether this lessening or inhibition of sepsis (infection) resulted from a promotion of phagocytosis.
Some light was thrown on this by Doctors Albert Berthelot and D.M. Bertrand in a research carried out in the Pasteur Institute, the results of which were published in August 1912. From these experiments the authors concluded that **allantoin was capable of strengthening the local resistance of the peritoneum against infection and of causing considerable leucocytosis.**
For it seemed possible that *if the local application of allantoin promoted a multiplication of phagocytes, it might also, if introduced in the blood, set up a general leucocytosis*, and that if this proved to be the case a controlling influence might be exerted on infective diseases, and especially perhaps in those types of cases where there is a controlled immunity."
-The Medicinal Uses of Comfrey by Dr. Charles MacAlister, M.D., F.R.C.P., written **1935**. (in 'Comfrey: Fodder, Food & Remedy')
    (The Pasteur Institute founded in 1887 in Paris, France is dedicated to the study of biology, microorganisms, diseases and vaccines. It is named after Louis Pasteur, who made breakthroughs in medicine, including pasteurization and vaccines for anthrax and rabies.) (Peritoneum is the membrane that lines the abdominal cavity.)

"***Allantoin and Leucocytosis:*** The first experiments were carried out in normal healthy individuals, to whom a grain of allantoin in solution was given by mouth at two-hourly intervals on an empty stomach. **The (white cell or leucocyte) counts were generally taken two hours after each of three successive doses.**
The polynuclear leucocyte cells were increased by from 5 to 15 percent, and it was noted by making subsequent counts that they had returned to normal in from sixteen to forty-eight hours.
Next came the administration of allantoin in cases of pneumonia commencing at the earliest possible periods after the onset of the disease, with the result that **a rapid increase in the number of leucocytes was generally observed.**
    A point to note in connection with the course of **pneumonia treated with allantoin** in 1914 (also observed by Mr. Green) is **the occurrence of a pseudo-crisis (not a real crisis) generally from twenty to thirty hours after the drug was given,** the temperature dropping to normal and remaining so for a few hours, but without much relief from symptoms. Ocassionally one or two such falls in temperature preceded (before) the true crisis.
        Mr. Green suggested that the pseudo-crisis might be explained by an abnormally rapid increase in the leucocytes causing a temporary check (reduction) to the multiplication of pneumococci bacteria and production of their toxins in the blood stream enabling the thermal (temperature) centre to regain its balance for a while. The symptoms are not abated (reduced) during the pseudo-crisis because the lung condition remains unchanged, and the disordered circulation in the lung does not enable many of the newly formed leucocytes to penetrate and attack its contained pneumococci bacteria.
That allantoin renders a useful service in pneumonia is indicated by the relatively low mortality (death) of patients treated with it as compared with the general run of cases treated on the usual therapeutic principals.
Allantoin may be useful in other infective conditions which might be benefited by the production of leucocytosis. It has been found serviceable (useful) in some cases of broncho-pneumonia, and in some septic infections, and even the common cold seems sometimes to have been adverted (avoided) when the drug has been given at the commencement (beginning).
Its employment in early stages of influenza and in other diseases associated with leucopaenia might be usefully investigated."
-The Medicinal Uses of Comfrey by Dr. Charles MacAlister, M.D., F.R.C.P., written **1935**. (in 'Comfrey: Fodder, Food & Remedy')
    (This was written before antibiotics and antivirals were in common use.)
    (A grain is a unit of measurement of mass, and in the troy weight, avoirdupois, and Apothecaries' system, equal to exactly 64.79891 milligrams.) (Polynuclear cells are multinucleate, i.e., containing multiple nuclei.)
    ('Pneumococcus' or the plural 'pneumococci' are bacterium associated with pneumonia and some forms of meningitis.)
    (Leucopaenia is abnormally low white blood cell count.)

"**Allantoin's utility as a promoter of healing in wounds, ulcers and sinuses has already been appreciated in this and other countries,** and it is generally recognized that the best way of applying it is on sterilized lint or gauze without waterproof coverings.

Ointments having lanoline (lanolin) bases which are capable of holding solutions of allantoin have proved useful.
The impression gained as to the cell-proliferating property of allantoin is that it acts like a hormone, *in the sense that such small quantities of it set going proliferations of considerable magnitude (great extent).* ***Solutions of allantoin should be freshly prepared in sufficient quantity for use during a period not exceeding twenty-four hours."***
-The Medicinal Uses of Comfrey by Dr. Charles MacAlister, M.D., F.R.C.P., written **1935**. (in 'Comfrey: Fodder, Food & Remedy')
    (Lanolin also called wool wax or wool grease, is a wax secreted by the skin glands of wool-bearing animals, especially sheep but also some goat and rabbit breeds.)

"As long ago as 1936 an article entitled **'Allantoin a New Granulation Tissue Stimulating Substance with Especial Emphasis on Allantoin in Ointment Form'** *was written by F.R. Greenbaum for 'American Journal of Surgery' Volume 34: 259-65, 1936. This author concluded: 'These reports demonstrate definitely the clinical value of allantoin ointment, and shows we are dealing with* **a remarkable drug of great promise and wide possibilities in the treatment of infected or slow healing wounds**.*' "*
-About Comfrey: The Forgotten Herb by G.J. Binding, M.B.E., Fellow of the Royal Horticultural Society, Wellingborough, Northhamptonshire, England: Weatherby Woolnough, **1974**, page 34.   ('American Journal of Surgery' is in Birmingham, Alabama.)

"Journal of the American Pharmaceutical Association, Volume NS 7, No. 11, November 1967 contained an article by Ronald Johnson who made these remarks about allantoin:
'Allantoin has been used many years by the medical profession to stimulate healthy granulation of tissues. *Recently a new group of Allantoin complexes had been developed called aluminium allantoinates. These salts can be incorporated into a hospital massage lotion and a hemorrhoidal suppository.'*
He also considered that it would bring prolonged healing and anti-irritant action."
-About Comfrey: The Forgotten Herb by G.J. Binding, M.B.E., F.R.H.S. (Fellow of the Royal Horticultural Society), Wellingborough, Northhamptonshire, England: Weatherby Woolnough, **1974**, page 33.
    (The Journal is now called 'Journal of Pharmaceutical Sciences' in New York, New York.)
    (Granulation is new connective tissue and blood vessels on the surface of a wound during healing.)
    (Suppository is a solid medicine inserted into rectum or vagina. It dissolves and has both local and systemic effects.)

"Another article entitled **'The Function and Application of Allantoins'** *by S.B. Mecca, B.S., is a clear indication of the power of allantoin to enhance all manner of lotions, creams and cosmetics.* **The small addition of 2 percent allantoin to various creams and lotions** *resulted in the following aids to the preparation.* **It proved anti-irritant, soothing, and aided the healing of the skin's surface.** *Mr. Mecca also wrote that allantoin increased the water-binding capacity of the skin, had an excellent moisturizing action, fine cleansing ability, prevented drying and chapping, and had the ability to soften hard skin and calluses."*
-About Comfrey: The Forgotten Herb by G.J. Binding, M.B.E., F.R.H.S. (Fellow of the Royal Horticultural Society), Wellingborough, Northhamptonshire, England: Weatherby Woolnough, **1974**, page 34.
    (This information is from S. B. Mecca, 1978, 'Allantoin and Its Derivatives' in 'Cosmetics & Toiletries' magazine, 93, pages 39-41. 'Cosmetics and Toiletries' has been in publication over 100 years. It is based Northbrook, Illinois.
    The article also mentions that allantoin removes necrotic/dead tissue with its cleansing action.)

"I am a gardener, not a doctor, and I am not going to risk having someone trust Comfrey on my advice and die, when he or she might have been saved by surgery. I cannot afford to risk the work that I have done in my own field, by writing of 'cures' I am not qualified to judge, which rest on hearsay, and on the enthusiasm of individuals who may well be justifed. **All I can do is use the work of Dr. Charles MacAllister, who died in 1940.**"
-Comfrey: Fodder, Food and Remedy by Lawrence D. Hills. New York: Rizzoli Universe Books: **1976**, page 178.
    (Dr. MacAlister is quoted several times above from: 'The Medicinal Uses of Comfrey', 1935.)

<u>Solubility, Stability, and Heat Sensitivity of Allantoin</u> (Useful information for making medicine.)

    The typical use of allantoin is between 0.1-0.5%, but may be increased to 2%.

    Allantoin Solubility: Allantoin is easily soluble in water.
    It is soluble to 0.5% in room temperature water.
    Solubility is best in water heated to 50-75 C (122-167 F) (up to 4.0%). Stirring helps.
    Allantoin is not soluble in cosmetic oils or other oils.
    Allantoin is slightly soluble in ethanol (ethyl alcohol). At 20 C (68 F) it dissolves to 0.1%.

    Allantoin Stability: Allantoin evaporates very slowly in pure water.

    Allantoin Heat Sensitivity: Boiling destroys allantoin. Water boils at 100 C (212 F).
    Allantoin is stable to 80 C (176 F) with prolonged heating. (Brief heating above that is OK.)

*"**Wiechowski (1910) states that allantoin is not stable in pure aqueous (water) solution.** He found a 0.1 percent allantoin solution after standing a few weeks in a closed flask, moulds being absent, free from allantoin (no allantoin). Under similar conditions I have noted some disappearance of allantoin, but neither so rapid nor so complete as Wiechowski reports.*

From the foregoing it seems without a doubt that there is a slow spontaneous destruction of allantoin. There is less destruction the more concentrated the solution.
While allantoin appears, therefore, to be relatively stable in neutral solutions, it is rapidly destroyed by alkalies. This is indeed well known and has recently been specially emphasized by Wiechowski."
-'Brief Notes Concerning Allantoin' by Maurice H. Givens, Physiology and Biochemistry, Cornell University Medical College, Ithaca, New York; Journal of Biological Chemistry, Volume 18, pages 417-424, **1914**.
(The type and purity of the water used by Wiechowski was not given.)
(An alkali is a chemical that neutralizes or bubbles with acids and turns litmus blue.)

"Whatever you do, do not boil Comfrey. The high temperature can break down the allantoin."
-Rodale's Illustrated Encyclopedia of Herbs edited by Claire Kowalchik and William H. Hylton. Emmaus, Pennsylvania: Rodale Press, **1998**, page 104.
(The boiling point of chemically/pharmaceutically made allantoin is 478 C = 892 F. Water boils at 100 C = 212 F.)

"Comfrey Root (Symphytum officinale, S. x uplandicum):
The healing allantoin is best extracted with heat and water, but you can use a decoction tincture method to capture it in a preserved form to use directly on the skin or mix into creams and liniments."
-'Autumn Roots, Barks and Berries: Day-Long Field Workshop' by Maria Noel Groves, Clinical Herbalist, Wintergreen Botanicals LLC, Allenstown, New Hampshire, www.WintergreenBotanicals.com, **2016**.

*"Properties and Stability: Allantoin is a heterocycle compound derived from purine. It is an odourless white powder, **soluble in water to 0.5%, very slightly soluble in alcohols, insoluble in oils and apolar solvents.***
*Allantoin is stable in the pH range of 3-8. **Allantoin is stable to 80 C (176 F) prolonged heating. The typical use of allantoin is between 0.1-0.5%, but may be increased to 2%.** Allantoin can be easily dissolved in aqueous (water) formulations at room temperature up to 0.5%. It can be incorporated in emulsions at 0.5-2.0% with temperature over 50 C (122 F)."*
-'Allantoin CTFA- Cosmetic Toiletry and Fragrance Association Data Sheet' by Akema Fine Chemicals, Coriano, Italy, manufactures raw materials for cosmetic and pharmaceutical industry, February 27 **2008**.
(Solvents are substances that dissolve a solute and form a solution. Solvents can be classified into 2 types: polar solvents such as water and non-polar {apolar} solvents such as pentane, hexane and gasoline {all found in petroleum oil}.) (An emulsion is a mixture of two or more liquids that are normally unmixable or unblendable. Homogenized milk is an example of an emulsion with liquid milk and fat.)

*"**Solubility: Allantoin is soluble in the aqueous (water) phase of cosmetic products, but is insoluble in the commonly used cosmetic oils.** Allantoin is best incorporated by addition to the aqueous phase and heating to 50 C (122 F), but can also be dissolved at 25 C (77 F) with stirring.*
Water at 20 C (68 F) dissolves allantoin to 0.4%. Water at 75 C (167 F) dissolves allantoin to 4.0%.
Ethanol (ethyl alcohol) at 20 C (68 F) dissolves allantoin to 0.1%. *The content of Allantoin in cosmetic preparations may be determined by HPLC (High Performance Liquid Chromatography) analysis.*
Stability: *In water-free formulations, allantoin can be stored virtually indefinitely under normal storage conditions.*
Aqueous solutions are stable in the range pH 3 to 8. Allantoin decomposes when heated for a long time in aqueous solution; strong alkalis have the same effect."
-'Allantoin Technical Sheet' by Ingredients to Die For, Austin, Texas, wholesale supplier of cosmetic materials, date unknown but **2018** or earlier.   (It says allantoin decomposes with prolonged heat but it did not say what temperature or for how long.)

1. Symphytum officinàle L. Comfrey. Healing-herb. Fig. 3547.
Symphytum officinale L. Sp. Pl. 136. 1753.
Roots thick, deep; stem erect, branched, 2°-3° high. Leaves lanceolate, ovate-lanceolate, or the lower ovate, pinnately veined, 3-10' long, acute or acuminate at the apex, narrowed into margined petioles, or the uppermost smaller and sessile, decurrent on the stem; petioles of the basal leaves sometimes 12' long; flowers numerous, in dense racemes or clusters; pedicels 2"-4" long; calyx-segments ovate or ovate-lanceolate, acute or acuminate, much shorter than the corolla; corolla yellowish or purplish, 6"-10" long; nutlets brown, shining, slightly wrinkled, 2" high.
In waste places, Newfoundland to Minnesota, south to Virginia and North Carolina. Naturalized or adventive from Europe. Native also of Asia. June-Aug. Back- or black-wort. Bruisewort. Knitback. Boneset. Consound. Gum-plant.
Symphytum tuberòsum L., with thickened tuberous roots, the nutlets granular-tuberculate, not shining, has been found in sandy meadows in Connecticut.

Symphytum Officinale. Healing-herb. 1913

'An Illustrated Flora of the Northern United States, Canada and the British Possessions, Volume 3: Gentianceae to Compositae (Gentian to Thistle)' by Nathaniel Lord Britton, PH.D., Sc.D., LL.D., Columbia University; and Honorable Addison Brown, A.B., LL.D., New York Botanical Garden. Published by Charles Scribner's Sons in New York, 1913.

Comfrey page 92.

Chapter 27

Making and Using Comfrey Medicine

Medicinal Gathering of Comfrey Leaves and Roots

For entertainment purposes only. See your medical practitioner about your health issues.

See subsection 'Comfrey Confused with Foxglove' in section 'Symphytum Genus Description'(Chapter 5, Volume 1). Foxglove is deadly when consumed. When not in flower, it looks somewhat like Comfrey. **Be sure you know what you are gathering.**

For more about digging Comfrey roots, see subsection 'Method of Digging Root Cuttings' in section 'Care of Comfrey Plant: Overview and How to Propagate' (Chapter 34).

"The green drug of Comfrey is proportionately much more active medicinally than the dried drug, and covers a greater therapeutic field."
-'National Association of Retail Druggists' with article 'The Cultivation of Medicinal Plants: Ideal Conditions for Comfrey, Stoneroot, Leptandra', Chicago, Illinois, Volume 23, November 9 **1916**.

"Gather Roots October to March. Gather leaves May to July.:
All Comfrey varieties were used, however, the roots being gathered during the winter months (October to March).
The leaves, also employed, were gathered in June and July during and after inflorescence (group or cluster of flowers).
> *Compare this with the Comfrey rhizome (root) which in the earliest months of the year (January to March) contains from 0.6 to 0.8 percent of allantoin.*
> *Analysed a couple of months later, it contains about 0.4 percent.*
> *In July the amount is still further diminished, and when the plant is in full growth practically none is to be found in the rhizome but it is discoverable in the terminal buds, leaves and young shoots.*
>> *This important fact may be regarded as evidence that the plant withdraws allantoin from its storehouse in the rhizome and utilizes it for purposes of cell-proliferation."*

-The Medicinal Uses of Comfrey by Dr. Charles MacAlister, M.D., F.R.C.P., **1935**. (in 'Comfrey: Fodder, Food & Remedy')

"Allantoin is also present in Comfrey roots, and in the temperate climates when the plant goes dormant in winter, this is returned to the roots for storage until growth begins in the spring.
It is recommended that roots an inch (2.5 cm) in diameter or under should be used for the maximum medicinal value, for elderly roots which can be losing vigour may be less effective."
-Comfrey: The Herbal Healer by Henry Doubleday Research Association, **1975**, 41 pages, page 21. (in 'Comfrey Report' book)
(Temperate climates are those without extremes of temperature and rain/snow. It is between the tropics and the polar regions.)

"In the spring the plants go into early flower, and the leaves are smaller, with heavy flower stems.
This is good Comfrey. In some respects this is the best medicinal Comfrey becase allantoin content is high in the flowering stage."
-Comfrey: Nature's Healing Herb & Health Food by Andrew Hughes. Japan: Sanyusha Publishing Co., Ltd, **1992**, page 154.

"Harvest by cutting off leaves to dry for tea, as soon as flower buds appear; feed also to animals as green chop.
Roots are collected in fall, or so early in spring that it has no leaves, used medicinally; wash, slice, dry, powder in blender; store in sealed, dark jar."
-The Organic Method Primer: A Practical Explanation: The How and Why for the Beginner and the Experienced by Bargyla and Gylver Rateaver. San Diego, California: The Rateavers, **1993**, page 163.

"Stability: This defines how long you can expect a gathered plant to stay reasonably strong or what characteristics it must retain to still have potency. **The rule of thumb is that green herbs are good for a year, and roots and barks for two year.**
Salves and oils, if they contain an antioxidant, are good for several years, and with few exceptions, tinctures last for years."
-Medicinal Plants of the Pacific West by Michael Moore. Santa Fe, New Mexico: Red Crane Books, **1993**, page 24.
> (This is stability of plants in general, not just Comfrey.)

"Fall is my favourite season. Our forests will soon be colourized, a brilliant last flash of colour before bedding down for the grey winter. **This is the time to collect herbs with medicinal roots, like ginseng, valerian (German: baldrian), and Comfrey (beinwell).** *Research at the 'University of Quebec' has shown that* **medicinal roots reach their largest size and highest potency in October and November, well after the above ground parts have died down with the killing frosts.**
Like their colourful arboreal (tree) cousins, herbs are programmed to withdraw nutrients from the leaves and stems and store

them in the roots for winter.
-'Fall is the Time to Harvest Roots to Make Home Remedies' by Conrad Richter, Richters Herbs, Goodwood, Ontario, Canada, (October 5 **1995**). First published in the German language weekly newspaper 'Kanada Kurier'.
(I searched for the 'University of Quebec' research report but could not find it. If you have it, could you send it to me.)

"Commercial Comfrey root harvest can be done with potato digging equipment, taking care to pile and cover the roots with tarps, or at least keep them from exposure to the drying effects of the sun.
Yields in dry weight for a field established for four year (or more) can be more than five ton per acre on four cuttings. With heavy irrigation, 6 cuttings are possible in some regions."
-'Comfrey Yields, Drying' by Richard Alan Miller, Commercial Herb Production and Marketing, Richters Herbs, Goodwood, Ontario, Canada, November 12 **1999**.

"In homeopathy, preparations of Symphytum officinale according to homeopathic pharmacopoeias are alcoholic extracts from the roots collected before the time of blossoming."
-'European Medicines Agency- Symphyti Radix: Committee for Veterinary Medicinal Products' by European Agency for the Evaluation of Medicinal Products, Veterinary Medicines Evaluation Unit, London, England, EMEA/MRL/649/99-FINAL, 5 pages, August **1999**.

"Comfrey contains many rare but important trace minerals that are important to the funciton of all living things. Herbal physicians long ago discovered this, and wisely understood that the most critical components were concentrated in the roots.
Therefore, Comfrey roots must be harvested before the plant begins to send up shoots, which is the reason the herbal states that March and April during a full moon are the preferred months for digging it."
-Sauer's Herbal Cures: America's First Book of Botanic Healing, 1762-1778 by William Woys Weaver. New York: Routledge, **2001**, pages 114-115.

"Medicinal plants should be harvested during the optimal season or time period to ensure the production of medicinal plant materials and finished herbal products of the best possible quality.
The time of harvest depends on the plant part to be used. Detailed information concerning the appropriate timing of harvest is often available in national pharmacopoeias, published standards, official monographs and major reference books.
However, it is well known that the concentration of biologically active constituents varies with the stage of plant growth and development.
Medicinal plants should be harvested under the best possible conditions, avoiding dew, rain or exceptionally high humidity. *If harvesting occurs in wet conditions, the harvested material should be transported immediately to an indoor drying facility to expedite (speed up) drying so as to prevent any possible deleterious (bad) effects due to increased moisture levels, which promote microbial fermentation and mould."*
-'WHO Guidelines on Good Agricultural and Collection Practices (GACP) for Medicinal Plants' by World Health Organization, Geneva, Switzerland, **2003**.

"Commercial Digging of Medicinal Roots in General (not just Comfrey):
Experience in digging roots in field plots shows that some degree of mechanization may be useful. For example, loosening roots with a tractor-pulled chisel plow *saves back muscles and knees, and allows deeper digging than if done by hand. However, a lot of hand work probably remains for sorting, washing and loading roots into the dryer.* **Other equipment recommended for handling roots includes a U-shaped bar to undercut roots or an L-shaped bar.** *These are sometimes used in the production of things like strawberry transplants, but probably won't go as deep as a chisel plow shank.*
The U-shaped bar was tested on field plots near Wichita (Kansas), on a sandy soil with moderate moisture in the fall. The bar did a nice job of cutting and lifting the roots, but the braces on the bar prevented it from going deep enough to completely uproot things like burdock. It did a nice job on the mallow roots, and even helped extract some of the licorice, which is a shallow, runner-type root. The bar was originally designed for sweet potato digging, and was fabricated locally."
-'Farming a Few Acres of Herbs: An Herb Grower's Handbook' by Rhonda Janke and Jeanie DeArmond, Department of Horticulture, Forestry and Recreation Resources, and David Coltrain, Department of Agricultural Economics, Kansas State University Agricultural Experiment Station and Cooperative Extension Service, www.oznet.ksu.edu, June **2005**.
    (The chisel plow has narrow, double-ended shovels or chisel points, mounted on long shanks. These points rip through the soil and stir it but do not invert and pulverize as much as the moldboard and disk plows.)

"Symphytum officinale: *This generous plant will give multiple leaf harvests in one season and* **abundant root harvests in the early spring and fall following two or three seasons of growth. Comfrey roots should be harvested in the spring or the fall when there is no aerial growth.** *The root grows up two feet (0.60 meters) in depth and* **can be dug by hand with a spading fork or mechanically using a modified potato digger or bed lifter.**
Prior to digging roots, it is helpful to mow down the aerial tops and loosen the soil by running a chisel shank or cultivator up the row on either side of the Comfrey."
-The Organic Medicinal Herb Farmer: The Ultimate Guide to Producing High-Quality Herbs on a Market Scale by Jeff Carpenter and Melanie Carpenter. White River Junction, Vermont: Chelsea Green Publishing, **2015**, pages 289.
    (A garden fork, spading fork or digging fork has a long handle and usually four short, sturdy tines/teeth. It is used to loosen, lift and turn soil. A potato digger is similar to a plow. A simple one is a double or lifting plow with long steel tines

at the back. The plow lifts soil and potatoes, then pushes them upon the tines so the potatoes are on top of the ground. There are two types of cultivators. One type has teeth/shanks that pierce the soil as they are dragged through it in a straight line. The other type has rotary/circular moving disks/teeth such as a rotary tiller. Cultivators stir and break up soil, used either before planting or after the crop has begun growing.)

"Harvesting Herbal Roots: Roots are the first thing people think of harvesting in the fall.
Once that first frost or two comes in, the plants shift their energy into the root so they can stay strong and survive the cold harsh winter to grow anew in the spring. **We harvest the roots after these first few frosts, until the ground is frozen solid (in places where there is a hard freeze).**
> The cold tells the roots to convert some of their complex starches into sugar, making them sweeter and generally giving them a more enjoyable flavor.

To clean roots (generally): Power spray with water, or clean in the sink using a vegetable brush to clean off the dirt."
-'Autumn Wildharvesting Guide: Herbal Roots' by The Herbal Academy, www.theherbalacademy.com, an online school of herbalism, October **2015**.   (If dirt is dried on the roots, you can soak them for a few minutes before cleaning them.)

"To obtain a high-quality efficacious (effective) herbal drug, the appropriate part of the medicinal plant must be harvested at the optimum stage of development, dried and stored at temperatures and conditions that do not decrease the active ingredients, and processed using a technique that maximizes phytochemical recovery.
It is well known that the concentration of required chemical constituents (active ingredients) is strongly influenced by its developmental stage of growth as well as the season (Pandey and Das, 2014). Developmental stage of plant directly influences the phytochemical concentration.
Phytochemical constituents are not evenly distributed throughout the plant. Maximum production of metabolites (active ingredients) depends on age and developmental stage of the plant.
Time and method of collection of different plant parts (not just Comfrey):
> **Root and rhizomes:** From perennials during autumn or winter following the second or third year's growth.
> **Leaves:** Collection should be made in dry weather whilst the plant is flowering. Leaves should be harvested before or at the time of initiation of flowering unless otherwise specified.
> **Flowers:** Flowers must be harvested when they have just opened or shortly afterwards to capture its aroma."

-'Harvesting and Post-Harvest Processing of Medicinal Plants: Problems and Prospects' by Ashok Pandey and Savita, Forest Research Institute, Indian Council of Forestry Research and Education (ICFRE), Dehradun, India; The Pharma Innovation Journal, New Delhi, India, Volume 6, No. 12, pages 229-235, **2017**.
(Polyphenols or phytochemicals are chemicals in plants with over 500 known.The include alkaloids, flavonoids, glycosides, phenolics {phenolic acid}, saponins, stilbenes, lignans, tannins, terpenes, anthraquinones, essential oils and steroids.)

Drying and Processsing Comfrey Roots

For information about harvesting and drying Comfrey leaves:
See section 'Harvesting Comfrey Leaves' (Chapter 37).
See sub-subsection 'Drying Comfrey Leaves' in subsection 'Comfrey Tea' in section 'Humans Eating Comfrey' (Chapter 32).
See subsection 'Comfrey as Hay' in section 'Comfrey Meal, Pellets, Hay and Silage' (Chapter 21).

"Wash the Comfrey roots soon after dug, break them shortish, spread them thin under cover, and without farther attention they will become dry enough to be ground in a coffee or corn mill."
-'The New England Farmer, and Horticultural Register' by Joseph Breck, Volume 23, New Series Volume 13, Boston, Massachusetts, **1845**. ('Symphytum or Comfrey, as Food for Men and Cattle' by Ezekiel Rich, Troy, New Hampshire, p. 10, July 10 1844.)

"Symphytum officinale L. Comfrey. Blackwort. Radix (root) Symphiti. Radix Consolidae Majoris.
A perennial, European herb often cultivated. Its root is spindle-shaped, branched, often more than 2.5 cm (0.98 inch) thick, and 3 decimeter (30 cm = 11.8 inch) long, externally smooth and brownish-black, internally white, fleshy, and juicy.
On drying it becomes wrinkled, of a firm, horny (hard) texture, and of a dark color within. It is almost inodorous (no odor), and has a sweetish, mucilaginous, feebly astringent taste."
-'The Dispensatory of the United States of America, 25th Edition' by Arthur Osol, George E. Farrar Jr., and Editor Horatio C. Wood Jr., Philadelphia, Pennsylvania, **1955**, page 1893. Based on the fifteenth revision (1955) of 'The United States Pharmacopeia', the tenth edition (1955) of 'The National Formulary', the 1953 edition of 'The British Pharmacopoeia', and the first edition (1951) of the 'International Pharmacopeia, Volumes I and II'.

"Comfrey root is more easily dried than Comfrey leaf, but this root must be washed and cut up by hand.
Though the roots do not contain the gummy proteins (as do the leaves), they vary in thickness from about a 1/4 inch to 2 1/2 inches (0.63 to 6.35 cm), so must be reduced to roughly the same size to avoid the problem of the small ones drying before the large. *The allantoin is present in the roots in winter, and dried Comfrey root is also used by herbalists, but it is very little easier than leaf."*
-Comfrey: Fodder, Food and Remedy by Lawrence D. Hills. New York: Rizzoli Universe Books: **1976**, page 118.

"Collect the roots of fully grown Comfrey plants in early spring and autumn. The roots are more easily removed on a rainy day- after the rain has stopped. Shake off the excess soil, and wash the material (in rain water which should always be collected during a downpour).
To dry the roots, *brush them whole with cold water to remove adhering soil particles, slice them lengthwise, and place them on clean, rust-free window screens or on trays of cheesecloth or fine mesh wire, or on aluminium trays or baking pans.*
To hasten (speed) drying, stir or turn the roots every other day. Allow 20 to 25 days to complete the drying process.
Comfrey roots lose between 3/4 to 4/5 of the water weight upon drying. Thus ten pounds (4.5 kg) of freshly gathered roots will end up around 2 pounds (0.90 kg) of dried material. *The least absorption of moisture invites mold and/or beetles. For root storage, use an air-tight glass jar. Label the container with the name and part of the herb and the date of collection."*
-Comfrey: What You Need to Know by Ben Charles Harris. New Canaan, Connecticut: Keats Publishing, Inc., **1982**, page 95.
(Lengthwise is parallel with a thing's length, i.e., in the direction of the longest side.) (Ordinary window screen is available in fiberglass and aluminium. Food grade mesh screens, either stainless steel or nylon, are available on the internet.)

"Comfrey, Symphytum officinale, root and leaf, standard processing specifications:
    Cut and Sifted: 3/8 inch (0.95 cm) screen.   Powder: 1/8 inch (0.31 cm) screen."
-The Potential of Herbs as Cash Crops: How to Make a Living in the Country by Richard Alan Miller. Berkeley, California: Ten Speed Press, **1992**, page 130.
('Cut and Sifted' or C/S means herb is cut into small pieces and then sifted to remove tiny bits and dust that result from cutting.)

"How to Dry Herbs Using a Dehydrator:
Preheat your dehydrator with the thermostat set between 95 and 115 F (35-46 C). *If your herbs are damp, be sure to gently blot them dry with a towel to remove as much moisture as you can.* **Place the herbs on the dehydrator trays in a single layer.** *Drying times will vary depending on the moisture content of your herbs.*
Leaves:
    You can leave small leaves on the stems, but remove larger leaves from thick stems to shorten the drying time. Tender herbs like dill, parsley, and oregano will dry more quickly than thick herbs like plantain and **Comfrey.** *Expect 1 to 4 hours for most herbs. Leaves, stems, and flowers are dry when they become brittle. The leaves and flowers easily crumble and the stems easily break.*
Roots:
    You'll want to cut roots into 1/4 to 1/2 inch (0.63-1.27 cm) pieces. They can take anywhere from 6 to 10 hours to dry. *Check them from time to time for dryness.* **Roots will be stiff when they are completely dry."**
-'Preserve Your Herbs Properly with a Dehydrator' by Jessica Lane, The 104 Homestead: Homestead Where You Live, https://104homestead.com, date unknown but **2013**-2018.

"Drying Roots: Roots are dried in the main drying shed and are much more tolerant of exposure to light and higher temperatures of up to 110 F (43.3 C) with some species. They also take longer to dry than leaves and blossoms.
Cut, Wash and Partially Dry (Par-Dry):
    Roots such as burdock, Comfrey, elecampane, and mallow, which are higher in moisture content and rich in mucilage (gooey) are quartered (cut), washed, and partially dried (in 100 to 110 F = 37.7 to 43.3 C) for thirty-six to forty-eight hours to remove at least 25 percent of their moisture.
    Comfrey roots are relatively soft and can be chopped easily with a field knife. It is helpful to quarter the roots before washing. **Unlike leaf and blossom crops, roots are all washed thoroughly before drying.**
Par-Drying and Milling (Grinding):
After this initial par-drying period, these roots can be milled into large pieces approximately a half-inch (1.27 cm) in diameter and spread out on racks again in the drying shed to complete their drying process.
    The reason we par-dry these type of roots is that they take so long to dry because of their high moisture and mucilage content. **If we didn't par-dry them, yeast and mold levels in the roots could possibly elevate because of lengthy times they sit on the drying racks while still moist.** *Comfrey root will get gummy if chipped (milled) when fresh.*
More Drying:
    After we par-dry and then mill (grind) these roots, we spread the milled root pieces out on drying racks and finish the drying process. Roots dry under optimum conditions in four to six days.
Testing Moisture Content of Roots:
    To determine acceptable moisture content for whole roots, we choose a handful of the largest pieces of dried roots from several racks and snap them in half by hand. *If they are still pliable and resistant to snapping, they need more time drying. If they easily snap in half with us using our fingers, they reached acceptable moisture content, given the fact that humidity levels in the drying shed are under 50 percent and the temperature is above 80 F (26.6 C).*
    This unscientific method for determining acceptable moisture content is also remarkably effective."
-The Organic Medicinal Herb Farmer: The Ultimate Guide to Producing High-Quality Herbs on a Market Scale by Jeff Carpenter and Melanie Carpenter. White River Junction, Vermont: Chelsea Green Publishing, **2015**, pages 222, 223 289.

"Peeling Comfrey Roots?: You do not have to peel the Comfrey roots.
We grind in a pretty heavy duty meat grinder on the coarsest setting, then dry. *Faster drying is best."*
-'Comfrey Roots' by IdahoHerbalist, Herb-Talk: The HomeGrown Herbalist Forum, School of Botanical Medicine, http://herb-talk.com, https://homegrownherbalist.net, Buhl, Idaho, May **2017**.

(You can peel Comfrey roots if you want to. But I do not believe it is necessary unless you are unable to get off all of the dirt. I looked at some dried Comfrey root I bought. It does not look like it was peeled.)

"It is good to give Comfrey roots a scrub with a brush to get them completely clean before processing, but you do not have to remove the black skin."
-'Comfrey, True (Symphytum officinale var patens)' by Richo Cech, Strictly Medicinal® Seeds, https://strictlymedicinalseeds.com, Williams, Orgeon, October 28 **2018**.

Making Comfrey Flour from Leaves and Roots

Some people coarsely grind clean, undried Comfrey root. Then they dry it.

"Home dried Comfrey can be ground to a green flour with a hand wheat mill as used by home breadmakers or the grinder attachment for a liquidizer (blender)."
-Comfrey: Fodder, Food and Remedy by Lawrence D. Hills. New York: Rizzoli Universe Books, **1976**, page 176.

Grind dry Comfrey leaves with a mortar and pestle:
Mortar and pestle are used to crush and grind substances into a fine paste or powder. A mortar is a bowl made of hard wood, metal, ceramic or hard stone. A pestle is a heavy, blunt club-shaped tool. The substance to be ground is put in the mortar, then the pestle is pressed and rotated on it to make small particles. This is good for making small amounts.

Grind dry Comfrey leaves or roots in a grain mill:
A grain mill or gristmill or flour mill is a machine that grinds cereal grain into flour. It can be manually operated or electric. Some brands will be able to grind dried Comfrey leaves or roots. You can contact the manufacturer, or look at their list of acceptable leaves and roots to see if some are similar to Comfrey in denseness and texture.

Grind dry Comfrey roots in a coffee grinder or spice mill:
A coffee grinder or coffee mill is used to grind whole coffee beans so they can be brewed. It should be able to grind dry Comfrey leaves and root. The grinding creates dust in the air. Be careful not to breath it.

Blenders or Food Processors:
Some blenders may be able to chop or grind the dry Comfrey leaves and roots.

Storing Dried Comfrey

"In today's market, the herbs can be found in many types of preparations, for instance syrups, infusions or herbal tea. Dry pieces of herb are usually stored at ambient (room) temperature without regard to relative humidity.
Since dried herbs have high prevalence of moulds, yeasts or coliforms, it is necessary to assess
appropriate storage conditions *to prevent growing of, or toxin production accompanying microflora.*
The purpose of this study was to determine the relationship between moisture content, water activity and glass transition temperature of Comfrey root samples at the temperature of 25 C (77 F) and thus obtain the critical moisture content or water activity level for safe storage.
Sorption isotherms (equal temperature) of Comfrey (Symphytum officinale L.) root samples with different particle size were obtained at 25 C. The shape of isotherm (lines) was similar to those of high-sugar-content foods and the particle size did not affect adsorption process in the 'aw' (water activity) range used in this study.
Blahovec-Yanniotis model was considered to give the best fit over the whole range of 'aw' tested. Various parameters describing the properties of sorbed water derived from GAB (Guggnheim-Anderson-de Boer equation)and Blahovec-Yanniotis models have been discussed.
DSC method was used to measure the glass transition temperatures (Tg) of Comfrey root samples in relation to water activity.
The safe moisture content was determined in 13.39 grams/100g in dried basis at 25 C. *Combining of the Tg line with sorption isotherm in one plot showed that the glass transition temperature concept overestimated the temperature stability for both root samples."*
-'Study of Moisture Adsorption Process in Comfrey (Symphytum Officinale L.) Roots at 25 C' by Libor Cervenka, Jana Kubinova, Leslaw Juszczak and Teresa Witzak; Pardubice, Czech Republic and Krakow, Poland; Scientific Papers of the University of Pardubice, Series A, Faculty of Chemical Technology, Volume 16, pages 5-18, **2010**.
(Sorption is absorption and adsorption considered as a single process.
Absorption is when molecules cross the surface and enter the inside of the material.
Adsorption is the accumulation of molecules on the outside of the material.)

*"**Testing Herbs for Dryness:** Herbs are sufficiently dry when leaves are crispy and crumble easily between the fingers.*
Storing Dried Herbs: Make sure herbs are completely dry to prevent mold growth during storage.
Avoid exposing to air, heat, and light. Airtight and vapor-proof containers will prevent herbs from absorbing moisture from the air

and other foods from absorbing the fragrance of the herbs.
Store in a cool, dry, dark place *such as cupboards or drawers away from stoves and sinks. Don't set them near the stove top or on a windowsill.* **Storing dried herbs in the refrigerator or freezer will maintain their freshness**, *but it creates other problems. If you take a container from a cold area to the warm kitchen, condensation may form, causing the dried herbs to absorb enough moisture in the jar to cause spoilage.*
Most dried herbs keep well for up to a year. *Judge their strength by their aroma. Store whole or crushed, but whole herbs are preferred because they hold their oils and retain their flavor longer."*
-'Lets Preserve: Drying Herbs' by Pennsylvania State College of Agricultural Science, Research and Extension Programs, Penn State Extension Home Food Preservation, extension.psu.edu/food/preservation, **2013**.

Dosages of Comfrey   (This is for entertainment purposes only. See your healthcare practitioner.)

"Comfrey, Symphytum officinale: The dosage of the tincture is 5 to 20 drops, 4 times a day.
Infusion is made out of either the entire fresh Comfrey plant or the powder of the entire plant:
    1 teaspoon steeped in a cup of hot water for one half hour, stir frequently, take 4 times a day."
-Natures Healing Agents: The Medicines of Nature or the Natura System by R. Swinburne Clymer, M.D. Philadelphia, Pennsylvania: Dorrance and Company, **1963**, page 124.

"Comfrey Dosages (Symphytum) (from various sources):
Do not use Comfrey (APA).
Do not use Comfrey root (JAD - James A. Duke, personal commentary).
2-4 grams root as tea 3 times/day (CAN).
2 teaspoons (= about 7.4 grams) root in hot tea (MAD).
2-4 ml liquid root extract (PNC).
2-4 ml liquid extract (1:1 in 25% ethanol) 3 times/day (CAN).
2-8 ml liquid leaf extract (1:1 in 25% alcohol) 3 times/day (CAN);
2-8 grams leaf in tea 3 times/day (CAN).
0.25-0.5 cup fresh leaf (PED).
6-12 grams dry leaf (PED).
9 grams dry leaf: 45 ml alcohol / 45 ml water (PED).
1-3 cups tea/day (5-10 grams herb) remembering Pyrrolizidine Alkaloids (PH2)."
-Handbook of Medicinal Herbs, second edition by James A. Duke with Mary Jo Bogenschutz-Godwin, Judi duCellier and Peggy-Ann K. Duke. Boca Raton, Florida: CRC Press, **2002**, page 215.
    APA = The American Pharmaceutical Association Practical Guide to Natural Medicines by Andrea Pierce. New York: Stonesong Press Book, Wm. Morrow & Co., Inc., 1999.
    CAN = Herbal Medicines: A Guide for Healthcare Professionals by C.A. Newall, L.A. Anderson and J.D. Phillipson. London, England: The Pharmaceutical Press, 1996.
    MAD = Lehrbuch der Biologischen Hilfmittel, Volumes 1-3 by G. Madaus. Georg Olms Verlag, Hildesheim, Germany, 1976. Reprint of 1938 Madaus.
    PED = Nutritional Herbology: A Reference Guide to Herbs by Mark Pederson. Indiana: Wendell W. Whitman Co., 1995.
    PH2 = PDR for Herbal Medicines (Physicians Desk Reference), 4th Edition. Montvale, New Jersey: Medical Economics Company, Inc., 2000. (Lists 600 herbs. It is mostly based on findings of the German Regulatory Authority for herbal medicine, called Commission E.)
    PNC = Potter's New Cyclopaedia of Botanical Drugs and Preparations by E.M. Williamson and F.J. Evans. Saffron Walden, C.W. Daniel Co., Ltd., Essex, England, 1988.

*"***Dosage:*** *The oil from Comfrey leaves and roots can be incorporated in creams and ointments or used in a compress.* **Ointments and other external preparations are typically made with 5% to 20% Comfrey.** *Comfrey should be applied topically on unbroken skin for less than 10 days or a maximum of 6 weeks per year in amounts at or below a daily dosage of 100 mcg of the unsaturated PAs. Although Comfrey has been used as a tea or taken in capsule form, it is not recommended for internal use because of its toxicity."*
-Professional's Handbook of Complementary and Alternative Medicines, 3rd Edition' by Charles H. Fetrow, Pharmd (Doctor of Pharmacy) and Juan R. Avila, Pharmd. Pennsylvania: Lippincott Williams & Wilkins (imprint of Wolters Kluwer), **2003**, pages 235-238.

*"***Dosages for Comfrey oral (unless otherwise stated) administration (adults)*** *for traditional uses recommended in standard herbal reference texts are given below*:*
    Dried root/rhizome: *2-4 grams as a decoction three times daily.*
    Root, liquid extract: *2-4 ml (1 : 1 in 25% alcohol) three times daily.*
    Ointment Symphytum root: *10-15% root extractive in usual type ointment basis applied topically (skin) 3 times daily.*
    Dried leaf: *2-8 grams or by infusion three times daily.*
    Leaf, liquid extract: *2-8 ml (1 : 1 in 25% alcohol) three times daily."*
-Herbal Medicines by Joanne Barnes, Linda A. Anderson and J. David Phillipson. London, England: Pharmaceutical Press, third

edition **2007**, pages 188, 189.
(* -British Herbal Pharmacopoeia, British Herbal Medicine Association, Keighley, England, 1983.)

"Typical daily doses of dried Comfrey leaf range from 5 to 30 grams (.176 to 1.05 ounces), but daily doses of the root are generally lower (0.5 to 10.0 grams, .017 to .352 ounces)."
-'Comfrey Toxicity Revisited' by Dorena Rode, Department of Animal Science, College of Agricultural and Environmental Sciences, University of California, Davis, California; Trends in Pharmacological Sciences: International Union Of Pharmacology, Volume 23, No. 11, pages 497-499, November **2002**.

"Comfrey, Symphytum officinale Dose:
    Root: 6-15 grams, decoct 1 teaspoon/cup (236 ml).
    Leaves: 3-9 grams, infuse 2 teaspoon/cup water.
    If acute, drink 1/2 to 1 cup tea every hour until condition lessens. Then drink 2 cups a day until problem is gone.
    Tincture: 10-30 drops, 1-3 times/day."
-Healing with the Herbs of Life: Hundreds of Herbal Remedies, Therapies, and Preparations by Lesley Tierra, L.Ac., Herbalist, A.H.G. New York: Crossing Press / Crown Publishing Group, **2003**.

"Dosage Symphyti radix (Comfrey root), Symphytum officinale: Ointments or other preparations containing up to 35% of root extract (1:2, ethanol 60% V/V), applied 3-4 times daily. Comfrey root preparations should be applied only to intact skin."
-'Pharmacognosy 2' by Dr. Gyorgyi Horvath, Professor Dr. Peter Molnar and Dr. Tímea Bencsik. Pecs, Hungary: University of Pecs, **2013**. (V/V = volume of solute per volume of solvent.)

*"**Although internal use of Comfrey was not advised (by 'European Medicines Agency'), dosages for oral administration for traditional uses were recommended in older standard herbal reference texts.** The recommended oral (unless otherwise stated) doses of 'The British Herbal Pharmacopoeia' (1974 and 1983) for the treatment of gastric and duodenal ulcer, colitis (inflamed bowel) and hematemesis (vomiting blood) were as follows:*
    Dried root/rhizome:
        2-4 grams (0.07- 0.14 ounces) (in 1983 edition) or
        2-8 grams (0.07- 0.28 ounces) (in 1974 edition) as a decoction (tea) three times daily.
    Root, liquid extract:
        2-4 ml (in the 1983 edition) or 2-8 ml (in the 1974 edition) (1:1 in 25% alcohol) three times daily."
-'European Medicines Agency- Assessment Report on Symphytum Officinale L., Radix' by European Medicines Agency, Committee on Herbal Medicinal Products (HMPC), London, England, 27 pages, May 5 **2015**.

"Comfrey (Symphytum officinale) Dosage and Administration:*
Cold extract, decoction, fluid extract, infusion, powder, and tincture. Comfrey should be for topical (skin) use only.
***Cold Extract:** 2 - 4 Tablespoons.*
***Decoctions:** 2 - 4 Tablespoons.*
***Fluid Extract:** 2 - 4 ml.*
***Infusion:** 4 - 8 Tablespoons.*
***Powder:** 1 Tablespoon.*
***Tincture:** 4 - 6 ml.*
***Herb/Ointments:** Ointments for external use should contain 5-20% of the dried Comfrey root.*
Duration of use should be no longer than 4-6 weeks per year. Adult dose is 3-4 times a day."
-'Herbal Remedies for Athletes: A Handbook' by Stephanie J Troscinski, CAT(C), ATC (Certified Athletic Therapist/Trainer), thesis for Masters of Science, Complementary Alternative Medicine, American College of Healthcare Sciences, Portland, Oregon. Published in Wainfleet, Ontario, Canada, August **2015**.
(* -'Herb 502: Advanced Herbal Materia Medica I' by D. Petersen, American College of Healthcare Sciences, Complementary Alternative Master's Program, Online Program Course Notes, Portland, Oregon, April 2012.)

Comfrey Medicine Overview (For entertainment purposes only. See your healthcare practitioner.)

This section includes: Powder, Extract for Tincture or Concentrated Powder, Fermentation, Syrup, Lozenge/Cough Drop, Bolus/Suppository, Oil Infusion, Ointment/Cream/Salve, Compress/Fomentation, Poultice, Baths/Soaks, Homeopathic, Flower Essences/Remedies.

For Comfrey tea (infusion or decoction), see subsection 'Comfrey Tea' in section 'Humans Eating Comfrey' (Chapter 32).

See subsection 'Solubility, Stability, and Heat Sensitivity of Allantoin' in section 'Comfrey Heals: Allantoin' (Chapter 26).
It has useful information for making medicine.

For an interesting overview of some of the ways of preparing Comfrey medicinally in the 1700s, see the writings by William Salmon in subsection 'Age of Enlightenment and Comfrey' in section 'History of Comfrey' (Chapter 14, Volume 1).

See subsection 'Comfrey Root Wine' in section 'Humans Eating Comfrey' (Chapter 32) for ways Comfrey root was used historically as a medicinal.

Do not use Comfrey externally on deep wounds. For insight on how to properly use Comfrey on deep cuts, see subsection 'Wounds' in section 'Personal/Clinical Observations of Healing' (Chapter 25). If you are treating a deep wound, be sure to contact your healthcare provider.

"Symphytum officinale: The Comfrey leaves are also used externally for purposes similar to those for which the root is employed, but not to any extent, being so far inferior to the root for all purposes."
-'The Botanic Pharmacopoeia: Comprising the Materia Medica, Doses and Preparations of the Medicines Employed in the Botanic Practice' by John G. Hatfield, National Association of Medical Herbalists, Birmingham, England, **1886**.

*"**Traditional Medicine:** Although they used to collect medicinal herbs in every family, there were some healers that specialised on herbal therapies and the profession was often inherited by their family lineage (bloodline) from one generation to the other.*
In making herbal medicines or gaining herbal extracts they used mostly water or alcoholic drinks: beer, fruit wines (made of grapes, raspberry, red-currants, rosehips, mulberry) and brandy (distilled from plums) as solvent.
They also used herbal oils and animal fats (lard, goose or duck fat, milk) in order to gain active agents. These fats promoted the smooth application of the powdered or pulped herbs on the skin surface.
The most popular ways of gaining extracts were maceration (cold soaking), scalding and preparation of decoctions, for which both fresh and dried herbal parts were used.
The most widely used medicinal forms were scaldings (majority of medicinal teas), decoctions (watery extracts used in medicinal baths and for washing injuries), alcoholic extracts made with alcohol of various concentration, creams, plasters, syrups and powdered medicines.
Herbal medicines were used both externally and internally. A part of them was freshly made on spot, but others were found in every household (mainly alcoholic extracts and creams made of them).
Medicines for internal usage were medicinal teas, and alcoholic extracts of certain plants.
Medicinal forms for external application contained plasters made of the leaves, bulbs or petals of certain herbs, decoctions used for poultice, bathing or enema, and embrocations (ointments/creams).
A well-known form of application was medicinal bath, steaming in the vapour of medicinal decoctions, or inhalation.
Healers were conscious of the advantageous and harmful effects (hot or poisonous) of herbs and herbal medicines and informed their patients on the proper quantities. In veterinary medicine, however, they more often applied herbs of strong effect than in human medicine."
-'Centuries of the Traditional Medicine in Hungary' by Peter Babulka, Budapest, Hungary, **2000**, page 12.
(Ethnomedicine is traditional medicine practiced by various ethnic groups, and especially by indigenous {native} peoples.)

Alkaloids in Comfrey Medicine Overview (including Harvest Time)

See subsection 'Hybrids, Hybrid Swarms, Introgression' in section 'Symphytum Species Overview' (Chapter 6, Volume 1).

*"**Russian Comfrey Leaf Alkaloids:** We have measured Pyrrolizidine Alkaloids in the leaves of the hybrid Russian Comfrey S. x uplandicum variety Bocking No. 14, grown at Bocking, Essex (England), and kindly supplied by Mr. Lawrence D. Hills of the 'Henry Doubleday Research Association'.*
    Leaves picked after the first 2 weeks (mid to late August) contained 0.115% of alkaloids (dry weight).
    By mid-September the level had fallen to 0.019% and remained between 0.019% and 0.022% until mid-October when growth ceased.
    Leaves cut April 13-28: alkaloids of the Bases + N-Oxides ranged from 0.048% to 0.222%.
Total alkaloid levels were 0.003% in the largest Russian Comfrey leaves and 0.049% in the smallest: a 16-fold variation.
The total amount of alkaloids per leaf remains fairly constant as the leaf grows heavier, so the percentage of alkaloids falls. My measurement show that the highest akaloid (percent) levels are in small, young leaves, especially early in the season. Protein extracted from Comfrey should not be harmful: a sample of Comfrey protein supplied by Mr. Hills proved, as expected, to be completely free of alkaloids.
The external use of Comfrey preparations should not be hazardous since the alkaloids are converted to toxic metabolites by liver enzymes only after being ingested."
-Toxic Pyrrolizidine Alkaloids in Comfrey' by A.R. Mattocks, Toxicology Research Unit, MRC Laboratories, Carshalton, Surrey, England; The Lancet, London, England, Volume 316, No. 8204, pages 1136-1137, November 22 **1980**, a letter.

*"**Russian Comfrey Leaf Alkaloids:** The total alkaloid content of leaves of Symphytum x uplandicum is variable but usually low.*
In the material studied here (voucher specimen No. CANB 286704 Australian National Herbarium, Canberra), we found 0.15% dry weight in young leaves, 0.01% in mature leaves and 0.05% in the bulk collection of intermediate-sized leaves."
-'The Alkaloids of Symphytum x uplandicum (Russian Comfrey)' by Claude C.J. Culvenor, John A. Edgar, John L. Frahn and Leslie W. Smith, Division of Animal Health, CSIRO, Animal Health Research Laboratory, Parkville, Victoria, Australia; Australian

Journal of Chemistry, Volume 33, No. 5, pages 1105-1113, **1980**.

"Protein extracted from Comfrey should not be harmful: a sample of Comfrey supplied by Mr. Lawrence D. Hills proved, as expected, to be completely free of alkaloids. A large percentage of the pyrrolizidine alkaloids are destroyed in the drying process, and as herbal preparations mainly use dried materials as a starting base, so there is no reason why Symphytum officinale should be restricted as a therapeutic agent."
-'Pharmacists and Comfrey' by Diane Wiesner, B.Pharm, MA, Ph.D., MPS, Principal of the NSW College of Natural Therapies, Sydney, Australia; Australian Journal of Pharmacy, Volume 65, pages 959-963, **1984**.

"Pyrrolizidine Alkaloids in Symphytum: Alkaloids + N Oxides (% dry weight):   Results from 6 different research reports.

	Fresh Leaves	Dried Leaves	Roots
Symphytum asperum	0.009%	0.059%	-
	0.01	-	-
Symphytum officinale	0.006	0.062	-
	-	-	0.07%
Symphytum x uplandicum	0.009	0.09	-
	-	0.013-0.062	-
	-	0.009-0.03 tea	-
	0.15 young leaf	0.22 young leaf	-
	0.05 intermediate	-	-
	0.01 mature leaf	0.05 mature leaf	-
	-	0.005 =undetectable	0.14-0.37 "

-Chemistry and Toxicology of Pyrrolizidine Alkaloids by A.R. Mattocks. London, England: Academic Press, **1986**, page 270.

"Symphytum officinale Alkaloids:
In an attempt to determine the cultural (cultivation) and environmental factors associated with production of the pyrrolizidine alkaloids in Comfrey (Symphytum officinale L.), 2 fields of the plants growing in Minneosta were studied through the 1986 and 1987 growing seasons. The previously established fields were under a balanced organic fertility program.
Using a modification of the procedure of *Molyneux et al., developed by J. Rana, University of Minnesota, Mineneapolis, the pyrrolizidine alkaloids in Comfrey tissue harvested at different times were extracted and quantified.
> *The first spring cutting of immature leaves (May 15, 1986) contained 0.026 percent pyrrolizidine alkaloids on a dry weight basis.*
> *Subsequent harvests during the growing season had no detectable pyrrolizidine alkaloids in leaf tissue (the minimum detectable quantity was 5 ppm {parts per million}).*

The data indicated harvest time was a critical factor in producing pyrrolizidine alkaloid-free Comfrey. Mature leaves have a lower alkaloid content than young leaves.
The quantity of alkaloids in the plant varies with the type of tissue.
> *Mature leaves have the lowest and the roots have the highest concentration of alkaloids. Generally, roots contain almost 10 times as much alkaloid as the leaf tissue."*

-'Growing Alkaloid-Free Comfrey' by Gary Steuart, Steuart Laboratories, Saint Paul, Minnesota; The Herb Spice and Medicinal Plant Digest, 'Massachusetts Cooperative Extension Service' and 'University of Massachusetts', Amherst, Volume 5, No. 4, page 9, Winter **1987**.
(* -'Chemistry of Toxic Range Plants: Determination Pyrrolizidine Alkaloid Content and Compositon in Senecio Species by Nuclear Magnetic Resonance Spectroscopy' by Russsel J. Molyneux, A. Earl Johnson, James N. Roitman and Mabry E. Benson, Journal of Agricultural and Food Chemistry published by American Chemical Society, Volume 27, No. 3, pages 494-499, 1979.)
> (260 ppm = 0.026%.) (5 ppm = 0.0005%.)

"Russian Comfrey (Symphytum x uplandicum Nyman) fresh leaf contains 0.01% to 0.15% of alkaloids, the major ones being echimidine and 7-acetyllycopsamine (24% and 32% of total alkaloid, respectively.
Symphytum officinale (Common Comfrey) contains less of the more toxic pyrrolizidine alkaloids, echimidine and symphytine, than does Russian Comfrey."
-'Comfrey: Assessing the Low-Dose Health Risk' by Peter J. Abbott, Toxicology Unit, Department of Community Services and Health, Canberra, Australia; Medical Journal of Australia, Volume 149, No. 11-12, pages 678-682, **1988**.

"The Cherokee harvested the Comfrey leaves during May through June when flowers had begun to bud. Leaves were highest in toxic alkaloids during this time.
The roots were gathered in autumn, cut into slices, and dried (J. Olds, 'The Encyclopeida of Organic Gardening', 1975). The dried roots were crushed, powdered, and placed on cuts, burns, and wounds to promote healing."
-'A Manual of Cherokee Herbal Remedies: History, Information, Identification, Medicinal Healing' by Patricia D. Schafer, Master's Thesis, Indiana State University, March **1993**.

"Symphytum officinale L.: In dried S. officinale leaves 0.02 to 0.18% and in the roots 0.25 to 0.29% alkaloids, respectively, their N-oxides were detected."
-'Medicinal Plants in Europe Containing Pyrrolizidine Alkaloids' by Erhard Thomas Roeder, Pharmazeutisches Institut der

Rheinischen Friedrichs-Wilhelms, University of Bonn, Germany, Pharmazie 50, pages 83-98, March **1995**.

"Symphytum tuberosum Alkaloids: S. tuberosum L. has been shown to contain 72 mg/kg PA (Pyrrolizidine Alkaloid) N-oxides in the leaf, and 180 mg/kg in the root.
S. tuberosum contains PA levels significantly lower than other Symphytum spp. (species) and relatively high levels of allantoin, 0.96% in the root and 0.98% in the leaf."
-'Using Herbs that Contain Pyrrolizidine Alkaloids' by Alison Denham, B.A., MNIMH (National Institute of Medical Herbalists), University of Central Lancashire, England; The European Journal of Herbal Medicine, Volume 2, No. 3, pages 27-38, **1996**.
>(**It seems that Symphytum tuberosum is the best species of Comfrey for medicinal purposes.** In the United States it is difficult to find. For more about S. tuberosum, see the subsection 'Tuberosum' in the section 'Details about Symphytum Species' {Chapter 9}.)

"Pyrrolizidine Alkaloids in Dried Symphytum Officinale (microgram/gram)
>*Root (1) 8320 microgram/gram*
>*Root (2) 1380*
>*Root (3) 2092*
>*Leaf (1) 22*
>*Leaf (2) 55*
>*Leaf (3) 15 "*

-'Analysis, Separation, and Bioassay of Pyrrolizidine Alkaloids from Comfrey- Symphytum officinale' by C.E. Couet, C. Crews and A.B. Hanley, Natural Toxins, Volume 4, page 163-167, **1996**.

"Man'ko et al. (*1970) reported that the pyrrolizidine alkaloid content of Symphytum officinale was highest in above-ground parts during flowering (0.06%), and in roots at the end of fruit formation (0.31%)."
'Comfrey and One of Its Constituent Alkaloids: Symphytine, Review of Toxicological Literature' by Raymond Tice, Ph.D., Integrated Laboratory Systems, Research Triangle Park, North Carolina, October **1997**.
(* -'Level of Alkaloids in Symphytum Officinale Dependent on the Phase of Plant Development' by I.V. Man'ko, B. K. Kotovskii et al, Rastitel'nye Resursy (Vegetable or Herbal Resources- Russian), Volume 6, No. 3, pages 409-411, 1970. Also cited in Chemical Abstracts, 74:61608, 1971. I was not able to get this. If you have an English translation, I would appreciate a copy.)

"Further studies would eventually show that there was essentially no toxic value of alkaloids in Comfrey leaf that had been cut from the plant prior to 10% flowering (much like Alfalfa).
>*If Comfrey is not cut at this stage of growth, then the leaf tends to become more root-like in both alkaloid content and texture.*

In some private studies conducted with Honda Corp. in Osaka, Japan, their feedlots were three 60-story buildings. Not being outside, the cattle were rampant with disease. When fed a 60% Comfrey / 40% Alfalfa pellet, however, virtually all diseases were eliminated."
-'A Farm Project: Comfrey Leaf and Root' by Richard Alan Miller, Oak Publishing Inc., Grants Pass, Oregon, and Northwest Botanicals, www.nwbotanicals.org, **1999**. (If you have information about the Osaka, Japan report, could you please send it.)

"Alkaloids are organic bodies, derived chiefly from plants in which they are believed to exist in combination with organic acids, forming salts (a safer, more soluble form of the alkaloid). These alkaloid salts are usually well-defined, colorless, odorless, crystalline and soluble. *Alkaloids may be unstable when heated.*
Alkaloids:
>*Most pure alkaloids are bitter, slightly alkaline, soluble in ether and chloroform, and often less readily in alcohol. In water they are comparatively insoluble.*

Alkaloid Salts:
>*On the other hand, the solubility of the alkaloid salts usually follow an opposite pattern; they are freely soluble in water and somewhat soluble in alcohol.* The ready solubility of the salts of alkaloids have caused them to be preferred to the alkaloids themselves for therapeutic uses.

Alkaloid Extracts:
>***For preparation of extracts with the highest levels of alkaloids,*** *water/vinegar/alcohol menstrua (solvent) having a 35 percent water, to 10 percent vinegar, to 55 percent alcohol content are recommended.*
>*Please keep in mind, that **while alkaloids and their salts have distinctive therapeutic properties of their own, they do not fully nor exactly represent the action of the whole plant from which they are derived.**"*

-The Herbal Medicine-Maker's Handbook: A Home Manual by James Green, Herbalist. New York, New York: Random House, Inc., **2000**, page 93.

"When Comfrey is dried, enzymes are released and much of the alkaloid is destroyed.
Harvest Time and Alkaloids:
>*From trials in Minnesota in 1987*, in an attempt to determine cultural and environmental factors associated with the production of PAs, **it was found that Comfrey, harvested at different times in the growing season, can be of varying PA amounts.***

In one trail in 1986, immature leaves contained 0.026% pyrrolizidine, on a dry weight basis. **A subsequent harvest during the**

growing season had no detectable PAs in the leaf (the minimum detectable quantity was 5 ppm {parts per million}. The data indicated, harvest time was a critical factor in producing PA free Comfrey, and that mature leaves have an even lower alkaloid content, than young leaves."
-How Can I Use Herbs in My Daily Life: Over 500 Herbs, Spices and Edible Plants: An Australian Practical Guide to Growing Culinary and Medicinal Herbs' by Isabell Shipard. David Stewart Publisher: **2003**.
(* The author did not give a reference about the 1987 Minnesota trials, but probably it was this: -'Growing Alkaloid-Free Comfrey' by Gary Steuart, Steuart Laboratories, Saint Paul, Minnesota; The Herb Spice and Medicinal Plant Digest, 'Massachusetts Cooperative Extension Service' and 'University of Massachusetts', Amherst, Volume 5, No. 4, page 9, Winter 1987.)

"As a result of a 1993 report by the 'Committee on Toxicity of Chemicals in Food' to the 'Food Advisory Committee' and the 'Ministry of Agriculture, Fisheries and Food' (United Kingdom), the health food trade voluntarily withdrew all Comfrey products, such as tablets and capsules, and advice was issued that the root and leaves should be labelled with warnings against ingestion. **It was considered that Comfrey teas contained relatively low concentrations of pyrrolizidine alkaloids and did not need any warning labels."**
-Herbal Medicines by Joanne Barnes, Linda A. Anderson and J. David Phillipson. London, England: Pharmaceutical Press, third edition **2007**, page 188.

"Symphytum officinale: *Contains alkaloids intermedine, lycopsamine, symphytine, echimidine and symglandine.*
　　　Total content of PAs (Pyrrolizidine Alkaloids) is about 0.5%."
-'Hepatic Veno-Occlusive Disease Associated with Toxicity of Pyrrolizidine Alkaloids in Herbal Preparations' by Zhe Chen and Ji-Rong Huo, Department of Digestive Disease, The Second Xiang-Ya Hospital of Central South University, Changsha, Hunan Province, China; The Netherlands Journal of Medicine, Volume 68, No. 6, pages 252-260, June **2010**.

"Pyrrolizidine Alkaloids:
Symphytum asperum:
　　　asperumine, echiumine, symlandine, symphytine, myoscorpine, echinatine, echimidine. (*Hartmann and Witte 1995)
Symphytum officinale:
　　　(7-acetyl)intermedine, (7-acetyl)lycopsamine, echimidine, symlandine, symviridine, myoscorpine, symphytine. (**Couet et al. 1996; ***Mei et al. 2005; ****Roder/Roeder 1995; *****Oberlies et al. 2004)
Symphytum tuberosum:
　　　amadoline, (7-acetyl)lycopsamine, symphytine, echimidine. (*Hartmann and Witte 1995)
Symphytum x uplandicum synonym Symphytum peregrinum:
　　　echimidine, (7-acetyl)intermedine, (7-acetyl)lycopsamine, uplandicine, symlandine, symviridine, myoscorpine, symphytine. (****Roder/Roeder 1995)"
-'Discussion Paper on Pyrrolizidine Alkaloids' by Joint FAO/WHO Food Standards Programme, Codex Committee on Contaminants in Foods, 5th Session, The Hague, The Netherlands, CX/CF 11/5/14, Agenda Item 9(f), 77 pages, March 21-25 **2011**. WHO= World Health Organization. FAO= Food and Agriculture Organization of United Nations. Codex Alimentarius Commission.
(* -Alkaloids: Chemical and Biological Perspectives, Volume 9, edited by S.W. Pelletier. Oxford, England: Pergamon Press, 1995. 'Chemistry, Biology and Chemoecology of the Pyrrolizidine Alkaloids' by T. Hartmann and L. Witte, pages 155-233.)
(** -'Analysis, Separation, and Bioassay of Pyrrolizidine Alkaloids from Comfrey- Symphytum officinale' by C.E. Couet, C. Crews and A.B. Hanley, Natural Toxins, Volume 4, page 163-167, 1996.)
(*** -'Mutagenicity of Comfrey (Symphytum Officinale) in Rat Liver' by N. Mei, L. Guo, P.P. Fu, R.H. Heflich and T. Chen, National Center for Toxicological Research, United States Food and Drug Administration, Jefferson, Arkansas; British Journal of Cancer, Volume 92, pages 873-875, 2005.)
(**** -'Medicinal Plants in Europe Containing Pyrrolizidine Alkaloids' by Erhard Thomas Roeder, Pharmazeutisches Institut der Rheinischen Friedrichs-Wilhelms, University of Bonn, Germany, Pharmazie 50, pages 83-98, March 1995.)
(***** -'Analysis of Herbal Teas Made from the Leaves of Comfrey (Symphytum Officinale): Reduction of N-oxides Results in Order of Magnitude Increases in the Measureable Concentration of Pyrrolizidine Alkaloids' by Oberlies, Kim, Brine, Collins, Handy, Sparacino, Wani and Wall (North Carolina and New Mexico); Public Health Nutrition Journal, Cambridge, England, Volume 7, Issue 7, pages 919-924, 2004.)

"Comfrey (Symphytum officinale) root pyrrolizidine alkaloids (0.05-0.08%)."
-'Symphyti Radix (Comfrey Root)' by ESCOP Monographs: The Scientific Foundations for Herbal Medicinal Products, European Scientific Cooperative on Phytotherapy, Exeter, United Kingdom, 16 pages, **2012**.

"As trained herbalists we learn how to harvest and prepare herbs for best effects.
Herbs with Pyrrolizidine Aalkaloids (PAs) traditionally are harvested and extracted in ways that would avoid the peak of concentration and extraction of PA's.
　　　They are more plentiful in annual herbs than perennial, more plentiful in the roots than in the leaves. This is why we don't see borage root used in herbal tradition, and it is usually the perennial leaves that are used instead.
PA's are used in the process of blooming and so this is why we harvest these herbs (boneset, borage, coltsfoot and Comfrey) for use while they are blooming, while flowers are on the plant or any time after.
Another way to obtain lower PA's is that we know that PA's are not very water soluble, and so drying these herbs and making herbal infusions is traditionally how they are used.

Tinctures of these herbs are made from fresh plant material, using low alcohol content, such as 40% alcohol, which is 80 proof. The resultant tincture tends to be 30-35% alcohol in its finished state, depending on juiciness of the herb."
-'Pyrrolizidine Alkaloids in Healing Herbs' by Heather Nic an Fhleisdeir, Academy Of Scottish Herbalism; Village Herbalist: Ancient Wisdom, Primal Healing and Modern Chemistry, http://celticherbalist.blogspot.com, Eugene, Oregon, August 8 **2013**.

"Unsaturated PA (Pyrrolizidine Alkaloids) occur in the plant families:
*Fabaceae, **Boraginaceae** (e.g., Borago officinalis, Symphytum officinale), and Asteraceae (e.g., Tussilago farfara = Coltsfoot).*
Unsaturated PA are water- and alcohol-soluble and so can be found in teas and tinctures.*"*
-'Pyrrolizidine Alkaloids: Key Points for Herbal Practitioners' by Helen Phillips M.Sc. Herbal Medicine, Diploma of Phytotherapy, MNIMH and Alison Denham, M.A., FNIMH (National Institute of Medical Herbalists); The Forager's Path, LLC, School of Botanical Studies, Flagstaff, Arizona, August **2016**.

"Total content of PAs (Pyrrolizidine Alkaloids) is:
    S. caucasicum 0.5%
    S. officinale (leaves: 0.02-0.18%; roots: 0.25-0.29%)
    S. peregrinum 0.2% alkaloids in the tops (leaves)."
-'Safety Issues Affecting Herbs: Pyrrolizidine Alkaloids" by Subhuti Dharmananda, Ph.D., Director, Institute for Traditional Medicine, Portland, Oregon, **2016**.

"From most toxic to least of PAs (Pyrrolizidine Alkaloids) is macrocyclic diesters > retronecine and heliotridine diesters > heliotridine monoesters > retroecine monoesters.
    Comfrey PAs are retronicene mono- and diesters. Symphytine and echimidine are derivatives of retronecine.
The most toxic alkaloid is echimidine. Only minute amounts are found in Symphytum officinale, but large amounts in Russian Comfrey and Prickly Comfrey leaves.
Symphytine (7-tiglylretronecine viridiflorate) is the major alkaloid of S. officinale."
-'Comfrey' by Glenn Axford, Our Common Roots: An Exploration of the Healing Power of Nature, Canada, www.ourcommonroots.com, date unknown but **2018** or earlier.

<u>Comfrey Powder</u>

See section 'Harvesting Comfrey Leaves' (Chapter 37).
See sub-subsection 'Drying Comfrey Leaves' in subsection 'Comfrey Tea' in section 'Humans Eating Comfrey' (Chapter 32).
See subsection 'Comfrey as Hay' in section 'Comfrey Meal, Pellets, Hay and Silage' (Chapter 21).
For concentrated Comfrey powder, see subsection 'Comfrey Extract to Make Tinctures and Concentrated Powder' in this section.

*"**Symphytum officinale:** The Comfrey leaves are also used externally for purposes similar to those for which the root is employed, but not to any extent, being so far inferior to the root for all purposes.*
Officinal preparation of the Comfrey root:
Pulvis Bistortae Composita: Compound Powder of Bistort. Restorative Powder.
Take of bistort, Comfrey, each, one ounce; *balmony, pond lily, cloves, each, half an ounce; pulverise and mix. This powder is especially designed for profuse menstruation, leucorrhoea, and weaknesses of the vagina and uterus. It forms one of the most valuable medicines for the weaknesses to which women are peculiarly liable, seldom requiring a prolonged employment to produce, even in the most severe and debilitating cases, a satisfactory change in the condition of the patient. It forms a most safe and reliable uterine astringent for hemorrhages, no less than for the mucous discharges from the female organs."*
-'The Botanic Pharmacopoeia: Comprising the Materia Medica, Doses and Preparations of the Medicines Employed in the Botanic Practice' by John G. Hatfield, National Association of Medical Herbalists, Birmingham, England, **1886**.

*"**Fineness of Powder:** To properly obtain the soluble constituents of drugs by the process of percolation, they should be so comminuted or divided that the menstruum may readily dissolve all soluble matter. To this end, different drugs are directed to be reduced to different degrees of fineness as experience has shown to be best suited to their nature.*
The 'United States Pharmacopoeia' has adopted the following standard for the fineness of powders:
    A very fine powder *should pass through sieve having 80 or more meshes to linear inch (2.54 cm). Equals No. 80 powder.*
    A fine powder *should pass through a sieve having 60 meshes to the linear inch. Equals No. 60 powder.*
    A moderately fine powder *should pass through a sieve having 50 meshes to the linear inch. Equals No. 50 powder.*
    A moderately coarse powder *should pass through a sieve having 40 meshes to the linear inch. Equals No. 40 powder.*
    A coarse powder *should pass through a sieve having 20 meshes to the linear inch. Equals No. 20 powder.*
Symphytum officinale root, Comfrey is Class D: No. 30 powder."
-'Fenners Twentieth Century Formulary and International Dispensatory: Working Formulas for the Official and Unofficial Preparations of All Countries' by B. Fenner, Westfield, New York, **1904**, pages 62, 63, 642.
    (Menstra or menstruum is a substance that dissolves a solid or holds it in suspension, i.e., a solvent.)

*"**Comfrey, Symphytum officinale, root and leaf, standard processing specifications:** Cut and Sifted: 3/8 inch (0.95 cm)*

screen. **Powder: 1/8 inch (0.31 cm) screen.**"
-The Potential of Herbs as Cash Crops: How to Make a Living in the Country by Richard Alan Miller. Berkeley, California: Ten Speed Press, **1992**, page 130.
('Cut and Sifted' or C/S means herb is cut into small pieces and then sifted to remove tiny bits and dust from the cutting.)

"*Comfrey Leaf (Symphytum officinale):*
   The powder of Comfrey leaf is green, almost inordorous (no odor), and has a mucilaginous, weakly astringent taste. Product should be of more or less uniform color with little or no 'browning' evident.
   As the flowers are edible, their inclusion is fine, except where high product uniformity is necessary.
Comfrey Root (Symphytum officinale):
   Comfrey root is almost inodorous, and has a sweetish, mucilaginous, feebly astringent taste.
   Powdered Comfrey root is greyish brown in color with many small dark brown specks of outer bark in it.
   It contains a mucilage which is water soluble."
-'A Farm Project: Comfrey Leaf and Root' by Richard Alan Miller, Oak Publishing Inc., Grants Pass, Oregon, and Northwest Botanicals, www.nwbotanicals.org, **1999**.

"**Powders are dusted on topically to treat skin conditions like rashes and chafing.**
Powdered herbs are sometimes made into a paste with a bit of water and then taken internally. You can also make your own capsules (empty size 00 capsules are available at health food stores).
To prepare an herbal powder: *Powder your herbs in a flour mill, spice mill or blender. I usually don't prepare more than I can use in a short period of time, because once an herb has been powdered, it will lose its potency more quickly.*
Store powdered herbs in a glass jar with a tightly sealed lid. Place the jar in a location where it is out of direct sunlight and away from excessive heat. Label the jar. **Herbs for powders: Comfrey promotes skin tissue regeneration.**"
-Growing 101 Herbs that Heal: Gardening Techniques, Recipes and Remedies by Tammi Hartung, Medical Herbalist. North Adams, Massachusetts: Storey Publishing, **2000**.

"**Dried powdered roots of Symphytum officinale, collected in spring or autumn, shows** parenchymatous cells containing mucilage and starch; fragments of cork tissue with dark brown cell walls; small groups of reticulate and bordered pitted vessels."
-'Comfrey Root Powder Specifications' by BiOrigins: Natural Products that Work, www.biorganics.co.uk, Fordingbridge, Hampshire, England, Madar Corporation, www.madarcorporation.co.uk, **2018**.
   (Parenchymatous cells make up the softer parts of leaves, stems, roots, etc.)
   (Reticulate means divided so as to resemble a net or network.)

Comfrey Extract to Make Tinctures and Concentrated Powder
See subsection 'Solubility, Stability, and Heat Sensitivity of Allantoin' in section 'Comfrey Heals: Allantoin' (Chapter 26).

"**The use of extracts (tinctures) of Comfrey root rather than root powder markedly decreases the PA (Pyrrolizidine Alkaloid) concentration.** *PA concentrations in the leaf are much lower than in the root.*
Given the low PA concentrations found in decoctions and tincture made from Comfrey leaf, it is likely that no hepatic damage whatsover results from their use, *and one can feel positive about prescribing them.*
Even if cumulative toxicity can develop from the low doses in Comfrey leaf teas, it may take more than the average lifetime to reach the toxic threshold.
One could advocate long-term usage for arthritis in elderly people in that it diminishes reliance on Non-Steroidal Anti-Inflammatory Drugs (NSAIDs) which are themselves an important cause of death due to perforation of occult duodenal ulcers."
-'Using Herbs that Contain Pyrrolizidine Alkaloids' by Alison Denham, B.A., MNIMH (National Institute of Medical Herbalists), University of Central Lancashire, England; European Journal of Herbal Medicine, Vol 2, No. 3, pages 27-38 (page 37), **1996**.

"**Extracts (Comfrey):** *Comfrey extracts are anti-inflammatory in vitro and in vivo, perhaps due to rosmarinic acid. Allantoin a well known dermatological (skin) agent. Comfrey aqueous (water) extract stimulates release of prostaglandin-like material from rat gastric mucosa. Two nonhepatotoxic PAs (Pyrrolizidine Alkaloids), platyphylline and sarracine, have been used for gastro-intestinal hypermotility and peptic ulceration.* **Yes, aqueous extracts increase survival time of mice with spontaneous tumors, and decrease tumor growth, and have antimutagenic activity. Is Comfrey more likely to cause, cure, or prevent cancer? This is what we really should be studying.**"
-Handbook of Medicinal Herbs, second edition by James A. Duke with Mary Jo Bogenschutz-Godwin, Judi duCellier and Peggy-Ann K. Duke. Boca Raton, Florida: CRC Press, **2002**, page 215.
   ('In Vitro' means done in laboratory equipment as opposed to in/on a living animal that is called 'In Vivo'.)

Definition of Extract and Tincture

Extract is a concentrated preparation of an herb made by removing the active constituents with solvent(s) such as ethanol (ethyl or grain alcohol) and/or water, then evaporating all or almost all of the solvent. Extracts may be tinctures or powders.

Alcohol is the most effective solvent for extracting herbal substances. Ethanol is a colorless, volatile, flammable liquid, produced by yeast fermentation of carbohydrates. Ethyl alcohols used medicinally include **vodka (40%=80 US proof)**, brandy (35-60%=70-120 US proof), and Everclear® grain alcohol (75.5%=151 US proof, 95%=190 U.S. proof). Alcohol 'proof' is a measure of the content of ethanol (alcohol). For example, 50% alcohol is 50/50 ethanol/water.

Isopropyl alcohol (isopropanol, 2-propanol, rubbing alcohol) is not ethanol and is not safe to consume. Do not use it for medicines you take internally.

A tincture is an alcoholic extract of a plant. To be called a tincture, the extract must be at least 25-60% ethanol (50-120 United States proof). Some tinctures are as high as 90% (180 US proof). Ethanol of 25% is the most common. A tincture can be made from just one or a combination of plants. Fresh or dried leaves, roots, bark, flowers, and berries can be used.

How to Make Comfrey Extract as Tincture

Overview of making herbal extract: Grind or finely chop herb(s) and/or root(s).
  Fresh leaves and flowers: fill jar 2/3 to 3/4 with herb.
  Dried leaves and flowers: fill jar 1/2 to 3/4 with herb.
  Fresh roots: fill jar 1/3 to 1/2 with roots.
  Dried roots: fill jar 1/4 to 1/3 with roots. Dried roots double in size when soaked in solvent.

Alcohol percent for standard tinctures is 40-50% (80-100 proof vodka) (all herbs, not just Comfrey). This is good for extracting both the alcohlic and water soluble substances. Pour alcohol and/or water mixture to top of jar. Add enough solvent to completely cover the herb or root. Put on cap, preferably plastic. The herb or root should move freely when shaking the jar. Label with date, name of herb(s), parts used, and percent alcohol. Put the jar in a dark place at room temperature. Shake daily or several times a week. Let sit for 4 to 8 weeks or more. The longer it sits, the stronger the tincture. Strain using a funnel with cheesecloth or similar cloth. Store in a dark glass bottle in a cool, dark place.

*"Dry Plant Tincture: **This is the classic one of grinding the herb, weighing the coarse powder, adding the solvent,*** putting the mixed gloop into a canning jar, and shaking it twice a day for two weeks. Then letting it set (sit) a day, pouring the standing tincture and squeezing the cloth-wrapped marc as dry as your wrists can make it. This is a fine and time-honored method, producing a full strength tincture; it has always been a proper alternative method to percolation."*
-Medicinal Plants of the Pacific West by Michael Moore. Santa Fe, New Mexico: Red Crane Books, **1993**, page 27.
  ('Marc' is the residue of the herb after extraction.)
  (Percolation is an extraction method where a solvent slowly goes through a powdered herb while it absorbs certain substances. The solvent drips out through a filter at the bottom. An example of percolation is making coffee in a coffee machine with a paper filter.)

*"**Toward the end of May, a distillate can be prepared in water from Comfrey leaves and root. An essence and extract of Comfrey is commonly made this way:**
  Take as much Comfrey as you want, and boil it in water until it becomes thick and mucusy. **Pour over this just enough brandy to cover it, then let it stand for a few days in warm sand.** (The sand is set over a charcoal stove or on top of a plate stove. This yields a constant, even, low heat necessary for separating and clarifying lighter components from the sediments.) **Carefully pour off the essence that has formed on the top,** and store away for future use.
  The dose should be twenty to forty drops at a time, taken frequently."*
-Sauer's Herbal Cures: America's First Book of Botanic Healing, 1762-1778 by William Woys Weaver. New York: Routledge, **2001**, pages 114-115.

*"Because different constituents of plants are soluble in different menstra, it's important to find a solvent that is effective for the constituents that you're working with. **The benefit of pure alcohol as a solvent is that it extracts balsams, camphors, resins, essential oils, alkaloids, and acrid and bitter constituents.** Homeopaths swear by pure alcohol.*
Alcohol also preserves your extracts and prevents decomposition, unlike plain water."
-'An Herbal Medicine-Making Primer' by Simon the Simpler, Yggdrasil Distro, 24 pages, **2010**.
  (Menstra or menstruum is a substance that dissolves a solid or holds it in suspension, i.e., a solvent.)

*"**Decoction Tincture Method: This is a variation on a dry tincture.** Some herbs, especially roots, extract better with heat (ginseng family). **Others do better with more water ('slimers' like Comfrey root** - for topical use - and high polysaccharide herbs and mushrooms). **Simmer your roots for 1-3 hours in a little water.***
When done, pour the hot mixture into a mason jar and add alcohol, *cap, and let sit (macerate) for at least 1 month, shaking daily (once it has cooled).* ***Ultimately you'll want at least 30% alcohol*** *to prevent your formula from growing mold and bacteria.*
  For example, per 1 ounce (28 gram) herb (weight), simmer in 3 ounces (88 ml) of water, then add 2 ounces (59 ml) of whole grain alcohol. Or, per 1 ounce herb (weight), simmer in 1 ounce (29 ml) of water, then add 4 ounces (118 ml) of high proof vodka.
Should keep for up to 10 years on the shelf."
-'Autumn Roots, Barks and Berries: Day-Long Field Workshop' by Maria Noel Groves, Clinical Herbalist, Wintergreen Botanicals

LLC, Allenstown, New Hampshire, www.WintergreenBotanicals.com, **2016**.

Many alkaloids dissolve poorly in water but readily dissolve in solvents such as ethanol (ethyl alcohol), ether, chloroform or ethylene dichloride.
Most alkaloids are present in raw plants in the form of salts of organic acids. These salts are usually freely soluble in water and ethanol (ethyl alcohol).

1. Water extracts acids, **alkaolids (some), alkaloid salts (most),** coloring matter, essential oils (some), glycosides, gums, mineral salts, mucilage, pectin, proteins, sugars, tannins.
2. Alcohol extracts acids (organic), acrid constituents, **alkaloids,** bitter constituents, chlorophyll, essential oils, glycosides, resins. Alcohol does not extract minerals, gums, or mucilage.
3. Infused oil extracts resins, essential oils, and flavonoids.
-Organic Growers School, Asheville, North Carolina, www.organicgrowersschool.org; Training, workshops, conferences, **2018**.

"*Comfrey leaf and root extracts have many constituents including allantoin, rosmarinic acid, triterpene saponins, silicic acid, and tannins,* believed to be the basis for its anti-inflammatory and wound healing activity.
However, Comfrey also contains several pyrrolizidine alkaloids (symphytine, echimidine, symglandine and lycopsamine) which are toxic and capable of causing sinusoidal obstruction syndrome (previously called veno-occlusive disease) and severe liver injury."
-National Institutes of Health, LiverTox, 'Clinical and Research Information on Drug-Induced Liver Injury', Bethesda, Maryland, https://livertox.nih.gov/Comfrey.htm, **2018**.
    (It did not say whether extracts are water and/or alcohol based. It did not say what species Comfrey was used.)

"*Tinctures Preparation and Ratios: There are three important numbers used to figure out how to prepare tinctures.*
    1. The weight of the plant material you want to tincture.
    2. The volume of fluid (ethanol and water) you need to prepare the tincture.
    3. The percentage of alcohol you are starting with (i.e., 50%, 95%).
The below ratios are standard and can be adjusted to the specific medicine you are preparing. The important part of this is to make sure you stay within the same system of measurement on each side of the equation, such as metric (grams: milliliters) or imperial (ounces: fluid ounces).
Alcohol proof into alcohol percentage.
    1. To change alcohol proof to a percentage, take the proof and divide it in half.
    2. This gives you the percentage of alcohol.
    3. Minus this number from 100, which is the amount of water.
    4. Example: 160 proof is 80% ethanol and 20% water.
    5. 160 divided by 2 = 80 (percentage of alcohol).
    6. 100 minus 80 = 20 (percentage of water).
*Herb weight : **Menstruum volume ratio** (i.e., 1:2, 1:5)*
    1. This is always a relationship of weight to volume (herb weight : liquid volume).
    2. Note on metric and imperial (American) systems:
        30 grams is about 1 ounce (weight), 30 milliliters is about 1 fluid ounce,
        1000 milliliters equals one liter, which is about 1 quart."
-'Herbal Pharmacy: Tinctures Preparation and Ratios' by Herbalist 7Song, Northeast School of Botanical Medicine, Ithaca, New York, www.7song.com, January **2018**.

Extracting Comfrey Constituents by Alcohol, Water and Other Solvents

"In 1965 Dr. A.H. Ward, analyst to the 'Henry Doubleday Research Association', **tested Russian Comfrey leaf tea made by two different methods (water or alcohol) to determine the allantoin:**
    1.** Allantoin content determined in the usual way after **alcoholic extraction: 0.083 percent.
    2.** Two heaped teaspoonfuls (4 grams) used to make four cupfuls (600 ml) of **Russian Comfrey leaf tea**, by pouring on the boiling water and steeping for 5 minutes. **Amount of allantoin extracted: 0.045 percent.
    3.** Four grams of the sample boiled with 600 ml of water for 5 minutes. **Amount of allantoin extracted: 0.046 percent.
These were leaf-only samples, for the major portion of the allantoin is in stems and midribs, *but it will be seen that the tea as drunk is approximately the same as the strength recommended for sores and gastric ulcers in the early editions of the 'British Pharmacopoeia'."*
-Comfrey: Fodder, Food and Remedy by Lawrence D. Hills. New York: Rizzoli Universe Books: **1976**, page 175.
    (A Pharmacopoeia is an official publication, containing a list of medicinal drugs with their effects and directions for use. The first edition of the 'British Pharmacopoeia' was in 1864.)

"**Ethyl alcohol is more selective in its extraction action, and therefore does not have as wide a solubility range as water.**
It is a good general solvent for extracting resins, balsams, camphors, essential oils, alkaloids and natural alkaloidal salts, glycosides, organic acids, chlorophyll, most coloring matter, nearly all the acrid and bitter constituents of a plant, uncrystallized

amorphous vegetable sugars.
Alcohol refuses, however, to extract gums, mucilaginous substances, starch, albuminous materials, or many mineral compounds. Some sugars, proteins, gums, mucilaginous substances and albuminous bodies are, however, capable of being mixed with dilute alcohol (a mixture of 50 percent water, 50 percent ethyl alcohol). **Mucilages (like gums) are expressly (very) soluble in water (they are best extracted in cold water) and insoluble (does not dissolve) in alcohol. Extraction of mucilaginous constituents must be done with as low an alcohol content as possible."**
-The Herbal Medicine-Maker's Handbook: A Home Manual by James Green, Herbalist. New York, New York: Random House, Inc., **2000**, pages 85, 95.

"**Mix an appropriate amount of alcohol into an existing extract being 'disturbed' by these components, and they will be thrown back out of solution.** They can then be removed, leaving the extract looking far more attractive.
    To illustrate this banishment (removal) of an offending (unwanted) component *(in this instance slimy mucilage by alcohol) from an herbal solution,* **make a strong water infusion of Comfrey root,** *and when it is cool, pour some into a small container, add to it a substantial quantity of pure 190-proof ethyl alcohol, and therein you will behold the outcast. Add more water to this, and the mucilage will dissolve back into solution.*
Often, however, the mucilaginous component of an herb is exactly what one wants to draw into solution. This is true in preparations of our most nourishing plants such as Comfrey, marshmallow, slippery elm, and cinnamon.
But if one wishes to preserve an extract of these mucilaginous plants for any length of time be well advised to use some alcohol as either a part of the extracting menstruum or as a later addition. Eighteen to 20 percent ethyl alcohol will be adequate, and usually this is not a sufficient amount to cause an eviction (removal) of the above-mentioned components."
-The Herbal Medicine-Maker's Handbook: A Home Manual by James Green, Herbalist. New York, New York: Random House, Inc., **2000**, page 85.   (Menstruum is a substance that dissolves a solid or holds it in suspension, i.e., a solvent.)

"To beginning students of herbal medicine-making who are faced with the task and awesome responsibiity of **formulating a suitable water-alcohol base menstruum (solvent) for making an extract,** I boldly suggest: When in doubt, **use 'dilute alcohol' as a menstruum.** This is a very practical solution.
There is sufficient amount of both water and alcohol to adequately dissolve the components of any plant, and there is enough alcohol to preserve most any solution. One hundred-proof vodka is a dilute alcohol (approximately 50 percent absolute ethyl alcohol, 50 percent water) which is available at any liquor store for your use.
For our purposes, merely divide the proof number by 2 to determine the percentage of absolute ethyl alcohol by volume contained in the product. An 80-proof vodka or gin, for example, is 40 percent absolute ethyl alcohol, 60 percent other liquid (water); a 150-proof rum is 75 percent absolute ethyl alcohol, 25 percent other liquid (water)."
-The Herbal Medicine-Maker's Handbook: A Home Manual by James Green, Herbalist. New York, New York: Random House, Inc., **2000**, page 98.

"**Solvent ranges for alcohol-created herbal tinctures using dry herbs:**
Every herb has an ideal solvent range: the most optimal alcohol percentage in which the plant constituents will extract.
If you use a lower percentage of alcohol than the herbal extraction requires, you may not extract the herb's full medicinal properties. If you use a higher percentage of alcohol, it can destroy the herb's efficacy. It is important to use the appropriate alcohol solvent ranges for the herbs you want to extract.
    Comfrey Leaf, Symphytum officinale: 50 to 65% alcohol.
    Comfrey Root, Symphytum officinale: 24 to 50% alcohol. "
-The Herbal Apothecary: 100 Medicinal Herbs and How to Use Them by J.J. Pursell. Portland, Oregon: Timber Press, **2015**, page 222.

"**Solvents and Materials Extracted:**
Water: Some alkaloids, gums (limited in action, but does soften), sugars, proteins/enzymes, vitamins (some), tannins, glycosides, saponins, bitter compounds, starch (that which is miscible/mixable), polysaccharides (in hot water), pectins.
Vinegar: Bitter compounds, glycosides, sugars, tannins.
Aqueous ethanol (50-70% alcohol to water): Limited alkaloids, bitter compounds, enzymes, glucosides, salts, sugar,   some tannins, vitamins.
Glycerin: Bitter compounds, glucosides, dilute saponins, tannins.
Absolute alcohol (100%): Some alkaloids, balsam, fats, volatile oils, resins sugars, some tannins, vitamins, waxes.
Lipids (fats, oils): Camphors, volatile oils, fat soluble vitamins, wax (when heated)."
-'Herbal Solvent Percentage and Ratio' by Rosemary Gladstar, The Science and Art of Herbalism: An Outstanding Herbal Homestudy Course, https://scienceandartofherbalism.com, Waitsfield, Vermont, **2018**.

    <u>**Comfrey Extract Formulas**</u>

"**Fluid Extract Hoarhound Compound:**
    Hoarhound, in No. 20 powder, red root, elecampane, spikenard, **Comfrey,** wild cherry, bloodroot, each, in No. 30 powder, 2 1/3 ounces av., alcohol 3, water 1, a sufficient quantity. **Make a fluid extract as directed in class C (for this group of herbs).** Fluid Extracts, Class C drugs require a menstruum of two measures of alcohol to one measure of

water, for preparing their fluid extracts.

Symphytum officinale root extract alone:
Fluid Extracts, Class D (includes Symphytum officinale root) drugs require diluted alcohol as a menstruum for preparing their fluid extracts: equal measures of alcohol and water.
The Drug (Symphytum officinale root), in No. 30 powder, 1000 grams (2.2 pounds).
Diluted alcohol, sufficient to make 1000 Cc (or 46 fluid ounces = 1.3 liters).
Moisten the drug with from 350 to 450 Cc (or 24 to 30 fluid ounces = 0.70 to 0.88 liters) of diluted alcohol and macerate for 24 hours in a covered vessel in a warm place. Transfer to the water-bath percolator, pack firmly, pour upon it sufficient diluted alcohol to saturate and cover the drug, and set in a warm place for two days. Then heat moderately and after one hour begin to percolate slowly, adding diluted alcohol to the drug and continuing the heat and percolation until 850 Cc. (or 40 fluid ounces = 1.18 liters) have passed, which reserve. Turn off the heat and continue the percolation with diluted alcohol until the drug is exhausted. Distil the alcohol (1/2 the measure) from this last portion, evaporate the residue to a soft extract, which dissolve in the reserved portion and add enough diluted alcohol to make 1000 Cc. (or 46 fluid ounces = 1.36 liters) of the fluid extract.
The alcohol remaining in the drug after percolation may be recovered by distillation.

Symphytum officinale root powder:
A moderately coarse powder should pass through a sieve having 40 meshes to the linear inch. Equals No. 40 powder.
A coarse powder should pass through a sieve having 20 meshes to the linear inch. Equals No. 20 powder.
Symphytum officinale root, Comfrey is Class D: No. 30 powder.

Percolation *is a process of pharmacy in which a liquid called a Menstruum, is caused to pass or filter through a powdered drug contained in a percolator, for the purpose of displacing and obtaining the soluble constituents of the drug.*
When a menstruum is poured upon a powdered drug contained in a percolator, and allowed to remain for some time, it absorbs or dissolves whatever constituents of the drug are soluble in the menstruum employed."

-'Fenners Twentieth Century Formulary and International Dispensatory: Working Formulas for the Official and Unofficial Preparations of All Countries' by B. Fenner, Westfield, New York, **1904**, pages 62, 63, 86, 637, 639, 642, 648.

('av.oz' refers to the avoirdupois measurement system of weights that uses pounds and ounces. Ounce {oz} = 28.35 gram = 16 dr {dram}. It is the system of weights used in the United States.)
(Menstra or menstruum is a substance that dissolves a solid or holds it in suspension, i.e., a solvent.)

"Extract, fluid, of Horehound, Compound:
    Horehound av.oz. 8 1/2,   **Comfrey av.oz. 4 1/4,**   *Senega av.oz. 2 1/8*
    Wild cherry av.oz. 2 1/8,   *Alcohol, water, each sufficient*
Mix the drugs, reduce to moderately coarse powder, and extract in the usual manner for fluid extract, using a mixture of one volume of alcohol and 3 of water as the menstruum."

-'The New Standard Formulary: Comprising in Part One All Preparations Official or Included in the Pharmacopoeias, Dispensatories or Formularies of the World' by A. Emil Hiss, Ph.G. and Albert E. Ebert, Ph.M, Ph.D.; G.P. Engelhard and Company, Chicago, Illinois, **1910**, page 155.

<u>Comfrey Fermentation</u>

"There are ways to ferment medicinal herbs and preserve them whole. Pickling *works well with fruits and vegetables, but some cultures use this method to preserve leafy herbs like grape leaves and roots like ginger.*
Lacto-fermentation *can also be used to quickly ferment herbal water infusions in just a few days, producing a very low-alcohol, fizzy brew. All you need for lacto-fermentation is water, sweetener, whey (the liquid part of your organic yogurt), and maybe some salt and other seasonings. As with out that it has a very limited range of solvency, not mixing well with resins or volatile or fixed oils. I would also like to point out from the do-it-yourself perspective that it is a highly processed substance and there are much simpler and more readily-available options."*
-'An Herbal Medicine-Making Primer' by Simon the Simpler, Yggdrasil Distro, 24 pages, **2010**.
    (See book 'Wild Fermentation' by Sandor Ellix Katz. White River Junction, Vermont: Chelsea Green, 2003.)

<u>Comfrey Syrup</u>

"Syrupus Marrubii Compositus Compound: Syrup of Hoarhound:
Preparation: Take of the bark of red-root, roots of elecampane, spikenard, and Comfrey, *bark of wild cherry, and leaves and tops of hoarhound, each, 16 troy ounces; bloodroot, 8 troy ounces. Grind and mix the articles together.*
Make a syrup after the process directed for Compound Syrup of Aralia, using the same menstruum and the proportional amount of sugar to produce 24 pints (= 12 quarts = 11.3 liters) of syrup. Each pint (2 cups = 0.47 liters) will contain the virtues of 4 ounces of the ingredients.
    How to make syrup *(from Syrup of Aralia method):*
First grind and mix the articles together, moisten with diluted alcohol and place in a percolator, cover with the same menstruum, and macerate for 2 days. Then gradually add diluted alcohol until 2 pints of percolate have been obtained, which retain and set aside. Continue the percolation until the drug is exhausted and distill or evaporate the alcohol from

it. Mix the solutions, add 12 pounds of refined sugar and water enough to make 16 pints of syrup, using a gentle heat to effect solution of the sugar.
In the earlier Dispensatories, this article was called Syrupus Aralia Compositus (Compound Syrup of Spikenard), but in consequence of the great improvement in the formula, and from the fact that this name has been now bestowed upon another preparation, the name of the article under consideration has been changed to avoid confusion.
Action, Medical Uses, and Dosage: *This is an elegant remedy of obstinate (stubborn) coughs of long standing and pulmonary (lung) affections generally. It has been called 'Pulmonary Balsam,' but is superior to the preparation bearing this name in past years. It is often employed advantageously in pulmonary and bronchial difficulties, combined with 1/4 part of fluid extract of queen's root. The dose of the syrup is 1/2 fluid ounce (14.7 ml), 3 or 4 times a day."*
-'Kings American Dispensatory, Volume 2, 19th Edition' by Harvey Wickes Felter, M.D. and John Uri Lloyd, Phr.M., Ph.D., The Eclectic Medical Institute, Cincinnati, Ohio, **1905**, pages 1878, 1894.

(One troy ounce is equal to 31.1034768 grams, or 1.0971 ounces avoirdupois which is the most common definition of ounce in the United States.)

"**Syrup, Strengthening.** *The following is credited to Thomsonian practice :*
   Comfrey ounce 4,   Elecampane ounce 2,   Horehound ounce 1,   Beth root, powder ounce 1/2
   Brandy pint 1 (= 2 cups = 0.47 liters),   Sugar pound 1 (= 453 grams),   Water quarts 3 (= 2.8 liters)
Boil the first three drugs with the water down to 3 pints, and add the remaining ingredients."
-'The New Standard Formulary: Comprising in Part One All Preparations Official or Included in the Pharmacopoeias, Dispensatories or Formularies of the World' by A. Emil Hiss, Ph.G. and Albert E. Ebert, Ph.M, Ph.D.; G.P. Engelhard and Company, Chicago, Illinois, **1910**, page 438.

"**Syrup for Weakly Patients. Take 1 pound (453 grams) each of the roots of elecampane, spikenard and Comfrey** *and 1/2 pound (226 grams) of boxwood flowers, bruise well together in a mortar, boil with 2 gallons (7.5 liters) for 1 hour, strain. And add while hot 1/2 ounce (14 grams) of golden seal, 2 ounces (56 grams) of dyspepsia powders, 4 ounces (113 grams) each of prickly-ash seed, acacia and slippery elm bark, all in fine powder, 8 pounds (3.6 kg) of sugar, 1 gallon (3.7 liters) of good Holland gin and 1/2 gallon (1.8 liters) of Madeira wine."*
-'The New Standard Formulary: Comprising in Part One All Preparations Official or Included in the Pharmacopoeias, Dispensatories or Formularies of the World' by A. Emil Hiss, Ph.G. and Albert E. Ebert, Ph.M, Ph.D.; G.P. Engelhard and Company, Chicago, Illinois, **1910**, page 448.

"*Syrups are delicious and easy to make. As with medicinal plants, a good guideline is to work honey at lower temperatures whenever possible.* **To make a medicinal syrup,** *make an herbal decoction with the plants you are interested in using. Simmer your tea, allowing half of the water to evaporate so that you are left with a very strong, dark liquid.
Keeping it on low heat, mix in honey so that your concoction is one part honey, one part herb-infused water. Once the honey is fully dissolved, take your syrup off the heat and bottle it up. Don't worry if it seems too thin or watery, it'll thicken as it cools. If you aren't canning it, be sure to store it in the refrigerator or somewhere cool, or you'll soon have a bottle of mead."*
-'An Herbal Medicine-Making Primer' by Simon the Simpler, Yggdrasil Distro, 24 pages, **2010**.

(Mead is an alcoholic drink of fermented honey and water.)

Comfrey Lozenges or Cough Drops

"*A lozenge or cough drop is a hard herbal candy.*
How to Make: *Cook honey or molasses to the 'hard-crack' stage on a candy thermometer. Then add your dry herbs for about 10 minutes of simmering or steeping. It won't matter if the honey cools off somewhat when the herbs are added if it was already at the 'hard-crack' stage of heat. The reason you add the herbs second is that it takes honey or molasses perhaps half an hour or more to get hot enough, and this might overcook the herbs. Spread this hot mixture (or place in small blobs) onto a buttered cookie sheet to cool. When the mixture is partially cool, it will be easy to score it with lines to facilitate breaking it up later.*
Cough: *Clove is especially effective when combined with 1 or 2 other potent 'cough herbs' among the ten essentials, such as slippery elm,* **Comfrey,** *onion, peppermint or garlic.* **Combine 2 to 4 of these herbs into a cough syrup or cough lozenges."**
-10 Essential Herbs: Everyone's Handbook to Health by Lalitha Thomas. Chino Valley, Arizona: One World Press, **1995**.

"**Lozenges are small, solid, flattened cakes of very finely powdered therapeutic herbal substance, prepared from a mass made with a base of sugar for its pleasant flavor and a mucilage prepared form slippery elm bark, or Comfrey root, or marshmallow root. Mucilages are thick, viscid (sticky), adhesive liquids that are produced by extracting by cold infusion the mucilaginous principles of slippery elm, marshmallow, or Comfrey.**
Lozenges are placed on the tongue and allowed to dissolve slowly. They are especially useful when the herbal ingredients are intended to come into contact with the mucous surface of the mouth, throat, and upper respiratory tract. The remedial action of lozenges is generally designed to be local rather than systemic (entire body).
Preparing a mucilage using a mucilaginous herb:
*1. Macerate (soften/separate) overnight 6 parts (6 grams) dried herb in 100 parts (100 ml) of water.
2. Press it forcibly through cotton muslin."*
-The Herbal Medicine-Maker's Handbook: A Home Manual by James Green, Herbalist. New York, New York: Random House,

Inc., **2000**, page 302.
(Muslin is cotton fabric of plain weave made in different degrees of fineness, used for sheets and other purposes.)

Comfrey Bolus or Suppository

" 'Suppository' is defined as something that is 'placed underneath', designed to penetrate and dissolve in a body cavity other than the mouth.
A 'bolus' is a 'rounded mass of anything', a large pill or tablet. Today, for all practical purposes, boluses and suppositories are considered to be the same thing.
It is a single dosage preparation, intended primarily for insertion into the vaginal or rectal cavity for local or systemic (whole body) action. Enemas and vaginal douches are extensions of the bolus. Herbs can be absorbed from the lower regions of the rectum and then can enter the general circulation, thus becoming systemic. We know now that rectal insertion can be an excellent method for administering herbs to infants, young children, and adults on correct occasions.
Medicinal herbs used in boluses can be strong astringents such as white oak bark, used to contract, firm and strengthen tissues and to reduce secretions and discharge; **demulcent and mild astringents such as Comfrey root or marshmallow root used to relax, soothe, protect and heal tissues;** anti-microbial tonics like goldenseal, echinacea, and chaparral, used to prevent and allay (reduce) infection while toning body tissues."
-The Herbal Medicine-Maker's Handbook: A Home Manual by James Green, Herbalist. New York, New York: Random House, Inc., **2000**, page 220.

"**Suppositories: Herbal suppositories are easy to make. They are helpful in providing direct application of herbs exactly where they are needed** and in supplying herbal medicine to those who are unable to take nourishment through regular means. Use suppositories to target day-to-day problems that occur in the lower body, such as in the rectum, lower intestines, and vagina, or in the upper body, such as in the nasal cavity.
Make suppositories using either a warm process or cold process.
    The warm process involves melting oil, adding the herbs, and pouring the mixture directly into molds.
    The cold process involves kneading and hand mixing the herbs and oil together and rolling the mixture into appropriate-sized suppositories."
-The Herbal Apothecary: 100 Medicinal Herbs and How to Use Them by J.J. Pursell. Portland, Oregon: Timber Press, **2015**, page 213.

"**To prepare a suppository,** melt cocoa butter on low heat, preferably in a double boiler, and stir in finely powdered herbs. The last thing you want to do is insert rough, scratchy plant material in a sensitive area, so **make sure the herbs are very finely powdered.** Add essential oils, if desired, and pour the herb-infused cocoa butter into a suppository mold, and let it cool. Wrap each suppository individually with waxed paper, and store it in the refrigerator.
Insert rectally or vaginally depending on need (Remove wax paper first). Use suppositories within 7-10 days, or keep in the freezer up to 6 months.
To make a bolus, mix a binding herb like marshmallow or Comfrey root with other herbs to form a stiff paste. Shape the paste into pieces the size of large pills. Boluses must be used immediately.
Herbs for suppositories and boluses include Comfrey root, marshmallow, Oregon grape, and cranesbill (geranium)."
-The Modern Herbal Dispensatory: A Medicine-Making Guide by Thomas Easley and Steven Horne. Berkeley, California: North Atlantic Books, **2016**, page 111.

Comfrey Oil Infusion
See subsection 'Solubility, Stability, and Heat Sensitivity of Allantoin' in section 'Comfrey Heals: Allantoin' (Chapter 26).

1. Water extracts acids, **alkaolids (some), alkaloid salts (most),** coloring matter, essential oils (some), glycosides, gums, mineral salts, mucilage, pectin, proteins, sugars, tannins.
2. Alcohol extracts acids (organic), acrid constituents, **alkaloids,** bitter constituents, chorophyll, essential oils, glycosides, resins. Alcohol does not extract minerals, gums, or mucilage.
3. Infused oil extracts resins, essential oils, and flavonoids.
-Organic Growers School, Asheville, North Carolina, www.organicgrowersschool.org; Training, workshops, conferences, **2018**.

Infusion is extracting chemical compounds from plants in a solvent such as water, oil or alcohol, by suspending it in the solvent over time. An infusion is also the name for the liquid that is created.
Herbal oils are plant extracts where the solvent is always vegetable oil with olive oil being the most common.

Cool infusion method: store in cool dark place.
Warm infusion method: put on low heat on stove, or place in sun.

If using fresh Comfrey leaves, wilt them first for 12-24 hours because too much water in the oil can cause it to go bad.
After putting cut Comfrey leaves in jar (push down leaves with medium pressure) and pouring in oil, use a knife or stick to

make sure the oil has reached all parts of the leaves.
In the below methods the time for infusing range from 3 days to 1 month. The longer you leave the infusion before straining, the stronger the infused oil. **The oil can be strained** with gauze, cheesecloth, loose-weaved cloth, metal strainer, bamboo strainer, or similar tool. If using cloth, a funnel is helpful.

*"**Herbal oils (real phytols or infused oils) are coming into fashion with certain aromatherapists.**
These include arnica (Arnica montana), calendula (Calendula officinalis, pot-marigold), centella (Centella asiatica, gotu kola or hydrocotyle), **Comfrey (Symphytum officinalis),** Devil's claw (Harpagophytum procumbens), echinacea (Echinacea purpurea), fenugreek (Trigonella foenumgraecum), lime blossom (Tilia sp.), meadowsweet (Filipendula ulmaria) and St John's wort (Hypericum perforatum). **Most are tea-like or alcoholic extracts, not essential oils, and have no aroma.**
All are well known as herbal remedies, usually taken internally or applied to burns or bruises, as poultices or compresses, but **many are potentially toxic orally and are sensitisers and should be given only at the advice of a qualified herbalist – not an aromatherapist.***
> *There is no toxicological evaluation for their aromatherapeutic application and their possible dermal irritation or sensitisation is often unknown, therefore their use should be restricted in pregnancy."*

-Aromatherapy Science: A Guide for Healthcare Professionals by Maria Lis-Balchin, BSc, PhD. London, England: Pharmaceutical Press, **2006**. (Chapter 7: 'The Safety Issue in Aromatherapy'.)
> (Phytol is a liquid alcohol found in cosmetics, shampoo, toilet soap, household cleaners and detergent.)
> (A chemical sensitizer causes allergic reactions after exposure.)

*"**Comfrey-infused olive oil is used to create healing salves. Before infusing leaves in oil, they must be fully dried.
Wet leaves will cause oils to become rancid,** and capped jars may burst under pressure. Hang leaves to dry in a shaded spot with good air circulation."*
-'Culturally and Economically Important Nontimber Forest Products of Northern Maine: Plant Profiles: Comfrey' by United States Department of Agriculture: Forest Service, Sustaining Forests, Northern Research Station, Madison, Wisconsin, May **2010**.

*"**Comfrey Root and/or Leaf Oil Infusion: Put 4 ounces (113 grams) of herb (finely chopped or ground) in a clean, dry quart (0.94 liter) jar and cover with oil of choice.** Oil level should be 1-2 inches (2.5-5.0 cm) above herb.
Method 1: Store in a cool dark place for 1 month, shaking mixture daily.
Method 2: Fill saucepan halfway with water, place on low heat and put jar in the water, let set for 3 days.
Method 3: Place jar of herbal oil in direct sunlight for 10 days.
Next: After the allotted time, strain herbs from oil with cloth or gauze and store in a clean, dry container. Comfrey infused oil or fine powder can be added to a suppository blend.
Comfrey infused oil can be used directly on skin or used to make a salve and mixed with other herbs if desired."*
-'Comfrey, Symphytum Officinale, Boraginaceae Family' by Sarah Heany, Highland Midwife Birth Services LLC (TM), Yakima, Washington, November **2011**.

*"**Herbal oils are often used as the base of other herbal creations such as salves, creams, scrubs and lotions.
Folk Method of Oil Infusion: Fill a mason jar halfway with dried, coarsely ground herb(s).** Add oil to reach the top of the jar and close the lid tightly. Place the jar in a paper bag and set it in a warm place.
Shake the jar several times per day for 1 or 2 weeks. Then strain the oil into another jar, secure the lid, and let it sit for a few more days. Strain the oil again with fine cheesecloth. Pour the herbal oil into a clean, dry jar, and store it in a cool, dark place."*
-The Herbal Apothecary: 100 Medicinal Herbs and How to Use Them by J.J. Pursell. Portland, Oregon: Timber Press, **2015**, page 199.

*"Graeme Little farms around 2500 Bocking #14 Comfrey plants on his 20 hectare French Island property in Victoria, Australia. **Once the Comfrey leaves are harvested, Graeme makes the oil, firstly grinding the root in a food processor, then adding olive oil, although he has experimented with almond and coconut oils.** Then the mixture is cooked using a double pan method over hot water for around 3 hours, stirring every 15 minutes. The mixture is then put through a juicer to leave a clear oil."*
-'The Comfrey Man' by Roger Clark, The Waterline News, www.waterlinenews.com.au, Grantville and Districts, Victoria, Australia, Volume 1, No. 12, August **2015**.

*"Basic formula for dry herb infused oils:
Comfrey root oil: 100 grams (3.5 ounces) dried and coarsely **ground Comfrey roots.** 500 ml (16.9 US fluid ounces) olive oil. The Comfrey oil is best used for speeding healing of broken bones, shallow wounds, scratches, diaper rash, abraded (scraped), or wind-burned skin, etc."*
-Making Plant Medicine by Richo Cech. Williams, Oregon: Herbal Reads, **2016**, page 85.

*"**How to Make Your Own Comfrey Oil Infusion:** If possible, use freshly dried herbs for this purpose. To get fresh dry Comfrey root: Dig the root when it is dry weather. Clean by hand or use some water and a vegetable brush. Brush the root gently. Chop finely; lay out on a paper bag overnight. To get freshly dry leaves: Harvest, wipe the dirt off with a towel and allow to dry whole overnight. The Comfrey roots should be broken down by chopping. Break up the Comfrey leaves by hand.
Method with Low Heat:*
> ***Create an herbal oil infusion by infusing 2 cups (0.47 liters) of cut Comfrey leaves in 4 cups (0.94 liters) of olive***

oil with a steady low heat (110 F degrees = 43 C) for two to three weeks. *Strain and pour into a clean, dry bottle.*
Method by Cold Infusion:
Ingredients: 8 ounces (236 ml) Comfrey leaf (70 percent), 4 ounces (118 ml) Comfrey root (30 percent), *extra virgin olive oil, to cover, approximately 16 ounces (473 ml). From 'Wildly Natural Skin Care'.*
Put all the herbs in a 16-ounce (473 ml) glass jar, cover with olive oil and cap and shake. **This can steep for 28 days.**
To strain, use a clean old shirt lined in a strainer, pour the mix through into a bowl and squeeze the shirt with herbs in it. The strained liquid is your Comfrey oil."
-Dr. Joseph Mercola, Chicago College of Osteopathic Medicine, Board Certified; New York Times bestselling author such as 'Fat for Fuel'; based in Illionis, www.mercola.com, **2018**.

Comfrey Ointment, Cream or Salve

"Comfrey ointment is used for skin conditions against which Allantoin, the healing principle in Comfrey appears to be most effective. The ointment sold by the H.D.R.A. (Henry Doubleday Research Association) which contains 1% of Allantoin."
-Comfrey: The Herbal Healer by Henry Doubleday Research Association, **1975**, 41 pages, page 3. (in 'Comfrey Report' book)

"Comfrey ointments or other preparation for external use only should contain up to 20% of the dried herb or equivalent amounts of extract."
-Rational Phytotherapy: A Physician's Guide to Herbal Medicine by V. Schulz, R. Hansel and V.E. Tyler. (Third edition, first English edition, translated by T.C. Telger.) Heidelberg, Germany: Springer Verlag, **1998**.

*"**Comfrey Salve:** To make a very simple salve,* **cut a few young Comfrey leaves, dig and clean some Comfrey roots, and gather some of the flowers.** *Chop all these up coarsely until you have about two cups and put this in an oven-proof dish with a cover. Add two cups of oil (olive, almond, grapeseed-it's your choice). Add two ounces of bee's wax. Cover and bake for about 40 minutes at 350 F degrees (177 C). Everything will melt down and mush together. Stir, then strain the mix through a fine sieve. Pour salve in small jars, cover tightly, and let cool.* **This pale green salve is as soothing and healing as the plant is prickly and itchy- another of Mother Nature's little mysteries in the herbal world.** *-Tanya Jackson, well known local herbalist."*
-'Comfrey Salve' by Tanya Jackson, The Plantsman: New Hampshire Plant Growers Association, edited by Robert Parker, University of New Hampshire Research Greenhouses, Durham, New Hampshire, page 28, June-July **1998**.

*"**The difference between an infused oil and ointment, salve or balm** is simply the addition of a solidifying agent, such as beeswax or cocoa butter. Making the preparation more solid allows you to apply it to skin more easily and keep it in place longer.* **Ointments** *have only a small amount of beeswax or cocoa butter added; they are almost the consistency of pudding and are easily spread on the skin.* **A salve** *is made firmer by the addition of more beeswax and it 'seals' better; it protects the skin from drying out or having excess moisture enter.* **Balms** *have even more beeswax and are firmer than salves; they are usually used on the lips or temples.* **Wonderful herbs for salves, ointments and balms: Comfrey salve will prevent scarring."**
-Growing 101 Herbs that Heal: Gardening Techniques, Recipes and Remedies by Tammi Hartung, Medical Herbalist. North Adams, Massachusetts: Storey Publishing, **2000**.

"The oil from Comfrey leaves and roots can be incorporated in creams and ointments or used in a compress. **Ointments and other external preparations are typically made with 5% to 20% Comfrey.**
Comfrey should be applied topically on unbroken skin for less than 10 days or a maximum of 6 weeks per year in amounts at or below a daily dosage of 100 mcg of the unsaturated PAs."
-Professional's Handbook of Complementary and Alternative Medicines, 3rd Edition' by Charles H. Fetrow, Pharmd (Doctor of Pharmacy) & Juan R. Avila, Pharmd. Pennsylvania: Lippincott Williams & Wilkins (imprint Wolters Kluwer), **2003**, page 235-238.

"How to Make Salves:
Step 1. *Prepare an infused oil. Strain.*
Step 2. *To each cup (236 ml) of herbal oil, add 1/4 cup (59 ml) of beeswax. Heat until the beeswax is completely melted. To check for consistency, place 1 tablespoon of the mixture in the freezer for just a minute or two. If it's too soft, add more beeswax; if too hard, add more oil.*
Step 3. *Remove from heat immediately and pour into small glass jars or tins. Store any extra salve in a cool, dark place. Stored properly, salves will last for months, even years."*
-Rosemary Gladstar's Herbal Recipes for Vibrant Health: 175 Teas, Tonics, Oils, Salves, Tinctures and Other Natural Remedies by Rosemary Gladstar. North Adams, Massachusetts: Storey Publishing, **2008**, page 384.
(For how to make infused oil, see subsection 'Comfrey Oil Infusion' in this section.)

Ointment or Cream from Comfrey Root

*"**A Comfrey ointment may be prepared by slowly simmering a tablespoonful of the dried root in a cup (236 ml) of unsalted lard, lanolin or suet for 10 minutes and straining.** Allow to cool before refrigerating. A few drops of tincture of Benzoin Compound may be incorporated into the cooled ointment."*

-Comfrey: What You Need to Know by Ben Charles Harris. New Canaan, Connecticut: Keats Publishing, Inc., **1982**, page 56.
(Compound Benzoin Tincture is a solution of benzoin resin in ethanol. It is applied to skin under an adhesive bandage. It protects skin from allergy to the adhesive and makes the bandage stay on longer.)

"*Comfrey Cream: To make an ointment to use externally, take 1 cup of finely cut Comfrey root and simmer in 1 cup of olive oil until it starts to soften. Cool and strain. Add 50 grams (1.7 ounces) of beeswax.*
Jasmine or orange blossoms may be added to the simmering mixture to give the cream a pleasant smell. The cream is used to relieve pain and aid healing of cuts, bites, sprains, arthritis, dry vaginal conditions, inflammation and neuralgia."
-How Can I Use Herbs in My Daily Life: Over 500 Herbs, Spices and Edible Plants: An Australian Practical Guide to Growing Culinary and Medicinal Herbs by Isabell Shipard. New York, New York: David Stewart Publisher / Simon & Schuster, Inc., **2003**.

"**Herbal Cream with Extract of Comfrey Root: Comfrey Gel:** *The root of Comfrey Symphyti radix (root) represents a very important medicinal herb raw material.* **Gel with 10% propylene glycol extract of Comfrey root (Symphyti Extractum fluidum 1: 7)** *containing mucus, tannins, saponosides, and allantoin improves epithelialization, drainage, tissue regeneration.* **The production process of Comfrey gel is performed in the following stages:**
    1. Purchase of high-quality herbal material Symphyti radix and quality control of parameters, as prescribed by Pharmacopeia selection of high-quality pharmaceutical raw materials according to manufacturer attest (certify formally).
    2. Grinding of Comfrey root and producing the liquid extract with propylene glycol with a method of percolation. The resulting extract is controlled to authentication, relative density, and microbiological safety.
    3. Liquid extract of Comfrey, after the processing procedure is incorporated into the semisolid base, and then filled into tubes. The final product, herbal gel from the roots of Comfrey, is sent to the control to laboratories.
    4. When a controlled product receives confirmation that corresponds to the standard quality, it is dispatched to the warehouse of final products and further distributed."
-Aromatic and Medicinal Plants: Back to Nature edited by Hany El-Shemy. London, England: IntechOpen Limited, **2017**. Chapter 16: From Medicinal Plant Raw Material to Herbal Remedies by Sofija M. Djordjevic.
(Propylene glycol is a synthetic organic compound. It is a sticky, colorless, almost odorless liquid with a slightly sweet taste. It is miscible with a broad range of solvents such as water, acetone, and chloroform. Miscible means it is uniform/ homogeneous when mixed together.)   (Extractum fluidum = latin for 'liquid extract'.)

"**The original Carpathian herbal remedies were either substituted by knowledge gathered from classical medicinal monographs or were identical to them.**
Medieval Celtic physicians (Scottish, Irish, and Welsh) often received education at the University of Padua, Italy, similar to their Ruthenian and Polish counterparts (Scottish Historical review, 1906). **Subsequent handwritten manuscripts** and surveys reported both smaller numbers of total medicinal plants, as well as those used for wound healing.
Comfrey root (Symphytum officinale) was used on multiple occasions to heal wounds associated with bone fractures, minced in animal fat. The type of fat used in this ointment varied depending on the location:
    in Silesian Beskids, pork fat was preffered,
    in Zywiec Beskids, goose lard was believed to possess the best curative properties,
    while in Sacz Beskids they were both considered as effective."
-'Botanical Provenance of Traditional Medicines From Carpathian Mountains at the Ukrainian-Polish Border' by Weronika Kozlowska, Charles Wagner, Erin M. Moore, Adam Matkowski and Slavko Komarnytsky from Wroclaw, Poland and North Carolina; Frontiers in Pharmacology, Lausanne, Switzerland, Volume 9, pages 295-311, April **2018**.
(The Carpathian Mountains or Carpathians form an arc across central and eastern Europe. It includes parts of Austria, Czech Republic, Hungary, Poland, Serbia, Slovakia, Ukraine and Romania.)
(Ruthenians and Ruthenes are Latin names formerly used in Western Europe for the ancestors of modern East Slavic peoples, especially Rus' people with a Ruthenian Greek Catholic religious background and Orthodox believers who lived outside Rus'.)
(Beskids or Beskid Mountains are mountain ranges in the Carpathians, going from Czech Republic in the west along the border of Poland with Slovakia up to Ukraine in the east.)

Ointment, Cream or Salve from Comfrey Leaves

"**Comfrey Salve or cream: Studies show that a 10% active ingredient cream, made from 25 grams (0.88 ounce) of fresh herb per 100 grams (3.5 ounce) of cream is very effective for sprains.** Commercial formulations are available. Look for a 0.4% allantoin solution. Homemade salves are possible, but the concentration of allantoin in them is more unpredictable. **How much do you use? A 5-20% ointment can be used several times per day.** Commission E recommends it not be used more than four to six weeks per year. Some herbalists suggest less than that, recommending three weeks at the most."
-Western Herbs for Martial Artists and Contact Athletes: Effective Treatments for Common Sports Injuries by Susan Lynn Peterson, Ph.D. Wolfeboro, New Hampshire: YMAA Publication Center, **2010**.

"**Herbal Healing Salve**: *If you are allergic to any of these ingredients, don't use them.*
2 ounces (56 grams) dried Comfrey leaves, 1 ounce (28 grams) dried calendula flowers, 2 cups (473 ml) olive oil, 1 ounce pure beeswax, 4 drops tea tree oil, 4 drops lavender oil, 1 ounce echinacea tea, 1 ounce goldenseal tea. **Heat the Comfrey leaves and calendula flowers in the olive oil over low heat for about 5 hours.** *Do not let the mixture bubble. If you can't*

lower the heat enough to keep from bubbling, turn it off and on. Strain off the olive oil while it is still warm. Put 1 1/4 cups (295 ml) of the olive oil in a pan; add the beeswax and heat it just enough to melt. Add the essential oils and stir. Pour in jar and store at room temperature. **Use for minor scrapes, burns, and other ouches.**
If you aren't growing Comfrey in your garden, give it a try. I have found it to be a wonderful herb for soothing scrapes and scratches. Just tear off a leaf and wrap the injured spot, and it will feel better in no time."
-Recipes From and For the Garden: How to Use and Enjoy Your Bountiful Harvest by Judy Barrett. Texas A&M University Press, **2014**, page 85.

"*Salve Master Recipe:* Since the only thing that differentiates one salve from another is the herb used, I use this master recipe as a base. **Materials:**
    yellow beeswax 40 gram (1.4 ounce) = 20%,   liquid oil, infused with herb 158 gram (5.5 ounce) = 78.5%
    Vitamin E oil 2 gram (0.07 ounce) = 1%,   essential oil blend 1 gram (0.03 ounce) = 0.5%
    double boiler, digital scale, spoon or stir stick, 4 tins, 2 ounce (56 gram) each
Yield: Makes enough to fill approximately four 2 ounce (56 gram) tins.
1. Heat the beeswax and infused oil in a double boiler until the beeswax is completely melted. Take care not to get the oil too hot, so that there is no damage to the oil.
2. Remove from heat, and add Vitamin E oil and essential oils, if desired.
3. Stir to mix well and pour into waiting containers. If using plastic jars, make sure the mixture isn't too hot before pouring into jars, as they can deform with too much heat.
All-Purpose Salve: This is my favorite salve. I call it my basic first-aid salve. It works really well at soothing and healing minor cuts and scrapes. **Simply follow the directions for the 'Salve Master Recipe' using infused liquid oil with the following ingredients.**
    Herbs: 1 part calendula blossoms, 1 part lavender blossoms, **1 part Comfrey leaf,** 1 part plantain leaf,
             1 part chickweed leaf.
    Oils: equal parts olive and grape seed.   *Essential Oils:* lavender or lavandin."
-'Beeswax Alchemy: How to Make Your Own Candles, Soap, Balms, Salves and Home Decor from the Hive' by Petra Ahnert. Beverly, Massachusetts: Quarry Books, **2015**.

"*Soothing Comfrey Cream:*
2 cups (473 ml) dried crushed Comfrey leaves, 2 cups organic olive oil, 1/2 cup (118 ml) beeswax pastilles, 1 ounce (28 grams) organic emulsifying wax (or another ounce of beeswax), 2 ounces (56 grams) shea butter or lanolin, 2 of 1000 mg vitamin E capsules, 5-6 drops essential oil of choice.
Loosely pack a 1 quart (0.94 liter) mason jar with Comfrey leaves. Fill with the olive oil, or enough to cover the Comfrey leaves, and allow to infuse for at least 30 days. Strain Comfrey leaves, yielding approximately 1 1/2 cups (354 ml) of infused oil. Place the 1 1/2 cup of infused oil in the top of double boiler which has been brought to a boil. Turn down heat to low simmer. Slowly add beeswax pastilles and emulsifying wax. I use an organic beeswax-emulsifying wax combination about 2/3 cup (157 ml). Blend in shea butter or lanolin until melted. Lanolin will produce a slightly greasier formula which can be easier to apply. Puncture vitamin E capsules and add oil to the mixture. Vitamin E is soothing to skin and is a natural antioxidant which prevents against oxidation and rancidity.
Add 5-6 drops of essential oil dependent on the use. Your choice, but my favorites are lavender to soothe tension and/or lemongrass to aid in healing ligament and muscle tears. Wintergreen is useful for sore muscles and chamomille aids in the reduction of swelling and bruising. Pour the melted mixture into containers with lids. Allow to solidify and cool before capping with lids. Label and date."
-'Comfrey Uses and Remedies' by Rebecca O'Bea, Kansas, The Herbal Academy of New England, an online school of herbalism, February **2015**.
    (A pastille is a type of sweet or medicinal pill made of thick, solidified liquid that is meant to be eaten by light chewing.)

"*Basic Healing Salve:* This is a basic salve for soothing minor irritations and injuries and promoting more rapid regeneration of tissue.
2 parts Comfrey leaf (key herb), 1 part calendula flower and/or yarrow flowering herb (supporting herb), 1 part plantain (supporting herb), 1 part myrrh (optional supporting herb for a more antiseptic effect). **Extract the herbs in olive oil, almond oil, or another quality fixed oil for 12-24 hours.** Strain. Add 1 ounce of beeswax for each 8 ounces of oil."
-The Modern Herbal Dispensatory: A Medicine-Making Guide by Thomas Easley and Steven Horne. Berkeley, California: North Atlantic Books, **2016**, page 162.
    (A fixed oil is a natural vegetable or animal oil that is nonvolatile such as avacado oil, coconut oil, lard, linseed oil, olive oil, palm oil, etc. It is a fatty oil.)

Comfrey Compress or Fomentation

Do not use Comfrey externally on deep wounds. See subsection 'Wounds' in section 'Personal/Clinical Observations of Healing' (Chapter 25).

The words 'compress', 'fomentation' and 'poultice' are sometimes used interchangeably. Definitions vary.

A compress is a cloth applied with pressure to skin and held in place for a period of time. The cloth is soaked with a medicinal liquid. It can be any temperature. If it is applied warm, then it is allowed to cool.

A fomentation is the same as a compress except it is always applied warm or hot, and it is kept warm/hot.

"**Compresses and poultices are applied topically to an injury or wound.** They should be prepared fresh each time they are needed. The herbs used for the compress or poultice will be determined by the condition being addressed.
To prepare a compress: Make a decoction or infusion with the selected herb. The herb and condition being treated will determine whether a cool or hot compress is necessary. Soak a clean cotton cloth in the strained liquid and gently squeeze out the excess. Apply the cloth over the affected area. Reapply fresh compresses as recommended, depending on the herbs used.
Soothing herbs for compresses and poultices: Comfrey for sports injury healing."
-Growing 101 Herbs that Heal: Gardening Techniques, Recipes and Remedies by Tammi Hartung, Medical Herbalist. North Adams, Massachusetts: Storey Publishing, **2000**.

"Fomentation is an externally applied application that helps to alleviate (reduce) pain, infection, swelling and inflammation. **They are strong infused herbal formulas made with water or vinegar. After saturating a cloth with the fomentation, you apply it externally where needed.** Depending on the problem, I typically recommend using it 10 to 30 minutes at a time. I've found it helpful in some cases to apply gentle heat over the saturated cloth."
-The Herbal Apothecary: 100 Medicinal Herbs and How to Use Them by J.J. Pursell. Portland, Oregon: Timber Press, **2015**, p. 204.

Comfrey Poultices

Do not use Comfrey externally on deep wounds. For more information about it, see subsection 'Wounds' in section 'Personal/Clinical Observations of Healing' (Chapter 25).

A poultice or cataplasm is a soft, moist, flexible mass, usually heated, put on skin either directly or with a cloth or plastic around it. The moist mass is usually an herbal medicine. It relieves pain/inflammation, stimulates circulation, draw out toxins and provides other forms of healing.

"**Comfrey, Ragwort, and Wood Sage Poultice:** Use equal parts of these three herbs and steep in boiling water. Apply poultice to external cancers and tumors. It is most beneficial and will give excellent results."
-Back to Eden: The Classic Guide to Herbal Medicine, Natural Foods and Home Remedies by Jethro Kloss. Wisconsin: Lotus Press, **1939**-1999, page 75.

"Mrs. J.A. of Bristol (southwest England): 'I have also two or three times cured bad sprains and bone bruises with Comfrey poultices. **Leaves sometimes give a rash whereas roots are more comforting and more potent, though much more messy to prepare.**"
-Comfrey: The Herbal Healer by Henry Doubleday Research Association, **1975**, 41 pages, page 23. (in 'Comfrey Report' book)

"**Comfrey (Symphytum officinale) is our most valuable plant ally for repairing wounds, while at the same time it soothes and softens tissue. It is probably the most healing mucilaginous remedy in our herbal materia medica,** having been used for centuries to treat external ulceration and all types of lesions and injuries ranging from small cuts and abrasions to large wounds and broken bones.
A Comfrey poultice quickens the repair of the normally slow healing process of torn cartilage, tendons, and ligaments. A common name 'knitbone' refers to Comfrey's historic use as a poultice for treating skeletal fractures in humans and animals. It heals bone tissue.
Its mucilage and tannins produce an astringent and contracting effect. By drawing a wound together at the surface, it reduces the need for stitching; and its generous allantoin content stimulates the regeneration of skin tissue, making the formation of scar tissue less likely."
-The Herbal Medicine-Maker's Handbook: A Home Manual by James Green, Herbalist. New York, New York: Random House, Inc., **2000**, page 280.
    (Materia medica {medical material} is Latin for the collected knowledge about the therapeutic properties of substances used for healing. The term is derived from the book by 1st century AD Greek physician Pedanius Dioscorides.)
    (A lesion is abnormal damage or change in body tissue caused by disease or trauma. An abrasion is superficial damage to the skin.)
    (A tendon is a band of tissue that connects muscle to bone. A ligament is an elastic band that connects bone to bone. Cartilage is soft, gel-like padding between bones.)

"**Fresh Comfrey Poultice:
Dig up Comfrey roots and scrub them well. Cut them into shorter lengths and put them in a blender with an equal amount of fresh Comfrey leaf.** Add just enough water to make it blendable, and blend until you have a gooey mess. Spread this onto a piece of gauze and apply to the body part affected, covering with a piece of muslin (cloth) or clingfilm (plastic wrap).

The gauze makes the poultice easier to remove. Replace poultice daily until healed.
The leaves can also be used on their own as a poultice, but the hairs on them can irritate the skin. *To avoid this, blend the leaves with a little water or pound them in a mortar and pestle, and then sandwich the leaf mush between two pieces of muslin before applying to the skin. The muslin will protect the skin from the irritating effects of the leaf hairs, and allow the juices to seep through."*
-Backyard Medicine: Harvest and Make Your Own Herbal Remedies by Julie Bruton-Seal and Matthew Seal. New York, New York: Skyhorse Publishing, **2009**. First published in Great Britain by Merlin Unwin Books, 2008 as 'Hedgerow Medicine'.
  (Muslin is cotton fabric of plain weave made in different degrees of fineness, used for sheets and other purposes.)

*"**Procedure for making Comfrey poultice from fresh whole plant:**
1. **Dig fresh Comfrey roots** and very thoroughly scrub them with a scrub brush to remove the black, slimy cortex (outer layer).
2. **Chop up the roots with an equal portion of fresh, green Comfrey leaf** and combine these in a blender, using only sufficient water to cause the mucilaginous goo to vortex (to mix).
3. Spread the fresh paste directly on the injured area and cover with a clean cloth."*
-Making Plant Medicine by Richo Cech. Williams, Oregon: Herbal Reads, **2016**, page 97.

*"**A plaster or poultice is a mixture of dried or fresh herbs moistened with water or oil and applied externally. It is similar to a compress, but plant parts are used rather than a liquid extraction.** Generally, a poultice implies a hot treatment, whereas a plaster is applied at room temperature. Poultices should be applied as hot as can be tolerated.
Smash or chop fresh herbs and apply them directly to the affected area.*
If you're using dried herbs, *boil them for up to 5 minutes or mix them with a small amount of boiling water.*
Apply the poultice or plaster in between layers of gauze to prevent messes. Use a gauze bandage covered with a warm cloth over the poultice.
Good herbs for making plasters and poultices include slippery elm, Comfrey leaf and root, *marshmallow, flaxseed, white oak bark, psyllium seeds, plantain, lily of the valley, pine gum, lobelia, calendula, yarrow, goldenseal, and aloe vera."*
-The Modern Herbal Dispensatory: A Medicine-Making Guide by Thomas Easley and Steven Horne. Berkeley, California: North Atlantic Books, **2016**, page 110.

Comfrey Flour Poultice

*"**Poultice: Plain white flour: 1 part. Comfrey flour: 2 parts.***
Mix the white flour with a little water (cold) in the dish, and add sufficient boiling water to make a runny consistency. Add the Comfrey flour, mix well, and add more boiling water as necessary to make a smooth mixture. This makes a smooth, easily spread mixture which is however, a poor heat retainer, and requires a very rapid technique.
It would probably be serviceable in dressing chronic lesions such as varicose ulcers, but Comfrey alone would be better."
-Comfrey: The Herbal Healer by Henry Doubleday Research Association, **1975**, 41 pages, page 19. (in 'Comfrey Report' book)

*"**Comfrey Flour Poutices:***
Equal parts of Comfrey flour and starch. *Mix the starch to a smooth paste with cold water. Add just sufficient boiling water to 'turn' the starch making a runny paste. Mix in the Comfrey flour, allowing it to take up surplus water to make a smooth paste. The consistency of this mixture is such that it may be moulded and applied direct to the part without the use of an intervening layer of material.*
It will lift on and off the skin easily and cleanly, and, if the technique is good, will retain the heat for some considerable time, provided sufficient of the constituents are used to allow the uptake of a considerable amount of boiling water, and that no time is lost in transit and in bandaging the preparation the affected part."
-Comfrey: Fodder, Food and Remedy by Lawrence D. Hills. New York: Rizzoli Universe Books: **1976**, page 86.

*"**Comfrey Poultice: Another way is to add to Comfrey flour in a dish sufficient boiling water to make a stiff paste.** Spread on lint or several layers of gauze. Cover top with a double layer of gauze.*
Apply to the part as hot as can be borne, the side with the double layer of gauze separating the poultice from the skin, so that the Comfrey makes contact through it.
A heated metal dish should be used to prepare the poultice, which should be carried to the patient between two heated metal plates. **This is a precaution against infection, for though Comfrey can inhibit bacterial growth, it is not a disinfectant.**"
-Comfrey: Fodder, Food and Remedy by Lawrence D. Hills. New York: Rizzoli Universe Books: **1976**, pages 179, 180.

Chopped Comfrey Leaf Poultice

"This column will be devoted largely to ways and means of preparing Comfrey leaves or roots as an internal medicine, or in the form of a poultice or fomentation in cases involving open wounds or ulcers, burns, insect bites, etc.
In cases of obstinate (slow healing) ulcers, gangrene, tumors, burns, open wounds, skin cancer, or inflammation caused by insect bites, the Comfrey leaves can be prepared for the poultice by putting them through a juicer.
However, as the Comfrey leaves contain no juice but a thick mucilaginous substance, like okra, the macerated (softened/

separated) leaves are gathered from the basket of the juicer following the operation, and not from the spout. The mass of triturated (pulverized) Comfrey leaves can then be spread on a cloth and applied to the infected area.
If no electric juicer is available, one can prepare the Comfrey leaves in a mortar with a pestle, similar to that used by druggists in compounding various medicinal substances.
Some of my friends have **prepared the Comfrey leaves for the poultice by placing them on a board, using a hammer to macerate them.** In making a large poultice, I would suggest that from 10 to 12 medium size leaves be prepared as indicated above."
-'Comfrey' by H.E. Kirschner, M.D., Let's Live Magazine, California, 8 pages, October/November/December **1958**.
>(Fomentation is the application of a hot moist substance to the outside of the body.)
>(Spices/herbs or other substances are put in a smooth, round bowl called a mortar. A pestle is a heavy, blunt tool used to finely grind these substances. Mortar and pestle have been used throughout history by pharmacists and herbalists.)

"**The Comfrey herb is also used in poultices to help heal swellings, inflammations, cuts and sores. To make such a dressing, let the leaves mush up in hot water,** squeeze out the excess liquid, and wrap several handfuls of the hot, softened foliage in a clean cloth. Apply the pad to the affected part- comfortably hot, but not scalding- and cover the area with a thick folded towel to keep the heat in. The moist warmth enhances the healing effect of the allantoin."
-'Comfrey for the Homestead' by Nancy Bubel, Mother Earth News, Kansas, May/June **1974**.

"***Fresh Comfrey Leaf Poultices:***
These are the traditional treatment for sores which fail to heal which were far more common in the pre-antiseptic past. *Allantoin made it one of the best weapons that herbalists had in their battles against injuries and disease. Its use is probably as old as using dock leaves against nettle stings.*"
-Comfrey: The Herbal Healer by Henry Doubleday Research Association, **1975**, 41 pages, page 35. (in 'Comfrey Report' book)

"*Mrs. P. (89 years old in 1966) of the 'Soil Association of South Africa' successful treatment of varicose ulcers using Comfrey poultices and Comfrey tea.* **Comfrey Symphytum leaf poultices were applied every twenty-four hours. The leaves were chopped up, a little boiling water added, a thin gauze was placed over the ulcer, and the poultice applied direct.** *Comfrey leaves were also infused by chopping up several leaves and adding boiling water and allowed to stand. A strong 8 ounce (236 ml) drink of this was given three times a day.*"
-Comfrey: The Herbal Healer by Henry Doubleday Research Association, **1975**, 41 pages, page 15. (in 'Comfrey Report' book)

"***A Comfrey poultice may be prepared by blending several large leaves, previously washed clean. The resulting okra-like, mucilaginous mass is placed on several thicknesses of cloth*** *(white shirt material will do) and applied to the area. No juicer? Macerate the leaves on a hand grater, or use a vegetable or meat grinder.*
Or place several layers on a board and beat with a hammer. A hot fomentation of leaves reduces the pain and swelling of bruises and sprains; cooled, it is also good to heal burns and fresh wounds.
The leaves may be substituted for Mallow or Hollyhock when they are indicated in external ulcers."
-Comfrey: What You Need to Know by Ben Charles Harris. New Canaan, Connecticut: Keats Publishing, Inc., **1982**, page 54.

Comfrey Root Poultice

"***If using the dried Comfrey rhizome (root), grind it and dissolve it in hot water to form a mucilage. Whatever you do, do not boil Comfrey. The high temperature can break down the allantoin.***"
-Rodale's Illustrated Encyclopedia of Herbs edited by Claire Kowalchik and William H. Hylton. Emmaus, Pennsylvania: Rodale Press, **1998**, page 104.
>(Boiling point of chemically/pharmaceutically made allantoin is 478 C = 892 F. Water boils at 100 C = 212 F.)

"***Comfrey Poultice: Use the dried roots of Comfrey, ground up as finely as possible.*** *The quantity used is dependent on the area of coverage desired. A midsized poultice will require about 100 grams (3.5 ounces) of dried root.*
Moisten the root powder with sufficient hot water to make a stiff paste. Spread the paste directly on the injured area to a thickness of approximately 1 inch (2.54 cm). *Cover the area with a clean cotton cloth.*
The poultice is preferably applied last thing before bed, left on all night, then scraped away and washed off in the morning. If the poultice is to be applied during the day, it may be secured with a long strip of cotton cloth. Repeat the procedure several times daily until the affliction is cured.
Comfrey poultices are excellent for repairing traumatic damage to bones, tendons, muscles, nerves, or spinal cord. Comfrey causes rapid cell proliferation, helps dissolve and remove dead tissue, and markedly speeds healing."
-Making Plant Medicine by Richo Cech. Williams, Oregon: Herbal Reads, **2016**, page 97.

"***Basic Poultice Formula:*** *Mix the following: 1 part plantain powder (key herb), 1 part calendula powder (key herb). This powder can be moistened with water or aloe vera juice and applied topically (to skin).*
It can also be modified as follows:
>*For an astringent effect to reduce the swelling of bites and stings, add 1 part white oak bark (key herb).*
>*For infection add 1/2 part echinacea (supporting herb), 1/2 part barberry (supporting herb), 1/2 part lobelia (catalyst).*

For an increased drawing action (to pull out pus, infection, or slivers), add one of the following as an additional supporting herb: 2 parts activated charcoal or fine clay (Redmond or other bentonite clay) to the base formula, 2 parts freshly crushed lily of the valley leaves, **2 parts Comfrey root powder,** or tincture of pine gum as the liquid.
For a greater styptic action (to reduce bleeding or oozing), add the following to the base formula: 1 part yarrow (supporting herb)."
-The Modern Herbal Dispensatory: A Medicine-Making Guide by Thomas Easley and Steven Horne. Berkeley, California: North Atlantic Books, **2016**, page 163.

Comfrey Baths and Soaks

Drawing Bath

"**Silk Soak:** *Submitted by Annette of Eugene, Oregon.*
Ingredients: *1 part sunflower seeds, 1 part corn starch, 2 parts finely ground oats, 2 parts milk powder, fragrance*
Instructions:
Coarsely grind the sunflower seeds in a blender or food processor. Add the other ingredients and mix thoroughly. Fragrance with oils or dry ingredients. **Because the sunflower seeds are so fragrant, earthy scents like nutmeg, almond and Comfrey work well.** *I prefer to use dry fragrances but oils work well too. This luxurious bath combines the skin soothing properties of oatmeal with softening milk and corn starch. The ground seeds release their natural oils when added to a hot bath.*"
-'Recipes and Information: Soap, Toiletries, Candles, Potpourri' by 'From Nature with Love', Dedicated to bringing you the best of nature from around the world, www.FromNatureWithLove.com, Oxford, Connecticut, October **2002**.

"Baths and soaks invigorate and stimulate circulation, or relax and soothe sore muscles. They can help to ease itching and skin eruptions and to hasten recovery from colds and fevers.
Add herbs to a bath by hanging 3-4 tea bags from the tap, or put a tea infuser in the water. A large gauze bag or a cotton sock will hold larger amounts of herbs. You can also make a strong decoction, strain it, and add the liquid to the bath.
The following are useful herbs for herbal baths and soaks:
Itching (hives, chicken pox, etc.): Yellow dock, burdock, Oregon grape, barberry, **Comfrey,** chickweed, lavender, vinegar (hot bath only), clay.
Swelling and inflammation: **Comfrey,** plantain, white oak bark, calendula, witch hazel, chamomile."
-The Modern Herbal Dispensatory: A Medicine-Making Guide by Thomas Easley and Steven Horne. Berkeley, California: North Atlantic Books, **2016**, page 336.

"**Drawing baths target the sebaceous glands (or oil ducts) in the skin. Drawing baths are useful for skin eruptive diseases, rashes, hives, pox, acne, and so forth. They can also be helpful for detoxification from heavy metals, especially the clay baths.**
A variety of herbs may be used in a drawing bath. You can also use Epsom salts and/or clay. **Herbs that are good for drawing baths include mucilaginous herbs like Comfrey leaf,** plantain, and various seaweeds and alterative herbs like Oregon grape, goldenseal, burdock, yellow dock, and red clover.
To make an herbal drawing bath, make a decoction of any of these herbs in a large pot containing 1-2 gallons (3.78-7.57 liters) of water. Use about 1/4 to 1/2 cup (2-4 United States fluid ounces = 59-118 ml) herbs per gallon. Simmer the herbs for at least 20-30 minutes; add this to the bath, then fill the tub with more water, adjusting the temperature to make the bath comfortable."
-The Modern Herbal Dispensatory: A Medicine-Making Guide by Thomas Easley and Steven Horne. Berkeley, California: North Atlantic Books, **2016**, page 339.

Sitz or Hip Bath

For a sitz or hip bath a person sits in water up to the hips (or lower). It relieves discomfort from hemorrhoids (piles), anal fissures, perianal fistulas, rectal surgery, episiotomy, uterine cramps, inflammatory bowel disease, pilonidal cysts and bladder/prostate/vaginal infections.
A sitz bath can be in a bathtub or large basin. It can be either warm or cool, or alternating between the two. Salt, baking soda, or vinegar may be added to the water. Warm baths reduce itching and discomfort from hemorrhoids and genital problems. Cool baths ease constipation, inflammation and vaginal discharge.
An infusion of Comfrey leaf or root can be added to the water. Comfrey does not cure infections.

"**Herbal Bath Recipe:** (Ingredients are enough for 4 brews.)
Uva Ursi - 1 cup (236 ml),   Shepherd's Purse - 1 cup,   Calendula - 1 cup,   Lavender - 1 cup
Comfrey leaves - 1 cup,   Comfrey root - 1/2 cup (118 ml),   garlic bulb (not just a clove!) - 4.
Mix 6 quarts (5.6 liters) of water with a small handful of each ingredient, a little less of the Comfrey root, and a full bulb of garlic (placed in the food processor with a bit of water to chop it up, but it is not necessary to peel it first). Bring to a boil and then let it simmer for 15 minutes. Let the brew sit for at least an hour before using it. The longer it sits, the stronger it gets. Then strain out

the herbs and **use it in a perineum spritzer (sprayer) or add at least 1 quart (0.94 liter) up to 1 gallon (3.78 liters) per bath to use it in the tub.** It is also ok to add sea salt to the tub, but not the spritzer. Use once or twice a day in post-partum (after giving birth) to help with healing and recovery. Brew can be refrigerated for up to a week."
-'Herbal Bath' by Shelley Snyder, A Time to Be Born, Natural Childbirth, Glenmoore, Pennsylvania, www.atimetobeborn.com, date unknown but between **2009** and 2018.
    (The perineum is the area between the anus and scrotum in a male and between the anus and vulva in a female.)

"**Comfrey Root: Comfrey root has a very long history of folk use for healing damaged skin, tissue, and broken bones.** It is highly mucilaginous. It is thought that allantoin and rosmarinic acid are the constituents mainly responsible for Comfrey's healing and anti-inflammatory actions.
Comfrey is indicated for topical use only. Use on broken skin or mucosa (mucous membrane) should be minimized but is reasonable for short durations (1 to 2 weeks at a time), and should not exceed 100 microgram of pyrrolizidine alkaloids with 1,2 unsaturated necine structure daily for a maximum of to 4 to 6 weeks annually.
Comfrey infusion may be added to a peri-rinse (perineum) or sitz-bath blend, or Comfrey oil or finely powdered herb may be added to a suppository blend."
-Botanical Medicine for Women's Health by Aviva Romm, CPM, Registered Herbalist of the American Herbalists Guild. London, England: Churchill Livingstone, **2009**. (Chapter 8: Vaginal Infections and Sexually Transmitted Diseases, pages 263-264.)

"**Sometimes herbs need to be applied directly to the mucous membranes** of the anus, vagina or urethra to combat irritation or infection. **My favorite method is the sitz bath.** Basically, you take a sock or cheese cloth and stuff it full of the herbs you need to apply to the delicate bits in question, draw a hot (but not too hot) bath, and toss it in. It's like sitting in a cup of tea!"
-'An Herbal Medicine-Making Primer' by Simon the Simpler, Yggdrasil Distro, 24 pages, **2010**.
    (Urethra is the duct that allows urine to come out of the body from the bladder.)

"**Sitz Bath Instructions:**
Try to soak in your sitz bath once a day for at least the first week after your birth. You can either soak in a few inches (2 inch = 5 cm) of water in your bath tub, or you can buy a plastic Sitz bath (available at pharmacies) that sits on the toilet.
To prepare your herbal infusion, place a couple of handfuls of the chosen herbal blend into a pot with a tight fitting lid. A handful is equal to about an ounce (28 grams). You will need 1 litre (4 cups) of boiled water for every 1 ounce of herb blend used. Pour the boiled water into the pot and replace the lid. Steep as directed in recipe. After straining the mixture, herbs can be discarded or composted.
When you are ready to soak, heat up a few litres, ensuring that the infusion does not boil. The temperature should be quite warm, but not so hot that it would be uncomfortable to sit in. Once it is at the desired temperature, pour the infusion into your bath tub or sitz bath. You can add cool water if necessary until you reach a comfortable temperature.
For greatest benefit, soak for at least 20 minutes, once or twice a day.
Recipe: 1 ounce Uva Ursi,   **1-2 ounces Comfrey**
       1 ounce Shepherd's Purse,   1 cup (236 ml) Sea Salt ,   1 bulb fresh Garlic
Simmer the herbs for 20-30 minutes in a big pot of water, then strain.
If you have fresh Comfrey from the garden, do not simmer, but use it raw by grinding in a blender, straining, then adding to the tea mixture. Blend the fresh garlic the same way, then add this and the Sea Salt to the tea. Keep in a covered container and add to warm water in your Sitz bath. If keeping for more than 2 days, keep in freezer."
-'Healing After Birth' by Willow Community Midwives, Penticton, British Columbia, Canada, www.willow-midwives.ca, May 24 **2013**.

"**A sitz bath is useful for hemorrhoids and for prostate, vaginal and uterine problems.**"
-The Modern Herbal Dispensatory: A Medicine-Making Guide by Thomas Easley and Steven Horne. Berkeley, California: North Atlantic Books, **2016**, page 336.

Homeopathic Comfrey **(Symphytum)** (For humans and animals.)

Homeopathy is an alternative medicine created in 1796 by Samuel Hahnemann, based on the doctrine of like cures like, meaning a substance that causes the symptoms of a disease in healthy people would cure similar symptoms in sick people. The preparations are manufactured using a process of homeopathic dilution, in where a substance is repeatedly diluted in alcohol or distilled water. This dilution continues past the point where no molecules of the original substance remain. Homeopaths select remedies by consulting books known as repertories, and by considering the patient's symptoms, physical/psychological state, and life history.

"**I have treated with success the (foot) rot in sheep by Homeopathic Comfrey** and Absinthium in alternating a few days, then give Arsenicum, Antimonium Crudum, and Sulphuris.
Contusions (bruises) are cured in a very short time by the external application of the tincture of Arnica diluted with half water. It is only in very severe cases, that the medicine should be employed internally.
If a bone has been affected along with the soft parts, or if the periostium has been injured, give Ruta Graveolens, and **Symphytum should be given internally,** several times in change, with Arnica wash externally.
Strain in the Shoulder:

This injury which in general is observed only in **oxen employed for drawing (pulling wagon or similar)**. When the disease has been occasioned by great efforts in drawing, by a false step or a slip: Rhus Toxicodendron; when external violence is the cause, Arnica is the remedy. **If bony parts are affected, we must have recourse to Symphytum internally.** Aconitum in inflammation.
Fractures:
Of the bones of the legs. After having reduced (set correctly in place) and adjusted the fracture, a strip of linen cloth is to he bound around the limb, over which two splints of wood or thick pasteboard are to be placed, which are to extend from four to six inches (10-15 cm) beyond the fracture, and be tied on snugly with a bandage; the bandage to be frequently moistened with Arnica water, and **Symphytum is to be given internally.** In 10 or 12 days the fracture is consolidated, if strictly attended to."
-'The Family Guide to Health and Husbandry: Containing Essays on Homeopathic and Other Medical Preparations for the Cure of Diseases of Men, Horses, Cattle and Sheep' by John Niesz, Canton, Ohio, **1857**
> (Foot rot, or infectious pododermatitis, is a bacterial hoof infection in sheep, goats and cattle.)
> (The periosteum is a membrane that covers the outer surface of all bones, except at the joints of long bones.)

"**Symphytum officinale 5C:** Pains across his epigastrium from one side to the other; worse opposite the spleen and in walking; when sitting pain is severe about the navel; griping pain; headache sometimes in the occiput (back of head), and again in top of head, occasionally in forehead; indefinable headache all over the head. Menses stopped; great deal of headache. Feeling of weight in the forehead constantly.
Considerable fever, which comes and goes often during the day. Often complains of coldness, cramp, and diarrhoea; nasal cavity sore, picking at the nose; rubs her eyes; inflamed ears, feels as if something was in them, stopped up, can't hear well, slight deafness; feeling miserably; generally weak and no desire or ability to be employed."
-'The Homeopathic Physician: A Monthly Journal of Homeopathic Materia Medica and Clinical Medicine, Volume 12, Philadelphia, Pennsylvania, April **1892**, page 137.
(Epigastrium is the upper abdomen immediately over the stomach. A griping pain is a sudden, sharp pain in stomach or bowels.)

"**Symphytum officinale 15C:** Alternately cold and feverish all day; after a few days continued coldness and desire to have on more clothing. General throbbing headache, all over the head; back, top, and frontal region. Itching of eyelids, disposition to rub them. Nose sore inside the alae (ala); wants to pick it. Pains across his epigastrium from one side to the other, worse opposite the spleen on walking; when sitting worse about the navel, griping pain about the umblicus often occurs. Cramp and diarrhea. Menses cease for a month when proving."
-'The Homeopathic Physician: A Monthly Journal of Homeopathic Materia Medica and Clinical Medicine, Volume 13, Philadelphia, Pennsylvania, **1893**, pages 289, 376, 383, 387, 441, 493.
> (The ala or wing of the nose is the lower surface, cartilaginous, that flares out.)

"**Symphytum Officinale Homeopathic Provings:** Consolida majoris. Comfrey. Gum plant. Healing herb. Tincture of root.
  1. **Guiding Symptoms of the Materia Medica**, Volume 10 by Constantin Hering. Philadelphia, Pennsylvania, 1878-1891.
  2. **Symptomen Codex**, Volume 2, by G.H.G. Jahr, page 1042. New York, 1848.
  3. **Provings and Clinical Observations of High Potencies** by M. Macfarlan. Philadelphia, Pennsylvania, 1894.
  4. **The Homoeopathic Physician** by M. Macfarlan, Philadelphia, Pennsylvania.
      Volume 12, 1892, page 137. Volume 13, 1893, pages 289, 376, 383, 387, 441, 493."
-'Index of Homeopathic Provings' by Thomas Lindsley Bradford, Philadelphia, Pennsylvania, original edition **1901**. Reprinted by Jain Publishing, Delhi, India, 2000. Adaptation by Jorg Wichmann, Fagus Publishing, Germany, 2017.
> (In 'homeopathic provings', a homeopathically prepared substance is tested on healthy volunteers, i.e., 'provers', to show through recording the effects, the state of health-disturbance that the substance creates.)

"**Homeopathic Symphytum Comfrey. Borraginaceae.**
  Facilitates union of fractured bones (Calcarea phosphorica).
  Lessens peculiar pricking pain; favors production of callous; when trouble is of nervous origin.
  Irritability at point of fracture; periosteal (periosteum) pain after wounds have healed.
  Mechanical injuries; blows, bruises, thrusts on the globe of the eye.
  Pain in eye after a blow of an obtuse body; snow ball strikes the eye; infant thrusts its fist into its mother's eye (to soft tissues around the eye, Arnica).
Relations- Compare: Arnica, Calendula, Calcarea phosphorica, Fluoricum acidum, Hep. (Hepar?), Silica.
Follows well: after Arnica, for pricking pain, and soreness of periosteum remaining after an injury."
-'Keynotes and Characteristics with Comparisons of Some of the Leading Remedies of the (Homeopathic) Materia Medica, Third Edition' by H.C. Allen, M.D., Boericke and Tafel, Philadelphia, Pennsylvania, **1910**, page 282.

"The results showed that by X-ray criteria none of the untreated controls or saline treated fractures were completely mineralized. Histological examination revealed there was still conspicuous presence of cartilage and also some fibrous tissue at the fracture site. **In the animals which received either of the two dilutions of Homeopathic Arnica and Symphytum, and irrespective of the mode of administration, both x-rays and histology showed completely mineralized new bone through the entire fracture site in 33% of the fractures.**"
-'Effect of Arnica Montana and Symphytum Officinalis on Bone Healing in Guinea Pigs' by M. Oberbaum, E. Yakovlev, D. Kaufman and S. Shoshan, British Homeopathic Journal, Volume 83, Issue 2, page 90, April **1994**.
> (Histology is the study of the microscopic anatomy of cells and tissues of plants and animals.)

"In homeopathy, preparations of Symphytum officinale according to homeopathic pharmacopoeias are alcoholic extracts from the roots collected before the time of blossoming.
In veterinary homeopathy 1:10 and 1:1000 dilutions of Symphytum officinale are intended for parenteral (injection) or oral (drops, tablets) use in all food-producing species.
The maximum recommended parenteral dose for large animals is 10 ml/animal. *Treatment may be repeated but a fixed dose schedule is not common in homeopathy."*
-'European Medicines Agency- Symphyti Radix: Committee for Veterinary Medicinal Products' by European Agency for the Evaluation of Medicinal Products, Veterinary Medicines Evaluation Unit, London, England, EMEA/MRL/649/99-FINAL, 5 pages, August **1999**.

*"**The objective of this study was to evaluate the influence of homeopathic treatment with Comfrey (Symphytum officinalis 6CH {Hahnemannian Centesimal scale}) on radiographic (X-ray or similar) bone density and area around titanium implants.***
The concentration used in this study was 6CH, which means that Comfrey was first dissolved in alcohol and then serial dilutions were made. ***In this case, 6CH has been diluted and succussed (vigorous shaking): one part of remedy to 99 parts of alcohol six successive times.***
   In homeopathy the remedies are not dispensed in milligrams or by weight; they are dosed according to their dilution factor (XH diluted 10 X, CH diluted 100 X, and MH diluted 1000 X). In a way, it is quite similar to the production of vaccine and allergen desensitization.
Material and methods: *Forty-eight rats were divided into two groups of 24 animals each: a control group (C) and a test group (SO). Each animal received one titanium micro-implant placed in the tibia. The animals in Group SO were subjected to **10 drops of Comfrey 6CH per day mixed into their drinking water** until the day of sacrifice. Eight animals of each group were sacrificed at 7, 14 and 28 days post-surgery, respectively.*
Results: *Subtraction images demonstrated that a significant difference existed in mean shade of gray at 14 days post-surgery between Group SO (mean 175.3 14.4) and Group C (mean 146.2 5.2). Regarding the area in pixels corresponding to the bone gain in Group SO, the differences observed between the sacrifice periods and groups were only significant at 7 days sacrifice between Group SO (mean 171.2 21.9) and Group C (mean 64.5 60.4).*
Conclusion: ***The present study showed that homeopathic solution (6CH) of Comfrey administration resulted in an increase in radiographic bone density and area of bone gain around titanium implants at 14 and 7 weeks postsurgery, respectively, demonstrated by subtraction radiography.***
These results suggest that the ingestion of Comfrey 6CH after implant surgery increased bone formation because subtraction radiography is capable of detecting small alterations in periimplant bone density (Engelke et al. 1990; Jeffcoat et al. 1992; Schou et al. 2003; Sakakura et al. 2007). ***This may be interpreted as an accelerated rate of bone maturation and mineral deposition.***
The effect of Comfrey on bone healing was evident in the initial healing period. In animals with the longest healing period (28 days), no parameter showed significant differences between the test and the control group. ***These results lead to the hypothesis that the action of Comfrey substances is evident only in the initial period of bone healing, accelerating the rate of mineral deposition until the hypothetical saturation level.****"*
-'Influence of Homeopathic Treatment with Comfrey on Bone Density Around Titanium Implants: A Digital Subtraction Radiography Study in Rats' by Celso Eduardo Sakakura, et al. (Sao Paulo, Brazil, and Arhus, Denmark); Clinical Oral Implants Research, Volume 19, Issue 6, pages 624-628, June **2008**.

*"**This study evaluated the effect of Symphytum officinale in homeopathic potency (6cH),** on the removal torque (twisting force) and radiographic bone density around titanium implants, inserted in rats tibiae (larger bone between knee and ankle). Implants were placed in male rat tibiae, and the animals randomized to two groups (Control and S. officinale 6cH treated), which were evaluated at 7, 14, 28 and 56 days post-implantation.*
Both removal torque and radiographic bone density evaluation showed that S. officinale 6cH treatment enhanced bone formation around the micro-implants, mainly at 14 days. At 56 days, the radiographic bone density was higher in the treated group.
We conclude that S. officinale 6cH enhances, principally at the early stages of osseointegration, bone formation *around titanium implants in rats' tibiae."*
-'Homeopathic Symphytum Officinale Increases Removal Torque and Radiographic Bone Density around Titanium Implants in Rats' by Rubens Spin-Neto, Marina Montosa Belluci, Celso Eduardo Sakakura, Gulnara Scaf, Maria Teresa Pepato and Elcio Marcantonio Jr., Brazil; Homeopathy: The Journal of the Faculty of Homeopathy, London, England, Volume 99, No. 4, pages 249-254, October **2010**.
   (Osseointegration is the structural and functional connection between bone and the surface of a load-carrying implant.)

*"**Homeopathy is a therapeutic method based on the empiric (from experience) law of similars with the hypothesis, that a given substance can cure in a diseased person the symptoms that it produces or causes in a healthy person.***
*The objective of this study was to investigate the effect of a commercial low potency homeopathic remedy Similasan® Arnica plus Spray on wound closure in a controlled, blind trial in vitro. We investigated the effect of an ethanolic preparation composed of equal parts of Arnica montana 4x, Calendula officinalis 4x, Hypericum perforatum 4x and **Symphytum officinale 6x.***
In this study we showed that the in vitro wound model used was sensitive enough to observe effects of substances at low

*potency homeopathic concentrations. **Its promoting wound filling effect could be related to the increased cell migration** without an increased mitotic activity of cells. It exerted in vitro wound closure potential in NIH 3T3 fibroblasts."*
-'A Homeopathic Remedy from Arnica, Marigold, St. John's Wort and Comfrey Accelerates in Vitro Wound Scratch Closure of NIH 3T3 Fibroblasts" by Hostanska et al., BioMed Central, Complementary and Alternative Medicine, London, England, 12:100, **2012**. (Mitotic is the usual method of cell division.)

"**Symphytum officinale, also known as Comfrey, knitbone and bone-set in herbal medicine, is a homeopathic first aid remedy.** The names precisely describe its sphere of genius, the mending of broken bones, and the treatment of bone diseases, ranging from inflammation of the bones, or osteitis (bone inflammation) to cancers of the bones or sarcomas (bone/connective tissue tumors).
Symphytum officinale (homeopathic remedy) is found to be widely used for treatment of injuries of bones, cartilage, tendons and periosteum. It helps in nonunion of fractures, with mending bones that are slow to heal, and where the wound has penetrated into the bones. It helps in alleviating the pain remaining in the periosteum or lining of the bones after the wound has healed.
Symphytum is recommended in wounds resulted with blows, falls and bruises, and even helps with the tendency to have such mishaps similar to Arnica.
The remedy can help with bruises to the tendons and bone healing. Symphytum helps healing of sprains and injuries from straining of muscles, bones and tendons similar to Ruta. Symphytum helps in treatment of fractures due to osteoporosis. Symphytum can help with diseases of the bones, such as inflammation of the bones, inflammation of the inferior maxillary bone and cancer of the bones or sarcomas (Olenev 2014)."
-'Role of Symphytum Officinale as an Osteoinducer in Long Bone Fracture Repair in Canine' thesis by Swarop Chandel for Master of Veterinary Science degree, Department of Veterinary Surgery and Radiology, Nagpur Veterinary College, India, **2014**.
(The periosteum is a membrane that covers the outer surface of all bones, except at the joints of long bones.)
(A tendon is a band of tissue that connects muscle to bone. A ligament is an elastic band that connects bone to bone. Cartilage is soft, gel-like padding between bones.)

"**The Medicines and Healthcare Products Regulatory Agency (MHRA) granted 'Biologische Heilmittel Heel GmbH' a 'Homeopathic Marketing Authorisation' for the homeopathic medicinal product Traumeel Ointment (Homeopathic Marketing Authorisation number: NR 08927/0019) on 13 June 2014.**
This product is available without prescription and can be bought from pharmacies. Traumeel Ointment is a **homeopathic medicinal product used within the homeopathic tradition to relieve symptoms such as pain and swelling which may result from minor injuries such as sprains, muscular strains and bruising.**
These indications are based on published Materia medica references and other bibliographic evidence.
*The active ingredients in Traumeel Ointment are Achillea millefolium 3X, Aconitum napellus 4X, Atropa belladonna 4X, Hepar sulfuris 9X, Matricaria recutita 2X, Mercurius solubilis Hahnemanni 9X, **Symphytum officinale 7X**, Bellis perennis 3X, Calendula officinalis 2X, Echinacea angustifolia 2X or Echinacea pallida 2X (or a mixture of both species), Echinacea purpurea 2X, Hamamelis virginiana 2X, Hypericum perforatum 9X and Arnica montana 4X."*
-'Traumeel Ointment: Homeopathic Marketing Number 08927/0019, UKPAR: United Kingdom Public Assessment Report' by Medicines and Healthcare Products Regulatory Agency, London, England, June 13 **2014**.
(I do not have any connection with this product or the company. It is an example of current regulations of homeopathic products in the United Kingdom.)

"**Homeopathy, the 'energy medicine'**, *is a branch of medical science based on the principle that disease can be cured by strengthening the body's defense mechanism with substances selected for their energy giving properties.*
Homeopathic Symphytum officinale:
Boericke in his Materia Medica indicates Symphytum in non-union of fractures in capital letters, whereas Allen states that it is an excellent remedy for fracture and mechanical injuries, facilitates union of fractured bones and favors production of callus. According to G. Vithoulkas' Materia Medica notes, Symphytum officinale promotes the repair of broken bones, especially when they heal slowly. It helps the slow repair of broken bones (stated in capital letters), lessens peculiar bone pain and favors production of callus.
We present four cases of patients with bone fractures, which were treated with the aid of the Homeopathic Remedy Symphytum officinale. *The Remedy was given to the patients after the proper alignment of the bone fragments was secured. In all of the cases, initially, potency used was 30CH, once daily for 10 days, and then 200CH, once a week for the next 3 weeks.*
Although in the first case the fracture was severely comminuted (many bone fragments), and in the third and fourth, the patient's compliance to the immobilization was very poor, the result of the treatment was excellent, both clinical and radiological (x-ray or similar).
From our experience, Symphytum officinale in homeopathic potencies is a powerful tool for the medical practitioner treating bone fractures. *It should be administered to the patient only after acceptable reduction of the fractured fragments has been secured (in order to avoid the possibility of mal-union).*
Even in very difficult cases or in patients with poor compliance, Symphytum can accelerate solid healing of the fractured bones."
-'Fracture Treatment with the Aid of the Homeopathic Remedy Symphytum Officinale: A Report of Four Cases' by Dionysis Tsintzas (Orthopaedic Surgeon) and George Vithoulkas (Homeopathic Practitioner), Greece; Clinical Case Reports and Reviews, Case Series, Volume 2, No. 5, pages 422-424, **2016**.

"Here is a list of conditions that homeopathic Symphytum officinale can help with:
> *1) **Injuries to bones, cartilage, tendons and periosteum**. Symphytum helps with the non-union of fractures, with mending bones that are slow to heal, and where the wound has penetrated into the bones. It helps with pain remaining in the periosteum or lining of the bones after the wound has healed.*
> *2) **Injuries to the eyes**. Symphytum helps with mechanical injuries to the eyes, such as blows from blunt objects. Examine Arnica and Ruta for this kind of injury as well. Think of Symphytum for injuries to the orbits of the eye.*
> *3) **Like Arnica, Symphytum helps with blows, falls and bruises**, and even helps with the tendency to have such mishaps. The remedy can help with bruises to the tendons and bones.*
> *4) **Like Ruta, Symphytum helps with sprains, & injuries from overlifting or straining of muscles, bones & tendons.***
> *5) **Symphytum should be thought of for fractures due to osteoporosis.***
> *6) **Symphytum can help with diseases of the bones**, such as inflammation of the bones, inflammation of the inferior maxillary bone (jaw).*

Here is a list of some of the sensations common to Symphytum: pain as if the bone is broken; gnawing, jerking, pricking, sticking, digging, burning, stitching, and bruised pains."
-'Symphytum Officinale: A Homeopathic First Aid Remedy' by Deborah Olenev, C.C.H., RSHom (NA), California, February **2016**.

*"Homeopathy, reiki, qi gong, polarity therapy, healing touch, acupuncture, and massage are all non-conventional energy healing modalities with applications for **fracture healing**.*
***Common over-the-counter homeopathic remedies** include arnica as an anti-trauma remedy for immediately after the fracture (not to be used if the person is unconscious),* **Symphytum (Comfrey) for pain relief and the joining of set bones,** *and Calcarea phosphorica for fractures that are difficult to heal.*
Low-potency homeopathic remedies (6x, 6c to 30x, 30c) are often used for self-help, as detailed in 'Homeopathic Self-Care' by Robert Ullman and Judyth Reichenberg-Ullman (Prima Publishing, 1997).
Homeopathy is a powerful medicine and when possible the best policy is to seek the advice of a professional homeopath."
-'How to Speed Fracture Healing' by Dr. Susan E. Brown, PhD, Center for Better Bones, E. Syracuse, New York, 11 pages, **2016**.

*"**Homeopathic Symphytum officinalis:***
Irritability with a sense of hurry. Anxious and vivid dreams with weeping. Intense headaches in the temples, forehead and extending to the neck. Nasal discharge, worse after eating and associated with flatulence and diarrhea. Improvement of dysmenorrhea (menstrual cramps) and premenstrual breast tenderness. Delayed menses (menstruation). Pain, aching, and soreness in the bones and joints of the body. Itching eruptions, especially on the digits (fingers/toes) of hands and feet.
David S. Riley's interest in the history, methodology, and results of homeopathic drug provings came from his study of homeopathy at the Hahnemann College of Homeopathy in Albany, California. The 69 drug homeopathic drug provings published here are the result of his investigation of the research methods that form part of the basis for prescribing in homeopathy."
-Materia Medica of New and Old Homeopathic Medicines by David S. Riley. Berlin, Germany: Springer, **2018**, Symphytum officinalis pages 227-230.

Comfrey Flower Essences and Remedies

Flower essences or remedies are homeopathic-like herbal infusions or decoctions made from flowers mixed with brandy. Flower remedies were first formulated by British physician, Edward Bach in the 1930's. They improve emotional and mental well being through energetic and vibrational influences.

*"**Indications (of need for Comfrey Flower Essence):***
Is depressed, has a poor memory, has few memories of early childhood, has no memories before age five, can't think clearly about some issues, is anxious, can't remember dreams, can't retain information, tests poorly, has difficulty accessing emotions, is uncoordinated, restless, anxious, repressed, has irregular heartbeat, high blood pressure.
Energetics of Comfrey:
Comfrey is an herb that has a strong presence. Just as Comfrey as an herb works to repair tissue and bone damage, Comfrey flower essence provides the same type of repair action on the nervous system. *This can have several beneficial effects on memory, coordination, reflex response, biofeedback and restoration of organ function, shut down due to repression. When memories have been suppressed by the unconscious mind, to protect the psyche from the pain of the event, the nervous system can close down or atrophy around this memory. Comfrey flower essence heals neural pathways that are shutdown, assisting one in gaining access to these memories.*
With the release of memories, feelings and information to the conscious mind, the individual is able to process and release the pain associated with the experience. *In order to have this kind of experience with Comfrey, a person must have done some initial preparatory work in psychotherapy, and have sufficient strength or support to endure the pain of the repressed energies.*
As the nervous system is the junction between the physical and the non-physical, its stimulation with Comfrey flower essence can increase one's ability to bring information from other levels into conscious awareness. This includes strengthening the ability to recall dreams, deepening meditation and journeying experiences or the developing of channeling abilities.
Energetically, Comfrey works with the deeper layers of the first three chakras, adding energy and rhythm to their functioning. *Small movements in these deeper layers can have dramatic and observable effects on the entire psyche. Most*

repressed information is buried in these deeper layers, while the upper layers contain mostly conscious information. The nervous system is stimulated to receive information from these deeper layers.

Comfrey's powerful awakening action demands understanding and respect. Use it later in the recovery process when a person has learned to recognize and process information arising from the deep subconscious, and has developed a reliable and trustworthy support network.

For animals as well as humans, healing from bone fractures is assisted by Comfrey, as the knitting of the tissues begins with the repair of the nervous system. Use Comfrey also to accelerate healing after surgery. Taking it several times per day starting one week before surgery to several weeks after is recommended."

-Stars of the Meadow: Medicinal Herbs as Flower Essences by David Dalton. Great Barrington, Massachusetts: Lindisfarne Books, **2006**.

"**Comfrey (Symphytum officinale) Flower Essence:**
Aids in the healing of wounds so deep and traumatic that they affect your soul's journey. Encourages feeling emotionally and physically safe, and engenders tenderness, self-compassion and joy during this deep healing process.
As the essence says of itself: I am comfort and solace for your soul during your deep healing process.
Flower Essence Symptom: Use Comfrey Flower Essence when you experience very deep emotional and psychological wounds that are usually from trauma, abuse and shock. The wounds could be recent or very old, from this lifetime or a past life.
Flower Essence Action: Comfrey Flower Essence magnetically reconnects soul parts that split off during trauma or shock and lovingly invites them back by creating a safe haven for their return. This includes soul parts that split off during this lifetime or previous lifetimes that did not come into this life with the soul at birth.
Comfrey essence also works through the root chakra to influence the body to rebuild cellular structures that support your continued healing and growth and with the heart chakra for gentle self-compassion."

-Tree Frog Farm: Flower and Tree Essences in Partnership with Nature, Lummi Island, Washington, www.treefrogfarm.com/store/flower-essences-tree-essences/comfrey-flower-essence.html, **2015**.

"**Flower Essences:**
Comfrey Flower Essence is a toner for nerves, and helps heal nerve misfunction. It also helps heal traumas from present or past lifetimes. It is a master healer and powerful grounding force.
Comfrey is a powerful tonic for the nervous system. It increases neurological response, and it invigorates the activity of the synapses between nerve cells.

> Any disease such as shingles, can be treated with Comfrey.
> If a person is trying to regain use of the nervous system after it has been in an atrophied state from, for instance, muscular degeneration or being in a wheelchair from an accident, this would be a powerful remedy to re-educate the body. Comfrey eases phantom limb pain because it heals nerve endings.

Comfrey can also be used when brain tissue has been damaged or destroyed. Comfrey does not rejuvenate brain tissue, but it allows dormant or atrophied portions of the brain to be used.

> Comfrey helps re-channel brain messages. In balancing the left and right brain, it increases physical coordination. It enables one to gain better control over the body processes.

Comfrey (Symphtum asperum) flower essence is for knowing that everything happens at the right time, and that where we are now, and how we are is precisely where we need to be for our continuing understanding and growth."

-'Comfrey' by Glenn Axford, Our Common Roots: An Exploration of the Healing Power of Nature, Canada, date unknown but **2018** or earlier. www.ourcommonroots.com   (Atrophy is gradual decline in effectiveness or vigor from lack of use or neglect.)

1884 Comfrey Apothecary Can / Tin

"Choice Botanic Drugs, Pressed"

"Parke, Davis & Co., Manufacturing Chemists Detroit, Michigan, U.S.A."

Hinged lid.
9" tall and 4 1/2" by 3 1/2"

Factory graphics on front.

1884 was the last patent date.

Before the 1930s: Symphytum Officinale Apothecary Jar

Eli Lilly Drug Company

8.5 inches tall by 4.5 inches
Weight: 1.8 pounds.

Ceramic painted green with gold lettering and art.

Eli Lilly and Company is an American pharmaceutical company in Indianapolis, Indiana. The company was founded in 1876 by Colonel Eli Lilly, a pharmaceutical chemist.

Chapter 28

Warnings and Negative Reactions to Comfrey

<u>Negative Reactions to Comfrey (Side Effects)</u> (in chronological order from older to recent)

"**Comfrey, Symphytum officinale- Possible Skin Reactions to Touching Leaf:**
    Allergic urticaria (hives): No.   Nonallergic urticaria (hives): No.
    Allergic contact dermatitis: No.   Irritant Reaction on skin: No.
    Photodermatitis (negative reactions to sun on skin): No.
    Mechanical: Yes. Dermatitis can be caused by cuts/abrasions from hairy appendages on the plants."
-'Phytodermatitis: Reactions in the Skin Caused by Plants', Safety and Health Assessment and Research for Prevention (SHARP) Report: 63-8-2001, Washington State Department of Labor and Industries, www.lni.wa.gov/sharp/derm, 12 pages, August **2001**.
    (Dermatitis is when the skin becomes red, swollen and sore, sometimes with small blisters, from direct irritation of the skin by an external agent or an allergic reaction to it.) (Hives are an allergic reaction with swollen, red patches of skin or mucous membrane and usually with intense itching. It is caused by contact with the agent.)

"**Adverse Reactions to Comfrey:**
CNS (Central Nervous System): *chills, fever.*
GI (Gastrointestinal): *abdominal pain, diarrhea, hematemesis (vomit blood), poor appetite, vomiting.*
Hepatic (Liver): *Hepatotoxicity (specifically veno-occlusive disease from pyrrolizidine alkaloids characterized by a nonthrombotic obliteration of small hepatic veins leading to cirrhosis and ultimately liver failure if the process is not arrested).*
Skin: *exfoliative dermatitis (skin peels), jaundice (yellowish pigmentation of skin, tissues, and body fluids caused by deposition of bile pigments).*
Other Reactions: *weight loss, cancer (several animal studies report hepatocellular adenomas and urinary bladder tumors caused by PAs in Comfrey [Hirono et al., 1978]), death.*"
-Professional's Handbook of Complementary and Alternative Medicines, 3rd Edition' by Charles H. Fetrow, Pharmd (Doctor of Pharmacy) & Juan R. Avila, Pharmd. Pennsylvania: Lippincott Williams & Wilkins (imprint Wolters Kluwer), **2003**, page 235-238.
    (The species of Comfrey was not given.)
    (In hepatic Veno-Occlusive Disease some small veins in the liver are blocked. It can be caused by high-dose chemo-therapy or after consuming certain plant alkaloids such as pyrrolizidine. Symptoms include weight gain due to fluid retention, increased liver size, and raised levels of blood bilirubin. Also called 'Sinusoidal Obstruction Syndrome'.)
    (For more about the Hirono et al. 1978 study, see subsection 'Scientific Studies Showing Dangers of Alkaloids in Comfrey' in section 'Alkaloids in Comfrey' {Chapter 30}.)

"**The features of Sinusoidal Obstruction Syndrome (Hepatic Veno-Occlusive Disease):**
Patients typically develop severe liver injury characterized by right upper quadrant pain, hepatomegaly (enlarged liver), and ascites (fluid in abdomen). Increased levels of alkaline phosphatase are common and, over time, advancement to the development of posthepatic portal hypertension with ascites and, eventually, liver failure occurs.
This presentation has been attributed to Comfrey's component pyrrolizidine alkaloids, *which create toxic intermediates that are injurious to sinusoidal endothelium, thus leading to the obstructive (blocked) process."*
-'Liver Injury Induced by Herbal Complementary and Alternative Medicine' by Victor J. Navarro, MD and Leonard B. Seeff, MD (Division of Hepatology, Einstein Healthcare Network, Philadelphia, Pennsylvania, and The Hill Group, Bethesda, Maryland; Clinics in Liver Disease, Pennsylvania, Volume 17, No. 4, pages 715-735, November **2013**.

"**Herbal Products in General, Not Just Comfrey: If you're taking a dietary supplement, follow the label instructions.** *Talk to your health care provider if you have any questions, particularly about the best dosage for you to take.*
If you experience any side effects that concern you, stop taking the dietary supplement, and contact your health care provider. *You can report serious problems suspected with dietary supplements to the United States 'Food and Drug Administration' and the 'National Institutes of Health' through the 'Safety Reporting Portal'.*
Keep in mind that although many dietary supplements (and some prescription drugs) come from natural sources, 'natural' does not always mean 'safe'.
    For example, the herbs Comfrey and kava can cause serious harm to the liver. *Also, a manufacturer's use of the term 'standardized' (or 'verified' or 'certified') does not necessarily guarantee product quality or consistency.*
Also consider the possibility that what's on the label may not be what's in the bottle. *Analyses of dietary supplements sometimes find differences between labeled and actual ingredients. For example:*
    1. An herbal supplement may not contain the correct plant species.
    2. The amounts of the ingredients may be lower or higher than the label states. *That means you may be taking less or more of the dietary supplement than you realize.*

3. The dietary supplement may be contaminated *with other herbs, pesticides, or metals, or even adulterated with unlabeled, illegal ingredients such as prescription drugs."*
-'Using Dietary Supplements Wisely', NCCIH Publication No. D426, by United States Department of Health and Human Services, National Institutes of Health, National Center for Complementary and Integrative Health, https://nccih.nih.gov, June **2014**.

Example 1 of Comfrey Tea Having Contaminants:
*"**Celestial Seasonings® recalls Comfrey leaf tea:** Celstial Seasonings recently concluded a voluntary consumer-level recall of their Comfrey tea, as a result of **accidental adulteration of the product by one of its suppliers.***
*The problem surfaced when a consumer drank a potent extract of Comfrey leaf for medicinal reasons. **The consumer was treating herself for a broken hip, and consumed the equivalent of around 15 cups (3.5 liters) of Comfrey tea at one sitting. She became ill, and later testing of the tea revealed traces of atropine-like alkaloids, the major active components of belladonna leaf (Atropa belladonna)** and several other Solanaceous (potato family) plants. Only slight traces (typically less than 0.1 mg/gram) were found in a few packages of the Comfrey tea. Atropine is not a natural component of Comfrey, however, and its presence indicated the likelihood that actual belladonna leaf had been accidentally mixed with the Comfrey by an Eastern European supplier.*
Technical Problems at Laboratories: *Celestial contracted outside consulting laboratories to test for the atropine, to assure the Food and Drug Administreation (FDA) of impartial results. **But the testing proved difficult for all 3 labs which attempted it, including FDA's own labs.***
Food and Drug Administration: *After three weeks of deliberation and several meetings with Celestial representatives, **FDA determined that the Comfrey situation was officially a 'Class 2' recall. FDA's decison states: 'The use of, or exposure to, the Comfrey tea, may cause temporary or medically reversible adverse health consequence.** Further, the probablity of serious adverse health consequence is remote."*
-'HerbalGram, Issue 2, Herb News' by Herb Research Foundation and American Botanical Council, Austin, Texas, 8 pages, Fall 1983-Winter 1984.   (Contamination of Comfrey products is rare.)

Example 2 of Comfrey Tea Having Contaminants:
*"A 30-year-old man visited a health-food store complaining of flatulence. Comfrey tea was recommended. He put 28 grams (0.98 ounce) into boiling water. He had several cups of the infusion, after which **he felt light-headed, agitated, and confused, and had difficulty in micturition (urinating).** He also complained of dry mouth and he had sinus tachycardia of 120/minute, dilated pupils, and a warm dry skin. He was admitted and his symptoms resolved over the next 24 hours.*
*An infusate of the Comfrey tea competitively blocked the contractile response of guinea pig ileum to acetylcholine and the tea contained a minimum of 4 mg atropine per 28 gram of leaf. **This is likely to have been due to contamination with Atropa belladona (deadly nightshade),** which was noted in a previous batch of Comfrey tea from a different supplier."*
-'Atropine as Possible Contaminant of Comfrey Tea' by P.A. Routledge and T.L.B. Spriggs, Welsh National Poisons Unit, Department of Pharmacology and Therapeutics, University of Wales College of Medicine, Cardiff, Wales; The Lancet, Volume 333, Issue 8644, Letter to the Editor, pages 963-964, April 29 1989.

This is a letter in response to the above letter about the Comfrey tea contamination:
*"**Such symptoms have also been observed in patients consuming contaminated burdock (Arctium) root, nettle (Urtica) tea, and mallow (Malva).***
*In two cases which required admission in Toronto, Canada, in 1981 and 1984, examination of purported Malva sylvestris, in packages recovered from patients, revealed **berries of Atropa belladonna (deadly nightshade). While it is indeed likely that Atropa belladonna (which has remarkable propensity for commingling with other plants) is responsible for the observed poisonings in Comfrey tea,** its presence cannot be confidently assumed from a hyoscyamine level as low as that observed in the Routledge and Spriggs' report (0.014%).*
Two other common agents of tropane alkaloid poisoning could be responsible for such levels of contamination, namely Datura stramonium (thornapple or Jimsonweed) and Hyoscyamus niger (black henbane)."
-'Atropine as Possible Contaminant of Comfrey Tea' by D.V.C. Awang and D.G. Kindack, Bureau of Drug Research, Health Protection Branch, Health and Welfare Canada, Ottawa, Ontario, Canada; The Lancet, Volume 334, Issue 8653, Letter to the Editor, page 44, July 1 1989.

*"**Herb-drug and herb-herb interaction and alteration of metabolizing enzymes:***
Both herbs and Western drugs require metabolizing enzymes for metabolism.** Herbs and herbal supplements are often used in combination with therapeutic drugs, and raise the potential of herb-drug interactions that may result in adverse effects. **It has been reported that a number of herbal extracts and herbal constituents can significantly alter the activity of phase I and phase II metabolizing enzymes, which may lead to herb-herb and herb-drug interactions.
> *The herbs that have been studied include kava extract and its kavalactones, Panax ginseng, St. John's Wort, G. biloba, goldenseal (Hydrastis canadensis), **Comfrey**, and tumorigenic pyrrolizidine alkaloids."*

-Nutraceutical and Functional Food Regulations in the United States and Around the World edited by Edited by Debasis Bagchi, PhD MACN CNS MAIChE, Department of Pharmacological and Pharmaceutical Sciences, University of Houston College of Pharmacy, Texas. London, England: Academic Press, **2014**.

*"**The use of HDS (Herbal and Dietary Supplements) is extensive and is largely uncontrolled neither by pharmacovigilance authorities nor by health practitioners.** Doctors are often confronted with an astonishing persuasion that HDS products must be effective and safe because they are 'pure' and 'natural'. This prevailing belief is fueled by the often long-lasting tradition of herbals in cultures with an apparently more humane view on health and well-being than modern Western societies. However, much of that is wishful thinking and rather reflects the deep desire of consumers to avoid some of the associated adverse effects Western medicine undoubtedly has.*
As compared to conventional pharmacologic agents, safety and efficacy of most HDS have not been rigorously tested, and some of them are less safe or even hazardous. *Thus, interrogating patients about their use is an essential diagnostic component when liver injury is suspected."*
-'Hepatotoxicity of Herbal and Dietary Supplements: An Update' by Felix Stickel (Switzerland) and Daniel Shouval (Israel); Archives of Toxicology, Berlin, Germany, Volume 89, pages 851-865, **2015**.

*"**Adverse drug reactions are classified as intrinsic/pharmacological or idiosyncratic**:*
Drugs or toxins that have an intrinsic/pharmacological hepatotoxicity *are those that have predictable dose-response curves where higher concentrations cause more liver damage, and well characterized mechanisms of toxicity, such as directly damaging liver tissue or blocking a metabolic process.*
Idiosyncratic injury occurs without warning, when agents cause non-predictable hepatotoxicity in susceptible individuals, which is not related to dose and has a variable latency/dormancy period. *This type of injury does not have a clear dose-response nor temporal (time) relationship, and most often does not have predictive models.*
Idiosyncratic drug reactions, also known as type B reactions, are drug reactions that occur rarely and unpredictably amongst the population. This is not to be mistaken with idiopathic, which implies that the cause is not known."
-Wikipedia®: The Free Encyclopedia, www.wikipedia.org, **2018**.

*"**Comfrey Adverse Reactions: Neither internal nor extensive topical use of Comfrey can be recommended because of numerous reports of liver toxicity.** Case reports show hepatic veno-occlusive disease and pulmonary (lung) hypertension related to Comfrey use. Infants are more susceptible to pyrrolizidine-alkaloid-related, veno-occlusive disease; therefore, the use of Comfrey in this population is contraindicated (not recommended).*
Hepatotoxic (Liver) Effects:
    Human poisonings with pyrrolizidine alkaloids are usually accidental *and may be caused by ingestion of contaminated flour, milk, certain goat products that are resistant to the alkaloids, honey produced by bees fed on pyrrolizidine-containing weeds, and consumption of certain herbal or bush teas. It also may be caused by Comfrey used in salads.*
Carcinogenicity (Cancer Causing):
    Symphytum officinale extract and components lasiocarpine and symphytine (alkaloids) are carcinogenic in rats, possibly via genotoxic mechanisms; however, an association of Comfrey consumption with cancer in humans is lacking. *A pyrrolizidine alkaloid-free liquid extract of Comfrey root was not mutagenic when tested by the bacterial reverse mutation assay."*
-'Comfrey' by Drugs.com® provides independent information on more than 24,000 prescription drugs, over-the-counter medicines and natural products. This material is provided for educational purposes only and is not intended for medical advice, diagnosis or treatment. Data sources include Micromedex®, Cerner Multum®, Wolters Kluwer® and others, July **2018**.

Individual Differences in Reactions to Alkaloids (Susceptibility)

    Alkaloids are organic compounds that contain basic nitrogen atoms. In addition to carbon, hydrogen and nitrogen, alkaloids may also contain oxygen, sulfur and, rarely, chlorine, bromine, and phosphorus.
    Most alkaloids have a bitter taste or are poisonous when drunk or eaten. Alkaloids in plants appear to have evolved in response to animals eating them; though, some animals have evolved the ability to detoxify alkaloids.
    Alkaloid-containing plants have been used throughout history. Many alkaloids are used today in medicine, usually in the form of salts. Many synthetic drugs are structural modifications of the alkaloids, designed to change the effect of certain drugs and reduce their unwanted side-effects.

    Pyrrolizidine Alkaloids (PA) are complex molecules that include a pyrrolizidine nucleus: a pair of linked pyrrole rings. Each pyrrole is a five-sided structure with four carbons and one nitrogen forming the ring. Pyrroles are incorporated into the chlorophyll molecule.

    There are hundreds of types of Pyrrolizidine Alkaloids, with numerous isomers (same chemical formula but different arrangement of atoms). PAs can be saturated which are not toxic, or unsaturated which are toxic. The toxicity of each depends on the chemical complexity of the associated necic acids. There are three main categories: macrocyclic diesters, branched diesters, and monoesters. PAs can be free bases or N-oxides.

*"**Many herbal poisonings (not just Comfrey) that are recognized as such involve people consuming what is by any measure unreasonable amounts of a plant, or people regularly consuming large numbers of herbs.** A disturbing aspect of the use of Comfrey-containing capsules is the ease with which toxicity can be produced by the 'recommended' doses. It is possible that the patients reported were receiving pyrrolizidine exposure from other sources.*

It is also possible that they represent part of a subgroup which for some reason is uniquely susceptible to the toxic actions of pyrrolizidines. Whatever the case, it is clear that a substantial fraction of the population is exposed to unnecessary risks in the use of Comfrey-pepsin preparations and other Comfrey-containing products."
-Toxicants of Plant Origin, Volume I: Alkaloids edited by Peter R. Cheeke. Boca Raton, Florida: CRC Press, **1989**. Chapter 3: 'Human Health Implications of Pyrrolizidine Alkaloids and Herbs Containing Them' by Ryan J. Huxtable, University of Arizona.

"Individual susceptibility (inter-individual differences) to PA (Pyrrolizidine Alkaloid) toxicity varies according to a number of factors such as age, liver function, genetic variation in microsomal enzymes and glutathione transferases, possible drug-drug interactions, and diet."
-'Using Herbs that Contain Pyrrolizidine Alkaloids' by Alison Denham, B.A., MNIMH (National Institute of Medical Herbalists), University of Central Lancashire, England; The European Journal of Herbal Medicine, Volume 2, No. 3, pages 27-38, **1996**.

"Many herbalists feel the Comfrey plant causes no threat to humans if consumed only as needed and avoided in prolonged high doses, while others have stopped using it altogether. It is important to realize that a constituent with a negative effect may be neutralized or greatly diminished when combined with other herbs in formulas.
To help make your own choice, those with a personal and/or parental history of alcoholism, hepatitis or mononucleosis, a history of drug use (recreational or otherwise) and caffeine users should all probably avoid using this Comfrey herb internally. However, it may safely be used externally by everyone. *Further, certain herb companies offer Pyrrolizidine Alkaloid-free (PA-free) Comfrey products."*
-Healing with the Herbs of Life: Hundreds of Herbal Remedies, Therapies, and Preparations by Lesley Tierra, L.Ac., Herbalist, A.H.G. New York: Crossing Press / Crown Publishing Group, **2003**.

*"**It has also been observed that cofactors can exacerbate (make worse) the PA (Pyrrolizidine Alkaloid) poisoning: liver damaging agents, bacterial or viral infections as well as medical drugs like barbiturates, or metals like copper, or mycotoxins like aflatoxins can increase the severity and likelihood of PA liver damage** (Yee, Kinser, et al.,2000; Newberne, Rogers, 1973; White, Mattocks, et al., 1973; Tuchweber, Kovacs, et al., 1974; Lin, Liu, et al., 1974: Bull, Culvenor, et al., 1968)."*
-'Plants Containing Pyrrolizidine Alkaloids: Toxicity and Problems' by Helmut Wiedenfeld, Pharmaceutical Institute, University of Bonn, Germany; Food Additives and Contaminants, Volume 28, No. 3, pages 282-292, **2011**.
(A cofactor contributes to the creation of a disease.)

*"**Tell all your health care providers about any complementary health approaches you use.** Give them a full picture of what you do to manage your health. This will help ensure coordinated and safe care.*
It's especially important to talk to your health care providers if you:
1. Take any medications (whether prescription or over-the-counter). Some dietary supplements have been found to interact with medications. For example, the herbal supplement St. John's wort interacts with many medications, making them less effective.
2. Are thinking about replacing your regular medication with one or more dietary supplements.
3. Expect to have surgery. Certain dietary supplements may increase the risk of bleeding or affect the response to anesthesia.
4. Are pregnant, nursing a baby, attempting to become pregnant, or considering giving a child a dietary supplement. Most dietary supplements have not been tested in pregnant women, nursing mothers, or children.
5. Have any medical conditions. Some dietary supplements may harm you if you have particular medical conditions. For example, by taking supplements that contain iron, people with hemochromatosis, a hereditary disease in which too much iron accumulates in the body, could further increase their iron levels and therefore their risk of complications such as liver disease."
-'Using Dietary Supplements Wisely', NCCIH Publication No. D426, by United States Department of Health and Human Services, National Institutes of Health, National Center for Complementary and Integrative Health, https://nccih.nih.gov, June **2014**.

<u>Age and Alkaloid Susceptibility</u>

*"**Children appear to be especially susceptible to VOD (Veno-Occlusive Disease) from the age of 2 to 14. One must use caution in the use of PA-containing (Pyrrolizidine Alkaloid) herbs (such as Comfrey) for children.***
Yet, there are cases where the use of Comfrey leaf (e.g., in colitis) carries a much smaller risk than drug treatment. But they should not be used if the child is already taking prednisolone or any other Cytochrome P450 enhancing drug."
-'Using Herbs that Contain Pyrrolizidine Alkaloids' by Alison Denham, B.A., MNIMH (National Institute of Medical Herbalists), University of Central Lancashire, England; The European Journal of Herbal Medicine, Volume 2, No. 3, pages 27-38, **1996**.

*"**Pregnancy and Breastfeeding:** Comfrey leaf contains considerably lower levels of the alkaloids than the roots and cultivated varieties are available that do not contain pyrrolizidine alkaloids.*
Internal Use of Comfrey:
   Plants containing pyrrolizidine alkaloids should not be used internally during pregnancy or lactation.
External Use of Comfrey:
   *Some sources extend this caution to topical use as well. In animal studies, however, the absorption of pyrrolizidine alkaloids through the skin is poor. **Comfrey ointments or salves are widely used for sore nipples in breastfeeding and for infants' skin rashes.** The external use of Comfrey has not been implicated in any liver toxicity. A number of leading herbalists recommend the topical use of Comfrey during pregnancy or lactation."*

-'Herbs and the Childbearing Woman: Guidelines for Midwives' by Cindy Belew, CNM (Certified Nurse Midwife), MS, San Francisco, California; Journal of Nurse-Midwifery: American College of Nurse-Midwives, Volume 44, No. 3, pages 231-252, May/June **1999**.

"Avoid Comfrey during pregnancy and breastfeeding as it may be hepatotoxic (liver damaging).
Toxins from Comfrey can be found in milk from grazing animals that have consumed Comfrey. Thus it is likely that Comfrey toxins would also be excreted in human breast milk."
-Natural Standard Herb and Supplement Guide: An Evidence-Based Reference by Catherine E. Ulbricht. Missouri: Mosby, **2010**, pages 250-254.

*"**Children aged between 1-14 years old are more at risk of PA (Pyrrolizidine Alkaloids) toxcity than adults,** even at low doses and when their lower body weight is taken into account. **Foetuses (fetuses) and babies are greatly at risk of PA toxicity,** even at low doses because of normal high levels of copper in their livers. **Pregnant and lactating (nursing) mothers are also at risk** as PA appears to cross the placental barrier and are excreted in milk."*
-'Pyrrolizidine Alkaloids: Key Points for Herbal Practitioners' by Helen Phillips M.Sc. Herbal Medicine, Diploma of Phytotherapy, MNIMH and Alison Denham, M.A., FNIMH (National Institute of Medical Herbalists); The Forager's Path, LLC, School of Botanical Studies, Flagstaff, Arizona, August **2016**.

Liver Function and Alkaloid Susceptibility

"Effect of dietary tansy ragwort (Senecio jacobaea), Comfrey (Symphytum officinale), bracken (Pteridium aquilinum) and alfalfa (Medicago sativa) on hepatic (liver) drug-metabolizing enzymes in rats were measured.
   Tansy ragwort and bracken increased (P less than 0.05) the activity of glutathione transferase and epoxide hydrolase.
   Comfrey and alfalfa increased (P less than 0.05) the activity of aminopyrine N-demethylase.
Comfrey (Symphytum officinale) was harvested in the prebloom vegetative stage. The plants were dried at 95 C (203 F), finely ground in a Wiley mill, and stored at room temperature. Alfalfa was included as a non-poisonous plant. The rats were fed the diet for 21 days ad lib (as much as they want) and were then killed and the livers removed.
   Weight gains were reduced (P < 0.05) in rats fed 1% tansy ragwort and 30% Comfrey.
   Comfrey and alfalfa stimulated the activity of aminopyrine N-demethylase but did not affect glutathione S-transferase or epoxide hydrolase activities.

Weight Gains and Liver Enzyme Activities of Rats (Enzyme activity = nmol product / min / mg protein)

Treatment	Daily Weight Gain	Aminopyrine N-demethylase	Glutathione S-transferase
Control: soy-corn	7.1 grams	7.05 +- 0.05	239 +-16
1% Tansy Ragwort	5.1	7.53 +- 0.29	296 +- 13
5% Comfrey	**6.9**	**8.75 +- 0.48**	**254 +- 18**
10% Comfrey	**6.4**	**9.26 +- 0.74**	**222 +- 20**
30% Comfrey	**3.3**	**8.64 +- 0.44**	**262 +- 24**
30% Alfalfa	5.5	9.74 +- 0.42	270 +- 23 "

-'Consumption of Poisonous Plants (Senecio jacobaea, Symphytum officinale, Pteridium aquilinum, Hypericum perforatum) by Rats: Chronic Toxicity, Mineral Metabolism, and Hepatic Drug-Metabolizing Enzymes' by B.J. Garrett, P.R. Cheeke, C.L. Miranda, D.E. Goeger and D.R. Buhler, Oregon State University, Corvallis, Oregon; Toxicology Letters, Amsterdam, Netherlands, Volume10, No. 2-3, pages 183-188, February **1982**.
   (In statistical hypothesis testing, the p-value or probability value is the probability that when the null hypothesis is true, the statistical summary would be greater than or equal to the actual observed results. Null hypothesis means there is no significant difference between specified populations. A small p-value, usually 0.05, indicates strong evidence against the null hypothesis. A large p-value, such as > 0.05, indicates weak evidence.)

"Taking Comfrey by mouth may increase the activity of the hepatic (liver) enzyme, aminopyrine N-demethylase."
-Natural Standard Herb and Supplement Guide: An Evidence-Based Reference by Catherine E. Ulbricht. Missouri: Mosby, **2010**, pages 250-254.

"Factors affecting alkaloid risk:
Liver health *(past or present viral or bacterial infection, alcohol intake, history of liver disease or alcohol dependency). Some viruses, bacteria and fungi act as hepatoxic (liver toxic) agents, working synergistically with PA and increasing the risk of hepatoxicity.*
Compromised liver function *(low levels of liver enzymes and microsomal enzymes, viral and bacterial infections, alcohol dependency) will encourage use of all pathways to deal with ingested PA, including the CYP450 route to toxic metabolite formation. Levels of liver enzymes vary hugely between individuals.*
High copper levels in the liver causes increased susceptibility to low-level PA toxicity."
-'Pyrrolizidine Alkaloids: Key Points for Herbal Practitioners' by Helen Phillips M.Sc. Herbal Medicine, Diploma of Phytotherapy, MNIMH and Alison Denham, M.A., FNIMH (National Institute of Medical Herbalists); The Forager's Path, LLC, School of Botanical Studies, Flagstaff, Arizona, August **2016**.
   (Synergistic means the interaction of two or more substances that create a combined effect greater than the sum of

their separate effects. In other words, the whole is greater than the sum of its parts.)

Kidney Function and Alkaloid Susceptibility

"Practice tips, guidelines and more for the nephrology professional:
Common botanical medicines that may be harmful to CKD (Chronic Kidney Disease) patients:
    Aloe, Buckthorn, Capsicum, Cascara, Chapparal, **Comfrey,** *Dandelion, Ephedra, Licorice, Mate, Nettle, Noni Juice, Pennyroyal, Rhubarb, Sassafras, Segrada, Senna.*
    Caution- Anticoagulant properties: Ginger, Gingko Biloba, Garlic, Ginseng, Feverfew."
-'Botanical Medicines and CKD Patients' by Catherine M. Goeddeke-Merickel, MS, RD, LD, Consultant and Nephrology Specialist, Lincoln, Nebraska; Rental Nutrition Forum: Renal Dieticians, Academy of Nutrition and Dietetics, Volume 28, No. 3, Supplement 1, page 2, Summer **2009**.
    (Nephrology is the medical practice about the physiology of the kidneys. It is the study of normal kidney function, kidney disease, and the treatment of kidney disease with diet, medication, dialysis and kidney transplants.)
    (Chronic Kidney Disease is long-term disease of the kidneys eventually leading to renal failure.)

*"***Nephrotoxicity of herbs and herbal preparations:*** The herbal treatments for complications caused by dialysis and chronic renal failure and any use of medicinal herbs may be inappropriate for the renal patient.*
Literature on herbs and the dialysis patient suggested the avoidance of *borage (Boragoofficinalis),* **Comfrey (Symphytum spp.),** *coltsfoot (Tussilagofarfara) and life root (Senecioaureus) because of their pyrrolizidine alkaloid content and hence hepatotoxic potential, and sassafras (Sassafras albidum) because of its safrole content."*
-'A Brief Study of Toxic Effects of Some Medicinal Herbs on Kidney' by Mohammad Asif, Department of Pharmacy, GRD (PG) Institute of Management and Technology, Uttarakhand, India; Advanced Biomedical Research, Wolters Kluwer - Medknow Publications, Mumbai, India, Volume 1, Issue 3, 9 pages, July-September **2012**.
(Nephrotoxicity is toxicity in the kidneys. It is a poisonous effect on kidney function due to toxic chemicals and/or medications.)

*"***Select Herbs That May Be Harmful to CKD (Chronic Kidney Disease) Patients:***
Alfalfa, Aloe, Aristolochic acid, Artemisia absinthium (wormwood plant), Autumn crocus, Bayberry, Blue cohosh, Broom, Buckthorn, Capsicum, Cascara, Chaparral, Chuifong tuokuwan (Black Pearl), Coltsfoot,* **Comfrey,** *Dandelion, Ephedra (Ma Huang), Ginger, Gingko, Ginseng, Horse chestnut, Horsetail, Licorice, Lobelia Mandrake Mate Nettle Noni juice Panax Pennyroyal Periwinkle Pokeroot Rhubarb Sassafras Senna St. John's wort, Tung shueh, Vandelia cordifolia, Vervain, Yohimbe."*
-'Mangement of Chronic Kidney Disease' by Chronic Kidney Disease Guideline Team, Guidelines for Clinical Care Ambulatory, Michigan Medicine, University of Michigan, Ann Arbor, Michigan, 27 pages, **2013**-2014.

Genetic Variation and Alkaloid Susceptibility: Cytochrome P450

    Cytochrome P450 (CYP) is a protein containing heme (an iron ion coordinated to a porphyrin) as a cofactor and, therefore, is a hemoprotein. CYPs use molecules as substrates in enzymatic reactions. They are the terminal (end) oxidase (reacts directly with oxygen) enzymes in electron transfer chains, categorized as P450-containing systems.

    Cytochrome P450 3A4 (CYP3A4) is an enzyme found in the liver and intestine. It oxidizes (combines with oxygen to make it inactive) foreign organic molecules (xenobiotics) such as toxins or drugs, so they can be removed from the body. While many drugs are made inactive by CYP3A4, some drugs are made more active by CYP3A4. Some drugs and some food such as grapefruit juice interfere with the activity of CYP3A4, either increasing or decreasing the oxidation by CYP3A4.

*"**Individual susceptibility to PA (Pyrrolizidine Alkaloid) toxicity must be determined by inter-individual diffferences in hepatocyte (liver cell) activation and detoxification of pyrroles by Cytochrome P450 enzymes (CYP)** as has been demonstrated with senecionine (alkaloid in the plant genus Senecio).*
CYP levels are influenced (increased or decreased) by intake of prescribed drugs."
-'Using Herbs that Contain Pyrrolizidine Alkaloids' by Alison Denham, B.A., MNIMH (National Institute of Medical Herbalists), University of Central Lancashire, England; The European Journal of Herbal Medicine, Volume 2, No. 3, pages 27-38, **1996**.
    (Pyrrolizidine alkaloids are complex molecules that include a pyrrolizidine nucleus: a pair of linked pyrrole rings. Each pyrrole is a five-sided structure with four carbons and one nitrogen forming the ring. Pyrroles are incorporated into the chlorophyll molecule.)

*"**Cytochrome P450 enzymes are essential for the metabolism of many medications.***
    Cytochrome P450 (CYP450) enzymes are essential for the production of cholesterol, steroids, prostacyclins, and thromboxane A2.
    They also are necessary for the detoxification of foreign chemicals and the metabolism of drugs.*
Although this class has more than 50 enzymes, six of them metabolize 90 percent of drugs, with the two most significant enzymes being CYP3A4 and CYP2D6. Genetic variability (polymorphism) in these enzymes may influence a patient's response

to commonly prescribed drug classes, including beta blockers and antidepressants.
Cytochrome P450 enzymes can be inhibited (decreased) or induced (increased) by drugs, resulting in clinically significant drug-drug interactions that can cause unanticipated adverse reactions or therapeutic failures. Interactions with warfarin, antidepressants, antiepileptic drugs, and statins often involve the cytochrome P450 enzymes.
Knowledge of the most important drugs metabolized by cytochrome P450 enzymes, as well as the most potent inhibiting and inducing drugs, can help minimize the possibility of adverse drug reactions and interactions."
-'The Effect of Cytochrome P450 Metabolism on Drug Response, Interactions, and Adverse Effects' by Tom Lynch, Pharm.D. and Amy Price, M.D., Eastern Virginia Medical School, Norfolk, Virginia; American Family Physician, Volume 76, No. 3, pages 391-396, August 1 **2007**.

"**Use Comfrey cautiously in patients taking cytochrome P450 3A4-inducing agents, which may increase the conversion of compounds in Comfrey to toxic metabolites.**"
-Natural Standard Herb and Supplement Guide: An Evidence-Based Reference by Catherine E. Ulbricht. Missouri: Mosby, **2010**, pages 250-254.

"**The PAs (Pyrrolizidine Alkaloids), which have minimal toxicity in their original form, are metabolized in the liver and can become toxic metabolites,** depending on the PA and on the particular condition of the liver enzymes.
The toxic metabolites, highly reactive smaller pyrroles (the dehydro- form of the alkaloids) resulting from action of microsomal enzymes in the liver, can act locally within the liver cells to cause damage at the chromosome level."
-'Safety Issues Affecting Herbs: Pyrrolizidine Alkaloids" by Subhuti Dharmananda, Ph.D., Director, Institute for Traditional Medicine, Portland, Oregon, **2016**.

"**Unsaturated PA (Pyrrolizidine Alkaloids) can be broken down and safely excreted via two 2 enzyme pathways in the liver: carboxylesterases and mono-oxygenases.**
If these two routes are saturated (e.g., high dose of PA, high concentration of macrocyclic PA, compromised liver function, or genetic tendency), **then PA are metabolised by CYP450 enzymes. This results in formation of pyrroles.**
High doses of toxic PA saturate the initial detoxification routes, thus initiating the CYP450 pathway and formation of pyrroles. If the dose is high enough (or the individual susceptible), acute poisoning occurs. Even in lower doses, if initial detoxification routes are saturated, cellular damage can occur and latent toxic metabolites bind to tissue."
-'Pyrrolizidine Alkaloids: Key Points for Herbal Practitioners' by Helen Phillips M.Sc. Herbal Medicine, Diploma of Phytotherapy, MNIMH and Alison Denham, M.A., FNIMH (National Institute of Medical Herbalists); The Forager's Path, LLC, School of Botanical Studies, Flagstaff, Arizona, August **2016**.
(Acute disease means it has an abrupt/rapid onset. It usually means an illness of short duration, rapidly progressive, and in need of urgent care. Acute is in reference to time as opposed to subacute or chronic. Chronic means lasting a long time, usually 3 months or more. Subacute is between acute and chronic.)

Drugs and Herbs that Increase Cytochrome P450 (CYP) Levels
Speeds up metabolism of some medicinal drugs so body can use them (this can be good or bad).
Speeds up detoxification of some foreign chemicals, and thereby increases breakdown by the liver
of Pyrrolizidine Alkaloids to toxic forms.

"*CYP3A4 (levels) are induced (increased):*
-by anticonvulsants (e.g., carbamazepine, phenobarbitone, phenytoin, primidone, topiramate).
-by the corticosteroid dexamethasone.
-by the anti-fungal grisefulvin.
-by rifampicin (antibiotic) which is used to treat tuberculosis.
Rifampicin is considered the strongest CYP inducer, and no PA-containing plants whatsoever should be used in combination with it.
Phenobarbitone also induces glutathione transferase and thus glutathione remanufacture after reduction.
Where CYP3A4 levels are induced (increased) by a drug, the rate of pyrrole formation (breakdown of pyrrolizidine alkaloids to toxic forms) will be higher (faster).
For example, oral contraceptives which are metabolized by CYP3A4 may be less effective if the woman is taking phenobarbitone as the oestrogens (estrogens) are metabolized more quickly. The 'British National Formulary' states that rifampicin (used to treat tuberculosis) renders (makes) oral contraceptives ineffective.
The case of VOD (Veno-Occlusive Disease) ascribed to (supposedly caused by) Symphytum which occurred in Britain involved a 14 year old boy suffering from Crohn's disease*. For part of the time, he was also taking prednisolone and sulphasalzine."
-'Using Herbs that Contain Pyrrolizidine Alkaloids' by Alison Denham, B.A., MNIMH (National Institute of Medical Herbalists), University of Central Lancashire, England; The European Journal of Herbal Medicine, Volume 2, No. 3, pages 27-38, **1996**.
(In hepatic Veno-Occlusive Disease some small veins in the liver are blocked. It can be caused by high-dose chemotherapy or after consuming certain plant alkaloids such as pyrrolizidine. Symptoms include weight gain due to fluid retention, increased liver size, and raised levels of blood bilirubin.)

(* -Details about this case are in subsection 'Scientific Studies Showing Dangers of Alkaloids in Comfrey' in section 'Alkaloids in Comfrey' {Chapter 30}.)

"Interactions of Pyrrolizidines with Other Drugs:
Pyrroles are inducing (increasing) agents; retrosine alkaloid gives rise to enhanced (increased) expression of hepatic (liver) CYPs 1A1, 1A2, 2E1, and 2B1/2 in rats, but it is not known whether the pyrroles induce their own metabolism during chronic administration nor whether pyrrolizidines, by analogy with paracetamol, are more toxic in patients with induced hepatic CYP 450 enzymes.
Chronic ingestion of pyrrolizidines in the form of herbal teas by humans is likely to lead to induction of the CYP450 enzymes, but whether this is specific to one or more of the isoforms has not yet been determined. Pyrrolizidine consumption is queried (asked) about the patient in cases in which such drug interactions may be important, e.g., the use of warfarin, phenytoin, cyclosporine, etc."
-'Pyrrolizidine Poisoning: A Neglected Area in Human Toxicology' by Michael J. Stewart and Vanessa Steenkamp, University of the Witwatersrand Medical School, Johannesburg, South Africa; Therapeutic Drug Monitoring Journal, New York, Volume 23, No. 6, pages 698-708, **2001**.

"Rates of CYP450 enzyme activity can vary significantly between individuals resulting in considerable variability of response to PA exposure. Some drugs and herbs induce (increase) CYP450 enzymes and could increase the risk of toxicity if taken alongside PA-containing herbs.
Factors that induce CYP450 enzyme pathway:
    Drugs: Griseofulvin (anti-fungal), rifampcin (antimycobacterial), carbamazepine / phenytoin / primidone / topiramate (anti-convulsants), phenobarbital (however, also induces glutathione S-transferase, encouraging detoxification of pyrrole ester), dexamethasone (corticosteroid) (lower potency glucocorticoids, for example, prednisolone have minimal effect on CYP3A4).
    Herbs: Hypericum perforatum (St. Johns Wort).
    DDT and Edrin: (organo-chlorine pesticides).
    Acute Systemic Hypoxia (not enough oxygen): (chronic respiratory or cardiac insufficiency increases CYP3A4 activity)."
-'Pyrrolizidine Alkaloids: Key Points for Herbal Practitioners' by Helen Phillips M.Sc. Herbal Medicine, Diploma of Phytotherapy, MNIMH and Alison Denham, M.A., FNIMH (National Institute of Medical Herbalists); The Forager's Path, LLC, School of Botanical Studies, Flagstaff, Arizona, August **2016**.

Drugs and Herbs that Decrease Cytochrome P450 (CYP) Levels
Slows down metabolism of some medicinal drugs (this can be good or bad).
Slows down detoxification of some foreign chemicals, and thereby decreases speed of breakdown by the liver
    of Pyrrolizidine Alkaloids to toxic forms.

"Drugs which suppress (decrease) CYP3A4 (levels) are the systemic antifungals ketoconazole and itraconazole, the oral contraceptive gestodene, chloramphenicol and cimetidine.
Some flavonoids, e.g., in grapefruit juice, are general inhibitors of Cytochrome P450s."
-'Using Herbs that Contain Pyrrolizidine Alkaloids' by Alison Denham, B.A., MNIMH (National Institute of Medical Herbalists), University of Central Lancashire, England; The European Journal of Herbal Medicine, Volume 2, No. 3, pages 27-38, **1996**.

Drugs Metabolized (Oxidized/Deactivated) by Cytochrome P450 (CYP)

"Other important drugs which are metabolized (deactivated) by CYP3A4 are cyclosporin (immunosuppressant), nifedipine (Adalat®), erythromycin (antibiotic), terfenadine (Triludan®), and ethinyloestradiol (estrogen).
If a person is taking any of these drugs, then the PAs (Pyrrolizidine Alkaloids) are competing with the drug for metabolism on a fixed number of sites. Drug-drug interactions are unlikely to be a problem in that competition for sites would reduce the rate of pyrrole formation."
-'Using Herbs that Contain Pyrrolizidine Alkaloids' by Alison Denham, B.A., MNIMH (National Institute of Medical Herbalists), University of Central Lancashire, England; The European Journal of Herbal Medicine, Volume 2, No. 3, pages 27-38, **1996**.

<u>Diet and Alkaloid Susceptibility: Glutathione</u>

Glutathione is an antioxidant that prevents damage to cellular components caused by reactive oxygen types such as free radicals, peroxides, lipid peroxides, and heavy metals. Glutathione is synthesized in the body from the 3 amino acids L-cysteine, L-glutamic acid, and glycine. Proteins are biomolecules of long chains of amino acids.

"Pyrroles (from the breakdown of Pyrrolizidine Alkaloids) can be broken down and safely excreted by glutathione and glutathione S-transferase in the liver.
There is substantial inter-individual (between individuals) genetic variation in levels of these compounds. Individual variations in levels of hepatic (liver) glutathione and glutathione S-transferase are also likely to be genetically influenced.
Low levels of glutathione have been found to increase toxic pyrrole formation *in rodent experiments."*
-'Pyrrolizidine Alkaloids: Key Points for Herbal Practitioners' by Helen Phillips M.Sc. Herbal Medicine, Diploma of Phytotherapy, MNIMH and Alison Denham, M.A., FNIMH (National Institute of Medical Herbalists); The Forager's Path, LLC, School of Botanical Studies, Flagstaff, Arizona, August **2016**.
(Pyrrolizidine alkaloids are complex molecules that include a pyrrolizidine nucleus: a pair of linked pyrrole rings. Each pyrrole is a five-sided structure with four carbons and one nitrogen forming the ring. Pyrroles are incorporated into the chlorophyll molecule.)
(Cytochrome P450 enzymes detoxify pyrroles but situations vary as to how this is done safely by the body. It can be detoxified too fast thereby creating health problems.)

Food with Glutathione

These foods with glutathione help detoxify pyrrolizidine alkaloids:
Fresh or frozen foods are good. Not in can, jar or bottle. The less processing and cooking, the better.
  Animal: red meat (beef, pork, etc.; pan fried, not processed), ham (boiled), poultry, chicken liver.
  The amount of glutathione is the same with and without fat / skin.
  Fruit: fresh fruit including blackberries, blueberries, cantaloupe, oranges, peaches, strawberries, watermelon.
  Vegetable: asparagus, avocado, broccoli and others in that family, garlic, leafy greens especially spinach, onions, parsley, potatoes (not sweet potatoes), nuts especially walnuts, squash (winter), tomatoes (fresh).
  Herbs and Supplements: curcumin/turmeric, milk thistle (Silybum marianum), vitamin C.
-Most of the above is from: 'Glutathione in Foods Listed in the National Cancer Institute's Health Habits and History Food Frequency Questionnaire' by Dean P. Jones, Ralph J. Coates, Elaine W. Flagg, John W. Eley, Gladys Block, Raymond S. Greenberg, Elaine W. Gunter and Bethany Jackson, United States Department of Health and Human Services; Nutrition and Cancer, Volume 17, No. 1, pages 57-75, **1992**.

*"**Seaweeds as Source of Polyphenols and Glutathione:** Seaweeds (marine macro-algae) are rich in bioactive compounds which have important pharmaceutical and biomedical values.*
Seaweeds are also reported to have natural assemblages of glutathione. *Seaweeds are rich in polysaccharides, vitamins, minerals and trace elements for which the consumption of marine algae has been increased globally, especially by the vegetarians.* ***Seaweeds are a potential reserve of naturally occurring antioxidants*** *and have been used since times immemorial in diets and formulations in many parts of the world."*
-'Glutathione Related Disorders: Do Seaweeds have Potential for Cure?' by Priyadarshini Rautray and Luna Samanta, India; Asian Journal of Biomedical and Pharmaceutical Sciences, Volume 6, No. 58, pages 20-26, **2016**.

Low Protein Diet Lowers Glutathione Levels

"Rhodes (*1954-1957) has produced evidence that, whatever the apparent nutritional status of the patient, veno-occlusive disease is invariably associated with lack of protein in some degree. Another instance of toxic liver injury, also occurring particularly in the age group one to six years in which veno-occlusive disease is commonest, is seen in the 'vomiting sickness of Jamaica'. ***The main clinical features of this condition, which also invariably occurs in relatively under-nourished subjects,*** *have been recently reviewed (**Stuart, Jelliffe, and Hill, 1955)."*
-'Veno-Occlusive Disease of the Liver' by K.L. Stuart and G. Bras, Departments of Medicine and Pathology, University College of West Indies, Jamaica; Quarterly Journal of Medicine, New Series XXVI, No 103, page 291-315, July **1957**.
(* -'Two Types of Liver Disease in Jamaican Children' by K. Rhodes, M.D. Thesis, University of St. Andrews, 1955; and West Indian Medical Journal, Jamaica, Volume 6, No. 1, pages 1-29, March 1957. I was unable to get this report. If you have a copy, could you send it to me.)
(** -'Acute Toxic Hypoglycaemia Occurring in the Vomiting Sickness of Jamaica: Clinical Aspects' by K.L. Stuart, M.D., M.R.C.P., M.R.C.P.E., D.T.M., and H.D.B. Jelliffe, M.D., M.R.C.P., D.C.H., D.T.M., and H.K.R. Hill, M.D., B.SC, A.R.I.C; The Journal of Tropical Pediatrics, London, England, Volume 1, Issue 2, pages 69-87, September 1 1955. I was unable to get a copy of this. Though there is a similar 1954 report that is shorter by the same authors.)

*"**Liver Disease in the West Indies: The next important factor is that of the nutritional background.** The economy of the majority of the populace in the West Indies is very, very poor, and, as a consequence, as Rhodes has stressed (Rhodes, 1952),* ***there is a deficiency of protein in the diet and the diet lacks variety.*** *According to Garrow (1954) even normal adult Jamaicans have a low albumin-globulin ratio and MCHC.*
This poor and inadequate dietary background probably plays a not inconsiderable part in the production of overt liver disease, for it is well known that the malnourished liver has reduced detoxicating powers and is more prone to injury."

-'Some Observations on Liver Disease in the West Indies' by Kenneth R. Hill, B.SC., M.D., M.R.C.P., A.I.R.C., Professor of Pathology, Royal Free Hospital, London, England; Transactions of the Royal Society of Tropical Medicine and Hygiene, London, England, Volume 53, No. 2, pages 217-237, March 31 **1959**.
(MCHC is Mean Corpscular Hemoglobin Concentration which is ratio of amount of hemoglobin to volume of red blood cells.)

"Glutathione status is extremely important in drug detoxification. Glutathione status depends on dietary protein (low protein means low glutathioine). Stuart and Bras state 'we believe that **malnutrition plays an important but subsidary (secondary) part, most likely as a contributory cause and in affecting the prognosis'** of Veno-Occlusive Disease. Many of the little children that they treated for VOD (Veno-Occlusive Disease) had a high carbohydrate and low protein diet. The people who suffered in Afghanistan** were on a starvation level diet.*
Glutathione levels drop 50% by overnight fasting."
-'Using Herbs that Contain Pyrrolizidine Alkaloids' by Alison Denham, B.A., MNIMH (National Institute of Medical Herbalists), University of Central Lancashire, England; The European Journal of Herbal Medicine, Volume 2, No. 3, pages 27-38, **1996**.
(* -'Veno-Occlusive Disease of the Liver' by K.L. Stuart and G. Bras, Departments of Medicine and Pathology, University College of West Indies, Jamaica; Quarterly Journal of Medicine, New Series XXVI, No. 103, pages 291-315, July 1957.)
(** -An Outbreak of Hepatic Veno-Occlusive Disease in North-Western Afghanistan' by Mohabbat, Younos, Merzad, Srivastava, Sediq, and Aram, The Lancet, London, England, Volume 308, No 7980, pages 269-271, August 7 1976.)

*"**Protein malnutrition (i.e., poor diet, erratic eating, fasting) and selenium deficiency also deplete glutathione levels, potentially making the individual more susceptible to PA poisoning.**"*
-'Pyrrolizidine Alkaloids: Key Points for Herbal Practitioners' by Helen Phillips M.Sc. Herbal Medicine, Diploma of Phytotherapy, MNIMH and Alison Denham, M.A., FNIMH (National Institute of Medical Herbalists); The Forager's Path, LLC, School of Botanical Studies, Flagstaff, Arizona, August **2016**.

Alcoholism and Low Glutathione Levels

*"**Patients chronically abusing ethanol (alcohol) are more susceptible to the hepatotoxic effects of paracetamol.** This could be due to an increased activation of the drug to a toxic metabolite or to a **decreased capacity to detoxify the toxic metabolite by conjugation with glutathione (GSH).** The urinary excretion of cysteine- plus N-acetyl-cysteine-paracetamol, the two major products of detoxification of the reactive metabolite of paracetamol, was not significantly higher in chronic alcoholics arguing against a substantially increased metabolic activation of paracetamol.*
***Chronic alcoholics had significantly lower plasma concentrations of GSH than healthy volunteers** before the administration of paracetamol, and plasma GSH reached lower concentrations in the alcoholics after paracetamol.*
*In a group of patients with alcoholic hepatitis intrahepatic GSH was significantly lower than in patients with chronic persistent hepatitis and patients with non-alcoholic cirrhosis, suggesting that **low plasma GSH in alcoholics reflects low hepatic concentrations of GSH.***
The data indicate that low GSH may be a risk factor for paracetamol hepatotoxicity in alcoholics because a lower dose of paracetamol will be necessary to deplete GSH below the critical threshold concentration where hepatocellular necrosis starts to occur."
-'Glutathione Deficiency in Alcoholics Risk Factor for Paracetamol Hepatotoxicity' by B.H. Lauterburg (Department of Clinical Pharmacology, University of Berne, Switzerland) and Maria E. Velez (Center for Experimental Therapeutics, Baylor College of Medicine, Houston, Texas; Gut: British Society of Gastroenterology, Volume 29, No. 9, pages 1153-1157, **1988**.
    (In a similar way, if alcoholism has created low levels of glutathione, then liver toxicity with Comfrey eating/drinking is more likely than in a healthy person with enough glutathione.)

*"**Alcoholics die from paracetamol (acetaminophen) poisoning at lower dosage partly because of microsomal enzyme induction but also because of poor nutrition.**"*
-'Using Herbs that Contain Pyrrolizidine Alkaloids' by Alison Denham, B.A., MNIMH (National Institute of Medical Herbalists), University of Central Lancashire, England; The European Journal of Herbal Medicine, Volume 2, No. 3, pages 27-38, **1996**.   (Similar health problems occur in alcohlics with intake of alkaloids.)

*"**Alcohol dependency can deplete glutathione levels and result in poor hepatic (liver) function, increasing risk of PA (Pyrrolizidine Alkaloid) toxicity even at low dose.**"*
-'Pyrrolizidine Alkaloids: Key Points for Herbal Practitioners' by Helen Phillips M.Sc. Herbal Medicine, Diploma of Phytotherapy, MNIMH and Alison Denham, M.A., FNIMH (National Institute of Medical Herbalists); The Forager's Path, LLC, School of Botanical Studies, Flagstaff, Arizona, August **2016**.

*"PAs-containing species induce apoptosis (cell death) in normal human hepatocytes (liver cells). In addition, **cells are susceptible to a greater degree of liver damage, when they are exposed to both PAs and alcohol.***
Therefore, individuals with preexisting liver injury or simultaneously misusing alcohol or a xenobiotic (foreign substance) that induce liver damage may be more susceptible to PA-induced hepatotoxicity. Our research suggests that

inflammation may play a role in the pathogenesis of PA-induced hepatotoxicity."
-'Pyrrolizidine Alkaloids Enhance Alcohol-Induced Hepatotoxicity In Vitro in Normal Human Hepatocytes' by M.G. Neuman, L.B. Cohen and V. Steenkamp, Ontario, Canada and Pretoria, South Africa; European Review for Medical and Pharmacological Sciences, Volume 21, Suppl 1, pages 53-68, March **2017**.

Sulfur and Selenium in Diet is Needed to Make Glutathione

"Other dietary factors which are important in glutathione status are intake of selenium and of Cruciferae (Brassicaceae or cabbage family) such as brussels sprouts (and broccoli, cauliflower, kale, turnip, radish, etc.)."
-'Using Herbs that Contain Pyrrolizidine Alkaloids' by Alison Denham, B.A., MNIMH (National Institute of Medical Herbalists), University of Central Lancashire, England; The European Journal of Herbal Medicine, Volume 2, No. 3, pages 27-38, **1996**. (Cruciferous vegetables have sulfur-containing compounds called glucosinolates.)

"Glutathione is the main sulphydryl compound in mammalian cells. **The important role that glutathione plays during the response to inflammatory agents and xenobiotics (foreign substance) has been reported by many authors. The present study indicates that lung GSH (reduced form of Glutathione) concentrations are sensitive to food and sulphur amino acid intake in a similar way to the concentration in liver.** *Glutathione concentrations in liver and lung fall when food intake or sulphur amino acid intake is inadequate.*
The ability to synthesize hepatic (liver) GSH may be compromised to a much greater degree than the ability to synthesize protein during dietary sulphur amino acid insufficiency."
-'Dietary Sulphur Amino Acid Adequacy Influences Glutathione Synthesis and Glutathione-Dependent Enzymes During the Inflammatory Response to Endotoxin and Tumour Necrosis Factor-Alpha in Rats' by Emma A.L. Hunter and Robert F. Grimble, Department of Human Nutrition, University of Southampton, England; Clinical Science, Volume 92, pages 297-305, **1997**.

*"***Factors affecting alkaloid risk: Glutathione status (alcohol intake, low dietary sulphur, protein and selenium intake, erratic eating habits).*** Low glutathione levels are linked to poor diet and alcohol dependency.*
People with low intake of sulphur-containing foods *(e.g., cruciferous vegetables, shellfish, eggs, red meat, dried apricots, Allium spp.)* ***and therefore lower levels of glutathione, are potentially more susceptible to PA poisoning."***
-'Pyrrolizidine Alkaloids: Key Points for Herbal Practitioners' by Helen Phillips M.Sc. Herbal Medicine, Diploma of Phytotherapy, MNIMH and Alison Denham, M.A., FNIMH (National Institute of Medical Herbalists); The Forager's Path, LLC, School of Botanical Studies, Flagstaff, Arizona, August **2016**.
(Allium spp. {species} include chives, garlic, leek, onion, scallion, shallot and others.)
(Some foods high in selenium: beans, Brazil nuts, eggs, liver, meat (red), mushrooms, oats, poultry, brown rice, seafood, sunflower seeds.)

Environment and Alkaloid Susceptibility

*"**Individuals whose drinking pipes are copper-lined are at more risk of PA-toxicity at low levels.**"*
-'Pyrrolizidine Alkaloids: Key Points for Herbal Practitioners' by Helen Phillips M.Sc. Herbal Medicine, Diploma of Phytotherapy, MNIMH and Alison Denham, M.A., FNIMH (National Institute of Medical Herbalists); The Forager's Path, LLC, School of Botanical Studies, Flagstaff, Arizona, August **2016**.

Drug and Herb Interactions with Comfrey

"Comfrey, Symphytum officinale, leaf or root:
Drug Class: None reported. Type of interaction: none. Remedies: Wound and many skin problems."
-Herbal Drug Interactions: Professional Reference Guide by Robert Rogers, B.Sc., Professional Member of American Herbalist Guild, with Royal Alexandra Hospital, Capital Health Authority Organization of Alberta, Canada. Guelph, Ontario, Canada: Mediscript Communications Inc., **2016**.

Caution When Using These Herbs with Comfrey

"Interactions: ***Eucalyptus may increase the risk of PA (Pyrrolizidine Alkaloid) toxicity*** *because of enzyme induction by eucalyptus. Avoid administration with Comfrey."*
-Professional's Handbook of Complementary and Alternative Medicines, 3rd Edition' by Charles H. Fetrow, Pharmd (Doctor of Pharmacy) & Juan R. Avila, Pharmd. Pennsylvania: Lippincott Williams & Wilkins (imprint Wolters Kluwer), **2003**, page 235-238.

"Based on clinical evidence topical (skin) Comfrey may offer anti-inflammatory effects. ***Caution is advised in patients taking anti-inflammatory herbs, such as oral licorice or topical Ginkgo biloba, due to possible additive (increasing) effects.***

The combination of pokeweed (Phytolacca americana) and Comfrey may result in additive effects. Although not well studied in humans, both herbs can precipitate human glycoproteins, agglutinate sheep red blood cells (SRBCs), and stimulate lymphocyte adherence to nylon fibers."
-Natural Standard Herb and Supplement Guide: An Evidence-Based Reference by Catherine E. Ulbricht. Missouri: Mosby, **2010**, pages 250-254.

"Traditional wisdom says that eucalyptus worsens the negative effects of borage, coltsfoot, Comfrey, hound's tooth, or Senecio species. We have no reliable research to that effect, but caution is advised."
-Western Herbs for Martial Artists and Contact Athletes: Effective Treatments for Common Sports Injuries by Susan Lynn Peterson, Ph.D. Wolfeboro, New Hampshire: YMAA Publication Center, **2010**.

"Comfrey may interact with the following herbs so caution is advised when taking these: oral licorice, Ginkgo biloba, kava, pokeweed (Phytolacca americana), alkanna, borage, butterbur, coltsfoot, forget-me-not, gravel root, hemp agrimony, hound's tongue, lungwort, Senecio species, Rosemary (Rosmarinus officinalis L.), sassafras (Sassafras albidum Nutt.), chamomile (Matricaria chamomile L.), pot marigold calendula (Calendula officinalis L.), cockscomb (Celosia cristata L.), plantain (Plantago lancolata L. and Plantago major L.), shepherd's purse (Capsella burse pastoris L.), and St John's wort (Hypericum perforatum L.)."
-'Herbal Remedies for Athletes: A Handbook' by Stephanie J Troscinski, CAT(C), ATC (Certified Athletic Therapist/Trainer), thesis for Masters of Science, Complementary Alternative Medicine, American College of Healthcare Sciences, Portland, Oregon. Published in Wainfleet, Ontario, Canada, August **2015**.

"Eucalyptus can increase the toxicity of herbs that contain hepatotoxic pyrrolizidine alkaloids (PAs). PAs can damage the liver. Herbs containing hepatotoxic PAs include alkanna, boneset, borage, butterbur, coltsfoot, Comfrey, forget-me-not, gravel root, hemp agrimony, and hound's tongue; and the Senecio species plants dusty miller, groundsel, golden ragwort, and tansy ragwort."
-Medline® (Medical Literature Analysis and Retrieval System Online) from the 'United States National Library of Medicine' is a bibliographic database of life sciences and biomedical information. It includes articles from academic journals about medicine, nursing, pharmacy, dentistry, veterinary medicine, and health care. https://medlineplus.gov, **2018**.

Do Not Use Comfrey with Drugs or Herbs that Have Alkaloids or Harm the Liver

"It is paramount (great importance) for clinicians (medical practitioners) to be aware of known or potential drug-herb interactions to adequately treat their patients. Borage does contain low concentrations of unsaturated pyrrolizidine alkaloids known to cause hepatotoxic (liver toxic) effects (e.g., Comfrey). **Therefore, do not use borage (or Comfrey) with other hepatotoxic drugs, such as anabolic steroids, phenothiazines (anti-psychotic), or ketoconazole (anti-fungal)."**
-'Herbal Medicinals: Selected Clinical Considerations Focusing on Known or Potential Drug-Herb Interactions' by Lucinda G. Miller, Pharm.D., BCPS, Department of Pharmacy Practice, Texas Tech University Health Sciences Center, Amarillo, Texas; Archives of Internal Medicine, American Medical Association®, Volume 158, pages 2200-2211, November 9 **1998**.

"Other PA (Pyrrolizidine Alkaloid) containing herbs may increase risk of toxicity.
Herbs that contain PAs include agrimony, alkanna, alpine ragwort, borage, colt's foot, dusty miller, golden ragwort, goundsel, gravel root, ground's tongue, hemp, petasties, and tansy ragwort. Avoid administration with Comfrey."
-Professional's Handbook of Complementary and Alternative Medicines, 3rd Edition' by Charles H. Fetrow, Pharmd (Doctor of Pharmacy) & Juan R. Avila, Pharmd. Pennsylvania: Lippincott Williams & Wilkins (imprint Wolters Kluwer), **2003**, page 235-238.

"Oral or absorbed Comfrey may have additive adverse effects on the liver when used in combination with hepatotoxic (liver damaging) herbs, such as kava or supplements.
Oral or absorbed Comfrey, in combination with other pyrrolizidine alkaloid-containing herbs, may increase total levels of pyrrolizidine alkaloid consumed, which increases the risk for toxicity.
Herbs containing pyrrolizidine alkaloids include alkanna (alkanet), borage, butterbur, coltsfoot, forget-me-not, gravel root (Joe Pye), hemp agrimony (holy rope), hound's tongue, lungwort, and Senecio species."
-Natural Standard Herb and Supplement Guide: An Evidence-Based Reference by Catherine E. Ulbricht. Missouri: Mosby, **2010**, pages 250-254.   (Senecio is a genus of the daisy family {Asteraceae} that includes ragworts and groundsels.)

"It has also been observed that cofactors can exacerbate (make worse) PA poisoning:
liver damaging agents, bacterial or viral infections but also medical drugs like barbiturates or metals like copper or mycotoxins like aflatoxins can increase the severity and likelihood of PA liver-damage (Yee et al. 2000; Newberne and Rogers 1973; White et al. 1973; Tuchweber et al. 1974; Lin et al. 1974: Bull et al. 1968)."
-'Toxicity of Pyrrolizidine Alkaloids to Humans and Ruminants' by Helmut Wiedenfeld and John Edgar, Phytochemistry Reviews: Proceedings of the Phytochemical Society of Europe, Volume 10, Issue 1, pages 137-151, March **2011**.

"Herbals with the potential to cause organ toxicity may cause further risk of toxicity when drugs with similar toxicity are administered concurrently, such as when the hepatotoxic herbal Comfrey is given with large and prolonged doses of acetaminophen."

-Readings in Advanced Pharmacokinetics: Theory, Methods and Applications edited by Ayman Noreddin. London, England: Intech Open, **2012**. ('Chapter 7: Pharmacokinetics and Drug Interactions of Herbal Medicines: A Missing Critical Step in the Phytomedicine/Drug Development Process' by Obiageri O. Obodozie, National Institute for Pharmaceutical Research and Development, Abuja, Nigeria, pages 127-156.)

    (Acetaminophen includes brands such as: Tylenol®, Mapap®, Ofirmev®, FeverAll®, Acephen®, Nortemp®, Jr. Strength Pain Reliever®, Little Remedies Fever and Pain®, Children's Pain Reliever®, Pain Relief Extra Strength®, and more.)

"Do not take Comfrey with medications that can harm the liver (Hepatotoxic drugs). Comfrey might harm the liver. Taking Comfrey along with medication that might also harm the liver can increase the risk of liver damage.
Some medications that can harm the liver include acetaminophen (Tylenol® and others), amiodarone (Cordarone®), carbamazepine (Tegretol®), isoniazid (INH®), methotrexate (Rheumatrex®), methyldopa (Aldomet®), fluconazole (Diflucan®), itraconazole (Sporanox®), erythromycin (Erythrocin®, Ilosone®, others), phenytoin (Dilantin®), lovastatin (Mevacor®), pravastatin (Pravachol®), simvastatin (Zocor®), and many others."
-'Comfrey' by WebMD.com®, the contents of the WebMD® site are for informational purposes only; it is not intended to be a substitute for professional medical advice, diagnosis, or treatment; always seek the advice of your physician or other qualified health provider with any questions you may have regarding a medical condition. WebMD® does not provide medical advice, diagnosis or treatment, July **2018**.

"What should I avoid while using Comfrey? Avoid using Comfrey together with other herbal/health supplements that can also harm the liver. This includes androstenedione, bishop's weed, borage, chaparral, DHEA, echinacea, garlic, germander, golden ragwort, hound's tongue, kava, licorice, niacin (vitamin B3), pennyroyal oil, red yeast, St. John's wort, and others."
-'Comfrey' by Drugs.com® provides independent information on more than 24,000 prescription drugs, over-the-counter medicines and natural products. This material is provided for educational purposes only and is not intended for medical advice, diagnosis or treatment. Data sources include Micromedex®, Cerner Multum™, Wolters Kluwer™ and others, July **2018**.

    Other Drugs that Should Not Be Used with Comfrey

Do Not Use Comfrey with Anticonvulsants:
*"I suspect that some undiscovered factors cause certain people to be more susceptible to certain herbs than others. Researcher and herb toxicity specialist Ryan J. Huxtable, Ph.D., who works at the 'Department of Pharmacology' at the University of Arizona, notes that **taking anticonvulsant drugs with the herbs Comfrey, coltsfoot or senecio increases the toxicity of certain compounds found in the herbs.**"*
-'Warnings of Herbs and Drug Interactions' by Julie Hatcher and Julie Munoz Jackson, based on the 'American Herbal Products Association Botanical Safety Handbook' and many other sources, 185 pages, various dates **1990**s to 2018.
    (Anticonvulsants are also known as antiseizure and antiepileptic drugs. Some are used for bipolar disorder and borderline personality disorder.)

*"**A study involving rats showed that phenobarbital induces (increases) the metabolism of PAs to their lethal metabolites***. The USP recognizes this as a possible drug interaction and also suggests that patients on any medications avoid taking Comfrey (**USP, 1998)."*
-Toxicology and Clinical Pharmacology of Herbal Products edited by Melanie Johns Cupp, PharmD, BCPS. Totowa, New Jersey: Humana Press, **2000**. Chapter 18: Comfrey by David Burch and Melanie Johns Cupp, page 210.
(* -'Hepatic Metabolism and Pulmonary Toxicity of Monocrotaline Using Isolated Perfused Liver and Lung' by Mark Lafranconi and Ryan J. Huxtable, Department of Pharmacology, School of Medicine, University of Arizona Health Sciences Center, Tucson, Arizona; Biochemical Pharmacology, Volume 33, No. 15, pages 2479-2484, 1984.)
(** -'The United States Pharmacopeia' {USP} is a compendium of drug information published annually by 'United States Pharmacopeial Convention', a nonprofit organization. USP is published in a combined volume with National Formulary as USP-NF.)

*"**Herb - Drug Interactions: Comfrey taken with Phenobarbital: increased metabolism of Comfrey producing a lethal metabolite from pyrrolizidine;** severe hepatotoxic effects. A study* showed that phenobarbital induced the metabolism of pyrrolizidine alkaloids to a lethal metabolite."*
-'Review of Abnormal Laboratory Test Results and Toxic Effects Due to Use of Herbal Medicines' by Amitava Dasgupta, Ph.D., Department of Pathology and Laboratory Medicine, University of Texas-Houston Medical School; American Journal of Clinical Pathololgy: Volume 120, pages 127-137, **2003**.
(* -'Hepatic Metabolism and Pulmonary Toxicity of Monocrotaline Using Isolated Perfused Liver and Lung' by W. Mark Lafranconi and Ryan J. Huxtable, Department of Pharmacology, School of Medicine, University of Arizona Health Sciences Center, Tucson; Biochemical Pharmacology, Volume 33, No. 15, pages 2479-2484, 1984.)
    (The 'Hepatic Metabolism' report did not test or mention Comfrey. It said: *"Monocrotaline is a pyrrolizidine alkaloid obtained from the seeds of Crotalaria spectabilis. Metabolism of monocrotaline was inducible (activated) with phenobarbital pretreatment."* This means it increased the body's metabolism of monocrotaline.)

*"Based on the carcinogenic (cancer causing) activity, **Comfrey taken by mouth or applied to the skin may have antagonistic (adverse or negative) effects to chemotherapeutic agents.** Caution is advised when taking concurrently (at same time) with*

other herbs or supplements with potential chemotherapeutic effects."
-Natural Standard Herb and Supplement Guide: An Evidence-Based Reference by Catherine E. Ulbricht. Missouri: Mosby, **2010**, pages 250-254.   (Chemotherapeutic or cytotoxic agents are used in chemotherapy to kill cancer.)

"Common Herb-Drug Interactions: Phenobarbital increases metabolism of Comfrey, producing a lethal metabolite from pyrrolizidine that results in severe hepatotoxicity."
-Use of Herbal Products and Potential Interactions in Patients With Cardiovascular Diseases' by Ara Tachjian MD, Viqar Maria MBBS and Arshad Jahangir MD, Rochester, Minnesota and Scottsdale, Arizona; Journal of the American College of Cardiology, Volume 55, No. 6, page 515-525, **2010**.   (Phenobarbital brand names: Luminal®, Solfoton® and others.)

"What other drugs will affect Comfrey? Carbamazepine; phenobarbital; phenytoin; or rifabutin, rifampin. This list is not complete. *Other drugs may interact with Comfrey, including prescription and over-the-counter medicines, vitamins, and herbal products.* ***Not all possible interactions are listed in this product guide."***
-'Comfrey' by Drugs.com® provides independent information on more than 24,000 prescription drugs, over-the-counter medicines and natural products. This material is provided for educational purposes only and is not intended for medical advice, diagnosis or treatment. Data sources include Micromedex®, Cerner Multum®, Wolters Kluwer® and others, July **2018**.

"Moderate Interaction with Comfrey: Be cautious with medications that increase the breakdown of other medications by the liver (Cytochrome P450 3A4 (CYP3A4) inducers). *Comfrey is broken down by the liver. Some chemicals that form when the liver breaks down Comfrey can be harmful. Medications that cause the liver to break down Comfrey might enhance (increase) the toxic effects of chemicals contained in Comfrey. Some of these medicines include carbamazepine (Tegretol®), phenobarbital, phenytoin (Dilantin®), rifampin, rifabutin (Mycobutin®), and others."*
-'Comfrey' by WebMD.com®, the contents of the WebMD® site are for informational purposes only; it is not intended to be a substitute for professional medical advice, diagnosis, or treatment; always seek the advice of your physician or other qualified health provider with any questions you may have regarding a medical condition.WebMD® does not provide medical advice, diagnosis or treatment, July **2018**.

"What Other Drugs Interact with Comfrey? *If your doctor has directed you to use this medication, your doctor or pharmacist may already be aware of any possible drug interactions and may be monitoring you for them. Do not start, stop, or change the dosage of any medicine before checking with your doctor, health care provider or pharmacist first.*
Comfrey has no known severe interactions with any other drugs.
Moderate interactions of Comfrey include: carbamazepine, phenobarbital, phenytoin, rifampin, rifabutin. Comfrey has no known mild interactions with any other drugs. This information does not contain all possible interactions or adverse effects. Therefore, before using this product, tell your doctor or pharmacist of all the products you use. *Keep a list of all your medications with you, and share this information with your doctor and pharmacist."*
-'Comfrey' by RxList.com®, informational purposes only; the content is not intended to be a substitute for professional medical advice, diagnosis, or treatment. Always seek the advice of your physician or other qualified health provider with any questions you may have regarding a medical condition, July **2018**.

Comfrey Affecting Laboratory Test Results

"Herbal medicines are used widely in the United States, and according to a recent survey, the majority of people who use herbal medicines do not inform their physicians about their use.
Herbal medicines can alter physiology, and these changes can be reflected in abnormal test results.
This can cause confusion in proper diagnosis.
    For example, kavakava can cause drug-induced hepatitis, leading to unexpected high concentrations of liver enzymes.
    Herbal medicines can alter test results by direct interference with certain immunoassays.
Drug-herb interactions can result in unexpected concentrations of therapeutic drugs.
    For example, low concentrations of several drugs (eg, cyclosporine, theophylline, digoxin) can be observed in patients who initiated self-medication with St John's wort.
Use of toxic herbal products:
    Such as ma huang (an ephedracontaining herbal product), Chan Su, and Comfrey may cause death. Other toxic effects of herbal medicines include cardiovascular toxic effects, hematologic toxic effects, neurotoxic effects, nephrotoxic effects, carcinogenic effects, and allergic reactions."
-'Review of Abnormal Laboratory Test Results and Toxic Effects Due to Use of Herbal Medicines' by Amitava Dasgupta, Ph.D., Department of Pathology and Laboratory Medicine, University of Texas-Houston Medical School; American Journal of Clinical Pathololgy: Volume 120, pages 127-137, **2003**.
    (An assay is a laboratory procedure to assess and measure the amount or functional activity of a biological substance. An immunoassay measures specific proteins or other substances through their properties as antigens or antibodies.)

*"**A herbal medicine can affect laboratory test results by one of three mechanisms:***
    1. Direct assay interference, most commonly with immunoassays due to cross-reactivity of a component present in the preparation. *For example, falsely elevated digoxin levels may be observed using fluorescence polariza-*

tion immunoassay (FPIA) for digoxin due to ingestion of the Chinese medicine chan su, lu-shen-wan or dan shen.
2. Physiological effects, either through toxicity or enzyme induction due to the use of herbal products. For example, kava-kava causes liver toxicity and elevated alanine aminotransferase (ALT), aspartate aminotransferase (AST) and bilirubin levels in healthy individuals.
3. A herbal product may contain undisclosed drugs or heavy metals."
-'How Herbal Remedies Affect Clinical Laboratory Test Results' by A. Dasgupta, Department of Pathology and Laboratory Medicine, University of Texas, Houston; National Medical Journal of India, Volume 20, No. 3, pages 109-113, May/June **2007**.
>(An enzyme inducer is a drug that increases the metabolic activity of an enzyme either by binding to the enzyme and activating it, or by increasing the expression of the gene coding for the enzyme.)

"Comfrey is a perennial plant whose leaves and roots are traditionally used for wound healing; repairing broken bones; and in the treatment of arthritis, gout, and psoriasis. However, there is no scientific evidence to support these claims.
Comfrey contains pyrrolizidine alkaloids, which are well-known hepatotoxins (liver toxins), and Russian Comfrey is even more toxic than European and Asian Comfrey due to a higher content of toxic alkaloids.
Many potentially toxic compounds enter the body through the gastrointestinal tract and are immediately transported via the portal vein to the liver. Therefore, the liver sustains the highest exposure to some toxins and is susceptible to damage.
Liver function tests are used for diagnosis of adequate liver function as well as to identify any potential injury. Liver function tests include enzymes, bilirubin, proteins, and coagulation factors. These analytes are chosen because they are readily released in large quantities following cellular injury.
Measurement of the serum or plasma activities of the enzymes aspartate aminotransferase (AST), alanine aminotransferase (ALT), alpha-glutamyl transferase (GGT), and alkaline phosphatase (ALP) are routinely performed to detect liver injury. These intracellular enzymes are released when hepatocytes are damaged, but none of these are specific to liver. GGT, along with ALP, also reflects injury to biliary cells. Total bilirubin and its conjugated and unconjugated forms are useful in differentiating cases of jaundice. **Various herbal products, such as kava, chaparral, Comfrey, pennyroyal, black cohosh, mistletoe, and green tea extract, can cause liver damage.**
Laboratory Test Results: Elevated liver enzymes/bilirubin and other tests showing liver damage: Kava, chaparral, **Comfrey,** germander, pennyroyal, lipokinetic, skullcap, valerian, mahuang."
-Accurate Results in the Clinical Laboratory: A Guide to Error Detection and Correction by Amitava Dasgupta. Netherlands: Elsevier, **2013**. (Chapter 7: Effect of Herbal Remedies on Clinical Laboratory Tests, pages 75-92)
>(Despite the authors statement, there are scientific studies about the healing properties of Comfrey. I'm not sure what the author means by 'Asian Comfrey', maybe Symphytum asperum or Symphytum caucasicum. I assume by 'European Comfrey' he means Symphytum officinale. The types and amounts of alkaloids in the different species of Comfrey is discussed in section 'Alkaloids in Comfrey: Overview, Types, Detection, Amounts, Food' {Chapter 29}.)

Heavy Metals and Comfrey

"**Accumulation of heavy metals in mixed agrocenosis of fodder plants consisting of inula, goat's-rue (Galega officinalis), nettle (Urtica sp.), and Comfrey (Symphytum officinale)** was studied in Kabardino-Balkarian Republic, Russia. A site characterized by low lead content (18 mg/kg), medium copper and zinc content (38 and 75 mg/kg) and high cadmium content (0.7 mg/kg) was chosen as a trial plot.
In total, heavy metal contents of green mass were within admissible levels with exceptions of leaves of nettle and rhizomes (roots) of Comfrey, which had high levels of cadmium and lead."
-'Accumulation of Heavy Metals in Mixed Agrocenosis of Fodder Plants' by A. Tamakhina, Kabardino-Balkarian State Agrarian Academy, Nal'chik, Kabardino-Balkarian Republic, Russia; Mezhdunarodnyi Sel'skohozyaistvennyi Zhurnal (International Agricultural Journal), No. 1, pages 55-57, **2008**. (I was unable to get this report. If you have English translation, could you send it.)

>Biocenosis is a self-sufficient community of naturally occurring organisms occupying and interacting within a specific biotope. A biotope is a portion of a habitat characterized by uniformity in climate and distribution of biotic {life} and abiotic components.
>**'Agrocoenosis' or 'agrobiocenosis'** is a Russian term meaning: "an association of organisms in the sowings and plantings of cultivated plants; one of the widespread forms of secondary biocenoses. Each species of cultivated plant forms its characteristic agrobiocenosis with a series of constant and dominant species.
>The agrobiocenosis is distinguished from the primary biocenoses by the incapacity for long-term independent existence as a result of the sharp weakening of the self-regulating processes; their temporary stability is supported by the activity of man. Another important property of an agrobiocenosis is the dominance of several herbivorous species of animals, primarily pests, particularly insects. A cultivated plant forms the energy base of the agrobiocenosis; this plant, together with the accompanying weeds, determines the composition of the animal population."
>-The Great Soviet Encylopedia, 1979, one of the largest Russian-language encyclopedias, published by Soviet Union from 1926-1990. (Inula is genus of 90 species of flowering plants in family Asteraceae, native to Europe, Asia and Africa.)

"Lead (Pb) is the number one heavy metal pollutant in the environment. The high cost and environmental concerns of conventional remediation technologies has led to **an emerging alternative technology for heavy metal remediation:**

phytoremediation. Using a hydroponic system, sand-grown Symphytum officinale L. plants were exposed to nutrient solutions with or without lead nitrate (Pb(NO3)2) and ethylenediamine tetraacetic acid (EDTA).

Using flame atomic absorption spectroscopy (to measure Pb content) and bovine serum albumin-protein precipitation (to measure polyphenol and tannin levels), a significant in vivo correlation between tannin level and Pb accumulation level was observed in roots of plants exposed to all Pb treatments.

Higher tannin containing-lateral roots accumulated significantly more Pb than lower tannin main roots.
Together, these findings demonstrate that S. officinale root tannins have the ability to chelate Pb.

This may be a mechanism to cope with Pb stress (adaptive tolerance). Despite the typical signs of Pb stress at root level (e.g. root growth inhibition, and degraded cytoplasms), **Comfrey shoots showed no signs of stress under any Pb treatments.**

Most importantly, since this chelation-based tolerance mechanism also influences the accumulation levels, the phytochemical composition of plants should also be considered when screening plants for phytoremediation.
Overall, since Symphytum officinale accumulated Pb predominately in the roots, it is most suited for rhizofiltration and phytostabilisation. Whilst chelating agents enhanced Pb accumulation in shoots, root levels were unexpectedly reduced compared to unchelated Pb treatments."
-'Investigations into Lead (Pb) Accumulation in Symphytum officinale L.: A Phytoremediation Study' by Lily Chin, Thesis for degree of Doctor of Philosophy in Plant Biotechnology, University of Canterbury, Christchurch, New Zealand, **2007**.

(Chelation is a type of bonding of ions and molecules to metal ions. It involves the formation of two or more coordinate bonds between a multiple bonded ligand and a single central atom. Usually these ligands are organic compounds, and are called chelants, chelators, chelating agents, or sequestering agents. A chelate is a molecular compound forming a complex of cations with organic compounds forming a ring structure. **Chelating agents convert these heavy metal ions into a chemically and biochemically inert {safe} form.**)

"Phytoremediation is a technology that uses plants to remove or render harmless pollutants in the environment *(Garbisu and Alkorta, 2001). Lead (Pb) is often considered as the target pollutant of remediation studies due to its widespread distribution, persistence, and toxicity to human health (Chen et al., 1997). Moreover, Pb is ranked the number one heavy metal pollutant (and number two of all hazardous substances) by the 'Agency for Toxic Substances and Disease Registry' (ATSDR, 2005).*
Reported correlations between tannin level and metal accumulation within plant tissues suggest that metal-chelating tannins may help plants to tolerate toxic levels of heavy metal contaminants.

This paper supports such correlations using a new method that demonstrated **the ability of plant tannins to chelate heavy metals,** *and showed that the relative levels of tannins in tissues were quantitatively related to lead chelation in vitro.*
Using this in vitro metal chelation method, we showed that **immobilised tannins prepared from lateral roots of Symphytum officinale L., that contained high tannin levels, chelated 3.5 times more lead than those from main roots with lower tannin levels.** This trend was confirmed using increasing concentrations of tannins from a single root type, and using purified tannins (tannic acid) from Chinese gallnuts.
This study presents a new, simple, and reliable method that demonstrates direct lead-tannin chelation. In relation to phytoremediation, it suggests that plant roots with more 'built-in' tannins may advantageously accumulate more lead."
-'**Lead Chelation to Immobilised Symphytum Officinale L (Comfrey) Root Tannins**' by L. Chin, D.W.M. Leung and H.H. Taylor, School of Biological Sciences, University of Canterbury, Christchurch, New Zealand; Chemosphere, Volume 76, No. 5, pages 711-715, **2009**.

Diagnosis of Herbal Hepatotoxicity

"Even when herbal poisoning is suspected, investigation is often not carried out properly.
The literature is full of confusing reports, because proper botanical identification was not done, proper chemical analysis was not done, or dose toxicity calculations were not made."
-'The Myth of Beneficent Nature: The Risks of Herbal Preparations' by Ryan J. Huxtable, PhD, College of Medicine, University of Arizona, Tucson, Arizona; Annals of Internal Medicine, Volume 117, No. 2, pages 165-166, July 15 **1992**.

"Diagnosis of drug-induced liver injury (DILI) remains a challenge and eagerly awaits the development of reliable hepatotoxicity biomarkers.
Several methods have been developed in order to facilitate hepatotoxicity causality assessments. These methods can be divided into **three categories: (1) expert judgement, (2) probabilistic approaches, and (3) algorithms or scales.**

The last category is further divided into general and liver-specific scales. The 'Council for International Organizations of Medical Sciences' (CIOMS) scale, *also referred to as the 'Roussel Uclaf Causality Assessment Method' (RUCAM), although cumbersome and difficult to apply by physicians not acquainted with DILI, is used by many expert hepatologists, researchers, and regulatory authorities to assess the probability of suspected causal agents."*
-'Causality Assessment Methods in Drug Induced Liver Injury' by Miren Garcia-Cortes, Camilla Stephens, M.I. Lucena, Alejandra Fernandez-Castaner and Raul J. Andrade, Campus Universitario de Teatinos, Malaga, Spain, on behalf of the Spanish Group for the Study of Drug-Induced Liver Disease; Journal of Hepatology: European Association for the Study of the Liver, Volume 55, No. 3, pages 683-691, September **2011**.

"Determining Cause of Liver Injury:

The diagnosis of herbal hepatotoxicity or herb induced liver injury (HILI) represents a particular clinical and regulatory challenge with major pitfalls for the causality evaluation. HILI causality assessment is challenging and is best achieved by the liver specific CIOMS scale, avoiding pitfalls commonly observed with other approaches.
 At the day HILI is suspected in a patient, physicians should start assessing the quality of the used herbal product, optimizing the clinical data for completeness, and **applying the 'Council for International Organizations of Medical Sciences' (CIOMS) scale for initial causality assessment.**
> This scale is structured, quantitative, liver specific, and validated for hepatotoxicity cases. Its items provide individual scores, which together yield causality levels of highly probable, probable, possible, unlikely, and excluded.

The CIOMS scale is preferred as tool for assessing causality in hepatotoxicity cases, compared to numerous other causality assessment methods, which are inferior on various grounds. Among these disputed methods are the Maria and Victorino scale, an insufficiently qualified, shortened version of the CIOMS scale, as well as various liver unspecific methods such as the ad hoc causality approach, the Naranjo scale, the World Health Organization (WHO) method, and the 'Karch and Lasagna' method.

LiverTox Database: The consideration of possible hepatotoxicity in various reports has been discussed by the 'National Institutes of Health' (NIH) in their recently released LiverTox database, covering a selected group of herbal and dietary supplement (HDS) products*.
> Among these are: Aloe vera, Black Cohosh, Cascara, Chaparral, Chinese and other Asian herbal medicines (Ba Jiao Lian, Chi R Yun, Ephedra, Jin Bu Huan, Sho Saiko To and Dai Saiko To, Shou Wu Pian), **Comfrey,** Fenugreek, Germander, Ginkgo, Ginseng, Glucosamine, Greater Celandine, Green Tea, Hoodia, Horse Chestnut, Hyssop, Kava, Margosa Oil, Milk Thistle, Noni, Pennyroyal, St John's Wort, Saw Palmetto, Senna, Skullcap, Usnic acid, Valerian, Yohimbine.

However, causality confirmation was surprisingly rare for individual cases of suspected herbal hepatotoxicity, which often were published as narrative and anecdotal reports without valid and transparent data collection that require stringent efforts for causality attribution."**

-'Herbal Hepatotoxicity: Challenges and Pitfalls of Causality Assessment' by Rolf Teschke, Christian Frenzel, Johannes Schulze and Axel Eickhoff (Germany); World Journal of Gastroenterology, Volume 19, No. 19, pages 2864-2882, May 21 **2013**.
(* -'LiverTox: Clinical and Research Information on Drug-Induced Liver Injury' by National Institutes of Health, Bethesda, Maryland, Drug Record: Herbal and Dietary Supplements, www.livertox.nih.gov)
(** -'Herbal Hepatotoxicity: A Critical Review' by Rolf Teschke, Christian Frenzel, Xaver Glass, Johannes Schulze and Axel Eickhoff {Germany}; British Journal of Clinical Pharmacology: British Pharmacological Society, Vol 75, No. 3, page 630-636, 2012.)
(Ad hoc means a solution designed for a specific problem, not generalizable, and not intended to be adapted to other purposes.)

"**Council for International Organizations of Medical Sciences' scale for liver injury:**
Causality assessment of suspected drug induced liver injury (DILI) and herb induced liver injury (HILI) is hampered by the lack of a standardized approach to be used by attending physicians and at various sub-sequent evaluating levels.
The aim of this review was to analyze the suitability of the liver specific 'Council for International Organizations of Medical Sciences' (CIOMS) scale as a standard tool for causality assessment in DILI and HILI cases.
> PubMed® database was searched for the following terms: drug induced liver injury; herb induced liver injury; DILI causality assessment; and HILI causality assessment.

The strength of the CIOMS lies in its potential as a standardized scale for DILI and HILI causality assessment.
> Other advantages include its liver specificity and its validation for hepatotoxicity with excellent sensitivity, specificity and predictive validity, based on cases with a positive reexposure test. This scale allows prospective collection of all relevant data required for a valid causality assessment. It does not require expert knowledge in hepatotoxicity, and its results may subsequently be refined. Weaknesses of the CIOMS scale include the limited exclusion of alternative causes and qualitatively graded risk factors."

-'Drug and Herb Induced Liver Injury: Council for International Organizations of Medical Sciences Scale for Causality Assessment' by Rolf Teschke, Albrecht Wolff, Christian Frenzel, Alexander Schwarzenboeck, Johannes Schulze and Axel Eickhoff (Germany); World Journal of Hepatology, Volume 6, No. 1, pages 17-32, January 27 **2014**.
> (PubMed®: United States National Library of Medicine, United States National Institues of Health. PubMed has 28 million citations for biomedical literature from MEDLINE®, life science journals, books. www.ncbi.nlm.nih.gov/pubmed)

PubChem CID	6440495
Structure	2D 3D Find Similar Structures
Molecular Formula	$C_{20}H_{31}NO_7$
Synonyms	comfrey [7-[(E)-2-methylbut-2-enoyl]oxy-4-oxido-5,6,7,8-tetrahydro-3H-pyrrolizin-4-ium-1-yl]methyl 2-hydroxy-2-(1-hydroxyethyl)-3-methylbutanoate 72698-57-8
Molecular Weight	397.5

Comfrey PubChem®
National Center for Biotechnology Information (2022).
PubChem Compound Summary for CID 6440495, Comfrey.
https://pubchem.ncbi.nlm.nih.gov/compound/Comfrey

"PubChem is an open chemistry database at the National Institutes of Health (NIH). 'Open' means that you can put your scientific data in PubChem and that others may use it.
PubChem mostly contains small molecules, but also larger molecules such as nucleotides, carbohydrates, lipids, peptides, and chemically-modified macromolecules. We collect information on chemical structures, identifiers, chemical and physical properties, biological activities, patents, health, safety, toxicity data, and many others.
PubChem records are contributed by hundreds of data sources. Examples include: government agencies, chemical vendors, and journal publishers."

Chapter 29

Alkaloids in Comfrey: Overview, Types, Detection, Amounts, Food

Warning about Alkaloids

Due to the controversial and negative reports about the alkaloids in Comfrey, scientists and health professionals recommend that pregnant women and nursing mothers do not use Comfrey medicinally. Contact a healthcare professional before using it with children or if you have special health problems. Use all herbs in moderation.

See the section above: 'Warnings and Negative Reactions to Comfrey' (Chapter 28).

It is best if you ask your healthcare professional about Comfrey if you have liver problems.
Do not take Comfrey internally if you are taking: acetaminophen (Tylenol®, Panadol® and others), amiodarone (Cordarone®), carbamazepine (Tegretol®), erythromycin (Erythrocin®, Ilosone® and others), fluconazole (Diflucan®), isoniazid (INHA®, isonicotinic acid hydrazide), itraconazole (Sporanox®), lovastatin (Mevacor®), methotrexate (Rheumatrex®), methyldopa (Aldomet®), phenytoin (Dilantin®), pravastatin (Pravachol®), simvastatin (Zocor®), and others.
If you are on medication such as carbamazepine (Tegretol®), phenobarbital (Luminal® and others), phenytoin (Dilantin®), rifampin (Rifadin®), or rifabutin (Mycobutin®), then Comfrey may interfere with their effectiveness or cause side effects.

"Comfrey appears to have assumed the attributes of a panacea (universal remedy). A single modern herbal book*, sold commonly in health food stores, recommends the use of Comfrey to alleviate some 47 ailments (including allergies, anemia, emphysema, and diabetes) and affected body organs (including bladder, lungs, kidneys, and pancreas)."
-'Ingestion of Pyrrolizidine Alkaloids: A Health Hazard of Global Proportions' by James N. Roitman, Natural Products Chemistry Research, Western Regional Research Center, Agricultural Research Service, U.S. Department of Agriculture, Berkeley, California, American Chemical Society, ACS Symposium Series, pages 345-378, October 25 **1983**.
(* -'Herbally Yours' by P.C. Royal. Provo, Utah: Biworld Publications, 1979.)

*"**A major source of exposure to Pyrrolizidine Alkaloids is the use of Comfrey-pepsin capsules and pills as digestive aids**. Some preparations are prepared from Comfrey root and some from Comfrey leaf, while others carry no information as to source. **There is a predictably wide variation in alkaloid content.***
*Thus, a preparation claiming to be from **Comfrey leaf** contained 40 mg/kg of pyrrolizidines, and 230 mg/kg of pyrrolizidine N-oxides, for a total alkaloid content of 270 mg/kg. A preparation claiming to be **Comfrey root**, on the other hand, held 400 mg/kg of pyrrolizidines, and 2500 mg of pyrrolizidine N-oxides, for a total alkaloid content of 2900 mg/kg.*
The use of Symphytum species is particularly to be discouraged. *The risk of occasional use of the leaves may be moderate or slight, but in the absence of any benefit, the risk is needless. Preparations containing Symphytum roots are so hazardous that there are grounds for banning their sale. Such a ban is enforced in Canada.*"*
-Toxicants of Plant Origin, Volume I: Alkaloids edited by Peter R. Cheeke. Boca Raton, Florida: CRC Press, **1989**. Chapter 3: 'Human Health Implications of Pyrrolizidine Alkaloids and Herbs Containing Them' by Ryan J. Huxtable, University of Arizona.
(* For information about Canada's regulation of Comfrey, see subsection 'Canada Bans Sale of Comfrey for Internal Use' in section 'Some Uses of Comfrey are Banned by Governments' {Chapter 31}.)

*"**The continued indiscriminate use of herbs as health foods remains a matter for concern, however. The lack of reports of toxicity from Comfrey despite claims of dietary use many years is not necessarily an indication of safety.***
The effects of pyrrolizidine alkaloids are cumulative (add up over time), and overt (obvious) damage may be delayed for a long time, thereby preventing an association with the ingestion of such herbs becoming clear. Doctors should be alerted to the dangers of some herbs, and should enquire more regularly into their patients' intake of traditional remedies and health foods."
-'Toxicity of Comfrey' by K.A. Winship, Department of Health, Medicines Control Agency, London, England; Adverse Drug Reactions and Toxicological Reviews, Oxford, England, Volume 10, No. 1, pages 47-59, **1991**.

"What then should a physician advise patients who wish to use herbs sensibly?
I have suggested the following guidelines for reducing the risk of herbal use:
    1. Do not take herbs if pregnant or attempting to become pregnant.
    2. Do not take herbs if you are nursing.
    3. Do not give herbs to your baby.
    4. Do not take a large quantity of any one herbal preparation.
    5. Do not take any herb on a daily basis.
    6. Buy only preparations when plants are listed on packet (no guarantee of safety or correctness but better than nothing).
    7. Do not take anything containing Comfrey."
-'The Myth of Beneficent Nature: The Risks of Herbal Preparations' by Ryan J. Huxtable, PhD, College of Medicine, University of

Arizona, Tucson, Arizona; Annals of Internal Medicine, Volume 117, No. 2, pages 165-166, July 15 **1992**.

"It is generally difficult to estimate the human cancer risk of a carcinogen (cancer causer) based on the results of carcinogenicity examinatiions of laboratory animals for the following reasons:
> The differences in life span and susceptibility to carcinogens between animals and human beings, multiple causal factors of human cancer, and the presence of factors inhibiting human carcinogenesis. Moreover, the dose and duration of exposure of human beings to a certain carcinogen cannot be accurately assessed.

It is now clear from epidemiological data that most human cancers are induced (caused) by environmental causes.
The possibility of cancer development by a certain carcinogen is increased when modifers, such as the habitual use of a specific food or medicine *or occupational conditions, provide favorable conditions for that development, even if the amount of carcinogen exposure is small.* **Thus, it is considered to be best to avoid the use of these plants as food or herbal remedies*.** *Reduction of the toal amount of carcinogens in the diet, including carcinogenic natural products, may eventually contribute to the prevention of cancer."*
-'Edible Plants Containing Naturally Occurring Carcinogens in Japan' by Iwao Hirono, Department of Pathology, Fujita Health University School of Medicine, Toyoake, Japan; Japanese Journal of Cancer Research, Tokyo, Volume 84, No. 10, pages 997-1006, October **1993**. (* The author includes Comfey in this group.)
> (Epidemiology is the study of pubic health. It monitors the incidence, distribution, cause, and control of diseases and other health factors in a population.)

"If the PA (Pyrrolizidine Alkaloid) content of a product is not standardized or stated, it is best not to use Comfrey or any other PA containing herbs. Because the content of PAs is highly variable and can vary naturally in the crude drug by a factor of approximately 10, only Comfrey products that have a declared content of PAs should be used therapeutically."
-Rational Phytotherapy: A Physician's Guide to Herbal Medicine by V. Schulz, R. Hansel and V.E. Tyler. (Third edition, first English edition, translated by T.C. Telger.) Heidelberg, Germany: Springer Verlag, **1998**.

*"***Hepatotoxic (liver toxic) effects of conventional drugs are widely acknowledged and most physicians are well aware of them, but the side effects of herbal preparations have not always reached public awareness. As with conventional drugs, interactions with other prescribed medication cannot be ruled out.*** Lack of awareness is even more profound in patients who take herbs mostly for self medication, with obvious dangers. First, the assumed diagnosis, for which the patient takes herbs, might be wrong. Second, the chosen medication might not be optimal, possibly delaying appropriate treatment.* **Toxicity symptoms and side effects may occur,** *but some people even increase the dose of the herbal product as a response to the onset of symptoms. These concerns may not hold for many herbal products that are distributed on the market but* **Comfrey definitely belongs to a potentially dangerous kind.**
One should follow the simple but effective rules offered by Ryan Huxtable*:
> *-do not use herbal drugs in infants and children;*
> *-avoid medication with herbs during pregnancy and while nursing;*
> *-do not take herbs on a regular basis and in large quantities;*
> *-***and lastly, beware of taking Comfrey**.*"*

-'The Efficacy and Safety of Comfrey' by F. Stickel and H.K. Seitz, Public Health Nutrition, United Kingdom, Volume 3, 4A, pages 501-508, December **2000**.
(* -'The Myth of Beneficient Nature: The Risks of Herbal Preparations' by Ryan J. Huxtable, Annals of Internal Medicine, Volume 117, pages 165-166, 1992.)

"Contraindications, Interactions, and Side Effects (Comfrey):
A. 'American Herbal Products Association':
> **Symphytum is Class 2a, 2b, 2c, 2d. Long term use of discourged.**

Effective July 1996, the 'American Heral Products Association' Board of Trustees recommends that all products with botanical ingredient(s) that contain toxic PAs (Pyrrolizidine Alkaloids), including Borago officinalis, display the following cautionary statement on the label, 'For external use only. Do not apply to broken or abraded skin. Do not use when nursing'.
> American Herbal Products Association's Botanical Safety Handbook edited by Zoe Gardner and Michael McGuffin. Boca Raton, Florida: CRC Press, 2013.

B. Commission E (Germany) reports Comfrey herb, leaf, and root permitted for external use only. *Skin should be intact and pregnant users should first consult physician. External dosage of pyrrolizidine alkaloids (PAs) maximum 100 g/day for a maximum 4-6 weeks/year. Internal use may cause severe hepatic (liver) damage. PAs are toxic to humans, with liver damage with cirrhosis and ascites, or seneciosis, or veno-occlusive disease (VOD) reported in almost all cases of severe or fatal intoxications, from intakes of 0.5 mg/kg to 3.3 mg/kg.*
> Adverse Effects of Herbal Drugs, Volumes 1, 2 and 3 edited by Peter A. G.M. DeSmet, Konstantin Keller, Rudolf Hansel and R. Frank Chandler. Berlin, Germany: Springer-Verlag, 1993.

C. Cautions that Comfrey PAs are genotoxic, carcinogenic, and hepatotoxic. *Because of the PAs, its use in pregnancy and lactation is to be avoided. Animal studies document placental transfer and secretion into breast milk of unsaturated PAs. Comfrey may speed up metabolism of other drugs (stimulates metabolism of aminopyrine-N-demethylase, a drug metabolizing enzyme).*
> Herbal Medicines: A Guide for Healthcare Professionals by C.A. Newall, L.A. Anderson and J.D. Phillipson. London, England: The Pharmaceutical Press, 1996."

-Handbook of Medicinal Herbs, second edition by James A. Duke with Mary Jo Bogenschutz-Godwin, Judi duCellier and Peggy-Ann K. Duke. Boca Raton, Florida: CRC Press, **2002**, page 215.

" *'American Herbal Products Association' Safety Ratings:*
Class 1: Herbs that can be safely consumed when used appropriately.
Class 2: Herbs for which the following use restrictions apply, unless otherwise directed by an expert qualified in their use:
    2a: *For external use only.*   **2b: Not to be used during pregnancy.**
    2c: Not to be used while nursing.   **2d:** *Other specific use restrictions as noted.*
Class 3: Herbs for which significant data exist to recommend the following labeling: 'To be used only under the supervision of an expert qualified in the appropriate use of this substance.'
Class 4: Herbs for which insufficient data are available to classify."
(Placenta develops in uterus during pregnancy. It provides oxygen and nutrients for the baby and removes waste.)

"The infant liver is not considered mature until two weeks of age, so it is at greater risk from exposure to toxic PAs (Pyrrolizidine Alkaloids). The risk of liver damage from toxic PAs in humans is currently considered highest for the developing fetus; *damage accumulates with every exposure and may not appear for decades.*
Toxic PAs have been detected in the milk of dairy animals as well as in research animals and honey. ***It is rational to assume that toxic PAs enter human milk.*** *Therefore, starting a baby's lifetime load of toxic PAs at birth seems unwise."*
-The Nursing Mother's Herbal by Sheila Humphrey, B.Sc. (Botany), R.N., IBCLC (International Board Certified Lactation Consultant®). Saint Paul, Minnesota: Fairview Press, **2003**. (Quote from 'HerbalGram, No. 108' by American Botanical Council, Austin, Texas, 84 pages, November 2015 to January 2016.)

"When the presence of pyrrolizidine alkaloids (alkaloids with a high risk of toxicity found in other plants such as Senecio and Crotalaria) was proved (symphytine, senkirkine, echimidine and lasiocarpine), *scientists began to study the toxicokinetics and toxicodynamics of these compounds."*
-'Symphytum Officinale: A Plant Being at the Border Between Phytotherapeutical Remedy and Poison' by C. Prisacaru, U.S.A.M.V. Iasi, Romania; Lucrai Scedilla tiinifice- Medicina Veterinara, Universitatea de Stilinte Agricole si Medicina Veterinara (Cedilla Scientific Works- Veterinary Medicine, University of Agricultural Sciences and Veterinary Medicine), Romania, Volume 48, No. 7, pages 63-71, **2005**.
    (Toxicokinetics is the rate a chemical will enter the body and what happens to it once it is in the body. Toxicodynamics are the dynamic interactions of a toxicant with a biological target {site of action} and its biological effects. A biological target can be binding proteins, ion channels, DNA or other receptors.)

"Comfrey: *Symphytum officinale and 35 other species in the genus Symphytum in Europe, Africa and Asia.*
Classification: *Liver poison, neurotoxin (neurological toxin), mutagen (gene mutation).*
Toxicity Class: *II: Moderately hazardous, III:Slightly hazardous.*
    (Comfrey is not Ia: Extremely hazardous, and is not Ib: Highly hazardous.)
Toxicity: *Symphytin LD50 mouse: 300 mg/kg i.p. (intraperitoneal= injected into abdomen).*
    Poisoning has been reported after ingestion (eating/drinking) of Comfrey.
Pharmacological Effects: *PAs (Pyrrolizidine Alkaloids) become metabolically activated in the liver. They can then alkylate proteins and DNA. As a result they cause liver damage (Veno-Occlusive Disease= VOD), and after prolonged intake they are mutagenic, teratogenic (congenital abnormalities and birth defects), and carcinogenic (especially causing liver tumours).*
First Aid: *Give medicinal (activated) charcoal to absorb the alkaloids; administer plenty of warm black tea, and sodium sulphate. Provide shock prophylaxis (keep patient warm and at a quiet place).*
(Consult your health care provider before applying any of these treatments.)
Clinincal Therapy (Hospital or Doctors Office): *In case of ingestion of large doses: gastric lavage (possibly with potassium permanganate), instillation of medicinal charcoal and sodium sulphate, electrolyte substitution, control of acidosis with sodium bicarbonate. Observe kidney/liver function and blood coagulation capacity. In case of VOD, apply therapies as for patients with liver cirrhosis. There are no specific antidotes. (In serious cases go to an Emergency Center immediately.)"*
-Mind-Altering and Poisonous Plants of the World: A Scientifically Accurate Guide to 1200 Toxic and Intoxicating Plants by Michael Wink and Ben-Erik Van Wyk. Portland, Oregon: Timber Press, **2008**.
    (Consult with your health care provider for the first aid and clinical treatment that you need.)
    (Neurotoxins are substances poisonous or destructive to nerve tissue by disrupting the normal function of nerve cells.)
    (A mutation is a change in hereditary material that involves either a change in chromosome structure or number, or a change in the nucleotide sequence.)
    (LD50 is the median Lethal Dose 50%. The value of LD50 for a substance is the dose required to kill half the members of a tested population after a specified test duration. It is used as an indicator of a substance's acute toxicity. A lower LD50 indicates increased toxicity. I think this method of scientific study is awful. There should be ways of testing substances without harming animals.)
    (Activated or medicinal charcoal is a fine black odorless, tasteless powder made from wood or other materials exposed to very high temperatures in an airless environment. It is then activated to increase its ability to adsorb substances in the digestive system.)
    (Sodium sulfate or sodium sulphate or sulfate of soda is the inorganic compound Na_2SO_4 as well as related hydrates. All are white solids highly soluble in water. It is a laxative and has nonmedical uses. Consult your health care provider

before using this.)

(Black tea is more oxidized than oolong, green and white teas. It is stronger in flavor than the less oxidized teas. All are made from leaves of the shrub Camellia sinensis.)

"Reflecting its former enormous popularity, Comfrey continues to be used both externally and internally by many people who are misinformed by advice provided in popular herbals. Believing that the use of herbs should be helpful, not harmful, to the consumer's health, we must conclude:
  1. The internal use of any species of Comfrey should be avoided.
  2. Comfrey root should never be used medicinally.
  3. Only the mature leaves of Symphytum officinale should be applied externally and then only to intact skin for limited periods of time.
  4. Comfrey should never be used by pregnant or lactating women or by young children.
  *5. Because so many other nontoxic yet effective treatments for minor skin ailments do not present the hazards associated with this herb, **Comfrey has little, if any, place in our modern materia medica.***"
-Tylers Herbs of Choice: The Therapeutic Use of Phytomedicinals, Third Edition by Dennis V.C. Awang. Boca Raton, Florida: CRC Press, **2009**.
  (Materia medica {medical material} is Latin for the collected knowledge about the therapeutic properties of substances used for healing. The term is derived from the book by 1st century AD Greek physician Pedanius Dioscorides.)

"We emphasize that we should avoid the fallacy of equating 'natural' with 'safe'.
  Poison ivy (Toxicodendron radicans L. Kuntze) is a perfect example.
Comfrey (Symphytum spp.) contains documented hepatocarcinogens (pyrrolizidine alkaloids) *that can **also be absorbed through topical applications, yet Comfrey still has very vocal adherents.***
(Awang et al. 1993; Betz et al. 1994; Brauchli et al. 1982; Culvenor et al. 1980; Furuya and Hikichi 1971; Gomes et al. 2007; Hirono et al. 1978, 1979; Jaarsma et al. 1989; Westendorf 1992)"
-Genetic Resources, Chromosome Engineering, and Crop Improvement: Medicinal Plants, Volume 6 edited by Ram J. Singh, University of Illinois, Urbana-Champaign, Illinois. Boca Raton, Florida: CRC Press, **2012**. ('Chapter 2: Medicinal Plants: Natures Pharmacy' by Ram J. Singh, Ales Lebeda and Arthur O. Tucker.)

"Direct and Indirect Hepatotoxins:
In conjunction with the hepatotoxic (liver toxic) profile of the specific drug, biological agent or herbal in question, dosing effects or individual patient susceptibility are major determinants in predicting the risk for the development of DILI.
Drugs that cause hepatotoxicity typically fit on a spectrum between those that are 'direct' hepatotoxins and those in which the toxicity is 'indirect'.
A predictable dose or exposure threshold marks direct hepatotoxins, when risk for liver injury rises quickly for most exposed individuals. Examples of herbals whose extracts are directly toxic when ingested at high exposure levels include Symphytum officinale (Comfrey), *Crotalaria, Heliotropium and Senecio.*
  *These plant species contain a number of different **pyrrolizidine alkaloids which, when ingested in high amounts, cause severe toxicity** through a mechanism of hepatocellular biotransformation into genotoxic pyrrole derivatives. The most common form of liver injury caused by these products is the sinusoidal obstruction syndrome that is marked by non-thrombotic obliteration of the hepatic sinusoids and terminal centrilobular hepatic veins."*
-'Scientific and Regulatory Perspectives in Herbal and Dietary Supplement Associated Hepatotoxicity in the United States' by Mark I. Avigan, Robert P. Mozersky and Leonard B. Seeff; Center for Drug Evaluation and Research, FDA, Silver Spring, Maryland and Office of Dietary Supplement Products, Center for Food Safety and Applied Nutrition, College Park, Maryland; International Journal of Molecular Sciences, Volume 17, No. 331, 30 pages, **2016**.
  (In hepatic Veno-Occlusive Disease some small veins in the liver are blocked. It can be caused by high-dose chemotherapy or after consuming certain plant alkaloids such as pyrrolizidine. Symptoms include weight gain due to fluid retention, increased liver size, and raised levels of blood bilirubin. Also called 'Sinusoidal Obstruction Syndrome'.)

Definition of Alkaloid and Pyrrolizidine Alkaloids

Alkaloids (alkali-like) are organic compounds that contain basic nitrogen atoms. In addition to carbon, hydrogen and nitrogen, alkaloids may also contain oxygen, sulfur and, rarely, chlorine, bromine, and phosphorus. Compared to other classes of natural compounds, alkaloids have great structural diversity. There is no simple, uniform classification.

Most alkaloids have a bitter taste or are poisonous when drunk or eaten. Alkaloids in plants appear to have evolved in response to animals eating them; though, some animals have evolved the ability to detoxify alkaloids. Examples of commonly known alkaloids include caffeine (in coffee), cocaine, nicotine (in tobacco), morphine, quinine (in cinchona tree bark), solanine (in potatoes), and theobromine (in chocolate).

Alkaloid-containing plants have been used by people throughout history. Many alkaloids are used today in medicine, usually in the form of salts. Many synthetic drugs are structural modifications of the alkaloids, designed to change the effect of certain drugs and reduce their unwanted side-effects.

Alkaloids are base (alkaline). They form salts by uniting with acids without setting hydrogen free. Only a few alkaloids are liquid; most of them and their salts can be crystallized into solids. Their salts are much more easily soluble in water than the alkaloids themselves.

Pyrrolizidine Alkaloids (PA) are complex molecules that include a pyrrolizidine nucleus: a pair of linked pyrrole rings. Each pyrrole is a five-sided structure with four carbons and one nitrogen forming the ring. Pyrroles are incorporated into the chlorophyll molecule.

There are hundreds of types of Pyrrolizidine Alkaloids, with numerous isomers (same chemical formula but different arrangement of atoms). **PAs can be saturated which are not toxic, or unsaturated which are toxic.** The toxicity of each depends on the chemical complexity of the associated necic acids. There are three main categories: macrocyclic diesters, branched diesters, and monoesters. PAs can be free bases or N-oxides.

"Alkaloids are organic substances other than certain simple amines and amino-acids associated with proteins which are produced by certain plants and which have a basic nitrogen atom.
Although alkaloids may be regarded as metabolic by-products, they frequently serve the plant through nitrogen reserve.
Alkaloids are widely distributed throughout the plant kingdom, being present in about 5 percent of all species, and occur in a wide variety of chemical forms.
Many of the alkaloids are pharmacologically active being either poisonous or having medicinal properties. Thus green potatoes produce a poisonous solanine, certain poppies- morphine, nux vomica- strychnine, tobacco- nicotine, and foxgloves- digitalis, and alkaloids are frequently the active component in herbal medicines."
'The Alkaloid Content of Comfrey' by Dr. D.B. Long, Ph.D, M.A., around **1970**. (In 'Comfrey: Fodder, Food & Remedy').

"Alkaloids are basic and form salts with acids, and in plants they are stored in the free state, usually as salts.
Certain alkaloids form salts with certain acids, such as meconic acid, quinic acid, veratric acid, aconitic acid, and tiglic acid.
Some other alkaloids are present in the plants as N-oxides, such as nupharidine, as glycosides such as solanine, amides such as piperine, and esters such as cocaine."
-The Alkaloids: Chemistry and Pharmacology, 49 volumes (1983-1997) by series editor Geoffrey A. Cordell. Quote from volume 39, edited by Arnold Brossi, **1990**, page 166. San Diego, California: Academic Press.

"Plant families containing Pyrrolizidine Alkaloids:
Boraginaceae, Orchidaceae, Compositae, Rhizophoraceae, Gramineae, Santalaceae, Leguminosae, Saptoceae."
-'Toxicity of Comfrey' by K.A. Winship, Department of Health, Medicines Control Agency, London, England; Adverse Drug Reactions and Toxicological Reviews, Oxford, England, Volume 10, No. 1, pages 47-59, **1991**.

"Over 200 PAs (Pyrrolizidine Alkaloids) have been found in plants, mainly from the borage family Boraginaceae (eg., Symphytum, Heliotropium); the aster family Compositae/Asteraceae (eg. Senecio); and the pea family Leguminosae/ Fabaceae (eg. Crotalaria) and an estimated 3% of the world's flowering plants may contain them."
-'Comfrey Update' by D.V.C. Awang, HerbalGram published by American Botantical Council, Austin, Texas, Volume 25, pages 20-23, **1991**.

"Over 160 pyrrolizidine alkaloids (PAs) have been isolated (in all plants). All are esters and vary substantially in toxicity according to their structure. They consist of a necine base and organic acid(s). The base is an amino-alcohol consisting of two 5-carbon rings with a nitrogen at position 4, a hydroxymethyl group at 9, and a hydroxyl group at 7.
Hepatotoxc PAs all have an unsaturated 1,2-double bond in the pyrrolizidine ring. Saturated PAs, which are also wide- spread, e.g., in Echinacea spp (species), are not hepatotoxic. The carboxylic acids, called necic acids, vary both in length and in the number of branches. PAs occur in plants either as free bases or as nitrogen-oxides (N-oxides)."
-'Using Herbs that Contain Pyrrolizidine Alkaloids' by Alison Denham, B.A., MNIMH (National Institute of Medical Herbalists), University of Central Lancashire, England; The European Journal of Herbal Medicine, Volume 2, No. 3, pages 27-38, **1996**.

"Pyrrolizidine alkaloids (PAs) are a class of phytochemicals found in more than 350 plant species, with the main suspect species being Heliotropium, Senecio, Crotalaria, and Symphytum. Many of these types of alkaloids have been shown to have important pharmacological properties, but many others have demonstrated severe toxicity.
Consumption of plants containing these alkaloids has been associated with potentially fatal hepatic veno-occlusive disease (Budd-Chiari Syndrome), carcinogenesis, and fibrotic lung disease."
-'Pyrrolizidine Alkaloids in Higher Plants: Hepatic Veno-Occlusive Disease Associated with Chronic Consumption' by Wendy Pearson, BSc in Agriculture; Department of Pharmacology and Toxicology, University of Western Ontario, Canada; Journal of Nutraceuticals, Functional and Medical Foods, Volume 3, No. 1, pages 87-96, **2001**.

"Pyrrolizidine Alkaloids (PAs) contain two fused five-membered rings in which a nitrogen atom is common to both rings. The precursor is ornithine. A hydrogen atom usually occurs opposite the nitrogen atom in the alpha position, with a hydroxymethyl substitute at the adjacent C-1 position.
Most pyrrolizidines have diester groups at C-1 and C-7: these may be non-cyclic, for example echimidine, or macrocy-

clic, for example senecionine. Toxicity appears to be mainly linked to the macrocyclic diester type (*Denham 1996).
> *PAs from Comfrey include symphytine, intermedidine and symlandine; these do not contain the macrocyclic diesters.*

The alkaloids also occur as water-soluble nitrogen oxides."
-The Constituents of Medicinal Plants: An Introduction to the Chemistry and Therapeutics of Herbal Medicine by Andrew Pengelly, BA, ND, DBM, DHom. Crows Nest, NSW, Australia: Allen & Unwin, **2004**, page 147.
(* -'Using Herbs that Contain Pyrrolizidine Alkaloids' by Alison Denham, B.A., MNIMH {Member of National Institute of Medical Herbalists}, University of Central Lancashire, England; European Journal of Herbal Medicine, Vol 2, No. 3, pages 27-38, 1996.)
> (In biochemistry, a precursor is a substance from which another is formed, usually by metabolic reaction.)
> (A diester is an organic compound with two ester functional groups. An ester is a compound derived from an acid in which at least one -OH group is replaced by an -O-alkyl group.)

"The PAs (Pyrrolizidine Alkaloids), which have minimal toxicity in their original form, are metabolized in the liver and can become toxic metabolites, depending on the PA and on the particular condition of the liver enzymes.
The toxic metabolites, highly reactive smaller pyrroles (the dehydro- form of the alkaloids) resulting from action of microsomal enzymes in the liver, can act locally within the liver cells to cause damage at the chromosome level."
-'Safety Issues Affecting Herbs: Pyrrolizidine Alkaloids" by Subhuti Dharmananda, Ph.D., Director, Institute for Traditional Medicine, Portland, Oregon, **2016**.

"There are three types of unsaturated PA, defined by their chemical complexity:
  1. **Monoester PA** have a single necic acid branch.
  2. **Open diester PA** have two necic acid branches.
  3. **Macrocyclic diester PA** have two branches of necic acid joined together to form a chemical ring.

In plant material, PA occur in a 'free-base' form and as N-oxides (with an attached oxygen).
The N-oxides are non-toxic and water-soluble but once ingested, N-oxides are converted into free-base PA, and are now therefore considered to contribute to total toxicity."
-'Pyrrolizidine Alkaloids: Key Points for Herbal Practitioners' by Helen Phillips M.Sc. Herbal Medicine, Diploma of Phytotherapy, MNIMH and Alison Denham, M.A., FNIMH (National Institute of Medical Herbalists); The Forager's Path, LLC, School of Botanical Studies, Flagstaff, Arizona, August **2016**.

"Alkaloids have a wide range of pharmacological activities including antimalarial (e.g. quinine), antiasthma (e.g. ephedrine), anticancer (e.g. homoharringtonine), cholinomimetic (e.g. galantamine), vasodilatory (e.g. vincamine), antiarrhythmic (e.g. quinidine), analgesic (e.g. morphine), antibacterial (e.g. chelerythrine), and antihyperglycemic activities (e.g. piperine).
Many have found use in traditional or modern medicine, or as starting points for drug discovery. Other alkaloids possess psychotropic (e.g. psilocin) and stimulant activities (e.g. cocaine, caffeine, nicotine, theobromine), and have been used in entheogenic rituals or as recreational drugs.
Alkaloids can be toxic too (e.g. atropine, tubocurarine). Although alkaloids act on a diversity of metabolic systems in humans and other animals, they almost uniformly evoke a bitter taste."
-Wikipedia®: The Free Encyclopedia, www.wikipedia.org, **2018**.

"Alkaloids are a diverse group of amino acid-derived and nitrogen-bearing molecules that display a wide range of roles in nature, where they occur in plants, microorganisms or animals.
In plants, alkaloids can be found in the form of salts of organic acids, mainly malate, acetate and citrate, or combined with other molecules, such as tannins. *Most alkaloids display basic (base or alkaline) properties and present a lipophilic character, being soluble in apolar organic solvents and alcohol.*
The Chemistry of Pyrrolizidine Alkaloid:
> PA are a group of alkaloids derived from ornithine that are distributed in plants of certain taxa, being also found in insects that uptake them for defense against predators. **They rarely occur in the free form as a pyrrolizidine base, being instead found as esters (mono-, di- or macrocyclic diesters)** formed by a necine base (amino alcohols) and one or more necic acids (mono- or dicarboxylic aliphatic acids), which are responsible for their structural diversity. They are usually found in the form of tertiary bases or Pyrrolizidine Alkaloids N-Oxides (PANO)."

-'Pyrrolizidine Alkaloids: Chemistry, Pharmacology, Toxicology and Food Safety' by Rute Moreira, David M. Pereira, Patrícia Valentao and Paula B. Andrade, REQUIMTE/LAQV, Laboratorio de Farmacognosia, Departamento de Quimica, Universidade do Porto, Portugal; International Journal of Molecular Sciences, Volume 19, No. 6, 1668, June **2018**.
> (Lipophilic means combining with or dissolving in lipids or fats.)

Some Alkaloids are More Toxic Than Others (in general, not just Comfrey alkaloids)

"Certain pyrrolizidine alkaloids, such as heliotrine, lassiocarpine and retrorsine, have long been known to cause chronic liver poisoning in animals, and the pathology has been well described. Progressive liver damage has been induced experimentally in rats by single doses of certain alkaloids.
Usually the most hepototoxic alkaloids are cyclic diesters, like retrorsine and senecionine. Higher doses of lasiocarpine and still more of heliotrine are needed to give similar effects, while supinine is even less hepatotoxic. Some alkaloids,

especially fulvine and monocrotaline, frequently cause lung as well as liver lesions.
A requirement for hepatotoxicity seems, inter alia (among other things), to be the allylic ester function. **Thus rosmarinine, platyphylline, and the amino-alcohol retronecine, among other alkaloids, are not hepatotoxic.**
The N-oxides display a toxicity which is similar to, but sometimes weaker than, that of the corresponding bases."
-'Toxicity of Pyrrolizidine Alkaloids' by A.R. Mattocks, Nature: International Journal of Science, Volume 217, pages 723-728, February 24 **1968**.

"The toxicity of alkaloids was sought in relation to the geography of alkaloid-bearing plants. Relative toxicity was inferred (implied/deduced) from the structure of the compounds, which exhibit regular structure-activity relationships.
The toxicity of alkaloids from tropical plants is much greater than those from temperate plants.
No significant differences were found between herbaceous and woody plants.
The alkaloid content of leaves tends to be a positive function of alkaloid toxicity.
The eco-geographical distribution of toxicity is interpreted in terms of plant defense. It is generally true that a given genus yields alkaloids of the same class and thus of the same toxicity index.
> **Boraginaceae family (that includes Symphytum Genus) has 'Mean Toxicity Index' of 2.00.** *Aceraceae family has Index of 1.00. Asteraceae or Compositae family has Index of 2.00. Amaranthaceae family has Index of 6.00.*
> *(The higher the number, the greater the toxicity.)"*

-'The Toxicity of Plant Alkaloids: An Ecogeographic Perspective' by Donald A. Levin and Billie M. York, Jr., Biochemical Systematics and Ecology, Volume 6, pages 61-76, **1978**.
(Herbaceous plants have no persistent woody stem above ground, i.e., above ground they die back completely in winter.)

"Individual pyrrolizidine alkaloids are metabolized at different rates and, therefore, exhibit somewhat differing toxicities.
The major pyrrolizidine alkaloids in Comfrey: symphytine; the isomeric alkaloids, lycopsamine/intermidine; 7-acetyllycopsamine; echimidine; and lasiocarpine, **all contain structural features which favour activation by oxidative dehydrogenation."**
-'Comfrey: Assessing the Low-Dose Health Risk' by Peter J. Abbott, Toxicology Unit, Department of Community Services and Health, Canberra, Australia; Medical Journal of Australia, Volume 149, No. 11-12, pages 678-682, **1988**.

"The most acutely (severely) toxic PAs include those in senecio and crotalaria.
The kinds of PAs found in Comfrey are generally less toxic. However, PAs of the type found in Comfrey must also be regarded as having the potential for liver damage due to chronic toxicity at surprisingly low levels.
Humans are believed to be more susceptible to PA poisoning than are the common laboratory animals (Culvenor, personal communication 1987). The most acutely toxic PAs are the macrocyclic diesters of unsaturated necine {aminoalcohol} bases, such as senecionine and monocrotaline.
*Noncyclic diesters are generally rather less toxic and monoesters notably much less so (*Mattocks 1986 book). However, all esterified 1,2-unsaturated necines ought to be regarded as having the potential for liver damage due to chronic toxicity at surprisingly low levels."*
-'Comfrey Update' by D.V.C. Awang, HerbalGram published by American Botanical Council, Austin, Texas, Volume 25, pages 20-23, **1991**.
(* -Chemistry and Toxicology of Pyrrolizidine Alkaloids by A.R. Mattocks. London, England: Academic Press, 1986.)

"Much of the quantitative data on the toxicity of pyrrolizidine alkaloids are from shorter-term lethality (harmfulness including death) studies. These data can give some indication about the relative toxicity among pyrrolizidine alkaloids.
There is considerable difference in the toxicity among pyrrolizidine alkaloids.
> *For example, the most potent alkaloids to which people are exposed, such as trichodesmine, senecionine, and senecipylline have have acute lethal toxicity (i.e., LD50) values of 25, 50, and 77 mg/kg, respectively, while* **for heliotrine, lycopsamine and heliotrine N-oxide the LD50 values are 296, >10,000 and c 5000, respectively (meaning lower toxicity).** *Monocrotaline is in the middle range of these toxins."*

-'Pyrrolizidine Alkaloids in Food' by Roger A. Coulombe Jr., Graduate Program in Toxicology and Department of Veterinary Sciences, Utah State University, Logan, Utah; Advances in Food and Nutrition Research, Volume 45, pages 61-99, **2003**.
> (LD50 is the median Lethal Dose 50%. The value of LD50 for a substance is the dose required to kill half the members of a tested population after a specified test duration. It is used as an indicator of a substance's acute toxicity. **A lower LD50 indicates increased toxicity**. I think this method of scientific study is awful. There should be ways of testing substances without harming animals.)

"Three Types of PAs: Among the PAs (Pyrrolizidine Alkaloids), **cyclic diesters are the most toxic, with non-cyclic diesters of intermediate toxicity, and the monoesters the least toxic.** *The amino alcohols are not toxic.*
The toxicity of the N-oxides, when first reduced to the basic alkaloid by bacteria in the gut, is of the same order as that of the basic alkaloid *(*Mattocks et al. 1988). N-oxides are, however, much more water-soluble and are subject to different pharmacokinetics when absorbed unchanged from the gut."*
-'Analysis of Pyrrolizidine Alkaloids in Crotalaria Species by HPLC-MS/MS in Order to Evaluate Related Food Health Risks' by Gertruida Magdalena Rosemann, University of Pretoria, South Africa, Phytomedicine Programme, Department of Paraclinical Science, Faculty of Veterinary Science, Ph.D. thesis, **2007**, page 8.
(* -'Pyrrolizidine Alkaloids: Environmental Health Criteria 80' by International Programme on Chemical Safety, United Nations

Environment Programme and World Health Organization, Geneva, Switzerland, www.inchem.org/documents/ehc/ehc/ehc080.htm, 1988.)
(Pharmacokinetics is the study of the time course {over time} of drug absorption, distribution, metabolism, and excretion.)

"The toxic pyrrolizidine alkaloids are a large, varied class of naturally occurring, stereochemically (spatial arrangement) diverse monoesters, diesters and macrocyclic diesters of 1-hydroxymethyl-7-hydroxy-1,2-dehydropyrrolizidines (e.g. esters of the diastereoisomeric necine bases heliotridine and retronecine) or their N-methylated, ringopened (otonecine) analogues (*Mattocks, 1986; IPCS, 1988; Stegelmeier et al., 1999; Prakash et al., 1999; Wiedenfeld et al., 2008). These alkaloids and their N-oxides occur as natural components of many herbal preparations, cooking spices and honey, and can contaminate food crops (e.g. cereals) and animal-derived food (e.g. milk, eggs) destined for the human food supply (IPCS, 1988; Colegate et al., 1998; Ahmad, 2001; Edgar, 2003)."
-'Safety Assessment of Food and Herbal Products Containing Hepatotoxic Pyrrolizidine Alkaloids: Interlaboratory Consistency and the Importance of N-Oxide Determination' by Y. Cao, S.M. Colegate and J.A. Edgar, CISRO Livestock Industries, Australian Animal Health Laboratory, Victoria Australia; Phytochemical Analysis, Volume 19, No. 6, pages 526-533, **2008**.
(* -Chemistry and Toxicology of Pyrrolizidine Alkaloids by A.R. Mattocks. London, England: Academic Press, 1986.)

"Alkaloids and reported intraperitoneal LD50 (mg/kg bodyweight) values obtained for the male rat unless otherwise stated (Cheeke and Shull, 1985 and World Health Organization, 1988). **(The lower the number, the greater the toxicity.):**

Alkaloid	LD50
Retrorsine	34
Senecionine	50 also quoted as 85
Heliosupine	**60**
Lasiocarpine	**72**
Seneciphylline	77
Jacobine	77 (mouse)
Riddelliine	105 (mouse)
Symphytine	**130 also quoted as 300**
Heleurine	140
Jaconine	168 (female rat)
Monocrotaline	175
Echimidine	**200**
Spectabiline	220
Senkirkine	220
Heliotrine	300
Echinatine	**350**
Supinine	450
Europine	>1000
Heliotridine	1200
Intermedine	**1500**
Lycopsamine	**1500**"

-'Committee on Toxicity (COT) of Chemicals in Food, Consumer Products and the Environment', Statement on Pyrrolizidine Alkaloids in Food, COT is a United Kingdom independent scientific committee that provides advice to 'Food Standards Agency', 'Department of Health' and other government agencies, October **2008**.
(I put in bold those alkaloids found in Comfrey.) (A lower LD50 indicates increased toxicity.)

"Data on toxicity of individual PAs (Pyrrolizidine Alkaloids) and their relative potencies are still very limited.
For endpoint mortality (death), an overview of available LD50 values after intraperitoneal (i.p. = intraperitoneal injection = injected into abdominal cavity) and intravenous injection (i.v. = injected into vein) is presented. Oral LD50 values are not available.
LD50-values of several PAs after i.p. injection in rats

PA	LD50 (mg/kg)	Observation period	Rat Strain and Sex
7-OAngeloylheliotridine	250	deaths hours after dose	Hooded Wistar, male
Cynaustine	260	3 days	Strain not specified, male
Echimidine	200	Not reported	Strain not specified, male
Echinatine	350	Not reported	Strain not specified, male
Europine	>1000	Not reported	Strain not specified, male
Heleurine	140	Not reported	Strain not specified, male
Heliosupine	60	Not reported	Strain not specified, male
Heliotridine	1500	Not reported	Hooded Wistar, male
Heliotrine	296	3 days	Hooded strain, male
Heliotrine	478	3 days	Hooded strain, female
Heliotrine-N-oxide	>5000	8 days	Hooded strain, male
Indicine	>1000	Not reported	Strain not specified, male
Intermedine	1500	Not reported	Strain not specified, male
Jacobine	138	3 days	Strain not specified, female
Jaconine	168	3 days	Strain not specified, female

PA	LD50 (mg/kg)	Observation Period	Species
Lasiocarpine	77	3 days	Hooded strain, male
Lasiocarpine	79	3 days	Hooded strain, female
Lasiocarpine-N-oxide	547	3 days	Hooded strain, male
Lasiocarpine-N-oxide	181	3 days	Hooded strain, female
Latifoline	125	3 days	Strain not specified, male
Lycopsamine	1500	Not reported	Strain not specified, male
Monocrotaline	175	Not reported	Strain not specified, male
Monocrotaline	109	4 days	Strain not specified, male
Monocrotaline	230	4 days	Strain not specified, female
Platyphylline	252	Not reported	Hooded Wistar, male
Retrorsine	153	7 days	Strain not specified, female
Retrorsine	34-38	7 days	Strain not specified, male
Retrorsine-N-oxide	250	7 days	Strain not specified, male
Rinderine	550	3 days	Strain not specified, male
Senecionine	85	Not reported	Strain not specified, male
Senecionine	50	7 days	Strain not specified, male
Seneciphylline	77	3 days	Strain not specified, male
Seneciphylline	83	3 days	Strain not specified, female
Senkirkine	220	Not reported	ACI, male
Spectabiline	220	3 days	Strain not specified, male
Supinine	450	Not reported	Strain not specified, male
Symphytine	130	Not reported	ACI, male
Symphytine	300	Not reported	Strain not specified, male
Triacetylindicine	164	4 days	Strain not specified, male

LD50-values of several PAs after i.v. injection

PA	LD50 (mg/kg)	Observation Period	Species
Heliotrine	274	7 days	Rat
Heliotrine	255	7 days	Mouse
Integerrimine	78	7 days	Mouse
Jacobine	77	7 days	Mouse
Lasiocarpine	88	5 days	Rat
Lasiocarpine	85	5 days	Mouse
Lasiocarpine	67.5	5 days	Hamster
Retrorsine	38	7 days	Rat
Retrorsine	59	7 days	Mouse
Retrorsine-N-oxide	834	7 days	Mouse
Riddelliine	105	7 days	Mouse
Senecionine	64	7 days	Mouse
Senecionine	61	7 days	Hamster
Seneciphylline	90	7 days	Mouse
Spartiodine	80	7 days	Mouse "

-'Pyrrolizidine Alkaloids in Herbal Preparations, RIVM Briefrapport 090437001/2014' by Rijksinstituut voor Volksgezondheid en Milieu (RIVM) (Dutch National Institute for Public Health and the Environment), Commissioned by Netherlands Food and Consumer Product Safety Authority (NVWA), 62 pages, all in Dutch except for English abstract and appendix, **2014**. If you have an English translation, I would appreciate a copy.
(These Pyrrolizidine Alkaloids are found in various species of plants, not just Comfrey. Comfrey contains only some of these PAs.)

"Among the diversity of secondary metabolites which are produced by plants as means of defence against herbivores and microbes, pyrrolizidine alkaloids (PAs) are common in Boraginaceae, Asteraceae and some other plant families. **Pyrrolizidine alkaloids are infamous as toxic compounds which can alkylate DNA and thus cause mutations and even cancer in herbivores and humans. Almost all genera of the family Boraginaceae synthesize and store this type of alkaloids.**"
-'Diversity of Pyrrolizidine Alkaloids in the Boraginaceae: Structures, Distribution and Biological Properties' by Assem El-Shazly and Michael Wink, Journal of Biodiversity, Basel, Switzerland, Volume 6, Issue 2, April 1 **2014**.
(Alkylation is the transfer of an alkyl group from one molecule to another. It may be transferred as an alkyl carbocation, a free radical, a carbanion or a carbene. In medicine, alkylation of DNA is used in chemotherapy to damage DNA of cancer cells.)

"***Bodi et al. subdivided the Pyrrolizidine Alkaloids according to their necine acid into monoesters, open-chain diesters and cyclic diesters,*** as Senecioneae species mainly produce cyclic diesters, Eupatorieae mainly monoesters and **Boraginaceae family mainly open-chain diesters and monoesters.**
Plant species, Structural features, Pyrrolizidine alkaloids:
  Senecioneae (Asteraceae family) mainly cyclic diesters:
    erucifoline, jacobine, retrosine, senecionine, seneciphylline, senecivernine, senkirkine.
  Crotalaria species (Fabaceae family) mainly cyclic diesters:
    erucifoline, jacobine, retrosine, senecionine, seneciphylline, monochrotaline, trichodesmine.

Eupatorieae (Asteraceae family) mainly mono- and open chain diesters: echimidine, lycopsamine, intermedine.
Boraginaceae family mainly mono- and open chain diesters:
    echimidine, lycopsamine, intermedine, europine, heliotrine, lasiocarpine.
Cyclic diesters (= macrocyclic diesters) (most toxic group):
    Erucifoline, Erucifoline-N-oxide, Jacobine, Jacobine-N-oxide, Monocrotaline, Monocrotaline-N-oxide, Retrorsine, Retrorsine-N-oxide, Senecionine, Senecionine-N-oxide, Seneciphylline, Seneciphylline-Noxide, Senecivernine, Senecivernine-N-oxide, Senkirkine, Trichodesmine.
Open chain diesters (= noncyclic diesters) (moderately toxic group):
    Echimidine, Echimidine-N-oxide, Lasoicarpine, Lasoicarpine-N-oxide.
Monoesters (least toxic group):
    Europine, Europine-N-oxide, Heliotrine, Heliotrine-N-oxide, Intermedine, Intermedine-N-oxide, Lycopsamine, Lycopsamine-N-oxide."
-'Pyrrolizidine Alkaloids: Impact of the Public Statements made by EMA and National Health Authorities on the Pharmaceutical Industry' by Katrin Lambrecht, Berlin for Wissenschaftliche Prufungsarbeit zur Erlangung des Titels (Scientific Exam Work to Obtain the Title) Master of Drug Regulatory Affairs, Rheinische Friedrich Wilhelms University, Bonn, **2016**, pages 19, 27, 34.
(* -'Determination of Pyrrolizidine Alkaloids in Tea, Herbal Drugs and Honey' by D. Bodi, S. Ronczka, C. Gottschalk, N. Behr, A. Skibba, M. Wagner, M. Lahrssen-Wiederholt, A. Preiss-Weigert and A. These, Department Safety in the Food Chain, Federal Institute for Risk Assessment, Berlin, Germany; Food Additives and Contaminants - Part A: Chemistry, Analysis, Control, Exposure and Risk Assessment, Volume 31, No. 11, pages 1886-1895, September 2014.)

"**This report presents four plant extracts, which were analyzed for their qualitative and quantitative content in PAs: Senecio vernalis (eastern groundsel), Petasites hybridus (butterbur), Tussilago farfara (coltsfoot) and Symphytum officinale (Common Comfrey).** While the last three are commonly used in phytotherapy, Senecio vernalis was analyzed mainly for its already known high content of PAs, in an attempt to emphasize the risk of using PAs in therapy, if any.
Senecio vernalis:
    The largest amount of PAs was found in Senecio: **494.86 mg/100 gram dw** (dw= dry material).
    The predominant alkaloid being senecionine (346.14 mg/100 g dw).
    Senecio, the extract with the highest concentration of PAs, exhibits the highest toxicity.
Petasites hybridus:
    For Petasites the total PAs concentration was **3.18 mg/100 gram dw** (dry material).
    The main PA for PET was senecionine (1.84 mg/100 g dw).
    The LC50 obtained for Petasites and Tussilago, extracts that have similar PAs concentrations, had similar values.
Tussilago farfara:
    For Tussilago the total PAs concentration was **3.17 mg/100 gram dw**.
    The main PA for Tussilago was the otonecine-type PA senkirkine (2.44 mg/100 g dw).
Symphytum officinale:
    For Symphytum the total PAs concentration was **157.56 mg/100 gram dw**.
    Symphytine was as the main alkaloid (117.92 mg/100 g dw).
Symphytum, the extract with the second PAs concentration 157.56 mg/100 gram dw, presented in both cases the lowest toxicity. This result could be explained by the fact that Symphytum officinale contained senecionine under the limit of detection, if any. Alternately, other components present in the extract could reduce the effect of existing PAs in Symphytum. Further studies are to focus on both hypotheses."
-'Toxicity of Plant Extracts Containing Pyrrolizidine Alkaloids Using Alternative Invertebrate Models' by O.C. Seremet, et al. (Romania, Greece, England); Molecular Medicine Reports, Volume 17, No. 6, pages 7757-7763, June **2018**.
(LC50 = Lethal Concentration 50. It is the amount of a substance in air or water. It is similar to LD50 which is Lethal Dose such as how many milligrams of a substance is ingested. LD50 is the dose required to kill half of a tested population after a specified test duration. It is a general indicator of a substance's acute toxicity. I think this type of testing is awful. Our society should have better ways to test substances that don't involve torturing animals.)

"From most toxic to least of PAs (Pyrrolizidine Alkaloids) is macrocyclic diesters > retronecine and heliotridine diesters > heliotridine monoesters > retroecine monoesters.
    Comfrey PAs are retronicene mono- and diesters. Symphytine and echimidine are derivatives of retronecine.
**The most toxic alkaloid is echimidine. Only minute amounts are found in Symphytum officinale, but large amounts in Russian Comfrey and Prickly Comfrey leaves.
Symphytine (7-tiglylretronecine viridiflorate) is the major alkaloid of S. officinale.**"
-'Comfrey' by Glenn Axford, Our Common Roots: An Exploration of the Healing Power of Nature, Canada, www.ourcommonroots.com, date unknown but **2018** or earlier.

"Your resource for searching databases on toxicology, hazardous chemicals, environmental health, and toxic releases."
-Toxnet: Toxicology Data Network, National Institutes of Health, Health and Human Services, United States National Library of Medicine, Bethesda, Maryland, https://toxnet.nlm.nih.gov, **2018**.

Metabolic Pathways of Pyrrolizidine Alkaloids

"Pyrrolizidine Alkaloid (PA) Metabolism: Ingested PA are absorbed and metabolized in the parenchymal cells (hepatocytes) of the liver, especially the cells of Zone 3 (centrilobular region) that have the highest activity of drug-metabolizing enzymes.
There are several pathways of PA metabolism, discussed in detail by Mattocks (*1986 book). These include hydrolysis, dehydration, N-oxidation and minor pathways such as hydroxylation and epoxidation.
    1. Hydrolysis, *whereby the side chain(s) are cleaved (split) from the necine nucleus,* **is a detoxification reaction, because neither the necine nor necic acid moities produced are toxic.** *Although hydrolysis of synthetic PA has been well studied (Mattocks, 1986) there is little evidence that this is an important pathway for plant PA.*
    2. Dehydrogenation of PA to yield pyrrole derivatives is the classic pathway of PA metabolism *(Mattocks, 1986). Pyrroles are very reactive and are strong alkylating agents; they cross-link strands of DNA and thus impair cell division and protein synthesis.*
    3. N-oxidation.
    4. Minor pathways such as hydroxylation and epoxidation."
-'Toxicity and Metabolism of Pyrrolizidine Alkaloids' by P.R. Cheeke, Oregon State University, Corvallis, Oregon; Journal of Animal Science: American Society of Animal Science, Volume 66, No. 9, pages 2343-2350, October **1988**.
(* -Chemistry and Toxicology of Pyrrolizidine Alkaloids by A.R. Mattocks. London, England: Academic Press, 1986.)

"After consumption, PAs may be excreted via the bowel or absorbed.
The N-oxides are water soluble, therefore, more likely to be excreted unchanged, but they are also absorbed, probably after metabolism by gut bacteria. *Absorption of N-oxides is more gradual than for free bases.*
Research on absorption is difficult because PAs are immediately metabolized in the liver. *After absorption, PAs travel to the liver. Detoxification of foreign compounds in the hepatocyte (liver cell) depends on making the compound more water soluble and thus more easily excreteted by the kidneys.*
Each class of compound is detoxified, often in stages, by the most appropriate chemical reactions available. This process may fail if the amount ingested exeeds the possible reaction rate(s).
The chemical reactions available in the hepatocyte vary in efficacy for each toxin, and there is also wide inter-individual (between individuals) variation in detoxification ability."
-'Using Herbs that Contain Pyrrolizidine Alkaloids' by Alison Denham, B.A., MNIMH (National Institute of Medical Herbalists), University of Central Lancashire, England; The European Journal of Herbal Medicine, Volume 2, No. 3, pages 27-38, **1996**.
    (A hepatocyte is a cell of the main parenchymal tissue of the liver. They are 70-85% of the liver's mass. These cells are involved in protein synthesis/storage, transformation of carbohydrates, synthesis of cholesterol/bile/phospholipids, and detoxification/excretion of unwanted substances.)

*"***The PAs are not toxic until they are metabolized in the liver. Dehydrogenation by P-450 enzymes forms toxic pyrrolic metabolites.** *These pyrrolic metabolites either undergo hydrolysis to pyrrolic alcohols or destroy surrounding tissues.*
Both the pyrrolic esters (primary metabolites) and the pyrrolic alcohols (secondary metabolites) have antimitotic (antimitosis) effects and are responsible for the damage to cells of the liver.*"
-Toxicology and Clinical Pharmacology of Herbal Products edited by Melanie Johns Cupp, PharmD, BCPS. Totowa, New Jersey: Humana Press, **2000**. Chapter 18: Comfrey by David Burch and Melanie Johns Cupp, page 211.
(* -'Comfrey: Assessing the Low-Dose Health Risk' by P.J. Abbott, Toxicology Unit, Department of Community Services and Health, Canberra, Australia; Medical Journal of Australia, Volume 149, No. 11-12, pages 678-682, 1988.)
(Mitosis is cell division that results in 2 daughter cells each having same number and kind of chromosomes as the parent.)

*"***Metabolism of Pyrrolizidine Alkaloids: There are two major routes of metabolism; one involves N-oxidation** *via the mixed-function oxidase system* **and the other dehydrogenation to dehydroalkaloids or pyrroles** *also by means of cytochrome (CYP) P450. The dehydroalkaloids and pyrroles are highly reactive.*
    P450-catalyzed alpha-carbon oxidation converts the 3-pyrrolinyl moiety of toxic pyrrolizidines into a pyrrolyl-containing metabolite, which undergoes spontaneous conversion to electrophilic products that react with nucleophiles on macromolecules. There are a number of other minor metabolites such as aldehydes, monocrotalic and monocrotic acids, and sulphur conjugates including those with N-acetylcysteine.
        The highly reactive dehydroalkaloids have half-lives of the order of seconds in aqueous (water) media. These dehydroalkaloids have four further reaction pathways available to them. All four of these pathways involve nucleophilic attack at position 7 of the fused ring system.
        (i) dehydrogenation to less toxic dehydropyrrolizidines.
        (ii) conjugation with glutathione (GSH) to form 7-glutathionyl-6, 7 dihydro-1-hydroxymethyl-5H-pyrrolizine (excreted in bile).
        (iii) alkylation of macromolecules (such as proteins or nucleic acids).
        (iv) release into the circulation leading to damage to the hepatic vasculature (endothelium) and other organs.
The relative contribution of these pathways determines both the sites at which toxicity is expressed and the degree of toxicity. *When deprived of its allylic group the pyrrolizidine nucleus loses its toxicity. Pyrrolizidine alkaloids are also metabolized to 6,7-dihydro-7-hydroxy-1-hydroxymethyl-5H-pyrrolizine.*
The toxicity and metabolism of pyrrolizidine alkaloids are markedly influenced by the availability of glutathione and taurine, to which they link to form nontoxic excretory products.
The progressive nature of PA-induced hepatotoxicity suggests that some pyrrolic adducts may be recycled, *reacting with*

new nucleophiles and causing further cellular damage. Pyrrolizidines form adducts with hemoglobin that are stable in red blood cells for 120 days."
-'Pyrrolizidine Poisoning: A Neglected Area in Human Toxicology' by Michael J. Stewart and Vanessa Steenkamp, University of the Witwatersrand Medical School, Johannesburg, South Africa; Therapeutic Drug Monitoring Journal, New York, Volume 23, No. 6, pages 698-708, **2001**.
> (An adduct is a product of a combination of two or more distinct molecules, resulting in a single reaction product containing all atoms of all components. It is a distinct molecular species.)

"Pyrrolizidine Alkaloids (PAs) are absorbed from the intestine and reach the liver via the portal vein. After hydrolytic cleavage, cytochromes and mixed function oxidases oxidize PAs to pyrrole derivatives, highly reactive electrophilic alkylating compounds that covalently bind to protein, RNA and DNA.
Because these reactions take place in the liver, it is the organ most affected."
-The 5-Minute Herb and Dietary Supplement Consult by Adriane Fugh-Berman. Philadelphia, Pennsylvania: Lippincott Williams & Wilkins, **2003**.

"Pyrrolizidine Alkaloid Metabolism/Mechanism of Action. PAs undergo three main metabolic pathways:
  1. PAs that are hydrolyzed to a carboxylic acid or N-oxidized to a N-oxide metabolite are non-toxic and soluble in water thus excreted via urine.
  2. PAs undergo biotransformation by CYP3A, its reactive metabolites. CYP3A oxidize the PAs, followed by the **dehydrogenation** of the necine ring. The phenomenon produces a dehydro-pyrrolizidine compound, **a toxic pyrrolic ester** that acts as an electrophile.
  Thus, CYP3A inducers could increase the susceptibility of PA-induced toxicity. CYP3A inhibitors could prevent toxic outcomes since inhibitors yield less dehydro-PAs.
  *3. **The excess of pyrrolizidine N-oxide metabolites*** can be further transformed into toxic epoxides and necine bases."
-'Pyrrolizidine Alkaloids Enhance Alcohol-Induced Hepatotoxicity In Vitro in Normal Human Hepatocytes' by M.G. Neuman, L.B. Cohen and V. Steenkamp, Ontario, Canada and Pretoria, South Africa; European Review for Medical and Pharmacological Sciences, Volume 21, Suppl 1, pages 53-68, March **2017**.

"Biological Activity of Pyrrolizidine Alkaloids: Pharmacokinetics:
Concerning PA pharmacokinetics, after oral ingestion these compounds are absorbed from the gastrointestinal tract. **Most of them, around 80%, are excreted in urine, feces and milk,** a few being able to pass the placenta due to their high lipophilicity. **Bioactivation occurs mostly in the liver and, for this reason, this organ is the most affected by toxicity. Other organs have been identified as targets, namely the lungs and kidneys.** The lung is the second most affected organ by the pyrroles formed after metabolic activation in the liver, since they can travel to the lungs through blood.
For PA to be excreted or exert toxicity, as with many xenobiotics (foreign substances), biotransformation must occur. There are three principal pathways for the metabolic activation of PA:
  1. Hydrolysis to produce necines and necic acids. Hydrolysis is an important detoxification route, promoting the clearance of these compounds. In the case of hydrolysis, liver microsomal carboxylesterases are involved.
  2. N-oxidation to form PANO (Pyrrolizidine Alkaloids N-Oxides). N-oxidation is also an important detoxification route, which allows the formation of PANO that can be conjugated for excretion. However, PANO can reverse back into PA and suffer oxidation into DHPA. This route is carried out mostly by cytochrome P-450 (CYP450) monooxygenases. In fact, the activity of these enzymes can partly explain the distinct susceptibility of different species to PA. The isoforms of CYP450 involved in the metabolism leading to DHPA are generally CYP3A and CYP2B.
  However, only retronecine-type and heliotridine-type PA are capable of suffering N-oxidation, otonecine-type PA being unable to generate PANO owing to their methylation in the nitrogen.
  3. Oxidation that leads to the formation of pyrrolic esters or dehydropyrrolizidine alkaloids (DHPA).
The balance between the formation of DHPA and the formation of detoxification compounds, such as necines, necic acids and PANO, is also important in explaining the distinct susceptibility of different species to these compounds."
-'Pyrrolizidine Alkaloids: Chemistry, Pharmacology, Toxicology and Food Safety' by Rute Moreira, David M. Pereira, Patrícia Valentao and Paula B. Andrade, REQUIMTE/LAQV, Laboratorio de Farmacognosia, Departamento de Quimica, Universidade do Porto, Portugal; International Journal of Molecular Sciences, Volume 19, No. 6, 1668, June **2018**.
> (Pharmacokinetics is the branch of pharmacology about the movement of drugs in the body.)

Solubility, Stability, and Heat Sensitivity of Alkaloids

"Protein extracted (removed) from Comfrey should not be harmful: a sample of Comfrey supplied by Mr. Hills proved, as expected, to be completely free of alkaloids. A large percentage of the pyrrolizidine alkaloids are destroyed in the drying process, and as herbal preparations mainly use dried materials as a starting base, so **there is no reason why Symphytum officinale should be restricted as a therapeutic agent."**
-'Pharmacists and Comfrey' by Diane Wiesner, B.Pharm, MA, Ph.D., MPS, Principal of the NSW College of Natural Therapies, Sydney, Australia; Australian Journal of Pharmacy, Volume 65, pages 959-963, **1984**.

"The alkaloids (not just of Comfrey) are fairly stable chemically, but the ester groups may undergo hydrolysis under alkaline

conditions. **Some alkaloids in plant material may decompose during drying** *(*Bull et al., 1968)*, but others appear to be stable under similar conditions *(**Pedersen, 1975; Birecka et al., 1980).*
The N-oxides of unsaturated pyrrolizidines are more readily decomposed by heat than the basic alkaloids, especially when dry. *However, the stability of the alkaloids and N-oxides in hot water as, for example, in cooking, is not known.*
Some pyrrolizidine alkaloids have a limited water solubility, unless neutralized with acid; but others (e.g., indicine), and all the N-oxides, are readily soluble.
According to Danninger et al. (***1983), in some species (Symphytum asperum), relatively long storage may lead to a reduction in the alkaloid content, *presumably because enzymes are released during drying."*
-'Pyrrolizidine Alkaloids: Environmental Health Criteria 80' by International Programme on Chemical Safety, United Nations Environment Programme and World Health Organization, Geneva, Switzerland,   www.inchem.org/documents/ehc/ehc/ehc080.htm, **1988**.
(*-The Pyrrolizidine Alkaloids: Their Chemistry, Pathogenicity and Other Biological Properties by L.B. Bull, C.C.J. Culvenor and A.T. Dick. Amsterdam, Netherlands: North Holland Publishing Co., 1968.)
(** -'Pyrrolizidine Alkaloids in Danish Species of the Family Boraginaceae' by E. Pedersen, Archive of Pharmacy and Chemistry Science Edition, Volume 3, pages 55-64, 1975. I was unable to get this report. If you have a copy, could you send it to me.)
(*** -'Toxicity of Pyrrolizidine Alkaloid-Containing Medicinal Plants' {Zur Toxizitat Pyrrolizidinalkaloid Haltiger Arzneipflanzen} by T. Danniger, U. Hagemann, V. Schmidt and P.S. Schoenhoefer, Pharmazeutische Zeitung, Volume 128, pages 289-303, 1983. I was unable to find this report. If you have a copy, could you send it to me especially if you have an English translation.)

*"***Alkaloids (not just of Comfrey) are organic bodies, derived chiefly form plants in which they are believed to exist in combination with organic acids, forming salts (a safer, more soluble form of the alkaloid).** *These alkaloid salts are usually well-defined, colorless, odorless, crystalline and soluble.* **Alkaloids may be unstable when heated.**
Solubility:
> *Most pure alkaloids are bitter, slightly alkaline, soluble in ether and chloroform, and often less readily in alcohol. In water they are comparatively insoluble (not soluble).*
> **On the other hand, the solubility of the alkaloid salts usually follow an opposite pattern; they are freely soluble in water and somewhat soluble in alcohol.** *The ready solubility of the salts of alkaloids have caused them to be preferred to the alkaloids themselves for therapeutic uses.*

For preparation of extracts with the highest levels of alkaloids, *water/vinegar/alcohol menstrua (solvent) having a 35 percent water, to 10 percent vinegar, to 55 percent alcohol content are recommended.*
Please keep in mind, that **while alkaloids and their salts have distinctive therapeutic properties of their own, they do not fully nor exactly represent the action of the whole plant from which they are derived.***"*
-The Herbal Medicine-Maker's Handbook: A Home Manual by James Green, Herbalist. New York, New York: Random House, Inc., **2000**, page 93.
> (Soluble means able to be dissolved in a liquid, especially water.)
> (Menstra or menstruum is a substance that dissolves a solid or holds it in suspension, i.e., a solvent.)

*"***When Comfrey is dried, enzymes are released and much of the alkaloid is destroyed.**
From trials in Minnesota in 1987, in an attempt to determine cultural and environmental factors associated with the production of PAs, **it was found that Comfrey, harvested at different times in the growing season, can be of varying PA amounts.**
In one trial in 1986*, immature leaves contained 0.026% pyrrolizidine, on a dry weight basis.
A subsequent harvest during the growing season had no detectable PAs in the leaf *(the minimum detectable quantity was 5 ppm {parts per million}. The data indicated, harvest time was a critical factor in producing PA free Comfrey, and that mature leaves have an even lower alkaloid content, than young leaves."*
-How Can I Use Herbs in My Daily Life: Over 500 Herbs, Spices and Edible Plants: An Australian Practical Guide to Growing Culinary and Medicinal Herbs' by Isabell Shipard. David Stewart Publisher: **2003**.
(* The author did not give a reference about the 1987 Minnesota trials, but probably it was this: -'Growing Alkaloid-Free Comfrey' by Gary Steuart, Steuart Laboratories, Saint Paul, Minnesota; The Herb Spice and Medicinal Plant Digest, 'Massachusetts Cooperative Extension Service' and 'University of Massachusetts', Amherst, Volume 5, No. 4, page 9, Winter 1987.)

*"***Various techniques of extraction of pyrrolizidine alkaloids (PAs) from Comfrey have been developed and compared.**
Different extraction media: methanol, ethanol, 1% methanolic solution of tartaric acid, 2.5% HCl solution, 5% CH3COOH solution, alkaline chloroform-methanol mixture at various temperatures (room temperature, 50-60 C {122-140 F}, solvent's boiling point) have been used in various extraction techniques (percolation, electric basket, ultrasonic water bath) and various extraction times. **The best result of PAs extraction from Comfrey was obtained using 1% methanolic solution of tartaric acid and electric basket technique at the temperature 100 +- 5 C (212 +- 41 F) for 2 hours.** *Possible applications of the method have been discussed.*
Another extraction method:
> **Extraction of PAs (pyrrolizidine alkaloids) from Symphytum officinale with 95% ethanol (alcohol) at room temperature (maceration):**
> *1 gram (0.035 ounce) samples of dried and pulverized Comfrey (Symphytum officinale) roots were placed in a 250 mL-in-volume round-bottom flask. 100 mL of ethanol were added and the samples were extracted by shaking in a wrist-action shaker for 18 hours at room temperature."*

-'Comprehensive Extraction of Pyrrolizidine Alkaloids from Plant Material' by Tomasz Mroczek, Jaroslaw Widelski and Kazimierz

Glowniak, Medical University of Lublin, Poland; Chemia Analityczna (Analytical Chemistry), Warsaw, Poland, Volume 51, pages 567-580, **2006**.

"Stability of Alkaloids:
*Alkaloids in general, are not very stable. **They normally undergo degradation or decomposition on being exposed to air, light, moisture and heat, besides chemical reagents.** A few typical examples of alkaloids vis-a-vis (in regard to) their stability are stated below, namely:*
An aqueous (water) solution of alkaloids undergo rapid decomposition or degradation as compared to their solid forms. *During the course of extraction of alkaloids followed by isolation, the solvent is preferably removed effectively by distillation under vacuum (or reduced atmospheric pressure) or by subjecting it to evaporation in a Rotary Thin-Film Evaporator under vacuum **so that the desired product is not exposed to excessive heat, thus avoiding decomposition.***
Acid salts of Alkaloids:
*In common practice the salts of alkaloids are prepared by using cold and dilute solutions of the mineral acid (not for human use). It may be pointed out that **the use of concentrated mineral acids, or heating an alkaloid even with a dilute acid under pressure may ultimately lead to profound changes in them."***
-'General Characteristics of Alkaloids', Pharmacognosy: Study of medicinal drugs derived from plants or other natural sources, www.epharmacognosy.com, July **2012**.

"Extracting (removing) substances from herbs:
*1. **Water extracts** acids, **alkaolids (some), alkaloid salts (most),** coloring matter, essential oils (some), glycosides, gums, mineral salts, mucilage, pectin, proteins, sugars, tannins.*
*2. **Alcohol extracts** acids (organic), acrid constituents, **alkaloids,** bitter constituents, chorophyll, essential oils, glycosides, resins. Alcohol does not extract minerals, gums, or mucilage.*
*3. **Infused oil extracts** resins, essential oils, and flavonoids."*
-Organic Growers School®, Asheville, North Carolina, www.organicgrowersschool.org; Training, workshops, conferences, **2018**.

*"**Alkaloids can be purified from crude extracts of these organisms by acid-base extraction.** Acid-base extraction is a procedure using sequential liquid–liquid extractions to purify acids and bases from mixtures based on their chemical properties. **Because of the structural diversity of alkaloids, there is no single method of their extraction from natural raw materials. Most methods exploit the property of most alkaloids to be soluble in organic solvents but not in water, and the opposite tendency of their salts.***
Most plants contain several alkaloids. *Their mixture is extracted first and then individual alkaloids are separated. Plants are thoroughly ground before extraction.* ***Most alkaloids are present in the raw plants in the form of salts of organic acids.*** *The extracted alkaloids may remain salts or change into bases.*
Base extraction *is achieved by processing the raw material with alkaline solutions and extracting the alkaloid bases with organic solvents, such as 1,2-dichloroethane, chloroform, diethyl ether or benzene.*
Then, the impurities are dissolved by weak acids; this converts alkaloid bases into salts that are washed away with water. If necessary, an aqueous solution of alkaloid salts is again made alkaline and treated with an organic solvent. The process is repeated until the desired purity is achieved.
In the acidic extraction, the raw plant material is processed by a weak acidic solution (e.g., acetic acid in water, ethanol, or methanol). *A base is then added to convert alkaloids to basic forms that are extracted with organic solvent (if extraction was performed with alcohol, it is removed first, and the remainder is dissolved in water). The solution is purified as described above."*
-Wikipedia®, The Free Encyclopedia, https://en.wikipedia.org, **2018**.

<u>Detecting Pyrrolizidine Alkaloids in Herbal Products</u>

"Preparations from Comfrey (Symphytum officinale and S. x uplandicum) root and leaf contain varying levels of the hepatotoxic (liver toxic) pyrrolizidine alkaloids (PAs).
Reference compounds for Comfrey are not commercially available, and there is currently no rapid extraction or analytical method capable of determining low levels in raw materials or as adulterants (contaminant) in commercially available extracts.
*A **Solid-Phase Extraction (SPE)** method was developed using an **Ergosil cleanup column that specifically binds the PAs**. With this method, powdered Comfrey root was extracted by sonication and shaking with basic chloroform.*
Percent recoveries of the PAs following Ergosil SPE had an overall average of 96.8%, with RSD (Relative Standard Deviation) of 3.8% over a range of 1.0 to 25.0 grams extracted in 100 mL. *Average precision of the method (n = 3 over 4 extraction concentrations) gave an overall RSD of 6.0% for the 5 alkaloids, with a range of 0.8% (5 g in 100 mL) to 11.2% (25 g in 100 mL).*
Recovery optimization testing showed that 1.0 gram (0.035 ounce) Comfrey root extracted in 100 mL yielded the greatest recovery (% dry weight) of the PAs, with an extraction efficiency and accuracy of 94.2%, and RSD of 1.7% (n = 9).
The unique properties of the Ergosil cleanup column provide rapid sample cleanup, volume reduction, and concentration of PAs from Comfrey extracts, and allow the eluant (extracted product) to be analyzed directly by traditional chromatographic methods."

-'A Rapid Cleanup Method for the Isolation and Concentration of Pyrrolizidine Alkaloids in Comfrey Root' by D.E. Gray, et al., Midwest Research Institute, Kansas City, Missouri; Journal of AOAC (Association of Official Analytical Chemists) International, Volume 87, No. 5, pages 1049-1057, September **2004**. (I only have the abstract. If you have all of the article, please send it to.)

"**The purpose of the current study was to develop a LC-MSn (Liquid Chromatography-Mass Spectrometry) method for the analysis of pyrrolizidine alkaloids (PAs) in Comfrey.**
      Published data presents an extensive list of PAs and their N-oxides present in Comfrey. However, standards are not commercially available for any of the PAs typically present in Comfrey.
Those PAs that are not stereoisomers were readily resolved on a C18 column using a water-acetonitrile gradient as the mobile phase. The use of a selective technique, LC-MS/MS, allowed us to identify groups of PAs and their N-oxides, as well as identify the number of PAs present in each group, including those that were not completely resolved chromatographically.
Published data presents an extensive list of PAs frequently found in Comfrey. In general, however, the presence or relative amounts of their N-oxides are not addressed. The N-oxide was the predominant form for each PA making it more important that the N-oxides be determined.
      The coupling of liquid chromatography (LC) with ion trap mass spectrometry (MS) provides a substantial gain in terms of selectivity and allows the simultaneous determination of PAs and their corresponding N-oxides in a single chromatographic run without the requirement of a reduction step.
This study presents a more extensive list of PAs detected in Comfrey root (Symphytum x uplandicum) compared to those routinely reported by other methods.
      Although it was possible to identify specific sets of PAs and N-oxides by mass spectrometry analysis, further studies, such as NMR (Nuclear Magnetic Resonance) or chiral separation strategies, are necessary to distinguish those PAs and N-oxides that are stereoisomers."
-'Simultaneous Analysis of Hepatotoxic Pyrrolizidine Alkaloids and N-oxides in Comfrey Root by LC-Ion Trap Mass Spectrometry' by Jorgelina C. A. Wuilloud, Samuel R. Gratz, Bryan M. Gamble and Karen A. Wolnik, United States Food and Drug Administration, Forensic Chemistry Center, Cincinnati, Ohio; The Analyst: Journal of the Royal Society of Chemistry, London, England, Volume 129, pages 150-156, **2004**.
(Chiral is a chemistry term meaning asymmetric in such a way that the structure and its mirror image are not superimposable.)

"**Two recent mass spectrometry-based reports concerning Senecio scandens yielded remarkably dissimilar (different) pyrrolizidine alkaloid constituents.**
In both studies, and in a related analysis of Senecio scandens and Tussilago farfara using micellar electrokinetic chromatography, **the presence of hazardous N-oxides of the alkaloids was either not considered or was inadequately considered. This raises concerns about the effectiveness of the methodologies used in these, and similar, studies in assessing the pyrrolizidine alkaloid content and the safety of food, food supplements and medicines for human use.**
Direct infusion-ESI MS and HPLC-ESI MS were used to analyse samples derived from liquid–liquid partitioning experiments and from strong cation exchange, solid-phase extraction of pyrrolizidine alkaloids and their N-oxides.
A preliminary LCMS analysis of commercially prepared extracts of Comfrey roots (Symphytum officinale and S. uplandicum s.l.) was used as a model to highlight the analytical importance of N-oxides in the safety assessment of pyrrolizidine alkaloid-containing medicinal herbs.
This study highlighted significant differences in the reported identification of pyrrolizidine alkaloids from the same plant species, and **clearly demonstrated the inadequacy of some procedures to include N-oxides in the assessment of pyrrolizidine alkaloid-related safety of food and herbal products.**"
-'Safety Assessment of Food and Herbal Products Containing Hepatotoxic Pyrrolizidine Alkaloids: Interlaboratory Consistency and the Importance of N-Oxide Determination' by Y. Cao, S.M. Colegate and J.A. Edgar, CISRO Livestock Industries, Australian Animal Health Laboratory, Victoria Australia; Phytochemical Analysis, Volume 19, No. 6, pages 526-533, **2008**.
      (sensu lato = s.l., sens. lat. In the broad sense. Used in taxonomy to clarify the scope of a taxon when it has been used to define more than one set of lower-level taxons. It includes all its subordinate taxa and/or other taxa that at other times are considered as distinct.)
      (N-oxide, or amine-N-oxide or amine oxide, is a chemical compound that contains functional group R3N+O-, an N-O coordinate covalent bond with 3 additional hydrogen and/or hydrocarbon side chains attached to N. Amine oxides are used as protecting group for amines and as chemical intermediates. They are readily metabolized and excreted if ingested.)

"**For the chemical standardization of botanicals and herbal preparations, the extraction of bioactive components or marker compounds is an important step.**
The methods found in the monographs of Pharmacopeias and other reports often use extraction methods that require significant volume of organic solvents and are rather tedious.
In previous PAs extraction studies, the use of methanol at temperature 50-60 C (122-140 F) for 12 hours was often recommended. **The best result of extraction of PAs from Comfrey was obtained using 1% methanolic solution of tartaric acid and electric basket technique at the temperature 100 +- 5 C (212 F) for 2 hours.**
Recently, simpler and more environmental friendly extraction methods have been developed. These include supercritical fluid extraction (SFE), microwave-assisted extraction (MAE), pressurized fluid extraction (PFE). Pressurized hot water extraction (PHWE) is another new extraction technique which can be applied for the isolation of bioactive or marker compounds from botanicals and medicinal plants.
Comfrey (Symphytum officinale): In the present study, PAs (pyrrolizidine alkaloids) such as lycopsamine, echimidine

and lasiocarpine were determined using electrospray liquid chromatography-mass spectrometry (LC-MS) with the method precision (relative standard deviation, RSD) <10%.
> Detection of lycopsamine, symviridine and their N-oxides could be confirmed with a newly developed method based on HPLC ion-trap and orbitrap MS with electrospray ionization interface. With LC-MS, quantitative analysis of lycopsamine in the botanical extract was carried out. The effect of extraction solvent was optimized by sonication and methanol: H2O (50:50) was selected. Then a rapid method based on pressurized hot water extraction (PHWE) was employed for the extraction of lycopsamine from Comfrey followed by the comparison with heating under reflux with the RSD ranging from 2.49% to 19.32%.

Our results showed a higher extraction efficiency for heating under reflux compared with PHWE. It was proposed that the lower extraction efficiency for PHWE was attributable to dissolved nitrogen from air which caused the reduction in the solubility of lycopsamine in the compressed hot solvent.

In this study, quantitative analysis of PAs in Comfrey was demonstrated. In addition, it was found that the use of subcritical water for extractions depended on the physical properties of the dissolved solutes and their tendency to degrade under the chosen extraction conditions."

-'Determination of Pyrrolizidine Alkaloids in Comfrey by Liquid Chromatography-Electrospray Ionization Mass Spectrometry' by Liu Feng, et al., (Department of Chemistry, National University of Singapore) (Universidad del Quindío, Columbia), Talanta: The International Journal of Pure and Applied Analytical Chemistry, Volume 80, No. 2, pages 916-923, **2009**.

"**Though Symphytum species have been the topic of investigations for several decades, the complete chemical profiling of plants belonging to these species has not been done. This is due primarily to the fact that many PAs in Symphytum species exist as isomeric mixtures, which are difficult to separate.**
> The two major compounds found in most of the Symphytum species are the diastereoisomers intermedine and lycopsamine. They are monoesters of the (+)-retronecine base and contain (-)-viridifloric acid (2S, 3S) and (+)-trachelanthic acid (2S, 3R) at the ninth position.

Enantioseparation of the pyrrolizidine alkaloid isomers intermedine and lycopsamine, isolated from Symphytum uplandicum, is discussed. The separatory power of two immobilized carbohydrate-based chiral HPLC (High Performance Liquid Chromatography) columns, Chiralpak IA and IC, in different chromatographic conditions is compared. The study demonstrated the importance of solvent and column selection while developing such chiral HPLC separation methods.

The baseline HPLC separation of the two alkaloid isomers in preparatory scale is reported for the first time. The optimized separations were achieved on a Chiralpak IA column with mobile phases of ACN/methanol (80:20) and methanol/methyl-t-butyl ether (90:10), both containing 0.1% diethylamine."

-'Chiral Stationary Phases for Separation of Intermedine and Lycopsamine Enantiomers from Symphytum Uplandicum' by Rahul S. Pawar, Erich Grundel, Eugene Mazzola, Kevin D. White, Alexander J. Krynitsky, Jeanne I. Rader, Office of Regulatory Science, Center for Food Safety and Applied Nutrition, United States Food and Drug Administration, Maryland; Journal of Separation Science, Germany, Volume 33, No. 2, pages 200-205, February **2010**.
(Enantiomers are each of a pair of molecules that are mirror images of each other. Enantioseparation is separation of enantiomers.)

"**Pyrrolizidine Alkaloid determination in plants: Several methods can be used for the qualitative and quantitative determination of PAs in plant material or its preparations:**
> **1. TLC: Thin-Layer Chromatography** (using the detection method of Dann / Mattocks [Dann, 1960; Mattocks, 1967]) is a quick, sure and easy method for a qualitative detection of PAs. Using TLC in a densitometric way (Bartkowski, Wiedenfeld, et al., 1997), it is also possible to have quantitative results; the detection limit is about 1 - 10 microgram.
> **2. LC-MS: Liquid Chromatography-Mass Spectrometry:** Different HPLC (High Performance Liquid Chromatography) methods are described. Obviously, this is the mostly used analytical method for the determination of PAs. A great benefit is the fact that PA-N-oxides can be analyzed as well as the free bases. Depending on the equipment, the detection limit is less than 0.1 microgram per injection.
> **3. GC-MS: Gas Chromatography-Mass Spectrometry:** Also in the case of GC a lot of methods are described. A GC problem is that N-oxides have to be reduced to the free bases before being analyzed. This reduction procedure is described intensively and doesn't seem to be a limiting factor. The detection limit is similar to that in LC.

All three methods can only be used accurately in case reference material is available. The reason is that the single PAs show different detector responses in the respective analyzing methods."

-'Plants Containing Pyrrolizidine Alkaloids: Toxicity and Problems' by Helmut Wiedenfeld, Pharmaceutical Institute, University of Bonn, Germany; Food Additives and Contaminants, Volume 28, No. 3, pages 282-292, **2011**.

> "**Problems with Thin-Layer Chromatography (TLC):**
> Lasiocarpine has only been found in studies of Comfrey root in Poland, and in leaves of Russian grown Symphytum officinale. Lasiocarpine has been shown to be strongly mutagenic and carcinogenic in rat studies.
> **However, the accuracy of identification by TLC (Thin-Layer Chromatography) has been called into question by some scientists, since it is a derivation of heliotridine and ALL other alkaloids in Comfrey are derivatives of retronecine; which itself is not hepatotoxic. Echimidine in fresh Comfrey has been misidentified in TLC.**
> One study of different ploidy races of Symphytum officinale shows echimidine content depends on the cytotype. When 2n = 24 or 40, echimidine is frequently absent, but at 2n=48 or higher it is almost always present."

-'Comfrey' by Glenn Axford, Our Common Roots: An Exploration of the Healing Power of Nature, Canada, www.ourcommonroots.com, date unknown but **2018** or earlier.

Detecting Pyrrolizidine Alkaloids in Humans

"**The present study describes the determination of two different types of hepatotoxic pyrrolizidine alkaloids (PAs) and also distinguishing the hepatotoxic PAs from non-toxic ones** by both in-source collision-induced dissociation high performance liquid chromatography mass spectrometry (CID-HPLC/MS) and HPLC/MS/MS (CID in the collision cell), using electrospray ionization. The mass spectra provided molecular ions and characteristic fragment ions, which could be used readily for a rapid identification of different types of PAs.
Applications of both in-source CID-HPLC/MS and HPLC/MS/MS analytical methods were successful for the determination of PAs in blood samples obtained from rats dosed with PAs and in the PA-containing plant.
The results demonstrated that the developed HPLC/MS methods with two different CID techniques **provided a very simple and rapid analysis for an unequivocal (no doubt) diagnosis of PA poisoning and for definitive identification of PAs in plants or herbal medicines.**"
-'Determination of Hepatotoxic Pyrrolizidine Alkaloids by On-Line High Performance Liquid Chromatography Mass Spectrometry with an Electrospray Interface' by Ge Lin, et al., (Chinese University of Hong Kong) (China Pharmaceutical University, Nanjing, The People's Republic of China); Rapid Communications in Mass Spectrometry, Volume 12, No. 20, pages 1445-1456, **1998**.

"**Until recently, there were considerable problems in detecting pyrrolizidine alkaloids in human blood or tissue samples, whereas a method for the identification of alkaloids in plants and herbal products was developed several years ago.**
As early as the 1960s, pyrrolizidines were identified using spectrophotometry and subsequent improvements were possible through thin-layer chromatography and gas chromatography-mass spectrometry.
>Betz et al. reported the development of a capillary gas chromatography technique and subsequent identification of different alkaloids via thin-layer chromatography.

More recently, a simple bioassay using a human cell line was developed to enable investigators to evaluate the mutagenic (gene mutation) potential of various alkaloids.
>Lin et al.* have established the first analytical method to detect pyrrolizidine alkaloids in blood samples using high performance liquid chromatography (HPLC) which **now provides the first tool for a definite diagnosis of pyrrolizidine alkaloid poisoning and for the identification of alkaloids in plants or herbal products.**"

-'The Efficacy and Safety of Comfrey' by F. Stickel and H.K. Seitz, Public Health Nutrition, United Kingdom, Volume 3, 4A, pages 501-508, December **2000**.
(* -'Determination of Hepatotoxic Pyrrolizidine Alkaloids by On-Line High Performance Liquid Chromatography Mass Spectrometry with an Electrospray Interface' by Ge Lin, et al., {The Chinese University of Hong Kong} {China Pharmaceutical University, Nanjing, The People's Republic of China}; Rapid Communications in Mass Spectrometry, Volume 12, No. 20, pages 1445-1456, 1998.)   (A bioassay determines the concentration or potency of a substance by its effect on living cells or tissues.)

"**There are no well-established methods available for the measurement of PAs (Pyrrolizidine Alkaloids) and their metabolites in body fluids,** although experimental applications of both in-source collision induced dissociation high-performance liquid chromatography mass spectrometry (CIDHPLC/MS) and HPLC/MS/MS analytical methods were successful for the determination of PA in blood samples obtained from rats treated with PA and in PA-containing plant material*.
Therefore, the diagnosis of PA-poisoning can only be made with chemical confirmation of the presence of toxic PA constituents in the herbal preparation that the patient consumed or, without chemical confirmation, when the herb in question is a known source of toxic PA."
-'Metabolism, Genotoxicity, and Carcinogenicity of Comfrey' by N. Mei et al., Division of Genetic and Reproductive Toxicology, National Center for Toxicological Research, United States Food and Drug Administration, Arkansas; Journal of Toxicology and Environmental Health, Part B Critical Reviews, Volume 13, No. 7/8, pages 509-526, **2010**.
(* -'Determination of Hepatotoxic Pyrrolizidine Alkaloids by On-Line High Performance Liquid Chromatography Mass Spectrometry with an Electrospray Interface' by G. Lin, K.Y. Zhou, X.G. Zhao, Z.T. Wang and P.P. But, Rapid Communications In Mass Spectrometry, 1998.)

"**Pyrrolizidine alkaloids (PAs) Induce (bring about) Liver Injury (PA-ILI) and is very likely to contribute significantly to Drug-Induced Liver Injury (DILI).**
In this study we used a newly developed ultra-high performance liquid chromatography-triple quadrupole-mass spectrometry (UHPLC-MS)-based method to detect and quantitate blood pyrrole-protein adducts in DILI patients.
Among the 46 suspected DILI patients, 15 were identified as PA-ILI by the identification of PA-containing herbs exposed. Blood pyrrole-protein adducts were detected in all PA-ILI patients (100%).
These results confirm that PA-ILI is one of the major causes of DILI and that blood pyrrole-protein adducts quantitated by the newly developed UHPLC-MS method can serve as a specific biomarker of PA-ILI."
-'Blood Pyrrole-Protein Adducts: A Biomarker of Pyrrolizidine Alkaloid-Induced Liver Injury in Humans' by Jianqing Ruan, Hong Gao, Na Li, Junyi Xue, Jie Chen, Changqiang Ke, Yang Ye, Peter Pi-Cheng Fu, Jiang Zheng, Jiyao Wang and Ge Lin (China, Hong Kong, Arkansas and Washington states: United States); Journal of Environmental Science and Health, Part C, Environmental Carcinogenesis and Ecotoxicology Reviews, Volume 33, Issue 4, **2015**.
>(Metabolic activation of Pyrrolizidine Alkaloids is catalyzed by hepatic {liver} cytochrome P450 and creates pyrrolic metabolites that bind to proteins to form pyrrole-protein adducts that lead to hepatotoxicity.)

Symptoms of Pyrrolizidine Alkaloid Poisoning (in general, not just from Comfrey)

"**Liver Toxicity Clinical Symptoms (from all PA containing herbs): The most common disease to be associated with the consumption of pyrrolizidine alkaloids (PAs) in humans is veno-occlusive disease** (a form of Budd-Chiari syndrome), which is characterized by the occlusion (blockage) of the small branches of the hepatic vein in the liver.
> The clinical symptoms include a dull ache in the upper half of the abdomen, hepatomegaly (enlarged liver), ascites (fluid in abdomen), and often pedal oedema (edema in feet and lower legs), and a fall in urinary output. Nausea and vomiting may be present, while jaundice and fever are rare.

The disease usually begins with an acute phase which, in severe cases, can lead to death as a result of hepatocellular (liver cell) failure. In less severe cases, the disease may enter a subacute and eventually a chronic phase. **In both acute and subacute phases, only slight changes in liver function test-results are observed.**
Children appear to be particularly vulnerable, and in a clinical study in Jamaica, children of up to six years of age accounted for 65% of the cases.
After a subacute presentation, patients either may recover or may enter the chronic phase which is characterized by cirrhosis."
-'Comfrey: Assessing the Low-Dose Health Risk' by Peter J. Abbott, Toxicology Unit, Department of Community Services and Health, Canberra, Australia; Medical Journal of Australia, Volume 149, No. 11-12, pages 678-682, **1988**.
> (Jaundice symptoms are yellowing of the skin or whites of the eyes from too much bilirubin pigment. It is usually caused by blockage of the bile duct, liver disease, or too much breakdown of red blood cells.)

"**Pyrrolizidine alkaloid related toxicity may be directed against different target organs.
The most frequently observed toxicity is against the liver.** The main liver injury caused by Comfrey and related pyrrolizidines is that of Veno-Occlusive Disease (VOD) also termed 'Stuart-Bras syndrome'. Pyrrolizidine-related VOD in Western countries has been recognized primarily in infants, who seem to be particularly susceptible.
Other known aetiologies (causes) for VOD are cytostatic drugs, bone marrow transplantation, oral contraceptives, systemic lupus erythematosus, and alcoholic hepatitis.
Diagnosis: The histologic (microanatomical) picture is essential for the diagnosis although false negative results may occur due to sample errors. Pathologically, a thrombotic (blood clot) form can be distinguished from a non-thrombotic variety and both entities can be observed in pyrrolizidine-induced VOD. **The clinical picture of VOD is often difficult to distinguish from other liver pathologies and sometimes remains unrecognized.**
A classification has been established separating acute and subacute from chronic signs, the latter being characterized by a subclinical appearance. **In general, the symptoms resemble those of Budd-Chiari syndrome** but 10% may be asymptomatic. In acute forms, a rapid progression to cirrhosis {scar tissue} of the liver, complicated by portal hypertension, and finally death from liver failure results in 20% of patients due to extensive hepatocyte necrosis. Chronic cases develop slowly and may only present with asthenia {debility}, fatigue, mild diarrhoea and signs of portal hypertension."
-'The Efficacy and Safety of Comfrey' by F. Stickel and H.K. Seitz, Public Health Nutrition, United Kingdom, Volume 3, 4A, pages 501-508, December **2000**.
> (Budd-Chiari syndrome is a very rare disease caused by occlusion {blocking} of the hepatic veins that drain the liver. It presents with abdominal pain, ascites, and liver enlargement. The acute syndrome presents with rapidly progressive severe upper abdominal pain, yellow discoloration of the skin and whites of the eyes, liver enlargement, enlargement of the spleen, fluid accumulation within the peritoneal cavity, elevated liver enzymes, and eventually encephalopathy {brain damage}. Ascites is abnormal buildup of fluid in the abdomen.)
> (Portal hypertension is an increase in the blood pressure in a system of veins called the portal venous system. Veins coming from the stomach, intestine, spleen, and pancreas merge into the portal vein, which then branches into smaller vessels and travels through the liver.)

"***Symphytum officinale*, Comfrey toxicity from Pyrrolizidine Alkaloids probably requires chronic consumption.:
Much delayed (weeks to months) symptoms: anorexia, depression, rough pelage, diarrhea, emaciation, constipation, ascites, edema of extremities (arms/legs), icterus (jaundice), hepatoencephalopathy, death.**"
-Small Animal Toxicology by Michael E. Peterson DVM MS and Patricia A. Talcott MS DVM PhD (Chapter 27: Household and Garden Plants by A. Catherine Barr, PhD, DABT). Pennsylvania: Saunders, **2012**, page 392.
> (Anorexia means loss of appetite and inability to eat. Pelage is hair, fur, wool, or other soft covering of a mammal. Ascites is the abnormal buildup of fluid in the abdomen. Edema is abnormal accumulation of fluid in interstitium, located beneath the skin and in cavities of the body; it manifests as swelling. Emaciation is extreme weight loss and thinness.)
> (Jaundice is a yellowish or greenish pigmentation of the skin and whites of the eyes due to high bilirubin levels. Hepatoencephalopahty or hepatic encephalopathy is a decline in brain function from liver disease because the liver can not adequately remove toxins from the blood.)

Alkaloids in Comfrey Develop in Roots and Flowers

"**Pyrrolizidine Alkaloids (PAs) are toxic secondary metabolites that are found in several distantly related families of the angiosperms (flowering plants).** The first specific step in PA biosynthesis is catalyzed by homospermidine synthase (HSS), *which has been recruited several times independently by duplication of the gene encoding deoxyhypusine synthase, an enzyme involved in the post-translational activation of the eukaryotic initiation factor 5A. HSS shows highly diverse spatio-temporal gene expression in various PA-producing species.*

Alkaloids in Comfrey Roots:
In Comfrey (Symphytum officinale; Boraginaceae), PAs are reported to be synthesized in the roots, with HSS being localized in cells of the root endodermis.

Alkaloids in Comfrey Flowers and Leaves:
*Here, we show that Comfrey plants activate a second site of HSS expression when inflorescences (flowers) start to develop. HSS has been localized in the bundle sheath cells of specific leaves. Tracer feeding experiments have confirmed that these young leaves express not only HSS but the whole PA biosynthetic route. **This second site of PA biosynthesis results in drastically increased PA levels within the inflorescences. The boost of PA biosynthesis is proposed to guarantee optimal protection especially of the reproductive structures.***"

-'Identification of a Second Site of Pyrrolizidine Alkaloid Biosynthesis in Comfrey to Boost Plant Defense in Floral Stage' by L.H. Kruse, T. Stegemann, C. Sievert, and D. Ober, Botanisches Institut und Botanischer Garten, Universitat Kiel (Botanical Institute and Botanical Garden, University of Kiel), Germany; Plant Physiology, Volume 174, No. 1, pages 47-55, **2017**.

(Endodermis is an inner layer of cells in the cortex of a root surrounding a vascular bundle.)

(An inflorescence is a group or cluster of flowers arranged on a stem that is composed of a main branch or a complicated arrangement of branches.)

"**Plant Defense:**
Pyrrolizidine alkaloids (PAs) are a typical class of plant secondary metabolites that are constitutively produced as part of the plant's chemical defense. *The biosynthesis of PAs is regarded as a constitutive defense mechanism that is coupled to plant growth. If no further biomass is produced, PA biosynthesis is switched off.*

Complex Patterns of PA Levels and Locations:
Therefore, the observed variability of PA levels between different parts of the plant, between individuals, and between samples harvested at different time points during the growing season shows a much more complex pattern than anticipated before.

Quantification of PA levels can always be only a glimpse of the metabolic state of the plant. The different developmental stages with their specific physiology including efficient transport processes and additional temporary sites of synthesis show unique PA levels, suggesting that this variability might have been selected in evolution.

Our data suggest that PA accumulation has to be understood as a highly dynamic system resulting from a combination of efficient transport and additional sites of synthesis that are only temporarily active.

In plant secondary metabolism it is a well-known phenomenon that the site of synthesis might be very different from the site of accumulation (storage).

Comfrey Roots:
For Symphytum officinale, studies on root cultures have shown that roots of S. officinale have the capacity to express the complete pathway of PA biosynthesis (*Frolich et al. 2007).

Therefore, roots have been interpreted as the exclusive site of PA biosynthesis, from where the alkaloids are distributed into the aerial parts of the plant *(Frolich et al. 2007).*

Comfrey Flowers and Leaves:
Recently, we have been able to show that S. officinale has, in addition to the roots, a second site of PA biosynthesis, viz. (that is), young leaves subtending developing inflorescences.

While roots are a well-established site of pyrrolizidine alkaloid biosynthesis, Comfrey plants (Symphytum officinale; Boraginaceae) have been shown to additionally activate alkaloid production in specialized leaves and accumulate PAs in flowers during a short developmental stage in inflorescence development.

PAs (in flowers) are almost exclusively accumulated in the ovaries, while petals, sepals, and pollen hardly contain PAs. High levels of PAs are detectable in the fruit (nutlets), but the elaiosome was shown to be PA free.

The absence of 7-acetyllycopsamine in floral parts while present in leaves and roots suggests that the additional site of PA biosynthesis provides the pool of PAs for translocation to floral structures.

We found only very low concentrations of PAs in pollen of S. officinale."

-'Specific Distribution of Pyrrolizidine Alkaloids in Floral Parts of Comfrey (Symphytum Officinale) and Its Implications for Flower Ecology' by Thomas Stegemann, Lars H. Kruse, Moritz Brutt and Dietrich Ober (Kiel, Germany and Ithaca, New York); Journal of Chemical Ecology, e-publication before print, July 28 **2018**.

(* -'Tissue Distribution, Core Biosynthesis and Diversification of Pyrrolizidine Alkaloids of the Lycopsamine Type in Three Boraginaceae Species' by Cordula Frolich, Dietrich Ober and Thomas Hartmann, Braunschweig and Kiel, Germany; Phytochemistry, Volume 68, pages 1026-1037, 2007.)

(Elaiosomes are fleshy structures attached to seeds. It is rich in lipids and proteins.)

(Subtend is a leaf or bract that extends under a flower to support it. A bract is a specialized leaf at the base of a flower.)

Research in 1970s and 1980s on Types and Percent of Alkaloids in Comfrey

For research articles in this book, some may not be relevant to the average person. However, I like to include them because there are researchers and scientists who are interested. These scientists can then find these articles and read the entire report.

"Studies on the alkaloid content and toxicity of Comfrey were co-ordinated between the Chemistry Department of the University of Exeter (Devon County, England), the Toxicology Unit of the Medical Research Council at Carshalton (south London, England), and the Michaelis Nutritional Research Laboratory at Harpenden (Hertfordshire County, England).
Comfrey root (as opposed to leaves) is known to have the highest concentration of alkaloid.
Pyrrolizidine alkaloid concentration in Comfrey leaf:
    S. peregrinum leaf is 0.03 percent.
    S. x uplandicum Bocking No. 4 leaf is .029 percent.
    S. x uplandicum Bocking No. 14 leaf is .024 percent.
    S. officinale leaf is .034 percent.
    S. asperum leaf is .062 percent."
-'The Alkaloid Content of Comfrey' by Dr. D.B. Long, Ph.D, M.A., around **1970**. (In 'Comfrey: Fodder, Food & Remedy').

"A new TLC- and HPLC-method (Thin-Layer Chromatography and High Performance Liquid Chromatography) for the detection and estimation of the main alkaloids from Symphytum roots, Symphytine, Echimidine, Lycopsamine, and Acetyllycopsamine in their genuine N-oxide form is described. The method provides characteristic finger-prints for the constituents of the Symphytum drugs of different origin and varying harvest seasons."
-'TLC and HPLC-Analysis of Pyrrolizidine-N-oxide Alkaloids of Symphyti Radix' by H. Wagner, U. Neidhardt and G. Tittel, Institut fur Pharmazeutische Atzneimittellehre der Universitat Munchen (Institute of Pharmaceutical Medicine at the University of Munich), Germany; Planta Medica: Journal of Medicinal Plant Research, Volume 41, pages 232-239, **1981**. (All is in German except a brief abstract is in English. If you have an English translation, I would appreciate a copy.)

"By means of Thin Layer Chromatography (TLC) in conjunction with mass spectrometry the pyrrolizidine alkaloid patterns derived from Symphytum asperum (Prickly Comfrey), several cytotypes of S. officinale (Common Comfrey) agg. (aggregate) and the artificial hybrids of the former taxa, were compared.
The obtained patterns were not essentially affected by variation in cytotype, harvesting times and location of plants.
Lycopsamine, acetyl-lycopsamine and symphytine or their isomers were generally found in the S. officinale cytotypes.
Echimidine and symphytine were found in S. asperum.
The interspecific (between species) hybrids contained all alkaloids mentioned.
    The definite lack of echimidine in the 2n = 40 cytotype proves that it is conspecific (same species) with S. officinale and does not belong to a hybrid-swarm S. asperum x S. officinale with 2n = 48."
-'Chemotaxonomical Investigations of the Symphytum Officinale Polyploid Complex and S. Asperum (Boraginaceae): The Pyrrolizidine Alkaloids' by H.J. Huizing, Th.W.J. Gadella, and E. Kliphuis, all from The Netherlands; Plant Systematics and Evolution, Volume 140, pages 279-292, **1982**.
(A cytotype is an individual of a species that has a different chromosomal factor to another, e.g., diploid versus haploid. Diploid cells contain two complete sets {2n} of chromosomes. Haploid cells have half the number of chromosomes {n}.)
(2n = 40 means 2 sets of 20 chromosomes in the plant cells. And 2n = 48 means 2 sets of 24 chromosomes.)
(Aggregate species represents a range of very closely related organisms.)
(A hybrid swarm is a group of hybrids that have survived past the first hybrid generation, with interbreeding taking place between hybrid individuals and also backcrossing with its parent type. These plants are highly variable, with genes and phenotypes of individuals ranging widely between the two parent types. Hybrid swarms blur the boundary between the parent taxa.)

"A number of recent studies of the pyrrolizidine alkaloids in various Symphytum species have succeeded in characterizing twelve mono- and di-ester alkaloids of retronecine and heliotridine.
Comfrey Root and Leaf:
We have examined twelve samples each of **Comfrey root and leaf** *sold as herbs in the United States. As determined by NMR (Nuclear Magnetic Resonance) spectroscopy,* **the total alkaloid content of the leaf samples was <0.005%; the roots, on the other hand, contained from 0.14-0.42%.**
Comfrey Root:
Examination of the root alkaloid mixture by capillary GC-MS (Gas Chromatography-Mass Spectroscopy) after derivatization allowed **identification of seven alkaloids; in all of the samples the first four peaks, lycopsamine, intermedine, and their 7-acetyl derivatives, accounted for 75% or more of the total alkaloids.**
    The same seven alkaloids have been found in Russian Comfrey, a cross between Symphytum officinale and S. asperum, sometimes called Symphytum x uplandicum."
-'Ingestion of Pyrrolizidine Alkaloids: A Health Hazard of Global Proportions' by James N. Roitman, Natural Products Chemistry Research, Western Regional Research Center, Agricultural Research Service, US Dept of Agriculture, Berkeley, California, American Chemical Society, ACS Symposium Series, pages 345-378, October 25 **1983**. (Species of Comfrey was not given.)

"Leaves of Russian Comfrey contain 0.01-0.15% alkaloid, of which the main component is echimidine (Culvenor et al., 1980). The consumption rate by Comfrey eaters varies from trivial levels up to 5 or 6 leaves/day. The high level implies 5-6 mg alkaloid/day."
-'Estimated Intakes of Pyrrolidizine Alkaloids by Humans: A Comparison with Dose Rates Causing Tumors in Rats' by C.C.J. Culvenor, CSIRO Division of Animal Health, Melbourne, Australia; Journal of Toxicology and Environmental Health, Volume 11, No. 4-6, pages 625-635, **1983**.

-'Phytochemistry, Systematics and Biogenesis of Pyrrolizidine Alkaloids of Symphytum Taxa' by Hindrik Jan Huizing, Thesis for Doctor of Philosophy, University of Groningen, Netherlands, 210 pages, **1985**.
(I was unable to get this. If you have a copy, could you please send it to me.)

"**Symphytum officinale: Analysis of commercial Comfrey (Symphytum officinale) root** by TLC (Thin-Layer Chromatography) shows three fully separated major spots. Because of the specificity of the sprays, this indicates the presence of at least **three different toxic pyrrolizidine alkaloids**. The spots have been identified: **di-ester symphytine, acetyl di-esters, mono-esters lycopsamine and intermedine.**
> The relative intensities of the spots indicate the acetyl di-esters 7-acetyllycopsamine and 7-acetylintermedine are by far the most prevalent. This is in agreement with reports in the literature.

The weight of unsaturated pyrrolizidine alkaloids per 10.0 gram of Comfrey (Symphytum officinale) root was calculated to vary from 1.3-8.3 mg. This is equivalent to a concentration of 0.013-0.083%.
> Values reported in the literature vary somewhat, but are similar to these findings: 0.036-0.046%, 0.07%, 0.14-0.37% and 0.23-0.38%. **Our results correspond to a concentration of 130-830 ppm (part per million).**

In the context of food contamination, these levels seem extremely high for chemicals known to be toxic and carcinogenic. Contamination of food with synthetic herbicides and pesticides is normally regulated at much lower levels, sometimes even in parts per billion (ppb).

The samples of Comfrey root (Symphytum officinale) examined contained an average of 0.15 mg of toxic pyrrolizidine alkaloids per capsule and about 0.5 mg per tea bag.
> According to the directions on the containers, two or three capsules are to be taken two or three times a day at mealtime. This leads to a consumption of 0.60 mg (four capsules) to 1.4 mg (nine capsules) per day.

The use of herbs as medicine has increased dramatically in recent years. It is based largely on a belief that the traditional and the natural must, inherently, be safer and healthier.
> **However, historical use does not, by itself, assure safety.** Many recent articles have reported the presence of natural toxins in medicinal herbs, and warnings have appeared about their use."

-'Pyrrolizidine Alkaloids: Testing for Toxic Constituents of Comfrey' by Vollmer, Steiner, Larsen, Muirhead and Molyneux, Journal of Chemical Education, Volume 64, Number 12, pages 1027-1030, December **1987**.

"**Pyrrolizidine Alkaloids:**
S. officinale	symphytine, **echimidine(?)**, lycopsamine, acetyllycopsamine, lasiocarpine, heliosupine N-oxide
S. peregrinum	lycopsamine
S. x uplandicum	intermedine, symphytine, echimidine, 7-acetyllycopsamine, 7-acetylintermedine, symlandine, uplandicine
S. asperum	asperumine, heliosupine N-oxide, echimidine, echinatine."

-International Programme on Chemical Safety {IPCS}: Pyrrolizidine alkaloids, Environmental Health Criteria 80, United Nations, World Health Organization, Geneva, Switzerland, **1988**.   (The author put the question mark after 'echimidine'.)

Research in 1990s on Types and Percent of Alkaloids in Comfrey

"The uses of **Comfrey (Symphytum officinale and Symphytum uplandicum)** in traditional medicine, its chemical composition (in particular, content of pyrrolizidine alkaloids), methods of detection and analysis, **human and animal toxicity** (including mechanism of toxicity, mutagenicity, carcinogenicity, teratogenicity, reproduction effects, foetotoxicity) are reviewed and discussed.
Symphytine Alkaloid: The symphytine content of Symphytum officinale roots is estimated to be between 0.23 and 0.38%; the leaves contain considerably less alkaloid.
Lasiocarpine Alkaloid: The lasiocarpine content of the roots of Symphytum officinale, determined by TLC (Thin Layer Chromatography), has been shown to be 0.0058%."

-'Toxicity of Comfrey' by K.A. Winship, Department of Health, Medicines Control Agency, London, England; Adverse Drug Reactions and Toxicological Reviews, Oxford, England, Volume 10, No. 1, pages 47-59, **1991**.

"**Pyrrolizidine alkaloids identified in Comfrey:** 7-acetyl echinatine, acetyl heliosupine, 7-acetyl intermedine, 7-acetyl lycopsamine, anadoline*, angelyl echimidine*, aspermine*, echimidine M, echinatine M, heliosupine, intermedine, lasiocarpine M C, lycopsamine M, myoscorpine, symlandine (7-angelyl-9-viridiflorylretronecine), symphytine (7-tiglylretronecine viridiflorate) M C, viridiflorine, 7-angelyl heliotridine, uplandicine (7-acetyl-9-echimidinylretronecine).
> (* Occurs in Symphytum species but has not been reported in S. officinale or S. x uplandicum.
> M = shown to be mutagenic. C = shown to be carcinogenic in rats.)

-'Toxicity of Comfrey' by K.A. Winship, Department of Health, Medicines Control Agency, London, England; Adverse Drug Reactions and Toxicological Reviews, Oxford, England, Volume 10, No. 1, pages 47-59, **1991**.

"**Symphytum officinale L.: In dried S. officinale leaves 0.02 to 0.18% and in the roots 0.25 to 0.29% alkaloids respectively, their N-oxides were detected.** The alkaloids include intermedine, lycopsamine, their 7-acetyl derivatives, symlandine, symviridine, myoscorpine, and symphytine.
The presence of the alkaloids (echinatine, heliosupine N-oxide, heliotrine, lasiocarpine, and viridiflorine) isolated and characterized by a Russian and Polish working group by means of paper chromatography could not be confirmed."

-'Medicinal Plants in Europe Containing Pyrrolizidine Alkaloids' by Erhard Thomas Roeder, Pharmazeutisches Institut der Rheinischen Friedrichs-Wilhelms, University of Bonn, Germany, Pharmazie 50, pages 83-98, March **1995**.

Research in 2000s on Types and Percent of Alkaloids in Comfrey

"As is often the case in the investigation of herbal drugs, reference standard compounds were unavailable from any commercial source. Thus, we have developed a preparatory scale, one-step, high-speed CounterCurrent Chromatography (CCC) procedure to isolate PAs (Pyrrolizidine Alkaloids) from an alkaloid extract of the roots of S. officinale. **Three pyrrolizidine alkaloids, symlandine, symphytine, and echimidine, were isolated from the roots of Symphytum officinale***.*

The structures were confirmed by several spectroscopic techniques including 2D NMR (Two-Dimensional Nuclear Magnetic Resonance spectroscopy) methods. **This is the first description of the separation of symlandine from its stereoisomer, symphytine.***"*

-'Isolation of Symlandine from the Roots of Common Comfrey (Symphytum officinale) Using Countercurrent Chromatography' by Nam-Cheol Kim, Nicholas H. Oberlies, Dolores R. Brine, Robert W. Handy, Mansukh C. Wani and Monroe E. Wall, Chemistry and Life Sciences, Research Triangle Institute, Research Triangle Park, North Carolina; Journal of Natural Products: American Society of Pharmacognosy, Volume 64, No. 2, pages 251-253, February **2001**.

(Isomers have the same chemical formula but different arrangement of atoms. Stereoisomers are two or more compounds differing only in the spatial arrangement of their atoms.)

"Comfrey species vary in their content of PAs (pyrrolizidine alkaloids). Comfrey (Symphytum spp. {species}) contains seven PAs: intermedine, lycopsamine, acetyl intermedine, acetyl lycopsamine, symlandine, symphytine and echimidine. **Symphytum officinale:**

However, nearly all (85-97%) of the PAs in the Comfrey grown in United States gardens (Symphytum officinale L.) are **retronecine monoesters.** *The remaining constituents are retronecine diesters.*
Symphytum x uplandicum:

By contrast, Russian Comfrey (Symphytum x uplandicum Nym.) **contains a high proportion of the slightly more toxic retronecine diester** *form of PA."*

-'Comfrey Toxicity Revisited' by Dorena Rode, Department of Animal Science, College of Agricultural and Environmental Sciences, University of California, Davis, California; Trends in Pharmacological Sciences: International Union Of Pharmacology, Volume 23, No. 11, pages 497-499, November **2002**.

*"***PAs (1, 2-dehydro-pyrrolizidine ester alkaloids) and their N-oxides in their native form have not been found to be significantly toxic. Their hazard arises from the formation of pyrrolic metabolites (dehydroalkaloids) in the liver.**

These dehydroalkaloids are highly reactive electrophilic alkylating agents capable of binding to nucleophilic centers in tissues or cross-linking DeoxyriboNucleic Acid (DNA), leading to hepatotoxicity and carcinogenicity. Chronic PA poisoning can result in hepatic veno-occlusive disease, hepato-splenomegaly, and emaciation.

Two authentic roots were analyzed (Symphytum x uplandicum and Symphytum officinale).
The data obtained for S. officinale root show that the total number of pyrrolizidine compounds (free bases and N-oxides) detected was 13. For S. x uplandicum root it was 22.
Also, it was observed that the amounts of intermedine, 7-acetylintermedine, and their isomers and N-oxides in S. officinale appear to be greater than for S. x uplandicum, based on relative peak areas.
In contrast, the amount of uplandicine N-oxide, symphytine isomers, and their N-oxides were lower for S. officinale than for S. x uplandicum.

The remaining pyrrolizidine compounds were not detected in the S. officinale material analyzed. These differences, while primarily attributed to the difference in species, may also arise from natural biological variation (e.g., age of the plant material, geographic region where it grew).

External use of Comfrey is of less concern than oral or systemic administration *given that the hepatic bioavailability should be minimal. In one study, the systemic bioavailability of PAs after dermal (skin) application was found to be 20- to 50-fold lower than when administered orally."*

-'Investigation of Pyrrolizidine Alkaloids and their N-Oxides in Commercial Comfrey-Containing Products and Botanical Materials by Liquid Chromatography Electrospray Ionization Mass Spectrometry' by Jorgelina C. Altamirano, Samuel R. Gratz and Karen A. Wolnik, United States Food and Drug Administration, Forensic Chemistry Center, Cincinnati, Ohio; Journal of AOAC International, Volume 88, No. 2, pages 406-412, March **2005**.

"Pyrrolizidine Alkaloids (PA) are constituents of various species of the Asteraceae, Boraginaceae and Fabaceae families. **PA with a 1,2-unsaturated necine structure show an acute toxic, carcinogenic, mutagenic and teratogenic effect.**
An appropriate analytical method was developed which is able to detect all toxic PA and their N-oxides in total. *A Gas Chromatographic Mass Spectrometric procedure was chosen which is characterized by high selectivity.* **A prerequisite for the GC determination is the reduction of the N-oxides to the alkaloid bases and clean-up of the raw alkaloid extract.**
Using Comfrey macerate (softened) the time response of the reduction was investigated by sampling at different time points of the process. **The PA mean content derived without reduction of the N-oxides is 0.43 mg/kg which means that nearly all PA are present in their N-oxide form.**
After 15 minutes under reduction conditions, the content increases to 38.5 mg/kg. Already after 15 minutes nearly all N-oxides

are transferred into the alkaloid bases."
-'Method Validation for the Determination of Toxic Pyrrolizidine Alkaloids and their N-oxides in Symphytum Officinale L. Using GC-MS' by C. Staiger, P. Ottersbach, M. Rudolph, G. Schulzki (PhytoLab, Vestenbergsgreuth, Germany and Merck Selbstmedikation GmbH, Darmstadt, Germany); Conference Paper in Planta Medica, Volume 75, Issue 9, page 6, July **2009**.

"**Symphytum officinale:**
   Symphytum officinale root contains 0.75-2.55% allantoin; about 0.3% alkaloids, including the pyrrolizidine alkaloids symphytine, echimidine, heliosupine, viridiflorine, echinatine, 7-acetyllycopsamine, 7-angelylretronecine viridiflorate, lasiocarpine, and acetylechimidine; the presence of lasiocarpine is questioned; lithospermic acid; 29% mucopolysaccharide that is composed of glucose and fructose; a gum consisting of L(-)-xylose, L-rhamnose, L-arabinose, D-mannose, and D-glucuronic acid; pyrocatechol tannins (2.4%); 0.63% carotene; glycosides, sugars; isobauerenol, b-sitosterol, and stigmasterol; steroidal saponins; triterpenoids; rosmarinic acid, and others.*
Symphytum asperum:
   S. asperum (roots and leaves) contains the pyrrolizidine alkaloids asperumine, echinatine, heliosupine, 7-acetyllycopsamine, and acetylechimidine.
Symphytum x uplandicum:
   S. x uplandicum (roots and leaves) contain symphytine, symlandine, echimidine; 7-acetyllycopsamine, 7-angelylintermidine, uplandicine, lycopsamine, and intermedine.
Comfrey leaves *(species not identified) also contain substantial quantities of allantoin, alkaloids (about 0.15%), and possibly other similar constituents as the root."*
-Leung's Encyclopedia of Common Natural Ingredients Used in Food, Drugs and Cosmetics, Third Edition by Ikhlas Khan and Ehab Abourashed. New Jersey: John Wiley & Sons, **2010**, pages 225-226.

"*Comfrey (Symphytum officinale) root pyrrolizidine alkaloids (0.05-0.08%) with 1,2-unsaturated necine ring structures are also present, almost entirely in the form of their N-oxides, the main ones being 7-acetylintermedine and 7-acetyllycopsamine together with smaller amounts of intermedine, lycopsamine and symphytine (*, **).*"
-'Symphyti Radix (Comfrey Root)' by ESCOP Monographs: The Scientific Foundations for Herbal Medicinal Products, European Scientific Cooperative on Phytotherapy, Exeter, United Kingdom, 16 pages, **2012**.
(* -'Pyrrolizidine Alkaloids from Symphytum Officinale L. and their Percutaneous Absorption in Rats' by J. Brauchli, J. Luthy, U. Zweifel and C. Schlatter, Basel, Switzerland, Experientia, Volume 38, No. 9, pages 1085-1087, 1982.)
   "**An analysis of a commercial sample of Symphyti radix (Comfrey root) originating from Poland with a total alkaloid content of 0.07% revealed the presence of **7 pyrrolizidine alkaloid-N-oxides: 7-acetyl intermedine, 7-acetyl lycopsamine as the main constituents and lycopsamine, intermedine, symphytine and traces of 2 further not yet identified alkaloids.** The percutaneous absorption of these alkaloids was investigated in rats, using a crude alcoholic extract of the plant corresponding to a dose of 194 mg alkaloid-N-oxides/kg body weight. The excretion of N-oxides in the urine during 2 days was in the range of 0.1-0.4% of the dose. The dermally absorbed N-oxides are not or only to a small extent converted to the free alkaloids in the organism. The oral application led to a 20-50 times higher excretion of N-oxides and free alkaloids in the urine.*"
(** -'Untersuchungen zum Pyrrolizidinalkaloidgehalt und Pyrrolizidinalkaloidmuster in Symphytum officinale L. Versuche zur Gewinnung pyrrolizidinalkaloidarmer Pflanzen bzw. Pflanzenteile' {'Investigations on Pyrrolizidine Alkaloid Content and Pyrrolizidine Alkaloid Patterns in Symphytum Officinale L. Experiments on the Production of Pyrrolizidine-intestinal or Plant Parts'} by R. Mutterlein R and C-G. Arnold, Pharm Ztg Wiss 138, pages 119-125, 1993.)

"*Pyrrolizidine Alkaloids in Comfrey Species:*

Symphytum aintabicum	Echimidine
Symphytum asperum	Acetylechimidine, aetyllyopsamine, asperumine, echimidine, echinatine, heliosupine, symphytine, symviridine
Symphytum bohemium	7-acetyllycopsamine, echimidine, lycopsamine, symphytine
Symphytum caucasium	Asperumine, echimidine, echinatine, heliotrine, lasiocarpine
Symphytum consolidum	Echimidine, symphytine
Symphytum grandiflorum	Echimidine, lycopsamine, symphytine
Symphytum ibericum	Echimidine, lycopsamine, symphytine
Symphytum officinale	7-acetylintermedine, 7-acetyllycopsamine, **echimidine**, echinatine, eliosupine, intermedine, lycopsamine, symphytine, symviridine, viridiflorine
Symphytum orientale	Anadoline, echimidine, symphytine
Symphytum peregrinum	7-acetylintermedine, 7-acetyllycopsamine, intermedine, lycopsamine, symphytine
Symphytum tanaiense	7-acetyllycopsamine, echimidine, lycopsamine, symphytine
Symphytum tuberosum	7-angeloylretronecine, 7-acetyllycopsamine, anadoline, echimidine, lycopsamine, symphytine
Symphytum x uplandicum	7-acetylintermedine, 7-acetyllycopsamine, echimidine, intermedine, lycopsamine, symlandine, symphytine, symviridine, uplandicine, uplandicine "

-Diversity of Pyrrolizidine Alkaloids in the Boraginaceae: Structures, Distribution and Biological Properties, by Assem El-Shazly and Michael Wink, Journal of Biodiversity, Basel, Switzerland, Volume 6, Issue 2, April 1 **2014**, page 200, Table 1.
   (Some studies state Symphytum officinale does not contain echimidine.)

"PA containing plants (Symphytum spp.) (Adapted from *Codex, 2011).

Species	Pyrrolizidine alkaloids
Symphytum asperum	Asperumine, Echiumine, Symlandine, Symphytine, Myoscorpine, Echinatine, Echimidine
Symphytum officinale	(7-acetyl)intermedine, (7-acetyl)lycopsamine, Echimidine, Symlandine, Symviridine, Myoscorpine, Symphytine
Symphytum tuberosum	Amadoline, (7-acetyl)lycopsamine, Symphytine, Echimidine
Symphytum x uplandicum	Echimidine, (7-acetyl)intermedine, (7-acetyl)lycopsamine, Uplandicine Symlandine, Symviridine, Myoscorpine, Symphytine"

-'Determination of Levels of Pyrrolizidine Alkaloids in Symphytum Asperum Lepech Growing in Selected Parts of Kenya' thesis by Shylock O. Onduso, Kenyatta University, Kenya, 119 pages, October **2014**.
(* -Codex, Joint FAO/WHO Food Standards Programme, Codex Committee On Contaminants In Foods: Discussion Paper On Pyrrolizidine Alkaloids, 2011.)

"*The optimised method of extraction for Pyrrolizidine Alkaloids (PAs) from Symphytum was found 50% aquous methanol extraction under sonication (sound) for one hour followed by solvent removal. PAs were separated by TLC (Thin-Layer Chromatography) with Hexane:Ethyl Acetate as mobile phase and Ehrlichs visualising reagent. Ehrlichs reagent was also used in their spectrophotometric determination with Monocrotaline as standard.*"
-'IPSAM Book of Abstracts' by Irish Plant Scientists Association Meeting, University College Cork, Ireland, April 28-29 **2014**. Includes article 'A Comparison of the Pyrrolizidine Alkaloids and Allantoin Content in Symphytum Officinale and Symphytum x Uplandicum, and Optimisation of Extraction and Detection Methods' by M. O'Keeffe and G. Levieille, Department of Biological Sciences, Cork Institute of Technology, Ireland. (I only have the abstract. If you have entire article, I would appreciate a copy.)

"Total content of PAs (Pyrrolizidine Alkaloids) is:
 S. caucasicum 0.5%
 S. officinale (leaves: 0.02-0.18%; roots: 0.25-0.29%)
 S. peregrinum 0.2% alkaloids in the tops (leaves)."
-'Safety Issues Affecting Herbs: Pyrrolizidine Alkaloids" by Subhuti Dharmananda, Ph.D., Director, Institute for Traditional Medicine, Portland, Oregon, **2016**.

Symphytym Officinale and Alkaloid Echimidine

"*Furuya and Araki* and Furuya and Hikichi** reported that **symphytine and echimidine, which are pyrrolizidine alkaloids, can be isolated from Symphytum officinale**.*"
-'Carcinogenic Activity of Symphytum Officinale' by I. Hirono et al, Japan National Cancer Institute Volume 61, No. 3, pages 865-869, September **1978**.
(* -'Studies on Constituents of Crude Drugs. I. Alkaloids of Symphytum Officinale Linn.' by T. Furuya and K. Araki, Chemical & Pharmaceutical Bulletin, Tokyo, Japan, Volume 16, No. 12, pages 2512-2516, December 1968.)
(** -'Alkaloids and Triterpenoids of Symphytum Officinale' by T. Furuya and M. Hikichi, Phytochemistry, Volume 10, Issue 9, pages 2217-2220. September 1971.)

"*The alkaloids lycopsamine, acetyl-lycopsamine and symphytine or their isomers were generally found in the Symphytum officinale cytotypes. **The alkaloids echimidine and symphytine were found in Symphytum asperum.** The interspecific (between species) hybrids contained all alkaloids mentioned.*
 ***The definite lack of echimidine in the 2n = 40 cytotype prove that it is conspecific (same species) with Symphytum officinale** and does not belong to a hybrid-swarm S. asperum x S. officinale with 2n = 48.*"
-'Chemotaxonomical Investigations of the Symphytum Officinale Polyploid Complex and S. Asperum (Boraginaceae): The Pyrrolizidine Alkaloids' by H.J. Huizing, Th.W.J. Gadella, and E. Kliphuis, all from The Netherlands; Plant Systematics and Evolution, Volume 140, pages 279-292, **1982**.
 (Chemotaxonomy uses phytochemical {biologically active compounds in plants} data to determine the proper taxonomic category. So far 33 groups of chemicals in plants have been found to be of taxonomic significance. Examples in Comfrey include alkaloids and triterpenoids.)
 (A cytotype is an individual of a species that has a different chromosomal factor to another, e.g., diploid versus haploid. Diploid cells contain two complete sets {2n} of chromosomes. Haploid cells have half the number of chromosomes {n}.)
 (2n = 40 means 2 sets of 20 chromosomes in the plant cells. And 2n = 48 means 2 sets of 24 chromosomes.)
 (A hybrid swarm is a group of hybrids that have survived past the first hybrid generation, with interbreeding taking place between hybrid individuals and also backcrossing with its parent type. These plants are highly variable, with genes and phenotypes of individuals ranging widely between the two parent types. Hybrid swarms blur the boundary between the parent taxa.)

"***Symphytum officinale s.s. (sensu stricto = in the strict/narrow sense) plants, however, do not contain echimidine.*** *As a consequence, the presence of echimidine in alkaloid extracts from the Symphytum x uplandicum (Russian Comfrey) plants, clearly indicates their hybrid character.*

In a communication by Furuya and Hikichi (*1971) in which the occurrence of echimidine in S. officinale s.l. (sensu lato = in the broad sense) has been reported, also the presence of a typical triterpenoid, isobauerenol and of phytosterols, beta-sitosterol and stigmasterol was mentioned.

From our previous work, however, we have assumed that Furuya and Hikichi might have used S. x uplandicum plants for their phytochemical survey instead of S. officinale plants."

-'Chemotaxonomical Investigations of the Symphytum Officinale Polyploid Complex and S. Asperum (Boraginaceae): Phytosterols and Triterpenoids' by H.J. Huizing, Th.M. Malingre, Th.W.J. Gadella, and E. Kliphuis, all from The Netherlands; Plant Systematics and Evolution, Volume 143, pages 285-292, **1983**.

(* -'Alkaloids and Triterpenoids of Symphytum Officinale' by Tsutomu Furuya and Manabu Hikichi, Kitasato University, Tokyo, Japan; Phytochemistry, Volume 10, pages 2217-2220, 1971.)

(sensu stricto = s.s., s. str., sens. str., sens. strict. In the strict/narrow sense. It is added after a taxon to mean it is being used in the sense of the original author, or without taxa which may otherwise be associated with it.

sensu lato = s.l., sens. lat. In the broad sense. Used in taxonomy to clarify the scope of a taxon when it has been used to define more than one set of lower-level taxons. It includes all its subordinate taxa and/or other taxa that at other times are considered as distinct.

More detail about this: After a species name has been first established by the author originally publishing the name, then specialist or later taxonomists may do more research on it and make changes based on new information. 'Taxonomic circumspection' is needed to classify properly.

Circumscription means the taxonomist must decide which specimens are included in the species described, and which are excluded. It is in this process of species description that the question of the 'sense / sensu' arises, because the taxonomist must explain his view of the proper circumscription.)

"Echimidine is considered to be the most toxic of the PAs found in Comfrey."
-'Zur Toxikologischen Beurteilung der Pyrrolizidin-Alkaloide in den Arzneipflanzen Symphytum officinale, Borago officinalis' (Toxicological Assessment of Pyrrolizidine Alkaloids in Medicinal Plants Symphytum officinale, Borago officinalis) by Julia Brauchli-Theotokis, Ph.D. Thesis, University of Zurich, Switzerland, **1987**, page 11. (I have the abstract. I would appreciate if anyone has an English translation of the thesis.)

(The source for the quote is: 'Comfrey Update' by D.V.C. Awang, HerbalGram, American Botanical Council, Austin, Texas, Volume 25, pages 20-23, Summer 1991.)

"In a chemotaxonomic study of the genus Symphytum pyrrolizidine alkaloids and triterpenes were used as chemotaxonomical markers. The occurrence of the pyrrolizidine alkaloids symphytine and (acetyl-)lycopsamine is very general for Symphytum taxa. **Echimidine is present in some S. officinale L. plants** *and in S. tanaicense Steven."*
-'Chemotaxonomy of the Symphytum Officinale Agg. (Boraginaceae)' by Jaarsma, Lohmanns, Gadella and Malingre, Plant Systematics and Evolution, 167, pages 113-127, **1989**.

"The alkaloid echidimine predominates in Symphytum asperum and is present in the artificial hybrids between S. officinale and S. asperum known as S. x uplandicum Nym. **Echimidine is not found in S. officinale plants.***'*
-'Chemotaxonomy of the Symphytum Officinale Agg. (Boraginaceae)' by Jaarsma, Lohmanns, Gadella and Malingre, Plant Systematics and Evolution, 167, pages 113-127, **1989**, page 113.

*"***The pyrrolizidine alkaloid (PA) content of 13 commercial Comfrey products sold in Canada was examined** *following an increased health concern about the consumption of PA-containing plant material, and particularly Comfrey, the use of which has been linked to a number of cases of human poisoning during the last decade. The samples were purchased so as to be representative of the variety of sources, shapes and forms of the commercially available products.*
Comfrey Root:
Total PA content of Comfrey root samples, determined by H-NMR spectroscopy, has indicated significant variation both quantitatively and qualitatively. The results of this investigation clearly show that the level of the highly toxic echimidine in Comfrey products can vary widely. *While some samples contained very little or no detectable amount, echimidine was the single most abundant component of the extracted PAS in at least 2 products.*

The occurrence of echimidine, probably the most toxic PA of Symphytum spp. {species}, has been evaluated by reverse-phase HPLC (RP-HPLC), TLC and GC-MS. Echimidine was detected in 9 of 13 samples analyzed, and was noted to be the major PA constituent of 1 root preparation (approximately 85% Symphytum as measured by RP-HPLC).
Comfrey Leaf: *Leaf preparations generally contained* **lower levels of echimidine,** *detected only by GC-MS, associated with lower overall quantities of extract.*
Of the 6 preparations (some root, some leaf) identified on the package label as being derived from the species Symphytum officinale, 3 were observed to contain echimidine at varying levels, likely an indication of inaccurate species designation since substantial echimidine content is usually indicative of Symphytum officinale hybridization."
-'Echimidine Content of Commercial Comfrey (Symphytum spp.- Boraginaceae)' by D.V.C Awang et al., Bureau of Drug Research, Health and Welfare Canada, Ottawa, Ontario, Canada; Journal of Herbs, Spices and Medicinal Plants, Volume 2, No. 1, pages 21-34, **1993**.

*"***Echimidine was not detected in any of the commercially available products {sold as Symphytum officinale} examined in the current study.** *This is an important point since Awang et al have demonstrated that the Comfrey of commerce is not*

always S. officinale. The absence of echimidine indicates that the products examined were not Prickly Comfrey {S. asperum} or Russian Comfrey {S. x uplandicum}. **This means the Comfrey was correctly labeled by the sellers."**
-'Determination of Pyrrolizidine Alkaloids in Commercial Comfrey Products (Symphytum sp.)' by Betz, Eppley, Taylor and Andrezejewski, Journal of Pharmaceutical Sciences, Volume 83, No 5, pages 649-653, May **1994**.

"**The alkaloids of Symphytum officinale are** the isomeric monoesters lycopsamine and intermedine, the isomeric diesters 7-acetyl lycopsamine and 7-acetyl intermedine, the isomeric diesters symphytine, symlandine and symviridine, and **rarely, echimidine.** They are all based on retronecine.
Some samples of S. officinale have been shown to contain echimidine but this is exceptional.*"
-'Using Herbs that Contain Pyrrolizidine Alkaloids' by Alison Denham, B.A., MNIMH (National Institute of Medical Herbalists), University of Central Lancashire, England; The European Journal of Herbal Medicine, Volume 2, No. 3, pages 27-38, **1996**.
(* -New Flora of the British Isles: Identification of Wild Vascular Plants of the British Isles edited by Clive Stace. England: Cambridge University Press, 1991, pages 645-658.)

"**Echimidine and Symphytine Alkaloids in Symphytum Officinale:**
The weight of the simple alkaloid extract was nearly equivalent for all three commercial samples (20 mg). **However, the concentrations of symphytine and echimidine varied considerably within these extracts.**
A genetically pure sample of Symphytum officinale does not produce echimidine.
   Therefore, the observation of echimidine in these teas could be indicative of varying levels of contamination of the
   Common Comfrey leaves with other species of Symphytum, such as **the hybrid S. x uplandicum (Russian Comfrey),
   which is known to contain echimidine.**
Also, it is possible that samples that have a higher relative concentration of the PAs may be contaminated with some root material, as the root is known to have a much higher concentration of PAs than the leaf. Alternatively, the observed variability in the concentration of symphytine and echimidine may be due to natural deviations among plant material grown and harvested under differing conditions."
-'Analysis of Herbal Teas Made from the Leaves of Comfrey (Symphytum Officinale): Reduction of N-oxides Results in Order of Magnitude Increases in the Measureable Concentration of Pyrrolizidine Alkaloids' by Oberlies, Kim, Brine, Collins, Handy, Sparacino, Wani and Wall (North Carolina and New Mexico); Public Health Nutrition Journal, Cambridge, England, Volume 7, Issue 7, pages 919-924, **2004**.

"**The herbal industry has the difficult task of having to prove that their products are free of the Pyrrolizidine Alkaloid echimidine.** According to Peter Child of 'Investigative Science Incorporated', Burlington, Ontario, Canada, a company that specializes in laboratory testing, **reference standards for echimidine are not available. This makes it very difficult for laboratories to properly test Comfrey."**
-'The Comfrey Ban: Canada' by Richters Herbs, Goodwood, Ontario, Canada, www.richters.com, February 21 **2004**.

"**The identity of the Comfrey species is important** because it is now recognized that, although Symphytum officinale contains toxic pyrrolizidine alkaloids (PAs), it usually does not contain large quantities of the highly toxic echimidine found in S. asperum (Prickly Comfrey) or S. x uplandicum (Russian Comfrey).
Originally, it was believed that S. officinale did not contain echimidine, but a *1989 chemotaxonomic study revealed that about one-fourth of the Common Comfrey samples examined did contain that alkaloid. This was generally in very small amounts, except for one 2n = 24 cytotype, which contained levels of echimidine comparable to concentrations found in S. asperum.
It is still reasonable to conclude that a high level of echimidine in a Comfrey sample indicates a species other than S. officinale."
-Tylers Herbs of Choice: The Therapeutic Use of Phytomedicinals, Third Edition by Dennis V.C. Awang. Boca Raton, Florida: CRC Press, **2009**.
(* -'Chemotaxonomy of the Symphytum Officinale Agg. {Boraginaceae}' by Jaarsma, Lohmanns, Gadella and Malingre, Plant Systematics and Evolution, 167, pages 113-127, 1989.)

"**Comfrey root** also consists of pyrrolizidine alkaloids with 1,2-unsaturated necine ring structures, almost entirely in the form of their N-oxides, the main ones being 7-acetylintermedine and 7-acetyllycopsamine together with smaller amounts of intermedine, lycopsamine and symphytine (Brauchli et al., 1982).
The total amount of pyrrolizidine alkaloids given by different authors varies from 0.013% to 1.2% based on the analytical methods used (Tittel et al., 1979; Brauchli et al., 1982; Neidhardt, 1982; Stengl et al., 1982; Gracza et al., 1985; Vollmer et al., 1987; Mutterlein and Arnold, 1993).
The pyrrolizidine alkaloids echimidine and symlandine are not found in S. officinale L. and can be used as indicators of possible adulteration (contamination) with other Symphytum species, such as S. uplandicum or S. asperum (Mutterlein and Arnold, 1993)."
-'Comfrey: A Clinical Overview' by Christiane Staiger in Germany, Phytotherapy Research, Volume 26, Issue 10, pages 1441-1448, February **2012**, page 1.

"**Some pyrrolizidine alkaloids are extremely toxic, the most toxic of which is echimidine,** which is found in Comfrey. **Only minute amounts are found in the leaves of Comfrey grown in the United States,** but very large amounts can be found in the leaves of Russian Comfrey as well as in the leaves of the closely related Prickly Comfrey (Symphytum asperum)."

-'Herbal Supplements' ebook, 5th edition for Continuing Education Credit by Leslie K. Kay-Getzinger, R.D. (Registered Dietician), M.S., OnCourse Learning Corporation, Brookfield, Wisconsin, March **2014**, page 32.
>(I am assuming by 'Comfrey grown in the United States', the author means Symphytum officinale. No references were given to these statements. It would be nice to know what 'very large' means in terms of mg or percent.)

"Alkaloids: Comfrey (Symphytum officinale) roots contain 0.2-0.4% pyrrolizidine alkaloids:
symphytine, lycopsamine/intermedine (diastereoisomers), acetyl-lycopsamine/acetyl-intermedine (diastereoisomers), myoscorpine, lasiocarpine, heliosupine, viridiflorine, echiumine, symlandine and **echimidine** *(*, **).*
A considerable proportion of the pyrrolizidine alkaloids may be present as their **N-oxides** *(***).*
Notable quantities of echimidine are often reported, apparently because other species that are morphologically very close, such as S. asperum Lepechin and S. x uplandicum Nyman, are mistaken for Symphytum officinale, which in theory contains no echimidine at all *(****).*
>*The presence of echimidine and symlandine may refer to the falsification (contamination) of the herbal substance with Symphytum peregrinum roots, since these two alkaloids are not present in the roots of Symphytum officinale (*****).*

The pyrrolizidine content of Symphytum varies with plant part, season, natural biological variation, and species. Small young leaves early in the season possess higher total alkaloid content than older leaves, and the roots contain greater concentrations of total pyrrolizidine alkaloids than above-ground plant parts *(******).*
In a study of commercial samples of **Common Comfrey (Symphytum officinale), analysis of the total pyrrolizidine alkaloid content demonstrated values ranging from 1380-8320 micrograms/gram root compared with 15-55 micrograms/gram leaf** *(*******).*
>*The major pyrrolizidine alkaloids were symphytine and symlandine along with lesser concentrations of echimidine, lycopsamine, and acetyl-lycopsamine. Analysis of* **Comfrey tablets** *indicated that total alkaloid concentrations are similar to the alkaloid content of Comfrey roots (******)."*

-'European Medicines Agency- Assessment Report on Symphytum Officinale L., Radix' by European Medicines Agency, Committee on Herbal Medicinal Products (HMPC), London, England, 27 pages, May 5 **2015**.
(* -'Herbal Medicines, 3rd Edition' by Barnes, Anderson and Phillipson, Pharmaceutical Press, London, page 188-190, 2007.)
(** -'Pyrrolizidine Alkaloids in Foods' by R.A. Coulombe, Advances in Food and Nutrition Research, Volume 45, page 61-99, 2003.)
(*** -'Adverse Effects of Herbal Drugs' by De Smet, Keller, Hansel and Chandler, Editors, Springer-Verlag, Berlin-Heidelberg, Germany, pages 194-205, 220-222, 1992.)
(**** -'The Complete German Commission E Monographs: Therapeutic Guide to Herbal Medicines' edited by Blumenthal, Busse, Goldberg, Gruenwald, Hall, Riggins, Rister and Klein, The American Botanical Council, Austin, Texas, page 116, 1998.)
(***** -'Pharmakognosie - Phytopharmazie, 7th Edition' by R. Hansel and O. Sticher, Springer-Verlag, Heidelberg, Germany, pages 915-920, 1138, 2004.)
(****** -'Medical Toxicology of Natural Substances: Foods, Fungi, Medicinal Herbs, Plants and Venomous Animals' by D.G. Barceloux, Wiley: Hoboken, New Jersey, pages 449-457, 2008.)
(******* -'Analysis, Separation, and Bioassay of Pyrrolizidine Alkaloids from Comfrey- Symphytum officinale' by C.E. Couet, C. Crews and A.B. Hanley, Natural Toxins, Volume 4, page 163-167, 1996.)

*"***The accuracy of identification by TLC (Thin-Layer Chromatography) has been called into question by some scientists***, since it is a derivation of heliotridine and ALL other alkaloids in Comfrey are derivatives of retronecine; which itself is not hepatotoxic. **Echimidine in fresh Comfrey has been misidentified in TLC. One study of different ploidy races of Symphytum officinale shows echimidine content depends on the cytotype. When 2n = 24 or 40, echimidine is frequently absent, but at 2n = 48 or higher it is almost always present.**"*
-'Comfrey' by Glenn Axford, Our Common Roots: An Exploration of the Healing Power of Nature, Canada, www.ourcommonroots.com, date unknown but **2018** or earlier.
>(Body cells and individuals are described by the number of chromosome sets {ploidy level} they have: monoploid: 1 set; diploid: 2 sets = 2n = 2x, triploid: 3 sets = 2n = 3x, tetraploid: 4 sets = 2n = 4x, pentaploid: 5 sets = 2n = 5x, etc.)

<u>Human Foods with Alkaloids</u> (in general, not just Comfey)

*"***Pyrrolizidine Alkaloids (PAs) which may find their way into human and animal food in Australia are derived mainly from the plants Heliotropium europaeum, Echium plantagineum, Symphytum spp. (species) and Crotalaria retusa.***
>**The Sympthytum spp. (Comfrey) are deliberately ingested while the remaining species are weeds in various grain crops.** *The main alkaloids involved in cases of human toxicity until 1988 were heliotrine from Heliotropium,* **echimidine from Symphytum,** *ridelliine and retrorsine from Senecio longilobus and crotananine and cronaburmine from Crotalaria nana.*

Some 13 families of the flowering plants contain PAs. *Only 6 of these families contain hepatotoxic PAs but they represent some 3% of all the species of flowering plants.*
>*The principal families involved are the Asteraceae (Compositae), Boraginaceae and Leguminaceae (Fabaceae), while the main genera are Senecio (Asteraceae), Crotalaria (Leguminaceae) Heliotroprium, Trichodesma and* **Symphytum (Boraginaceae)**. *In Australia Echium plantagineum (Boraginaceae) is also an important PA-containing species.*

Apart from the deliberate use of herbal remedies and nutritional supplements containing PAs, humans can become

inadvertently (accidentally) exposed through consumption of contaminated food. The foods that have been found to contain PAs include grains, honey, milk, offal and eggs. It is still unknown whether there are residues of PAs in meat.
The target organ for PA toxicity in both experimental animals and humans is the liver. In animals, this toxicity is manifested as anti-mitotic (stops cell division) activity *leading to extensive fibrosis, nodular regeneration, parenchymal megalocytosis and cancer, while in humans, the major effects are hepatocellular injury, cirrhosis and veno-occlusive disease.*
There is no evidence from the significant poisoning outbreaks that have occurred that PAs cause liver cancer in humans. *Further research on the mechanisms of PA-induced hepatotoxicity may clarify the apparent differences in species specificity. At this time, the major toxicological endpoint for humans is considered to be veno-occlusive disease."*
-'Pyrrolizidine Alkaloids in Food: A Toxicological Review and Risk Assessment', Technical Report Series No. 2 by Professor Alan Seawright, Australia New Zealand Food Authority (ANZFA), Canberra BC, Australia, November **2001**.

"In addition to the many well-known major nutrients (protein, fat, carbohydrate and fiber) and minor nutrients (vitamins, minerals and nonessential compounds), foods contain thousands of naturally present toxic plant compounds.
> *Some are carcinogenic (cancer causing) in animals, and thus may be potentially carcinogenic in people. Many of these compounds are commonly termed 'nature's pesticides' because they are often toxic to predators such as insects and animals, thereby conferring a competitive advantage to the plant that produces them.*

Although these chemicals are in every meal we eat, they have received little attention compared to that given to the relatively minor residues of synthetic chemicals such as polychlorinated biphenyls (PCBs) and pesticides.
> **Our food contains greater than 10,000-fold more natural toxins than the synthetic kind, and in terms of metabolic reactions, our bodies are not able to distinguish between the two.**
> *Despite the popular notion equating 'natural' and 'healthy,' it is clear that natural toxins pose a far greater health risk than that posed by synthetic chemicals in our foods.*

One important and well-known class of naturally occuring chemicals in foods is the pyrrolizidine alkaloids. Pyrrolizidine alkaloids are a large group of compounds; more than 350 pyrrolizidine alkaloids have been isolated from over 6,000 plant species. The majority of pyrrolizidine alkaloids are toxic.
> **Probably the most popular and widely used pyrrolizidine alkaloid-containing medicinal herb is Comfrey (Symphytum).** *Since Greek and Roman antiquity, this herb has been part of the official pharmacopeia of many cultures, and has been used as a 'cure-all' for a variety of ailments.*

This is not intended to be an exhaustive review of all aspects of pyrrolizidine alkaloids. For that, I recommend Robin Mattock's book 'Chemistry and Toxicology of Pyrrolizidine Alkaloids' (1986)."
-'Pyrrolizidine Alkaloids in Food' by Roger A. Coulombe Jr., Graduate Program in Toxicology and Department of Veterinary Sciences, Utah State University, Logan, Utah; Advances in Food and Nutrition Research, Volume 45, pages 61-99, **2003**.

"Unsaturated PA have also been found in foodstuffs including milk, eggs, meat, bee pollen and honey. This is because the animal has consumed PA-containing plants, and the bees have visited PA-containing plants. *The topic is discussed in a report for the 'European Food Safety Authority' by Mulder et al.*.* **Vegans, for example, have a much smaller risk of chronic low-level PA exposure."**
-'Pyrrolizidine Alkaloids: Key Points for Herbal Practitioners' by Helen Phillips M.Sc. Herbal Medicine, Diploma of Phytotherapy, MNIMH and Alison Denham, M.A., FNIMH (National Institute of Medical Herbalists); The Forager's Path, LLC, School of Botanical Studies, Flagstaff, Arizona, August **2016**.
(* -'Occurrence of Pyrrolizidine Alkaloids in Food' by P.P.J. Mulder, P.L. Sanchez, A. These, A. Preiss-Weigert and M. Castellari {Netherlands, Germany, Spain}; EFSA {European Food Safety Authority} supporting publication EN-859, 116 pages, 2015.)

"To obtain a comprehensive overview of the Pyrrolizidine Alkaloid (PA) content in animal- and plant-derived food from the European market, and to provide a basis for future risk analysis, a total of 1105 samples were collected in 2014 and 2015. **These comprised milk and milk products, eggs, meat and meat products,** *herbal teas, and herbal food supplements collected in supermarkets, retail shops, and via the internet.*
> **Only 2% of animal derived products, in particular 6% of milk samples and 1% of egg samples, contained PAs.** *Determined levels in milk were relatively low, ranged between 0.05 and 0.17 microgram L^{-1} and only trace amounts of 0.10-0.12 microgram kg^{-1} were found in eggs. No PAs were detected in other animal-derived products.*

The results revealed that in animal products, PAs were only detectable in trace amounts and only in a few cases. Contamination of milk, eggs, and meat products with significant levels of PAs seems to be rare in the European Union.
This is likely due to a combination of the situation that animal feed is rarely highly contaminated and the fact that metabolic processes in the animals lead to an efficient reduction of the ingested PAs."
-'Occurrence of Pyrrolizidine Alkaloids in Animal- and Plant-Derived Food: Results of a Survey Across Europe' by Patrick P.J. Mulder, Patricia Lopez, Massimo Castelari, Dorina Bodi, Stefan Ronczka, Angelika Preiss-Weigert and Anja These (The Netherlands, Spain, Germany), Food Additives and Contaminants: Part A, Volume 35, No. 1, pages 118-133, **2018**.

Foods with Alkaloids: *"Members of the spinach family (Amaranthaceae) and the brassicas (mustard family Brassicaceae) (cabbage, broccoli, brussels sprouts) are high in oxalates (oxalic acid, an alkaloid), as are sorrel and umbellifers like parsley. Spinach 0.97g/100g, cabbage 0.10g, broccoli 0.19g, brussels sprouts 0.36g."*
-Wikipedia®: The Free Encyclopedia, www.wikipedia.org, **2018**.

Honey and Alkaloids (in general, not just Comfrey)

"Honey has also been found to contain PAs (Pyrrolizidine Alkaloids).
Honey from Senecio jacobaea has been shown to contain between 300 and 3200 micrograms of PAs per kilogram. Senecio vernalis honey is reported to contain between 500 and 1000 microgram of PAs per kilogram while Echium plantagineum honey has been shown to contain between 270 and 950 micrograms per kilogram."
-'Transfer of Pyrrolizidine Alkaloids into Eggs: Food Safety Implications' by John Edgar and Leslie Smith, CSIRO Animal Health, Victoria, Australia; American Chemical Society, ACS Symposium Series, Washington, DC, pages 118-128, **1999**.

"Some Pyrrolizidine Alkaloid-Containing Plants Reported to Contribute to Honey Production in Various Countries: Plant genera producing pyrrolizidine alkaloids are significant contributors to honey production in many countries. *Many of these plants are not native to the countries under which they are listed but are inadvertently introduced weeds or plants deliberately naturalized for use as herbs and for their flowers. The latter may become widely distributed throughout the countryside as garden escapes.*
The frequent lack of natural predators in areas where such introductions have become established often results in a plant density far greater than that in their native areas and displacement of more desirable native species.
Common examples of this phenomenon are the following: common borage (Borago officinalis); Paterson's curse (Echium plantagineum); tansy ragwort (Senecio jacobaea); and, **Comfrey (Symphytum officinale).**

Country	Genus
Albania	Senecio
Argentina	Echium
Austria	Myosotis (Mouse's Ear, Forget-Me-Not)
Australia	Echium, Ageratum, Heliotropium
Brazil	Senecio, Eupatorium (Aster family genus)
Burma	Chromolaena (Aster family genus)
Canada	Borago
Denmark	Borago
Egypt	Borago
Finland	Borago
Germany	Borago, Petasites (butterbar), Myosotis
India	Crotalaria (Fabaceae family genus), Senecio, Ageratum (whiteweed, Aster familly genus)
Italy	Echium, Senecio, Borago, Myosotis, Cynoglossum (Borage family), Petasites, Tussilago (coltsfoot)
Lithuania	**Symphytum**
Mexico	Eupatorium (Aster family genus), Senecio
Morocco	Echium
Netherlands	Tussilago
New Zealand	Echium
Nigeria	Ageratum, Chromolaena
Poland	Echium, Tussilago
Portugal	Echium
Senegal	Crotalaria
Somalia	Eupatorium
South Africa	Echium, Ageratum
Spain	Echium
Switzerland	Myosotis, Senecio
Thailand	Chromolaena (Eupatorium)
Taiwan	Ageratum
Turkey	Myosotis
Ukraine	**Symphytum**
United Kingdom	Senecio, Borago, Myosotis
Uruguay	Echium
USA	Senecio, Borago
USSR (Former)	**Echium, Symphytum, Borago, Cynoglossum**
Venezuela	Crotalaria
Yugoslavia	Echium
Zimbabwe	Senecio"

-'Honey from Plants Containing Pyrrolizidine Alkaloids: A Potential Threat to Health' by Edgar, Roeder and Molyneux, Journal of Agricultural and Food Chemistry, (50), pages 2719-2730, **2002**.
(Genus is a taxonomic rank that is above species and below family. In binomial nomenclature, the genus name is the first part of the binomial species name. Genera is the plural form of genus.)

"The 'German Federal Health Bureau' (1992) and the 'International Programme on Chemical Safety' (World Health Organization) (1988) both concluded that **the level of PAs present in honey may contribute to chronic liver disease and liver tumours."**

-'Toxicity Profile of Pyrrolizidine Alkaloid-Containing Medicinal Plants: Emphasis on Senecio Species' by Manuela G. Neuman (Toronto, Ontario, Canada) and Vanessa Steenkamp (Pretoria, South Africa), International Journal of Biomedical and Pharmaceutical Sciences, Volume 3, Special Issue 1, pages 104-108, **2009**.

"Pyrrolizidine alkaloids (PAs) are a class of naturally occurring compounds produced by many flowering plants around the world. Their presence as contaminants in food systems has become a significant concern in recent years. **For example, PAs are often found as contaminants in honey through pollen transfer.**
A validated method was developed for the quantification of four pyrrolizidine alkaloids and one pyrrolizidine alkaloid N-oxide in plants and honey grown and produced in British Columbia (western Canada).
Based on the findings in this single-laboratory validation, this method is suitable for the quantitation of lycopsamine, senecionine, senecionine N-oxide, heliosupine and echimidine in Common Comfrey (Symphytum officinale), tansy ragwort (Senecio jacobaea), blueweed (Echium vulgare) and hound's tongue (Cynoglossum officinale) and for PA quantitation in honey. **It found that PA contaminants were present at low levels in British Columbia honey.**
In Canada, and British Columbia specifically, there are a limited number of PA producing plants. In the Lower Mainland and Fraser Valley, the main plants are Senecio jacobaea (tansy ragwort), **Symphytum officinale (Common Comfrey),** and Senecio vulgaris (common groundsel), while more northeastern regions such as Nicola Valley and the Okanagan, Echium vulgare (blueweed) and Cynoglossum officinale (hound's tongue) are prevalent PA-containing plants."
-'Quantification of Pyrrolizidine Alkaloids in North American Plants and Honey by LC-MS' by Elizabeth M. Mudge, A. Maxwell, P. Jones and Paula N. Brown, Food Additives and Contaminants, Part A, Volume 32, Issue 12, **2015**.

"**Honey: Unsaturated PA (Pyrrolizidine Alkaloids) have been found in honey collected by bees from Pyrrolizidine Alkaloid-containing flowers.** Daily consumption of 15-25 grams (0.529-0.88 ounces) of honey is enough for chronic toxicity*. Now that the problem has been identified, then efforts are being made to reduce the level of PA in honey.
Children and the elderly are of particular concern with regard to chronic overload of dietary PAs, as both tend to consume higher-than-average quantities of honey.
Using honey as a therapeutic medium should also be considered a potential source of PA. A daily amount of 10 mg of **bee pollen** gives on average, 15 micrograms PA which is sufficient to give risk of chronic toxicity."
-'Pyrrolizidine Alkaloids: Key Points for Herbal Practitioners' by Helen Phillips M.Sc. Herbal Medicine, Diploma of Phytotherapy, MNIMH and Alison Denham, M.A., FNIMH (National Institute of Medical Herbalists); The Forager's Path, LLC, School of Botanical Studies, Flagstaff, Arizona, August **2016**.
(* -'Pyrrolizidine Alkaloids in Food: A Spectrum of Potential Health Consequences' by J.A. Edgar {Australia}, S.M. Colegate {United States}, M. Boppre {Germany} and R.J. Molyneux {United States}; Food Additives and Contaminants: Part A, Volume 28, No. 3, pages 308-324, 2011.)

"Comfrey PAs are almost exclusively accumulated in the ovaries, while **Comfrey petals, sepals and pollen hardly contain PAs.** High levels of PAs are detectable in the fruit, but the elaiosome was shown to be PA free."
-'Specific Distribution of Pyrrolizidine Alkaloids in Floral Parts of Comfrey (Symphytum Officinale) and Its Implications for Flower Ecology' by Thomas Stegemann, Lars H. Kruse, Moritz Brutt and Dietrich Ober (Kiel, Germany and Ithaca, New York); Journal of Chemical Ecology, e-publication before print, July 28 **2018**.
(Elaiosomes are fleshy structures attached to seeds.)
(See sub-subsection 'Comfrey Flower and Pyrrolizidine Alkaloids' in subsection 'Comfrey Flower' in section 'Symphytum Genus Description' {Chapter 5, Volume 1}.)

Eggs and Alkaloids (in general, not just Comfrey)

"**PAs (Pyrrolizidine Alkaloids) occur widely in agricultural production systems throughout the world and can enter the human food chain as contaminants. Grain, milk and honey are among the products that can sometimes be contaminated by PAs.**
A natural episode of PA poisoning in chickens caused by a lapse (failure) in the quality control of the grain component of their feed, has provided evidence that **PAs can also be transferred into eggs**. The concentrations of PAs detected in the eggs exceeded the levels deemed tolerable for herbal medicines in Germany.
The likely cause of PA poisoning among three flocks of layer chickens was considered to be Heliotropium europaeum seeds present in wheat at an estimated concentration of 0.6% by weight. Also present as contaminants in the grain, but in much smaller quantities, were seeds of yellow Burr weed (Amsinckia spp.) and Sheepweed (Buglossoides arvensis) which are also possible sources of PAs. Even higher concentrations of PAs have previously been recorded in milk and honey. **While there is no evidence of chronic health problems caused by PAs in these products,** they and other products, including grain and meat, warrant (require) monitoring for PAs to ensure safe levels of dietary intake are not being exceeded.
It is unusual for PA-containing plants to be intentionally used as foods. Examples of such limited use include Comfrey (Symphytum spp. {species}) that is sometimes eaten in salads and Borage (Borago officinalis) which is occasionally used as a cucumber flavored garnish.
The principal route of human exposure to PAs in food is via contaminated staple foods.
As expected from a consideration of the levels of PAs present in these products relative to the amount required

to cause acute toxicity, there have been no confirmed cases of acute human poisoning from these sources."
-'Transfer of Pyrrolizidine Alkaloids into Eggs: Food Safety Implications' by John Edgar and Leslie Smith, CSIRO Animal Health, Victoria, Australia; American Chemical Society, ACS Symposium Series, Washington, DC, pages 118-128, **1999**.

"PA have been found in eggs, as a consequence of PA contamination in the hen's grain but were not found when hens were deliberately fed PA-containing plants.
The risk or incidence rate of PA-contaminated eggs has not been fully evaluated*."
-'Pyrrolizidine Alkaloids: Key Points for Herbal Practitioners' by Helen Phillips M.Sc. Herbal Medicine, Diploma of Phytotherapy, MNIMH and Alison Denham, M.A., FNIMH (National Institute of Medical Herbalists); The Forager's Path, LLC, School of Botanical Studies, Flagstaff, Arizona, August **2016**.
(* -'European Medicines Agency- Public Statement on the Use of Herbal Medicinal Products Containing Toxic, Unsaturated Pyrrolizidine Alkaloids (PAs)' by European Medicines Agency, Committee on Herbal Medicinal Products (HMPC), London, England, 24 pages, November 24 2014.)

Milk, Meat, Grain and Alkaloids (in general, not just Comfrey)

"PAs (Pyrrolizidine Alkaloids) have for example been shown to pass into the milk of cows, goats and rats. In the case of cows, experimental exposure to a PA-containing plant, Senecio jacobaea, has been reported to lead to levels of PAs between 470 and 835 micrograms per liter (1.05 quart) of milk."
-'Transfer of Pyrrolizidine Alkaloids into Eggs: Food Safety Implications' by John Edgar and Leslie Smith, CSIRO Animal Health, Victoria, Australia; American Chemical Society, ACS Symposium Series, Washington, DC, pages 118-128, **1999**.

*"**Milk:** Lipophilic (attracted to fats) toxic PA have been found in milk*. (Lipophilic alkaloids are not in Comfrey.)*
***Meat**, especially liver, may contain toxic tissue-bound PA ingested by the animal before its slaughter. PA are not destroyed by cooking*.*
***Wheat:** There is some concern that there may be residual low levels of PA-containing plants in wheat that are not fully eradicated (removed) by the grain-filtering processes used by countries in the West**."*
-'Pyrrolizidine Alkaloids: Key Points for Herbal Practitioners' by Helen Phillips M.Sc. Herbal Medicine, Diploma of Phytotherapy, MNIMH and Alison Denham, M.A., FNIMH (National Institute of Medical Herbalists); The Forager's Path, LLC, School of Botanical Studies, Flagstaff, Arizona, August **2016**.
(* -'European Medicines Agency- Public Statement on the Use of Herbal Medicinal Products Containing Toxic, Unsaturated Pyrrolizidine Alkaloids (PAs)' by European Medicines Agency, Committee on Herbal Medicinal Products (HMPC), London, England, 24 pages, November 24 2014.)
(** -'Pyrrolizidine Alkaloids in Food: A Spectrum of Potential Health Consequences' by J.A. Edgar {Australia}, S.M. Colegate {United States}, M. Boppre {Germany} and R.J. Molyneux {United States}; Food Additives and Contaminants: Part A, Volume 28, No. 3, pages 308-324, 2011.)

Symphytum Officinale 1880

'Illustrations of the British Flora: A Series of Wood Engravings, with Dissections, of British Plants, Drawn by W.H. Fitch and W.G. Smith. Forming an Illustrated Companion to Mr. Bentham's Handbook and other British Floras'.
London, England, 1880 and other editions.

Symphytum page 171.

Chapter 30

Alkaloids in Comfrey: Scientific Studies and Various Perspectives

<u>Scientific Studies Showing Dangers of Alkaloids in Comfrey</u>

With Some Rebuttals: Rebuttal is a statement that a claim or criticism is not true. It is an argument or proof to contradict or disprove something previously stated.

"The citing of earlier papers proceeds on the assumption of certainty, which, as we have seen, is not always as sure as it is portrayed. According to authors Bone and Pembery: Mohabbat, Culvenor and Hirono research reports are flawed to some extent, yet their conclusions are interpreted as sure.
In particular, as mentioned above, Hirono's paper is regularly quoted as proof of the carcinogenicity (cancer causing) of Comfrey.
In 1978, Lawrence Hills of the 'Henry Doubleday Research' Institute published an article, reported in the British Medical Journal* 1979, which concluded that until further research clarified the long-term health hazards of Comfrey ingestion, 'no human being or animal should eat, drink, or take Comfrey in any form.'
The report by Pembery was subsequently published in 1982 (1983) in which the foreword by Hills revoked the 1978 report, and Comfrey was declared safe. Many post 1982 papers on Comfrey quote the 1978 warning and fail to acknowledge the 1982 revocation."
-'In Defence of Comfrey' by Margaret Whitelegg, MNIMH. Paper presented to 'Department of Health' and 'Ministry of Agriculture, Fisheries and Food', 'National Institute of Medical Herbalists', United Kingdom, January 1993. Published in European Journal of Herbal Medicine, Exeter, England, Volume 1, No. 1, pages 11-17, **1994**.
(* -'Letter from Chicago: Henry Doubleday Research Association', The British Medical Journal, London, England, page 598, March 3 1979.)
(** -The Safety of Comfrey: Report from the Henry Doubleday Research Association by J.A. Pembery B.Sc., Special Advisor, Research Chemist, Aldgate Press: London, England, 20 pages, 1983.)

The following scientific reports are discussed: (in chronological order from older to recent)
1. 'Toxicity of Pyrrolizidine Alkaloids' by Mattocks, **1968**.
2. 'Studies on Constituents of Crude Drugs. I. Alkaloids of Symphytum Officinale Linn.' by Furuya, **1968**.

3. 'An Outbreak of Hepatic Veno-Occlusive Disease in North-Western Afghanistan' by Mohabbat, **1976**.
4. 'Carcinogenic Activity of Symphytum Officinale' by Hirono, **1978**.

5. 'The Alkaloids of Symphytum x uplandicum (Russian Comfrey)' by Culvenor, **1980**.
6. 'Structure and Toxicity of the Alkaloids of Russian Comfrey (Symphytum x uplandicum, Nyman), a Medicinal Herb and Item of Human Diet' by Culvenor, **1980**.
7. 'Toxic Pyrrolizidine Alkaloids in Comfrey' by Mattocks, **1980**.
8. 'Comfrey and Liver Damage' by Roitman, **1981**.
9. 'Mutagenic Effects of Aqueous Extracts of Symphytum Officinale L. and of Its Alkaloidal Fractions' by Furmanowa, **1983**.
10. 'Hepatic Venocclusive Disease Associated with Consumption of Pyrrolizidine Containing Dietary Supplements' by Ridker, **1985**.
11. 'Toxicity of Comfrey-Pepsin Preparations' by Huxtable, **1986**.
12. 'Veno-Occlusive Disease of the Liver Secondary to Ingestion of Comfrey' by Weston, **1987**.
13. 'Comfrey: Assessing the Low-Dose Health Risk' by Abbott, **1988**.
14. 'Hepatic Veno-Occlusive Disease in Newborn Infant of a Woman Drinking Herbal Tea' by Roulet, **1988**.

15. 'Comfrey Herb Tea-Induced Hepatic Veno-Occlusive Disease' by Bach, **1989**.
16. 'Studies on the Effect of an Alkaloid Extract of Symphytum Officinale on Human Lymphocyte Cultures' by Behninger, **1989**.
17. 'Comfrey Herb Tea and Hepatic Veno-Occulusive Disease' by Ridker, **1989**.

18. 'Hepatic Veno-Occlusive Disease Associated with Comfrey Ingestion' by Yeong, **1990**.
19. 'The Effects of Comfrey Derived Pyrrolizidine Alkaloids on Rat Liver' by Yeong, **1991**.
20. 'CPMP Listing of Herbs and Herbal Derivatives Withdrawn for Safety Reasons: Herbal Drugs with Serious Risks' by Committee for Proprietary Medicinal Products, **1992**.
21. 'Hepatic and Pulmonary Complications of Herbal Medicines' by Miskelly, **1992**.
22. 'Hepatocyte Membrane Injury and Bleb Formation Following Low Dose Comfrey Toxicity in Rats' by Yeong, **1993**.
23. 'Determination of Pyrrolizidine Alkaloids in Commercial Comfrey Products (Symphytum sp.)' by Betz, **1994**.
24. 'Medicinal Plants in Europe Containing Pyrrolizidine Alkaloids' by Roeder, **1995**.

25. 'Pyrrolizidine Alkaloids in Human Diet' by Prakash, **1999**.
26. 'Acute Hepatitis After Ingestion of Herbs' by Shad, **1999**.

27. 'Pyrrolizidine Poisoning: A Neglected Area in Human Toxicology' by Stewart, **2001**.
28. 'Systematic Review: Hepatotoxic Events Associated with Herbal Medicinal Products' by Pittler, **2003**.

29. 'Analysis of Herbal Teas Made from the Leaves of Comfrey (Symphytum officinale): Reduction of N-oxides Results in Order of Magnitude Increases in the Measureable Concentration of Pyrrolizidine Alkaloids' by Oberlies, **2004**.
30. 'Mutagenicity of Comfrey (Symphytum Officinale) in Rat Liver' by Mei, **2005**.
31. 'Comparison of Gene Expression Profiles Altered by Comfrey and Riddelliine in Rat Liver' by Guo, Mei, et al., **2007**.
32. 'Bioactive Compounds in Food' by Gilbert, **2008**.
33. 'Severe Pulmonary Hypertension Possibly Due to Pyrrolizidine Alkaloids in Polyphytotherapy' by Gyorika, **2009**.

34. 'Metabolism, Genotoxicity, and Carcinogenicity of Comfrey' by Mei, **2010**.
35. 'Toxicity of Pyrrolizidine Alkaloids to Humans and Ruminants' by Wiedenfeld, **2011**.
36. 'Herbal Hepatotoxicity: A Tabular Compilation of Reported Cases' by Teschke, **2012**.
37. 'Pyrrolizidine Alkaloids in Medicinal Plants from North America' by Roeder, **2015**.
38. 'The Effect of Aqueous Leaf Extract of Symphytum Officinale (Common Comfrey) on the Liver of Adult Wistar Rats' by Ezejindu, **2015**.
39. 'The Comparative Toxicity of a Reduced, Crude Comfrey (Symphytum Officinale) Alkaloid Extract and the Pure, Comfrey-Derived Pyrrolizidine Alkaloids, Lycopsamine and Intermedine in Chicks' by Browna, **2016**.

1. **'Toxicity of Pyrrolizidine Alkaloids'** by A.R. Mattocks, Toxicology Research Unit, MRC Laboratories, Carshalton, Surrey, England; Nature: International Journal of Science, Volume 217, pages 723-728, February 24 **1968**.
*"**A new class of metabolite, with a pyrrole-like structure, has been demonstrated in the tissues of animals poisoned by Pyrrolizidine Alkaloids. There is some correlation between the degree of hepato-toxicity and the amount of 'pyrrole' found in the liver.** Evidence has been found of the types of reactions such metabolites might undergo with tissue constituents. Certain Pyrrolizidine Alkaloids, such as heliotrine, lasiocarpine, and retrorsine. have long been known to cause chronic liver poisoning in animals, and the pathology has been well described.*, ***
The 'metabolic pyrroles' are partly excreted in urine, but some are also bound strongly to the tissues of the liver and to a decreasing extent, the lungs and other organs for 48 hours or more after being formed.
With the exception of rosmarinine, there is a rough correlation between the hepatotoxicity of the alkaloids and the amounts of pyrroles to which they give rise in vivo.
The following hypothesis is consistent with these results: **The alkaloids themselves are not hepatotoxic. A proportion of the alkaloid (depending on its structure) is metabolized in the liver (by a process amounting to dehydrogenation) to a pyrrole-like derivative.**"
(* -'The Action of Retrorsine on Rat's Liver' by James Davidson, Department of Pathology, University of Edinburgh, Scotland; The Journal of Pathology, Volume 40, Issue 2, pages 285-295, March 1935.)
(** -'The Acute Toxic Effects of Heliotrine and Lasiocarpine, and their N-oxides, on the Rat' by L.B. Bull, A.T. Dick and J.S. McKenzie, Division of Animal Health and Production, CSIRO, Animal Health Research Laboratory, Victoria, Australia, The Journal of Pathology, Volume 75, Issue 1, pages 17-25, January 1958.)
(Correlation is a mutual connection between two or more things.)
('In Vitro' means done in laboratory equipment as opposed to in/on a living animal that is called 'in vivo'.)
(This report is one of the first about Pyrrolizidine Alkaloids. Comfrey is not mentioned.)

2. **'Studies on Constituents of Crude Drugs. I. Alkaloids of Symphytum Officinale Linn.'** by T. Furuya and K. Araki; Kitasato University, Japan; Chemical and Pharmaceutical Bulletin, Volume 16, No. 12, pages 2512-2516, **1968**.
"Preliminary investigation disclosed that the alkaloids were present mainly in the form of N-oxides. After reduction of the aqueous (water) acid solution of total base, the reduced bases were investigated.
Two pyrrolizidine alkaloids, symphytine (I), a new compound, and echimidine (II) have been isolated from the dried roots of Symphytum officinale. *Both have a retronecine nucleus esterified on the 7-hydroxyl group with angelic acid.*
The other esterifying acid of symphytine is l-viridifloric acid (III), *identical with the authentic sample. The structure of symphytine, a diastereoisomer of echiumine (retronecine esterified with angelic acid and trachelanthic acid), has been confirmed. A large amount of allantoin has also been isolated from the dried roots and leaves."*

   Problems with Above Research (Rebuttal)

   Rebuttal i. "***The Japanese work (by Furuya) in 1968 showed** that the root of Symphytum officinale, Common Comfrey, contained pyrrolizidine alkaloids of the type found in ragwort (Senecio squalidus) which had caused liver damage in experimental rats. The quantity present was very small but the risk was of these adding up to danger.
   At the Henry Doubleday Research Association committee meeting held before the Annual General Meeting in 1971, Dr. Dennis Long (Ph.D. from the Michaelis Nutritional Foundation) was able to announce that **the Medical Research***

Council had found that Comfrey tea held 0.028% of the alkaloids, and his own tests had shown 0.020 percent. Assuming that men had the same sensitivity to alkaloids as experimental rats, anyone weighing 160 pounds would need to drink 102,000 cups of Comfrey tea one after the other to produce toxic symptoms. If the same amount was necessary to induce chronic toxicity, then, at five cups a day, this would take 56 years.

> Arthritics average it (taking Comfrey) for 20-30 years, no one had yet achieved 56. Today, our vigorous vegetarian member over 80, Miss Aughton, who eats her own Comfrey as spinach about three times a week in summer and dries it for tea in winter, is unlikely to live to 113, but she might. If she misses her Comfrey for as long as three weeks on holiday (vacation) the arthritis is back, and she feels safer taking it than orthodox anti-arthritic drugs.

There have been generation after generation of 'North Country' (England) women keeping arthritis at bay by drinking 'knitbone' (Comfrey) tea and using the healing leaves for injuries. Surely, if there were a genuine danger to any livestock or human being there would be some record of something suffering, especially from the 1880-1890 period when it was fed in such quantities to horses."

-Fighting Like the Flowers: An Autobiography: The Life Story of Britain's Best-Known Organic Gardener by Lawrence D. Hills. Bideford, Devon, England: Green Books, 1989, page 215.

Rebuttal ii. *"In their paper, Furuya and Araki also report on a pharmacological test with rats showing symphytine alkaloid to have a LD50 of about 300 mg/kg; that is an intravenous (in vein) injection of 300 mg of the purified alkaloid per kg of rat tissue caused death in approximately 50 percent of the experimental animals.*
Thus in the case of Comfrey tea if it be assumed that normal methods of infusion (steeping) could extract just over half the alkaloid that was extracted in 8 hours in a Soxhlet apparatus in the laboratory, each cup of tea could contain 100 micrograms of alkaloid.
At this level the consumer could never attain the lethal dose of 300 milligrams/kg (1 kg= 2.2 pounds) tissue found necessary to produce the acute reaction in rats. **Even to consume this quantity it would take a 150 pound man drinking 4 cups of tea per day a total of 140 years.**
Furthermore it is known that to produce chronic reactions sub-lethal doses over a prolonged period are necessary. Normally such sub-lethal doses would need to be of a much higher order particularly as sensitivity to the alkaloid decreases with age."
-'The Alkaloid Content of Comfrey' by Dr. D.B. Long, Ph.D, M.A., around 1970. (In 'Comfrey: Fodder, Food & Remedy').
> (A Soxhlet extractor is used when the desired compound has limited solubility in a solvent, and the impurity is insoluble in that solvent.)

Rebuttal iii. *In this report by Furuya, he writes:* **"Symphytum officinale Linn. is widely distributed in Europe and now cultivated in Japan. It is called Russian Comfrey, which generally indicates the dried roots, and in Japan the fresh or dried leaves or occasionally the dried roots."**
Symphytum officinale is Common Comfrey, not Russian Comfrey, so which species was tested? In the late 1950s Russian Comfrey (Symphytum x uplandicum) was brought to Japan from Australia that got it from England. Japan does not have any native Symphytum species. So most likely the species tested is Symphytum x uplandicum, not Symphytum officinale as is stated in the title of the report.

3. 'An Outbreak of Hepatic Veno-Occlusive Disease in North-Western Afghanistan' by Mohabbat, Younos, Merzad, Srivastava, Sediq, and Aram, The Lancet, London, England, Volume 308, No. 7980, pages 269-271, August 7 **1976**.

"Following a 2-year period of severe drought a very large number of patients with massive ascites (fluid in abdomen) and emaciation (starvation) were observed in north-western Afghanistan. Clinico-pathological study showed that these were typical cases of hepatic veno-occlusive disease.
The outbreak was caused by consumption of bread made from wheat contaminated with seeds of Heliotropium plants, which were shown to contain pyrrolizidine alkaloids. *Examination of 7200 inhabitants from the affected villages showed evidence of liver disease in 22.6%.*
Clinical improvement was observed in thirteen cases after 3 to 9 months of supportive hospital treatment, and in three cases liver biopsies showed almost complete disappearance of initial abnormalities."

> (In hepatic Veno-Occlusive Disease some small veins in the liver are blocked. It can be caused by high-dose chemotherapy or after consuming certain plant alkaloids such as pyrrolizidine. Symptoms include weight gain due to fluid retention, increased liver size, and raised levels of blood bilirubin. Also called 'Sinusoidal Obstruction Syndrome'.)

Problems with Above Research (Rebuttal)

Rebuttal i. *"Appraisal of the toxicology shows some* **discrepancies (inconsistencies) in the figures used to calculate the level of alkaloids in the wheat flour.**
For instance, the average sample of wheat seeds contained forty Heliotropium seeds per kilogram (2.2 pounds) of wheat. These were included in the flour when it was made. The total alkaloid content of seed samples varied from 0.72% to 1.49%, whereas the alkaloid content of seed samples varied from 0.50% to 0.186%.
This would mean that the samples of wheat flour examined would contain upward from 10% Heliotropium seeds, which is in contrast to their average figure for contaminated wheat of 40 seeds (300 mg) per kilogram.

This drastically affects the conclusions which can be drawn about the chronic toxicity of the alkaloids. Certainly the authors claim that their estimate of a consumption of 1.46 grams of alkaloid over two years is a 'very conservative estimate'."
-'The Safety of Comfrey: Report from the Henry Doubleday Research Association' by J.A. Pembery B.Sc., Special Advisor, Research Chemist, Aldgate Press: London, England, 20 pages, 1983, page 7.

Rebuttal ii. *"In a subsequent edition of 'The Lancet', C. Anderson* of University College, London, points out that **two contrasting figures occur in this paper by Mohabbat et al, one leading to a figure of 2 mg per day consumption of alkaloid, and the other figure for contaminated wheat leading to a figure of 1300 mg alkaloids per day.** There is also the point that the alkaloids in Heliotropium are probably considerably more toxic than those in Comfrey. **If one extends toxicity figures of Comfrey alkaloids from juvenile (young) rats to humans, one arrives at a figure of the order of 1,000 cups of root tea presenting a risk of liver damage.**
Therefore long term consumption of Comfrey root products is not recommended.
Comfrey leaf tea presents quite a different picture (it is much safer)."*
-'The Safety of Comfrey: Report from the Henry Doubleday Research Association' by J.A. Pembery B.Sc., Special Advisor, Research Chemist, Aldgate Press: London, England, 20 pages, 1983, page 17.
(* -'Comfrey Toxicity in Perspective' by Clare Anderson, The Lancet, London, England Volume 317, Issue 8235, page 1424, June 27 1981.)

Rebuttal iii. *"Lancet correspondence on the potential hepatotoxicity of the herb Comfrey has referred to a study in Afghanistan where wheat was contaminated with seeds of Heliotropium, which contain pyrrolizidine alkaloids, and where liver disease resulted. In referring to this work, others have quoted only one of the analyses- that of total alkaloids in the Heliotropium seeds themselves.*
However, samples of contaminated wheat flour were also analysed and total alkaloid content varied around 0.186%, representing a daily intake of approximately 1300 mg alkaloids in 700 grams (1.54 pounds) flour.
This is in marked contrast to the estimated intake of 2 mg alkaloids daily, extrapolated from the other analysis based on seeds."
-'Comfrey Toxicity in Perspective' by Clare Anderson, The Lancet, London, England Volume 317, Issue 8235, page 1424, June 27 1981.
(Extrapolate means to extend a method or conclusion, especially one based on statistics, to an unknown situation by assuming that existing trends will continue or similar methods will be applicable.)

4. 'Carcinogenic Activity of Symphytum Officinale' by I. Hirono et al, Japan National Cancer Institute Volume 61, No. 3, pages 865-869, September **1978**.
*"The carcinogenicity (cancer causing) of **Symphytum officinale L., Russian Comfrey,** used as a green vegetable or tonic, was studied in **inbred ACI rats**. Three groups of 19-28 rats each were fed Comfrey leaves for 480-600 days {1.3 to 1.6 years}; four additional groups of 15-24 rats were fed Comfrey roots for varying lengths of time. A control group was given a normal diet.*
Hepatocellular (liver cell) adenomas (benign tumor) were induced (created) in all experimental groups that received the diets containing Comfrey roots and leaves.
The incidence of the liver tumors tended to be higher in groups fed a diet containing Comfrey root than in those fed a diet containing Comfrey leaf, despite the fact that the percentage of Comfrey flour is higher in the leaf diet than in the root diet.
*Furthermore, rats became seriously affected when roots were fed in a concentration of 4% in 2 groups of rats (groups V-I and 2), whereas **rats tolerated well the feeding of the Comfrey leaves in concentrations as high as 33%**. Therefore, it may be evident that feeding of the Comfrey roots was much more toxic than was feeding of the leaves.*
The incidence of the hemangioendothelial sarcoma of the liver was low in this study.
*However, the same type of tumor was predominant in rats fed Petasites or colts foot. **Carcinogenic activity of Comfrey may be assumed to be weaker than that of Petasites or coltsfoot."***
(The author defines 'Symphytum officinale' as 'Russian Comfrey'. They are different species.)
(Carcinogenicity means its ability or tendency to cause cancer. Adenoma is a benign, noncancerous epithelial tumor. Sarcoma is a cancerous tumor.)

Problems with Above Research (Rebuttal)

Rebuttal i: *"**At the start of Hirono's study the rats were from one month to six weeks old, at which age they had been weaned, but were very vulnerable to hepatoxic (liver toxic) effects from pyrrolizidine alkaloids.***
The maximum time of administration of the diet containing Comfrey leaf was 600 days (1.3 to 1.6 years) which represents a very large proportion of the life of a laboratory rat.
The control group in this study received the basal diet without Comfrey but, unfortunately, comparative survival of the control animals was not reported.
It is clear that Comfrey root is not a suitable item of diet for an extended period of time, even at a level of 0.5% of the basal (base) diet. This, however, does not mean that it is unsafe to use Comfrey root preparations for occasional medical applications.

The effect of the diet containing dried Comfrey leaves is less marked, and at the lowest level, of the 28 rats which were fed a diet containing 8% dry weight Comfrey, only one showed a liver tumour at 600 days. A level of 4% to 6% dried Comfrey leaf in the basal diet would be unlikely to produce any tumours.
The total amount of dried Comfrey represents about 240% of the estimated final bodyweight of the rat. It is difficult to envisage (see) a situation in which a human subject would assimilate (eat) an equivalent amount of Comfrey over a similar time span, or an equivalent time span in their lives."
-'The Safety of Comfrey: Report from the Henry Doubleday Research Association' by J.A. Pembery B.Sc., Special Advisor, Research Chemist, Aldgate Press: London, England, 20 pages, 1983, pages 10, 11.

    (In the wild, rats live to about 1 year {365 days}. In a home as a pet, rats live about 1.8 years {21.6 months = 657 days}. Only 5% live to 3 years old {36 months = 1095 days}.)

Rebuttal ii: Two Toxicological Studies (1978: Hirono, 1980: Culvenor):
"My quotations are from the December 1989 and February 1990 issues of 'Medi-Herb' newsletter, published in Queensland (Australia) (titled 'Comfrey- The Facts'):
'A group of well-meaning scientists actively lobbied the Australian government to have Comfrey restricted. The basis for their concern was just **two toxocological studies, both of which have doubtful relevance to normal human use. The arguments generally used were related to pyrrolizidine alkaloids (PA), not Comfrey itself,** *and their theme was the pyrrolizidine alkaloids should be entirely eliminated from human diet and human medicine. Their zeal saw Comfrey in some states of Australia receive a higher poisons classification than arsenic, hemlock, belladona and strychinine.*
As far as pyrrolizidine alkaloid carcinogenesis is concerned, an important part of the argument rests on the disputed identity of the lesions reported as hepatoma (malignant liver tumor).
It is noteworthy that there has never been a link with PA intake in cancer or in veterinary studies.' "
-Comfrey: Nature's Healing Herb & Health Food by Andrew Hughes. Japan: Sanyusha Publishing Co, 1992, page 72.

Rebuttal iii: *"In one study ('Carcinogenic Activity of Symphytum Officinale') rats were fed green leaves of Senecio jacobea (Oxford Ragwort) and Comfrey in their diet.*
At 5% Comfrey leaves there was no sign of toxicity, but at 1% Ragwort leaves in the diet there were many signs of toxicity, including changes in liver enzyme activity.
Even 20% Comfrey leaf in the diet did not cause the liver enzyme changes caused by 1% Ragwort.
Comfrey PAs (pyrrolizidine alkaloids) are therefore much less toxic to the liver than those of Ragwort. In fact there are no recorded cases of livestock poisoning due to Comfrey."
-Comfrey: Nature's Healing Herb & Health Food by Andrew Hughes. Japan: Sanyusha Publishing Co, 1992, page 72.

Rebuttal iv: *"The Hirono study does not satisfy many of the criteria demanded for a rigorous assessment of carcinogenicity.*
Rats were fed Comfrey leaf from 8 to 33% of their diet, thus all test levels exceeded the 5% maximum normally recommended for such trials. Test levels for the root were 0.5 to 4%.
The maximum time of administration of the diet containing Comfrey leaf was 600 days, which represents a very large proportion of the life of a laboratory rat. The comparative survival of the control group of animals was not reported.
The fact that rats could be fed 33% Comfrey leaves in their diet and still survive to old age is testimony to relatively low toxicity. How many drugs could survive such scrutiny?"
-'The Safety-in-Use of Comfrey and Comfrey Products: Results of a Research Survey' by the Society for the Promotion of Nutritional Therapy (SPNT), Linda Lazarides, Sussex County, United Kingdom, 8 pages, 1993-1994, page 7.

Rebuttal v: *"Yet even here in 'Hirono et al' when researchers used the whole plant, the results of the trial are extremely dubious (questionable).*
Bone* and Pembery** again question various aspects of the paper. PA levels were never measured, toxicity was admitted to vary.
Diets of Comfrey at certain levels are protein-deficient.
600 days is a long time in the life of a laboratory rat: according to Pembery 80% on a basal diet survive sixty days. The comparative control group survival was not indicated.
Pembery again extrapolated figures and argues that **the average rat consumed 24 times its own body weight at which level only one rat showed toxic symptoms.**
Bone's main criticism of the paper is the misleading nature of the title 'Carcinogenic Activity of Symphytum Officinale': in the text the Comfrey is referred to as Russian Comfrey, not officinale, so it is unclear which Comfrey is on trial.
But more importantly the tumours in all but three of the rats were benign (no cancer); the plant appeared to be hepatotoxic (liver toxic), but not carcinogenic (cancer causing).
Metastases were not mentioned, perhaps indicating that the three malignant tumours were of low malignancy.
Malignancy occurred in the lower dosage groups, therefore **a dose-response relationship was not evident.**
Abbott*** suggests that the relatively high doses that were used leave open the possibility of secondary effects that influenced the development of the cancers.
Yet this Hirono paper is cited frequently in subsequent literature as concrete evidence of the carcinogenic

properties of Comfrey in animals."
-'In Defence of Comfrey' by Margaret Whitelegg, MNIMH. Paper presented to 'Department of Health' and 'Ministry of Agriculture, Fisheries and Food', 'National Institute of Medical Herbalists', United Kingdom, January 1993. Published in European Journal of Herbal Medicine, Exeter, England, Volume 1, No. 1, pages 11-17, 1994.
(* -'Studies in Materia Medica, Part I: Symphytum Species' by K.M. Bone, School of Herbal Medicine, Tunbridge Wells, England, undated 1983 or 1984 probably. I was unable to find this report. If you have a copy, could you send it to me.)
(** -'The Safety of Comfrey: Report from the Henry Doubleday Research Association' by J.A. Pembery B.Sc., Special Advisor, Research Chemist, Aldgate Press: London, England, 20 pages, 1983, pages 10, 11.)
(*** -'Comfrey: Assessing the Low-Dose Health Risk' by P.J. Abbott, Toxicology Unit, Department of Community Services and Health, Canberra, Australia; Medical Journal of Australia, Volume 149, No. 11-12, pages 678-682, 1988.)
(Metastasis is secondary malignant growths at a distance from a primary cancer site.)

Rebuttal vi: This rebuttal is based on what I found when looking for a definition of an 'ACI rat'. As shown by the examples below, these rats are bred to be prone to having tumors. So how is this kind of test going to be accurate when comparing to humans?
a. "ACI rats, an inbred line derived from August and Copenhagen strains, is unique for its susceptibility to estrogen-induced mammary tumors.
While the ACI rat model is ideal to study estrogen-induced mammary tumorigenesis, associated problems arising with pituitary tumors do not provide a sufficient window to utilize this model for several other studies such as intervention."
-'Mammary Tumor Induction in ACI Rats Exposed to Low Levels of 17Beta-Estradiol' by S. Ravoori, M.V. Vadhanam, S. Sahoo, C. Srinivasan and R.C. Gupta, University of Louisville, Kentucky and University of Kentucky, Lexington, Kentucky; International Journal of Oncology, Volume 31, No. 1, pages 113-120, July 2007.
***b. "ACI rats will grow transplantable Morris hepatomas** 3924A which can be used as a model for the treatment of liver cancer (Yang et al 1995)."*
-MGI: Mouse Genome Informatics, The Jackson Laboratory, Bar Harbor, Maine; Inbred Strains of Rats: ACI, www.informatics.jax.org/inbred_strains/rat/docs/ACI.shtml, 2018.

Rebuttal vii: *"Japanese Experiments: groups of rats were fed over a prolonged period, large amounts of chopped Comfrey, both root and leaf. The rats would not normally eat Comfrey so 'other food' was mixed with it.*
As far as what the other rats were fed on, the control diets, the data does not say, or what the food that was mixed with the Comfrey, but there was cancer in the control rats as well as those on Comfrey.
These experiments were completely out of balance and would prove nothing. If anyone will try such unbalanced types of action disaster will result. We see it every day, people who think they can live solely on canned foods, devoid of enzymes, or diets that are in excess in protein or fats. We see it constantly in the heart-artery blockage syndrome.
These experiments with Comfrey prove nothing more than a balanced diet is a must for good health."
-'Comfrey: School of Natural Science' by Dr. Frederick John Steed. Ph.D. (Natural Science), NSc., D.N., Australia, 2 pages, 1997. Founder of Natural Science correspondence course about healthy food and herbal medicine.

5. 'The Alkaloids of Symphytum x uplandicum (Russian Comfrey)' by Claude C.J. Culvenor, John A. Edgar, John L. Frahn and Leslie W. Smith, Division of Animal Health, CSIRO, Animal Health Research Laboratory, Parkville, Victoria, Australia; Austrailian Journal of Chemistry, Volume 33, No. 5, pages 1105-1113, **1980**.
"Eight pyrrolizidine alkaloids (PAs) were identified in the leaves of Symphytum x uplandicum.
They were echimidine, symphytine, lycopsamine, intermedine, and the 4 new alkaloids 7-acetyllycopsamine, 7-acetylintermedine, symlandine {7-angelyl-9-viridiflorylretronecine} and uplandicine {7-acetyl-9-echimidinylretronecine}.
> ***The total alkaloid content of leaves of S. x uplandicum is variable but usually low.*** *In the material studied here (voucher specimen No. CANB 286704 Australian National Herbarium, Canberra),* ***we found 0.15% dry weight in young leaves, 0.01% in mature leaves and 0.05% in the bulk collection of intermediate-sized leaves***.

Species of the family Boraginaceae are characterized by alkaloids in which the main esterifying acid is of the cc-isopropylbutyric type, commonly trachelanthic and viridifloric acids. These acids may occur in several diastereoisomeric forms, allowing a number of possible isomers (same chemical formula but different arrangement of atoms) for each gross structure.
Identification of these alkaloids requires careful attention to the physical separation of closely related isomers and the choice of methods that distinguish and characterize such isomers. *Direct spectral methods may be insufficient, although in principle they could be refined to achieve complete identifications."*
(See subsection 'Australia Bans Sale of Comfrey for Internal Use' in section 'Some Uses of Comfrey are Banned by Governments' {Chapter 31}.) (Comfrey does contain pyrrolizidine alkaloids. This is one of the early research papers on it.)

6. 'Structure and Toxicity of the Alkaloids of Russian Comfrey (Symphytum x uplandicum, Nyman), a Medicinal Herb and Item of Human Diet'by Claude C.J. Culvenor, M. Clarke, J.A. Edgar, J.L. Frahn, M.V. Jago, J.E. Peterson and L.W. Smith, CSIRO, Division of Animal Health, Victoria, Australia; Experientia, Switzerland, Volume 36, No. 4, pages 337-379, **1980**.
*"****Eight pyrrolizidine alkaloids of hepatotoxic (liver toxic) type have been indentified in leaves of Symphytum x uplandicum. The combined alkaloids exhibit chronic hepatotoxicity in rats using intraperitoneal injection.***
Extraction: *Young leaves of a clone of Russian Comfrey regarded as Bocking No.4 were immersed in methanol immediately*

after collection and chopped into small pieces after soaking for 1 day. The extraction was completed in the usual way, and the alkaloid fraction recovered after reduction of N-oxide with zinc dust.

14-Day-Old Rats:
Total Comfrey alkaloids, lasiocarpine, platyphylline or saline were given 3 times weekly i.p. (intraperitoneal injection = injected into abdominal cavity) to groups of hooded rats, both male and female, commencing when the rats were 14 days old (2 weeks).

Lethal Dose 50:
Echimidine has an acute LD50 of 200 mg/kg by i.p. (intraperitoneal) injection in rats and repeated injections at the rate of 20 mg/kg, thrice weekly, produce moderate chronic liver damage within 18 weeks. The LD50 of symphytine alkaloid for mice is approximately 300 mg/kg by i.p. injection. These considerations suggest that the acute LD50 of the total alkaloid fraction of Russian Comfrey should be about 550 mg/kg for rats."

Liver Disease:
These results suggest that under certain conditions of administration of pyrrolizidine alkaloids (PAs) impairment of liver function may be sufficient to cause death before the changes that characterize the histology of the chronic disease become apparent.

The morphological changes induced (created) directly by the alkaloids and the reparative (repair) responses to tissue damage either have insufficient time to develop or may possibly be suppressed by the continued presence of the alkaloid.
These deaths are clearly distinguishable from the acute deaths with massive hepatic necrosis (liver cell death) that occur within about 1 week of the administration of high doses of alkaloid. They are more closely related to the chronic disease and are indicative of selective interference with parenchymal (functional part) cell function that is sustained for periods of some weeks without loss of cell viability.

Humans Eating Alkaloids:
The susceptibility of humans to pyrrolizidine alkaloid poisoning is now well established although only one estimate is available of a low-level intake leading to toxicity.

In an outbreak involving wheat contaminated with seed of Heliotropium popovii subsp. gillianum (in Afghanistan), the intake of alkaloid which caused severe liver disease in 2 years was estimated at about 2 mg/person/day or approximately 30-40 microgram/kg/day.

The rate of ingestion of Russian Comfrey alkaloids may exceed this level since several leaves may be eaten per day and the level of alkaloid in the material available to us was approximately 1 mg/leaf.

Root and leaf of Symphytum officinale have recently been shown to be carcinogenic in rats when fed at rates down to 0.5% and 8% of diet, respectively."

(See subsection 'Australia Bans Sale of Comfrey for Internal Use' in section 'Some Uses of Comfrey are Banned by Governments' {Chapter 31}.)
(I wonder if any of the chemicals used in the processing ended up being injected into the rats.)
(Hepatoxic means causing injury to the liver. Intraperitoneal is injecting a substance into the abdominal cavity.)
(LD50 or median Lethal Dose 50 is the dose required to kill half of a tested population after a specified test duration. It is a general indicator of a substance's acute toxicity. I think this type of testing is awful. Our society should have better ways to test substances that don't involve torturing animals.)
(For more about the Afghanistan poisoning, see number 3.)
(For more about Symphytum officinale and carcinogenic effects, see number 4.)

Problems with Above Research (Rebuttal)

Rebuttal i. "The alkaloids were administered intra-peritoneally (I.P.) either as a single dose to two week old baby rats, or as multiple doses commencing at two weeks of age.
Although it may be argued that injection of the alkaloid into the body gives more certain delivery than oral administration, **it does not seem logical to test a drug administered in this way when it is normally eaten.**
Also it must be remembered that two week old rats, still unweaned, are expected to be more vulnerable than adult rats to the effect of pyrrolizidine alkaloids.
One must question the relevance of projecting results based on I.P. injections in baby rats to the effect on adult humans consuming the plant foliage (leaves).
At the very low concentration of alkaloid found in Comfrey leaves the chances of acuumulation sufficient to cause significant liver damage are very remote.
Many of the most valuable drugs in medicine are toxic in high concentrations."
-'The Safety of Comfrey: Report from the Henry Doubleday Research Association' by J.A. Pembery B.Sc., Special Advisor, Research Chemist, Aldgate Press: London, England, 20 pages, 1983, pages 12

Rebuttal ii. "**This is where the false extrapolations are made. Arguing from the intraperitoneal injection of alkaloids into 3-4 weeks-old baby rats to the ingestion of Comfrey leaves by animals and humans.**
Culvenor says that leaves (no size or age of leaf stated, an important factor that determines the percentage of alkaloid in the leaf) contain an estimated one milligram of total alkaloids."
-Comfrey: Nature's Healing Herb & Health Food by Andrew Hughes. Japan: Sanyusha Publishing Co, 1992, page 60.
(Extrapolation is estimating, beyond the original observation range, the value of a variable on the basis of its relationship with another variable.)

Rebuttal iii. *"In 'The Safety of Comfrey' (by Henry Doubleday Research Association) we read, 'One must question the relevance of projecting results based on intraperitoneal injections in baby rats to the effect on adult humans consuming the plant foliage.'*
The report shows that it would require the alkaloids from 19,880 leaves of 100 grams (.22 pounds) each (28 times the body weight of a man-sized rat) to produce the result produced in a baby rat, *even if such a projection were valid, which for a number of reasons is not.* **Rats do not necessarily predict the results in primates, man or monkey, especially when the usage of Comfrey by man is the whole leaf, not the extracted alkaloid.** *How unscientific can one get!"*
-Comfrey: Nature's Healing Herb & Health Food by Andrew Hughes. Japan: Sanyusha Publishing Co, 1992, page 68.

Rebuttal iv. *"Toxicity studies of Symphytum x uplandicum leaf alkaloids, adapted from Culvenor et al Experientia 36 377 (1980):*
To cause death the alkaloid dose for rats is 284 mg/kg (1 kg= 2.2 pounds) which is the equivalent human dose of 66,300 leaves per day. The dose needed for death is impossibly high.
How could anyone possibly consume 66,300 leaves at one sitting- more than a person's body weight in Comfrey. This equivalent oral human dose of leaves is based on the fact that a leaf consistently contains 0.33 mg of alkaloids whether it is old (large) or young (small).
Once something is in print in a journal it is often quoted in a superficial way as fact."
-Comfrey: Nature's Healing Herb & Health Food by Andrew Hughes. Japan: Sanyusha Publishing Co, 1992, page 73, 74.

Rebuttal v. *"This applies to all experiments with isolated alkaloids:* **to give alkaloids in isolation and injected intraperitoneally into animals, cannot reflect the effects in humans of the entire plant taken orally.**
This reflects one of the central tenets (principles) of herbal medicine, that an isolated chemical of a plant, while useful for certain indications, cannot define the action of the whole herb, where **the herb is more than the sum of the individual parts, its constituents working synergistically to create its healing effects."**
-'In Defence of Comfrey' by Margaret Whitelegg, MNIMH. Paper presented to 'Department of Health' and 'Ministry of Agriculture, Fisheries and Food', 'National Institute of Medical Herbalists', United Kingdom, January 1993. Published in European Journal of Herbal Medicine, Exeter, England, Volume 1, No. 1, pages 11-17, 1994.
  (Synergistic means the interaction of two or more substances that create a combined effect greater than the sum of their separate effects.)

Rebuttal vi. *"***Very young rats, 1-2 weeks of age, were highly sensitive to the alkaloids heliotrine and lasiocarpine,*** and male rats remained considerably more sensitive than adults to heliotrine until they were about 8 weeks old. Young rats have also been reported to be more susceptible than adults to retrorsine (Schoental, 1959), isatidine (Schoental, 1955) and monocrotaline (Schoental and Head, 1955)."*
-'Factors Affecting the Chronic Hepatotoxicity of Pyrrolizidine Alkaloids by Marjorie V. Jago, CSIRO (Commonwealth Scientific and Industrial Research Organisation) Division of Animal Health, Animal Health Research Laboratory, Melbourne, Victoria, Australia; The Journal of Pathology, Volume 105, No. 1, pages 1-11, 1971.
(In Culvenor's study the rats began treatment at 14 days old. It would have been more realistic to test adult rats.)

Rebuttal vii. *"Comfrey illustrates the fallacy (faulty reasoning) and intentional harmfulness of industrial 'science'. The 'Food and Drug Administration' here in the United States banned the use of Comfrey in commercial pharmaceuticals a couple years ago after concluding it is unsafe for human consumption.*
This farcical (ridiculous) analysis is based on a study in which the alkaloids found in the plant were extracted, concentrated in huge amounts, and injected into rats who then died of liver failure.
This is farce (absurd) first because rats are not humans, and therefore rat testing does not reflect human consumption of a given substance.
It is also a misleading conclusion because the alkaloids found in any plant (carrots, tomatoes, coffee, or potatoes, for example) will cause liver failure in mammals when heavily concentrated.
My point in mentioning all this is that Comfrey has an undeserved negative reputation based on some very shaky research."
-'Edible Medicinal and Utilitarian Plants, Volume I: Weeds and Common Plants' by Rowan Walking Wolf, Ph.D. and Harun Highmountain, M.D., published by Yggdrasil Distro, 2009.

Rebuttal viii. *"The Experiment on Rats: First I might say that we humans are not rats and do not act bodily the same. Our growth is so different - rats mature in weeks, humans in 20 years. Our needs are so far apart.*
The CSIRO extracted by means of chemicals the active ingredients from Comfrey. Then over a period of time injected this material into rats at a rate far above that which would be used by any person. The imbalance of materials brought a state of disease to the rats, some more than others."
-'Comfrey: School of Natural Science' by Dr. Frederick John Steed. Ph.D. (Natural Science), NSc., D.N., Australia, 2 pages, 1997. Founder of Natural Science correspondence course about healthy food and herbal medicine.

7. 'Toxic Pyrrolizidine Alkaloids in Comfrey' by A.R. Mattocks, Toxicology Research Unit, MRC Laboratories, Carshalton,

Surrey, England; The Lancet, London, England, Volume 316, No. 8204, pages 1136-1137, November 22 **1980**, a letter.
"We have measured Pyrrolizidine Alkaloids in the leaves of the hybrid Russian Comfrey S. x uplandicum variety Bocking No. 14, grown at Bocking, Essex (England), and kindly supplied by Mr. Lawrence D. Hills of the 'Henry Doubleday Research Association'.

*Twelve long-established plants were cut back to ground level in **early August**, 1979. The largest (over 18 cm long = 7 inch) and shortest (under 15 cm = 5.9 inch) were separately pooled, air-dried in the dark at 15-25 C (59-77 F) degrees, milled, and extracted exhaustively with methanol; the intermediate sized leaves were discarded.*
The alkaloids were isolated and analysed by spectrophotometry. Subsequently (after cutting in early August) the total new leaf growth from pairs of the same plants were harvested and analysed.

*Leaves picked after the first 2 weeks **(mid to late August) contained 0.115% of alkaloids (dry weight)**.*
By mid-September the level had fallen to 0.019% *and remained between 0.019% and 0.022% until mid-October when growth ceased.* ***Leaves cut April 13-28: alkaloids of the Bases + N-Oxides ranged from 0.048% to 0.222%.***
Total alkaloid levels were 0.003% in the largest Russian Comfrey leaves and 0.049% in the smallest: a 16-fold variation.
The total amount of alkaloids per leaf remains fairly constant as the leaf grows heavier, so the percentage of alkaloids falls and alkaloids in the form of N-oxides are progressively converted to bases.
The base and N-oxide forms of pyrrolizidine alkaloids are likely to have similar toxicity when taken by mouth.
The amounts of alkaloids found are consistent with the range of values previously reported.

Thus Culvenor et al. found 0.01-0.15% total alkaloids in different samples of dried Comfrey leaves, and Pederson up to 1.97 parts per 1000 (0.2%); however, the age and size of leaves were not recorded.
My measurement show that the highest akaloid (percent) levels are in small, young leaves, especially early in the season. Moreover, protein extracted from Comfrey should not be harmful: a sample of Comfrey protein supplied by Mr. Hills proved, as expected, to be completely free of alkaloids.
The external use of Comfrey preparations should not be hazardous since the alkaloids are converted to toxic metabolites by liver enzymes only after being ingested."

8. **'Comfrey and Liver Damage'** by James N. Roitman, Natural Products Chemistry, Western Regional Research Center, United States Department of Agriculture, Berkeley, California; The Lancet, Volume 317, Issue 8226, page 944, April 25 **1981**, a letter.
"We have analysed a series of dried commercial Comfrey samples (12 root, 12 leaf) and a fresh sample of mature Symphytum asperum (Prickly Comfrey) leaf for pyrrolizidine alkaloids (alkaloid + alkaloid N-oxide).
The alkaloid content of all the dried leaf samples was below the limit detectable by NMR (Nuclear Magnetic Resonance) (0.005%). *The fresh leaf sample of S. asperum contained 0.01%, and the 12 root samples ranged from 0.14 - 0.37%, all calculated on a dry weight basis.*
To estimate how much alkaloid is consumed by drinking a cup of Comfrey root tea, *an infusion (tea) was prepared according to directions on the package of a locally purchased sample; 250 ml (1.05 cup) boiling water was added to 1.5 teaspoonsful (8.7 grams) of chopped root. After separation of the gelatinous (jellylike) residue by centrifugation, the supernatant (clear) liquid was worked up and analysed by NMR;* **it contained 8.5 mg alkaloid. Were some or all of the gelatinous residue consumed, the amount of alkaloid could be as much as 26 mg/cup."**

(Except for the Symphytum asperum, the species of Comfrey for the other tests was not given.)
(It woud have been better if the gelatinous residue from Comfrey tea had been tested, rather than being guessed at.)

Problems with Above Research (Rebuttal)

Rebuttal i. *"When Roitman made up Comfrey 'root tea', he analysed the clear solution and found that 33% of the available alkaloid had dissolved, being 8.5 mg.*
He then referred to a paper by Mohabbat et al* published in 'The Lancet' in 1976, which discusses the outbreak of veno occlusive disease in north western Afghanistan, and the result of eating wheat flour contaminated with Heliotropium seeds. *Taking the figure of 2 mg consumption of alkaloid per day, as calculated from one set of figures in the paper, he concludes that Comfrey root tea presents an unacceptable risk if it contains at least four times as much alkaloid per cup.*
From this he also concludes that the leaves, although containing a much lower level of alkaloid, he himself having been unable to detect any at all in the leaf tea, also presents an unacceptable risk.
Two contrasting figures occur in this paper by Mohabbat et al, *one leading to a figure of 2 mg per day consumption of alkaloid, and the other figure for contaminated wheat leading to a figure of 1300 mg alkaloids per day."*
-'The Safety of Comfrey: Report from the Henry Doubleday Research Association' by J.A. Pembery B.Sc., Special Advisor, Research Chemist, Aldgate Press: London, England, 20 pages, 1983, page 17   (* See number 3.)

Rebuttal ii. *"In order to justify his warning about leaves, even though he could detect no alkaloid in leaf tea, Roitman quoted the results of Hirono et al*, although the rats in that study must have consumed several times their own bodyweight of Comfrey leaves before liver damage occurred.*
How many cups of Comfrey leaf tea must a person consume in order to consume the equivalent of many times their own bodyweight of leaves? In the appendix to 'Comfrey: Fodder, Food & Remedy' by L.D. Hills, Dr. Denys Long (D. B. Long, Ph.D.) has calculated that **it would take 140 years drinking four cups (0.94 liters) of Comfrey leaf tea a day, in order to run the risk of alkaloid poisoning.**"

-'The Safety of Comfrey: Report from the Henry Doubleday Research Association' by J.A. Pembery B.Sc., Special Advisor, Research Chemist, Aldgate Press: London, England, 20 pages, 1983, page 17   (* See number 4.)

9. 'Mutagenic Effects of Aqueous Extracts of Symphytum Officinale L. and of Its Alkaloidal Fractions' by Mirostawa Furmanowa, Joanna Guzewska and Bozena Betdowska, Department of Pharmaceutical Botany, The Medical Academy, Warsaw, Poland; Journal of Applied Toxicology, Volume 3, No. 3, pages 127-130, **1983**.

"*Aqueous (water) solutions of three alkaloid fractions obtained from infusions of Symphytum officinale L. root* were tested for their antimitotic (stops cell division) and mutagenic (gene mutation) activity in meristematic cells of the lateral roots of Vicia fuba L., var minor (field bean).

Lasiocarpine, a proven carcinogen, served as a positive control. **Mutagenic effects were induced (created)** by lasiocarpine, **by the alkaloidal fraction I (41.9 mg) and by diluted infusions from Radix symphyti (Comfrey root).** Alkaloidal fraction II was 19.0 mg.

Fraction III (16.9 mg) had only antimitotic effect. *The lasiocarpine content of roots of Symphytum officinale (determined by thin-layer chromatography) was 0.0058%."*

(Infusion is when a tea bag or ball is put in a cup of hot water and allowed to steep for a few minutes before drinking.)
(Meristem is undifferentiated cell tissue in plants, found where growth takes place.)

10. 'Hepatic Venocclusive Disease Associated with the Consumption of Pyrrolizidine Containing Dietary Supplements' by Paul M. Ridker, Seitaro Ohkuma, William V. McDermott, Charles Trey and Ryan J. Huxtable, (Boston, Massachusetts and Tucson, Arizona), Gastroenterology, Volume 88, Number 4, pages 1050-1054, **1985**.

"*Venocclusive disease, a form of Budd-Chiari syndrome, was diagnosed in a 49-year-old woman.* The patient had portal hypertension associated with obliteration (destruction) of the smaller hepatic venules. A liver biopsy specimen showed centrilobular necrosis (tissue death) and congestion.

Analysis of food supplements the woman regularly consumed showed the presence of pyrrolizidine alkaloids. The major source was a powder purporting to contain ground Comfrey root (Symphytum species). *We calculated that during the 6 months before the woman was hospitalized, she had consumed a minimum of 85 mg of pyrrolizidine alkaloids (15 micrograms/kg body weight/day).*

The clinical and analytic findings were consistent with chronic pyrrolizidine intoxication, indicating that low-level, chronic exposure to such alkaloids can cause venocclusive disease. For the 6 months before admission she had consumed 1 quart/day (0.94 liters) of a herbal tea known as MU-16 (unknown herbs).

Each packet of MU-16 tea contained eight 6-gram (0.21 ounce) tea bags. The earlier sample, obtained from the patient, contained 80 nmol pyrrolizidine/gram dry weight of tea. A later sample, purchased by us, **contained 23.9 nmol/gram pyrrolizidine alkaloids, and 3.0 nmol/gram pyrrolizidine N-oxides. Our analyses suggest that she had, therefore, consumed between 12.9 and 38.4 micro-mol of total pyrrolizidines (from MU-16 tea).** *This is equivalent to between 0.49 to 1.45 microgram/kg body weight/day.*

In addition, for the 4 months before admission, she had taken two capsules of 'Comfrey-pepsin pills' with each meal (6 capsules a day).

Each Comfrey-pepsin pill contained 400 mg of a white powder. The powder **contained 107 nmol/gram pyrrolizidine alkaloids and 757 nmol/gram pyrrolizidine N-oxides.** Her daily consumption of pyrrolizidines from Comfrey-pepsin was, therefore, 2.07 micro-mol/day (14.1 microgram/kg body weight/day).

Over a 4-month period, she had consumed around 250 micro-mol, or 1700 microgram/kg body weight, of total *pyrrolizidines from the Comfrey-pepsin capsules.*

The total pyrrolizidine consumption we can establish for this patient is relatively low.
It is possible that she had other sources of exposure, and it is probable that she had been consuming pyrrolizidine containing supplements for longer than the period we could establish."

(Portal hypertension is an increase in the blood pressure in a system of veins called the portal venous system. Veins coming from the stomach, intestine, spleen, and pancreas merge into the portal vein, which then branches into smaller vessels and travels through the liver.)

Problems with Above Research (Rebuttal)

Rebuttal i: *"For six months, she had taken MU-16, which also contained Pyrrolizidine Alkaloids.* **The researchers failed to investigate. The Pyrrolizidine Alkaloids in the MU-16 tea could have been the toxic Macrocyclic diesters or another alkaloid.** *Yet, Comfrey was implicated in this incident. There is also evidence that Comfrey-pepsin tablets contain higher amounts of Pyrrolizidine Alkaloids than other Comfrey containing products.* **Comfrey was not the only source of purrolizidine alkaloids in her diet."**
-'Comfrey, Symphytum Officinale: The Much Debated Healer' by Mimi Kamp, medicinal herbalist, botanical illustrator and photographer from southwestern United States, probably 2017.

Rebuttal ii. **"The authors admit: 'It is possible that she had other sources of exposure.'**
Yet conclude 'To our knowledge, this is the first report of veno-occlusive disease in any human after the use of a preparation claiming to be made from Comfrey."

-'In Defence of Comfrey' by Margaret Whitelegg, MNIMH. Paper presented to 'Department of Health' and 'Ministry of Agriculture, Fisheries and Food', 'National Institute of Medical Herbalists', United Kingdom, January 1993. Published in European Journal of Herbal Medicine, Exeter, England, Volume 1, No. 1, pages 11-17, 1994.

11. 'Toxicity of Comfrey-Pepsin Preparations' by Ryan J. Huxtable, Jurg Luthy and Ulrich Zweifel (Tucson, Arizona and Switzerland); The New England Journal of Medicine, Volume 315, No. 17, page 1095, October 23 **1986**, letter.
"To the Editor: **We wish to draw attention to the health risks of a food supplement containing Symphytum that is widely available in the United States. Numerous brands of Comfrey–pepsin capsules and tablets are sold** *in herbal and health-food stores as a digestive aid. Users take them daily over long periods or even permanently.*
We have analyzed the pyrrolizidine content of two brands of Comfrey-pepsin capsules, one brand purporting to be prepared from the leaves and the other from the roots of the Comfrey plant.
> *The brand made from the roots contained 520 to 590 mg per capsule, with a pyrrolizidine content of 400 mg/kg and an N-oxide content of 2500 mg/kg (total pyrrolizidines, 2900 mg/kg). The brand made from the leaves contained 40 mg/kg of free alkaloids and 230 mg/kg of N-oxides (total pyrrolizidines, 270 mg/kg).*

These alkaloids are hepatotoxic, and some of them are demonstrated carcinogens.
A person consuming two capsules per meal for six months would receive a total of 162 mg of alkaloids from the 'leaf' preparation and 1740 mg from the 'root' preparation."

12. 'Veno-Occlusive Disease of the Liver Secondary to Ingestion of Comfrey' by C.F.M. Weston, B.T. Cooper, J.D. Davies and D.F. Levine (Bristol and Cornwall, England), British Medical Journal, Volume 295, page 183, July 18 **1987**.
*"***Case report: A 13 year old boy was admitted in July 1986 for investigation of hepatomegaly (enlarged liver) and ascites (fluid in abdomen).**
> *Three years earlier Crohn's disease (inflammatory bowel) had been diagnosed from radiographs (X-rays) showing consistent changes in the terminal ileum (part of small intestine) and colon and from histological (tissue) studies of the colon. He was treated with prednisolone (synthetic glucocorticoid) and sulphasalazine/sulfasalazine (anti-inflammatory drug) with benefit.*

At his parents' request these drugs were discontinued, and he was treated with acupuncture and Comfrey root, prescribed by a naturopath. *Up to 1986 he had been regularly (for 2-3 years) given a herbal tea containing Comfrey leaf. The exact quantities of leaves given and frequency of administration are unknown.*
An exacerbation (worsening) of his inflammatory bowel disease in 1984 required a further course of prednisolone. *In June 1986 he presented with fatigue, diarrhoea, and weight loss, and a few weeks later developed fever, abdominal pain, and swelling. He was taking prednisolone and sulphasalazine. He had never taken azathioprine (immunosuppressive drug).*
> *Percutaneous (through skin) liver biopsy showed the thrombotic (stationary blood clot) variant of hepatic veno-occlusive disease (some of the small veins in the liver are obstructed). He was treated with spironolactone (reduces fluid build up), salt restriction, and bed rest with a good response.*
> *His bowel disease remained relatively inactive with treatment with prednisolone and sulphasalazine; at the time of writing he was back at school and tolerably well on his medication.*

Risk Factors:
Malnutrition and poor health may be risk factors in Jamaica* (previous Pyrrolizidine Alkaloid poisoning, but not from Comfrey) so our patient may have been susceptible to hepatic veno-occlusive disease from Comfrey because of his underlying inflammatory bowel diseases. *Major hepatic vein thrombosis (blood clots) but not veno-occlusive disease has been described in patients with colitis**.*
Known or suspected causes of hepatic veno-occlusive disease are *systemic lupus erythematosus, alcoholic hepatitis, immune deficiency, azathioprine (in renal transplant recipients), radiotherapy, chemotherapy (especially in bone marrow transplant recipients), and pyrrolizidine alkaloids.*
> **Only two cases of hepatic veno-occlusive disease as a result of pyrrolizidine alkaloid ingestion have been described in Britain, however, and both patients had ingested imported herbal teas.**

Comfrey and Alkaloids:
> **The Common Comfrey, Symphytum officinale,** *a native British plant, contains at least nine potentially hepatotoxic pyrrolizidine alkaloids in its leaves and roots.* **These alkaloids are less toxic than those in other plants** *-for example, senecios (a genus of the daisy family Asteraceae)- which may explain why only a few cases of hepatic veno-occlusive disease caused by ingestion of Comfrey are known (G. Nicholson and C.C.J. Culvenor, personal communications), and only one has been published***. This second published case is the first to result from a native British plant."*

(* -'Veno-Occlusive Disease of the Liver' by K.L. Stuart and G. Bras, Departments of Medicine and Pathology, University College of the West Indies, Jamaica; Quarterly Journal of Medicine, New Series XXVI, No. 103, pages 291-315, July 1957.)
(** -Pathology of the Liver edited by R.N.M. McSween, P.P. Anthony and P.J. Scheuer. Edinburgh, Scotland: Churchill Livingstone, 1979. Article: Vascular Disorders by G. Bras and K.H. Brandt, pages 315-334.)
(*** -'Hepatic Venocclusive Disease Associated with the Consumption of Pyrrolizidine Containing Dietary Supplements' by Ridker, Ohkuma, McDermott, Trey and Huxtable, Gastroenterology Volume 88, Number 4, pages 1050-1054, 1985. This report is discussed in this list of various research.)
(Prednisolone is a corticosteroid hormone that is also made by the adrenal glands. It is used to treat arthritis, immune system disorders, skin/eye problems, breathing problems, and allergies. It slows the immune system's response so

symptoms such as pain and swelling are reduced.)
(Sulphasalazine/sulfasalazine is an anti-Inflammatory drug used to treat ulcerative colitis and rheumatoid arthritis. It does not cure the problem but it does decrease symptoms.)
(The authors do not give the names of the reports about hepatic veno-occlusive disease in Britain from herbal teas. One possibility is: 'Herbal Tea Induced Hepatic Veno-Occlusive Disease: Quantification of Toxic Alkaloid Exposure in Adults' by Kumana, Ng, Lin, Ko, W and Todd, Gut {British Society of Gastroenterology, British Medical Association}, London, England, Volume 26, No. 1, pages 101-104, January 1985. Four young Chinese women took daily doses of an unidentified 'Indian' herbal tea as treatment for psoriasis.
Another possibility is: 'Hepatic Venocclusive Disease Associated with the Consumption of Pyrrolizidine Containing Dietary Supplements' by Ridker, Ohkuma, McDermott, Trey and Huxtable; Gastroenterology, Volume 88, Number 4, pages 1050-1054, 1985. These reports are discussed in this list of various research.)

Problems with Above Research (Rebuttal)

Rebuttal i: *"Although at the time all other known factors were ruled out, further investigation shows that the medications he was taking could have been the major culprits (causes) in this case. Prednisolone can cause abdominal pain, gastrointestinal upset, and damage the liver. This liver damage (brought on through hypokalimic alkalosis) would have weakened his liver and made him more susceptible to liver toxicity. Sulfasalazine can case headache, nausea, vomiting, gastric distress, and hepatitis. This indicates that people with a history of liver toxicity should not use Prednisolone and Sulfasalazine together."*
-'Comfrey, Symphytum Officinale: The Much Debated Healer' by Mimi Kamp, medicinal herbalist, botanical illustrator and photographer from southwestern United States, probably 2017.

Rebuttal ii: *"The authors concede (admit) that the patient may have been more susceptible to hepatic veno-occlusive disease because of underlying bowel disease causing malnutrition, but they conclude that 'the only possible causal factor in this patient was Comfrey'. The drugs are not considered as possible factors."*
-'In Defence of Comfrey' by Margaret Whitelegg, MNIMH. Paper presented to 'Department of Health' and 'Ministry of Agriculture, Fisheries and Food', 'National Institute of Medical Herbalists', United Kingdom, January 1993. Published in European Journal of Herbal Medicine, Exeter, England, Volume 1, No. 1, pages 11-17, 1994.

13. 'Comfrey: Assessing the Low-Dose Health Risk' by P.J. Abbott, Toxicology Unit, Department of Community Services and Health, Canberra, Australia; Medical Journal of Australia, Volume 149, No. 11-12, pages 678-682, **1988**.
"While minimal direct evidence exists of liver toxicity in humans as a result of the consumption of Comfrey, this may be due to a long latency period and subclinical symptoms, or to a lack of recognition of the symptoms of veno-occlusive disease, particularly if these are mild.
> *From the data that are available, it must be concluded that the potential exists for pyrrolizidine alkaloid-induced liver toxicity after long-term exposure to Comfrey, albeit (although) with mild effects, even at modest levels of intake.*

On the basis of the present data, it is not possible to determine a 'safe' level of long-term Comfrey exposure to avoid veno-occlusive disease.
> *An additional concern is the greater susceptibility of children to pyrrolizidine alkaloid-induced liver damage, as indicated by the relatively-high number of cases of cirrhosis to be reported in this age-group in exposed communities. The 'safe' dosage level in this age-group is likely to be much lower.*

With regard to carcinogenicity (cancer causing), although the evidence at present suggests that Comfrey should be regarded as a potential human carcinogen, it seems likely that the normally-low levels of exposure, the risk would be very small.
On the basis of the data that are available currently, the small but significant long term risk that is associated with the consumption of Comfrey justifies the need to limit its intake. *This is being achieved by controls under various state Poisons Acts (of Australia), but also requires further education on the potential dangers of naturally-occurring chemicals of plant origin."*
> (Latency is a time interval between the stimulation and response. In other words, it is a time delay between the cause and the effect of some physical change in the system.)
> (Subclinical, when referring to a disease, means it is not severe enough to give readily observable symptoms.)

Problems with Above Research (Rebuttal)

Rebuttal i: *"A research paper on the Comfrey (Symphytum) species appears in which the author (Abbott, 1988) demonstrates the dangerous presence of pyrrolizidine alkaloids in Comfrey and the reasons for access to the 'poisonous' herb being restricted to pharmacists and medical practitioners.*
Although the author does acknowledge that: 'It seems likely that at the normally low levels of exposure, the risk of taking Comfrey would be very small', *this passing reference to dosage appropriateness is lost to the discourse on the dangers of 'naturally-occurring chemicals which are found in plants':*
> *A discourse which employs the languages of pharmacology and phytochemistry to convey (give) the toxicity of pyrrolizidine alkaloids, with only two international cases and one Australian case of actual suspected Comfrey poisoning cited in the article.*

The primary basis for the Abbott article on Comfrey is to support the argument for what has been perceived by

herbalists and naturopaths as a drastic regulatory move which prevents prescription of Comfrey (and other such toxic plant substances) by Herbal Medicine practitioners or lay usage.
As Mendel (*2001) points out in her research, the 'promotion of science as the ultimate authority' is a key rhetorical tool in elite medical discourse in Australia."
-'Herbal Medicine and Risk Constructions: Representations in Australian Print Media' by Monique Renae Lewis, Media Studies Ph.D. Thesis, Southern Cross University, Lismore, New South Wales, Australia, December 2011, page 199.
(* -'Risk and Evidence: Political and Philosophical Hegemonies in Australian Health Care' by J. Mendel, School of Social and Workplace Development, Southern Cross University, Lismore, Australia, 2001.)

14. 'Hepatic Veno-Occlusive Disease in Newborn Infant of a Woman Drinking Herbal Tea' by Michel Roulet MD, Ricardo Laurini MD, Laurent Rivier PhD, and Andre Calame MD, Departments of Pediatrics and Pathology and the Institute of Legal Medicine, University of Lausanne, Switzerland; The Journal of Pediatrics, Volume 112, No. 3, pages 433-436, March **1988**.
"*We report a newborn infant with fatal hepatic vaso-occlusive disease. Pyrrolizidine alkaloids were identified in the herbal medicine bought by the mother at a pharmacy and consumed daily during the entire pregnancy.*
Veno-occlusive disease of the liver is a rare disorder seen mainly in children who have consumed toxic pyrrolizidine alkaloids. **The hepatotoxicity seems to depend on their chemical structure, on the total dose ingested, and in particular on the susceptibility of the individual.**
In our patient the Pyrrolizidine Alkaloid concentration was relatively low (0.60 mg/kg dry weight), but the maternal consumption was prolonged. **The daily consumption during the entire pregnancy of a single cup of the tea analyzed here represents a toxic amount of senecionine (a Pyrrolizidine Alkaloid)**, because the effects of pyrrolizidine alkaloids in the body are cumulative.
The absence of hepatic (liver) damage in the mother suggests that the fetal liver may be more sensitive. This hypothesis is supported by the fact that young animals are known to be more sensitive, and most of the cases of epidemic poisoning have been in infants and children.
Coltsfoot Tea:
In our search for the source, the exact composition of the incriminated tea was obtained from the manufacturer; **of the 10 different plants in this preparation, leaves of Tussilago farfara (horsefoot, coltsfoot, or coughwort) made up 9% (weight/weight). This plant contains various pyrrolizidine alkaloids, such as senecionine, senkirkine, and tussilagine.** Its long-term administration is responsible for liver carcinoma in animals, and its use as herbal tea in humans is discouraged. It is classified by United States Food and Drug Administration as an herb of undefined safety.
Toxic agents other than pyrrolizidine alkaloids, especially in this mother with a previous history of use of cannabis and hallucinogenic mushrooms, must be considered. However, mycotoxins lead to another kind of hepatic lesion."

Comment 1: This report never mentions Comfrey or Symphytum, yet it is referred to in some articles as a reason why Comfrey should not be used. However, it is safer to not use Comfrey internally during pregnancy or breast feeding.
The Pyrrolizidine Alkaloid senecionine is not found in Comfrey.

Comment 2: "*Most cases of pyrrolizidine alkaloid poisoning in the scientific literature have involved third-world epidemics among people who consumed contaminated grain over a long period. Most clinical cases have involved Senecio, Heliotropium, and Crotalaria species.*
Infant Case: A 5-day old infant with jaundice, liver enlargement and abdominal swelling. Liver biopsy revealed veno-occlusive disease. Patient died at 27 days of life.
Mother consumed an herbal expectorant tea daily throughout pregnancy. Tea purported to contain 9% coltsfoot (Tussilago farfara) by weight.
They isolated senecionine from the tea sample, but not senkirkine. This calls into question the identification of Tussilago. Coltsfoot was not clearly enough identified as the offending agent.
The absence of senkirkine suggests that coltsfoot was not present, or at least was not the cause of death in the infant.
Senecionine, one of the most toxic of the pyrrolizidine alkaloids, is present in a wide variety of plants, including Petacites spp. (species) a possible adulterant of commercial coltsfoot."
-'Symphytum: Comfrey, Coltsfoot, and Pyrrolizidine Alkaloids' by Paul Bergner, Medical Herbalism: Journal for the Clinical Practitioner, Oregon, Volume 1, No. 1, pages 1, 3-5, (around 1990).

15. 'Comfrey Herb Tea-Induced Hepatic Veno-Occlusive Disease' by Nancy Bach M.D., Swan Thung M.D., and Fenton Schaffner M.D., The American Journal of Medicine, Volume 87, Issue 1, pages 97-99, July **1989**.
"*A 47-year-old non-alcoholic woman was well until 1978 when she began to have vague complaints of abdominal pain, fatigue, and allergies. She consulted a homeopathic doctor who recommended Comfrey tea. She began consuming as many as 10 cups (2.3 liters) of tea per day in addition to taking Comfrey pills by the handful, which continued for more than one year.*
Four years later, in 1982, her serum (blood) aminotransferase activities were noted to be twice the normal values. By 1986, she had developed ascites and a workup of her liver disease was undertaken. Results of paracentesis, computerized tomographic scan, and magnetic resonance imaging of the abdomen, gynecologic examination, and sigmoidoscopy were all negative except for the presence of ascites. **She was hospitalized in December 1986 for massive ascites, hyponatremia (low blood sodium), and confusion.**

This case emphasizes the importance of questioning patients about all medications including home remedies, herbal teas, vitamins, and pills, and warning them of potential risks of these products. Products sold in health food stores are not regulated by the government in terms of either safety or efficacy (effectiveness), and thus are a potential health hazard. In particular, consumption of Comfrey and it's teas should be avoided."

(Aminotransferase is an enzyme that catalyzes a reaction between an amino acid and alpha-keto acid. It helps synthesize amino acids to form proteins.)   (Ascites is abnormal buildup of fluid in the abdomen.)

Article Agreeing with Above Research:

"Bach et al (Am J Med 1989; 87: 97-99) are to be commended for recognizing Comfrey herb tea as a source of toxic pyrrolizidine alkaloids in a patient with hepatic veno-occlusive disease.
In our original description of this entity in 1985*, an extensive biochemical analysis for pyrrolizidine alkaloids was undertaken to precisely define the dietary source of these hepatotoxic compounds.
Because our patient proved to have two sources of pyrrolizidine alkaloids- MU16 herb tea and Comfrey-pepsin capsules- we** and others*** have argued that the diagnosis of hepatic veno-occlusive disease should not be made without chemical confirmation of the presence of toxic alkaloids in herbs that a patient has consumed.
Elegant studies by Huxtable et al**** and Awang*****, however, have clearly demonstrated the presence of several different pyrrolizidine alkaloids in Comfrey samples, often at surprisingly high levels.
We now believe, as case reports of Comfrey-induced hepatic veno-occlusive disease become more common (Am J Med 1989; 87: 97-99, *,******), that the diagnosis can be made without chemical confirmation when the herb in question is a known source of these toxic alkaloids."
-'<u>Hepatotoxicity Due to Comfrey Herb Tea: Letter to the Editor</u>' by Paul M. Ridker, M.D. (Brigham and Women's Hospital, Harvard Medical School, Boston, Massachusetts) and William V. McDermott, M.D. (New England Deaconess Hospital, Harvard Medical School, Boston, Massachusetts); The American Journal of Medicine, Volume 87, page 701, December 1989.
(* This is number 10 above. -'Hepatic Venocclusive Disease Associated with the Consumption of Pyrrolizidine Containing Dietary Supplements' by Paul M. Ridker, Seitaro Ohkuma, William V. McDermott, Charles Trey and Ryan J. Huxtable, {Boston, Massachusetts and Tucson, Arizona}, Gastroenterology, Vol 88, Number 4, pages 1050-1054, 1985.)
(** -'Comfrey Herb Tea and Hepatic Veno-Occulusive Disease' by Paul M. Ridker and William V. McDermott, {Department of Medicine, Brigham and Women's Hospital, and Department of Surgery, New England Deaconess Hospital, Harvard Medical School, Boston, Massachusetts}; The Lancet, London, England, Volume 333, Issue 8639, pages 657-658, March 25 1989.)
(*** -'New Aspects of the Toxicology and Pharmacology of Pyrrolizidine Alkaloids' by Ryan J. Huxtable, Department of Pharmacology, University of Arizona, Tucson, Arizona; General Pharmacology: The Vascular System, Volume 10, Issue 3, pages 159-167, 1979.)
(**** See number 11 above. -'Toxicity of Comfrey-Pepsin Preparations' by Huxtable, Luthy and Zweifel, The New England Journal of Medicine, Volume 315, No. 17, page 1095, October 23 1986.)
(***** -'Comfrey' or 'Herbal Medicine: Comfrey' by D.V.C. Awang, Health Protection Branch, Canada; Canadian Pharmaceutical Journal, Volume 120, pages 101-104, February 1987.)
(****** See number 12 above. -'Veno-Occlusive Disease of the Liver Secondary to Ingestion of Comfrey' by C.F.M. Weston, B.T. Cooper, J.D. Davies and D.F. Levine (Bristol and Cornwall, England), British Medical Journal, Volume 295, page 183, July 18 1987.)

Problems with Above Research (Rebuttal)

Rebuttal i: *"We have read with interest the letter of Ridker and McDermott (Am J Med 1989; 87: 701) on the hepatotoxicity of herbal teas made from Comfrey. However, we disagree strongly with their conclusion that a diagnosis of pyrrolizidine poisoning can be made without chemical confirmation.* Such a practice would return us to the situation of dangerous confusion that has so often surrounded the question of herbal safety.
Even if one limits the question to Comfrey (Symphyturn species), four major objections to the Ridker-McDermott suggestion can be made.
1. Comfrey Labels May Not Be Accurate
  First, several plant species and varieties of Symphytum are involved in the Comfrey trade. *A recent Canadian study* that examined 13 commercial Comfrey products revealed the unreliability of commercial labeling of herbal products. Although all products were labeled either simply 'Comfrey' or 'Comfrey Symphyturn officinale', six were shown not to be S. officinale (Common Comfrey) because they contained the pyrrolizidine alkaloid echimidine, indicative of Symphytum asperum and its hybrids, such as Symphytum x uplandicum (Russian Comfrey); Russian Comfrey has long been recognized to be the most common commercial Comfrey in Britain**.*
  In view of differing alkaloidal profiles and different pyrrolizidine alkaloid toxicities, a simple 'Comfrey' label is obviously not a sufficient indication of toxic potential.
2. Herbal Products Can Be Contaminated
  Second, the poor quality control in the herbal industry *frequently leads to adulteration of plant products. Poisonings have occurred as a result of consumption of deadly nightshade (Atropa belladonna) sold*

commercially as Comfrey***.

3. Botanical Identification Can Be Difficult

Third, with herbal products containing chopped or powdered roots, leaves, and herbage, or mixtures of substances, botanical identification is nearly impossible. *This situation holds, for example, with the commonly used Comfrey-pepsin capsules, which probably present the highest current risk of pyrrolizidine poisoning in the United States.*

4. Alkaloids Vary Depending on Individual Plant

Fourth, even within a defined botanical species, there is marked variation in alkaloid content, *depending on the time of year the material was collected, the location, the conditions of growth, maturity, and the part of the plant used. Thus, with S. officinale (Common Comfrey), there is an up to 10-fold greater level of alkaloid content in root as compared to leaf, while young leaves of S. x uplandicum (Russian Comfrey) may contain up to 15 times more alkaloid than old leaves****."*

-'Pyrrolizidine Poisoning: Letter to the Editor' by Ryan J.Huxtable, Ph.D. (College of Medicine, University of Arizona, Tucson, Arizona) and Dennis V.C. Awang, Ph.D. (Bureau of Drug Research, Health and Welfare Canada, Ottawa, Ontario, Canada; The American Journal of Medicine, Volume 89, page 547, October 1990.

(* -'Chemotaxonomy and the Regulation of Commercial Plant Products: Identity and Standardization' by D.V.C., Presentation to the 57th Congress of the French Canadian Association for the Advancement of the Sciences, Montreal, Quebec, Canada, May 15-19 1989.)

(** -Flora of the British Isles, Second Edition by A.R. Clapham, T.G. Tutin and E.F. Warburg. England; Cambridge University Press, 1962, first edition 1952.)

(*** -'Clinical Curio: Hallucinations in Elderly Tea Drinkers' by E.J. Galizia, Consultant Anaesthetist, Bath, England; British Medical Journal, Volume 287, Issue 6397, page 979, October 1 1983.)

(**** See number 6 above. -'Structure and Toxicity of the Alkaloids of Russian Comfrey {Symphytum x uplandicum, Nyman}, a Medicinal Herb and Item of Human Diet'by Claude C.J. Culvenor, M. Clarke, J.A. Edgar, J.L. Frahn, M.V. Jago, J.E. Peterson and L.W. Smith, CSIRO, Division of Animal Health, Victoria, Australia; Experientia, Switzerland, Volume 36, No. 4, pages 337-379, 1980.)

Rebuttal ii: *"It was not determined exactly what species of Symphytum or indeed, if it was even a Symphytum species at all through current tests. She also had symptoms of abdominal pain and fatigue before using Comfrey. These are both symptoms of veno-occlusive disease that she may have had before consuming the plant. This is not determined in this case."*

-'Comfrey, Symphytum Officinale: The Much Debated Healer' by Mimi Kamp, medicinal herbalist, botanical illustrator and photographer from southwestern United States, probably 2017.

Rebuttal iii. *"In this case it is not clear what the woman first presented with, nor how long before the development of the liver abnormalities she stopped taking Comfrey. In a case of addiction such as this, deleterious effects are nevertheless counted as condemning the plant taken in normal, moderate doses."*

-'In Defence of Comfrey' by Margaret Whitelegg, MNIMH. Paper presented to 'Department of Health' and 'Ministry of Agriculture, Fisheries and Food', 'National Institute of Medical Herbalists', United Kingdom, January 1993. Published in European Journal of Herbal Medicine, Exeter, England, Volume 1, No. 1, pages 11-17, 1994.

16. 'Studies on the Effect of an Alkaloid Extract of Symphytum Officinale on Human Lymphocyte Cultures' by Cornelia Behninger, Gudrun Abel, Erhard Roder, Viktor Neuberger and Waltraud Goggelinann, Federal Republic of Germany; Planta Medica: Society for Medicinal Plant and Natural Product Research, Volume 55, pages 518-522, **1989**. All is in German except the abstract is also in English.

*"**An alkaloid extract of Symphytum officinale was investigated for its chromosome-damaging effect in human lymphocytes (white blood cells) in vitro.***

*In concentrations of 1.4 microgram/ml and 14 microgram/ml the alkaloids had no effect, **in concentrations of 140 microgram/ml and 1400 microgram/ml the alkaloids induced Sister Chromatid Exchanges (SCE) as well as chromosome aberrations (abnormalities).***

Additionally, the influence of rat liver enzymes (S9) was tested. The SCE-inducing capacity and the clastogenic (mutagenic) effect of Symphytum alkaloids was increased by simultaneous application of S9-Mix."

(Sister Chromatid Exchange is the exchange of genetic material. Sister chromatids are identical copies formed by DNA replication of a chromosome. The rate of exchange is used as a mutagenic test of substances.)

17. 'Comfrey Herb Tea and Hepatic Veno-Occulusive Disease' by Paul M. Ridker and William V. McDermott, (Department of Medicine, Brigham and Women's Hospital, and Department of Surgery, New England Deaconess Hospital, Harvard Medical School, Boston, Massachusetts); The Lancet, London, England, Volume 333, Issue 8639, pages 657-658, March 25 **1989**.

*"**In at least two patients the commercially available herb Comfrey was thought to be the source of hepatotoxic pyrrolizidine alkaloids.*,*****

Hepatotoxicity is related to host susceptibility, total ingested dose, and route of exposure. Dose information is difficult to obtain because techniques of harvesting and brewing greatly affect alkaloid content, as does the use of either leaves or roots.

Carcinogenic activity is also reported for pyrrolizidine-containing compounds: **Comfrey, in particular, has been associated with hepatocellular tumours in rats owing primarily to the symphytine content.*****

As there is no method available to measure pyrrolizidine metabolites in body fluids (as of 1989), **the diagnosis of pyrrolizidine alkaloid poisoning** can only be made by exclusion of veno-occlusive disease in all patients with hepatic failure; recognition of the pathognomonic histological changes in hepatic biopsy specimens; and **analysis for the presence of pyrrolizidine alkaloids in herbal preparations that the patient has been exposed to.******

> Physician and consumer awareness of herbal toxicities is extremely limited, and the incidence of hepatic veno-occlusive disease and of pyrrolizidine poisoning may be grossly underestimated."

(* See number 12 above. -'Veno-Occlusive Disease of the Liver Secondary to Ingestion of Comfrey' by C.F.M. Weston, B.T. Cooper, J.D. Davies and D.F. Levine (Bristol and Cornwall, England), British Medical Journal, Vol 295, page 183, July 18 1987.)

(** See number 10 above. -'Hepatic Venocclusive Disease Associated with the Consumption of Pyrrolizidine Containing Dietary Supplements' by Paul M. Ridker, Seitaro Ohkuma, William V. McDermott, Charles Trey and Ryan J. Huxtable, (Boston, Massachusetts and Tucson, Arizona), Gastroenterology, Volume 88, Number 4, pages 1050-1054, 1985.)

(*** See number 4 above. -'Carcinogenic Activity of Symphytum Officinale' by I. Hirono et al, Japan National Cancer Institute Volume 61, No. 3, pages 865-869, September 1978.)

(**** -'New Aspects of the Toxicology and Pharmacology of Pyrrolizidine Alkaloids' by Ryan J. Huxtable, Department of Pharmacology, University of Arizona, Tucson, Arizona; General Pharmacology: The Vascular System, Volume 10, Issue 3, pages 159-167, 1979.)   (Pathognomonic menas a sign or symptom characteristic of a particular disease.)

> **Comment i:** This article mentions 3 research reports about the health problems with Comfrey. Those reports all have rebuttals as to the problems with the research.
> There are no new test findings or other new information in this article. It is a compilation of health problems caused by pyrrolizidine alkaloids in various herbs.

18. 'Hepatic Veno-Occlusive Disease Associated with Comfrey Ingestion' by Yeong, Swinburn, Kennedy and Nicholson, Auckland Hospital, New Zealand; Journal of Gastroenterology and Hepatology, Volume 5, No. 2, pages 211-214, March **1990**.

"*A 23 year old man presented with hepatic veno-occlusive disease and severe portal hypertension (high blood pressure) and subsequently died from liver failure.*

> Light microscopy and hepatic angiography showed occlusion (blockage) of sublobular veins and small venous radicles of the liver, associated with widespread haemorrhagic necrosis (tissue death) of hepatocytes (liver cells).

He had, however, a striking 'binge-type' eating pattern whereby he would eat large quantities of a particular food such as grapes or cashew nuts for days and weeks on end.

> **The patient's unusual diet with its emphasis on large quantities of fruit and vegetables may have resulted in a degree of protein malnutrition. Individual susceptibility may be influenced by dietary and nutritional factors, with a protein deficient diet playing a contributory role.**

The patient had been on a predominantly vegetarian diet and, **prior to his illness, took Comfrey leaves** which are known to contain hepatotoxic pyrrolizidine alkaloids.

Comfrey is widely used as a herbal remedy, but so far has only been implicated in two other documented cases* of human hepatic veno-occlusive disease.

> In the 1-2 weeks before the onset of symptoms, he ate young Comfrey leaves which were steamed and eaten as a vegetable. **The quantity of Comfrey he ingested was reported to be 4-5 leaves every day for 1-2 weeks.**

A possible causal association of Comfrey and this patient's veno-occlusive disease is suggested by the temporal (time) relationship of the ingestion of Comfrey to his presentation, the histological changes in the liver and the exclusion of other known causes of the disease.

The relative rarity of recognized hepatotoxicity in humans may be due to a number of factors. These include the variability of alkaloid concentration in commercial preparations and possible deterioration with storage. Mild forms of the disease may well be reversible and under-diagnosed, particularly if morphologic variations of the disease are not well recognized.

Marked individual variations in dosage susceptibility have been found with other types of pyrrolizidine alkaloids.* "

(* -'Herbal Tea Induced Hepatic Veno-Occlusive Disease: Quantification of Toxic Alkaloid Exposure in Adults' by Kumana, Ng, Lin, Ko, W and Todd, Gut {British Society of Gastroenterology, British Medical Association}, London, England, Volume 26, No. 1, pages 101-104, January 1985. Four young Chinese women took daily doses of an unidentified 'Indian' herbal tea as treatment for psoriasis.)

> **Problems with Above Research (Rebuttal)**
>
> **Rebuttal i:** *"In one case report of 'death due to Comfrey', the plant had not even been in season at the time the fresh leaves were alleged (said) to have been eaten. In the New Zealand (sometimes referred to as Australian) case, Comfrey was not even in season during the 7-14 day period in which the New Zealand man was reported to have eaten the fresh leaves.* The plants would have died back at least two to three months prior to the patient's admission to hospital in August 1985 (winter in New Zealand or Australia)."

-'The Safety-in-Use of Comfrey and Comfrey Products: Results of a Research Survey' by the Society for the Promotion of Nutritional Therapy (SPNT), Linda Lazarides, Sussex County, United Kingdom, 8 pages, 1993-1994, pages 1, 8.

Rebuttal ii: *"The man presented with a three-month history of initial influenza-like symptoms followed by continued malaise and night sweats. Three weeks before admission he noticed peripheral edema (swelling) and abdominal distension.*
In the one to two weeks before the onset of symptoms he ate four to five steamed young Comfrey leaves as a vegetable every day. *The authors suggest that the patient's protein deficient diet could have played a contributory role;* ***they attributed Comfrey as a possible cause due to the temporal (time) sequence of events.***
In a separate review of potential risk to consuming Comfrey published in the 'Australian Medical Journal'*, the author declined to consider this case in his report because 'there is some controversy surrounding this case.' "
-'Safety Issues Affecting Herbs: Pyrrolizidine Alkaloids" by Subhuti Dharmananda, Ph.D., Director, Institute for Traditional Medicine, Portland, Oregon, 2016.
(* -'Comfrey: Assessing the Low-Dose Health Risk' by P.J. Abbott, Toxicology Unit, Department of Community Services and Health, Canberra, Australia; Medical Journal of Australia, Volume 149, No. 11-12, pages 678-682, 1988.)

Rebuttal iii: *"A 23 year old man was diagnosed with veno-occlusive disease after going to the hospital for a fever and malaise. Diuretics (increases urine) were administered and a shunt employed to relieve liver congestion.* ***He died of liver failure seven days after installing the shunt. The liver was not biopsied to see if Pyrrolizidine Alkaloids or their metabolites were present.***
His friends reported that he had consumed Comfrey 1-2 weeks prior to his hospitalization. It was not determined what species or if indeed it was Comfrey at all. It was also not determined if there were other Pyrrolizidine Alkaloid sources present in his diet. *Also, it was not determined if his habit of binge eating and consumption of marijuana caused deterioration of the hepatic (liver) cells leading to this problem."*
-'Comfrey, Symphytum Officinale: The Much Debated Healer' by Mimi Kamp, medicinal herbalist, botanical illustrator and photographer from southwestern United States, probably 2017.

Rebuttal iv: *"**The details of the case are again not clear.** It is not stated whether he ingested Comfrey one to two weeks before the onset of the initial flu-like symptoms, or before the more recent oedema (edema) and distension. The authors suggest that the patient's protein deficient diet could have played a contributory role and admit that 'marked individual variations in dosage susceptibility have been found with other PAs.'*
Abbott* declines to consider this case in his report because 'there is some controversy surrounding this case.' "
-'In Defence of Comfrey' by Margaret Whitelegg, MNIMH. Paper presented to 'Department of Health' and 'Ministry of Agriculture, Fisheries and Food', 'National Institute of Medical Herbalists', United Kingdom, January 1993. Published in European Journal of Herbal Medicine, Exeter, England, Volume 1, No. 1, pages 11-17, 1994.
(* -'Comfrey: Assessing the Low-Dose Health Risk' by Peter J. Abbott, Toxicology Unit, Department of Community Services and Health, Canberra, Australia; Medical Journal of Australia, Volume 149, No. 11-12, pages 678-682, 1988.)

Rebuttal v: *"**There has been a documented case of veno-occlusive disease in a recent New Zealand Coroner's Report, which implicated (implied) the use of Comfrey as being associated directly with the death of a young New Zealander.** The Comfrey allegedly (supposedly) was consumed while he was living on a farming community near Bredbo in New South Wales, Australia.*
Since there is some controversy surrounding this case, it will not be discussed further here."
-'Comfrey: Assessing the Low-Dose Health Risk' by Peter J. Abbott, Toxicology Unit, Department of Community Services and Health, Canberra, Australia; Medical Journal of Australia, Volume 149, No. 11-12, pages 678-682, 1988.

19. 'The Effects of Comfrey Derived Pyrrolizidine Alkaloids on Rat Liver' by M.L. Yeong, S.P. Clark, J.M. Waring, R.D. Wilson and S.J. Wakefield; Wellington, New Zealand; Pathology: The Journal of the Royal College of Pathologists of Australasia, Volume 23, No. 1, pages 35-38, January **1991**.
*"**Three groups of young adult rats were fed pyrrolizidine alkaloids derived from Russian Comfrey to study the effects of the herb on the liver.***
Group I animals received a single dose of 200 mg/kg body weight, Group II 100 mg/kg three times a week for 3 weeks and Group III 50 mg/kg three times a week for 3 weeks. ***All rats showed light and electron-microscopic evidence of liver damage, the severity of which was dose dependent.***
> *There was swelling of hepatocytes and hemorrhagic necrosis of perivenular cells. There was a concomitant loss of sinusoidal lining cells with disruption of sinusoidal wall and the sinusoids were filled with cellular debris, hepatocyte organelles and red blood cells. Extravasation of red blood cells was evident. Terminal hepatic venules were narrowed by intimal proliferation, and in Group II and III, reiculin fibres radiated from these vessels.*

These appearances have been described in veno-occlusive disease due to pyrrolizidine alkaloids from other plant sources such as Senecio and Crotalaria.
The safety of Comfrey, a widely used herb, in relation to human consumption requires further investigation."
(For transgenic mice and rats, see subsection 'Flawed Experimental Design' in section 'Alkaloids in Comfrey' {Chapter 30}.)

Problems with Above Research (Rebuttal)

Rebuttal i: *"There is imprecision: Yeong describes his work on rats as 'Three groups of young adult rats were fed PAs* *derived from Russian Comfrey to study the effects of the herb on the liver.' **The effects of the herb are not those of just the PAs.***"

(Comfrey herb is the total plant, not just the alkaloids. But the researcher treats them as the same thing.)

-'In Defence of Comfrey' by Margaret Whitelegg, MNIMH. Paper presented to 'Department of Health' and 'Ministry of Agriculture, Fisheries and Food', 'National Institute of Medical Herbalists', United Kingdom, January 1993. Published in European Journal of Herbal Medicine, Exeter, England, Volume 1, No. 1, pages 11-17, 1994.

20. 'CPMP Listing of Herbs and Herbal Derivatives Withdrawn for Safety Reasons: Herbal Drugs with Serious Risks' by Committee for Proprietary Medicinal Products (CPMP), Commission of the European Communities, Brussels, Belgium, 5 pages, August 15 and October 26 **1992**.

"*1. Herbal drugs with serious risks without any accepted benefit; not acceptable for revision:*
  Symphytum all species, internal use. Parts: herb, leaf, root.
  Reason: contains pyrrolizidine-alkaloids with genotoxic, carcinogenic and hepatotoxic properties. No benefit proven.
*2. **Drugs with toxic principles**, where a more detailed discussion concerning benefit/risk ratio is necessary.*
  Drugs with pyrrolizidine-alkaloids where a use is accepted under special precautions/labelling:
  Symphytum officinale L., external use. Parts: herb, leaf, root.
  Restrictions: Use only on unbroken, intact skin. Use during pregnancy requires medical advise. Use not longer than 6 weeks per year. Temporarily Tolerable Dose (TTD) 100 microgram PA/day. For these drugs a limitation of the toxic principle and a strict definition of the conditions of use is necessary."

(There were 34 herbs in the first category. Two herbs plus Symphytum were in the second category: Tussilago farfara L. {coltsfoot} leaf and Petasites hybridus L. {butterbur} rhizome.)

(In May 2004 the name 'Committee for Proprietary Medicinal Products' was changed to 'Committee for Medicinal Products for Human Use' {CHMP}. It is the European Medicines Agency's committee responsible for human medicines. For more about their regulations, see subsection 'European Union Comfrey Medicinal Regulations' in section 'Some Uses of Comfrey are Banned by Governments' {Chapter 31}.)

21. 'Hepatic and Pulmonary Complications of Herbal Medicines' by F.G. Miskelly and L.I. Goodyer, Charing Cross Hospital, London, England; Postgraduate Medical Journal: The Fellowship of Postgraduate Medicine, London, England, Letters to the Editor, Volume 68, No. 805, pages 935-936, November **1992**.

"*A 77 year old woman presented with tiredness, anorexia and weight loss for 6 months, cough with green sputum for 3 months and dark urine but normal stools for a few weeks. She consumed approximately 6 units of alcohol per week. On examination she was moderately jaundiced but apyrexial. She had no hepatosplenomegaly or ascites. Her chest was clear on auscultation.*

The patient, despite repeated questioning about medications, admitted taking three types of herbal remedies for the previous 6 months: *BFC for 'wasting diseases', Bowel Tonic for 'peristalsis' and Nervine for anxiety. Each was taken as 1/2 a level teaspoon, three times daily, for 6 days out of seven.*

Along with seven other herbs, Comfrey root and skullcap were both present in the BFC, Comfrey 6 parts in 27 and skullcap 1 part in 27.

Both Comfrey and skullcap may produce an acute hepatitis but only Comfrey is known to produce pulmonary lesions, and then only in rats. This suggests that the pulmonary lesions were evidence of endothelial hyperplasia and due to Comfrey."

(In the United Kingdom, one unit of alcohol is defined as 10 ml {8 grams} of pure 100% alcohol. One unit of alcohol is equal to 25 ml of spirits with ABV {Alcohol by Volume} 40%.)

Problems with Above Research (Rebuttal)

Rebuttal i: The other 7 herbs used by the patient were not given. This article can not be properly evaluated because without the other herb information, there is no way of knowing if any of them could have been all or part of the cause of the disease. The diagnosis is not approached scientifically. Comfrey is only a possibility.

22. 'Hepatocyte Membrane Injury and Bleb Formation Following Low Dose Comfrey Toxicity in Rats' by M.L. Yeong, S.J. Wakefield and H.C. Ford, Department of Pathology, Wellington School of Medicine, Wellington South, New Zealand; International Journal of Experimental Pathology, Volume 74, pages 211-217, **1993**.

"*Comfrey, a popular herbal remedy, contains hepatotoxic pyrrolizidine alkaloids and has been implicated in recent human toxicity. Although alkaloids from other plant sources have been extensively researched, studies on the hepatotoxic effects of Comfrey alkaloids are scant (rare).*

The effects of high dose Comfrey toxicity have been studied and the present investigation was undertaken to identify changes associated with relatively low dose toxicity. *Eight young adult rats were dosed weekly for six weeks with 50 mg/kg of Comfrey derived alkaloids.*

  The animals were dissected one week after the last dose and the livers examined by light and electron microscopy. Changes at the light microscopic level showed vascular congestion, mild zone 3 necrosis (tissue death) and loss of

definition of hepatocyte cellular membranes. Extensive ultrastructural abnormalities were identified in the form of endothelial sloughing and the loss of hepatocyte microvilli.
A striking finding was florid bleb formation on the sinusoidal (capillary) borders of hepatocytes (liver cells). Many blebs were shed into the space of Disse and extruded to fill, and sometimes occlude, sinusoidal lumina. Platelets were frequently found in areas of bleb formation. There was evidence of late damage in collagenization of Disse's space.
Hepatocyte bleb formation is known to occur under a variety of pathological conditions but there is little to no information in the literature on the effects, if any, of bleb formation on fibrogenesis and the microcirculation and its role in the pathogenesis of liver disease.
The pyrrolizidine alkaloids of Comfrey may serve as an experimental tool to study the process of bleb formation and the intimate relationship between hepatocyte and sinusoidal injury in the liver."
(A bleb is a hemispherical protrusion from a cell's surface, which may be filled with fluid or supported by a network of microfilaments. It is like a small blister or fluid-filled cyst.)
(Fibrogenesis is a mechanism of wound healing and repair. Prolonged injury causes deposits of extracellular matrix proteins and fibrosis. Fibrosis is thickening and scarring of connective tissue.)
(For transgenic mice & rats, see subsection 'Flawed Experimental Design' in section 'Alkaloids in Comfrey' {Chapter 30}.)

23. 'Determination of Pyrrolizidine Alkaloids in Commercial Comfrey Products (Symphytum sp.)' by Betz, Eppley, Taylor and Andrezejewski, Center for Food Safety and Applied Nutrition, Food and Drug Administration, Washington, DC; Journal of Pharmaceutical Sciences, Volume 83, No 5, pages 649-653, May **1994**.

"*In the past, Comfrey was one of the most popular herbal teas in the world. Fortunately, as its dangers have become known, its popularity has declined, but it is still available commercially in several forms.*
The unsaturated PAs (Pyrrolizidine Alkaloids) are toxic because they are very rapidly converted to the corresponding pyrroles by the mixed-function oxidases of the liver, resulting in cellular destruction or abnormal growth patterns.
Accumulation of this cellular damage results in a syndrome known as **Hepatic Veno-Occlusive Disease (HVOD)**. *Even if acute intoxication does not occur, the likelihood of increased incidence of liver cancer must be considered.*
Bulk ground Comfrey (Symphytum officinale) root and leaf material were purchased from a local dealer *in bulk botanicals. Hot water infusions (tea) of the bulk root and leaf were prepared and then analyzed. In this method, 2 grams (about 1 teaspoon) of root or leaf was added to 250 ml (1.05 cup) of hot water, and the mixture was allowed to steep for 5 minutes. The resulting infusion was carefully decanted, cooled to room temperature, and extracted three times with 250 ml of chloroformammonium hydroxide (99:1).*
Lack of standards *prevented us from determining which, if any, of the compounds was symphytine and prevented identification of all of the PAs known to occur in Symphytum species.*
  The lowest level of an individual PA found in any product was 0.1 ppm (parts per million).
  The PA content of commercially available Comfrey products varies considerably.
  The highest PA levels were found in the authentic powdered root.
  Levels of intermedine and lycopsamine were approximately 4 times lower in authentic leaf than in root, whereas levels of acetyllycopsamine and acetylintermedine (the 7-acetyl derivatives) were roughly 8 and 11 times lower, respectively.
A person consuming two 250-mg capsules of the bulk Comfrey root used in this study three times a day would receive approximately 1.8 mg/day of alkaloid (excluding symphytine).
Consumption of a similar amount of bulk Comfrey leaf in capsule form would provide approximately 0.3 mg/day."
(Decant means slowly pouring a liquid from one container to another, trying not to disturb the sediment on the bottom.)

24. 'Medicinal Plants in Europe Containing Pyrrolizidine Alkaloids' by Erhard Thomas Roeder, Pharmazeutisches Institut der Rheinischen Friedrichs-Wilhelms, University of Bonn, Germany, Pharmazie 50, pages 83-98, March **1995**.

"**Medicinal plants in Europe of the family Boraginaceae containing pyrrolizidine alkaloids:**
  Alkanna tinctoria (Dyer's Alkanet)
  Anchusa officinalis (Common Bugloss or Alkanet)
  Borago officinalis (Borage or Starflower)
  Cynoglossum officinale (Houndstongue or Houndstooth)
  Heliotropium arborescens (Garden Heliotrope)
  Lithospermum officinale (Common Gromwell or European Stoneseed)
  Myosotis scorpioides (True Forget-Me-Not or Water Forget-Me-Not)
  Symphytum asperum (Prickly Comfrey)
  Symphytum caucasicum (Caucasian Comfrey)
  Symphytum officinale (Common Comfrey)
  Symphytum tuberosum (Tuberous Comfrey)
  Symphytum x uplandicum (Russian Comfrey)
Medicinal plants in Europe of the family Asteraceae containing pyrrolizidine alkaloids:
  Eupatorium cannabinum (Hemp-Agrinomy or Holy Rope)
  Adenostyles alliariae (Hedge-Leaved Adenostyles)
  Emilia sonchifolia (Lilac Tasselflower or Cupid's Shaving Brush)
  Petasites hybridus (Butterbur)

Petasites spurius (Wooly Butterbur)
Senecio aureus L. (Golden Ragwort)
Senecio bicolor (Dusty Miller)
Senecio doronicum (Chamois Ragwort)
Senecio jacobaea = Jacobaea vulgaris (Stinking Willie)
Senecio nemorensis (Broad-Leaved Ragwort)
Senecio vulgaris (Common Groundsel or Old-Man-in-the-Spring)
Tussilago farfara (Coltsfoot)"

25. 'Pyrrolizidine Alkaloids in Human Diet' by A.S. Prakash, T.N. Pereira, P.E. Reilly and A.A. Seawright (Australia); Mutation Research Journal: Genetic Toxicology and Environmental Mutagenesis, Elsevier, Netherlands, Volume 443, No. 1-2, pages 53-67, July **1999**.
"PAs (Pyrrolizidine Alkaloids) in medicinal plants:
PAs have been identified in traditional herbal medicines of South America, Sri-Lanka and China.
Of the herbal remedies containing pyrrolizidine alkaloids, Comfrey has received the most attention. *Studies have shown the presence of toxic PAs in fresh leaves, commercial Comfrey preparations and in Comfrey-pepsin capsules.*
    Following an episode of Comfrey-pepsin related poisoning*, the sale of Comfrey products for internal use was banned in the United States and in Canada**.
    Nevertheless, Comfrey leaves and extracts continue to be used in poultices, creams and ointments for topical application. Comfrey leaves are consumed in salads, particularly in Europe, North America, Japan and Australia.
Carcinogenicity: *While there is no evidence of cancer in the literature concerning domestic animals (livestock) exposed to PAs, studies carried out under laboratory conditions have been able to produce PA-induced cancer in rodents.*
    Some of the plant species known to cause cancer in rodents are S. longilobus, Petasites japanicus Maxim, Tussilago farfara L., **Symphytum officinale******, Farfugium japonicum, Ligularia dentata and S. cannabifolis.*
    Further, individual PA compounds such as monocrotaline, heliotrine, lasiocarpine, clivorine, petasitenine and riddelliine have also been shown to be carcinogenic in experimental animals."
(* -'Comfrey Herb Tea-Induced Hepatic Veno-Occlusive Disease' by Nancy Bach M.D., SwanThung M.D., and Fenton Schaffner M.D., The American Journal of Medicine, Volume 87, Issue 1, pages 97-99, July 1989. See number 13.)
(** See section 'Some Uses of Comfrey are Banned by Governments'.)
(*** -'Carcinogenic Activity of Symphytum Officinale' by I. Hirono et al, Japan National Cancer Institute Volume 61, No. 3, pages 865-869, September 1978. See number 4.)

26. 'Acute Hepatitis After Ingestion of Herbs' by J.A. Shad, C.G. Chinn and O.S. Brann, Department of Internal Medicine, Naval Medical Center, San Diego, California; Southern Medical Journal, Birmingham, Alabama, Volume 92, No. 11, pages 1095-1097, November **1999**.
"Case 2: A 69-year-old man who had been taking 14 tablets per day of a mixed herb prepartion for 6 weeks had pruritus (itchiness) and nausea. These symptoms became more intense and were followed by anorexia (not eating), weight loss, and jaundice. His only medication was metoprolol tartrate (Lopressor) for hypertension. The herbal medications were discontinued, and the patient's symptoms, as well as liver function abnormalities, progressively improved. He was asymptomatic (no symptoms) 8 weeks later, and results of laboratory tests were normal.
Ingredients of Herbal Product in Case 2: *Alfalfa, bee pollen, black cohosh, burdock root, capsicum,* **cascara sagrada, chaparral,** *chickweed,* **Comfrey leaf,** *dandelion root, fennel seed, ginger root, hawthorne berries, juniper berry, kelp, licorice root, mullein, papaya leaf, pau d'arco, rose hips and safflower."*

    Comment: The levels of Pyrrolizidine Alkaloids are low in Comfrey leaf as opposed to the root. It is not clear which herb or combination of herbs caused the health problems. However, there are later articles that use this case report as proof that Comfrey is bad for health.

    -'**Cascara Sagrada-Induced Intrahepatic Cholestasis Causing Portal Hypertension:** Case Report and Review of Herbal Hepatotoxicity' by A. Nadir A, D. Reddy, D.H. Van Thiel; American Journal of Gastroenterology, Volume 95, No. 12, pages 3634-3637, December 2000.

    *"***Chaparral*** (Larrea tredentata, Larrea divaricata): The chaparrals (chapparral, creosote bush, greasewood, hediondilla) are a group of closely related wild shrubs found in the deserts of the American southwest and Mexico. Chronic ingestion of chapparral may be associated with acute or chronic hepatotoxicity. In December 1992, the United States issued a public warning about four cases of hepatitis."*
-'Medical Toxicology' edited by Richard C. Dart, MD, PhD. Philadelphia, Pennsylvania: Lippincott, Williams and Wilkins, page 1687, 2004.

27. 'Pyrrolizidine Poisoning: A Neglected Area in Human Toxicology' by Michael J. Stewart and Vanessa Steenkamp, University of the Witwatersrand Medical School, Johannesburg, South Africa; Therapeutic Drug Monitoring Journal, New York,

Volume 23, No. 6, pages 698-708, **2001**.

"Pyrrolizidine poisoning in humans is regarded by most clinical toxicologists as of little relevance. However, a number of individual case studies in the West and some severe cases of mass poisoning by contaminated grains have led to increased interest in these alkaloids.
The increasing use of herbal remedies, some of which contain toxic pyrrolidines, suggests that the incidence of pyrrolizidine poisoning is likely to increase.
In this review the authors describe the chemistry and metabolism of pyrrolizidine alkaloids, the salient features of pyrrolizidine poisoning, and the methods available for detection of these compounds in human fluids."

28. 'Systematic Review: Hepatotoxic Events Associated with Herbal Medicinal Products' by M.H. Pittler and E. Ernst, Complementary Medicine, Peninsula Medical School, Universities of Exeter and Plymouth, Exeter, England; Alimentary Pharmacology and Therapeutics, Oxford, England, Volume 18, pages 451-471, September 1 **2003**.
"Chaparral (Larrea tridentata) is used for a variety of conditions such as upper respiratory tract infections. A number of cases implicating chaparral as the cause of severe liver damage have been reported previously and have been summarized in the light of a further report.
A herbal combination product containing chaparral, Comfrey and Cascara sagrada was also associated with a case of acute hepatitis.* *Liver function stabilized in both patients after discontinuing these preparations.*
Pyrrolizidine-containing HMPs (Herbal Medicinal Products):
Veno-occlusive disease (the main liver injury induced by pyrrolizidine alkaloids) is characterized by portal hypertension and noncirrhotic ascites, and often progresses to hepatic failure.
Hepatotoxic events and fatalities associated with pyrrolizidine alkaloids have mainly been observed after exposure to four plant species, Crotalaria, Heliotropium, Senecio and Symphytum.**
    Comfrey (Symphytum officinale), *used externally to treat inflammatory disorders of the joints, wounds, distorsions and haematomas, has been associated with liver damage and fatalities.***, *****
    These cases occurred before 1990, and it seems that none have been published since."
(* See number 26. -'Acute Hepatitis After Ingestion of Herbs' by J.A. Shad, C.G. Chinn and O.S. Brann, Department of Internal Medicine, Naval Medical Center, San Diego, California; Southern Medical Journal, Birmingham, Alabama, Volume 92, No. 11, pages 1095-1097, November 1999.)
(** See number . -'Pyrrolizidine Alkaloids in Higher Plants: Hepatic Veno-Occlusive Disease Associated with Chronic Consumption' by Wendy Pearson, BSc in Agriculture; Department of Pharmacology and Toxicology, University of Western Ontario, Canada; Journal of Nutraceuticals, Functional and Medical Foods, Volume 3, No. 1, pages 87-96, 2001.)
(*** See number 18. - 'Hepatic Veno-Occlusive Disease Associated with Comfrey Ingestion' by Yeong, Swinburn, Kennedy and Nicholson, Auckland Hospital, New Zealand; Journal of Gastroenterology and Hepatology, Volume 5, No. 2, pages 211-214, March 1990.)
(**** -'The Efficacy and Safety of Comfrey' by F. Stickel and H.K. Seitz, Public Health Nutrition, United Kingdom, Volume 3, No. 4A, pages 501-508, December 2000. This article did not do original research or report a new case history. It gives references to other research that I include in this section. It is an overview.)

29. 'Analysis of Herbal Teas Made from the Leaves of Comfrey (Symphytum Officinale): Reduction of N-oxides Results in Order of Magnitude Increases in the Measureable Concentration of Pyrrolizidine Alkaloids' by Oberlies, Kim, Brine, Collins, Handy, Sparacino, Wani and Wall (North Carolina and New Mexico); Public Health Nutrition Journal, Cambridge, England, Volume 7, Issue 7, pages 919-924, **2004**.
*"**Pyrroles and N-Oxides:**
    PAs (Pyrrolizidine Alkaloids) may be metabolised to either pyrroles, possibly responsible for the hepatotoxicity, or N-oxides, possibly a detoxification process. Studies on the structure-activity relationships responsible for either of the above mechanisms have been inconclusive.*
    Some proponents (for use) of Comfrey have maintained that consumption of tea made from Comfrey leaves may not be a risk since the PAs are not particularly soluble in water, and thus the subsequent tea should contain only the more water-soluble N-oxides.
Reduction of N-Oxides:
    By reducing N-oxides prior to analysis, we observed an order of magnitude increase in the measurable concentration of symphytine alkaloid. This demonstrates the relatively high level of total PAs that may result from drinking Comfrey herbal tea.
        Comfrey leaves (10 grams = 0.35 ounces) were added to 1 liter (1.05 quart) of hot (90 C = 194 F) water, and the mixture was allowed to steep for 5 minutes. The resulting solution was decanted and passed through cheesecloth, allowed to cool to room temperature, and extracted three times with 1 liter of chloroform-ammonium hydroxide (99:1). The PAs were visualised in this fraction via thin layer chromatographic analysis using Dragendorff's reagent.
The N-oxide derivatives of the PAs are more hydrophilic (strong affinity for water) than the native free bases, yet, in vivo (in the living body), these N-oxides can be reduced to the native PAs in the gut.
The reduction procedure produced a larger amount of PAs from the aqueous tea as evidenced by a higher concentra-

tion of symphytine measured in the vendor A material (14.5 versus 110 microgram).
In fact, the zinc dust reduction step resulted in a 10-fold increase in the amount of symphytine measured in the tea. Thus the total PA content of the tea may be underestimated substantially.

Determing Total PAs in Comfrey Tea:
Thus, to most accurately determine the total concentration of PAs in teas made from Comfrey leaves, procedures that account for the N-oxides should be utilised. These N-oxide derivatives of PA are apparently not produced during the tea-making process.

Government Regulations:
In Germany, consumption of total PAs with 1,2-unsaturated necine moieties (part of a molecule), such as seen in the structures of both symphytine and echimidine, is limited to 1 microgram daily, although special consideration is given for **Comfrey tea, which is limited to a maximum dose of 10 microgram daily."**

30. 'Mutagenicity of Comfrey (Symphytum Officinale) in Rat Liver' by N. Mei, L. Guo, P.P. Fu, R.H. Heflich and T. Chen, National Center for Toxicological Research, United States Food and Drug Administration, Jefferson, Arkansas; British Journal of Cancer, Volume 92, pages 873-875, **2005**.

"**Comfrey is a rat liver toxin and carcinogen** that has been used as a vegetable and herbal remedy by humans. **There is little known about the mechanism of tumour induction (creation) by Comfrey.**
Although induction of hepatic (liver) tumours has been associated with the pyrrolizidine alkaloids (PAs) that are present in Comfrey, and PAs are genotoxic and carcinogenic by binding to liver DNA in humans and animals (*Prakash et al, 1999; Fu et al, 2004), **a comprehensive study of Comfrey mutagenesis has not been conducted.**
This inspired us to investigate the mutagenicity of Comfrey in rat liver, a target tissue for its carcinogenesis, by using a transgenic Big Blue® rat mutational model (Dycaico et al, 1994).

Groups of six 6-week-old male Big Blue rats were fed either a basal diet or the 2% Comfrey root diet. The animals were killed after 12 weeks of treatment. **Mutant frequencies (MFs)** were determined for the liver 'cII' gene of the rats treated with Comfrey. The MF for rats fed Comfrey was $146 \pm 15 \times 10^{-6}$, which was significantly greater than the MF for control rats, $30 \pm 16 \times 10^{-6}$ ($P<0.001$, Anova, Holm-Sidak test).

Although we encountered no overt signs of liver toxicity in our relatively short term study, the liver histology of rats fed Comfrey for prolonged periods is quite similar to that produced by some hepatotoxic PAs (Schoental, 1968; Hirono et al, 1976, 1977).

Liver cell necrosis (cell death), haemorrhage (bleeding), bile duct proliferation, and liver cirrhosis are frequently encountered even in rats from experimental groups that have no tumours. **This suggests that the liver tumours in Comfrey-treated rats might be induced by the PAs present in Comfrey.**

In conclusion, treatment of transgenic Big Blue® rats with Comfrey induced mutations in the liver 'cII' gene. This result suggests that Comfrey induces liver tumours by a genotoxic mechanism. The mutational spectrum from Comfrey-treated rats suggests that PAs in the plant are responsible for mutation induction and tumour initiation in rat liver."

(* -'Pyrrolizidine Alkaloids in Human Diet' by A.S. Prakash, T.N. Pereira, P.E. Reilly and A.A. Seawright, Australia; Mutation Research Journal: Genetic Toxicology and Environmental Mutagenesis, Elsevier, Netherlands, Volume 443, No. 1-2, pages 53-67, July 1999.)

(Transgenic is an organism that contains genetic material into which DNA from an unrelated organism has been artificially introduced. Transgenic Big Blue® rat is sold by Stratagene, La Jolla, California, exclusively in the homozygous form with 30-40 copies of the shuttle vector per genome. The Big Blue® mouse and rat transgenic systems are based on the bacterial lacI gene. Big Blue® rats were generated by microinjection of lambda LIZ phage DNA into fertilized eggs of Fischer 344 rats. For more about transgenic mice and rats, see subsection 'Flawed Experimental Design' in section 'Alkaloids in Comfrey'.)

(Histology is the study of the microscopic anatomy of cells and tissues of plants and animals.)

(Proliferation of liver bile ducts is due to various liver injuries and disease. The term 'ductular reaction' is preferred. There is an increase in the number of bile duct structures with inflammatory cell infiltrates and periportal fibrosis.)

(Cirrhosis is a chronic disease of the liver with degeneration of cells, inflammation, and fibrous thickening of tissue.)

31. 'Comparison of Gene Expression Profiles Altered by Comfrey and Riddelliine in Rat Liver' by Lei Guo, Nan Mei, Stacey Dial, James Fuscoe and Tao Chen, National Center for Toxicological Research, Food and Drug Administration, Jefferson, Arkansas; Fourth Annual MCBIOS Conference, Computational Frontiers in Biomedicine, New Orleans, Louisiana, February 1-3 2007. Published in BMC Bioinformatics, Volume 8, Supplement 7, S22, November 1 **2007**.

"**Groups of 6 Big Blue Fisher 344 rats were treated with riddelliine at 1 mg/kg body weight by gavage five times a week for 12 weeks or fed a diet containing 8% Comfrey (Symphytum officinale) root for 12 weeks.**
Animals were sacrificed one day after the last treatment and the livers were isolated for gene expression analysis. The gene expressions were investigated using Applied Biosystems Rat Whole Genome Survey Microarrays and the biological functions were analyzed with Ingenuity Analysis Pathway software.

Although there were large differences between the significant genes and between the biological processes that were altered by Comfrey and riddelliine, there were a number of common genes and function processes that were related to carcinogenesis. There was a strong correlation between the two treatments for fold-change alterations in expression of drug metabolizing and cancer-related genes.

Conclusion: Our results suggest that the carcinogenesis-related gene expression patterns resulting from the treatments of comfrey and riddelliine are very similar, and Pyrrolizidine Alkaloids contained in Comfrey are the main active components responsible for carcinogenicity of the plant."

(Riddelliine is a chemical compound classified as a pyrrolizidine alkaloid. It was first isolated from Senecio riddellii and is also found in Jacobaea vulgaris, Senecio vulgaris, and others plants in the genus Senecio. **Riddelliine is not found in Comfrey.**) (Gavage is administration of food or drugs by force through a tube going down the throat to the stomach.)

32. 'Bioactive Compounds in Food' edited by John Gilbert (York, England) and Hamide Z. Senyuva (Ankara, Turkey). Oxford, England: Blackwell Publishing Ltd, **2008**, pages 16-17.

"Comfrey (Symphytum officinale) has long been a popular herb in Europe and the United States.

Preparations of Comfrey in the form of dried leaves, dried root, and root powder tablets and capsules, often mixed with other herbs, have been sold with active promotion of the plant's supposed healing and digestive properties.

Herbal preparations of Symphytum, Tussilago, Borago, and Eupatorium, in the form of leaf, root powders, tablets and root extract tinctures, **sold in the United Kingdom in 1994** were surveyed for pyrrolizidine alkaloid content (Ministry of Agriculture, Fisheries and Food, 1994).

Comfrey Tablets:

Comfrey (Symphytum) tablets contained up to 5000 mg/kg and root powders up to 8300 mg/kg of pyrrolizidine alkaloids, giving estimated potential intakes in excess of 35 mg/day.

Comfrey and Borage Leaf Tea:

Comfrey and borage (Borago) leaf preparations intended for consumption as teas contained less than 100 mg/kg total pyrrolizidine alkaloids.

About 50% of the total acetyllycopsamine and symphytine (alkaloids) but only about 5% of the lycopsamine (alkaloid) were extracted into the water on brewing Comfrey leaf teas, possibly due to binding of the more polar lycopsamine to the plant tissue.

Comfrey Leaf and Root:

A survey of Comfrey leaf and root products sold in the United States in 1989 showed them to contain up to 1200 mg/kg of pyrrolizidine alkaloids (*Betz et al., 1994).

Teas prepared from Comfrey root and leaf showed preferential extraction of acetyllycopsamine and acetylintermedine over lycopsamine and intermedine.

Later studies confirmed that pyrrolizidine alkaloids were present in many Comfrey preparations sold in the United States (**Altamirano et al., 2005). In one case where the N-oxides were reduced prior to determination, the level of symphytine measured increased from 0.1 to 1 mg/L (***Oberlies et al., 2004)."

(* -'Determination of Pyrrolizidine Alkaloids in Commercial Comfrey Products {Symphytum sp.}' by Betz, Eppley, Taylor and Andrezejewski, Center for Food Safety and Applied Nutrition, Food and Drug Administration, Washington, DC; Journal of Pharmaceutical Sciences, Volume 83, No 5, pages 649-653, May 1994.)

(** -'Investigation of Pyrrolizidine Alkaloids and their N-Oxides in Commercial Comfrey-Containing Products and Botanical Materials by Liquid Chromatography Electrospray Ionization Mass Spectrometry' by Jorgelina C. Altamirano, Samuel R. Gratz and Karen A. Wolnik, United States Food and Drug Administration, Forensic Chemistry Center, Cincinnati, Ohio; Journal of AOAC International, Volume 88, No. 2, pages 406-412, March 2005.)

(*** -'Analysis of Herbal Teas Made from the Leaves of Comfrey (Symphytum Officinale): Reduction of N-oxides Results in Order of Magnitude Increases in the Measureable Concentration of Pyrrolizidine Alkaloids' by Oberlies, Kim, Brine, Collins, Handy, Sparacino, Wani and Wall (North Carolina and New Mexico); Public Health Nutrition Journal, Cambridge, England, Volume 7, Issue 7, pages 919-924, 2004.)

33. 'Severe Pulmonary Hypertension Possibly Due to Pyrrolizidine Alkaloids in Polyphytotherapy' by Sandor Gyorika and Hans Stricker, Division of Pulmonary Medicine, Ospedale San Giovanni, Bellinzona, Switzerland and Division of Angiology, Ospedale La Carita, Locarno, Switzerland; Swiss Medical Weekly, Volume 139, No. 13-14, pages 210-211, **2009**.

*"**We report the case of a 66 year old woman in Switzerland** with known arterial hypertension, non-insulin dependent diabetes, moderate adiposity (BMI 33 kg/m2) and mild renal insufficiency who came to the emergency room in November 2006 because of progressive dyspnoea (difficulty breathing).*

***At follow-up, when explicitly asked for the use of alternative and complementary medicines, the patient reported that she used a mixture of several herbs to make a tea,** which she described as 'excellent for her health'. In the months prior to hospitalisation she had drunk between one and one and a half litres (1.05-1.58 quarts) a day of this tea.*

*At the next visit she brought a bag full of herb packages. **Our patient took a mix of nine different herbal remedies.***

One of them, Comfrey, is made from an herb named Symphytum officinale.

***All herbs were carefully checked in the internet for possible relationship with Pulmonary Hypertension (PH) and one of the compounds, Comfrey, was found to contain pyrrolizidine alkaloids (PAs),** which have been associated with hepatic veno-occlusive disease and possibly with PH in the literature.*

We told the patient to immediately stop her herbal remedies because of their potential severe side effects.

Although not proven, we believe the PH of our patient to be possibly caused by the prolonged use of large quantities of boiled herbal remedies containing Comfrey.

***PA are present in more then three hundred plants and therefore probably in many herbal remedies.** Heath described the* *case of a young African who died due to primary PH suspected of having ingested the seeds of crotalaria laburnoides (another PA containing plant)."*

Problems with Above Research (Rebuttal)

Rebuttal i: The other 8 herbs used by the patient were not given. This article can not be properly evaluated because without the other herb information, there is no way of knowing if any of them could have been all or part of the cause of the disease. And it would have been even better to know the percentages of each herb in the mixture, and whether they were leaf or root. If possible, a brand and product name would have been good to know.
The diagnosis is not approached scientifically. As stated at the end of the article, there are many herbs that contain Pyrrolizidine Alkaloids. Comfrey is only a possibility.

34. 'Metabolism, Genotoxicity, and Carcinogenicity of Comfrey' by N. Mei et al., Division of Genetic and Reproductive Toxicology, National Center for Toxicological Research, United States Food and Drug Administration, Arkansas; Journal of Toxicology and Environmental Health, Part B Critical Reviews, Volume 13, No. 7/8, pages 509-526, **2010**.
"Although there are no epidemiological data regarding the carcinogenicity (cancer causing) of Comfrey, there are a number of cases that implicate (imply / suggest) human consumption of Comfrey in the development of liver diseases.
Comfrey, however, produces hepatotoxicity in livestock* and humans and carcinogenicity in experimental animals. Comfrey contains as many as 14 Pyrrolizidine Alkaloids (PAs), including 7-acetylintermedine, 7-acetyllycopsamine, echimidine, intermedine, lasiocarpine, lycopsamine, myoscorpine, symlandine, symphytine, and symviridine.
    The PAs in Comfrey are retronecine mono- and diesters, a class with lower toxicity than heliotridine monoester and macrocyclic diesters found in other plants**.
    The PA content of Comfrey is less than 1% and variably dependent on the plant part ***, ****.
Pyrrolizidine alkaloids are biologically and toxicologically inactive and require metabolic activation to exert these effects. These compounds undergo metabolic activation to yield the corresponding pyrrolic metabolites that react with cellular macromolecules, including proteins and DNA, to exert toxicity.
The mechanisms underlying Comfrey-induced (created) genotoxicity and carcinogenicity are still not fully understood.
The available evidence suggests that the active metabolites of PA in Comfrey interact with DNA in liver endothelial cells and hepatocytes, resulting in DNA damage, mutation induction, and cancer development.
Genotoxicities attributed to Comfrey and riddelliine (a representative genotoxic PA and a proven rodent mutagen and carcinogen) are discussed in this review.
    Both of these compounds induced similar profiles of 6,7-dihydro-7-hydroxy-1-hydroxymethyl-5H-pyrrolizine (DHP)-derived DNA adducts and similar mutation spectra. Further, the two agents share common mechanisms of drug metabolism and carcinogenesis.
Overall, Comfrey is mutagenic in liver, and PAs contained in Comfrey appear to be responsible for Comfrey-induced toxicity and tumor induction."
(* The author does not give a reference to information about Comfrey and liver damage in livestock.)
(** -'Comfrey Toxicity Revisited' by Dorena Rode, Department of Animal Science, College of Agricultural and Environmental Sciences, University of California, Davis, California; Trends in Pharmacological Sciences: International Union Of Pharmacology, Volume 23, No. 11, pages 497-499, November 2002.)
(*** -'Medicinal Plants in Europe Containing Pyrrolizidine Alkaloids' by Erhard Thomas Roeder, Pharmazeutisches Institut der Rheinischen Friedrichs-Wilhelms, University of Bonn, Germany, Pharmazie 50, pages 83-98, March 1995.)
(**** -'The Efficacy and Safety of Comfrey' by F. Stickel and H.K. Seitz, Public Health Nutrition, United Kingdom, Volume 3, No. 4A, pages 501-508, December 2000.)
    (Epidemiology is the study of pubic health. It monitors the incidence, distribution, cause, and control of diseases and other health factors in a population.)
    (Riddelliine is a chemical compound classified as a pyrrolizidine alkaloid. It was first isolated from Senecio riddellii and is also found in Jacobaea vulgaris, Senecio vulgaris, and others plants in the genus Senecio. **Riddelliine is not found in Comfrey.**)

35. 'Toxicity of Pyrrolizidine Alkaloids to Humans and Ruminants' by Helmut Wiedenfeld (Bonn, Germany) and John Edgar (North Ryde, Australia); Phytochemistry Reviews: Proceedings of the Phytochemical Society of Europe, Volume 10, Issue 1, pages 137-151, March **2011**.
"PA (Pyrrolizidine Alkaloid) intoxication in humans is not only related to the amount and the duration of the exposure but also to age and gender: males react more sensitively than females and foetuses and children (especially neonates/ newborns or infants) show the highest sensitivity for PA poisoning*.
    In 2003 it was shown that the daily uptake of approximately 7 micrograms PA (from a herbal tea containing Comfrey) during pregnancy did not show a toxic effect in the mother's liver but damaged the foetal liver in this way that the new born child died after 2 days**)."
(* -International Programme on Chemical Safety {IPCS}: Pyrrolizidine alkaloids, Environmental Health Criteria 80, United Nations, World Health Organization, Geneva, Switzerland, 1988.)
(** -'Veno-Occlusive Disease in a Foetus Caused by Pyrrolizidine Alkaloids of Food Origin' by R. Rasenack, C. Muller, M.

Kleinschmidt, J. Rasenack and H. Wiedenfeld, Friedrich-Wilhelm University, Bonn, Germany; Fetal Diagnosis and Therapy, Switzerland, Volume 18, No. 4, pages 223-225, July-August 2003.)

Problems with Above Research (Rebuttal)

Rebuttal i. Quote from above article about Veno-Occlusive Disease in a fetus:
"History of the family indicated tea as possible source for pyrrolizidine alkaloids, but the ingested teas were free of alkaloids.
However, a herbal mixture which was used daily for cooking contained significant amounts of different pyrrolizidine alkaloids. **The pyrrolizidine alkaloids found are observed in Heliotropium (genus in the Borage family with 325 species) and Comfrey (genus in Borage family with 35-40 species).** *The patient identified these herbs as ingredients of the herbal mixture which was imported from Turkey."*
(It is not known whether the Heliotropium or the Comfrey or both combined caused the liver problem. It is possible that Comfrey was not a significant factor. And it is not known what percentage of each was used.)

36. 'Herbal Hepatotoxicity: A Tabular Compilation of Reported Cases' by Rolf Teschke, Albrecht Wolff, Christian Frenzel, Johannes Schulze and Axel Eickhoff, Germany; Liver International: Official Journal of the International Association for the Study of the Liver, Oxford, England, Volume 32, No. 10, pages 1543-1556, November **2012**.

"Herbal hepatotoxicity is a field that has rapidly grown over the last few years along with increased use of herbal products worldwide. **To summarize the various facets of this disease, we undertook a literature search for herbs, herbal drugs and herbal supplements with reported cases of herbal hepatotoxicity.**
A selective literature search was performed to identify published case reports, spontaneous case reports, case series and review articles regarding herbal hepatotoxicity. **A total of 185 publications were identified and the results compiled. They show 60 different herbs, herbal drugs and herbal supplements.**
Based on stringent causality assessment methods and/or positive re-exposure tests, causality was highly probable or probable for Ayurvedic herbs, Chaparral, Chinese herbal mixture, Germander, Greater Celandine, green tea, few Herbalife® products, Jin Bu Huan, Kava, Ma Huang, Mistletoe, Senna, Syo Saiko To and Venencapsan®.
   In many other publications, however, causality was not properly evaluated by a liver-specific and for hepatotoxicity-validated causality assessment method such as the scale of CIOMS (Council for International Organizations of Medical Sciences).
Comfrey: Symphytum officinale, Symphytum asperum, Symphytum uplandicum:
Ridker et al., 1985*, Weston et al., 1987**, Bach et al., 1989***, Ridker and McDermott, 1989****, Miskelly and Goodyer,1992*****."
(*See number 10. **See number 12. ***See number 15. ****See number 17. *****See number 21.)
(See subsection 'Diagnosis of Herbal Hepatotoxicity' in section 'Warnings and Negative Reactions to Comfrey' {Chapter 28}.)

Problems with Above Research (Rebuttal)

Rebuttal i. The authors state that in the Comfrey case reports the 'causality was not properly evaluted by a liver-specific and for hepatotoxicity-validated cauality assessment'. This is a common problem that other authors mention in some of the rebuttals.
Five case reports were found for Comfrey in this literature search. All of them are discussed in this section with all having rebuttals or comments about them.

37. 'Pyrrolizidine Alkaloids in Medicinal Plants from North America' by E. Roeder, H. Wiedenfeld and J.A. Edgar (Germany and Australia); Die Pharmazie, Eschborn, Germany, Volume 70, pages 357-367, **2015**.

"***Symphytum asperum Lepech. Prickly Comfrey:***
   Distribution: Native to Europe, naturalized in southern provinces of Canada. Canadian Aboriginal peoples use this plant. Plant can cause veno-occlusive symptoms leading to liver cirrhosis and liver tumor (*Munro 2013).
   Symphytum asperum contains: intermedine, O3-acetylintermedine, lycopsamine, O3-acetyl-lycopsamine, echimidine, symlandine, symviridine, myoscorpine, symphytine *(Culvenor et al. 1980; Roeder et al. 1992).*
Symphytum officinale L., synonym S. consolida L., Common Comfrey:
   Distribution: Introduced and naturalized herb from Europe. It occurs in many parts of Canada.
   Cherokee Indians take it to treat dysentery (bad diarrhea) and as a gastrointestinal aid. It is also used as a gynaecological aid and taken to treat heartburn in pregnancy and for 'flooding' after birth. It is also taken as laxative infusion 'costiveness' (constipation) in pregnancy and used as an orthopedic aid against sprains and bruises. An infusion (tea) of roots in water is used against gonorrhea (**Moerman 2009).
   Symphytum officinale causes: venoocclusive symptoms, liver cirrhosis, and death (Munro 2013).
   It contains: echinatine, asperumine, intermedine, O3-acetyl-intermedine, anadoline, lasiocarpine, heliosupine, lycopsamine, O3-acetyl-lycopsamine, uplandicine, echimidine, echiumine, symlandine, symviridine, myoscorpine, symphytine *(Furuya et al. 1968, 1971; Pedersen 1975; Culvenor et al. 1980; Resch et al. 1982; Huizing et al. 1985, Roeder et al. 1992; Kim et al. 2001; Wuilloud et al. 2004; Liu et al. 2009)*

Symphytum x uplandicum Nyman, synonym S. peregrinum Ledeb., Russian Comfrey:
It is a hybrid generated from Symphytum officinale L. and Symphytum asperum Lepech.
Distribution: It is widely distributed in the United States and Canada, and used as a trial forage crop in Lethbridge (city), Alta (Alberta province), and Vancouver Island (province of British Columbia, Canada).
This plant contains PAs, which cause veno-occlusive symptoms, liver cirrhosis, and death (***Ridker, et al. 1989; Altamirano et al. 2005). **Roots and leaves contain: intermedine, O3-acetyl-intermedine, lycopsamine, O3-acetyl-lycopsamine, uplandicine, echimidine, symlandine, symviridine, myoscorpine, and symphytine** *(Culvenor et al. 1980; Roeder et al. 1992)."*

(* -'Munro 2013': The author does not give any other information about this reference.
  1. One relevant report I found: -'Correlation of Structural Class with No-Observed-Effect Levels: A Proposal for Establishing a Threshold of Concern' by I.C. Munro, R.A. Ford, E. Kennepohlt and J.G. Sprenger, CanTox Inc., Mississauga, Ontario, Canada, and Research Institute for Fragrance Materials, Inc., Hackensack, New Jersey; Food and Chemical Toxicology, Volume 34, No. 9, pages 829-867, September 1996.
  "The relationship between chemical structure and toxicity was explored through the compilation of a large reference database consisting of over 600 chemical substances tested for a variety of endpoints resulting in over 2900 No-Observed-Effect Levels {NOELs}."
  2. Another possible reference: CBIF: Canadian Poisonous Plants Information System, Government of Canada, www.cbif.gc.ca. This database was started by G.A. Mulligan and D.B. Munro in 1983.
  3. *"Work of Munro et al. 1996: Compiled existing toxicity data on 613 substances {industrial, food, consumer, environmental and agricultural chemicals, pharmaceuticals} and determined their No-Observed-Effect Levels.*
  Allocated each substance to one of three broad structural classes:
  Class I –simple structures, low likelihood of toxicity
  Class III –more complex structures suggesting significant toxicity
  Class II –structures in between I and III."
  -'Threshold of Toxicological Concern: Draft Opinion of the Scientific Committee' by European Food Safety Authority, Stakeholder Consultative Platform, Brussels, Belgium, November 17-18 2011.)
(** -Native American Medicinal Plants: An Ethnobotanical Dictionary by D.E. Moerman. Portland, Oregon: Timber Press, 2009.)
(*** -'Comfrey Herb Tea and Hepatic Veno-Occulusive Disease' by Paul M. Ridker and William V. McDermott, (Department of Medicine, Brigham and Women's Hospital, and Department of Surgery, New England Deaconess Hospital, Harvard Medical School, Boston, Massachusetts); The Lancet, London, England, Volume 333, Issue 8639, pages 657-658, March 25 1989.)

38. 'The Effect of Aqueous Leaf Extract of Symphytum Officinale (Common Comfrey) on the Liver of Adult Wistar Rats' by Ezejindu, Udemezue, Anyabolu, Chukwujekwu, Anike, Obialor, Akingboye and Ihim, Nigeria; International Journal of Innovative Research and Review, Volume 3, No. 2, pages 76-82, April-June **2015**.
"The aim of this study is to investigate the effect of aqueous (water) extract of Symphytum officinale leaf (Common Comfrey) on the liver of adult wistar rats.
  Twenty adult wistar rats of an average weighing 219 grams were used for the study. They were divided into four groups of five animals each. Group A served as the experimental control and were orally administered 0.3 ml of distilled water; the experimental groups B, C & D orally received 0.4 ml, 0.6 ml and 0.8 ml of aqueous extract of Symphytum officinale leafs for twenty eight (28) days.
Twenty four hours after the last administration, the animals were weighed and weights were recorded. The animals were sacrificed (killed) under the influence of chloroform vapour and dissected. The liver organ were harvested, weighed and trimmed down to a size of 3 mm x 3 mm and fixed in 10% formalin for histological studies.
The results of this study revealed that consumption of Symphytum officinale in low dose or small amount had no effect on the histological appearance of the liver but when consumed in high dose or excessive it induced mild distortion of histological liver appearance which includes mild central vein hypertrophy (enlargement), increased cellularity and periportal fibrosis of liver cells."
  (The Wistar rat is an outbred albino rat developed at the Wistar Institute, Philadelphia, Pennsylvania in 1906 for use in biological and medical research.)
  (Histology is the study of the microscopic anatomy of cells and tissues of plants and animals.)
  (Fibrosis is thickening and scarring of connective tissue.)

39. 'The Comparative Toxicity of a Reduced, Crude Comfrey (Symphytum Officinale) Alkaloid Extract and the Pure, Comfrey-Derived Pyrrolizidine Alkaloids, Lycopsamine and Intermedine in Chicks' by Browna, Stegelmeiera, Colegatea, Gardnera, Pantera, Knoppela and Hallc (Logan, Utah); Journal of Applied Toxicology, Volume 36, pages 716-725, May 1 **2016**.
"**Comfrey (Symphytum officinale),** *a commonly used herb, contains DeHydro-Pyrrolizidine Alkaloids (DHPA) that, as a group of bioactive metabolites, are potentially hepatotoxic (toxic to liver), pneumotoxic (toxic to lungs), genotoxic (mutates genes) and carcinogenic (causes cancer).*
How extract was produced:
  Dry, powdered root of Common Comfrey (Symphytum officinale) *was purchased from 'Starwest Botanicals' (Cordova, California) or 'Take Herb' (Alhambra, California). The crude alkaloid extract of the S. officinale powdered root was treated in the usual way with zinc and sulphuric acid to reduce most of the N-oxides to the free base DHPA forms.*

A crude extract of the Comfrey root alkaloids (including reduction of N-oxides to their free base forms) and the monoester DHPAs lycopsamine and intermedine were isolated from a reduced, crude alkaloidal extract of the powdered Comfrey as previously described (*Colegate et al., 2014).

Experiment:
The objective of this work was to compare the toxicity of a crude, reduced Comfrey root alkaloid extract to purified lycopsamine and intermedine that are major constituents of S. officinale.
Male, California White chicks were orally exposed to daily doses of 0.04, 0.13, 0.26, 0.52 and 1.04 mmol (millimole= 1/1000th of a mole) lycopsamine, intermedine or reduced Comfrey extract per kg bodyweight (BW) for 10 days. After another 7 days chicks were euthanized (killed).

Results:
Based on clinical signs of poisoning, serum (blood) biochemistry, and histopathological analysis the reduced Comfrey extract was more toxic than lycopsamine and intermedine.
This work suggests a greater than additive effect of the individual alkaloids and/or a more potent toxicity of the acetylated derivatives in the reduced Comfrey extract. It also suggests that safety recommendations based on purified compounds may underestimate the potential toxicity of Comfrey.

Authors' statements of limits to this research:
The Comfrey extract used in this study is somewhat different than what would be expected from natural exposure. In the assessment of risks presented to human health by dietary supplements, medicinal herbs, nutraceuticals or new functional foods, **the application of data acquired using various in vitro (in laboratory) or in vivo (live animal) models is a major challenge with respect to extrapolation to the human situation.**
Therefore, models are continually changed or modified in an attempt to establish meaningful indications of toxicity. In vivo toxicity assessments of the pro-toxic DHPAs are complicated by considerable variations in species-, gender- and age-related susceptibilities.
Additionally, assessment of entire products, crude extracts or purified components presents potential complications associated with any intrinsic (essential part of) additive or protective effects of the whole plant or crude extract relative to the purified components."

(* -'Heterozygous p53 Knockout Mouse Model for Dehydropyrrolizidine Alkaloid-Induced Carcinogenesis' by A.W. Brown, B.L. Stegelmeier, S.M. Colegate, K.E. Panter, E.L. Knoppel and J.O. Hall; Journal of Applied Toxicology, Volume 35, No. 12, pages 1557-1563, December 2015.)

(In the International System, a 'mole' is the base unit used in representing an amount of a substance, equal as many atoms, molecules, ions, or other elementary units as the number of atoms in 0.012 kilogram of carbon-12. The number is 6.0221×10^{23}, or Avogadro's number.)

(Histopathology is the study of changes in tissues caused by disease.)

(A nutraceutical is food with health-giving additives that has medicinal benefit.)

(Extrapolate means to extend a method or conclusion, especially one based on statistics, to an unknown situation by assuming that existing trends will continue or similar methods will be applicable.)

Scientific Studies and Indirect Evidence Showing Low Risk of Comfrey (in chronological order)
Before eating or drinking Comfrey, consult your healthcare provider.

A. Indirect Evidence of Low Toxicity:
"One group of alkaloids known as the pyrrolizidines has come into prominence more recently. This group is hepatoxic, causing in the liver either acute reactions with massive necrosis (that is, total destruction of tissues) or slow chronic symptoms of wasting with the development of extensive liver tumours according to the level and duration of ingestion. Comfrey belongs to the plant family of the Boraginaceae which includes the Heliotropium species with their high alkaloid content.
Nevertheless experience gained at over many years with feeding cattle and horses on Comfrey in different parts of the world has failed to produce any evidence of an acute reaction.
Equally well there is an absence of any direct evidence of liver tumours of the chronic reaction in Comfrey-fed animals having been observed in slaughter houses.
However, from this negative indirect evidence it cannot be decisively concluded that Comfrey does not present the toxic hazard of a chronic reaction because cattle bred for meat are slaughtered early in life when at their prime and long before chronic reaction develops later in life following the slow accumulation of alkaloid."
-'The Alkaloid Content of Comfrey' by Dr. D.B. Long, Ph.D, M.A., around **1970**. (In 'Comfrey: Fodder, Food & Remedy').

B. Unstable Alkaloid and Healthy Animals
"*Alkaloid in Comfrey is Unstable:*
The alkaloid proved to be unstable and was very easily oxidized and destroyed during extraction.
It would appear conceivable that a considerable portion of the alkaloid would be destroyed in preparation and cooking of the Comfrey flour.
With fresh herbage such as leaves in salads it is possible to eat only a relatively small quantity and with its high water content the amount of alkaloid thus actually eaten by man would be very small, and **there would naturally be present**

in such Comfrey herbage catalytic enzymes which would hasten the alkaloid destruction.

No Liver Problems:
Rats fed on a diet containing a high proportion of Comfrey flour over a long period were examined after death anatomically, biochemically and histologically for symptoms of a chronic reaction of the liver to the alkaloid. None of these symptoms were found in the experimental rats.

Healthy Livestock:
Livestock which may consume larger amounts of herbage frequently only eat it when wilted, and thus at a time when enzymatic breakdown could well have begun.

Certainly prolonged and extensive use of Comfrey herbage as a feeding stuff for animals has failed to reveal any deleterious effects, *but rather that of considerable benefit to the health of the livestock.*

Furthermore it must also be remembered that many other species of plant considered safe for foodstuffs actually contain toxic alkaloids but in amounts too small to be harmful."

-'The Alkaloid Content of Comfrey' by Dr. D.B. Long, Ph.D, M.A., around **1970**. (In 'Comfrey: Fodder, Food & Remedy').

C. Healthy Rats with Comfrey Diet

"*In 1978 Cheeke and Carlsson of Oregon State University* fed rats with diets containing various plant materials as potential sources of protein, the species tested being alfalfa, Chenopodium (goosefoot), Comfrey, Amaranthus (amaranth), Sudan grass and Atriplex (saltbush, orache). These diets were fed for 21 days groups of six male rats whose initial body weight was approximately 125 grams (4.4 ounces).* **Of the six diets only alfalfa and Comfrey gave normal weight gains** *when compared with diets containing conventional protein sources.*

Those rats fed the Comfrey diet, which contained 38% dried Comfrey leafmeal, gained an average of 4.8 grams (0.169 ounces) per day and consumed an average of 19.9 grams (0.701 ounces) of the diet each day. *This represents 7.56 grams (0.266 ounces) of dried Comfrey leaf meal per day, or 158 grams (5.57 ounces) of Comfrey leaf meal per rat over the duration of the experiment, or 127% of the initial body weight.*

No abnormalities in these rats were reported and, in fact, they did best of the six test groups on experimental diets.
It is essential that all information about control groups should be adequately documented, if one is to comment on the survival of test groups of animals."

-'The Safety of Comfrey: Report from the Henry Doubleday Research Association' by J.A. Pembery B.Sc., Special Advisor, Research Chemist, Aldgate Press: London, England, 20 pages, **1983**, page 10.
(* -'Evaluation of Several Crops as Sources of Leaf Meal: Composition, Effect of Drying Procedure, and Rat Growth Response' by P.R. Cheeke and R. Carlsson, Nutrition Reports International, Volume 18, No. 4, pages 465-472, 1978. I was unable to get this report. If you have a copy, could you please send it to me.)

D. Comfrey Helps Wound Healing

"**The crude extract of Symphytum officinale (Comfrey) afforded (helped) the cicatrization process by increasing at first the number of fibroblast and, in a later phase, the number of collagen fibers** *in experimental lesions produced in rats.*
The number of blood vessels was also increased *at the 7th day of treatment.*
On the experimental edema (swelling) induced (created) by carrageenin in rat's paws, the crude extract of Comfrey at doses of 150 and 300 mg/kg per os (by mouth) showed no effect.
Analgesic (pain killing) effect was seen with doses of 300 mg/kg per os (by mouth)."

-'Wound Healing and Analgesic Effect of Crude Extracts of Symphytum Officinale in Rats' by R.S. Goldman, P.C. de Freitas and S. Oga; Fitoterapia: The Journal for the Study of Medicinal Plants, Volume 56, No. 6, pages 323-330, **1986**.
(Cicatrization is wound healing with scar formation.)
(A fibroblast is a cell in connective tissue that produces collagen and other fibers.)
(Carrageenin is a colloidal extract from carrageen seaweed and other red algae.)

"**Other reports on Comfrey fed directly to animals or applied topically report favourable effects, for example, *Goldman et al report wound healing and analgesic properties of crude extracts of Symphytum officinale in rats.**"
-'In Defence of Comfrey' by Margaret Whitelegg, MNIMH. Paper presented to 'Department of Health' and 'Ministry of Agriculture, Fisheries and Food', 'National Institute of Medical Herbalists', United Kingdom, January 1993. Published in European Journal of Herbal Medicine, Exeter, England, Volume 1, No. 1, pages 11-17, 1994.   (* The above report.)

E. Healthy Consumers of Comfrey

"**Volunteers were sent a questionnaire regarding how long participants had been using Comfrey and the amount and form in which the herb was taken. At the same time liver function tests were performed at local hospital laboratories.**
There was considerable variation in the amount and way Comfrey was taken.

Actual alkaloid intake was impossible to determine because of the unknown origin or age of leaves used but, if the average value of about 1.0 mg alkaloid base per gramme of dry leaf is used as a guide, then intake would be around 1-10 mg alkaloid base per day, (i.e., 0.015 - 0.15 mg kg^{-1} d^{-1}). Acute toxic effects are found at around 30 mg alkaloid/kg body weight and carcinogenic effects at about 16 mg kg^{-1} d^{-1}.

Results of liver function tests in the 29 volunteers were found to be within the normal range *of the local laborato-*

ries with the exception of a slight elevation in bilirubin level in two and AST (aspartate aminotransferase) in one sample. Alpha-foetoprotein levels were normal in the seven volunteers tested. **There is no evidence of liver injury in this small sample, even for those who have been regularly taking the herb for 20 years."**
-'Comfrey and Liver Damage' by P.C. Anderson and A.E.M. McLean, Laboratory of Pharmakinetics and Toxicology, School of Medicine, University College, London, England; Human Toxicology Journal (now called Human and Experimental Toxicology), Volume 8, No. 1, pages 55 and 68-69, **1989**. Presented at Joint Meeting of the 'British Toxicology Society' and 'Institute of Biology', Oxford, England, September 22-23, 1988.

"Clinical Studies: There was no evidence of liver damage in a group of 29 people who had regularly consumed Comfrey (Anderson and McLean, 1989).
Twenty-nine volunteers responded to mailed questionnaires regarding duration, amount, and form of Comfrey used. At the same time, liver function tests (bilirubin, transaminase, and GGT) were performed on the volunteers.
Most volunteers (21/29) had used Comfrey for 1-10 years *(mean intake 3.0 grams dry leaf/day); 5/29 used it for 11-20 years (mean intake 2.6 g dry leaf/day); and 3/29 used it for 21-30 years (mean intake 11 g dry leaf/day)."*
-'Comfrey and One of Its Constituent Alkaloids: Symphytine, Review of Toxicological Literature' by Raymond Tice, Ph.D., Integrated Laboratory Systems, Research Triangle Park, North Carolina, October 1997.

"Another report, is of trials by Dr. Clare Anderson, from the Laboratory of Pharmakinetics and Toxicology, School of Medicine, University College, London, with testing long-term Comfrey consumers, who then submitted for liver function tests. All were found to have perfectly fit livers!
One of these long-term Comfrey consumers was Emsie du Plessis. **She had eaten quantities of raw Comfrey since 1960.** Her story, she wrote at 78 years of age: 'A few years ago, I felt a little below par, not really sick, but not my usual self. I went to the doctor, and he examined me and found that I was a bit anemic and that I should go to the hospital for a full blood test. I decided to postpone my visit to the hospital for 3-4 weeks, and to **step up my Comfrey eating to 3-4 leaves every day. When I returned to my doctor, he said my blood was perfect.** If the alkaloid damaged my liver, during that time, then I would like to think that allantoin is one of the world's most powerful, natural healers.' "
-How Can I Use Herbs in My Daily Life: Over 500 Herbs, Spices and Edible Plants: An Australian Practical Guide to Growing Culinary and Medicinal Herbs by Isabell Shipard. New York, New York: David Stewart Publisher / Simon & Schuster, Inc., 2003.
  (Anemia is when there are not enough healthy red blood cells to carry adequate oxygen to the body's tissues.)

F. Safer Use of Comfrey
"One statement that appeared in 'The Lancet' of June 27, 1981 is worthy of special note:
'People who consider the benefit of Comfrey to outweigh the perhaps slight risk involved may like to know that large mature leaves contain the lowest concentration of alkaloids.
The external use of Comfrey preparations should not be hazardous since the alkaloids are converted to toxic metabolites by liver enzymes only after being ingested.' "
-Comfrey: Nature's Healing Herb & Health Food by Andrew Hughes. Japan: Sanyusha Publishing Co., Ltd, **1992**, page 62.
('The Lancet', London, United Kingdom is a weekly peer-reviewed medical journal founded in 1823. This quote is from the Lancet article 'Comfrey Toxicity in Perspective' by Clare Anderson, Volume 317, Issue 8235, June 27 1981, page 1424. Also from the same article: *"In addition, **cooking will ensure some of the soluble alkaloids are lost in the vegetable water.**"*)

G. Ranking Cancer Hazards
"Dr. Bruce Ames, a respected scientist in the fields of carcinogenicity and mutagenicity has recently published an article in the journal 'Science', entitled 'Ranking Possible Carcinogenic Hazards' in Science Journal, April 1987. An index was developed called **HERP- Human Exposure Dose / Rodent Potency Dose.**'"
-Comfrey: Nature's Healing Herb & Health Food by Andrew Hughes. Japan: Sanyusha Publishing Co., Ltd, **1992**, page 75.

"**HERP- Human Exposure Dose / Rodent Potency Dose:**
(The higher the percent, the more carcinogenic the food or environment is.)

Hazard %	Daily Human Exposure	Carcinogen Dose 70 kg person (154 pounds)
0.03%	**Comfrey Herb Tea (1 cup=236 ml)**	**Symphytine Pyrrolizidine Alkaloids**
0.03%	Peanut Butter (32 grams= 1.12 oz)	Aflatoxin mycotoxin (mold toxin)
0.06%	Diet Coke (12 ounces= 354 ml)	Saccharin artificial sweetener
0.07%	Brown Mustard (5 grams= .17 oz)	Allyl isothiocyanate (organosulfur compound)
0.10%	Mushroom, raw (15 grams)	Hydrazines N_2H_4
0.10%	Basil (1 gram=.03 ounce dry leaf)	Estragole (a phenylpropene)
0.20%	Root Beer- Natural (12 ounces)	Safrole (a phenylpropene) from Sassafras tree
0.60%	Air, Conventional Home	Formaldehyde
2.80%	Beer (12 ounces= 354 ml)	Ethyl Alcohol
4.70%	Wine (8.4 ounces= 250 ml)	Ethyl Alcohol

This review discusses reasons why animal cancer tests cannot be used to predict absolute human risks. Such tests, however, may be used to indicate that some chemicals might be of greater concern than others.
Possible hazards to humans from a variety of rodent carcinogens are ranked by an index that relates the potency of each carcinogen in rodents to the exposure in humans."
-'Ranking Possible Carcinogenic Hazards' by Bruce N. Ames, Renae Magaw, and Lois S. Gold (Berkeley, California), Science Journal: American Association for the Advancement of Science, Vol 236, Issue 4799, page 271-280, April 17 1987.

H. Herb Practitioner Surveys: 1993 and 2014

i. "Questionnaire about Comfrey Use:

In 1993-1994 30,000 questionnaires were distributed to the public, through outlets most likely to reach Comfrey consumers: 'Green Farm' magazine, 'Nutritional Therapy Today', 'Henry Doubleday Research Association' newsletter, 'Living Earth' magazine, 'Herbs' magazine, and by direct mailing to the members of the 'National Institute of Medical Herbalists' and the 'General Council and Register of Consultant Herbalists'.

More than 1,000 questionaires were returned. 44.9% of consumers report consumption of fresh Comfrey leaves, 29.5% report consumption of tablets or capsules made from Comfrey root, and 25.5% report consumption of tablets or capsules made from Comfrey leaf. Many report consuming both fresh Comfrey and tablets/capsules.

Data on Comfrey tea was not requested in the questionnaire because Comfrey tea has not been banned (in the United Kingdom). Most people having begun Comfrey consumption 10 to 40 years ago. 3% of consumers have used Comfrey for more than 40 years (maximum 60 years). Consumption habits were mainly occasional, or 'as required'.

Health Benefits: *Although no effects which could be linked with long or short-term Comfrey toxicity were reported by any users, two people developed a skin allergy after Comfrey use. In contrast, many health benefits were reported by Comfrey users.*

Most striking was the number of people reporting relief from arthritic pain and various other types of pain such as cystitis, angina and pain from bone fractures. Reports of relief from gastrointestinal problems were also numerous, e.g., irritable bowel syndrome, gastric ulcers, diverticulitis. Reports of rapid healing of fractures and wounds ranked next.

Comfrey Ban:

There was much anger at the ban on Comfrey products, which was seen by many as 'senseless'.
It is suggested that the three known anecdotes of toxic reactions to Comfrey are likely to be due either to idiosyncratic factors or to contamination of the Comfrey crops or products with unknown toxins."
-'**The Safety-in-Use of Comfrey and Comfrey Products: Results of a Research Survey**' by the Society for the Promotion of Nutritional Therapy (SPNT), Linda Lazarides, Sussex County, United Kingdom, **1993-1994**, pages 1, 2.

ii. "Herbal Practitioners Use Comfrey: The survey aimed to assess how often and in what ways herbal practitioners use Comfrey (Symphytum officinale L.) externally in everyday practice.

A 2-sided A4 (8.27" x 11.69") survey was sent to all United Kingdom members of the 'National Institute of Medical Herbalists', the 'College of Practitioners of Phytotherapy' and the 'Association of Master Herbalists' with viable practice addresses (number = 598).

Results of Survey:

239 herbalists responded, of whom 179 (75%) reported regularly using Comfrey, in 15% of their consultations. It was most commonly prescribed as a cream for tendon, ligament and muscle problems, for fractures, and for wounds, the indications for which it was also perceived to be most effective.

Comfrey was rated least effective for haemorrhoids, varicose veins and boils and was considered to carry the greatest risk when prescribed for ulcers, wounds and boils.

Practitioner experience suggests that Comfrey can be used safely and effectively externally for certain indications."
-'**External Use of Comfrey: A Practitioner Survey**' by R. Frost, S. O'Meara and H. MacPherson, Department of Health Sciences, University of York, England; Complementary Therapies in Clinical Practice, Elsevier, Netherlands, Volume 20, No. 4, pages 347-355, November 2014.

I. Proper Doses of Herb

"*The question of dosage is central to any discussion of PA (Pyrrolizidine Alkaloid) toxicity; there is a huge disparity (difference) between dosage levels in research findings and dosage levels used by herbal practitioners.*
Many drugs have toxic effects at high doses but are widely prescribed at a recommended dosage. A classic exampe is acetaminophen (Paracetamol) which, at high doses, causes similar hepatocyte (liver) tissue damage to PAs."
-'**Using Herbs that Contain Pyrrolizidine Alkaloids**' by Alison Denham, B.A., MNIMH (National Institute of Medical Herbalists), University of Central Lancashire, England; The European Journal of Herbal Medicine, Volume 2, No. 3, pages 27-38, **1996**.

J. Alkaloids Excreted

"The concern over the health effects of Comfrey is based on the toxic pyrrolizidine alkaloid constituents. Unsaturated pyrrolizi-

dine alkaloids are metabolically activated to toxic compounds in the liver by mixed function oxidases.
In experimental animal studies, pyrrolizidine alkaloids are almost completely excreted within 24 hours."
-'Comfrey and One of Its Constituent Alkaloids: Symphytine, Review of Toxicological Literature' by Raymond Tice, Ph.D., Integrated Laboratory Systems, Research Triangle Park, North Carolina, October **1997**.

> **There still may be problems:** *"A concern that has arisen during the past two decades is that persons who take an herbal preparation regularly for many months or years might experience a cumulative (additive) effect from the PAs.* ***Although these ingredients do not accumulate in the liver or elsewhere in the body, the prolonged use of amounts that cause no acute reactions may increase the chances for developing liver disease, including liver cancer.***"
> -'Safety Issues Affecting Herbs: Pyrrolizidine Alkaloids" by Subhuti Dharmananda, Ph.D., Director, Institute for Traditional Medicine, Portland, Oregon, 2016.

K. Sometimes Comfrey Use is OK
"*A small number of case reports linking Comfrey intake with* **Hepatic Veno-Occlusive Disease (HVOD)** *have appeared in the past 20 years.*
> *In every case, there were numerous complicating circumstances, including ingestion of other potential hepatotoxins or the presence of other diseases that might predispose a patient to liver problems.*

Roots of Comfrey and coltsfoot are much higher generally in PAs (Pyrrolizidine Alkaloids); thus, **leaves and flowers are mostly used (and have been traditionally) to increase safety further.**
In addition, a species of Comfrey known as Symphytum uplandicum (Russian Comfrey) has been shown to contain much higher levels of PA than Symphytum officinale*.
Therefore, the belief that Comfrey is absolutely contraindicated for internal use is greatly oversimplified, and both Comfrey and coltsfoot remain as important and safe internal and external medications.
> Internally, Comfrey and coltsfoot are used for gastrointestinal (GI) tract inflammations, particularly in patients with peptic ulcer or gastritis. The two herbs can also soothe an inflamed pharynx and help to relieve coughs that stem from a variety of causes.
> Topically, Comfrey is most often used, primarily to help heal patients with wounds and ulcers. Up to 5 ml of tincture three times per day or 3 cups of tea made from PA-free Comfrey or coltsfoot daily is completely safe.

If PA content is unknown, patients should not use the products for more than 3 months consecutively internally. Short-term (a few days) use internally is also generally safe.
Creams or poultices can be used indefinitely with no ill effects.
Some possible signs of overdose: *jaundice (liver bile pigments in blood), dark urine, itching, edema (fluid retention), elevated (blood) serum enzyme levels.*"
-'**Misunderstood Toxic Herbs**' by Eric Yarnell, N.D., Journal of Alternative and Complementary Therapies, New Rochelle, New York, pages 6-11, February **1999**.
(* -'Echimidine Content of Commercial Comfrey {Symphytum spp.- Boraginaceae}' by D.V.C Awang et al., Bureau of Drug Research, Health and Welfare Canada, Ottawa, Ontario, Canada; Journal of Herbs, Spices and Medicinal Plants, Volume 2, No. 1, pages 21-34, 1993.)   (I do not recommend using coltsfoot unless your healthcare practitioner recommends it.)

L. Comfrey is Good for Health
"*Another Australian, who valued HDRA (Henry Doubleday Research Association) research, was* **Foster Savage** *who I had the opportunity to know him, personally, when he settled in Nambour (Queensland, Australia).*
He writes: '**I was perhaps responsible for 95% of the Comfrey in Australia, having introduced the plant to this country in 1954, and having used the plant in great quantities, since then; I am, perhaps, competent to speak about it and to make a few comments on the remarks about Comfrey made by the CSIRO (Commonwealth Scientific and Industrial Research Organisation) scientist.**
To say that two leaves, eaten daily –over a couple of years– will cause serious disease, is simply not true. In our house, we have eaten 70 leaves, or thereabouts, daily, for 24 years: in the form of Comfrey tea, liquidised in a vitamiser (blender) as a green drink, and in salads.
I also fed Comfrey to my farm animals.
> Knowing the power of Comfrey to restore a worn out animal quickly, and make her milk again, I once bought an old cow at the Dandenong Market, when farming in Victoria (Australia). It had been discarded by some farmer, as worn out.
> I put her on Comfrey, giving her 90 pounds (40.8 kg) of wilted Comfrey (wilted to increase the cow's intake of Comfrey's extraordinary nutrients), and 90 pounds made a pretty big heap, about 4 feet (1.2 meter) high.
> **This poor, old, creature took to the Comfrey, without hesitation. She was starving for minerals and her instincts gave her a craving for Comfrey.** When she began to eat, she would eat off the heap of leaves for a couple of hours, then sit down for an hour or so. Later, she would continue eating, until every leaf was gone.

If Dr. Culvenor's* words were true, imagine the poison she would be taking into her body, with this quantity of Comfrey daily. If Comfrey attacked the liver, then this cow would have died, because she was in a worn out condition.
> Instead, she doubled her milk output, within a week, and in a fortnight (2 weeks), trebled (tripled) it. The remarkable thing, was that the cream that settled overnight, was some 3/4 inch (1.9 cm) thick and the separation of cream from the

milk was so perfect, that the cream could be lifted off, with none remaining.
I fed Comfrey to calves, as much as they could eat, again with only gratifying results.
I fed pigs, entirely on Comfrey and grain, as much Comfrey as they could eat, and the quality of those pigs was legendary in the district. The fame of Comfrey spread far and wide, for my farm was visited by 6,000 farmers from around Australia and from overseas.
Finally, I well remember the enthusiastic remarks of the butcher who regularly killed our Comfrey-fed calves. He told us that he had never before, seen such healthy livers, that, mind you, after being reared on a herb that was supposed to cause liver diseases!'"
-**How Can I Use Herbs in My Daily Life:** Over 500 Herbs, Spices and Edible Plants: An Australian Practical Guide to Growing Culinary and Medicinal Herbs by Isabell Shipard. New York, New York: David Stewart Publisher / Simon & Schuster, **2003**.
(* -'Structure and Toxicity of the Alkaloids of Russian Comfrey {Symphytum x uplandicum, Nyman}, a Medicinal Herb and Item of Human Diet'by Claude C.J. Culvenor, M. Clarke, J.A. Edgar, J.L. Frahn, M.V. Jago, J.E. Peterson and L.W. Smith, CSIRO, Division of Animal Health, Victoria, Australia; Experientia, Switzerland, Volume 36, No. 4, pages 337-379, 1980.)

M. Comfrey Root is Not Mutagenic
*"***With regard to safety, the absence of genotoxic effects was demonstrated*** in the bacterial reverse mutation assay (Ames test)* **for a pyrrolizidine alkaloids (PA)-free Comfrey root liquid extract** *(*Benedek et al., 2010).*
  The extract was investigated for its ability to induce (create) gene mutations in Salmonella typhimurium strains TA 98, TA 100, TA 102, TA 1535 and TA 1537 with and without metabolic activation using the mammalian microsomal fraction S9 mix. Reference mutagens were used to check the validity of the experiments.
The Comfrey root extract showed no biologically relevant increases in revertant colony numbers of any of the five tester strains, neither in the presence nor in the absence of metabolic activation. ***In conclusion, the fluid Comfrey root extract was not mutagenic (cause gene mutation) in the bacterial reverse mutation assay."***
-**'Comfrey: A Clinical Overview'** by Christiane Staiger (Germany), Phytotherapy Research, Volume 26, Issue 10, pages 1441-1448, February **2012**, page 1446.
(* -'Absense of Mutagenic Effects of a Particular Symphytum Officinale L. Liquid Extract in the Bacterial Reverse Mutation Assay' by B. Benedek, A. Ziegler, and P. Ottersbach, Phytotherapy Research, Volume 24, pages 466-468, 2010.)

N. Comfrey Shown to Be Effective
"Comfrey has a long tradition as a medicinal plant. In general, ***the effects of Comfrey extracts can be described as pain relieving, anti-inflammatory and callus (bone repair) formation promoting.***
To date, the activity-determining constituents and mechanisms of action of the medicinal plant are only partly known. However, in accordance with the modern approach of evidence-based medicine, ***Comfrey extract creams have demonstrated their efficacy (effectiveness) and tolerability (acceptability) in a number of muscle and joint injuries, such as acute myalgia (muscle pain) in the back area, and in blunt injuries.***
Comfrey herb has also been shown to be efficacious in wound healing. Comfrey root has also proven to be efficacious in activated osteoarthritis, and equivalent or more efficacious in distortions (sprains) *compared with topical diclofenac (Voltaren)."*
-**'Comfrey: A Clinical Overview'** by Christiane Staiger (Germany), Phytotherapy Research, Volume 26, Issue 10, pages 1441-1448, February **2012**, page 1446.
  (A callus is a mass of tissue, at first uncalcified, that appears between and around the broken ends of bones.)

O. Herbs and Liver Disease: Case Study Problems
*"**This review deals with herbal hepatotoxicity, identical to Herb Induced Liver Injury (HILI), and critically summarizes the pitfalls associated with the evaluation of assumed HILI cases.***
Analysis of the relevant publications reveals that several dozens of different herbs and herbal products have been implicated to cause toxic liver disease, but ***major quality issues limit the validity of causality attribution (cause and effect).***
  Causality assessment of herbal hepatotoxicity is a major clinical and regulatory challenge based on low product quality, poor case data presentation and use of insufficient causality algorithms.
Though the production of herbal drugs is under regulatory surveillance and quality aspects are normally not a matter of concern, ***low quality of the less regulated herbal supplements may be a critical issue*** *considering product batch variability, impurities, adulterants (contaminants) and herb misidentifications.*
Regarding case data presentation, essential diagnostic information is often lacking, *as is the use of valid and liver specific causality assessment methods that also consider alternative diseases.*
  At present, causality is best assessed by using the Council for International Organizations of Medical Sciences scale (CIOMS) in its original or updated form, which should primarily be applied prospectively (before) by the treating physician when evaluating a patient rather than retrospectively (after) by regulatory agencies."
-'Herbal Hepatotoxicity: A Critical Review' by Rolf Teschke, Christian Frenzel, Xaver Glass, Johannes Schulze and Axel Eickhoff (Germany); British Journal of Clinical Pharmacology: British Pharmacological Society, Volume 75, No. 3, pages 630-636, **2012**.

P. Most Herbs are Safe if Used Properly

"Problems with Case Studies:
Many chemicals can cause hepatotoxicity. This is true of pharmaceuticals, industrial chemicals, and compounds in herbs. The current authors posit (propose) that, despite concerns raised in the conventional literature, there is no increase in herbal hepatotoxicity out of proportion to their use, there is no 'hidden epidemic' of herbal hepatotoxin.
Much of the purported (supposed) herb-induced hepatotoxicity appears in poor-quality case studies that fail to provide any analysis of the product(s) involved and thus fail to: (1) rule out adulteration and contamination; (2) provide important information about patients that might suggest alternative causes of liver disease; and (3) use existing tools for assessing causality*.
Most of the research discussed here supports the case that herbs can be associated with idiosyncratic liver toxicity but that there are very few situations in which the intrinsic hepatotoxicity of herbs have been shown. There is always a rare risk of idiosyncratic hapatotoxicity with any herb, medication, or chemical, and this cannot be the basis for restriction of use of these substances. Only those herbs with proven, consistent track records of problematic activity should be considered for regulatory control.
Safe Use of Herbs:
> *Unsaturated-Pyrrolizidine-Alkaloid (UPA) containing herbs should generally be avoided for long-term internal use, during pregnancy and lactaton (breast feeding), and by people with concomitant (at the same time) liver and/or kidney disease.*
> *Topical (skin) use is generally safe and many studies of Symphytum show it to be effective for treating ankle sprains and promoting wound healing.*
> *Short-term use of extracts with very low UPAs is acceptable in generally healthy patients, particularly with Symphytum officinale, which has proven to be one of the most effective herbs for healing peptic ulcers. Extracts tested and shown to be UPA-free can be used long-term safely."*

-'**Hepatotoxicity of Botanicals**' by Eric Yarnell, N.D. and Kathy Abascal, Journal of Alternative and Complementary Therapies, New Rochelle, New York, Volume 20, No 3, pages 136-144, June **2014**.
(* -'Herbal Hepatotoxicity: A Critical Review' by Rolf Teschke, Christian Frenzel, Xaver Glass, Johannes Schulze and Axel Eickhoff (Germany); British Journal of Clinical Pharmacology: British Pharmacological Society, Volume 75, No. 3, pages 630-636, 2012.)

> ("***Adverse drug reactions are classified as intrinsic/pharmacological or idiosyncratic.*** *Drugs or toxins that have an intrinsic/pharmacological hepatotoxicity are those that have predictable dose-response curves where higher concentrations cause more liver damage, and well characterized mechanisms of toxicity, such as directly damaging liver tissue or blocking a metabolic process.*
> *Idiosyncratic injury occurs without warning, when agents cause non-predictable hepatotoxicity in susceptible individuals, which is not related to dose and has a variable latency/dormancy period. This type of injury does not have a clear dose-response nor temporal relationship, and most often does not have predictive models."*
> -Wikipedia®: The Free Encyclopedia, www.wikipedia.org, 2018.)

More Alkaloid Research is Needed

"The controversy over Comfrey deserves closer inspection.
It appears to be damned by its association with the effects of other PA (Pyrrolizidine Alkaloid) containing plants, by the effects of its alkaloids on laboratory animals and by certain cases of hepatotoxicity through ingestion of the plant by humans.
Yet I would argue that the case against Comfrey is by no means proven in the scientific literature."
-'In Defence of Comfrey' by Margaret Whitelegg, MNIMH. Paper presented to 'Department of Health' and 'Ministry of Agriculture, Fisheries and Food', 'National Institute of Medical Herbalists', United Kingdom, January 1993. Published in European Journal of Herbal Medicine, Exeter, England, Volume 1, No. 1, pages 11-17, **1994**.

"The information currently available is not sufficient to permit an accurate assessment of the risks or potential therapeutic benefits of Comfrey. A more precise and credible (convincing) measure of health benefits would ensure appropriate use by herbalists and medical practitioners.
Research to date has often been flawed by the use of inappropriate animal models and faulty experimental design. Correct botanical identification and analysis of the plant material for PA (Pyrrolizidine Alkaloid) content and profile is essential."
-'Comfrey Toxicity Revisited' by Dorena Rode, Department of Animal Science, College of Agricultural and Environmental Sciences, University of California, Davis, California; Trends in Pharmacological Sciences: International Union Of Pharmacology, Volume 23, No. 11, pages 497-499, November **2002**.

Proper Botanical Identification is Needed

"Comfrey Plants are Sometimes Wrongly Identified: Comfrey leaves of commerce in Britain are collected from wild plants for use internally against colds and bronchitis and as a poultice or tea to alleviate sprains and associated inflammations (Ministry of Agriculture and Fisheries, 1941; Clapham, Tutin and Warburg, 1962).
Since extensive hybridisation (cross breeding) of S. officinale (Common Comfrey), particulary with S. asperum (Prickly Comfrey) occurs, it is virtually certain that most commercial material consists of leaves of S. x uplandicum (Russian

Comfrey), even if labelled S. officinale.
It is possible to distinguish whole plants of S. officinale, S. tuberosum and S. x uplandicum by morphological characteristics. Since commercial materials consists of the separated leaves or roots only, the present work was undertaken to show histological (microscopic structure) similarities and variation within the leaves of the three species: S. officinale L. (the leaf anatomy of which was described earlier by Fell and Peck, 1961), S. x uplandicum Nyman, and S. tuberosum L."
-'British Medicinal Species of the Genus Symphytum' by K.R. Fell and Janet M. Peck, Pharmacognosy Research Laboratories, University of Bradford, England; Planta Medica: Journal of Medicinal Plant and Natural Product Research, Volume 16, No. 2, pages 208-216, May **1968**.

"Much of the literature on Comfrey makes no reference to botanical name, so that doubt arises as to which species is meant."
-'Comfrey Symphytum spp. as a Forage Crop' by J.C. Forbes, A.D. McKelvie, and P.J.C. Saunders, North of Scotland College of Agriculture, Aberdeen, United Kingdom; Herbage Abstracts, Volume 49, No. 12, pages 523-539, **1979**.

"All Comfrey species investigated have been found to contain hepatotoxic pyrrolizidine alkaloids (PAs), but the literature on the subject is confused due to a glaring lack of attention to proper botanical identificaiton of the various Symphytum species studied."
-The Honest Herbal: A Sensible Guide to the Use of Herbs and Herbal Products by V.E. Tyler, Ph.D. Philadelphia, Pennsylvania: G.F. Stickley Co., **1981**. Now called 'Tyler's Honest Herbal'.

"A widespread lack of attention to proper botanical identification of Symphytum species by herbal investigators from different disciplines has led to much confusion and perpetuated (keep repeating) serious errors in the literature.
> *The 'British Medical Journal', for example, is very misleading in stating that at least nine PAs are present in the leaves and roots of Symphytum officinale. The reference cited*, in fact, lists the tabulated results of eight investigations covering, variously, whole plant, root, and herb, from four different geographic locations.*

Perhaps the most damaging account in this area is the publication from Japan which reports on the constituent of 'Symphytum officinale Linn.' **. The authors' statement that this species 'is called Comfrey or Russian Comfrey' is responsible for the widespread- and fallacious (false) - claim that that species contains echimidine.
> *It is quite obvious that these workers were dealing with a S. asperum (Prickly Comfrey) hybrid.*

-'Comfrey Update' by D.V.C. Awang, HerbalGram published by American Botanical Council, Austin, Texas, Volume 25, pages 20-23, Summer **1991**.
(* -'Plant Sources of Hepatotoxic Pyrrolizidine Alkaloids' by L.W. Smith and C.C.J. Culvenor, CSIRO Division of Animal Health, Parkville, Victoria, Australia; Journal of Natural Products, Volume 44, No. 2, pages 129-152, 1981.)
(** -'Studies on Constituents of Crude Drugs. I. Alkaloids of Symphytum Officinale Linn.' by T. Furuya and K. Araki from Kitasato University in Japan; Chemical & Pharmaceutical Bulletin, Volume 16, No. 12, pages 2512-2516, 1968.)

"Awang has shown that about half of the Comfrey products available in Canada contained echimidine, in spite of the fact that they were labeled as consisting of Common Comfrey. This inattention to proper taxonomic identification by suppliers creates a particularly dangerous situation. In addition, Comfrey roots contain about ten times the concentration of PAs (Pyrrolizidine Alkaloids) found in leaves, rendering the roots unsuitable for any therapeutic application. 'Health and Welfare Canada' has long refused to register any Comfrey root products for medicinal application."
-Tylers Herbs of Choice: The Therapeutic Use of Phytomedicinals, Third Edition by Dennis V.C. Awang. Boca Raton, Florida: CRC Press, **2009**.

"What is a voucher and why is it important in research? As a preserved specimen of an identified taxon deposited in a permanent and accessible storage facility, the voucher serves as the supporting material for published studies of the taxon and ensures that the science is repeatable. Vouchers are crucial in authenticating the taxonomy of an organism, as a tool for identifying localities of the taxon, and for additional taxonomic, genetic, ecological, and/or environmental research.
> *A voucher can be broadly defined as a representative sample of an expertly identified organism that is deposited and stored at a facility from which researchers may later obtain the specimen for examination and further study. As such, a voucher is a specimen that has been specifically collected and accessioned to support a research project (e.g., genetic analysis of a taxon) or activity (e.g., a floristic survey of a park). For plants, a voucher typically consists of a herbarium specimen, a pressed and dried sample of an individual containing aboveground structures (leaves, stems, flowers, and/or fruits) and below ground structures when possible."*

-Why Vouchers Matter in Botanical Research' by Theresa M. Culley, Editor-in-Chief, Applications in Plant Sciences, Department of Biological Sciences, University of Cincinnati, Ohio; Applications in Plant Sciences: Official Publication of the Botanical Society of America, Volume 1, No. 11, 5 pages, November **2013**.
> (A taxon is a taxonomic group of any rank, such as a species, family, or class.)
> (A voucher example: 'Voucher Specimen No. CANB 286704 Australian National Herbarium, Canberra'.)
> (Accession is a new addition to a plant collection, usually in reference to an herbarium.)

Examples of Wrong Identification of Comfrey Species

"Two pyrrolizidine alkaloids, symphytine (I), a new compound, and echimindine (II) have been isolated from the dried roots of Symphytum officinale. **Symphytum officinale Linn. is widely distributed in Europe and now cultivated in Japan. It is called Russian Comfrey,** *which generally indicates the dried roots, and in Japan the fresh or dried leaves or occasionally the dried roots."*
-'Studies on Constituents of Crude Drugs. I. Alkaloids of Symphytum Officinale Linn.' by T. Furuya and K. Araki from Kitasato University in Japan; Chemical & Pharmaceutical Bulletin, Volume 16, No. 12, pages 2512-2516, **1968**.
(Symphytum officinale is Common Comfrey. Symphytum x uplandicum is Russian Comfrey.)

"Yields of **Russian Comfrey (Symphytum asperum)** *sown in rows 50 cm apart..."*
-'Productivity and Composition of Russian Comfrey as a Fodder Crop in Egypt' by A. El-Bassousy, F. El-Sabban, A.M. Makky, M.S. El-Danasoury and M.A. El-Ashry, Animal Production Research Institute, Agriculture Research Center, Cairo, Egypt; Agricultural Research Review, Volume 53, No. 7, pages 51-57, **1975**.
(This is an example of wrong identification of Comfrey. Russian Comfrey is Symphytum x uplandicum. Prickly Comfrey is Symphytum asperum.)

"The carcinogenicity (cancer causing) of **Symphytum officinale L. (Russian Comfrey),** *used as a green vegetable or tonic, was studied in inbred ACI rats."*
-'Carcinogenic Activity of Symphytum Officinale' by I. Hirono et al, Japan National Cancer Institute Volume 61, No. 3, pages 865-869, September **1978**.
(Symphytum officinale is Common Comfrey. Symphytum x uplandicum is Russian Comfrey.)

"Carcinogenicity of Comfrey, Symphytum officinale L.:
S. officinale L. is a herb of the family Boraginaceae. This plant is called Comfrey or Russian Comfrey *and is cultivated for use in Japan as a green vegetable or tonic."*
-'Edible Plants Containing Naturally Occurring Carcinogens in Japan' by Iwao Hirono, Department of Pathology, Fujita Health University School of Medicine, Toyoake, Japan; Japanese Journal of Cancer Research, Tokyo, Volume 84, No. 10, pages 997-1006, October **1993**.   **(S. officinale L. is Common Comfrey, not Russian Comfrey.)**

"The first Canadian action was taken in 1982, when the 'Health Protection Branch of Health and Welfare Canada' introduced an amendment to 'Canada's Food and Drug Regulations' which prohibits the sale, for medicinal purposes, of any products containing echimidine (Canada Gazette, 30 March 1988).
The intent of this legislation is to have more careful attention paid to identification of botanical species by the herbal industry, *and to alert the Canadian public to the potential danger of PA consumption.*
There was no intent to underestimate the relative potential danger of echimidine-free S. officinale. Both root and leaf of 'S. officinale' have been shown to be carcinogenic in rats (Hirono et al. 1978), though here again **there is species confusion because the authors equate Common Comfrey and Russian Comfrey!"**
-'Comfrey Update' by D.V.C. Awang, HerbalGram published by American Botantical Council, Austin, Texas, Volume 25, pages 20-23, Summer **1991**.

"To differentiate it from other Comfreys, this species (Symphytum officinale) may be known as Common Comfrey, Quaker Comfrey, and Cultivated Comfrey."
-'Pharmacognostical Studies of Roots of Symphytum Officinale' by Manpreet Kaur and Hayat M. Mukhtar, Punjab, India, International Journal of Pharmaceutical Sciences Review and Research, Volume 45, No. 2, Article No. 27, pages 146-148, July-August **2017**.
(The author is correct about Symphytum officinale being called 'Common Comfrey'. However, 'Quaker Comfrey' is Russian Comfrey. 'Cultivated Comfrey' is a vague term with no certain usage. Many species of Comfrey are cultivated.
The above authors used this article as their source for the incorrect information: 'Potential of Russian Comfrey as an Animal Feedstuff in Uganda' by Bareeba, Odwongo and Mugerwa, Department of Animal Science, Faculty of Agriculture and Forestry, Proceedings of the First Uganda Pasture Network Workshop, Makerere University, Kampala, Uganda, Africa, 1987.
Unfortunately, errors like this get repeated by referring to other research articles that have incorrect information. I believe the authors are using Symphytum officinale because they got the roots through an authenticated botanical voucher system.)

Flawed Experimental Design

"The alkaloids appear to be more toxic to male rats than female rats, and more toxic to young rats than to mature rats.
Usually young rats are three times more susceptible (likely to be harmed) to the alkaloid than adults, that is one third the dose will produce the same effects as in the adult.
There is known resistance to alkaloids in rabbits, sheep, guinea pigs and Japanese quail."
-'The Safety of Comfrey: Report from the Henry Doubleday Research Association' by J.A. Pembery B.Sc., Special Advisor, Research Chemist, Aldgate Press: London, England, 20 pages, **1983**, page 9.

(Results of experiments vary with the age, sex and type of animal species being tested. These tests may not apply to humans.)

"It should be noted that consumption of Comfrey is usually at lower levels than in toxicity research.
Studies need to be conducted that involve the normal or low intake of Comfrey for a proper evaluation of the health hazard to people or farm animals. Comfrey can be used externally as a medicinal herb for the allantoin content and as a crop for composting, mulching, or green manuring."
-'Comfrey' by Teynor, Putnam, Doll, Kelling, Oelke, Undersander and Oplinger, 'Alternative Field Crops Manual', University of Wisconsin: Cooperative Extension, University of Minnesota: Center for Alternative Plant & Animal Products, and Minnesota Cooperative Extension, February 1992, updated November **1997**, page 6.

"Most of the research which has focused on Comfrey toxicity up to this date is often inadequate, flawed due to experimental animal models which are not applicable to humans and also faulty in experimental design.
The concept of using different animal species is contentious (controversial) because animal species vary considerably in their susceptibility (ability to be harmed) to PA (Pyrrolizidine Alkaloid) toxicity, and therefore it would be of more benefit to carry out toxicity testing on several animal species.
Attaining differential dose responses would also be a step forward in learning about safety in greater detail."
-'Is Comfrey Safe for Human Use' by Usmaan Hafiz, Elective Report for Advanced Topics in Pharmacy, Reading School of Pharmacy, University of Reading, Berkshire, England, 16 pages, date unknown but between **2007** and 2010, page 13.

"The evidence available thus far pertaining to Comfrey toxicity is not conclusive.
> *At first glance previous cases and studies do point towards a relationship between Comfrey ingestion and hepatotoxicity (liver toxicity), however, a deeper insight into the evidence reveals key unanswered questions and raises other relevant issues.*

Liver Function Tests: *Firstly, assessing hepatic (liver) function in the clinical setting normally involves monitoring serum (blood) concentration of certain proteins.*
> For example, elevations in aspartate aminotransferase (AST) may be reflective of liver pathology, gamma-glutamyl-transferase (GGT) and bilirubin levels can elevate with choleostasis, and alpha-fetoprotein (AFP) is a specific marker for liver cancer (*Rode 2002).

However, these liver markers (test results) are not necessarily elevated in every case of VOD (Veno-Occlusive Disease).
> A study by Anderson et al., (**1989) determined serum concentrations of AST, GGT and bilirubin in 29 long-term Comfrey users and AFP in a subgroup of seven Comfrey users. It can be noted that the cohort (group) is rather small and thus makes it difficult to ascertain (determine) risk, nonetheless, it is intriguing that AST, GGT, bilirubin and AFP levels were determined to be **in the normal range, especially after prolonged consumption of Comfrey leaf** (0.5 to 25 grams a day for 1 to 30 years).

This can suggest that not all those who ingest Comfrey (the study looked at long-term users) are subject to liver toxicity however, a larger cohort (group) and prospective (future looking) focus would have given this line of argument a more substantial basis.
Clinical Trials Needed: Furthermore, systematic toxicity testing or alternatively, clinical trials are yet to be performed for Comfrey ingestion. The restrictions on internal use of Comfrey are based upon evidence from a few cases and from studies which involved highly purified PA (Pyrrolizidine Alkaloid) administration to rodents (Rode 2002).
> *Furthermore, it can be argued that the toxicity which presents itself in humans occurs not due to Comfrey ingestion, but due to the consumption of a variety of other plants which contain PAs (Rode 2002)."*

-'Is Comfrey Safe for Human Use' by Usmaan Hafiz, Elective Report for Advanced Topics in Pharmacy, Reading School of Pharmacy, University of Reading, Berkshire, England, 16 pages, date unknown but between **2007** and 2010, page 7.
(* -'Comfrey Toxicity Revisited' by Dorena Rode, Department of Animal Science, College of Agricultural and Environmental Sciences, University of California, Davis, California; Trends in Pharmacological Sciences: International Union Of Pharmacology, Volume 23, No. 11, pages 497-499, November 2002.)
(** -'Comfrey and Liver Damage' by P.C. Anderson and A.E.M. McLean, Laboratory of Pharmakinetics and Toxicology, School of Medicine, University College, London, England; Human Toxicology Journal (now called Human and Experimental Toxicology), Volume 8, No. 1, pages 55 and 68-69, 1989. Presented at Joint Meeting of the 'British Toxicology Society' and 'Institute of Biology', Oxford, England, September 22-23, 1988.)

"Also widely circulated but never confirmed is the notion that Symphytum spp. (species) may contain the noxious diester PA, lasiocarpine- following a USSR report, which based identification of the alkaloid solely on paper chromatography! See **Awang 1987 for an evaluation of the evidence."*
-'Symphytum / Smeerwortel' (Dutch for Comfrey) by Maurice Godefridi, Kruidwis website for Herbalists Education, Belgium, February 20 **2014**.
(* -'Level of Alkaloids in Symphytum Officinale Dependent on the Phase of Plant Development' by I.V. Man'ko and B.K. Kotovskii, et al, Rastitel'nye Resursy {Vegetable Resources}, Volume 6, No. 3, pages 409-411, 1970, as cited in Chemical Abstracts: American Chemical Society, Volume 74, 61608, 1971. I could not find this report. If you have a copy, could you send it.)
(** -'Comfrey' or 'Herbal Medicine: Comfrey' by D.V.C. Awang, Health Protection Branch, Canada; Canadian Pharmaceutical Journal, Volume 120, pages 101-104, February 1987.)

"Rats are a standard experimental animal for cancer bioassay and toxicological research for chemicals. Although the

*genetic analyses were behind (less than) mice, rats have been more frequently used for toxicological research than mice. This is partly because **rats live longer than mice and induce (create) a wider variety of tumors**, which are morphologically similar to those in humans. **In addition, there are a number of chemicals that exhibit marked species differences in the carcinogenicity. These compounds are carcinogenic in rats but not in mice.***
Such examples are aflatoxin B1 and tamoxifen, both are carcinogenic to humans. Therefore, negative mutagenic/carcinogenic responses in mice do not guarantee that the chemical is not mutagenic/carcinogenic to rats or perhaps to humans.
To facilitate research on in vivo (live animal) mutagenesis and carcinogenesis, several transgenic rat models have been established. *In general, the transgenic rats for mutagenesis are treated with chemicals longer than transgenic mice for more exact examination of the relationship between mutagenesis and carcinogenesis.*
Transgenic rat models for carcinogenesis are engineered mostly to understand mechanisms underlying chemical carcinogenesis."
-'Transgenic Rat Models for Mutagenesis and Carcinogenesis' by Takehiko Nohmi, Kenichi Masumura and Naomi Toyoda-Hokaiwado, Division of Genetics and Mutagenesis, National Institute of Health Sciences,
Tokyo, Japan; Genes and Environment: Official Journal of the Japanese Environmental Mutagen Society, Volume 39, No. 11, 32 pages, **2017**.
> (Transgenic is an organism that contains genetic material into which DNA from an unrelated organism has been artificially introduced. Laboratory animals are genetically engineered to give specific results that researchers want. This does not seem to accurately reflect what will happen in human beings.)

Inaccurate Statements Made Based on Poor Understanding of Research

"Toxicology and Herbs: In recent years there have been reports of death and poisonings attributed to the use of medicinal plants such as Comfrey and chaparral. Herbalists have a responsibility to determine the inaccuracies in such reports. Literature concerning poisonous plants is replete with (full of) misinformation and erroneous reporting. These same mistakes continue to plague the reporting of poisoning by medicinal plants."
-'Phytochemical Investigation and Biological Activity of Leaves Extract of Plant BoswelliaSerrata' by T. Susan Srujana, Konduri Raveendra Babu and Bodavula Samba Siva Rao; Khammam, Andhra Pradesh, India; The Pharma Innovation Journal, New Delhia, India, Volume 1, No. 5, pages 22-46, **2012**.

Early research states something is true but later it is found questionable or false:
One huge problem is that a researcher will make a statement in an article that something is absolutely true. Then future researchers find it is not true or at the very least it is questionable.
However, later scientists and other people report this first research as an absolute truth. Sometimes decades go by with paper after paper repeating the same questionable or inaccurate statement as totaly true. This leads to misunderstandings and falsehoods about Comfrey.

> *"Once something is in print in a journal, it is often quoted in a superficial way as fact."*
> -Comfrey: Nature's Healing Herb & Health Food by Andrew Hughes. Japan: Sanyusha Publishing, **1992**, pages 73, 74.

> *"Yet this *Hirono paper is cited frequently in subsequent literature as concrete evidence of the carcinogenic properties of Comfrey in animals."*
> -'In Defence of Comfrey' by Margaret Whitelegg, MNIMH. Paper presented to 'Department of Health' and 'Ministry of Agriculture, Fisheries and Food', 'National Institute of Medical Herbalists', United Kingdom, January 1993. Published in European Journal of Herbal Medicine, Exeter, England, Volume 1, No. 1, pages 11-17, **1994**.
> (* For excerpts of why the results of the Hirono study are questionable, see number 4 in this section.)

Early article or book is misquoted many times:
I found in some papers that typographical and other errors made by some authors referencing the first research, is then repeated with the exact same mistakes by a third or fourth author. Some researchers never read the orginal article but instead repeat second- and third-hand information that may not be what the first author actually wrote.
This same problem occurs with references at the end of articles or books. Some researchers list the exact same research about Comfrey for the exact same reasons as other researchers. When I check the reference, it is not what the reference says.

> **An example.** The orginal article is about roots, then the later article that references it changes it to leaves. Then an even later article references the second article and also refers to leaves.:
> **1. Orginal article:**
> *"The <u>roots</u> of the Common Comfrey from the Ukraine contain tannins of the pyrogallic group (2.4%)."*
> -'Chemical Study of the Root of Common Comfrey (Symphytum officinale L.)' by G.V./H.V. Makarova, K.N. Zarai'ska and Y.H. Borisyuk.; Farmatsevtychnyi Zhurnal (Pharmaceutical Journal- Russia/Ukraine), Vol 20, No. 5, pages 41-43, **1966**.
> **2. Later article that references original article:**
> *"Tannin, probably responsible for the astringency of Comfrey leaf infusions, is reported to be 2.4% of dried <u>leaves</u>."*
> -'Comfrey' by D.V.C. Awang, Health Protection Branch, Health and Welfare Canada; Canadian Pharmaceutical Journal,

Volume 120, pages 101-104, February **1987**.
3. Much later article references second article:
"Known active constituents of Comfrey: Tannins may compose 2.4% of dried <u>leaves</u>."
-The 5-Minute Herb and Dietary Supplement Consult by Adriane Fugh-Berman. Philadelphia, Pennsylvania: Lippincott Williams & Wilkins, 2003.

Various articles condemn Comfrey unfairly. Example of false conclusions:
"Experiments have shown that one cup (236 ml) of tea brewed from Comfrey <u>root</u> contains approximately 8.5 mg of toxic alkaloids*. These levels of consumption, then, are 4 to 13 times as great as the amount of pyrrolizidine alkaloids accidentally consumed in contaminated wheat by a population of 7,200 people in Afghanistan. Within two years, 23 percent of that population were found to have experienced severe liver damage.*,**
The regular consumption of cups of Comfrey tea, one must conclude, is a dangerous practice."
-'Comfrey in the Chinese Materia Medica' by Robert Anderson, Mills College, Oakland, California; Asian Medicine Newsletter, International Association for the Study of Traditional Asian Medicine, New Series, Volume 2, pages 7-11, July **1992**.
(* -'Comfrey and Liver Damage' by James N. Roitman, Natural Products Chemistry, Western Regional Research Center, United States Department of Agriculture, Berkeley, California; The Lancet, Volume 317, Issue 8226, page 944, April 25 1981, a letter.)
(** -'An Outbreak of Hepatic Veno-Occlusive Disease in North-Western Afghanistan' by Mohabbat, Younos, Merzad, Srivastava, Sediq, and Aram, The Lancet, London, England, Volume 308, No. 7980, pages 269-271, August 7 1976.)

The statements by this author are a good example of how misunderstood research is repeated as absolute truths, and then wrong conclusions are made.
1. The author makes the statement that drinking 'Comfrey tea' is dangerous. The only conclusion the author should have made is that drinking 'Comfrey <u>root</u> tea' is dangerous. Comfrey leaf tea is not part of this research.
2. Please read the reports and rebuttals for the two reports mentioned. They are numbers 3 and 8 in this section. There is disagreement about how much alkaloids were in the wheat in Afghanistan so these conclusions are suspect.
3. The poisoning in Afghanistan was not from Comfrey. This Afghanistan population was suffering from malnutrition after a 2 year drought. Liver damage could be caused by more than just alkaloids.

It is common for statements about Comfrey to be made with no references or incomplete references that can not be verified: Sometimes references are given to books or articles about Comfrey that themselves give no references as to their source for that statement.

Many times I have followed references back over 3 or 4 referrals from one article to another article to another article, with in the final article, there is no source for the statement. So it is impossible to determine if it is accurate or not. It is about as scientific as saying: 'I heard it through the grapevine'.

If an important or unusual statement is made in a book or article, I always go back through references to find the final source. If I can't find the source, most of the time I disregard that statement. Sometimes I have added them to the book so it can be compared to other statements on that topic.

Giving just a web site address is not a good way to reference information because addresses change or no longer exist. Giving a web site address with the name of the article or book included is OK. The more information, the better. Sometimes references are hard to understand, with not enough information given to find the original source.

Some articles or books condemn Comfrey based on problems with other herbs:
Another problem is that some articles or books make statements that Comfrey is bad for health, and then the references that are supposed to back up that statement are not about Comfrey but about another herb that has alkaloids.
Sometimes these references about alkaloids in other herbs are types of alkaloids that are not even in Comfrey.
Or if Comfrey does have that alkaloid, the amounts are significantly less in Comfrey.
Some authors make blanket statements that all alkaloids are bad, regardless of any other factors.

Some articles or books condemn Comfrey when problems are caused by herb mixtures:
There are case histories of people getting sick when they consumed large quantities of herb mixtures over long periods of time. Sometimes these herb mixtures contain 10 or more different herbs. However, there are articles that blame only Comfrey for the health problems when it is very possible for other herbs in the mixture to have caused the problems. At the very least, the negative health effects could be caused by a combination of all the herbs, not just Comfrey.

Example of Comfrey in Herb Mixture:
-'Acute Hepatitis After Ingestion of Herbs' by J.A. Shad, C.G. Chinn and O.S. Brann, Department of Internal Medicine, Naval Medical Center, San Diego, California; Southern Medical Journal, Birmingham, Alabama, Volume 92, No. 11, pages 1095-1097, November 1999.   (See number 25 in this section.)

<u>Putting Alkaloids in Perspective</u>

"Possible Long Term Effects: There is increasing recognition that the hepatotoxic (liver toxic) pyrrolizidine alkaloids

(not just Comfrey but all similar herbs) may be of greater importance as a cause of human disease than the presently known outbreaks of poisoning would indicate.
The chronic and progressive character of their effects, including carcinogenicity (cancer causing) means that in the type of hazard they present, they resemble the mycotoxins rather than the main body of alkaloids.
> *Disease is likely to be induced (created) in man or animals by their ingestion over long periods of time in plants or foods in which they are present at low concentrations.*

More Education Needed:
> *In the interests of public health, present knowledge of plant species containing hepatotoxic pyrrolizidine alkaloids should be diffused as widely as possible and brought particularly to the people who are most at risk.*
> *The purpose of this compilation of plant sources is to make the information more readily available to the health-oriented investigators and educators who have this task.*

Mycotoxins More Dangerous than Alkaloids: *The early warnings by Schoental* about the hazard of pyrrolizidine alkaloids in medicinal herbs were probably overshadowed to a large extent by the discovery of the greater and more ubiquitous (everywhere) hazard of the mycotoxins."*
-'Plant Sources of Hepatotoxic Pyrrolizidine Alkaloids' by L.W. Smith and C.C.J. Culvenor, CSIRO Division of Animal Health, Parkville, Victoria, Australia; Journal of Natural Products, Volume 44, No. 2, pages 129-152, **1981**.
(* -'Kwashiorkor-Like Syndrome and Other Pathological Changes in Rats as a Result of Feeding with Senecio Alkaloids {Isatidine}' by R. Schoental, Voeding, Volume 16, pages 268-285, 1955.)
> (A mycotoxin is a toxic substance produced by a fungus and especially a mold.)
> (The long-term effects of low doses of Comfrey still need further research.)

"Tea, almonds, apples, pears, mustard, radishes and hops, to list only a few items, all contain substances which, if extracted, can be shown to be poisonous when tested under conditions similar to those used in Comfrey experiments. Must we then ignore our experience of the usefulness and wholesomeness of these foods because controlled trials and scientific evidence have not been published to establish their safety?"
-'Comfrey as a Medicine' by F. Fletcher Hyde, FNIMH, National Institute of Medical Herbalists, Exeter, England, Press Release, undated but probably **1982**.

"I would suggest that many common foods would have to be banned if specific constituents were given in large doses to rats, or humans for that matter.
A few examples are: glucosinolates in cabbage, myristicin in parsley, safrole in aniseed and cocoa.
Other examples are given in the 'United States Academy of Sciences' publication 'Toxicants Occurring Naturally in Food', such as the 150 distinct chemicals in the potato, including the solanine alkaloids, arsenic, tannins, nitrate and oxalic acid.
It is toxicologically axiomatic (self evident) that if almost any one of these were tested in experimental animals by today's standards of safety evaluation, it would be shown to be toxic. The human organism can readily tolerate small amounts of many different chemical substances, even though any one of them might not be tolerated in a somewhat larger amount.
When viewed along side the potential and real dangers of salt, alcohol, cigarettes, nitrites, aspirin, and the commonly used pharmaceuticals, the current legislative measures against Comfrey are unduly harsh."
-'Pharmacists and Comfrey' by Diane Wiesner, B.Pharm, MA, Ph.D., MPS, Principal of the NSW College of Natural Therapies, Sydney, Australia; Australian Journal of Pharmacy, Volume 65, pages 959-963, **1984**.

"Over and above the scientific arguments that advocate (support) a reduced intake of Comfrey, the question of an individual's freedom of choice to determine an acceptable risk must be considered. Unfortunately, public perceptions of acceptable risk rarely coincide with the real risk, and can be influenced greatly by the perceived benefit.
*An example of this is **the widespread misconception that naturally occurring chemicals intrinsically, essentially) are less harmful than are the synthetic (man made) chemicals** that are used in food. In some cases, the limited toxicological data on naturally-occurring chemicals can hide the potential hazards, and the health risks in these cases cannot be ignored."*
-'Comfrey: Assessing the Low-Dose Health Risk' by Peter J. Abbott, Toxicology Unit, Department of Community Services and Health, Canberra, Australia; Medical Journal of Australia, Volume 149, No. 11-12, pages 678-682, **1988**.

*"**A. What dose is toxic?** It is not easy to identify a toxic dose of pyrrolizidine alkaloids for several reasons:*
> *1) Pyrrolizidine alkaloids vary widely among themselves in potential toxicity.*
>> *One alkaloid can be six or more times as toxic as another, and some are not toxic at all.*
> *2) Plants often contain more than one alkaloid, and proportions may vary from plant to plant and time to time.*
>> **Comfrey (S. officinale) contains at least nine alkaloids of varying degrees of toxicity.**
> *3) The alkaloid content of plants may vary widely from one plant part to another.*
>> *A test sample showed that the pyrrolizidine alkaloids in a commercial sample of Comfrey root (S. officinale) were 10.7 times the level in the leaf.*
> *4) The content can vary from season to season. The amount of alkaloids in Russian Comfrey*
>> *(S. uplandicum) leaves can vary at least up to 16 times depending on time of year and size of the leaf.*
> *5) The dose may also vary widely depending on age.*
> *6) There may be differences between one person and another. Malnutrition may be a predisposing factor.*
B. When is toxicity known? *What of the argument that if Comfrey were toxic, it would be known by now after centuries of use?*
> ***Pyrrolizidine alkaloid poisoning was only discovered in 1954.** It is insidious (gradually harmful), and can take*

from 2-13 weeks for onset of symptoms, even after stopping ingestion. In the various epidemics and single cases around the world, none of those who consumed them suspected the plants as the cause.

C. Is the Problem Common?
Should we really be concerned when tons of Comfrey root are consumed each year in the United States alone, and we can only find 2 cases of Comfrey poisoning in the scientific literature?

Possibly so- the extent of the problem could be much wider than assumed. Most significant is that hepatic (liver) veno-occlusive disease has been misdiagnosed as viral hepatitis or cirrhosis. The most serious risk of pyrrolizidine alkaloid poisoning in herbal practice in the United States appears to be the use of Comfrey root for chronic digestive conditions. Such treatment assumes long-term use, and patients are often weakened from the disease.

D. Public Health Hazards are Greater with Conventional Drugs:
In perspective, the public health hazards of pyrrolizidine alkaloid poisoning pale (are much less) when compared with to those of conventional drugs. Adverse drug reaction costs the United States about 5 billion a year. **Ten to twenty thousand people a year die from gastric complications of non-steroidal anti-inflammatory drugs alone (NSAIDs such as aspirin, ibuprofen, naproxen)**, and recent study shows that overly aggressive treatment of high blood pressure with drugs may cause 25,000 heart attacks a year. Nevertheless, the hypocrisy of regulatory agencies should not prevent clinicians from protecting their patients from potential harm."

-'Symphytum: Comfrey, Coltsfoot, and Pyrrolizidine Alkaloids' by Paul Bergner, Medical Herbalism: Journal for the Clinical Practitioner, Oregon, Volume 1, No. 1, pages 1, 3-5, around **1990**.

"With herbal preparations as well as with modern pharmaceuticals, the physician must undertake a therapeutic cost-benefit analysis. What health benefit does the drug provide for the patient, and is this sufficiently important to make unwanted side-effects acceptable?
Aspirin, the contemporary descendant of an old herbal medication (willow bark), may be taken to illustrate this principle.
The analgesic (pain killer), antipyretic (fever reducer), and anti-inflammatory benefits of aspirin are sufficiently valued to make transient (temporary) gastritis (upset stomach) an acceptable side-effect for many patients, although not for those with a tendency to ulcer formation.
In the last two decades, however, we have become aware that aspirin can cause encephalopathy (brain disease) and fatty metamorphosis of the liver (Reye's syndrome) leading to permanent impairment and death in infants and children.
This sinister (alarming) adverse (negative) reaction required a dramatic re-evaluation of the cost-benefit status of aspirin, which is now contraindicated (not recommended) for the treatment of infants and children suffering from viral infections such as chicken pox."

-'Comfrey in the Chinese Materia Medica' by Robert Anderson, Mills College, Oakland, California; Asian Medicine Newsletter, International Association for the Study of Traditional Asian Medicine, New Series, Volume 2, pages 7-11, July **1992**.

"While for orthodox medicine the laboratory experiments lead to a 'universal knowledge' , i.e., they are accessing the 'pure' effects of a substance on the body, for herbalists, the context is all-important.
The rigour (thoroughness/strictness) of the laboratory finding may be more or less demonstrable, but the relevance of the findings to the effect of the whole herb is questionable.
One can find similar cases of abuse or harmful effects of orthodox medicines throughout the medical literature, but this does not necessarily lead to withdrawal or limitation of the drug.
Despite claims that the rigour of science is being applied to the investigation of the Comfrey plant, there is much uncertainty within the debate, and the uncertainty is often ignored. The levels of PAs (Pyrrolizidine Alkaloids) in the leaves and roots of Comfrey are never the same in any two experiments, indeed never the same in any two leaves."

-'In Defence of Comfrey' by Margaret Whitelegg, MNIMH. Paper presented to 'Department of Health' and 'Ministry of Agriculture, Fisheries and Food', 'National Institute of Medical Herbalists', United Kingdom, January 1993. Published in European Journal of Herbal Medicine, Exeter, England, Volume 1, No. 1, pages 11-17, **1994**.
(Orthodox medicine is a system where medical doctors and healthcare professionals such as nurses/pharmacists treat disease with drugs, radiation and surgery. Also called allopathic medicine, conventional medicine, mainstream medicine, Western medicine.) (Context is the circumstances that form the setting of an event or idea, and in terms of which it can be fully understood.)

"Is Comfrey Safe?
Herb safety issues are complex, extending far beyond the realm of the reliable, objective scientific method.
The questions raised and answers given often need to be deconstructed (reduced to parts) in the light, when assessing not only the toxicological facts, but also the political and sociological context within which the discussion is occurring.
Unfortunately, there is a common assumption that might be paraphrased as 'Herbs are natural, and therefore safe.' This is clearly not the case. If 'safe' is taken to mean completely risk-free, then the answer must be no, Comfrey is not safe.
For example, as with all herbs and drugs, there is a small chance that an allergic response will occur.
The following quotation comes from Dorena Rode, author of an insightful review on Comfrey toxicity:
'One might expect that new toxicity research or an unacceptable number of adverse reactions prompted these recent actions, but neither is the case.'*
The very few specific reports of human toxicity related to Comfrey all come from the period between 1980 and 1990, when a number of cases of veno-occlusive disease were reported. There is no question about the diagnoses.
However, it is important to note that in these cases, the connection with Comfrey was not considered in the context of other contributing factors.

For example, concomitant (at the same time) illness, the use of prescription or over-the-counter hepatotoxic drugs (like acetaminophen, for example), and impaired nutritional status clearly increase the likelihood that PA-containing herbs will cause hepatotoxicity.
With minimal epidemiological data, what insights can be garnered (gained) from the laboratory research into toxicity?
  As with many statements about herbal toxicity, the evidence proffered (given) comes primarily from rodent studies that utilized high levels of purified PAs.
  No systematic toxicity testing or clinical trials of Comfrey have been performed. Although PA poisoning in humans does occur, it is most commonly a consequence of consuming plants other than Comfrey.* "

-Medical Herbalism: The Science and Practice of Herbal Medicine by David Hoffmann, FNIMH, AHG. Rochester, Vermont: Healing Arts Press, **2003**, pages 188-190.

(* -'Comfrey Toxicity Revisited' by Dorena Rode, Department of Animal Science, College of Agricultural and Environmental Sciences, University of California, Davis, California; Trends in Pharmacological Sciences: International Union Of Pharmacology, Volume 23, No. 11, pages 497-499, November 2002.)

(** -'The Efficacy and Safety of Comfrey' by F. Stickel and H.K. Seitz, Public Health Nutrition, United Kingdom, 3-4A, pages 501-508, December 2000.)

(*** -Natural Toxicants in Feeds, Forages, and Poisonous Plants by Peter Cheeke. Danville, Illinois: Interstate Publishers, 1998.)
  (The scientific method is a procedure used in science since the 1600s. It involves systematic observation, measurement, and experiment, and the formulation and testing of various hypotheses.)
  (Epidemiology is the study of pubic health. It monitors the incidence, distribution, cause, and control of diseases and other health factors in a population.)

"In Comfrey's case, due to a configuration of events, the research that today is used to cite Comfrey as a toxic plant was and still is mis-applied.
  *1. The few cases of hepatotoxicity that were 'reported' for Comfrey use in late 1980s and early 1990s **mainly involved individuals who had pre-existing liver-centered illnesses, and/or who who were users of hepatotoxic drugs.***
  *2. **Comfrey toxicity is mainly theoretical and is essentially based only upon animal studies.** Purified PA fractions (not Comfrey) were fed to rats, who then developed hepatotoxic reactions. High administered PA to body weight ratio, and the fact that not all animals process PAs the same, factor into the alarmist result.*
  *3. **Where do humans fit in? The fact is we don't really know,** but the most likely the situation with Comfrey is not as dire as some would suggest.*
  ***Some perspective is needed.** How many individuals die each year of liver-related pharmaceutical/over-the-counter poisonings? Hundreds, if not thousands. How many individuals have succumbed to the rational use of only Comfrey, with no pharmaceutical/prior disease history… zero!*
  *4. **Comfrey's overall usage history is so empty of toxicity report,** it is strange that an apparent concentration of cases developed when they did. With some speculation it is not far-fetched to suggest that the few individuals supposedly affected by 'Comfrey toxicity' were stricken, despite, not because of Comfrey."*

-Herbal Medicine Trends and Tradition: A Comprehensive Sourcebook on the Preparation and Use of Medicinal Plants by Charles W. Kane. Tucson, Arizona: Lincoln Town Press, **2009**.

*"**What is a poison? A poison is substance which has a harmful effect on a living system.***
  Paracelsus (1493-1541) was one of the first to distinguish between the therapeutic and toxic properties of substances. He thought the only difference between a medicine and a poison was the dose. Very few substances are actually classed as a 'poison'. Harmful chemicals are not necessarily poisons.
We are exposed to potentially toxic substances every day without immediate harm. Our bodies can usually safely metabolize toxins if we are exposed to them in small amounts. It is only when we overwhelm our body and reach the toxic dose of a substance that life threatening results occur.
  That is, all substances have a potential toxicity. All herbs can therefore be harmful, but most would have to be ingested in impossible amounts to cause harm. Herbs which have a high toxicity, such as Gelsemium and Aconitum, can be used safely and effective if taken in a small, therapeutic dose.
Thus, the primary determinate of the safety of a substance is the dose, not the herb, which makes the poison. A correct use of semantics and a correct understanding of these terms are crucial to avoid confusion and misinformation."

-'Phytochemical Investigation and Biological Activity of Leaves Extract of Plant BoswelliaSerrata' by T. Susan Srujana, Konduri Raveendra Babu and Bodavula Samba Siva Rao; Khammam, Andhra Pradesh, India; The Pharma Innovation Journal, New Delhia, India, Volume 1, No. 5, pages 22-46, **2012**.

'Drug Induced Liver Injury' in general, not just from Comfrey:
"**The diagnosis of Drug Induced Liver Injury (DILI) is based primarily on the exclusion of alternative causes.** To assess the frequency of alternative causes in initially suspected DILI cases, we searched the Medline® database with the following terms: drug hepatotoxicity, drug induced liver injury, and hepatotoxic drugs.
  A total of 2,906 cases of initially assumed DILI were analyzed in these 15 publications, with **diagnoses missed in 14% of the cases due to overt alternative causes. In another 11%, the diagnosis of DILI could not be established because of confounding variables.**
  Alternative diagnoses included hepatitis B, C, and E, CMV, EBV, ischemic hepatitis, autoimmune hepatitis, primary biliary cirrhosis, primary sclerosing cholangitis, hemochromatosis, Wilson's disease, Gilbert's syndrome, fatty liver, non

alcoholic steatohepatitis, alcoholic liver diseases, cardiac and thyroid causes, rhabdomyolysis, polymyositis, postictal state, tumors, lymphomas, chlamydial and HIV infections.
In conclusion, alternative diagnoses are common in primarily suspected DILI cases *and should be excluded early in future cases, requiring a thorough clinical and causality assessment."*
-'Drug Induced Liver Injury: Accuracy of Diagnosis in Published Reports' by R. Teschke, C. Frenzel, A. Wolff, A. Eickhoff and J. Schulze (Germany); Annals of Hepatology, Volume 13, No. 2, pages 248-255, **2014**.
(Medline® {Medical Literature Analysis and Retrieval System Online} from the 'United States National Library of Medicine' is a bibliographic database of life sciences and biomedical information. It includes articles from academic journals about medicine, nursing, pharmacy, dentistry, veterinary medicine, and health care.)

'Drug Induced Liver Injury' in general, not just from Comfrey:
"Drug-induced liver injury (DILI) is common and nearly all classes of medications can cause liver disease. Most cases of DILI are benign (not cancerous or seriously harmful), and improve after drug withdrawal. It is important to recognize and remove the offending agent as quickly as possible to prevent the progression to chronic liver disease and/or acute liver failure.
There are no definite risk factors for DILI, but preexisting liver disease and genetic susceptibility may predispose certain individuals. *Although most patients have clinical symptoms that are identical to other liver diseases, some patients may present with symptoms of systemic hypersensitivity. Treatment of drug and herbal-induced liver injury consists of rapid drug discontinuation and supportive care targeted to alleviate unwanted symptoms."*
-'Drug-Induced Liver Injury' by Stefan David, M.D. and James P. Hamilton, M.D., Division of Gastroenterology and Hepatology, The Johns Hopkins University School of Medicine, Baltimore, Maryland; US Gastroenterology and Hepatology Review, Volume 6, pages 73-80, January 1 2010.

'Drug Induced Liver Injury' in general, not just from Comfrey:
"Drug-Induced Liver Injury (DILI) (Drug-Induced Hepatotoxicity) is an uncommon, but potentially fatal, cause of liver disease that is associated with prescription medications, OTC drugs, and herbal and dietary supplements (HDS). DILI has two types: intrinsic and idiosyncratic.
Intrinsic DILI *refers to liver toxicity induced by a drug in a predictable and dose-related manner (e.g., acetaminophen [APAP]).* **Idiosyncratic DILI***, which occurs less frequently, is associated with a less consistent dose-toxicity relationship and a more varied presentation.*
Patient, environmental, and drug-related factors may play a role in pathogenesis of DILI. In United States, antibiotics and antiepileptic drugs are the most common drug classes associated with DILI, but HDS are on the rise as a cause."
-'Drug-Induced Liver Injury: An Overview' by Donna Lisi, PharmD, BCPS, BCPP, Clinical Pharmacist, Somerset, New Jersey; United States Pharmacist: Pharmacist's Resource for Clinical Excellence, Volume 41, No. 12, pages 30-34, December 2016.

'Attribution of causation (of tissue damage) in individual cases is replete (abundant) with problems. Teschke and his team have investigated these problems in cases of hepatotoxicity and drug-induced liver disease attributed to consumption of medicinal plants (*).*
-'Pyrrolizidine Alkaloids: Key Points for Herbal Practitioners' by Helen Phillips M.Sc. Herbal Medicine, Diploma of Phytotherapy, MNIMH and Alison Denham, M.A., FNIMH (National Institute of Medical Herbalists); The Forager's Path, LLC, School of Botanical Studies, Flagstaff, Arizona, August **2016**.
(* -'Drug Induced Liver Injury: Accuracy of Diagnosis in Published Reports' by R. Teschke, C. Frenzel, A. Wolff, A. Eickhoff and J. Schulze (Germany); Annals of Hepatology, Volume 13, No. 2, pages 248-255, 2014.)

<u>Comfrey with Low or No Alkaloids</u>

*"**The variations in PA (Pyrrolizidine Alkaloid) concentration of both wild-collected and commercial samples suggest that this is an area where cultivation may be appropriate to develop low PA / high allantoin varieties.**"*
-'Using Herbs that Contain Pyrrolizidine Alkaloids' by Alison Denham, B.A., MNIMH (National Institute of Medical Herbalists), University of Central Lancashire, England; The European Journal of Herbal Medicine, Volume 2, No. 3, pages 27-38, **1996**.
(Cultivation is the planting, tending, improving, propagating and harvesting of plants.)

"Preparations derived from Comfrey are well known in traditional medicine for their use in the treatment of patients with blunt trauma, such as broken limbs, distortions, contusions, or pulled muscles and tendons. Anti-inflammatory, analgesic, anti-edematous, and wound-healing properties have been noted with Symphytum extracts.
Applicability in the traditional sense is especially well documented for topical (skin) preparations, as has been demonstrated in several clinical trials. The effects of Symphytum were officially approved in a German Commission E monograph. Topical Symphytum preparations have a very good safety profile.
Comfrey Cultivation and Cultivars:
Concerns about the possible content of pyrrolizidine alkaloids with 1,2-unsaturated necine structure can be overcome through **the use of plant material derived from cultivars that are devoid of alkaloids in the aerial (above ground) parts.**
Herbal raw material was exclusively obtained by means of a controlled cultivation of the Symphytum officinale hybrid

Symphytum x uplandicum Nyman (synonym Symphytum peregrinum auct. non Ledeb., Russian Comfrey), which does not include alkaloids in the aerial parts."
-'Topical Symphytum Herb Concentrate Cream Against Myalgia: A Randomized Controlled Double-Blind Clinical Study' by Kucera, Barna, Horacek, Kalal, Kucera, and Hladikova, Advances in Therapy Journal, Voluem 22, No 6, page 681-692, November-December **2005**.

(I do not know what 'controlled cultivation' means in this context. If you have more information, please let me know.)

"Due to contrasting toxicity of the different types of PAs (Pyrrolizidine Alkaloids) and the variation in distribution of PAs amongst the Comfreys species, it can be established that research relating to toxicity and safety which focuses on one Comfrey species may not accurately portray the results that would be obtained if the tests were subject to an alternative species of Comfrey.

This alludes (refers) to the fact that one Comfrey species may be safer to consume than an alternative of the same species, yet it seems both are capable of causing harmful affects if consumed in excessive amounts."
-'Is Comfrey Safe for Human Use' by Usmaan Hafiz, Elective Report for Advanced Topics in Pharmacy, Reading School of Pharmacy, University of Reading, Berkshire, England, 16 pages, date unknown but **2006**-2018, page 9.

"Extracts from the Caucasian species of Comfrey: Symphytum asperum (Prickly Comfrey) and Symphytum caucasicum have been used in folk medicine in the treatment of some kinds of disorders, mainly fractures and wounds.

The aforenamed extracts contain allantoin, claimed to be a cell proliferation-stimulating agent responsible for the wound-healing properties of Symphytum, and, on the other hand, hepatotoxic pyrrolizidine alkaloids which strongly restrict internal use of Comfrey extracts.
In the present investigation, we obtained allantoin- and toxic pyrrolizidine alkaloids-free composition containing crude polysaccharides and novel biopolymer from S. asperum roots:

The main constituent of S. asperum and S. caucasicum high-molecular (>1000 kDa) fractions was found to be **a new caffeic acid-derived polymer, namely poly[3-(3,4-dihydroxyphenyl)glyceric acid] (PDGA)** *or poly[oxy-1-carboxy-2-(3,4-dihydroxyphenyl)ethylene] [10-13]. It is a super-gigantic caffeic acid-derived polymer.*

We attempted to appraise its pharmacological properties in 'in vitro' (anticomplementary and antioxidant assays) and in vivo experiments (mouse excisional wound and skin burn models).
PDGA exhibited marked antioxidant and anticomplementary activity in contrast with polysaccharides, which displayed no detectable anticomplementary and antioxidant efficacy. *Besides, ointment, containing 2.5% crude polysaccharides and PDGA was found to have pronounced wound healing properties, by efficacy not yielding to 2.5% allantoin ointment.*
The obtained results allow assuming with high degree of reliability that wound healing activity of Comfrey preparations could be associated not only with allantoin but also with PDGA."
-'Allantoin- and Pyrrolizidine Alkaloids-Free Wound Healing Compositions from Symphytum Asperum' by V. Barbakadze, et al., I. Kutateladze Institute of Pharmacochemistry, Tbilisi, country of Georgia; Bulletin of the Georgian National Academy of Sciences, Volume 3, No. 1, **2009**.

('In Vitro' means done in laboratory equipment as opposed to in/on a living animal that is called 'In Vivo'.)
(Masses of proteins are expressed in daltons. For example, a protein with a molecular weight of 64,000 g mol-1 has a mass of 64,000 daltons or 64 kDa.) (Anticomplemenatary reduces action of an immunological complement. It is a substance that combines with a complement component and neutralizes its action by stopping its joining with an antibody.)

"The Common Comfrey or Symphytum officinale L. (Boraginaceae) is well-known for its wound-healing properties. **At the site of a dermal wound, activation of complement is one of a complex series of processes. Since complement factors upon activation exert important immuno-regulatory effects, modulation of complement activity may promote the healing of wounds. It was established that an ethanol (30%) extract of Symphytum officinale roots strongly inhibited both the classical and alternative pathway of complement."**
-'Inhibition of Complement Activity by High Molecular Compounds of Symphytum Officinale' by F.M. van den Dungen, A.J.J. van den Berg, C.J. Beukelman, H.C. Quarles van Ufford, H. van Djk and R.P. Labadie; Research Centre for Natural Products and Phytopharmaceuticals, University of Utrecht, Department of Pharmacognosy, Utrecht, The Netherlands; Planta Medica, Volume 57, Supplement Issue 2, pages A62-A63, 1991.

*"***Comfrey (Symphytum officinale L.) root** *is traditionally used for the topical treatment of contusions, strains and sprains. Besides allantoin and rosmarinic acid, which are discussed as pharmacologically active principles, the drug contains pyrrolizidine alkaloids (PAs) known for their hepatotoxic, carcinogenic and mutagenic properties.*
The topical herbal medicinal products Kytta-Salbe® f and Kytta-Plasma® f contain a PA-free liquid extract from Comfrey root as active substance.
The aim of this study was to demonstrate the absence of genotoxic effects of this special extract in the bacterial reverse mutation assay (Ames test). Briefly, Comfrey root liquid extract was investigated for its ability to induce gene mutations in Salmonella typhimurium strains TA 98, TA 100, TA 102, TA 1535 and TA 1537 with and without metabolic activation using the mammalian microsomal fraction S9 mix.
Comfrey root fluid extract showed no biologically relevant increases in revertant colony numbers of any of the five tester strains, neither in the presence nor in the absence of metabolic activation. **In conclusion, the Comfrey root fluid extract contained in Kytta-Salbe® f and Kytta-Plasma® f was not mutagenic in the bacterial reverse mutation assay."**
-'Absence of Mutagenic Effects of a Particular Symphytum Officinale Liquid Extract in the Bacterial Reverse Mutation Assay' by

B. Benedek, A. Ziegler and P. Ottersbach, Phytotherapy Research 24, pages 466-468, **2010**.
(Genotoxic means a toxic agent that damages DNA molecules in genes, causing mutations, tumors, etc.)

"Literature on Comfrey often focuses on the content of pyrrolizidine alkaloids (PA) of the raw plant material.
*It is important to note that **fully licensed medicinal products available today contain depleted or PA-free processed extracts. In fact, using these approved and safe products, pyrrolizidine alkaloids are no longer of clinical significance.** Still, some authors recommend a restriction of the duration of treatment, also with externally applied Comfrey preparations.*
In Germany, the restriction limiting application to 4-6 weeks/year applies only to preparations containing more than 10 micrograms, but less than 100 micrograms pyrrolizidine alkaloids (daily allowance).
*The application of modern (PA-free) preparations results in far below the daily allowance of 10 micrograms. As a consequence, **there are no restrictions in Germany on these PA-free products as regards the duration of treatment***.*
The absence of genotoxic effects was demonstrated in the bacterial reverse mutation assay (Ames test) for the PA-free liquid extract used in the above-mentioned clinical trials and studies**."
-'Comfrey Root from Tradition to Modern Clinical Trials' by Christine Staiger, Wiener Medizinische Wochenschrift (Vienna Medical Weekly, Austria), 163, pages 58–64, **2013**.
(* -Bundesgesundheitsamt. Bekanntmachung uber die Zulassung und Registrierung von Arzneimitteln {Abwehr von Arzneimittel-risiken- Stufe II}. Bundesanzeiger No. 111 {Federal Health Office. Notice on the authorization and registration of medicinal products: Defense against drug risks- Level II. Federal Gazette.}, page 4805, June 5 1992.)
(** -'Absense of Mutagenic Effects of a Particular Symphytum Officinale Liquid Extract in the Bacterial Reverse Mutation Assay' by B. Benedek, A. Ziegler and P. Ottersbach, Phytotherapy Research 24, pages 466-468, 2010.)

"Comfrey is a medicinal plant, extracts of which are traditionally used for the treatment of painful inflammatory muscle and joint problems, because the plant contains allantoin and rosmarinic acid. However, its medicinal use is limited because of its toxic pyrrolizidine alkaloid (PA) content.
PAs encompass more than 400 different compounds that have been identified from various plant lineages. To date, only the first pathway specific enzyme, homospermidine synthase (HSS), has been characterized. *HSS catalyzes the formation of homospermidine, which is exclusively incorporated into PAs. HSS has been recruited several times independently in various plant lineages during evolution by duplication of the gene encoding deoxyhypusine synthase (DHS), an enzyme of primary metabolism.*
Here, we describe the establishment of RNAi knockdown hairy root mutants of HSS in Symphytum officinale. *A knockdown of HSS by 60-80% resulted in a significant reduction of homospermidine by about 86% and of the major PA components 7-acetylintermedine N-oxide and 3-acetylmyoscorpine N-oxide by approximately 60%. The correlation of reduced transcript levels of HSS with reduced levels of homospermidine and PAs provides 'in planta' support for HSS being the central enzyme in PA biosynthesis. **Furthermore, the generation of PA-depleted hairy roots might be a cost-efficient way for reducing toxic by-products that limit the medicinal applicability of S. officinale extracts.**"*
-'Reduction of Pyrrolizidine Alkaloid Levels in Comfrey (Symphytum officinale) Hairy Roots by RNAi Silencing of Homospermidine Synthase', by Kruse, Stegemann, Jensen-Kroll, Engelhardt, Wesseling, Lippert, Ludwig-Muller, and Ober (Germany and United States); Planta Medica, Volume 85, pages 1177-1186, Stuttgart, Germany, and New York, (2019).

705. Symphytum tuberosum.

Symphytum Tuberosum 1880

'Illustrations of the British Flora: A Series of Wood Engravings, with Dissections, of British Plants, Drawn by W.H. Fitch and W.G. Smith. Forming an Illustrated Companion to Mr. Bentham's Handbook and other British Floras'.
London, England, 1880 and other editions.

Symphytum page 171.

Chapter 31

Some Uses of Comfrey are Restricted by Governments

<u>Overview of Legal Restrictions by Governments</u> (in chronological order from older to recent)

"**Symphytum officinale L. = Symphytum consolida Gueldenst. ex. Ledeb.:**
    France: Comfrey root permitted for external use only (toxicological categories pd:-ht/ae/wa:1 sa/ti:1).
    Sweden: Comfrey is classified as a drug, which must be registered as pharmaceutical specialty.
-Adverse Effects of Herbal Drugs, Volume 2, edited by Peter A. G.M. DeSmet, Konstantin Keller, Rudolf Hansel and R. Frank Chandler. Berlin, Germany: Springer-Verlag, **1993**. (I do not know what the French codes mean. If you do, please let me know.)

"*Pyrrolizidine Alkaloid Toxicity: Given the widespread consumption of Pyrrolizidine-Alkaloid-containing herbs, the number of people who have been clearly shown to have developed VOD (Veno-Occlusive Disease) after taking herbal remedies is so small that they could be dismissed as an idiosyncratic (unique to an individual) reaction.*
However, the legislators may wish to drastically curtail (reduce) over-the-counter (OTC) sales because the significance of the research into Pyrrolizidine Alkaloid (PA) toxicity lies in the fact that the damage to liver cells is silent. *No clinical indication or abnormality in liver function tests would be expected until after serious damage has occurred.*
    Many pieces of research demonstrate a dose dependant response to PA.
    In vitro (laboratory) research also suggests that there is a threshold to tissue damage. Beneath an undetermined level, at which detoxificaiton mechanisms are overpowered, one can argue that no tissue damage occurs.
Usefulness of Herbs: Discussions in scientific journals advocating (supporting) the removal of medicinal plants from sale generally minimise their usefulness.
    Whereas, quoting form the NIMH Journal of 1942*, **Comfrey is 'the most useful plant in the Pharmacopoeia,** *in particular for gastro-intestinal conditions, e.g., duodenal ulceration, colitis, for chronic bronchitis, bronchiectasis and tuberculosis, and externally for wounds, brusing, and slow to heal fractures.' "*
-'Using Herbs that Contain Pyrrolizidine Alkaloids' by Alison Denham, B.A., MNIMH (National Institute of Medical Herbalists), University of Central Lancashire, England; The European Journal of Herbal Medicine, Volume 2, No. 3, pages 27-38, **1996**.
(* -'Modern Practice Monographs', The Herbal Practitioner {now called Journal of Herbal Medicine}: Official Journal of National Institute of Medical Herbalists, Exeter, England, Volume 3, pages 7-9, 1942. If you have this, could you please send me a copy.)

"***Austria:*** *In Austria, medicinal preparations of Comfrey must be registered with the 'Federal Ministry of Health and Environmental Protection', and can be sold only in pharmacies.*
Sweden: *A similar situation exists in Sweden, where Comfrey products must be registered.*
Belgium, United Kingdom, Indonesia: *Sale for medicinal use is prohibited in Belgium, the United Kingdom, and Indonesia.*
France, Germany: *In France, as in Germany, the root is permitted for external use only."*
-Toxicology and Clinical Pharmacology of Herbal Products edited by Melanie Johns Cupp, PharmD, BCPS. Totowa, New Jersey: Humana Press, **2000**. Chapter 18: Comfrey by David Burch and Melanie Johns Cupp, page 212.

"*United Kingdom:*
    In the United Kingdom, Comfrey is only available as a tea, unless prescribed by a medicinal herbalist.
Germany: In Germany, Comfrey is not permitted for internal use, and external use is limited to a maximum daily dose of 100 micro-grams per day of PAs (pyrrolizine alkaloids) for a total period of 4 to 6 weeks, per year.
France: In France, Comfrey is not permitted for internal use, and external use is permitted only on intact skin."
-Health Canada, Marketed Health Products Directorate, Ottawa, Canada, 04-100842-849, January 21 **2004**.

"*Regulatory Status of Comfrey:*
    Canada: Prohibited substance (oral use).
    United Kingdom: General Sale List, Table B.
    United States: Negative monograph in the USP DI®. 100% Pyrrolizidine-Alkaloid-free Comfrey is a 'New Dietary Ingredient' (NDI) that may be permitted in dietary supplements. External products with PA-containing Comfrey have the following warning: 'For external use only. Do not apply to broken or abraded skin. Do not use when nursing.' "
-Herbal Drugs and Phytopharmaceuticals: A Handbook for Practice on a Scientific Basis edited by Max Wichtl. Boca Raton, Florida: CRC Press, **2004**. Symphytum Radix: Comfrey Root, pages 590-592.
    (According to the 'Medicines Act' of the United Kingdom: Medicinal products that can, with reasonable safety, be sold without involving a pharmacist or prescription are known as 'General Sale List' medicines. Medicinal products that are sold only from pharmacies with a prescription given by a practitioner are 'Prescription Only Medicines'. If a medicine is neither of those, then it is a 'Pharmacy Medicine'.)
    (USP DI = United States Pharmacopoeia®: Drug Information for the Healthcare Professional®. It includes

11,000 generic and brand-name drugs dispensed in the United States and Canada.)

"Regulations and recommendations have been introduced in some countries in attempts to limit human exposure to pyrrolizidine alkaloids.

Germany:
Regulations introduced in Germany by the 'Federal Health Bureau' (Germany Federal Health Bureau, 1992) limit the tolerable pyrrolizidines in herbal medicines to levels providing less than 1 microgram per day orally based on an assessment of genotoxic carcinogenicity, reduced to 0.1 microgram per day when used for over 6 weeks. Similar limits are likely to be adopted across Europe (*Roeder, 2000).

New Zealand:
'Food Standards Australia New Zealand' (FSANZ) has set a provisional exposure level of 1 microgram/kg (1 kg = 2.2 pounds) body weight per day and advised heavy consumers not to eat honey from Echium plantagineum (purple viper's-bugloss or Paterson's curse) every day (**FSANZ, 2001).

United Kingdom:
Herbal preparations containing Comfrey root were removed voluntarily from the market in the United Kingdom (Ministry of Agriculture, Fisheries and Food, 1994) and in the United States (Food and Drug Administration)."

-'Bioactive Compounds in Food' edited by John Gilbert (United Kingdom) and Hamide Z. Senyuva (Turkey), United Kingdom, Oxford: Blackwell Publishing Ltd, **2008**, pages 16-17.
(* -'Medicinal Plants in China Containing Pyrrolizidine Alkaloids' by E. Roeder, Pharmazie, Vol 55, No. 10, pages 711-726, 2000.)
(** -'Pyrrolizidine Alkaloids in Food: A Toxicological Review and Risk Assessment', Technical Report Series No. 2 by Professor Alan Seawright, Australia New Zealand Food Authority (ANZFA), Canberra BC, Australia, November 2001.)

"**Japan:**
In Japan, **Comfrey (Symphytum spp.) and products thereof cannot be marketed,** in accordance with Article 6.2 of the 'Food Sanitation Law'.
For precaution, the decision was made by the 'Ministry of Health, Labour and Welfare' in June 2004, due to incident of adverse effects reported overseas, to ban the sale of any Comfrey and Comfrey containing foods for consumption.

South Africa:
The South African **'Ministry of Agriculture, Forestry and Fisheries' reiterated in 2008 that the sale of Comfrey (Symphytum spp.) as a foodstuff had been prohibited** by the 'Regulations Relating to the Prohibition of the Sale of Comfrey, Foodstuffs Containing Comfrey and Jelly Confectionery Containing Konjac', which were published in the 'Government Gazettee' on 10 October 2003, to prohibit the use of Comfrey as a foodstuff (*,**DAFF, 2008).
Regulations as regards to Comfrey containing products with medicinal claims were under development at the time."

-'Discussion Paper on Pyrrolizidine Alkaloids' by Joint FAO/WHO Food Standards Programme, Codex Committee on Contaminants in Foods, 5th Session, The Hague, The Netherlands, CX/CF 11/5/14, Agenda Item 9(f), 77 pages, March 21-25 **2011**. WHO= World Health Organization. FAO= Food and Agriculture Organization of United Nations. Codex Alimentarius Commission.
(* -'Warning Against Sale and Use of Banned Comfrey' by 'Department of Agriculture, Forestry and Fisheries' {DAFF}, South Africa, 2008.)
(** -'Regulations Prohibiting Comfrey and Comfrey in Foodstuffs and Jelly Confectionery Containing Konjac' by 'Department of Agriculture, Forestry and Fisheries' {DAFF}, Food Safety Notification No. G/SPS/N/ZAF/16, Comments Deadline 13 March 2003, South Africa.)

"**United States:**
According to the **'Dietary Supplement Health and Education Act of 1994', herbal remedies sold in the United States are classified as food supplements.**
Manufacturers of herbal remedies are not allowed by law to claim any medical benefit from these products, but at the same time they are not under surveillance of the United States Food and Drug Administration (FDA).

Germany:
In Germany, however, **German Commission E has some control over marketing of herbal supplements** because the commission publishes monographs prepared by an interdisciplinary committee using historical information; chemical, pharmacological, clinical, and toxicological study findings; case reports; epidemiological data; and unpublished manufacturers' data. **If an herbal supplement has an approved monograph, it can be marketed.**

European Union:
European Directive 2004/24/EC, released in 2004 by the European Parliament and also by the 'Council of Europe', provides the basis for regulation of herbal supplements in the European market.
This directive requires that authorization be obtained from the national regulatory authorities of each European country in which herbal medicines are to be released in the market and that these products must be safe. The safety of a supplement is established based on published scientific literature, and when the data on safety are not sufficient, this is communicated to consumers.
In Europe, there will be two kinds of herbal supplements in the future: (1) herbal supplements with well established safety and efficacy and (2) traditional herbal supplements that do not have a recognized level of efficacy but are relatively safe.

Australia:
The Australian government also created a 'Complementary Medicine Evaluation Committee' in 1997 to address

regulatory issues regarding herbal remedies.

Canada:
> *In Canada, the federal government implemented a policy in 2004 to regulate natural health products and naturopaths. Many traditional Chinese medicine practitioners, homeopaths, and Western herbalists are concerned that this policy will eventually affect their access to the products they need to practice effectively."*

-Accurate Results in the Clinical Laboratory: A Guide to Error Detection and Correction by Amitava Dasgupta. Netherlands: Elsevier, **2013**. (Chapter 7: Effect of Herbal Remedies on Clinical Laboratory Tests, page 75.)

"Norway:
> ***Comfrey is classified as a drug, prescription only*** *(FOR 1999-12-27 nr 1565).*
> *The Norwegian cosmetics regulation (FOR 1995-10-26 nr 0871, Annex 3), which is in force until 11th June 2013, states that **Symphytum officinale is allowed in cosmetics provided it does not contain pyrrolizidine alkaloids."***

-'Risk Profile Symphytum Officinale Extracts', CAS No. 84696-05-9, www.mattilsynet.no, Statens tilsyn for planter, fisk, dyr og naeringsmidler (State supervision of plants, fish, animals and food), Brumunddal, Norway, March 11 **2013**.

"The Netherlands:
> *Pyrrolizidine alkaloids (PAs) are natural toxins occurring in a wide variety of plants, including herbs. When excessive amounts are ingested, these toxins are harmful to humans because PAs are carcinogenic and may cause severe liver damage.*
> ***The quantity of PAs present in herbal preparations may not exceed 1 microgramme per kilo (1 kg = 2.2 pounds). The 'Dutch National Institute for Public Health and the Environment' (RIVM)*** *has investigated whether this limit value is still in accordance with the latest scientific insights. This is indeed the case for herbal teas and food supplements containing herbs.*
> ***From a scientific point of view, it is even possible to adopt a slightly less strict limit value (5 microgrammes per kilo).*** *Because PAs are genotoxic carcinogens, however, it is advisable to minimize exposure to these toxins."*

-'Pyrrolizidine Alkaloids in Herbal Preparations, RIVM Briefrapport 090437001/2014' by Rijksinstituut voor Volksgezondheid en Milieu (RIVM) (Dutch National Institute for Public Health and the Environment), Commissioned by Netherlands Food and Consumer Product Safety Authority (NVWA), 62 pages, all in Dutch except for English abstract and appendix, **2014**. If you have an English translation, I would appreciate a copy.

"Belgium:
> ***Based on the Belgium recommended limit for PAs (Pyrrolizidine Alkaloids) in herbs at 1 ppm (parts per million), it would require ingestion of 1 kilogram (2.2 pounds) of an allowed herb to yield 1 mg (the smallest amount cited as toxic for Comfrey ingestion),*** *so the limit set would appear to be quite safe, in terms of any potential for acute reactions. Normal daily intake of herbs is usually only a few grams, so there is a large margin of safety; a 10 gram dose would yield a maximum of 10 micrograms of PAs from an allowed herb.*

Germany:
> ***The German recommended limit, just 1 to 10 micrograms per day, is consistent with the Belgium proposal based on PA content."***

-'Safety Issues Affecting Herbs: Pyrrolizidine Alkaloids" by Subhuti Dharmananda, Ph.D., Director, Institute for Traditional Medicine, Portland, Oregon, **2016**.

"United States:
> *In 2001, **'United States Food and Drug Administration' sent an advisory letter to manufacturers of dietary supplements** requesting that they remove all Comfrey products intended for consumption from the market (*FDA, 2001).*

Germany:
> *In 2011 Germany, the 'Federal Institute for Risk Assessment' (BfR) conducted a risk assessment for DHPAs (dehydro-pyrrolizidine alkaloids) and concluded that exposure should be kept as low as possible **limiting tolerable daily intake to 0.007 microgram of unsaturated pyrrolizidine alkaloids per kg bodyweight (BW)** (**BfR, 2011).*

United Kingdom:
> *In 2008, the 'United Kingdom Committee on Toxicity of Chemicals in Food, Consumer Products and the Environment' released a statement on DHPAs in food, which supported that of the BfR, **limiting daily oral exposures to less than 0.007 micrograms DHPAs kg-1 BW (bodyweight)** (***Committee on Toxicity of Chemicals in Food, 2008).*

New Zealand:
> *The 'Food Standards Australia New Zealand Authority' recommends a somewhat higher tolerable exposure of 1 microgram DHPAs kg-1 BW per day based exclusively on hepatotoxicity as opposed to potential carcinogenicity (****FSANZ, 2001)."*

Miscellaneous Organizations:
> *The 'World Health Organization', 'Dutch National Institute for Public Health and the Environment', and 'European Food Safety Authority' have all conducted similar reviews with similar concerns and recommendations (*****RIVM 2005/2014; ******EFSA 2011; *******WHO 2011)."*

-'The Comparative Toxicity of a Reduced, Crude Comfrey (Symphytum officinale) Alkaloid Extract and the Pure, Comfrey-derived Pyrrolizidine Alkaloids, Lycopsamine and Intermedine in Chicks' by Browna, Stegelmeiera, Colegatea, Gardnera, Pantera, Knoppela and Hallc, Journal of Applied Toxicology, Volume 36: 716-725, May 1 **2016**.

(* -'FDA Advises Dietary Supplement Manufacturers to Remove Comfrey Products From the Market' by Christine Lewis, Ph.D., 'Office of Nutritional Products, Labeling and Dietary Supplements', 'Center for Food Safety and Applied Nutrition', Food and Drug Administration, College Park, Maryland, July 6 2001.)

(** -'Chemical Analysis and Toxicity of Pyrrolizidine Alkaloids and Assessment of the Health Risk Posed by their Occurrence in Honey' by Bundesinstitut fur Risikobewertung (BfR) (Federal Institute for Risk Assessment), BfR Opinion No. 038/2011, August 11 2011.)

(*** -'Committee on Toxicity (COT) of Chemicals in Food, Consumer Products and the Environment', Statement on Pyrrolizidine Alkaloids in Food, COT is a United Kingdom independent scientific committee that provides advice to 'Food Standards Agency', 'Department of Health' and other government agencies, October 2008.)

(**** -'Pyrrolizidine Alkaloids in Food: A Toxicological Review and Risk Assessment', Technical Report Series No. 2 by Professor Alan Seawright, Australia New Zealand Food Authority (ANZFA), Canberra BC, Australia, November 2001.)

(***** -'Pyrrolizidine Alkaloids in Herb Preparations, RIVM Briefrapport 090437001/2014' by Rijksinstituut voor Volksgezondheid en Milieu {RIVM} {Dutch National Institute for Public Health and the Environment}, Commissioned by Netherlands Food and Consumer Product Safety Authority {NVWA}, 62 pages, all in Dutch except for English abstract and appendix, 2014. If you have an English translation, I would appreciate a copy.)

(****** -'Scientific Opinion on Pyrrolizidine Alkaloids in Food and Feed' by EFSA Panel on Contaminants in the Food Chain {CONTAM}, European Food Safety Authority {EFSA}, Parma, Italy; EFSA Journal, Volume 9, No. 11, 2406, 134 pages, 2011.)

(******* -'Discussion Paper on Pyrrolizidine Alkaloids' by Joint FAO/WHO Food Standards Programme, Codex Committee on Contaminants in Foods, 5th Session, The Hague, The Netherlands, CX/CF 11/5/14, Agenda Item 9(f), 77 pages, (March 21-25 2011). WHO= World Health Organization. FAO= Food and Agriculture Organization of the United Nations. Codex= Codex Alimentarius Commission.)

Australia Bans Sale of Comfrey for Internal Use

In June 1984 Australia's 'National Health and Medical Research Council' (NH and MRC) placed Comfrey on the 'Poison Schedule 1' in the 'Uniform Poisons Standard' based on the recommendation of the 'Poisons Schedule Committee of Victoria'.
Then it was only available through pharmacists, by a doctor's prescription.
(Other plants on this list are arsenic, belladonna, hemlock and strychnine.)

'Poison Schedule 1: Comfrey (Symphytum) being preparations and admixtures for internal use of comminuted (pulverized) leaves or dried and powdered root or any part of the dried plant.'

'Schedule 1: Substances which are of such danger to life as to warrant them being available only from:
    i) medical practitioners, dentists, veterinary surgeons, or pharmacies.
    ii) wholesale dealers, to authorised persons or to be prescribed classes of persons.'

This decision was based on these 3 reports:

    A. 'The Alkaloids of Symphytum x uplandicum (Russian Comfrey)' by Claude C.J. Culvenor, John A. Edgar, John L. Frahn and Leslie W. Smith, Division of Animal Health, CSIRO, Animal Health Research Laboratory, Parkville, Victoria, Australia; Austrailian Journal of Chemistry, Volume 33, No. 5, pages 1105-1113, 1980.
(For details about this report, see number 5 in subsection 'Scientific Studies Showing Dangers of Alkaloids in Comfrey' in section 'Alkaloids in Comfrey'.)

    B. 'Structure and Toxicity of the Alkaloids of Russian Comfrey (Symphytum x uplandicum, Nyman), a Medicinal Herb and Item of Human Diet' by Claude C.J. Culvenor, M. Clarke, J.A. Edgar, J.L. Frahn, M.V. Jago, J.E. Peterson and L.W. Smith, CSIRO, Division of Animal Health, Victoria, Australia; Experientia, Switzerland, Volume 36, No. 4, pages 337-379, 1980.
"Eight pyrrolizidine alkaloids of hepatotoxic type have been indentified in leaves of Symphytum X uplandicum. The combined alkaloids exhibit chronic hepatotoxicity in rats using intraperitoneal injection."
        (Hepatoxic means causing injury to the liver. Intraperitoneal is injecting a substance into the abdominal cavity.)
(For details about this report, see number 6 in subsection 'Scientific Studies Showing Dangers of Alkaloids in Comfrey' in section 'Alkaloids in Comfrey'.)

    C. 'Carcinogenic Activity of Symphytum Officinale' by I. Hirono et al, Japan National Cancer Institute, Volume 61, No. 3, pages 865-869, September 1978.
"The carcinogenicity of Symphytum officinale L., Russian Comfrey, used as a green vegetable or tonic, was studied in inbred ACI rats. Three groups of 19-28 rats each were fed Comfrey leaves for 480-600 days (1.3 to 1.6 years); four additional groups of 15-24 rats were fed Comfrey roots for varying lengths of time. A control group was given a normal diet. Hepatocellular adenomas were induced in all experimental groups that received the diets containing Comfrey roots and leaves. Hemangioendothelial sarcoma of the liver was infre-

quently induced."
(Carcinogenicity means its ability or tendency to cause cancer. Adenoma is a benign, noncancerous epithelial tumor. Sarcoma is a cancerous tumor.)
(For details about this report, see number 4 in subsection 'Scientific Studies Showing Dangers of Alkaloids in Comfrey' in section 'Alkaloids in Comfrey' {Chapter 30}.)

"Toxins Document:
A CSIRO (Commonwealth Scientific and Industrial Research Organization) scientist (Culvenor) recently took part in an international conference in Tashkent (Uzbekistan) called to discuss the environmental hazard caused by pyrrolizidine alkaloids.
Dr. Claude Culvenor, coordinator of the pasture associated toxins program at the 'Division of Animal Health' (CSIRO, Australia), *visited Tashkent at the invitation of the 'World Health Organisation'.*
He contributed to a task group responsible for preparing an environmental health criteria document on human health hazards from this substance. ***In Australia, pyrrolizidine alkaloid occurs in heliotrope, Paterson's curse, and Comfrey."***
-'CoResearch: CSIROs Staff Newspaper', Commonwealth Scientific and Industrial Research Organisation, Melbourne, Victoria, Australia; No. 298, December **1986** and January 1987. Comfrey article page 9.   (This is the entire article.)

"The regular use of Comfrey as part of the diet or for medicinal purposes may be a potential health risk as a result of the presence of naturally-occurring pyrrolizidine alkaloids. *The majority of these alkaloids are hepatotoxic in both animals and humans, and some have been shown to induce (create) tumours in experimental animals.*
In this article, the toxic properties of pyrrolizidine alkaloids are reviewed briefly, with particular reference to their presence in Comfrey. The acute and long-term health risks at the normally-low levels of Comfrey consumption are evaluated and discussed. ***On the basis of the data that are available currently, the small but significant long-term risk that is associated with the consumption of Comfrey justifies the need to limit its intake.***
   This is being achieved by controls under various state Poisons Acts (of Australia), but also requires further education on the potential dangers of naturally-occurring chemicals of plant origin."
-'Comfrey: Assessing the Low-Dose Health Risk' by Peter J. Abbott, Toxicology Unit, Department of Community Services and Health, Canberra, Australia; Medical Journal of Australia, Volume 149, No. 11-12, pages 678-682, **1988**.
   (Acute disease means it has an abrupt/rapid onset. It usually means an illness of short duration, rapidly progressive, and in need of urgent care. Acute is in reference to time as opposed to subacute or chronic. Chronic means lasting a long time, usually 3 months or more. Subacute is between acute and chronic.)

"In Australia, Comfrey (Symphytum spp. {species}) is included in Appendix C *(Substances, other than those included in Schedule 9, of such danger to health as to warrant prohibition of sale, supply and use) of the Poisons Standard 2010, for therapeutic or cosmetic use except for dermal (skin) use, for which it is included in Schedule 5 of the Poisons Standard (*DOHA, 2010).*
The 'Australia New Zealand Food Authority' (ANZFA) became 'Food Standards Australia New Zealand' (FSANZ) on 1 July 2002. ***The Joint Food Standards Code (ANZJFSC), which applies to both Australia and New Zealand, lists PA-containing plant species under Schedule 1:***
   'Prohibited Plants and Fungi of Standard 1.4.4.: The most important are the following plants: Borago officinalis; Crotolaria spp; Echium plantagineum; Echium vulgare; Heliotropium spp; Senecio spp; ***Symphytum asperum; Symphytum officinale; and Symphytum x uplandicum'.***
   A plant or fungus, or a part or a derivative of a plant or fungus listed in Schedule 1, or any substance derived there from, must not be intentionally added to food or offered for sale as food."
-'Discussion Paper on Pyrrolizidine Alkaloids' by Joint FAO/WHO Food Standards Programme, Codex Committee on Contaminants in Foods, 5th Session, The Hague, The Netherlands, CX/CF 11/5/14, Agenda Item 9(f), 77 pages, March 21-25 **2011**.
WHO= World Health Organization. FAO= Food and Agriculture Organization of United Nations. Codex Alimentarius Commission.
(* -Australian Government, Poisons Standard 2010, Department of Health and Ageing {DOHA}, Therapeutic Goods Administration, Federal Register of Legislative Instruments, F2010L02386. Standard for the Uniform Scheduling of Medicines and Poisons No. 1, 368 pages. Also: Explanatory Statement F2010L02386ES, 4 pages.)

Response to Decision to Put Comfrey on Poison Schedule in Australia

"The 'National Herbalists Association of Australia' has received a list of 69 references from the Victorian Government. *A preliminary check indicates that: 30 relate to other plant species, 21 apparently refer to liver or cancer generally and may not be relevant, 14 relate to studies of individual alkaloids or groups of alkaloids, 3 are Russian papers and copies of translations have been requested,* ***one relates to carcinogenic activity of Symphytum officinale in rats*.***
There do not appear to have been tests for the level of pyrrolizidine alkaloids in Symphytum officinale extracts, tinctures or tablets. *As these are the products which are affected by the new scheduling, and as there may well be negligible (extremely low) pyrrolizidines in them, it would have been more appropriate for the Department of Health to have had these checked."*
-'Pharmacists and Comfrey' by Diane Wiesner, B.Pharm, MA, Ph.D., MPS, Principal of the NSW College of
Natural Therapies, Sydney, Australia; Australian Journal of Pharmacy, Volume 65, pages 959-963, **1984**.

(* -'Carcinogenic Activity of Symphytum Officinale' by I. Hirono et al, Japan National Cancer Institute Volume 61, No. 3, pages 865-869, September 1978.)   (Victoria is a state in southeast Australia.)

"The 'National Health and Medical Research Council' recommended in 1984 that Comfrey be placed in Schedule I of the 'Standard for Uniform Scheduling of Drugs and Poisons'; thus, it recommended that Comfrey be available only from pharmacists, medical practitioners or, in isolated communities, other licensed individuals.
Criticism of the decision was vigorous and generally fell into one of four categories:
   1. Disbelief that a natural herb could be anything but safe, in view of its long history of use.
   2. Attempts to discredit the scientific evidence on the toxicity of pyrrolizidine alkaloids in general.
   3. Attempts to show that Comfrey (particularly Symphytum officinale) was less toxic than were other pyrrolizidine alkaloid-containing plants.
   4. Concerns that the professional integrity of herbalists was in question.
The controversy serves to highlight the difficulty of arriving at a consensus (general agreement) of opinion about the assessment of risk to human health, as well as to question the role of governments in regulating naturally-occurring toxic chemicals."
-'Comfrey: Assessing the Low-Dose Health Risk' by Peter J. Abbott, Toxicology Unit, Department of Community Services and Health, Canberra, Australia; Medical Journal of Australia, Volume 149, No. 11-12, pages 678-682, **1988**.

"The responsibility for the decision rests fundamentally with the 'Commonwealth Scientific and Industrial Research Organization' (CSIRO, Canberra, Australian Capital Territory, Australia), its decision based explicitly (clearly only) on the work of Dr. Claude Culvenor. CSIRO News File of March 29, 1978 carries following statement:
   'Dr. Culvenor's research group at Parkville (Melbourne, Victoria, Australia) has been studying these compounds, called pyrrolizidine alkaloids, which occur in such pasture weeds as ragwort (Jacobaea vulgaris), Paterson's curse (Echium plantagineum) and heliotrope (Heliotropium)- the last two weeds being from the same plant family as Comfrey.
   At least four of these alkaloids are known to be carcinogens, and it is probable that the type found in Comfrey is also carcinogenic, according to Dr. Culvenor.
   While it is unlikely anybody eating Comfrey in small quantities would suffer serious effects, its regular use as a green vegetable could cause chronic liver damage and worse.
   Plants in the same family (borage family, Boraginaceae) have caused human poisonings in the USSR, Africa, India and Afghanistan after their accidental consumption in bread over a period of one to two years.
   The evidence of these outbreaks, considering the amount of the alkaloid we have measured in Comfrey, **suggests that daily consumption of two young leaves of the Comfrey plant over a similarly lengthy period will lead to serious disease**.'
Dr. Claude Culvenor cites the case of human poisoning in Afghanistan*, where wheat flour containing a high percentage of seeds of Heliotropium mixed with it was used for bread, with consequently a high level of alkaloids.
The report by Dr. Culvenor reveals that he has very limited knowledge about the history of the hybrid Comfrey, Symphytum peregrinum (Russian Comfrey), which we use, and its many medical and therapeutic uses, or about its parent plants, Symphytum officinale and asperimum.
There is no excuse for this kind of ignorance. Dr. Culvenor even questions the validity of its history of nearly 2000 years."
-Comfrey: Nature's Healing Herb and Health Food by Andrew Hughes. Japan: Sanyusha Publishing Co., Ltd, **1992**, pages 54, 55, 60, 65.
(* -An Outbreak of Hepatic Veno-Occlusive Disease in North-Western Afghanistan' by Mohabbat, Younos, Merzad, Srivastava, Sediq, and Aram, The Lancet, London, England, Volume 308, No 7980, pages 269-271, August 7 1976.)
   (Created in 1916, 'Commonwealth Scientific and Industrial Research Organisation' {CSIRO} is an Australian federal government agency responsible for scientific research. Based in Canberra, Australia with 50 locations in Australia, France, Chile and United States. Research areas include: agriculture/food, data, energy, land/water, mineral resources, manufacturing, ocean/atmosphere and health/biosecurity.)
   (The U.S.S.R. also known as the Soviet Union was a state in Eurasia that existed from 1922 to 1991. It is now Russia and 12 independent nations.)
   (Heliotropium is a genus of flowering plants in Borage family with 250-300 species, also known as heliotropes.)

"In March 1978 there was a news release a few days before the 'Organic Festival' held at Brighton, Tasmania, Australia. I was a speaker at that time and a question from the audience asked about 'this Comfrey story in the newspapers'.
The news item was as a result of an experiment by the CSIRO which concluded that Comfrey was 'a dangerous herb and could cause liver cancer'. The details of this experiment were not available, I learned*.
It seemed evident this was a scare tactic. This CSIRO Report or press release was enough to frighten the ignorant and fearful but not enough to upset the enlightened.
   At that time conventional medicine was losing many to the so-called alternative methods of healing. It was evident that this was the beginning of an effort to stop people medicating themselves."
-'Comfrey: School of Natural Science' by Dr. Frederick John Steed. Ph.D. (Natural Science), NSc., D.N., Australia, 2 pages, **1997**. Founder of Natural Science correspondence course about healthy food and herbal medicine.
(* -'Structure and Toxicity of the Alkaloids of Russian Comfrey (Symphytum x uplandicum, Nyman), a Medicinal Herb and Item of

Human Diet' by Claude C.J. Culvenor, M. Clarke, J.A. Edgar, J.L. Frahn, M.V. Jago, J.E. Peterson and L.W. Smith, CSIRO, Division of Animal Health, Victoria, Australia; Experientia, Switzerland, Volume 36, No. 4, pages 337-379, 1980. The press release was sent out before the research report was available to the public.)

"*This problem with Comfrey started back in *1978 following sensational headlines in newspapers.*
    Headlines and topics could be seen as herbs that damage the liver which seem to imply many herbs: 'Popular Herb is a Killer.' - 'Scientist Warns Herb is a Killer' - 'Health Drink Causes Cancer, Declares a CSIRO Expert' - 'Comfrey is a Killer' - 'Be Careful With Herbs, They Can Kill You' and so forth.
In 1984 Comfrey was prohibited for internal usage in Australia where it was placed on the 'Poisons Schedule' by the 'Poisons Advisory Bureau', through the 'National Health and Medical Research Council' (NH and MRC). It then only became available by a doctor's prescription through a pharmacist."
-'Comfrey is Safe and Highly Beneficial' by Michelle Honda, Ph.D., M.H., R.N.C.P., D.Sc., Doctor of Holistic Science, Natural Health Blog, Ontario, Canada, **2016**.
(* -'Carcinogenic Activity of Symphytum Officinale' by I. Hirono et al, Japan National Cancer Institute, Volume 61, No. 3, pages 865-869, September 1978.)

Canada Bans Sale of Prickly (S. asperum) and Russian Comfrey for Internal Use

"**The first Canadian action was taken in 1982**, when the 'Health Protection Branch of Health and Welfare Canada' introduced an amendment to 'Canada's Food and Drug Regulations' which **prohibits the sale, for medicinal purposes, of any products containing echimidine** (Canada Gazette, 30 March 1988).
    Echimidine, considered to be the most toxic of Comfrey PAs (Pyrrolizidine Alkaloids) (*Brauchli-Theotokis 1987), is not found in Common Comfrey (Symphytum officinale L.). However, it is present in Prickly Comfrey (S. asperum Lepechin) and its hybrids with S. officinale (**Huizing, Gadella, and Kliphuis 1982), including Russian Comfrey (S. x uplandicum Nyman), which is the most commonly encountered commercial Comfrey in Britain (***Clapham, Tutin and Warburg 1962).
The intent of this legislation is to have more careful attention paid to identification of botanical species by the herbal industry, and to alert the Canadian public to the potential danger of PA consumption.
There was no intent to underestimate the relative potential danger of echimidine-free Symphytum officinale. Both root and leaf of S. officinale have been shown to be carcinogenic in rats (****Hirono et al. 1978), though here again there is species confusion because the authors equate Common Comfrey and Russian Comfrey!"
-'Comfrey Update' by D.V.C. Awang, HerbalGram published by American Botanical Council, Austin, Texas, Volume 25, pages 20-23, 1991.
    (* -'Zur Toxikologischen Beurteilung der Pyrrolizidin-Alkaloide in den Arzneipflanzen Symphytum officinale, Borago officinalis' {Toxicological Assessment of Pyrrolizidine Alkaloids in Medicinal Plants Symphytum officinale, Borago officinalis} by Julia Brauchli-Theotokis, Ph.D. Thesis, University of Zurich, Switzerland, 1987. I have the abstract. I would appreciate if anyone has an English translation of the thesis.)
    (** -'Chemotaxonomical Investigations of the Symphytum Officinale Polyploid Complex and S. Asperum {Boraginaceae}: The Pyrrolizidine Alkaloids' by H.J. Huizing, Th.W.J. Gadella, and E. Kliphuis, all from The Netherlands; Plant Systematics and Evolution, Volume 140, pages 279-292, 1982.)
    (*** -Flora of the British Isles, Second Edition by A.R. Clapham, T.G. Tutin and E.F. Warburg. England; Cambridge University Press, 1962, first edition 1952.)
    (**** For more about the Hirono et al. 1978 study, see subsection 'Studies Showing Dangers of Alkaloids in Comfrey' in section 'Alkaloids in Comfrey' {Chapter 30}.)

"The 'Health Protection Branch' of the 'Ministry of National Health and Welfare of Canada' has issued a 'Report of the Expert Advisory Committee on Herbs and Botanical Preparation' (1986): Four basic categories were determined:
    1. Herbs and botanical preparation unacceptable for use in or as food.
    2. Herbs and botanical preparations generally accepted as food.
    3. Herbs and botanical prepartions acceptable as food under specified conditons.
    4. Herbs and botanical prepartions generally used for medicinal purposes.
Of particular note is a recommendation to consider establishing a subcategory of drugs to include 'Folklore Medicines'.
It was suggested that this might follow Sweden's lead. In 1978 Sweden established marketing and regulatory guidelines for 450 natural remedies. These regulations are not as stringent as those for standard drug products."
-'HerbalGram, Issue 9' by American Botanical Council, Austin, Texas, 12 pages, Spring **1986**. Includes 'Canadian Panel Makes Recommendations on Herb Usage' article.
(-'Report of the Expert Advisory Committee on Herbs and Botanical Preparations to the Health Protection Branch, Health and Welfare Canada' by J.L. Blackburn. Ottawa, Canada: Published by Minister of National Health and Welfare, 1986.)

"**In the fall of 1987, the Canadian Health Protection Branch moved to ban the sale of certain types of Comfrey leaf (Symphytum officinale and related species). The Canadians have previously banned the sale of all Comfrey root several years ago.**
    The Canadian ban does not affect the status of all Comfrey leaf products. Only those products that contain a

variety of Comfrey known as 'Russian Comfrey' are being banned; *the 'Common Comfrey' variety is reportedly not being affected as it contains considerably lower levels of Pyrrolizidine Alkaloids.*
Comfrey leaf and root are still sold in the United States, although one major United States manufacturer has recently removed Comfrey root from some of its herbal formulas and has discontinued the sale of Comfrey root capsules altogether."
-'HerbalGram, Issue 15' by American Botanical Council, Austin, Texas, 24 pages, Winter 1988. Includes 'Canada Bans Comfrey Leaf' article.

"Examination of Comfrey products available in Canada (*1988) revealed that none was designated or labelled as Russian Comfrey or by its Latin binomial, Symphytum x uplandicum. *Products were labelled as either simply 'Comfrey' or Symphytum officinale (Common Comfrey).*

> *However, just about half (6) of all products analyzed (13) were found to contain echimidine, and must therefore have derived from S. asperum or a hybrid of that species, probably S. x uplandicum. Three of the six echimidine-containing products were specifically labelled as Symphytum officinale.*

'Health and Welfare Canada' has for many years refused to register Comfrey root products for any medicinal application, *in recognition of the much greater risk presented by root material as compared to leaf.*

> *Comfrey root has been consistently observed to contain roughly ten times the concentration of PA (pyrrolizidine alkaloids) found in leaves**, ***.*
>
> *Manufacturers have been advised that the inclusion of Comfrey root in herbal preparations is no longer acceptable."*

-'Symphytum / Smeerwortel' (Dutch for Comfrey) by Maurice Godefridi, Kruidwis website for Herbalists Education, Belgium, February 20 2014.
(* -'International Congress on Natural Products Research' by Awang, Fillion, Girard and Kindack, Park City, Utah, July 17-21 1988.)
(** -'Toxicity of Pyrrolizidine Alkaloids' by A.R. Mattocks, Nature: International Journal of Science, Volume 217, pages 723-728, February 24 1968.)
(*** -'Comfrey and Liver Damage' by James N. Roitman, Natural Products Chemistry, Western Regional Research Center, United States Department of Agriculture, Berkeley, California; The Lancet, Volume 317, Issue 8226, page 944, April 25 1981.)

"The Canadians were off to a great start. **The Canadian equivalent of our 'Food and Drug Administration', the 'Health Protection Branch' (HPB), set up an expert panel to evaluate the regulation of botanicals and make recommendations.** Their report, released over two years ago, was heralded as a most promising approach.

> A major industrialized nation was actually considering rational regulation of folk medicines. Realizing that folk medicines will always be used, Canada actually seemed prepared to admit that there is a legimate class of products that lie somewhere between foods and drugs.

Yes, many herbs could fall into this category, but so do some other common substances currently regulated as foods. Coffee, tea and prune juice are obvious examples of 'foods' that are actually used as 'drugs', at least within the functional definition of a drug, which centers on a substance's intended use.
The advisory panel suggested that a new category be established for folk drugs, which could be labeled with known physiological effects, along with a disclaimer that the claims are based on cultural, historical and folk information, not necessarily proven by modern laboratory experimentation.

> This would allow manufacturers of herbal products to label their products with information about known folk uses and physiological effects, without submitting them to the drug approval process. Because botanical folk medicines are low profit and not patentable, they are not serious candidates for formal drug approval.

Canada 'Health Protecion Branch' has decided to play FDA instead. On March 11, 1989, the HPB published two lists of botanicals: one listing botanicals which cannot be sold as food (or food supplements) and one which can be sold only with cautionary labeling.

> **Those on the first list are thus removed from the food category, with no category established to provide any more rational regulatory status- they are essentially banned.** Plants on the other list could still be sold as food, but only with this warning: 'Caution: Do not consume (common name of plant)(Latin name) during pregnancy'.

The inclusion of plants with small amounts of carcinogens are particularly problematic, since neither Canada nor the United States have reached a rational regulatory stance on natural carcinogens in food. The ban on Comfrey leaf is a good example.

> **Comfrey leaf** does indeed contain mutagenic alkaloids (so does coffee). According to the classic risk assessment article by Dr. Bruce Ames, a cup of Comfrey leaf tea is about as dangerous as a peanut butter sandwich, a third as dangerous as a raw mushroom and a tenth as risky as a beer. Surveys by Dr. Farnsworth and others have shown that nearly every plant studied contains carcinogens.

Now it appears that in Canada, herbs are to be held to a different standard of safety- a higher standard-than other foods.
They are to be held to a much tougher standard than drugs, in which some risk is accepted if substantial benefit is present.

> Make no mistake about it, this is not a consistent move by HPB. There will be no attempt to 'ban' mustard powder, which is more acutely toxic than golden seal (Hydrastis canadensis), mushrooms or coffee. 'Morbidity and Mortality Weekly Report', published by the Center for Disease Control, rarely reports toxic reactions to herbal products despite increasingly widespread usage, while toxic reactions are common with Over-the-Counter (OTC) and prescription drugs.

Here are some highlights of the list of botanicals banned from food use:
Barberry root, Bloodroot, Coltsfoot, **Common comfrey,** Common wormwood, European mistletoe berries, European pennyroyal oil, Goldenseal root, Horse chestnut, Lobelia, Mountain grape root, Mugwort, **Prickly Comfrey,** Yohimbe."

-'HerbalGram, Issue 20' by American Botanical Council, Austin, Texas, 52 pages, Spring **1989**. Includes 'Canadian FDA Ignores Advisory Panel; Bans Many Folk Medicines' article.

"**'Second Report of the Expert Advisory Committee on Herbs and Botanical Preparations'*** by J.L. Blackburn to 'Health Protection Branch', Health Canada, Ministry of Health, Canada, **October 1993**.
    The 'Expert Advisory Committe' was established in 1984 to assess the safety concerns associated with the sale of herbs and botanicals as foods. Its work resulted in a recommendation that a class of Traditional Medicines be identified, an action that has subsequently been adopted in Canada. **This work classified 64 plants as adulterants in foods and 7 others as requiring labeling which contraindicates in pregnancy. This list was later modifed (**Welsh, 1995),** and the current classifications are reported in the '***Botanical Safety Handbook'."
-American Herbal Products Association's Botanical Safety Handbook edited by Zoe Gardner and Michael McGuffin. Boca Raton, Florida: CRC Press, 2013.
(* -'Second Report of the Expert Advisory Committee on Herbs and Botanical Preparations to the Health Protection Branch, Health and Welfare Canada'. Ottawa, Canada: Published by Health Canada, 1993.)
(** -Letter to 'Canadian Health Food Association' from 'Health Protection Branch, Health Canada' from F.W. Welsh, February 6 1995. If you have a copy of this, could you please send it to me.)
(*** -American Herbal Products Association's Botanical Safety Handbook edited by Zoe Gardner and Michael McGuffin. Boca Raton, Florida: CRC Press, 2013.)

"**Analysis of 'Second Report of the Expert Advisory Committee on Herbs and Botanical Preparations'*** by J.L. Blackburn to 'Health Protection Branch', Health Canada, Ministry of Health, Canada, October 1993.
    The Committee held a two-day session in Ottawa, Canada on June 22-23, 1993, and worked through the summer through correspondence. It reviewed the classification of the six herbs and botanical preparations on the priority list.
    Their mandate was to address six herbs: Common Comfrey (Symphytum officinale); common wormwood (Artemisia absinthium); feverfew Tanacetum parthenium); goldenseal root (Hydrastis canadensis); Oregon grape root (Mahonia aquifolium, M. nervosa); St. John's wort (Hypericum perforatum).
The Committee felt that the toxicity of all herbs and botanical preparations listed in Appendix VI was well documented, and presented the following synopsis of **the rationale for maintaining Comfrey on the list of adulterants: 'the Symphytum species contain pyrrolizidine alkaloids that have been shown in experimental animals to produce a hepatotoxic effect.'** The fact that the Committee was reconvened is, undoubtedly, due to the great outcry from the public when it felt its right to use herbal medicine was threatened.
The Canadian system of allowing herbs to be sold as OTC (Over-the-Counter = no prescription) drugs with label claims for their traditional uses allows Canadian consumers to choose these products with some rational basis for their uses and potential health benefits. The Canadian system is consistent with the World Health Organization's 'Guidelines for the Assessment of Herbal Medicines' and with international trends in regulating herbs for their well-documented historical uses."
-'Herbs as Traditional Medicines in Canada', HerbClip®, Herbal Regulation in Canada by American Botanical Council, 4 pages, January 12 1996. HerbClip provides summaries and reviews of articles covering research, regulation, marketing and responsible use of medicinal plants.
(-'Report of the Expert Advisory Committee on Herbs and Botanical Preparations to the Health Protection Branch, Health and Welfare Canada' by J.L. Blackburn. Ottawa, Canada: Published by Minister of National Health and Welfare, 1986.)
(* -'Second Report of the Expert Advisory Committee on Herbs and Botanical Preparations to the Health Protection Branch, Health and Welfare Canada'. Ottawa, Canada: Published by Health Canada, 1993.)

"**The Canadian 'Food and Drug Act' and findings of an 'Expert Advisory Committee on Herbs and Botanical Preparations' were consulted to provide an overview of the issues regarding herbal product regulation in Canada.**
Herbal products not registered as drugs in Canada are sold as foods and are exempt from the drug review process that evaluates product efficacy and safety. This places the public at risk of unwanted effects from the use of herbal products that are adulterated with other substances and of forgoing effective conventional therapy.
    Moreover, consumers are exposed to a plethora of information portraying herbal products as harmless.
    Some progress has been made to address these concerns by facilitating the registration of herbal products as drugs. **Most herbal products that were evaluated were unsafe or ineffective, or no information was available to evaluate their efficacy.**"
-'Herbal Products in Canada: How Safe are They?' by Anita Kozyrskyj, BScPhm (Bachelor of Science in Pharmacy), MSc; Canadian Family Physician, Volume 43, pages 697-702, April **1997**.

In December 2003, 'Health Canada' ('Sante Canada'), the governmental agency responsible for national public health, banned all products containing Comfrey (Symphytum spp. {species}) because of toxicity reports about liver damaging pyrrolizidine alkaloids.

> "**On December 12th, 2003, 'Health Canada' announced that Canadian manufacturers must pull all Comfrey-containing products (for external or internal use) from the market. Canadian herbalists are actively opposing this unjustified action that will prevent consumers from using products that are effective and safe.**
> We ask the 'Natural Health Products Directorate' and 'Health Canada' to modify this regulatory action in a way that will protect Canadian consumers without depriving them of effective and harmless products.

We see no evidence of toxicity for the external use of Comfrey in the documents presented by 'Health Canada'. After examining the proof provided by 'Health Canada' to justify this rash (without due consideration) decision, we see no reason to classify Comfrey as a toxic product when used externally, no matter the species. Canadian herbalists, now preparing their documents of evidence to support their request for Natural Health product licenses, would never form such an erroneous (wrong) conclusion based on the findings. We are also asking 'Health Canada' to give natural health product manufacturers the same chance as companies whose products have drug identification numbers (DIN). Those companies have a grace period to prove that their products are labeled correctly (external use only) and do not contain any other species of Comfrey than Symphytum officinale."
-'The Future of Common Comfrey: Traditional Herbalists are Worried: Petition' by Le Guilde des Herboristes (The Herbalists Guild), Quebec, Canada, **2003**.
(This petition was signed by herbalists and others concerned about this drastic action.)

This ban does not affect the sale of Comfrey seeds and roots for planting in Canada. It is legal to grow Comfrey, both personally and commercially.

On January 21, 2004 the Canadian ban was clarified:
"Banned:
Two species of Comfrey (Prickly = Symphytum asperum, and Russian = Symphytum uplandicum) are prohibited pursuant to section 8 of the 'Food and Drugs Act' and section C.01.038 of the 'Food and Drug Regulations'. These particular species have been prohibited since 1988. **Also prohibited is a specific PA (pyrrolizidine alkaloid), echimidine, which is toxic to the liver and is a constituent of Prickly and Russian Comfrey.**
Not Banned:
Other Comfrey species, including Common Comfrey (Symphytum officinale), are not prohibited."
-Health Canada, Marketed Health Products Directorate, Ottawa, Canada, 04-100842-849, an unsigned letter, January 21 **2004**.

"**The herbal industry has the difficult task of having to prove that their products are free of the Pyrrolizidine Alkaloid echimidine.** According to Peter Child of 'Investigative Science Incorporated', Burlington, Ontario, Canada, a company that specializes in laboratory testing, **reference standards for echimidine are not available. This makes it very difficult for laboratories to properly test Comfrey.**"
-'The Comfrey Ban: Canada' by Richters Herbs, Goodwood, Ontario, Canada, www.richters.com, February 21 **2004**.

"**Elements of the 'Natural Health Products Regulations', in effect since 2004, have proven to be an effective means to achieve a flexible, risk-based regulatory approach to consumer health products.**
Regulated under the 'Natural Health Products Regulations', natural health products include vitamins and minerals, herbal remedies, homeopathic medicines, traditional medicines such as traditional Chinese medicines, probiotics and other products like amino acids and essential fatty acids. Similar to non-prescription drugs, natural health products also include products such as pain relievers, toothpastes and anti-dandruff shampoos."
-'A Framework for Consumer Health Products', Health Canada, updated March 2018.

"**Canada: Consultations with various stakeholders led to the creation (in 2004) of the 'Natural Health Products Regulation' that addresses the availability and safety of natural health products.**
All natural health products must have a product license and manufacturers, packagers, labelers, and importers must have site licenses. Licensing requires specific labeling and packaging requirements, cGMP (Current Good Manufacturing Practice), and evidence of safety and efficacy."
-Drug-Induced Liver Disease edited by Neil Kaplowitz and Laurie DeLeve. London, England: Academic Press, 2013. (Chapter 35: 'Hepatotoxicity of Herbal and Dietary Supplements' by L.B. Seeff, F. Stickel and V. Navarro, pages 631–657.)

"**Postmarket surveillance, particularly Adverse Reactions (ARs), forms an integral part of the ongoing safety evaluation for Natural Health Products (NHPs).** ARs can be related to many factors, including inherent toxicity, misuse, hypersensitivity, NHP-drug interactions, or product quality.
Canada's 'Natural Health Products Regulations' mandate (official order) NHPs to be licensed.
Although the vast majority of NHPs are considered to be low risk, **there have been some serious, life-threatening adverse reactions associated with the use of several NHPs that have required regulatory action** (Health Canada Advisory *2002, **2003). This has provided the impetus for developing a regulatory framework in Canada.
This article will briefly outline the following:
(i) **Canadian Natural Health Products Regulations (NHPR)**, including Health Canada's risk-based approach to noncompliant NHPs (Natural Health Products);
(ii) **the challenges of postmarket surveillance of NHPs** from the perspective of clinicians and federal regulators;
(iii) **Health Canada's postmarket initiatives** to strengthen the vigilance of NHPs and promote the safe use of NHPs."
-'Postmarket Surveillance of Natural Health Products in Canada: Clinical and Federal Regulatory Perspectives' by Mano Murty, Health Canada, 'Marketed Biologicals, Biotechnology and Natural Health Products Bureau', Marketed Health Products Directorate, Ottawa, Ontario, Canada; Canadian Journal of Physiology and Pharmacology, Vol 85, No. 9, pages 952-955, Sept **2007**.
(* -'Health Canada requests recall of certain products containing Ephedra / ephedrine', Health Canada Advisory, 2002.)

(** -'Health Canada advises consumers not to use or ingest the herb Comfrey or health products that contain Comfrey'. Health Canada Advisory 2003-101, December 12, 2003.)
(Postmarket surveillance is monitoring the safety of a drug, herb or medical device after it has been released on the market to consumers or patients.)

"**The overall objective of this study is to review the natural health product laws and regulations for Canada's primary Natural Health Products (NHP) trading partners,** namely, Australia, China, France, Germany, Hong Kong, India, United States, United Kingdom, and the European Union and identify priority areas where policy research should be focussed and then propose strategies to address these selected policy research areas.
Health Canada's new NHP regulations came into effect in 2004. The major components of the Canadian regulatory framework for NHPs that were used for research comparison purposes were the following: Product Licensing including Standards of Evidence, Site Licensing, Good Manufacturing Practices and Adverse Reaction Reporting."
-International Regulation of Natural Health Products by John Robert Harrison. Boca Raton, Florida: Universal Publishers, **2008**.

"**In March 2009, 'Health Canada' introduced The new 'Risk-Based Approach' (RBA) for Natural Health Products (NHP),** which outlines proposed modifications to product and site licensing (Health Canada 2009). **The RBA was developed in response to a regulatory review where stakeholders felt that NHPs should be regulated proportionally to their risk level** and that the regulatory requirements were too onerous (difficult) for this product type, given the low risk nature of these products and their history of safe use (Health Canada 2008).
The RBA envisions two classes of product licenses, which depend on the degree of certainty associated with a product's safety, quality or health claims.
Class I Products and/or claims for which there are readily available, authoritative and high-quality sources of evidence (called Pre-Cleared Information) relating to their efficacy, safety and quality.
Class II Products and/or claims considered a higher risk due to lack of existing evidence (e.g., a novel product)."
-'Regulation of Natural Health Products in Canada' by Yoshinori Mine and Denise Young, Department of Food Science, University of Guelph, Ontario, Canada; Food Science and Technology Research, Volume 15, No. 5, pages 459-468, **2009**.

"'**Health Canada' has advised Canadian consumers not to use or to ingest the herb Comfrey (Symphytum officinale) or any health products that contain Comfrey.**
As a precaution, consumers were advised not to topically (to skin) apply Comfrey-containing products to broken skin. This advisory applied to both approved and unapproved products (*Health Canada, 2003)."
-'Discussion Paper on Pyrrolizidine Alkaloids' by Joint FAO/WHO Food Standards Programme, Codex Committee on Contaminants in Foods, 5th Session, The Hague, The Netherlands, CX/CF 11/5/14, Agenda Item 9(f), 77 pages, March 21-25 **2011**.
WHO= World Health Organization. FAO= Food and Agriculture Organization of United Nations. Codex Alimentarius Commission.
(* -'Health Canada advises consumers not to use or ingest the herb Comfrey or health products that contain Comfrey'. Health Canada Advisory 2003-101, December 12, 2003.)

United States Regulates Sale of Comfrey for Internal Use

USA 1990: FDA Tests Comfrey Products

"**The Food and Drug Administration (FDA) has asked its district offices to collect samples of products made with the common herb Comfrey (Symphytum officinale and related species) to determine the levels of pyrrolizidine alkaloids (PAs) in these products.**
PAs in Comfrey and other herbs are suspected of being responsible for various types of adverse toxic reactions in several animal test systems, including hepatotoxicity and carcinogenicity.
The May 1990 FDA assignment said, '**In order for FDA to complete a toxicological risk assessment to support a regulatory decision on Comfrey,** a survey of commercial samples to detennine potential levels of exposure to pyrrolizidine alkaloids needs to be conducted.' "
-'HerbalGram, Issue 23' by American Botanical Council, Austin, Texas, 52 pages, Summer **1990**. Includes article 'FDA Collecting Information on Herb Products to Assess Safety: Comfrey Main Item for Review' by Mark Blumenthal.

USA 1994: Dietary Supplement and Health Education Act

The '**Dietary Supplement Health and Education Act of 1994' (DSHEA)** is a statute of United States Federal legislation which defines and regulates dietary supplements. **Supplements are effectively regulated by the Food and Drug Administration (FDA) for 'Good Manufacturing Practices'** under 21 CFR Part 111.
Under the act, supplement manufacturers do not need to receive FDA approval before marketing dietary supplements that were marketed in the United States before 1994.
Manufacturers of 'New Dietary Ingredients' (NDI) must provide reasonable evidence of safety, or reasonable expectations of safety, and must be reviewed (not approved) by the FDA prior to marketing.

"**Dietary supplements were defined in a law passed by Congress in 1994 called the 'Dietary Supplement Health and**

Education Act' (DSHEA). According to DSHEA, a dietary supplement is a product that:
   1. Is intended to supplement the diet.
   2. Contains one or more dietary ingredients (including vitamins, minerals, herbs or other botanicals, amino acids, and certain other substances) or their constituents
   3. Is intended to be taken by mouth, in forms such as tablet, capsule, powder, softgel, gelcap, or liquid.
   4. Is labeled as being a dietary supplement.
Herbal supplements are one type of dietary supplement. *An herb is a plant or plant part (such as leaves, flowers, or seeds) that is used for its flavor, scent, and/or potential health-related properties. 'Botanical' is often used as a synonym for 'herb'. An herbal supplement may contain a single herb or mixtures of herbs.*
The law requires that all of the herbs be listed on the product label.
Federal Trade Commission:
   Once a dietary supplement is on the market, the FDA monitors product information, such as label claims and package inserts. The Federal Trade Commission (FTC) is responsible for regulating product advertising; it requires that all information be truthful and not misleading."
-'Using Dietary Supplements Wisely', NCCIH Publication No. D426, by United States Department of Health and Human Services, National Institutes of Health, National Center for Complementary and Integrative Health, https://nccih.nih.gov, June 2014.

"According to the 'Dietary Supplement Health and Education Act of 1994', herbal remedies sold in the United States are classified as food supplements.
Manufacturers of herbal remedies are not allowed by law to claim any medical benefit from these products, but at the same time they are not under surveillance of the United States Food and Drug Administration (FDA)."
-Accurate Results in the Clinical Laboratory: A Guide to Error Detection and Correction by Amitava Dasgupta. Netherlands: Elsevier, 2013. (Chapter 7: Effect of Herbal Remedies on Clinical Laboratory Tests, page 75)

*"**The term 'dietary supplement' means a product (other than tobacco) intended to supplement the diet** that bears or contains one or more of the following dietary ingredients:*
a vitamin; a mineral; an herb or other botanical; an amino acid; a dietary substance for use by man to supplement the diet by increasing the total dietary intake; or a concentrate, metabolite, constituent, extract, or combination of any ingredient described."
-**United States Department of Health and Human Services, National Institutes of Health, Office of Dietary Supplements, Dietary Supplement Health and Education Act of 1994,** Public Law 103-417, 103rd Congress.

*"In 1994, Congress passed the 'Dietary Supplement Health Education Act' (DSHEA), **defining 'dietary supplement' and 'New Dietary Ingredient' (NDI).***
Current Dietary Ingredients:
   Regarding the NDI, Congress stipulated that all dietary supplements sold before 1994 were to be considered safe and could therefore remain on the market without the manufacturer having to file an NDI notification.
New Dietary Ingredients:
   The Food and Drug Administration must receive notification of any supplement with a new ingredient(s) marketed after 1994, showing information regarding the manufacturer, the manufacturing process, and the product's safety. The NDI notification must be received 75 days before marketing of the product."
-'Scientific and Regulatory Perspectives in Herbal and Dietary Supplement Associated Hepatotoxicity in the United States' by Mark I. Avigan, Robert P. Mozersky and Leonard B. Seeff; Center for Drug Evaluation and Research, FDA, Silver Spring, Maryland and Office of Dietary Supplement Products, Center for Food Safety and Applied Nutrition, College Park, Maryland; International Journal of Molecular Sciences, Volume 17, No. 331, 30 pages, 2016.

USA 1996: AHPA Recommends Restricting Comfrey to External Use Only

*"**Founded in 1982, the American Herbal Products Association (AHPA) is comprised of more than 350 member companies,** consisting primarily of domestic and foreign companies doing business as growers, processors, manufacturers and marketers of herbs and herbal products as foods, dietary supplements, cosmetics, and non-prescription drugs, and also including companies that provide expert services to the herbal trade.*
AHPA's mission is to promote the responsible commerce of herbal products to ensure that consumers continue to enjoy informed access to a wide variety of herbal goods."
-American Herbal Products Association: The National Trade Association and Voice of the Herbal Products Industry, Silver Spring, Maryland, www.ahpa.org, 2018.

*"Comfrey has remained commercially available in the United States, though in 1993 the **'American Herb Products Association' (AHPA) alerted its members to restrict Comfrey's use to external applications.***
*It issued the following recommendation **(adopted July 1996):***
   *'**AHPA recommends that all products with botanical ingredients which contain toxic pyrrolizidine alkaloids bear the following cautionary statement on the label:***
      For external use only. Do not apply to broken or abraded skin. Do not use when nursing.' "
-'Safety Issues Affecting Herbs: Pyrrolizidine Alkaloids" by Subhuti Dharmananda, Ph.D., Director, Institute for Traditional Medicine, Portland, Oregon, 2016.

"The **American Herbal Products Association (AHPA)** took the initiative to restrict the sale of Comfrey products for internal use by issuing a policy that all commercial Comfrey products sold by AHPA members should contain the following warning:
'The product is intended for External Use Only. It should not be used by nursing mothers nor applied to abraded skin.'"
-'Variability in Comfrey PA Content' by Ginger Webb, HerbalGram published by American Botanical Council, Austin, Texas, Issue 38, page19, **1996**.

"*AHPA Trade Requirements: Pyrrolizidine Alkaloids (adopted July 1996; revised July 2010):*
Products with botanical ingredients that contain toxic pyrrolizidine alkaloids are not offered for sale for internal use and bear the following cautionary statement on the label:
For external use only. Do not apply to broken or abraded skin. Do not use when nursing.
   Including but not limited to:
   Alkanna tinctoria (alkanet); Arnebia euchroma, Anchusa officinalis (bugloss); Borago officinalis (borage); Crotalaria spp., Cynoglossum spp., Erechtites hieraciifolia, Eupatorium cannabinum (hemp agrimony); Eupatorium purpureum (Joe Pye), Gynura segetum, Heliotropium spp., Lithospermum officinale (European gromwell); Packera candidissima, Petasites spp. (e.g., butterbur); Pulmonaria spp. (e.g., lungwort); Senecio jacobaea (European ragwort); Senecio vulgaris (groundsel herb); **Symphytum spp. (Comfrey);** and Tussilago farfara (coltsfoot). Borage seed oil is specifically exempt from the above label requirement."
-'AHPA Code Of Ethics, Business Conduct and Trade Regulations' by American Herbal Products Association, Silver Spring, Maryland, March 2017.

"**Drastic restriction of PAs (Pyrrolizidine Alkaloids), such as avoidance of any amount in an herb intended for internal use as suggested by AHPA (American Herbal Products Association), may not be justified.**
There is a great diversity of alkaloid structures among the pyrrolizidine group. Some of the pyrrolizidine alkaloids, such as farfugine and tussilagine, are considered non-toxic.
The range of amounts of PAs in plants (and different parts of the same plant) is also quite large: from less than 0.001% to 0.1.2%, more than a 1,000-fold range among samples with reported levels.
This variability in amounts and toxicity makes it difficult to rationally suggest that no pyrrolizidines, including the toxic ones, are acceptable in herbs used medicinally as internal remedies."
-'Safety Issues Affecting Herbs: Pyrrolizidine Alkaloids" by Subhuti Dharmananda, Ph.D., Director, Institute for Traditional Medicine, Portland, Oregon, 2016.

USA 1998: USP Discourages Internal Use of Comfrey

"**The USP (United States Pharmacopeia) discourages the internal use of Comfrey because of studies showing hepatotoxicity (liver toxicity) and carcinogenicity (cancer causing), and no studies have shown the benefit of taking Comfrey orally (USP, 1998).**
The Delaney Clause of the 'Food, Drug and Cosmetic Act' establishes no tolerance for carcinogens in foods. **Even so, Comfrey is considered a dietary supplement, and can be sold if labeled as such.**"
-Toxicology and Clinical Pharmacology of Herbal Products edited by Melanie Johns Cupp, PharmD, BCPS. Totowa, New Jersey: Humana Press, 2000. Chapter 18: Comfrey by David Burch and Melanie Johns Cupp, page 212.
(United States Pharmacopeia {USP} is a compendium of drug information published annually by the United States Pharmacopeial Convention, a nonprofit organization. USP is published in a combined volume with National Formulary as USP-NF.)

USA 2001: FDA Bans Sale of Comfrey for Internal Use

"*FDA and FTC Action Comfrey:*
On July 6 2001 the **Food and Drug Administration (FDA)** sent a letter to industry trade associations and herbal nonprofits to inform their members about products containing the herbal ingredient Comfrey (Symphytum spp.{species}, Boraginaceae) and its link to potentially serious health hazards when ingested.
FTC:
In a parallel move on July 6 2001, as part of its 'Operation Cure.All'*, the **Federal Trade Commission (FTC)** took enforcement action against a manufacturer for marketing Comfrey products for both oral and external use. The action and advisory follow recent FTC findings of Comfrey products for sale on the Internet.
The key to the FTC Comfrey action was advertising claims by the company that their Comfrey product was safe, i.e., false advertising from a regulatory perspective.
   The FTC action, part of an ongoing internet sweep in search of fraudulent health claims, involved an injunction against one company, 'Christopher Enterprises' ** of Springville, Utah, to stop selling Comfrey as a dietary supplement and an ointment for open wounds. According to the FTC, 'Christopher Enterprises' agreed to stop making the health and safety claims that the agency challenged. In addition, the company agreed to place an extensive warning label on any Comfrey product intended for external use.
   FTC also took actions against another company, 'Western Botanicals' *** in Fair Oaks, California.
PA-Free Comfrey: Curiously, the FDA and FTC letters did not recognize the availability of PA (Pyrrolizidine Alkaloid) free

Comfrey extracts, *which are sold by at least two companies in the United States and are available as a raw material in Europe. The PAs are said to be removed by a resin bed ion-exchange process, although HerbalGram has not been able to locate independent analyses to verify this claim.*
However, as most herb industry watchers know, there is little Comfrey sold in the United States: relatively few products are offered since most leading manufacturers discontinued Comfrey in mid-1980s due to concerns about potential toxicity."
-'HerbalGram, Issue 53' by 'American Botanical Council', Austin, Texas, and 'Herb Research Foundation', 84 pages, Summer 2001. Includes 'FDA and FTC Act on Comfrey' article.
(* -'Latest FTC Case in Operation Cure.All Focuses on Safety Risks of Comfrey Products Promoted Via Internet' by Federal Trade Commission, Washington, DC, July 6 2001.)
(** -'Marketer of Herbal Supplement Products Agrees to Settle Charges that Safety and Health Benefit Claims were False' by Federal Trade Commission, Washington, DC, December 6 2001.)
(*** 'FTC Charges Western Botanicals with Unsubstantiated Comfrey Claims' by Natural Products Insider: An Informa Business, July 16 2001.)

In 2001 the United States 'Food and Drug Administration' (FDA) banned the sale of Comfrey products for internal use, and put a warning label ('External Use Only') on those intended for external use such as creams and ointments. 'Safety Alert and Advisory' was sent to: 'American Botanical Council', 'American Herbal Products Association', 'Council for Responsible Nutrition', 'Consumer Healthcare Products Association', 'National Nutritional Foods Association', 'Utah Natural Products Alliance', 'American Association of Oriental Medicine', 'American College of Acupuncturists and Traditional Medicine'.

Safety Alert and Advisory:
'The Food and Drug Administration (FDA) is issuing this letter to communicate to you our concern about **the marketing of dietary supplements that contain the herbal ingredient Comfrey (Symphytum officinale = Common Comfrey, S. asperum = Prickly Comfrey, and S. x uplandicum = Russian Comfrey).** *These plants are a source of pyrrolizidine alkaloids that present a serious health hazard to consumers when they are ingested.*
Under the 'Federal Food, Drug, and Cosmetic Act' (the Act), as amended by the 'Dietary Supplement Health and Education Act' of 1994, the manufacturer bears the primary responsibility for ensuring that its dietary supplement products are safe.
FDA believes that the available scientific information is sufficient to firmly establish that dietary supplements that contain Comfrey or any other source of pyrrolizidine alkaloids are adulterated under the Act.
***The agency strongly recommends that firms marketing a product containing Comfrey or another source of pyrrolizidine alkaloids remove the product from the market and alert its customers to immediately stop using the product.'* "**
-Lewis, C. J. 'FDA Advises Dietary Supplement Manufacturers to Remove Comfrey Products From the Market'. **July 6 2001** (Christine J. Lewis, Ph.D., Director, 'Office of Nutritional Products, Labeling, and Dietary Supplements', 'Center for Food Safety and Applied Nutrition', Food and Drug Administration, College Park, Maryland.)

"*The 'Consumer Healthcare Products Association' established a voluntary program for manufacturers with the same recommendation as AHPA (American Herbal Products Association) (adopted March 2001).* **On July 6, 2001, the United States FDA (Food and Drug Administration) took official action to remove Comfrey from all dietary supplements.**"
-'Safety Issues Affecting Herbs: Pyrrolizidine Alkaloids" by Subhuti Dharmananda, Ph.D., Director, Institute for Traditional Medicine, Portland, Oregon, 2016.

"**On July 6, 2001, the FDA sent a letter to eight trade associations representing the botanical dietary supplement industry advising manufacturers to remove Comfrey products from the market.**
It stated that the scientific evidence was sufficient for the FDA to consider products that contain Comfrey or other sources of pyrrolizidine alkaloids to be adulterated under the Federal Food, Drug, and Cosmetic Act and that the FDA was prepared to use its authority and resources to remove these products from the market.
In tandem (at the same time), the **Federal Trade Commission (FTC)** took action against two companies that were marketing Comfrey-containing products. Both firms agreed to injunctions prohibiting them from marketing these products for internal use or for use on open wounds, and requiring the following warning on the label of products sold for external use:
> **Warning: External Use Only. Consuming this product can cause serious liver damage. This product contains Comfrey. Comfrey contains pyrrolizidine alkaloids, which may cause serious illness or death. This product should not be taken orally, used as a suppository, or applied to broken skin. For further information contact the Food and Drug Administration, https://www.fda.gov/.**

Perhaps these actions will deter others from selling Comfrey-containing products intended for ingestion. If not, the FDA and/or the FTC will have the costly task of identifying these products and taking action on a case-by-case basis."
-Herbal Medicine: Chaos in the Marketplace by Rowena K Richter. New York: Haworth Herbal Press, 2003.

"**July 6, 2001 Letter from Food and Drug Administration:**
FDA also believes that manufacturers need to take adequate steps to identify and report adverse events, especially adverse events that may include liver disorders, associated with any product that contains an ingredient that may contain pyrrolizidine alkaloids.
> *FDA recommends that firms promptly notify FDA's MedWatch program of reports of adverse events associated with the use of products containing any source of pyrrolizidine alkaloids.*"

-Dietary Supplement Labeling Compliance Review (Understanding and complying with dietary supplement labeling requirements of the Food and Drug Administration) by James L. Summers. New Jersey: John Wiley & Sons Incorporated, 2008.
>(MedWatch is the Food and Drug Administration's 'Safety Information and Adverse Event Reporting Program'. Founded 1993, this system of voluntary reporting allows information to be shared with the medical community or general public.)

"The 'Food and Drug Administration' recently sent out a warning letter to supplement manufacturers, trade associations, and practitioners stating that the herb Comfrey can pose a serious health threat and should not be used as a dietary supplement. The herb contains pyrrolizidine alkaloids, which are known hepatotoxins and which have been shown to cause veno-occlusive disease in animals. It is also a suspected carcinogen.
The 'Federal Trade Commission' has gone a step further, taking legal action against an herbal supplement marketer. The FTC has filed charges against 'Western Botanicals' in US District Court concerning its oral Comfrey products and ordered the company to add a warning label to all topical products, instructing consumers not to use them orally, as a suppository, or on broken skin."
-'A New FDA Warning on Comfrey' by Anonymous, Contemporary OB/GYN, Iselin, New Jersey, Volume 46, Issue 10, page 120, October **2001**.

"**The FDA's (Food and Drug Administration) Christine Lewis, who wrote the letter,** says it's not clear there is a consensus against internal use of the herb, noting that AHPA (American Herbal Products Association) represents only a portion of a large industry. **Mark Blumenthal, president of the American Botanical Council, one of eight groups that received the FDA letter, says there are flaws in it:**
>'The agency failed to distinguish Russian and Prickly Comfrey, which contain liver-toxic chemicals, from Common Comfrey, which lacks them. The agency failed to point out that Comfrey's root has about 10 times the alkaloids as its leaf, and didn't acknowledge that alkaloid-free Comfrey extracts are available.

'Our concern is the unsafe ingredients out there,' FDA's Christine Lewis replies. There are, she adds, 'probably quite a few details here, but generally, the products are not safe.'
James Duke, former chief of the United States Department of Agriculture's medicinal plants research laboratory, thinks otherwise. Duke, who grows Comfrey in his garden, says:
>'**Comfrey is an excellent herb externally and very unlikely to cause problems internally.** Comfrey does contain traces of pyrrolizidine alkaloids, and anyone eating it like spinach three times a day for 10 years might have a problem, but nobody in their right mind is going to do that.' "
-'Herb Industry Welcomes Curbs on Comfrey' by Judy Packer-Tursman, Washington Post, Washington, D.C., www.washingtonpost.com, July 17 **2001**.

"**The Food and Drug Administration (FDA) told the American Herbal Products Association (AHPA) that marketing Comfrey products that are free of Pyrrolizidine Alkaloids (PA) would not be objectionable;** however, if those products are chemically altered to remove the PA, they would be considered New Dietary Ingredients (NDI) and would require a 75-day notice prior to marketing.
In a July 6, 2001, letter circulated to AHPA members, FDA advised the industry to cease marketing supplements that contain PA or Comfrey (Symphytum officinale, S. asperum and S. x uplandicum) because they are potentially harmful when ingested.
On July 31, AHPA responded to clarify that some Comfrey cultivars contain very low PA levels, and there are processing technologies that can remove PA altogether. AHPA requested that FDA delineate its position on PA-free products and provide a specific minimum level of PA for such products.
On Oct. 3, FDA responded to AHPA's letter, saying that as safety concerns with Comfrey are associated with PA:
>'We would not necessarily object to the marketing of dietary supplement products containing dietary ingredients that had been processed to remove the pyrrolizidine alkaloids. It would expect such a product to be completely free of the pyrrolizidine alkaloids and not simply have been processed to reduce the quantity.'
>FDA stated that manufacturers should ensure through validated analytical methods and good manufacturing practices that **all batches of products are 100 percent PA-free.**"
-'FDA, AHPA Clarify PA-Free Comfrey Guidelines' by Natural Products Insider: Provides Information for Executives in the Global Nutrition Industry, An Informa Business, www.naturalproductsinsider.com, November 8, **2001**.

USA 2006: Adverse Effects from All Dietary Supplements Must be Reported

"**In 2006, Congress passed the 'Dietary Supplement and Non-Prescription Drug Protection Act'***. Prior to its passage, manufacturers of dietary supplements and over-the-counter drugs were not required to notify Food and Drug Administration of adverse events regarding their products.
After passage of this law, the requirement to report serious adverse events came into being."
-'Scientific and Regulatory Perspectives in Herbal and Dietary Supplement Associated Hepatotoxicity in the United States' by Mark I. Avigan, Robert P. Mozersky and Leonard B. Seeff; Center for Drug Evaluation and Research, FDA, Silver Spring, Maryland and Office of Dietary Supplement Products, Center for Food Safety and Applied Nutrition, College Park, Maryland; International Journal of Molecular Sciences, Volume 17, No. 331, 30 pages, 2016.
(* -'Dietary Supplement and Non-Prescription Drug Act', Public Law 109-462, United States Government Printing Office, Washington, DC, 2006.)

USA 2007: Good Manufacturing Practices

"In 2007, the FDA (Food and Drug Administration) required a modification to the cGMPs (Current Good Manufacturing Practices) regarding manufacturing, packaging, labeling, and holding operations for dietary supplements and a 'Final Rule' was published on 22 June 2007."
-Drug-Induced Liver Disease edited by Neil Kaplowitz and Laurie DeLeve. London, England: Academic Press, **2013**. (Chapter 35: 'Hepatotoxicity of Herbal and Dietary Supplements' by L.B. Seeff, F. Stickel and V. Navarro, pages 631–657.)

USA 2007, 2008: Regulations and Pyrrolizidine Alkaloids

"The 'Food and Drug Administration' (FDA) issued a letter to industry in 2001 communicating concern about the safety of supplement products containing Comfrey (Symphytum spp. {species}) because of its PAs (Pyrrolizidine Alkaloids).

FDA further recommended that firms immediately stop marketing Comfrey-containing supplements and alert consumers to stop using the products. Finally, FDA urged manufacturers to identify and report any adverse events, including liver disorders, which had been associated with Comfrey and other ingredients containing PAs (FDA, 2001).

In the United States, animals presented for slaughter that show signs of PA-related disease, are condemned and not allowed to enter the food supply (*USDA, 2007)."
-'Discussion Paper on Pyrrolizidine Alkaloids' by Joint FAO/WHO Food Standards Programme, Codex Committee on Contaminants in Foods, 5th Session, The Hague, The Netherlands, CX/CF 11/5/14, Agenda Item 9(f), 77 pages, March 21-25 2011. WHO= World Health Organization. FAO= Food and Agriculture Organization of United Nations. Codex Alimentarius Commission.
(* -'Multi-Species Disposition Basics with a Public Health Focus' by United States Department of Agriculture, Food Safety and Inspection Service, Center for Learning, Public Health Veterinarian Training, 115 pages, 2007-2012.)

"Amount of Pyrrolizidine Alkaloids in Various Species of Comfrey:
To the extent that some members of the Comfrey plant family do not contain PAs (Pyrrolizidine Alkaloids), then the concerns FDA (Food and Drug Administration) has about a Comfrey-containing dietary supplement product or ingredient depends upon the exact species identified.

The FDA ruling banned the internal use of Symphytum officinale (Common Comfrey), S. asperum (Prickley Comfrey), and S. x uplandicum (Russian Comfrey), as well as any other plant/substance containing pyrrolizidine alkaloids.

While FDA did not examine the safety of other Comfrey species such as Symphytum tuberosum L. (Tuberous Comfrey), *which is suggested to contain negligible amounts of pyrrolizidine alkaloids, FDA would rely on the presence of the pyrrolizidine alkaloid to determine whether this species was permitted to be used as a dietary ingredient.*

Only after such sampling and analysis of this biomarker of toxicity would FDA be in a position to provide an opinion on whether its use in dietary supplements would not present a significant or unreasonable risk of illness or injury under the conditions of use recommended in labeling."
-Bioactive Compounds in Food: Edited by John Gilbert (United Kingdom) and Hamide Z. Senyuva (Turkey), United Kingdom, Oxford: Blackwell Publishing Ltd, **2008**, page 619.

USA 2011: New Dietary Ingredients

"More recently (2011), the FDA (Food and Drug Administration) has developed draft guidance for industry regarding 'New Dietary Ingredients' (NDI) notification.
Needless to say, there continues to be controversy about the 'Act' (Dietary Supplement and Education Act of 1994), including the question of what actually constitutes a dietary supplement and what represents a 'New Dietary Ingredient'*."
-Drug-Induced Liver Disease edited by Neil Kaplowitz and Laurie DeLeve. London, England: Academic Press, 2013. (Chapter 35: 'Hepatotoxicity of Herbal and Dietary Supplements' by L.B. Seeff, F. Stickel and V. Navarro, pages 631–657.)
(* -'Dietary Supplements: Regulatory Issues and Implications for Public Health' by Bryan E. Denham, Ph.D., JAMA: Journal of the American Medical Association, Volume 306, No.4, pages 428-429, July 27 2011.)

USA 2011: Comfrey Still Sold Under 'Dietary Supplement Act'

"Various herbal products such as kava, chaparral, Comfrey, pennyroyal, mistletoe and green tea extract can cause liver damage. Despite known toxicity and warnings issued by the authorities, these herbal supplements are still available (2011) in health food stores because herbal supplements are sold in the United States under the 'Dietary Supplement Act of 1994', under which, as long as no medical claim is made, a herbal supplement can be sold as a dietary supplement.

When the United States Food and Drug Administration (FDA) banned the use of phenylpropanolamine, all over-the-counter cold medications containing phenylpropanolamine were removed from all stores within 1 week.

By contrast, despite FDA caution against the use of kava in 2003, this product is freely available in health food stores in Houston, Texas where the author lives.

Because abnormal function tests may be an early indication of liver damage due to certain herbal supplements, *laboratory scientists, clinical chemists and pathologists are the first healthcare professionals who can intervene to advise clinicians to investigate what herbal supplements a patient is taking.*

Abnormal liver function tests in the absence of negative serology for hepatitis (liver damage), cytomegaly (herpes viral infection) and Epstein Barr virus are a strong indication of drug- or herb-induced liver damage."
-Effects of Herbal Supplements on Clinical Laboratory Test Results by Amitava Dasgupta. Germany: De Gruyter Inc, **2011**.
(Serology is the diagnostic examination of blood serum, especially the response of the immune system to pathogens or introduced substances.)
(Herbal supplements can interact with Western drugs so there is possibiity of drugs not working as indicated. See section 'Warnings and Negative Reactions to Comfrey'.)
(Herbal products in the United States must have this warning: 'This statement has not been evaluated by the Food and Drug Administration. This product is not intended to diagnose, treat, cure or prevent disease.')
(As of 2018, a store such as Walmart® sells online: Comfrey cream, extract, oil, powder {leaf/root}, salve, tea {leaf}, and tincture {leaf/root}. The ban never included live Comfrey for gardening or for feeding to animals. And it did not ban external use of Comfrey for humans.)

USA 2013: Comfrey Extracts and Safety

"*'American Herbal Products Association' States Some Comfrey is Safe:*
Evidence of toxicity of one specific extract of an herb cannot be extrapolated to another extract of the same herb.
Concerns about the safety, including evidence of carcinogenic (cancer causing) activity, of one extract of a source herb cannot be automatically transferred to any other extract of the same herb. **This is because, as discussed above, there are many different and unique substances that can be manufactured through variations in extraction processes that start with a single herbal source ingredient.**
Recognizing this scientific fact, the 'United States Senate Committee on Appropriations' recorded the following statement in its Report to accompany appropriations for the fiscal year ending September 30, 2014 for 'Departments of Labor, Health and Human Services, and Education':
'National Toxicology Program' (NTP): The Committee urges NTP to be highly precise when describing the results of its studies on particular extracts of an herbal species to avoid any possible confusion about the relevance of such studies to other extracts of the species.
As an example, Comfrey (Symphytum officinale) root is known to contain pyrrolizidine alkaloids (PAs), which are in turn known to have mutagenic (gene mutation) properties.
But there are in the marketplace Comfrey root extracts that have been manufactured with an extraction process that removes the PAs. A study conducted on one such PA-free Comfrey root extract showed that this specific extract was 'not mutagenic in the bacterial reverse mutation assay.'
Any extrapolation of safety concerns about whole Comfrey root to this extract is therefore scientifically unsound and in fact inaccurate. *Any such extrapolation of research on one extract of an herbal ingredient in which the specific compound responsible for a known or observed toxicity has not been identified to another unique extract of that same herb is also scientifically unsound.*"
-'Comments of the American Herbal Products Association before the United States of America Department of Health and Human Services, National Institutes of Health, National Toxicology Program, Office of the Report on Carcinogens, on Request for Information on Nominations: Ginkgo Biloba, Goldenseal, and Kava', October 18 **2013**.
(Extrapolate means to extend a method or conclusion, especially one based on statistics, to an unknown situation by assuming that existing trends will continue or similar methods will be applicable.)

USA 2015, 2016: Supplement Claims and Safety

"*'Good Manufacturing Practices' and Supplement Claims:*
All supplements, including herbs, vitamins, minerals, etc., must conform to Federal regulations that control their manufacture, labeling, and advertising. *In order to sell an herbal supplement, a manufacturer must meet many different Federal (and sometimes state) regulations, and must also adhere to state and local health and business regulations.*
Since supplements are legally classified as a specifically defined type of food, all supplements are required to be manufactured to the same high standards that are required of all foods. *These mandated Good Manufacturing Practices (GMP)* establish basic guidelines to assure that supplements are manufactured under sanitary conditions that result in properly identified products that are not contaminated or adulterated, and that are fit for consumption. Any supplement that does not conform to these basic guidelines is subject to regulatory action by FDA.*
In addition, all supplement products are required by law to provide certain information about their formulation.
Like foods, supplements must provide consumers with nutritional information. Unlike foods, supplements must state the quantity of each of the contained ingredients, or of the 'proprietary blends' that make up a product.
All herbal products are required to identify the parts used of each of the plant ingredients, and to label them with their commonly accepted names.
One of the areas of the most detailed Federal regulation of supplements is in the area of product claims, whether on product labels or in advertising. *The Food and Drug Administration specifies exactly what kind of claims are allowed, and prohibits the use of any statement that would brand the product as a drug.***
Herbal supplements are not allowed to make statements regarding prevention, cure, mitigation or treatment of diseases. Instead, their claims are limited to statements that are legally defined as 'statements of nutritional support' or 'structure/function statements'.

-'Answers to FAQs about Herbal Products' by American Herbal Products Association (AHPA), Silver Spring, Maryland, December **2015**.
(* -'Current Good Manufacturing Practice in Manufacturing, Packing, or Holding Human Food', Title 21, Code of Federal Regulation, Section 110, Government Printing Office, Washington, DC, 2000.)
(** -'Certain Types of Statements for Dietary Supplements', Title 21, Code of Federal Regulation, Section 101.93, Government Printing Office, Washington, DC, Revised February 7, 2000.)   (FAQ = Frequently Asked Questions.)

"**Safety Evaluations of Dietary Supplements:** *In the United States, the risk of hepatotoxicity linked to the widespread use of certain herbal products has gained increased attention among regulatory scientists.*
Based on current United States law, all dietary supplements sold domestically, including botanical supplements, are regulated by the Food and Drug Administration (FDA) as a special category of foods.
Under this designation, regulatory scientists do not routinely evaluate the efficacy of these products prior to their marketing, despite the content variability and phytochemical complexity that often characterizes them.
Nonetheless, there has been notable progress in the development of advanced scientific methods to qualitatively and quantitatively measure ingredients and screen for contaminants and adulterants in botanical products when hepatotoxicity is recognized.
Pre-Marketing Safety Evaluation:
The Food and Drug Administration (FDA) does not routinely perform pre-marketing safety evaluation of dietary supplements and does not register all marketed supplements.
Post-Marketing Safety Evaluation:
FDA regulatory scientists often do not have available for review product ingredient content or exposure outcome measurements at the onset of an investigation that is tasked to assess **a post-marketing adverse event, such as supplement-associated liver injury**. *Compounding this challenge in the evaluation of herbal-related safety signals, FDA scientists together with non-agency regulatory analysts must take into account the content variability and phytochemical complexity that typically characterizes different preparations and batches of many botanical products.*"
-'Scientific and Regulatory Perspectives in Herbal and Dietary Supplement Associated Hepatotoxicity in the United States' by Mark I. Avigan, Robert P. Mozersky and Leonard B. Seeff; Center for Drug Evaluation and Research, FDA, Silver Spring, Maryland and Office of Dietary Supplement Products, Center for Food Safety and Applied Nutrition, College Park, Maryland; International Journal of Molecular Sciences, Volume 17, No. 331, 30 pages, **2016**.

<u>United Kingdom Bans Some Uses of Comfrey</u>

"*The Japanese work* brought Comfrey into 'Poisonous Plants in Britain and Their Effects on Animals and Man'**, Great Britain Ministry of Agriculture, Fisheries and Food, H.M.S.O. (Her Majesty's Stationery Office) **1984** with the verdict:
'*The carcinogenic (cancer causing) response followed continous high dosing over long periods, and it is unlikely that human consumption of Comfrey in much smaller amounts could cause liver damage, and no examples of poisoning by the plant have been reported.*' "
-Fighting Like the Flowers: An Autobiography: The Life Story of Britain's Best-Known Organic Gardener by Lawrence D. Hills. Bideford, Devon, England: Green Books, 1989, page 216.
(* -"**Two pyrrolizidine alkaloids, symphytine, a new compound, and echimidine have been isolated from the dried roots of Symphytum officinale.**" -Studies on Constituents of Crude Drugs. I. Alkaloids of Symphytum Officinale Linn.' by T. Furuya and K. Araki from Kitasato University in **Japan**; Chemical & Pharmaceutical Bulletin, Volume 16, No. 12, pages 2512-2516, 1968. For more information about this research, see number 2 in subsection 'Scientific Studies Showing Dangers of Alkaloids in Comfrey' in section 'Alkaloids in Comfrey' {Chapter 30}.)
(** -Poisonous Plants in Britain and Their Effects on Animals and Man by Marion Cooper and Anthony W. Johnson for 'Ministry of Agriculture, Fisheries, and Food'. London, England: H.M. The Stationery Office, 1984.)

"*The 'Committee on Toxicity' (COT) last reviewed PAs (Pyrrolizidine Alkaloids) in 1992, focussing on Comfrey, a herb which at the time was available in tablet and capsule form as well as for tea and infusions. The recommendations of that review of Comfrey were as follows:*
*1. **The public should be warned of the potential dangers associated with the consumption of Comfrey** and products containing Comfrey. This advice applies equally to commercial and home-grown Comfrey and preparations made from it.*
*2. **Concentrated forms of Comfrey such as tablets and capsules should no longer be available.***
*3. **The public should be advised against the ingestion of Comfrey** root and leaves, and of teas and infusions made from Comfrey root.*
*4. **Comfrey teas and tinctures may continue to be available to the public. However, this recommendation should not be construed as an endorsement of these products.***
The COT advice was subsequently endorsed by the 'Food Advisory Committee', 'Department of Health' (DH) and 'Ministry of Agriculture Fisheries and Food' (MAFF) Ministers accepted the committees advice and action was taken to implement it."
-'Committee on Toxicity (COT) of Chemicals in Food, Consumer Products and the Environment', Statement on   Pyrrolizidine Alkaloids in Food, COT is a United Kingdom independent scientific committee that provides advice to 'Food Standards Agency', 'Department of Health' and other government agencies, October 2008.

"Comfrey Tablets and Capsules Banned by MAFF:
*At a meeting dated 2 March **1993** UK (United Kingdom) **'Ministry of Agriculture, Fisheries and Food' (MAFF) Minister Nicholas Soames MP (Member of Parliament), on the advice of the 'Committee on Toxicity' (COT) and 'Food Advisory Committee' (FAC), asked the health food trade and industry to discontinue the sale of all tablets and capsules containing Comfrey,** and to withdraw existing supplies from shelves. Representatives of relevant associations agreed voluntarily to do so. The grounds for the ban were that the content of pyrrolizidine alkaloids (PAs) in Comfrey could constitute a hazard to health.*
Herbal Practitioners and Consumers Not Represented:
*In the course of this meeting of 2 March, which was **convened with four days' notice**, the **'Society for the Promotion of Nutritional Therapy' (SPNT),** which represents consumers and practitioners of nutritional therapy (which may include the use of Comfrey products) expressed deep concern that **the government had not held any consultation procedure with consumers prior to announcing its decision.***

The Society had not been informed that the COT was investigating Comfrey, and had not been given the opportunity to present any material which could be used in an assessment of the safety of Comfrey. In the course of the meeting of 2 March the Society's request to delay any ban on Comfrey until it had been given such an opportunity was denied.
Objections to Comfrey Ban:
The Society was also concerned that by making the Comfrey ban a voluntary one, the government avoided having to subject its arguments to Parliamentary scrutiny.

*After the ban, the SPNT subjected the COT's and FAC's reports to close scrutiny. **Many errors and inaccuracies were found, and it became clear that the criteria used as the basis for the ban did not stand up to this scrutiny. Epidemiological evidence against Comfrey appeared to have been fabricated (made up) by referring to poisonings by plants other than Comfrey.***
Banning is Not Only Way to Regulate Comfrey:
***Should the government consider lifting the ban,** Society submits that the government can answer safety concerns by means of cautions on labels, or by setting upper limits for the content of pyrrolizidine alkaloids in Comfrey products, as is the custom in other countries such as Germany. The Society would also not object to the restriction of Comfrey sales to trained, registered nutritional therapists and medical herbalists, for professional use only."*
-'The Safety-in-Use of Comfrey and Comfrey Products: Results of a Research Survey' by the Society for the Promotion of Nutritional Therapy (SPNT), Linda Lazarides, Sussex County, United Kingdom, 8 pages, **1993**-1994, pages 1, 2.

*"In 1988 the **British 'Ministry of Agriculture, Food and Fisheries' (MAFF)** had withdrawn licenses for products containing mistletoe, squaw weed, broom and sassafras. Griggs (*1997) explains that the **later moves (in the 1990s) to proscribe (legally forbid) Comfrey** produced challenges, to which the herbal profession had to respond vigorously to protect other interests.*

The pyrrolizidme alkaloids of Comfrey also occur in other plants such as ragwort, borage and coftsfoot, and even in whole plant form, the alkaloids can pose dangers. For example, farmers know that stock may be at risk from grazing ragwort.

***However, Comfrey had always been regarded as one of the safest and most effective medicinal plants,** particularly therapeutic as a wound healer and soother of the mucous membrane. Herbalists believed that if Comfrey, or any other medicinal plants, had any dangerous side effects, then they would have been revealed by centuries of widespread use.*

***The herbalists defence of Comfrey was based on the theory that evidence of toxic constituents within a plant does not automatically prove that the whole plant is toxic;** a plant was more than a mere physical dilution of alkaloids (Griggs, 1997). Supporting evidence for this theory came in the form of tea, almonds, apples, pears, mustard, radishes and hops, which are safe to consume, but have potentially toxic constituents (Whitelegg; cited in Griggs, 1997).*
**Eventually, MAFF's 'Committee on Toxicology' allowed the continuing use of Comfrey leaf tea or tincture, as a food, as well as mistletoe and broom. However, MAFF called for a voluntary ban on chaparral, and the herbalists decided to concede (stop resisting the ban on) Comfrey as well, whilst retaining Comfrey tea."*
-'Evolution of Natural Medicine and Biomedicine and Their Future Roles in Health Care' by Peter Thomas Sherwood, dissertation for Doctor of Philosophy, School of Health Sciences, St. Albans, Victoria, Australia, 2004.
(* -New Green Pharmacy: Story of Western Herbal Medicine by Barbara Griggs. London, England: Vermillion, 1997. Also called 'Green Pharmacy: The History and Evolution of Western Herbal Medicine' by Barbara Griggs. Rochester, Vermont: Healing Arts Press, 1997.)

"Internal Use of Comfrey Banned, Except Comfrey Tea is OK:
***As a result of a 1993 report** by the 'Committee on Toxicity of Chemicals in Food' to the 'Food Advisory Committee' and the 'Ministry of Agriculture, Fisheries and Food' (United Kingdom), **the health food trade voluntarily withdrew all Comfrey products, such as tablets and capsules, and advice was issued that the root and leaves should be labelled with warnings against ingestion. It was considered that Comfrey teas contained relatively low concentrations of pyrrolizidine alkaloids and did not need any warning labels.**"*
-Herbal Medicines by Joanne Barnes, Linda A. Anderson and J. David Phillipson. London, England: Pharmaceutical Press, third edition 2007, page 188.

*"**United Kingdom: The 'Herbal Medicines Advisory Committee'* was established in 2005** to advise on the safety, quality, and efficacy of herbal medicinal products for human use. Although there is no requirement for proof of efficacy, pharmacological effects must be plausible and supported by long-standing use and experience.*

Regulation through European Union:
The herbal medicinal product must be eligible under the 'Traditional Herbals Medicine Products Directive' 2004/24/EC**.
Prior to this, regulation of the herbal industry was covered by the 1968 'Medicines Act'.
Adverse Reactions:
> Under the current act, suspected adverse reactions to herbal medicines are reported voluntarily, but are compulsory for manufacturers with registration of a product under the 'Traditional Herbal Medicines Registration Scheme' that has published guidelines for retailers, wholesalers, importers, and manufacturers on its requirements."

-Drug-Induced Liver Disease edited by Neil Kaplowitz and Laurie DeLeve. London, England: Academic Press, **2013**. (Chapter 35: 'Hepatotoxicity of Herbal and Dietary Supplements' by L.B. Seeff, F. Stickel and V. Navarro, pages 631–657.)
(* -The United Kingdom 'Herbal Medicines Advisory Committee' advises on the safety and quality of herbal medicines when there is an application for registration, marketing authorisation or product license.)
(** -The 'European Directive on Traditional Herbal Medicinal Products', Directive 2004/24/EC was established by the 'European Parliament and Council' on 31 March 2004 to provide a simplified regulatory approval process for traditional herbal medicines in the European Union.)

> *"How to apply for a Traditional Herbal Registration (THR) to market a herbal medicine (remedy) in the United Kingdom (UK). You must apply for a traditional herbal registration (THR) before you can market a herbal medicine in the UK. **A THR is only granted if the medicine is used for minor health conditions where medical supervision is not required (e.g. a cold).***
>> *If your Traditional Herbal Medicine claims to treat major health conditions, you need to apply for a marketing authorisation (License to market a medicine in the United Kingdom) before you can place it on the market.*
> **You must include scientific evidence with your application relating to the safety, quality and traditional use of the herbal product.** *You must also show evidence that the herbal medicinal product has been traditionally used to treat the stated condition for a minimum of 30 years, 15 years of which must have been in the European Union (EU).*
> **If you're a herbal practitioner, you don't need a licence to supply herbal medicines you create on your premises to patients following one-to-one consultations."**

-'Apply for a Traditional Herbal Registration (THR)' by Medicines and Healthcare Products Regulatory Agency, London, England, updated September **2016**.

Germany Bans Some Uses of Comfrey

*"**Regulation of Pyrrolizidine Alkaloids:***
Different national and European legislations setting limits for pyrrolizidine alkaloids in herbal medicinal products have been enacted. This work discusses the recent regulatory developments and their effects on the pharmaceutical industry.
BfArM (Bundesinstitut fur Arzneimittel und Medizinprodukte), i.e., the 'Federal Institute for Drugs and Medical Devices', Germany, 1992.
The graduated plan regarding pyrrolizidine alkaloid containing plants in medicinal products was published by the BfArM. *Namely for the following medicinal plants: Alkanna, Anchusa, Borago, Brachyglottis, Cineraria, Cynoglossum, Erechthites, Eupatorium (except E. perfoliatum), Heliotropium, Lithospermum, Petasites, Senecio,* **Symphytum** *and Tussilago.*
> **The exposure of pyrrolizidine alkaloids should not exceed the following limits:**
> **- 100 microgram/day for topical (skin) applications (maximum 6 weeks per year)**
> **- 1 microgram/day for internal use (maximum 6 weeks per year)**
> **- 0.1 microgram/day for internal or 10 microgram/day for topical medicinal products**
>> **without therapeutic indications or without restriction of intake to 6 weeks."**

-'Pyrrolizidine Alkaloids: Impact of the Public Statements made by EMA and National Health Authorities on the Pharmaceutical Industry' by Katrin Lambrecht, Berlin for Wissenschaftliche Prufungsarbeit zur Erlangung des Titels (Scientific Exam Work to Obtain the Title) Master of Drug Regulatory Affairs, Rheinische Friedrich Wilhelms University, Bonn, Germany, 2016.
(1 milligram = 1,000 micrograms.)

*"**Today (1996), the most accurate information available on the safety and efficacy of herbs and phyto-medicinals is found in the German medical and pharmaceutical literature.** Natural drugs have long been popular in Germany. The subject is taught in German medical schools, and all aspiring physicians must pass an examination regarding their use. German pharmacy students also receive extensive instruction in pharmaceutical biology (pharmacognosy).*
Traditionally Used Herbs:
German law requires 'reasonable' proof of efficacy for these long-used natural drugs, making product research and development economically feasible.
In 1978, the German 'Federal Health Agency' convened a commission *of reputable scientists and practitioners to examine all available information on the safety and efficacy of commonly used herbs.*
Other countries, notably Britain and France, allow sales of herbal products with traditional claims of value; that is, if an herb had long been used to cure headache, the manufacturer could continue to sell it as a drug labeled for that purpose.
Monographs (German Commission E) have now been published on about 300 different herbs."
-'What Pharmacists Should Know about Herbal Remedies: Pharmacists Can Help Patients Differentiate the Useful Herbs from the Harmful Ones' by Varro E. Tyler, Ph.D., Purdue University School of Pharmacy and Pharmacal Sciences, West Lafayette,

Indiana; Journal of the American Pharmaceutical Association, Volume NS36, No. 1, pages 29-37, January **1996**.
(The 'American Society of Pharmacognosy' defines pharmacognosy as 'the study of the physical, chemical, biochemical and biological properties of drugs, drug substances or potential drugs or drug substances of natural origin as well as the search for new drugs from natural sources'.)
(The 'Federal Health Agency' (Bundesgesundheitsamt) was a German federal government agency and research institution for public health. It was dissolved 1994, and its responsibilities transferred to: 'Federal Institute for Drugs and Medical Devices', 'Robert Koch Institute', and 'Federal Institute for Health Consumer Protection and Veterinary Medicine'.)

"*Government Regulations of Pyrrolizidine Alkaloids:*
In Germany, consumption of total PAs (Pyrrolizidine Alkaloids) with 1,2-unsaturated necine moieties (part of a molecule), such as seen in the structures of both symphytine and echimidine, is limited to 1 microgram daily, although special consideration is given for Comfrey tea, which is limited to a maximum dose of 10 microgram daily."
-'Analysis of Herbal Teas Made from the Leaves of Comfrey (Symphytum Officinale): Reduction of N-oxides Results in Order of Magnitude Increases in the Measureable Concentration of Pyrrolizidine Alkaloids' by Oberlies, Kim, Brine, Collins, Handy, Sparacino, Wani and Wall (North Carolina and New Mexico); Public Health Nutrition Journal, Cambridge, England, Volume 7, Issue 7, pages 919-924, **2004**.

German Commission E Monographs

"*The* **'*German Commission E*'** *is a scientific advisory board of the 'Bundesinstitut fur Arzneimittel und Medizinprodukte' (the German equivalent of the 'Food and Drug Administration' FDA), formed in 1978.*
The commission gives scientific expertise for the approval of substances and products previously used in traditional, folk and herbal medicine.
The commission became known beyond Germany in the 1990s for compiling and publishing 380 monographs evaluating the safety and efficacy of herbs for licensed medical prescribing in Germany.
The monographs were published between 1984 and 1994 *in the Bundesanzeiger; they were not updated since then but are still considered valid.*"
-Wikipedia®: The Free Encyclopedia, www.wikipedia.org, 2018.

"*Symphytum officinale L. :*
Comfrey root (Symphyti radix).
Comfrey leaf (Symphyti folium).
Comfrey herb (Symphyti herba) = *all parts that grow above ground.*
Composition of Drug: The drug contains allantoin and rosmarinic acid.
Uses: External: Bruising, pulled muscles and ligaments, sprains.
Contraindications: None known. Application should only occur on intact skin. During pregnancy use only after consultation with a physician.
Side Effects: None known.
Dosage: Unless otherwise prescribed: Ointments and other preparations for external application with 5-20 percent dried drug or equivalent preparations. The daily applied dosage should not exceed 100 microgram (mcg) of pyrrolizidine alkaloids with 1,2-unsaturated necine structure, including their N-oxides.
Mode of Administration: Comminuted (pulverized) herb and other galenical (natural) preparations for external use.
Duration of Administration: Not more than 4-6 weeks per year.
Action: Anti-inflammatory."
-'**Commission E Monograph** (Phytotherapy), Comfrey Herb and Leaf (Symphytii Herba-Folium)', published July 27, **1990** by the 'Bundesanzeiger', the official publication of the Federal Republic of Germany.

"*In the German market there are specially cultivated varieties (cultivars) of Comfrey that do not contain PAs (Pyrrolizidine Alkaloids) (*Schilcher, 1997).*
Nearly all products for internal use made from Comfrey root were removed from the German market in 1992 *in accordance with section 8 of the 'Second Medicines Act' (2 AMG 76)** of 1976 (*Schilcher, 1997).*
Medicinal Actions:
In almost all of the 'Approved Commission E Monographs' and in some of the 'Unapproved Monographs', the **'*Actions' (Wirkungen) section refers to pharmacological actions carried out in laboratory conditions,** *either experimentally in vitro, or in vivo in test animals, and in animal organs.*
These actions usually do not refer to observations based on human clinical trials and are meant to help describe the potential activity of the herbal drug. Professor Schilcher has translated this section as 'Medicinal Actions', a term that implies (to us) medicinal applications, which we are not sure we are ready to accept. Hence, we use only the word 'Actions', qualifying it here as pharmacological."
-'Commission E Monographs: Explanation of Monograph Sections' by American Botanical Council, Austin, Texas, **1999**.
(* -H. Schilcher, December 30 1997, 'Personal communication'. Heinz Schilcher: University Professor Dr. rer. nat. {'Doctor rerum naturalium' which is same as Ph.D.}, Marx-Zentrum Pharmacy, Munich, Germany.)
(** -Medicinal Products Act {Arzneimittelgesetz - AMG}, Germany.)

"The book 'The Complete German Commission E Monographs' documents how phytotherapy in Germany has become rational, responsible, and well integrated with conventional medicine. Herbal medicines comprise 30% of all drugs sold in German pharmacies.
Good Overview about Herbs:
Certainly worth studying, the 'Commission E Monographs' detail which herbs are approved or disapproved, along with their uses, dosages, contraindications, adverse effects, drug interactions, and pharmacologic actions.
Limitations of the Monographs:
    1. Toxic side effects are less extensively documented here than in 'The Lawrence Review of Natural Products' or 'Ellenhorn's Medical Toxicology'. Although some 'Commission E Monographs' state or imply that certain herbs can kill us, others omit mentioning the possible fatal reactions. Thus, no matter how this book excels, it is not the final source.
    2. Many monographs are briefer than terse (few words). All lack literature references, which denudes (removes) their value as believable wisdom. Such opacity (lack of clarity) is perplexing.
    3. Also missing is exactly which brands were tested."
-'The Complete German Commission E Monographs: Book Review' by Alan T. Marty, M.D., Newburgh, Indiana; JAMA: Journal of the American Medical Association, Chicago, Illinois, Volume 281, No. 19, pages 1852-1853, May 19 **1999**.
(The Complete German Commission E Monographs: Therapeutic Guide to Herbal Medicines edited by Blumenthal, Busse, Goldberg, Gruenwald, Hall, Riggins, Rister and Klein. Austin, Texas: The American Botanical Council, 1998.)

"This book is a translation of all 380 monographs on medicinal herbs published by the German Commission E, an expert multidisciplinary group within the German federal health agency charged with evaluating the safety and efficacy of herbal medications.
Each monograph provides easily accessible and necessary information for patients and health care providers.
Limitations: The text lacks references to the primary research, making it impossible for clinicians to independently assess the weight of the evidence for a particular indication."
-'The Complete German Commission E Monographs: Book Review' by Michael Nathan, M.D. and Robert Scholten, MSLIS, Massachusetts; Annals of Internal Medicine: American College of Physicians, Volume 130, No. 5, page 459, March 2 **1999**.

European Union: Comfrey Medicinal Regulations

The 'European Medicines Agency' (EMA) is a 'European Union' agency that evaluates medicinal products. Prior to 2004, it was known as the 'European Agency for the Evaluation of Medicinal Products' or 'European Medicines Evaluation Agency' (EMEA). The EMA, based in London, England, is meant to harmonize, but not replace, existing national medicine regulatory bodies. It is intended to have just one approval process for all of the European Union rather than separate processes for each nation. As of 2018, the European Union is the source of one-third of new drugs brought to the world market each year.

"**European Union 1992 and 2011:**
Internal Use of Comfrey Root:
    Due to the **'List' of herbs and herbal derivatives with serious risks dated 1992** prepared by 'Committee for Proprietary Medicinal Products' (CPMP) and other measures made by national medicine authorities, **there is no product containing Symphyti radix (Comfrey root) for oral use as herbal medicine in Europe***.
External Use of Comfrey:
    Only some preparations for cutaneous (skin) use are available as well with limited content of unsaturated pyrrolizidine alkalkaloids (PA).
Based on 'Public statement on the use of herbal medicinal products containing toxic, unsaturated pyrrolizidine alkaloids (PAs)' prepared and issued by **'Committee on Herbal Medicinal Products' (HMPC) (EMA/HMPC/893108/2011)** the following recommendations were taken into account for the purpose of the monograph on Symphyti radix.
    Because of their known involvement in human poisoning and their putative (generally regarded) carcinogenicity, exposure to toxic, unsaturated PAs should be kept as low as practically achievable."
-'**European Medicines Agency- Comfrey Root Symphytum Officinale L., Radix- Herbal Medicine Summary for the Public**' by European Medicines Agency, Committee on Herbal Medicinal Products (HMPC), London, England, 2 pages, Dec 4 2015.
(* -'Committee for Proprietary Medicinal Products' {CPMP} 'Listing of Herbs and Herbal Derivatives Withdrawn for Safety Reasons: Herbal Drugs with Serious Risks', 1992.)

"**European Union 2004 and 2011:**
Current regulations are based on the 'Traditional Herbals Medicine Products Directive' 2004/24/EC announced on 31 March 2004. The intent was that all traditional medicines in health food shops and pharmacies must be formally registered and the products approved before they could be sold.
    Acceptable products for licensure were considered to be those whose use was 'plausible on the basis of long-standing use and experience' and whose quality and safety could be guaranteed. The duration required to prove their safety was at least 30 years in all, and at least 15 years in the EU.
The Directive allowed the passage of 7 years from its announcement for manufacturers to gather the necessary information on

their products and **on 1 May 2011, the requirement that herbal medicines and their ingredients be registered with evidence of safety went into effect.**
Based on this Directive, no herbals from China would meet the specifications, and they are therefore currently banned in the European Union."
-Drug-Induced Liver Disease edited by Neil Kaplowitz and Laurie DeLeve. London, England: Academic Press, 2013. (Chapter 35: 'Hepatotoxicity of Herbal and Dietary Supplements' by L.B. Seeff, F. Stickel and V. Navarro, pages 631–657.)

"**The 'European Directive on Traditional Herbal Medicinal Products' (THMPD), formally the Directive 2004/24/EC amending, as regards traditional herbal medicinal products, Directive 2001/83/EC on the Community code relating to medicinal products for human use**, was established by European Parliament and Council on 31 March 2004 to provide a simplified regulatory approval process for traditional herbal medicines in European Union (EU). Previously, there was no formal EU wide authorisation procedure, so each EU member state regulated these types of products at the national level. **Under this regulation, all herbal medicinal products are required to obtain an authorisation to market within the EU.** Those products marketed before this legislation came into force can continue to market their product until 30 April 2011, under the transitional measures defined in the 'Traditional Herbal Medicinal Products Directive'. Once this time limit has expired, **all herbal medicinal products must have prior authorisation before they can be marketed in the EU.**"
-Wikipedia®: The Free Encyclopedia, www.wikipedia.org, 2018.

"**This is a summary of the scientific conclusions reached by the 'Committee on Herbal Medicinal Products' (HMPC) on the medicinal uses of Comfrey root.** The HMPC conclusions are taken into account by EU (European Union) Member States when evaluating applications for the licensing of herbal medicines containing Comfrey root.
Comfrey Root Extract:
This summary covers Comfrey root medicines containing a specific herbal preparation, which is obtained by ethanol extraction (a technique used to extract compounds from plant material by dissolving them in alcohol). **Herbal medicines containing this Comfrey root preparation are available in semi-solid forms (such as creams or ointments) to be applied to the skin.**
Sprains and Bruises:
The HMPC concluded that, on the basis of its long-standing use, **these Comfrey root medicines can be used for the relief of symptoms of minor sprains and bruises.** These medicines should only be used in adults and should not be taken for longer than 10 days. The HMPC conclusions on the use of Comfrey root medicines for the relief of symptoms of minor sprains and bruises are **based on their 'traditional use' in these conditions. This means that, although there is insufficient evidence from clinical trials, the effectiveness of these herbal medicines is plausible**, and there is evidence that they have been used safely in this way for at least 30 years (including at least 15 years within the EU). Moreover, the intended use does not require medical supervision.
Clinical Trials:
In its assessment, the HMPC considered the well documented use of Comfrey root for the relief of symptoms of minor sprains and bruises. The HMPC also noted 4 clinical studies carried out with a different Comfrey preparation (not covered by this summary). These studies suggested a reduction in swelling and pain in patients using this herbal preparation for sprains and bruises. However, since the exact composition of the herbal preparation used in these studies is not known, **these clinical data were not taken into account, and the HMPC conclusions on the use of Comfrey root medicines are based on their long-standing use.**"
-'European Medicines Agency- Comfrey Root Symphytum Officinale L., Radix- Herbal Medicine Summary for the Public' by European Medicines Agency, Committee on Herbal Medicinal Products (HMPC), London, England, 2 pages, Dec 4 **2015**.

"**Symphytum officinale Root:**
With regard to the registration application of Article 16d (1) of Directive 2001/83/EC as amended Symphytum officinale L., radix (Comfrey root).** The daily exposure has to be below 0.35 micrograms for adults.
Extract and Ointment:
Liquid extract prepared by extraction with ethanol 65% (V/V =volume of solute per volume of solvent) followed by partial evaporation and adjustment to a DER 2:1 (with respect to mass of the starting plant material). It is a **10% liquid extract in ointment base** that has been on the market for more than 30 years.
Semi-solid dosage forms for cutaneous (skin) use. The pharmaceutical form should be described by the European Pharmacopoeia full standard term. The amount of pyrrolizidine alkaloids has to be specified in the given product.
Dosage:
Adults and elderly: Single dose for semi-solid dosage forms: apply a thin layer, 2 times daily. The use in children and adolescents under 18 years of age is not recommended. Not to be used for more than 10 days. **Not to be applied to broken or irritated skin.** Avoid contact with the eyes or mucous membranes.
Pregnancy and Breastfeeding:
Safety during pregnancy and lactation has not been established. Studies with isolated pyrrolizidine alkaloids from the plant part in animals have shown reproductive toxicity. In the absence of sufficient data, the use during pregnancy and lactation is not recommended.
In the 'British Herbal Pharmacopoeia' (1974 and 1983), fresh Symphytum root is indicated externally for the treatment of ulcers, wounds, fractures and hernia."

-'**European Medicines Agency- European Union Herbal Monograph on Symphytum Officinale L., Radix, Final**' by European Medicines Agency, Committee on Herbal Medicinal Products (HMPC), London, England, 6 pages, May 5 **2015**.

"United Kingdom:
*In United Kingdom, 'Medicine Control Agency' (now 'Medicines and Healthcare Products Regulatory Agency' MHRA) recently **included Comfrey in a list of herbs under consideration for restriction to physician prescription only****.*
Germany:
The use of Comfrey root in Germany is limited to external products. The 'Commission E' suggest the external use of Comfrey root (crushed root, extracts, the pressed juice of the fresh plant for semi-solid preparations and poultices) in case of bruising, pulled muscles and ligaments, and sprains. According to the 'Commission E', the daily dose should not exceed more than 100 microgram pyrrolizidine alkaloids with 1,2 unsaturated necine structure, including its N-oxides. The duration of treatment should not be longer than 4-6 week per year**.
France:
In France, the only indication that may be claimed for the Comfrey root is as follows:
as an adjunct in the emollient and anti-pruriginous treatment of skin disorders, and as a trophic protective agent for cracks, bruises, frostbite and insect bites.***"

-'**European Medicines Agency- Assessment Report on Symphytum Officinale L., Radix**' by European Medicines Agency, Herbal Medicinal Products Committee (HMPC), London, England, 27 pages, May 5 **2015**.
(* -'Comfrey Toxicity Revisited' by Dorena Rode, Department of Animal Science, College of Agricultural and Environmental Sciences, University of California, Davis, California; Trends in Pharmacological Sciences: International Union Of Pharmacology, Volume 23, No. 11, pages 497-499, November 2002.)
(** -The Complete German Commission E Monographs: Therapeutic Guide to Herbal Medicines edited by Blumenthal, Busse, Goldberg, Gruenwald, Hall, Riggins, Rister and Klein. Austin, Texas: The American Botanical Council, page 116, 1998.
(*** -Pharmacognosy, Phytochemistry, Medicinal Plants, 2nd Edition by J. Bruneton. Paris, France: Lavoisier Publishing Inc., pages 839-841, 1999.)

(An emollient softens and soothes skin. Prurigo is a chronic inflammatory skin disease with itchy small, solid, round bumps. Prurigo is a skin disease that causes hard, itchy lumps on skin. Trophic pertains to food/nourishment in the sense of growth. The opposite is atrophy.)

*"**Dosage Pyrrolizidine Alkaloids:***
*Based on the HMPC (Herbal Medicine Products Committee) 'Public statement on **the use of herbal medicinal products containing toxic, unsaturated pyrrolizidine alkaloids (PAs)**' (EMA/HMPC/572844/2009) **the acceptable daily intake is maximum 0.007 microgram/kg/day (kg= 2.2 pounds) for cutaneous (skin) preparations**.*
*Generally for adults the calculation is done with a body weight of 50 kg (110 pounds). **Therefore the daily dosage would be: 0.007 microgram/kg/day x 50 kg body weight = 0.35 microgram/person/day for short-time use only (maximum 2 weeks)**. The use is restricted to intact skin.*
Pregnancy and Children:
The public statement does not exclude the use of the preparations with a lower limit on the PAs content (0.014 microgram PA/day) in children and during pregnancy, however, taking into consideration the general rule that there are not adequate data on the use of the preparation mentioned in the monograph in these populations for more than 30 years, the use is not recommended."

-'**European Medicines Agency- Assessment Report on Symphytum Officinale L., Radix**' by European Medicines Agency, Committee on Herbal Medicinal Products (HMPC), London, England, 27 pages, May 5 **2015**.

*"**European Medicines Agency Issues Consumer-Friendly Herb Summaries:***
*The European Medicines Agency (EMA), which is responsible for the evaluation of medicines used within the European Union (EU), has announced that **it will regularly publish consumer-friendly summaries of the findings of its 'Committee on Herbal Medicinal Products' (HMPC)** on its website. The main purpose of the HMPC is to review all available scientific data on the use of specific herbal medicines, including information on safety and effectiveness, and to issue conclusions in EU herbal monographs on **how to use these medicines responsibly**."*

-'HerbalGram, No. 108' by American Botanical Council, Austin, Texas, 84 pages, November **2015** to January 2016.

*"**Comfrey Root, Symphytum Officinale L. Radix:***
*This summary is not intended to provide practical advice on how to use medicines containing Comfrey root. **For practical information about using Comfrey root medicines, patients should read the package leaflet that comes with the medicine or contact their doctor or pharmacist.***
Applying Comfrey Root to Skin:
*This summary covers Comfrey root medicines containing a specific herbal preparation, which is obtained by ethanol extraction (a technique used to extract compounds from plant material by dissolving them in alcohol). **Herbal medicines containing this Comfrey root preparation are available in semi-solid forms (such as creams or ointments) to be applied to the skin.***
*The HMPC (Committee on Herbal Medicinal Products) concluded that, on the basis of its long-standing use, these Comfrey root medicines can be **used for the relief of symptoms of minor sprains and bruises**. These medicines should only be used in adults and should not be taken for longer than 10 days.*

Traditional Use:
> The HMPC conclusions on the use of Comfrey root medicines for the relief of symptoms of minor sprains and bruises are based on their 'traditional use' in these conditions. This means that, **although there is insufficient evidence from clinical trials, the effectiveness of these herbal medicines is plausible and there is evidence that they have been used safely in this way for at least 30 years** (including at least 15 years within the European Union). Moreover, the intended use does not require medical supervision.

Side Effects:
> At the time of the HMPC assessment, no side effects had been reported with these medicines.
> Comfrey root contains substances known as pyrrolizidine alkaloids, which have toxic effects on the liver when taken by mouth. **No significant risk is expected when Comfrey root medicines are used on the skin for short periods.** However, the amount of pyrrolizidine alkaloids has to be specified in each Comfrey root medicine and patients should not be exposed to more than 0.35 microgram of pyrrolizidine alkaloids per day."

-'European Medicines Agency- Summary for the Public, Herbal Medicine: Comfrey Root, Symphytum Officinale L. Radix' by European Medicines Agency, Committee on Herbal Medicinal Products (HMPC), London, England, 2 pages December 4 **2015**.

Symphytum 1644

'Cruydt-Boeck Remberti Dodonaei Volghens Sijne Laetste Verbeteringhe' by Rembert Dodoens, Antwerp, Belgium. Edited and translated into Dutch by Carolus Clusius, 1644. First published 1554.
Symphytum pages 12, 101, 192-198, 243, 498, 551, 557, 979, 983.

Rembert Dodoens, 1517-1585. He worked as a physician in the Flemish town Malines (Mechelen, Belgium) and became court physician to Holy Roman Emperor Maximilian II and his son. In 1582 Dodoens became professor of Medicine and Pharmacologie / Botany at Leiden University, Netherlands.
He is considered the father of botany. In the early 1500s it was believed that all plants had been completely described by Dioscorides in his 'De Materia Medica'. Then began a botanical Renaissance. Botanical knowledge was expanding, especially by unknown plants found in the Americas, and the discovery of printing and wood-block illustration.

'Cruydt-Boeck' focused on medicinal herbs, which made this a pharmacopoeia (official list of medicinal drugs). It is one of the most important botanical works of the late 16th century. It was used as a reference book for 2 centuries.
It was translated into French in 1557 by Charles de L'Ecluse as 'Histoire des Plantes'. Into English in 1578 by Henry Lyte as 'A New Herbal, or Historie of Plants'. And into Latin in 1583 as 'Stirpium Historiae Pemptades Sex'. It was used by John Gerard in 1597 for his 'Herball'.

Comfrey with Purple Flowers.

Comfrey with Purple Flowers 1710

'Botanologia, The English Herbal, or History of Plants with Names, Species, Descriptions, Places, Times, Qualities, Specifications, Preparations, Virtues and Uses' by William Salmon, M.D., London, England, 1710. It is 1,375 pages. Comfrey is on pages 210-213.

Other Herbals:
John Gerard's 'Herball or Generall Historie of Plantes' (1597)
John Parkinson's 'Theatrum Botanicum' (1640)
Nicholas Culpeper's 'Complete Herbal and English Physician' (1681)
Elizabeth Blackwell's 'Curious Herbal' (1737)
William Withering's 'Botanical Arrangement of British Plants' (1787)

Chapter 32

Humans Eating Comfrey

Humans Eating Comfrey Overview

These ideas are for entertainment purposes only.
Please consult your health care advisor about the safety of eating Comfrey.

Comfrey as Food in 1400s:
"A retrospective view of Spanish agriculture and the range of species cultivated during the last 500 years would clearly show the considerable change that has taken place regarding the nature of crops. These changes are evident not only through the gradual incorporation of American flora into the Iberian (Spain) and island agricultural landscape (potato, maize, sunflower, beans, tomato, American cotton plants, avocados, custard apple, tobacco, etc.), but also through the loss of quite a few cultivated species **during the centuries prior to Columbus's voyage in 1492.**
Neglected Species:
> Widely different species have lost much of their importance, been marginalized or even completely forgotten. Some remain in the wild state, growing in ditches and on boundaries of cultivation, as a testimony to their past agricultural use, and they even behave as weeds of other crops. Others have disappeared completely from Spanish agricultural flora.

Horticultural Species:
> **This is perhaps the group with the largest number of marginalized species, especially horticultural species which may be called bitter. The species involved are mainly consumed as greens (boiled, cooked in butter or oil or fresh in the form of salads).**
> Some current gastronomies in Europe (and also in America because of the export of the crop and traditional consumption patterns) even use them preferentially as a garnish for meat.
> There are others which are very flavoursome and which are difficult to separate from their categorization as spices or aromatic plants.
> These include Amaranthaceae: Amaranthus lividus (blite); Apiaceae: Foeniculunt vulgare (fennel), Pastinaca sativa (parsnip), Smyrnium olusatrutn (alexanders or alisander); Asteraceae: Taraxacum (icinale (dandelion), Silybum marianum (holy, milk thistle or lady's thistle), Cichorium intybus (chicory, succory or witloop), Scolymus maculatus (spotted golden thistle), Scolymus hispanicus (Spanish salsify, golden thistle or Spanish oyster plant), Tragopogon porrifolius (salsify or vegetable oyster), Scorzonera hispanica (scorzonera or black salsify); Boraginaceae: Borago officinalis (borage), **Symphytum officinale (Comfrey)**; Brassicaceae: Eruca vesicaria (rocket, garden or salad rocket), Nasturtium officinale (summer or green watercress), Lepidium sativum (cress), Armoracia rusticana (horse-radish); Polygonaceae: Rumex acetosa (sorrel) and other species of the genus; Portulacaceae: Portulaca oleracea (purslane); and Chenopodiaceae: Atriplex hortensis (orache), Chenopodium album (goosefoot or fat-hen)."

-Neglected Crops: 1492 From a Different Perspective coordinated by Dr. J.E. Hernandez Bermejo (Spain) and Dr. J. Leon (Costa Rica) with contributions from 31 authors from 9 countries. United Nations Food and Agriculture Organization, Plant Production and Protection Series, No. 26, in collaboration with the Botanical Garden of Cordoba, Spain, 1994.

Comfrey as Food in 1824:
"The leaves are frequently employed to give a 'grateful' flavour to cakes and panada (thick sauce or paste made with bread crumbs, milk, and seasonings), and when boiled are esteemed by many a very great delicacy."
-'The Universal Herbal; or, Botanical, Medical, and Agricultural Dictionary; Containing an Account of All the Known Plants in the World, Arranged According to the Linnean system. Specifying the uses to which they are or may be applied, whether as food, as medicine, or in the arts and manufactures, with the best methods of propagation, and the most recent agricultural improvements' by Thomas Green, London, England, **1824**, page 641.

Comfrey as Food in 1838:
"The leaves of this Comfrey plant were formerly used to impart a flavour to cakes and panada, and **the young shoots are said to be good and nutritious food."**
-'The British Flora Medica, or, History of the Medicinal plants of Great Britain, Volume 1' by Benjamin Herbert Barton and Thomas Castle, London, England, **1838**. (Symphytum officinale, pages 211-215.)

Comfrey as Food in 1841:
"**Comfrey: In Bailey's Dictionary*** I find that this plant is called an excellent wound healer. I consulted this old dictionary in preference to any other, because this vegetable was considered by people in former days, as **an excellent esculent (edible). I tried it myself dressed in the same way as spinach**, and found it sustain its character. It is free from that flavour which spinach possesses, that renders it to some palates not nice till a taste is acquired for it it.
Can you tell me why Comfrey is not more in use as an esculent? Is it because it is old-fashioned and superseded by spinach? It has this advantage for cottagers, that it is perennial; and if it will throw a second and greater crop of leaves after the

first and successive cuttings, it must be valuable.
I find that it thickens from the root, and grows more bushy each year. I carried soil from a broken-up garden, that dates itself from the time of Elizabeth, in which this Comfrey grew; and that soil was afterwards removed twice: but not withstanding every disturbance, and diggings without any attention paid to it, this esculent showed an unconquerable tenacity for life, and in consequence this year, attracted my attention, and I had it cooked.
-Your Constant Reader of the Old School, Carnarvon (I'm not sure if this is in Wales or western Australia)
Editors Response: *Comfrey is Symphytum officinale, a plant belonging to the Borage tribe, and, like all others of that natural assemblage, is perfectly free from unwholesome qualities. It was formerly regarded as a vulnerary, that is, a plant capable of staunching wounds (stop flow of blood), a merit to which it has no claim.*
It has probably gone out of cultivation because its leaves become harsh and coarse unless very young. If gathered while tender, they are certainly a good perennial substitute for spinach; and the young shoots, blanched by being forced to grow through heaps of earth, are eaten like asparagus. *It would be a good plant for a cottage garden, provided the cottager is made to understand that he must never let it grow more than a foot high, if he means to eat it. The worst of it is, that it runs about very much by its creeping underground stems."*
-'The Gardeners Chronicle: A Stamped Newspaper of Rural Economy and General News', by Professor Lindley, Great Britain, Weekly- No. 1: January 2 to No. 52: December 25, **1841**. Comfrey in No. 17: April 24, page 262.
(* -'An Universal Etymological English Dictionary' by Nathan (Nathaniel) Bailey. First published in London, England in 1721. It was the most popular English dictionary of the 1700s. Etymology is the study of the history of words.)

Comfey as Food in 1877:
"**The young Comfrey leaves, when boiled,** *form a tolerable vegetable, and are not unfrequently eaten by country people where the plant abounds.*
Another person wrote: It should be observed that some days since we had a few of the green leaves, from the main plant, boiled, with a view of testing the qualifications of the Comfrey for a pot herb, and **we must confess that we found it very agreeable, as much so as our most delicate greens and spinach."**
-'Forage Plants and Their Economic Conservation by the New System of Ensilage: Part I: Caucasian Prickly Comfrey' by Thomas Christy, Jun., F.L.S. (Fellow of the Linnean Society), Christy & Co., London, England, **1877**, page 3, 6.

Comfey as Food in 1917:
"**The Comfrey Field Allotments Glasnevin, Ireland**
In connection with a fete in aid of the funds of the Irish Counties War Hospital (World War 1), Glasnevin, the above society held a most successful exhibition of produce in the grounds of the Claremount Institution, on Saturday, the 15th September 1917. Consisting entirely of vegetables, the exhibition was highly creditable, and reflected great credit on those responsible for the organisation of the show and on the numerous competitors who enthusiastically came forward with their entries.
Potatoes, Celery, Parsnips, Carrots, Onions, Beetroot, Cabbages, Marrows and such like crops of food value were most favoured. We will be very glad if the Secretaries of other **Allotment Societies*** *will kindly favour us with a report of any shows or other competitions in connection with the Allotment Movement."*
-'Irish Gardening: A Monthly Journal Devoted to the Enhancement of Horticulture and Arboriculture in Ireland', Volume XII, January to December 1917. Published by Irish Gardening Limited, Dublin, Ireland. Comfrey: October **1917**, No. 140, page 151.
(*In the United Kingdom an allotment is a small parcel of land rented to individuals to grow food. Most are 10 rods, an ancient measurement equal to 302 square yards or 253 square metres. Land is owned by local government or an association. Allotments go back 1000 years to when Saxons would clear woods which would be held in common.)
(I am assuming Comfrey would be one of the plants grown in a 'Comfrey Field Allotment', for food and/or medicine.)

Comfrey as Food in 1919:
"*Symphytum officinale Linn. Boragineae. Boneset. Comfrey: Europe and adjoining Asia.*
The leaves, when young, form a good green-vegetable and are not infrequently eaten by country people. They are sometimes used to flavor cakes and other culinary preparations. *The blanched stalks form an agreeable asparagus."*
-'Sturtevants Edible Plants of the World' by Dr. E. Lewis Sturtevant, Director of New York Agricultural Experiment Station, Geneva, New York, page 637, June **1919**.

Comfrey as Food in 1952:
"**The young shoots of Comfrey are especially nutritious. Arab peasants eat of them freely.**
But above all the herb is famed for its pecular powers upon the bones and ligaments. It has the power of aiding the body in the speedy and firm uniting of fractured surfaces."
-The Complete Herbal Handbook for Farm and Stable by Juliette de Bairacli Levy. London, England: Faber and Faber, **1952**.

> "*The Arabic words filaha, 'cultivation, tillage', and by extension 'agriculture, farming, husbandry', and fallah,* **'husband-man, tiller of the soil, peasant, farmer'**, *are derived from the verbal form falaha meaning 'to cleave, split', and in particular, 'to plough, till, cultivate the land'. Another principle that emerges clearly from the 'Books of Filaha' is that of diversity.*
> **A very large number of crops and plant species were cultivated, to be eaten or otherwise utilized, including many that have since fallen out of general use or been marginalized in southern Europe such as** *rocket, purslane, sorrel, dandelion, alexanders, scorzonera or black salsify, spotted golden thistle, milk thistle,* **Comfrey,**

Spanish salsify, vetches, cow-peas, spelt, pearl millet, sorghum, lotus tree, service tree or sorb, azarole, and the hackberry or nettle tree."
-'An Introductory Survey of the Arabic Books of Filaha and Farming Almanacs' by A. H. Fitzwilliam-Hall, The Filaha Text Projects: The Arabic Books of Husbandry, www.filaha.org, October 2010.

Comfrey Food in Japan in 1968:
"***Comfrey, called 'Miracle Grass', is being extensively used not only for stockfood, but also for human food.*** A tasting party (**1968** in Tokyo, Japan) to try 'Comfrey-fed meat', pork and chicken from animals raised on Comfrey was sponsored by Yomiuri Shimbun (one of the big 3 Japanese daily newspapers). The meat of pig and chicken fed on Comfrey has a high nutritional value, but is lean with a delicate flavor.
In addition to 'Comfrey meat' cooked by the chef, Mr. Takeo Kisawa, of the (Tokyo Prince) hotel, there are more than 20 kinds of dishes using Comfrey such as Comfrey Tempura (deep fried leaf), Comfrey Oden (Comfrey rolls like cabbage rolls cooked in special water-based mixture with rice-wine sake and soy sauce), Ramen (noodles), etc."
-Comfrey: Nature's Healing Herb & Health Food by Andrew Hughes. Japan: Sanyusha Publishing Co., Ltd, 1992, page 28.

Man Ate Comfrey for 28 Years:
"I (Andrew Hughes, Ringwood, Victoria, Australia) am writing this on my eighty-ninth birthday (in 1991), after taking some 85 grams (3 ounces) or more of green leaf of Comfrey in tablet form every day for 28 years, a routine adopted by my wife and myself. Back in 1978 we were **taking the equivalent of approximately 135 grams (4.7 ounces) of green Comfrey leaf a day with no side effects.** We reduced it to 85 grams as an appropriate average amount, considered diet- and cost-wise."
-Comfrey: Nature's Healing Herb & Health Food by Andrew Hughes. Japan: Sanyusha Publishing Co., Ltd, **1992**, page 5.

"***Since as much as 90% of total plant alkaloids in Comfrey may be in the form of highly water-soluble N-oxides, boiling Comfrey leaf as a vegetable should reduce toxicity, because most of the alkaloids would be discarded with the water.***"
-'Comfrey Update' by D.V.C. Awang, HerbalGram published by American Botanical Council, Austin, Texas, Volume 25, pages 20-23, Summer **1991**.

"***Non-Traditional and Local Food Plants under Investigation at the Nursery in Russia:***
Non-traditional food plants - species from the following genera:
    Arctogrostis, Astragalus, Arundo, Ammi, Heracleum, Baptisia, Asclepias, Vicia, Angelica, Trifolium, Hedysarum, Chamaenerion, Urtica, Cimicifuga, Lupinus, Rhaponticum, Chenopodium, Alcea, Malva, Daucus, **Symphytum**, *Pastinaca, Rheum, Glycyrrhiza, Silphium, Trigonella, Oxytropis, etc.*
The 'Nursery for the Introduction of Food, Crop and Medicinal Plants' was created in 1823. Our nursery was the recipient of all collections of economic plants from different regions. In years with a long spring, or changing frosts and warm weather, or with a cold and wet summer and early winter, nearly 25% of species can die."
-'The Introduction Nursery for Food, Crop and Medicinal Plants at the Komarov Botanical Institute of the Russian Academy of Sciences: Its Role in the Conservation of Biodiversity' by K.G. Tkachenko, I.A. Pautova and M.M. Korobova; Botanic Gardens Conservation International (BGCI), www.bgci.org, Volume 2, Number 8, July **1997**. BGCI provides a global voice for all botanic gardens, championing and celebrating their inspiring work. We are the world's largest plant conservation network, open to all.
    (Symphytum can survive under many weather extremes.)

"***Analysis of beta-carotene content of fresh Comfrey leaves that were blanched/unblanched, oven or solar dried were carried out.*** Storage studies on beta-carotene stability, sieve analysis and adsorption capacity of Comfrey powder were also carried out.
    The fresh Comfrey leaf was washed and excess water allowed to drain for 20-30 minutes. After drying it was followed by milling in a Sanyo food mix blender. About 50 grams (1.7 ounce) of leaves were broken by adjusting a Sanyo blender to the 'blending' operation for two minutes. The blender was then adjusted to the 'liquefying' operation and the vegetable processed for another 2 minutes. The vegetable was further milled to increase the surface area by adjusting the blender to the 'flash' position also for 2 minutes.
Processing of blanched Comfrey:
    Whole and chopped Comfrey (both without the midribs) were steam-blanched. The sublot of whole Comfrey leaves without the midrib was steam blanched for one minute while another was steam blanched for three minutes. These were quickly cooled with clean cold water while still in the blancher.
Processing of unblanched, oven-dried Comfrey:
    The midribs were removed and the vegetable dried on stainless steel wire mesh trays of the Fessmann (Zurich, Switzerland) air circulating oven, with drying at 60 degrees C (140 F) for 3 hours.
Processing of sun-dried Comfrey:
    The midribs were removed and the vegetable dried in a direct solar drier for 24 hours with mean temperature recorded as 25 degrees C (77 F).
Preparation of cooked Comfrey powder:
    The unblanched, and oven dried powders were cooked as follows: About 10 grams (0.3 ounce) of powder was fried in 20 grams (0.7 ounce) of fat and 50 mls (1.7 fluid ounce) of water added and cooking done for 5 minutes from boiling for one sublot of the sample and for 7 minutes for another sublot of the sample. These were then allowed to cool at room temperature (25 C = 77 F) for 15 minutes and analyzed for beta-carotene content.

The beta-carotene content of whole unblanched, oven dried (60 C = 140 F) Comfrey powder was significantly higher (49.03+-0.14mg/ 100 grams) than solar dried Comfrey powder (34.3+-1.9mg/ 100 grams) by 30% with $P<0.05$.
The content of whole steam blanched, oven dried powder was higher than that of chopped steam blanched powder. The loss of beta-carotene after 5 minutes of cooking Comfrey powder was 4.5% and 21.7% after 7 minutes of cooking.
After 2 weeks of storage of Comfrey powder in 200 gauge (2 mil) black polythene bags *at 25 degrees C under CO_2 (carbon dioxide), the loss was 27.8% (i.e., from 49.03+-0.14 mg/ 100 grams to about 35.4+-0.24 mg/ 100 grams) and fell gradually to about 55% after five weeks of storage."*
-'Assessment of the Potential of Comfrey (Symphytum Asperrimum) as a Source Of Vitamin A in a School Feeding Programme' by Joyce Violet Chania, Thesis Master of Science in Applied Human Nutrition, Department of Food Technology and Nutrition, University of Nairobi, Kenya, March **1998**.   (Carotene is a vitamin A precursor.)

(Absorption is when molecules cross the surface and enter the inside of the material.
Adsorption is the accumulation of molecules on the outside of the material.)
(Blanching is scalding or parboiling in water or steam to stop enzymatic action in food. Parboil is boiling briefly as a preliminary or incomplete cooking procedure.)

Overview of Many Ways to Eat Comfrey

*"**The Stalks and Roots of the Common Comfrey (Symphytum officinale) are very farinaceous: the stalks have been blanched and eaten** like those of Angelica, and we have no doubt of the tuberculated roots being at least as good as those of the Stachys palustris (Marsh All-Heal)."*
-'Loudon's Gardeners Magazine' or 'The Gardeners Magazine and Register of Rural and Domestic Improvement' by John Claudius Loudon, London, England, Volume 5, **1829**. Symphytum pages 442, 546.
(Farinaceous means a food containing starch or having a mealy texture.)

"Symphytum officinale (Comfrey) has been a traditional dietary and medicinal herb in Europe for hundreds of years (Grieve, 1978). **The hybrid Symphytum x uplandicum (Russian Comfrey) has also been consumed on an increasing scale in recent years and is the Comfrey most used in Australia and probably in the United States as a salad item.**
No toxic effects attributable to consumption of Comfrey have been reported. Leaves of Russian Comfrey contain 0.01-0.15% alkaloid, of which the main component is echimidine (Culvenor et al., 1980). **The consumption rate by Comfrey eaters varies from trivial levels up to 5 or 6 leaves/day.** *The high level implies 5-6 mg alkaloid/day."*
-'Estimated Intakes of Pyrrolidizine Alkaloids by Humans: A Comparison with Dose Rates Causing Tumors in Rats' by C.C.J. Culvenor, CSIRO Division of Animal Health, Melbourne, Australia; Journal of Toxicology and Environmental Health, Volume 11, No. 4-6, pages 625-635, **1983**.

"We have often suggested using fresh Comfrey in a salad with other fresh, green vegetables. When a transportation strike occurs and folks will not be able to obtain some of the 'foods' we are accustomed to eating, Comfrey may save lives. It should be grown in every garden. It is a must. **A wonderful drink can be made in a blender with fresh Comfrey, pineapple juice, and a little fresh peppermint. This is nourishing as well as healing.** *A person suffering from asthma, chronic or acute, can be relieved by this drink. During the winter,* **dried Comfrey may be made into a tea.***"*
-School of Natural Healing: Herbal Reference Guide by John R. Christopher. Colorado: Nutri Books Corp, **1996**.

*"**Comfrey is occasionally used as an ingredient of soups and salads. It is listed by the 'Council of Europe'* as natural source of food flavouring (category N4).** This category indicates that although Comfrey is permitted for use as a food flavouring, insufficient data are available to assess toxicity."*
-Herbal Medicines by Joanne Barnes, Linda A. Anderson and J. David Phillipson. London, England: Pharmaceutical Press, third edition **2007**, page 188.
(* -Natural Sources of Flavourings, Volume 2 {Report No. 2, 'Blue Book'} by Committee of Experts on Flavouring Substances. Strasbourg, France: Council of Europe Publishing, 2007.)

(In the 'Natural Sources of Flavourings' books, flavorings were classified into four categories: N1 to N4. Category N4 is 'Plants, animals and other organisms, and parts of these or products thereof, and preparations derived there from, not normally consumed as food items, herbs or spices in Europe, which contain defined active principles requiring limits on use levels. These source materials and preparations are not considered to constitute a risk to health in the quantities used provided that the limits set for the active principles are not exceeded.'
Symphytum officinale L.: CoE No. 441, pub 1. I could not find this Symphytum publication. If you have it, could you send it to me please.)

*"**Symphytum officinale L. (Boraginaceae), Comfrey.** Friuli Venezia-Giulia (region of Italy). Local name: Concuardie. Wild.* **The young Comfrey leaves are gathered in spring, boiled, and added to meat and old bread to make meatballs; or they are boiled and consumed fried with other greens.** *A similar use of these leaves was recorded earlier in the same area."*
-'The Importance of a Taste: A Comparative Study on Wild Food Plant Consumption in Twenty-One Local Communities in Italy' by M.P. Ghirardini, et al., (School of Life Sciences, University of Bradford, United Kingdom), (University of Gastronomic Sciences, Bra/Pollenzo, Italy), (Dept of Social Sciences, University of Wageningen, Netherlands); Journal of Ethnobiology and

***"Food: For vegetarians, numerous recipes are available for different food with Comfrey: salads, souffles, soups, bread, rolls and root beverages. The consumption of food with Comfrey is widespread** (*Council of Europe 2008)."*
-'Risk Profile Symphytum Officinale Extracts', CAS No. 84696-05-9, www.mattilsynet.no, Statens tilsyn for planter, fisk, dyr og naeringsmidler (State supervision of plants, fish, animals and food), Brumunddal, Norway, March 11 **2013**.
(* -Active Ingredients Used in Cosmetics: Safety Survey by Council of Europe, Committee of Experts on Cosmetic Products. Strasbourg, France: Council of Europe Publishing, 2008. Symphytum officinale extracts, Monograph No. 38, pages 369-374.)

*"**The use of wild greens is an important issue in gastronomic (food) ethnobotany** as in some parts of the world, wild greens have been widely used to supplement human nutrition.*
It is interesting that many toxic wild vegetables, such as buttercups Ranunculus spp. (species) and Comfrey Symphytum grandiflorum, are used and sold in the market of Kutaisi (Georgia, country in the Caucasus region of Eurasia).
> *Raw buttercups contain protoanemonin which is very pungent, and Symphytum species contain pyrrholizidine (PA) alkaloids (e.g., Rode 2002; Roitman 1981).*

Prolonged cooking probably removes most of these toxins, but there is a lack of studies focused specifically on the alimentary (food) use of Comfrey after longer cooking.*"*
-'Comfrey and Buttercup Eaters: Wild Vegetables of the Imereti Region in Western Georgia, Caucasus' by L. Luczaj, B. Tvalodze and D. Zalkaliani, (Department of Botany, Institute of Biotechnology, University of Rzeszow, Kolbuszowa, Poland) (Kutaisi Botanic Garden, Kutaisi, Georgia); Economic Botany, Volume 71, No. 2, pages 1-6, May **2017**.
(Ethnobotany is study of a region's plants and their practical uses through traditional knowledge of a local culture and people.)

<u>Reasonable Eating of Comfrey Leaves</u>

Use Comfrey in moderation. Consult your healthcare provider.

*"**You can go nutty about something like Comfrey or dandelions. But as a component in food, these things are good.***
Some people were urging on everybody to feed their children, chickens, horses and cows on Comfrey, until another gentleman said, 'Look, be careful!'.
> *Once a nut starts urging nutrition on someone, they are going to do it. They get their blenders down and start drinking green glue. It's stupid! **Of course it is possible, under certain conditions, to damage the liver. So there has been a note of caution sounded.***

Nobody has found that Comfrey will kill you; we are already certain it won't. Everybody I know eats Comfrey and a few borage leaves, and we put borage leaves in our drinks.
> ***The main thing is, don't go to your garden and eat Comfrey as your main food, like a lot of those people were doing.*** *It is not the complete food; nothing is. Everything you do like that is stupid.*

If you eat a hundred things, you are not very likely to die of it; and you will get everything you ever need. *The point is, in a varied diet you add a component where that component was short."*
-'Introduction to Permaculture: Permaculture Design Course Series, Pamphlets 1 to 14' by Bill Mollison at 'The Rural Education Center', Wilton, New Hamshire, **1981**; published by Yankee Permaculture, Sparr, Florida.

*"In conclusion, the research shows that although there are alkaloids of the liver damaging type in Comfrey, there are such small quantities in the leaves that **reasonable consumption of mature leaves, or of Comfrey tea, even over a number of years, is unlikely to cause any problems.** If people take Comfrey for its nutritional or medicinal benefits, they may continue to do so without any undue concern. Probably they should be more circumspect about continued assimilation of Comfrey root products, although the occasional topical application should cause no problems."*
-'The Safety of Comfrey: Report from the Henry Doubleday Research Association' by J.A. Pembery B.Sc., Special Advisor, Research Chemist, Aldgate Press: London, England, 20 pages, **1983**, page 14.

*"**The dried Comfrey leaves can be cooked in water and consumed as a cooked green vegetable.***
*Consequently, two plants in a city yard can supply a family with all of its **cooked green vegetable food** throughout the year. Other green leaf vegetables such as spinach, New Zealand spinach, Swiss chard, mustard, amaranth, turnip tops, and beet tops produce lower yields and for much shorter periods of time than Comfrey. Comfrey will grow in partially shaded areas that are typical of city yards. However, the hairs on the Comfrey leaves give it a different (furry) texture than that of other leaf vegetables.*
Longer cooking reduces the leaf furriness.*"*
-'Comfrey: A Controversial Crop' by Robert G. Robinson, University of Minnesota, Agricultural Experiment Station, Minnesota Report MR-191, Item No. AD-MR-2210, **1983**, pages 4, 5.

*"**An additional problem is that a segment of the population uses herbal preparations not occasionally or moderately, but in vast quantities each day.***
Numerous reports have appeared of people literally drinking liters (quarts) of herbal teas per day for prolonged periods; levels at which anything would represent a hazard. A related problem is the great variety of herbs that may be consumed.
> ***In most herb-using cultures, herbs are taken episodically (on and off) to treat illness or occasionally for diag-***

*nostic purposes or used in small amounts for gustatory (taste) pleasure. This includes the adding of spices to food, or **the occasional cup of Comfrey**, etc. The frequency and amount of usage maximize the chance of ill effects."*
-Toxicants of Plant Origin, Volume I: Alkaloids edited by Peter R. Cheeke. Boca Raton, Florida: CRC Press, **1989**. Chapter 3: 'Human Health Implications of Pyrrolizidine Alkaloids and Herbs Containing Them' by Ryan J. Huxtable, University of Arizona.

Eating Regularly:
*"This means that to get the benefits of Comfrey, it must be taken regularly and consistently. The effect of a drug may be immediate, but it is not long-lasting. **The effect of Comfrey is long-lasting because it is built into the very cells of the body.** It penetrates to every part of the body and brain, improving both the structure and function of each part."*
-Comfrey: Nature's Healing Herb and Health Food by Andrew Hughes. Japan: Sanyusha Publishing Co., Ltd, **1992**, page 46.

*"Both Comfrey roots and leaves are reported to be used for medicinal purposes. **Comfrey is occasionally used as an ingredient of soups and salads. It is listed by the 'Council of Europe' as natural source of food flavouring** (category N4). This category indicates that although Comfrey is permitted for use as food flavouring, insufficient data are available to assess toxicity*.*
The herb has long been used as a cooked green vegetable in early spring, and the fresh, young leaves have been added to salads. The widespread suffering caused by the Irish potato famine of the 1840s motivated Henry Doubleday, an Englishman, to fund research into Comfrey's potential as a nutritional food crop.
Farmers have valued Comfrey as a nutritious fodder for cattle**."
-'European Medicines Agency- Assessment Report on Symphytum Officinale L., Radix' by European Medicines Agency, Committee on Herbal Medicinal Products (HMPC), London, England, 27 pages, May 5 **2015**.
(* -'Herbal Medicines, 3rd Edition' by Barnes, Anderson and Phillipson, Pharmaceutical Press, London, pages 188-190, published 2007.)
(** -'The Gale Encyclopedia of Alternative Medicine, Volume 1' edited by J.L. Longe, Thomson Gale, Michigan: Gale Group, pages 526-527, 2005.)

Eat Small or Large Comfrey Leaves?

Eat Small Leaves Because Tender

*"**The Comfrey leaves, when young, form a good green vegetable,** and are not unfrequently eaten by country people where the plant abounds; **when fully grown they become coarse and unpleasant in taste.** They are sometimes used to flavour cakes and other culinary preparations."*
-'The Useful Plants of Great Britain: Part I, August' by John E. Sowerby & C. Pierpoint Johnson, London, England, **1862**, page 182.

*"In Britain and the United States, a great many gardeners **eat Comfrey as a spinach-like vegetable**, slightly bitter in flavor and **best picked small when the leaves are not more than six inches (15.2 cm) long**. This is not the way to maximum yield, but those who have half-a-dozen plants often cut one back hard for compost, or liquid manure, and pick this continually for a fortnight (2 weeks), then let it grow and change over to another one.*
Others let the Comfrey leaves grow larger, but the stems are too stringy for normal eating."
-Comfrey: Fodder, Food and Remedy by Lawrence D. Hills. New York: Rizzoli Universe Books: **1976**, page 172.

*"**It's important that one consider young early leaves, about four to five inches (10.1-12.7 cm) long, for purposes of food or herb tea.** There are a few cases, of course, where the larger leaves are used for internal purposes.*
There are several ways to let Comfrey leaves serve you as a worthwhile food source. The uncooked leaves may be included whole or sliced thin in a mixed vegetable salad."
-Comfrey: What You Need to Know by Ben Charles Harris. New Canaan, Connecticut: Keats Publishing, Inc., **1982**, page 63.

*"**I feel strongly that one is perfectly secure in eating the early three- to five-inch (7.6-12.7 cm) leaves but the later larger leaves should be used for external purposes."***
-Comfrey: What You Need to Know by Ben Charles Harris. New Canaan, Connecticut: Keats Publishing, Inc., **1982**, page 21.

*"**Local Food Uses for Symphytum caucasicum M. Bieb. Boraginaceae:** Armenia: **Young Comfrey leaves are used in salads and soups, adult leaves as spinach** (Grossheim 1952; Tsaturyan and Gevorgyan 2007)."*
-Ethnobotany of Caucasus by Ketevan Batsatsashvili, et. al. (Chapter: Symphytum caucasicum M. Bieb. Boraginaceae, pages 683-688). Cham, Switzerland: Springer, **2017**.

Eat Large Leaves Because Less Alkaloids

*"**We have measured pyrrolizidine alkaloids in leaves of the hybrid Russian Comfrey S. x uplandicum, variety Bocking No. 14,** grown at Bocking, Essex, England, and kindly supplied by Mr L.D. Hills of the Henry Doubleday Research Association. **Total alkaloid levels were 0.003% in the largest Comfrey leaves and 0.049% in the smallest: a 16-fold variation.***

Subsequently the total new leaf growth from pairs of the same plants were harvested and analysed.
Leaves picked after the first 2 weeks contained 0.115% of alkaloids (dry weight) while by mid-September the level had fallen to 0.019% and remained between 0.019% and 0.022% until mid-October (when growth ceased).
The total amount of alkaloids per leaf remains fairly constant as the leaf grows heavier, so the percentage of alkaloids falls. My measurement show that the highest akaloid levels are in small, young leaves, especially early in the season."
-'Toxic Pyrrolizidine Alkaloids in Comfrey' by A.R. Mattocks, The Lancet, Surrey, United Kingdom, Volume 316, No. 8204, 1136-1137, November 22 **1980**, a letter.

Ways to Eat Comfrey

These recipes are for entertainment purposes only.
Please consult your health care advisor about the safety of eating Comfrey.

"Our Canadian members of the 'Henry Doubleday Research Association' have winters with no green vegetables at all. **They now dig up young Comfrey before the frost, pot them up in tubs and keep them indoors.** They grow and force like chicory, to use Comfrey in place of lettuce, cabbage or spinach."
-Compost, Comfrey and Green-Manure: 'Henry Doubleday Research Association' First Gardeners Report by Lawrence D. Hills. Braintree, Essex, England: Henry Doubleday Research Association, **1959**.
> ('Forcing' roots means putting them in a somewhat warm, dark place in order to force them into rapid and early growth. The roots are covered completely except for the crown. The leaves that grow are pale. Or you can grow the Comfrey inside in pots in a sunny window.)

"**The cooking Comfrey is Russian Comfrey Bocking No. 4.**
Like cabbage, it doubles as poultry green food but is richer in protein, calcium and phosphorus, and **its season goes on until November, but begins in March and April when the first leaves are ready to pick.**"
-Comfrey Report Number Two: For Gardeners, Farmers and All Comfrey Growers in All Countries, with Analysis, Yields and Cultivation Methods, and the Results of Seven Years More Work Since Our First Report by Henry Doubleday Research Association, Braintree, Essex, England, 45 pages, March **1963**.

Raw Comfrey Leaves and Flowers in Salad

"One phase which appeals to the writer is **the use of Comfrey's flowers, which are edible.** *The blooms develop at the top of the stems with umbels of small bell-like flowers.*
These flowers contain a certain amount of honey, and if served on salads with parsley, on open-faced sandwiches or floating in soups, *Comfrey adds another facet to the enjoyment of nature's products supplied for our benefit.*"
-'Comfrey: The Cinderella of Plants' by Maria Wilkes, Herbarist: The Herb Society of America Journal, Volume 33, pages 47-50, **1967**. Also in 'Herbs for Use and for Delight: An Anthology from the Herbarist' by Daniel J. Foley, Herb Society of Ameria. Gloucester, Massachusetts: Peter Smith Publisher Inc., 1974.
(Umbel is a flower cluster where stalks grow from a common center and form a flat or curved surface. An example is parsley.)

"**Comfrey, Symphytum officinale,** *a healing herb, its very name implies a knitting together.* **Use its young leaves sparingly, raw in salads or cook them as spinach, cutting the leaves before the plant blooms.**"
-Joy of Cooking by Irma S. Rombauer and Marion Rombauer Becker. Indianapolis, Indiana: The Bobbs-Merrill Company, Inc., page 581, **1976**.

"**Symphytum officinale L. is a herb of the family Boraginaceae. It is cultivated for use in Japan as a green vegetable or tonic. The fresh leaves are used in salads, and their juice is used as a drink. Sliced roots are also eaten.**"
-'Edible Plants Containing Naturally Occurring Carcinogens in Japan' by Iwao Hirono, Department of Pathology, Fujita Health University School of Medicine, Toyoake, Japan; Japanese Journal of Cancer Research, Tokyo, Volume 84, No. 10, pages 997-1006, October **1993**.

"**Many herb lovers use Comfrey leaves as a vegetable, like spinach, in salads or cooked.** *As the leaves are rather rough and hairy, it is best to chop the leaves finely, when adding to a tossed salad or tucked in a sandwich.* **Some Comfrey connoisseurs (food experts) eat the leaves with lemon juice, because the lemon is said to release the Comfrey's calcium.**"
-How Can I Use Herbs in My Daily Life: Over 500 Herbs, Spices and Edible Plants: An Australian Practical Guide to Growing Culinary and Medicinal Herbs by Isabell Shipard. New York, New York: David Stewart Publisher / Simon and Schuster, Inc., **2003**.

"*Only a few countries were lucky enough, at the end of the nineteenth century, to have* **ethnographers who would write down local plant lore in great detail, providing Latin names. Jozef Ludovit Holuby (1836-1923), a Slovakian Renaissance-type scholar,** *was a prime example of such an ethnographer.*
Holuby lists a variety of children's snacks:
> *Sorrel (Rumex acetosa L.) leaves and stalks, Lamiaceae and* **Comfrey (Symphytum officinale L.) flower nectar,**

stalks of Tragopogon orientalis L., tubers of Lathyrus tuberosus L. and Chaerophyllum bulbosum L. and immature fruits of mallow Malva rotundifolia L., not counting numerous species of wild berries."
-Pioneers in European Ethnobiology edited by Ingvar Svanberg and Lukasz Luczaj. Uppsala, Sweden: Uppsala University, October **2014**. Chapter: 'A Pastor who Loved Blackberries, Jozef Ludovit Holuby, 1836-1923' by Rastislava Stolicna.
   (Ethnography is the description of the customs of peoples and cultures.)
   (Slovakia is a country in central Europe bordered by Poland to the north, Ukraine to the east, Hungary to the south, Austria to the west, and the Czech Republic to the northwest.)

Comfrey Cooked Alone

*"**When young, and before it blossoms, Comfrey is very delicious, and it looks so tempting, with its broccoli-like heads, that I gathered some and had them cooked as a dish of broccoli, and really found them very good.**
As they come in early in April, they may prove a valuable addition to our spring vegetables."*
-'Ensilage: A System for the Preservation in Pits of Forage Plants and Grasses, Independent of Weather: A Collection of Facts and Statistics on the Cheapest Mode of Providing Winter Food for Dairy Cattle, Sheep, Horses, Etc.', by Thomas Christy, F.L.S. (Fellow of Linneaen Society), London, England, **1883**. Caucasian Prickly Comfrey, pages 57-63.

*"**In picking the Comfrey leaves to use fresh,** cut them down to the base, collecting them from the second row up, which leaves the basal (bottom) leaves to absorb the soil splatter and the possibility of slugs or snails. **Wash the leaves, cut them in strips and place in an enamel or glass saucepan,** cover tightly and put over a low flame for a very few minutes. Do not boil. There should be enough moisture to let them cook enough without burning. However, it might be necessary to add a tablespoon of water before putting on the fire.
Keep any liquor (liquid) which may remain for gravies or soups. Season with herbal salt and any dressing preferred. **the flavor of Comfrey is a blend of endive and asparagus.**"*
-'Comfrey: The Cinderella of Plants' by Maria Wilkes, Herbarist: The Herb Society of America Journal, Volume 33, pages 47-50, **1967**. Also in 'Herbs for Use and for Delight: An Anthology from the Herbarist' by Daniel J. Foley, Herb Society of Ameria. Gloucester, Massachusetts: Peter Smith Publisher Inc., 1974.

*"**Comfrey Cabbage:** Gather a sufficient quantity of fresh Comfrey leaves, selecting only those that are young and tender. Wash and remove the stalks. Cut up the leaves, put a knob of butter or vegetable oil into a pan, and add the leaves slowly- there should be enough juice to cook them; this will gradually appear, and as it does the rest of the leaves may be added.
Add salt to taste whilst cooking, and for those persons who like aromatic caraway seeds, a teaspoonful, stirred into the cooking Comfrey cabbage makes the dish so much more enjoyable. **Cooking should only take ten minutes or less** if only a small amount is being prepared."*
-About Comfrey: The Forgotten Herb by G.J. Binding, M.B.E., F.R.H.S. Wellingborough, Northhamptonshire, England: Weatherby Woolnough, **1974**, page 59.
   (A knob of butter is a lump of butter. For sauteing add just enough butter to cover the bottom of the pan. It is about 12 to 25 grams, or one to two tablespoons of butter.)

*"**Fresh Comfrey as a green vegetable:** Pick the leaves, wash them, shake off surplus water, and put straight into a dry enamel saucepan where the wet on the leaves and **their juice is usually enough to cook them tender in ten minutes**, with the lid on over a low gas or electric hot-plate setting. If there seems too little water, add a tablespoonful or so, but **boiling- as with cabbage- wastes protein and destroys the flavour,** which is a blend of endive and asparagus."*
-Comfrey: Fodder, Food and Remedy by Lawrence D. Hills. New York: Rizzoli Universe Books: **1976**, page 172.

> *"**Comfrey cooked this way is a spinach-like vegetable,** but it can make vegetarian dishes or a change for those on 'milk-fish-rice' diets for gastric troubles."*
> -Down to Earth Fruit and Vegetable Growing by Lawrence D. Hills. London, England: Faber & Faber, **1960**, page 54.

*"**Comfrey leaves may be steamed (1 or 2 mintues) alone or with other vegetables in a double boiler in as little water as possible.** Heat only long enough to soften the tissue- the less heating time the better. **The leaves should definitely not be boiled, or cooked too long.** Boiling is the inevitable destruction and loss of vitamins and minerals. Be sure to use the remaining liquid as a drink between meals, a soup or stew ingredient, a liquid in bread-making or reheating other vegetables."*
-Comfrey: What You Need to Know by Ben Charles Harris. New Canaan, Connecticut: Keats Publishing, Inc., **1982**, page 64.

*"Young Comfrey shoots were often covered with straw and blanched (made pale) so that they could be cooked much like cardoons. **The young Comfrey leaves were also eaten as cooked greens,** but when the plant becomes too large, the leaves toughen and become prickly.
Indeed, Comfrey was once a common kitchen garden herb, and it is evident from Sauer's herbal that it was already well established in colonial America by the 1760s."*
-Sauer's Herbal Cures: America's First Book of Botanic Healing, 1762-1778 by William Woys Weaver. New York: Routledge, **2001**, pages 114-115.
   (Cardoon is a tall thistle-like southern European plant related to the globe artichoke.)

Comfrey Soup and Stew

"Soy Soup: Melt 3 tablespoons butter in the top of a double boiler; **add 1/2 cup (118 ml) soy flour,** blend thoroughly, and cook a few minutes. Slowly stir in 2/3 cup (157 ml) hot water, and simmer 15 minutes till thick, stirring to prevent sticking to pan. Add 2 cups (473 ml) stock, place pan over boiling water, and bring to the boil. Add a pinch of curry powder, 1/2 teaspoon brown sugar, seasoning, and **some chopped parsley, chives, Comfrey or spinach leaves.** Finally add 2 teaspoons soy sauce."
-The Tradition of Australian Cooking by Anne Gollan. Canberra, Australia: Australian National University Press, page 184, **1965**.

"There are many ways of cooking Comfrey as a green vegetable (all of which require very little fuel), for fast cooking is essential. The amino acid analysis incidentally solves a mystery many vegetarians have noticed when making **Comfrey soup. This can form a jelly as though it were beef stock, explained by the presence of lysine, the amino acid in gelatin."**
-Comfrey Report: The Story of the World's Fastest Protein Builder and Herbal Healer, Conservation Gardening and Farming Series: Series C by Lawrence D. Hills. England: Henry Doubleday Research Association, **1975**, page 38.

"Fresh Comfrey and Nettle Soup:
About 12 leaves Comfrey, large handful Stinging Nettle tips, 1 small onion sliced, 1 medium potato peeled and sliced, 7 fluid ounces (207 ml) water, 1 egg yolk, 1 teaspoonful Marmite (brown food paste), salt and pepper to taste, 1/2 pint (1 cup = 236 ml) creamy milk, whipped cream to garnish. Wash Comfrey and Stinging Nettle tips, and **remove middle rib from Comfrey leaves** and stalky bits from Nettles. Put into a saucepan the Comfrey, Stinging Nettles, onion, potato, a little salt and the water. Cover with lid, simmer gently till everything is quite soft. Put through electric blender or fine sieve, add Marmite. Beat egg yolk and milk together and add to puree."
-Comfrey: Fodder, Food and Remedy by Lawrence D. Hills. New York: Rizzoli Universe Books: **1976**, page 173.

"Over the years many Comfrey devotees have incorporated the finely ground leaves- their choice is the small-sized ones- in soups, stews and casseroles, omelettes, hamburger and even sandwiches.
> *When taken with soups or stews, the leaves, now cut into small pieces or diced and slivered, should be stirred into the cooking food long enough to be softened adequately- and just before the soup or stew is served."*

-Comfrey: What You Need to Know by Ben Charles Harris. New Canaan, Connecticut: Keats Publishing, Inc., **1982**, page 65.

"Comfrey soup is easily made:
Try blending or chopping finely 3 Comfrey leaves with half a cup (118 ml) of eggplant (aubergine), 2 or 3 carrots or tomatoes, or any other greens, plus a little garlic, some herbs and 1 pint (2 cups = 473 ml) of water. Then simmer for 10 minutes or so. **Alternatively, blend about 6 Comfrey leaves** with 1 tomato, a quarter of a cucumber, some celery, lettuce, avocado, garlic, fresh herbs and ground roasted sesame.seeds. Chill to serve as a consomme or heat throug for a soup."
-Self Reliance: A Recipe for the New Millennium by John Yeoman. East Meon, England: Permanent Publications, **2003**.
> (A consomme is a clear soup made from richly flavored stock.)

"Comfrey will also bind together croquettes and patties, in the absence of egg- a valuable economy tip. It will also thicken sauces and fortify stews."
-Self Reliance: A Recipe for the New Millennium by John Yeoman. East Meon, England: Permanent Publications, **2003**.
> (A croquette is a small roll of chopped vegetables, meat, or fish, fried in breadcrumbs.)

"Recipes in a Middle English (1150 to 1470 AD) cookbook called the 'Liber cure Cocorum' (Book of Cookery) contains a recipe describing how to get back at a cook. In medieval cookbooks recipes for punishing the cook by spoiling or manipulating his food are quite rare. But this is to be expected, given that it was neither in the interest of cooks to give readers any ideas for pranks, nor in the interest of upper-class households to admit to a diners' revolt.
It is in books on magic and, yes, books on warfare, that we find such recipes listed, and the picture they paint of a cook under attack is not a pretty one:
> He was faced with chickens, pieces of meat, peas or beans made to jump out of the pot with the help of such unsavory additives as quicksilver (mercury), vitriol (sulfate), and saltpeter, or **with the pieces of meat in his pot sticking together in one big lump because somebody had poured in Comfrey powder."**

-Food in Medieval Times by Melitta Weiss Adamson. Westport, Connecticut: Greenwood Press, **2004**.
> ("The rootes of Comfrey stamped are so glutenatiue, that it will sodder or glew together meate that is chopt in peeces seething in a pot, and make it in one lumpe." -'The Herball, or, Generall Historie of Plantes Gathered by John Gerarde of London, Master in Chirurgerie' by Gerard, 1730 pages, published in London, England in 1633, pages 805-807.)
> (Mercury is very toxic. Never put it in food. Do not touch it.)
> (Saltpeter or niter is a mineral form of potassium nitrate. Do not put it in food. Do not put sulfate in food.)

Baked Comfrey

*"Comfrey au Gratin is made by putting a layer of cooked rice in the bottom of a Pyrex® dish, then a layer of cooked

Comfrey leaf with some grated cheese and a dab of margarine (or butter), then more rice and another thin Comfrey layer and finish with rice, cheese and margarine (or butter), leaving half an inch (1.2 cm) below the rim to allow a filling of milk. Bake in a fairly hot oven for half an hour and serve."
-Compost, Comfrey and Green-Manure: 'Henry Doubleday Research Association' First Gardeners Report by Lawrence D. Hills. Braintree, Essex, England: Henry Doubleday Research Association, **1959**.

"**A beta-carotene rich biscuit (betaCRB) was made using Comfrey** and a relatively low beta-carotene biscuit (betaCLB) without Comfrey was also made and their beta-carotene content determined. **The Comfrey rich biscuit had a significantly higher beta-carotene content (1060 microgram/100 grams) than the Comfrey free biscuit, which had (170 microgram/100 grams).** The biscuits were baked at 'Mealz Jamaica Restaurant' in Nairobi, Kenya under high standards of hygiene. The main equipments used were **a blender to macerate (cut and soften) the fresh Comfrey leaves,** a weighing scale, a dough mixer, baking trays, biscuit rings to standardize the size of the biscuits, oven and cooling racks for the biscuit preparation.
Biscuit Recipe:
Sugar 75 grams (2.6 ounce), Shortening 75 grams, Wheat flour 100 grams (3.5 ounce), Baking powder 1/8 teaspoon, Water 50 ml (1.69 fluid ounce), **Comfrey leaf 100 grams**.
The mixture was kneaded to a soft consistency. The biscuits were baked for 60 minutes at 150 degrees Centigrade (302 F)."
-'Assessment of the Potential of Comfrey (Symphytum Peregrinum) as a Source of Vitamin A for Malnourished Children (8-16 Years): A Case Study of Kirigiti Girls Approved School in Kiambu, Kenya' by Wambui Gatigwa, Thesis for Master of Science in Applied Human Nutrition, University of Nairobi, **2002**.

"**Comfrey cakes** are a tasty way to use up leftovers. Simply **steam or simmer the washed leaves** for 5 minutes or so until soft, then chop and saute them briskly with diced onions. Then mix with an equal amount of mashed cooked pulses (beans, lentils, peas) or grains or rice or potato or even porridge. Add some bean sprouts if available, and herbs, and a dash of soy or Worcester sauce. **Form into cakes and bake or grill** for about 15 minutes till brown both sides Serve with a tomato sauce, made by mixing tomato puree with cream or yoghurt, plus seasoning and a dash of Worcester sauce."
-Self Reliance: A Recipe for the New Millennium by John Yeoman. East Meon, England: Permanent Publications, **2003**.

"**Stuffed Comfrey leaves also offer endless variations.** Parboil for a minute or so, until they're flexible, the **largest leaves you can find.** Then make a stuffing from 8 ounce (227 grams) of cooked pulses (beans, lentils, peas), 1 chopped onion, 1 tablespoon of tomato puree, fresh chopped herbs, 1 tablespoon of chopped nuts, some raisins or chopped dried fruit, plus 2 tablespoons uncooked rice or buckwheat. Stuff the leaves and fold into packets fastened with a toothpick, leaving room for the grain to swell. Place upright in a casserole, cover with stock and **bake in a medium oven** for 45 minutes. Use the drained cook water as soup."
-Self Reliance: A Recipe for the New Millennium by John Yeoman. East Meon, England: Permanent Publications, **2003**.
(Parboil is partially cooking by boiling. A medium oven is 180 C = 356 F.)

"**Try this farmhouse casserole recipe, adapted for Comfrey: Lay chopped Comfrey leaves in a dish,** cover with a layer of grated cheese, then breadcrumbs, then repeat till the dish is loosely full. Finish with breadcrumbs mixed with grated cheese. Fill with stock to the top layer **Bake 40 minutes in a medium oven till golden on top.**"
-Self Reliance: A Recipe for the New Millennium by John Yeoman. East Meon, England: Permanent Publications, **2003**.

Comfrey Bread, Pasta and Flour

"**Comfrey root, cleaned, broken, dried, and ground in a coffee or corn mill, is a very mucaliginous and nutricious flour,** resembling slippery-elm, oatmeal, arrow-root and sago (from tropical palm trees).
It has much the appearance of good, light-colored rye meal, and is peculiarly easy of digestion.
    A lady who could keep no other food on her stomach, was sustained three months on pudding of Comfrey flour, made by pounding the dried roots in a mortar. Another, whose voice failed by disease, had it soon restored by the green Comfrey root cut and simmered with molasses.
I had 4 pounds (1.8 kg) of Comfrey root ground at a corn mill, to use with other flour or meal in gruel for family colds, coughs, and bowel complaints. **I would recommend to use about one-fourth part of Comfrey meal with three-fourths of wheat, corn meal, barley or buckwheat for bread, pastry or dough-nuts; but not with rye.**
    With the other ingredients, you may use it for custards instead of eggs, probably to great advantage. I would, for mixing with my own food, gladly exchange two pounds (0.90 kg) of good wheat flour for one pound (0.45 kg) of Comfrey meal."
-'The New England Farmer, and Horticultural Register' by Joseph Breck, Volume 23, New Series Volume 13, Boston, Massachusetts, **1845**. ('Symphytum or Comfrey, as Food for Men and Cattle' by Ezekiel Rich, Troy, New Hampshire, page 10, July 10 1844.)
    (Gruel is made from a cereal grain such as oats, wheat or rye flour that is boiled in water or milk. It is a thinner than porridge and may be more often drunk than eaten.)

"**In Japan the following receipe for Comfrey bread is becoming popular,** modified from 'Doris Grant Loaf' by Andrew Hughes: 10 1/2 oz (310 ml) whole wheat flour, 3 1/2 ounce (103 ml) whole corn flour, **3/4 ounce (22 ml) Comfrey flour,** 1/2 ounce (14 ml) skim milk powder, honey or molasses, salt and yeast. Mix the Comfrey flour and milk powder thoroughly through the wheat and corn flour. **The Comfrey gives the mixture an excellent 'break', and it rises quickly and well. One must be careful not to**

fill the bread tin high because it rises more with Comfrey. It makes a dark loaf that keeps moist and is very tasty."
-Comfrey Report: The Story of the World's Fastest Protein Builder and Herbal Healer, Conservation Gardening and Farming Series: Series C by Lawrence D. Hills. England: Henry Doubleday Research Association, **1975**, page 38.

*"**Mix Comfrey flour, soya flour and wholemeal flour in equal quantities** with sunflower seed oil, stirring for about ten minutes in moderate heat. This can be kept in a jar for a week as a nourishing and first-class protein **as a base for gravies, soups and stews.** It can be added to the various nut savory mixtures available at the health stores to make rissoles and to scramble eggs. **Tomato is the best contrasting flavor to off-set the Comfrey."***
-Comfrey Report: The Story of the World's Fastest Protein Builder and Herbal Healer, Conservation Gardening and Farming Series: Series C by Lawrence D. Hills. England: Henry Doubleday Research Association, **1975**, page 78.

*"**Comfrey has more mucilage than even marshmallow, so its roots and leaves are excellent, dried and stored, as a flour** which can be mixed with wheat flour (or equal parts of soy and wheat flour) plus water to make a pastry, intriguingly green but nutritious."*
-Self Reliance: A Recipe for the New Millennium by John Yeoman. East Meon, England: Permanent Publications, **2003**.

"Comfrey Pasta:
Ingredients: *One onion, garlic to taste, 5 mushrooms (button or freshly picked ceps), **20 or so leaves of Comfrey,** pinch of nutmeg, some oil for frying (prefer rapeseed oil), olive oil or butter to flavour, black pepper, basil, pasta for two.*
Method: *Fry up the onions until soft then add garlic. Add the mushrooms, and once softened **add the Comfrey and cook until wilted.** Chuck (put) in the remaining ingredients. Serve with pasta and garlic bread and a smile."*
-'Comfrey' by Dave Hamilton, The Urban Guide to Becoming Self-Sufficientish: Urban Homesteading on a Budget, www.selfsufficientish.com, October 21 **2016**.
    ('Cep' mushroom is Boletus edulis, Boletus pinophilus, Boletus reticulatus also known as Porcini in Italy and United
    Kingdom, Penny bun in United Kingdom, Cepe in France, Steinpiltz in Germany, and King bolete in the United States.)

Vegetables/Fruit and Comfrey

"Confetti Breadfruit (Hawaiian American):
    3 1/2 pounds (1.5 kg) (or 7 cups = 1.6 liters) breadfruit, cooked and mashed
    4 tablespoons butter or margarine, 1 1/2 to 2 teaspoons salt, 3/4 to 1 teaspoon pepper
    2 cups (0.47 liters) fresh Comfrey, chopped and steamed
    1 cup (0.23 liters) diced sweet red pepper, 1 cup heated milk
Wash green, unripe breadfruit. Do cutting under running water to prevent hands and knife from getting very sticky from sap. Cut off stem and cut breadfruit into quarters, cutting out the core as you would an apple. Cut quarters in half. Cover with lightly salted water and boil one hour until soft. Drain and mash, and add remaining ingredients. Mix well and serve. Serves 8-10."
-'Use of Tropical Vegetables to Improve Diets in the Pacific Region' by Stacy K. Evensen and Bluebell R. Standal, Research Series 028, HITAHR: College of Tropical Agriculture and Human Resources, University of Hawaii, 35 pages, June **1984**.
    (Breadfruit or Artocarpus altilis is a flowering tree in the mulberry and jackfruit family {Moraceae}. It is grown in 90
    countries throughout south and southeast Asia, the Pacific Ocean, Caribbean, Central America and Africa. Its name
    comes from the texture of the cooked ripe fruit that is similar to baked bread.)

*"**The best way to deal with coarse leaves, such as those of stinging nettles and Comfrey,** I have found, is to boil up a saucepanful of potatoes, **adding the leaves when the potatoes are just cooked.**
Cutting the Comfrey leaves up with the potatoes is sufficient to remove the roughness and sting, and the result is a surprisingly tasty variation on the traditional British dish known as 'bubble-and-squeak'."*
-Forest Gardening: Cultivating an Edible Landscape by Robert Hart. White River Junction, Vermont: Chelsea Green Publishing Company, **1991**, page 29.
    ('Bubble and squeak' is a traditional British breakfast made from boiled potatoes and cabbage. In modern times it is
    made with leftover vegetables from a roast dinner.)

Frying and Dry-Roasting Comfrey

*"To Make Clary-Sage or Comfrey Leaf Fritters. Take your clary-leaves, cut off the stalks, dip them one by one in a batter made with milk and flour, your butter being hot, fry them quick. **This is a pretty heartening dish for a sick or weak person; and Comfrey-leaves do the same way."***
-'The Art of Cookery, Made Plain and Easy' by Mrs. Hannah Glasse, London, England, page 217, **1796**.
(Clary sage or Salvia sclarea is a biennial or short-lived perennial. Used as a medicinal herb and is grown to make essential oil.)

*"Cooked Bean Sprouts: Fry a little onion and garlic in butter or oil. **Add bean sprouts,** and any additions that are to hand such as cut up zucchini, celery, capsicum (pepper), mushrooms, a few chopped lettuce, silver beet (chard), Chinese cabbage, or **Comfrey leaves,** herbs, etc., then moisten with chicken stock or soy sauce, add seasonings, and serve."*

-The Tradition of Australian Cooking by Anne Gollan. Canberra, Australia: Australian National University Press, page 183, **1965**.

"In Germany, Comfrey leaves are dipped in batter and fried before eaten."*
-Toxicants of Plant Origin, Volume I: Alkaloids edited by Peter R. Cheeke. Boca Raton, Florida: CRC Press, **1989**. Chapter 3: 'Human Health Implications of Pyrrolizidine Alkaloids and Herbs Containing Them' by Ryan J. Huxtable, University of Arizona. (* -Wild Flowers of the United States, Volume 2, Part 2 by H.W. Rickett. New York: McGraw-Hill, 1975.)

"The very young mucilaginous Comfrey leaves deep fry, as do wild mallow leaves, particularly well in tempura."
-Self Reliance: A Recipe for the New Millennium by John Yeoman. East Meon, England: Permanent Publications, **2003**.
    (Tempura is a Japanese dish of seafood or vegetables dipped in batter and deep-fried.)

*"**Stuffed Comfrey Leaves:**
Batter: 1 small egg, 50 ml (1.6 ounces) beer, 25 grams (0.88 ounces) buckwheat flour, 10 grams (0.35 ounces) melted butter, 1/4 teaspoon Herbamare® Original (organic herb seasoning salt).
Filling: 5-6 sprigs of marjoram, 200 grams (7.0 ounces) fresh cheese, 1/4 teaspoon Herbamare® Original, freshly ground pepper.
Other Ingredients: 40 Comfrey leaves, approximately 10 cm (3.9 inches) long, from the garden.* Olive oil for frying.
*1. Mix all the ingredients for the batter together and leave to rest for 30 minutes.
2. For the filling, chop the marjoram and mix with the remaining ingredients.
3. Shake the Comfrey leaves, do not wash if possible or if necessary clean with a soft brush. **Spread 1 teaspoon of the filling on the hairy underside of half of the Comfrey leaves. Cover with the undersides of the remaining leaves.**
4. Preheat a serving plate in the oven at 80 C (176 F). **Dip your pre-prepared Comfrey leaves in the batter and shallow fry in very hot oil** on both sides for 3 to 4 minutes. Lay on kitchen paper and keep warm on the plate in the oven. Season with pepper and serve."*
-'Naturally Fresh and Delicious: Healthy Cooking with A. Vogel' translated by Irene E. Robbie. Switzerland: A. Vogel Publishing House, **2009**.   (I do not have any connection with A. Vogel or their products.)
    (A sprig is a small stem with leaves or flowers, taken from a plant.)

*"**Comfrey Root Deep-Fried in Beer Batter:** For 4 persons. Preparation time is 20 minutes. Cooking time is 30 minutes.
Ingredients: 2 pounds (0.9 kg) Comfrey root, 4 ounces (113 gram) flour, 3/4 cups (177 ml) beer, 1 pinch salt, egg, juice of lemon, salt and pepper from the mill, nutmeg, peanut oil for deep frying.
Preparation: Bring a large pan of water to the boil, as Comfrey must be completely covered while cooking, add the lemon juice and salt lightly. Wash and peel Comfrey, cut in about 2 inch (5 cm) long pieces and place immediately into the prepared cooking water, so they do not discolour and get dark.
    Cook Comfrey for about 10-15 minutes. Remove from water and dry with kitchen paper, set aside. Sprinkle with pepper from the mill and the nutmeg. Beat egg and add the flour and the beer, stirring well. Season with a little salt.
    Turn Comfrey in the beer batter and deep fry until golden brown."*
-Schafer-Recipes: 333 Cooking Recipes from All Over the World by Hans J. Schafer. Hamburg, Germany: Tredition, **2010**.

*"**Dry-roasting certain herbs will enhance their fire or yang characteristics.** Herbs like bupleurum are prepared in this way. **Western herbs that are dry-roasted for added flavor include dandelion, chicory, and Comfrey roots.**
In the raw state these herbs have a cooling energy, but when roasted they develop a slightly warmer energy.
When a little honey is added to the herb as it is stirred and dry-roasted in a wok, it adds a tonic effect through its sweetness."*
-The Modern Herbal Dispensatory: A Medicine-Making Guide by Thomas Easley and Steven Horne. Berkeley, California: North Atlantic Books, **2016**, page 120.
(In dry roasting heat is applied to dry food without adding oil or water. It is used with nuts and seeds. Food is stirred as it roasts so it evenly cooks. Dry roasting can be done in a frying pan or wok. A wok is a bowl-shaped frying pan used in Chinese cooking.)

*"**Comfrey Pancakes:** Make batter: Sift 200 grams (7 ounces) of plain flour. Add a pinch of salt and pepper. Add a knob of butter and half a pint (1 cup = 236 ml) of milk, stir and add egg. You can leave the mixture to stand but this is not always necessary.
Chop two Comfrey leaves finely and add to a pan of hot oil. Pour in the mixture so that it just covers the bottom of the pan. Fry each side until golden."*
-'Comfrey' by Dave Hamilton, The Urban Guide to Becoming Self-Sufficientish: Urban Homesteading on a Budget, www.selfsufficientish.com, October 21 **2016**.

Comfrey Green Drinks

*"**Formula for Dr. Kirschner's Therapeutic 'Green Drink':** 15 almonds, 4 pitted dates, 5 teaspoonfuls sunflower seeds. Soak nuts/seeds overnight in water. Fill the liquefier (blender) above the blades with unsweetened pineapple juice (approximately 8 ounces = 236 ml). Place softened nuts, seeds and dates in the pineapple juice and liquify. Pour this mixture into a pitcher.
Next, take four large handfuls of green leaves- such as alfalfa, parsley, Comfrey, mint, spinach, beet greens, watercress, kale, chard, and if obtainable, such nutritious wayside weeds as filaree, malva, lamb's quarter and dandelion. Do not use the stems. Liquify the 'greens' in 8 ounces of unsweetened pineapple juice. Then put the two mixtures together in the liquifier and

triturate (crush, grind) for a few moments. Place in referigeration."
-Nature's Healing Grasses by H.E. Kirschner, MD. California: H.C. White Publications, page 114. First printing **1960**, 16th print 1980.

"These were invented by the late Dr. H.R. Kirschner in the United States as a way of taking allantoin through the summer. **Take about four large Comfrey leaves, plus a teacupful of cold water and put them through the electric liquidizer (blender).** Then pour through a sieve into a glass, and the result is a bright green fluid that can be drunk easily.
Bocking No. 14 is too bitter to make an attractive drink, which may be because of its higher potash. Bocking No. 4 or a Bocking Mixture is far pleasanter liquidized. As the protein breaks down fast, liquidized Comfrey is always made fresh."
-Comfrey: Fodder, Food and Remedy by Lawrence D. Hills. New York: Rizzoli Universe Books, **1976**, page 174.
(I think the writer may have meant Dr. H.E. Kirschner, MD, who wrote 'Live Food Juices: For Vim, Vigor, Vitality' in 1972.)

"**Putting four young small Comfrey leaves in a blender with orange, pineapple or tomato juice** makes a palatable (agreeable) green drink. This may be strained, but I drink the entire glassful. **Since reducing my coffee consumption, an ample pinch of peppermint or orange mint with a teaspoon of dried Comfrey makes a healthful substitute.** Add to a cup (236 ml) of boiling water and steep 5 minutes.
The English add a teaspoon of dried Comfrey per cup to their regular tea, using milk, sugar or honey, if desired.
The Comfrey roots may roasted with dandelion roots for coffee."
-The Herbalist: A Publication of The Herb Society of America, No. 44. Boston, Massachusetts: **1978**. Article: 'Quaker or Russian Comfrey' by Rosella F. Mathieu.

"**Nutritive:** As Comfrey contains a large percentage of protein, it may be regarded as a nutritive food.
When fresh Comfrey leaves are liquidized (put in a blender) in a base of some fruit juice, such as unsweetened pineapple juice, the result is a nutritious drink that has been know to relieve many ailments. In some instances, other herbs have been added to the drink, such as alfalfa, parsley, dandelion, mint, kale or other healthy food."
-Modern Encyclopedia of Herbs by Joseph M Kadans, N.D., Ph.D. New York, New York: Simon & Schuster, Inc., **1993**, page 102.

"**Blended Green Drink:** Combine the following ingredients in a blender with 2 cups (473 ml) of filtered or spring water. Blend at medium or high speed until thoroughly liquefied, then strain and serve. Several handfuls of fresh young wheat or barley shoots, or 1 teaspoon of barley grass or algae powder, 1/4 cup (59 ml) of alfalfa sprouts, 1/2 cup (118 ml) of spinach, chard, kale, beet greens, or a mixture of them all. Suggested additions: **2 leaves of fresh Comfrey (remove the center rib),** 4-6 pieces of fresh rosemary, 1-2 stems of fresh basil, several stems of fresh mint, a few sprigs of fresh cilantro."
-Viral Immunity: A 10-Step Plan to Enhance Your Immunity Against Viral Disease Using Natural Medicines by J.E. Williams, O.M.D. (Doctor of Oriental Medicine), **2002**.

Comfrey Root with Sweeteners

"**Comfrey Root Sweets: 1 tablespoonful grated raw Comfrey root,** 1 tablespoonful desiccated (dry) coconut, 1 teaspoonful cocoa, 1 tablespoonful honey. Mix the four ingredients together until gluey. Allow to stand for a while, and take pieces about the size of peanuts and roll them in the coconut."
-About Comfrey: The Forgotten Herb by G.J. Binding, M.B.E., F.R.H.S. Wellingborough, Northhamptonshire, England: Weatherby Woolnough, **1974**, page 62.

"**Candied Comfrey Root: Wash Comfrey roots and boil them slowly till tender,** drain off but retain the water. Soak roots for one hour in syrup made of 1 pound (0.45 kg) of cane sugar to 1 1/2 pints (= 3 cups = 709 ml) of the water the roots were boiled in, and add a little lemon juice. Then boil over a low heat till the syrup candies. Drain off the syrup, and dry the roots in a warm place."
-About Comfrey: The Forgotten Herb by G.J. Binding, M.B.E., F.R.H.S. Wellingborough, Northhamptonshire, England: Weatherby Woolnough, **1974**, page 62.

"**Comfrey Root Marmalade: Grate equal quantities of washed raw Comfrey root** and lemon or orange peel and leave them in a covered dish with an equal quantity of sugar for a few days Then cook with grated apple and sugar to a standard marmalade recipe."
-Comfrey Report: The Story of the World's Fastest Protein Builder and Herbal Healer, Conservation Gardening and Farming Series: Series C by Lawrence D. Hills. England: Henry Doubleday Research Association, **1975**, page 78.
(Marmalade is a preserve made from citrus fruit, especially bitter oranges, prepared like jam.)

Comfrey Root Wine

For entertainment purposes only. If done incorrectly, it could make you sick.

"**To Boil Comfrey-Roots: Directions for the Sick: Take a pound (0.45 kg) of Comfrey-roots, scrape them clean, cut them into little pieces,** and put them into three pints (6 cups = 1.4 liter) of water, let them boil till there is about a pint (2 cups = 0.47

liter), then strain it. When it is cold put it into a sauce-pan; if there is any settling at the bottom, throw it away; mix it with sugar to your palate, **add half a pint (1 cup = 0.23 liter) of mountain wine** and the juice of a lemon, let it boil.
Then pour it into a clean earthen pot, and set it by for use. Some boil it with milk, and it is very good where it will agree, and is **reckoned a very great strengthener."**
-'The Art of Cookery, Made Plain and Easy' by Mrs. Hannah Glasse, London, England, page 270, **1796**.

"To Make Plague-Water:
Roots: Of each of these a pound (0.45 kg). Angelica, dragon, maywort, mint, rue, carduus (plumeless thistle), origany (marjoram), winter-savory, broad thyme, rosemary, pimpernell, sage, fumitory, coltsfoot, scabeous, borage, saxifraga, betony, liverwort, germander, gentian root, dock root, butterbur root, piony root.
Flowers: Wormwood, succory, hysop, agrimony, fennel, cowslips, poppies, plaintain, setsoyl, vocvain, maidenhair, motherwort, cowage, golden-rod, gromwell, dill.
Leaves, Seeds and Berries: A good handful of each of these things. Hart's tongue, horehound, fennel, melilot, St. John's wort, **Comfrey,** feverfew, red rose leaves, wood-sorrel, pellitory of the wall, heart's safe, centaury, sea-drink, bay berries, juniper berries. One ounce (28 grams) of nutmegs, one ounce of cloves, and half an ounce (14 grams) of mace.
Pick the herbs and flowers, and shred them a little. Cut the roots, bruise the berries, and pound the spices fine; take a peck of green walnuts and chop them small; **mix all these together, and lay them to steep in sack lees, or any white wine lees, if not in good spirits (20% or more alcohol); but wine lees are best.**
Let them lie a week, or better; be sure to stir them once a day with a stick, and keep them close covered, then still (distill) them in an alembic (alchemical still) with a slow fire, and take care your still does not burn.
The first, second, and third running is good, and some of the fourth. Let them stand till cold, and then put them together."
-'The Art of Cookery, Made Plain and Easy' by Mrs. Hannah Glasse, London, England, page 378-379, **1796**.
('Plague Water' is the name for a variety of medicinal waters meant to fight the plague. Most recipes had 20 or more species of leaves and roots. They were steeped in white wine or brandy, and then distilled.)
(The Black Death or Great Plague was one of the most devastating pandemic diseases with 75 to 200 million people dying in Eurasia. The peak in Europe was from 1347 to 1351.)
(The Great Plague of London, England, from 1665 to 1666, killed 20% of London's population.)
(Plague and other outbreaks of disease through a large population occur throughout all of human history. This recipe is for entertainment and historical purposes only.)
(A peck is 8 quarts = 2 gallons, or 1/4 bushel, or 554.8 cubic inches or 9.09 liters.)
('Lees' are deposits of yeast and other particles that precipitate to bottom of a tank of wine after fermentation and aging.)

"Comfrey, Symphytum officinale Properties and Uses:
The root of Comfrey contains a large amount of mucilage, which makes the dried article dense and horny. **When treated with water, or wine, it yields a moderate tonic power** with this mucilage; and makes a mild remedy for recent and old coughs, sub-acute dysentery and diarrhea, simple forms of leucorrhea, spitting of blood, and other pulmonary affections.
Comfrey root is rarely used alone, but makes a good soothing addition to more tonic agents; and has much merit, when used in the fresh state, as an application to bruises, and irritable ulcers.
> **Two ounces (59 ml) boiled in a pint (2 cups = 473 ml) of water, and then a gill of wine added, may be taken in doses of a fluid ounce (29 ml) or more three times a day."**
-'The Physio-Medical Dispensatory: A Treatise on Therapeutics, Materia Medica and Pharmacy in Accordance with the Principles of Physiological Medication' by Wm. H Cook, M.D., Physio-Medical College/Institute, Cincinnati, Ohio, **1869**.
(A gill or teacup is a unit of measurement for volume equal to 1/4 pint. One United States gill = 4 fluid ounces = 1/2 cup =118 ml.)

"**Pharmaceutical Preparations for Comfrey, Symphytum officinale: Compound Wine:**
Boots of Comfrey, convallaria, and aralia racemosa, each, one ounce (28 gram); cocculus palmatus (not frasera, as is generally used), camomile, and gentiana ochroleuca, each, half an ounce (14 gram). **Crush, and macerate (soften) for twenty-four hours with sherry wine;** transfer to a percolator, and add wine till two quarts (1.8 liters) in all have been used; then add water till two quarts of tincture have been obtained.
> This is a mild and valuable tonic for female difficulties, loss of appetite, nervousness, and insufficient menstruation.
> **Dose, half a fluid ounce (14.7 ml) or more three times a day."**
-'The Physio-Medical Dispensatory: A Treatise on Therapeutics, Materia Medica and Pharmacy in Accordance with the Principles of Physiological Medication' by Wm. H Cook, M.D., Physio-Medical College/Institute, Cincinnati, Ohio, **1869**.
(I'm not sure what 'boots' means above, but 'boot stage' is when plant growth begins to concentrate on seed head development rather than leaf tissue.)

"**But few medicinal wines, except those already noted, are now used in medicine.
Wine was once a favorite vehicle for exhausting medicinal principles and for the administration of medicine, but has now given place to more stable and uniform alcoholic liquids."**
-'Fenners Twentieth Century Formulary and International Dispensatory: Working Formulas for the Official and Unofficial Preparations of All Countries' by B. Fenner, Westfield, New York, **1904**, page 1238.

"**Vinum Symphyti Compositum: Wine of Comfrey Compound:**
Preparation: Take of Comfrey root, Solomon's-seal, helonias root, each, in coarse powder, 1 ounce (28.3 gram); chamomile

flowers, calumba root, gentian root, cardamom seeds, sassafras bark, each, in coarse powder, 1/2 ounce (14.17 gram); sherry wine, 4 pints (8 cups = 1.89 liter); boiling water, a sufficient quantity. Place the herbs in a vessel, cover with boiling water, and let the compound macerate (soften) for 24 hours, keeping it closely covered; then add the sherry wine. Macerate for 14 days; express and filter. **This preparation is sometimes called 'Restorative Wine Bitters',** but is much superior to the article formerly known by this name.
Action, Medical Uses, and Dosage:
This is a most valuable tonic in all diseases peculiar to females, especially leucorrhoea (vaginal discharge), amenorrhoea (no menstruation), weakness of the back, etc. The dose is from 1/2 to 2 fluid ounces (14.7-59.1 ml), 3 or 4 times a day."
-'Kings American Dispensatory, Volume 2, 19th Edition' by Harvey Wickes Felter, M.D. and John Uri Lloyd, Phr.M., Ph.D., The Eclectic Medical Institute, Cincinnati, Ohio, **1905**.

"**Wine of Comfrey, Compound: 'Restorative Wine Bitters' or 'Vinum Symphyti Compositum':**
   Comfrey (av.oz.) avoirdupois ounce 1/4
   Solomon's seed av.oz. 1/4, Helonias av.oz. 1/4, Roman chamomile gr. (grain) 55
   Columbo gr. 55, Gentian gr. 55, Cardamom seed gr. 55, Sassafras gr. 55
   Sherry wine, to make fluid ounces 16 (473 ml)
Reduce the drugs to moderately coarse powder, and extract by percolation with the wine so as to obtain 16 fluid ounces of percolate. -Eclectic modified. This is a tonic for diseases peculiar to females."
-'The New Standard Formulary: Comprising in Part One All Preparations Official or Included in the Pharmacopoeias, Dispensatories or Formularies of the World' by A. Emil Hiss, Ph.G. and Albert E. Ebert, Ph.M, Ph.D.; G.P. Engelhard and Company, Chicago, Illinois, **1910**, page 537.
('av.oz' refers to the avoirdupois measurement system of weights that uses pounds and ounces. Ounce {oz} = 28.35 gram = 16 dr {drams}.)   (A grain is a measurement of mass, and in troy weight, avoirdupois, and Apothecaries' system, equal to exactly 64.79891 milligrams.)

"**Comfrey Root Wine: Clean and peel the Comfrey roots** and cut in pieces four or five inches (10.1-12.7 cm) long. Boil four or five roots in a gallon (3.7 liters) of water until they are tender and add three pounds (1.3 kg) of sugar to each gallon of the strained liquor. Boil this syrup for three quarters of an hour, pour in a pan, and when it is luke warm float a slice of toast with yeast spread on it on the top. Leave covered for ten days, stirring now and then, and **put in a cask or jar to ferment.** When it has stopped working, close tightly and leave for six months before bottling."
-Comfrey Report: The Story of the World's Fastest Protein Builder and Herbal Healer by Lawrence D. Hills, written **1975**, page 77
If you are not an expert at wine making, I would not try this, just in case you create something not drinkable that could make you sick.

Comfrey Beer and Ale

For entertainment purposes only. If done incorrectly, it could make you sick.

Gruit, grut or gruyt is an herb mixture used for bittering and flavoring beer that was popular before hops became the primary herb. In the Middle Ages, Comfrey root was used to flavor beer.

"**The Gruit: It is not known exactly when Europeans added herbs to ale for flavor.**
As the Middle Ages neared their end and the Renaissance began, the gruit or grout was the term used to describe the herbs and spices added to the brew.
This could be added to flavor the drink, for medicinal value, as a preservative, or to cover the bad flavor of a spoiled brew. The ingredients of a gruit differed from brewer to brewer. As certain locations became famous for the flavor of their ale, the ingredients of the gruit and their ratios became secret, as in modern formulas for liqueurs or modern soft-drinks.
Most brewers used the herb ground ivy as the foundation for their gruit. Ground ivy has the Latin name of Glechoma hederacea, and is also known by the names of alehoof and creeping jenny. Ground ivy is bitter and aromatic, and it was the most commonly used herb prior to the introduction of hops. The herb also has some preservative qualities. It should not be confused with English ivy, which is the ivy found growing on old buildings. English ivy is poisonous.
Other herbs used in the gruit were: alecost, bog myrtle, buckbean, carduus, **Comfrey,** coriander, elecampagne, eyebright, horehound (called mountain hops in Germany), marjoram, mugwort, pennyroyal, sage, woodruff (famous as an addition to May wine), wood sage, and yarrow.
This is by no means all of the herbs used in gruits, and **the herbs could be used alone or in combinations**."
-Libations of the Eighteenth Century: A Concise Manual for the Brewing of Authentic Beverages from the Colonial Era of America, and of Times Past by David Alan Woolsey. Irvine, California: Universal Publishers, pages 19-20, **1997**.
       (Libation means pouring out wine or other liquid in honor of a deity. Informally, it means beverage.)
       (Ale is a type of beer brewed using warm fermentation, that gives it a sweet, fruity, full-bodied taste.)

"**Borage is a relative of Comfrey (Symphytum officinale)** and looks much like it, though its leaves are not nearly so dark a green. Formerly, it was frequently used in herbal practice but is only now becoming commonly known to this new generation. **Borage is a very refreshing plant, and I can understand its frequent use in ale in the middle ages.**"

-'Sacred and Herbal Healing Beers' by Stephen Harrod Buhner. Boulder, Colorado: Brewers Publications, **1998**, page 324.
(If borage can be used in ale, then so can Comfrey.)

"*Before hops, there was Grout. If you are German, there was Gruit. Grout was the blend of spices and herbs the brewer used to make their beer interesting.* Fermented malt water is very sweet and bland. Because of this, brewers generally used at least one bitter herb. If they didn't, they chose herbs that tasted good in a sweet medium. **There aren't a lot of grout recipes, because this was the brewers secret.** Incidently, most period brewers were women. It was cooking, and therefore considered womens work, hence the term 'Alewife'. Even monasteries would sometimes hire a woman to come in and help the brothers make beer. Once brewing became a profitable buisness, instead of housework, well...
Bitter herbs: Ground Ivy (Alehoof, Creeping Jenny), Buckbean, Carduus, Centaury, Nettle, Wood Sage, Wormwood, Germander.
Less bitter or sweet herbs: Juniper berries, Sweet Gale, Sweet Woodruff, Lavender, Tansey, Alecost, **Comfrey,** Dandelion, Elecampane, Eyebright, Hyssop, Mugwort, Pennyroyal, Sage, Coriander seed, Cloves, Seville orange or any orange peel, Cinnamon, Vanilla, Ginger, Cherries, Raspberries, etc."
-'Making Beer, Period Beer and Ale. Use of Hops in Medieval Beverages' by Mark S. Harris, Stefan's Florilegium, www.florilegium.org, and Society for Creative Anachronism, California, www.sca.org, September **2015**.

Comfrey Tea (Infusion or Decoction)

See subsection 'Comfrey Extract to Make Tinctures and Concentrated Powder' in section 'Making and Using Comfrey Medicine' (Chapter 27).

"The roasted Comfrey roots are used with dandelion and chicory roots for making coffee."
-'Comfrey' by BotanicEye: Herbs and Botanics, https://botanicseye.com, Ireland, (date unknown, **2019** or earlier). BotanicEye aims to be a one-stop information site for all matters herbal.

Overview of Extracting Substances from Comfrey

1. Water extracts acids, **alkaolids (some), alkaloid salts (most),** coloring matter, essential oils (some), glycosides, gums, mineral salts, mucilage, pectin, proteins, sugars, tannins.
2. Alcohol extracts acids (organic), acrid constituents, **alkaloids,** bitter constituents, chorophyll, essential oils, glycosides, resins. Alcohol does not extract minerals, gums, or mucilage.
3. Infused oil extracts resins, essential oils, and flavonoids.
-Organic Growers School, Asheville, North Carolina, www.organicgrowersschool.org; Training, workshops, conferences, **2018**.

"*Amounts of Alkaloid in Comfrey Tea: Thus in the case of Comfrey tea if it be assumed that normal methods of infusion (steeping) could extract just over half the alkaloid that was extracted in 8 hours in a Soxhlet apparatus in the laboratory, each cup of tea could contain 100 micrograms of alkaloid.*
At this level the consumer could never attain the lethal dose of 300 milligrams/kg (1 kg = 2.2 pounds) tissue found necessary to produce the acute reaction in rats.
Even to consume this quantity it would take a 150 pound man drinking 4 cups of tea per day a total of 140 years.
Furthermore it is known that to produce chronic reactions, sub-lethal doses over a prolonged period are necessary. Normally such sub-lethal doses would need to be of a much higher order particularly as sensitivity to the alkaloid decreases with age."
-The Alkaloid Content of Comfrey by Dr. D.B. Long, Ph.D, M.A., around **1970**. (In 'Comfrey: Fodder, Food & Remedy'.)

"In 1965 Dr. A.H. Ward, analyst to the 'Henry Doubleday Research Association', tested **Russian Comfrey leaf tea made by two different methods (water or alcohol) to determine the allantoin:**
1. Allantoin content determined in the usual way after alcoholic extraction: 0.083 percent.
2. Two heaped teaspoonfuls (4 grams) used to make four cupfuls (600 ml) of **Russian Comfrey leaf tea,** by pouring on the boiling water and steeping for 5 minutes.
Amount of allantoin extracted: 0.045 percent.
3. Four grams of the sample boiled with 600 ml of water for 5 minutes.
Amount of allantoin extracted: 0.046 percent.
These were leaf-only samples, for the major portion of the allantoin is in stems and midribs, but it will be seen that the tea as drunk is approximately the same as the strength recommended for sores and gastric ulcers in the early editions of the British Pharmacopoeia."
-Comfrey: Fodder, Food and Remedy by Lawrence D. Hills. New York: Rizzoli Universe Books: **1976**, page 175.
(The first edition of 'British Pharmacopoeia' was 1864. It is the national pharmacopoeia of United Kingdom that is published each year. It states the quality standards for medicinal substances. It is used by pharmaceutical researchers and manufacturers.)

Drying Leaves for Comfrey Tea

See sub-subsection 'Dry Small Amounts of Comfrey Leaf' in sub-section 'Comfrey as Hay' in section 'Comfrey Meal, Pellets, Hay and Silage' (Chapter 21).
See subsection 'Drying Comfrey Roots' in section 'Making and Using Comfrey Medicine' (Chapter 27).

There are similarities between drying tobacco leaf and drying Comfrey leaf.

"The 'Henry Doubleday Research Association' has been selling Comfrey tea for many years, and hundreds of members have made it. **The usual method is to gather the Comfrey leaves only, spread them out in the sun, ideally on inch (2.5 cm) mesh wire netting** *supported above the ground so there is free air circulation under it, and then to crisp it off in an airing cupboard.*
The Comfrey leaves are large and some gardeners snip the thick midrib out of the leaf and dry the outer leaf separately, because it finishes quicker. *Crisping can be done in an electric oven turned low, but the door must be left ajar because of the steam. Rub it through a seive until it is broken small and store in a screw top jar or in a polythene (plastic) bag sealed with Celotape (cellophane tape).*
Comfrey should not be heated above boiling point (212 F = 100 C), for this will smash the allantoin. Commercial driers usually keep the temerpature below 120 F (48.8 C)."
-Comfrey: The Herbal Healer by Henry Doubleday Research Association, **1975**, 41 pages, page 27. (in 'Comfrey Report' book)

*"****Shredding and Drying Comfrey Leaf: An ordinary compost shredder can be used.*** *If Comfrey foliage goes through a shredder, it can be dried in an ordinary drier as at Southery (Norfolk County, England), and there is rather more allantoin in the stems than in the leaf.* **A batch or continuous process drier can be used at the low temperature that tea needs if it is dried for medicinal purposes."**
-Comfrey: Fodder, Food and Remedy by Lawrence D. Hills. New York: Rizzoli Universe Books, **1976**, page 118.
   (One example of a 'Kitchen Compost Shredder' is the 'Green Cycler Platinum 2' by Ecotonix®. It is a food scrap shredder and recycling system. I do not have any connection with this company or their products. If you shred Comfrey, you will need a fine mesh screen to hold the pieces for drying. If you have any information about a low to mid-priced machine that would be good for shredding fresh Comfrey leaves so they can be dried faster, let me know.)

*"****There is big demand for Comfrey tea, and the fundamental difficulty is the hand labour.*** *Comfrey has thick stems and midribs which hold in both moisture and the protein, and can seal some of the moisture inside.* **Therefore the only satisfactory way of drying for the Comfrey tea trade is to grow it to the 'fountain of leaves stage', before it has inch-thick (2.5 cm) solid stems, and either pick the leaves by hand, or mow and sort them out, discarding the stems.**
   One way of handling the crop is to spread these leaves on inch mesh wire netting supported at a convenient height off the ground to allow free air circulation all round.
A basic problem is that too much heat can coagulate the gummy proteins like egg white and prevent the escape of further moisture, apart from the fact that if the drying temperature goes much above 180 degrees F (82 C), the allantoin breaks down, and the medicinal value is lost.
If the tea is to be green, attractive and tasty, it must be dried like any other herb, and faster than tobacco."
-Comfrey: Fodder, Food and Remedy by Lawrence D. Hills. New York: Rizzoli Universe Books: **1976**, page 117.
   (Coagulate means to change from a liquid to a solid or semi-solid.)

*"****In the leaf drying process, great care must be taken to dry at low temperature (not more than 50 degrees C = 122 F) so as not to change the protein structure, and not to lose the volatile elements so necessary in a health and medicinal food, allantoic acid in particular.****"*
-Comfrey: Nature's Healing Herb & Health Food by Andrew Hughes. Japan: Sanyusha Publishing Co., Ltd, **1992**, page 23.
   (Allantoic acid is an organic crystalline acid obtained by hydrolysis of allantoin.)

*"****Comfrey, Symphytum officinale, Leaf: Dehydration between 85 and 120 F (29.4-48.8 C).***
      Drying method: *Dryer. Most difficult to dry.* ***Packaging:*** *Polypropylene sack."*
-The Potential of Herbs as Cash Crops: How to Make a Living in the Country by Richard Alan Miller. Berkeley, California: Ten Speed Press, **1992**, page 110.

*"****Herbs must never be dried in sunlight;*** *instead, they should be hung from hooks or nails in shaded areas with adequate ventilation until both the tops and the bottoms are brittle-dry.* **The end point of drying botanicals has always been to finish with herbs that resemble the living plant in color and texture.**
Canning jars and cleaned reused jars are the optimum storage containers. Coffee cans, plastic bags, even paper bags may also be appropriate. Storage should be in a cool, dark area. **And, most importantly, label everything.** *It is best to include when and where you picked the plant as well, even listing preparation and dosage directions."*
-Medicinal Plants of the Pacific West by Michael Moore. Santa Fe, New Mexico: Red Crane Books, **1993**, page 23.

*"****You read in most herb books to dry in the shade. But one of the things about drying in a greenhouse is you're blocking the ultraviolet rays that you would get outside.*** *It's the ultraviolet rays that are so detrimental to the quality of medicinal herbs. If you were just to take herbs and hang them out in the sun, they would definitely turn brown, and you would have a very poor quality product.*

We try to maintain a temperature of about 110 to 115 Fahrenheit (43.3-46.1 C) and that is all through solar gain, primarily during the months of June, July, August and September.
One of the other important things you need for drying in the greenhouse is air circulation. *You need to be very concerned about air circulation in any drying house so we have four fans located about seven feet (2.1 meter) high, located in the four corners of the greenhouse. Because you have so much moisture in there, you need to circulate that moisture around and move the air.* **A 'wet' herb like Comfrey or coltsfoot with a lot of natural moisture will simply steam without air circulation and will cook right in the greenhouse."**
-'Richters® Fifth Commercial Herb Growing Conference' edited by Helen Snell and Conrad Richter. Transcripts from November 4 **2000**, Richters: The Herb Specialists, Goodwood, Ontario, Canada.

"**Drying Medicinal Plants:**
When medicinal plant materials are prepared for use in dry form, **the moisture content of the material should be kept as low as possible** in order to reduce damage from mould and other microbial infestation. Information on the appropriate moisture content for particular medicinal plant materials may be available from pharmacopoeias or other authoritative monographs.
Medicinal plants can be dried in a number of ways:
> in the open air (shaded from direct sunlight); placed in thin layers on drying frames, wire-screened rooms or buildings; by direct sunlight, if appropriate; in drying ovens/rooms and solar dryers; by indirect fire; baking; lyophilization (freeze drying); microwave; or infrared (radiant heat) devices.

When possible, temperature and humidity should be controlled to avoid damage to the active chemical constituents.
The method and temperature used for drying may have a considerable impact on the quality of the resulting medicinal plant materials. For example, shade drying is preferred to maintain or minimize loss of colour of leaves and flowers; and lower temperatures should be employed in the case of medicinal plant materials containing volatile substances.
The drying conditions should be recorded.
In the case of natural drying in the open air, medicinal plant materials should be spread out in thin layers on drying frames and stirred or turned frequently. In order to secure adequate air circulation, the drying frames should be located at a sufficient height above the ground. Efforts should be made to achieve uniform drying of medicinal plant materials and so avoid mould formation.
Drying medicinal plant material directly on bare ground should be avoided. If a concrete or cement surface is used, medicinal plant materials should be laid on a tarpaulin or other appropriate cloth or sheeting. Insects, rodents, birds and other pests, and livestock and domestic animals should be kept away from drying sites.
For indoor drying, the duration of drying, drying temperature, humidity and other conditions should be determined on the basis of the plant part concerned (root, leaf, stem, bark, flower, etc.) and any volatile natural constituents, such as essential oils.
If possible, the source of heat for direct drying (fire) should be limited to butane, propane or natural gas, and temperatures should be kept below 60 C (140 F). If other sources of fire are used, contact between those materials, smoke and medicinal plant material should be avoided."
-'WHO Guidelines on Good Agricultural and Collection Practices (GACP) for Medicinal Plants' by World Health Organization, Geneva, Switzerland, **2003**.

"**When drying large quantities of herbs for commercial sale, growers should use a forced-air dryer** to preserve their color, flavor, oil content, and medicinal properties. To achieve this, good air circulation within the dryer is important, as it reduces drying time and allows the use of lower temperatures, both of which can prevent the degradation of chemical constituents during the drying process.
> The dryer should have well-spaced racks to ensure that all sides of the plant receive sufficient air flow and the plant material dries evenly. **The shelves should be constructed of food grade screens or covered with an acceptable food grade material.**

In eastern North Carolina, existing **forced air tobacco dryers (kilns) can be used to dry herbs.** Or a dryer can be constructed from a new or existing shed outfitted with a heater, fan, and dehumidifier.
What temperature should I use? Drying temperatures and times differ by plants, plant parts, and ambient conditions. In general, most herbs should be dried at low temperatures, ideally around 90-100 F.
> In the mountains, expect herbs to dry at somewhere around 90-105 F. In the Piedmont and Coastal areas (hilly and flat areas) of North Carolina, temperatures may need to be around 110-130 F. Typically, the higher the ambient humidity, the higher the temperature that is required to dry the material thoroughly.

Depending on the plant material and your buyer's specifications, the final product should be between 8-12% moisture. This means that approximately 2/3 of the fresh weight will be lost in the drying process."
-'An Introduction to Herb Dryers. Dryers for Commercial Herb Growers: A Construction Guide' by Jeanine Davis and Amy Hamilton, North Carolina State University, Mills River, North Carolina, October **2008**.

"**Information we keep on the harvest and drying logs:**
> Specificity of drying conditions is often noted so we can capture successes and replicate them in the future.

Pull Comfrey leaves off the stems prior to drying to speed up the drying process. We dry Comfrey leaf differently than other leaf crops because Comfrey leaf has a higher water content and is prone to browning.
> **Comrey leaves have a high water content and can easily bruise and turn brown.**
> Browning is often caused by rough handling and by drying too rapidly in high heat.

Comfrey leaf can also easily compost if not processed quickly after harvest. To process lay out leaves in a single layer with

minimal overlap. **Good airflow is very important. Begin drying leaves with fans at lower temperatures of 80 to 90 F (26.6 to 32.2 C). Then gradually raise the heat after the leaves begin to lose their moisture, and finish drying at temperatures of no more than 100 F (37.7 C)."**
-The Organic Medicinal Herb Farmer: The Ultimate Guide to Producing High-Quality Herbs on a Market Scale by Jeff Carpenter and Melanie Carpenter. White River Junction, Vermont: Chelsea Green Publishing, **2015**, pages 34, 289.

Making Comfrey Tea (Infusion or Decoction)

Infusion: When a tea bag or ball is put in a cup of hot water and allowed to steep for a few minutes before drinking. It extracts vitamins and volatile ingredients from soft ingredients like leaves, flowers, etc.
Decoction: Used to extract the mineral salts and bitter principles from hard material such as roots, bark, seeds and wood. They need boiling for at least 10 minutes, and then are allowed to steep for a number of hours.

"**Comfrey Root Tea: To prepare the mucilage the sliced Comfrey root should be put into cold water, care being taken to prevent the liquor (liquid) from boiling, which causes great clots to be formed;** and it should not be performed in an iron vessel. When the mucilage is dissolved it should be removed from the fire, and after being sweetened with honey or sugar, it may be taken a glassful at a time."
-'The British Flora Medica, or, History of the Medicinal plants of Great Britain, Volume 1' by Benjamin Herbert Barton and Thomas Castle, London, England, **1838**. (Symphytum officinale, pages 211-215.)

"**Comfrey Leaf Tea:** In the vicinity of Williams College, Massachusetts and in some places in the neighboring states of Vermont and New York, the green leaves of Comfrey are often used as pot-herbs (cooked greens). It is found in many New England gardens and used much in syrups to nourish the feeble, parry (ward) off consumption (tuberculosis).
The liquor (liquid), produced by boiling the dried Comfrey leaves, is of a dark color, moderately tonic, restringent and exhilarating; much like the black tea, but more nutritious. Of course, a very good substitute, for people who will still use hot and exciting drinks."
-'The New England Farmer, and Horticultural Register' by Joseph Breck, Volume 23, New Series Volume 13, Boston, Massachusetts, **1845**. ('Symphytum or Comfrey, as Food for Men and Cattle' by Ezekiel Rich, Troy, New Hampshire, page 10, July 10 1844.)
(Restringent is a medical term meaning binding, drying, constricting and tending to stop the flow of body fluids, especially blood or diarrhea.)
(Black tea is more oxidized than oolong, green, and white teas. It has a stronger flavor. All four teas are made from leaves of the shrub Camellia sinensis. Over 90% of tea sold in the West is black tea.)

"**Comfrey Root Tea:** Miss Lupton of Rhodesia (now Zimbabwe, Africa) has reported **success with Comfrey tea made from Comfrey root against a variety of complaints:**
'Take 1 large or smaller Comfrey roots, scrape and clean, then place in 2 pints (4 cups = 946 ml) of cold water to steep for 6 hours. Do not throw away, but boil it and the roots, then simmer for 5 minutes. Remove from stove and allow to cool, then place in fridge to infuse all night. Next morning remove roots from Comfrey tea, bottle and keep in fridge. Take 1 wineglassful 3-4 times daily. This can also be used as a lotion for wounds, rashes, etc., or for dressings for same.
Cold water is a good drawing agent. For dried, dehydrated Comfrey roots I would soak them for a minimum of 24 hours before boiling.' "
-Comfrey: The Herbal Healer by Henry Doubleday Research Association, **1975**, 41 pages, page 21. (in 'Comfrey Report' book)
(As an apothecary measure, wineglass / wineglassful / cyathus vinarius was defined as 1/8 pint or 2 fluid ounces or 59 ml {2 1/2 fluid ounces in the imperial system}. Before 1800 it was 1 1/2 fluid ounces or 44 ml. These measurements are no longer relevant to current capacity of wineglasses.)

"**Comfrey Leaf Tea: There are many recipes for Comfrey tea, and the usual one is equal parts of Comfrey tea and ordinary tea in a normal pot.**
A herbalist's one is as follows: 'Pour 1 1/2 cupfuls of cold water into a pan, add 3 teaspoonfuls of dried Comfrey, bring to the boil, simmer for five minutes, then cover and allow to cool. Strain and drink as required."
-Comfrey: The Herbal Healer by Henry Doubleday Research Association, **1975**, 41 pages, page 27. (in 'Comfrey Report' book)

"**Comfrey Leaf Tea: Put leaves including stems through hand mincer (fine chopping such as with meat grinder).** Cover resulting pulp with cold boiled water- let it stand for 1 hour- press through fine wire sieve using wooden spoon to press all juice out. Use without any further dilution."
-Comfrey: The Herbal Healer by Henry Doubleday Research Association, **1975**, 41 pages, page 33. (in 'Comfrey Report' book)

"**Comfrey Leaf Tea: The best flavour comes from the quickest drying, and snipping the Comfrey leaf midribs out hastens the process. Two heaped teaspoonfuls (4 gram) used to make four cupfuls (600 ml) of tea,** by pouring on the boiling water and steeping for 5 minutes. Amount of allantoin extracted- 0.045 percent.
These were leaf-only samples, for **the major portion of the allantoin is in stems and midribs,** but it will be seen that the tea as drunk is approximately the same as the strength recommended for sores and gastric ulcers in the early editions of the British Pharmacopoeia."

-Comfrey: Fodder, Food and Remedy by Lawrence D. Hills. New York: Rizzoli Universe Books, **1976**, page 175.
    (The first edition of the 'British Pharmacopoeia' was in 1864.)

"Boil six large Comfrey leaves in two pints (4 cups) of water. Let it stand cooling for four hours after it has come to a boil, then strain the liquid into a 2 pound (0.9 kg) jam jar and fit on a tight cover. It will keep for several days and tastes quite pleasant. Take half a teacupful night and morning."
-Comfrey: Fodder, Food and Remedy by Lawrence D. Hills. New York: Rizzoli Universe Books, **1976**, page 179.

"About Tisanes (Herbal Tea) and Other Infusions: From time immemorial various plants, less stimulating than regular tea or coffee, have been used the world over as restoratives. **Some of the homegrown herbs which, singly or in combination, may become interesting beverages are the fresh or dried leaves** *of angelica, bergamot,* **Comfrey,** *hyssop, lemon verbena, mints, linden, orange, lemon, wintergreen and elderberry; the seeds of anise and fennel.*
There is a good general rule for quantiy per cup of water in preparing these infusions:
For strong herbs, allow: 1/2 to 1 tablespoon fresh material, 1/4 to 1/2 teaspoon dried material. *For mild herbs, allow twice the above amounts. Never use a metal pot. Before straining and serving, steep for 3 to 10 minutes in water brought to a rolling boil. Serve with honey or lemon."*
-Joy of Cooking by Irma S. Rombauer and Marion Rombauer Becker. Indianapolis, Indiana: Bobbs-Merrill Company, page 40, **1976**.

"The gardeners partake of the strength of the Comfrey, too. You can cut off a big leaf like those I collected, and **dry it for tea, which has a clean, deep flavor that goes very well with chamomile or lemon balm.**
You can also use small leaves and the ends of stems as a healthful addition to your salads."
-Enchanted Garden: Alan Chadwick's Organic Method of Gardening by Tom Cuthbertson. London, England: Rider & Company / Hutchinson & Co. Publishers Ltd, **1978**, page 122.

"Comfrey Tea by Cool Infusion:
Suspend 1 part (by weight) of the herb in cloth or paper towel in 32 parts of water at room temperature for at least six hours, preferably overnight, *squeezing out the excess tea from the herb packet when finished.*
It is best to moisten the dry herb first before suspending it; gravity does the rest. The substances that dissolve in the water are heavier than water, sink to the bottom of the jar, and set up a slow displacement current. The water containing more solubles is heavier than the water containing less solubles, so there is always a rise towards the top of weaker tea, always a draw downwards from the suspended herb.
It is a very efficient method and, as it uses no heat, the least altering to plant constituents of any tea process.
    The cold infusion preparation is somewhat more efficient at extracting dense storage compounds, since the extended submersion in water allows both enzyme activity and reabsorption of water by hydrolysis and digestion, making some poorly soluble constituents dissolvable.
When preparing tea, make no more than a day's worth at at time."
-Medicinal Plants of the Pacific West by Michael Moore. Santa Fe, New Mexico: Red Crane Books, **1993**, page 24.

"Comfrey Root Tea: A decoction of 1 part dry Comfrey roots to 10 parts of water is a suitable alternative. A daily dose of 5-10 grams (0.176-0.352 ounces) of dry roots was traditionally taken in the form of tea but internal use is now obsolete because of liver damage."
-Medicinal Plants of the World: An Illustrated Scientific Guide to Important Medicinal Plants and Their Uses by Ben-Erik Van Wyk and Michael Wink. Portland, Oregon: Timber Press, **2004**, page 314.
    (A decoction is used to extract the mineral salts and bitter principles from hard material such as roots, bark, seeds and wood. They need boiling for at least 10 minutes and then are allowed to steep for a number of hours.)

"Standard Infusion Dosage for Comfrey (Symphytum officinale):
4-8 ounces (113-226 grams) up to 3 times daily for no more than 6 weeks a year. *We recommend the internal use of Comfrey only for broken bones that don't seem to be healing together naturally or with the assistance of safer herbs."*
-The Modern Herbal Dispensatory: A Medicine-Making Guide by Thomas Easley and Steven Horne. Berkeley, California: North Atlantic Books, **2016**, page 217.   (It did not say whether it was for Comfrey leaf or root, though probably it is for leaf.)

Comfrey Protein Concentrates
See subsection 'Protein' in section 'Nutritional Value of Comfrey' (Chapter 19, Volume 1).

Leaf protein concentrate is a concentrated form of proteins found in the leaves of plants for animal and human food. It is potentially the cheapest, most abundant source of protein.

"Plants as Unconventional Protein Foods: In the United States the whole unfractionated pulp from relatively fibre-free leaves such as Comfrey, and the laminae removed from sugar beet by rubber flails, are being used as pig feed.
Unlike the original forage, the 'whey' has physical porperties that make its economical drying easy.
In spite of this, and in spite of the advocacy of Hartman, Akeson and Stahmann (*1967), Hollo and Koch (1971) and Kohler and Bickoff (1971), it seems unlikely that it is worth drying in countries where the whey from milk is not fully used."

-The Biological Efficiency of Protein Production edited by J.G.W. Jones. Cambridge, England: Cambridge University Press, **1973**, page 113.
(* -'Leaf Protein Concentrate Prepared by Spray-Drying' by Grant Henry Hartman, Walter R. Akeson and Mark A. Stahmann; Journal of Agricultural and Food Chemistry, Volume 15, No. 1, pages 74-79, 1967.)

"**Protein from Plant Cells: The cell of a plant contains thousands of substances** and is constantly engaged in the manufacture of new substances. **The basic reaction is the production of sugar. The sugar is the energy source for the many metabolic reactions, each one regulated by an enzyme which is a protein.**
   In effect if the cells are caught young, **green leaf can provide a valuable selection of proteins,** with sugar and with water according to the age of the cells, and with representative samples of most organic substances required in body chemistry. **They may also contain substances which are not desirable in human body chemistry, including many toxins which are likely to be proteins.**"
-World Protein Resources by Allen Jones. New York: John Wiley & Sons, **1974**. (Chapter 22: 'Green Leaf Protein', page 209-216)

"**Comfrey is an interesting potential source of protein.** Comfrey is seen as a potential replacement for fish meal in South America as more fish meal is used to enrich bread, and as a replacement for oilseeds in Africa when oilseeds are improved to become diet for humans, since **100 tons of Comfrey supply 3.5 tons of dry protein** which is three times the yield of oilseeds on a area basis.
   There is a claim that Comfrey is individual in its tryptophane content of 0.64%, much more than in other vegetables and twice the content of cheese. Conversely, Comfrey lacks methionine (**0.58% against 0.66%** for cheese) and isoleucine (**1.15% against 1.28%** for cheese or beans).
   The ratio of protein to fibre in Comfrey being about 3.4, whereas that of most green leaf is from 1.5-2.5, although green maize (corn) and lucerne (alfalfa) can show a ratio of from 5:7 if cut at the correct stage."
-World Protein Resources by Allen Jones. New York: John Wiley & Sons, Inc., 1974. (Chapter 22: 'Green Leaf Protein', pages 209-216.)
(Percent of methionine and isoleucine for Comfrey and cheese are close. I'm not sure why he says it is a big difference.)

"**This book discusses the technology and economics of protein on a global scale and describes the quality and availability of protein sources.** It consists of 34 chapters and is arranged in 5 parts.
Part 1, 'Introduction to Proteins' deals with protein chemistry, proteins in human diets and methods of protein preservation.
Parts 2, 3 and 4, entitled 'Animal Proteins', 'Vegetable Proteins', and 'Microbial Proteins', respectively, discusses these groups as present and potential future sources of proteins.
Green leaf protein and the use of cultivated crops, grasses and Comfrey as potential sources of protein are considered in Chapter 22.
Part 5, 'Protein Economics', examines the variations in production, distribution and consumption of proteins, world protein supply and demand, and future sources of proteins.
The text is well supported with tables and statistical data."
-'World Protein Resources' by Allen Jones, Medical and Technical Publishing Co. Ltd., Lancaster, England, 381 pages, 1974, book review.

"**Problems with Expressing Juice from Comfrey:**
   A problem that had to be faced at the outset was the impossibility of expressing juice from Russian Comfrey by the methods commonly employed for leaf protein production from grass or lucerne (alfalfa). Juice expression is prevented by the large quantities of mucilage or gum that exude from the tissues when they are macerated (softening by soaking); this difficulty has necessitated the addition of relatively large volumes of water to the macerated leaf, mixing thoroughly before extraction and, finally, recovering the extracted solids from the resulting dilute liquor.
   Gumminess is an inherent characteristic of minced or macerated Russian Comfrey and no way has been found to overcome this barrier to extraction processes except by adding water and thereby diluting the juice.
More Soluble Solids; Coagulation Stage is Important:
   The results obtained are quite unlike those obtained during the production of leaf protein from grass or lucerne. The extraction differs from leaf protein extraction from lucerne (expression of juice) mainly in the respect that **(a) a much greater quantity of soluble solids is extracted, a high proportion of which are non-protein and (b) a greater effectiveness exists at the coagulation stage in separating protein from non-protein components.**
Crude Protein:
   Notwithstanding the latter effect, coagulated protein from Russian Comfrey contains a lower percentage of crude protein than coagulated protein from lucerne.
Non-Protein Parts of Comfrey:
   At present, no information is available concerning the non-protein components in the coagulum from Russian Comfrey; if they include a considerable amount of energy sources that are available to non-ruminants, then the protein concentrates may represent a very good non-ruminant feed which would be more balanced in protein and energy than is found in protein concentrate from lucerne. However, if the non-protein components are mainly polyphenols and non-available carbohydrates, this could not apply.
Supernatant Liquors:

Approximately a third of the crop dry matter remains in soluble form in the supernatant liquors remaining after coagulation. Such a substantial by-product would need to be utilised in a commercial process and may be suitable for yeast or mycelial protein production by processes that have already been evaluated for the 'whey' from lucerne leaf protein production (Worgan & Wilkins, 1977). **The supernatant liquors from Russian Comfrey were shown to support fairly rapid growth of Penicillium spp (species).**

This fraction can also be expected to contain the allantoin which is thought to be responsible for at least some of the curative properties of Russian Comfrey (Titherley, 1976) and a concentrate of allantoin might be obtainable from it.

Extraction Process:

Extracts were prepared from fresh Russian Comfrey plants by successive extractions with water and NaOH (sodium hydroxide, lye). *Almost 55% of crop dry matter and 69% of crop crude protein were extracted. The extracts were then acidified to produce precipitates that contained 21.7% of crop dry matter and 52.6% of crop crude protein.*

Protein Amounts:

The crude protein in these precipitates was found to be over 90% true protein; it is suggested that such concentrates may prove to be valuable as non-ruminant feeds.

The solid residues, containing 11-12% of crude protein, may be usable as feed for beef cattle."

-'The Preparation of Protein Concentrates from Symphytum Asperrimum (Donn) (Russian Comfrey) for Non-Ruminant Feeds and Human Foods' by L.G. Plaskett, Biotechnical Processes, Ltd., Devon, England; Food Chemistry, Volume 7, No. 2, pages 109-116, **1981**.

(Supernatant is the liquid lying above a solid residue after crystallization, precipitation, centrifugation, or other process.)

(Sodium hydroxide, also known as lye and caustic soda, is an inorganic compound with the formula NaOH. It is a white solid ionic compound. It is a highly caustic base and alkali that decomposes proteins at ordinary room temperature and may cause severe chemical burns.)   (Symphytum asperrimum is Prickly Comfrey, not Russian Comfrey.)

"Protein extracted from Comfrey should not be harmful: a sample of Comfrey supplied by Mr. Lawrence D. Hills proved, as expected, to be completely free of alkaloids. A large percentage of the pyrrolizidine alkaloids are destroyed in the drying process, and as herbal preparations mainly use dried materials as a starting base, so there is no reason why Symphytum officinale should be restricted as a therapeutic agent."

-'Pharmacists and Comfrey' by Diane Wiesner, B.Pharm, MA, Ph.D., MPS, Principal of the NSW College of Natural Therapies, Sydney, Australia; Australian Journal of Pharmacy, Volume 65, pages 959-963, **1984**.

"Consumption of protein extracted from Comfrey should be safe, since no pyrrolizidine alkaloids have been detected in such material."

-'Comfrey' or 'Herbal Medicine: Comfrey' by D.V.C. Awang, Health Protection Branch, Health and Welfare Canada; Canadian Pharmaceutical Journal, Volume 120, pages 101-104, February **1987**.

"Always I thought of Comfrey as food for a hungry world, as Henry Doubleday did, but every ton of the 12 tons of dry matter in the 100 ton an acre crops common in Africa needs 500 tons of water."

-Fighting Like the Flowers: An Autobiography: The Life Story of Britain's Best-Known Organic Gardener by Lawrence D. Hills. Bideford, Devon, England: Green Books, **1989**, page 182.

LIST OF CULINARY VEGETABLES.

Asparagus - - - 648	Celery :—		Onion - - - 88. 185	
Marshall's dwarf Prolific - - - 614	Bailey's Gigantic - 76		O'xalis Déppei - 614, 648	
	Kentucky - - - 76		Parsnep - - - 79	
Beans - - - - 74	Law's Giant - - 76		Peas - - - 74. 375	
Mazagan - - 614	Manchester Giant - 76		Milford Marrow - - 614	
Cabbage :—	Perkins's Large - 76		Potatoes - - - 329	
Mammoth - - 330	Russian Pink - - 76		Sourmillier - - 331	
Carrot :—	Seymour's Red Solid - 76		Radish :—	
Altringham - - 80	Superb White - 76		Java - - - 86	
Altringham long Orange 28	Siberian - - - 76		Rhubarb - 75. cult. 174	
Early Horn cult. 27. 80	Comfrey - - - 648		Victoria - - 648	
New White Altringham 28	Cucumber - 85. 53. 90. cult.		Scarlet Runners - - 211	
Surrey long - - 28	262. 563		Sea-kale - 75. cult. 269	
White or Belgian - 80	Mushrooms - 227. cult. 328. 526			

List of Culinary Vegetables includes Comfrey 1841

'The Gardeners Magazine, and Register of Rural and Domestic Improvement', New Series Volume VII, by J.C. Loudon, F.L.S., H.S., London, England, 1841. Contents xiii and page 648.

"Comfrey {Symphytum officinale) is recommended as a perennial spinach plant; and the young shoots, blanched by being forced to grow through loose soil, as a substitute for asparagus. {'The Gardeners Chronicle: A Stamped Newspaper of Rural Economy and General News', by Professor Lindley, Great Britain, 1841. Comfrey in No. 17: April 24, page 262.}"

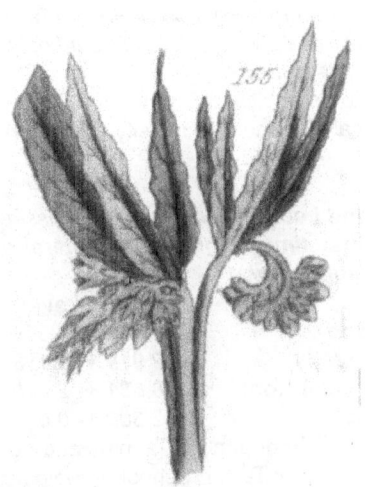

Symphytum officinale 1862
Plate 13, Figure 155

'The Useful Plants of Great Britain: Part I, August' by John E. Sowerby (Illustrator) and C. Pierpoint Johnson (Description), London, England, 1862.

Chapter 33

Miscellaneous Uses of Comfrey

<u>Tannins in Comfrey</u>

Tanning turns the hides of animals into leather. It changes the protein structure of the hide, making it more durable. Traditionally, tanning used tannin, an acidic chemical.

Tannin is an astringent, polyphenolic biomolecule that binds to and precipitates proteins and other organic compounds including amino acids and alkaloids.
Tannic acid is a specific form of tannin, a type of polyphenol. It is weakly acidic. 'Tannic acid' and 'tannin' are sometimes used interchangeably but they are not the same thing.

"Comfrey Root:
The roots of the Common Comfrey from the Ukraine contain tannins of the pyrogallic group (2.4%), glycosidic substances, steroidal saponins, carotene (0.63%), and allantoin (0.8%). A new method for isolating allantoin is proposed."
-'Chemical Study of the Root of Common Comfrey (Symphytum officinale L.)' by G.V./H.V. Makarova, K.N. Zarai'ska and Y.H. Borisyuk.; Farmatsevtychnyi Zhurnal (Pharmaceutical Journal- Russia/Ukraine), Volume 20, No. 5, pages 41-43, **1966**. And in Chemical Abstracts, Volume 66, No. 12, 49229h, 1967. If you have an English translation of this, could you send it to me.

"Comfrey Root and Leaf:
According to Morton*, Comfrey (Symphytum officinale L.) has considerable tannin in the leaves and the infusion of the dried leaves is very astringent. Comfrey root possesses 2.4% tannin...used in Germany for tanning leather."
-'Symphytum Peregrinum Ledeb., Boraginaceae, Comfrey, Russian Comfrey, Quaker Comfrey' by James A. Duke, Handbook of Energy Corps, unpublished, Purdue University, Indiana, **1983**.
(* -'Is There a Safer Tea?: Popular Tisanes Require Pharmacognostic Study' by Julia F. Morton, Morris Arboretum Bulletin, University of Pennsylvania, Philadelphia, Pennsylvania, Volume 26, No. 2, pages 24-30, 1975. And a short version in Lloydia: Quarterly Journal of Biological Science, Lloyd Library of Natural History, Cincinnati, Ohio, Volume 37, No. 4, pages 643-644, 1975. If you have the Morris Arboretum article, I would appreciate a copy.)

*"Comfrey Root: Symphyti radix is the fresh or dried root section of Symphytum officinale (synonym: Comfrey) which is a plant of the Boraginaceae family. Mucilaginous substances (up to 30% of the dry weight), **tannins (4 to 6% dry weight)** and silicic acid (about 4%) are contained in the Comfrey plant."*
-'European Medicines Agency- Symphyti Radix: Committee for Veterinary Medicinal Products' by European Agency for Evaluation of Medicinal Products, Veterinary Medicines Evaluation Unit, London, England, EMEA/MRL/649/99-Final, August **1999**.

*"Comfrey Root: On a dry weight basis, lateral (side) root tissue from 3-month-old **Symphytum officinale** plants contained eight times more extractable polyphenols and 13 times more tannins than main root tissue. **Lateral roots contained on average 3.9% (dry weight) tannins.**"*
-'Lead Chelation to Immobilised Symphytum Officinale L (Comfrey) Root Tannins' by L. Chin, D.W.M. Leung and H.H. Taylor, School of Biological Sciences, University of Canterbury, Christchurch, New Zealand; Chemosphere, Volume 76, No. 5, pages 711-715, **2009**.

"Comfrey Root and Leaf: Tannins content in tissues: The concentration of extractable tannins (on a dry weight basis) was approximately three times higher in the lateral roots than in the main roots, leaf blades and petioles (leaf stalk), which did not differ significantly.

<u>Tissue</u>	<u>Tannin Level</u> (mg Tannic Acid g^{-1} dry weight, 3-4 month S. officinale)
Leaf Blade	0.913 +- 0.389
Petiole (stalk)	2.919 +- 0.113
Lateral (side) Root	8.681 +- 2.620
Main Root	2.300 +- 0.615 "

-'Correlation Between Endogenous Tannins and Lead Accumulation in Roots of Symphytum Officinale L.' by Lily Chin, David W.M. Leung and H. Harry Taylor, School of Biological Sciences, University of Canterbury, Christchurch, New Zealand; Australasian Journal of Ecotoxicology, Volume 15, Issue 1, pages 5-10, January **2009**.

"Comfrey Root: Chemical composition of Comfrey roots (Symphytum officinale) revealed tannins 2.4%."
-'Anti-Inflammatory and Antinociceptive Effect of Symphytum Officinale Root' by Oliviu Vostinaru, Simona Conea, Christina Mogosan, Claudia Crina Toma, Corina Claudia Borza and Laurian Vlase in Romania; Romanian Biotechnological Letters, University of Bucharest, 9 pages, **2017**.

"Comfrey Leaf: Comfrey, Symphytum officinale: Tannin plant (leaf) 80,000 - 90,000 ppm."
-'Natural Anti-Irritant Plants' by Manfred Axterer, Cornelia Muller and Anthony C. Dweck (Symrise GmbH and Co. KG, and Dweck Data) from Dweck Data: Consultants on Natural Products to the Cosmetic, Toiletry and Pharmaceutical Industry, United Kingdom, **2017**.   (80,000 - 90,000 ppm = 8-9%.)

*"The parts of the herb **(Symphytum officinale)** under the ground, especially the root, encloses around 0.6 to 0.7 percent of allantoin and approximately **4.0 to 6.5 percent tannin**. On the other hand, **Comfrey leaves contain** lesser amount of allantoin (approximately 0.3 percent), but more of **tannin (anything between 8.0 to 9.0 percent)**."*
-'A Practical Guide for Nutritional and Traditional Health Care' based on the book 'A Modern Herbal' by Mrs. Grieves, www.Herbs2000.com, **2018**.

*"Professor J. A. Barral, Perpetual Secretary of the 'Central Society of Agriculture in France' ('Societe Centrale de l'Agriculture en France') writes in the Journal de l'Agriculture, of 7th October, 1876 (No. 391): **Tanners and Dyers both use Comfrey; a glue is made from it and used in the preparation of the wool, when mixed with goats hair, without which it could not be spun**."*
-'Forage Plants and Their Economic Conservation by the New System of Ensilage: Part I: Caucasian Prickly Comfrey' by Thomas Christy, Jun., F.L.S. (Fellow of the Linnean Society), Christy and Co., London, England, **1877**, page 9.

*"**A strong decoction has been used on the Continent (Europe) for tanning leather, and in Angora (Turkey) a sort of glue is got from the Common Comfrey, which is used for spinning the famous fleeces of that country**."*
-A Modern Herbal: The Medicinal, Culinary, Cosmetic and Economic Properties, Cultivation and Folk-Lore of Herbs, Grasses, Fungi, Shrubs and Trees with their Scientific Uses by Mrs. M. Grieve. New York: Dover Publications, 1971. First published **1931**.
   (A decoction is used to extract the mineral salts and bitter principles from hard material such as roots, bark, seeds and wood. They need boiling for at least 10 minutes and then are allowed to steep for a number of hours.)

<u>Leather and Tanning</u>   If you have a recipe for tanning leather with Comfrey, please let me know.

*"**Leather: This most useful Comfrey plant**, which deserves more attention than it has hitherto received, has also been **employed in tanning, and has made leather not only more durable than any other method, but it also remained always pliable and elastic**."*
-'Crosbys Merchants and Tradesmans Pocket Dictionary' by a London Merchant and Several Experienced Tradesmen, printed for Crosby and Co. Stationers Court, London, England, **1808**.

*"**Leather: Tabernamontan (Tabernaemontanus), in his 'German Herbal'**, relates a curious fact, which, if not exaggerated, would be of great value in the important process of **tanning, and rendering leather water-proof**.
   He boiled, in a pailful of water, ten pounds (4.5 kg) of the fresh Symphytum officinale Comfrey root, dug out in November, till 1/2 of liquor was evaporated; with this decoction, when cool, **he repeatedly dressed the leather** which, thus prepared, became not only more durable than by any other method, but it always remained pliable and elastic.
M. Dorffurth, an apothecary of Wittenberg, in Germany, also employed these Comfrey roots in his experiments on tanning, with considerable success. **After drying, and reducing them to powder, or cutting the fresh Comfrey roots into small pieces, he infused them in a proportionate quantity of water**, frequently stirring the mass, till it acquired the consistence of treacle.
   It was then allowed to stand at rest several days, till the fibrous and woody part had subsided, when the clear fluid was poured off, or passed through a basket lined with straw.
   By dropping diluted oil of vitriol into this liquor, he precipitated the mucilaginous part, which was again filtered and rendered fit for another process of tanning, after depriving it of its acidity, by means of a lye mace of common pot-ash."*
-'A New Medicinal Economical and Domestic Herbal Containing a Familiar and Accurate Description of Upwards of Six Hundred of British Herbs, Shrubs and Trees, Etc.' selected from the works of Linnaeus, Bechstein, Withering, Dambourney and Barthollet, published by R. Parker Stationer and Bookseller in Blackburn, England, **1808**, page 46.
   (Treacle is a thick, sticky dark syrup made from partly refined sugar, i.e., molasses.)
   (In chemistry, vitriol is an archaic name for a sulfate.)
   (I'm not sure what the author means by 'mace'. The herbs mace and nutmeg come from different parts of the fruit of nutmeg trees. The inner seed is made into nutmeg while the outer layer is made into mace.)

***Tanning Leather: Tabernamontan (Tabernaemontanus), a German writer, highly extols Comfrey's superiority in tanning leather.** He boiled ten pounds (4.5 kg) of Comfrey root in four gallons (15 liters) of water till one-half was consumed; with this decoction he repeatedly dressed the leather, which, thus prepared, became not only more durable than by any other method, but it uniformly remained pliable and elastic.
Another experimentalist, after making a strong infusion of the Comfrey root, allowed it to stand for several days, and when the woody and fibrous parts had subsided, poured off the liquor, and by dropping into it diluted oil of vitriol, he precipitated the mucilaginous part, which was again filtered and rendered **serviceable for tanning**, by neutralizing its acidity with a ley (lye) formed of common potash (potassium)."*
-'The British Flora Medica, or, History of the Medicinal plants of Great Britain, Volume 1' by Benjamin Herbert Barton and Thomas Castle, London, England, **1838**. (Symphytum officinale, pages 211-215.)
      (Jacobus Theodorus {Jacob Diether}, called Tabernaemontanus, 1525-1590, was a physician, botanist and herbalist.

He wrote 'Neuw Kreuterbuch' in 1588 {also called 'Eicones Plantarum' , 1590}. He is the father of German botany.)
(Ley is an archaic word for lye. Lye is an alkaline liquid made by leaching wood ashes. It is also potassium or sodium hydroxide which is called caustic soda. Be very careful with lye. Experts only...)

"Leather-dressing compositions. S.C. McNally, Britain, 9741, July 5, 1915.
A composition for preserving leather, etc., or for use as a substitute for blacking, consists of grease, lard, and a decoction of the root of the Symphytum officinale or Comfrey plant. *Definite proportions are specified."*
-'Chemical Abstracts, Volume 11, Issues 1-6' by American Chemical Society, Easton, Pennsylvania, January-April **1917**, page 106.   (I was not able to find more information about this. I think it is an abstract of a patent.)

Spinning   (If you have details about how Comfrey is used in spinning, let me know.

*"**Spinning: The natives of Angora (part of Turkey), who possess the finest breed of goats in the world, prepare from the Symphytum officinale Comfrey-roots a kind of glue, that enables them to spin the fleece into a very fine yarn,** from which camblets (camlet or camelot woven fabric) and shawls are manufactured.* ***The Germans have lately employed the same mucilage for correcting the brittleness of flax, and roughness of wool in spinning:*** *this preparation neither soils the fingers nor the yarn, and may be preserved in a fresh state for many days, in close wooden boxes."*
-'A New Medicinal Economical and Domestic Herbal Containing a Familiar and Accurate Description of Upwards of Six Hundred of British Herbs, Shrubs and Trees, Etc.' selected from the works of Linnaeus, Bechstein, Withering, Dambourney and Barthollet, published by R. Parker Stationer and Bookseller in Blackburn, England, **1808**, page 46.

*"**Spinning: The natives of Angora (part of Turkey) prepare from Comfrey root a kind of glue, which they use to spin the celebrated fleeces of their country into fine yarn,** from which camblets (camlet or camelot woven fabric) and shawls are manufactured.* ***It has also been employed for correcting the brittleness of flax and the roughness of wool.****"*
-'The British Flora Medica, or, History of the Medicinal plants of Great Britain, Volume 1' by Benjamin Herbert Barton and Thomas Castle, London, England, **1838**. (Symphytum officinale, pages 211-215.)
   (Flax fiber is extracted from the outside of the stem of the plant. It is the strongest of natural cellulosic fibers.)

Comfrey is a Natural Dye

*"**Common Comfrey: A decoction of the stalks with leaves and flowers, gives to wool prepared by a solution of bismuth, a fine and permanent brown colour. But the most useful part of the Comfrey, is its viscid (sticky) and mucilaginous root,** which may be classed among the neglected treasures of the vegetable kingdom. These roots are, at present chiefly employed by colour-makers, who, by means of a decoction made of them,* ***extract the beautiful crimson colour from gum-lac.****"*
-'A New Medicinal Economical and Domestic Herbal Containing a Familiar and Accurate Description of Upwards of Six Hundred of British Herbs, Shrubs and Trees, Etc.' selected from the works of Linnaeus, Bechstein, Withering, Dambourney and Barthollet, published by R. Parker Stationer and Bookseller in Blackburn, England, **1808**, page 46.
   (Bismuth is a chemical element with symbol Bi. It a lustrous brownish-silver metal. It is used in cosmetics, pigments and pharmaceuticals.) (Lac is a resinous substance secreted as a protective cover by the lac insect. People use it to make varnish, shellac, sealing wax, and dyes.)

*"**Scarlet of Gum-Lacque:** The red part of the gum lacque may be used for the dieing of scarlet. I tried gum-lacque with weak lime water, with a decoction of the heart of agaric,* ***with a decoction of Comfrey root, recommended in an ancient book of physic.*** *In all these the water leaves a part of the die, and it still passes too full of colour, and it ought to be evaporated to get all the dye; this evaporation I wanted to avoid,* ***therefore I made use of mucilaginous or slimy roots, which of themselves gave no colour, but whose mucilage might retain the colouring parts,*** *so that they might remain with it on the filter.*
The great Comfrey root has as yet the best answered my intention:
I use it dry and in a gross powder, *putting half a dram (1/2 dram = 1.84 ml) to each quart (0.94 liter) of water, which is boiled a quarter of an hour (15 minutes). Passing it through a hair sieve immediately extracts from it a beautiful crimson tincture. Put the vessel to digest in a moderate heat for twelve hours shaking it seven or eight times to mix it with the gum that remains at the bottom, then pour off the water that is loaded with colour in a vessel sufficiently large, that three-fourths may remain empty and fill it with cold water.*
Then pour a very small quantity of strong solution of Roman alum on the tincture; the mucilaginous or slimy die precipitates itself, and if the water which appears on the top appears still coloured, add some drops of the solution of alum to finish the precipitation, and this repeat till the water becomes as clear as common water.
When the crimson mucilage or slime is all sunk to the bottom of the vessel, draw off the clear water, and filter the remainder; after which, dry it in the sun.
If the first mucilaginous water has not extracted all the colour of the gum lacque, (which is known by the remaining being of a weak straw colour) repeat the operation until you separate all the die the gum-lacque can furnish; and as it is reduced to powder when dry, the quantity to be used in the die is more exactly ascertained than by evaporating it to the consistence of an extract."
-'The Diers Assistant in the Art of Dying Wool and Woollen Goods: Extracted from the Philosophical and Chymical Works of Those Most Eminent Authors Ferguson, Dufay, Hellot, Geoffery, Colbert, and that Reputable French Dier Mons. De Julienne' by

James Haigh, Leeds, England, translated from French, **1813**, pages 137-138.

*"The Comfrey root has been applied to several important purposes. **A strong decoction of the plant will dye wool of a brown colour. The beautiful crimson colour obtained from gum-lac is extracted by means of a decoction of this root.**"*
-'The British Flora Medica, or, History of the Medicinal plants of Great Britain, Volume 1' by Benjamin Herbert Barton and Thomas Castle, London, England, **1838**. (Symphytum officinale, pages 211-215.)
>(A decoction extracts mineral salts and bitter principles from hard material such as roots, bark, seeds and wood. It needs boiling for at least 10 minutes and then is steeped for a number of hours.)

*"**Dye: The leaves of Comfrey produce a brown dye in wood mordanted with iron.**"*
-Rodale's Illustrated Encyclopedia of Herbs edited by Claire Kowalchik and William H. Hylton. Emmaus, Pennsylvania: Rodale Press, **1998**, page 105.
>(A mordant sets dyes on fabrics or tissue by forming a coordination complex with the dye which then attaches to it.)

*"**Boil fresh Comfrey leaves for golden fabric dye.**"*
-'The Complete Book of Vegetables, Herbs & Fruit' by Biggs, McVicar and Flowerdew. London, England: Kyle Cathie Limited, **2004**, page 364.

*"**Now, Comfrey has revealed yet another marvellous characteristic: it produces a soft green dye on wool. Green is said to be a difficult colour to get with simple plant dyes.**
>It is generally recommended to combine a yellow plant dye with the complex blue dyes of indigo or woad (Isatis tinctoria) to get a true green. However, I had been looking for a way to get more muted green, and without having to undertake the complex fermentation steps needed to extract the blue dye present in indigo or woad.*

I borrowed Jenny Dean's 'Wild Colour: How to Grow, Prepare and Use Natural Plant Dyes' * from the local library. I have been disappointed with many books and web pages on natural dyeing, as the information is often unclear, or they use toxic mordants. Jenny Dean's book is quite different: there are fantastic descriptions of various dye extraction techniques, there is no use of the toxic mordants, and the amount of mordant is kept to an absolute minimum.
In 'Wild Colour' I found a description of the use of Comfrey to produce a soft green. *I went out and harvested some Comfrey and followed the dyeing instructions in the book with some of my alum/cream of tartar treated sock wool.*
Process steps:
Chop Comfrey leaves and flower stalks *up into a pot and cover with boiling water. Leave to steep overnight. Bring to the boil and simmer for 1 hour. Strain solution to remove plant material. Add pre-soaked, alum/cream of tartar-treated yarn. Keep just under a simmer for 45 minutes to 1 hour. Rinse skein and dry.* ***The green colour was immediately apparent on the yarn.***"
-Holda's String, www.holdasstring.com, Fleece / Fiber / Yarn, South Island, New Zealand, November 24 **2015**.
(* -Wild Color: The Complete Guide to Making and Using Natural Dyes by Jenny Dean and Karen Diadick Casselman. New York, New York: Potter Craft {part of Random House}, 2010.) (Skein is a loosely coiled length of yarn or thread wound on a reel.)

>*"I am sure I tried dyeing wool with Comfrey a couple of years ago, getting a beige that faded.* ***Jenny Dean's book 'Wild Colour' says Comfrey gives sage green.*** *As I had 1.5 kg (3.3 pounds) Comfrey leaves, I took out three 50 grams (0.11 pound) skeins of wool. I gave them all a hot soak with washing up liquid and a few hot rinses, then premordanted three skeins in a pot with a cupful of iron water added - vinegar and water in which rusty nails had been soaking - which turned the wool a pale rusty orange. The other three were premordanted in copper water - same vinegar and water mix which had had bits of copper piping soaking in it for months. These turned palest green.*
>***The Comfrey leaves were simmered for an hour and left to cool overnight.*** *Next day, I simmered one skein with each mordant all together in the brown dye bath.* ***The results are greenish. Comfrey on alum mordant gave the most herbal green.***"
>-'A Trial of Comfrey Plant Dye on Wool with Alum, Iron and Copper Mordants' by Fran Rushworth, Wool: Tribulations of Hand Spinning and Herbal Dyeing. http://wooltribulations.blogspot.com, June 13 2014.

*"**Making Comfrey dye is easy: Simmer the Comfrey leaves**, about a grocery bag full (3.5 gallon = 13.2 liter), in about a gallon (3.7 liter) of water for about 1/2 hour, then remove the plant material. Submerge wet, alum mordanted wool into the bath and simmer in an iron container for another 1/2 hour or so. **Olive green should be the result.**
Using brass, glass, stainless steel containers will not yield the same results; probably a watery yellow will emerge, but who knows? **Natural dyeing is more art than science.**"*
-'Plants for Dyeing: Comfrey' by KatKnit, Dances with Wools: Knitting, Spinning, Dyeing and Related Fiber Arts, https://danceswithwool.wordpress.com, May 31 **2016**.

*"**Local Handicraft and Other Uses for Symphytum caucasicum M. Bieb. Boraginaceae:**
Armenia: An extract of the plant can stimulate plant growth.
Azerbaijan: **A dye solution is prepared from Comfrey root to obtain red and violet color and used for dyeing wool yarn as well as products made of wool***."*
-Ethnobotany of Caucasus by Ketevan Batsatsashvili, et. al. (Chapter: Symphytum caucasicum M. Bieb. Boraginaceae, pages 683-688). Cham, Switzerland: Springer, **2017**.
(* -'Dye Plants of Azerbaijan' by M.A. Qasimov. Baku, Azerbaijan: Azerbaijan State Publishing House, 1980. If you have an

English copy of this report, could you please send it to me.)

Comfrey Soap   Be careful when making soap. Lye is dangerous. Wear safety gear. Work in a well-ventilated area.

"*A Very Healing Treatment Soap: Neem Oil Soap: Submitted by Kathleen of Burlington, Washington.*
Ingredients:
    18 ounce (532 ml) canola oil, 18 ounce olive oil, 5 ounce (147 ml) virgin coconut oil
    4 ounce (118 ml) palm oil, 4 ounce neem oil,   2 ounce (59 ml) sesame oil (to be reserved to mix with herbs and clay)
    12 ounce (354 ml) of liquid (**distilled water herb tea with marshmallow root, Comfrey root, calendula petals**)
    6 1/2 ounce (192 ml) of lye (sodium hydroxide)
Herbs: 1 tablespoon Tulsi, 1 teaspoon Tumeric, 1 teaspoon basil, **1 teaspoon Comfrey root**
    1 tablespoon neem powder, 1 tablespoon Bentonite Clay, 1 tablespoon Fullers Earth
    1 tablespoon each of essential oil of Sandalwood and Myrrh
Instructions: Follow basic soap making instructions: Make sure the herb tea is cold before adding lye, cool to 100 degrees F (37 C), add to warmed oils, (about 110 degrees F = 43 C) stir to a light trace. Add herbs and clay mixed with the two ounces of sesame oil and blend well, then stir in essential oils. This recipe makes 1 dozen 5 ounce bars. This has a powerful odor while making, but cures out to be a very pleasant, fresh smelling bar. I can't seem to keep enough of this on hand!"
-'Recipes and Information: Soap, Toiletries, Candles, Potpourri' by 'From Nature with Love', Dedicated to bringing you the best of nature from around the world, www.FromNatureWithLove.com, Oxford, Connecticut, October **2002**.

"**Soap Colorants:** *In general, organic pigments are most effective when used to infuse the oils, rather than the water.*
The fastest way to infuse the oils with organic pigments is to mix together a portion of the oils and the organic pigment in a pan on the stove. Do not allow the oil-pigment mixture to heat over 145 F (62.7 C). Keep the mixture warm for 10 minutes to an hour, depending on the intensity of the color desired, and then strain the organic pigment material (herbs, flowers, etc.) from the oil. Next, add 200 IU natural d-gamma tocopherol (vitamin E) per pound (0.45 kg) of oil to greatly prolong the life of organic pigment.
Green: 1/2 teaspoon dried Comfrey. *Comfrey is an organic pigment that is beneficial to skin such as demulcent (relieves irritation), humectant (preserve moisture), or vulnerary (wound healing).*"
-'Soap Colorants' by Kerri Mixon, Pallas Athene Soap and Natural Skin Care, Spring Valley, California; The Handcrafted Soap Maker: Journal of the Handcrafted Soap Makers Guild, Saratoga Springs, New York, Summer **2004**.

"**Colorants in Soapmaking:** *There are a number of herbs, spices, plant derivatives and clays that can be used to color soap.*
    Comfrey Leaf - Deep Sage Green - Add 1 tablespoon per pound (0.45 kg) soap at trace.
    Comfrey Root - Light Milky Brown."
-'Colorants in Soapmaking Manual' by Erica D. Pence, Certified Advanced CP/HP Soapmaker and Certified Soapmaking Teacher, Pence Enterprises Inc., Bath Alchemy Lab, www.bathalchemylab.com, Greensboro, North Carolina, **2015**.

"**Comfrey and Spirulina Multicolored Cubes:** *Makes 9 bars. Spirulina is a dark-green algae that grows in lakes. It is rich in essential fatty acids, such as Omega-3s, -6s, and -9s, that are great for the skin.* **Comfrey leaves contain allantoin, which is said to protect the skin and promote new cell growth.** *The addition of refreshing rosemary and peppermint essential oils cools the skin and awakens the mind.*
Mold and Special Tools: 9-bar silicone cube mold, heating pad.
Lye-Water Amounts: 5.1 ounces lye (5% superfat), 11.5 ounces distilled water, 2 teaspoons sodium lactate (optional).
Oil Amounts: 8.5 ounces palm oil (23%), 9.2 ounces coconut oil (25%), 1.9 ounces avocado butter (5%),
    11.1 ounces olive oil pure (30%), 0.7 ounce castor oil (2%), 5.6 ounces rice bran oil (15%).
Essential Oil Blend: 1.1 ounces rosemary essential oil, 0.3 ounce peppermint essential oil- second distill.
Colorant and Additive Amounts: 1 tablespoon Comfrey powder dispersed into 1 tablespoon rice bran oil.
    1 teaspoon spirulina powder dispersed into 1 tablespoon rice bran oil.
    1 teaspoon rose clay dispersed into 1 tablespoon distilled water.
    1 teaspoon alkanet root powder dispersed into 1 tablespoon rice bran oil.
Note: *This mold creates beautiful bars, but it can be tricky to remove them if sodium lactate is not included.*"
-Pure Soapmaking: How to Create Nourishing Natural Skin Care Soaps by Anne-Marie Faiola. North Adams, Massachusetts: Storey Publishing, **2016**.

"**Comfrey Soap Making: Comfrey will color your soap anywhere from a light lime green to a dark forest green** depending on which Comfrey soap making method you decide to use. **Comfrey topics in this article:**
    Infusing Comfrey into your liquid oils,   5% Comfrey infused oil soap (light),   15% Comfrey infused oil soap (medium),
    35% Comfrey infused oil soap (dark),   Adding Comfrey powder to your soap at trace,
    Adding 1 teaspoon Comfrey powder to soap at trace (light), Adding 2 teaspoons Comfrey powder to soap at trace (medium)
    Adding 4 teaspoons Comfrey powder to your soap at trace (dark),   Adding Comfrey powder to the lye solution,
    How will your super-fat percentage affect Comfrey in soap?   How will the gel phase affect Comfrey in soap?
Infusing Comfrey into Your Oil:
    When working with natural colorant botanical powders, like Comfrey, I like to infuse the colorant into one of my liquid oils. *After the infusion is complete, I then drain out the actual plant matter so that it is not added to my soap. This*

helps avoid any scratchy feeling that can sometimes occur when adding botanical powders to your product. **Comfrey is one of those nicer botanicals that don't produce an overly scratchy product when it is added directly to soap.**
Comfrey Soap Recipe:
   Olive Oil = 10.5 ounces (310 ml) (50% of total oils)
   Coconut Oil (type = 76 Degree) = 4.2 ounces (124 ml) (20% of total oils)
   Palm Oil Refined/Bleached/Deoderized (RBD) = 4.2 ounces (20% of total oils)
   Castor Oil = 2.1 ounces (62 ml) (10% of total oils),   Lye = 82 grams (2.8 ounces),   Water = 8 ounces (236 ml)
Our first step is to create the Comfrey infusion to use in our Comfrey soap making experiments.
I am infusing the Comfrey at the rate of 2 tablespoons Comfrey powder to 1 cup (236 ml) of oil.
   If you want a darker soap color, of course, you can always add more Comfrey to increase the 2 tablespoon Comfrey to 1 cup of oil ratio. If you want a lighter color than what is shown in the following Comfrey soap making experiments, then decrease the amount of Comfrey in your infusion.
For consistencies sake all Comfrey infused oil used in this Comfrey soap making batch was infused for exactly 2 hours. **The longer you infuse your oil, the darker the color will be.** *Keep that in mind if you decide to adjust the hue of your bars!"*
-'Comfrey Soap Making' by Soap Making Resource LLC, www.soap-making-resource.com, supplies the soap and cosmetic manufacturing industry, Lancaster, Pennsylvania, date unknown but 2008-**2017**. (I do not have a connection with this company.)

<u>**Comfrey Toiletries and Cosmetics**</u>   See section 'Comfrey Heals: Allantoin' (Chapter 26).

"Another German writer, M. Reuss, mentions the root of the Comfrey, Symphytum officinale, among those plants, from which good starch and hair powder may be prepared."
-'A New Medicinal Economical and Domestic Herbal Containing a Familiar and Accurate Description of Upwards of Six Hundred of British Herbs, Shrubs and Trees, Etc.' selected from the works of Linnaeus, Bechstein, Withering, Dambourney and Barthollet, published by R. Parker Stationer and Bookseller in Blackburn, England, **1808**, page 47.

"A more elegant method of improving the complexion is to use the juice of Comfrey root; the root is macerated (softened) and the juice pressed out as needed. It has an astringent action and would be quite pleasant on the skin."
-Irish Country Cures by Patrick Logan. Dublin, Ireland: Talbot Press, **1972**, page 77.

"Yeast Face Mask: 1 teaspoon honey, 1 tablespoon brewer's yeast, **1 teaspoon Comfrey infusion (Symphytum officinale)**
1 teaspoon milk or yoghurt, 1 teaspoon Marigold infusion (Calendula officinalis), 1 teaspoon skin oil
This can be a favourite proprietary oil or Sunflower oil (Helianthus annuus), Avocado (Persea americana), Olive oil (Olea europaea) or Peanut (Arachis hypogaea) oil. Combine the honey with a few drops of very hot water. This will thin the honey down and make it easier to use. Blend in the yeast, then add the milk or yoghurt and the herb infusions. Stir until it becomes a thick paste. Pat your face with the oil, and then spread a layer of the paste. Allow to set for approximately 15 minutes, then wash off with tissues and splash with tepid (lukewarm) water."
-The Encyclopedia of Herbs and Herbalism edited by Malcolm Stuart. New York: Grosset and Dunlap, **1979**, page 108.

"Healing Clay Facial: Submitted by Jennifer of Hersey, Michigan.
Ingredients: *1 cup (236 ml) French Clay (or one better suited to your skin type)*
     1 tablespoon finely ground Comfrey, *1/2 tablespoon Calendula powder,   1/2 tablespoon finely ground Mullein*
Instructions: *Mix herbs and clay together very well with a mortar and pestle. Store in a sterile glass jar. Use 1 tablespoon per application.* **Wet with water or a tea infused with one of the above herbs."**
-'Recipes and Information: Soap, Toiletries, Candles, Potpourri' by 'From Nature with Love', Dedicated to bringing you the best of nature from around the world, www.FromNatureWithLove.com, Oxford, Connecticut, October **2002**.

"Hair Rinse: Submitted by Tiffany of Cartersville, Georgia.
Ingredients: *1 heaping tablespoon of each herb, 1 quart (0.94 liter) boiling water, 1/4 cup (59 ml) apple cider vinegar*
For dry hair, try the following herbs: chamomile flowers, nettle, Comfrey root, elder flowers.
For oily hair: lemon peel, lemongrass, peppermint. For general care: nettle, elder flowers, rosemary leaves, burdock root.
Instructions: *Make an infusion of the herbs, steeping at least 30 minutes. Strain and add vinegar. Pour mixture over hair, catch the run-off in a bowl and pour over hair repeatedly. You will get the benefits of the herbs, while the vinegar seals the hair cuticle, making the hair soft, shiny, and less tangled. Use after shampooing."*
-'Recipes and Information: Soap, Toiletries, Candles, Potpourri' by 'From Nature with Love', Dedicated to bringing you the best of nature from around the world, www.FromNatureWithLove.com, Oxford, Connecticut, October **2002**.

*"***Medicinal botanicals of proven and potential dermatologic significance by therapeutic uses:**
     Comfrey (Symphytum officinale): *Bruises/contusions, mucocutaneous pain, stomatitis/gingivitis.*
Some of the herbs with published human studies treating dermatologic conditions with topical formulations include *almond,* **allantoin and Comfrey,** *aloe, anise, bitter orange, black nightshade, black seed, camptotheca, cayenne, curcumin, date palm, echinacea, german chamomile, horse chestnut, lemon balm, neem, oat, onion, oregon grape, pomegranate, St. John's wort, tea tree, oolong tea, and western medicinal herbal mixtures.*
Herbs in Cosmeceuticals: *Comfrey (Symphytum officinale) is approved by the German 'Commission E' to treat blunt injuries*

due to the activity of triterpene saponins, tannins, and silicic acid as well as allantoin.

Allantoin has been extracted from the Comfrey root and leaves but is now commercially manufactured. *Allantoin is an antipholgistic, antioxidant, and soothing keratolytic that has antitrichomonal effect and induces cell proliferation. It is listed in the 'Food and Drug Administration' over-the-counter monograph as a safe and effective skin protectant at 0.1% (0.5%) to 2.0%*.*

Allantoin- and/or Comfrey-based products are used to treat wounds, ulcers, burns, dermatitis, psoriasis, impetigo, and acne. *When formulated with surfactant and benzalkonium chloride it is an effective hand sanitizer and onychomycosis therapy."*

-Cosmetic Formulation of Skin Care Products, Cosmetic Science and Technology Series, Volume 30 edited by Zoe Diana Draelos and Lauren A. Thaman. New York, New York: Taylor and Francis Group, **2006**.
(* -'Skin Protectant Drug Products for Over-the-Counter Human Use: Final Monograph' by Food and Drug Administration, Department of Health and Human Services, Federal Register, Rules and Regulations, Rockville, Maryland, Volume 68, No. 107, June 4 2003.)   (Onychomycosis or tinea unguium is a fungal infection of the nail.)

*"**Allantoin and Dry Skin: The Panel recommended that ingredients allantoin,** cocoa butter, dimethicone, glycerin, petrolatum, and shark liver oil **be included in the monograph as active ingredients for symptoms of dryness. Allantoin and Wound Healing: The Panel classified the ingredients allantoin, live yeast cell derivative, and zinc acetate as Category III skin protectants for wound-healing based on the lack of effectiveness data.** No additional data were submitted for allantoin or zinc acetate to support a 'wound healing' use. **Drug products containing certain active ingredients offered over-the-counter (OTC) for certain uses:** Allantoin- wound healing claims only.*
Skin protectant active ingredients: Allantoin with concentration of 0.5 to 2 percent."

-'Skin Protectant Drug Products for Over-the-Counter Human Use: Final Monograph' by Food and Drug Administration, Dept of Health & Human Services, Federal Register, Rules & Regulations, Rockville, MD, Vol 68, No 107, June 4 2003.

"Symphytum officinale in Cosmetic Products in Denmark:

INCI Name	CAS No.	Chemical name	Function	Products/Type	Rank
S. officinale extract	84696-05-9	Extract rhizomes/roots.	Soothing/antidandruff	2 / All	11.5
S. officinale extract	84696-05-9	Extract rhizomes/roots.	Soothing/antidandruff	1 / Balsam	13.0
S. officinale extract	84696-05-9	Extract rhizomes/roots.	Soothing/antidandruff	1 / Shampoo	10.0"

-'A Survey and Health Assessment of Cosmetic Products for Children: Survey of Chemical Substances in Consumer Products, No. 88' by Pia Brunn Poulsen and Anders Schmidt, Force Technology; Danish Ministry of the Environment, Environmental Protection Agency, Denmark, 220 pages, **2007**.
(INCI is 'International Nomenclature Cosmetic Ingredients' that is used in declarations of contents for cosmetic products in European Union.)   (A CAS Registry Number is a unique numerical identifier assigned by the 'Chemical Abstracts Service' to every chemical substance described in the open scientific literature.)
('Products' above refers to number of products that contain Comfrey for that purpose out of 208 products tested.)
('Rank' is the average order of the substance in the list of constituents on the product label. The ranking is an indication of the relative concentration of the constituents in the product. A low number indicates the substance is a main constituent.)

*"**After Shave:** Aloe Vera, Chamomile, Calendula, **Comfrey**, Althea, Slippery Elm, Fenugreek, Cucumber, Oats, Milk Thistle, Bay Laurel.*
Astringents: *Agrimony, Arnica, Oak Bark, Bistort, Bayberry, Witch Hazel, Myrrh, Plantain, Rhatany, Tormentil, Myrtle, Lemon, Raspberry, Rose, Rosemary, Sandalwood, **Comfrey**, Yarrow, Nettle, Horsechesnut, Corn Flower, Hawthorn.*
Eczema: *Cleavers, **Comfrey**, Fig, Burdock, Red Clover, Golden Seal, Nettle, Yellow Dock, Sarsaparilla, Pansy, Pine, Thyme, White Willow.*
Emollients: *Aloe Vera, **Comfrey**, Althea, Slippery Elm, Fenugreek, Blue Mallow, Oats, Flax, Orange Flowers, Quince Seed, Various Sea Weeds, Elderflower, Cucumber.*
Hemorrhoids: *Bistort, Witch Hazel, **Comfrey**, Horsechestnut, Bayberry Bark, Butchersbroom.*
Dry Hair: *Chamomile, Red Clover, Quince Seed, Horsetail, **Comfrey Root**, Elderflower, Orange Blossom, Peach Leaves, Rosemary, Sage, Basil, Southern Wood.*
Hair, Split Ends: *Horstail, **Comfrey**, Fenugreek, Quince, Rosemary, Echinacea, Lavender, Olive, Basil.*
Healing: *Aloe, **Comfrey**, Horsetail, Yarrow, Pansy, Rose Buds.*
Nails, to Strengthen: *Horsetail, Calendula, Aloe, **Comfrey**.*
Dry Skin: *Aloe Vera, **Comfrey**, Apple, Chamomile, Red Clover, Dandelion, Elder Flowers, Fennel, Quince, Marsh Mallow Root, Slippery Elm Bark, Sea Weeds, Licorice Root, Oats, Orange Blossom, Orange Peel, Citrus Bioflavonoids, Pansy, Peach, Evening Primrose, Yarrow, Parsley, Violet, Cleavers, Capsicum, Arnica, Ginseng.*
Soothing: *Marsh Mallow, Echinacea, Corn Flower, Apple, Colts Foot, **Comfrey**, Calendula, Slippery Elm, Chamomile, St. John's Wort, Cucumber, Red Poppy, Sage, Licorice, Hawthorn, White Lily Pond.*
Sunburn: *Aloe Vera, Calendula, Lemon, Capsicum, Nettle, Slippery Elm Bark, **Comfrey**, Sea Weeds, Witch Hazel.*
Varicose Veins: *Golden Seal, Capsicum, Mullein, Horsechestnut, **Comfrey**, Calendula, Marsh Mallow Root, Hawthorn, St. John's Wort, Witch Hazel, Bayberry, Bistort.*
Wounds: *Milk Thistle, Chamomile, Plantain, **Comfrey**, Chickweed, Golden Seal, St John's Wort, Arnica, Calendula, Aloe Vera, Yarrow."*

-'Botanicals in Cosmetics: Reference Chart' by Vege-Tech Botanicals: Nature and Science, certified organic specialties, www.vegetch.com, Glendale, California, **2011**.

"Dry Skin Type and Their Care: *Low level of sebum and prone to sensitivity. Has a parched look, feels tight. Chapping and cracking are signs of extremely dry, dehydrated skin.*
Suitable herbal skin care for dry skin: *Aloe Vera, Olive Oil, Calendula, Comfrey."*
-'General Review on Herbal Cosmetics' by A. Fathima, Sujith Varma, P. Jagannath and M. Akash, National College of Pharmacy, Manassey, Kozhikode, Kerala State, India; International Journal of Drug Formulation and Research, India, Volume 2, Issue 5, pages 140-165, September- October **2011**. (Sebum is an oily secretion of the sebaceous glands.)

"Sweet Yarrow Facial Mask: *If your skin is dry or irritated, you'll find this combination of yarrow, honey and egg soothing and nourishing: 1 tablespoon yarrow flowers, 2 tablespoons honey, 1 egg yolk, 1 tablespoon yogurt, dry milk powder. Pour 1 cup boiling water over the yarrow flowers. Simmer for one minute, turn off the heat, stir and cover. Steep for about 10-15 minutes. Strain, reserving the liquid. In a separate bowl, combine honey, egg yolk and yogurt. Stir in 1 teaspoon of the yarrow infusion. Add enough milk powder to make a paste.*
Herbal Highlights: *Customize your mask by adding herbs like these:*
To moisturize: ***Comfrey leaf,*** *elder flowers, red clover blossoms, yarrow flowers.*
To cleanse and tone: *Chamomile (German),* ***Comfrey root, Comfrey leaf,*** *elder flowers, lady's mantle, lemongrass, lovage root, nettle leaf, peppermint leaf, rosemary leaf, sage leaf, horsetail, yarrow flowers."*
-'Herb Savvy: Herbal Facial Masks' by Frontier Natural Products Co-Op®: Quality Products for Natural Living, www.frontiercoop.com, Norway, Iowa, around **2011**.

"Soothing Summertime Powder: *This blend helps ease inflammation and discomfort from summer heat by soothing skin irritations.*
Ingredients: *1 cup arrowroot powder or cornstarch, 1/2 cup French green clay*
      1/4 cup each powdered ***Comfrey root or leaf,*** *chamomile flowers, and Calendula petals*
Directions: *Grind up your herbs if needed. Please note that roots can be very hard and difficult to grind so use caution. Mix the powdered herbs and powder base in a bowl, whisking to break up any clumps. Pass the powdered herbs and base through a strainer and finish by bottling up!*
Comfrey leaves and root, Symphytum officinale:
    Regenerative, emollient (softening/soothing), astringent, anti-inflammatory, vulnerary (wound healer)."
-'Keep Cool this Summer with a Homemade Floral Body Powder' by Angela Justis, The Herbal Academy, www.theherbalacademy.com, an online school of herbalism, July **2017**.

"Belif®: The True Cream Moisturizing Bomb:
Comfrey leaf, a natural detoxifier and moisturizing, *make this cream a must for leaving skin more supple and hydrated."*
-'12 of the Best Moisturizers for Protecting Your Face Against the Cold' by Nikki Brown, StyleCaster; Business Insider®: Financial and Business News, www.businessinsider.com, New York, January 16 **2018**.   (I do not have a connection with this product.)

"Cosmetically speaking, the demulcent quality of Comfrey will help ensure a wrinkle free complexion and would be beneficial in skin creams. I use it as a hair conditioner and de-tangler. Because the tea made from the root makes a dark brown color, it keeps my hair brown with frequent usage."
-'The Thymekeeper: A Sea of Comfrey' by Mari Marques-Worden, UTE Country News: Putting the Unity Back in Community, Divide, Colorado, Volume 10, No. 9, September **2018**.

<u>Smoking and Vaporizing Comfrey</u>

"Herbal Smoking Mixtures: *All are 100% free from nicotine.*
The herbs used in the recipes have traditional uses for cleaning the lungs.
American Indian Herbal: *Damiana, strawberry leaf,* ***Comfrey leaf,*** *coltsfoot, uva ursi, peppermint."*
-The Potential of Herbs as Cash Crops: How to Make a Living in the Country by Richard Alan Miller. Berkeley, California: Ten Speed Press, **1992**, pages 191-192.

"A composition for use as a tobacco substitute and as an aid in the cessation of tobacco use, *containing a Vebascum thapsus (great mullein) component, an algae component, a Medicago sativa (alfalfa, lucerne) component, and a* **Symphytum officinale component,** *together with other optional components.*
    The Symphytum officinale component may be dried root of Symphytum Officinale. *The composition may contain approximately equal quantities of each of the components by weight.*
Use of the composition of the present invention as a tobacco substitute, in cigarettes or pipes, **produces a diminished desire for tobacco.** *The composition may further include cigarette paper, and may be formed into a cigarette.*
It will provide the user with a decreased craving or desire for tobacco, and further **provides healing benefits to ameliorate (reduce) the deleterious (bad) effects of tobacco use.** *Such healing benefits include healing of the lungs and general strengthening of the immune system."*
-'Herbal Composition as Substitute for Tobacco: United States Patent 6497234' by Pamela Coy-Herbert, Durango, Colorado, December 24 **2002**.

*"**Natural Herbs to Smoke and Vaporize:** Best temperatures for vaporizing herbs: **Comfrey: 150 C to 175 C (302 F to 347 F)**."*
-'The Best Natural Smoking Herbs' by Nikki Harmony, Harmony Herbals: Ancient Wisdom Meets the Modern World, http://harmonyherbals.net, Victoria, Canada, November 20 **2013**.

*"**Mishma Smoking Blend: A traditional Native American smoking blend:** Smoking this blend creates a sense of peace and tranquility, useful for meditation or the offering of prayers. **Contains:** Mullien leaf, damiana, catnip, blackberry leaf, California poppy, **Comfrey leaf**, raspberry leaf, strawberry leaf, spearmint and spearmint oil."*
-'Mishma Smoking Blend' by Grandfather's Spirit®: Art Inspired by the Ancestors, Native American Culture, www.grandfathersspirit.com, United States, **2018**.   (I do not have a connection with this product or the company.)

Making Paper with Comfrey

*"**An Enquiry Concerning the Materials that May be Used in Making Paper:** By Mr. Guettard, of the Royal Academy of Sciences, and Physician to his Serene Highness the Duke of Orleans, France.*
Neither ought we to entertain any doubt about the use to which we might put the stalks of hemp and flax; and I think we have reason to hope, that one day an advantageous use may be found, for the different kinds of down, not only of the cotton of which, it is perhaps very singular to have entertained any suspicion, but also of the thistle, the trumpet-wood, and the wad.
All that remains, therefore, in order to fulfil our hope of this down as well as of the others, is to find out some method, perhaps very easy and simple, and perhaps for that very reason the more difficult to invent.
If, for example, when the materials are ready to be beaten, instead of simple water, we should substitute a gummy or mucilaginous water like that in which have been boiled the parings of leather, roots of marsh-mallows, the great Comfrey, or such substances; *the paste by these means would be endued with a kind of glue, which might be an expedient by means of which the parts would cohere more strongly.*
*Perhaps it would be sufficient to prepare in this manner the water of the tub in which the paste is diluted when it comes from under the pestle. If not withstanding this preparation, the paste should not have body enough, perhaps by substituting compression in the room of immersion, which is the ordinary method of **forming the sheets of paper**, we should be able to render the parts of the paste more coherent; and **I imagine this is the method which must be taken with that cottoneous substance that owes its origin to the conferva of Pliny**.*
*The heaps formed by the re-union of the different feet of this Comfrey plant, are already of a certain thickness, and not easily torn; **so that in extending the paste made of this plant, we might give what thickness we would, to every sheet,** and the compression would afterwards do the rest."*
-'Select Essays Containing: The Manner of Raising and Dressing Flax and Hemp;.....And an Enquiry Concerning the Materials that May be Used in Making Paper; With Valuable Dissertations on Other Useful Subjects' by Tobias George Smollett, Philadelphia, Pennsylvania, **1777**.
>(In the middle ages, 1275-1325 AD, the word used for Comfrey in Middle English was 'cumfirie' and 'conferye'. In Anglo-French the word was 'cumfirie'. In Old French it was 'confire'. Those words come from Latin words '**conferva**' and 'confervere'. 'Conferva' appears in Pliny's 'Natural History', 77 AD encyclopedia. Pliny wrote: *"The roots be so glutinative that they will solder or glew together meat that is chopt in pieces, seething in a pot, and make into one lump."*)

Wildlife and Fire Protection

*"In the correspondence respecting **this most valuable Comfrey plant**, one important use has been overlooked, namely, as **a plant affording the best cover for game, especially in dry summers, when there is a scarcity of root crops**. Game of all kinds harbour in it, and rabbits cannot eat it down owing to its rapid growth. The plant is a very sightly object on the farm or game preserve."*
-'Forage Plants and Their Economic Conservation by the New System of Ensilage: Part I: Caucasian Prickly Comfrey' by Thomas Christy, Jun., F.L.S. (Fellow of the Linnean Society), Christy & Co., London, England, **1877**, page 17.

"Ornithology was born in ancient Greece, when Aristotle and other writers studied and sought to identify birds. 'Birds in the Ancient World from A to Z' gathers together the information available from classical sources, listing all the names that ancient Greeks gave their birds and all their descriptions and analyses.
Page 53:
>*Aelian (NA 4.47) is hopelessly confused when he claims that Chlorion is the male and Chlorion (q.v.) the female name of one and **the same bird, which builds its nest from Comfrey, hairs and wool** but is a summer migrant.*
>*According to Aristotle's 'History of Animals', the Chloris eats worms and grubs (592b16-17), has received its name because its underparts are greenish yellow, is the size of a Crested Lark, **lays four or five eggs, and makes its nest from Comfrey with a bedding of hairs and wool (615b32–6a2) in a tree** where it is parasitised by the Cuckoo.*

Page 161:
>*HA 616a1-3 says that **Blackbird nests are made from Bulbous Comfrey**, with a bedding of hair and wool; in fact, any combination of dry grass, roots, stalks and moss is in use, cemented on the inside with mud and damp leaves and lined with dry grass."*

-'Birds in the Ancient World from A to Z' by W. Geoffrey Arnott, Professor of Greek at the University of Leeds, and Fellow of the British Academy; Routledge, Taylor & Francis Group, London, England, **2007**, pages 53 and 161.
(In late summer I had a part of my Comfrey garden that I had not cut at all. It was full of tall and dense leaves and stalks. In the middle of 1 Comfrey plant I found a bird's nest. There were no eggs so the eggs must have hatched successfully.)

"Aside from its healing qualities, Graeme further emphasises the huge potential for **Comfrey in land management.** While Comfrey is widely recognised as a fertiliser, a Landcare trial discovered the plant is a very useful green fodder crop in the hotter months and also **provides fire protection for fences.** Another significant benefit is that during the first year, roots break through clay soil acting as an aerator and dispersing water from waterlogged pastures."
-'The Comfrey Man' by Roger Clark, The Waterline News, www.waterlinenews.com.au, Grantville and Districts, Victoria, Australia, Volume 1, No. 12, August **2015**.

"**Faunal Associations of Comfrey: For North America, little is known about floral-faunal (plant-animal) relationships for this Comfrey plant.** According to Muller (1873/1883) in Germany, nectar-seeking long-tongued bees are the primary pollinators of the flowers, particularly bumblebees and Anthophorine bees (Anthophora species); sometimes bumblebees steal nectar by chewing holes near the corolla bases of the flowers. Muller also reported that honeybees, Halictid bees, and a Syrphid fly (Rhingia species) would also steal nectar from the corolla holes that were created by bumblebees."
-'Common Comfrey, Symphytum officinale, Borage family, Boraginaceae' by Dr. John Hilty, Illinois Wildflowers, www.illinoiswildflowers.info, **2017**.

"Some of our most abundant species are often treated as 'weeds' when they appear in the garden. Yet they can be **extremely beneficial to wildlife** - providing food for nectar-loving insects and shelter for minibeasts - and also helpful to the gardener; Common Comfrey has many uses, for example. **Try leaving wilder areas in your garden, such as patches of clover in your lawn, nettles near the compost heap, and Common Comfrey by the pond, and see who comes to visit...**"
-'Comfrey Symphytum officinale' by Ulster Wildlife part of The Wildlife Trusts, www.ulsterwildlife.org, Belfast, Ireland, **2018**.

Feed Worms

"**I regularly put Comfrey leaves in my worm bins,** especially those for which I have no other waste at the moment. The Comfrey is a very fast source of food for them as it rots in just a few days, and Comfrey does contain lots of nutrients. **Comfrey is particularly useful for bins about to be harvested** since I can still feed the worms, but don't need to worry about having all kinds of undigested stuff in the castings at harvest.
Generally, for the last month before harvesting a bin I either feed nothing new to the worms, or just a small bit of something like Comfrey. **For my bins (27 gallon tubs =102 liter), I put in about 2 quarts (1.9 liter) of Comfrey leaves per bin as a feeding.**"
-'Comfrey for the Worms' by commenters, **2015**. Provides experience for home renovation and design, connecting homeowners and home professionals with tools, resources and vendors. www.houzz.com/discussions/2812068/comfrey-for-the-worms

"If you've followed the blog for a number of years, you've likely seen some mentions of **my favorite plant for vermicomposting - Comfrey!** What I love about it - in comparison to most of the more typical weeds - is the abundance of lush foliage. **A perfect 'green manure' for any composting system – including one with worms!**
I use Comfrey a lot as a sort of 'slow food' in my outdoor systems (eg. vermicomposting trenches, vermicomposting planter, backyard composters). I just toss it in - maybe chop it up a bit if I'm feeling motivated - and usually cover it up."
'Comfrey as Sole Food Source for Worms' by Bentley Christie, creator of Compost Guy, and Worm Composting Canada, May **2019**. www.redwormcomposting.com/fun-stuff/comfrey-as-sole-food-source-for-worms/
(Vermicomposting is making compost by providing organic waste as food to earthworms and then collecting their excrement.)
(Green manure is a growing crop, such as clover or grass, that is plowed under the soil to increase fertility.)

As Energy Source (Biofuel)

Bioethanol is ethyl alcohol produced from corn, sorghum, wheat, straw, sugar cane, cornstalks, willow, sawdust, grasses, jerusalem artichokes, and other sources. These biomass products have carbohydrate known as cellulose, hemi-cellulose and lignin. It is treated with acids or enzymes to reduce the size. Cellulose and hemi-cellulose are broken down or hydrolysed into sucrose sugar that is fermented into ethanol. Lignin is used as fuel for the ethanol production boilers.

Biogas is produced by the breakdown of organic matter from raw materials such as agricultural waste, manure, silage, sewage sludge, straw, municipal waste, plant material or food waste. In anaerobic digestion microorganisms break down biodegradable material in the absence of oxygen. Biogas is made up mostly of methane and carbon dioxide. A mesophilic biodigester operates between 20-40 C / 68-104 F.

"**Bioethanol or Alcohol from Comfrey: Comfrey has a reported potential of 247 Metric Ton / hectare (MT/ha) of green fodder,** but the average is usually less than that figure, about 237 MT/hectare reported from England. Australia claims up to 250 MT/hectare green fodder, with 33% protein based on dry material. **Such biomass would have the energetic equivalent of 30**

to 40 barrels of oil per hectare. According to 'U.S. Oil Week' (September 17, 1979), an Oregon company, 'Western Comfrey' **refines the plant into a 25% protein cattle feed** selling for $210 a ton. It is estimated to produce 2 to 2.5 gallon alcohol/bushel with 44% protein cattle feed as a byproduct. Agri-Fuels of Portland, Oregon put up $1.5 million to build a distillery to turn out 1 million gallons fuel grade alcohol along with tons of rich cattle feed."
-'Symphytum Peregrinum Ledeb., Boraginaceae, Comfrey, Russian Comfrey, Quaker Comfrey' by James A. Duke, Handbook of Energy Crops, unpublished, Purdue University, Indiana, **1983**.

> (There are 20 hundredweight in a ton, producing a 'short ton' of 2000 pounds and a 'long ton' of 2240 pounds. A short ton is used in United States. A long ton is used in Great Britain and countries that used to be territories of it. The metric ton or tonne is the 'International System of Units' that is used worldwide. It is 1,000 kg which equals 2204 pounds.)
> ('U.S. Oil Week' was a trade publication published by Capitol Publications, Inc., Alexandria, Virginia. I was unable to find the article or any information about 'Western Comfrey' company. If you have it, could you please send it to me.)

"Biogas from Comfrey: The cost of establishing the energy farm is based on a published figure for Comfrey (cost of initial establishment, spread over 10 years, with interest on the outstanding balance at 12% = $461/hectare/year). The likely cost is about $2.00 to $2.50 per GJ (gigajoule). If the initial planting cost could be halved, it would cost only $1.40 to $1.80 per GJ, at which level these energy sources would be competitive.
As feedstock to anaerobic digestion, perennial crops at $2.25/GJ gross thermal value would give a feedstock cost of $3.75/GJ of **product gas** (without transport and storage) and a probable minimum gas cost of $8.25 per GJ.
These figures are based also on harvesting cost of $75.00/hectare*."
-'Symphytum Peregrinum Ledeb., Boraginaceae, Comfrey, Russian Comfrey, Quaker Comfrey' by James A. Duke, Handbook of Energy Crops, unpublished, Purdue University, Indiana, **1983**.
(* -Energy in Biomass from Europe edited by Wolfgan Palz and P. Chartier. London, England: Elsevier Applied Science Publishers Ltd., 1980.)   (GJ or gigajoule is equal to one billion joules. Six gigajoules is chemical energy of combusting 1 barrel or 42 gallons or 159 liters of crude oil.)

"In 1991 the Matanuska-Susitna Borough, Alaska became interested in commercial ethanol production from farm products which could be grown in the area: fodder potatoes, fodder beets, and **a perennial herb, Comfrey.**
Ethanol Plan: Variety tests and field trial plots of the most promising crops would have been grown each year. These crops were European varieties of stockfeed potatoes, sugarbeet/mangel hybrids, and the perennial herb, Comfrey. This would have provided materials for tests with both laboratory size (1 quart per hour) (0.94 liter) and pilot-plant commercial size (10 gallons per hour) (37.8 liter) fermentation and distillation units.
What finally killed the ethanol project was the the storage requirement for any commercial scale production. Of the three potential crops, potatoes, hybrid beets and Comfrey, only the beets might have been suitable for unprotected winter storage.
> **Comfrey, at 87 percent moisture content, could only have been ensiled, which would have used up a good part of the carbohydrates needed for ethanol production.**

The investors promoting Comfrey thought it would take 20,000 acres, at more than 20 tons per acre, to support a commercial ethanol plant, so **the sheer volume of material to be harvested and stored was beyond reason.** Therefore, the energy budget for the ethanol project was never calculated."
-'Small-Scale Bioenergy Alternatives for Industry, Farm and Institutions: A User's Perspective' edited by Richard Folk, Proceedings of the National Bioenergy Conference, Coeur d'Alene, Idaho, March 12-18 **1991**.

"Bioethanol or Alcohol from Comfrey:
Various green energy crops are available for the production of renewable energy vectors such as second generation bioethanol. The efficiency of the energy recovery potential of these lignocellulosic crops depends on the crop husbandry, their content of main components (cellulose, hemicelluloses, lignin, ash) and on the second generation bioethanol production process.
The 9 lignocellulosic crops investigated in this study are miscanthus (silvergrass, Miscanthus x giganteus), switchgrass (Panicum virgatum L.), fescue (Festuca arundinacea), fiber sorghum (Sorghum bicolor L.), fiber corn (Zea mays L.), cocksfoot-alfalfa mixture (Dactylis glomerata L. - Medicago sativa L.), **Common Comfrey (Symphytum officinale L., la consoude),** jerusalem artichoke (aerial part) (Helianthus tuberosus L.) and hemp (Cannabis sativa L.).
> The samples came from different energy crop trials and their content in cellulose, hemicelluloses and lignin was quantified using the Van Soest method. The ash content has also been quantified.

The lignocellulosic crops with the best theoretical potential to produce second generation bioethanol based on their biochemical composition (with a yield of 100% for the hydrolysis and the fermentation of cellulose and hemicelluloses) are in decreasing order miscanthus (silvergrass), switchgrass, fescue, fiber sorghum, fiber corn and hemp. On one hand, these crops are composed of high amounts of cellulose and hemicelluloses and, on the other hand, the lignin and ash concentration are low.
A principal component analysis showed that commeniloid monocotyledonous and dicotyledonous lignocellulosic crops formed two differentiated categories.

Content of Main Chemical Constituents in Lignocellulosic Crops (kg /100 kg-1 Dry Matter)

Lignocellulosic Crop	Number	Cellulose	Hemicelluloses	Lignin	Ash	Other
Miscanthus/Silvergrass	9	48.4	26.1	8.8	3.5	13.1
Switchgrass	7	40.1	30.3	7.2	5.5	17.0
Fescue	26	34.0	25.2	4.3	9.6	27.0
Fiber Sorghum	16	29.7	25.1	3.2	9.5	32.5
Fiber Corn	21	26.3	23.8	3.4	5.4	41.1

Cocksfoot - Alfalfa	9	27.0	12.9	4.3	9.4	46.4
Hemp	15	47.5	6.4	8.0	8.8	29.4
Jerusalem Artichoke	14	36.5	1.2	10.1	10.9	41.3
Comfrey	**4**	**22.4**	**9.6**	**6.9**	**14.2**	**46.9**
"
-'Cellulose, Hemicelluloses, Lignin, and Ash Contents in Various Lignocellulosic Crops for Second Generation Bioethanol Production' by Bruno Godin, et al., Walloon Center for Agricultural Research, Belgium; Biotechnology Agronomy Society and Environment (BASE), Volume 14, S2, pages 549-560, **2010**.

(Lignocellulose or lignocellulosic biomass is composed of carbohydrate polymers {cellulose, hemicellulose}, and an aromatic polymer {lignin}.) **(Comfrey was rated low for second generation bioethanol. It is low in cellulose and hemiculluloses. It is high in ash/minerals.)**

"Biogas from Comfrey: The study observed the effects of Comfrey on the anaerobic digestion process of swine feces. Swine feces was employed as main material for biogas production and Comfrey as addition material. The trial was conducted in self-manufactured anaerobic equipment. Effect of different material ratios during mesophillic anaerobic fermentation was studied through **measuring CH4 (methane) content, pH, biogas production.**
Result: Among TS (Total Solids) ratios of Comfrey : swine feces: 1:10, 2:10, 3:10, 4:10, the CH4 contents of 1:10, 2:10 were first up to 40%, whose total trend was better than those of other two treatments and two controls.

The pH value of 1:10, 2:10 was 6.6-7.7, 6.6-7.6 respectively, whose total trend was better than those of other two treatments and two controls. **The average biogas production (756.20 mL/day) of 1:10 was the highest. The total biogas production (43,540 mL) of 2:10 was the highest.**
Conclusion: Proper Comfrey can be used to promote the fermentation. The fitting curve predicted the best TS (Total Solids) ratio of Comfrey : swine feces is 1.40:10-1, 48:10."
-'Effects of Comfrey on Mesophilic Anaerobic Digestion of Swine Feces' by L. Tao, Q. Ling, S. YanQiu and D. YuanFang, College of Mechancial and Electronic Engineering, Northwest A&F University, Yangling, Shaanxi, China; Journal of Northwest A&F University - Natural Science Edition, Volume 39, No. 4, pages 177-181, **2011**.

"Biogas from Comfrey: A further alternative use that I am also considering is the use of Comfrey as part of the feedstock for a small-scale biogas plant. Due to its low Carbon : Nitrogen ratio, around 10:1, it would make a good feedstock alongside vegetable waste, or a higher carbon crop grown specifically to feed it - especially the forage part of maize.
The digestate from this process can be used as a replacement for compost, saving some fertiliser costs, or further used to produce ethanol. I have done a little research and whilst a **household sized methane digester** is cheap and feasible on a do-it-yourself basis, I have not gone into costs of a slightly larger scale plant, and done very little work on the ethanol production stage because obviously I could not do this without first producing the digestate.
My personal use of methane would be to power a generator to pump irrigation water. This means I would only need to produce methane during summer, which fortunately for me is the time when ambient temperatures are ideal for methane production. Consequently I would be using a **batch process rather than an all year round constant flow digester.** Batch production is cheaper and the equipment less complex although less efficient."
-'Comfrey' by Old McDonald in Portugal: Farming, Gardening, Wildlife and Good Food, http://oldmcdonaldinportugal.blogspot.com, November 23 **2012**. (Digestate is the material left over after anaerobic digestion of biodegradable matter. Anaerobic digestion produces two main products: digestate and biogas.)

"Selection of potential biomass species for risk assessment:

Latin Name	Dutch Common Name	First, Second or Third Generation Energy Crop
Symphytum peregrinum	Russische smeerwortel	**Second Generation Cellulose / Ethanol, Solid / Pellet Fuel.**"

-'Horizon Scanning and Environmental Risk Analyses of Non-Native Biomass Crops in the Netherlands' by J. Matthews, R. Beringen, M.A.J. Huijbregts, H.J. Van der Mheen, B. Ode, L. Trindade, J.L.C.H. Van Valkenburg, G. Van der Velde and R.S.E.W. Leuven; Reports Environmental Science, Department of Environmental Science, Institute for Water and Wetland Research, Radboud University, Nijmegen, The Netherlands, December 31 **2015**.

(First Generation biofuels are produced from food crops by removing oil for biodiesel or producing bioethanol through fermentation. Second Generation biofuels are from non-food crops such as wood, organic waste, food crop waste and other biomass crops. Third Generation are specially engineered energy crops such as algae.)

"Bioethanol Production: Lignocellulosic biomass has the potential to offer a cleaner alternative as a renewable source for fuel production. The present work aimed to use two plants, Symphytum officinale L. (Common Comfrey) and **Panicum virgatum L. (switchgrass) to produce 5-hydroxymethylfurfural (HMF)** using metal chloride catalysis in two ionic liquids, 1-butyl-3-methylimidazolium chloride ([BMIM]Cl) or 1-ethyl-3-methylimidazolium chloride ([EMIM]Cl). Pre-treatments were used to increase sugar availability, and **two types of treatments were found to be suitable for HMF production.**

First, the 0.5 M (molar) sulfuric acid hydrolysis yielded 230 +- 23 mg of sugars per gram of hydrolysed Comfrey, and 425 +- 13 mg of sugars per gram of hydrolysed switchgrass. Second, the methanol extraction yielded 300 +- 60 mg of sugars per gram of extracted Comfrey, and 202 +- 16 mg of sugars per gram of extracted switchgrass.
The yield of HMF produced was improved from <1% using untreated biomass, to 6.04% and 18.0% using methanol extracts of Comfrey and switchgrass, respectively."
-'Transformation of Symphytum Officinale L. and Panicum Virgatum L. Biomass to 5-Hydroxymethylfurfural for Biofuel Production' by Alexandrine Martel, Thesis for degree Master of Science in Chemical Sciences, Laurentian University, Sudbury,

Ontario, Canada, **2016**.

Protection During a Journey

"*Comfrey was one of the most popular and widely used herbs of the last two centuries; people had faith in the plant, used it, and experience miraculous healing.* **It was held in such high esteem that it was believed that even wearing or carrying Comfrey could guard and protect a person on a journey.** *In my bookshelf, I have more books on Comfrey than any other individual herb.*"
-How Can I Use Herbs in My Daily Life: Over 500 Herbs, Spices and Edible Plants: An Australian Practical Guide to Growing Culinary and Medicinal Herbs by Isabell Shipard. New York, New York: David Stewart Publisher / Simon & Schuster, Inc., **2003**.
    (Another source said Comfrey roots are better for this than the leaves.)

"***Comfrey:*** *Gender: Cold.* ***Planet:*** *Saturn.* ***Element:*** *Air.*
Ritual Uses and Basic Power is for Protection: *To ensure your safety while traveling, carry some Comfrey. Put some in your luggage to ensure its safety.* **Many nomadic peoples of the northern countries put Comfrey leaves in their shoes to avoid blisters, but also to protect against illness and misfortune while traveling.**
Relocation Spell: *Tuck bits of Comfrey root into possessions and furniture before loading them onto the moving van, to ensure their safe arrival at their destination.*"
-'Herbalpedia®: Comfrey' by Maureen Rogers, The Herb Growing & Marketing Network, Silver Spring, Pennsylvania, www.herbalpedia.com, **2006**.

"***Comfrey, Symphytum officinale:***
Gender: Feminine. ***Planet:*** *Saturn.* ***Element:*** *Water.* ***Powers:*** *Safety during travel, Money.*
Magical Uses: *Wear or carry it for protection and to ensure safety during travel. Tuck some into your suitcase to prevent them from being lost or stolen. Also used in money spells.*"
-'Celtic Book of Shadows' by Zyanya, Site Manager / Administrator; Herbs, Health, Magic, Planets, Recipes, http://celticbookofshadows.wikidot.com, **2008**.

Another use during travel: "**In Travelling, if you go by Water, as it is not easy to have Rice got ready for you in the Morning, furnish yourself beforehand with small Pills of 'Ti hoang' (Comfrey),** *and as soon as you awake swallow three or four Drachms of them in a Cup of warm Water.*
These Pills are called Ti hoang, because the Ti hoang is the principal of the five small Ingredients of which they are composed; but for want of these Pills you may take the Ti hoang by itself."
'The General History of China Containing a Geographical, Historical, Chronological, Political and Physical Description', Vol 4, by Jean-Baptiste Du Halde, London, England, page 77-78, **1739**. Translated from French. (Drachms= 60 grains=64.79891 milligrams)

Improve Finances with Comfrey Leaves

"**Plants that Increase Wealth:** *alfalfa, cloves, almonds, basil, iris, pine nut, chamomile, cinnamon, cinquefoil, clover,* **Comfrey***, dill, sorrel, elder, fenugreek, flax, ginger, honeysuckle, jasmine, apple, marjoram, mint, myrtle, patchouli , periwinkle, poppy, sesame, verbena.*"
-'Magic Helpers: Amulets and Talismans- Kabbalah' by Rabbi David Azulai, Morocco, expert on Jewish Mysticism, date unknown but around **2015**.   (Another source said Comfrey leaves are better for this than the roots.)

Astrology and Comfrey

Culpeper lists about 80 herbs that are in the sign of Saturn with Comfrey being listed there.
"*To Saturn, which is a malignant (causes harm) planet, diurnal (daytime cycle), masculine, and very cold; a friend to Mars, and an enemy to the rest, and answereth to the spleen of the microcosm: yet some ascribe it to the head, as also Jupiter and Mars.* **Some say also, that Saturn ruleth the right ear, also the bones, fundament (anus), and the retentive faculties, cold and dry, in the whole body; and the bladder with the Moon.**
> *Of sicknesses; the leprosy, cankers, quartan ague, palsy, consumption, black jaundice, iliac passion, dropsy, catarrh, gout in the feet, and serophulus; as also apoplexies, tooth-ach, all melancholic diseases, cold and dry, trembling, vain fears, fancies, gout, dog-like appetite, hemorrhoids, broken bones, dislocations, deafness, pains in the bones, ruptures (if he be in Leo or Scorpio, or in an evil aspect to Venus), the chin-cough, pain in the bladder, all long diseases, melancholic madness, fear or grief; he governs the memory also, which is cold and dry; and the hearing also.*"
-'Culpeper's English Physician and Complete Herbal to Which are Now First Added Upwards of One Hundred Additional Herbs' by Nicholas Culpeper, **1652-1653**, page 17.

"***Comfrey, a herb of Saturn under Capricorn,*** *is mentioned in herbal records which can be traced back almost to the birth of Christ. When Dioscorides wrote his famous 'Materia Medica' in A.D. 200, he praised the virtues of Comfrey.*"
-About Comfrey: The Forgotten Herb by G.J. Binding, M.B.E., F.R.H.S. Wellingborough, Northhamptonshire, England: Weatherby

Woolnough, **1974**, page 21.
    (Materia medica {medical material} is Latin for the collected knowledge about the therapeutic properties of substances used for healing. The term is derived from the book by 1st century AD Greek physician Pedanius Dioscorides.)

"The winter solstice or hibernal solstice, also known as midwinter, is an astronomical phenomenon marking the day with the shortest period of daylight and the longest night of the year.
In the Northern Hemisphere this is the December solstice and in the Southern Hemisphere this is the June solstice."
-Wikipedia®: The Free Encyclopedia, www.wikipedia.org, 2018.

The Capricorn zodiac dates are December 22 to January 19. Capricorn is the earth element and cardinal (beginnings, changes) ruled by the planet Saturn. It is symbolized by the goat. **In the body it concerns the bones, joints (particularly the knees), and skin. Comfrey,** Horsetail, Ivy, Mint, Poppy, Sassafras, Slippery Elm, and Woodruff are aligned with Capricorn. Capricorn people are self-disciplined and practical with a focus on self-sufficiency and material success.

*"**The Saturnine nature of Comfrey (Symphytum officinale) is readily reflected by its preference for shady places that are cool and damp.** The leaves of the plant are rough from numerous stiff hairs while the veins of the leaf have a blackish tinge that produces a characteristic shadowy complexion to the foliage.*
    Though the plant can endure sunny locations, the sheer intensity of its growth in a few seasons soon produces a dense thicket. The concentration of foliage rapidly creates a sombre darkness in which few other plants flourish. One of the ancient folk names of the plant was blackewoort.
In any plant Saturn rules the roots, *which delve down into the earth providing foundation and stability to the aerial parts. The roots are also the part that endures when cold and darkness predominate in winter.*
Accordingly Comfrey develops strong roots, *so that its unrestrained growth, once established, makes it hard to dig out of the ground. However its foliage very quickly rots down to form an excellent black compost, so that harvesting the leafy stems is less arduous to the problem of digging out a Comfrey patch.* **Saturn is of course linked to death and decomposition.**"
-'Comfrey, Symphytum Officinale: The Embodiment of a Saturnine Herb' by Dylan Waren-Davis, researcher of lost European metaphysical teachings upon which Western herbal knowledge is based. SkyScript: The Mountain Astrologer, Skyscript.co.uk, and 'Traditional Astrologer Magazine', Issue 11, Winter **1996**.

*"**Traditional astrology holds that herbs and plants come under the rulership of the planets,** and have curative powers over the parts of the body associated with those planets.* **By using the system based on the Planetary Hours, Sue harvests the Comfrey by the hour associated with its planetary rulership, the hour of Saturn.**
By combining the ritual of planetary hours and herbalism, Sue creates a salve which includes not only the freshest organic herbs available but herbs harvested with the intention for healing. By watching for these times, it is possible to ensure that **plants are picked at the moment when their curative powers are at their height.**
Symbolic understanding to the interconnectedness of life enhances the healing process by adding an additional dimension to the often overlooked magical means by which healing can take place."
-'Astrology and Herbalism' by Sue J. Morris, Sue's Salves: Something for Your Skin, https://suesalves.com, Mill Hall, Pennsylvania, **2014**. (I do not have any connection with these products.) (Planetary hours are an ancient system where 7 planets have rulership over each day and parts of the day. The day and hour for the planets keep changing so an astrological 'Table of Hours' is used. The 7 planets are Jupiter, Mars, Mercury, Moon, Saturn, Sun, Venus.)

*"**Esoteric uses of Comfrey: Magickal uses include money, safety during travel, and any Saturnian purpose. Use for workings involving stability, endurance, and matters relating to real estate or property.** Put some Comfrey in your luggage to help prevent loss or theft. Wear for travel safety and protection. Use the Comfrey root in money spells and incenses."*
-'Comfrey' by BotanicEye: Herbs and Botanics, https://botanicseye.com, Ireland, (date unknown, **2019** or earlier). BotanicEye aims to be a one-stop information site for all matters herbal.

<u>Increase Fertility and Help Pregnancy</u>

*"**Special Allies for Pregnant Women over Forty: The single most important herb for pregnant women over forty is Comfrey (Symphytum uplandica hybrids).** The leaves of the mature plant contain an abundance of constituents beneficial to mother and babe, including generous amounts of minerals, allantoin, proteins, and many vitamins.*
Regular use of the leaf infusion, at least a quart (0.94 liter) a week, promotes a safe delivery by:
    Strengthening uterine muscles and preparing them to work easily and well.
    Strengthening perineal tissues so they become resistant to tearing.
    Strengthening uterine ligaments so the uterus does not prolapse.
    Strengthening the bladder and increasing resistance to bacterial infection.
    Strengthening the vagina and helping to promote an environment hostile to infection.
    Providing easily assimilated minerals to prevent eclampsia and other complications.
    Helping the bones of the pelvis flex and open during birth.
    Increasing iron in the blood and thus forestalling post-partum hemorrhage.
Some people feel that Comfrey is not safe to use during pregnancy. Some people feel Comfrey is not safe to use

internally at all. I disagree.
The roots of Comfrey do contain compounds that are best avoided during pregnancy. *(As do all parts of the wild plant.)*
In fact, I rarely use Comfrey root because of the possibility of liver congestion, and I strongly caution those who have had hepatitis, chemotherapy, or alcohol problems to strictly avoid comfrey root. Yet even these people can benefit from use of comfrey leaf infusions."
-'Fertility After Forty' by Susun S. Weed, www.susunweed.com, Woodstock, New York, **2000**. Author of the book 'Wise Woman Herbal for the Childbearing Year', safe remedies for pregnancy, childbirth, lactation and newborns, Ash Tree Publishing.

"**Fertility Potion:** *Time this spell to coincide with either the New or Full Moon, whichever suits your reproductive cycles best.* **Make an infusion with spring water and Comfrey.** *Combine with equal parts brandy. Pour into a single glass, which the man and woman drink together. Drink it in bed; make love immediately afterwards. Repeat three nights in a row. If it doesn't work, wait until next month's corresponding moon phase before attempting this method again."*
-'Herbalpedia®: Comfrey' by Maureen Rogers, The Herb Growing & Marketing Network, Silver Spring, Pennsylvania, www.herbalpedia.com, **2006**.
(New Moon is the first lunar phase, when moon and sun have same ecliptic longitude. The moon is not visible to unaided eye.)

"**Aphrodisiacs by signature** *(appearance) also appeared relatively frequently in ballads, pornographic literature and jests. In 'The Womans Brawl'* the protagonist Doll complained:*
Thou unnatural clown thou, feeble Dick thou: As I am Honest Woman Neighbor, I went like a fool and made him a Caudle with Turkey-eggs, and afterwards a Tanzey with new-laid eggs, from a Hen trod by a Game Cock, **put in Comfrey and Clary,** *and fed him Lamb-stones, cavior, and potatoe-pies; yet he could do no more good to a woman than a Boy of a year old.' "*
-Aphrodisiacs, Fertility and Medicine in Early Modern England by Jennifer Evans. Woodbridge, Suffolk, England: Royal Historical Society, Boydell Press, **2014**.
(* -The Womans Brawl, Or Billingsgate Against Turn-Mill-Street, London, England, 1680. A chapbook from Pepys Ballad Collection. Chapbook is a pamphlet with tales, ballads or tracts, sold by peddlers.) (Caudle is a thick, sweet hot drink, usually with alcohol.)

Comfrey, Minnesota
Comfrey is a town in Brown and Cottonwood counties in the state of Minnesota. The population was 382 at the 2010 census. **The town was platted (mapped) in 1902. It was named after the plant called Comfrey.**

Comfrey Postage Stamps
Ukraine printed a Symphytum officinale postage stamp in August 2017. It is under the category of 'Medical and Melliferous Plants'. Melliferous means yielding or having to do with honey. The denomination is 5.00 UAH (Ukaine Hryvnia).

Tobacco Drying 1884
I could not find antique images of Comfrey leaves drying. Comfrey leaves are similar to tobacco leaves.

'Report on the Culture and Curing of Tobacco in the United States', by Joseph Buckner Killebrew, Department of the Interior, Government Printing Office, Washington, D.C., 1884.

Field scaffold for small scale production, page 152.
Drying shed for large scale production, page 248.

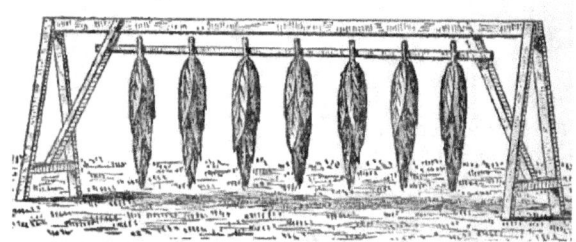

PART F

CULTIVATION AND PRODUCTIVITY OF COMFREY

Chapter 34

Care of Plant Overview and How to Propagate

General Care of Plant

"Chas. Drake, Cattle Inspector on Tambracherry Coffee Estates in Wynaard, Malabar Coast (southern part of India's western coast), October 1876:
Thus far I am glad to be able to report most favourably on the progress of the **Prickly Comfrey** roots I brought out with me. **I have had them planted on a low marshy soil, in ridges three feet (0.9 meter) apart, taking care, previously, to have the soil broken up two or three feet (0.6-0.9 meter) deep, and at subsoil of the ridges making a good coating of cattle manure mixed with jungle soil.**
By this cultivation the roots will not only have considerable depth of soil to grow in, but in the event of having a dry season, the manure being placed at fair depth under the top soil, will tend to make it moist for a very considerable time.
I was greatly surprised at the quick germinating qualities of these Comfrey roots, which in several instances had not been planted more than forty-eight hours, **at about three to four inches (7.6-10.1 cm) below the surface,** and eight had appeared in that time one inch (2.5 cm) above the surface.
I also found after a voyage of six weeks from England, on opening the case, that the roots had germinated a little.
It will be invaluable in my opinion, here on Coffee estates, as a standard food for cattle. Grass is often difficult to obtain during some seasons of the year. To a dairy farmer it would be an acquisition. I confidently expect to get here, a crop every two months, if not more frequently."
-'Forage Plants and Their Economic Conservation by the New System of Ensilage: Part I: Caucasian Prickly Comfrey' by Thomas Christy, Jun., F.L.S. (Fellow of the Linnean Society), Christy & Co., London, England, **1877**, page 20.

"**The Symphytum asperrimum (Prickly Comfrey)** has of late years again received considerable attention, and **wishing to test its value in New Zealand, I obtained roots of the different kinds from England, France, Australia and other places,** and having got them, subjected them to test culture. **They have grown and thriven well,** and I have no doubt that this climate and the conditions it will find here, will suit it very well. I believe it will be a very valuable plant for using as fodder for cows to increase their milk, for feeding bullocks, horses, and sheep.
There are several varieties have been sent me, some are more vigorous growers than others, they can be easily brought here in wardian, or such-like cases, with very few failures. **My last consignment was a case containing 1,000 small roots, they were 4 months before I could get them.** I had them put in the ground by common labourers, and yet over 800 of them are now growing, they have had no watering during warm weather, or other artificial care, as I wanted to try what they would do if planted out and left to themselves. The result has been that they stood the driest, hottest, coldest, and most windy weather, and grew through it all, so that, bearing this rough treatment without any digging, manuring, hoeing, or other cultivating, and yet growing vigorously.
They must be regarded as able to stand unfavourable conditions well, and if, with this treatment, they prove that they can keep a large number of live stock to the area upon which they grow, and make them improve in condition quickly, **they may then be regarded as a useful and good addition to our fodder plants, but as I never think much of any plant until it has had five or six years testing,** it is too soon yet to say much in favour of this plant."
-'Further Observations Upon Certain Grases and Fodder Plants', Transactions and Proceedings of the Royal Society of New Zealand, Volume 11, August **1878**. 'Botany' by S. M. Curl, M.D., pages 409-410.
(Wardian case was an early type of terrarium or sealed container for plants. It was used in the 1800s to protect plants imported to Europe from overseas. Without this protection most plants died from exposure during long sea journeys.)

"**The three main maxims for successful Comfrey growing are: 'Keep it Clean'; 'Keep it Cut'; 'Keep it Fed'.**"
-Russian Comfrey: A Hundred Tons an Acre of Stock or Compost for Farm, Garden or Smallholding by Lawrence D. Hills. London England: Faber and Faber, Limited, **1953**, page 67.   ('Clean' means free of weeds, especially do not let grass grow near it.)

From 'Catalogue of Messrs. Suttons' of Reading, Berkshire, England in 1878:

"The holes filled with rotted dung would make absolutely sure that however bad the soil, the customer would get a crop and establish his Comfrey plants, even on soils of superlatively dry and sandy nature.
This forage plant, introduced into this country a few years since, is rapidly increasing in favor. We believe Comfrey to be a valuable plant for giving a supply of green food in hot, dry seasons. The long roots, which penetrate a great distance into the ground, enable it to obtain moisture beyond the reach of ordinary plants."
-Russian Comfrey: A Hundred Tons an Acre of Stock or Compost for Farm, Garden or Smallholding by Lawrence D. Hills. London England: Faber and Faber, Limited, **1953**, page 33.

"Mr. Stephenson (of Yorkshire County, England, 1942) attempted to establish plants in his stallion paddock so that it could be grazed, but the result was a heavy fall in yield and the Comfrey merely struggled along as a weed; an experiment that is still tried again and again. **Comfrey is only a crop when it is kept clean, cut and manured."**
-Russian Comfrey: A Hundred Tons an Acre of Stock or Compost for Farm, Garden or Smallholding by Lawrence D. Hills. London England: Faber and Faber, Limited, **1953**, page 45.

"Lady Eve Balfour's most important discovery was the way in which **allowing the plant to run up to stem and flower before cutting would cripple the yield for the rest of the season. The next season's cut is unimpaired.**
A bull was then tethered on the plot; it ate the Comfrey greedily, but by trampling and close grazing it reduced the crop to the weed condition in which it is found on many farms."
-Russian Comfrey: A Hundred Tons an Acre of Stock or Compost for Farm, Garden or Smallholding by Lawrence D. Hills. London England: Faber and Faber, Limited, **1953**, page 46.
>(Lady Evelyn Barbara Balfour, 1898-1990, was a British farmer, educator, and a founding figure in the organic movement. She studied agriculture at University of Reading, Bershire County, England.)

In 1952: *"**This grower found that Comfrey was very easy to keep clean, no weeds grew under the plants, so no hand hoeing was needed; a clean up of the middles between early and late cuts was all that was required.**
The main problem on the mixed farm is that, though the crop is useful at both ends of the year, the main cuts have often to be left to waste in June and July with a flush of grass."
-Russian Comfrey: A Hundred Tons an Acre of Stock or Compost for Farm, Garden or Smallholding by Lawrence D. Hills. London England: Faber and Faber, Limited, **1953**, page 50.

*"**Garden-scale crops yield more than in the field**, but their product is just as much help to the national larder in terms of home-produced meat."*
-Russian Comfrey: A Hundred Tons an Acre of Stock or Compost for Farm, Garden or Smallholding by Lawrence D. Hills. London England: Faber and Faber, Limited, **1953**, page 51.   (Larder is a room or large cupboard for storing food.)

"The Comfrey bed should be in full sun, never shaded by trees or hedges, because either will rob the crop of the manure it needs."
-Fertility Without Fertilizers: A Basic Approach to Organic Gardening by Lawrence D. Hills. New York: Universe Books, **1975**.

"Comfrey Climate Requirements:
5.3-8.7 pH, 19.7-106.2 inches (50-269 cm) annual precipitation (rain), 42.8-77.2 degrees F (6-25 C)."
-'Comfrey Leaf: A New Animal Food Supplement' in 'The Encyclopedia of Alternative Agriculture' ebook by Dr. Richard Alan Miller, https://richardalanmiller.com, Agricultural Consultant and Researcher, Oregon and Washington, **1992**.

*"**Comfrey is cultivated from rootstock.** Roots from an older field are quartered and cut into 3 to 5 inch (7.6-12.7 cm) lengths, by hand.* **They are planted 1.5 to 2.0 inches (3.8-5.0 cm) deep and one foot (30.4 cm) apart in rows. Some recommended planting 4 inches (10.1 cm) deep, but this can lead to rotting before emergence.** *Plant in 17 to 20 inch (43-50 cm) furrows. Some 6 to 10 inches (15.2-25.4 cm) of rootstock can be taken from an established field every fourth year, with one acre reseeding five.* **Comfrey likes a heavy irrigation which it sets up its root system the first year, probably as much as a five-day rotation on well drained soils.** *Comfrey should be cut before 10% of the crop goes to flower."*
-'Comfrey Yields, Drying' by Richard Alan Miller, Commercial Herb Production and Marketing, Richters Herbs, Goodwood, Ontario, Canada, November 12 **1999**.

*"**Quaker Comfrey, Symphytum peregrinum, also called Russian Comfrey,** is a perennial used for green manure or as forage.*
***Fertilization**: 60 pounds (27 kg) Nitrogen/acre. Apply P (Phosphourus) & K (Potassium) according to soil test recommendations.*
***Soil pH Range**: 6.0-6.5.*
***Approximate Planting Date**: Fall or early spring. Root cuttings in rows 3 feet (0.9 meter) apart in prepared seedbed.*
***Harvesting**: Cut to a 2 inch (5 cm) stubble when leaves reach a length of 18-24 inches (0.45-0.60 meters)."*
-'The Mid-Atlantic Nutrient Management Handbook' by Land Grant Universities in Delaware, Maryland, Pennsylvania, Virginia and West Virginia, and USDA's Cooperative State Research, Education and Extension Service (CSREES), working with Environmental Protection Agency Region III. Date unknown but **2005** to 2018.
>(The 'Cooperative State Research, Education, and Extension Service' {CSREES} is an agency of the 'United States Department of Agriculture' {USDA}. Its mission is to 'advance agriculture, the environment, human health and well-being, and communities' by supporting research, education, and extension programs at land-grant universities and other

organizations it partners with. Also called 'Cooperative Extension' or 'County Extension'. All counties in the United States have an office. They provide free or low cost soil testing as well as free agriculture information.)

*"**The thesis dealt with the evaluation methods of reproduction of Comfrey and evaluation of the impact of environmental conditions to the growth of Comfrey. Further it dealt with germination tests and their statistical evaluation** while respecting the specific work with seeds. They were evaluated seed characteristics, which are reflected in the growth of these plants arising both in soil and in a laboratory environment.*
*Literary part gave information about **biology, characteristics, morphology, prevention and regulation of Comfrey** even more its cultivation, nutrition of the grassland and seed characteristics, germination and dormancy.*
In the final section of work was monitored coverage, botanical images, representation of herbaceous species growing near Comfrey. The study also discussed vernalization, size of Comfrey leaves, growth and development depending on the content of nutrients (soil fertility) and water in the soil and on the number of cuts to its growth. There were designed a suitable ways of reproduction and growing of Comfrey."
-'Biology, Seed Characteristics and Growth Establishment of Comfrey (Symphytum Officinale L.)' by Jaroslava Kovarova, Master Thesis, University of South Bohemia in Ceske Budejovice, Department of Crop Production and Agroecology, Czechoslovakia, **2016**. All in Czech except abstract is also in English.   (If you have an English translation, I would appreciate a copy.)
   (Vernalization is influencing a plant's flowering process or seed germination by exposure to prolonged cold. After vernalization, plants have the ability to flower or germinate, but may require additional seasonal cues or time. The term is named after Professor Jarowitsch.)

<u>**General Cultivation of Prickly Comfrey**</u> (in chronological order)
See subsection 'Prickly Comfrey as a Forage Crop' in section 'Comfrey as Food and Medicine for Livestock' (Chapter 20).

Prickly Comfrey is Symphytum asperum or Symphytum asperrimum. It is native to Caucasus Mountains located at the border of Europe and Asia, between Black Sea and Caspian Sea and occupied by Russia, Georgia, Azerbaijan and Armenia. Prickly Comfrey was popular in England in 1800s. Its popularity declined with the introduction of Russian Comfrey in the early 1900s. This information is good for all species of Comfrey.

*"**Prickly Comfrey** is conveniently propagated by divisions of the root; the root may be cut into small pieces, as the smallest bit will produce a plant. **The best time for planting the root-cuttings is February or the beginning of March;** on good soils it should be planted in rows 2 to 24 feet (0.6-7.3 meters) apart, and 15 to 18 inches (38.1-45.7 cm) apart in the rows.*
*If the plantation is made early in the year and the cuttings strike root and get well established before genial (pleasant) spring weather sets in, it will yield two good cuttings the first season; and in the succeeding years, if it is kept fairly clean and occasionally manured, **it will give three or four cuttings furnishing an abundance of green food from April to October.**"*
-'The Journal of the Royal Agricultural Society of England, Second Series, Volume 7', London, England, **1871**. Includes article 'XV: On the Composition and Nutritive Value of the Prickly Comfrey (Symphytum asperrimum)' By Dr. Augustus Voelcker, F.R.S. (Fellow of the Royal Society), pages 387-389.

*"After the ground has been properly prepared, **the Prickly Comfrey roots may be subdivided and planted in all but the dry season of the year. In Europe this plant cannot be cultivated from seed.***
   *Mr. Henry Doubleday, of Coggeshall, Essex, England, took a great deal of pains to test the fertility of the seed. He says, 'I would not hybridze or have any inferior variety near them; I sowed the seed under all conditions, and under glass, and in the open ground; I only obtained a score or two (20 to 40) on a considerable space of ground under glass, and kept very moist a few more. **My opinion is, that there are a few fertile seeds, but certainly not one percent; but as the root cuttings are so successful, we have ample means of increase.'****
*I began thus: I forked my ground ten inches (25.4 cm) deep, I opened a small trench and forked the bottom six inches (15.2 cm) deeper, I then put on the bed **a good covering of stable and farm manure**. I mixed this well with the soil having **scattered a little bone dust upon it**.*
*I next planted the Comfrey root cuttings or sets, two feet (0.6 meter) apart, in the rows, and the rows two feet apart also; but I see already that the ground coud be worked more advantageously to the plants, if the plants and rows were **three feet (0.9 meter) apart** instead of two, and where space of ground is no object, I should recommend it. I intend to do so in the future.*
   *If this latter suggestion be adopted, **1,000 root-cuttings will occupy a quarter of an acre (1,011 square meters).***
*My ground being ready, I took out a handful of soil where each set was to be planted, put a little moist sand in the hold and planted a root-cutting in it three inches (7.6 cm) deep. I selected moist ground. I keep the ground clean and free form weeds; **in two or three weeks after planting, leaves like Foxglove appeared.***
Mr. Kinard Edwards, at Hinckley, in Leicestershire (in midlands of England) wrote:
   *'**The first year as much as twenty tons may be obtained; the second year fifty; and every year after, eighty to one hundred tons**. But to do this it will be necessary to lay on a heavy amount of manure, as in this respect Comfrey is no exception to the rule which demands an equivalent being returned to the soil to keep up its fertility.*
   *I have some cuttings rising to a height of five feet (1.5 meters), each plant averaging 10 to 12 pounds (4.5-5.4 kg) to the cut. It may be cut with a hook, tied up in bundles, and so carried to the stall or farmstead as required, day by day; and **for amateurs and cottagers having a horse, cow, or pigs to feed, few crops will be found so useful or more easily cultivated.**'*

A few hundreds of root cuttings will suffice to make a start, as every spring the roots may be raised and divided into twelve parts, and twelve times the area of ground planted."
-'Prickly Comfrey: Its History, Cultivation, Extraordinary Production, and Uses: A Letter Addressed to His Excellency Sir Hercules Robinson, President of the Agricultural Society of New South Wales' by Arthur T. Holroyd, Sydney, Australia, **1876**, pages 6, 8.
(* In the following report from 1881, the author says that Prickly Comfrey seeds are very fertile.)

"***Prickly Comfrey,*** *of which I have had some growing for the last nine months, not long enough, however, for me to speak positively as to its adaptability to this climate of San Mateo County, California.*
I would say, first of all, to those who feel enchanted by the reports of the enormous yields per acre we have heard about it, try it for yourselves by all means, but don't give it a half trial by planting and then neglecting it. For a plant that is a large producer must of necessity be a large feeder, consequently the proper plant food must always be abundantly supplied in the shape of good stable manure.
This, coupled with thorough cultivation, and a fair amount of moisture in the soil will, I have no doubt, make the plant yield a large amount of fodder, for, like all broad-leaved plants, it obtains a large portion of nourishment from the air.
I think the plants ought to be four feet (1.2 meter) apart each way, *instead of two feet (0.60 meter) each way, as I have seen it stated they should be."*
-'Pacific Rural Press', Volumes 15 and 16, San Francisco, California, 853 pages, every other week from January 5 to December 28 **1878**. Article: 'Ensilage, Corn Fodder, Mangolds and Comfrey' by Robert Ashburner, Baden Farm, San Mateo County, California, written December 21 1877, printed January 5 1878.

"We read in the 'London Farmer' magazine some notes of ***Prickly Comfrey trials by Herr A. Ruef, made in a humous, clayey, unmanured soil, in the raw, unfriendly climate of the mountainous districts of lower Austria.***
In May, 1877, 20 Prickly Comfrey settings (roots) were planted, at 18 inches (0.45 meter) distance apart. Not one of these failed; all were cut three times in the year, *the hight of their growth each time being two feet (0.60 meter).*
In the following year the plants had made such headway by the beginning of May that they were then almost fit for cutting, while the neighboring lucerne (alfalfa) crops stood only nine inches (0.22 meter) high.
In this season the Comfreys were cut five times, the hight of the crops ranging from 75 to 62 centimeters (2.4 to 2.0 feet), and the weight of the green stuff obtained from the 20 plants at each cutting varying from 49 pounds (22 kg) in June to 29 pounds (13 kg) in October. ***Altogether the five crops from the 20 Comfrey plants yielded 184 pounds (83 kg) of green food.***
The attention bestowed on the plantation was confined to simply loosening the earth round the Comfrey roots and extirpating (removing) the weeds, no manure of any kind being employed.
Herr Ruef believes that in fair average seasons a yield of 72 hundred weight per acre could be obtained, even under these unfavorable circumstances. Both calves and full-grown stock devoured the produce greedily."
-'Pacific Rural Press', Volumes 17-18, San Francisco, California, 843 pages, every other week from January 4 to December 27 **1879**. Article: 'Prickly Comfrey' by editors, March 1 1879.

"***Prickly Comfrey (Symphytum asperrimum) can be propagated from seeds, root cuttings, crown cuttings and stem-cuttings; the seeds and roots being best.***
    The seeds may be sown in autumn, winter or spring. If planted early they may not appear till spring. *They should be barely covered with soil.*
    The roots may be cut in pieces an inch (2.5 cm) long, and the larger ones may also be split in two or more pieces and set almost any time if the ground is sufficiently wet; but best from February first till April. *I speak with a view to my own experience in latitude 31 (Mississippi).*
The plant is perennial and requires little cultivation after once getting rooted. For large crops it must be manured whenever the yield falls off too much. The ground should be well broken and as deeply as possible with convenience. ***The distance at which the cuttings or plants must be set will depend on the quality of the soil and the preparation.***
    Two feet (0.60 meter) by one and a half feet (0.45 meter) would require 14,520 plants per acre.
    2 1/2 feet (0.76 meter) by 1 1/2 feet (0.45 meter), 11,600 plants.
    2 feet (0.60 meter) by 2 feet, 10,890 plants.
    2 1/2 feet (0.76 meter) by 2 feet (0.60 meter), 8,712 plants.
This is as close as I would advise to plant; and on very rich land I would have the plants three by three feet (0.91 meter).
Mr. Ashburner of Virginia recommended setting the cuttings very deep in the soil, and following his instructions in a heavy soil, many of my cuttings never got out. Four or five inches (10.1-12.7 inch) may do in very light soil, but in very heavy soil one inch (2.5 cm) is much better."
-'The Farmers Book of Grasses and Other Forage Plants for the Southern United States' by D.L. Phares, A.M. (Master of Arts), M.D., Professor of Biology, A&M College of Mississippi, Starkville, Misssissippi, **1881**, pages 21-23.
    (The 31st Parallel North is a circle of latitude 31 degrees north of the Earth's equator. It crosses Africa, Asia and North America. In the United States it goes through Alabama, Florida, Georgia, Louisiana, Mississippi, and Texas.)

"***Symphytum asperrimum, Prickly Comfrey:***
Number 2152 from France. Received through Mr. W. T. Swingle, February 13, 1899.
A coarse, rank-growing (vigorous), perennial herb, with purple flowers in nodding, one-sided clusters, and large, rough leaves. It is a native of the Caucasus.
It has been extensively tried in this country (United States). Although it will produce a great bulk of forage on rich or

swampy soils its cultivation is not recommended. It has been recommended for waste, swampy lands in Florida. Prickly Comfrey does not compare with the clovers, alfalfa, or cowpeas in feeding value, and where the latter can be grown it is not advisable to plant it. It is propagated from the roots. These are set out in rows 1 1/2 to 2 feet (0.45-0.60 meter) apart, the plants 16 to 20 inches (40.6-50.8 cm) apart in the row. At first of slow growth, **Prickly Comfrey will in the course of two years yield from 3 to 6 tons of cured (dried) forage per acre.**"
-'Inventory No. 5 of Foreign Seeds and Plants Imported by the Department of Agriculture, and for Distribution Through the Section of Seed and Plant Introduction, Numbers 1901-2700' by United States Department of Agriculture, Division of Botany, S.P.I. 10, Washington, DC, November 3 **1899**.

"**It is expedient to create plantations of Prickly Comfrey of multipurpose use (forage, medicinal harvest) in the mountain zone of central North Caucasus.**
An agrocoenosis of this plant on typical mountain soils is characterized by high aftergrowth, stable productivity, self-renewal, and early spring regrowth.
Uniform illumination of the herbage makes it possible to include Prickly Comfrey in a mixture with other forage grasses."
-'Biological Productivity and Stability of a Prickly Comfrey Agrocoenosis in the Mountain Zone of Central North Caucasus' by A. Ya. Tamakhina, Russian Agricultural Sciences, Volume 34, No. 1, pages 21-24, 2008, and Doklady Rossiiskoi Akademii Sel'skokhozyaistvennykh Nauk (Reports of the Russian Academy of Agricultural Sciences), No. 1, pages 19-21, **2008**.

(The North Caucasus or Ciscaucasia is the northern part of the Caucasus region between the Sea of Azov and Black Sea on the west and the Caspian Sea on the east, within European Russia.)

(Biocenosis is a self-sufficient community of naturally occurring organisms occupying and interacting within a specific biotope. A biotope is a portion of a habitat characterized by uniformity in climate and distribution of biotic {life} and abiotic components.

'Agrocoenosis' or 'agrobiocenosis' is a Russian term meaning *"an association of organisms in the sowings and plantings of cultivated plants; one of the widespread forms of secondary biocenoses. Each species of cultivated plant forms its characteristic agrobiocenosis with a series of constant and dominant species. The agrobiocenosis is distinguished from the primary biocenoses by the incapacity for long-term independent existence as a result of the sharp weakening of the self-regulating processes; their temporary stability is supported by the activity of man.*

Another important property of an agrobiocenosis is the dominance of several herbivorous species of animals, primarily pests, particularly insects. A cultivated plant forms the energy base of the agrobiocenosis; this plant, together with the accompanying weeds, determines the composition of the animal population." -The Great Soviet Encyclopedia, 1979, one of the largest Russian-language encyclopedias, published by the Soviet Union from 1926 to 1990.)

"**In the mountain regions of the central part of North Caucasus belonging to the zone of risky agriculture** *(high probability of hail, low degree of soil improvement, pronounced temperature fluctuations),* **it is expedient to create agroecosystems characterized by productive longevity.**
Of practical interest in connection with this is Prickly Comfrey (Symphytum asperum L.) found in the Caucasus from the lowlands to the high mountain belt.
This earliest nontraditional forage plant surpasses many leguminous and gramineous grasses in rate of growth. It is well adapted to growth and development under conditions of a cold climate, surplus of atmospheric moisture, and various soils. It ensiles well in a pure form as well as in a mixture with other silage crops.
Its major shortcoming is the difficulty of seed production owing to nonuniform ripening and rapid shedding of seeds.
The years of the investigation differed in temperature conditions.

In 2002-2004 it was close to the normal values: the average annual temperature was 8.3-8.4 C (46.9-47.1 F); 2005 was hot with a long period of high temperatures that sometimes reached 35 C (95 F), which is not typical for mountain conditions. In 2002 precipitation was closest to normal values (794 mm= 31.25 inches). In 2003-2004 there was 2-2.8% more precipitation than in 2002.

The Prickly Comfrey coenosis is characterized by more or less uniform illumination (light) of all layers of the herbage. **By the first cut a spread-out, rough mass forms, within which most weed plants die,** except Mentha arvensis Soll. (corn/field mint) and Artemisia vulgaris L. (mugwort).

Consequently, this forage plant is an edificator (improver) of the phytocoenosis creating unfavorable conditions for growth of other species.
The laboratory germination rate of Prickly Comfrey seeds is 55-60%, under natural conditions it is much less, 20-30%.
This is due to dropping in a radius of 0.6-1.0 meter (1.96-3.28 feet) from the base of the stem of the mother plant and accumulation of viable seeds in the soil. A large part of the seed specimens is located at a distance of 20-30 cm (7.8-11.8 inch) from the mother plant.

Under favorable conditions sprouts appear after a week. Juvenile individuals are characterized by high energy of growth of above- and underground organs. Seed individuals stay for a long time (2 years and more) in a juvenile state.

On average during the year the plantation yielded 47.3 ton/hectare green mass on gray forest soil and 11.6% less on alluvial meadow soil. The yield of green mass of second cut in the case of sufficient moisture was 45-55% and under the 2005 drought conditions, 33% of the first. On both soil types the underground part of the community reached 51-52% of total biomass.
The second direction of using this plant is medicinal harvest. The wide-row band planting on the plantation makes it possible to procure Prickly Comfrey roots and rhizomes annually, selective digging up at a distance of 5 meter (16.4 feet) in the row with preserving part of the root. The productivity of medicinal raw material (air dry weight) on gray forest soil was 7-8 cwt/hectare (cwt= hundredweight= 112 pounds) and on alluvial meadow soil 6.2-7.5 cwt/hectare.

Already in the second growth year of the plantation, the profitability of the medicinal harvest reached 321%, in 2004= 327%, and in 2005= 332%. The one-time expenses for establishing the plantation per hectare were completely paid back in the second year of its operation with the use of just selective medicinal harvesting, i.e., without consideration of forage use.
The results obtained attest to the expediency of creating and operating Prickly Comfrey plantations in a monoculture as well as in a mixture with other grasses in the mountain zone of central North Caucasus."
-'Biological Productivity and Stability of a Prickly Comfrey Agrocoenosis in the Mountain Zone of Central North Caucasus" by A. Ya. Tamakhina, Russian Agricultural Sciences, Volume 34, No. 1, pages 21-24, 2008, and Doklady Rossiiskoi Akademii Sel'skokhozyaistvennykh Nauk (Reports of the Russian Academy of Agricultural Sciences), No. 1, pages 19-21, **2008**.
(Monoculture is the cultivation of a single crop.)

<u>Collecting Pods and Seeds</u>
The following seed germination rates are for controlled conditions by a grower:
Most Common Comfrey seeds will grow. Almost always Russian Comfrey seeds will not grow.
From what I have read, about half of Prickly Comfrey seeds will grow.

My experience with Common Comfrey: On each Common Comfrey flowerstalk the flowers start blooming from the bottom of the stalk going gradually to the tip. Seeds become mature in the same pattern so you collect seeds from an individual flowerstalk over several weeks. In other words, seeds do not mature all at the same time.
In June (usually) start checking plants for seed development. The seedpods created from the flower heads swell, and you see green seeds growing. Check plants at least once a week.
Once the seeds turn blackish-brown, they are ready to be picked. If you wait too long, the seeds fall to the ground. Once they start maturing, it is best to collect seeds every few days.
You can collect the entire pod. This is the easiest way, and you are less likely to have seeds fall to the ground while picking. There are usually 4 seeds per pod. The most any pod can have is 4 seeds but sometimes not all of them develop.

"Scabrous (Rough) Comfrey seed ripens very unevenly, and is easily shattered, therefore picking is difficult."
-'Seminar on New Ensilage Forage Plants' by V.S. Sokolov and P.F. Medvedev, Botanical Institute V.L. Komarov, Academy of Sciences, Leningrad; Botanicheskiy Zhurnal (Botanical Journal), Moscow, Russia, No 9, pages 1404-1406, **1963**.

"Collecting the Comfrey seeds may present a minor problem. **Ordinarily they are gathered when nearing maturity or when the seed pods have opened.**
But Comfrey's seeds, those brownish-black nut-like kernels which form when the flowers shed their petals, are not easily available. They remain 'hidden', Comfreyites (Comfrey lovers) tell me, 'and **pop out when we're not ready for them'**. One may gently cover the flowering heads either when fresh or as the suspended herbs hang to dry, with cheesecloth or other cloth, or with plastic bags. Or spread out a newspaper under the flowering herb, suspended and drying, and you may collect a few falling seeds. An outdoor seed drop will mean a new plant which is easily transplanted to your favorite spot."
-Comfrey: What You Need to Know by Ben Charles Harris. New Canaan, Connecticut: Keats Publishing, Inc., **1982**, page 96.

"Many perennials produce seeds readily, often in papery capsules or pods. **Gather from plants with the best characteristics of the form to ensure good-quality seedings.** Seedheads can ripen quickly, so watch them closely and gather the seeds before they are dispersed. Choose a dry day to ensure that the seeds are not damp and at risk of rot. Always label bags of seeds when you gather them to avoid confusion later."
-Plant Propagation by the American Horticultural Society, edited by Alan Toogood. New York, New York: DK Publishing, Inc., **1999**, page 150.

"In Common Comfrey (Symphytum officinale), the flower buds and flowers in bloom point downward. The calyxes with the ripening fruits are located on the upper surface of the scorpioid cyme. The flowers are arranged in a double scorpioid cyme. Note hairs on the stem, leaves, pedicels, and sepals.
Flowering starts along the underside of the scorpioid cyme. The flowers that are the first to come into bloom are also the first to form ripe fruits.
If flowering takes place over a longer period of time, it can happen that the fruits that formed the earliest have already dispersed, while the fruits of the flowers along the middle of the scorpioid cyme are still present in the calyx and the flowers at the end of the scorpioid cyme are still flowering."
-A Manual for the Identification of Plant Seeds and Fruits by R.T.J. Cappers and R.M. Bekker, Netherlands: Barkhuis Publishing, **2013**, page 69.   (A cyme is an inflorescence in which each floral axis terminates in a single flower. An inflorescence is a group of flowers arranged on a stem that is composed of a main branch.)

Drying Comfrey Pods and Seeds

My experience with Comfrey: Some just-picked Comfrey pods may need a few days or a week to dry to make it easier for the seeds to be pulled out as the inside can be sticky.

The lower the humidity, the faster they dry. Do not dry in the sun. Dry in a place with good ventilation. You can stir once or twice a day to improve drying.

After removing from the pods, let the seeds dry for a month before you store them. (Or you can plant them right away.) Make sure they are thoroughly dry before putting them in an enclosed bag or jar because they might grow mold and you could lose all of your seeds.

I could not find any specific information about drying Comfrey seeds so the below paragraphs in this subsection are about seed drying in general.

*"**Seeds should go into storage as dry as possible.** Green seeds or seeds that have accumulated moisture will heat up when heaped in a pile. Moisture also speeds up the seed's metabolism, causing it to use up its stored nourishment too quickly.*
***Seeds of most vegetables may simply be spread on newspapers in a dry, well-ventilated place.** Change the papers once or twice if the seeds were damp at first."*
-The New Seed Starters Handbook by Nancy Bubel. Emmaus, Pennsylvania: Rodale Press, **1988**, page 202.

*"The crops usually need additional curing time before they are ready for processing. The seed must be dry and hard enough to withstand processing, and **the plant material it's attached to must be brittle enough to easily shatter and break away from the seed.** This time also gives seed that is still immature at harvest additional time to ripen.*
***Be sure to allow for enough air flow as the crop cures.** Depending on your scale and type of crop, this may entail such things as **placing your seed pods in a shallow container and stirring them daily** or hanging plants or seed heads upside down from the ceiling.*
***How to process seed: When your crop is dry enough, you can separate the seed from the non-seed material that it's mixed with.** This non-seed material may include leaves, stems, and pods from the seed crop, called chaff, as well as dirt, stones, and weed seeds.*
The first step in this process is to thresh the crop. Next, depending on how important it is that there is no non-seed material mixed with the seed, the crop is usually cleaned through cycles of winnowing and screening. Often, between threshing and cleaning, or between cycles of winnowing and screening, the seed is given time to dry further.
How to dry seed using ambient (room) conditions
***Dry cleaned seed in thin layers (around 1/4 inch thick = 0.6 cm) on plywood, window screen, sheet pans, or any hard, non-stick surface.** Stirring seed helps ensure even drying. Placing fans near seeds will also facilitate drying. Seeds can become damaged when the ambient temperature rises above 95 F (35 C) and dark colored seeds can become overheated when exposed to direct sunlight, especially at temperatures at or above 80 F (27 C). Therefore, take care to dry your seed in a location that is warm, but not too warm, and **dry them out of the sun** if necessary."*
-'A Seed Saving Guide for Gardeners and Farmers' by Organic Seed Alliance, Port Townsend, Washington, www.seedallliance.org, April **2010**.

Number Dry Seeds Per Gram

According to one Comfrey grower: After the Symphytum officinale seeds have dried thoroughly for a month, **100 seeds should weigh about 1 gram (.035 ounce).**

"Symphytum officinale: Average 300 seeds per gram."
-'Comfrey Seed Symphytum Officinale' by Bristol Botanicals Ltd: High grade botanicals to businesses, healthcare professionals and educational/research establishments, www.bristolbotanicals.co.uk, England, **2019**.

"Symphytum officinale: Seeds per gram: 150 seeds."
-'Comfrey Symphytum Officinale' by The Seed Collection: Providing gardeners with quality herb, vegetable and flower seeds, www.theseedcollection.com.au, Upper Ferntree Gully, Victoria, Australia, **2019**.

"Symphytum officinale and Symphytum x uplandicum: Seed density: 90-100 seeds/gram."
-'Richters® ProGrowers Info' by Richters: The Herb Specialists, www.richters.com, Goodwood, Ontario, Canada, 1997-**2019**. Includes market and culture of herbs.

"Symphytum officinale, True Comfrey: Average 90 seeds per gram."
-'Medicinal Herb Seeds Per Gram' compiled over the last 30 years by herbalist Richard A. (Richo) Cech, Strictly Medicinal® Seeds, https://strictlymedicinalseeds.com, Williams, Orgeon, **2019**.

Storing Seeds

*"**The genetic vigor of your saved seeds has already been determined by thier parentage.** Environmental conditions affecting the parent plant during seed formation- temperature, available moisture, weed competition, nutrient supply, and so forth- help to determine the physiological vigor of the seeds.*
*Although improvements can't be made, **the vigor of seeds can be drastically reduced by poor storage conditions.***

In fact, William Crocker and Lela Barton, in their excellent book 'Physiology of Seeds'*, go so far as to say that **storage conditions are more influential than age of seeds in determining viability- the ability of the seeds to germinate.**
Keep in mind that the stored seed is alive, with its life processes barely humming. Even in its dormant state, it reacts with its environment. Your aim when storing seeds, then, should be to keep their metabolism operating at the lowest possible level, to keep the seeds on 'hold'.

Heat urges the seeds into premature internal activity, which uses up their stored food supply. **Seeds stored at home at temperatures between 32 and 41 F (0 to 5 C), not freezing but cold enough to retard enzyme activity, usually keep well.** Ideally, temperature should fluctuate as little as possible. If you do want to try keeping some seeds in your freezer, be sure that they are good and dry. High moisture in frozen seeds will spoil them.

Moisture revs up seed metabolism. In some cases, it can be even more damaging then heat. Seeds that have gotten damp and then been redried suffer irreparable damage.

Seeds to be stored for long periods of time will fare best if moisture is kept around 4 to 6 percent but no lower than 1 to 2 percent, or the embryo may be damaged.

Stored seeds are safest in securely closed containers such as the following: cans with metal lids, screw-top glass jars or individual envelopes sealed in a large glass jar, plastic or metal film containers, vitamin bottles."
-The New Seed Starters Handbook by Nancy Bubel. Emmaus, Pennsylvania: Rodale Press, **1988**, page 204.
(* -'Physiology of Seeds: An Introduction to the Experimental Study of Seed and Germination Problems' by W. Crocker and Lela V. Barton. Massachusetts: Waltham Press, 1953.)

"Seeds must be stored in a cool, dry place; humidity and warmth cause seeds to deteriorate and die.
A good place to store seeds is in the refrigerator at 41 F (5 C). Place dry seeds in labeled paper packets in an airtight, plastic or glass container. A little dessicant, such as silica gel, placed in the container will remove excess moisture. Avoid opening the container unnecessarily."
-Plant Propagation by the American Horticultural Society, edited by Alan Toogood. New York, New York: DK Publishing, Inc., **1999**, page 150.

Propagation by Seeds

If you have experience growing Comfrey from seed and would like to add to the ideas presented here, please contact me.

Common Comfrey (Symphytum officinale), Prickly Comfrey (Symphytum asperum) and others create viable seeds (capable of sprouting) that store well for several years.
Planted seeds take about 3 years to reach the size of a normally planted root.

Russian Comfrey does not usually produce seeds, and in the rare occasion that it does they are usually sterile. Therefore, all propagation of Russian Comfrey is from root cuttings.

"Comfrey seed germination was rapid and best on peat in water or watered loam."
-'Studies on the Cultivation and Allantoin Content of Comfrey (Symphytum officinale)' by D. Fijalkowski and M. Seroczynska, Instytut Biologii Uniwersytetu (University Institute of Biology), Lublin, Poland; Herba Polonica (Polish Herbs), Volume 23, No. 1, pages 47-53, **1977**. (I was not able to get this report. If you have an English translation of it, could you please send it to me.)
(Peat is partially decayed vegetation or organic matter in areas called marshes, peatlands, bogs, moors or mires. Loam is soil of equal amounts of clay, sand and silt. It retains water and nutrients in the right amount.)

"Foraging Ants Help Comfrey Seeds Germinate:
The objective of this study was to quantify preferences of ants for seeds of different plant species and to test if these preferences were caused by foraging strategies or by historical constraints.
We compared seed removal rates of ten different ant-dispersed plant species found in temperate forests, along forest edges and in grassland. **We found that seeds with larger elaiosomes had significantly higher removal rates.** We found that plants of a certain habitat type were in general not preferred by ants of the same habitat type. These results demonstrate that seed preferences of ants are mainly determined by foraging strategies and not by historical constraints. **The most attractive seeds were those of Symphytum officinale L.,** the least attractive those of A. ursinum and Reseda lutea L.
Symphytum officinale:
Seed mass: 11.50 mg, Elaiosome mass: 1.74 mg, Mass ratio (elaiosome/seed): 0.152.
Seed volume: 127.8 mm3 (cubic millimeters), Eliosome volume: 6.4 mm3, Volume ratio (elaiosome/seed): 0.050."
-'Seed Dispersal by Ants: Are Seed Preferences Influenced by Foraging Strategies or Historical Constraints?' by M. Peters, R. Oberrath and K. Bohning-Gaese, Zoologisches Forschungsinstitut und Museum Alexander Koenig, Bonn, and Institut fur Zoologie/Tierphysiologie, Kopernikusstr, Aachen, and Institut fur Zoologie, Johannes-Gutenberg-Universität Mainz, Germany; Flora, Volume 198, pages 413-420, **2003**.
(Elaiosomes are fleshy structures attached to seeds of many plant species. It is rich in lipids and proteins. Elaiosomes attract ants, which take the seed to their nest and feed to their larvae. Then the ants take the seed to their waste area, which is rich in nutrients from ant frass {excrement} and dead bodies, and the seeds germinate. This type of seed dispersal is called myrmecochory.)

"Comfrey Seed Germination Requirements: Seeds germinate rapidly in moist soils of open sites. They germinate especially well in gardens, peat and loam. They are also able to germinate in water*."
-'Common Comfrey: Symphytum Officinale L.- Alaska Natural Heritage Program' by Timm Nawrocki, University of Alaska Anchorage, Alaska Center for Conservation Science, October **2010**.
(* -'Weeds of California and Other Western States', Volume 1 by J. DiTomaso and E. Healy, University of California, Agriculture and Natural Resources Communication Services, Oakland, California, 834 pages, 2007.)

Online translation from Czech to English:
*"**Laboratory emergence of Symphytum officinale seeds on germinators is not recommended.** It is a way of unnecessarily lengthy and inefficient, especially when you want to plant in October, when there is significantly less sunshine.*
The better way is to plant the Comfrey seeds in the germinates in the soil, where they will start to germinate more easily. Dormancy plays a major role in cultivation.
*In monitoring the seed of Comfrey in the soil, the seeds from the previous year of harvesting were seized. **Seeds planted in the same year a few months after harvest did not germinate.** The only exception was two plants of 72 seeds, which is negligible.*
It follows that I recommend using seeds from last year of harvest when growing from seeds.
The effect of outdoor Jarovization (vernalization) was confirmed. Comfrey stood in the soil waiting for a favorable period until spring, *when then began to grow. The effect in the laboratory environment was also confirmed. In the laboratory environment, the influence of dormancy was more apparent.*
*Plants, which germinated and grew in the same autumn (November), in winter stopped their development. After some time (after 5 months, April), they continued to grow in the spring, adding two new, germinating Comfrey plants (5 plants in total). **However, this is a small proportion of the sown seeds.**"*
-'Biology, Seed Characteristics and Growth Establishment of Comfrey (Symphytum Officinale L.)' by Jaroslava Kovarova, Master Thesis, University of South Bohemia in Ceske Budejovice, Department of Crop Production and Agroecology, Czechoslovakia, **2016**. All in Czech except abstract is also in English.
(If you have an English translation, I would appreciate a copy. The above is from an online translator.)
(Vernalization is influencing a plant's flowering process or seed germination by exposure to prolonged cold. After vernalization, plants have the ability to flower or germinate, but may require additional seasonal cues or time. The term is named after Professor Jarowitsch.)

*"**True Comfrey Seed:** Sow the seed just under the surface and tamp in securely.*
Sown directly in warm soils, germination usually occurs within 30 days.
A 30 day period of cold, moist refrigeration followed by planting in warm conditions will speed germination appreciably.
Grow the seedlings out in pots for about 3 months, then transplant to the garden.
You can also direct-seed into a fertile bed in the spring, as soon as the soil can be worked."
-'Comfrey, True (Symphytum officinale var patens)' by Richo Cech, Strictly Medicinal® Seeds, https://strictlymedicinalseeds.com, Williams, Orgeon, **2017**.

*"**Symphytum Seeds:** For best results, sow immediately onto a good soil-based compost. Cover the seeds with fine grit or compost to approximately their own depth.*
*They can be sown at any time, and **germination can sometimes be quicker if kept at 15 to 20 degrees C (59-68 F).***
We sow most seeds in an unheated greenhouse and wait for natural germination, as many seeds have built-in dormancy mechanisms, often waiting for natural spring germination, hence giving them a full season of growth."
-Plant World Seeds: 3500 Seed Varieties, www.plant-world-seeds.com, Newton Abbot, Devon, England, **2018**.
(I do not have any connection with this company.)

*"**Starting Comfrey from Seed:***
If choosing to start your Comfrey from seed, it is critical that you select 'True' or 'Common' Comfrey. These are the seeds which are capable of reproducing. The other variety of Comfrey, 'Russian Comfrey', is sterile, which means that the seeds do not grow.
Comfrey requires a winter 'chilling period' in order to germinate, and it is common to wait two years after sowing seed before seeing signs of germination."
-'Growing Comfrey' by AJ and Rebecca, Growing Organic: Whole World Organics approach to all areas of life, https://growingorganic.com, Delta, Colorado, April 9 **2018**.

*"**Comfrey Seed Germination:***
In Australia sow Comfrey seeds early autumn (March) or during spring (September to November).
Sow Comfrey seeds 5 mm (0.19 inch) deep, and spacing plantings about 50 cm (19.6 inch) apart to provide enough room for their large leaves to grow.
Comfrey can be slow to germinate, taking from 10 to 28 days to emerge. *Cold stratification can improve germinate rates. To do this mix the Comfrey seeds with drained damp sand in a zip lock plastic bag and refrigerate for a week prior to sowing."*
-'Comfrey Growing Advice' by Succeed Heirlooms: Quality, old-fashioned, open-pollinated heirloom vegetable seed varieties, www.succeedheirlooms.com.au, Brisbane, Queensland, Australia, **2018**.
(Australia is in the southern hemisphere so seasons are the opposite of the northern hemisphere such as the United

States and Europe. If it's summer in Australia, it is winter in the United States and Europe.)

"Sow 3 mm (0.11 inch) deep in late summer to autumn in cold frame. Sowing spring requires 6-8 weeks cold stratification. Pot up into individual pots, grow on for first winter before planting out after last frosts, in full sun to partial shade. Best grown in open sunny site in deep rich soil or compost material."
-'Comfrey Seed Symphytum Officinale' by Bristol Botanicals Ltd: High grade botanicals to businesses, healthcare professionals and educational/research establishments, www.bristolbotanicals.co.uk, England, **2019**.

*"**Symphytym Seed Germination Tips** for Symphytum officinale, Symphytum popovii, Symphytum tuberosum, Symphytum x uplandicum 'Axminster Gold': **Sow at 20 C (68 F)**. If seed does not germinate within 3 months, try 4 C (39 F) for 1 to 2 months, then 20 C again."*
-'Symphytum Popovii' by Ontario Rock Garden and Hardy Plant Society, **2019**. Focus on alpine plants, related hardy perennials, creation of rock gardens, suitable habitats for all hardy plants, and support for the planting and care of rock and hardy plants in botanic and public gardens. www.onrockgarden.com

Starting Seeds Indoors or in Greenhouse / Cold Frame

You can experiment with soaking some of your Comfrey seeds overnight, for 24-48 hours before sowing them. If soaking 48 hours, rinse the seeds after 24 hours so slime does not develop.

Germination is better when planted in a warm greenhouse, sunny house window or under a grow light. You can use a horticulture heat pad under a seed tray. You can even use a poultry incubator. **Adjust thermostat for 70 to 80 degrees F (21.1 to 26.6 C).** In good conditions germination should take place in about 10 days.

Sow the seeds 1/8 inch (0.317 cm) deep. Keep soil or seedling mixture moist but not soggy. You can use potting soil, seedling mixture, coconut coir and other materials. Peat alone is probably too acidic. Germination rates tend to be low.

Comfrey grows slowly at first so it is better to start indoors in trays or pots. Sow in individual cell flats (3" x 3" x 2") or (3" x 2" x 2") or (2" x 2" x 2") or (1" x 1" x 2") in seed starter trays. Or sow 2 to 3 seeds in a small pot. (1 inch = 2.5 cm, 2 inch = 5.0 cm, 3 inch = 7.6 cm.) **Grow until seedlings are at least 3 inches across or for a few months or more.**

You can start your indoor seeds late winter, early spring or in the summer, depending on how large you want your plants to be when you plant them outside after your last spring frost date. In much of the United States the last spring frost date is some time in March through May. Seeds can be started at any time as long as you have the facilities to give them the environment they need.

Then put outside to harden off. Hardening off is the process of preparing plants started from seed indoors for the change in environment they encounter when moved outdoors. It takes one to two weeks. It is a transitional period in which plants are left outside during daylight hours only, and where they are shaded and protected from wind and rain. They are then brought back indoors at night.

Plant seedlings at least 2 weeks after the last frost date in spring up until about September, depending on your climate. Plant about 8 inches (20 cm) apart. Transplant later to about 2-3 feet (0.6-0.9 meters) apart in all directions. Water well and keep moist until established but do not overwater. The ground should drain well. Soil that is constantly soggy will kill the seedlings.

Keep the seed trays or pots that have not germinated. Take the seeds and soil medium out of the trays/pots, and plant them outside for overwinter stratification. (Or stratify them in the refrigerator. See below.) Usually the seeds germinate the following spring. Mark where you planted them.

*"**Prickly Comfrey will grow from seeds**; but, after seeding, the plant will give very little produce that year. I have sowed seeds. **I have had the best success from sowing them in a wooden box, such as is used in stables for stopping horses' feet, a little mould being put on the top of the dung.**"*
-'The Gardeners Chronicle and Agricultural Gazette for 1871' published in Covent Garden, WC, London, England, 1269 pages, January 7 to December 30 **1871**. Article: 'Notices to Correspondents', January 21 1871, pages 59-60.

*"**Equal parts of vermiculite, milled sphagnum moss, and perlite combine to make a good seed starting medium.** This provides a spongy, friable (easily crumbled) seedbed that promotes good root development. Perlite, despite its plastic appearance, is a natural product, a form of 'popped' volcanic ash, and while I do not like to use it alone, as I do vermiculite in some cases, it promotes good drainage in seed-starting mixes. The moss must be milled sphagnum, which is very fine, not peat moss, which is too coarse for small seeds and tends to dry and crust."*
-The New Seed Starters Handbook by Nancy Bubel. Emmaus, Pennsylvania: Rodale Press, **1988**, page 11.
(Horticultural or agricultural vermiculite is a natural mica mineral. It increases the water holding capacity and air space of soil.)

"*Comfrey Culture: Comfrey is relatively easy to germinate, just keep the seed bed moist and warm (e.g. with the seed box placed above an electric heating outlet) and 15 to 20% should germinate in 8 to 9 days.*
Comfrey is a sporadic germinator, meaning that seedlings appear here and there over a 4-6 week period and not in one flush of growth as is the case with many other herbs and vegetables. *Total germination is rarely more than 50%.*
Fungus: It is worth noting that if you sow seeds outside in garden soil the pathogenic fungi are usually kept in check in healthy, balanced soils by other, harmless micro-organisms and your seedlings should be fine.
However, often *it is better to start seedlings in containers in sterilized soil because medicinal herbs tend to be slow to establish* and in the garden they can easily be overwhelmed by competing weeds."
-'Coltsfoot, Comfrey, Ginseng and Goldeseal Culture' by Inge Poot, Richters® Herbs, www.richters.com, Goodwood, Ontario, Canada, January 9 **2002**.

"*Common comfrey (Symphytum officinale) Seedlings:* Within a relatively short time I had achieved about 30% to 40% successful germination, and they were doing very well. However, this afternoon I noticed that most of them appeared to be drooping, but on closer inspection I found that the lower section of the stems had a 'pinched' appearance.
Answer- Damping Off: This sounds like typical damp-off. To avoid the infection in the first place, use sterilized pots (soak in diluted household bleach) and a soil-less mix such as Pro-Mix as a sowing medium.
Before sowing, wet the medium with boiling water and sow as soon as cool. Cover with clean plastic, held off the soil by some sterile sticks -or recycle some lidded clear plastic food container as a seeding pot.
Try to rescue any seedlings still alive in your planting by spraying the soil with 3-5% hydrogen peroxide, sprinkling with powdered cinnamon, or spraying with strong chamomile tea, or even 70% isopropyl (rubbing) alcohol."
-'Elecampane Tolerance to Wind, and Comfrey Germination' by Inge Poot, Richters®: The Herb Specialists, www.richters.com, Goodwood, Ontario, Canada, February 28 **2006**.

"*Starting seeds inside:* Your first task is choosing a starting medium. *As a general rule, start seeds for perennials in a compost-based soil mix. If you are concerned about their getting a damping-off fungus, many perennial seedlings can be started successfully in a cold frame.*"
-The Plant Propagator's Bible: A Step-by-Step Guide to Propagating Every Plant in Your Garden by Miranda Smith. Emmaus, Pennsylvania: Rodale Inc., **2007**, page 24.
(Damping-off is a fatal fungal disease of seedlings. It is most common when seeds are germinated in cool, wet soil with poor drainage, and there is low light with little air circulation. Crowded seedlings encourage damping off. Seedlings become infected at or just below the soil line. Soil temperatures below 68 F or 20 C before seed germinates increases chances of damping off.)
(A cold frame is a transparent-roofed enclosure, built low to the ground, used to protect plants from weather that is too cold or wet. The top admits the sun and prevents heat escape. It functions as a miniature greenhouse.)

"*Moisture, Heat and Germination:* I tried various methods to germinate Comfrey seeds, and it seems as if the seeds are impartial to (don't care) the media in which they're sowed on for germination.
However, *the trick seemed to be keeping them constantly moist,* which also makes them susceptible to fungus growing on the seed case, and in which case they won't sprout.
And also to keep the germinating temperature as hot as possible. I place the covered container under full afternoon sun for about four hours a day."
-'Growing Conditions: Comfrey' by Skyfiery or Casey, Asea Aranion: Healing and Death; Gardens, plants and seeds, https://aseaaranion.wordpress.com, Singapore, March 7 **2011**.

"*Comfrey seed germination question* from Jonah:
'I sowed Comfrey seed I bought this spring, and it has been three weeks and so far only one seed germinated. *I soaked the seed for 24 hours and sowed them in a standard peat/humus/perlite potting mix.* The seed was sowed and covered 1/8th of an inch (0.3 cm). I kept them moist and my apartment temperature is about 20 degrees C (68 F). I covered them with another seed plug tray. Do you have any suggestions? Perhaps they need higher humidity, and I cover them with a plastic dome? More heat with a heating pad?'
Answer: You are doing well, Jonah, and with a bit more time, you should see more germination of the Comfrey seeds. *Sometimes, germination takes as long as 30 days.*
I would recommend that you not cover the seed tray with another seed tray, as there should be some exposure to light."
-'Comfrey Seed Germination' by Ginny Cotterill, Richters Herbs, www.richters.com, Goodwood, Ontario, Canada, Oct 22 **2012**.

"*Seeds are prepared for seeding by soaking them in lukewarm herbal tea, milk or whey for 3 to 48 hours.*
Herbs, such as yarrow, chamomile, thyme, calendula, **Comfrey**, valeriana, oak bark and nettle are used. This procedure is claimed to shorten germination time by 1-2 weeks and to fortify germinating plants.
The small greenhouse has wood heated warm beds that are used for preparing seedlings, which are planted in the compost produced from the vegetable garden.
To prevent fungal infections, the compost is sterilized at 80 C (176 F) in an oven. Nettle, garlic and horsetail preparation are also used to water the seedlings and prevent infection.
Seedlings are planted outside after 6-8 weeks of raising and fortified with a diluted nettle preparation.

Seeding time for plants is based on the biodynamic moon calendar and weather conditions."
-'Agroecological System Analysis of the Krishna Valley in Hungary' by Katalin Rethy, Master of ScienceThesis for Organic Agriculture and Agroecology, Farming Systems Ecology Group, Wageningen University, Wageningen, The Netherlands, **2013-2014**.   (Whey is the watery part of milk that remains after the formation of curds.)

"**Comfrey Germination:** *Success will be around 20% in 8-9 days. Keep the soil moist and warm.* **Sow spring or autumn in a cold frame.** *When they are large enough to handle, prick the seedlings out into individual pots and* **grow them on in the greenhouse for their first winter.**
Plant them out into their permanent positions in late spring or early summer, after the last expected frosts."
-'Comfrey Symphytum Officinale' by Rory Turnbull, 'Permies: Homesteading and Permaculture All the Time', https://permies.com by Paul Wheaton, Missoula, Montana, **2014**.

"*Sow the Comfrey seed just under the surface and tamp in securely.* **Keep warm and moist until germination, which takes approximately 10 days in standard greenhouse culture.**
Germination may take as long as 30 days if you do not have a greenhouse."
-'Permies: Homesteading and Permaculture All the Time', https://permies.com by Paul Wheaton, Missoula, Montana, from a post 'Getting Comfrey Seeds to Sprout', quoting Horizon Herbs®, www.horizonherbs.com, **2016**.

"*Alternatively you can sow Comfrey seeds in pots or trays using a good quality compost such as John Innes 'Seed and Potting'* and **place under a protective environment such as a greenhouse. Germination will be considerable quicker, usually from 10 days onwards.**"
-'How to Grow Comfrey from Seed' by Simon Eade, Garden of Eaden: For Growing Plants, Beautiful Gardens and Propagation; UK's No. 1 Gardening Blog, https://gardenofeaden.blogspot.com, March **2017**.
  (John Innes compost is four different formulas of growing media/medium developed in the 1930s at the John Innes
   Horticultural Institution, Norwich, Norfolk, England.
   Sowing of seeds and rooting cuttings. No. 1: Sowing of large seeds and pricking out.
   No. 2: Potting up and potting on. No. 3: Final potting.)

A successful Comfrey seed germination: One person put his dry seeds in the freezer for 1 month. Then soaked his seeds in water for 48 hours at 65 F (18.3 C). Then he put them on top of a moist paper towel on a dish and covered it with plastic wrap. He put the dish on top of his water heater that was warm, but not hot. He kept the paper towel moist.
In 8 days, 26 out of 35 Comfrey seeds germinated. -**2017**.

"**Starting Comfrey seed indoors (recommended):** **Sow in medium individual cell flats or two seeds to a small pot.**
Grow on until plants are at least 3 inches (7.6 cm) across when they should be placed outside to harden off.
  Transplant to well prepared bed placing plants at least 2 feet (0.6 meter) apart with three feet (0.9 meter) between rows.
  Comfrey plants can be located slightly closer together if grown as ground cover but they will most likely fill the space at this distance in about three years. Water well and keep moist until established.
For those who want all the seeds to germinate. Keep the seed tray or pots that have not germinated. *When transplanting plants dig these 'empty' pots into areas between the plants that you would like to fill and leave them for overwinter stratification.* **In most cases the seeds will germinate for the following spring. Mark where you planted them so they don't get weeded out by mistake.**"
-'Comfrey, Symphytum Officinale: How to Grow' by Floral Encounters: Direct from Our Organic Farm, www.floralencounters.com, Hightstown, New Jersey, **2018**.

"**Comfrey (Symphytum officinale) Seed: Sowing rate:** *600-900 grams / hectare (indirect). Or 50-70 grams / 1000 plants.*
Planting: *Although Comfrey seed can be sown direct, the high cost of seed dictates* **indirect sowing in plugs or seedbed and transplanting to field.** *Sow Symphytum officinale in plugs or seedflats 12-14 weeks before sale.*
Seeds to saleable plants: *12 weeks.*"
-'Richters® ProGrowers Info' by Richters: The Herb Specialists, www.richters.com, Goodwood, Ontario, Canada, 1997-**2019**. Includes market and culture of herbs.
  (Hectare is a square with 100 meter sides, or 10,000 m2. There are 100 hectares in one square kilometer. An acre is
   0.405 hectare, and one hectare is 2.47 acres.)

"**Comfrey (Symphytum officinale) Seed: Preculture from March to April or sow outdoors in autumn or early spring.**
The seeds may require a cold period to germinate. Sow 0.5 cm (0.19 inch) deep.
Germination should occur within 3-4 weeks. If the seeds do not germinate, store them in the fridge for 2-4 weeks and try again."
-'Benific Herbs' by Wiebke Rost; Pflanzenkunst: Garden and Art Blog, https://pflanzenkunst.wordpress.com, Germany, **2019**.
  (Preculture is the opposite of direct sowing of seeds. In preculture, seeds are placed in a culture dish to achieve pre-
   germination. The seeds are kept in the culture until seedlings appear. These are then planted.)

Direct Sowing of Seeds in Fall
You can direct sow in the fall so the seeds are naturally stratified over the winter.

"The Symphytum officinale seeds may be obtained from wild plants. It flowers from June to August, and the brown, shining, and somewhat wrinkled nutlets (seeds), which are present in great numbers, may be gathered in September or later.
These seeds may be planted at once, that is, in early fall, carried through the winter in a coldframe, and transplanted to their permanent position in the spring.
The seeds may be carried over until spring and sowed in their permanent place after all danger of frost is over, in beds of finely pulverized soil, covered with light soil, and firmed; then thinned out when necessary.
The plants will be firmly rooted before cold weather, and the roots may be harvested the succeeding year."
-'National Association of Retail Druggists' with article 'The Cultivation of Medicinal Plants: Ideal Conditions for Comfrey, Stoneroot, Leptandra', Chicago, Illinois, Volume 23, November 9 **1916**.

Direct Sowing of Seeds in Spring

Spring sowing of Comfrey seeds after last expected frost is somewhat less successful than fall sowing. It is best if soil temperature is between 68-80 F (20 to 26.6 C) degrees but should be above 40 degrees F (4.4 C). Sow the seeds 1/8 inch (0.317 cm) deep and 1 inch (2.54 cm) apart. Pat the soil around it lightly. Keep the soil damp but not wet. These take 20-30 days to germinate.

Comfrey plants like full sun to partial shade in well drained soil. They prefer loam soil but sandy is OK if it is kept moist. If you want more Symphytum officinale (Common Comfrey) to grow from your plants, then let the plants grow seed so it spreads. Or you can cut off the flower stalks before they go to seed if you do not want them to spread.

*"**Comfrey seed is easily grown by direct sowing outdoors from March to June.** You will need a large, prepared seed bed in a sunny position away from smaller plants. For best performance the soils should be moist, fertile and well drained.*
Sow Comfrey seed thinly at a depth of 1 cm (0.39 inch) in rows 30 cm (11.8 inch) apart.
Germination is variable, but keep the seedbed moist, and you can expect the seedlings to emerge anywhere from 4-8 weeks.
When large enough to handle, thin seedlings out to just 1 plant every 60 cm (23.6 inch).
Water as needed during dry periods, and keep the bed weed free. Enrich the soil regularly with well-rotted farm compost or garden compost."
-'How to Grow Comfrey from Seed' by Simon Eade, Garden of Eaden: For Growing Plants, Beautiful Gardens and Propagation; UK's No. 1 Gardening Blog, https://gardenofeaden.blogspot.com, March **2017**.

*"**True Comfrey, Symphytum officinale, seeds are best planted in very early spring in outdoor conditions,** or given a good 30 days moist refrigeration before planting in the greenhouse or receptive garden bed. Seedlings grow fast and in time send down a good taproot. You really have to grow them for 2 years before adequate root yield can be obtained.*
It always makes sense to plant perennials in pots and then transplant out. It is more saving of the seed and gives better control and better overall results. If you don't have a reasonable place for growing plants in pots (sunroom, greenhouse) then direct seeding in a clean seedbed is also a workable choice."
-'Growing True Comfrey from Seed' by Richo Cech, Richo's Blog, Strictly Medicinal® Seeds, https://blog.strictlymedicinalseeds.com, Williams, Oregon, 2017.

*"**Comfrey seed germination can be a little erratic.** Some seeds need to have cold stratification (cold moist period) before they will germinate while most of the seeds seem happy with just a period of cold. All our seeds have been cold treated but not stratified.*
Germination rates for spring sowing tend to be around 80% for such seeds.
Comfrey often grows fairly slowly to begin with so for these reasons we recommend starting indoors in pots, but direct seeding is an option."
-'Comfrey, Symphytum Officinale: How to Grow' by Floral Encounters: Direct from Our Organic Farm, www.floralencounters.com, Hightstown, New Jersey, **2018**.

Cold, Moist Stratification (Optional)

Stratification means treating seeds to simulate or imitate natural conditions that some species of seeds must experience before germination occurs. It is an embryonic dormancy phase where these species will not sprout until the dormancy is broken.
Comfrey seed does not require stratification to sprout, however, in many cases you get a higher rate of germination if you stratify your seeds.

Stratification is a cold and moist period of 1-2 months or more. It mimics the conditions of winter in northern climates.

Put seeds in a 'medium' such as moist sand, vermiculite, coir, potting soil, soil-less potting mix, or paper towel. A paper towel is least preferred because it does not contain growth hormones to encourage germination. However, it does work. Peat moss alone is too acidic. Seeds can be mixed together with a small amount of 'medium'. Or you can put 1 to 2 seeds each in a starter cube/plug.

Put the 'medium' or the cubes/plugs in a baggie, cloth bag, plastic container or glass jar with lid. It should be damp but not soggy. Moisture must be added. If mold forms, it means it has too much water.
Kelp has naturally occurring growth hormones that help seeds germinate. Add a little to the germination medium. It is also good for seedlings to prevent damping off.

Put the bag or jar in your refrigerator (usually 40 degrees F = 4.4 C). Do not put them in your freezer when moist. Keep there for 30-60 days with 30 days probably being the best. After 3-4 weeks, check seeds to see if they have tiny white buds. After stratifying, transfer the loose seeds to starter cubes/plugs or outside.

"Germinate Comfrey Seed:
Germination depends on how fresh the seed is, and if they were fertilized. Comfrey has a difficult flower to fertilize, and they normally have low germination even for professionals. If you are direct seeding, then expect low germination.
In my own testing, I start them inside using cold stratification. I soak them in water overnight and then put them into a large ziplock plastic bag filled 25% Milled Peat Moss (anti-fungal medium).
In about 4 weeks I check the seeds for germination, you'll see little white nubs on the ones that are ready to plant. The others can be put back in for another 2 weeks.
After that I soak them in near boiling water and then repeat. The last ones rarely germinate but sometimes I might get 5%. **By doing it this way versus direct seeding I get over 40% germination.**
Another idea: **You could also try to put them in warm to hot water for a day or so after the stratification process."**
-'Need Help Getting Comfrey to Germinate from Seed' by The Survival Podcast Forum, The Survival Podcast: Dedicated to modern survivalism, sustainability, alternative energy and thriving in changing economic times, https://thesurvivalpodcast.com, June **2015**.

"**Is cold stratifying worth it for Comfrey?** For many plants it is essential or they just won't germinate. For Comfrey, we don't think so. **We have consistently germinated about 80% of our seeds without it,** for us that is enough.
For every seed there are always some that don't germinate no matter what. **If you are looking for higher rates, then it might be worth your while** but it takes time and patience that many people just don't have."
-'Comfrey, Symphytum Officinale: How to Grow' by Floral Encounters: Direct from Our Organic Farm, www.floralencounters.com, Hightstown, New Jersey, **2018**.

Propagation by Stem Cuttings

You can create a new Comfrey plant by cutting off a medium-sized leaf with the stem and putting it in some potting soil. Keep it moist and in a few weeks roots will develop. Then leaves grow from the base of the stem. You may want to experiment by putting a plastic bag over it to create a mini-greenhouse.

Or cut off a section of stem and cover all of it with a few inches of soil or compost. In a few weeks roots appear. **This works better with solid stems rather than hollow stems.**

" 'The Solid-Stem Symphytum Asperrimum or Caucasian Prickly Comfrey' by Henry Doubleday, 1876:
'This **Solid Stem Prickly Comfrey** branches out much more than the old varieties, so that though the plants be placed three feet (0.9 meter) from each other they soon cover the ground.
The crowns and stem cuttings blossom the first year, but as a rule the root cuttings generally not till the second year, but the latter produce an amazing quantity of leaves forming a head of great beauty from the graceful wavy curve of its long leaves. No other cultivated plant produces the enormous weight per acre of such valuable food.' "
-'Forage Plants and Their Economic Conservation by the New System of Ensilage: Part I: Caucasian Prickly Comfrey' by Thomas Christy, Jun., F.L.S. (Fellow of the Linnean Society), Christy & Co., London, England, page 11, **1877**.

"**There are two varieties of the Comfrey plant, one with a hollow and the other with a solid stem.**
The latter is an excellent food for stock of all kinds; especially does it increase the quantity and improve the quality of cows' milk. It grows with marvelous rapidity and luxuriance. Land which yields eight tons of grass per acre gives from sixty to a hundred and fifty tons of Comfrey. The plant is four or five feet (1.2-1.5 meters) high when near flowering, and the leaves attain a length of three feet (0.9 meters). The flowers abound in honey.
The solid stem is like a succulent root, and the plant is easily propagated by cuttings from this stem, containing a couple of eyes each. When once well rooted, it will go on producing from fifteen to twenty years.
The fodder may be cut six or even eight times a year; and if the leaves are stacked green, or partially dried, with a little salt between the layers, they keep well through the winter."
-'Popular Miscellany: A Drought-Proof Fodder-Plant', Popular Science Monthly (founded May 1872 in the United States to give

scientific knowledge to educated layman), Volume 13, page 763, October **1878**.

"*In certain plants of Prickly Comfrey: Symphytum asperrimum the root and **flower stems grew almost solid- so much so that cuttings from the flower stems could be propagated like geraniums**.*"
-'Ensilage: A System for the Preservation in Pits of Forage Plants and Grasses, Independent of Weather: A Collection of Facts and Statistics on the Cheapest Mode of Providing Winter Food for Dairy Cattle, Sheep, Horses, Etc.', by Thomas Christy, F.L.S. (Fellow of Linneaen Society), London, England, **1883**. Caucasian Prickly Comfrey, pages 57-63.
 (How to propagate geraniums: Cut the stem just above a node which is a swollen part. On the cut piece, make another cut just below a node, so the length from the leafy tip to the node is 4 to 6 inches. Remove all leaves except those on the tip. Put the cutting in a container of damp, sterile potting soil with the leaves above the soil. Water and place in a bright location out of direct sun. Water when the soil feels dry. After a week or two, it should have started to form roots.)

"***Rooting Media for Stem Cuttings:***
Materials into which cuttings are inserted must give them support and be sterile, water-retentive, and well aerated: mixtures of peat and fine grit, perlite, vermiculite, or sand are among the most popular.
Protecting Cuttings:
Cuttings taken from the topgrowth of perennials are usually soft or semi-ripe, and it is essential that their tissues remain turgid (well supplied with water). In dry air or in wind, water will be lost from stem and leaf surfaces and the cutting will rapidly wilt, so a sheltered, humid growing environment is essential.
Taking Cuttings from Stems:
Stem, stem-tip, and basal stem cuttings can all be used to propagate perennials. Take the material where possible from the younger, more vigorous shoots. Nonflowering shoots are always preferable. The softer the growth, the faster it will root, but the more vulnerable the cutting will be to pests and diseases and adverse conditions.
With nearly all plants, the lower cut is made just below a leaf joint, where nautral growth hormones (auxins) are more active in the initiation of roots. *A hormone rooting powder or gel helps; most plants root well but more slowly without it.*"
-Plant Propagation by the American Horticultural Society, edited by Alan Toogood. New York, New York: DK Publishing, Inc., **1999**, page 154.

Propagation by Root Cuttings

Overview

Power-tilling in spring or summer can be used if you want a fast way to spread Comfrey. Any place where a root is cut off from the original plant, a new plant will form. The root can stay in its original position or the tiller might move it to a new location. Any roots left on top of the soil will dehydrate and die. So you need to replant exposed roots right away. Do not power-till in winter.
(A power tiller is a set of 4 blades or tines mounted on 2 wheels. It is powered by gasoline or electricty. Each blade alternates in opposite directions as it digs up the soil.)

Comfrey patch expansion also works by using a shovel going down only vertically to break roots in a large plant. You don't have to move any roots; you just need to cut them off from the main plant.
These methods are only useful if you have enough room in your patch for new plants to grow.
It is a good way to revitalize an old Comfrey patch. See subsection 'Older Comfrey Plots and How to Restore Them' in section 'Productivity and Farm Economics of Comfrey' (Chapter 36).

"***Agricultural Comfreys have always been increased by division or root cuttings.*** *No other farm crop is propagated vegetatively (grafted fruit trees do not concern the general farm), and the tubers of the potato are about the most costly 'seed' per acre on the farm.*
This Comfrey patch should be prepared by digging thoroughly to remove the roots of all perennial weeds, *especially couch grass, creeping thistle, dock, creeping buttercup, dandelions, and convolvulus (bindweed, morning glory).*"
-Russian Comfrey: A Hundred Tons an Acre of Stock or Compost for Farm, Garden or Smallholding by Lawrence D. Hills. London England: Faber and Faber, Limited, **1953**, pages 21, 118.

"*The ideal agricultural Comfrey plant is one whose flower-stems when cut are solid, without a hollow in the centre, and that has consistently given heavy cuts.* **Select a good Comfrey plant or two at the end of the season (autumn) when they are in full growth. It should be labeled so that it can be identified when you dig up the plant in March.**
Get as much root as possible. It will be necessary to go down at each side as though lifting a fruit tree, to get the main taproots out. It will resemble a giant and very fangy dock (plant)."
-Russian Comfrey: A Hundred Tons an Acre of Stock or Compost for Farm, Garden or Smallholding by Lawrence D. Hills. London England: Faber and Faber, Limited, **1953**, page 124.
 (Dock roots are large, yellow with a forking taproot. Comfrey root is not yellow.)

"*The Reverend E. Highton of Bude (Cornwall County, England) bought 200 plants from Messrs. Sutton in 1875 (they sold it*

between 1875 and 1896) carried out the directions on the packet to the letter, and cut at the rate of 60 tons an acre the year of the planting.
He then dug his plants up and kept increasing them till he had a quarter of an acre, maintaining his yield all the time.
He reports that house cows and pigs throve on it; by feeding it ad lib to his horses. He reduced the oats ration from 6 to 3 quarts (5.6 to 2.8 liters) a day without loss of condition."
-Russian Comfrey: A Hundred Tons an Acre of Stock or Compost for Farm, Garden or Smallholding by Lawrence D. Hills. London England: Faber and Faber, Limited, **1953**, page 34.   (Ad lib means as much as they want.)

From "Catalogue of Messrs. Suttons" of Reading, Berkshire, England in 1878:
"Comfrey is cultivated by dividing the roots, and spring and autumn are the best seasons for planting. Holes should be dug 24 to 30 inches (0.6-0.76 meters) apart each way and filled with well-rotted stable dung.
The cuttings should then be deposited and covered over with earth, leaving the crowns 1 1/2 to 2 inches (3.8-5.0 cm) under ground. It is very important to keep the ground clean and free from weeds."
-Russian Comfrey: A Hundred Tons an Acre of Stock or Compost for Farm, Garden or Smallholding by Lawrence D. Hills. London England: Faber and Faber, Limited, **1953**, page 33.

Rapid Increase in Area of Land Planted with Comfrey

"To multiply your Comfrey plants, chop through the plants about 3 inches (7.6 cm) below ground level between April and September, when they are growing well, and replant the cutoff portion.
The is the best garden method of propagation, for **after the third spring the left-in roots will grow rapidly, making it possible to multiply the size of the bed by five times."**
-Fertility Without Fertilizers: A Basic Approach to Organic Gardening by Lawrence D. Hills. New York: Universe Books, **1975**.

"It is possible to identify clones by flower colour and habit, and with adequate yield recording, like milk recording dairy cows, **it would be possible to multiply a single plant to thousands in the course of a few years.**
These thousands would need re-recording to make certain that the particular plant was not a product of a specially favoured position or soil condition."
-Comfrey: Fodder, Food and Remedy by Lawrence D. Hills. New York: Rizzoli Universe Books, **1976**, page 70.

"Over the past few growing seasons, we have multiplied our number of Comfrey plants several times over, but have not had to do so by ordering more from an online nursery.
> Fortunately for us and the land, the prolific taproots of this plant not only dynamically accumulate nutrients from the subsoil, but also are easy to dig up, cut into 1 1/2 to 2 inch (3.8-5.0 cm) sections, and then plant into the ground elswhere, becoming new Comfrey plants themselves!

The number of root cuttings you are able to divide off of the dug up roots will depend upon how loose and workable your soil is around your Comfrey plant.
In a well-developed veggie garden, we divided a couple Comfrey plant root systems into well over 100 root cuttings, which we then planted by our fruit and nut trees in our tree belts, in our hugelkulture beds, and wherever else we wanted to condition soil, build organic matter, provide food for bees, feed livestock, and have a ready source of herbal medicine at the ready."
-'Divide and Conquer: Multiplying Comfrey by Root Cuttings, Episode #029' by Grant Shadden; ABC Acres: A Guest Ranch Experience, https://abcacres.com, Tim and Sarah, Hamilton, Montana, **2018**.
> (Hugelkulture is creating a small mound or hill from decaying wood and other biomass plant material. This is used as a raised bed for garden plants.)

"From one half of a year old plant, I've obtained 42 offsets (roots), and a year later the original plant is as good as those you've not split."
-'How to Propagate Comfrey: Propagating Bocking 14 Comfrey' by John Harrison, Allotment and Gardens: Grow Your Own - Allotment - Gardening Help, https://www.allotment-garden.org, Penygroes, Gwynedd, Wales, 2018.

Plant Hormones and Comfrey Root Cuttings

"Methods are described for the in vitro propagation of Symphytum x uplandicum Nyman from bud, root and stem explants. Highest shoot numbers were produced from root explants > 4 mm (0.157 inch) in diameter, cultured vertically with their distal cut surface on the medium.
The most suitable medium for shoot production was Murashige and Skoog's (MS) with 0.3 mg l-1 6-benzylaminopurine (BAP). These shoots developed roots on MS medium without hormones and were successfully transplanted into pots.
Subculturing shoots onto MS medium containing BAP, kinetin (K), 6-alpha, alpha-(dimethylallylamino)-purine or gibberellic acid (GA3) at 0.5, 1.0, 5.0, 10.0 or 30.0 mg l-1 failed to stimulate outgrowth of axillary buds in culture.
In vitro propagation from root explants was also achieved with Symphytum asperum Lepech., S. officinale L., and S. x uplandicum cultivars 'Bocking 1', 'Bocking 2', 'Bocking 4' and 'Bocking 17', but not with S. bohemicum, S. grandiflorum

DC, S. tuberosum L., or S. x uplandicum 'Bocking 7' and 'Variegatum'."
-'In Vitro Propagation of Symphytum Species' by P.J.C.Harris, C.G.Grove and A.J.Havard, Henry Doubleday Research Association, Coventry, Great Britain, and Department of Biological Sciences, Coventry Polytechnic, Great Britain; Scientia Horticulturae, Volume 40, Issue 4, pages 275-281, November **1989**.

>(In Vitro means made to occur in a laboratory vessel or other controlled experimental environment rather than within a living organism.)
>
>(Explant means to take living material from an animal or plant and place it in a culture medium. Distal means situated away from the point of origin or attachment. The axillary or lateral bud is an embryonic shoot located in the axil of a leaf. Axil is the angle between the upper side of a leaf or stem and the supporting stem or branch.)
>
>(Murashige and Skoog medium is a plant growth medium used in the laboratories for cultivation of plant cell culture.)
>
>(Benzylaminopurine is a phytohormone that increases plant growth, development, setting blossoms and stimulating fruit by enhancing cell division.)

"Callus derived from Symphytum officinale L. regenerants was cultured in the presence of various phytohormones. The growth rate of callus was stimulated by all phytohormones at various concentrations.

>With 1-naphthaleneacetic acid no organ differentiation could be observed. With indole-3-butyric acid at low concentrations only roots were formed, whereas 6-benzylaminopurine, kinetin and zeatin at various concentrations induced either root or shoot formation or the simultaneous regeneration of both.

Fructans: Minor amounts of fructans were formed at high 6-benzylaminopurine-, zeatin- and at all indole-3-acetic acid-concentrations. The concentration of 1-naphthaleneacetic acid had no influence on the fructan content. Highest rates of fructan synthesis occurred at low zeatin-concentrations up to 1.5 mg/liter. Only zeatin at all concentrations induced the synthesis of polyfructans, whereas appreciable amounts of oligofructans were formed under the influence of all other phytohormones."
-'The Influence of Phytohormones on Growth, Organ Differentiation and Fructan Production in Callus of Symphytum Officinale L.' by D. Haass, A.A. Abou-Mandour, W. Blaschek, G. Franz and F.C. Czygan, Institut fur Pharmazeutische Biologie, Regensburg and Wurzburg, Germany; Plant Cell Reports, Volume 10, Issue 8, pages 421-424, October **1991**.

>(Callus is a mass of unorganized plant parenchyma cells that cover a plant wound.)
>
>(Fructan is a polymer of fructose molecules. Fructose is a sugar in fruit and honey.)
>
>(See sub-subsection 'Fructans' in subsection 'Fat, Protein, Carbohydrate, Fiber, Ash, Miscelleaneous' in section 'Nutritional Value of Comfrey' {Chapter 19, Volume 1}.)

"With the objective to evaluating the influence of 3-Indolebutyric acid (IBA, a plant hormone to promote root growth) on the initial growth of Comfrey (Symphytum officinale L.) plants, a medicinal specie, an experiment was conducted in a greenhouse. Rhizome segments obtained from plants 90 days old were submitted to five IBA concentrations (0, 0.246, 0.492, 0.738, 0.984 mM {milliMolar}) and three incubation periods (6, 12 and 18 hours). The experimental design used was a randomized block, under a 5 x 3 x 3 factorial scheme.
After 42 days, total and partitioned dry matter in leaves, petioles, roots and rhizomes of Comfrey was evaluated. **No IBA effect was observed on dry matter of total and partitioned in leaves and petioles; however, the increase in IBA concentration promoted a gain of root and rhizome biomass being higher when 0.984 mM IBA treatment was used during 6 hours incubation."**
-'Influence of 3-Indolebutyric Acid on the Initial Growth of Comfrey (Symphytum Officinale L.)' by A.H.F. Castro and A.A. de Alvarenga, Departamento de Biologia / Setor de Fisiologia Vegetal da Universidade Federal De Lavras (Department of Biology / Department of Plant Physiology, Federal University of Lavras), Brazil; Ciencia e Agrotecnologia (Science and Agrotechnology), Volume 25, No. 1, pages 96-101, **2001**.

>(3-Indolebutyric acid {IBA} is a white to light-yellow crystalline solid, with the molecular formula $C_{12}H_{13}NO_2$. It is a plant hormone in the auxin family and is an ingredient in many commercial horticultural plant rooting products.)

Methods of Digging Root Cuttings

Definitions and Overview

A garden fork, spading fork or digging fork has a long handle and usually four short, sturdy tines/teeth. It is used to loosen, lift and turn soil.

A spade has a narrower and smaller blade than a shovel. A spade blade is flat or almost flat with a straight edge. The handle is shorter than a shovel and may be T or D shaped at the end. A spade is used to slice through roots and soil. A shovel is good for breaking up and turning soil.

A potato digger is similar to a plow. A simple one is a double or lifting plow with long steel tines at the back. The plow lifts the soil and potatoes, then pushes them upon the tines so the potatoes are on top of the ground.

A bed lifter or undercutter bar is a machine that is lowered down into the soil to lift root crops out of the soil. It loosens up the soil above it as the blade passes below. It is a mechanical means of replacing the pitchfork.

There are two types of cultivators. One type has teeth/shanks that pierce the soil as they are dragged through it in a straight line. The other type has rotary/circular moving disks/teeth such as a rotary tiller. Cultivators stir and break up soil, used either before planting or after the crop has begun growing.

"**Root-Set: This is a section cut off of a tap root,** and if placed in soil with the small end downwards will throw a large numer of heads, which do not always bloom the first year, but yield a large crop.
Crown-Set: This is taken from the root of the plant near the surface of the ground, and the smallest piece forms a crown-set that blooms at once."
-Russian Comfrey: A Hundred Tons an Acre of Stock or Compost for Farm, Garden or Smallholding by Lawrence D. Hills. London England: Faber and Faber, Limited, **1953**, page 125.

(A taproot is a large, central, and dominant root from which other roots sprout laterally. In Comfrey there is not just one main taproot. There are several main roots that divide into medium sized roots. The root rarely goes straight down.)
(The crown is where the plant stem meets the roots. There are great changes here in vascular structure, as the root system gets transformed to the shoot system. It is also called root crown, root collar and root neck. The top of the crown is level with the soil surface or just a little below it.)

"**The easiest method of vegetative propagation for perennials is by division.** It is the method most commonly used by gardeners for rejuvenating an old plant while providing extra plants, and commercially for propagating many garden perennials in large numbers. **Plants are divided in autumn or early spring,** when they are not in active growth.
If necessary, perennials can be divided at any time, except during hot, dry periods and freezing winter weather.
Early summer division of some perennials works well. All plants that are divided in summer should be watered thoroughly until they establish.
The secret of successful division at any time is always to have more root than shoot, to cut away excess foliage, and to keep the divisions moist and sheltered until established.
Divide plants with a spreading rootstock early in spring, just as the new growth is breaking. **Lift the plant with a fork,** inserting it well away from the crown to avoid damaging the roots.
Shake the roots free of loose soil. Divide the plant into smaller pieces by chopping through the woody center with a spade. **Pull the divisions into smaller pieces with your hands.**
The exposed roots should never be allowed to dry out. Replant the divided sections immediately, to the same depth as before or slightly deeper. Firm in lightly and water throughly."
-Plant Propagation by the American Horticultural Society, edited by Alan Toogood. New York, New York: DK Publishing, Inc., **1999**, page 148.

(In gardening, division is a method of asexual plant propagation. The plant is cut up into two or more pieces.)
(If you can not replant right away, place the roots in some soil until you are ready to plant.)

Garden Method of Digging Roots

"Select a good Comfrey plant or two at the end of the season (autumn) when they are in full growth. It should be labeled so that it can be identified when you dig up the plant in March.
Get as much root as possible. It will be necessary to go down at each side as though lifting a fruit tree, to get the main taproots out. It will resemble a giant and very fangy dock (plant)."
-Russian Comfrey: A Hundred Tons an Acre of Stock or Compost for Farm, Garden or Smallholding by Lawrence D. Hills. London England: Faber and Faber, Limited, **1953**, page 124.

"**In March or April, drive a sharp spade through the main root of each Comfrey plant about three inches (7.6 cm) below the surface, and lift off the severed crowns.**
Divide these with a sharp knife into sections, each with a growing point and a section of the thick, brown root."
-Comfrey: Fodder, Food and Remedy by Lawrence D. Hills. New York: Rizzoli Universe Books, **1976**, page 110.

"They have begun to dig up some big tough Comfrey plants that are growing in abundance here.
When the gardeners have dug up several plants, **they divide each clumped crown into many small units, cutting through each one with a knife, singling out pieces of crown and roots.**
They don't seem to care if the roots break or even if they get any of the new foliage that's starting to sprout out of the dormant crowns. While dividing the Comfrey, the two gardeners work very quickly and brusquely (roughly), **prying the plants up with their forks,** then cutting large and small crown sections, like so many pieces of a messy pie.
If there are some new leaves on a segment of a crown, they leave them, but they give them no special treatment."
-Enchanted Garden: Alan Chadwick's Organic Method of Gardening by Tom Cuthbertson. London, England: Rider & Company / Hutchinson & Co. Publishers Ltd, **1978**, page 70.

"**Comfrey Planting Material:**
You can plant out with crown divisions or root cuttings best done in the spring when the soil has warmed.
A crown division can be obtained from simply putting a spade through the center of a mature Comfrey plant and transplanting the divided sections. For our beds, I divided two year old plants into quarters, sometimes sixths, and these

established very well in the first year. It's best not to harvest the leaf biomass in the first year in order to allow a deep root system to develop. However if you use large divisions, you can start harvesting in July."
-'Comfrey: Its History, Uses and Benefits' by Paul Alfrey, Permaculture Magazine: Earth Care, People Care, Future Care; Hampshire, England, www.permaculture.co.uk, March 3 **2016**.

"Split a Comfrey plant into halves:
Choose a strong mature plant and start by loosening the soil to one side with a fork. **Insert a spade across the middle of the plant and cut straight down, splitting the plant in two.** *Dig up half and fill the hole left with compost or a compost and soil mix. The half of the plant left in the ground will recover and be as productive as the other plants in the Comfrey bed in a year's time.* **Take the lifted half of the plant and wash off the soil adhering to the roots.** *I found it easiest to use a plastic box and the spray nozzle on the hosepipe but a bucket of water would be fine.*
Split the half Comfrey plant into root and crown offsets:
I take the cleaned half plant into the potting shed for this part. You'll need a sharp knife for this part. **Start by cutting off the leaves leaving a very short stalk,** *about 1/2 inch or 1 cm. At this point the leaves will demand more water than the root can supply and will not feed the root. Removing them allows the root time to recover and develop leaves of a size it can support.* **Cut across the plant about 2-3 inches or 5 to 7.5 cm from the top.** *This will give you your crown offsets. Take the top slice of the plant and split into pieces using your knife."*
-'How to Propagate Comfrey: Propagating Bocking 14 Comfrey' by John Harrison, Allotment and Gardens: Grow Your Own - Allotment - Gardening Help, https://www.allotment-garden.org, Penygroes, Gwynedd, Wales, **2018**.

Farm Method of Digging Roots

"The farm method of increase of Comfrey is best carried out in March or April, as soon as the plants can be seen starting into growth. Take a beet-lifter along the rows, to cut the roots about six inches (15 cm) below the surface, first one side and then the other.
Then lift them from the soil by hand and cart them off to a shed or barn where they can be cut up, with sacks spread beneath the cutters. The best time to carry out the operation is in the third year; either the spring following the big cut from fully-established plants, or in the autumn about October, after a cut."
-Russian Comfrey: A Hundred Tons an Acre of Stock or Compost for Farm, Garden or Smallholding by Lawrence D. Hills. London England: Faber and Faber, Limited, **1953**, page 69.
(A beet-lifter or beet puller is a plow designed to gently plow up sugar beets.)
(When the crown is cut off, the rest of the roots remain in the soil undisturbed, so the plant recovers quickly.)

"A potato digger would be run along the Comfrey rows taking up the roots to about ten inches (25 cm) deep.
These roots would be cut up in 1 1/2 inch (3.8 cm) long sections, bagged and sent by road or rail to customers as easily as seed potatoes.
On a large scale, harrow the Comfrey field after the crowns have been removed, or spade back the soil over the cut surfaces on a small plot. *Then after six weeks new growing points will appear, and if the operation is carried out in early April, there will be a cut of Comfrey ready in July. The crowns will cut up to provide 4 to 8 new plants as originally bought. The original plot misses two cuts but the following year it will be as good as ever."*
-Comfrey: Fodder, Food and Remedy by Lawrence D. Hills. New York: Rizzoli Universe Books, **1976**, pages 110, 219.
(A harrow tool breaks up and smooths out the surface of the soil.)

"Comfrey roots should be harvested in the spring or in the fall when there is no aerial (above ground) growth.
The root can be dug by hand with a spading fork or mechanically using a modified potato digger or bed lifter.
Prior to digging roots, it is helpful to mow down the aerial tops and loosen the soil by running a chisel shank or cultivator up the row on either side of the Comfrey."
-The Organic Medicinal Herb Farmer: The Ultimate Guide to Producing High-Quality Herbs on a Market Scale by Jeff Carpenter and Melanie Carpenter. White River Junction, Vermont: Chelsea Green Publishing, **2015**, page 289.

Size of Root Cuttings (Offsets)

"The offsets are sections of root about three inches (7.6 cm) long, each with a growing point which should go just below the surface. Slice the roots into fragments, each consisting of a growing point and a length of the brown or black rootstock from three to six inches (7.6-15.2 cm) long. **If growth is advanced, the foliage should be trimmed back to about an inch (2.5 cm) of stalk;** *they are a handicap to the young crowns when first planted because of transpiration."*
-Russian Comfrey: A Hundred Tons an Acre of Stock or Compost for Farm, Garden or Smallholding by Lawrence D. Hills. London England: Faber and Faber, Limited, **1953**, pages 62, 69.
(An offset is part of the main plant that is removed and planted elsewhere.)
(Transpiration is when water is carried from roots to small pores on the underside of leaves, where it changes to vapor and is released to the air.)

"Digging Comfrey Roots:
Whole plants can be dug up with as much root as possible. ***First cut off the tops to make as many plants with growing points as possible, and then slice up the roots into 2 to 3 inch (5.0 - 7.6 cm) long sections.***
The root cutting bed will become full of broken root ends which will grow, and therefore those who wish to increase Russian Comfrey Bocking No. 4, No. 14 or any other variety should keep each to its own root cutting bed."
-Comfrey: Fodder, Food and Remedy by Lawrence D. Hills. New York: Rizzoli Universe Books, **1976**, pages 111, 112.

"For quick and good results we have established that an offset should not be less than 6-8 cm (2.37-3.15 inches) in length, by a diameter of not less than 1 cm (.39 inch) at the thinner end.
If the diameter is 2-3 cm (.78-1.18 inches), the length can be less, but small offsets take too long to begin to get their nutrition from the soil. They must wait until they have a root structure, and food reserves in small offsets are insufficient.
Commercially mature plants in the 3rd or 4th growing season are best for cutting offsets."
-Comfrey: Nature's Healing Herb and Health Food by Andrew Hughes. Japan: Sanyusha Publishing Co., Ltd, **1992**, page 153.

Care and Storing of Root Cuttings

If you need to temporarily store root cuttings in a bag or box, it is much better to have the roots too dry than too wet.
Roots that are too wet quickly rot. Roots that are too dry can last days or weeks depending on how dry it is and on the size of the root. Of course, planting sooner is always best.
For storage of several months, one method is putting whole plants, with the leaves/stalks cut off, in slightly moist sawdust and/or peat moss mixture at 35 to 40 F (1.6-4.4 C) degrees.
The ideal is a temperature-controlled storage room of the type used by farmers who grow vegetables and perennials. Or you could store them in a root cellar.
If you have experience with storing live Comfrey, let me know the method you use.

"Root sets, if in winter, place on the ground in some warm sheltered, temporary situation. Arrange so that every 100 sets shall cover an area of two feet (0.6 meter) long by two feet broad.
After so doing, cover the roots to a depth of 2 inches (5 cm), with rich garden loam soil, well manured; in a short time the sets will appear above ground, having thrown out a mass of white fibrous roots."
-'Forage Plants and Their Economic Conservation by the New System of Ensilage: Part I: Caucasian Prickly Comfrey' by Thomas Christy, Jun., F.L.S. (Fellow of the Linnean Society), Christy & Co., London, England, **1877**, page 24.

"The bare roots are, however, sensitive to frost and also should not be allowed to get dry and flabby.
If they arrive when they cannot be planted for some days, lay them along a shallow trench and cover with soil, or spread them out and heap straw or strawy dung over them. They should be planted 4,840 to the acre."
-Russian Comfrey: A Hundred Tons an Acre of Stock or Compost for Farm, Garden or Smallholding by Lawrence D. Hills. London England: Faber and Faber, Limited, **1953**, page 62.

"If the Comfrey root sets are not fresh and vigorous, keep them moist in a box of light compost in a closed frame until new growth is started, before planting them out."
-Goat Husbandry by David Mackenzie. London, England: Faber and Faber Ltd., **1967**, page 315.

"Do not let Comfrey roots dry out:
The essential is to get the offsets (root cuttings) with their growing points upwards and below the surface, for ***if the roots stand out of the ground they can dry out and die very easily- the only time the crop is really at risk."***
-Comfrey: Fodder, Food and Remedy by Lawrence D. Hills. New York: Rizzoli Universe Books, **1976**, page 105.

"Storing Root Cuttings (Temporary Bed): Take out a furrow two inches (5.0 cm) deep along a line, and lay the Comfrey root sections flat along it about an inch (2.5 cm) apart, *so their upper cut surfaces will grow new shoots while they root from their lower ends. There is no need to take trouble to keep the root cuttings the same way that they grew.*
Nursery Beds for Comfrey Roots:
Root cuttings planted in the spring should be ready for their permanent positions by the following spring, or by autumn if the bed has been well manured. The root cuttings rows should be a foot (30 cm) apart, and they will need at least one hoeing to keep the weeds down between the rows, before they are covered with deep-litter compost for maximum growth."
-Comfrey: Fodder, Food and Remedy by Lawrence D. Hills. New York: Rizzoli Universe Books, **1976**, pages 111, 112.
(This method can be used to store roots for days, weeks or months.)

Comfrey root is similar to carrots so I am using that storage information:
*"***Keep carrots cold and moist: 32 to 40 F (0 to 4.4 C) degrees, and 90 to 95 percent humidity. They keep 4-5 months.***
Cut off the tops (leaves and stalks). Our favorite carrot storage method is simple:
Spread an inch-thick (2.5 cm) layer of damp sawdust in a carton, box or can. Arrange the carrots side by side in this bedding. They can touch, but should be kept in a single layer, not piled on each other. Put a 1-inch thick layer of damp sawdust over the first course of carrots. Press another tier of lined-up carrots into that and continue to alternate carrots and sawdust until

the container is full, ending with a covering of damp sawdust. If your storage area is dryer than you would like it to be, cover the carton with damp newspaper. **If you have no sawdust, you can use damp sand, peat moss, wild moss or leaves.*"*
-Root Cellaring: Natural Cold Storage of Fruits and Vegetables by Mike and Nancy Bubel. North Adams, Massachusetts: Storey Publishing, **1991**, pages 61-62.

"Wilted Comfrey root cuttings should be soaked in cold water until they become firm before planting.
Root cuttings develop buds about 3 to 6 weeks after planting, while crown divisions emerge in about 10 days."
-'Comfrey' by Teynor, Putnam, Doll, Kelling, Oelke, Undersander and Oplinger, 'Alternative Field Crops Manual', University of Wisconsin: Cooperative Extension, University of Minnesota: Center for Alternative Plant & Animal Products, and Minnesota Cooperative Extension, February 1992, updated November **1997**, page 3.
    (Soak a wilted root cutting a few hours at most so it does not rot.)

Comfrey root is similar to carrots and other root vegetables so I am using that storage information:
*"***Specific Cooling Methods for 'Toppped' Root Vegetables:*** *Hydro-Cooling is very fast cooling. Uniform cooling in bulk if properly used. Package Icing for fast cooling. Water-ice contact. Forced-Air Cooling is much faster than room cooling. Container venting required.*
Recommended Temperature and Relative Humidity, and Approximate Storage Life:

Carrot, bunched	*32 F = 0 C*	*98-100%*	*10-14 days*
Carrot, topped	*32 F = 0 C*	*98-100%*	*6-8 months*

Moisture Loss from Vegetables: *Carrot: Medium loss if no tops. Root crops with tops have a high rate of moisture loss.*
Vegetables Classified According to Chilling Injury Susceptibility: *Beet, Carrot, Parsnip, Radish, Turnip: Not susceptible."*
-Knott's Handbook for Vegetable Growers by Donald N. Maynard and George J. Hochmuth, University of Florida. Hoboken, New Jersey: John Wiley & Sons, Inc., **2007**, pages 424, 426, 430, **442**.
('Topped' means the leaves and stalk are cut off. 'Bunched' means leaves are still on root.)
(100% relative humidity means air is holding as much moisture as it can at that temperature. It does not mean that it is raining.)

"Temporary and New Comfrey Beds: Root cuttings are a great way to plant out large areas of Comfrey. The cuttings should be grown on in small pots *with 50% compost, 50% river sand mix, kept moist, and planted out in the spring as soon as the first leaves emerge and the soil has warmed.*
If you are planting large numbers of root cuttings, *you can plant directly into the beds by creating 'nests' in the straw, adding two cupped handfuls of the above mentioned potting mix and plant the cuttings into this. Keep them moist like a wrung out sponge, and the success rate will be very close to 100%."*
-'Comfrey: Its History, Uses and Benefits' by Paul Alfrey, Permaculture Magazine: Earth Care, People Care, Future Care; Hampshire, England, www.permaculture.co.uk, March 3 **2016**.
    (Both of these methods can be used if you can't plant to a permanent location right away.)

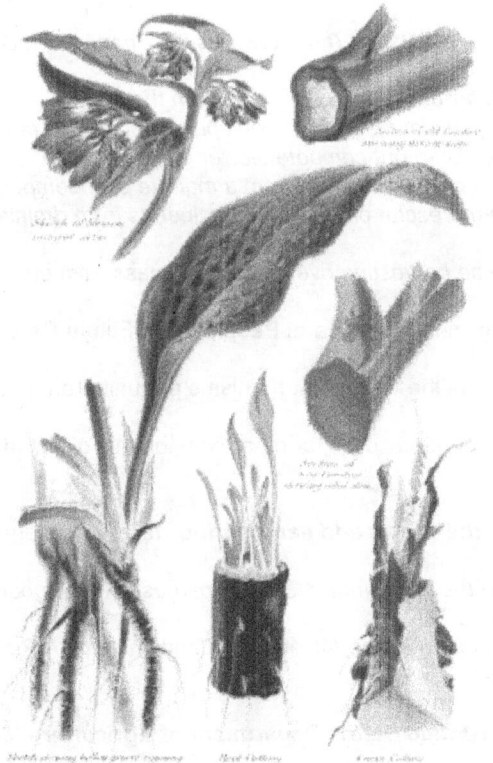

Prickly Comfrey 1877

'Forage Plants and Their Economic Conservation
by the New System of Ensilage:
Part I: Caucasian Prickly Comfrey'
by Thomas Christy, Jun., Fellow of the Linnean Society,
Christy & Co., London, England, 1877.

Top left: *"Sketch of blossom."*

Top right: *"Section of old Comfrey showing hollow stem."*

Big Comfrey leaf with roots: *"Sketch showing hollow groove running from the leaf down the stem to the root."*

Middle right: *"Section of new Comfrey showing solid stem."*

Bottom middle: *"Root cutting."*

Bottom right: *"Crown cutting."*

Chapter 35

Planting, Soil Fertilization, Water, Weeds, Disease

<u>Location of Planting</u>

"I believe Russian Comfrey has great possibilities as a spare corner crop or on land not in rotation or around the hedgerows of the organic farm."
-Fertility Farming by Newman Turner. Austin, Texas: Acres USA, **1951**, page 54.
    (A hedgerow is a hedge of shrubs or low trees growing along a bank, especially one bordering a field or lane.)

    *"**The place for Russian Comfrey is in the odd corner near the compost heap and the farmyard, now growing weeds, where it can be cut quickly for green fodder or compost.** It is surprising how little work with scythe and fork will fill a cart to take out to a 'threadbare' field in a dry summer. When it is not required as feed, half an hour with a scythe on an odd Saturday morning will add a ton of material to the compost heap."*
    -Fertility Farming by Newman Turner. Austin, Texas: Acres USA, 1951, page 56.
    (Scythe is a tool with a long, curving blade fastened at an angle to a handle used to cut grass, grain, etc., by hand.)

*"**Comfrey will grow where alfalfa will not grow, almost anywhere from cold snowy regions to semi-tropical climates,** yielding 60-140 tons of high grade green leaf per acre every year, and go on growing in the same place throughout one's working life."*
-Comfrey: Nature's Healing Herb and Health Food by Andrew Hughes. Japan: Sanyusha Publishing Co., Ltd, **1992**, page 11.

*"Comfrey is a perennial. **There is no known limit to the number of years one plant will grow in the same place without deterioration if given proper cultivation.**
Planting Comfrey is more like planting a tree than planting grass. It is there to stay, so it must have a permanent place, and preferably a place to itself. Comfrey needs special care, and a large plantation for farming or commercial purposes (say, 1/4 acre and upwards, more than 1,000 plants) should be given a clear open space where mechanical cultivation is possible."*
-Comfrey: Nature's Healing Herb and Health Food by Andrew Hughes. Japan: Sanyusha Publishing Co., Ltd, **1992**, page 150.

*"**Choosing the Site for Comfrey:** We're growing for biomass and want the Comfrey plants to receive as much light as possible. Accordingly, **we lay out our beds on an east to west axis (we're in the northern hemisphere).**
Irrigation is necessary if you want to get good yields from the Comfrey plants so picking a place with access to irrigation is of paramount importance.
Rainfall:*
    ***In areas of low rainfall,** using the gradient of the land to channel precipitation (rain) towards your beds will reduce the water needs of your plants.*
    ***In areas of high rainfall** with a high water table you should consider diverting water away from the beds.*
Once established, Comfrey is difficult to get rid of, so choose a site where you want it to stay. Don't plant Comfrey in any area you cultivate, as the broken root pieces quickly establish into new plants and can out compete slower growing crops.
***Position Comfrey downhill from where you expect leachate to be present,** i.e., downhill from a manure pile, compost heap, outside toilet, animal pen etc., can provide **passive fertility** to the plants and rescue otherwise lost minerals from draining away with the subsoil ground water.*
***Grow the Comfrey where you want to use it.** As you'll see later we may be harvesting over 1/4 ton biomass from our patch and don't want to be carrying that over long distances."*
-'Comfrey: Its History, Uses and Benefits' by Paul Alfrey, Permaculture Magazine: Earth Care, People Care, Future Care; Hampshire, England, www.permaculture.co.uk, March 3 **2016**.
(Biomass is the amount of living matter in a given habitat, expressed either as the weight of organisms per unit area or as the volume of organisms per unit volume.)
(Leachate is liquid that, in passing through material, extracts soluble or suspended solids, or other components of the material.)

*"**Symphytum species, Comfrey, is good in dry shade:**
In dry shade situations, there is usually a period of some moisture in the winter and early spring. Take advantage of this period by planting new starts. This moist time helps the plants settle in.
Greatest success comes when the new plants receive some water through the hot summer, so they can establish a good root system. This root system helps the plants to survive drought in coming years."*
-'Tough Shade Situations: Recommendations for Dry, Damp and Deep Shade' by Laura Altvater, Portland Nursery: A Passion for Plants. A Nursery for Plant People., https://portlandnursery.com, Oregon, **2018**.

*"**Growing Comfrey plants requires a climate in hardiness zone USDA (United States Department of Agriculture) 3-9, although some ornamental varieties are only hardy to zone 5.**"*

Read more at Gardening Know How: What Is Comfrey: Information For Growing Comfrey Plants
https://www.gardeningknowhow.com/edible/herbs/comfrey/growing-comfrey-plants.htm
-'What is Comfrey: Information for Growing Comfrey Plants' by Amy Grant, Gardening Know How: Passionate about helping you with your garden, www.gardeningknowhow.com, Bedford, Ohio, April 5 **2018**.
> (Examples of ornamental Comfrey include Symphytum rubrum and Symphytum grandiflorum Goldsmith.)

USDA Agriculture Research Service, Hardiness Zones, https://planthardiness.ars.usda.gov. *"The **USDA Plant Hardiness Zone Map** is the standard by which gardeners and growers can determine which plants are most likely to thrive at a location. The map is based on the average annual minimum winter temperature, divided into 10-degree F zones."*
You can enter your zip code and find your hardiness zone.

Average annual extreme minimum temperature:
- Zone 3 is -40 to -30 F = -40 to -34.4 C.
- Zone 4 is -30 to -20 F = -34.4 to -28.9 C.
- Zone 5 is -20 to -10 F = -28.9 to -23.3 C.
- Zone 6 is -10 to 0 F = -23.3 to -17.8 C.
- Zone 7 is 0 to 10 F = -17.8 to -12.2 C.
- Zone 8 is 10 to 20 F = -12.2 to -6.6 C.
- Zone 9 is 20 to 30 F = -6.7 to -1.1 C.

Timing of Digging and Planting

Astronomical Seasons: Spring begins on the spring equinox (March 20 in the northern hemispshere). Summer begins on the summer solstice (June 21). Fall (autumn) begins on the fall equinox (September 22-23). Winter begins on the winter solstice (December 21).

A **solstice** occurs when the sun appears to reach its most northerly or southerly excursion relative to the celestial equator on the celestial sphere. Two solstices occur annually, around **June 21 and December 21**. (Excursion is a movement outward and back or from a mean position or axis.)

An **equinox** is the moment the plane of the Earth's equator passes through the center of the Sun, which occurs twice each year, around **March 20 and September 22-23**. It is the point when the center of the visible sun is directly over the equator.

"Plant Comfrey in the autumn or spring like any herbaceous subject, but gardeners who can water in drought can plant in summer and gain a better yield next season than by waiting for the farmer's planting times."
-Compost, Comfrey and Green-Manure: 'Henry Doubleday Research Association' First Gardeners Report by Lawrence D. Hills. Braintree, Essex, England: Henry Doubleday Research Association, **1959**.
> (Herbaceous plants, i.e., herbs, are plants that have no persistent woody stem above ground. They can be annuals, biennials or perennials.)

"Today is a 'root day' according to the Biodynamic calendar, perfect time to transplant.
To transplant Comfrey, dig them up along with as much of their roots as possible. Then break apart the roots into smaller sections, or make 2 inch (5 cm) cuttings. Then simply stick them in the ground and water.
Plant by autumn so they have time to establish before winter."
-'Digging Up Comfrey' by Sean Dixon-Sullivan, Next Succession: Ecologists and Farmers. Pioneers of a Restored Planet, www.nextsuccession.com, Restoration sites and farms around the world, September **2016**.
> (Biodynamic agriculture is similar to organic garden/farm methods but also includes esoteric/mystical concepts based on the writings of Rudolf Steiner, 1861-1925. Soil fertility, plant growth and livestock are ecologically interrelated tasks.)

"Roots are associated with the Zodiac Earth Signs: Taurus, Virgo, and Capricorn.
The Full Moon and Waning Moon:
*During this phase, energy is drawn down, and after the moon has peaked, the light starts to decrease, **good for root growth**.*
***Plant root crops** such as beets, carrots, onions, and potatoes. This is also a good phase to transplant any type of plant."*
-'Using the Moon Phases and the Zodiac to Guide Your Gardening' by Melissa Keyser, Quarter Moon Living: Dedicated to rewilding, the old ways, and slow, intentional living, https://quartermoonliving.com, Sacramento, California, **2018**.
> (The full moon is when the moon appears completely illuminated from Earth's perspective. Waning of the moon is when a progressively smaller part of its visible surface is illuminated, so that it looks like it is decreasing in size. The moon starts to wane right after the full moon.)

*"**As the moon makes its monthly journey in the sky, it passes through one of the 12 constellations that make up the signs of the Zodiac and that vary in their fruitfulness.***
- *Cancer tops the list of fruitful signs: plant, transplant, graft and bud.*
- *Scorpio is almost as good: plant vine crops such as pole beans or cucumbers.*
- *Pisces favors plants requiring strong root development, which is practically every plant.*
- ***Taurus is semifruitful: good for root and leaf crops.***

Capricorn is less fruitful: good for root crops and tubers.
Libra is good for lettuce, cabbage and corn.
Third Quarter of Moon (decreasing or waning):
Plant biennials, perennials, and bulb and roots plants. *Plant trees, shrubs, berries, artichoke, beets, carrots, chicory, onions, parsnips, potatoes, radishes, rhubarb, rutabagas, strawberries, turnips, winter wheat, grapes, etc.*
Dig root crops for seed in the third quarter of the moon. *They will keep longer and are usually drier and better."*
-'Planting by Moon Signs' by Billy E. Warrick, Ph.D., worked 32 years for Cooperative Extension Service in west Texas; Soil, Crop and More Information: Compiling Information for Family and Friends, www.soilcropandmore.info, **2018**.

Best Time to Dig Up Comfrey Roots is Late Fall or Early Winter

See subsection 'Medicinal Gathering of Comfrey Leaves and Roots' in section 'Making and Using Comfrey Medicine' (Chapter 27).

"It is easier for a root cutting to develop shoots than a stem cutting to develop roots. **Root cuttings are best taken from a plant when it is dormant, in mid- to late autumn or early winter.** Plants with thick roots such as Papaver orientale, **Symphytum**, and Verbascum can be propagated by this method. The thinner the roots, the longer they should be."
-Plant Propagation by the American Horticultural Society, edited by Alan Toogood. New York, New York: DK Publishing, Inc., **1999**, page 158
(It is best to dig roots when plants are dormant, however, Comfrey is such a vigorous plant that it can be dug up any time the ground is not frozen. **The disadvantage of digging roots in late fall or early winter is that at this time the roots are more likely to die if planted outdoors because they do best if they have time to grow roots/leaves before very cold weather arrives.** If digging when dormant, planting in a coldframe or greenhouse is best, then transplanting them outside in the spring.)

"Weeds in your garden have probably demonstrated how easily some plants propagate themselves from a tiny piece of root. **The common herbs Comfrey** and milk thistle, for example, can turn into weeds if you make the mistake of moving pieces of the roots around with a tiller.
Take root cuttings in late fall or early winter, when the plants are dormant. At this time, the carbohydrate levels in the root are high enough to see it through its dormant period. This food supply will sustain it through the rooting period as well, making it much easier for the plant to form roots and shoots in time for the coming season."
-The Plant Propagator's Bible: A Step-by-Step Guide to Propagating Every Plant in Your Garden by Miranda Smith. Emmaus, Pennsylvania: Rodale Inc., **2007**, page 96.

"As the cold weather takes hold and beats back the perennials and biennials, their leaves begin to change color and fall away. Sap, energy, vitality, and nutrients are sent down to the plants' roots, and **autumn is also the prime time for harvesting almost any medicinal root (although spring harvests may also be serviceable).**"
-'Autumn Roots, Barks and Berries: Day-Long Field Workshop' by Maria Noel Groves, Clinical Herbalist, Wintergreen Botanicals LLC, Allenstown, New Hampshire, www.WintergreenBotanicals.com, **2016**.

"**Division of Comfrey succeeds at almost any time of the year.** Simply use a spade to chop off the top 7 cm (2.7 inch) of root just below the soil level. The original root will regrow, and you will have a number of root tops, each of which will make a new plant. These can either be potted up or planted out straight into their permanent positions."
-'Comfrey' by BotanicEye: Herbs and Botanics, https://botanicseye.com, Ireland, (date unknown, **2019** or earlier). BotanicEye aims to be a one-stop information site for all matters herbal.

Best to Plant Comfrey in Spring or Early Fall

Comfrey roots can be planted any time the soil is not frozen. It is very hardy. It is best to plant Comfrey in your garden or field. However, it can grow in a pot, at least 5 gallons (18.9 liters) for best development.

"**I think the best time to dig and replant Comfrey is about the time of early plowing in the spring**; that the said root-seed might be well preserved through winter in a cellar, for planting in spring, and that by properly dividing the said root cap (crown), and by strict economy, double the quantity of ground dug over, may be replanted by the seed (root) produced- you may dig annually."
-'The New England Farmer, and Horticultural Register' by Joseph Breck, Volume 23, New Series Volume 13, Boston, Massachusetts, **1845**. ('Symphytum or Comfrey, as Food for Men and Cattle' by Ezekiel Rich, Troy, New Hampshire, page 10, July 10 1844.)

"**Spring is the most proper time for planting in England, but no month comes amiss with it unless mid-winter, when the frost might kill the fresh planted root.**"
-'Prickly Comfrey: Its History, Cultivation, Extraordinary Production, and Uses: A Letter Addressed to His Excellency Sir Hercules Robinson, President of the Agricultural Society of New South Wales' by Arthur T. Holroyd, Sydney, Australia, **1876**, page 7.

"**There are two planting seasons; September, October and November,** the last month to be avoided in Scotland and the

colder parts of the North Country, **and March, April or May in the spring."**
-Russian Comfrey: A Hundred Tons an Acre of Stock or Compost for Farm, Garden or Smallholding by Lawrence D. Hills. London England: Faber and Faber, Limited, **1953**, page 62.   (This is true in the northern hemisphere.)

"Then plant the offsets (roots) 2 feet apart with their growing points just peeping (above the soil). **Ideally, this should be done between March and August, so that the plants become established for a full cutting season in the following year."**
-Down to Earth Gardening by Lawrence D. Hills. London, England: Faber & Faber, **1967**, 1975, page 219.

"The optimum time of Comfrey planting is in April or as early as the soil can be tilled, but the crop can be planted throughout the growing season."
-'Comfrey: A Controversial Crop' by Robert G. Robinson, University of Minnesota, Agricultural Experiment Station, Minnesota Report MR-191, Item No. AD-MR-2210, **1983**, page 3.

"Comfrey offsets (roots) can be planted at almost any time of the year, depending on the local climate. From early spring until mid-autumn is generally suitable. *At the stage of early spring the plant is just breaking into growth, and the allantoin in the plant is up in the new-forming buds, and the new plants will get away to a quick start."*
-Comfrey: Nature's Healing Herb and Health Food by Andrew Hughes. Japan: Sanyusha Publishing Co, **1992**, pages 154, 157.

Latest Planting is Early Fall So Roots Can Develop

"When the Comfrey offsets (roots) are planted so late in the year, **many of them do not take deep root;** *in the thaw they rise up with the frost columns, and are exposed on the surface with the melting frost. They dry up and die."*
-Comfrey Report No. 3: Feeding Dairy Cattle in Japan' by Meiji Milk Producing Co., Tokyo, Japan for Henry Doubleday Research Association, Braintree, Essex, England, 31 pages, July **1964**.

**"Comfrey can be planted at any time of the year except December and January, when the roots are fully dormant and planting can mean heavy losses in hard weather.
A start in early September gives the plants time to establish growth before winter.**
Leave the plants till the following July or August before cutting them with shears 2 inches (5 cm) above the ground, which will prevent their flowering in the first year. It is best to let what would be the second cut die down on the plants during their first season, to build up strength for the next, as with newly planted rhubarb."
-Fertility Gardening: The Organic Way to Make Your Garden Grow by Lawrence D. Hills. London, England: Cameron & Tayleur Books Limited, **1980**.

"Comfrey transplants and crown cuttings can be planted as late as October, but root cuttings should preferably be planted before September. *The plants must establish and grow before winter in order to produce a high yield the next year. Plantings are often made in a checkerboard arrangement to allow cross cultivation for weed control. Rows 3 to 4 feet (0.9-1.2 meter) apart are common. Closer spacings such as 30 inches (0.7 meter) probably produce higher yields."*
-'Comfrey: A Controversial Crop' by Robert G. Robinson, University of Minnesota, Agricultural Experiment Station, Minnesota Report MR-191, Item No. AD-MR-2210, **1983**, page 3.

"When to Plant Comfrey: Ideally plant out Comfrey offsets in March, April and May or in September. *September plantings may not show until the following spring.*
Late planted offsets in pots *can be moved into the greenhouse or a cold frame in late January, early February and will be well along for planting into the ground in early March."*
-'Planting, Cultivating, Harvesting and Problems of Comfrey' by John Harrison, Allotment and Gardens: Grow Your Own - Allotment - Gardening Help, https://www.allotment-garden.org, Penygroes, Gwynedd, Wales, **2018**.

Dig Roots Starting Third Year

"The farm method of Comfrey increase is best carried out in March or April, as soon as the plants can be seen starting into growth. **The best time to carry out the operation is in the third year; either the spring following the big cut from fully-established plants, or in the autumn about October, after a cut."**
-Russian Comfrey: A Hundred Tons an Acre of Stock or Compost for Farm, Garden or Smallholding by Lawrence D. Hills. London England: Faber and Faber, Limited, **1953**, page 69.

"The owner will wish to increase the Comfrey bed. This is best left until the third spring after planting to make certain that the crop has been allowed the minimum time to become established, *and will yield enough to pay when kept clean, cut and fed."*
-Comfrey: Fodder, Food and Remedy by Lawrence D. Hills. New York: Rizzoli Universe Books, **1976**, page 110.

__Temperature__   See 'Location of Planting' in this section.

*"Comfrey seems almost independent of moisture, still growing when coffee plants, cinchona (quina), and maize (corn) have died for lack of rain; and, **once established, no heat appears to have much effect upon it**. Singular to say, **Comfrey seems almost as independent of cold**, for it flourishes in Saint Petersburg (Russia), and in north Britain gives great satisfaction. It dies down when the severe frosts commence, but **no degree of cold appears to hurt its roots**."*
-'New Commercial Plants with Directions How to Grow Them to the Best Advantage, No. 1 (**1878**) and No. 3 (1880)' by Thomas Christy, F.L.S. (Fellow of Linnaean Society), London, England. Number 1 includes Caucasian Prickly Comfrey pages 14-16. Number 3 includes Prickly Comfrey pages 11-14. Includes 'Caucasian Prickly Comfrey' and 'Russian Comfrey' advertisements.

*"**The Comfrey plant apparently survives a wide extreme of climate from the cold winters of Canada, to the tropical climate of Nakuru, Africa**."*
-'Quaker Comfrey' by Milton D. Miller and Lloyd Harwood, Agronomy Progress Report No. 2, Agricultural Experiment Station, Agricultural Extension Service, University of California, Davis, March **1958**.
   (Nakuru is in Kenya. Kenya is in east Africa, 69 miles or 111 kilometers north of the equator.)

*"**Growth of Comfrey starts at around 10 C (50 F), and is best between this & 27 C (80.6 F), over which, growing declines**."*
-Comfrey Report No. 3: Feeding Dairy Cattle in Japan' by Meiji Milk Producing Co., Tokyo, Japan for Henry Doubleday Research Association, Braintree, Essex, England, 31 pages, July **1964**.

*"**The relationship between temperature and stem elongation** has been investigated for representative herbaceous plants sampled from populations of seven species which were growing under semi-natural conditions in the Botanical Garden of the University of Padua, Italy.*
   The growth of the floral stem was measured for the following species: *Galanthus nivalis L., Corydalis cava Schw. et Krt., Anemone nemorosa L.,* ***Symphytum tuberosum L.,*** *Allium ursinum L., Aegopodium podagraria L., and Campanula rapunculoides L.*
Optimum Mean Temperature:
   *The temperature ranges between 0 C and 30 C (32 and 86 F) which gave the best correlations with stem elongation have been calculated for the entire period of this process of growth. The optimum mean temperature increases from the start of the process to ripening, with the exception of a period **just before flowering when all plants seem to require lower daily temperatures.***
From charts for Symphytum tuberosum:
   Stem growth starts slowly in mid-March. In early April the rate of growth goes up quickly. This new rate stays the same through April. In early May the growth stops due to anthesis (flowering).
   Temperature in February varies from 2-3 C (35.6-37.4 F). Temperature in March varies from 3 to 16 C (37.4-60.8 F). In April froam 10 to 23 C (50-73.4 F).
Temperature and Light:
   With Corydalis cava and Symphytum tuberosum, the oscillations (up and down) of the minimal daily temperature which take place during March, may represent a strong disturbing element to growth.
   Personal observations have revealed that only Symphytum tuberosum and Campanula rapunculoides have strongly photoperiodic requirements.
The results of this analysis have also indicated that unknown factors other than temperature are affecting stem elongation."
-'Optimum Mean Temperature for a Plant Growth Calculated by a New Method of Summation' by Giovanni Abrami, Hortus Botanicus, University of Padua, Italy; Ecology: Ecological Society of America, Columbus, Ohio, Volume 53, No. 5, pages 893-900, September **1972**.

*"Ecology: **Comfrey is suitable for the temperate and subtropical regions**. It will grow and produce where many other forage plants will not. Almost any soil that allows deep root penetration (to about 2.5 meters = 8.2 feet) will grow Comfrey. The crop needs lots of water, and will stand flooding.*
Manures and fertilizers, particularly potash (potassium), should be added to the soil. Comfrey will grow in partial shade.
The above-ground foliage will stand 15 F (-9.4 C) of frost for a short time, and the roots will stand winter temperatures of -40 C (-40 F).
*In the Crop Diversification Matrix, **Comfrey is reported to range from the Boreal Moist Forest Life Zone to the Warm Temperate Moist Forest Life Zone,** tolerating annual biotemperature of 6-15 C (42.8-59.0 F), annual precipitation of 5-11 decimeters (19.68-43.30 inches), and pH of 5.3-6.8."*
-'Symphytum Peregrinum Ledeb., Boraginaceae, Comfrey, Russian Comfrey, Quaker Comfrey' by James A. Duke, Handbook of Energy Corps, unpublished, Purdue University, Indiana, **1983**.

Frost Resistance of Leaves

"Monsieur Petin, Member of the St. Marcellin Agricultural Society (southeastern France), writes to the 'Journal d'Agriculture Progressive' (Journal of Progressive Agriculture):
'On the 1st of May, 1875, I received fifty-two plants from England, and set (planted) them on the 3rd. In eight days the leaves

appeared, and I remarked very vigorous vegetation.
The Comfrey stands the cold of Isere (a department/area in the Auvergne-Rhone-Alpes region in eastern France, altitude 1,465 metres = 4,806 feet), even the late frosts of April *-and produces three kilogrammes (6.6 pounds) per head (plant). It is planted at a metre (3 feet) apart every way, which gives 10,000 per hectare (2.471143 acres), and consequently 30,000 kilogrammes (66,165 pounds) of green fodder for cutting, and 120,000 kilos (264,660 pounds) a year!'"*
-'Forage Plants and Their Economic Conservation by the New System of Ensilage: Part I: Caucasian Prickly Comfrey' by Thomas Christy, Jun., F.L.S. (Fellow of the Linnean Society), Christy & Co., London, England, **1877**, page 20.

*"**The Comfrey plant is used to very much colder winters than those in Britain; the last cut will stand until late November in a normal year.** It is also moderately shade-tolerant."*
-Russian Comfrey: A Hundred Tons an Acre of Stock or Compost for Farm, Garden or Smallholding by Lawrence D. Hills. London England: Faber and Faber, Limited, **1953**, page 59.

"The Comfrey plants start growth very early in the spring, and during the mild winter of 1958 a clip (cut) was actually made in the third week of March (Saskatoon, Saskatchewan, Canada).
The plant leaves have survived up to 15 degrees F (-9.4 C) of frost for short periods (night frost). However, prolonged temperatures below freezing froze the leaves black. *Temperatures have only gone down to about 0 degrees F (-17.7 C) in the area, and there has been no winter injury of any kind to the plants."*
-Comfrey: Fodder, Food and Remedy by Lawrence D. Hills. New York: Rizzoli Universe Books, **1976**, page 66.

*"**Comfrey has its own defenses against light frosts. The hairy upper surfaces of the leaves keep most of the dew off the leaves,** and this protects those surfaces from any frost as light as two or three degrees (Fahrenheit? which equals -16.6 to -16.1 C). Because there is less frost, and because all of the tiny little particles of ice are held up in the air, the melting and evaporating processes occur much more quickly than they can on the lettuce leaves."*
-Enchanted Garden: Alan Chadwick's Organic Method of Gardening by Tom Cuthbertson. London, England: Rider & Company / Hutchinson & Co. Publishers Ltd, **1978**, page 90.
(The author is based in Covelo, California so I assume his temperature is Fahrenheit, even though the book is published in England.)

> *"**Plants may appear to have hair, but the technical term for plant hair is trichomes.** Trichomes are not the same as our hair, but insofar as the definition of hair is that it is an outgrowth of the epidermis, then trichomes are for all practical purposes, a kind of hair. Unlike animal hair, though, trichomes are often living cells.*
> **Trichomes can be insulating by keeping frost away from leaf cells. They can help reduce evaporation by protecting the plant from wind and heat.**
>
> *In many cases, trichomes protect plants from herbivorous insects that may want to feed on them. And in some cases, if the trichomes are especially stiff or irritating, they may protect a plant from larger herbivores."*
> -'Plant Hair' by Michelle Ross, A Moment of Science, Indiana Public Media, https://indianapublicmedia.org, Indiana University, Bloomington, Indiana, May 23, 2018.
> (See subsection 'Comfrey Leaf Description' in section 'Symphytum Genus Descripton' {Chapter 5, Volume 1}.)

*"**Comfrey produces large amounts of foliage from late May until hard frosts in October or November.**"*
-'Comfrey' by Teynor, Putnam, Doll, Kelling, Oelke, Undersander and Oplinger, 'Alternative Field Crops Manual', University of Wisconsin: Cooperative Extension, University of Minnesota: Center for Alternative Plant & Animal Products, and Minnesota Cooperative Extension, February **1992**, updated November 1997.
(A hard freeze is at least four consecutive hours of air temperature below 25 degrees Fahrenheit = -4 C.)

Cold Hardiness of Entire Plant

Comfrey is dormant in winter: the leaves die but the roots live. In extremely cold climates such as USDA Hardiness Zone 2 or colder, you can mulch your plants to protect them from the cold.

*"**Prickly Comfrey (Symphytum asperrimum) is a perennial herb native to the Caucasus region.** The plant has a large taproot 8 or 9 feet deep (2.4-2.7 meters); stems 2 to 4 feet high (0.6-1.2 meters); leaves oblong, large, rough, sometimes a foot (0.3 meter) or more long; flowers tubular, bright-blue, nodding in one-sided clusters. **The plant is hardy, withstanding the winters in Ontario (east-central Canada) and succeeding well in most of the United States.**"*
-'Forage Plants and Their Culture: Rural Text-Book Series' by Charles V. Piper, M.S., Bureau of Plant Industry, Department of Agriculture, United States; edited by L.H. Bailey, New York, **1916**.
> (The Caucasus Mountains are located at the border of Europe and Asia, between the Black Sea and Caspian Sea and occupied by Russia, Georgia, Azerbaijan, and Armenia.)
> (Hardy means robust and capable of enduring difficult conditions. It survives outside in the winter very well.)

"Comfrey is hungry for nitrogen; it is rich in potash (potassium); it likes a slightly alkaline soil at pH 7.2 and loves rich compost.
Given these conditions you can produce top grade Comfrey all the year round where the soil temperature is above 10

degrees C (50 F)."
-Miracle Grass: Comfrey by Andrew Hughes published in **1966** in Japanese. Part of it is in: 'Comfrey: Nature's Healing Herb & Health Food' by Andrew Hughes, Japan: Sanyusha Publishing Co., Ltd, 1992. And part is in 'Comfrey Report No. 1: The 1954 Research Results' by Lawrence Hills published in 1955. (If you have translation in English, I would appreciate having a copy.)

"Realizing that the original Symphytum peregrinum (Russian Comfrey) comes from the cold Caucasus mountains, one is not surprised to learn that is has adapted very well to the cold climate even of Hokkaido, Japan, with a growing season of some 5 months at most, and **has survived through the deep winter freeze to a depth of 20 cm (7.8 inches) or more, where the snow lies deep for five or six months."**
-'Miracle Grass: Comfrey' by Andrew Hughes, **1966**, published in Japanese.
> (Hokkaido is the second largest island of Japan, and the largest and northernmost prefecture/territory. It is located at the boundary of a cool-temperate zone and subarctic zone.)

"Comfrey will survive even through severe winters. If conditions are suitable, plants may be given vinyl shelters, and will continue to provide winter feed, even without artificial heat if the climate is not too cold."
-Comfrey: Nature's Healing Herb and Health Food by Andrew Hughes. Japan: Sanyusha Publishing Co., Ltd, **1992**, page 157.
> (A row cover is any transparent or semi-transparent, flexible material such as fabric or plastic sheeting, that is put over plants to protect them from wind and cold.)

"Even through Comfrey is a perennial, it will not be able to survive winter outside in a container or pot. When plants are growing in the ground, soil provides insulation from the extremes of winter air temperature, which can get well below freezing or even into below 0 F (-17 C) temperatures.
If you want to leave the Comfrey in the pot, you could bring it into your greenhouse. Or you can put it somewhere cold that will not get below freezing.
It will go dormant for winter, but the crown and roots will survive to grow again next year. **You will need to water periodically even when its dormant, possibly about once a month, to make sure the plant doesn't desiccate (dry out) during winter."**
-'Comfrey Over Winter' by Sarah B., Extension: Ask an Expert, https://ask.extension.org, Made up of groups and individual experts from Land-Grant universities and Extension offices across the United States. A part of the Cooperative Extension System, Decemer 4 **2015**.

Heat Sensitivity of Comfrey
If you live in a very hot region, then Comfrey does better if you provide around 40% shade,
either natural or artificial such as shade cloth.

"Hot and Dry Climates: Comfrey is a little difficult to establish. Many young plants have died due to excessive heat.
It is also attacked by local insect pests such as grasshoppers and caterpillars.
However, once it is established, its roots go several feet (2 feet = 0.6 meter) deep, and it will remain a hardy perennial for years. Well established plants thrive on raw farm and sewerage manure."
-'Nature Study: Comfrey in the College Farm' by Gift Siromoney, M.A., M.Sc., Ph.D., F.S.S., India; Madras Christian College Magazine, Chennai, Tamil Nadu, India, No. xiv, pages 21-26, **1976**.
> (Tamil Nadu is tropical and prone to drought when monsoons fail. The climate ranges from dry sub-humid to semi-arid. The annual rainfall is about 94.5 cm = 37.2 inches.)

"Comfrey does have a weakness when it comes to sun scorch, but it will recover quickly unless the temperature goes over 110 degrees (43.3 C). For periods of extreme heat, you can spread a light mulch of straw over the soil of the Comfrey bed to protect the young plants.
> In general, you don't want to put mulches, such as straw or wood chips, in the garden. They don't contribute much to the soil. In fact, they use up a lot of nitrogen as they decompose, owing to their high carbon content. The plants can make much better use of that nitrogen, so you try to help them to make a living mulch.
> You time their planting out and their growth so that when the hot weather comes, the plants make a foliage cover for the bed. A live cover is much healthier than a dead one like straw. Moist straw can encourage fungi and snails to multiply, and if there's a thick layer of it, some of the sucking and chewing insects can start breeding in it.

An exception about mulching is made in the case of the bed with the young Comfrey plants, however. They grow at varying rates, and in this early scorch season that we get in Covelo, California, the first leaves can be burnt to such an extent that the plants miss a whole season of production.
> But since Comfrey is very resistant to fungus and insects, you can use a light mulch of straw, spread around the plants that are still too small to provide a living canopy when the scorch season starts.

Straw protects the soil from the sun and at the same time slows the growth of weeds, thus helping the Comfrey to get a strong start. And the cover keeps the soil from seizing up, too, so you don't have to do much mechanical cultivation with the hoe. You can also water less often and more heavily to encourage the deep mineral-mining root system that is one of the most valuable assets of the Comfrey plant.
The leaves of the plant do droop a little during the heat of the day, but this is a natural state of relaxation, as it is for lettuce. They'll perk up and look fine by dark."

-Enchanted Garden: Alan Chadwick's Organic Method of Gardening by Tom Cuthbertson. London, England: Rider & Company / Hutchinson & Co. Publishers Ltd, **1978**, page 90.

"*Comfrey Culture:* **Comfrey does well in Florida gardens, where it grows year round** *and tolerates cold weather.*
Since it is a perennial, cut it back yearly in January or February to reduce the thatch and encourage new succulent leaf growth. Start Comfrey any time of the year, although spring is best, using root or crown cuttings that are 2 to 6 inches (5 to 15 cm) long. Place them 2 to 4 inches (5 to 10 cm) deep in furrows spaced 3 feet (0.9 meter) apart."
-'Comfrey Symphytum Peregrinum L.' by James M. Stephens, Horticultural Sciences Department, University of Florida, Institute of Food and Agricultural Sciences (IFAS) Extension, Gainesville, Florida, publication HS587, First published May 1994. Revised September **2015**.

Photoperiod (Day Length)

Photoperiod is the interval in a 24-hour period during which a plant or animal is exposed to light. Photoperiodism is the physiological reaction of organisms to the length of day or night.

At the equator the daylight period lasts about 12 hours, regardless of season. The middle latitudes have significant change in day length through the year but it is never 24-hour dark or 24-hour sun like polar regions. An example in the middle latitudes is 14 hours sun in the summer and 10 hours sun in winter.

"**The development of the root and shoot system of Symphytum officinale** *always begins with the formation of a rape (radicle) with 6-8 leaves on its epicotyl.* **After this stage, development is determined by the length of the day.**
If the day is shorter than 15 hours, *the subterraneous (underground) organs grow thicker.*
Flower formation needs a day length of at least 12 hours.
Shoot growth needs a day length of at least 14 hours.
The shorter the day length, the more leaves are formed.
Starch Storage:
In Symphytum officinale, starch is always stored until 6-8 leaves are formed by the young plant. After this stage it is only stored when the days are longer than 14 hours.
When the days are shorter, the amount of starch is reduced, partly during the thickening of the subterraneous organs.
Fructosans are stored and reduced independent of day length."
-'Annual Developmental Cycle of Roots and Shoots in Symphytum Officinale L.' by Karin Staesche, Institut fur Spezielle Botanik und Pharmakognosie der Universitat Tubingen (Institute of Special Botany and Pharmacognosy of the University of Tubingen), Germany; Planta, Berlin, Germany, Volume 71, No. 3, pages 268-282, September 1966. All in German except this abstract is also in English.

(See sub-subsection 'Fructans' in subsection 'Fat, Protein, Carbohydrate, Fiber, Ash, Miscelleaneous' in section 'Nutritional Value of Comfrey' {Chapter 19, Volume 1}.)
(Radicle is the first part of a seedling to emerge from the seed. Epicotyl is the seedling stem above the stalks of the first 2 seed leaves.)

"**Staesche (1966) studied the development of the root and shoot system of Symphytum officinale, as well as the storage and consumption of carbohydrates in different organs over the year and under different day-length conditions.** *She showed that* **under short-day conditions** *many large ground leaves are formed, that the root becomes very thick and that shoot elongation is inhibited.*
Under long-day conditions *shoot elongation is stimulated and almost no thickening of subterranean organs occurs.*
The storage and consumption of carbohydratres is dependent upon the day length, *as is the development of the cormus.*"
-Plant Carbohydrates I: Intracellular Carbohydrates edited by F.A. Loewus and Widmar Tanner. Berlin, Heidelberg, Germany: Springer-Verlag, 1982, page 440.   (Cormus or corm is a rounded underground storage organ.)

"**In root-layers of Symphytum officinale, development as well as storage and consumption of carbohydrates is determined by day length, in a manner similar to that in plants developed from seeds.**
Root-layers differ in the following points:
   1. **Flowers are always formed after 16-19 leaves, even at a day length of 12 hours at which 26-29 leaves usually appear before flowers are formed.**
   2. In cultures kept at temperatures of at least +10 C (50 F) fructosans are stored in the young shoot-born roots, while the amount of fructosans is reduced in the buds, in the subterraneous shoot parts and in the old root pieces.
The old root piece remains a living part in the root-system of the layer and takes part in the renewed storage of starch just like the primary root of plants developed from seeds."
-'Development of Root-Layers of Symphytum Officinale L.' by Karin Staesche, Institut fur Spezielle Botanik und Pharmakognosie der Universitat Tubingen (Institute of Special Botany and Pharmacognosy of the University of Tubingen), Germany; Planta, Berlin, Germany, Volume 75, No. 4, pages 352-357, **1967**.

"Near the equator the day length is equal, with sunset about 6 p.m., and Comfrey can be cut all the year round, provided the water supply is kept up.
Comfrey in Kenya kept two cows in milk where one cow starved before, providing the high protein part of the ration, fed with Napier Fodder, a kind of fodder sugar cane, to balance up the starch equivalent. ***Again in the sub-tropical North Island of New Zealand,*** *the daylength and the moisture scored over the 100 tons an acre for the Southery Strain (Russian Comfrey)."*
-Comfrey: Fodder, Food and Remedy by Lawrence D. Hills. New York: Rizzoli Universe Books, **1976**, pages 45, 47.
(Southery is a town in Norfolk County, England. Lawrence D. Hills researched Comfrey there from 1948 to 1951.)

"In order to evaluate the influence of photoperiod on the accumulation of allantoin in Comfrey (Symphytum officinale L.), a well known medicinal plant, an experiment was conducted during July and August, 1998. Cuttings obtained from 90 day old plants were submitted to **four photoperiods (8, 12, 16 and 20 hours).**
After 60 days, allantoin content in roots and rhizomes was evaluated. ***The results showed that increases in photoperiod promoted an increment in the average content of allantoin in roots (0.06%, 0.303%, 1.213% and 4.78%).***
On the other hand, in rhizomes, allantoin accumulation decreased (9.65%, 7.14%, 0.55%) when the photoperiod was increased from 8 to 12 and 16 hours, respectively, stabilizing on a 20-hour photoperiod (0.53%). *Plants cultivated under field conditions presented 2.55% and 2.63% allantoin content in rhizomes and roots, respectively.*
Based on the fact that in Comfrey the roots are considered to be sites of allantoin synthesis, the results demonstrated that photoperiod could influence both the synthesis of allantoin in these organs as well as its accumulation in the rhizomes."
'Influence of Photoperiod on the Accumulation of Allantoin in Comfrey Plants' by A.H.F. Castro, M. Young, A.A. de Alvarenga and J.D. Alves, Departamento de Biologia / Setor de Fisiologia Vegetal da Universidade Federal De Lavras (Dept of Biology / Department of Plant Physiology, Federal University of Lavras), Brazil; Revista Brasileira de Fisiologia Vegetal (Brazilian Journal of Plant Physiology), Volume 13, No. 1, pages 49-54, **2001**.

"Alterations in boimass production, leaf area and chlorophyll content were ***evaluated during the initial growth of Symphytum officinale plants under photoperiods (daylight) of 8, 12, 16 and 20 hours.***
After 60 days, total and factionated (divide into groups) dry matter (leaves, petioles, roots and rhizomes), leaf area, leaf area ratio, leaf weight ratio, specific leaf area and chlorophyll content were evaluated.
Photoperiod increased the accumulation of total and fractioned (divided) dry matter up to 16 hours, followed by a decrease when the photoperiod was increased from 16 and 20 hours. *Photoperiod decreased the leaf area ratio, specific leaf area and chlorophyll a and b contents. No alterations were observed for leaf weight ratio and chlorophyll a:b ratio."*
-'Influence of Photoperiod on the Initial Growth of Comfrey (Symphytum Officinale L.) Plants' by A.H.F. Castro and A.A. de Alvarenga, Departamento de Biologia / Setor de Fisiologia Vegetal da Universidade Federal De Lavras (Department of Biology / Department of Plant Physiology, Federal University of Lavras), Brazil; Ciencia e Agrotecnologia (Science and Agrotechnology), Volume 26, No. 1, pages 77-86, **2002**.

<u>Soil Type and Soil Depth for Root Growth</u>

There are 3 main particle sizes in soil: clay, sand, and silt.
Clay: Tiny particles (less than 0.002 mm) that cling together reducing water and nutrient transfer. It is hard and easily compacted.
Sand: Coarse particles (2.0 to 0.05 mm) that allow water and nutrients to leach out too quickly.
Silt: Particle size (0.05 to 0.002 mm) is between clay and sand.

Loam: Soil of equal amounts of clay, sand and silt. It retains water/nutrients in right amount.
Clay Loam: A fine-textured soil that breaks into clods or lumps that are hard when dry.
Sandy Loam: Soil with a lot of sand but enough clay for good structure and fertility.
Silty Loam: It is two-thirds silt with the rest split equally between sand and clay.

Chalk, limestone or calcium carbonate are alkaline/basic (pH 7.1 or higher). Chalky soils do not hold water. It reduces the absorption of iron by plants. These soils can be fertile depending on other conditions.

Peat is partially decayed vegetation or organic matter in areas called marshes, peatlands, bogs, moors or mires. Peaty soil has a high amount of organic matter because its acidic nature slows decomposition. There tend to be few nutrients but it is good for growth if fertilizer is added. It holds water and needs drainage.

*"**As to the place where Prickly Comfrey is to be planted, any soil seems to suit it.** I have, where I am at present, two spots where it is planted; one, an open sunny place on almost a sand; the other, a stiff clay against a north wall; and they seem to do nearly equally well in both. **Any soil but chalk suits it.***
My notion is that the plant derives a good deal of nourishment from its decayed leaves in autumn, which I have always left."
-'Prickly Comfrey: Its History, Cultivation, Extraordinary Production, and Uses: A Letter Addressed to His Excellency Sir Hercules Robinson, President of the Agricultural Society of New South Wales' by Arthur T. Holroyd, Sydney, Australia, **1876**, page 7.

*"**I have chiefly grown Prickly Comfrey (Symphytum asperrimum) upon a stiff clay,** what is commonly called a good 'pipe and tile earth'; but I have also grown it upon land of a lighter description, but not what is called a light soil, and **have found very***

little difference in its growth upon the two or three different soils.
In all cases I have found that good cultivation and well manuring tells more than the difference of soil."
-'Ensilage: A System for the Preservation in Pits of Forage Plants and Grasses, Independent of Weather: A Collection of Facts and Statistics on the Cheapest Mode of Providing Winter Food for Dairy Cattle, Sheep, Horses, Etc.', by Thomas Christy, F.L.S. (Fellow of Linneaen Society), London, England, **1883**. Caucasian Prickly Comfrey, pages 57-63.

"On poor stony soil, such as the Yorkshire, England wold limestone-fragment filled corner used by Mr. E. V. Stephenson, with good manuring Comfrey will thrive; on sands a hard pan will defeat them, and they will need feeding on a considerable scale to produce a bulk of fodder.
The soils which rule Comfrey out altogether are thin soils over rock (including chalk), pure peats, and land with a high water table. On clays, loams, and sandy loams Comfrey are at their best, *and there is no objection to a moderate flood risk, provided the land is well drained otherwise.*
*They need at least four feet foot-run (1.2 meter), and **a neutral soil (pH); pH 6.0 is about as low as they can safely go.*** *A soil test should be made to show the lime dressing required to bring it to pH 7.0 or above."*
-Russian Comfrey: A Hundred Tons an Acre of Stock or Compost for Farm, Garden or Smallholding by Lawrence D. Hills. London England: Faber and Faber, Limited, **1953**, page 59.
(Wold is high, open, uncultivated land or moor.) (Water table is the level below which the ground is saturated with water.)

"Clay soils are ideal for Comfrey, and they will do well on sand, but thin soils over chalk or peaty ones over rock are unsuitable, *because the roots need room to get down into subsoil."*
-Fertility Without Fertilizers: A Basic Approach to Organic Gardening by Lawrence D. Hills. New York: Universe Books, **1975**.
(Subsoil is the layer of soil under the surface soil or topsoil. It has less organic matter and humus than topsoil.)

"In field trials with Comfrey plants from root cuttings or seedlings, growth was most vigorous on rich garden soil, or alluvial soil, followed by brown soil, rendzinas and pseudo-podzolic soils. *Dry root yields showed a similar trend and varied from 5 to 12 ton/hectare. The highest average allantoin content was found in roots from rendzinas (1.71%) and the lowest in those from pseudo-podzolic soils (0.88%)."*
-'Studies on the Cultivation and Allantoin Content of Comfrey (Symphytum officinale)' by D. Fijalkowski and M. Seroczynska, Instytut Biologii Uniwersytetu (University Institute of Biology), Lublin, Poland; Herba Polonica (Polish Herbs), Volume 23, No. 1, pages 47-53, **1977**. (I was not able to get this report. If you have an English translation of it, could you please send it to me.)
(Alluvium is loose soil or sediments, which has been eroded by water and redeposited in a non-marine setting. It is made up of a variety of materials, including fine particles of silt and clay and larger particles of sand and gravel.)
(Brown earths or soils have reasonable natural fertility that grow deciduous woodland and are sometimes used for farming. They are usually located in regions with a humid temperate climate.)
(Rendzina soils are humus-rich shallow soils usually formed from carbonate or sometimes sulfate parent material. They are often found in karst and mountainous regions.)
(Podzol, Podosol or Spodosol soils are typical soils of coniferous {pines + others} or boreal {taiga or snow} forests. They are also typical soils of eucalypt forests and heathlands in southern Australia. In Western Europe podzols develop on heathland.)

"Comfrey is adaptable to many soils, but prefers moist, fertile soils.
Thin soils over rock will give a poor crop, but on light sands and loams, this crop will be productive if adequate nutrients are present.
Comfrey productivity is not very sensitive to soil pH, but highest yields occur on soils with a pH of 6.0 to 7.0."
-'Comfrey' by Teynor, Putnam, Doll, Kelling, Oelke, Undersander and Oplinger, 'Alternative Field Crops Manual', University of Wisconsin: Cooperative Extension, University of Minnesota: Center for Alternative Plant & Animal Products, and Minnesota Cooperative Extension, February **1992**, updated November 1997, pages 2, 4.

"Optimum soil conditions for Comfrey call for pH 7.2."
-Comfrey: Nature's Healing Herb and Health Food by Andrew Hughes. Japan: Sanyusha Publishing Co., Ltd, **1992**, page 156.

"Best Plants for Problem Clay Soils:
Symphytum officinale (Comfrey): *Height 2 to 3 feet (0.6 to 0.9 meter), USDA Zone 5-10.*
Flowers: White, pink, mauve in late spring to summer. Has bell-shaped flowers and large hairy leaves.
Likes full sun or part shade, Water amount: average."
-'Best Plants for Problem Clay Soils: Perennials' by William T. Kemper Center for Home Gardening, www.gardeninghelp.org, Missouri Botanical Garden®, Saint Louis, Missouri, **2019**.

Soil Depth Needed for Root Growth
See subsection 'Comfrey Root Description' in section 'Symphytum Genus Description' (Chapter 5, Volume 1).

"All species of Comfrey require a deep soil, not necessarily a good one, but deep. *They owe their reputation for growing and producing on land where nothing else will grow to their long and powerful roots.*

Their vertical search is large, so **they can tap subsoil water on dry soils,** *draw up calcium, phosphorus and potassium salts, not to mention trace elements. Moreover their suction power is considerable;* **the roots have a vigour far beyond that of any other plant."**
-Russian Comfrey: A Hundred Tons an Acre of Stock or Compost for Farm, Garden or Smallholding by Lawrence D. Hills. London England: Faber and Faber, Limited, **1953**, pages 59, 60.

"Comfrey shares with alfalfa the distinction of being one of the deep rooted fodder plants. *It does not go as deep as alfalfa, but it needs when mature to be able to get down into the subsoil.* **This is one secret of its success on mountain sides and high lands; it does not do well on low lying areas where drainage is not good.**
It is only when the main roots get down to the subsoil that the plant reaches its maximum in food value, in vigor of growth and palatability, *because here lie the rich microelements that have leached down through the top soil. Only the deep growing main roots can reach these. This is one of the secrets of alfalfa too. And with the depth of the root comes resistance to disease and insect pests."*
-Miracle Grass, Comfrey by Andrew Hughes published in 1966 in Japanese. Part of it is in: 'Comfrey: Nature's Healing Herb & Health Food' by Andrew Hughes, Japan: Sanyusha Publishing, **1992**. Part is in 'Comfrey Report No. 1: The 1954 Research Results' by Lawrence Hills published in 1955. (If you have a translation in English, I would appreciate a copy.)

How to Plant Comfrey Roots

Overview

"Home Cultivation and Treatment of Symphytum Asperrimum, or Prickly Comfrey Sets:
On the receipt of the Crown sets (cuttings) place them at once in the field, if there is no frost. **In spring and summer if planted in the field at once they will thrive,** *but in dry hot weather water them a few times.*
In planting on a moor (peaty, poor drainage) or boggy land, it is only necessary to make a hole with a bar, and drop in a large set. It does well on a loam and a stiff clay soil, and taps down deep into a sandy soil.
The root-sets should be planted perpendicularly (vertical), with the thin end downwards.
The best site to select for planting is that nearest to where the cattle will be fed, to save carrying such a weight of forage further than necessary, and to ensure its being cut regularly. If planted in woods for harbouring game (hunted wildlife), the sets may be placed closer together, and the flower stems will rise to five or six feet (1.5 to 1.8 meter) high."
-'Forage Plants and Their Economic Conservation by the New System of Ensilage: Part I: Caucasian Prickly Comfrey' by Thomas Christy, Jun., F.L.S. (Fellow of the Linnean Society), Christy & Co., London, England, **1877**, page 24.

"Comfrey has several serious limitations. For one thing, it is quite difficult to get a satisfactory stand, since cuttings have to be used. **The first attempt to get a stand here at Davis, University of California, with a spring planting, resulted in only a 20% stand establishment (survival rate).**
Plantings made in November of 1958 emerged the following May only a few days earlier than the plantings made in February 1959. The plantings made in April emerged in June. The early planting date had a higher plant mortality (death), which was probably caused by severe decay of root cuttings.
There was an apparent viability advantage of whole root cuttings over the crown, and half root cuttings. The crown cuttings did emerge faster. **The cuttings planted three inches and six inches (7.6 and 15.2 cm) deep** *emerged about the same time, while those planted nine inches (22.8 cm) deep emerged about two weeks later.*
Gibberellic acid (plant hormone) had no effect on emergence or survival.
The most successful method of establishing Comfrey has been by using transplants. *The live plants may be taken from cuttings started in the greenhouse or from an established plant in the field."*
-'Quaker Comfrey in 1961' by M.D.M. and R.T. Edwards, Agronomy Notes, University of California Cooperative Extension Service, Davis, California, pages 22-23, February 28 **1961**.

(A 20% survival rate is extremely low. Most survival rates are around 100%. For a planting in April to emerge in June, then it must have been a very small root cutting that was not presprouted. They did not give size of cuttings.)
(I'm not sure what they mean by 'whole root cuttings' vs 'half root cuttings'.)
(Planting 9 inches deep is much too deep. Even 6 inches deep is usually too deep unless the soil is very sandy so there is risk of the root drying out.)

"The home and small plantation: In this case it is possible to open up the soil for each plant to a depth and width of 30 cm (11.8 inches) each way and place 5-8 kg (11-17.6 pounds) of poultry manure compost in the hole. *This basic manuring is ideal for getting a plant established and into quick growth.*
If poultry manure compost is not available, use other forms of well rotted dung, cattle or stable manure, pig, night-soil, fish meal, etc. If the manure is fresh and not composted, mix it well with straw and soil, and put it at the bottom of the hole, and it will activate there.
Cover this base with soil mixed 50% x 50% with compost, and plant the offset (root) vertically, budding (large) end up, about 3 cm (1.18 inch) below the surface. Cover the hole with a mulch of straw, leaves or dried grass cuttings, and **the offset will break through with leaves in about 8-10 days."**
-Comfrey: Nature's Healing Herb and Health Food by Andrew Hughes. Japan: Sanyusha Publishing Co., Ltd, **1992**, page 152.

(Night soil is human feces. Using unprocessed human feces as fertilizer is a risky practice because it may contain disease-causing pathogens.)

(How deep you plant the root depends on soil type and rain. Plant deeper in sandy soils and dry weather. I have clay soil and plant mine 2 inches deep; 3 inches is OK too.)

"The cultivation of a high-performance variety of Symphytum officinale L. selected especially for the minimization of the content of pyrrolizidine alkaloids exclusively requires vegetative propagation to maintain its properties. Therefore, the development of effective methods for its vegetative propagation is necessary. **The use of aerial (above ground) plant parts as propagative material is not effective. The aim of this study was to examine the vegetative propagation of Comfrey by root cuttings.** In April and July 2013 root cuttings of 1.0, 2.0 and 3.0 cm (0.39, 0.78, 1.18 inch) length and of <0.5, 0.5,1.0 and 1.5 cm (<0.196, 0.196, 0.393 and 0.590 inch) thickness were prepared from potted mother plants, cultivated in a greenhouse and planted in 9 cm (3.54 inch) pots.
After 42 days (6 weeks), all root cuttings that were at least 2.0 cm long and 0.5 cm thick showed roots. In addition, 87% of 1.0 cm long and 0.5 cm thick root cuttings also showed roots."
-'Vegetative Propagation of Symphytum officinale L (Comfrey) by Root Cuttings' by R. Kadner and W. Junghanns, Journal of Medicinal and Spice Plants, Germany, Volume 20, No. 1, pages 7-11, March **2015**. (I was only able to get an abstract of this article. If you have the full article, could you send it to me please.)

Spacing Between Roots, and Number Per Acre

An acre is an area of 1 chain by 1 furlong {66 by 660 feet}, which is equal to 1/640 of a square mile, 43,560 square feet, 4,047 square meter, or 40% of a hectare.

For 1,000 square feet (92.9 square meter) with plants 3 feet (0.9 meter) apart from each other in all directions, then you need 111 plants. For 1,000 square feet with plants 2 feet (0.6 meter) apart, you need 250 plants.
For 100 square feet (9.29 square meter) with plants 3 feet apart, then you need 11 plants. For 100 square feet with plants 2 feet apart, then you need 25 plants.

"**If the Prickly Comfrey plants and rows were three feet (0.9 meter) apart** instead of two, and where space of ground is no object, I should recommend it. If this suggestion be adopted, **1,000 root-cuttings will occupy a quarter of an acre (1,011 square meters).**"
-'Prickly Comfrey: Its History, Cultivation, Extraordinary Production, and Uses: A Letter Addressed to His Excellency Sir Hercules Robinson, President of the Agricultural Society of New South Wales' by Arthur T. Holroyd, Sydney, Australia, **1876**, pages 6, 8.

"Although Prickly Comfrey seeds, it rarely germinates in this country (England). The only successful way of propagating it, is by dividing the roots in spring or autumn.
5,000 sets (roots), planted one yard (3 feet = 0.9 meters) apart, are sufficient for the cultivation of an acre of land."
-'New Commercial Plants with Directions How to Grow Them to the Best Advantage, No. 1 (**1878**) and No. 3 (**1880**)' by Thomas Christy, F.L.S. (Fellow of Linnaean Society), London, England. Number 1 includes Caucasian Prickly Comfrey pages 14-16. Number 3 includes Prickly Comfrey pages 11-14. Includes 'Caucasian Prickly Comfrey' and 'Russian Comfrey' advertisements.

"**The distance at which the Prickly Comfey cuttings or plants must be set will depend on the quality of the soil and the preparation.**
    Two feet (0.60 meter) by one and a half feet (0.45 meter) would require 14,520 plants per acre.
    2 1/2 feet (0.76 meter) by 1 1/2 feet (0.45 meter), 11,600 plants.
    2 feet (0.60 meter) by 2 feet, 10,890 plants.
    2 1/2 feet (0.76 meter) by 2 feet (0.60 meter), 8,712 plants.
This is as close as I would advise to plant; and on very rich land I would have the plants three by three feet (0.91 meter)."
-'The Farmers Book of Grasses and Other Forage Plants for the Southern United States' by D.L. Phares, A.M. (Master of Arts), M.D., Professor of Biology, A&M College of Mississippi, Starkville, Misssissippi, **1881**, pages 21-23.

"Following a botanical description of Symphytum peregrinum (Russian Comfrey), an account is given of results of experiments on some aspects of cultivation and use.
A close spacing of 40 cm (15.7 inch) between rows and 40 cm between plants in the row gave the highest yields, but a spacing of 62.5 cm x 35-40 cm (24.6 x 13.7-15.7 inches) was better because it allowed mechanical cultivation.
Square-pocket planting with a spacing of 62.5 x 62.5 cm was suitable on land likely to be heavily infested with weeds, but yields were 3.12% lower than when spacing was 62.5 x 62.5 cm."
-'Cultivation of Comfrey' by W. Doring, Inst Acker- u PflBau, University of Halle-Wittenber, East Germany; Deutsche Landwirtschaft (German Agriculture), Volume 10, No. 2, pages 62-66, **1959**. (I do not have this report. If you have it, could you please send it to me.)

"**The standard spacing for farm plots of Comfrey is three feet apart each way, or 4,840 to the acre.** The reason for this wide spacing is because a Comfrey plant which is four years old and good for six more years of productive youth can be

eighteen inches (0.45 meter) across, and though the plants look wide through their first two years, **this leaves room for between row cultivations to keep down the weeds which are at their worst in the early years.**"
-Comfrey: Fodder, Food and Remedy by Lawrence D. Hills. New York: Rizzoli Universe Books: **1976**, page 105.

"In a trial in the Komi ASSR (Autonomous Soviet Socialist Republic) reported by Vavilov and Kondrat'ev (1975), **Symphytum asperum gave the highest yield at the widest plant spacing (60 x 60 cm or 70 x 50 cm) (23.6 x 23.6 inch, 27.5 x 19.6 inch).** The lower plant density was more than compensated for by the much greater individual plant weight.
In the Netherlands, however, Van der Zweerde (1965) **found that the reverse was true for Symphytum x uplandicum, in that a 45 x 45 cm (17.7 x 17.7 inch) spacing gave higher yields than a 90 x 90 cm (35.4 x 35.4 inch) spacing, except in very dry weather.**
Similar results were obtained in Bulgaria by Shchereva et al. (1965), who **found an even higher yield at 20 x 20 cm (7.8 x 7.8 inch) than at 45 x 45 cm spacing.** Lee, Kang and Han (1969) in Korea obtained a similar effect in the first year, but in the second year they observed a reduced yield only at the lowest density."
-'Comfrey Symphytum spp. as a Forage Crop' by J.C. Forbes, A.D. McKelvie, and P.J.C. Saunders, North of Scotland College of Agriculture, Aberdeen, United Kingdom; Herbage Abstracts, Volume 49, No. 12, pages 523-539, **1979**.

"**Standard spacing recommended throughout the world and based on experience of intense cultivation is 60-90 cm (2-3 feet) each way between plants for farm use, the distance determined by the machine used for inter-row cultivation.**
This is standard stockfeed cultivation, allowing for mechanical operation of manuring and cultivating, where the Comfrey can be cut in strips, cutting only enough each day for that day's needs, cutting the crop completely over 28 days.
For those who plan to cut the leaves when smaller, say every 21 days, when the protein content is higher and the fiber less, planting can be reduced to 50 to 80 cm (19.6-31.5 inches) between the rows each way to suit small machine cultivation. This method of straight line cultivation makes for easy cultivation by small farm machines."
-Comfrey: Nature's Healing Herb & Health Food by Andrew Hughes. Japan: Sanyusha Publishing Co., Ltd, **1992**, page 152.

"**Comfrey, Symphytum officinale: Plant spacing for Comfrey is eight inches (20.3 cm) within the rows and twenty-eight inches (71.1 cm) between the rows.**"
-The Organic Medicinal Herb Farmer: The Ultimate Guide to Producing High-Quality Herbs on a Market Scale by Jeff Carpenter and Melanie Carpenter. White River Junction, Vermont: Chelsea Green Publishing, **2015**, pages 288.

Depth of Planting Roots, and Vertical vs Horizontal

"Mr. Ashburner of Virginia recommended setting the Comfrey root cuttings very deep in the soil, and following his instructions in a heavy soil, many of my cuttings never got out.
Four or five inches (10.1-12.7 inch) may do in very light soil, but in very heavy soil one inch (2.5 cm) is much better."
-'The Farmers Book of Grasses and Other Forage Plants for the Southern United States' by D.L. Phares, A.M. (Master of Arts), M.D., Professor of Biology, A&M College of Mississippi, Starkville, Misssissippi, **1881**, pages 21-23.

"**These Comfrey offsets are short sections of root with a growing point which should just stick out of the soil; on the farm they are planted quickly with a dibber.** In the garden you need an early variety. The best is Russian Comfrey Bocking No. 14 which is used to feed race-horses and in most seasons gives an April cut for the early foals."
-Compost, Comfrey and Green-Manure: 'Henry Doubleday Research Association' First Gardeners Report by Lawrence D. Hills. Braintree, Essex, England: Henry Doubleday Research Association, **1959**.
    (A dibber, dibble or dibbler is a pointed wooden stick for making holes in the ground so that seeds, seedlings, bulbs or roots can be easily and quickly planted. They have a variety of designs such as straight, T-handle, trowel and L-shape.)

"**There is no need to take trouble to keep the root cuttings the same way (vertical, horizontal, etc.) that they grew.**"
-Comfrey: Fodder, Food and Remedy by Lawrence D. Hills. New York: Rizzoli Universe Books, **1976**, pages 111, 112.

"**If there is foliage on the Comfrey crown he leaves it sticking up, but if there isn't any he buries the whole root and crown.** He seems indifferent as to whether there is foliage or not. He does the whole planting procedure so quickly that it seems almost careless. He doesn't make any preparation of the soil before he sticks the roots into it. At the same time he firms the soil around the roots after they are planted.
There are a few foot pieces left over and these are stuck into the Comfrey bed at random."
-Enchanted Garden: Alan Chadwick's Organic Method of Gardening by Tom Cuthbertson. London, England: Rider & Company / Hutchinson & Co. Publishers Ltd, **1978**, page 71.
    (The crown is where the plant stem meets the roots. There are great changes here in vascular structure, as the root system gets transformed to the shoot system. It is also called root crown, root collar and root neck. The top of the crown is level with the soil surface or just a little below it.)
    (Comfrey is very hardy. It will survive under almost any condition. The more care you take to prepare the soil, the better your Comfrey plants will grow, but you don't have to do anything special unless you want to.)

"**Comfrey root cuttings should be laid flat (horizontal) and covered with soil.**

The cuttings should be planted at a depth of 2 to 4 inches (5 to 10 cm) according to soil texture and expected soil moisture. A 4 inch depth is commonly used, but 2 inches is adequate with irrigation. Very small cuttings should not be planted as deeply as the longer cuttings. Young transplants should be planted upright with the crowns about 2 inches deep."
-'Comfrey' by Teynor, Putnam, Doll, Kelling, Oelke, Undersander and Oplinger, 'Alternative Field Crops Manual', University of Wisconsin: Cooperative Extension, University of Minnesota: Center for Alternative Plant & Animal Products, and Minnesota Cooperative Extension, February **1992**, updated November 1997, page 3.

> (**It doesn't matter whether you plant root cuttings horizontally or vertically. Crown cuttings should be planted with the buds facing up.** I prefer an inch or so of soil over the top of it. All of the large leaves should be removed before planting to help reduce moisture loss through evaporation.)

"Horizontal or vertical placement of Comfrey cuttings in the pots had no influence on rooting rate."
-'Vegetative Propagation of Symphytum officinale L (Comfrey) by Root Cuttings' by R. Kadner and W. Junghanns, Journal of Medicinal and Spice Plants, Germany, 20(1):7-11, March **2015**. (I was only able to get an abstract of this article. If you have the full article, could you send it to me please.)

<u>Fertilization of Comfrey Plant</u>   It is best to get a soil test before adding fertilizers to your garden or farm land.

For how to use Comfrey as a fertilizer, see subsection 'Grow-It-Yourself Fertilizer' in section 'Garden Uses of Comfrey: Compost, Fertilizer, Potting Mix' (Chapter 17, Volume 1).

Primary Nutrients:
    Nitrogen (N): Good for leaf development. Yellowing of leaves may mean a nitrogen deficiency.
    Phosphorus (P): Improves the growth of roots, flowers, seeds, and fruits. It is important in spring.
    Potassium (K): Increases root and stem development.
Secondary Nutrients:
    Calcium (Ca): The primary function is to provide structural support to cell walls.
    Magnesium (Mg): Involved in photosynthesis and is an enzyme activator for plant growth.
    Sulfur (S): An essential building block in chlorophyll development and protein synthesis.
Micronutrients:
    Essential for plant growth. They include boron (B), chloride (Cl), copper (Cu), iron (Fe), manganese (Mn), molybdenum (Mo), nickel (Ni), zinc (Zn) and others.

*"Prepare the Comfrey bed by digging out perennial weed roots with great care, especially the long white roots of couch grass, and **giving a lasting dressing of bone meal at the rate of 1 pound (453 grams) a square yard (0.836 square meter), plus compost or manure for a well-fed start.**"*
-Fertility Without Fertilizers: A Basic Approach to Organic Gardening by Lawrence D. Hills. New York: Universe Books, **1975**.
> (Bone meal is ground animal bones. The Nitrogen-Phosphorus-Potassium ratio varies depending on the source: 3-15-0 to 2-22-0.)

"Comfrey response to very high fertilizer levels was better in the second year than in the first year."
-'Comfrey Symphytum spp. as a Forage Crop' by J.C. Forbes, A.D. McKelvie, and P.J.C. Saunders, North of Scotland College of Agriculture, Aberdeen, United Kingdom; Herbage Abstracts, Volume 49, No. 12, pages 523-539, **1979**.

*"A Comfrey bed, like the ground for asparagus or rhubarb, should be thoroughly dug to remove the roots of perennial weeds, especially convolvulus (bindweed family), couch grass and docks. **It then needs some lasting sources of nitrogen and potassium dug in, starting with 1 pound a square yard (540 gram per square meter) of coarse bone meal.**
Feathers, wool shoddy from the stuffing of old mattresses, and human hair, if you can get it from a barber, are far more lasting than manure, especially on sandy soil. Failing these, use manure.
It is a good idea to prepare your bed in the late autumn for planting in the spring."*
-Fertility Gardening: The Organic Way to Make Your Garden Grow by Lawrence D. Hills. London, England: Cameron & Tayleur Books Limited, **1980**.
> (Wool shoddy is recycled wool from materials that are not felted, that is of better quality and longer staple than mungo that is short staple wool. A wool staple is a cluster or lock of wool fibers and not a single fiber.)

*"**Comfrey Disadvantages:**
Comfrey has some disadvantages compared with other forage crops. **Comfrey requires the addition of nitrogen fertilizer to produce a high yield and protein content, while alfalfa produces high yields and protein content without addition of nitrogen fertilizer.** Alfalfa and other forages can be established more cheaply than Comfrey, since it is usually planted as root cuttings, especially at the close spacing needed for maximum yield.
Comfrey Advantages:
Advantages of Comfrey are that it is very winter hardy in northern environments and could stabilize soil on erodible lands. It also produces fresh forage at a time of the year (spring, fall) when forage may be short from other sources."*
-'Comfrey' by Teynor, Putnam, Doll, Kelling, Oelke, Undersander and Oplinger, 'Alternative Field Crops Manual', University of

Wisconsin: Cooperative Extension, University of Minnesota: Center for Alternative Plant & Animal Products, and Minnesota Cooperative Extension, February **1992**, updated November 1997, page 6.

"**The simple principle to remember is that the growth of Comfrey leaf needs high nitrogen**, whereas the production of fruit (seeds) calls for higher phosphate, and the formation of roots (tubers, potatoes, etc.) calls for higher potash (potassium). **So feed a higher balance of nitrogen to the Comfrey to promote leaf growth.**"
-Comfrey: Nature's Healing Herb & Health Food by Andrew Hughes. Japan: Sanyusha Publishing Co., Ltd, **1992**, page 154.

"**It is necessary to pay special attention to regular after-manuring (after cutting leaves), which should be done after every cut.** Plants should also have organic micro-element spraying. Recently some very valuable organic micro-element seaweed-based products that orginated with Comfrey growers have been put on the market. The spraying of the leaves of Comfrey with one of these increases the solids of the leaves.
It has already been demonstrated by more than one Comfrey grower in Japan that Comfrey treated with these organic micro-elements or grown in soil treated this way is not seriously affected by nematodes."
-Comfrey: Nature's Healing Herb and Health Food by Andrew Hughes. Japan: Sanyusha Publishing Co., Ltd, **1992**, page 155.
  (I put Azomite® and/or dried, granulated seaweed in my soil. Azomite is a mined product that has minerals and trace elements. Seaweed such as kelp has minerals, trace elements and enzymes.)

"Starting with a foundation of feathers, old mattress hair, dead carcases, farmyard manure dug in, or the best compost and an alkaline soil. Mushroom compost can be useful as it contains some lime as of course does calcified seaweed, and **seaweeds** will form part of an annual feeding programme for the future.
Comfrey is a nitrogen hungry plant so this needs to be supplied in early spring, usuallly in the form of fish, blood and bone, or hoof and horn which tends to last longer. Pigeon or poultry manure is ideal.
As Comfrey produces more potassium than any other plant, this should be provided for with an occasional dose of **wood ashes or rock potash**, whilst it is also available in the seaweed meal."
-Comfrey: Symphuo Symphytum: A Multi-Purpose Herb by Philip Clarke. Edinburgh, England: The Pentland Press, **1997**, page 7
  (Seaweed has 60 trace / micro minerals and some nitrogen, potassium, phosphate, and magnesium.)
  (Wood ashes from hardwood trees such as oak and maple contain around 3% potassium, 15% calcium, and micronutrients. It increases the alkalinity of soil. Scatter it lightly.)
  (Rock potash is produced by crushing rock and contains 10-12% potassium oxide. It is effective for a long time because it is very slow to form soluble forms of potassium. It best applied in autumn or early winter.)

"Growing Comfrey in the Garden: **Feed Comfrey plants every year or two with a high-fertility soil improver or high-nitrogen fertilizer.** Grass clippings applied in spring, with shredded prunings added in autumn, can be an effective feeding schedule."
-Rodale's Illustrated Encyclopedia of Organic Gardening edited by Pauline Pears. Emmaus, Pennsylvania: Rodale Press, **2002**, page 206.

"**Come the winter when the Comfrey plants have died back, spreading a mulch of manure or compost will get them off to a good start next year. Every three or four years, lime if the soil is going acid.**
  Throughout the growing season you can spread manure, even fresh poultry manure, between the plants.
  Alternatively a good couple of handfuls of chicken manure pellets per plant with about 250 grams (8 ounces) of fish, blood and bone fertiliser in March and July to really get them set up.
If the Comfrey plants are not looking too well, it may be a nitrogen deficit. They're a very leafy and greedy plant. Give around 100 grams (3.5 ounces) per square metre (10.7 square feet) of high nitrogen fertiliser like 'Prilled Urea' or 'Nitro-Chalk'®."
-'Planting, Cultivating, Harvesting and Problems of Comfrey' by John Harrison, Allotment and Gardens: Grow Your Own - Allotment - Gardening Help, https://www.allotment-garden.org, Penygroes, Gwynedd, Wales, **2018**.
(Prilled Urea is a nitrogen fertilizer manufactured by the reaction of ammonia and carbon dioxide. It contains 46% nitrogen.)
(Nitrochalk® is a trademarked chemical fertilizer that contains calcium carbonate and ammonium nitrate. It is 26% nitrogen, half is in the ammonical form and half is nitrate. Some nitrogen is immediately available and some later. It is also used under other trade names.)

Manure

Most animal manure consists of feces. These include farmyard manure (FYM) or farm slurry (liquid manure). Farmyard manure also contains plant material, usually straw, that has been used as bedding for animals. Farm slurry is produced by intensive livestock rearing where concrete or slats are used, instead of straw.

Manure from different animal species has varying nutrients. Manure increases soil fertility by adding organic matter and nutrients, such as nitrogen, that are utilised by bacteria, fungi and other organisms in the soil.

One of the problems of manure is that it can have weed seeds that will grow in your Comfrey patch. The heat of composting kills weed seeds but reduces the amount of nitrogen. Straw is better than hay to use for bedding for animals because straw has fewer weed seeds than hay. The most common seed is grass.

You can put used cat litter lightly around Comfrey plants, but try to keep it off the leaves. I would not do this around plants that you are harvesting for your own consumption. If your cat has a contagious illness, I would not do it then.

"**The Reverend E. Highton's (of Bude, Cornwall, England) success in growing a heavy Prickly Comfrey produce is plainly attributable to high manuring with good fat farmyard-dung** in preparing his land for the Comfrey sets (roots to plant), and the point cannot be too strongly enforced that, unless the soil be naturally rich in deep black mould, the extraordinary produce the roots are generally expected to yield can only be realized by a similar course being adopted, or by correspondingly large applications of artificial manures being made."
-'The Journal of the Royal Agricultural Society of England, Second Series, Volume 18', London, England, **1882**. Includes article: 'On Green or Fodder Crops Not Commonly Grown, which Have Been Found Serviceable for Stock-Feeding' by Joseph Darby.

"**Manuring to obtain and increase this Comfrey yield, particularly on poor soil, is necessary.**
Those who have poultry have no problems; they can fork on very roughly four ounces a square yard (113 gram per 9 square feet or 0.83 square meter) of fresh droppings in the spring between the plants, and give another application between cuts about July.
Give lime at the rate of two or three ounces (56 or 85 grams) a square yard in November, and leave it to Nature.
The 'burning' effect of the excessive and rapidly available nitrogen in fresh poultry manure cannot harm Comfrey as the roots are down out of harm's way, and they can take their nitrogen crude.
Potassium:
Where poultry manure is used exclusively for a long period, such as five years, balancer dressings of **potash (potassium)** are essential, transposed on this small scale to one ounce (28 gram) of sulphate of potash (muriate of potash only if the other is unobtainable) to the square yard.
The true organic gardener, and those who dislike buying what they can supply from the garden, can use wood ashes.
-Russian Comfrey: A Hundred Tons an Acre of Stock or Compost for Farm, Garden or Smallholding by Lawrence D. Hills. London England: Faber and Faber, Limited, **1953**, page 119.

"**Stable manure** was 10 kg (22 pounds) per offset (root) which equals 17 tons per 'are' or 22 tons (United Kingdom ton) per acre.
Chemical fertilizer was 30 grams per offset which equals 50 kg per 'are' or 445 pounds per acre. N-P-K (Nitrogen-Phosphorous-Potassium) = 2:3:5.
Carbonate of lime (limestone, calcium carbonate) was 300 grams per offset which is 500 kg per 10 'are' which equals 2 tons per acre.
There were 1,710 offsets per 10 'are' which equals 6,800 offsets per acre."
-Comfrey Report No. 3: Feeding Dairy Cattle in Japan' by Meiji Milk Producing Co., Tokyo, Japan for Henry Doubleday Research Association, Braintree, Essex, England, 31 pages, July **1964**, page 9.
('Are' is 120 square yards = 1080 square feet =100 square meters.) (United Kingdom ton is 2,240 avoirdupois pounds =1,016 kg.)

"Manuring:
Comfrey responds well to dressings of dung and compost, and 'Hoof and Horn' meal, once it is well established, but in its first year in a fertile soil no manuring is required; on a hungry patch, a barrowload of compost per pole will be sufficient."
-Goat Husbandry by David Mackenzie. London, England: Faber and Faber Ltd., **1967**, page 315.
('Pole' or 'perch' or 'rod' is 30.5 square yards = 3 yards x 10 yards = 25.5 square meters. One acre = 160 square poles or square rods. One ton per acre = 1 stone per pole. One stone = 14 pounds = 6.3 kg.)

"**Another excellent Comfrey manure and compost heap activator is pigeon manure.** Comfrey will take it. Scatter the droppings between the rows, ideally in showery weather or when you can hose the beds, and the plants will leap ahead. Use it in layers not more than half an inch (1.27 cm) thick in your compost heap, or to feed your Comfrey bed.
The Comfrey crop will take crude sewage sludge and liquid farmyard manure even when it is dormant, so tanker sewage sludge as sprayed on pastures and stubbles by many local authorities in Britain and the United States could go on at the rate of between 6,000 and 12,000 gallons (22,712 and 45,424 liters) an acre. This would have to be low in metals because of the risk of their building up to toxic levels in the soil as the years went by.
Alternatives would be farm slurry, and methane generator slurry, which would hold far more humus."
-Comfrey: Fodder, Food and Remedy by Lawrence D. Hills. New York: Rizzoli Universe Books: **1976**, pages 163, 164, 226.

"**For Comfrey in the spring and again about June, if possible, scatter pigeon or poultry manure on the surface and fork it in, or any other fresh manure you have.**
A 2 to 4 inch (5 to 10 cm) thick coat of spent mushroom compost is also an excellent weed suppressor and food while the bed is young, and watering with **Household Liquid Activator** at the rate of roughly 1 quart (0.94 liter) a plant is a useful way of providing cheap nitrogen for a crop that needs this faster than compost can supply it."
-Organic Gardening by Lawrence D. Hills. Middlesex, England: Penguin Books Ltd., **1977**, page 57.
(For 'Household Liquid Activator', see below, 'Urine as Fertilizer'.)

"**Apply manure regularly every spring, always on the surface or lightly forked in; you can use fresh poultry or pigeon manure, sewage sludge, even cesspool pumpings, for Comfrey can take its nitrogen crude, but never cover the crowns**

with pigeon manure.
If you can obtain a straw or peat based deep litter compost to spread on the surface, this is ideal.
Comfrey grows faster than any other plant in your garden and needs a good supply of nitrogen to keep it gathering the other plant foods and trace elements, since it is not a legume and therefore cannot fix its own supply. Compost contains too little nitrogen to keep pace with its greed."
-Fertility Gardening: The Organic Way to Make Your Garden Grow by Lawrence D. Hills. London, England: Cameron & Tayleur Books Limited, **1980**.

"**The preference for poultry manure lies in the excellent blending of high nitrogen and calcium, both needed in large quantities by Comfrey.** Its value lies not only in the nitrogen-rich droppings, but in the fact that Vitamin B12 develops in the deep litter, in the fungus growths so relished by the chickens that scratch in the straw.
Stable and cow-shed manure, full of straw and stacked to mature, is also top class. Pig manure, especially straw based, is high quality too. The application of these will restore nature's cycle, returning to the soil what has been taken from it.
The period of maturing compost can be reduced to 6 weeks (1 1/2 months) instead of the usual 3 months, making the turning of compost unnecessary in the process.
One of the micro-element concentrates made from kelp (seaweed) can do this for the farmer and add the necessary micro-elements to the compost at the same time."
-Comfrey: Nature's Healing Herb & Health Food by Andrew Hughes. Japan: Sanyusha Publishing Co., Ltd, **1992**, page 157.
    (Most writers about growing Comfrey say raw manure is fine, i.e., it does not need to be composted first.)

"**Comfrey is a high-protein forage that, unlike legumes, obtains all of its nitrogen from the soil.**
Older plantings with leaves showing a lighter green color will usually require broadcasting or sidedressing of nitrogen fertilizer. **Recommended rates vary from 40 to 100 pounds (18 to 45 kg) Nitrogen per acre depending on soil organic matter. Barnyard manure is an excellent nutrient source for Comfrey.**
Productive Comfrey, like silage corn, removes relatively large amounts of potassium, phosphorus, and calcium from the soil. Comfrey productivity is not very sensitive to soil pH, but highest yields occur on soils with pH of 6.0 to 7.0. **Soils testing in medium range should receive about 25 pounds (11 kg) P2O5 (phosphorus) & 120 pounds (54 kg) K2O (potassium) per acre.**"
-'Comfrey' by Teynor, Putnam, Doll, Kelling, Oelke, Undersander and Oplinger, 'Alternative Field Crops Manual', University of Wisconsin: Cooperative Extension, University of Minnesota: Center for Alternative Plant & Animal Products, and Minnesota Cooperative Extension, February **1992**, updated November 1997, page 4.
    (A legume is a plant, fruit or seed in the family Fabaceae. Legumes are grown for their grain seed called pulse, for livestock forage and silage, and as soil-enhancing green manure. These include alfalfa, beans, carob, clover, indigo, lentils, mesquite, mimosa, peanuts, peas, soyeans and tamarind.)
    (Soil pH is a measure of the acidity or basicity {alkalinity} of a soil. pH is the negative logarithm base 10 of the activity of hydronium ions {H+} in a solution. It normally falls between 3 and 10, with 7 being neutral. Acid soils have a pH below 7 and alkaline soils have a pH above 7. Soil pH affects many chemical processes.)

"For crops such as angelica, burdock, elecampane, **Comfrey** and some of the 'mint family' plants such as nettles and skullcap **that prefer a richer soil with more readily available nitrogen, we follow the soil test recommendation data forms that that the soil labs give for 'field crops'.**
Incidentally, I have never seen 'medicinal herb crop' listed on these soil test recommendation data forms.
The fertility recommendations from the soil testing labs for field crops generally recommend one to one-and-a-half tons per acre of composted chicken manure."
-The Organic Medicinal Herb Farmer: The Ultimate Guide to Producing High-Quality Herbs on a Market Scale by Jeff Carpenter and Melanie Carpenter. White River Junction, Vermont: Chelsea Green Publishing, **2015**, page 119.
    (The 'Cooperative State Research, Education, and Extension Service' {CSREES} is an agency of the 'United States Department of Agriculture' {USDA}. Its mission is to 'advance agriculture, the environment, human health and well-being, and communities' by supporting research, education, and extension programs at land-grant universities and other organizations it partners with. Usually, this agency is called 'Cooperative Extension' or 'County Extension'. And specifically for each state such as 'North Carolina Cooperative Extension Office'. All counties in the United States have an office. They provide free or low cost soil testing as well as free agriculture information.)

Fishmeal or Dried Blood Fertilizers

Fish meal is made from fish not used for human consumption; some is made from bones and internal organs left over from processing fish. It is powder or cake made by drying and often cooking, and then grinding. Fish meal is **Nitrogen 9.6%**, Phosphorus 3%, Potassium 0%.

Blood meal is dry powder made from blood. It is used as a high-nitrogen organic fertilizer and a high protein animal feed. **Nitrogen 13.25%**, Phosphorus 1.0%, Potassium 0.6%. It is one of the highest non-synthetic sources of nitrogen. It usually comes from cattle or hogs as a slaughterhouse by-product. Blood meal is a concentrated source of nitrogen so follow directions on the package.

"If dried blood or fishmeal are used on the Comfrey bed, they should be applied between May and July, when the soil is warm enough for them to break down fast instead of going mouldy and being wasted, as they would in winter."
-Fertility Gardening: The Organic Way to Make Your Garden Grow by Lawrence D. Hills. London, England: Cameron & Tayleur Books Limited, **1980**.

Urine as Fertilizer

"The mixture of two parts water to one of urine (human) which is known to H.D.R.A. (Henry Doubleday Research Association) members as 'H.L.A.' or 'Household Liquid Activator' because it is the best and cheapest of all compost heap activators."
-Comfrey: Fodder, Food and Remedy by Lawrence D. Hills. New York: Rizzoli Universe Books: **1976**, page 163.
(The average urine of someone living in a western developed country has Nitrogen - Phosphorus - Potassium 11 - 1 - 2.5.)

*"**Household Liquid Activator is a mixture of 1 part urine to about 3 parts water.** Organizing one's family affairs to provide this useful mixture has more advantages than cash saving and simplicity.*
Like all mammals we pass the surplus potash (potassium) from out bodies in our urine, *and this is why there is only a trace of this vital plant food in sewage sludge. It has all gone down to the sea in the effluent (liquid waste), where it grows algae no one wants on its way through out polluted rivers.*
Our bodies need relatively little potassium compared with the quantity there is in our food, and this should be returned quickly and naturally to the soil from our livestock and ourselves, so that the same potassium molecules are used over and over again.
Because the carbon-nitrogen ratio is high enough, the use of diluted urine on compost heaps involves no smell from ammonia wasted into the air, *and the form in which nitrogen is present favours some remarkably good cellulose-breaking-down bacteria, while conditions in a compost heap are unfavourable to the bacteria concerned with human disease."*
-Organic Gardening by Lawrence D. Hills. Middlesex, England: Penguin Books Ltd., **1977**, page 24.

" 'Household Liquid Activator' can be used, though not on clay soils, because the accumulatoion of salt can produce a sodium clay which is permanently sticky."
-Fertility Gardening: The Organic Way to Make Your Garden Grow by Lawrence D. Hills. London, England: Cameron & Tayleur Books Limited, **1980**.
    (I would think that Household Liquid Activator could be used in a clay soil if it was spread around a lot rather than fertilizing the same piece of land over and over. It would also depend on how much rain there is.)

*"**Comfrey Production: Urine can also be used for the production of Comfrey (Symphytum officinale).***
The Comfrey plant is a fast growing plant that is able to take up huge amounts of nutrients (particularly Nitrogen) that are accumulated in the leaves. *Comfrey can be harvested almost monthly and due to the lack of fiber in the plant, the leaves break down quickly to a thick black nutrient-rich liquid after harvest. Comfrey can be used as follows:*
    1. As a compost activator that adds Nitrogen to the heap and helps to heat the compost heap.
    2. As a liquid fertilizer (Comfrey tea) produced by rotting the leaves down in rainwater for around 4-5 weeks.
    3. As a mulch or side dressing with a 2-inch (5 cm) layer of Comfrey leaves around the plants that will slowly break down and release the plant nutrients.
The production of Comfrey offers an additional safety barrier in the system, *and it allows the urine nutrients to be stored in the Comfrey plants."*
-'Urine as Liquid Fertilizer in Agricultural Production in the Philippines: A Practical Field Guide' by Robert Gensch, Analiza Miso and Gina Itchon for Sustainable Sanitation Alliance, published by Xavier University Press, Cagayan de Oro City, Philippines, **2011**.

*"**A most excellent Comfrey feed is undiluted urine applied at a rate of approximately 500 ml (16.9 United States fluid ounces) per plant twice per growing season.**"*
-'Comfrey: Its History, Uses and Benefits' by Paul Alfrey, Permaculture Magazine: Earth Care, People Care, Future Care; Hampshire, England, www.permaculture.co.uk, March 3 **2016**.

*"**Human urine provides an excellent source of nitrogen, phosphorous, potassium and trace elements for plants,** and can be delivered in a form that's perfect for assimilation. With a constant, year-round and free supply of this resource available, more and more farmers and gardeners are making use of it.*
Urine is 95% water. The other 5% consists of urea (around 2.5%), and a mixture of minerals, salts, hormones and enzymes. *It is a blood bproduct, but despite containing some bodily waste it is non-toxic.*
    The average urine from a healthy adult will release 11 grams nitrogen/urea, 1 grams phosphorus/super-phosphate and 2.5 grams potassium. The normal range for a 24-hour urine output is 800 to 2000 milliliters (27.0-67.6 United States fluid ounce) per day with a normal fluid intake of about 2 liters (2.11 quart) per day. That's an average of 1400 ml (47.3 US fluid ounce) per person, per day!
Fresh human urine is sterile and so free from bacteria. Only when it is older than 24 hours the urea turns into ammonia, which is what causes the distinctive smell. Antibiotics, vitamin supplements and other medications will end up in your urine, but in such minute quantities as to be negligible, especially when diluted in water.

Some plants such as Symphytum x uplandicum- Comfrey 'Bocking 14' can handle neat (undiluted) urine. Following cutting the plants, you can provide each plant with approximately 700 ml (23.6 US fluid ounces) of undiluted urine.
-'Biological Fertiliser: Human Urine', www.balkep.org, The Balkan Ecology Project is a permaculture-inspired, grassroots project based in south eastern Europe, Bulgaria; http://balkanecologyproject.blogspot.com/2016/02/taking-piss.html, February 11 **2016**.

Lime Fertilization

Agricultural lime (agricultural limestone, aglime, garden lime) is a soil additive made from pulverized limestone or chalk. The primary active component is calcium carbonate.
Further classification includes high calcium, argillaceous (clayey), silicious, conglomerate, magnesian, dolomite, and others. Additional chemicals vary depending on the mineral source and may include calcium oxide, magnesium oxide and magnesium carbonate.
Unlike quicklime (calcium oxide) and slaked lime (calcium hydroxide), agricultural lime does not need burning in a kiln; it only requires milling.

Dolomite lime contains magnesium along with calcium carbonate while calcitic lime only contains calcium carbonate. Usually, dolomite is 50% calcium carbonate and 40% magnesium carbonate.

Agricultural lime increases the pH of soil (the lower the pH, the more acidic the soil) and provides calcium and magnesium. It improves the uptake of major plant nutrients on acid soils.
Check your soil pH before applying lime. Different animal manures affect the soil in various ways.

Gypsum is calcium sulfate, a naturally occurring mineral. It is useful in loosening up compacted soil such as clay, especially when impacted by heavy machinery. One of its main uses is to remove excess sodium from soil while it adds calcium and sulfur.

"The autumn lime-scattering should be at the rate of from four to six ounces (113-170 grams) per square yard (0.836 square meter)."
-Russian Comfrey: A Hundred Tons an Acre of Stock or Compost for Farm, Garden or Smallholding by Lawrence D. Hills. London England: Faber and Faber, Limited, **1953**, page 119.

"In a London, England garden 4 ounces (113 grams) to the square yard (0.836 square meter) of lime every autumn is not too much."
-Compost, Comfrey and Green-Manure: 'Henry Doubleday Research Association' First Gardeners Report by Lawrence D. Hills. Braintree, Essex, England: Henry Doubleday Research Association, **1959**.

Carbonate of lime (limestone, calcium carbonate) *was 300 grams per offset (root) which is 500 kg per 10 'are' which equals 2 tons per acre. There were 1,710 offsets per 10 'are' which equals 6,800 offsets per acre."*
-Comfrey Report No. 3: Feeding Dairy Cattle in Japan' by Meiji Milk Producing Co., Tokyo, Japan for Henry Doubleday Research Association, Braintree, Essex, England, 31 pages, July **1964**, page 9.
    (An 'are' is 120 square yards.)   (In the United Kingdom the ton is 2,240 avoirdupois pounds = 1,016 kg.)

"Every other autumn give the Comfrey bed 1/2 pound (8 ounces or 226 grams) of lime a square yard (0.836 square meter) dug in with the autumn clean-up."
-Down to Earth Gardening by Lawrence D. Hills. London, England: Faber & Faber, **1967**, 1975, page 219.

"**In the late autumn, dig between the Comfrey, and fork in 8 ounces (226 grams) of slaked lime (calcium hydroxide) a square yard (250 gram a square meter) every other autumn.**"
-Fertility Gardening: The Organic Way to Make Your Garden Grow by Lawrence D. Hills. London, England: Cameron & Tayleur Books Limited, **1980**.
    (Calcium hydroxide is also called slaked lime, hydrated lime, caustic lime, builders lime and pickling lime. It is an inorganic compound made when calcium oxide {lime or quicklime} is mixed, or slaked with water.)
    (**For organic farming and gardening not all sources of lime are permitted. Mined limestone {calcium carbonate} and dolomite {magnesium carbonate} are allowed. Quicklime {calcium oxide} and slaked lime {calcium hydroxide} are not.** Mined gypsum is permitted; calcium sulfate that is a by-product of superphosphate manufacture or from reclaimed drywall is not.)

Chemical Fertilizers

"Comfrey will do well only in soil that is rich in humus, rich in active aerobic bacteria, fungi and micro elements. **Comfrey does not like artificial fertilizers,** and these should be used if ever, only with the greatest care. To do well it must find Vitamin B12 in the soil, and this will be there only if it has the right conditions for its production."

-'Miracle Grass, Comfrey' by Andrew Hughes published in **1966** in Japanese. Part of it is in: 'Comfrey: Nature's Healing Herb & Health Food' by Andrew Hughes, Japan: Sanyusha Publishing Co., Ltd, 1992.
And part is in 'Comfrey Report No. 1: The 1954 Research Results' by Lawrence Hills published in 1955.
(If you have a translation of it in English, I would appreciate having a copy.)

"The average yield of Comfrey grown during 1959 to 1962 and fertilized with 30 and 60 kg (66-132 pounds) nitrogen per hectare was 28 to 30 tons per hectare. Yield was highest, 60 tons per hectare, during the second year of growth and thereafter declined. Crude protein ranged from 15.6 to 16.4% and crude fibre from 24.3 to 25.8%."
-'Effect of Nitrogen (N) Top-Dressing on Yield Quality of Fodder Comfrey (Symphytum peregrinum)' by by S. Tabin, S. Berbec and H. Wrebiakowski, Kat Szcaegol Uprawy Roslin Wydz Rol WSR, Lublin, Poland; Annales University Mariae Curie-Sklodowska, Volume 21, pages 139-153, **1966**. (I do not have this report. If you have an English translation, could you send it.)
> (A hectare is a square with 100 meter sides, or 10,000 m2. There are 100 hectares in one square kilometer. An acre is 0.405 hectare, and one hectare is 2.47 acres.)

"In trials during 1967-69, length and number of leaves of Russian Comfrey increased with fertilizer level. Leaf number and yield also increased as distance between plants increased.
Optimum fertilizer applications for maximum yields were 30,000 kg f.y.m. (farm yard manure) + 300 kg urea + 300 kg triple superphosphate + 240 kg KCl (potassium chloride) and 1500 kg lime/hectare."
-'Effects of Planting Density and Amount of Fertilizer on the Growth and Yield of Russian Comfrey (Symphytum peregrinum)' by K.S. Lee, C.J. Kang and I.K. Han, Livestock Expo Statn, Suwon, Korea; Nong-sa Si-hem Yen-ku Po-ko= Research Report to Office of Rural Development, Suwon, Korea, Volume 12, No. 4, pages 83-88, **1969**. (I do not have this report. If you have an English translation, could you please send it to me.)
> (Urea is an inexpensive form of nitrogen fertilizer with an NPK {Nitrogen-Phosphorus-Potassium} ratio of 46-0-0. Triple Superphosphate was one of the first high-analysis phosphorus fertilizers that became widely used in the 20th century. It is calcium dihydrogen phosphate and as monocalcium phosphate. It is NPK 0-45-0 with 15% calcium. Potassium chloride is a commonly sold fertilizer with NPK 0-0-60.)

"Do not waste compost on the Comfrey bed. There is no record of anyone securing a good yield that way. Comfrey needs its nitrogen faster than compost releases it. **Calcium nitrate is in theory the ideal fertilizer,** but in practice produces poor Comfrey yields, while fresh poultry droppings are the best."
-Comfrey Report: The Story of the World's Fastest Protein Builder and Herbal Healer, Conservation Gardening and Farming Series: Series C by Lawrence D. Hills. England: Henry Doubleday Research Association, **1975**, page 85.
> (Calcium nitrate is the only water soluble source of calcium available for plants. Diseases such as blossom end rot are controlled by it. It is usually applied dissolved in water, but can be used as a solid on top of the soil. Calcium nitrate is produced by applying nitric acid to limestone and then adding ammonia. It is a 'double salt' because it is two nutrients high in sodium. It is an artificial fertilizer.)

"I have never seen a good Comfrey crop that was fed with chemical nitrogen, though in theory, Nitro-chalk® (calcium nitrate) is ideal. It is in fact the research stations who insist on feeding it chemically who often have the lowest yields. Both plots (University of California, Davis)...**four top dressed treatments of 1726 pound (782 kg) per acre of elemental nitrogen (as ammonium nitrate) were applied."**
-Comfrey: Fodder, Food and Remedy by Lawrence D. Hills. New York: Rizzoli Universe Books: **1976**, pages 91, 163.
(Nitrochalk® is a trademarked chemical fertilizer that contains calcium carbonate and ammonium nitrate. It is 26% nitrogen, half is ammonical form and half is nitrate. Some nitrogen is immediately available and some later. It is used under other trade names.)

"Doring in 1959 in West Germany obtained a higher dry matter yield of Symphytum x uplandicum with 75 kg (165 pounds) Nitrogen, 62.5 kg (137 pounds) P2O5 and 90 kg (198 pounds) K2O per hectare (4.57 tons/hectare) than with the phosphate and potassium fertilizers alone (3.66 tons/hectare). Increasing the nitrogen level to 100 or 125 kg/hectare produced no further yield increases.
Symphytum x uplandicum:
> In Poland the average fresh yield of Symphytum x uplandicum in the second, third and fourth years was 24.6 tons/hectare with application of 30 kg (66 pounds) nitrogen/hectare and 33.4 tons/hectare with application of 60 kg (132 pounds) nitrogen/hectare with 50 kg (110 pounds) P2O5 and 120 kg (264 pounds) of K2O/hectare supplied for both treatments (Tabin, Berbec and Wrebiakowski, 1966).

Symphytum asperum:
> Symphytum asperum dry matter yield increased form 3.90 tons/hectare with no fertilizer to 4.80 tons with 30 kg (66 pounds) Nitrogen, 45 kg (99 pounds) P2O5 and 60 kg (132 pounds) K2O/hectare. **Doubling the rates of phosphate and potassium application increased the yield** to 5.30 tons/hectare (Vavilov and Kondrat'ev, 1975).

Evidence that Comfrey is particularly responsive to potassium fertilizers was obtained by Chubarova, Vorob'ev and Rybnikova (1970) who obtained a higher dry matter yield of Symphytum x uplandicum with 100 kg (220 pounds) Nitrogen, 60 kg (132 pounds) P2O5 and 218 kg (480 pounds) K2O/hectare (4.40 tons/hectare) than with 120 kg (264 pounds per hectare each of Nitogen, P2O5 and K2O (3.74 tons/hectare)."
-'Comfrey Symphytum spp. as a Forage Crop' by J.C. Forbes, A.D. McKelvie, and P.J.C. Saunders, North of Scotland College of Agriculture, Aberdeen, United Kingdom; Herbage Abstracts, Volume 49, No. 12, pages 523-539, **1979**.

*"Fertilization of the soil is necessary to keep the Comfrey fields producing well, **the fertilizer being best applied at end of April or beginning of May, at rate of 400 kg/hectare (881 pounds/ 2.47 acres) of a 5-20-20 (N-P-K) formula.** Soil should be adjusted to pH 6.5."*
-'Symphytum Peregrinum Ledeb., Boraginaceae, Comfrey, Russian Comfrey, Quaker Comfrey' by James A. Duke, Handbook of Energy Corps, unpublished, Purdue University, Indiana, **1983**.

Mineral Fertilization

"The high ash (mineral) content of Comfrey, approximately 14 percent of the dry matter, is 35 percent potash (potassium), 14 percent calcium and only 5 percent phosphorus. The general routine at the end of the season should be lime and potash if required, on the surface and worked in between the rows with a cultivator about late November after the last cut. In February or early March before the shoots are up, F.Y.M. (FarmYard Manure) or poultry manure again, and scuffle it in. A summer dressing (about June, when the crop is growing most rapidly) of F.Y.M. or pig manure is advisable."
-Russian Comfrey: A Hundred Tons an Acre of Stock or Compost for Farm, Garden or Smallholding by Lawrence D. Hills. London England: Faber and Faber, Limited, **1953**, page 66.

*"Grasses wither in summer owing to high temperature and low precipitation. Dallisgrass belongs to the southern grass (Panicoidea or Eragrostoidea) but **Russian Comfrey is not influenced by the climatic condition.** Dallisgrass is recognized to be a suitable grass for grazing land and **Comfrey is a useful crop to soiling.***
***The effects of trace elements** on them were studied in this paper.*
*Both plants were cultivated in pot or earthern pipe which were filled with granite residual soil. **Ammonium sulfate (or sodium nitrate), calcium superphosphate and potassium sulfate were applied** 1.2 kg (2.6 pounds) per one 'are' (100 square meters) as N, P_2O_5 and K_2O respectively.*
Trace elements were applied in the following doses:
4 kg (8.8 pounds) of $CuSO_4$, 2 kg (4.4 pounds) of $CoCl_2$, 6 kg (13.2 pounds) of $MnCl_2$, 2 kg of Na_2MoO_4, 2 kg of $ZnCl_2$ and 6 kg of H_2BO_3 per hectare each.
The growth of Russian Comfrey was higher by the application of ammonium sulfate than of sodium nitrate, and increased 30% by Co (Cobalt), Mo (Molybdenum) and B (Boron) but decreased by Mn (Manganese).
Trace element contents of Russian Comfrey were much higher by the application of sodium nitrate than of ammonium sulfate except B. Especially Co content by the application of the former fertilizer was five times more than by the latter.
Application of trace elements increased the content of them in plants, especially Co and Mo.
The content of moisture, crude protein, crude fiber, N free extract and crude fat of Russian Comfrey increased by any kind of trace elements.
Though trace elements promoted the growth and increased the yield of plants, the influences of them were different by the kind of plant, soil and the application method of trace element. The contents of trace elements in plant were different by the kind of nitrogenous fertilizer, plant, soil and the climatic conditions. The contents of moisture, crude protein, crude fiber, crude fat, N free extract and ash were effected by the kinds of plants and trace elements."
-'Effect of Trace Elements for Dallisgrass and Russian Comfrey' by M. Ikeda, K. Kurozumi, J. Tsubota and H. Matsumura, Department of Animal Husbandry, University of Hiroshima, Fuku-yama, Japan; Journal of Japanese Society of Grassland Science, Volume 10, Issue 2, pages 100-104, **1964**.

> (Granite is a naturally occurring earth mineral, present in large deposits across the western United States. Soils that have significant amounts of decomposed granite present challenges to growing plants. However, granite dust is recommended to improve clay soils.)
>
> (An 'are' is a unit of area in the metric system, equal to 100 square meters which is the equivalent of 0.0247 acre.)

"Ikeda et al. (1964) obtained **increased yields of Symphytum x uplandicum in garden pots following the addition of cobalt, molybdenum or boron to the soil, but not using copper, manganese or zinc.**"*
-'Comfrey Symphytum spp. as a Forage Crop' by J.C. Forbes, A.D. McKelvie, and P.J.C. Saunders, North of Scotland College of Agriculture, Aberdeen, United Kingdom; Herbage Abstracts, Volume 49, No. 12, pages 523-539, **1979**.
(* -'Effect of Trace Elements for Dallisgrass and Russian Comfrey' by M. Ikeda, K. Kurozumi, J. Tsubota and H. Matsumura, Department of Animal Husbandry, University of Hiroshima, Fuku-yama, Japan; Journal of Japanese Society of Grassland Science, Volume 10, Issue 2, pages 100-104, 1964.)

*"**The biology of Symphytum asperum flowering and fruit bearing was studied in the Komi SSSR, Russian automomous republic.** The effect of soil type, plant density, mineral fertilizers, time of fertilization and leaf feeding with microelements on S. asperum seed productivity was studied.*
S. asperum seed growing in the central taiga subzone of the Komi SSSR should be based on obtaining seeds from thyrses.
***Central inflorescences produced the best seeds.** Seed plots should be located on mineral soils, but not on turf soils. The optimal area of plant nutrition is 70 x 50 cm (27.5 x 19.6 inch).*
***Total mineral fertilization should be conducted in seed plots not in early spring, but after the plants started growing when the soil gets dry.** It is necessary to carry out leaf feeding with Mo (molybdenum), Cu (copper), Zn (zinc) and Basalts or a mixture of these microelements at the beginning of plant budding before plant rows close."*

-'Effect of Growing Conditions on Seed Production by Symphytum Asperum Lepech' or 'Effect of Cultivation Conditions on Symphytum Asperum Seed Productivity' by Y.M. Frolova and N.P. Frolova, Rastitel'nye Resursy, Leningrad, Russia, Volume 19, No. 3, pages 336-341, **1983**. (I could not find this article. If you have an English translation, could you please send it to me.)
>(Taiga, or boreal or snow forest, is a plant community of coniferous forests with mostly pines, spruces and larches.)
>(A thyrse is a type of inflorescence where the main axis grows indeterminately, and branches have determinate growth.)
>(A mineral soil is derived from minerals or rocks with little humus or organic matter.)
>(Basalt is an igneous rock formed from rapid cooling of magnesium-rich and iron-rich lava.)

Rock Powders and Seaweed as Fertilizers

"**Powdered Rock:** After the third year's gathering of leaves and roots of the same Comfrey plants, it is advisable to feed the soil in late fall not only with the usual composted material but with **a thin layer of crushed, ground or powdered rock (granite, for example) over and around the Comfrey beds.**
Seaweed: If you are fortunate enough to live on or near the coast, gather and **thoroughly dry all kinds of seaweeds in April or May, cut them up as finely as possible and incorporate them, i.e., dig them well into the soil.** And last, to put the beds to sleep for the winter, spread a blanket of one-foot (0.3 meter) lengths of seaweed over the entire flower and garden space, and cover with leaves or grass cuttings.
Dig well-rotted manure between the rows when dressing them for winter, spread a layer of compost and leaves."
-Comfrey: What You Need to Know by Ben Charles Harris. New Canaan, Connecticut: Keats Publishing, Inc., **1982**, page 97.

"**Comfrey plants should be sprayed with a seaweed spray** using Maxicrop® or SM3® extract 4 times a year. This will ensure good healthy roots able to resist infection especially from nematodes, promoting the vigour and growth rate of the plants as well as increasing solid content of the leaves, minerals, proteins, carbohydrates and fats by up to 20%."
-Comfrey: Symphuo Symphytum: A Multi-Purpose Herb by Philip Clarke. Edinburgh, England: Pentland Press, **1997**, page 14
>(Maxicrop® is a company that sells seaweed-based liquid or powdered organic fertiliser sold as a bio-stimulant. Chase SM3® is 15% seaweed solids that contain natural plant growth stimulants and trace elements. I do not have any connection with either of these companies.)

Timing of Fertilization

"**The idea is to apply a high nitrogen general fertilizer when the Comfrey plants are growing strongly, not in winter when they are dormant,** and of these dried sewage sludge and dried blood are both excellent, the last at two ounces (56 grams) to the square yard (0.83 square meter) as it is both richer and more expensive."
-Russian Comfrey: A Hundred Tons an Acre of Stock or Compost for Farm, Garden or Smallholding by Lawrence D. Hills. London England: Faber and Faber, Limited, **1953**, page 120.

"**The autumn lime-scattering** should be at the rate of from four to six ounces (170 grams) per square yard.
Ten to fifteen tons of farm-yard manure to the acre turned under in the autumn or spring, with any liming or potash (potassium) balancing fitted in during the alternative period."
-Russian Comfrey: A Hundred Tons an Acre of Stock or Compost for Farm, Garden or Smallholding by Lawrence D. Hills. London England: Faber and Faber, Limited, **1953**, pages 61, 119.

"**Manure always in the spring when the plants are beginning to grow,** for the plant foods will wash out on a sandy soil before the Comfrey is feeding again, if this goes on in autumn."
-Grow Your Own Fruit & Vegetables by Lawrence D. Hills. London, England: Faber and Faber Limited: **1971**, page 50.

"**For Comfrey as much as 2 pounds (0.9 kg) a square yard (0.83 square meter) of dried sludge in the spring and again about August is not too much.** Those who are vegans, and will not buy manure for their Comfrey can use compost or even chemicals like **nitro-chalk®**, but they must expect a lower yield."
-Grow Your Own Fruit & Vegetables by Lawrence D. Hills. London, England: Faber and Faber Limited: **1971**, page 46.
(Nitrochalk® is a trademarked chemical fertilizer that contains calcium carbonate and ammonium nitrate. It is 26% nitrogen, half is ammonical form and half is nitrate. Some nitrogen is immediately available and some later. It is used under other trade names.)

Water

"Instead of carefully watering the soil around each piece of **just planted Comfrey root** with a handheld hose, they hook up a single sprinkler, push its long metal stake into the soil at an angle and adjust the water flow so that it spreads into a wide fan over the Comfrey bed from the far end to almost the middle.
They walk away and leave the sprinkler running for a while, then come back when that part of the Comfrey bed is soaked and beginning to get puddly."
-Enchanted Garden: Alan Chadwick's Organic Method of Gardening by Tom Cuthbertson. London, England: Rider & Company /

Hutchinson & Co. Publishers Ltd, **1978**, page 71.

*"**Loosen Hard Crust on Top of Soil Before Watering:** You don't have to worry much about the hard surface choking the Comfrey plants- Symphytum is so strong it can't possibly get throttled. But it isn't good to let that crust form on top of the soil and prevent air and water from getting down into the soil.*

If we break up the crust well, we'll be able to do the deep, slow watering that Comfrey likes. There's a garden saying you can use for situations like this: 'Water with the hoe, not the hose."

All you have to do to the rough-tilthed bed is to make a quick run up and down with the knife hoe, punching through the crust like this, and it will open the bed as much as it needs.

Comfrey has Strong Roots to Get Water:

The young Comfrey roots need plenty of air and water exchange to get started, but soon each plant will send down a big taproot and spread its foliage out so as to cover the soil completely. **The plant keeps its own continuum of moisture** *and moderation of atmosphere, even though it's less than a month old!*

Fast Growing Foliage Keeps in Soil Moisture:

That's the kind of strength Comfrey has. **Not just the strength to push up through poor soil and drive a root down into the heaviest clay hardpan, but also the ability to make a quick cover that will protect that delicate life-sustaining zone an inch (2.5 cm) above the soil surface and a couple of inches (2 inch = 5.0 cm) down below.** *The whole bed will be covered with foliage like that in a few weeks.*

Watering Technique:

There, just that much cultivation will be enough for the Comfrey. **As long as the watering technique is good, the plants will get along fine. The best watering implement for this bed is the oscillator sprinkler or back-and-forth sprinkler as some people call it.** *The oscillator makes very fine drops that spread evenly over a much wider area than the small patch made by the rose hand-held sprinkler. You can set the sprinkler so it will cover a large area, then leave it for up to a half hour before you come back and move it. You can keep your eye on the bed until you see it start to puddle, then move the oscillator."*

-Enchanted Garden: Alan Chadwick's Organic Method of Gardening by Tom Cuthbertson. London, England: Rider & Company / Hutchinson & Co. Publishers Ltd, **1978**, page 81.

(Soil tilth is its physical condition, especially its suitability for growing crops. It includes the formation and stability of aggregated soil particles, moisture content, degree of aeration, rate of water infiltration and drainage. Soil with good tilth has large pore spaces for air infiltration and water movement.)

(A 'rose' head sprinkler has lots of small holes on the top or cap.)

How Much Water is Needed (Right Amount versus Too Much)

If Comfrey roots sit in water for a long time, they will rot and die. So make sure your soil has good drainage.

*"**Both Comfrey plots (University of California, Davis) were irrigated with 3 inches (7.6 cm) of water per irrigation about every two weeks or as called for by weather conditions.***

Four top dressed treatments of 1726 pound (782 kg) per acre of elemental nitrogen (as ammonium nitrate) were applied."

-Comfrey: Fodder, Food and Remedy by Lawrence D. Hills. New York: Rizzoli Universe Books: **1976**, page 91.

*"**In Hong Kong the growth of a trial crop of Symphytum x uplandicum was retarded by the onset of the wet season, and finally the plants died, the crown and roots having rotted** (Hong Kong Department of Agriculture, /Fisheries and Forestry, 1959). In a garden pot experiment Tabin and Berbec (*1968)* **obtained a higher yield of Symphytum x uplandicum over three years with the soil at 60% of field capacity than at 45% or 75% of field capacity."**

-'Comfrey Symphytum spp. as a Forage Crop' by J.C. Forbes, A.D. McKelvie, and P.J.C. Saunders, North of Scotland College of Agriculture, Aberdeen, United Kingdom; Herbage Abstracts, Volume 49, No. 12, pages 523-539, **1979**.

(* -'Influence of Calcium at Different Soil Humidity Levels on the Yield and Quality of Green Herbage of Fodder Comfrey: Symphytum Peregrinum' by S. Tabin and S. Berbec; Annales Universitatis Mariae Curie-Sklodowska, Sectio E Agricultura, Lublin, Poland, Volume 23, pages 301-321, 1968. All in Polish. If you have an English translation, I would appreciate a copy.)

(Comfrey roots are sensitive to soil that is constantly wet. Too dry is better than too wet soil. This is true for storing roots.)

(Field Capacity is the amount of water held in the soil after excess water has drained away. This takes place 2 to 3 days after rain or irrigation in permeable soil of uniform structure. Medium textured soil such as fine sandy loam, silt loam and silty clay loam have the highest water holding capacity. Coarse soil such as sand, loamy sand and sandy loam have the lowest. The volumetric soil moisture amount remaining at field capacity is about 15 to 25% for sandy soil, 35 to 45% for loam soil, and 45 to 55% for clay soil.)

*"**Comfrey thrives in heavy irrigation while it sets up its root system the first year.**
Use up to a 5-day rotation on well-drained soils."*

-'Comfrey Leaf: A New Animal Food Supplement' in The Encyclopedia of Alternative Agriculture by Dr. Richard Alan Miller, Agricultural Consultant and Researcher, USA, **1992**.

*"**Copious (a lot of) water is a help in good Comfrey growth, but water-logged soil is an enemy.** We found too that in a long dry spell before the spring rains come, followed by the rainy season, plants continued their growth, even if slower, in all fields."*

-Comfrey: Nature's Healing Herb & Health Food by Andrew Hughes. Japan: Sanyusha Publishing Co., Ltd, **1992**, page 140.

Very Drought Tolerant

"The thermometer has rarely been below 95 F (35 C) degrees in the shade, and frequently over 105 F (40.5 C). **We have had a long drought of nearly three months,** *followed by almost incessant rain for a fortnight (2 weeks), and in consequence the crops have suffered severely; but, at a time when everything else was burned up,* **the Prickly Comfrey (Symphytum asperrimum) continued to afford abundant green fodder for my horses and stock.** *I may mention that my land is not particularly good, and that the Prickly Comfrey has had no attention beyond being manured when the sets (root cuttings) where first put out, and being kept clear from weeds. -C.E. Ashburner, Richmond, Virginia, September 28, 1876."*
-'Forage Plants and Their Economic Conservation by the New System of Ensilage: Part I: Caucasian Prickly Comfrey' by Thomas Christy, Jun., F.L.S. (Fellow of the Linnean Society), Christy & Co., London, England, **1877**, pages 19, 26.

"In a paper on the progress of agriculture in Natal, South Africa, Dr. P.M. Sutherland, surveyor-general of that colony, speaks of **the advantages possessed by the Caucasian Prickly Comfrey (Symphytum asperrimum) as a fodder-plant, in regions characterized by annually-recurring seasons of drought.** *His remarks will doubtless be of interest to farmers settled in some of our United States and Territories where like meteorological conditions exist."*
-'Popular Miscellany: **A Drought-Proof Fodder-Plant**', Popular Science Monthly (founded May 1872 in the United States to give scientific knowledge to educated layman), Volume 13, page 763, October **1878**.
    (Today Caucasian/Caucasicum Comfrey and Prickly Comfrey are considered two different species.)

"Having a long tap root, Prickly Comfrey is little affected by prolonged droughts. *In 1879, we had a long drought terminating the last day of May. Pastures were parched, streams and springs dried up; but the Comfrey flourished, and on the first day of June we commenced cutting the fourth time since March.*
At such times it is extremely valuable for all farm animals; but especially for hogs and cows.
It may be cut six or eight times a year; and if the seasons are very favorable perhaps ten times in my locality (Mississippi)."
-'The Farmers Book of Grasses and Other Forage Plants for the Southern United States' by D.L. Phares, A.M. (Master of Arts), M.D., Professor of Biology, A&M College of Mississippi, Starkville, Misssissippi, **1881**, page 23.

"The result of experiments made by Professor Buckman several years since, he says:
'Having procured (received) a few sets of **Prickly Comfrey** *with roots attached, we planted them in a plot on the cold clay of the Forest Marble, previously slightly manured. These sets grew rapidly, and we were soon enabled to divide them, until we had as much as a quarter of an acre of ground occupied. The crop was enormous, and this too upon land of very medium quality.*
But we have since then been trying its growth on light sandy soil, and can report that, all through a season of drought, the thick deep roots of the Comfrey have drawn up the moisture which rises hygrometrically in our sand bed, and the result has been a succession of green leaves when surface plants were an utter failure."
-'The Journal of the Royal Agricultural Society of England, Second Series, Volume 18', London, England, **1882**. Includes article: 'On Green or Fodder Crops Not Commonly Grown, which Have Been Found Serviceable for Stock-Feeding' by Joseph Darby.
    (Forest Marble is a geological formation in England that is a flaggy {rock that splits into layers}, sandy, carbonate
    {limestone}, and mudstone.)

"In the south western part of Japan, the regular feed crops wither in summer due to high temperature and little precipitation.
Russian Comfrey does not wither because of its deep root, and can be grown for a period of seven or eight months- i.e., from April or May to November or December, *making it a very suitable feed crop in the dairy farming region of Japan."*
-'On the Ascorbic Acid Content of Russian Comfrey' by M. Ikeda, S. Uchimura and E. Matsui, Journal of the Faculty of Fisheries and Animal Husbandry, Hiroshima University, Japan, Volume 4, pages 103-109, **1962**.

"The only problem is drought, for though Comfrey can draw on subsoil moisture, Rhodesian (now Zimbabwe, Africa) droughts can go on for years. **Three years without rain can kill Comfrey."**
-Comfrey: Fodder, Food and Remedy by Lawrence D. Hills. New York: Rizzoli Universe Books, **1976**, page 148.

"Darby (1882) described S. asperum as a drought-tolerant species, but **in the USSR drought is reported to have adverse (negative) effects both on yield (Vavilov, Dotsenko and Dotsenko, 1974) and on longevity (Tanfil'ev, 1975).**
Yields of Symphytum x uplandicum are seriously reduced by drought (New South Wales {Australia} Department of Agriculture, 1959; Van der Zweerde, 1965; Lachance, 1968)."
-'Comfrey Symphytum spp. as a Forage Crop' by J.C. Forbes, A.D. McKelvie, and P.J.C. Saunders, North of Scotland College of Agriculture, Aberdeen, United Kingdom; Herbage Abstracts, Volume 49, No. 12, pages 523-539, **1979**.

"Forage Systems and Permaculture Zone Two:
Zone two is not fully mulched. It may contain main crop gardens. They shouldn't be in that little area of annual garden. Here grow crops that you much use, much store, maybe only have a single harvest, maybe only visit three times to fully harvest.
When we come to the period of summer drought, we look to greens- to Comfrey, cleavers, and any amount of chard.
There are gardeners at home who grow more chard for their chickens than they grow for themselves."

-'Introduction to Permaculture: Permaculture Design Course Series, Pamphlets 1 to 14' by Bill Mollison at 'The Rural Education Center', Wilton, New Hamshire, **1981**; published by Yankee Permaculture, Sparr, Florida.

"Food Plants for Dry Regions of the Tropics:
Crop plants for arid regions are those that survive and produce in spite of aridity. However, in almost all of these crops, seeds must be germinated or cuttings must be rooted under conditions of almost normal water availability. Therefore, when one speaks of tolerance of dry conditions one is talking mostly about the drought tolerance of the growing or mature plant.
Symphytum officinale, Comfrey, Degree of Tolerance = 1. Rated from 0 (no tolerance) to 3 (high tolerance)."
-'Dryland Farming: Crops and Techniques for Arid Regions' by Randy Creswell and Dr. Franklin W. Martin, ECHO Technical Note, ECHO: Fighting World Hunger, www.echonet.org, North Fort Myers, Florida, **1998**.
    (The tropics are areas on earth that are near the equator.)

*"Irrigation: Comfrey will produce more biomass if irrigated and **in dry climates it's essential to irrigate. Comfrey plants wilt very fast in hot conditions and will stop photosynthesising at this point. Twenty litres (21 quarts) per square meter (1.19 square yard) per week of drought should be more than adequate.** The beauty of biological systems are that, if managed properly, each year the soils improve and the ability of the soil to store water will improve over time.*
We use passive irrigation diverting water from a mountain stream into the paths around the beds. The paths fill with water, we raise the level by blocking the low points with sacks of sawdust, and the water is drawn throughout the soil via capillary action."
-'Comfrey: Its History, Uses and Benefits' by Paul Alfrey, Permaculture Magazine: Earth Care, People Care, Future Care; Hampshire, England, www.permaculture.co.uk, March 3 **2016**.
    (Capillary action or wicking is the ability of a liquid to flow in narrow spaces without the assistance of, or even in opposition to, external forces like gravity.)

Weeds

*"**When the Comfrey plants are established, weeds cannot survive under it,** and the cut is greatly increased by a mulch of manure or compost. Experiments have been tried with **sawdust mulching between the rows,** for the fewer weed seeds that are brought to the surface by hoeing, the better.*
***The Comfrey will, even in the first year, put on so much growth in three weeks that weeds rarely have a chance to set seed,** and the hoeing is only to prevent them taking advantage of temporary daylight to flower and seed."*
-Fertility Farming by Newman Turner. Austin, Texas: Acres USA, **1951**, page 56.
    (Sawdust is acidic so it is good for acid-loving plants. It needs nitrogen to decompose so will take it away from your plants. Therefore, use small to moderate amounts of sawdust.)

"Comfrey will need hoeing shallowly for the first few months after planting, but they are essentially a 'no-digging' crop. The swift growth and constant cutting prevent much weed-seed formation, once the stored seed in the surface layer has germinated."
-Russian Comfrey: A Hundred Tons an Acre of Stock or Compost for Farm, Garden or Smallholding by Lawrence D. Hills. London England: Faber and Faber, Limited, **1953**, page 118.

*"After establishment the Comfrey crop was cultivated with inter-row (between row) equipment, but **perennial weeds, such as creeping thistle, couch grass, and lucerne (alfalfa) from the previous crop, tended to be troublesome."***
-'Russian Comfrey' by L.A. Willey and R.L. Knight, Journal of the National Institute of Agricultural Botany, Cambridge, England, Volume 9, No. 2, pages 139-144, **1962**.

*"**Clean up the Comfrey bed in the autumn and dig between the plants if there are perennial weeds starting between them. On a garden scale, it is possible to mulch a Comfrey bed with leaf mold** as though it were a rhubarb bed, to conserve plant foods and avoid any winter weed growth, with liming about every third or fourth year except on chalky soils."*
-Fertility Without Fertilizers: A Basic Approach to Organic Gardening by Lawrence D. Hills. New York: Universe Books, **1975**.
(For leaf mold see subsection 'Potting Mixture with Comfrey Leaf Mold' in section 'Garden Uses of Comfrey' {Chapter 17, Vol 1}.)

*"**The only difference on a garden scale is that at two foot (0.6 meter) apart each way spacing means that the Comfrey plants crowd together faster and reduce the need for weeding."***
-Comfrey: Fodder, Food and Remedy by Lawrence D. Hills. New York: Rizzoli Universe Books: **1976**, page 162.

"Comfrey must be kept free of grasses, otherwise it will suffer from nitrogen deficiency."
-Comfrey: Nature's Healing Herb & Health Food by Andrew Hughes. Japan: Sanyusha Publishing Co., Ltd, **1992**, page 150.

"Using Comfrey for Grass Exclusion:
*Question: We have an ongoing problem with **invasive Bermuda grass** from outside the garden area. I attended a class recently in which the grower had planted Comfrey all around the borders of his garden because he said that it suppressed grass from infiltrating.*
*Answer: **Comfrey can itself be invasive;** so you may be replacing one evil with another, depending on how you look at the*

situation. Once established, Comfrey is hard to get rid of; so a decision to grow it must be made carefully.
All of the major varieties of Comfrey (Common, Russian Bocking No. 4, Russian Bocking No. 14) have the same effect: they form dense patches that crowd out virtually everything else."
-'Using Comfrey for Grass Exclusion' by Conrad Richter; Richters®: The Herb Specialists, www.richters.com, Goodwood, Ontario, Canada, October 20 **2004**.

"**Quack Grass and Other Undesirable Perennials:** Quack grass is one of the most difficult weeds to manage organically in a perennial garden and requires special consideration. For this it is best to have a strategy of exclusion, removal and tolerance (which could be applied to any weed pest). If at all possible it should be removed prior to planting during the site preparation phase. In this garden we removed as much as possible by running a drag (drag harrow or cultivator) across the soil pulled by a tractor. **Once the garden is planted, quack grass can be excluded from smaller garden spaces by using rhizome barriers. In larger areas its progress can be impeded by using plants such as Comfrey, rhubarb or daffodils but in most cases it is not likely to ever be completely eliminated.**"
-'Edible Forest Garden Permaculture for the Great Lakes Bioregion: Background, Development and Future Plans for The Michigan State University Student Organic Farm Edible Forest Garden' by Jay Tomczak, East Lancing, Michigan, July **2007**.
(Quack grass, couch grass, quitch, twitch or witchgrass is botanically Elytrigia repens or Agropyron repens. It is an invasive grass that spreads by creeping rhizomes.)

"**In our Comfrey nursery bed, we have observed over the past season that it has suppressed the weed growth in the bed rather effectively,** as we planted the root cuttings about a foot (30 cm) apart. Part of this is due to the fact that Comfrey, in its hardiness, begins to grow early in the spring, and its quick growth will fill in a space nicely before something else can fill its niche. With this observation in mind, we decided to use this characteristic of Comfrey to our advantage in some of our food hedge planting projects this season.
In order to maintain the edge between the trees and shrubs we planted and the pasture, we are planting 1 1/2 inch (3.8 cm) root cuttings of Comfrey on 1 foot centers along our plantings. They will reduce weeding requirements and keep the pasture from encroaching back into the hedge planting line, which would cause competition for water and nutrients with our newly planted edible trees and shrubs. By using the Comfrey as a boundary defense, we will conquer the pasture's charge in the battle for the edge."
-'Divide and Conquer: Multiplying Comfrey by Root Cuttings, Episode #029' by Grant Shadden; ABC Acres: A Guest Ranch Experience, https://abcacres.com, Tim and Sarah, Hamilton, Montana, **2018**.

"**Preparing The Comfrey Bed: Once established, Comfrey will overshadow and overwhelm most weeds but it can be taken over by couch grass in the early years. So a thorough weeding and cleaning of the soil is the first task.
Don't forget to carefully dig up any perennial weeds** like docks and dandelions prior to rotovating, or you'll have a mass of them growing back from bits of roots left in the soil.
Deep digging and adding organic matter, especially rotted manure or compost, prior to planting will certainly get your Comfrey off to a good start when planted out. It makes it easier for the roots to get down deep into the soil to get at the minerals."
-'Location and Preparing the Comfrey Bed' by John Harrison, Allotment and Gardens: Grow Your Own - Allotment - Gardening Help, https://www.allotment-garden.org, Penygroes, Gwynedd, Wales, **2018**.

Mechanical and Chemical Control of Weeds

"**Mr. Stephenson cultivates both ways with a Howard rotary hoe, and keeps his crop clean, which is the secret of success. He has tried again and again to establish it in pasture and failed, for grass is the worst weed in Comfrey,** and no attempt to remove the need for weeding by sowing a legume has ever suceeded. There are plenty of good deep rooting herbs to add as seeds to a ley mixture, and in grass chicory is better than Comfrey.
-Comfrey Report Number Two: For Gardeners, Farmers and All Comfrey Growers in All Countries, with Analysis, Yields and Cultivation Methods, and the Results of Seven Years More Work Since Our First Report by Henry Doubleday Research Association, Braintree, Essex, England, 45 pages, March **1963**.
(The powered rotary hoe was invented by Arthur Clifford Howard in 1912 in New South Wales, Australia. He found that ground could be mechanically tilled without soil-packing occurring, as was the case with normal ploughing.)
(Ley is grassland or pastureland that is grazed.)

"**A successful programme combining mechanical and chemical weed control in medicinal Comfrey (Symphytum officinale) is described.
A pre-crop emergence spray** with Elbanil® (chlorpropham) + Trakephon® (buminafos), effective for 6-7 weeks, was followed by **a post-emergence spray** with Satecid 65 WP® (propachlor), also effective for 6-7 weeks, and finally by Probanil® (chlorophram + propazine), effective for 8-10 weeks.
Mechanical control comprised raking after sowing but before seedling emergence, followed by 3 hoeings (before Satecid 65 WP®), 2-3 weeks later, and finally after a further 3-4 weeks."
-'Weed Control in Comfrey Production' by K. Zschau, Institut fur Pflanzenschutzforschung (Institute for Crop Protection Research) Kleinmachnow, Germany; Gartenbau (Horticulture), Volume 30, No. 7, pages 203-204, **1983**.
(Pre-emergence is before a seedling grows out of the ground. Post-emergence is the stage between the

emergence of a seedling and crop maturity.)

*"1. Mechanical Cultivation: Comfrey is an excellent weed competitor due to its rapid and dense growth. Weeds may become established between Comfrey plants under a multiple-cut harvesting regime. As a result, **two cultivations per year are often required.** Rototilling between plants is an effective method for destroying weeds.*
2. Chemical Control: Comfrey has usually been grown without herbicides. No herbicides are labeled for use on this crop in the Upper Midwest of the United States."
-'Comfrey' by Teynor, Putnam, Doll, Kelling, Oelke, Undersander and Oplinger, 'Alternative Field Crops Manual', University of Wisconsin: Cooperative Extension, University of Minnesota: Center for Alternative Plant & Animal Products, and Minnesota Cooperative Extension, February 1992, updated November **1997**, page 4.

Fertilization Helps Comfrey Outgrow Weeds

*"**The labour saving method is to dung (manure) it well so it grows fast enough to suppress most weeds.** Trial ground experiments in 1961 with Dalapon® selective weedkiller to destroy the grass and leave the Comfrey, showed that crude poultry droppings thick spread were cheaper and more effective."*
-Comfrey Report Number Two: For Gardeners, Farmers and All Comfrey Growers in All Countries, with Analysis, Yields and Cultivation Methods, and the Results of Seven Years More Work Since Our First Report by Henry Doubleday Research Association, Braintree, Essex, England, 45 pages, March **1963**.
(Dalapon® is a selective herbicide used to control perennial grasses.)

*"**The better Comfrey is fed, the faster it grows with less weeding to do, and spreading a two-inch (5 cm) coat of deep-litter compost between the plants combines weed suppressing and ample feeding.**"*
-Grow Your Own Fruit & Vegetables by Lawrence D. Hills. London, England: Faber and Faber Limited: **1971**, page 46.
(Deep litter is repeated spreading of straw, hay or sawdust in a coop or barn. More is added as the bedding gets dirty.)

*"**The better Comfrey is fed, the less weeding there is to do.** On poor soils a second feed about June helps the bed, and about every other year give it 8 ounces (226 grams) a square yard (0.836 square meter) of slaked lime in the autumn to keep the soil neutral rather than acid from the heavy manuring."*
-Fertility Without Fertilizers: A Basic Approach to Organic Gardening by Lawrence D. Hills. New York: Universe Books, **1975**.   (Check your soil pH before applying lime. Different animal manures affect the soil in various ways.)

Pests and Disease

*"**Diseases and pests:** During the eight years that the Russian Comfrey crop has been in observation, **we found only a few times a few aphids on the crop.** Once was observed 'earth fleas'; against this it is immediately sprayed. Furthermore, no parasites were observed.*
Strange (*1959) shares that the crop is affected by nematodes and by the parasite Alectra asperrima."
-'Verslag van een Proef met het Gewas Kaukasische Smeerwortel Uuitgevoerd in de Jaren 1953-1960' (Report on a Trial with Russian Comfrey in 1953-1960) by H. Van Der Zweerde, Verslagen Instituut voor Biologisch en Scheikundig Onderzoek van Landbouwgewassen (Institute for Biological and Chemical Research on Field Crops and Herbage), Netherlands, No. 35, 12 pages, **1965**. All in Dutch. (If you have an English translation of this, please send it to me.)
(* -'A Comparison Between Russian Comfrey and Lucerne' by Richard Strange, Grassland Research Station, Department of Agriculture, Kenya; East African Agricultural Journal, Volume 24, pages 203-205, 1959.)
(An aphid is a tiny insect that feeds by sucking sap from plants.)
(Alectra asperrima also called Glossostylis asperrima is annual herb up to 36 cm tall {14 inches} found only in Africa.)

*"**Grasshoppers, slugs, cutworms and pyrethrum eelworms (nematode) have been reported as the worst pests for Comfrey.** Several nematodes attack Comfrey: Meloidogyne hapla, Meloidogyne javanica and several species in East Africa. **The following fungi have been found on Comfrey:** Corticium solani, Pleospora herbarum, Stemphylium botryosum (leaf spot) and Sclerotium rolfsii."*
-'Symphytum Peregrinum Ledeb., Boraginaceae, Comfrey, Russian Comfrey, Quaker Comfrey' by James A. Duke, Handbook of Energy Corps, unpublished, Purdue University, Indiana, **1983**.

*"**Comfrey has very few pests and minimal disease pressure.** Comfrey rust or powdery mildew can sometimes be an issue, but proper crop rotations and good cultivation practices are recommended to avoid any pest and disease issues that may arise."*
-The Organic Medicinal Herb Farmer: The Ultimate Guide to Producing High-Quality Herbs on a Market Scale by Jeff Carpenter and Melanie Carpenter. White River Junction, Vermont: Chelsea Green Publishing, **2015**, page 289.

"Symphytum officinale, Common Comfrey, Plant Parasites:

Organ	Parasitic Mode	Stage	Taxonomic Group	Parasite
leaf	vagrant	summer	Aphididae	Brachycaudus helichrysi

stem	borer		Agromyzidae	*Melanagromyza symphyti*
flower	gall		Cecidomyiidae	*Contarinia symphyti*
flower	gall		Cecidomyiidae	*Dasineura symphyti*
flower	gall		Tingidae	*Dictyla echii*
flower	gall		Tingidae	*Dictyla humuli*
leaf	down		Erysiphales	*Golovinomyces asperifoliorum*
leaf	down		Peronosporales	*Peronospora symphyti*
leaf	leaf spot		Entylomatales	*Entyloma serotinum*
leaf	gall		Aphididae	*Aphis symphyti*
leaf	gall		Cecidomyiidae	*Dasineura foliumcrispans*
leaf	gall		Chytridiales	*Synchytrium trichophilum*
leaf	miner		Agromyzidae	*Agromyza abiens*
leaf	miner		Agromyzidae	*Agromyza ferruginosa*
leaf	miner		Agromyzidae	*Agromyza myosotidis*
leaf	miner		Agromyzidae	*Phytomyza medicaginis*
leaf	miner		Agromyzidae	*Phytomyza pulmonariae*
leaf	miner		Coleophoridae	*Coleophora pennella*
leaf	miner		Coleophoridae	*Coleophora pulmonariella*
leaf	miner		Gracillariidae	*Dialectica imperialella*
leaf	miner		Gracillariidae	*Dialectica scalariella*
leaf	miner		Sciaridae	*Phytosciara macrotricha*
leaf	pustule		Blastocladiales	*Physoderma speciosum*
leaf	pustule		Chytridiales	*Synchytrium jaapianum*
leaf	pustule	aecia	Pucciniales	*Puccinia symphyti-bromorum*
leaf	pustule	uredinia telia	Pucciniales	*Melampsorella symphyti*
leaf	vagrant	summer	Aphididae	*Brachycaudus cardui*
leaf	vagrant		Aphididae	*Brachycaudus mordvilkoi*

Parsitic Mode:
Borer: larva living internally, almost no outwards signs.
Down: 0.5-2 mm high fungal down.
Film: very thin cover of fungal tussue.
Gall: swelling and/or malformation.
Grazer: feeding at the outside of the plant.
Leaf spot: discoloured, often necrotic, generally not galled, sign of a fungus infection.
Miner-borer: larve initially makes a mine, lives as a borer later.
Pustule: plug of fungal tissue, generally brown-black and < 2 mm.
Stripe: longitudinal line of fungal tissue in a grass leaf.
Vagrant: (aphids, mites) living freely on the plant, at higher densitiy causing malformations.
-'Symphytum Officinale, Common Comfrey: Plant Parasites' by Dr. Willem N. Ellis, Plant Parasites of Europe: Leafminers, Galls and Fungi, https://bladmineerders.nl, Amsterdam, The Netherlands, **2019**. Supported by Naturalis Biodiversity Center and EIS.
(Aecium {plural aecia} is a reproductive structure found in some plant pathogenic rust fungi that produce aeciospores. Aecia may also be referred to as 'cluster cups'.) (Uredinia telia is a reddish pustulelike structure that is formed in tissue of a plant infected by a rust fungus and produces urediniospores.)

Comfrey Rust (Fungus: Melampsorella symphyti)

"The Comfreys have no insect pests; the records of the Royal Horticultural Society Entomology Department reveal no instance of these plants being attacked by anything during the long period covered by the Host Plant Index at Wisley (Surrey County, England). **There is, however, a disease, the rare Comfrey Rust, which is found on the wild Symphytum officinale.** This is Melampsorella symphyti. The fungus resembles very small butterfly eggs occurring as dark yellow or orange patches on the underside of leaves.
The treatment recommended by the N.A.A.S. (National Agricultural Advisory Service in England) mycologists is to burn straw on the plot in winter after the crop has gone down, in the same way as one would for the Peppermint and Mint Rusts. The strength of Russian Comfrey is such that very powerful fungicides could be employed to destroy the Rust."
-Russian Comfrey: A Hundred Tons an Acre of Stock or Compost for Farm, Garden or Smallholding by Lawrence D. Hills. London England: Faber and Faber, Limited, **1953**, pages 73, 74.

"**Genus Melampsorella Schroet. Spore Stages:**
 0: Spermagonia amphigenous, producing small rounded orange spots, superficial, subcuticular, hemispherical, with flat bases, devoid of ostiolar filaments.
 I: Aecia cylindrical, orange-yellow; peridium colorless, irregularly erumpent; peridial cells with verrucose walls. Aeciospores in chains with intermediate cells, from globoid to ellipsoid or ovoid, warted; walls colorless; contents orange-yellow. Aecia on Abies (fir trees) in summer.

II: Uredia *subepidermal, covered by peridium; peridium thin, hemispherical, rupturing centrally at the apical pore; peridial cells smooth, polygonal. Urediospores yellow or orange-yellow, produced singly, on very short pedicels, ellipsoid or subgloboid, echinulate.*
III: Teliospores *intraepidermal, unicellular, occasionally dividing longitudinally into 2 cells, in varying number in each epidermal cell, globoid or, following reciprocal pressure, polyhedral; walls smooth, colorless, thin; contents colorless or pale yellow.*
Teliospores develop in the spring and immediately germinate into 4-celled IV: Basidia.
Uredio and teliospores developing on diffuse mycelium overwinter in the host's roots.
Two species are known: Melampsorella symphyti ranging throughout Europe and Transcaucasia, *and Melampsorella cerastii, in Europe, Asia, and North America.*
Melampsorella symphyti produce teliospores in spring and aecia spores in summer.
Aecia are on local mycelium infecting single leaves of current year, and developing soon after infection (within a month), uredio and teliospores on species of Symphytum."
-'Cryptogamic Plants of the USSR, Volume IV: Fungi and Rust Fungi' by V.F. Kuprevich and V.G. Transhel,
Academy of Sciences of the USSR, Komarov Institute of Botany, Moscow and Leningrad, Russia, **1957**, pages 53-303.
    (There are five spore stages:
    0: Pycniospores=Spermatia- These are haploid gametes in rusts.
    I: Aeciospores=Aecium- They are non-repeating, asexual spores that infect the primary host.
    II: Urediniospores=Uredium- They are repeating vegetative spores that re-infect the primary host. They are abundant, red/orange, and a visible sign of rust disease.
    III: Teliospores=Telium- These spores are the overwintering stage of the life cycle.
    Later they germinate to produce IV: Basidia=Basidum.)
    (The Caucasus Mountains are located at the border of Europe and Asia, between the Black Sea and Caspian Sea and occupied by Russia, Georgia, Azerbaijan, and Armenia.)
    (Mycelium is the vegetative part of fungus. It is a network of fine white filaments.)

"Rust, Melampsorella symphyti (Bulak), occurred in both years (1957, 1958) on Russian Comfrey variety Newman Turner."
-'Russian Comfrey' by L.A. Willey and R.L. Knight, Journal of the National Institute of Agricultural Botany, Cambridge, England, Volume 9, No. 2, pages 139-144, **1962**. (The other Russian Comfrey varieties grown in this experiment were Bocking No. 4, 12, 14, 17, Stepenson and Webster. Other types of Comfrey were not grown.)

"Comfrey rust (Melampsorella symphyti) is a disease of the wild Symphytum officinale. Symphytum asperum (Prickly Comfrey) does not contract the disease. So the more S. officinale there is in a hybrid, the higher the risk of rust.
This looks like an orange-yellow powder on the underside of the leaves, which is the spore stage of one of the awkward types of rust that get right inside the plant.
If this is observed, cut down the plants at once, either using it or dumping it in the trash bin. Then give a tonic such as dried blood. Continued cutting exhausts the fungus in the end."
-Fertility Without Fertilizers: A Basic Approach to Organic Gardening by Lawrence D. Hills. New York: Universe Books, **1975**.
    (Blood meal or dried blood is a high-nitrogen organic fertilizer. Nitrogen 13.25%, Phosphorus 1.0%, Potassium 0.6%. It is one of the highest non-synthetic sources of nitrogen.)

"Research to find a cure for Comfrey Rust still continues (1976), *and among the remedies tried have been Equisitum (Equisetum, horsetail, snake grass) preparations, decoctions of infected leaves, heavy composting, hot water treatment as for Mint Rust, flame gunning, and almost all of the fungicides.*
The rust is visible at the uredospore stage, *when the undersides of the leaves are covered with orange powder, starting usually in June, and spreading until in late July it loosens and blows on the wind.*
Treatment:
On a garden scale it is best to dig up the attacked plant and dump it in your dustbin (garbage can), but **cutting the foliage as soon as the spores are detected and watering with a strong solution of Jeyes® Fluid (disinfectant), or a phenol-based farm disinfectant can cure a starting infection."**
-Comfrey: Fodder, Food and Remedy by Lawrence D. Hills. New York: Rizzoli Universe Books: **1976**, page 72.
    (Jeyes® Fluid is manufactured in Preston, Lancashire, England. They sell various types. It contains 4-Chloro-M-Cresol, Tar Acids, (Poly)Alkyphenol Fraction. It damages skin and is harmful if swallowed so be careful.)
    (Phenol or carbolic acid is one of the oldest antiseptics. It is bacteriostatic {stops reproduction} at concentrations of 0.1% to 1% and is bactericidal / fungicidal at 1% to 2%. The bactericidal activity is increased by warm temperatures. Its strength is decreased by lipids {fatty acids}, soaps, and cold temperatures. Use as directed on the bottle. Phenol damages skin and is toxic to cats and babies. Household phenol disinfectants include Birex®, Dettol®, Lysol®, Meytol®, Pine Sol®, ProSpray®, Wex-cide® and others.)

"Comfrey rust fungus (Melampsorella symphyti) appears to be the only important disease of the crop and attacks S. asperum and S. officinale as well as S. x uplandicum. The mycelium perennates in the rhizome, and uredospores are produced on the leaves in early summer (*Wilson and Henderson, 1966).
There appear to be differences between clonal selections of S. x uplandicum in their susceptibility to the disease (Hills, 1976)."
-'Comfrey Symphytum spp. as a Forage Crop' by J.C. Forbes, A.D. McKelvie, and P.J.C. Saunders, North of Scotland College of

Agriculture, Aberdeen, United Kingdom; Herbage Abstracts, Volume 49, No. 12, pages 523-539, **1979**.
(* -British Rust Fungi by Malcolm Wilson and D.M. Henderson. Cambridge, England: Cambridge University Press, 1966. If you have the section about Comfrey rust Melampsorella symphyti, could you please send it to me.)
(Perennate means it lives a few years with an annual cycle of dormancy and activity.)

"Comfrey has no pests, though some caterpillars will eat it occasionally, and **only one serious disease: Comfrey Rust. The uredosore stage of Melampsorella symphyti becomes visible as an orange powder on the undersides of the leaves, usually in May and June**; in July the spores blow like orange smoke, which is how the disease spreads from the wild Comfrey. **If you notice the bright orange powder, pick off affected leaves and dump them in your dustbin (trash can)**, though they are safe in a compost heap or liquid manure butt (barrel). This is the policy most Comfrey growers adopt, and **if they observe more than the old rusty leaf, they mark the plant, cut it every four weeks instead of every six, and give it extra feeding**. This rust is one of the awkward type, like the mint rust (Puecinia menthae), that gets right inside the roots and stems, and though the hot water treatment used for strawberry virus has also been tried with mint rust, you cannot get all the Comfrey roots in a bed into water at 110 F (43 C).
If there are too many rusted Comfreys in a bed, it pays to kill them out and start again with fresh stock."
-Fertility Gardening: The Organic Way to Make Your Garden Grow by Lawrence D. Hills. London, England: Cameron & Tayleur Books Limited, **1980**.

"Almost all Comfrey in the United States has grown from that one stock (from Elmer Deetz of Oregon). **Fortunately, Comfrey rust (Melampsorella symphyti), a disease of the wild S. officinale, did not cross with them so American Comfrey is rust-free.**"
-Fighting Like the Flowers: An Autobiography: The Life Story of Britain's Best-Known Organic Gardener by Lawrence D. Hills. Bideford, Devon, England: Green Books, **1989**, page 104.   (I do not know if Comfrey in the United States is still rust free.)

"Diseases have not been a serious problem with Comfrey in the United States. **Comfrey rust fungus (Melampsorella symphyti) overwinters in roots and reduces yield of old plantings in Great Britain.** This disease problem has not spread to the United States due to plant quarantine regulations on the importation of roots or plants.
Insects have not been reported to be a problem with Comfrey in the United States."
-'Comfrey' by Teynor, Putnam, Doll, Kelling, Oelke, Undersander and Oplinger, 'Alternative Field Crops Manual', University of Wisconsin: Cooperative Extension, University of Minnesota: Center for Alternative Plant & Animal Products, and Minnesota Cooperative Extension, February **1992**, updated November 1997, page 4.

"**Comfrey's only disease is Melampsorella symphiti, a rust not common to well-fed plants or those grown on good soils.** It can be alleviated with the addition of wood ash sprinkled amonst the plants; wood ash is a good source of potash (potassium)."
-Comfrey: Symphuo Symphytum: A Multi-Purpose Herb by Philip Clarke. Edinburgh, England: Pentland Press, **1997**, page 14.

"**Melampsorella symphyti Bubak:** [0, I on Abies=Fir trees, on needles.]
II on Symphytum asperum, S. officinale, S. tuberosum; II, III on S. x uplandicum (= S. officinale x asperum).
'Wilson and Henderson' Flora of 1966, page 46. Great Britain including the Channel Islands and Ireland including northern Ireland. The spore stages are indicated by the conventional symbols 0, I : II, III.
Spore stages known, but not recorded in Britain, are often included in square brackets [ ] to encourage the search for them."
-'A Checklist of the Rust Fungi of the British Isles' by D.M. Henderson, British Mycological Society, Royal Botanic Gardens Kew, Richmond, Surrey, England, **2000**, page 7.

"**Comfrey rust spores appear as orange or orange yellow dust or spots.** These are found on the underside of leaves in early summer. Plants are stunted and have pale coloured leaves. Cut back any plant that shows signs of the disease, **treat with a high potassium amendment such as compost or manure or wood ash** if you have this available. This will help the plants to 'grow away' from the disease. Badly affected plants should be removed and destroyed."
-Comfrey for Gardeners: A Garden Organic Guide by Garden Organic. Coventry, England: no date but around **2015**, page 17.

"**Comfrey Diseases and Problems: Comfrey is remarkably trouble free in Britain. The only serious disease it suffers from is Comfrey rust and the Bocking 14 clone is exceptionally resistant to that.** Comfrey rust is common in the Common Comfrey grown by herbalists. Rust is a fungal disease that manifests as small orange spots under the leaves, spreading to cover them and eventually sending out spores to infect other plants.
Treatment:
If picked up really early, the leaves can be cut off at ground level and disposed of in the dustbin rather than compost. With luck they'll grow back clean. **Sterilising by watering with armillatox or Jeyes® fluid used to be a recommended treatment following the leaf cut** but as this is not an approved use for the products I cannot recommend it. **A systemic fungicide** is also worth trying but treated plants should not be used to feed livestock for the rest of the year."
-'Planting, Cultivating, Harvesting and Problems of Comfrey' by John Harrison, Allotment and Gardens: Grow Your Own - Allotment - Gardening Help, https://www.allotment-garden.org, Penygroes, Gwynedd, Wales, **2018**.
   (Armillatox® is a castor oil soap-based fungicide of 'high boiling tar acids' that kills moss. It is manufactured in Alfreton, Derbyshire, England. It is harmful on skin or if swallowed.)   (I have never had Rust on my Common Comfrey plants.)

Other Fungus Disease

*"The Aecidium fungus is said to be one condition (phase) of the Uredo and Puccinia. It is one of the rarest of British (fungal) plants. Aecidia grow within the tissues of plants, and in these positions they form minute spherical balls, filled with chains of **whitish or yellowish, semi-transparent, generally spherical, spores.***
*In the process of growth the immersed spheres burst through the cuticle of the host plant- generally from the **under side of the leaves, and often through the stem**. As the Aecidium cups mature they burst at the exposed apex, and the fractured part turns back so as to give the little fungus growths the form of minute cups filled with spores.*
Each cluster of cups is surrounded by a large pallid (pale) disease patch**, which has been caused by the exhaustion of the vital material of the leaf by the spawn of the fungus which at first grew within. **The fungus growth has also caused the leaf to become torn.
***Aecidium asperifolii, Pers., has been found in Europe on** Cynoglossum officinale, L.; Borrago officinalis, Tour.; Anchusa officinalis, L.; A. arvensis, Bieb.; Nonnea pulla, D.C.; **Symphytum officinale, L.; Symphytum tuberosum, L.;** Cerinthe minor, L.; O. alpina, Kit.; Echium vulgare, L.; Pulmonaria officinalis, L.; P. tuberosa, Schrk.; and Lithospermum arvense, L."*
-'Diseases of Field and Garden Crops Chiefly Such as are Caused by Fungi' by Worthington George Smith, F.L.S., M.A.I., Royal Horticultural Society, London, England, **1884**.
> (Aecidium asperifolii is also known as Puccinia recondita. Puccina wheat leaf rust is a fungus that affects wheat, barley and rye. In temperate zones the pathogen overwinters. Puccinia triticina causes black rust, P. recondita causes brown rust, and P. striiformis causes yellow rust.)

*"At the Regional Plant Introduction Station, Geneva, New York, in 1957-1958, **a severe leaf spot on the lower leaves of Comfrey (Symphytum peregrinum) was found to be caused by Stemphylium sp.**, with conidia similar to those of Stemphylium botryosum (Pleo-spora herbarum), but it was non-pathogenic to red and white clover, lucerne (alfalfa) and birdsfoot trefoil (Lotus corniculatus). **This is the first report of Stemphylium on Comfrey.**"*
-'Stemphylium Species on Comfrey' by S.W. Braverman, Plant Disease Reporter Journal, Volume 43, No. 9, pages 1050, **1959**.
> (If you know more about this fungus on Comfrey, please let me know.)
> (Stemphylium is a genus that comprises imperfect fungi of the order Moniliales, with dark greenish-brown spores closely resembling those of Alternaria genus but borne singly rather than in chains and is often included in Alternaria genus or sometimes by Macrosporium genus. It is often found growing in soil, wood, and decaying vegetation. Some species found on leaves are plant pathogens.)

*"**Mildew, Erisyphe spp. (species), occurred in all Russian Comfrey varieties in 1957 (none in 1958), more particularly on Bocking No. 17 and Websters varieties.**"*
-'Russian Comfrey' by L.A. Willey and R.L. Knight, Journal of the National Institute of Agricultural Botany, Cambridge, England, Volume 9, No. 2, pages 139-144, **1962**.
(Russian Comfrey varieties grown in this experiment were Bocking No. 4, 12, 14, 17, Newman Turner, Stepenson and Webster. Other types of Comfrey were not grown.) (Erysiphe is a genus of fungi in Erysiphaceae family that causes powdery mildew.)

*"**Russian Comfrey** was harvested at weekly intervals from eight to 12 weeks after planting. Harvesting was stopped at **12 weeks of growth** because the plants had become very coarse. **They were also heavily infested with bacterial blight caused by Cladosporium sp. (species).**"*
-'Potential of Russian Comfrey as an Animal Feedstuff in Uganda' by Bareeba, Odwongo and Mugerwa, Department of Animal Science, Faculty of Agriculture and Forestry, Proceedings of the First Uganda Pasture Network Workshop, Makerere University, Kampala, Uganda, Africa, **1987**.
> (Cladosporium is a genus of fungi of some of the most common indoor and outdoor molds. They produce olive-green to brown or black colonies. They are found on living and dead plant material. Some species are plant pathogens, others parasitize other fungi. Cladosporium spores are wind-dispersed.)
> (I was unable to find anything more about this fungus. If you have information, contact me.)

*"**Comfrey powdery mildew (Erisyphe polygoni) is seen as white patches or a white layer on leaves and stems.** Mildew is more common on wild Comfrey (Common Comfrey) and tends to occur on older leaves at the end of the season when the plants are less vigorous. **Strong, vigourous plants will be less susceptible to disease**, so aim to keep your plants well fed, adequately spaced, and regularly cut and watered."*
-Comfrey for Gardeners: A Garden Organic Guide by Garden Organic. Coventry, England: no date but around **2015**, page 11.
> (Powdery mildew is a fungal disease that affects a wide range of plants. It is caused by different species of fungi in the order Erysiphales. Infected plants have white powdery spots on leaves and stems with lower leaves the most affected. Powdery mildew grows well in environments with high humidity and moderate temperatures such as in greenhouses.)

Pyrethrum Eelworm (Pratylenchus destructor) (Nematodes)

Pratylenchus is a genus known as lesion nematodes. They are parasitic on plants and cause root lesion disease. They are migratory endoparasites that feed and reproduce in the root and then move, unlike cyst or root-knot nematodes that stay in one place. They usually feed on the cortex (outer layer) of the root.

Signs of disease are necrotic (dead) lesions (wounds) of the roots. Above ground the plant is stunted, chlorotic (yellow), and wilted. Plants wither and die.

Pratylenchus is affected by soil moisture, minerals, temperature, aeration, organic matter, and pH. They are more common in sandy soils and land with many weeds. It survives in a wide range of conditions. Moist temperate soils are ideal for breeding and migrating. In very dry conditions they are dormant. They are inactive when soil temperatures are below 59 F (15 C) and are fully active above 68 F (20 C).

"The 'Long Rains' in Kenya, Africa in 1955 failed, and there was nothing green but the Comfrey field, which had got it roots down ten feet (3 meters) to water. His whole crop dried out. **The pyrethrum eelworm, Pratylenchus destructor, one of the migratory species, found that it could attack Comfrey and slaughtered it.** *"*
-Comfrey: Fodder, Food and Remedy by Lawrence D. Hills. New York: Rizzoli Universe Books, **1976**, page 47.

"Where suddenly a good plant falls ill and dies, look at the roots. The chances are 10 to 1 that it is due to nematodes, and if you want to save your crop, you will treat the soil at once. *As soon as it is treated, look to the soil structure for permanent rectification (correction) organically and with* **micro-elements**. *If this is done, there will be no need to inject chemical poisons into the soil.* **It took up to 4 years to get a clean crop by completely organic treatment of the soil. The problems of nematodes in Japan is due in large part to the over-chemicalization of agriculture.**
But we have confirmed over the subsequent years of experience with our own plantation, that **properly used organic methods do reduce the nematode problem to an almost negligible factor.**"
-Comfrey: Nature's Healing Herb and Health Food by Andrew Hughes. Japan: Sanyusha Publishing Co, **1992**, page 21, 156, 158.

"This plot we have here is planted very much closer than recommended and presents a solid block of green today with the older plants obviously very heavy. The effluent (waste) from the pigs is simply spread directly onto the plot, and **there is no evidence of nematode infestation.**
I am familiar with the havoc nematodes can create in Comfrey, as I planted years ago on tobacco land and lost most of that planting, the roots of the plants becoming a black mass. *I did note however that amongst this black mass, the plants were still making an effort at regrowth.* **I feel, with you, that this heavy manuring ensures the nematode has no chance to compete.** *It's quite remarkable, the evenness of the stand with no evidence of blank spaces."*
-Comfrey: Nature's Healing Herb and Health Food by Andrew Hughes. Japan: Sanyusha Publishing Co., Ltd, **1992**, page 111.
(Manure increases the beneficial bacteria in the soil, thereby reducing nematode problems.)

"Plants should also have organic micro-element spraying. It has already been demonstrated by more than one Comfrey grower in Japan that **Comfrey treated with these organic micro-elements or grown in soil treated this way is not seriously affected by nematodes."**
-Comfrey: Nature's Healing Herb and Health Food by Andrew Hughes. Japan: Sanyusha Publishing Co., Ltd, **1992**, page 155.
(I put Azomite and dried seaweed in my soil. Azomite is a mined product that has minerals and trace elements. Seaweed such as kelp has minerals, trace elements and enzymes.)

"Mr. Strange* further observed that Russian Comfrey was very susceptible to nematodes *and was also weakened by the plant parasite 'Alectra asperrima'."*
-East Africa's Grasses and Fodders: Their Ecology and Husbandry by Joseph G. Boonman. Netherlands: Kluwer Academic Publishers, **1993**, page 302.
(* -'A Comparison Between Russian Comfrey and Lucerne' by Richard Strange, Grassland Research Station, Department of Agriculture, Kenya; East African Agricultural Journal, Volume 24, pages 203-205, 1959.)
(Alectra asperrima also called Glossostylis asperrima is an annual herb up to 36 cm tall {14 inches} found only in Africa.)

"1,2-Dehydropyrrolizidine alkaloids (Pyrrolizidine Alkaloids = PA), known to be nematotoxic (toxic to nematodes) in vitro (in laboratory), *represent a class of secondary plant metabolites from hundreds of plant species worldwide.*
Pot experiments with the commercially available PA-containing plants Ageratum houstonianum, Borago officinalis, Senecio bicolor, and **Symphytum officinalis demonstrate that Meloidogyne hapla (northern root-knot nematode) is not per se repelled by these plants** *as all species were infested with nematodes. However, the development of Meloidogyne hapla juveniles was completely suppressed on Ageratum houstonianum and Senecio bicolor. Ten days after inoculating 400 J2 (second stage juveniles) per plant, all PA- and non-PA-plant species exhibited clear symptoms of nematode infestation.*
> **The degree of galling varied significantly, Symphytum officinalis being the most** *and Ageratum houstonianum the least affected one.* **Significantly more J2 had penetrated the roots of Lycopersicon esculentum and Symphytum officinalis** *than the roots of Senecio bicolor and Ageratum houstonianum.*
> **The proportion of juveniles that had already reached the J3-stage (third stage juveniles) was three to ten times higher for roots of Borago officinalis, Symphytum officinalis** *and Lycopersicon esculentum than for roots of Ageratum houstonianum and Senecio bicolor."*
-'Effects of Plants Containing Pyrrolizidine Alkaloids on the Northern Root-Knot Nematode Meloidogyne Hapla' by Tim C. Thoden, Johannes Hallmann & Michael Boppre, Germany; European Journal of Plant Pathology, Vol 123, No. 1, page 27-36, **2009**.
(Galls are abnormal outgrowths of plant tissue caused by parasites such as viruses, fungi, bacteria, insects and mites.)

Bacterial Disease

*"**Symphytum officinale** (Common Comfrey) is grown as quality silage crop. **In 1979 Mycoplasma-Like Organisms (MLO) were first found in the phloem of wilted plants, showing stunt, yellow edge, distortion of leaves and retardation of growth.** Some MLO bodies were rounded or pleomorphic, 60-600 nm (nanometer) diameter, bounded by a 9-12 nm thick bilaminar limiting membrane or occasionally by a trilaminar one. Electron-densed ribosome-like particles and DNA-like strands occurred in the large rounded cells. MLO in division phase were sometimes found.*
Symptoms could be eliminated by tetracycline treatment of the roots. *No other pathogens were found."*
-'On the Comfrey Stunt Disease I. Pathogen Identification' by S.H. Xu, K.Q. Mang and C.X. Huang, Institute of Microbiologica Academia Sinica, Beijing, China; Acta Phytopathologica Sinica, Volume 13, No. 2, pages 5-8, **1983**. (I was unable to get this report. If you have English translation, could you please send it. If you have more information about this disease, let me know.)
(Mycoplasma is a genus of bacteria that lack a cell wall around their cell membrane. Without a cell wall, they are unaffected by many common antibiotics such as penicillin or other beta-lactam antibiotics that target cell wall synthesis.)
(In vascular plants, phloem is tissue that transports soluble organic compounds from photosynthesis to other parts of the plant.)
(The nanometre or nanometer is a unit of length in the metric system, equal to one billionth of a meter = 0.000000001 meter.)

*"**Bacteria were also isolated from lesions on Rudbeckia and Symphytum, but it is thought that these are probably 'low-grade' pathogens or secondary invaders of damaged / senescing tissue.**
Symphytum:*
Symptoms: *Brown angular spots / larger lesions.*   **Pathogen:** *Pseudomonas viridiflava.*
Severity: *3.2 (3 = very obvious lesions, unacceptable for retail). 2 samples from 10 retail nurseries in England. Priority: Low."*
-'Bacterial Diseases of Herbaceous Perennials' by Dr. Steven J Roberts, Project Number: HNS 178, Annual Report, Plant Health Solutions, Warwick, England; Grower Nurseries, Henry Doubleday Research Association, Ryton, Coventry, England, March **2011**.
(Lesion is an area with damage from injury or disease, such as a cut, wound, ulcer or tumor.)   (Senesce means to grow old.)
(Pseudomonas viridiflava is a fluorescent, gram-negative, soil bacterium that is pathogenic to plants.)

Insect Damage

*"**Insect resistance: Symphytum asperrimum (Prickly Comfrey) and Symphtum peregrinum (Russian Comfrey) (father and child) appear to share the high degree of resistance to these two enemies of the farmer, virus and insect.**
Like Alfalfa when the roots get down into the subsoil and begin to feed on the micro-elements, **the resistance of Comfrey is strongest when mature.** This is clear, that Comfrey does not need spraying to keep off insect pests."*
-Miracle Grass, Comfrey by Andrew Hughes published in 1966 in Japanese. Part of it is in: 'Comfrey: Nature's Healing Herb & Health Food' by Andrew Hughes, Japan: Sanyusha Publishing Co., Ltd, **1992**. Part is in 'Comfrey Report No. 1: The 1954 Research Results' by Lawrence Hills published in 1955. (If you have a translation in English, I would appreciate having a copy.)

*"**Lace Bugs: Numerous occurrences of Dictyla humuli (lace bugs) were observed in the year 2000 on Common Comfrey (Symphytum officinale) plants growing in Slowinski National Park, Poland. Adults of this pest hibernate near Common Comfrey plants, and then live and develop on the plants during the vegetative season."***
-'Dictyla Humuli F.: A Pest of Common Comfrey' by A. Korcz and P. Olejarski, Instytut Ochrony Roslin w Poznaniu (Institute of Plant Protection in Poznan), Poland; Ochrona Roslin (Plant Protection), Volume 45, No. 5/6, pages 12-13, **2001**.
    (If anyone knows more about this insect damaging Comfrey, please let me know.)
    (Dictyla is a genus of lace bugs in family Tingidae. There are 80 species. Full name: Dictyla Humuli Fabricius, 1794. There are 140 North American species of lace bug. Most have a specific host preference. Adults are 1/8 to 3/8 inch {0.317 to 0.952 cm} long. All have wings and thorax sculptured with an intricate pattern of veins that resembles lace. Nymphs are dark and pointed at both ends. They are most often found on underside of leaves.)

*"**Beetles:** In the hop gardens of Miklavz, Slovenia, the **Longitarsus anchusae beetle** was found to be a very common species. **It feeds on the Common Comfrey (Symphytum officinale L.) and on the Tuberous Comfrey (Symphytum tuberosum L.),** both of which were present in the area."*
-'International Hop Growers Convention: Proceedings of the Scientific Commission' by Dr. Elisabeth Seigner, Tettnang, Germany, June 2007. Article: 'Hop Flea Beetle (Psylliodes attenuatus Koch) in Slovenia' by M. Rak Cizej and L. Milevoj.
    (Longitarsus anchusae Beetle: 1.5-2.4 mm. Shiny black. Scarce in England; small numbers of old records from Scotland and Ireland, none from Wales. Now found only in south and east England. Habitat: Various. Host plant: Various Boraginaceae. Overwintering: Probably as adults. Food: Adults on leaves, larvae at roots.)

*"**Aphids and Symphytum:** Apterous and alate viviparous female, oviparous female and alate male of **Macrosiphum symphyti sp. nov. (new species) living on Symphytum asperum (Boraginaceae)** are described and illustrated.
Macrosiphum symphyti sp. nov. (new species) is closely related to Macrosiphum funestum (Macchiati). These two species have different host plants: M. funestum lives mostly on Rubus spp., but has also been recorded from Galium sp., Geranium robertianum, while M. symphyti lives on Symphytum asperum."*
-'A New Species of the Genus Macrosiphum (Hemiptera: Aphididae) Living on Symphytum Asperum (Boraginaceae) from

Georgia' by Shalva Barjadze and Nino Chakvetadze, Tbilisi, Georgia; Annales Zoologici: Museum and Institute of Zoology, Polish Academy of Sciences, Poland, Volume 58, No 3, pages 551-556, **2008**.
    (Aphids are small sap-sucking insects of the superfamily Aphidoidea, also called greenfly and blackfly. There is very little information about this aphid: Macrosiphum symphyti Barjadze & Chakvetadze, 2008.)

*"**The Scarlet Tiger Moth (Callimorpha dominula, formerly Panaxia dominula):***
This species is present in most of Europe and in the Near East (Turkey, Transcaucasus and northern Iran). These moths prefer damp areas (wet meadows, river banks, fens and marshes), but they also can be found on rocky cliffs close to the sea.
They mainly feed on Comfrey (Symphytum officinale), *but also on a number of other plants (Urtica, Cynoglossum, Fragaria, Fraxinus, Geranium, Lamium, Lonicera, Myosotis, Populus, Prunus, Ranunculus, Rubus, Salix and Ulmus species).*
    The Scarlet Tiger Moth (family Arctiidae) occurs in continental Europe, western Asia and southern England. It is a day-flying moth, noxious-tasting, with brilliant warning colour in flight, but cryptic at rest. The moth is colonial in habit, and prefers marshy ground or hedgerows. ***The preferred food of the larvae is the herb Comfrey (Symphytum officinale).*** *In England it has one generation per year. "*
-Wikipedia®: The Free Encyclopedia, www.wikipedia.org, **2018**.

    *"**Scarlet Tiger Moth: This widely distributed, distinctly local species is polyphagous on herbaceous plants, but most abundant on Comfrey (Symphytum officinale),*** *green alkanet (Pentaglottis sempervirens) and hounds-tongue (Cynoglossum officinale).*
In areas where colonies are established, older larvae sometimes feed in the spring on ornamental herbaceous plants and young trees, including ash (Fraxinus excelsior), blackthorn (Prunus spinosa), elm (Ulmus), flowering cherry (Prunus), oak (Quercus), rowan (Sorbus aucuparia) and willow (Salix).
Although causing slight damage to the leaves, this attractive insect is not of pest status and specimens found on cultivated plant should not be destroyed. *The young larvae feed briefly in the late summer or early autumn, before hibernating, and complete their development in the following spring. Fully grown larvae are about 40 mm (1.57 inch) long and mainly black, marked prominently with white and bright yellow.* ***The spectacular, bright red, yellow, white and black adults occur in June and July."***
-Pests of Ornamental Trees, Shrubs and Flowers: A Colour Handbook, Second Edition by David V Alford Bsc, Ph.D. United Kingdom: CRC Press, 2012.

*"**Aphid: Aphis symphyti Schrank:** Apterae are pale yellowish green to dark green, rarely lemon yellow, with dark head and siphunculi; BL 1.3-2.0 mm. Alatae have secondary rhinaria distributed III 8-12, IV 1-4, V 0-2 (cf. gossypii).*
Living scattered under leaves of Symphytum officinale, or when numerous extending up the stems and in the influorescences *(*Stroyan 1984). Also recorded from various other Boraginaceae, but many of these records may apply to Aphis gossypii, with which there is wide overlap in many features of morphology.*
Recent molecular work suggests a close relationship with Aphis frangulae (Cocuzza et al. 2009, Cocuzza and Cavalieri 2014).
Throughout Europe (except Scandinavia and Iberian peninsula).
Monoecious holocyclic, with alate males; ***the life cycle is only known to be completed on Symphytum officinale.***"
-'The Aphids: Systematic Treatment of Aphid Genera' by Roger Blackman, Natural History Museum, London, England, and author Victor Eastop; Aphids on the Worlds Plants: An Online Identification and Information Guide, www.aphidsonworldsplants.info, **2019**.
(* -'Handbooks for the Identification of British Insects, Volume 2, Part 6: Hemiptera, Homoptera, Aphididae' by Royal Entomological Society, St. Albans, England. 'Aphids: Pterocommatinae and Aphidinae' by H.L.G. Stroyan. London, England: Dramrite Printers Limited, 1984.)   (The Iberian Peninsula is a mountainous region in Spain and Portugal.)

    Virus Disease

*"**Virus resistance: Symphytum asperrimum (Prickly Comfrey) and Symphtum peregrinum (Russian Comfrey) (father and child) appear to share the high degree of resistance to these two enemies of the farmer, virus and insect.***
But Symphytum officinale is one type cultivated for stock and for medicine that suffers from one of the virus diseases, *and lacks something of the vitality of the other two. Like Alfalfa when the roots get down into the subsoil and begin to feed on the micro-elements,* ***the resistance of Comfrey is strongest when mature."***
-Miracle Grass, Comfrey by Andrew Hughes published in 1966 in Japanese. Part of it is in: 'Comfrey: Nature's Healing Herb & Health Food' by Andrew Hughes, Japan: Sanyusha Publishing Co., Ltd, **1992**. Part is in 'Comfrey Report No. 1: The 1954 Research Results' by Lawrence Hills published in 1955. (If you have a translation in English, I would appreciate having a copy.)

*"**In July 2004, symptoms of a virus-like disease that consisted of yellow spots, rings and chlorotic or yellow line patterns, were observed in the foliage of wild Symphytum tuberosum L.*** *(family Boraginaceae) growing in the Botanical Garden of Bologna University, Italy. Symptomatic leaves were collected and tested for virus by using the protein A-sandwich enzyme-linked immunosorbent assay (PAS-ELISA) technique.*
    The polyclonal antisera tested were against either one of the nepoviruses strawberry latent ringspot, tobacco ringspot, tomato ringspot, tomato black ring, cherry leaf roll, arabis mosaic, or cucumber mosaic virus, tobacco streak virus or Alfalfa Mosaic Virus (AMV: PVAS 92, American Type Culture Collection, Manassas, VA, USA and AMVVinca minor,

DiSTA collection).
PAS-ELISA revealed only the presence of Alfalfa Mosaic Virus (AMV). *Electron microscope examination of ultrathin sections of small fragments of symptomatic leaves confirmed the presence of aggregated and scattered AMV particles in the cytoplasm of parenchymatous cells, but not in the vascular tissue or the vacuoles. Large aggregates consisted of many bacilliform particles packed side-by-side to form long bands across the cytoplasm.*
The nature of these ultrastructural modifications suggests that the S. tuberosum isolate of AMV could be included in group 2A of the 4 groups described by Hull et al. (1970). **No virus infection has been reported in S. tuberosum and, on the basis of our results, it appears that this wild plant can be considered a new natural host for AMV."**
-'The Occurrence of Alfalfa Mosaic Virus in Symphytum Tuberosum L.' by M.G. Bellardi and A. Benni, Dipartimento di Scienze e Tecnologie Agroambientali (DiSTA), Patologia Vegetale, Universita degli Studi di Bologna, Italy; Journal of Plant Pathology, Volume 87, No. 1, page 75, March **2005**.

Unwanted Deer, Rabbits and Groundhogs Eating Comfrey
See subsection 'Rabbits' in section 'Livestock / Pet Species and Comfrey' (Chapter 22).

I have never had any problems with deer or wild rabbits eating my Comfrey. However, other people have told me they sometimes have problems with deer eating it. Comfrey is considered deer-resistant. However, deer are ruminants just like goats, and my goats love it. They certainly won't kill it. If you have serious problems with this, you may want to fence it in but that would rarely be needed.
I had some Comfrey leaves eaten in a way that had not happened in 10 years. On about 20 plants all the leaves not touching the ground were eaten. The stalks were not eaten. Then a few weeks later I saw a groundhog (woodchuck, Canada marmot) in my garden which I had never seen in my garden before. So I suspect those were eaten by this groundhog. They are mostly herbivores.

*"**If deer are hungry enough, they will eat almost anything.** However, there are a number of woody and herbaceous plants that deer usually don't find appealing. The best long-term strategy may be to note which plants they seem to favor or avoid in your yard or neighborhood and plant accordingly. Remember, that in severe conditions all plants are vulnerable to deer browsing.*
Deer Resistant Herbs:
Allium spp. (chives), Anethum graveolens (dill), Angelica archangelica (angelica), Borago officinalis (borage), Foeniculum vulgare (fennel), Hyssopus officinalis (hyssop), Levisticum officinale (lovage), Marrubium vulgare (horehound), Matricaria recutita (chamomile), Melissa officinalis (lemon balm), Mentha x piperita (peppermint), Mentha spicata (spearmint), Nepeta x faassenii (catmint), Ocimum basilicum (sweet basil), Origanum vulgare (oregano), Perilla frutescens (perilla), Petroselinum crispum (parsley), Rosmarinus officinalis (rosemary), Ruta graveolens (common rue), Salvia officinalis (common sage), Sanguisorba spp. (burnet), Santolina chamaecyparissus (lavender cotton), Satureja montana (winter savory), **Symphytum spp. (Comfrey species),** *Tanacetum parthenium (feverfew), Tanacetum vulgare (tansy), Teucrium chamaedrys (wall germander), Thymus spp. (thyme), Verbascum spp. (mullein)."*
-'Plants Not Favored by Deer' by Laura G. Jull, Assistant Professor of Horticulture, College of Agricultural and Life Sciences; Publication A3727, Board of Regents of University of Wisconsin, University of Wisconsin-Extension, Cooperative Extension, Madison, Wisconsin, **2001**.

*"**Wildlife experts never use the term 'deer-proof'. They know white-tailed deer will eat almost any kind of plant if they're hungry enough and, even if well fed, will sample plants they're not supposed to like.** One reason it's difficult to compile a list of deer-resistant plants is that deer appetites seem to vary from one deer herd to another.*
Individual deer also may have varying preferences and needs. Deer have been known to develop a taste for a species of plant they previously ignored, and the increased nutritional needs of pregnant or lactating does may lead them to consume plants they otherwise would avoid.
Observers rate many herbs as deer-resistant, because of their strong aromas and flavors:
> *Angelica, Anise Hyssop, Basil, Catmint, Chamomile, Chives,* **Comfrey,** *Dill, Fennel, Lamb's ears, Lavender, Lavender Cotton, Lemon balm, Mint, Mullein, Oregano, Parsley, Rosemary, Sage, Thyme."*

-'Deer Gardening' by Suzanne Wilson, Missouri Department of Conservation, https://mdc.mo.gov, Missouri Conservationist Magazine, April **2003**.

*"**The following is a list of landscape plants rated according to their resistance to deer damage.** The list was compiled with input from nursery and landscape professionals, Cooperative Extension personnel, and Master Gardeners in northern New Jersey. Success of any of these plants in the landscape will depend on local deer populations and weather conditions.*
> *Realizing that* **no plant is deer proof,** *plants in the 'Rarely Damaged', and 'Seldom Rarely Damaged' categories would be best for landscapes prone to deer damage.*
> *'Plants Occasionally Severely Damaged' and 'Frequently Severely Damaged' are often preferred by deer and should only be planted with additional protection such as the use of fencing, repellents, etc.*

Categories: *Rarely Damaged, Seldom Severely Damaged, Occasionally Severely Damaged, Frequently Severely Damaged.*
> **Comfrey, Symphytum officinale: Occasionally Severely Damaged."**

-'Landscape Plants Rated by Deer Resistance' by Pedro Perdomo and Peter Nitzsche, Morris County Agricultural Agents, and David Drake, Ph.D., Extension Specialist in Wildlife Management; Rutgers Cook College, Rutgers Cooperative Research and

Extension Bulletin E271, New Jersey Agricultural Research Station, New Brunswick, New Jersey, **2004**.

"Deer damage is usually identified by the torn or jagged appearance of branches or twigs compared to the clean-cut feeding damage caused by rabbits and squirrels. If they are hungry enough and food is scarce enough, deer will eat almost anything. However, there are a number of plants that deer don't find particularly palatable (tasty). Using these plants in your landscape is often the most cost-effective, least time consuming, and most aesthetically pleasing solution.
Deer-Resistant Herbs:
Angelica archangelica (Angelica), Artemisia absinthum (Artemisia), Ocimum basilicum (Basil), Borago officinalis (Borage), Nepeta x faassenii (Catmint), Matricaria spp. (Chamomile), Allium schoenoprasum (Chives), **Symphytum x rubrum (Comfrey),** *Anethum graveolens (Dill), Foeniculum vulgare (Fennel), Tanacetum parthenium (Feverfew), Teucrium chamaedrys (Germander), Hyssopus officinalis (Hyssop), Stachys byzantina (Lamb's Ear), Lavandula angustifolia (Lavender), Melissa officinalis (Lemon Balm), Mentha spp. (Mint), Verbascum spp. (Mullein), Origanum vulgare (Oregano), Petroselinum spp. (Parsley), Rosmarinus officinalis (Rosemary), Ruta graveolens (Rue), Salvia officinalis (Sage), Satureja montana (Savory), Tanacetum coccineum (Tansy), Thymus spp. (Thyme)."*
-'Plants Not Favored by Deer' by The Morton Arboretum®: The Champion of Trees, www.mortonarb.org, Lisle, Illinois, **2018**.

Gopher Problems

I have never had a problem with gophers or other underground mammals eating the Comfrey roots. I am in western North Carolina. All the references I found about Comfrey and gophers were in California. In particular in Santa Barbara county (central-southern California coast), Fresno county (central valley California), Butte county (Sacramento Valley), Mendocino county (north California coast), and San Mateo county (San Francisco Bay area).

If you live in California, you may want to plant small Comfrey roots in pots first so that gophers do not eat them. Large roots or whole plants should be OK to plant directly.

"We noticed that the gophers are quite partial to the Prickly Comfrey plant (in San Mateo County, California)."
-'Pacific Rural Press', Volumes 17-18, San Francisco, California, 843 pages, every other week from January 4 to December 27 **1879**. Article: 'A Visit to Baden Farm', San Mateo County, California, October 11, 1879.

"Here in California, Comfrey roots are eaten by the gophers who will harvest hundreds of short root pieces and then stash them in little pockets which then sprout into a jungle of Comfrey. Whether this happens with any other rodents, I don't know, but I do know that gophers will store Bermuda grass with the same results. This hasn't stopped us from growing Comfrey."
-'Comfrey Tendencies' by Keith Johnson, www.ibiblio.org, permaculture@listserv.oit.unc.edu, October 12 **1997**.

"Pocket Gophers: In some areas of the West Coast of California with loamy or sandy soils, gophers can make gardening a nightmare. In the short term, the only reliable solutions that I know of are trapping the varmints or fencing them out with wire cages. However, gophers prefer the soft, succulent growth provided by annuals and young perennials, and once a perennial is established, gophers become less of a problem. In addition, gophers seem to avoid forest, so a densely-planted food forest free of grass should eventually have little gopher activity. Unfortunately, gophers love clover, and I've heard of a mature fig that was toppled and turned into a hedge by gophers after clover was planted around it.
In any case, it can't hurt to increase habitat for barn owls and gopher snakes."
-West Coast Food Forestry: A Permaculture Guide by Rain Tenaqiya, Ukiah, California, **2005**.

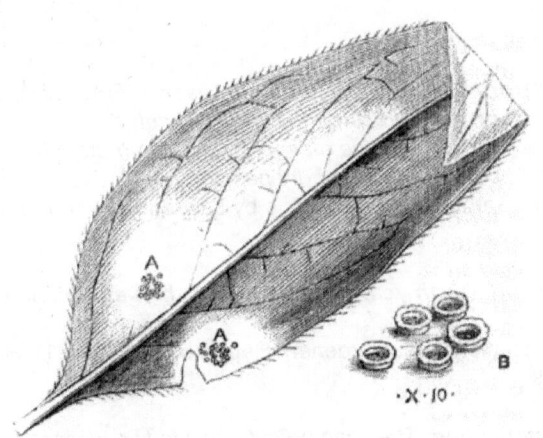

FIG. 70.
Leaf of Tuberous Comfrey, *Symphytum tuberosum*, L., invaded by *Æcidium asperifolii*, Pers., natural size.
Æcidium cups at B enlarged 10 diameters.

Leaf of Tuberous Comfrey 1884
Invaded by Aecidium asperifolii, Pers.

'Diseases of Field and Garden Crops Chiefly Such as are Caused by Fungi' by Worthington George Smith, F.L.S., M.A.I., Royal Horticultural Society, London, England, 1884.

"The under side of a leaf of Tuberous Comfrey, Symphytum tuberosum, L., is illustrated, at Fig. 70.

Two groups of the cups belonging to Aecidium asperifolii, Pers., are shown at AA. Each cluster of cups is surrounded by a large pallid disease patch, which has been caused by the exhaustion of the vital material of the leaf by the spawn of the fungus which at first grew within.

The fungus growth as also caused the leaf to become torn. Five of the little Aecidium cups are shown, at B."

Chapter 36

Productivity and Farm Economics of Comfrey

Rate of Growth

"In 3-year trials in Lithuania, Symphytum officinale (Common Comfrey) was propagated by root suckers. **Growth was most vigorous during flower bud formation and early flowering.**
Mean leaf productivity at flowering was 353.3 gram/plant (12.46 ounce), and root productivity at flowering and at the end of vegetation was 505 and 1106 gram/plant (17.81-39.01 ounces), respectively."
-'Growth and Development of Comfrey in Cultivation' by A.A. Pyasyatskene and Ya. A. Vaichyunene, Lekarstvennye Rasteniya - Narodnomu Khozyaistvu Tezisy Respublikanskogo Soveshchaniya Kaunas (Theses of the Republican Conference of Kaunas), Lithuania, page 58, September 16-17 **1986**.

"**One of the truly astonishing facts about Comfrey is its rate of growth. Leaves of a mature plant grow to 90 cm (35.4 inches) long and 20-25 cm (7.8-9.8 inches) wide in 28 days at the height of the season, if cultivated properly in the right location. This means 3 cm (1.18 inch) a day,** almost a 'Jack and the Beanstalk' story.
Our experience in Japan confirms what has been done in England, Australia, Canada, New Zealand and Kenya in yield per acre. Cut every 28 days (which means cut off completely, large and small leaves, about 2-3 cm {0.78-1.18 inch} above the ground), one can get 4 1/2 kg (9.9 pounds) of green leaf from well-established plants at each cut in the main growing season if properly cultivated. Three kilograms (6.6 pounds) is a fair average yield, and depending on soil treatment this can be raised to optimum."
-Miracle Grass, Comfrey by Andrew Hughes published in 1966 in Japanese. Part of it is in: 'Comfrey: Nature's Healing Herb & Health Food' by Andrew Hughes, Japan: Sanyusha Publishing Co., Ltd, **1992**. Part is in 'Comfrey Report No. 1: The 1954 Research Results' by Lawrence Hills published in 1955. (If you have a translation in English, I would appreciate having a copy.)

"**The taproot can grow to six feet (1.8 meter) within three years.**"
-'Comfrey Leaf: A New Animal Food Supplement' in The Encyclopedia of Alternative Agriculture by Dr. Richard Alan Miller, Agricultural Consultant and Researcher, USA, **1992**.

"**The Comfrey root shows remarkable growth, and are over 300 grams (0.66 pounds) in total weight about 100 days after planting, and root branches reach 40 in number.**"
-Comfrey Report No. 3: Feeding Dairy Cattle in Japan' by Meiji Milk Producing Co., Tokyo, Japan for Henry Doubleday Research Association, Braintree, Essex, England, July 1964, page 29.   (Roots planted were 'main root sections'. Size was not given.)

"**Comfrey vegetative growth does not cease with the start of flowering, and the plant will add new stems continuously during the growing season.** The plant will grow rapidly after harvest and flower again. Unlike annual crops, the leaves do not readily wilt during extended periods of drought due to its deep root system. This crop is also very frost resistant."
-"Comfrey' by Teynor, Putnam, Doll, Kelling, Oelke, Undersander and Oplinger, 'Alternative Field Crops Manual', University of Wisconsin: Cooperative Extension, University of Minnesota: Center for Alternative Plant & Animal Products, and Minnesota Cooperative Extension, February 1992, updated November **1997**, page 2.

First, Second, Third and Fourth Years

"**The trial plot, with autumn or early-spring planting, should produce from 20 to 30 tons the first season (first year), 40 to 60 the second year, and 80 to 100 the third year, though the fourth season may produce the biggest weight.**
A really dry spring and summer on a new-planted plot may hold back the yield, but once the main roots are down drought will not affect it, and even the water content of a big crop is an asset, carted to the beasts in a year of dry pastures."
-Russian Comfrey: A Hundred Tons an Acre of Stock or Compost for Farm, Garden or Smallholding by Lawrence D. Hills. London England: Faber and Faber, Limited, **1953**, page 63.

"**The second year Comfrey starts to move (grows fast), and the third and fourth year it grows faster than any plant in the English garden.**"
-Compost, Comfrey and Green-Manure: 'Henry Doubleday Research Association' First Gardeners Report by Lawrence D. Hills. Braintree, Essex, England: Henry Doubleday Research Association, **1959**.

"**The first year the Comfrey plants need until July to grow and get established.** They can be cut down to two inches (5 cm) from the ground, and in three weeks the young leaves will be ready to pick.
Their second season starts in early spring, and they can be picked hard till the peas come in, cutting back any that run to stem or stop leaf production. Then let them rest and grow stout, solid stems with small bell flowers the colour of purple carbon paper. These can be cut every six weeks for poultry or compost.

If you want Comfrey for cooking through the whole season, have two rows, one resting and one for picking."
-Comfrey Report Number Two: For Gardeners, Farmers and All Comfrey Growers in All Countries, with Analysis, Yields and Cultivation Methods, and the Results of Seven Years More Work Since Our First Report by Henry Doubleday Research Association, Braintree, Essex, England, 45 pages, March **1963**.

(Peas are a cool season crop that like temperatures from 13 to 18 C = 55 to 64 F. In temperate areas they are usually planted in March or April. Depending on variety, they are ready to harvest in 11 to 15 weeks, or about 3 to 4 months.)

*"**A Comfrey field builds up to maximum yield in its third or fourth year and keeps in full production for about 12 years.**"*
-Comfrey Report: The Story of the World's Fastest Protein Builder and Herbal Healer by Lawrence D. Hills, **1975**, page 51.
(Comfrey can be rejuvenated to last indefinitely. See subsection 'Older Comfrey Plots and How to Restore Them' in this section.)

*"**The third year should see the Comfrey crop in full production, and the plants should be meeting (touching) in rows on good land, or with ample feeding.**
The only chance the weeds should get to grow is between cuts, so any soil cultivation should be shallow and fitted in then."*
-Comfrey: Fodder, Food and Remedy by Lawrence D. Hills. New York: Rizzoli Universe Books, **1976**, page 106.

Production Per Plant or Small Area (Yield)
If you have experience with yields and weights of dried / fresh Comfrey roots and leaves, please let me know.

Pounds of Fresh Comfrey Leaves Per Plant or Small Area

"Monsieur Petin, Member of the St. Marcellin Agricultural Society (southeastern France), writes to the 'Journal d'Agriculture Progressive' (Journal of Progressive Agriculture):
The Comfrey produces three kilogrammes (6.6 pounds) per head (plant).
It is planted at a metre (3 feet) apart every way, which gives 10,000 per hectare (2.47 acres). Consequently 30,000 kilogrammes (66,165 pounds) of green fodder for cutting, and 120,000 kilogrammes (264,660 pounds) a year!' "
-'Forage Plants and Their Economic Conservation by the New System of Ensilage: Part I: Caucasian Prickly Comfrey' by Thomas Christy, Jun., F.L.S. (Fellow of the Linnean Society), Christy & Co., London, England, **1877**, page 20.

"The Reverend E. Highton, of Bude, Cornwall, England writes about Prickly Comfrey:
'With reference to the weight, I find this note in my diary of April 24th, 1880:
I cut a head of Prickly Comfrey weighing 6 1/4 pounds (2.8 kg), an average-sized head, *fifty to the square rod. A few days after this, each head weighed considerably more, but I believe that as great a weight of fodder is obtained in the course of the year when **cut before it runs to flowering stalk**, and the animals eat it much better.*
At a low computation, I should think an average of 4 pounds (1.8 kg) a head for each of four cuttings, is got here, which would give 16 pounds (7.2 kg) a head for the year, *or nearly 60 tons per acre.' "*
-'The Journal of the Royal Agricultural Society of England, Second Series, Volume 18', London, England, **1882**. Includes article: 'On Green or Fodder Crops Not Commonly Grown, which Have Been Found Serviceable for Stock-Feeding' by Joseph Darby.
(A rod is a linear measure for land equal to 5 1/2 yards or 5.029 meters. A 'square rod' is a square with each side measuring a rod. It is equal to 160th of an acre or 30 1/4 square yards or 25.29 square meters).

*"The best Comfrey yield in 1958 was a poultry farmer in New Zealand who got 124 tons per acre by mulching heavily with deep-litter compost, suppressing weeds, and **growing plants up to 23 pounds (10.4 kg) in weight each**."*
-Compost, Comfrey and Green-Manure: 'Henry Doubleday Research Association' First Gardeners Report by Lawrence D. Hills. Braintree, Essex, England: Henry Doubleday Research Association, **1959**.

"A Comfrey bed holding 3 dozen plants (36 plants), 2 feet (0.6 meter) apart, takes up about 14 feet (4.26 meter) of garden length if your beds are 13 feet (3.96 meter) wide.
These should average 18 pounds (8 kg) each of foliage between April and November, *roughly 5 cwt (560 pounds = 254 kg) of foliage to use **and the equivalent in minerals of 1 cwt (112 pounds = 50.8 kg) of chemical potato fertilizer**, but the with organic flavour and quality in crops grown with it."*
-Down to Earth Fruit and Vegetable Growing by Lawrence D. Hills. London, England: Faber & Faber, **1960**, page 56.
(cwt= hundredweight= 112 pounds)

*"**Gardeners have found that a dozen established plants yield about 2 cwt (224 pounds) of foliage a summer in four to five cuts with the last about October.**"*
-Down to Earth Gardening by Lawrence D. Hills. London, England: Faber & Faber, **1967**, 1975, page 219.

*"**Gardeners growing Comfrey are not concerned with tons an acre but how many pounds of Comfrey they get from a dozen plants.**
Mr. Harold Kirkman of Southport (Merseyside County, England) planted a dozen Russian Comfrey Bocking No. 14 in the spring of 1959 and harvested as follows in 1960: April 18th= 18 pounds, June 5th= 88 pounds, July 24= 52 pounds,*

*September 18th= 63 pounds, **total= 221 pounds (100 kg).**"*
-Comfrey Report: The Story of the World's Fastest Protein Builder and Herbal Healer, Conservation Gardening and Farming Series: Series C by Lawrence D. Hills. England: Henry Doubleday Research Association, **1975**, page 84.

*"Gardeners who grow Comfrey are not concerned with tons per acre but rather with how many pounds they get from a dozen or so plants. **Under good growing conditions a dozen plants can yield more than 200 pounds (90.7 kg).**"*
-Fertility Without Fertilizers: A Basic Approach to Organic Gardening by Lawrence D. Hills. NY: Universe Books, **1975**.

*"**This one Comfrey bed, measuring 150 by 5 feet (45.7 x 1.52 meter), will produce at least a (United States) ton (2000 pounds) of useful foliage in a year.**"*
-Enchanted Garden: Alan Chadwick's Organic Method of Gardening by Tom Cuthbertson. London, England: Rider & Company / Hutchinson & Co. Publishers Ltd, **1978**, page 72.
  (The author is based in California, so I assume it is US ton {2000 pounds}, not British ton {2240 pounds}.)

*"Compost, Carbon, Organic Matter, Fodder and Cover Crops: **Russian Comfrey:***
53 roots per 100 square feet *(9.29 square meter).*
Possible 'Grow Biointensive'® Yield Pounds/100 Square Foot.
Biomass, wet: 92 / 220 / 339 pounds (41.7 / 99.7 / 153.7 kg). 6-month yield.
Estimates based on our experience and research. Use lowest figure if you are a beginning gardener; middle if a good one; highest if an excellent gardener with a good soil and climate."
-How to Grow More Vegetables and Fruits, Nuts, Berries, Grains and Other Crops, 6th Edition, by John Jeavons, Ecology Action of Midpeninsula. Berkeley, California: Ten Speed Press, page 100, **2002**. First published 1974. A Primer on the Life-Giving Sustainable 'Grow Biointesive'® Method of Organic Horticulture.

*"**2017 Russian Comfrey 'Bocking 14' Biomass Records: Site Overview:***
Shipka Bulgaria, Temperate climate, Latitude 41. Elevation 580 meter (1902 feet).
Average annual rainfall: 588 mm (58.8 cm = 23 inch).
42 plants. *60-70 cm apart (1.9-2.2 feet).* **Bed 13 m2 (square meters) = 140 square feet.**

Cut	Date	Flowering	Days Between Cuts	Weight All	Weight Per Plant
1	May 8 2017	yes	since last season	20.4 kg	0.49 kg
2	June 5 2017	partial	27	25.8	0.62
3	July 17 2017	partial	42	23.5	0.56
4	August 29 2017	no	43	22.5	0.53
5	October 9 2017	no	41	7.08	0.16
			Total	99.28 kg	2.2 kg (4.8 pound)

Fertilizing:
  The initial input of 20 liter (5.2 gallons) of mature compost per meter (3.2 feet) length of bed and 70 gram (2.4 ounce) of wood ash per meter length was applied when we established the bed in the spring of 2015.
  In 2016 and 2017 the only fertility inputs were the trimmings from the pathways between and around the garden beds. We mowed this section after each cut, and each time emptied approximately four 30 liter (7.9 gallons) mower bags of trimmings onto the Comfrey beds.
Mulch Production:
  The bed produced enough Comfrey leaf mulch to cover 18.5 m2 (199 square feet) with a 10-15 cm (3.9-5.9 inch) deep layer of leaves.
Casual observations of organisms on the Comfrey plants:
  Many spiders in undergrowth. Ladybirds (ladybugs). Many species of bees. Carpenter bees were one of many species of solitary bees that feed from the Comfrey flowers as well as Honey bees. For higher yields the plants can be cut before flowering. We did not carry out this practice to allow the bees to forage."
-'How Much Comfrey Can You Grow on 13 m2? Comfrey Trial Results Year 2: 2017' by Balkep: The Balkan Ecology Project: A Permaculture-Inspired, Grassroots Project, https://balkanecologyproject.blogspot.com, www.balkep.org, southeastern Europe, Bulgaria; October 9 **2017**.

Pounds of Dried Comfrey Leaves and Roots Per Part of an Acre

"I will now state some facts with regard to the production of Comfrey, root and herb.
Comfrey Root Fresh and Dried:
In April, 1841, I dug from 18 square feet (1.67 square meter) of ground, *of moderately good tilth or heart, besides the said seed caps (crowns) dug earlier, one large bushel of green Comfrey root, of two year's growth.*
In this proportion, an acre would yield 2420 bushels green Comfrey root. **This when wash, dried and ground, weighed ten pounds (4.5 kg); which would be 24,200 pounds (10,976 kg) to the acre.**
Comfrey Leaf Dried:
  As to the Comfrey herb for hay: **At two cuttings on 15 square feet (1.39 square meter) of soil, in 1842, the second year from planting, July 15th and September 1st, I obtained four pounds (1.8 kg) of good well-*

*dried hay, better I think than good clover; which would be **5 tons and 1616 pounds to the acre.** It is, I think, after the roots become considerable in the ground, capable of producing larger crops of hay than this."*
-'The New England Farmer, and Horticultural Register' by Joseph Breck, Volume 23, New Series Volume 13, Boston, Massachusetts, **1845**. ('Symphytum or Comfrey, as Food for Men and Cattle' by Ezekiel Rich, Troy, New Hampshire, page 10, July 10 1844.)
(A United States bushel is equal to 64 United States pints {1 pint = 2 cups} equal to 35.2 liters, used for dry goods. A British bushel is equal to 8 Imperial gallons equal to 36.4 liters, used for dry goods and liquids.)
(One acre = 43,560 square feet.)
(There are 20 hundredweight in a ton, producing a 'short ton' of 2000 pounds and a 'long ton' of 2240 pounds. A short ton is used in United States. A long ton is used in Great Britain and countries that used to be territories of it. The metric ton or tonne is the 'International System of Units' that is used worldwide. It is 1,000 kg which equals 2204 pounds.)
(For the dried Comfrey leaf hay, it is a United States ton at 2,000 pounds. So that is 10,000 pounds + 1616 pounds = 11,616 pounds per acre {5268 kg}.)

"Compost, Carbon, Organic Matter, Fodder and Cover Crops: **Russian Comfrey:**
53 roots per 100 square feet (9.29 square meter).
Possible 'Grow Biointensive'® Yield Pounds/100 Square Foot.
   Biomass, air-dry: 10 / 27 / 37 pounds (4.5 / 12.2 / 16.7 kg). 6-month yield. *Estimates based on our experience and research. Use lowest figure if you are a beginning gardener; middle if a good one; highest if an excellent gardener with a good soil and climate.*
Possible 'Grow Biointensive'® Yield Pounds/Plant
   Dry: .06-.23 pounds (0.027 - 0.10 kg) (27-104 gram).
Average United States Yield Pounds/100 Square Foot
   Biomass, air-dry: '62.6' pounds (28.3 kg) world high, *12-month season. From United States Department of Agriculture, Agricultural Statistics 2000, 1998 data, Washington, DC, U.S. Government Printing Office. Numbers in quotes are approximations from other data, because official data are not available for this crop."*
-How to Grow More Vegetables and Fruits, Nuts, Berries, Grains and Other Crops, 6th Edition, by John Jeavons, Ecology Action of Midpeninsula. Berkeley, California: Ten Speed Press, page 100, **2002**. First published 1974. A Primer on the Life-Giving Sustainable 'Grow Biointesive'® Method of Organic Horticulture.

"A crop adaptation study of 25 species of medicinal and culinary herbs and essential oil crops was established in 2003 at the Crop Diversification Center North, Edmonton, Alberta, Canada.
Comfrey Symphytum officinale L (Boraginaceae):
Comfrey was grown as a perennial under rain fed conditions with supplementary irrigation at seedling establishment. Comfrey transplanted into the field on June 17 at a density of 33,333 kg (73,486 pounds) per hectare in single row plot. Harvesting was done between September 28 and September 30.
Comfrey had 100 percent establishment. Winter survival was excellent, 100 percent plant survival over three years. No flowers were observed.
Yield of Roots and Aerial Parts of Comfrey, CDC North, Alberta, Canada

Year	Roots dry weight kg per hectare	Aerial parts kg per hectare
2003	3215 kg (7087 pound)	2935 kg (6470 pound)
2004	1328 (2927 pound)	
2005	8247 (18181 pound) "	

-'Alberta 2007 Specialty Crop Report' by Chuanliang Su, Crop Statistician, Alberta Agriculture and Food, Economics and Competitiveness Division, Statistics and Data Development Unit, Edmonton, Alberta, Canada, **2007**.
   (Hectare is a square with 100 meter sides, or 10,000 m2 {square meters}. There are 100 hectares in one square kilometer. An acre is 0.405 hectare, and one hectare is 2.47 acres.)

"Dried Comfrey Leaf:
Five to six hundred pounds (226 to 272 kg) of dried Comfrey leaf per one-eighth-acre bed (multiple cuttings per season). **Moisture ratio for Comfrey leaf is 6:1 fresh:dried.**
Dried Comfrey Root:
One hundred fifty pounds (68 kg) of dried Comfrey root per one-eighth-acre bed. **Moisture ratio for Comfrey root is 3:1, fresh:dried."**
-The Organic Medicinal Herb Farmer: The Ultimate Guide to Producing High-Quality Herbs on a Market Scale by Jeff Carpenter and Melanie Carpenter. White River Junction, Vermont: Chelsea Green Publishing, **2015**, page 289.
(Based on above, there is 4,000 to 4,800 pounds = 1814 to 2177 kg of dried Comfrey leaf per acre. And 1,200 pounds = 544 kg of dried Comfrey root per acre. These amounts are very different from the article from 1845.)

Production Per Acre (Yield)

Tons or Pounds of Fresh Comfrey Leaves Per Acre

*"**Russian Comfrey grown at Ottawa, Canada and cut 4 times per year yielded 19.3 tons green forage per acre** containing 2.06 tons dry matter, of which 21-22% was crude protein and 10% crude fibre. Cultural practices are described."*
-'Russian (Quaker) Comfrey: Symphytum Peregrinum' by R.W. Robertson, Forage Notes, Volume 5, No. 1, pages 32-33, **1959**. (I was unable to get this report. If you have a copy, could you please send it to me.)

"
	Russian Comfrey Leaf (50 tons/acre)	**Soybean** (15 cwt/acre=1680 pounds/acre)
Dry matter	124 cwt (13,888 pounds= 6299 kg)	13.5 cwt (1,512 pounds= 685 kg)
Crude Protein	3,808 pounds	314 pounds
Oil	336 pounds	225 pounds
Carbohydrates	5,500 pounds	488 pounds
Fibre	1,680 pounds	60 pounds

From 'Feeds and Feeding' by Professor Frank B. Morrison, **1959**."
-Comfrey: Fodder, Food and Remedy by Lawrence D. Hills. New York: Rizzoli Universe Books: 1976, page 86.
        (cwt= hundredweight= 112 pounds = 50.8 kg)

"1958 Quaker Comfrey Leaves Yield Per Acre (Russian Comfrey):
University of California, Davis; R.T. Edward's Records:

Harvest Date	Yield Green	Yield Dry	Protein Dry Weight
June 3	104,544 pound = 52.2 US tons	7,318 pound	21.17%
July 1	40,656 pound = 20.3 US tons	4,066	24.15%
August 6	46,464 pound= 20.2 US tons	5,576	22.31%
November 11	39,638 pound = 19.8 US tons	7,144	15.74%
Total	231,352 pound=104,939kg=115.6 ton	24,104 pound	Average 20.84% "

-'Quaker Comfrey in 1961' by M.D.M. and R.T. Edwards, Agronomy Notes, University of California Cooperative Extension Service, Davis, California, pages 22-23, February 28 **1961**.

"**The Russian Comfrey trial** had row width of 30 inches (76.2 cm) with 36 inches (91.4 cm) between the plants. The trial was planted from sets (roots) on 6th April 1956, 2nd April 1957, and autumn of 1957.
The trial, which was grown on a gault clay soil, received a basal dressing of four cwts per acre of complete fertiliser (Nitroten 9 : Phosphorus 9 : Potassium 15). In 1957, eight cwts per acre of a similar fertiliser was applied in the spring, and also four cwts per acre, supplemented with three cwts of NitroChalk®, after both the June and the August cuts.
In 1958, the spring application was four cwts per acre, supplemented by four cwts per acre of fertiliser (9 : 6 : 18) applied after both the June and August cuts.
In this Cambridge, England trial the total annual yields of the best clones, Russian Comfrey Bocking No. 12 and Bocking No. 14, were about 48 tons per acre.
Bocking No. 12, which is from a survival of a planting said to have been made many years ago at Abbey Farm, Coggeshall, England, gave good yields in both 1957 and 1958.
Bocking No. 14, a selection from Stephenson's, gave the highest yield in 1958.
In the trial reported here the yields of Symphytum asperum (Prickly Comfrey) were much below those of the Russian Comfrey clones."
The trial crop removed the equivalent of more than five hundredweights (560 pounds = 254 kg) of Muriate of Potash (60%) in the produce of two years."
-'Russian Comfrey' by L.A. Willey and R.L. Knight, Journal of the National Institute of Agricultural Botany, Cambridge, England, Volume 9, No. 2, pages 139-144, **1962**.
        (The Gault Formation is a geological formation of stiff blue clay. It is a glutinous/sticky marine deposit found in southeastern England and France.)   (Basal dressing means broadcasting when sowing or planting so it is uniformly distributed over the entire field and mixed in the soil.)
        (Nitrochalk® is a trademarked chemical fertilizer that contains calcium carbonate and ammonium nitrate. It is 26% nitrogen, half is in the ammonical form and half is nitrate. Some nitrogen is immediately available and some later. It is also used under other trade names.)

"*Representative Comfrey Leaf Yields from Various States and Foreign Countries:*

State or Nation	Yield: Tons/Acre		Spacing	Manure	Fertilizer	Cuts per Year
	Green	Dry			N-P-K Pounds/Acre	
South Korea	36.0	5.4	18 inches	13 tons/acre	120-54-107	6
Vermont	37.0	3.8	36	50	0-0-0	4
Netherlands	29.6	3.6	18	13	128-108-192	4-5
Wisconsin	33.7	3.2	18x48	Heavy	0-0-0	4
England	25.6	3.1	30x36	0	180-120-255	4-5
West Germany	22.3	2.4	16x24	0	68-24-25	2-4
Poland	14.9	2.1	16	0	54-0-0	3-4
Kenya	21.2	1.7	36	5	0-29-0	7

Many strains of Comfrey are sold, and there are large differences in yield. The English study included eight strains of

Comfrey. The two poorest yielded only 41 and 68 percent as much as the best strain."
-'Comfrey Yields and Forage Value' by Richard H. Hart, United States Department of Agriculture, Agricultural Research Service Plant Physiology Institute, Beltsville, Maryland, CA-NE-2, December **1972**.
> (The type of Comfrey species was not given. And other information such as soil type, sun exposure, water input, size of planted roots, and age of plants was not given.)

"In six years of trials carried out at Felin, near Lublin, Poland, of **three ecotypes of Symphytum officinale L. from different parts of Poland and two clones derived from reciprocal crosses between two of them, clone A (female Tarnow x male Lodz) gave the highest mean yield of green fodder, 230 q/ha (q/hectare), and was followed by the Lodz ecotype and clone B, both with 200 q/hectare.**
Clone A showed its superiority particularly in years when yields were high, but in three years of low yields was outyielded by three of the other four."
-'The Yield of Some Ecotypes of Common Comfrey (Symphytum Officinale L.)' by S. Tabin, S. Berbec and T. Bobrzynski, Akademia Rolnicza (Agricultural Academy), Lublin, Poland; Hodowla Roslin, Aklimatyzacja Nasiennictwo (Plant Breeding, Acclimatization and Seed Production), Volume 17, No. 6, pages 505-511, **1973**.
> (Ecotype or ecospecies is a genetically distinct geographic variety, population or race within a species, which is adapted to specific environmental conditions.)
> (I think for q/ha that 'q' is quintale or 100 kg = 220 pounds. Ha is hectare which is a square with 100 meter sides, or 10,000 m2 {square meter}. There are 100 hectares in one square kilometer. An acre is 0.405 hectare, and one hectare is 2.47 acres.)

"Yields of Russian Comfrey *(Symphytum asperum) {as quoted from report}* **sown in rows 50 cm (19.68 inches) apart,** *increased with increasing cutting interval with a total fresh yield after 2 years of* **55.0 ton/feddan (0.42 hectare) when cut a 5-week intervals** *compared with 13.5 when cut every 2 weeks.*
Only plots cut every 4 or 5 weeks continued to grow throughout the second year. For all cutting intervals vegetative growth rate was highest during the warmer months and was much reduced during the colder months.
On a DM (Dry Matter) basis, content of CP (Crude Protein) was 22.87%, CF (Crude Fiber) 9.37%, ether extract 5.36%, NFE (Nitrogen-Free Extract) 37.21%, ash 25.19%, Ca (Calcium) 2.92% and P (Phosphorus) 1.00%; carotene content of fresh Comfrey averaged 290.4 mg/kg."
-'Productivity and Composition of Russian Comfrey as a Fodder Crop in Egypt' by A. El-Bassousy, F. El-Sabban, A.M. Makky, M.S. El-Danasoury and M.A. El-Ashry, Animal Production Research Institute, Agriculture Research Center, Cairo, Egypt; Agricultural Research Review, Volume 53, No. 7, pages 51-57, **1975**.
> (Russian Comfrey is not Symphytum asperum. Asperum is Prickly Comfrey.)

"No account has appeared in the English language of the substantial amount of research on Comfrey which has been carried out in the USSR and other eastern European countries. *This article is intended to meet the need for a critical review of the scientific literature on the crop, with particular reference to its yield and chemical composition. Aspects reviewed include taxonomy; breeding (principally Russian Comfrey, Symphytum x uplandicum after crosses),* **yields of Symphytum x uplandicum, Symphytum asperum (Prickly Comfrey) and Symphytum officinale (Common Comfrey) as recorded in trials in different parts of the world,** *comparisons of spp. (species), strains and selections and yield in relation to stand age, water supply, fertilizers, plant population and cutting regime; chemical composition, including CP (Crude Protein), ether extract, NFE (Nitrogen-Free Extract), vitamins, alkaloids, minerals; digestibility and palatability. Much of the literature cited is from the USSR."*
-'Comfrey Symphytum spp. as a Forage Crop' by J.C. Forbes, A.D. McKelvie, and P.J.C. Saunders, North of Scotland College of Agriculture, Aberdeen, United Kingdom; Herbage Abstracts, Volume 49, No. 12, pages 523-539, **1979**.

> *"***The highest individual record for Symphytum x uplandicum** *is 259.3 tons green herbage/hectare obtained by Heitman and Miller in California in 1958 (Hills, 1976).*
> ***For Symphytum asperum,*** *a yield of 186.7 tons/hectare was recorded from a second-year crop in Uzbekistan (Vavilov and Kondrat'ev, 1975).*
> *A record yield of 130.5 tons green herbage/hectare was reported for* **Symphytum officinale** *in Romania by Popescu, Pitis and Casanova (1971).*
> *From the range of values in the Tables 1-3,* **it can be concluded that a reasonable expectation of yield from a crop of Comfrey in a temperate climate would be 30 to 60 tons green herbage and 3.5 to 7 tons dry matter/hectare per year."*
> -'Comfrey Symphytum spp. as a Forage Crop' by J.C. Forbes, A.D. McKelvie, and P.J.C. Saunders, North of Scotland College of Agriculture, Aberdeen, United Kingdom; Herbage Abstracts, Volume 49, No. 12, pages 523-539, **1979**.
>> (One hectare is a square with 100 meter sides, or 10,000 m2. There are 100 hectares in one square kilometer. An acre is 0.405 hectare, and one hectare is 2.47 acres.)

> **"Medvedev (1974) grew five Symphytum species and hybrids in Leningrad Province, USSR and measured their yield each year from the second to the ninth year of growth.**
> *In the second year S. asperum and S. x uplandicum were much more productive than the others.*

In later years the differences were less marked, but S. asperum was consistently higher yielding than S. x uplandicum.
S. caucasicum was the least productive species in the second year but by the ninth year was the most productive. In a later study in the same area by Medvedev and Sidorava (1976), S. asperum again outyielded S. x uplandicum."
-'Comfrey Symphytum spp. as a Forage Crop' by J.C. Forbes, A.D. McKelvie, and P.J.C. Saunders, North of Scotland College of Agriculture, Aberdeen, United Kingdom; Herbage Abstracts, Volume 49, No. 12, pages 523-539, **1979**.

"S. officinale, S. asperum and S. x uplandicum have all been grown as forage crops, but **Forbes et al. (1979) reported that S. x uplandicum is apparently the only species in cultivation outside the U.S.S.R. and Eastern Europe.** The high cost and practical difficulties of establishing, maintaining and utilizing Comfrey as a perennial forage crop (Forbes et al., 1979) have led these authors to conclude that it has little value in conventional agriculture."
-'In Vitro Propagation of Symphytum Species' by P.J.C.Harris, C.G.Grove and A.J.Havard, Henry Doubleday Research Association, Coventry, Great Britain, and Department of Biological Sciences, Coventry Polytechnic, Great Britain; Scientia Horticulturae, Volume 40, Issue 4, pages 275-281, November **1989**, page 276.

"**We estimated the yield would be about 1,000 kg (2204 pounds, 1.1 United States ton) from 1 tan (1/4 acre) from plants that less than 3 month before had been mere root sections, 8-10 cm (3.14-3.93 inches) long and 1 cm (0.39 inches) wide.**
Every root had struck, 100% fertility, and the vigor of the plants was phenomenal. Each plant had been given 10 kg (22 pounds) of ordinary farm cow manure compost."
-Comfrey: Nature's Healing Herb & Health Food by Andrew Hughes. Japan: Sanyusha Publishing Co, **1992**, page 120.
  (In Japanese units of measurement, a 'tan' is 300 tsubo, 991.7 square meters or 10,670 square feet. A 1/4 acre is 10,890 square feet.)

"**The highest known figures for (fresh) yield for S. officinale in recent years is from Holland (Netherlands) in 1936: 36 tons to the acre (9,000 kg per 1/4 acre),** S. asperrimum (Prickly Comfrey) gave 50 tons to the acre (12,500 kg per 1/4 acre) in Denmark in 1940."
-Comfrey: Nature's Healing Herb & Health Food by Andrew Hughes. Japan: Sanyusha Publishing Co, **1992**, page 142.

Tons or Pounds of Dried Comfrey Leaves Per Acre
See subsection 'Dry Matter' in section 'Nutritional Value of Comfrey' (Chapter 19, Volume 1).

"Prickly Comfrey is propagated from the roots. These are set out in rows 1 1/2 to 2 feet (0.45-0.60 meter) apart, the plants 16 to 20 inches (40.6-50.8 cm) apart in the row. At first of slow growth, **Prickly Comfrey will in the course of two years yield from 3 to 6 tons of cured (dried) forage per acre.**"
-'Inventory No. 5 of Foreign Seeds and Plants Imported by the Department of Agriculture, and for Distribution Through the Section of Seed and Plant Introduction, Numbers 1901-2700' by United States Department of Agriculture, Division of Botany, S.P.I. 10, Washington, DC, November 3 **1899**.

"1958 Quaker Comfrey Leaves Yield Per Acre (Russian Comfrey):
University of California, Davis; R.T. Edward's Records:

Harvest Date	Yield Green	Yield Dry	Protein Dry Weight
June 3	104,544 pound = 52.2 US tons	7,318 pound	21.17%
July 1	40,656 pound = 20.3 US tons	4,066	24.15%
August 6	46,464 pound= 20.2 US tons	5,576	22.31%
November 11	39,638 pound = 19.8 US tons	7,144	15.74%
Total	231,352 pound=104,939kg=115.6 ton	**24,104 pound**	Average 20.84% "

-'Quaker Comfrey in 1961' by M.D.M. and R.T. Edwards, Agronomy Notes, University of California Cooperative Extension Service, Davis, California, pages 22-23, February 28 **1961**.

"Representative Comfrey Leaf Yields from Various States and Foreign Countries:

State or Nation	Yield: Tons/Acre Green	Dry	Spacing	Manure	Fertilizer N-P-K Pounds/Acre	Cuts per Year
South Korea	36.0	5.4	18 inches	13 tons/acre	120-54-107	6
Vermont	37.0	3.8	36	50	0-0-0	4
Netherlands	29.6	3.6	18	13	128-108-192	4-5
Wisconsin	33.7	3.2	18x48	Heavy	0-0-0	4
England	25.6	3.1	30x36	0	180-120-255	4-5
West Germany	22.3	2.4	16x24	0	68-24-25	2-4
Poland	14.9	2.1	16	0	54-0-0	3-4
Kenya	21.2	1.7	36	5	0-29-0	7 "

-'Comfrey Yields and Forage Value' by Richard H. Hart, United States Department of Agriculture, Agricultural Research Service Plant Physiology Institute, Beltsville, Maryland, CA-NE-2, December **1972**.
> **(Based on the above Green/Dry tons per acre, fresh green Comfrey leaves dry to a little over 10% of the weight of fresh leaves.)**

"Dry-weight yields of Comfrey, Symphytum officinale, are estimated to be more than 1.2 ton/acre/cutting, with up to 6 cuttings per year in this region (Grants Pass, Oregon).
In full production, Comfrey leaf yields should be in excess of 6 ton/acre dry-weight, with 5 to 6 cuttings."
-'Comfrey Leaf: A New Animal Food Supplement' in 'The Encyclopedia of Alternative Agriculture' ebook by Dr. Richard Alan Miller, https://richardalanmiller.com, Agricultural Consultant and Researcher, Oregon and Washington, **1992**.

"Yields in dry weight of Comfrey leaf (Symphytum officinale) for a field established for four years or more can be five tons per acre on four cuttings. With heavy irrigation, up to six cuttings are available in some regions."
-'A Farm Project: Comfrey Leaf and Root' by Richard Alan Miller, Oak Publishing Inc., Grants Pass, Oregon, and Northwest Botanicals, www.nwbotanicals.org, **1999**.

"Yield and energy value of dry biomass of perennial herbs was measured in the mountain zone of Kabardino-Balkarian Republic, Russia in May 2002. **Rhizomes (roots) of goat's-rue, inula, Comfrey and nettle were planted** in 2002 as pure cultures and in a mixture. **Yield of dry hay in 2005 was** 11.0, 9.9, **8.7** and 16.6 **tons/hectare** for goat's-rue, inula, **Comfrey** and nettle, respectively.
Energy value of 1 kilogram of dry biomass was 16.5, 12.9, **10.7** and 9.1 **MJ (megajoule)** for goat's-rue, inula, **Comfrey** and nettle, respectively. Yield and energy value of mixed planting were 13.6 tons/hectare and 12.6 MJ, respectively. Detailed data are presented in a table and a graph."
-'Effectiveness of Multi-Purpose Grasses in Fodder Production' by A.Ya. Tamakhina, Kabardino-Balkarian State Agrarian Academy, Russia; Kormoproizvodstvo (Plant Breeding), No. 10, pages 2-4, **2009**.

"Comfrey, Symphytum officinale Leaf: Yield: Dry leaf 12 tons / hectare (years 2 to 5)."
-'Richters® ProGrowers Info' by Richters: The Herb Specialists, www.richters.com, Goodwood, Ontario, Canada, 1997-**2019**. Includes market and culture of herbs.
(Hectare is a square with 100 meter sides, or 10,000 m2. There are 100 hectares in one square kilometer. An acre is 0.405 hectare, and one hectare is 2.47 acres.) (So 0.405 x 12 tons = **4.86 United States tons dried leaf**.)

Tons or Pounds of Fresh Comfrey Root Per Acre

"Comfrey Root Fresh and Dried:
> *In April, 1841, I dug from 18 square feet (1.67 square meter) of ground,* of moderately good tilth or heart, besides the said seed caps (crowns) dug earlier, **one large bushel of green Comfrey root, of two year's growth. In this proportion, an acre would yield 2420 bushels green Comfrey root.**
> *This when wash, dried and ground, weighed ten pounds (4.5 kg); which would be 24,200 pounds (10,976 kg) to the acre."*

-'The New England Farmer, and Horticultural Register' by Joseph Breck, Volume 23, New Series Volume 13, Boston, Massachusetts, **1845**. ('Symphytum or Comfrey, as Food for Men and Cattle' by Ezekiel Rich, Troy, New Hampshire, page 10, July 10 1844.)
> (A United States bushel is equal to 64 United States pints {1 pint = 2 cups} equal to 35.2 liters, used for dry goods. A British bushel is equal to 8 Imperial gallons equal to 36.4 liters, used for dry goods and liquids.)
> (The weight of a bushel depends on the density of the product. Barley bushel = 48 pounds or 21 kg. Beet bushel = 60 pounds or 27 kg. Carrot bushel = 50 pounds or 22 kg. Onions = 57 pounds. Parsnips = 50 pounds. Potatoes = 60 pounds. Turnips = 55 pounds or 25 kg. Wheat = 60 pounds. **So perhaps a fresh Comfrey root bushel weighs around 55 pounds = 25 kg. Based on the author's production, that would be 2420 bushel x 55 pounds = 133,100 pounds = 60,500 kg per acre. This amount is much higher than the article below from 1916 so perhaps my guess at the weight per bushel is wrong.**)
> (One acre = 43,560 square feet = 4,046 square meters. 43,560 square feet divided by 18 square feet = 2420 x 1 bushel= 2420 bushels per acre. In other words, the author's math is correct.)

"The root of Comfrey (Symphytum officinale) is the part used, and should be collected late in October, or at least before the ground freezes. The root is quite large, very fleshy and juicy, and **an acre should yield between 10,000 and 12,000 pounds (4,535 - 5,443 kg) of fresh root.**
The water contents of fresh roots is about 75 percent, hence four pounds (1.8 kg) of fresh root usually represent about one pound (0.45 kg) of dry root."
-'National Association of Retail Druggists' with article 'The Cultivation of Medicinal Plants: Ideal Conditions for Comfrey, Stoneroot, Leptandra', Chicago, Illinois, Volume 23, November 9 **1916**.

Tons or Pounds of Dried Comfrey Root Per Acre

*"**Comfrey root: 0.8 ton dry yield per acre.** The yields per acre are estimates based on information from the 'United States Department of Agriculture' trade magazines, especially 'Chemical Marketing Reporter', and the author's background in marketing as a regional wholesaler."*
-The Potential of Herbs as Cash Crops: How to Make a Living in the Country by Richard Alan Miller. Berkeley, California: Ten Speed Press, **1992**, page 11.
>('Chemical Marketing Reporter' was published from 1972-1996. Its name changed to 'Chemical Market Reporter' and was published from 1996-2006. Then its name was changed to 'ICIS Chemical Business', www.icis.com. It covers American, European and Asian markets.)

*"**Comfrey, Symphytum officinale Root: Yield: Dry root 1,800 kg / hectare** (year 3, 4 or 5)."*
-'Richters® ProGrowers Info' by Richters: The Herb Specialists, www.richters.com, Goodwood, Ontario, Canada, 1997-**2019**. Includes market and culture of herbs.
>(Hectare is a square with 100 meter sides, or 10,000 m2. There are 100 hectares in one square kilometer. An acre is 0.405 hectare, and one hectare is 2.47 acres.)
>(1,800 kg = 3,968 pounds = 1.98 United States ton = 1.77 Imperial ton. So 0.405 x 1,800 kg = 1 acre = 729 kg/ acre. Then **729 kg = 0.8 United States ton = 0.7 Imperial ton dried root per acre**.)

Comparing Comfrey to Other Crops

*"Comparison of Russian Comfrey with crops grown in adjacent trials for yield of **Dry Matter in cwts per acre**:*

	Russian Comfrey	Kale	Timothy Hay (S.48)	Perennial Ryegrass (S.24)
1957	53.8	63.2	67.9	-
1958	55.5	68.0	75.8	89.8

The Crude Protein content of Russian Comfrey resembled that of lucerne (alfalfa) and on the basis of Crude Protein production Russian Comfrey may well outyield kale.
The chemical composition of the green Comfrey material was similar to that of mangel (fodder beet) leaves, at least, for the main constituents (*Woodman, 1952).
In comparison with kale the Comfrey Crude Fibre content was lower and the ash content much higher. **The ash (minerals), however, while containing relatively high quantities of calcium and phosphate (phosphorus), also contained a large quantity of silica and an exceptionally high proportion of potash (potassium)."**
-'Russian Comfrey' by L.A. Willey and R.L. Knight, Journal of the National Institute of Agricultural Botany, Cambridge, England, Volume 9, No. 2, pages 139-144, **1962**.
(* -'Rations for Livestock' by H.E. Woodman, Bulletin No. 48, Ministry of Agriculture and Fisheries, Her Majesty's Stationery Office, London, England, 15th Edition, 1952. First published January 1921.) (cwt= hundredweight= 112 pounds)
>(Timothy is available as a hay strain and a pasture strain. The hay strain is more stemmy and less persistent. S.51 = hay type, S.48 = pasture type.) (Perennial ryegrass Aberystwyth S. 24 = dense hay type.)

*"**Yields of the best strain of Comfrey were 83, 76, and 62 percent of the yields of kale, timothy, and ryegrass, respectively. Comfrey yields were 81 percent of red clover yields in Wisconsin and 74 percent of alfalfa yields in Kenya."***
-'Comfrey Yields and Forage Value' by Richard H. Hart, United States Department of Agriculture, Agricultural Research Service Plant Physiology Institute, Beltsville, Maryland, CA-NE-2, December **1972**.   (The type of Comfrey species was not given.)

*"Russian Comfrey was tested in southern Norway in two trials in the Stavanger region (59 degrees northern latitude) and in the Oslo region (60 degrees northern latitude). **Russian Comfrey was outyielded by both leys and annual fodder crops (Italian ryegrass, Westerwold ryegrass, fodder rape, fodder turnips, marrow stem kale).**
The crude protein content was somewhat higher, but the digestibility much lower than in grass. Early and frequent cutting raised the protein content and digestibility, but lowered the dry matter yields and plant vigour.
New selections from England (Bocking 4 and 14) did not give higher yields than the progenies (offspring) of Russian Comfrey cultivated in Norway since about 1925.
Growing of Russian Comfrey is not recommended because of low yields, high costs of establishment and a content of toxic alkaloids."*
-'Russian Comfrey: Yield and Quality' by M. Pestalozzi and N. Skaland, Statens Forskingsstasjon (National Research Institute) Saerheim Klepp, and Norges Landbrukshoegskole (Norwegian Agricultural High School), Inst. for Plantekultur (Institute for Plant Culture), Norway; Forskning og Forsok i Landbruket (Research and Progress in Agriculture), Volume 37, No. 1, pages 37-44, **1986**.   (I do not have this report. If you have a copy, could you please send it to me.)
>(Ley farming is growing grass or legumes in rotation with grain or tilled crops as a soil conservation measure.)

Alfalfa (Lucerne) Overview

Alfalfa or lucerne (Medicago sativa) is a perennial flowering plant that usually lives 4 to 8 years. It is in the legume family Fabaceae that helps add nitrogen to the soil. It is grown for grazing, hay, pellets, silage, green manure and as a cover crop. It is high in protein and digestible fiber. It grows about 3.28 feet or 1 meter tall with a deep root system that makes it drought tolerant. It is usually cut 3 to 4 times a year.

"Prickly Comfrey (Symphytum asperrimum) afforded the first green fodder of the season on April 24th, being ahead of lucerne. It is not in the least injured by the severe winter."
-'Experimental Work of the Agricultural Department of the University of Tennessee (Report of the Experimental and Other Work of the School of Agriculture and Botany of the University of Tennessee for Session of 1880-1881)' by John M. McBryde, Professor of Agriculture, Horticulture and Botany, Knoxville, Tennessee, **1881**.

"With well-adapted alfalfa varieties such as Caliverde and Lahontan, we regularly harvest 12 tons of alfalfa per acre at Davis, University of California. **Comfrey has been no more productive than alfalfa.**
>Popular reports, however, have indicated yields of up to 100 tons per acre. Certainly nowhere in the state have farm advisors ever recorded any such records for Comfrey.

You will note that the protein content of Comfrey appears to be equivalent to if not just a little bit better than alfalfa. **Although admittedly the crop may produce as much dry matter as alfalfa, Comfrey does contain more water, on a green weight basis.** Thus, the grower must handle more water In a green chop operation with Comfrey. Moreover, spacing plants on a 3-foot (0.9 meter) spacing causes special harvesting problems as contrasted with harvesting green chop alfalfa."
-'Quaker Comfrey in 1961' by M.D.M. and R.T. Edwards, Agronomy Notes, University of California Cooperative Extension Service, Davis, California, pages 22-23, February 28 **1961**.

"Lucerne Forage: In Vitro Dry Matter Digestibility (IVDMD)

Canopy Age	2 weeks	4 weeks	6 weeks	8 weeks	10 weeks
IVDMD %	75.5%	67.9	63.5	58.1	54.9 "

-'Estimation of Lucerne Forage Quality by Means of Morphological and Meterological Data' by L. Kratchunov and T. Naydenov, Forage Institute, Bulgaria; European Journal of Agronomy, Volume 4, No. 2, 263-267, **1995**.
>(Canopy is the above ground, leafy part of the plant.)
>('In Vitro Digestible Dry Matter' {IVDDM} or 'In Vitro Dry Matter Digestibility {IVDMD}. Digestible Dry Matter {DDM} is digestible part of a feed. 'In Vitro' means done in laboratory equipment as opposed to in/on a living animal that is called 'In Vivo'.)
>(To compare to the IVDMD or IVDDM of Comfrey, see subsection 'Digestibility of Leaf' in section 'Nutritional Value' {Chapter 19, Volume 1}.)

"Lucerne (Medicago sativa or Alfalfa) is often referred to as the queen of forages because of its ability to provide consistently high yields of high quality. Lucerne is the most widely grown forage legume in the world with 30 million hectares, 85 percent of which is in the United States, Commonwealth of Independent States (previously USSR), Argentina, Canada, China, and Italy (Frame et al., 1998).
It produces the greatest yield of protein per hectare of any of the temperate crops, including grains and oilseeds (Barnes and Sheaffer, 1995). It is most frequently used in conjunction with forage maize (corn) in dairy systems, because the protein of the lucerne complements the high energy maize.
It is rarely grazed as a pasture crop because of the risk of bloat (gas in stomach) and possible death. It is generally harvested for hay or silage.
Crude Protein (grams/kg Dry Matter basis): Lucerne, forage 129-324. Protein values for lucerne are dependent upon the growth stage at which it is harvested and are generally around 200 grams/kg (Dry Matter basis), but have been reported to range from 129 to 324 g/kg (Spedding and Deikmahns, 1972)."
-'Practical Production of Protein for Food Animals' by Stephen A Chadd, W. Paul Davies and Jason M. Koivisto, Royal Agricultural College, Cirencester, Gloucestershire, England; Protein Sources for Animal Feed Industry, Food and Agriculture Organization of United Nations (FAO), Expert Consultation and Workshop Bangkok, Thailand, April 29 to May 3 **2002**.
>(A hectare is a square with 100 meter sides, or 10,000 m2. There are 100 hectares in one square kilometer. An acre is 0.405 hectare, and one hectare is 2.47 acres.)
>**(200 grams/kg = 200 grams/1000 grams = 20% Crude Protein of Dry Matter for Lucerne.)**

Alfalfa fertilization needs so can compare to Comfrey: "Average nutrient removal (pounds per ton) by established Alfalfa at harvest (based on moisture content common at harvest):
>Nitrogen 55.0, Phosphorus (P2O5) 10.0, Potassium (K2O) 60.0, Calcium 30.0, Magnesium 4.6, Sulphur 8.0, Zinc 0.006, Copper 0.14, Manganese 1.8, Iron 1.8, Boron 0.02.

On established alfalfa stands, nitrogen fertilization is not required because Rhizobium meliloti bacteria convert nitrogen gas from the air into a plant usable form of nitrogen. To ensure establishment of these bacteria in the alfalfa roots, alfalfa seed needs to be inoculated with the bacteria prior to planting."
-'Fertilizer Management for Alfalfa' by David D. Tarkalson, Extension Soil Specialist and Charles A. Shapiro, Extension Soil Specialist, University of Nebraska- Lincoln Extension, Institute of Agriculture and Natural Resources, G1598, **2005**.

"Alfalfa Fresh: 24% Average Dry Matter, Beef Magazine 2009.
Alfalfa (hay): 90.2%, University of California Cooperative Extension 2009.
Alfalfa Silage: 30%, Beef Magazine 2009."
-'Average Dry Matter Percentages for Various Livestock Feeds' by California Certified Organic Farmers (CCOF), May **2015**. Data from 'University of California Cooperative Extension', 'Dairy One' and 'Beef Magazine'.

Comfrey versus Alfalfa: Yield and Land Use

"In comparison with the yield of other green soiling plants, according to Mr. Thomas Christy, it stands thus in England for cut green yields:

Good Grass Land	8 tons to the acre
Lucerne (Alfalfa)	**40 tons to the acre**
Rye Grass	50 tons to the acre
Vetches	20 tons to the acre
Prickly Comfrey	**80 to 120 tons to the acre**

-'Prickly Comfrey: Its History, Cultivation, Extraordinary Production, and Uses: A Letter Addressed to His Excellency Sir Hercules Robinson, President of the Agricultural Society of New South Wales' by Arthur T. Holroyd, Sydney, Australia, **1876**, page 8.

(A soiling crop or plant is cut green and fed to livestock immediately with no processing.)
(Thomas Christy from England promoted Prickly and Russian Comfrey in the mid to late 1800s.)
(Vetches or Vicia is a genus of 140 species of flowering plants part of the legume family, i.e., Fabaceae.)

"During a late call on Mr. G. Hunziker, at **Cloverdale, California,** to whose interesting experiments we have already had occasion to refer, we were shown a patch of Prickly Comfrey, which was set out last April, and has been cut eight times since then.
From the results obtained, however, we should say that **with proper attention Prickly Comfrey would be likely to yield largely a most desirable green fodder the year round, on land which could not be relied on for alfalfa."**
-'Pacific Rural Press', Volumes 15 and 16, San Francisco, California, 853 pages, every other week from January 5 to December 28 **1878**. Article: 'Prickly Comfrey, Russian River Flag', January 19 1878.

"Professor Buckman has recently informed me that **Prickly Comfrey (Symphytum asperum)** *was always a favourite plant of his, and remains so still, as he obtained three pickings from his roots last season.*
As a soiling plant, however, he expresses a great preference for lucerne (alfalfa), *from which, in the same time, he obtained as many as four good cuttings. The Professor here touches on a point which is well worthy of attention.*
The two crops named by him can never come into competition, for, wherever lucerne thrives, there cannot be the slightest doubt of its being by far the most valuable. Still there are numerous places for which this invaluable forage-plant would be ill-adapted, and where Prickly Comfrey would do well.
Sir Thomas Acland sends me the following:
'*I have found that Prickly Comfrey is a very valuable addition to our food at all times of the year.* Once established it requires no trouble. **It may be planted in any odd corners of the farm; under trees, or on the sides of old ditches and wet places where hardly anything else will grow.**'
The bull's-eye is fairly hit in the above statement; for nothing can be more true than that the special value of Comfrey consists in being so well suited to plant odd spots, which, if not thus occupied, would only produce weeds."
-'The Journal of the Royal Agricultural Society of England, Second Series, Volume 18', London, England, **1882**. Includes article: 'On Green or Fodder Crops Not Commonly Grown, which Have Been Found Serviceable for Stock-Feeding' by Joseph Darby.

"**Russian Comfrey was inferior to lucerne (alfalfa) in dry-matter yield, % dry matter, % crude protein and digestible crude protein.** Soil contamination of the herbage was considerably greater on Comfrey than on lucerne. However, Comfrey had a lower fibre content than lucerne."
-'A Comparison Between Russian Comfrey and Lucerne' by Richard Strange, Grassland Research Station, Department of Agriculture, Kenya; East African Agricultural Journal, Volume 24, pages 203-205, **1959**.

"**A good deal of interest in Russian Comfrey has been shown by Kenya (east Africa) farmers** during the last few years (1950s), and it was therefore decided to obtain some information regarding the agricultural value of this plant at Kitale, where the **altitude is 6,200 feet (1889 meters), and the soil is a free draining medium sandy loam.** Lucerne was used as a yardstick as it competes with Comfrey under these conditions.
The lucerne was established from seed in continuous lines 3 feet (0.9 meters) apart, and the **Comfrey from roots at 3 feet x 3 feet spacing.** The plots were kept clean weeded, and the herbage was harvested simultaneously in all plots at intervals of about six weeks during the growing season.
Analysis of Lucerne and Russian Comfrey Herbage (percent of dry matter):

	August 1955		*August 1956*	
	Lucerne	*Comfrey*	*Lucerne*	*Comfrey*
Ash	*10.36%*	*19.73%*	*9.59%*	*22.05%*
Crude Protein	*30.52*	*23.38*	*20.17*	*17.31*
Ether extract	*3.36*	*4.08*	*2.61*	*2.54*
Crude fibre	*27.76*	*9.90*	*19.90*	*12.61*
Carbohydrate	*28.00*	*42.91*	*47.73*	*45.49*
Phosphorus	*0.29*	*0.34*	*0.33*	*0.47*
Calcium	*1.43*	*1.35*	*0.96*	*1.56*
Silica	-	-	*0.67*	*5.78*

The high silica content of the Comfrey is not due to soil contamination as the samples had been washed prior to analysis. **The dry matter content of fresh lucerne and Russian Comfrey, 'mean' of all cuts: Lucerne 24.9%, Comfrey 14.2%."**
-'A Comparison Between Russian Comfrey and Lucerne' by Richard Strange, Grassland Research Station, Department of Agriculture, Kenya; East African Agricultural Journal, Volume 24, pages 203-205, **1959**.
(Statistical 'mean' is determined by adding all data in a population and dividing the total by the number in the population.)

"***In an investigation on the growth and chemical composition of 4 varieties of Russian Comfrey on chalky loam,*** *the crop was laborious to grow since it had to be propagated by off-sets (root cuttings) and needed continual weeding.* ***It yielded considerably less than lucerne (alflafa),*** *and had a lower DM (Dry Matter) percentage but about the same CP (Crude Protein) content. The trial was ended by a serious attack of rust."*
-'Russian Comfrey' by S.P. Mcclean, Bridget's Experimental Husbandry Farm, United Kingdom; Experimental Husbandry, No. 10, pages 46-51, **1964**. (I do not have this report. If you have a copy, could you please send it to me.)
(Rust is a fungus. See subsection 'Pests and Disease' in section 'Planting, Soil, Fertilization, Water, Disease' {Chapter 35}.)

"***Quaker Comfrey (Symphytum x uplandicum Nym., Russian Comfrey)*** *has been promoted for years as a high-yielding protein rich forage crop. Although considerable research has been done in other countries, almost none has been reported from the United States. Therefore, we established field trials on a Codorus silt loam (Fluvaquentic Dystrochrept) at Beltsville, Maryland and on an Archerson sandy clay loam (Aridic Argiustoll) at Cheyenne, Wyoming.*
   At Beltsville, three cultivars of Comfrey under two cutting schedules were compared to orchardgrass (Dactylis glomerata L.) at three nitrogen rates, and to three cultivars of alfalfa (Medicago saliva L.) alone and in mixture with orchardgrass.
   At Cheyenne, two cultivars of Comfrey at three nitrogen rates were compared with alfalfa.
   Forage dry matter yield was determined at both locations; protein concentration and in vitro dry matter digestibility were determined at Beltsville, and vitamin B12 concentration was determined at Cheyenne.
Comfrey yields at both locations were about half those of alfalfa or orchardgrass at the same nitrogen rate. Closer spacing of Comfrey probably would have increased yield. Protein concentration of Comfrey was less than that of alfalfa or orchardgrass except at very high nitrogen rates.
In vitro dry matter digestibilities of Comfrey, alfalfa, and orchardgrass were 37, 62, and 61%, respectively. Contrary to some reports, **Comfrey forage did not contain detectable amounts of vitamin B12.** *(The bioassay used was capable of detecting amounts of B12 as small as 3 ppb {parts per billion}.) Moisture content of fresh Comfrey averaged 85% and was consistently higher than that of alfalfa or orchardgrass.*
In view of its low yield and digestibility in these tests, high cost of establishment and weed control, and low palatability (tastiness) to some animal species, Quaker comfrey is not recommended as a forage crop."
-'Forage Yield and Quality of Quaker Comfrey, Alfalfa, and Orchardgrass' by R.H. Hart, A.J. Thompson III, J.H. Elgin Jr., and J.E. McMurtrey III, Agronomy Journal, Volume 73, No. 4, pages 737-742, July **1981**.

"***Cuttings of 'Bocking Mixture' Quaker Comfrey*** *were provided by Colton Marshall, Portland, Oregon from England into Canada and thence to the United States.* ***As the name implies, this is a mixture of many clones, which differ from one another noticeably in vigor and somewhat in general morphology.*** *Cuttings were also obtained from Merry Gardens, Camden, Maine, but as the study progressed it appeared that these also were 'Bocking Mixture'.*

	Beltsville, Maryland	*Cheyenne, Wyoming*
Planting Date	*May 22, 1974*	*July 18, 1975*
Spacing	*90 x 90 cm (3 x 3 feet)*	*75 x 100 cm (2.5 x 3.3 feet)*
Plants/Treatment	*12*	*16*
Harvest Stage	*First Bloom 1974-1975*	*Full Bloom*
	Full Bloom 1976	
Fertilization *(kg/hectare/year)*		
Nitrogen *(ammonium nitrate)*	*112 (first cut 1974-76)*	*66 (first cut 1976)*
		0 (first cut 1977-78)
	224 (second cut 1975-76)	*133 (second cut 1976)*
		0 (second cut 1977-78)
	448 (third cut 1975-76)	*200 (third cut 1976),*
	? (fourth cut 1975)	*0 (third cut 1977-78)*

Phosphorus *(superphosphate)*	50	50
Potassium *(muriate of potash)*	93	93
Irrigation	None	2.5 cm/week (1 inch/week)

Comfrey yields at both locations were below maximum because of the wide spacing between plants. We used this spacing because it was recommended by several purveyors (sellers) of Comfrey planting stock, but this recommendation was evidently intended to reduce planting costs rather than maximize production.

Shtereva et al. reported green forage yield 10 times as large from **Comfrey spaced 20 cm x 20 cm (.65 feet) as from that spaced 90 cm x 90 cm (3 feet) at one location in Bulgaria. Optimum spacing would seem to be about 40 cm x 40 cm (1.3 x 1.3 feet), and this might increase yields in the first few years after planting by 50 to 100% over yields at 90 cm X 90 cm. Thus, if we had planted at the optimum spacing and received maximum increase in yield, Comfrey yields would have been nearer to but still less than alfalfa or orchardgrass yields in most cases.*

-'Forage Yield and Quality of Quaker Comfrey, Alfalfa, and Orchardgrass' by R.H. Hart, A.J. Thompson III, J.H. Elgin Jr., and J.E. McMurtrey III, Agronomy Journal, Volume 73, No. 4, pages 737-742, July **1981**.

(* 'Investigatons into a New Fodder Crop: Russian Comfrey' by R. Shtereva, I. Popov, V. Petkov and N. Kamishev, Rasteniev'dni Nauki {Plant Science}, Bulgaria, Volume 2:6, pages 113-118, 1965.)

(**'Bocking Mixture'** is Webster Strain of Russian Comfrey before it has been divided into Bocking Numbers.)

Comfrey versus Amaranth

"Cheeke and Bronson (*1979) found that amaranth leaves and stems were higher in hemicellulose and ash and lower in Acid Detergent Fiber (ADF) than alfalfa (Medicago sativa L.).
They also found a greater amount of protein bound to the cell wall constituents in amaranth than in alfalfa and Comfrey (Symphytum officinale L.)."

-'Forage Nutritive Value of Various Amaranth Species at Different Harvest Dates' by Byron B. Sleugh, Kenneth J. Moore, E. Charles Brummer, Allan D. Knapp, James Russell and Lance Gibson (Kentucky and Iowa); Crop Science, Volume 41, pages 466-472, **2001**.

(* -'Feeding Trials with Amaranthus Grain, Forage and Leaf Protein Concentrations' by P.R. Cheeke and J. Bronson, Proceedings of Second Amaranth Conference, Rodale Research Center, Kutztown, Pennsylvania, September 13-14 1979. Pages 5-11, printed by Rodale Press, Emmaus, Pennsylvania.)

Comfrey versus Timothy Grass Yield

Timothy (meadow cat's-tail, common cat's tail) (Phleum pratense) is a perennial grass native to most of Europe except for the Mediterranean area. It is naturalized in North America. It is used as a fodder crop and dried as hay.

*"**Yield data for Comfrey are most useful when they include comparative data for well-known forage crops. For example, Chubarova (*1974) obtained from Symphytum asperum in Moscow Province (Russia) an average dry matter yield over six years of 5.23 tons/hectare, similar to that obtained from timothy (Phleum pratense).** However, Kalinina, D'yakanova and Laidinen (1970) found that in Karelia (part of Russia and Finland) Symphytum asperum out-yielded timothy."*

-'Comfrey Symphytum spp. as a Forage Crop' by J.C. Forbes, A.D. McKelvie, and P.J.C. Saunders, North of Scotland College of Agriculture, Aberdeen, United Kingdom; Herbage Abstracts, Volume 49, No. 12, pages 523-539, **1979**.

(* -'A Study of Promising Perennial Silage Plants' by G.V. Chubarova; Sbornik Nauchnykh Rabot, Vsesoyuznyi Institut Kormov, Russia, Volume 9, pages 168-175, 1974. In Russian.)

Comfrey and Farm Economics

*"**Varietal Improvement of Herbs Needed:**
Conrad Richter presented at the 'First Richters Commercial Herb Growing Conference', in 1997 (1996), on improvement of strains of medicinal herbs grown commercially in Canada.*
He suggested the following herbs, for which few if any improved varieties exist, for targeted research: *ginseng, echinacea, borage, evening primrose, feverfew, goldenseal, catnip, St. Johnswort, valerian, milk thistle, foxglove, chamomile, angelica, sheep sorrel, burdock, **Comfrey**, and nettle."*

-'ATTRA Herb Production in Organic Systems' by Katherine L. Adam, Agriculture Specialist, National Center for Appropriate Technology, National Sustainable Agriculture Information Service, Butte, Michigan, www.attra.ncat.org, **2005**.

International Comfrey Production and Marketing

*"**Marketing: Comfrey is traditionally sold to processors for milling and to manufacturers.** The current United States domestic tea markets for Comfrey leaf are more than 400 ton, with world markets greater than 4,000 ton.*
Prices vary, depending on availability and volume of sale. Most current sales to small processors is usually in 2 to 4-ton

lots. A typical mid-sized wholesaler will use up to 20 ton, depending on statewide 'Food and Drug Administration' recommended compliance. This is changing, and less and less Comfrey is sold for retail distribution.
Most Comfrey now goes to manufactured cosmetics and topical (skin) applications.
The price for Comfrey leaf has increased to about $1.50 per pound in 2-ton quantities. Comfrey root can fetch up to $2.80 per pound, again depending on availability. These two markets are limited, and becoming more so because of concerns about Pyrrolizidine Alkaloids and other potential and 'unknown' carcinogenic.
There is, however, another very large and potentially important new market as an animal food supplement.
Recent studies with Comfrey leaf have shown that it contains several essential amino acids missing in alfalfa. When it is combined with alfalfa in a 60/40% ratio, it constitutes a 'whole food' for feedlot cattle. The potential future markets for this crop in combination with other currently produced feeds can become very significant, with availability.
With a pellet-combination price of $400/ton, the demand within five years for a new animal feed could be more than 50,000 acres. This would represent more than 300,000 ton.
Mixed with birdseed, the exotic bird (ostrich) markets are growing more than 500% per year, while the racetrack industry cannot get enough of these specialty feed mixes.
Now estimate this market to be in excess of 4,000 ton/month just for the cattle food markets of Japan alone."
-'Comfrey Leaf: A New Animal Food Supplement' in 'The Encyclopedia of Alternative Agriculture' ebook by Dr. Richard Alan Miller, https://richardalanmiller.com, Agricultural Consultant and Researcher, Oregon and Washington, **1992**.

"Several species of Comfrey are in cultivation, but two, Symphytum x uplandicum, the Russian Comfrey, and Symphytum officinale, the Common Comfrey, are the most important for the commercial herb market."
-'Richters® First Commercial Herb Growing Conference' edited by Rita Berzins and Conrad Richter. Transcripts from October 26 **1996**, Richters: The Herb Specialists, www.richters.com, Goodwood, Ontario, Canada.

"1999 World and Domestic Volumes of Symphytum officinale:
Estimate world use for dried Comfrey leaf 80,000 ton. Domestic (United States) food and drug use is 800 ton.
1999 Sources of Supply of Comfrey: *South America, Germany, Bulgaria, Brazil. Oregon and California."*
-'A Farm Project: Comfrey Leaf and Root' by Richard Alan Miller, Oak Publishing Inc., Grants Pass, Oregon, and Northwest Botanicals, www.nwbotanicals.org, **1999**.

*"**Comfrey for Dry Raw Material Market (Bulk Wholesale), United States Dollar, Year 2000:***

Cost to Establish	Cost to Maintain	Pounds per Acre: Dry Weight Yield	Prices per Pound	Gross Revenue, $ per Acre:			
				Year 1	Year 2	Year 3	Year 4
$1,000 per acre	$400 per year	16,000 pounds	$0.25-0.75	$0	$600	$2,000	$4,000

Depending on the material and markets, most wrapped bales are packed at 200 pounds (90.7 kg) and then strapped. These can then be put onto pallets and shrink-wrapped for better transport. **Sometimes the bale may be strapped with flat wire to hold shape and integrity. Comfrey and other large production herbs are shipped this way.**"
-'Getting Started: Important Considerations for the Herb Farmer' by Richard Alan Miller, published by Richters®: The Herb Specialists, Goodwood, Ontario, Canada, 27 pages, **2000**.

*"**The dried Comfrey material of commerce is imported from Bulgaria, Poland, Romania, and Hungary. Some of the supply is also cultivated in the United Kingdom and in the United States.**"*
-Herbal Drugs and Phytopharmaceuticals: A Handbook for Practice on a Scientific Basis edited by Max Wichtl. Boca Raton, Florida: CRC Press, **2004**. Symphytum Radix: Comfrey Root, pages 590-592.

"Important European Medicinal and Aromatic Plants in Trade:
Comfrey leaf and root, Symphytum officinale produced by: **Bosnia and Herzegovina, Croatia, Hungary, Poland, Romania.**"
-'Medicinal Plants and Extracts: Market News Service' by Division of Market Development, International Trade Centre, The Joint Agency of the World Trade Organization and the United Nations, Geneva, Switzerland, www.intracen.org, Quarterly Bulletin, March **2011**.
('Market News Service' provides market trends, price information, market news, botanical product specifications, meetings and trade shows, directory of global herb trade associations, and special features such as company profiles and regulations. It covers major international markets in Africa, North and South America, Europe and South-East Asia. Latin America, Caribbean and Pacific countries are covered from time to time.)

"The importance of the various subspecies in the Nairobi, Kenya market of 'African Leafy Vegetables':
Common Comfrey

Daily Share Entire Market	Popularity Ranking in Entire Market in 2006
1.21%	18 out of 21 products on market

Weekly Gross Sales in Kenya Shilling (Kshs currency):   1,061.4 in 2006   682.5 in 2001
Results indicate that consumers demand particular African Leafy Vegetable species due to nutritive aspects associated with them."
-'The Effect of Market Development On-Farm Conservation of Diversity of African Leafy Vegetables around Nairobi' by Charity Irungu, J. Mburu, P. Maundu and M. Grum; Limuru and Nairobi, Kenya; International Journal of Humanities and Social Science, Volume 1, No. 8, July **2011**.

*"**Natural Can Be Expensive:** In the minds of many consumers, preservatives extracted from plants are preferable to those made in factories. Yet because plants usually contain only minuscule (parts per million) amounts of preservatives and other cosmetic ingredients, isolating the chemicals can be cost prohibitive.*
Back in 1983, Steinberg worked at a chemical manufacturing plant. **He recalls that a customer wanted to buy 1 kilogram (2.2 pounds) of the cosmetic ingredient allantoin extracted from Comfrey root.**
> '**I told him he would have to send me a certified check for 15 million dollars,** *and I would get him one kilo of naturally obtained allantoin from Comfrey root,' says Steinberg. 'The guy went nuts, and I said, Do you know how many millions of pounds of Comfrey root I would have to buy.'*

Steinberg notes that at that time, the price of one kilogram of synthetic allantoin was $4.00."
-'Are Natural Preservatives Better' by Inform: International News on Fats, Oils and Related Materials; AOCS: American Oil Chemists Society, Urbana, Illinois, Volume 24, No. 2, February **2013**. 'The Preservative Wars' by Laura Cassiday, pages 70-77.

*"**Comfrey, Symphytum officinale is commonly imported from Hungary.**"*
-The Organic Medicinal Herb Farmer: The Ultimate Guide to Producing High-Quality Herbs on a Market Scale by Jeff Carpenter and Melanie Carpenter. White River Junction, Vermont: Chelsea Green Publishing, **2015**, page 289.

*"**In the global cosmetics market, herbal ingredients are estimated to have a 6% share of the market,** and are exhibiting the strongest growth, between 8% and 12%. In terms of geography, the global herbal medicines/supplement market is divided among Germany (28%), Asia (19%), Japan (17%), France (13%), rest of Europe (12%), and North America (11%).*
***Among preferred botanicals used in cosmeceuticals are** grape seed, bilberry, acerola, baobab, turmeric, ginkgo biloba, white and green tea, red clover, soy, tomato, **Comfrey**, papaya, rosemary, wheat, evening primrose oil, sweet potatoes, carrots, olives, flax, aloe vera, coffee plant, centella asiatica, avocado and passion fruit."*
-'Sub-Sector and VC Analysis of Medicinal Plants' by Md. Altaf Hossain, Value Chain Specialists, International Fund for Agricultural Development (IFAD), a specialised agency of the United Nations, https://asia.ifad.org, 84 pages, **2015**.

*"**Diversify your crops.** On our farm we often think about crops and their marketability in the following categories: 'Core Crops', 'Specialized/Challengers', and 'Contracted Crops'.*
> *'Core Crops' are, for the most part, the tried-and-true crops that are easy to grow and easy to market. They make up a large portion of what we grow and sell, season to season.*
> *'**Core Crops' include** anise hyssop leaf/flower, ashwagandha root, burdock root, calendula flower, **Comfrey Leaf/Root (Symphytum officinale),** echinacea root/tops, garlic, lemon balm leaf/flower, nettle leaf, milky oat tops, peppermint leaf, tulsi leaf/flower and valerian root."*

-The Organic Medicinal Herb Farmer: The Ultimate Guide to Producing High-Quality Herbs on a Market Scale by Jeff Carpenter and Melanie Carpenter. White River Junction, Vermont: Chelsea Green Publishing, **2015**, page 245.

*"The industry report '**Allantoin Market Size By Application (Cosmetics, Pharmaceuticals, Oral hygiene), Industry Analysis Report, Regional Outlook (United States, Germany, China, Brazil), Growth Potential, Price Trends, Competitive Market Share and Forecast, 2016 – 2023**' by Global Market Insights, Inc. says allantoin market size is poised to cross United States dollar 670 million by 2023.*
> *The product allantoin is broadly used as an elementary constituent in skin treatment creams and lotions, which are used in skin diseases including eczema, psoriasis, keratosis (noncancer skin growth) and xerosis (dry skin). It is also preferred in the oral hygiene applications owing to its superior cell proliferation properties.*

The global allatoin market size is negatively influenced by ban executed on Comfrey extracts, as the product is extracted from Comfrey plants. *The United States, Europe and Australia has strict regulatory bans on these products, which may hamper the overall industry growth over the estimated time frame. Additionally, limited consumer awareness about the product in the Middle East and Africa and Latin America shall obstruct industry growth.*
However, booming product demand across cosmetic formulations due to its value-added properties will open up industry growth prospects."
-'Allantoin Market to Hit $670 Million by 2023' by Global Market Insights, Inc.: Insights to Innovation, A global market research and consulting service provider, www.gminsights.com, Ocean View, Delaware, March 30 **2017**.

*"**The global allantoin market is forecast to grow at a CAGR (Compound Annual Growth Rate) of 5.76% during the period 2017-2021. The latest trend gaining momentum in the market is the use of allantoin in scar treatment and tattoo industry.** The growing use of allantoin-based products in scar treatments, as a result of advanced treatment methods and formulations containing allantoin as one the key ingredient, is significantly driving the market growth.*
Further, the report states that one of the major factors hindering the growth of this market is the medical complications with oral consumption. Allantoin in Comfrey was earlier popular as an oral medicine for treating renal (kidney) problems. Comfrey is a class of flowering plant that is widely used in herbal medicines and cosmetic products.
Comfrey is a rich source of allantoin and is used as very good cell proliferant and is used in skin repair, scar treatment, bone growth, and hair growth.
Owing to the remarkable benefits of Comfrey, it is widely used in chemical drugs and herbal medicines."
-'Global Allantoin Market: Analysis, Technologies & Forecasts to 2021: Growing Use of Allantoin in Scar Treatment & in Tattoo Industry' by PR Newswire for Research and Markets®: The World's Largest Market Research Store, www.researchandmarkets.com, Dublin, Ireland, September 1 **2017**.

"Global Comfrey Root Market Segmentation:
On the basis of nature, the global Comfrey root market has been segmented as:
Organic Comfrey root, Conventional Comfrey root.
On the basis of source, the global Comfrey root market has been segmented as:
Symphytum officinale, Symphytum asperum, Symphytum x uplandicum, Others.
On the basis of form, the global Comfrey root market has been segmented as: Powder, Liquid.
On the basis of end use, the global Comfrey root market has been segmented as:
Pharmaceuticals: Ointment, Cream, Salve, Others.
Cosmetics and Personal Care: Skin Care, Hair Care, Others.
On the basis of distribution channel, the global Comfrey root market has been segmented as:
Supermarkets/Hypermarkets, Convenience Stores, Specialty Stores, Pharmaceuticals, Online.
Global Comfrey Root Market Participants:
Some of the market participants operating in the global Comfrey root market identified across the value chain includes Herbo Nutra, Green Heaven India, Vital Herbs, Kshipra Biotech Private Limited, Azafran Innovacion Limited, Madhu Fitness Care, XTZ, Green Harbal Health Care, Admark Herbals Limited, Himalaya Herbal Health Product, Pashan Ventures, Verma Wellness Care, Sky Healthcare Private Limited, Ravi Wellness Products, Enjoy Life Care among the other Comfrey root players. (® for all names listed.)
Opportunities for Participants in the Comfrey Root Market:
The market potential for the Comfrey root market is expected to grow, owing to the health benefits of the Comfrey root. The existing consumer awareness of Comfrey root serves as an opportunity for the market participants of Comfrey root."
-'Comfrey Root Market: Global Industry Analysis, Size, Share, Growth, Trends and Forecast 2018 - 2028' by Transparency Market Research, www.transparencymarketresearch.com/comfrey-root-market.html, Albany, New York, **2018**.
(I do not have any connection with this company. For $5,795 in 2018 they provide a report on the international Comfrey root market. Other companies provide this service.)

"**The Comfrey Root industry report is a complete market study** which ensures availability of insightful data and strategic information. **Market statistics** such as global Comfrey Root market size, estimated and forecasted data for next 5 years with breakdown at regional and country level is covered in the report.
The market segmentation by type and application along with manufacturers profile and share gives a clear understanding of the market dynamics. The analysis on sales along with the study of latest developments in the past years - covering product launches, mergers and technological innovation is covered in the report. Single User License: $3,480.00."
-'Global Comfrey Root Market 2021 by Manufacturers, Regions, Type and Application, Forecast to 2026' by GIR (Global Info Research), Mumbai, India, August 2021, www.decisiondatabases.com/ip/5715-comfrey-root-industry-analysis-report
(I do not have any connection with this company.)

Usefulness Varies with Number of Tons Per Acre

"**At fifty tons an acre**, about the yield of the Soil Association's unmanured plot, a low figure often achieved and quoted, only the fifty-ton crop of kale is ahead on dry matter, and Symphytum peregrinum (Russian Comfrey) is over 850 pounds (385 kg) more than the best pasture grass yield of crude protein. Good lucerne (alfalfa) is still leading on fibre, and maize (corn) and kale on carbohydrates.
At this figure then the Comfrey crop pays the pig-and-poultry man, giving the green fodder vitamins and vegetable protein with the least possible fibre, at the lowest labour cost per ton.
It also pays the race-horse breeder and riding-stable proprietor with the concealed subsidy of improved condition.
On the diary farm, except where the pastures are late starting on a cold, heavy clay, or dried up from July onwards in bad years, orthodox fodder and grazing crops at their best still lead."
-Russian Comfrey: A Hundred Tons an Acre of Stock or Compost for Farm, Garden or Smallholding by Lawrence D. Hills. London England: Faber and Faber, Limited, **1953**, page 55.
('Soil Association' is a United Kingdom charity for healthy, humane and sustainable food, farming and land use. Based in Bristol, England, it is the UK's largest organic certification organization.)

"**With between seventy-five and a hundred tons an acre, the yield of established and well-grown Comfrey for the past seventy years since Henry Doubleday, the picture is completely altered.** The labour cost has remained constant, so has the first cost of planting; the only increase to set against double the food production is that of harvesting the greater bulk, and in petrol (gasoline) or T.V.O. (Tractor Vaporising Oil), and labour and machinery running-time costs. This is a small figure.
It becomes worth reorganizing the farm programme to fit the yield, taking rather more trouble with silage-making to fit in the summer cuts, either as a mixture or alone, and working out the stocking system that will best utilize the bulk of feed."
-Russian Comfrey: A Hundred Tons an Acre of Stock or Compost for Farm, Garden or Smallholding by Lawrence D. Hills. London England: Faber and Faber, Limited, **1953**, page 55.

Usefulness Varies with Type of Environment

"The farm economics of different countries determine the livestock that can be fed at a profit.
In Britain *it pays for pigs, calf rearing, race horses and cinchillas (rodents raised for their fur). It is also used for human food, and as greenstuff for cage birds and poultry.*
**In countries where grass is not the cheapest cattle food, it is a valuable fodder for beef and dairy farmers.*
  The Japanese *are using the high production of the crop for cattle and as a vegetable for human consumption in a temperate climate, where the major problem is too little land and too many people. Japan has a climate relatively near that of England. The Japanese experiments were carried out 75 miles west northwest of Tokyo, in the Nagano Prefecture (County) in a valley 2,000 feet (609 meter) above sea level with high mountain ranges on both sides.*
It is precisely under these conditions with small fields and steep slopes that Comfrey can be of greatest value to modern farmers needing maximum production."
-Comfrey Report No. 3: Feeding Dairy Cattle in Japan' by Meiji Milk Producing Co., Tokyo, Japan for Henry Doubleday Research Association, Braintree, Essex, England, 31 pages, July **1964**.

Best Farm Comfrey is Solid Stem and Magenta (Purplish) Flowers

"It should be said that **there is no record of any Comfrey of agricultural value with yellow, cream or white flowers, and bright blue is associated with the lower yields. The hollow stem is another bad sign; solid stems, magenta pink flowers and lack of wings (on the flower stem) are good ones,** *but yield per acre from the strain in question is the only sure guide."*
-Russian Comfrey: A Hundred Tons an Acre of Stock or Compost for Farm, Garden or Smallholding by Lawrence D. Hills. London England: Faber and Faber, Limited, **1953**, page 39.

Intercropping (Growing Comfrey with Other Crops)
For Comfrey with fruit trees/bushes, see subsection 'Fruit and Comfrey' in section 'Garden Uses of Comfrey' (Chapter 18, Vol 1).

"I once, by way of experiment, had some cabbage plants put among the Prickly Comfrey, but the cabbage very soon gave up their competition in assimilating the nutriment of the soil."
-'The Gardeners Chronicle and Agricultural Gazette for 1871' published in Covent Garden, WC, London, England, 1269 pages, January 7 to December 30 **1871**. Article: 'Notices to Correspondents', January 21 1871, pages 59-60.

"The rather upright growth of the Comfrey plant and the necessity for fairly frequent cutting to provide young leafy herbage, prevent it covering the ground rapidly, and as a crop, therefore, it is rather difficult to establish and liable to serious weed infestation. **The two most successful methods of establishment proved to be in rows two feet (0.6 meter) apart, and close-planted and intersown with Italian rye grass."**
-Russian Comfrey: A Hundred Tons an Acre of Stock or Compost for Farm, Garden or Smallholding by Lawrence D. Hills. London England: Faber and Faber, Limited, **1953**, page 47.
  (Italian Rye Grass {Lolium multiflorum} also called Annual Ryegrass is native to temperate Europe. It is a herbaceous annual, biennial or perennial grown for silage and a cover crop. It naturalizes and can become a noxious weed in agricultural areas and an invasive species in native habitats.)

"But there is no need to leave the inter-row soil idle in the cold winter, when Comfrey is dormant, if a crop of cabbage or other quick growing winter plants are needed."
-Miracle Grass, Comfrey by Andrew Hughes published in **1966** in Japanese. Part of it is in: 'Comfrey: Nature's Healing Herb & Health Food' by Andrew Hughes, Japan: Sanyusha Publishing Co., Ltd, 1992. Part is in 'Comfrey Report No. 1: The 1954 Research Results' by Lawrence Hills published in 1955. (If you have a translation in English, I would appreciate having a copy.)

"Though research workers and others who know the Comfrey crop only in theory may feel that the ideal system would be to grow a weed suppressing and nitrogen fixing legume between the plants, this has so far always produced a problem by making row crop cleaning impossible and allowing a buildup of grass. If Comfrey is growing well enough to pay, there is no room for anything in between it, which would produce two poor crops instead of one good one."
-Comfrey Report: The Story of the World's Fastest Protein Builder and Herbal Healer, Conservation Gardening and Farming Series: Series C by Lawrence D. Hills. England: Henry Doubleday Research Association, **1975**, page 51.

*"**It is also possible to grow clover inter-row in season without harm to the Comfrey.** In fact, the nitrogen-fixing habit of the clover will help the nitrogen-hungry Comfrey."*
-Comfrey: Nature's Healing Herb and Health Food by Andrew Hughes. Japan: Sanyusha Publishing Co., Ltd, **1992**, page 150.
  (The most common cultivated clovers are the small white/Dutch/Ladino clover {Trifolium repens} and red clover {Trifolium pratense} that is 20-80 cm {8-31 inches} tall.)

*"**Fertility Patches:** This a technique developed in England for increasing the organic content and nutrients of soil, especially nitrogen and phosphorus, and is perfect for sandy, low nutrient soils such as those we have on the Coast (Pacific Northwest of

United States and Canada).
The suggested planting is up to one third of the area under production be grown in alternating rows of Comfrey and alfalfa (lucerne) for up to 6 years. *Alfalfa does not grow that well in this area, so white clover and lupines (of the pea family) might be a better option.* **Several harvests a year can then be mulched or incorporated into the soil of any areas that are in intensive annual vegetable production."**
-'An Introduction to Permaculture Sheet Mulching: Experimental Strategies and Techniques for the Pacific Northwest Coast' by Harold Waldock, 'Edible Landscape Creations' and 'Vancouver Permaculture Network', British Columbia, Canada, August **2001**.
(Comfrey should be planted as a permanent crop because it is hard to kill once it is established.)

"Competition between plants in mixed agricultural crop communities was studied using intercrops and monocultures (single crop) of the non-traditional feed crops elecampane (horse-heal, elfdock), Rough Comfrey (Prickly Comfrey), eastern goat's rue and dioecious (2 sexes) nettle.
Plant communities were established in spring 2002 on soils typical of **Central Northern Caucasian mountain-forest regions in the Russian Republic of Kabardiono-Balkariya** *(i.e., grey forest soils and alluvial meadow soils).*
Two harvests were carried out each year, one in late June and one in mid-August. *Yields were higher for crops grown on grey forest soils than those grown on alluvial meadow soils.*
In 2002 yields of Dry Matter (DM) were higher for intercrops on both soil types, *and were higher than for monocultures of nettle and goat's rue* **but lower than yields of elecampane and Comfrey grown in monocultures.**
In 2003 the DM yield of intercrops exceeded that of goat's rue grown in monoculture, and in 2004-2005 exceeded that of elecampane, Comfrey and goat's rue monocultures.
For elecampane-Comfrey mixes, Comfrey was more aggressive and competitive on grey forest soils, and elecampane on alluvial meadow soils.
Overall trial results indicated that creation of dioecious nettle / eastern goat's rue / elecampane / Rough Comfrey intercrops was effective means of providing animal feeds for farms in mountain-forest zones of Central Northern Caucasus."
-'Phytocenotic Competition in Mixed Agricultural Plant Communities of Non-Traditonal Feed Crops' by A. Tamakhina, Kabardino-Balkarskaya Gosudarstvennaya Akademiya, Russia; Mezhdunarodnyi Sel'skokhozyaistvennyi Zhurnal (International Agricultural Journal), Russia, No. 5, pages 53-54, **2007**.
(Caucasus Mountains are at border of Europe and Asia, between Black Sea and Caspian Sea and occupied by Russia, Georgia, Azerbaijan and Armenia.) (Alluvial soil is fine-grained fertile soil deposited by water flowing over flood plains or in river beds.)

*"***Gray/Grey Forest Soil** *is a type of soil that forms primarily in a continental, moderately humid climate in forests, usually hardwood, with grassy vegetation. Gray forest soils form on loesslike cover loams, carbonate moraines, and other parent materials that are usually rich in calcium and have an elutriate water regime. Gray forest soils are mostly acidic, especially in the A2B horizons and the upper part of the B horizon; alkaline or neutral soils are typical of the lower part of B horizon. Gray forest soils have relatively good physical properties and are biologically active and fertile. Gray forest soils are found primarily in the USSR, in the northern part of the forest steppe, where they form a discontinuous band from the Carpathians to Transbaikalia. They are also found in Canada and adjacent regions of the United States. The soils are used for the cultivation of grains, industrial crops, vegetables, and fruits."*
-The Great Soviet Encylopedia, 1979, one of the largest Russian-language encyclopedias, published by the Soviet Union from 1926 to 1990.

"If you follow the Richland Electric Cooperative lines through the rolling open hills of Richland County to Ash Ridge near **Viola, Wisconsin,** *you will discover someone who is* **on the cutting edge of green farming and agroforestry- Mark Shepard.** *Influenced by a book written by Russell Smith titled 'Tree Crops', Mark began to formulate his idea to transition from the usual traditional annual crops like soybeans and corn to* **a perennial crop ecosystem.** *This bio-diverse system is made up of multi-use crops designed to enhance and work with each other while working on good land as well as marginal land.*
Other poly-culture plants: *Adding to the already diverse mix of plants on his farm are* **Comfrey,** *daffodils, iris, wormwood, pears, grapes currants, roses, elderberries, hybrid poplar."*
-'Profits Not Problems' by Doug Wallace, NRCS Lead Forester; Inside Agroforestry, USDA National Agroforestry Center, University of Nebraska, Lincoln, Nebraska, Volume 19, Issue 2, page 5, **2011**.

Pasture and Comfrey

"Where it is intended to have a permanent Comfrey meadow, seeds might be sown along with oats, *after a well-manured root crop, at the rate of 6 pounds (2.7 kg) per acre, in the month of March or April.*
On deep, rich arable land, where heavy crops of white and green crops are annually grown, it might not be so profitable to convert it into a Comfrey meadow, but as it will grow on the poorest of land, it would undoubtedly be the most profitable crop for waste lands, *and will yield food for an additional number of cattle, where formerly it yielded little or nothing."*
-'Transactions of the Highland and Agricultural Society of Scotland, Fourth Series, Volume 14', Edinburgh, Scotland, **1882**. With the article: 'The Cultivation of Prickly Comfrey, and Its Use as a Fodder Plant' by David W. Wemyss, Newton Bank, St. Andrews, Fife Council, Scotland, pages 264-267.

"Seeds for Hedgerow Herbal Mixture for Cows, Goats and Horses: 1 pound (453 gram) Comfrey, 1 pound chicory, 1 pound sheep's parsley, 1 pound burnet, 1 pound ribgrass (plantain), 1 pound sweet clover, 1 pound kidney vetch, 1 pound caraway, 1/4 pound yarrow. Total is 8 1/4 pounds (3742 gram) of seeds an acre."
-Herdsmanship: A Guide for the Herd Owner, Herdsman and Cowman by Newman Turner. Austin, Texas: Acres USA, **1952**, page 84.   (A hedgerow is a hedge of shrubs or low trees growing along a bank, especially one bordering a field or lane.)

*"**Many attempts have been made to supply the nitrogen Comfrey needs by growing a legume, such as crimson clover, betwen the plants, but this always fails,** for the Comfrey either destroys the legume by shading it, or is reduced to a shadow of its original yield by the grass that the clovers encourage.*
Grass is the worst weed in Comfrey, *robbing it of the nitrogen it needs for speed, and most of the account of poor Comfrey yields from research stations come from attempts to grow it in pasture.*
Chicory is a far better deep-rooted mineral miner; add seed to the pasture or to the pasture mixture."
-Fertility Without Fertilizers: A Basic Approach to Organic Gardening by Lawrence D. Hills. New York: Universe Books, **1975**.

"The cultivation of the introduced species Heracleum sosnowskyi, Sida hermaphrodita, Silphium perfoliatum, Galega orientalis, **Symphytum asperum,** *Malva meluca and M. crispa on reclaimed peat soils under seasonal flooding in the western Ukraine gave high fresh fodder yields and showed good regrowth after harvesting and grazing. They are recommended for reclamation of peat-bog soils to improve grasslands in the forest zone of the western Ukraine."*
-'Enrichment of Forest Grasslands with Fodder Plants' by M.D. Latyshenko; Nauchnye Trudy, Ukrainskaya Sel'skokhozyaistvennaya Akademiya, Volume 146, pages 109-112, **1975**. In Russian. If you have English translation, I would appreciate a copy.)

*"**Comfrey can usefully contribute calcium, phosphorus and copper as well as protein to a grass sward (upper layer of soil),** at least in the first two years. **Latyshenko (*1975) has recommended the sowing of Symphytum asperum and other species to improve the quality of grasslands on peaty forest soils."***
-'Comfrey Symphytum spp. as a Forage Crop' by J.C. Forbes, A.D. McKelvie, and P.J.C. Saunders, North of Scotland College of Agriculture, Aberdeen, United Kingdom; Herbage Abstracts, Volume 49, No. 12, pages 523-539, **1979**.
(* -'Enrichment of Forest Grasslands with Fodder Plants' by M.D. Latyshenko; Nauchnye Trudy, Ukrainskaya Sel'skokhozyaistvennaya Akademiya, Volume 146, pages 109-112, 1975. In Russian.)

*"**Productivity and Botanical Composition of Meadows (Grassland Farming):** Alopecuretum pratensis (grass) v. Bromus inermis (smooth brome grass) community grows in the similar habitats as Phalaroidis arundinacea (reed canary grass). **Both communities occupy large areas and develop in flooded areas with fluctuating soil water conditions.***
    Higromesophytes and mesohigrophytes (plants needing a moderate amount of water) are abundant in these communities forming grass swards. Such plant communities are widespread in west and middle Europe, and along river valleys reach to the north and east as far as Krasnoyarsk land (in Siberia, Russia).
In these communities the diagnostic species were: *Bromopsis inermis, Calamagrostis epigejos, and Carex praecox. As well as diagnostic species of communities Persicaria amphibia,* **Symphytum officinale,** *Veronica longifolia were found."*
-'Integrating Efficient Grassland Farming and Biodiversity' edited by R. Lillak, R. Viiralt, A. Linke and V. Geherman, Proceedings of the 13th International Occasional Symposium of the European Grassland Federation, Grassland Science in Europe, Volume 10, Tartu, Estonia, August 29-31 **2005**.
    ('Alopecuretum pratensis' may mean 'Alopecurus pratensis', the meadow foxtail perennial grass.)
    (Diagnostic species or indicator species have a distinct concentration of abundance in a particular habitat or plant community.)   (Comfrey grows well in moist soil but does not like soggy soil. If soil is too wet, roots will rot.)

<u>Age of Comfrey Plot and Yield</u>

"I should add that **Symphytum asperrimum (Prickly Comfrey)** *is a perennial, and that* ***it does not appear to have declined at all in 14 years: the longest experience I have had of it in one place.***
It only requires a little care to keep it clean. But from its rapid growth it gives weeds a poor chance."
-'The Gardeners Chronicle and Agricultural Gazette for 1871' published in Covent Garden, WC, London, England, 1269 pages, January 7 to December 30 **1871**. Article: 'Notices to Correspondents', January 21 1871, pages 59-60.

*"**The yield of Russian Comfrey increases steadily until it is between five and seven years old, but at ten years after planting it begins to go back,** a decline that varies with the soil and manurial treatment."*
-Russian Comfrey: A Hundred Tons an Acre of Stock or Compost for Farm, Garden or Smallholding by Lawrence D. Hills. London England: Faber and Faber, Limited, **1953**, page 73.

"The average crude protein content of the old Russian or Quaker Comfrey (Symphytum peregrinum) stand increased from 20.81% in 1958 to 25.41% in 1967. Reasons may have been age, yield, nitrogen fertilization, or others.
Over this nine-year period, the yield per acre of dry matter decreased from 12.0 to 9.4 tons, indicating longevity."
-'Plant Leaf Protein with Emphasis on Comfrey' by Hubert Heitman and Milton D. Miller with Edward Johnson (agronomic phases of study), Sergio E. Gyarzun of University of Chile (swine digestion trial), Bob D. Wilson (rat study) and James T. Elings (Exten-

sion Animal Scientist); California Experiment Station and Agricultural Extension Service, University of California, **1969**.

"*When the yields from a set of plots are measured over a number of years, the effect of stand age is unfortunately confounded (made less clear) with the effect of weather and probably also with management in different years. In general, however, the highest yields are obtained in the second or third year of growth.*
    In crops grown from seed, as is usual with Symphytum asperum in the USSR, the first year yield is generally very small. In crops grown from root or rhizome cuttings, as is usual in Symphytum x uplandicum, the first year yield is appreciable, but almost always lower than in the second year.
In Moscow Province (Russia), Vavilov, Dotsenko and Dotsenko (*1974) in the fourth year of utilization of Symphytum asperum obtained yields 18-23% greater than in the third year. In trials in Poland the maximum yield of Symphytum officinale was not achieved until the sixth year (**Tabin, Berbec and Bobrzynski, 1973).
Vavilov and Kondrat'ev (***1975) give an instance of a thirty-year-old crop of Symphytum asperum yielding 55-60 tons green herbage/hectare."
-'Comfrey Symphytum spp. as a Forage Crop' by J.C. Forbes, A.D. McKelvie, and P.J.C. Saunders, North of Scotland College of Agriculture, Aberdeen, United Kingdom; Herbage Abstracts, Volume 49, No. 12, pages 523-539, **1979**.
(* -'Effect of Age and Weather Conditions on Productivity of Perennial Silage Plants' by P.P. Vavilov, A.I. Dotsenko and R.A. Dotsenko; Biologicheskie Osnovy Povyzheniya Urozhainosti Sel'skokhozyaistvennykh Kul'tur, Vypusk 2, Moscow, Russia, pages 5-9, 1974. In Russian. If you have in English the parts about Comfrey, I would appreciate a copy.)
(** -'The Yield of Some/Several Ecotypes of Common Comfrey {Symphytum Officinale L.}' by S. Tabin, S. Berbec and T. Bobrzynski, Akademia Rolnicza {Agricultural Academy}, Lublin, Poland; Hodowla Roslin, Aklimatyzacja Nasiennictwo {Plant Breeding, Acclimatization and Seed Production}, Volume 17, No. 6, pages 505-511, 1973. In Polish. I do not have this report. If you have it, could you please send it to me, especially if you have an English translation.)
(*** -'New Fodder Crops' {Novye Kormovye Kul'tury} by P.P. Vavilov and A.A. Kondratev; Referativnyi Zhurnal, Moscow, Russia, 350, 1975. In Russian. If you have this in English, I would appreciate a copy.)
    (A hectare is a square with 100 meter sides, or 10,000 m2. There are 100 hectares in one square kilometer. An acre is 0.405 hectare, and one hectare is 2.47 acres.)

Older Comfrey Plots and How to Restore Them

"**The yield of Russian Comfrey increases steadily until it is between five and seven years old, but at ten years after planting it begins to go back,** *a decline that varies with the soil and manurial treatment. The main roots die of old age and rot off in black decay, while the growing points on the outsides spread into a kind of 'fairy ring' or 'moon crater' of vigour, with a bare hollow in the middle. There is greater tendency to run to stem and flower.*
After ten years, three-quarters of the area of each plant can be dug up, *taking the circumference of the circles so that smaller clumps are left at reasonable intervals for cultivation.*
The offsets (roots) from these will grow to full vigour when replanted on fresh land. The land is then well dunged and scuffled between the rows and plants, and **with steady cutting the roots regain strength and the yield is restored.**"
-Russian Comfrey: A Hundred Tons an Acre of Stock or Compost for Farm, Garden or Smallholding by Lawrence D. Hills. London England: Faber and Faber, Limited, **1953**, page 73.
    (A fairy ring, fairy circle or elf circle is a naturally occurring ring or arc in soil.)
    (Scuffle is scraping or brushing the surface of the soil back and forth.)

"**As Comfrey plants age they die out in the middle,** *looking rather like coral atolls or moon craters.* **This is because instead of making a main tap root like a tree out of woody material, Comfrey grows soft ones like horse radish.**
Only the layers immediately under the bark of a tree root take part in the two-way traffic of minerals and water and waste products. Tree roots stay alive with hemicellulose and lignins in long strong fibers.
Those of root vegetables have also a two-way traffic layer on the outside, but in the centre they have stored starches and proteins ready to provide energy and materials for the towering flower and seedhead. **Comfrey roots come between the two.**
The 'crater effect' on ageing Comfrey plants comes from the main roots wearing out and rotting off, *just as the root hairs are dying and decaying in about 3 to 4 days.*
The normal routine as the plot ages is to walk along the rows with a hand fork after the April or May cut, dig up any of the outer crater rim that has extended far enough to interfere with row crop cultivation, and replant the pieces in the empty middles, *making sure the growing points are well below the surface.*
The fungi which break down the worn-out roots do not attack living Comfrey."
-Comfrey: Fodder, Food and Remedy by Lawrence D. Hills. New York: Rizzoli Universe Books: **1976**, page 109.
    (An atoll is a ring-shaped reef, island, or chain of islands.)

"*Old Comfrey plants gradually come to resemble moon craters or coral atolls in shape as the main roots die out from the middle and replacements grow outwards. How long this takes depends on the quality of the soil, but if the result is merely a fall in yield,* **take up some plants in March, and transplant the offsets in the hollow centres.**"
-Fertility Gardening: The Organic Way to Make Your Garden Grow by Lawrence D. Hills. London, England: Cameron & Tayleur Books Limited, **1980**.

Chapter 37

Harvesting Comfrey Leaves

For nutrition and harvest time, see sub-subsection 'Nutrition and Digestibility of Leaves/Stems and Maturity' in subsection 'Digestibility of Comfrey Leaf', in section 'Nutritional Value of Comfrey' (Chapter 19, Volume 1).
See sub-subsection 'Dry Small Amounts of Comfrey Leaf' in sub-section 'Comfrey as Hay' in section 'Comfrey Meal, Pellets, Hay and Silage' (Chapter 21).
See subsection 'Method of Digging Root Cuttings' in section 'Care of Comfrey Plant: Overview & How to Propagate' (Chapter 34).
See subsection 'Drying Comfrey Roots' in section 'Making and Using Comfrey Medicine' (Chapter 27).

Overview of Harvesting Leaves

*"Your correspondent asks **'How often Prickly Comfrey can be mown'**:*
*'I do not think it can be mown. By far the most practical, as well as economical, mode of 'taking' its leaves is not to take the whole of them, as I did till 'within a few years past, but to **pluck the four or five most forward leaves from each crown**.*
A tender hand will be better for a glove, but a hand ordinarily exposed will not need one.
The weight of each leaf will be some what more than an ounce (28 gram). *By this method of treatment, the leaves that are left will be allowed to expand to their full size.* ***When the whole crown was taken at once, it was reproduced in the short space of 10 or 12 days in the summer time, and in a fortnight (2 weeks) or a little more in late spring and autumn.***
The leaves begin to show themselves in April. They may be plucked about the first week in May; and have lasted (this year and last year they did) to the end of October."
-'The Gardeners Chronicle and Agricultural Gazette for 1871' published in Covent Garden, WC, London, England, 1269 pages, January 7 to December 30 **1871**. Article: 'Notices to Correspondents', January 21 1871, pages 59-60.

*"It has been noticed that the **Prickly Comfrey** plant thrives in all kinds of soil and aspect.* ***The leaves, as they reach maturity, are torn off without injury to the coming crop.*** *It is a hardy and free grower, the roots taking firm hold of the soil. After being once established they are difficult to eradicate, and **the leaves, which are most abundant, can be gathered from the beginning of May to the first frosts without injury to the plant**."*
-'The Southern Planter and Farmer, Devoted to Agriculture, Horticulture and Rural Affairs' by L.R. Dickinson, Richmond, Virginia, No. 1, January **1876**. Article: 'Symphytum Asperrimum' by C.E. Ashburner, Englishman in Henrico County, Virginia, page 55.

*"**It is advisable not to cut the leaves too low, say three inches (7.6 cm) up. The yield is largest if cut just before the flower opens.** The plant likes clay, loam or any deep soil, and the roots will tap down eight feet (2.4 meter) to moisture.*
The roots sometimes globe, and hold half-a-pint (1 cup = 236 ml) of gummy water.
When preserved in tanks (silage) or dried into hay, it is the richest fodder known.
It is estimated on a clay soil to produce from 60 to 120 tons an acre per annum (year), when the plants are established, and has been known to yield much above this."
-'Forage Plants and Their Economic Conservation by the New System of Ensilage: Part I: Caucasian Prickly Comfrey' by Thomas Christy, Jun., F.L.S. (Fellow of the Linnean Society), Christy & Co., London, England, **1877**, page 25.

*'From 'Catalogue of Messrs. Suttons' of Reading, Berkshire, England in **1878**:*
> *'**When the Comfrey leaves have grown from 18 to 24 inches (0.45 to 0.60 meter) high, they should be cut and given to the stock in a fresh green condition. In about six weeks, a second cutting will be ready and a succession of cuttings can be obtained through the summer and autumn.***
> ***As many as five heavy cuttings, each 20 tons per acre, or 100 tons per acre in one season, have been obtained by good management.*** *If it is cultivated for one or two heavy cuttings, the stems should be allowed to grow to 4 or 5 feet (1.2 or 1.5 meter), and **it may be cut with an ordinary hook, tied up in bundles and conveyed to the homestead as required**. We recommend it especially for small occupations, as few crops can be more easily grown or prove so useful to those whose livestock consists of a horse, cow, and a few pigs."*

-Russian Comfrey: A Hundred Tons an Acre of Stock or Compost for Farm, Garden or Smallholding by Lawrence D. Hills. London England: Faber and Faber, Limited, **1953**, page 33.

*"**All Comfrey cuts should be at two inches (5 cm) from the ground. The daily cut system is the most widely used but others are possible, so long as the plot is cut a minimum of five times a season.** These cuts can be close together, especially in June, July and August. **In October and November the growth slows, and it will be noticed that sun and warmth are of more importance to the yield than drought or wet.**"*
-Russian Comfrey: A Hundred Tons an Acre of Stock or Compost for Farm, Garden or Smallholding by Lawrence D. Hills. London England: Faber and Faber, Limited, **1953**, page 65.

*"**Leaf and stem harvesting reduces a plant's ability to gather energy, and hence its ability to grow and compete.** It may set a plant back in the successional marathon, allowing others to take the lead and direct the successional sequence. Intensive harvesting may spur growth of new or neighboring plants by giving them more sun or water or open up sites for weeds to establish. **Some crops can respond to cutting quickly, barely skipping a beat. Such fast-recovering plants include mints (Mentha spp.), Comfreys (Symphytum spp.), and stinging nettle (Urtica dioica), for example.**"*
-Edible Forest Gardens: Volume Two: Design & Practice by Dave Jacke and Eric Toensmeier. White River Junction, Vermont: Chelsea Green, **2005**, page 439.   (sp.= single species) (spp.= more than 1 species)

*"As I sink my hands into mass after mass of fallen leaves, I am always happy to see creepy crawlies. **Despite many folks' odd antipathy to spiders, the arachnids are in fact a very helpful generalist predator in the garden.** Spiders will eat just about anything that moves, so they keep insect population explosions from getting out of hand.*
But spiders hate bare soil, so they are often absent from conventional agricultural situations. *Mulching is the best way to attract spiders to your garden, but **having perennial plants around is also a good bet. Comfrey seems to be especially attractive, even more so if you let the winter-killed leaves lie on the ground rather than 'cleaning' them up.***
One study* in Switzerland found 240 spiders for every square meter (10.7 square feet) of soil beneath Comfrey leaves."
-'Spiders and Comfrey' by Anna Hess and Mark Hamilton, The Walden Effect, Virginia and Ohio, November **2009**. www.waldeneffect.org/blog/Spiders_and_comfrey/
(*'Uberwinterung von Arthropoden im Boden und an Ackerkrautern Kunstlich Angelegter Ackerkrautstreifen' {Overwintering of Arthropods in the Soil and on Field Weeds, Artificially Created Strips of Field Weed} by Hans-Martin Burki and Ariane Hausammann; Haupt, Bern Report, Agrarokologie, Band 7, 158 pages, 1993. In German.)

*"**Q: I was wondering when to stop harvesting my Comfrey for the season?** It is mid September in zone 5b, and I have a lot of Comfrey to harvest but heard somewhere you should stop at a certain time to allow the plant to store up energy for the winter. I have about 60-90 days to first frost....will this be enough time if I harvest now?*
A1: I usually stop now, not for the sake of the plants, but for the sake of the spiders. Hundreds of them can over-winter in the big Comfrey leaves.
A2: That's a good point! I always feel so bad when I see people doing the 'full fall cleanup' leaving nothing for the predators and prey to hang out in over the winter."
-"Comfrey: When to Stop Harvesting for Season' by Permies: Homesteading and Permaculture, All the Time; created by Paul Wheaton, Missoula, Montana, **2019**. https://permies.com/t/70587/Comfrey-stop-harvesting-season
(I have found spiders in my Comfrey plants in February when previous temperatures have been down to 12 degrees F {-11 C}. They get as close to the ground as they can under the debris so they don't freeze to death.)

Tools Used for Harvesting Comfrey Leaves

*"Comfrey should be cut when it is about half-grown, as stock like it better then, and it springs up again quicker; besides, when cut at that time, four or five crops may be taken in one year. But if cut just before the flower-buds open, it should not be cut closer to the crown than two inches (5 cm). **It is usual to reap Comfrey with a sickle, the crop from one root being as much as a man can get his arm round when set close. However, it may be mown with a strong scythe.**"*
-'Transactions of the Highland and Agricultural Society of Scotland, Fourth Series, Volume 14', Edinburgh, Scotland, **1882**. With the article: 'The Cultivation of Prickly Comfrey, and Its Use as a Fodder Plant' by David W. Wemyss, Newton Bank, St. Andrews, Fife Council, Scotland, pages 264-267.
(A sickle is a short-handled tool with a semicircular blade used for cutting grain or trimming plants. A scythe is a long-handled tool with a long, curving blade fastened at an angle to the handle. It is used to cut grass, grain, and weeds.)

*"**On a small scale, Comfrey cutting can be by scythe, hook, beet-topping knife, or motor scythe.***
The usual procedure is for the pigman to take out a slit-open bag or two stitched together, and drag them after him, cutting until he has enough to carry away on his shoulders; the same applies to poultry.
Shears can be used, cutting two inches (5 cm) from ground level; or a sickle (bagging-hook, reaping-hook), or on a larger scale, the type of motor scythe used for cutting rough grass in orchards.
Gathering is a matter of a garden fork and a barrow (wheel barrow)."
-Russian Comfrey: A Hundred Tons an Acre of Stock or Compost for Farm, Garden or Smallholding by Lawrence D. Hills. London England: Faber and Faber, Limited, **1953**, pages 64, 118.
(A scythe is a tool with a long, curving blade fastened at an angle to a handle used to cut grass, grain, etc., by hand.)
(A beet-topping knife has a wide blade with hook on the end. It is used to harvest sugar beets and chop off the leaves.)
(Motor Scythe is a mower with a short reciprocating knife attached to a garden tractor mechanism.)

*"Comfrey can be cut at soil level, and this makes cutting very efficient. **1,200 to 2,000 kg (2645 to 4409 pounds) of Comfrey leaf were cut by hand in an hour.** It is possible to cut in the rain, because water scarcely adheres to the leaf and stem."*
-Comfrey Report No. 3: Feeding Dairy Cattle in Japan' by Meiji Milk Producing Co., Tokyo, Japan for Henry Doubleday Research Association, Braintree, Essex, England, 31 pages, July **1964**, page 13.

*"**Comfrey leaves were harvested by hand with a hedge clipper.** Harvesting continued from April 26 until November 16. Early*

in the season harvesting interval was 10 to 20 days, but from August on yields dropped and harvesting intervals became longer."
-'Plant Leaf Protein with Emphasis on Comfrey' by Hubert Heitman and Milton D. Miller with Edward Johnson (agronomic phases of study), Sergio E. Gyarzun of University of Chile (swine digestion trial), Bob D. Wilson (rat study) and James T. Elings (Extension Animal Scientist); California Experiment Station and Agricultural Extension Service, University of California, **1969**.

*"**In late summer or early fall two gardeners approach the Comfrey bed with a wheelbarrow and two pairs of long-bladed shears.** They begin cutting down the plants with their shears. They slide the open blades of the shears around the base of each plant, then lift the shears up off the soil a few inches (2 inch = 5 cm) and snip the blades shut, so that every leaf falls away, and **the crown of the plant is left with a clump of short stems sticking out of it.***
The cutting goes very quickly, even though the gardeners stop every minute or so to pick up all the leaves they've cut, and put them into the wheelbarrow."
-Enchanted Garden: Alan Chadwick's Organic Method of Gardening by Tom Cuthbertson. London, England: Rider & Company / Hutchinson & Co. Publishers Ltd, **1978**, page 134.

*"Medvedev and Sidorova (1976) found that **in both S. asperum and S. x uplandicum the yield was strongly influenced by the cutting regime (system) in the previous year.***
Yields were 30-35% greater from plots cut only once in the previous year than from plots cut three times.
A similar effect on S. x uplandicum yield was noted by Chubarova, Vorob'ev and Rybinkova (1970)."
-'Comfrey Symphytum spp. as a Forage Crop' by J.C. Forbes, A.D. McKelvie, and P.J.C. Saunders, North of Scotland College of Agriculture, Aberdeen, United Kingdom; Herbage Abstracts, Volume 49, No. 12, pages 523-539, **1979**.

*"**The best form of harvest is to cut Comfrey leaf at 6 inches (15 cm) from the ground with a side-bar cutter, attempting not to bruise the leaf as this darkens the final color of the dried leaf. Some can use a flail-chop if the end use is for cattle.** The Comfrey is laid out in wide windrows (a row of cut crop raked together to dry), avoiding leaf stacking and compaction. Let the leaf come to a 50 percent sun-cure, and then pick it up using a draper (conveyor pick-up) or other conveyor-type delivery to wagons. The wagons should be taken to large dehydration facilities for final drying. These facilities can be hop kilns, tobacco dryers, and plywood kilns. Grain dryers are too small. Basically, forced warm air shafts work best.*
Once the Comfrey leaf is dried, it is usually put into 180-pound (81.6 kg) rectangular bales, wrapped in burlap (much like hops). Since it is quite light, the 'cube' is bulky, so dimension of the bale are designed for stacking in a warehouse.
Comfrey root harvests can be done with potato digging equipment."
-'Comfrey Leaf: A New Animal Food Supplement' in 'The Encyclopedia of Alternative Agriculture' ebook by Dr. Richard Alan Miller, https://richardalanmiller.com, Agricultural Consultant and Researcher, Oregon and Washington, **1992**.
    (A flail chopper cuts/shreds leaves and stalks into small pieces so it can be fed to livestock or used as mulch.)
    (Potato digger is a machine that digs up potatoes, removes tops/dirt from tubers, then throws tubers on top of the field.)

*"**Harvesting Comfrey:** For years I happily harvested a bed of around 24 Comfrey plants using ordinary garden shears until I discovered the **long-handled horizontal bladed shears.** They save a lot of bending or kneeling when harvesting.*
For larger patches, a scythe or sickle will pay for itself in time saving."
-'Planting, Cultivating, Harvesting and Problems of Comfrey' by John Harrison, Allotment and Gardens: Grow Your Own - Allotment - Gardening Help, https://www.allotment-garden.org, Penygroes, Gwynedd, Wales, **2018**.

*"**EA:** Thought I'd share how I'm stacking a fair bit of function in the later summer harvest of our **Comfrey rhizome barrier** between our gardens and the neighbors lawn. Lots of layers of yield coming from this, it's such an exciting plant to work with!*
    *R: Just a note that **with a European scythe with a ditch blade, you could slash all of that Comfrey easily, without having to bend down or exert much effort at all.** I mention the ditch blade explicitly because **it's meant to handle even smaller woody plants and weeds rather than just trimming fine grasses** (many blades are thinner and more fragile, meant only to mow lawn). I mention European explicitly because it's a much lighter scythe, and the American scythe is extremely heavy and doesn't work with your body to make cutting easy.*
EA: I use a scythe quite a bit and often to harvest Comfrey. *The problem is that in this scenario there is material standing tall and a fair amount laying flat. **I wanted to get it really cleanly cut this time so the hori was the perfect tool for me in this particular case.** But yeah, the Euroopean scythe is a wonderful tool!"*
-'Comfrey:Leaf Harvest and Stacking Function' by Edible Acres, August **2018**, Finger Lakes area, New York, from Reddit Permaculture. https://www.reddit.com/r/Permaculture/comments/9c2vho/comfrey_leaf_harvest_and_stacking_function/ www.edibleacres.org
(A hori or hori-hori is also called a 'soil knife', 'weeding knife' or 'leisure knife'. It is a heavy serrated multi-purpose steel blade for garden digging or cutting. The blade is sharp on both sides and comes to a somewhat sharp point at the end. Hori means 'to dig' in Japanese.)

*"**The Japanese Kama - 'Mow where your scythe can't go!'** I actually prefer to use a Japanese kama for harvesting certain grains, over the European sickle. The short, comparatively straight blade of the kama, is much easier to control, in my opinion, than the western, hook-shaped sickles. Harvest grains or tall grass by grabbing a handful of stems and then cut a safe distance underneath with a pull stroke. It can also be used as a grass sickle and cut with a forward slicing motion, like a scythe blade.*
Long-Handled Japanese Kama: *This super-sharp Japanese kama has a 41 inch (104 cm) long oak handle. The laminated blade is 8.5 inches (21.6 cm) long and 2 inches (5 cm) wide. It is great for reaching down into tight spots and hooking and*

cutting weeds, larger stemmed grasses, and brambles. I use it for managing a native prairie planting on a very steep slope. Unlike the kama above, the tang of the blade is angled to make mowing while upright a little easier. **It cuts Comfrey like it isn't even there.** I do not recommend it for any plant woodier than blackberry brambles."
-'Sickles' by One Scythe Revolution, Winona, Minnesota, (February 2022). The Japanese Kama and European sickles.
https://onescytherevolution.com/sickles.html

Timing and Frequency of Harvests

"The morphological and biological characters of Symphytum asperum are described. Growth was rapid in the initial stages. Symphytum asperum gave 2-3 cuts in a season, and herbage yields were high, with high nutrient contents, especially Digestible Crude Protein. **Contents of Crude Protein, carotene and mineral nutrients were highest at the early-bud stage. It was recommended that plants for silage should be cut at the end of July** when they give high herbage yield and have a high sugar content."
-'Okopnik na Sakhaline' {Comfrey in Sakhalin, Russia} by N.G. Khrushkova and A.A. Odegova; Trudy Sakhalinskogo Kompleksnogo Nauchno-Issledovatel'skogo Instituta, Russia, Volume 23, pages 175-179, **1971**.

"Plant Profiles: Comfrey, Symphytum officinale:
When to harvest: Herbalist Natalia Bragg says that it is possible to get two or three harvests of Comfrey leaves in a summer, and **recommends harvesting leaves just before flowers open.** She explained that the plants will become about four feet (1.2 meter) tall, flop over, and then begin to regrow. **Harvest plants after 10 am in order to give time for dew to dry off plants.**

Tips for Sustainable Harvesting/Management: Comfrey is a plant that responds well to cutting. **Natalia Bragg makes her last Comfrey leaf harvest in September, before the frost, and then cuts stems down to the ground to promote hearty growth the following season.**"
-'Culturally and Economically Important Nontimber Forest Products of Northern Maine: Plant Profiles: Comfrey' by United States Department of Agriculture: Forest Service, Sustaining Forests, Northern Research Station, Madison, Wisconsin, May **2010**.

Timing According to First Cut of Season

"Effect of cutting regime (system) on yield:
The average dry matter yield of S. x uplandicum over two years in West Germany was significantly affected by the stage of growth at which the first cut of the season was taken. The annual yield:
- 3.81 tons/hectare from 5 or 6 cuts beginning at the rosette stage.
- 5.44 tons/hectare from 3 or 4 cuts beginning just before flowering.
- 8.00 tons/hectare from 3 cuts beginning when the plants were in full flower (*Doring, 1959).

With S. asperum in Sakhalin (Russian island in Pacific Ocean), however, the date of the first cut had no effect on the total annual yield (**Khrushkova and Odegova, 1971)."
-'Comfrey Symphytum spp. as a Forage Crop' by J.C. Forbes, A.D. McKelvie, and P.J.C. Saunders, North of Scotland College of Agriculture, Aberdeen, United Kingdom; Herbage Abstracts, Volume 49, No. 12, pages 523-539, **1979**.
(* -'Cultivation of Comfrey' by W. Doring, Inst Acker- u PflBau, University of Halle-Wittenber, East Germany; Deutsche Landwirtschaft {German Agriculture}, Volume 10, No. 2, pages 62-66, 1959.)
(** -'Okopnik na Sakhaline' {Comfrey in Sakhalin, Russia} by N.G. Khrushkova and A.A. Odegova; Trudy Sakhalinskogo Kompleksnogo Nauchno-Issledovatel'skogo Instituta, Russia, Volume 23, pages 175-179, 1971.)
> (Hectare is a square with 100 meter sides, or 10,000 m2. There are 100 hectares in one square kilometer. An acre is 0.405 hectare, and one hectare is 2.47 acres.)

Timing According to Height

"**Comfrey is ready to cut when it is 1 foot (0.3 meter) in height.** Lop it off 1 inch (2.54 cm) from ground level with shears or a knife. Cutting every six weeks gives the highest yield, but most gardeners make their cuts to fit with the crops that need the most potash (potassium)."
-Down to Earth Gardening by Lawrence D. Hills. London, England: Faber & Faber, **1967**, 1975, page 219.

Timing According to Month

First Cut for Established Comfrey Plants

I am in western North Carolina in the mountains at 3200 feet (975 meter). **I start cutting in early April** which is before the Comfrey flower stalks start developing. I prefer cutting with a sickle.

First Cut for Comfrey Planted that Spring

"**The first year after planting, the Comfrey plants need until July to grow and get established,** then they can be cut down to 2 inches (5 cm) from the ground, and in 3 weeks the young leaves will be ready to pick (for humans to eat).
Their second season starts in early spring, and they can be picked hard till the peas come in as a green vegetable to beat spring droughts, cutting back any that run to stem or that stop leaf production.
Then let them rest and grow stout, solid stems with small ball flowers the colour of purple carbon paper."
-Down to Earth Fruit and Vegetable Growing by Lawrence D. Hills. London, England: Faber & Faber, **1960**, page 55.

"We harvested that Comfrey foliage yesterday afternoon **(in late summer or early fall)** to give us a high-quality additive **for the green matter that is going into a pea trench today. The Comfrey plants were all strong enough to take their first harvest (planted in spring);** in fact, they will recover just enough to withstand the shock of the first frost easily, then go through winter with minimal damage to the crown and root system. They will produce at a terrific rate next year, so fast that we'll have to cut the foliage at least three times to keep the plants from running to bloom.
If the crop had been cut earlier this summer, though, much of it wouldn't have been mature enough, and there might have been some dying back of the younger crowns. Also, a good deal of foliage would have grown after the harvest. That foliage would freeze, which is not only wasteful but also hard on the crown.
If the crop were cut later (in late fall), there might have been weak, newly cut stems sticking up at the time of the first killing frost, and that could have killed back some of the crowns. Even the plants that made it through the winter after a too-late cutting would be weak next year, and subject to a takeover by weeds."
-Enchanted Garden: Alan Chadwick's Organic Method of Gardening by Tom Cuthbertson. London, England: Rider & Company / Hutchinson & Co. Publishers Ltd, **1978**, page 137.

(The author is in Covelo, California that is very dry in late summer and early fall. His pea trench is for next spring's sweet-pea crop though in some areas peas can be grown in the fall too.)

Cutting in June or July

"**About June or July the plants will throw up flower stems among their leaves, and this is when it pays to mow,** setting the knife to cut about two inches (5 cm) above the ground.
Flowering early weakens the crop, and cutting puts its strength into building up bigger plants for next season."
-Comfrey: Fodder, Food and Remedy by Lawrence D. Hills. New York: Rizzoli Universe Books, **1976**, page 106.

"**Despite root reserve depletion from intensive harvesting (4 or 5 harvests per year), Comfrey showed high longevity in these trials.** No plants died, and there were no diseased plants. The lower yield of the 1975 planting that was harvested in 1979 was due to lack of fertilizer nitrogen rather than to any weakening of aged plants.
More than half of the annual yield was produced in the first two harvests. The second harvest was greater than the first because of slow regrowth in the spring on plots harvested the previous year.
This suggests that harvesting only in June and July might permit root reserves to increase in late summer, result in higher first cutting yields the following year, and maintain productivity at a sustainable level."
-'Comfrey: A Controversial Crop' by Robert G. Robinson, University of Minnesota, Agricultural Experiment Station, Minnesota Report MR-191, Item No. AD-MR-2210, **1983**, page 4.

"More than half of the annual yield in Minnesota trials was produced in the first two cuttings during June and July. Chinese* researchers found that the greatest annual yields were obtained from **three cuttings, which start at full bloom in mid-June**."
-'Comfrey' by Teynor, Putnam, Doll, Kelling, Oelke, Undersander and Oplinger, 'Alternative Field Crops Manual', University of Wisconsin: Cooperative Extension, University of Minnesota: Center for Alternative Plant & Animal Products, and Minnesota Cooperative Extension, February **1992**, updated November 1997, page 4.
(* -'Optimum Time for Cutting Symphytum Peregrinum' by C.Y. Zhao and Z.J. Chen, Liaoning Institute of Animal Husbandry and Veterinary Medicine, Liaoning, China, 6:1921, 1981. I was unable to get this report. If you have a copy, could you please send it.)

Last Cut is in September or October

"**The final cut of Comfrey being taken in mid-September.**"
-'Cultivation of Comfrey' by W. Doring, Inst Acker- u PflBau, University of Halle-Wittenber, East Germany; Deutsche Landwirtschaft (German Agriculture), Volume 10, No. 2, pages 62-66, **1959**.

"**When Comfrey goes dormant in winter, it returns all the potash (potassium) in its foliage to store in the roots, which is why it pays to let the last cut die down naturally.** This gives a starting stock for the spring."
-Grow Your Own Fruit & Vegetables by Lawrence D. Hills. London, England: Faber and Faber Limited: **1971**, page 50.

"**Most growers make a final cut in early October or equivalent in their latitude,** and leave the last three weeks of growth to increase the size of the first cut in spring when they need the food more urgently."
-Comfrey: Fodder, Food and Remedy by Lawrence D. Hills. New York: Rizzoli Universe Books, **1976**, page 109.

Cutting Prebloom versus Full Bloom

"Further interesting information about **Prickly Comfrey** is found in the recently published **pamphlet of Messrs. Thomas Christy and Co., of London, England.** From it I glean that the plants spread in size of crown each time they are cut, and the forage may be taken all year round to within three inches (7.6 cm) of the ground. **The yield is largest if cut before it opens into flower.** It likes moist, boggy places, and the roots will tap down many feet in open soil to moisture."
-'Prickly Comfrey: Its History, Cultivation, Extraordinary Production, and Uses: A Letter Addressed to His Excellency Sir Hercules Robinson, President of the Agricultural Society of New South Wales' by Arthur T. Holroyd, Sydney, Australia, **1876**, page 8. (Thomas Christy from England promoted Prickly and Russian Comfrey in the mid to late 1800s.)

"**Plant should be cut as soon as it shows for flower. When allowed to flower it weakens the root, and it runs into stem.**"
-'Forage Plants and Their Economic Conservation by the New System of Ensilage: Part I: Caucasian Prickly Comfrey' by Thomas Christy, Jun., F.L.S. (Fellow of the Linnean Society), Christy & Co., London, England, **1877**, page 24

"Mr. W. Stevens (Sir Thomas Acland's chief local manager), of Broadclyst (Devonshire, England) has furnished me with the following particulars on the growth of Prickly Comfrey at the home farm, Killerton (garden and estate):
>'We get three cuttings a year, and the Comfrey crop produced is a heavy one. **The cattle like it best just as it begins to come into flower. Growers will find it important to attend to this, as, if allowed to stand too long, the stems get hard and tough, and much is wasted.**'

Probably the true reason of the repugnance (dislike) of stock to Comfrey, in some instances at least, may be that it is not gathered at the right time. At all events there is very conflicting evidence as to the likings and dislikings of the same kind of animals for the produce of the plant."
-'The Journal of the Royal Agricultural Society of England, Second Series, Volume 18', London, England, **1882**. Includes article: 'On Green or Fodder Crops Not Commonly Grown, which Have Been Found Serviceable for Stock-Feeding' by Joseph Darby.

"Taking into consideration yield, nutrient content, and subsequent growth, **the most suitable Comfrey cutting management was 5 times per year beginning after the completion of basal-leaf formation, but before flowering, the final cut being taken in mid-September.**"
-'Cultivation of Comfrey' by W. Doring, Inst Acker- u PflBau, University of Halle-Wittenber, East Germany; Deutsche Landwirtschaft (German Agriculture), Volume 10, No. 2, pages 62-66, **1959**.

"You have to see the use of Symphytum from a wider perspective than the rabbit's in order to understand **why we cut back the stems before they flower.** The part of the Comfrey plant which is most useful to the garden and to us is the leafy foliage. **This foliage will grow best if the plant is prevented from blooming. Since you propagate Comfrey by root division, flowering is a waste of the plant's strength.** By keeping the flower stems cut back, we provide for a continuing growth of strong foliage, and we encourage the original root system to mine deeply into the soil; as far as three feet (0.9 meter) down; those roots will bring up great supplies of potassium, phosphorus and calcium, as well as many trace minerals, collecting them from the depths below where the roots of most plants can reach."
-Enchanted Garden: Alan Chadwick's Organic Method of Gardening by Tom Cuthbertson. London, England: Rider & Company / Hutchinson & Co. Publishers Ltd, **1978**, page 119.

"**Delaying harvest of Comfrey until full bloom, rather than cutting at prebloom, has increased yields of Quaker Comfrey 50%* and yields of Prickly Comfrey 26%**.**
Therefore, at Beltsville, Maryland in 1976 we split all Comfrey plots, and half of each was cut at first bloom as in 1974 and 1975, whereas the other half was cut at full bloom. Four cuttings were obtained at first bloom, but only three at full bloom; alfalfa and orchardgrass were cut four times. Yields of Comfrey at both cutting schedules and of orchardgrass increased with increasing nitrogen. **Comfrey cut at full bloom produced almost 1.5 times as much forage as Comfrey cut at first bloom.**"
-'Forage Yield and Quality of Quaker Comfrey, Alfalfa, and Orchardgrass' by R.H. Hart, A.J. Thompson III, J.H. Elgin Jr., and J.E. McMurtrey III, Agronomy Journal, Volume 73, No. 4, pages 737-742, July **1981**.
(* -'Research on the Value of Comfrey for Cultivation' by W. Doring, Kuhn-Arch., Volume 72, pages 335-375, 1958. I was unable to get this report. If you have a copy, could you send it to me please.)
(** -'On the Introduction of Symphytum Asperum into Belorussia' by H.N. Smolski and I.I. Checkalonskay, Rastit Resur {Rastitel'nye Resursy= 'Vegetable Resources' translated from Russian}, Volume 6, pages 223-227, 1970.
>"In field trials near Minsk, Russia, S. asperum {Prickly Comfrey} **yielded from 2 cuts** 850, 960 and 920 hkg {hundreds kilogram} air-dry matter/hectare in the 2nd, 3rd and 4th year of growth, respectively, when the first cut was taken at the stage of bud formation; **higher yields than these were obtained when the first cut was taken at flowering rather than at bud formation.**")

(**In the 'Forage Yield' article it is not clear whether this is just one or several cuts per season. If just one cut, then these results make sense.** However, total yield is higher with several or many cuts done at prebloom throughout the season.)

"**The research studied the dynamic pattern of amino acids in 6 kinds of forages in different growth periods** in Yellow River beach regions. Six kinds of forages (Alfalfa, White clover, Ryegrass, Red fescue, Cup plant, Common Comfrey) were selected to determine and analyze the contents of crude protein and amino acids in 4 different growth periods in Yellow River

beach regions. Analyzing synthetically the above index and yield, the author considered the optimal harvesting time for Alfalfa, White clover, Ryegrass, Red fescue and Cup plant was the flowering stage, however **the optimal reaping stage of Common Comfrey was the maturing stage."**
-'Dynamic Analysis of Contents of Crude Protein and Amino Acids in 6 Kinds of Forages in Different Growth Periods' by L. TaiYu, N. FuRong, L. QingHua, L. MenYun, W. YanLing and S. SuFang, College of Animal Husbandry and Veterinary, He'nan Agricultural University, Zhengzhou, China; Journal of Northwest A&F University, Natural Science Edition, Volume 37, No. 1, pages 11-16, **2009**.   (By 'maturing stage' I assume they mean the prebloom stage.)

Number of Cuts per Year (Or Days Between Cuts)

"Taking into consideration yield, nutrient content, and subsequent growth, **the most suitable Comfrey cutting management was 5 times per year beginning after the completion of basal-leaf formation, but before flowering, the final cut being taken in mid-September.** A cultivation area is envisaged which allows the cutting of sufficient fodder for pigs and poultry daily and an interval of 3-5 weeks before the same plant is cut again. **The 2-year average yield of Comfrey (from 5 cuts per year) was lower** than the yield of Jerusalem artichoke, lucerne (alfalfa), fodder rye + maize, or fodder rye + millet, but the high protein and low fibre content of the fodder compensated for the lower yield."
-'Cultivation of Comfrey' by W. Doring, Inst Acker- u PflBau, University of Halle-Wittenber, East Germany; Deutsche Landwirtschaft (German Agriculture), Volume 10, No. 2, pages 62-66, **1959**.
(I do not have this report. If you have a copy, could you please send it.) (A basal leaf grows from the lowest part of the stem.)

"For highest protein content, Comfrey plants should be cut every 21 days, thus increasing the yield and raising the number of cuts in one year from 8 or 10 to 13 or 14, depending on the season and the location of the plantation. But the soil must be properly conditioned to get such results. By cutting mature leaves every 20-28 days the plant is kept in a constant process of nitrogen absorption for 6 or 8 or 10 months of the year, and the rate of take-up of nitrogen continues, reaching its peak in the middle of summer and early autumn."
-'Miracle Grass, Comfrey' by Andrew Hughes published in **1966** in Japanese. Part of it is in: 'Comfrey: Nature's Healing Herb & Health Food' by Andrew Hughes, Japan: Sanyusha Publishing Co., Ltd, 1992. Part is in 'Comfrey Report No. 1: The 1954 Research Results' by Lawrence Hills published in 1955.
(If you have a translation of it in English, I would appreciate having a copy.) **(This needs a lot of manure put on frequently.)**

"In a field trial at Jaboticabal, Brazil, Symphytum sp. (species) was planted 1 September 1976 at 0.80 meters (2.62 feet) between plants in the row, in rows 1 meter (3.28 feet) apart. The plants were cut after 21, 28, 35, 42 or 49 days growth twice during the experimental period January to May 1977. There were significant differences in yields of DM (Dry Matter) and CP (Crude Protein). **The best results were obtained by cutting after 35 or 49 days of growth. Yield of DM and CP at these 2 dates were 2.70 and 0.76 tons/hectare, and 3.42 and 0.93 tons/hectare (total of 2 cuts), respectively."**
-'DM Production and Nutritive and Chemical Composition of Comfrey (Symphytum sp.) Cut at Different Stages' by V. Favoretto, J. Ariki and I. Arcaro Junior, Faculdade de Ciencias Agrarias e Veterinarias, Universidade Estadual (Faculty of Agrarian and Veterinary Sciences, State University), Jaboticabal, Sao Paulo, Brazil; Cientifica (Science), Vol 6, No. 3, pages 471-476, **1978**.
   (Hectare is a square with 100 meter sides, or 10,000 m2. There are 100 hectares in one square kilometer.
   An acre is 0.405 hectare, and one hectare is 2.47 acres.)

PRICKLY COMFREY, WHEN CUT.

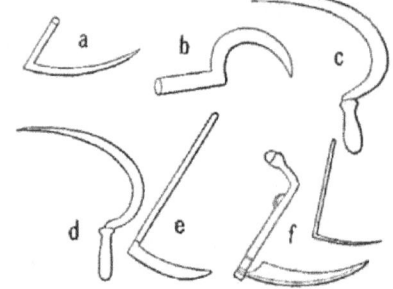

FIG. 1. VARIOUS FORMS OF SICKLES AND SCYTHES.

Prickly Comfrey, When Cut 1876

'Prickly Comfrey: Its History, Cultivation, Extraordinary Production, and Uses: A Letter Addressed to His Excellency Sir Hercules Robinson, President of the Agricultural Society of New South Wales' by Arthur T. Holroyd, Sydney, Australia, 1876. Comfrey page 9

'The New International Encyclopaedia', Volume 16, edited by Gilman, Peck and Colby; New York: Dodd, Mead and Company, **1905**. Reaper, page 740.

"Reapers: The first implement used for reaping was the reaping hook or sickle, dating from the Stone and Bronze ages (Fig. 1, a, b, c, d). Records of this implement are found in Egyptian history B.C. 1400 to 1500. The earliest form of the sickle had a slightly curved blade with straight handle.
The scythe followed the sickle and was apparently introduced by the Romans, by whom it was employed mainly for cutting grass. Pliny, in his writings, distinguishes between the sickle and the scythe, and Crescenzio described both in 1548. At first the scythe was intermediate in construction between the sickle and the modern scythe, as in the Hainault scythe (Fig. 1, f)."

Chapter 38

How to Get Rid of Unwanted Comfrey

Invasiveness of Comfrey

"Comfrey is a plant determined to live and grow. **These fragments will grow, even if half rotten on the outer portion. If there is the least bit of viable interior cells, they proliferate and eventually push a shoot to the soil sufrace.**
It spreads madly, so plant in a large container, or shield its roots from other plants by sheet metal sunk 1 foot (0.3 meter) down."
-The Organic Method Primer: A Practical Explanation: The How and Why for the Beginner and the Experienced by Bargyla and Gylver Rateaver. San Diego, California: The Rateavers, **1993**, page 163.
(I don't think a 1 foot deep piece of sheet metal will stop Comfrey roots from spreading, but it will slow it down. **The blue-flowered Symphytum Hidcote Comfrey that I sell is very invasive by the roots. I poured a concrete floor for it by my house.** It won't grow under the house since there is no light or water. Then I made a 3-high cinder block wall on 3 sides. It does grow between the cinder blocks but I keep an eye on it, and dig out any roots escaping the container. If you pour concrete walls, there needs to be some ways for rain to drain. Fill the container with good soil.)

"Their roots are extremely presistent, however, so the plants are hard to remove once in place, and **most Comfrey species will 'walk' around the garden to at least some degree.**
Russian Comfrey (Symphtyum x uplandicum) will not walk, however, and stays in its place."
-Edible Forest Gardens: Volume One: Vision & Theory by Dave Jacke and Eric Toensmeier. White River Junction, Vermont: Chelsea Green, **2005**, page 187.

"A chunk the size of one finger joint is enough to found a new plant, so if a gardener digs or tills into the root system, the plant will spread wherever cuttings land. **Power-tilling recklessly through a single large plant is enough to sow Comfrey sucessfully and persistently across a surprising expanse of garden. Thus one caveat (warning) about Comfrey: Don't dig or till close to it.** When I wanted to eliminate a plant, sheet mulch did the job."
-Gaia's Garden: A Guide to Home-Scale Permaculture by Toby Hemenway. White River Junction, Vermont: Chelsea Green Publishing Company, **2009**, page 128.
(**Power-tilling can be used if you want to spread Comfrey on purpose.** Any place where a root is cut off from the original plant, a new plant will form. The root can stay in its original position or the tiller might move it to a new location. Any roots left on top of the soil will dehydrate and die. So if you want to expand your Comfrey patch, you need to replant exposed roots. The Comfrey patch expansion idea also works by using a shovel going down vertically to break roots in a large plant. You don't have to move any roots; you just need to cut them off from the main plant.)

" **'Invasive Plant' as classified by the 'United States Department of Agriculture' per 'Executive Order 13112'** is defined as a plant that is:
1. Nonnative (or alien) to the ecosystem under consideration, and
2. Whose introduction causes or is likely to cause economic or environmental harm or harm to human health.
The problem is that there is little to no scientific method for validating which species are truly causing 'economic or environmental harm'. There also seems to be no accurate method for determining exactly what qualifies as 'native'.
The term 'native' when referring to plants and their habitats is a fairly subjective term and can carry with it a certain degree of controversy.
Since plants have been dispersed far and wide both naturally and anthropogenically (by humans) since the dawn of humans, **it can be challenging to determine exactly which species qualify as 'native' and which ones were 'introduced'. Here we are referring to species that existed in their native environments in North America before European colonization.**
Many species of medicinal plants that are grown or wild-harvested and marketed today are considered invasive plants. This list includes Comfrey."
-The Organic Medicinal Herb Farmer: The Ultimate Guide to Producing High-Quality Herbs on a Market Scale by Jeff Carpenter and Melanie Carpenter. White River Junction, Vermont: Chelsea Green Publishing, **2015**, pages 59, 60.
(Wild-harvesting or wild-crafting is foraging or collecting plants from their natural habitat, usually for food or medicine. It applies to uncultivated plants wherever they may be found. There are ethical considerations such as protecting endangered species, overharvesting of commonly held resources, and if on private property preventing theft.)

Symphytum Asperum (Prickly Comfrey) Invasiveness

"**Prickly Comfrey (S. asperum) is a tracked non-native plant in Alaska, and it is considered noxious in California(*, **).**"
-'Common Comfrey: Symphytum Officinale L.- Alaska Natural Heritage Program' by Timm Nawrocki, University of Alaska Anchorage, Alaska Center for Conservation Science, October **2010**.

(* -AKEPIC Database: Alaska Exotic Plant Information Clearinghouse Database, 2010. http://akweeds.uaa.alaska.edu)
(** -The Plants Database, National Plant Data Center, Natural Resources Conservation Service, United States Department of Agriculture, Baton Rouge, Louisiana, 2010. http://plants.usda.gov)

"**Symphytum asperum has been shown to escape from cultivation, but most sources describe escapes as limited or ephemeral (short lived) populations.** *Descriptions of significant damage are lacking. However, NOBANIS (European Network on Invasive Alien Species) lists Symphytum asperum as invasive in Sweden, and* **it is listed as a noxious weed in California.**
Symphytum officinale is naturalized across much of the United States and Canada, but no species of Symphytum are included on Midwest or Mid-Atlantic invasive lists.
The native range in the Caucasus is a climate match for Illinois as is the naturalized distribution in Northern Europe. However, occurrences in California, New Zealand, and Western Europe do not match.
Does this plant displace native plants and dominate (overtop or smother) the plant community in areas where it has established? There is no evidence of it dominating the plant community. Plants may spread somewhat aggressively by creeping rhizomes and by self-seeding, *and can be quite invasive in the garden.*
Moreover, once planted, Comfrey can be very difficult to dig out because any small section of root left behind can sprout a new plant. Easily propagated by seed, root cuttings or division.
Though there are reports of self-seeding, there are not estimates of the number of viable seeds, except here: '**The production of ripe nutlets (seeds) varies considerably in different plants,** *even fertile plants may have many flowers which produce only 1, 2, or 3 (and often 0) nutlets which are able to germinate.*' "
-'Plant Risk Evaluator: PRE Evaluation Report: Symphytum asperum- Illinois' by PlantRight and Chicago Botanic Garden, https://pre.ice.ucdavis.edu, Online database about invasive risk of non-native plants, July **2017**.
(Other species of Comfrey are not listed.)   (NOBANIS: 'The European Network on Invasive Alien Species' is a gateway to information on land and water alien and invasive species in North and Central Europe. www.nobanis.org)
(A noxious weed has been designated by agricultural authority as one injurious to agricultural or horticultural crops, natural habitats or ecosystems, or humans or livestock.)

"***Symphytum asperum Lepechin (Prickly or Rough Comfrey):***
This plant can be weedy or invasive according to the authoritative sources noted below.
Noxious Weed Information: California B list (noxious weeds). *Lower 48 States, Alaska, Canada: Introduced."*
-United States Department of Agriculture, Natural Resources Conservation Service, Plants Database, www.plants.usda.gov, **2018**.   (Other species of Comfrey are not listed.)

"***Symphytum asperum (Boraginaceae, Angiosperms):***

Country	Introduction Year	Status	Frequency	Invasiveness
Austria (AT)	<1910	Not established	Local	Not known
Denmark (DK)		Established	Rare	Not invasive
Estonia (EE)	1925	Established	Rare	Not known
Germany (DE)		Not known	Not known	Not known
Iceland (IS)	1945	Not established	Rare	Not known
Ireland (IE)		Not known	Not known	Not invasive
Latvia (LV)	<1950	Established	Rare	Not invasive
Netherlands (NL)		Established	Not known	Not known
Norway (NO)	1865	Established	Not known	Potentially invasive
Sweden (SE)	<1844	Established	Local	Invasive "

-NOBANIS: 'The European Network on Invasive Alien Species' is a gateway to information on land and water alien and invasive species in North and Central Europe. Participating countries: Austria, Belarus, Belgium, Czech Republic, Denmark, Estonia, Finland, Germany, Greenland, Iceland, Ireland, Latvia, Lithuania, the Netherlands, Norway, Poland, Slovakia, Sweden and the European part of Russia. www.nobanis.org, **2019**.

Symphytum Officinale (Common Comfrey) Invasiveness

"*In the past,* **Common Comfrey (Symphytum officinale)** *was cultivated extensively in Japan for medicinal use, as well as for use as a vegetable and feed of domestic animals in the 1960s. In recent years (early 1990s), however, due to the reduced demand for Common Comfrey production,* **the plant has tended to remain in crop fields and to behave like a weed.**
In a study of one field, it was noted that the individuals of Common Comfrey which exhibited various sizes and growing stages formed a dominant population. Within this population the height and area occupied by each plant at the reproductive stage tended to be larger than at the vegetative stage.
Common Comfrey plants which emerged in the field were derived from buried root fragments, which were of two types: fragments with or without a crown. *The former deveoped to the reproductive stage, while the later remained in the vegetative stage and in some of them, the top of the leaf bud did not sprout on the soil surface.*
**Plants emerging from former ridges were relatively larger than those emerging from former furrows. The position of the root fragments under former ridges were shallower than those under former furrows."*
-'Characteristics of Vegetative Propagation of Common Comfrey' by M. Nemoto, K. Shibuya and M. Saigusa, National Institute of

Agro-Environmental Sciences, MAFF, Tsukuba, Ibaraki, Japan; Weed Research, Tokyo, Japan, Volume 40, No. 3, pages 203-208, **1995**.

"*Symphytum Officinale:*

Naturalised	Enviro Score	Australia Rating	Prohibited for Sale
Victoria (southeast Australia)	X	4	None

Available for Sale: New South Wales (east Australia), Queensland (northeast Australia), Tasmania (Australia island), Victoria.
  Environmental score: H Significant environmental weed, **X Environmental weed**, S Sleeper.
  4: *Naturalised and known to be a major problem at 3 or fewer locations within a State or Territory.*"
-'WWF (World Wildlife Fund) National List of Naturalised Invasive and Potentially Invasive Garden Plants' (2004-**2006**). This national list has been derived from the Western Australia Department of Agriculture Plants Database' to identify plants documented as being weedy here or overseas and that have also been cultivated at some stage by the Australian garden industry.

"*Symphytum officinale Invasiveness Rank 48 (Weakly Invasive):* The invasiveness rank is calculated based on a species' ecological impacts, biological attributes, distribution, and response to control measures.
The ranks are scaled from 0 to 100, with 0 representing a plant that poses no threat to native ecosystems and 100 representing a plant that poses a major threat to native ecosystems. **Specific ecosystem impacts caused by Common Comfrey are largely unknown. While this species may reduce the nutrients and moisture available for native species, it is unlikely to have any major impacts on ecosystem processes.** (*, **)."
-'Common Comfrey: Symphytum Officinale L.- Alaska Natural Heritage Program' by Timm Nawrocki, University of Alaska Anchorage, Alaska Center for Conservation Science, October **2010**.
(* -AKEPIC Database: Alaska Exotic Plant Information Clearinghouse Database, 2010. http://akweeds.uaa.alaska.edu)
(** -The Plants Database, National Plant Data Center, Natural Resources Conservation Service, United States Department of Agriculture, Baton Rouge, Louisiana, 2010. http://plants.usda.gov)
  (The Alaska Invasive Ranking System is a climate screening and evaluation of ecological impacts, biological characteristics and dispersal ability, ecological amplitude and distribution, and feasibility of control.)

"*Symphytum officinale (Common Comfrey)* *(The higher the number, the more invasive.)*
  Ecological Impact: 16 (out of 40 points).   Biological Characteristics and Dispersal Ability: 12 (out of 25 points).
  Ecological Amplitude and Distribution: 13 (out of 25 points).   Feasibility of Control: 7 (out of 10 points).
  Invasiveness 48 (out of 100 points) 'Weakly Invasive'
  Pacific Maritime: Yes.   Interior Boreal: Yes.   Arctic Alpine: Yes"
-'Invasiveness Ranking of 50 Non-Native Plant Species for Alaska' by Nawrocki, Klein, Carlson, Flagstad, Conn, DeVelice, Grant, Graziano, Million and Rapp, University of Alaska Anchorage, Alaska Center for Conservation Science, April 8 **2011**.

"*Invasive neophytes,* such as *Lupinus polyphyllus Lindl., Calystegia sepium (L.) R.Br.* and **Symphytum officinale L., were mainly rare in agricultural habitats, although most of them favour agricultural and ruderal habitats** (e.g. Hamet-Ahti et al. 1998). The establishment and spread of invasive neophytes may be hindered by the **harsh climate in Finland** and high proportion of forests in Finnish agricultural landscapes (e.g. Luoto 2000).
  Neophyte: *An alien plant introduced to Finland after the early 17th century (Hamet-Ahti et al. 1998). An alien plant introduced to Central Europe after 16th century, both deliberately or accidentally (e.g. Pysek et al. 2004).*"
-'Plant Invasions in Boreal Agricultural Habitats: The Effect of Environmental Conditions, Species Traits, and the Impact on Native Diversity' by Miia Jauni, Doctoral Thesis, Department of Agricultural Sciences, Agriculture and Forestry of the University of Helsinki, Finland, August **2012**.   (A ruderal species is a plant that is first to colonize disturbed land.)

"*Symphytum officinale* **is listed as invasive in Norway and Sweden by NOBANIS.**"
-'Plant Risk Evaluator: PRE Evaluation Report: Symphytum asperum- Illinois' by PlantRight and Chicago Botanic Garden, https://pre.ice.ucdavis.edu, Online database about invasive risk of non-native plants, July **2017**.
  (NOBANIS: 'The European Network on Invasive Alien Species', www.nobanis.org)

"*Symphytum Officinale:* Naturalised in some parts of south-eastern and eastern Australia (i.e., in Victoria and in some coastal and sub-coastal districts of New South Wales). Also sparingly naturalised in the cooler parts of south-eastern Queensland and possibly naturalised in south-western western Australia.
Comfrey (Symphytum officinale) is regarded as an environmental weed in Victoria (Australia)."
-Queensland Government, Weeds of Australia, Biosecurity, Queensland Edition, Department of Employment, Economic Development and Innovation (DEEDI), **2018**.

"*Symphytum officinale* (Boraginaceae, Angiosperms):

Country	Introduction year	Status	Frequency	Invasiveness
Finland (FI)		Established	Common	Potentially invasive
Iceland (IS)	1967	Not established	Rare	Not known
Norway (NO)	<1800	Established	Not known	Invasive
Sweden (SE)	1400	Established	Local	Invasive

-NOBANIS: 'The European Network on Invasive Alien Species' is a gateway to information on land and water alien and invasive species in North and Central Europe. Participating countries: Austria, Belarus, Belgium, Czech Republic, Denmark, Estonia, Finland, Germany, Greenland, Iceland, Ireland, Latvia, Lithuania, the Netherlands, Norway, Poland, Slovakia, Sweden and the European part of Russia. www.nobanis.org, **2019**.

Symphytum Tuberosum (Tuberous Comfrey) Invasiveness

"Symphytum tuberosum is a spreading species with yellow flowers. It grows from large rhizomes, and often spreads to form large patches in gardens and the wild. It is native to parts of the United Kingdom and surrounding region. The roots are sometimes ground and roasted to make a non-caffeinated coffee substitute like Taraxacum or Cichorium intybus roots."
-Pacific Bulb Society, www.pacificbulbsociety.org, April **2015**. A volunteer-written encyclopedia about flower bulbs. This includes both cold hardy and tender bulbs, and all the bulbs in between.

"*Symphytum tuberosum (Boraginaceae, Angiosperms):*

Country	Status	Frequency	Invasiveness
Ireland (IE)	Not known	Not known	Not invasive "

-NOBANIS: 'The European Network on Invasive Alien Species' is a gateway to information on land and water alien and invasive species in North and Central Europe, www.nobanis.org, **2019**.

Symphytum x Uplandicum (Russian Comfrey) Invasiveness

"*Symphytum x Uplandicum:*

Naturalised	Enviro Score	Australia Rating	Prohibited for Sale
Sleeper	S	2	None

Available for Sale: New South Wales (east Australia), Queensland (northeast Australia), Tasmania (southern Australia island).
    Environmental score: H Significant environmental weed, X Environmental weed, **S Sleeper**.
    2: Naturalised and known to be a minor problem warranting control at 3 or fewer locations within a State or Territory."
-'WWF (World Wildlife Fund) National List of Naturalised Invasive and Potentially Invasive Garden Plants' (2004, updated **2006**). This national list has been derived from Western Australia Department of Agriculture Plants Database' to identify plants documented as being weedy here or overseas and that have also been cultivated at some stage by Australian garden industry.

"*The territory of the Upper Volga basin is 203,500 square km (78571 square miles). It is situated in the* **central part of European Russia** *and comprises the following administrative regions: Ivanovo, Kostroma, Yaroslavl, Vladimir and Tver.*
There are also some examples of local invasives: *Galega orientalis,* **Symphytum x uplandicum in Ivanovo region,** *Zizania latifolia in Korstroma region, Valisneria spiralis in Yaroslavl region, Anisantha tectorum, Linaria canadensis in Vladimir region."*
-'Invasive Species in the Flora of the Upper Volga Basin' by E.A. Borissova; In: Neobiota: From Ecology to Conservation (Book of Abstracts), 4th European Conference on Biological Invasions, Vienna, Austria, September **2006**, BfN-Skripten 184, page 85.
    (Ivanovo is a Russian city northeast of Moscow.)

"*Risk classifications of non-native biomass crops:*
Five other species were **classified as medium risk** *(Andropogon gerardii, Asclepias syriaca, Fallopia sachalinensis var. igniscum, Miscanthus sacchariflorus and* **Symphytum x uplandicum***).*
Ecological risk scores and classification of non-native biomass crops for the Netherlands:

	Total Risk Score	Distribution in Netherlands	Risk Classification
Symphytum x uplandicum	10 (Watch list)	Widespread	B3 = medium risk

    Total Risk Score *(ISEIA = Belgium Invasive Species Environmental Impact Assessment).*
    Risk Classification *(BFIS list system = Belgian Forum Invasive Species).*
    B3 = Medium risk = Except when assisted by man, the species doesn't colonise remote places. Natural dispersal rarely exceeds more than 1 kilometer (0.62 mile) per year. The species can however become locally invasive because of a strong reproduction potential.
Reproductive Capacity of Russian Comfrey *(Dutch names: Bastaardsmeerwortel, Smeerwortel hybride):*
    In United Kingdom, most reproduction is thought to occur vegetatively, however varying degrees of fertility in naturalised populations suggest that **some sexual reproduction may occur** *(*NNSS, 2015). What originally was an ornamental species in many countries of Europe has often escaped and established outside cultivation.*
    In European part of Russia, *Symphytum x uplandicum was seen to grow individually or in small groups. Large stands were found in forest clearings. Plants were very tall at this location (up to 1.7 meter = 5.5 feet) and abundantly flowered. Plants were observed to flower twice per year (**Borissova, 2006). Symphytum x uplandicum has been recorded in pine-birch-spruce forest, herbaceous spruce forest and herbaceous spruce forest in Russia (**Borissova, 2006).*
Cultivated Range:
    Symphytum x uplandicum has been cultivated as a crop for hundreds of years (*NNSS, 2015). Limited information is available on the cultivated range of S. x uplandicum. However, the plant is cultivated in the United Kingdom, the United States of America and Canada.
Non-Native Range:

Symphytum x uplandicum is locally invasive in the Ivanovo region of the upper Volga basin, Russia (**Borissova, 2006). The plant has been recorded in the Czech Republic, Denmark, Germany, Finland, France, Great Britain, Ireland, Lithuania, the Netherlands, Norway, Portugal (The Azores), Sweden, Austria, Belgium, Luxembourg, New Zealand, Australia, Japan, Brazil (*NNSS 2015; EPPO Global Database; United States Department of Agriculture http://plants.usda.gov/core/profile?symbol=SYUP; GBIF www.gbif.org/species/2926068).

Distribution in the Netherlands:

The distribution of Symphytum x uplandicum in the Netherlands is poorly known due to issues with identification and hybridization. Symphytum x uplandicum exhibits two chromosome numbers (2n = 36 and 2n = 40). Plants with chromosome number 2n = 40 are able to backcross with native S. officinale. The plants of this hybrid group are difficult to identify and may be labelled as either Symphytum officinale or Symphytum asperum (***Gadella, 1978). Moreover, confusion may occur with other species present in the plant trade such as Symphytum caucasium and Symphytum grandiflorum. Most records identified as Symphytum x uplandicum occurred in southern half of Netherlands.

Ecological Risk Assessment with the ISEIA Protocol:

The expert team allocated Symphytum x uplandicum a 'high' ecological risk classification to the categories dispersion potential and invasiveness, and colonisation of high value conservation habitats, a 'medium' risk classification to the category adverse impacts on native species and a 'likely' risk classification to the category alteration of ecosystem functions.

Consensus scores for potential risks of Russian Comfrey (Symphytum x uplandicum) in the current situation in the Netherlands, using the ISEIA-protocol:

ISEIA Section	Risk	Consensus Score
Dispersion potential or invasiveness	High	3
Colonization of high value conservation habitats	High	3
Adverse impacts on native species	Medium	2
Alteration of ecosystem functions	Likely	2
Ecological risk score		**10** "

-'Horizon Scanning and Environmental Risk Analyses of Non-Native Biomass Crops in the Netherlands' by J. Matthews, R. Beringen, M.A.J. Huijbregts, H.J. Van der Mheen, B. Ode, L. Trindade, J.L.C.H. Van Valkenburg, G. Van der Velde and R.S.E.W. Leuven; Reports Environmental Science, Department of Environmental Science, Institute for Water and Wetland Research, Radboud University, Nijmegen, The Netherlands, December 31 **2015**.

(* -'Russian Comfrey, Symphytum officinale x asperum = S. x uplandicum' by R.V. Lansdown; NNSS: Great Britain Non-Native Species Secretariat, www.nonnativespecies.org, York, England, 2011.)

(** -'Invasive Species in the Flora of the Upper Volga Basin' by E.A. Borissova; In: Neobiota from Ecology to Conservation, 4th European Conference on Biological Invasions, Vienna, Austria, September 2006, BfN-Skripten 184, page 85.)

(*** -'Variation and Hybridization in Some Taxa of the Genus Symphytum' (Variatie en Hybridisatie bij enkele Taxa van het Genus Symphytum) by Th.W.J. Gadella, Department of Population and Evolutionary Biology {Vakgroep Populatie en Evolutiebiologie, Utrecht, The Netherlands}; Gorteria: Tijdschrift voor de Floristiek, de Plantenoecologie en het Vegetatie-Onderzoek van Nederland {Journal for Floristics, Plant Ecology and Vegetation Research in the Netherlands}, Volume 9, No. 4, pages 88-93, 1978.)

"*Symphytum x uplandicum (Boraginaceae, Angiosperms):*

Country	Introduction year	Status	Frequency	Invasiveness
Austria (AT)	<1880	Not established	Common	Not known
Czech Republic (CZ)	<1908	Established	Rare	Not invasive
Denmark (DK)		Established	Common	Not invasive
Finland (FI)		Established	Common	Potentially invasive
Germany (DE)		Not known	Not known	Not known
Ireland (IE)		Not known	Not known	Not invasive
Lithuania (LT)	1994	Established	Local	Not known
Netherlands (NL)		Established	Not known	Not known
Norway (NO)	1873	Established	Not known	Invasive
Sweden (SE)	1700's	Established	Local	Not known "

-NOBANIS: 'The European Network on Invasive Alien Species' is a gateway to information on land and water alien and invasive species in North and Central Europe. Participating countries: Austria, Belarus, Belgium, Czech Republic, Denmark, Estonia, Finland, Germany, Greenland, Iceland, Ireland, Latvia, Lithuania, the Netherlands, Norway, Poland, Slovakia, Sweden and the European part of Russia. www.nobanis.org, **2019**.

Plow Field in Winter, Then Grow Grass (How to Kill Comfrey)

The first step is removing as much of the Comfrey roots as you can. It is much easier to remove the roots if you do it when the ground is wet such as after a good rain. Or use a garden hose to get it wet.

"Large-scale Comfrey extermination is better effected by exploiting the 'resentment' of the plants at winter disturbance. Plough not in November even, but in December or January.

If there are large roots, as from a good well-grown field (an agricultural folly to destroy it), take up the main clumps for pig-feed

as if they were mangolds (large fodder beets). The large roots were carted off and fed to pigs.
Leave the ground rough, and let the frost get at the roots; deep ploughing at this time of year not only kills exposed roots, but the broken ends rot in the land to a considerable extent.
The efficiency of the kill depends on the plant; our native Symphytum officinale (Common Comfrey) is the most resistant, and Symphytum peregrinum (Russian Comfrey) the most easily destroyed.
After leaving to the frost, the land was sown down to a ley. **It was grazed by sheep and these animals, though slower as exterminators, completed the clearing of stray fragments in two years without damaging the pasture as rooting pigs would have done."**
-Russian Comfrey: A Hundred Tons an Acre of Stock or Compost for Farm, Garden or Smallholding by Lawrence D. Hills. London England: Faber and Faber, Limited, **1953**, page 71.
(Ley farming is growing grass or legumes in rotation with grain or tilled crops as a soil conservation measure.)

"**The simplest way is for the farmer who wishes to convert his Comfrey bed back to pasture:**
He should plough (plow) and cultivate in December or January, the only time of the year when Comfrey plants are damaged by machinery, and sow to a four-year ley in the spring. *The grass will slowly destroy his Comfrey, though a few fragments will persist in the hedge bank.*
Ploughing and harrowing in summer should never be used to try to kill Comfrey, and rotavation is still worse. All these often-tried answers merely spread the broken roots wider and mean thousands of small plants which grow into tall strong perennial weeds in arable land. **Grass is always the worst weed in Comfrey, and in time it will kill the remains of the crop.** *In a lawn, however, it can persist, because it withstands mowing as well as it does cutting."*
-Comfrey: Fodder, Food and Remedy by Lawrence D. Hills. New York: Rizzoli Universe Books, **1976**, pages 112, 113.
(Rotavators® are bigger and more powerful than cultivators. Rotavator blades dig deeper into the ground about 228 mm / 9 inches compared to a cultivator that is 51-76 mm / 2 to 3 inches deep. Rotavator is a trademark name.
I don't know which kind of plow the writer meant. Three of the most common types of plows: Moldboard plow cleanly cuts through soil and turns it over. It is used to break up a fallow or hay field with 200-300 mm / 7.8-11.8 inch depth. Chisel plow is used for deep tillage by moving soil below the surface, and minimizing disturbance of top layers with 15-46 cm / 6-18 inches deep. Disc plows use discs instead of a wedge-shaped blade. It is good in sticky or rocky soil.)

Let Pigs Dig Up and Eat the Comfrey Roots

"*The simplest method to remove Comfrey from for a small area is undoubtedly the one used by German smallholders and peasant farmers.* **They wait until May when the first Comfrey cut is really tall, about four feet (1.2 meter), and turn on the pigs unringed. These stay until October; they eat every fragment of crown and snout up the shoots that spring from the buried roots until these are exhausted.**
If the ground is hard, plough and cultivate, and the snouts will look after the fragments. The ploughing to allow further snouting is best done after the land is already cleared of foliage; by cutting deep, a great many can be brought to the surface. Even ringed they can do a great deal of damage."
-Russian Comfrey: A Hundred Tons an Acre of Stock or Compost for Farm, Garden or Smallholding by Lawrence D. Hills. London England: Faber and Faber, Limited, **1953**, page 70.
(Smallholding is a small farm. In developing countries, they support a single family with a mixture of cash crops and subsistence farming. In developed countries, smallholdings are not usually self-sufficient but provide some income and a rural lifestyle.)

"*In Germany, where Russian Comfrey is a favourite smallholder's crop, the custom is to leave the patch down about seven years, and then fold pigs on it.* **The pigs are not rung and eat every bit of root they can scent, after they have cleared the foliage. This is the only way of getting rid of it economically,** *and it enables it to be used as a farm crop like lucerne (alfalfa), outside the normal rotation."*
-Fertility Farming by Newman Turner. Austin, Texas: Acres USA, **1951**, page 57.

"**The best way to destroy Comfrey is to turn on the pigs, ideally tethering dry sows or the boar on the worst places, and they should clear the plot in six months.** *Those who have pigs can use these as ground clearers. In areas where it is desired to kill out wild Comfrey, or when a poorly-sited crop or a change in farm policy demands that a plot should be destroyed, then* **pigs are the cheapest answer. They are extremely fond of Comfrey root, which they hunt by scent like truffles, and in three months they will leave no trace of the crop."**
-Comfrey: Fodder, Food and Remedy by Lawrence D. Hills. New York: Rizzoli Universe Books, **1976**, pages 104, 112.
(Truffles are the fruiting body of an underground Ascomycete fungus.)

Let Chickens or Other Poultry Kill the Comfrey

"*In Kenya and Rhodesia (Zimbabwe) and Zambia (all African countries), where Comfrey plants wear out fast with the constant growth cycles, lasting about twelve years,* **the favourite method to get rid of Comfrey is to run chickens on the old plot after removing the crowns for a new planting, in the dry season.** *The birds peck up every blade of leaf as fast as it appears, and the roots die out before the rains.* **The chicken answer is for tropical countries only."**

-Comfrey: Fodder, Food and Remedy by Lawrence D. Hills. New York: Rizzoli Universe Books, **1976**, page 112.
> (I think chickens might be able to kill Comfrey any location if their area of free ranging is small. It might take a year but eventually the roots would not get any nutrition and would die.)

"If you should need to kill Comfrey (what a sad thought!), **ducks and chickens, pastured on it, will kill it or knock it so far back that it might as well be dead.** *This was not an intentional discovery...."*
-'Comfrey Question' by Chamoisee, Idaho; Gardening and Plant Propagation Forum, Homesteading Today, www.homesteadingtoday.com, Forums, articles and media, June 27 **2012**.
> (Whether or not poultry will kill Comfrey growing in pasture depends on how long the birds are on the pasture. Rotational grazing works best if you don't want to kill it.)

"If you really want to kill Comfrey, get a couple of ducks. I was told by various people that Comfrey, once well established, was virtually indestructible. But then my ducks discovered the stuff. Before I knew it, **the ducks had stripped the Comfrey patches bare...right down to the earth.** *I immediately fenced the areas off away from them and hoped for the best, but alas, they had managed to do the impossible as a year later not so much as a single leaf reappeared."*
-'Comfrey Types and Facts' by LibertyBelle, The Survival Podcast Forum, https://thesurvivalpodcast.com, Helping you live a better life, if times get tough or even if they don't, December 2 **2014**.

Kill Comfrey by Sheet Mulching or Covering with Plastic

"Comfrey Out of Control:
Question: *How does one get rid of a patch of Comfrey which has overtaken the garden including two elderly but still magnificent and much loved rhododendron bushes?*
Answer: *Comfrey roots penetrate deeply, as much as 1 metre (3 feet), and since each small piece can develop into another plant,* **Comfrey can be difficult to eradicate. The key is persistence.** *But not of the sweaty, backbreaking kind. Cut back the plants to ground level.* **Cover the area with a strong plastic sheeting, at least 4 mil in thickness. Extend the plastic 60 cm (24 inches) past the outer edges of the Comfrey patch.** *Then weigh down the plastic with 10 centimetres (4 inches) of clean soil. In a year you can remove the sheeting and the roots will be dead from lack of light."*
-'Comfrey Out of Control' by Richters Staff; Richters®: The Herb Specialists, www.richters.com, Goodwood, Ontario, Canada, April **1998**. (Some people prefer black plastic sheeting; others like clear plastic sheeting. You can use a tarp or anything similar.)

"If you have Comfrey on your property and you want to get rid of it, do not try to dig it out or rototill near it. If even a tiny piece of root breaks off and is left in the ground, it will re-grow.
The best way to get rid of Comfrey is to sheet mulch over it. It will not send out shoots or runners to get around the mulch. Just lay some cardboard and other mulching materials on top of it to rob it of sunlight and you'll be all set."
-'The Awesomeness of Comfrey' by Jocelyn Durston, The Farm for Life Project, Maple Ridge, British Columbia, Canada, **2014**.
> (With sheet mulching the soil is covered with a thin layer of slowly decomposing material known as the weed barrier, typically cardboard. This suppresses weeds by blocking sunlight. Sheets of newspaper, clothing or rugs can be used instead of cardboard. For this to work with Comfrey, you need to have it extend at least 2 feet (0.6 meter) in all directions from the center of the Comfrey plant. Experiment with it, and let me know your results.)

"Killing Comfrey:
1. Stick a shovel or fork under the Comfrey plant and pull out the bulk of the root ball.
> *By yanking the bulk of the plant out, you are stressing the plant and forcing it to push up new leaves. Thus, the plant will use all of it's final energy in an effort to re-establish itself above the soil.*

2. Get a large black plastic pot or black plastic mixing tub.
> *I've got a black tub that I use for mixing small batches of concrete. They are cheap enough at Home Depot® or Lowes®.* **Anything that seals out the light would seem to work.**
> **Why black plastic? It's a heat sink.** *The summer sun beating on a black plastic pot would make it very hot underneath. A parked car with the windows rolled up can get up to 150 F degrees = 65 C.*

3. Block all sunlight:
> *With your hands or a little trowel, dig a bit of shallow trench around the base of the plant so that you can situate your pot or tub slightly below soil level. Place the pot or tub over the Comfrey and nestle it firmly into the trench. Then backfill up against the exterior sides of your pot. It shouldn't allow any light to get to any emerging plant leaves.*

4. Put a rock or a couple of bricks on top, *so your pot doesn't get toppled.*

5. Wait 4 months or more.
Deprived of sunlight and over-head water, and subjected to the hot summer sun, the three Comfrey plants I've tried this on all seem to have died. **No Round-Up® or salt or anything chemical is being used."**
-'Killing Comfrey' by Marco Banks, Los Angeles, California, **2017**. Permies: Homesteading and Permaculture All the Time, https://permies.com by Paul Wheaton, Missoula, Montana.

*"***Heat and Dry Out Soil: The only way I've known to kill Comfrey is to dry it out, and this can take all summer. Chop it down, lay clear plastic (vented) or heavy paving stones over it, and be religious about not watering in that area.**

Vented because you're wanting to turn the area into a mini-greenhouse, heating that soil enough to dry it out, and you can't dry it out unless it's vented. Solarizing the soil is for another purpose, not this...exactly."
-Permaculture Research Institute, https://permaculturenews.org, **2018**.
(I have not tried this method. This would work better in a dry climate that gets little rain. Comfrey is very drought tolerant.)

Dig, Dig, Dig (This method is a lot of work.)

*"**Here is how to kill Comfrey: Mark where it is growing in the autumn, then in December or January dig it up with as much of the root as possible, leaving the ground rough.** The disturbance and the frost at its dormant period will rot off the remaining roots of the Russian species.*
*Our native Symphytum officinale (Common Comfrey) is tougher, and so are some of its natural hybrids with the other species. These will probably appear again in late spring or early summer. **Dig up again, going down as much as eight inches (20 cm); if necessary do this three or four times, and the buried roots will die of exhaustion, but only if the leaves are not allowed to develop and send reinforcements of carbohydrates down to the buried fragments.***
A summer's careful weeding will destroy the toughest, and the fragments should be burnt, or in the case of a good variety, planted where they are required.
Never hoe any Comfrey, the buried root will merely grown again, and the chopped-off fragment may have enough root on it to make a further plant. A weed is a plant in the wrong place."
-Russian Comfrey: A Hundred Tons an Acre of Stock or Compost for Farm, Garden or Smallholding by Lawrence D. Hills. London England: Faber and Faber, Limited, **1953**, page 117.

*"**Management: Common Comfrey can be difficult to remove due to the potential for vegetative regeneration from root fragments. Digging is required to remove the plant and the large network of roots.** Populations in Glacier Bay National Park, Alaska, persisted after multiple years of manual removal efforts.*
Mowing plants before they produce seeds can prevent populations from spreading."
-'Common Comfrey: Symphytum Officinale L.- Alaska Natural Heritage Program' by Timm Nawrocki, University of Alaska Anchorage, Alaska Center for Conservation Science, October **2010**.

Use Salt to Kill Comfrey Roots

You can use regular table salt (sodium chloride), rock salt (sodium chloride), or Epsom® salt (magnesium sulfate). Rock salt is used to melt snow and ice.
Be careful when working with salt. It is a corrosive and will kill other plants besides just Comfrey. Salt pulls the moisture out of everything. It prevents the proper flow of magnesiuim and potassium in the plant.
Salt is used to kill tree stumps. In that situation you drill holes 3 to 4 inches (7.6 to 10.1 cm) deep into the stump, and then **pour the salt in the holes.** Then add just enough water to make the salt moist.
Or instead of putting salt and then water in the hole, you can create **a salt solution of 2 parts salt to 1 part water.** You pour the solution into the holes. Cover the holes with soil, or a plastic tarp to keep out rain. You may have to do this once every few weeks for a few months. I have not tried this method but it seems that it should work with Comfrey roots. Let me know your experience with it. I prefer the black plastic method.

Rot Roots with Too Much Water

Comfrey roots are sensitive to constant, water-logged soil. If they stay very wet all the time, the roots get soft and die. I have not tried this or read anyone using this method but it makes sense.
If there is a way you can put a barrier around the Comfrey plants that will hold water, then in a few months (maybe more) the roots will die. You would keep filling with water so the soil is always soggy. If you try this, tell me about your experiences.

Use Herbicides to Get Rid of Comfrey

Be careful with all herbicides and read the instructions. I prefer not to use herbicides. But I include this information for those that do use them so they do it most effectively with the least environmental damage.

Use Herbicide Ammonium Sulphamate (Sulfamate®)

*"**A better and safer chemical killer is ammonium sulphamate,** which was formerly used for fireproofing railway sleeper carriage curtains. Roughly speaking it is sulphate of ammonia 'made crooked', and **the greedier a weed is for nitrogen, the more it takes and the harder it falls. Though the chemical is expensive, it has the great advantage of becoming sulphate of ammonia in about six weeks, and washing harmlessly from the soil.***
Mix up a solution of 1 pound (.45 kg) in a gallon (3.7 liter) of water and put this on with a perforated can over 100

square feet (9.3 square meters), that is a square ten feet (3 meters) each way.
This is for killing out the mess of small broken roots that attempting to kill by summer cultivation will bring.
If the bed is a normal one, with plants at intervals, cut them first, wait till growth has started again, say a week, and then water the clumps carefully, giving them about a quart of the solution each, but of course they vary in size- if they are eighteen inches (45 cm) across, they take more, but if they are growing as well as this, it is cheaper to learn how to use the crop properly.
Ammonium sulfamate should only be put on during the spring and summer. The first week of September is about the latest date for using in Britain, for if it is watered on in winter when the roots are not feeding, you will waste this expensive but safe and effective weedkiller.
It kills by bad diet, like poisoning a rat with a diet of white bread, but more quickly, and in six weeks the land can be sown or planted again. The test is where there is a rapid growth on the cleared ground of chickweed and groundsel from suppressed seeds. This killer is also effective for horse radish, equisetum (horsetail), convolvulus and all perennial weeds."
-Comfrey: Fodder, Food and Remedy by Lawrence D. Hills. New York: Rizzoli Universe Books, **1976**, pages 113, 114.

(Ammonium sulfamate is a white crystalline solid readily soluble in water. It is a commonly used broad-spectrum herbicide sold under these trade names®: Amcide®, Amicide®, Amidosulfate®, Ammate®, Ammate X-NI®, AMS®, Fyran 206k®, Ikurin®, Sulfamate®, and Root-Ou®t. Some of those brands are available in Great Britain and Australia; none seem to be available in the United States except as small amounts for laboratory use.

It is useful in killing tough woody weeds, tree stumps and brambles. It is also used as a compost accelerator that helps break down tough and woody weeds in the compost heap.

Ammonium sulfamate is considered slightly toxic to humans and animals, making it appropriate for the home garden, professional and forestry uses. It is generally accepted as safe for use on land that will be used for growing fruit and vegetables. It is considered environmentally friendly because it degrades to non-toxic residues.

European Union's pesticide legislation 'unlicensed' herbicides containing ammonium sulfamate, thereby banning it starting 2008. It is still available for use as a compost accelerator. Be careful with all herbicides and read the instructions.)

<u>**Use Herbicide Sodium Chlorate**</u> (Be careful with all herbicides and read the instructions.)

"Comfrey is a long-lived perennial, lasting up to forty years, and it is very hard to kill. The roots have so much depth from which to come up that no method of cultivation will destroy it, and **the use of sodium chlorate is about the only way to get rid of it.**"
-Fertility Farming by Newman Turner. Austin, Texas: Acres USA, **1951**, page 56.

"**Large-scale control of Symphytum officinale with sodium chlorate is uneconomic,** but for treatment of isolated plants, 5-10 grams of granular formulation or 0.25 liter of a 5% spray solution are recommended."
-'Common Comfrey (Symphytum Officinale) and Its Control' by A. Becker, Germany; Gesunde Pflanzen: Pflanzenschutz, Verbraucherschutz, Umweltschutz (Healthy Plants: Plant Protection, Consumer Protection, Environmental Protection), Volume 17, No. 6, pages 115-118, **1965**.

"**On a garden or plot scale Comfrey can be easily killed with Sodium Chlorate.**
Cut through the crowns about an inch (1 inch = 2.5 cm) below the ground and spread the white crystals on the creamy cut surface of the root. This will soak down the sap channels and should kill out the roots completely as it will those of a tree stump."
-Comfrey: Fodder, Food and Remedy by Lawrence D. Hills. New York: Rizzoli Universe Books, **1976**, page 113.

(Sodium chlorate is an inorganic compound {$NaClO_3$} that is a white crystalline powder soluble in water. It is used as a non-selective herbicide that is phytotoxic to plants. It also kills through root absorption. Sodium chlorate is used to kill plants such as bamboo, canada thistle, johnson grass, morning glory, ragwort, and St John's wort. The herbicide is used on non-crop land for spot treatment.

Sodium chlorate is highly volatile when exposed to heat. It stays in the soil for up to a year. Whereas, ammonium sulphamate breaks down in 6 to 8 weeks. When it rains, sodium chlorate travels more than ammonium sulphamate so it might harm other plants. Be careful with all herbicides and read the instructions.)

<u>**Other Herbicides to Kill Comfrey**</u> (Be careful with all herbicides and read the instructions.)

"**Roots remaining after ploughing up an old Comfrey stand can be killed by treating individual plants with 6 grams Agrosan® per plant.**"
-'Cultivation of Comfrey' by W. Doring, Inst Acker- u PflBau, University of Halle-Wittenber, East Germany; Deutsche Landwirtschaft (German Agriculture), Volume 10, No. 2, pages 62-66, **1959**. (I do not have this report. If you have a copy, could you please send it to me.)
(Agrosan® is a broad-spectrum concentrated, non-iodine sanitizer that provides effective germicidal action on a broad range of micro-organisms, including gram-positive and gram-negative bacteria. It is also listed as a fungicide and herbicide.)

"***Weed Species and Herbicides for Control:***
MCPA (2-methyl-4-chlorophenoxyacetic acid): MCPA is very similar in use to 2,4-D. It is used for the control of herbaceous broad-leaved weeds. Resistant species are: Common Comfrey.
2,4,5-T (2,4,5-trichlorophenoxyacetic acid): 2,4,5-T can be used for the selective control of broad-leaved weeds in turf and for

some emersed aquatics of nongrass species, but it is used principally for the control of woody plants. 2,4,5-T is generally used in the ester formulation although there are sodium and triethanolamine salts. There is also a new diamine formulation.

2,4-D and 2,4,5-T are also combined in a formulation called 'brush killer'.

Resistant species are: Common Comfrey."

-'Herbicide Manual for Noncropland Weeds: Agriculture Handbook No. 269' by R.S. Dunham, Agriculture Research Service, United States Department of Agriculture, with Bureau of Yards and Docks, Department of Navy, Washington, DC, March **1965**.

(MCPA, 2-methyl-4-chlorophenoxyacetic acid, is a powerful, selective, widely used phenoxy herbicide. It controls broadleaf weeds, including thistle and dock, and deciduous trees, in cereal crops and pasture. Clovers are tolerant at moderate application levels.)

(2,4-D or 2,4-Dichlorophenoxyacetic acid is an organic compound that is a systemic herbicide which selectively kills most broadleaf weeds by causing uncontrolled growth. Most grasses are relatively unaffected.)

(2,4,5-Trichloro-phenoxyacetic acid is a synthetic auxin, a chlorophenoxy acetic acid herbicide used to defoliate broad-leafed plants. It was developed in late 1940s and was widely used until being phased out, starting in late 1970s due to toxicity concerns.)

"A note is given on the biology of Symphytum officinale, together with observations on its control. **The weed is not particularly susceptible to such growth-regulators as 2,4-D (2,4-Dichlorophenoxyacetic acid), MCPA (2-methyl-4-chlorophenoxyacetic acid), or 2, 4, 5-T (2,4,5-Trichlorophenoxyacetic acid), and regenerates after relatively high doses of amitrole.**

On arable land and grassland, clumps can be eradicated by spot-treatment with mecoprop (methyl-chlorophenoxypropionic acid, MCPP) + 2, 4, 5-T, *while several-year clover or lucerne (alfalfa) leys maintain suppression of dense infestatons.*

Large-scale control of Symphytum officinale with sodium chlorate is uneconomic, *but for treatment of isolated plants, 5-10 grams of granular formulation or 0.25 liter of a 5% spray solution are recommended."*

-'Common Comfrey (Symphytum Officinale) and Its Control' by A. Becker, Germany; Gesunde Pflanzen: Pflanzenschutz, Verbraucherschutz, Umweltschutz (Healthy Plants: Plant Protection, Consumer Protection, Environmental Protection), Volume 17, No. 6, pages 115-118, **1965**.   (Ley is grassland or pastureland that is grazed.)

(Mecoprop® is a common general use herbicide found in many household weed killers and 'weed-and-feed' lawn fertilizers. It is primarily used to control broadleaf weeds. It is often used in combination with other chemically related herbicides such as 2,4-D, dicamba, and MCPA.)

"Close grazing and rooting by hogs through the summer and late into the fall is an effective control. **Comfrey is reported to be tolerant of 2,4-D and 2,4,5-T and susceptible to atrazine, sodium chlorate, and ammonium sulfamate."**

-'Comfrey: A Controversial Crop' by Robert G. Robinson, University of Minnesota, Agricultural Experiment Station, Minnesota Report MR-191, Item No. AD-MR-2210, **1983**, page 3.

(2,4-D {2,4-Dichlorophenoxyacetic acid} kills broadleaf plants but not grasses. 2,4,5-T {2,4,5-Trichlorophenoxyacetic acid} kills broadleaf plants. Atrazine kills broadleaf plants. Sodium chlorate is a non-selective herbicide that is phytotoxic to all plant parts. Ammonium sulfamate {ammonium sulphamate} is broad-spectrum herbicide.)

"When an area is to be replanted with a different crop, repeated tillage is usually used to remove the Comfrey plants. Deep moldboard plowing should be done in September or October, and then followed by tillage with a field cultivator, which will expose roots to the drying and freezing conditions of winter.

Herbicides could be used in removing the old plants. **Glyphosate was sprayed in June at Rosemount, Minnesota and killed stems and leaves. However, plants grew again from the roots to produce a full stand by September."**

-'Comfrey' by Teynor, Putnam, Doll, Kelling, Oelke, Undersander and Oplinger, 'Alternative Field Crops Manual', University of Wisconsin: Cooperative Extension, University of Minnesota: Center for Alternative Plant & Animal Products, and Minnesota Cooperative Extension, February **1992**, updated November 1997, page 3.

(Glyphosate {N-(phosphonomethyl)glycine} is a broad-spectrum systemic herbicide and crop desiccant {drying agent}. It is an organophosphorus compound. Monsanto® brought it to market for agricultural use in 1974 under the trade name Roundup®.)

"Tillage both multiplies the Comfrey population and prolongs its emergence period. Even without soil disturbance, plants emerge well into the growing season complicating post-emergence application timings.

Roundup Ready® (RR) corn and soybean were planted by the producers and post-emergence herbicides were evaluated for Comfrey control. Unless otherwise noted, the glyphosate (Roundup®) rate in all years was 0.75 pounds ae/a (acid equivalent per acre), and all other herbicides were applied at normal use rates.

The 2001 site was **moldboard plowed** *prior to planting RR corn. No tillage was done in the 2002 and 2003 trials where RR corn and RR soybean were planted, respectively.*

2001 Herbicide Treatments

Persistent rains in 2001 prevented the farmer from planting the field until the end of May. At that time, Comfrey was 12 to 30 inches (0.30 to 0.76 meters) tall and flowering. The field was moldboard plowed on May 26 to obtain a level playing field. RR corn was planted in 38-inch (0.96 meter) rows on May 28.

Post-emergence herbicides *were applied on July 6. Comfrey was 2 to 11 inches (0.05 to 0.27 meters) tall and had not flowered. On July 11, the second part of split treatments was applied.*

In 2001, dicamba (broad spectrum herbicide) alone, dicamba premixed with diflufenzopyr, and dicamba tank mixed with glyphosate gave 80% or more control of treated Comfrey plants 30 days after application.

The premix of primisulfuron plus dicamba, halosulfuron alone, mesotrione plus dicamba and glyphosate alone (single and sequential treatments) gave 60 to 80% control of treated plants. Clopyralid (selective herbicide used for control of broadleaf weeds, especially thistles and clovers) and mesotrione alone had little effect on Comfrey.
Due to continued emergence of Comfrey after herbicide application, Comfrey levels in September were greater than when treatments were applied.

2002 Herbicide Treatments
In 2002, RR corn was no-till planted on May 29 following a May 16 broadcast application of paraquat (a burndown herbicide) and acetochlor (a preemergence treatment to control annual broadleaf and grass weeds). **Surprisingly, paraquat had only a temporary effect on Comfrey and did not 'burn down' the treated plants adequately.**
Post-emergence treatments were made when Comfrey had started flowering on June 17 and again on July 2.
On October 10, glyphosate (1.5% v/v solution of a 3-pound ae/gallon formulation) was applied preharvest between corn rows with a backpack sprayer fitted with a single TK 3 tip. **Even in a no-till system, Comfrey emerged well into the growing season but less so than in 2001 when tillage is done.**
This year dicamba, halosulfuron, primisulfuron plus dicamba, mesotrione and 2,4-D failed to give acceptable Comfrey suppression. Glyphosate at 0.75 pounds ae/a in single and sequential applications and dicamba plus diflufenzopyr at the highest labeled rate for corn gave 90% or more Comfrey control.
Nearly all the 2002 treatments that contained glyphosate had less than 10% Comfrey abundance (0-100 scale) on August 20 but only those with two glyphosate applications had less than 15% Comfrey abundance at the end of the season.
Comfrey control with the preharvest glyphosate application was nearly 100% when evaluated in May 2003.
Sequential application of glyphosate in RR corn in 2002 also resulted in less than 10% Comfrey abundance the year after application and sequential use of dicamba plus diflufenzopyr had 15% abundance, the best of all non-glyphosate-based treatments.

2003 Herbicide Treatments
In 2003, RR soybeans were no-till planted on May 30. Paraquat or glyphosate with metolachlor was applied as a burndown/pre-emergence treatment prior to planting. **We observed that preplant glyphosate noticeably reduced Comfrey vigor but paraquat did not. Glyphosate applied in-crop gave 81 to 97% Comfrey control in late July with only slight advantage to sequential applications and no greater control from rates above 0.75 pounds ae/a.**
Glyphosate was applied alone as a preharvest treatment to mature soybean on October 6 and after harvest alone and in combination with either tribenuron and rimsulfuron plus thifensulfuron (Basis®), or with tribenuron and clorimuron + thifensulfuron (Synchrony®) on October 30.

2004 Herbicide Treatments
In May 2004, the preharvest treatments were nearly free of Comfrey (0 to 2%) and even the postharvest applications were as effective as in-crop application of glyphosate in reducing Comfrey abundance.

Comfrey can be conquered.
The formula for success is to:
1) plan to use a no-till system and glyphosate resistant crops for two seasons,
2) plant Comfrey-infested fields last to delay the burndown application of glyphosate as long as possible,
3) apply a low rate of a soil-active herbicide with the burndown treatment,
4) plant 3 days after applying the burndown treatment,
5) apply 0.75 pounds ae/a of glyphosate when Comfrey is well into the flowering stage (probably mid to late June) and
6) consider preharvest glyphosate applications if necessary."
-'Conquering Comfrey' by Jerry D. Doll, Extension Weed Scientist, Department of Agronomy, University of Wisconsin, Madison, Wisconsin at 'North Central Weed Science Proceedings', Volume 59, No. 178, **2004**.
(Be careful with all herbicides and read the instructions.)

"Glyphosate Toxicity:
1. Research shows glyphosate is toxic to water fleas at extraordinarily low levels, well within the levels expected to be found in the environment. These findings throw serious doubt on glyphosate's safety.
2. Previous research has shown that Roundup® is toxic to human DNA even when diluted to concentrations 450-fold lower than used in agricultural applications.
3. 'Inactive' ingredients such as solvents, preservatives, and surfactants contribute to toxicity in a synergistic manner, and ethoxylated adjuvants in glyphosate-based herbicides have been found to be 'active principles of human cell toxicity'.
4. Cell damage and even cell death can occur at the residual levels found on Roundup®-treated food crops, as well as lawns and gardens where Roundup® is applied for weed control.
5. Liver, embryonic and placental cell lines are adversely affected by glyphosate at doses as low as 1 ppm (parts per million). GM (genetically modified) corn can contain as much as 13 ppm of glyphosate, and Americans eat an average of 193 pounds (87 kg) of GM foods annually."
-'Roundup® and Glyphosate Toxicity Have Been Grossly Underestimated' by Dr. Mercola; Mercola: Take Control of Your Health, https://articles.mercola.com, July 30 2013. Sharing groundbreaking and up to date natural health information and resources with the public. Mercola.com also includes pets, fitness, recipes, newsletter and blog.
(Use of any synthetic herbicide involves risk.)